Classics in Mathematics

Joseph L. Doob

Classical Potential Theory and Its Probabilistic Counterpart

Springer Science+Business Media, LLC

Born in Cincinnati, Ohio on February 27, 1910, Joseph L. Doob studied for both his undergraduate and doctoral degrees at Harvard University. He was appointed to the University of Illinois in 1935 and remained there until his retirement in 1978. Doob worked first in complex variables, then moved to probability under the initial impulse of H. Hotelling, and influenced by A.N Kolmogorov's famous monograph of 1933, as well as by Paul Lévy's work. In his own book *Stochastic Processes* (1953), Doob established martingales as a particularly important type of stochastic process. Kakutani's treatment of the Dirichlet problem in 1944, combining complex variable theory and probability, sparked off Doob's interest in potential theory, which culminated in the present book.

(For more details see:
http://www.dartmouth.edu/~chance/Doob/conversation.html)

Joseph L. Doob

Classical Potential Theory and Its Probabilistic Counterpart

Reprint of the 1984 Edition

Springer

Joseph L. Doob
University of Illinois
Department of Mathematics
101 West Windsor Road #1104
Urbana, IL 61802
USA
e-mail: doob@math.uiuc.edu

Originally published as Vol. 262 of the
Grundlehren der mathematischen Wissenschaften

Cataloging-in-Publication Data applied for

Die Deutsche Bibliothek - CIP-Einheitsaufnahme

Doob, Joseph L.:
Classical potential theory and its probabilistic counterpart / J. L. Doob. - Reprint of the 1984 ed. - Berlin; Heidelberg; New York; Barcelona; Hong Kong; London; Milan; Paris; Singapore; Tokyo: Springer, 2001
(Classics in mathematics)
ISBN 978-3-540-41206-9 ISBN 978-3-642-56573-1 (eBook)
DOI 10.1007/978-3-642-56573-1

Mathematics Subject Classification (2000): 31-XX, 60J45

ISSN 1431-0821
ISBN 978-3-540-41206-9

Originally published by Springer-Verlag Berlin Heidelberg New York in 2001

Printed on acid-free paper SPIN 10786705 41/3142ck-5 4 3 2 1 0

J. L. Doob

Classical Potential Theory and Its Probabilistic Counterpart

Springer Science+Business Media, LLC

J. L. Doob
Department of Mathematics
University of Illinois
Urbana, IL 61801
U.S.A.

AMS Subject Classifications: 31-XX, 60J45

Library of Congress Cataloging in Publication Data
Doob, Joseph L.
Classical potential theory and its probabilistic counterpart.
(Grundlehren der mathematischen Wissenschaften; 262)
Bibliography: p.
1. Potential, Theory of. 2. Harmonic functions. 3. Martingales (Mathematics) I. Title. II. Series.
QA404.7.D66 1983 515.7 83-12446

Originally published by Springer-Verlag New York Inc. in 1984

Typeset by Asco Trade Typesetting Ltd., Hong Kong.

9 8 7 6 5 4 3 2 1

ISBN 978-3-540-41206-9 ISBN 978-3-642-56573-1 (eBook)
DOI 10.1007/978-3-642-56573-1

Contents

Chapter VI
The Fundamental Convergence Theorem and the Reduction Operation

Chapter VII
Green Functions

Chapter VIII
The Dirichlet Problem for Relative Harmonic Functions

Chapter XVI
Subparabolic, Superparabolic, and Parabolic Functions on a Slab ... 285

Chapter XVII
Parabolic Potential Theory (Continued) ... 295

Chapter XVIII
The Parabolic Dirichlet Problem, Sweeping, and Exceptional Sets ... 329

Chapter X

Conditional Brownian Motion 668

Part 3

Chapter I

Lattices in Classical Potential Theory and Martingale Theory 705

Chapter II

Brownian Motion and the PWB Method 719

Introduction

Potential theory and certain aspects of probability theory are intimately related, perhaps most obviously in that the transition function determining a Markov process can be used to define the Green function of a potential theory. Thus it is possible to define and develop many potential theoretic concepts probabilistically, a procedure potential theorists observe with jaundiced eyes in view of the fact that now as in the past their subject provides the motivation for much of Markov process theory. However that may be it is clear that certain concepts in potential theory correspond closely to concepts in probability theory, specifically to concepts in martingale theory. For example, superharmonic functions correspond to supermartingales. More specifically: the Fatou type boundary limit theorems in potential theory correspond to supermartingale convergence theorems; the limit properties of monotone sequences of superharmonic functions correspond surprisingly closely to limit properties of monotone sequences of supermartingales; certain positive superharmonic functions [supermartingales] are called "potentials," have associated measures in their respective theories and are subject to domination principles (inequalities) involving the supports of those measures; in each theory there is a reduction operation whose properties are the same in the two theories and these reductions induce sweeping (balayage) of the measures associated with potentials, and so on.

The purpose of this book is to develop this correspondence between potential theory and probability theory by examining in detail classical potential theory, that is, the potential theory of Laplace's equation, together with the corresponding probability theory, that is, martingale theory. The joining link which makes this correspondence especially perspicuous is the Brownian motion process, so this process is studied as needed. In order to carry through this program it is necessary to study parabolic potential theory, that is, the potential theory of the heat equation, and the corresponding process of space time Brownian motion. No knowledge of potential theory is presupposed but it is assumed that the reader is familiar with basic probability concepts through conditional expectations. The necessary lattice theory, analytic set theory and capacity theory are covered in the Appendixes.

Thus this book on the one hand contains an introduction to classical and parabolic potential theory and on the other hand contains an introduction to martingale theory, including a smattering of the general theory of stochastic processes and of Markov process theory. There is cross referencing between the nonprobabilistic and probabilistic aspects of the work, and the linking of classical and parabolic potential theory with martingale theory, by Brownian motion and space time Brownian motion, is examined in depth.

One natural criticism of this project is that there is no reason to treat the very special potential theories of the Laplace and heat equations rather than general axiomatic potential theory. Another criticism is that there is no reason to treat potential theory other than as a special subhead of Markov process theory. In the author's opinion, however, classical potential theory is too important to serve merely as a source of illustrations of axiomatic potential theory, which theory in turn is too important in its own right to be left to the probabilists. To learn potential theory from probability is like learning algebraic geometry without the geometry.

It would be quite impossible to cover all those parts of modernized classical potential theory which are relevant to the purpose of this book. Thus there are striking gaps. For example the treatment of energy is skimpy, and Dirichlet spaces and the concept of bounded mean oscillation are not even mentioned in the text. The emphasis is on the Dirichlet problem and related topics; these are treated in considerable depth. The treatments of classical and parabolic potential theories are sometimes separated, sometimes together, but the notation is designed to exhibit the parallelism of the two theories: dots in the notation distinguish parabolic from classical concepts, thereby muddling eyes but saving brains. And the martingale theory notation is designed to point out to readers the corresponding potential theory notation.

Only the part of Markov process theory needed for the relevant discussion of Brownian motion and conditional Brownian motion is covered. In this book a stochastic process is a specified family of random variables, frequently coupled with a filtration to which the family is adapted, but the measure space of the process is left unspecified and there is no translation operator. Thus in a discussion of Brownian motion from a varying initial point the measure space on which the process is defined may vary with the initial point. This definition of a process may not be best for general Markov process theory but is convenient in the special context of this book; it implies for example that no matter how or on what measure space a process is defined, if it has the properties of a Brownian motion (continuous sample functions and the correct distributions of independent increments) then it is a Brownian motion. In a traditional song, a child finds an object which looks smells and tastes like a peanut so the child concludes that the object is a peanut. As stochastic processes are sometimes defined, with special properties demanded of the measure space on which the process random variables are defined, this simple logic is invalid. However the point of view of this book makes it essential in discussing Brownian motion to prove certain invariance properties, for example that two Brownian motion processes in N space, with a common initial point and variance parameter, have the same probability of hitting an analytic set. This fact is not trivial and such questions are treated.

There is nothing very novel in this book. Potential theorists may find the treatment of reductions on boundary sets of interest, as well as the use of iterated reductions to obtain limit theorems. Correspondingly, probabilists may find the new supermartingale crossing inequalities and the technique of iterated reductions of supermartingales of interest. A new domination principle for supermartingales illustrates the fact that classical potential theory still suggests interesting probability results.

The author thanks Bruce Hajek, Naresh Jain and John Taylor for helpful comments on various chapters and, finally, thanks his typist: usually faithful, sometimes accurate.

Notation and Conventions

$\mathbb{R}^N$ is N dimensional Euclidean space, $\mathbb{R} = \mathbb{R}^1$, and $\mathbb{R}^+$ is the set $[0, +\infty[$ of positive reals. $\overline{\mathbb{R}}$ is the set $[-\infty, +\infty]$ of extended reals and $\overline{\mathbb{R}}^+$ is the set $[0, +\infty]$ of positive extended reals.

$\mathbb{Z}$ is the set of integers, $\mathbb{Z}^+$ is the set $0, 1, 2, \ldots$, and $\mathbb{Z}_n^+$ is the set $0, \ldots, n$.

The boundary of an unbounded subset of $\mathbb{R}^N$ contains the adjoined point ∞ of the one point compactification of $\mathbb{R}^N$ unless some other compactification has been specified. This boundary relative to the one point compactification of $\mathbb{R}^N$ will be called the Euclidean boundary of the set.

If ξ is a point of $\mathbb{R}^N$ and A is a subset of $\mathbb{R}^N$ the distance between ξ and A is written $|\xi - A|$.

$B(\xi, \delta)$ is the ball, in whatever metric space is specified, of center ξ and radius δ, specifically in $\mathbb{R}^N$: $B(\xi, \delta) = \{\eta : |\eta - \xi| < \delta\}$.

l_N refers to N dimensional Lebesgue measure.

If A and B are subsets of a space the set of points in A but not in B is denoted by $A - B$.

"Positive" means "≥ 0" and monotone concepts are to be taken in the wide sense, so that for example a constant function from $\mathbb{R}$ into $\mathbb{R}$ is both monotone increasing and monotone decreasing.

If D is an open subset of $\mathbb{R}^N$ the notation $\mathbb{C}^{(k)}(D)$ refers to the class of functions from D into $\mathbb{R}$ which are continuous together with their derivatives of order $\leq k$.

Limit concepts for a function f at a point do not involve the value of f at the point. Thus $\lim_{\eta \to \xi} f(\eta) = \alpha$ means that f is near α in small deleted neighborhoods of ξ.

The notation for a sequence frequently uses a dot for the index set; unless otherwise identified the index set is $\mathbb{Z}^+$, so that $A_{\bullet} = \{A_0, A_1, \ldots\}$.

The set on which a function f satisfies some set S of conditions is frequently denoted by $\{S\}$. Thus if f is a function from $\mathbb{R}$ into $\mathbb{R}$ the positivity set of f is $\{f \geq 0\}$.

The book is divided into three Parts. Section 1.II.3 is Section 3 of Chapter II of Part 1; in any Part, Section II.3 is Section 3 of Chapter II of that part; in any Chapter, Section 3 is Section 3 of that Chapter, and so on.

Part 1

Classical and Parabolic Potential Theory

(dimensionality number $N > 1$ in Chapters I–XIII)

Chapter I

Introduction to the Mathematical Background of Classical Potential Theory

1. The Context of Green's Identity

In this chapter some of the mathematical ideas of classical potential theory are introduced, under simplifying assumptions. The basic space is Euclidean N space $\mathbb{R}^N$. For a ball $B(\xi, \delta)$ in $\mathbb{R}^N$

$$l_{N-1}(\partial B(\xi, \delta)) = \pi_N \delta^{N-1}, \qquad \pi_N = 2\pi^{N/2}\Gamma\left(\frac{N}{2}\right)^{-1}$$
$$l_N(B(\xi, \delta)) = \frac{\pi_N \delta^N}{N}. \tag{1.1}$$

An application of the Lebesgue dominated convergence theorem shows that if u is a locally l_N integrable function on an open subset D of $\mathbb{R}^N$ and if $\delta > 0$, the function

$$\xi \mapsto \int_{B(\xi, \delta)} u \, dl_N$$

is continuous on the set $\{\xi : |\xi - \partial D| > \delta\}$.

A bounded open set for which Green's identity (1.2) is true will be called *smooth*. Since a ball is obviously smooth in this sense, as is the set between two concentric balls, and since these sets are the only smooth sets for which the details of this chapter will be used later, a precise description of smooth sets is omitted. The Laplacian of a function u, that is, the sum of the unmixed second partial derivatives of u, will be denoted as usual by Δu. If B is a smooth open subset of $\mathbb{R}^N$, if $\mathbf{D_n}$ is the partial derivation operator on ∂D in the direction of the exterior normal, and if u and v are functions in $\mathbb{C}^{(2)}(\bar{B})$, Green's identity is

$$\int_B (u \, \Delta v - v \, \Delta u) \, dl_N = \int_{\partial B} (u\mathbf{D_n} v - v\mathbf{D_n} u) \, dl_{N-1}, \tag{1.2}$$

which for $v \equiv 1$ reduces to

$$\int_B \Delta u \, dl_N = \int_{\partial B} \mathbf{D}_{\mathbf{n}} u \, dl_{N-1}. \tag{1.3}$$

The right side of (1.3) is the flux of the vector field grad u out through ∂B.

2. Function Averages

The unweighted averages of a function u over $\partial B(\xi, \delta)$ and over $B(\xi, \delta)$ will be denoted by $L(u, \xi, \delta)$ and $A(u, \xi, \delta)$, respectively; that is

$$L(u, \xi, \delta) = \frac{1}{\pi_N \delta^{N-1}} \int_{\partial B(\xi, \delta)} u \, dl_{N-1}, \qquad A(u, \xi, \delta) = \frac{N}{\pi_N \delta^N} \int_{B(\xi, \delta)} u \, dl_N, \tag{2.1}$$

assuming that the necessary measurability and integrability conditions are satisfied. If u is defined, l_N measurable and locally l_N integrable on an open subset D of $\mathbb{R}^N$, the function $A(u, \cdot, \delta)$ is defined and continuous on the set $\{\xi : |\xi - \partial D| > \delta\}$ (see Section 1). In this case according to Fubini's theorem the average $L(u, \xi, r)$ is defined for l_1 almost every r in the interval $]0, \delta]$ and

$$A(u, \xi, \delta) = N\delta^{-N} \int_0^\delta L(u, \xi, r) r^{N-1} l_1(dr). \tag{2.2}$$

Define

$$\gamma_N(r) = \begin{cases} c_N \exp[-(1-r)^{-1}] & \text{if } 0 \le r < 1 \\ 0 & \text{if } r \ge 1, \end{cases} \tag{2.3}$$

choosing c_N so that $\pi_N \int_0^\infty r^{N-1} \gamma_N(r) l_1(dr) = 1$, and under the preceding hypotheses on u, define $A_\delta u$ for $\delta > 0$ by

$$\begin{aligned} A_\delta u(\xi) &= \delta^{-N} \int_{\mathbb{R}^N} \gamma_N\left(\frac{|\xi - \eta|}{\delta}\right) u(\eta) l_N(d\eta) \\ &= \pi_N \int_0^\infty r^{N-1} \gamma_N(r) L(u, \xi, \delta r) l_1(dr) \qquad (|\xi - \partial D| > \delta). \end{aligned} \tag{2.4}$$

The function $A_\delta u$ is infinitely differentiable, $A_\delta 1 \equiv 1$, and if u is continuous $\lim_{\delta \to 0} A_\delta u = u$ locally uniformly on D.

3. Harmonic Functions

A *harmonic function* is defined as a (finite-valued) continuous function u, defined on a nonempty open subset of $\mathbb{R}^N$, satisfying

$$u(\xi) = L(u, \xi, \delta) \quad \text{if } \bar{B}(\xi, \delta) \subset D. \tag{3.1}$$

The stated continuity condition is stronger than necessary. In fact, if it is weakened to l_N measurability and local integrability, (2.4) implies that $A_\alpha(u, \xi) = u(\xi)$ for $|\xi - \partial D| > \alpha$, so that u is infinitely differentiable and therefore harmonic, and we have incidentally shown that harmonic functions are infinitely differentiable. The class of harmonic functions on D is trivially linear.

If u is harmonic on D, (2.2) yields

$$u(\xi) = A(u, \xi, \delta) \qquad (|\xi - \partial D| > \delta). \tag{3.2}$$

Conversely (3.2) implies (3.1) if u is l_N measurable and locally integrable because under (3.2) the function u must be continuous and (2.2) then yields (3.1).

Equation (1.3) with $B = B(\xi, \delta)$ reduces to

$$\int_{B(\xi,\delta)} \Delta u \, dl_N = \pi_N \delta^{N-1} \frac{d}{d\delta} L(u, \xi, \delta). \tag{3.3}$$

Hence a function u with continuous second partial derivatives is harmonic if and only if u satisfies Laplace's equation $\Delta u = 0$, and it follows that harmonicity is a local property. If u is harmonic all its partial derivatives are harmonic because they satisfy Laplace's equation also. According to (1.3) we can conclude that if u is harmonic on D the flux of its gradient out of any smooth subdomain of D vanishes, and conversely, if u has continuous second partial derivatives and if the flux of its gradient out of every ball B with closure in D vanishes, then u is harmonic. (It is easy to see that it is sufficient here if u has continuous first partial derivatives.)

$N = 2$. The real and imaginary parts of a (complex) analytic function $f = u + iv$ are harmonic. In fact, u and v satisfy Laplace's equation because they satisfy the Cauchy–Riemann equations. Alternatively, the Cauchy integral formula yields (3.1).

4. Maximum-Minimum Theorem for Harmonic Functions

Recall our convention that in the absence of another stated topology the space $\mathbb{R}^N$ with its usual Euclidean topology is supposed and is compactified by a point at infinity, denoted by ∞. This point ∞ is not included in $\mathbb{R}^N$ but is included in the boundary of every unbounded subset of $\mathbb{R}^N$.

Theorem. *Let u be harmonic on the open subset D of $\mathbb{R}^N$.*

(a) *If D is connected and if u attains its supremum or infimum at a point ξ of D then u is identically constant.*
(b) *The supremum and infimum of u are limits of u along sequences of points approaching ∂D.*

(c) *If u has a continuous extension to $D \cup \partial D$, the supremum and infimum of the extension are attained on ∂D.*

A typical implication of this theorem is the fact that a harmonic function on D with limit 0 at every boundary point must vanish identically. To prove (a), suppose that u attains, say, its infimum α at a point ξ of D. Since (continuity of u) the set $\{\eta : u(\eta) > \alpha\}$ is open, the harmonic function average property (3.2) implies that u is identically α on $B(\xi, \delta)$ for $\delta < |\xi - \partial D|$. It follows that the set of points of D at which $u = \alpha$ is open. Since continuity of u implies that this set is also closed relative to D, this set is D if D is connected. Thus part (a) of the theorem is true and parts (b) and (c) follow easily.

5. The Fundamental Kernel for $\mathbb{R}^N$ and Its Potentials

If $f \in \mathbb{C}^{(2)}]a, b[$ the function u defined by $u(\xi) = f(|\xi|)$ on the domain $\{\xi : a < |\xi| < b\}$ is harmonic if and only if (denoting $|\xi|$ by r)

$$\Delta u = f''(r) + \frac{N-1}{r} f'(r) = 0. \tag{5.1}$$

Thus if G is defined by

$$G(\xi, \eta) = \begin{cases} \log \dfrac{1}{|\xi - \eta|} & \text{if } N = 2, \\ \dfrac{1}{|\xi - \eta|^{N-2}} & \text{if } N > 2, \end{cases} \qquad (= +\infty \text{ if } \xi = \eta) \tag{5.2}$$

the function $G(\xi, \cdot)$ is harmonic on $\mathbb{R}^N - \{\xi\}$. The function kernel G is the fundamental kernel of classical potential theory. For $D = \mathbb{R}^N$ we shall sometimes write G_D instead of G when $N > 2$ to match later notation for Green's functions.

If $|\xi - \eta| < r < s$, equation (1.3) with $B = B(\xi, s) - \bar{B}(\xi, r)$ and $u = G(\cdot, \eta)$ reduces to

$$0 = \pi_N s^{N-1} \frac{d}{ds} L(u, \xi, s) - \pi_N r^{N-1} \frac{d}{dr} L(u, \xi, r) \tag{5.3}$$

so that $s^{N-1}\, d/ds\, L(u, \xi, s)$ does not depend on s and since

$$\lim_{s = |\zeta| \to +\infty} L(u, \xi, s)/G(0, \zeta) = 1,$$

$$L(G(\cdot, \eta), \xi, r) = \begin{cases} -\log r & \text{if } N = 2 \\ r^{2-N} & \text{if } N > 2 \end{cases} \qquad (|\xi - \eta| < r). \tag{5.4}$$

On the other hand since $G(\cdot, \eta)$ is harmonic on $\mathbb{R}^N - \{\eta\}$, the harmonic function average property yields

$$L(G(\cdot, \eta), \xi, r) = G(\xi, \eta) \quad \text{when } |\xi - \eta| \geq r. \tag{5.5}$$

(The fact that this evaluation is valid when $|\xi - \eta| = r$ follows from an easy continuity argument.)

If μ is a measure of Borel subsets of $\mathbb{R}^N$ the function $G\mu$ defined by

$$G\mu(\xi) = \int_{\mathbb{R}^N} G(\xi, \eta)\, \mu(d\eta) \tag{5.6}$$

is the *potential of* μ. We shall discuss the convergence of this integral later. It is clear however that if $\mu(\mathbb{R}^N) < +\infty$, the integral converges absolutely at every point not in the closed support A of μ and thereby defines a continuous function on $\mathbb{R}^N - A$. The function is harmonic on this domain because it has the harmonic function average property there.

6. Gauss Integral Theorem

Let D be a smooth domain, and let μ be a signed measure supported by a compact set A not meeting ∂D. Then

$$\mu(A \cap D) = -\frac{1}{\pi'_N} \int_{\partial D} \mathbf{D}_{\mathbf{n}_\eta} G\mu(\eta) l_{N-1}(d\eta), \tag{6.1}$$

where

$$\pi'_N = \begin{cases} 2\pi & \text{if } N = 2, \\ (N-2)\pi_N & \text{if } N > 2. \end{cases} \tag{6.2}$$

If A does not meet D the function $G\mu$ is harmonic on a neighborhood of $\bar{D}$ so (6.1) is true because according to Section 3 the flux out of D of the gradient of a harmonic function vanishes. The potential of the projection of μ on $\mathbb{R}^N - D$ is covered by this remark so from now on we suppose that $A \subset D$. If (6.1) is true for one choice of D, it is true for every smooth domain D_1 containing $\bar{D}$ if $D_1 - \bar{D}$ is smooth because $G\mu$ is harmonic on a neighborhood of $\bar{D}_1 - D$ so the flux out of $D_1 - \bar{D}$ of the gradient of $G\mu$ vanishes. Observe that (6.1) is trivially true if D is a ball with center ξ and if $A = \{\xi\}$, and therefore, in view of the remark in the last sentence, (6.1) is true whenever A is a singleton,

$$\frac{1}{\pi'_N} \int_{\partial D} \mathbf{D}_{\mathbf{n}_\eta} G(\xi, \eta) l_{N-1}(d\eta) = -1 \quad \text{if } \xi \in D. \tag{6.3}$$

Integrating (6.3) with respect to μ yields (6.1).

See Section 7 for an extension of this theorem allowing the support of μ to meet ∂D.

7. The Smoothness of Potentials; The Poisson Equation

In the following theorem the coordinates of a point ξ in $\mathbb{R}^N$ are denoted by $\xi^{(1)}, \ldots, \xi^{(N)}$. We shall use the inequalities

$$\left|\frac{\partial G(\xi,\eta)}{\partial \xi^{(i)}}\right| \le N|\xi-\eta|^{1-N}, \qquad \left|\frac{\partial^2 G(\xi,\eta)}{\partial \xi^{(i)}\partial \xi^{(j)}}\right| \le N(N+1)|\xi-\eta|^{-N}, \qquad N \ge 2. \tag{7.1}$$

Theorem. *Let μ be a signed measure on $\mathbb{R}^N$ given by an l_N measurable density f relative to l_N, $d\mu = f\,dl_N$.*

(a) *If f is bounded and vanishes near ∞ then $u = G\mu$ is in class $\mathbb{C}^{(1)}(\mathbb{R}^N)$ and*

$$\frac{\partial u(\xi)}{\partial \xi^{(i)}} = \int_{\mathbb{R}^N} \frac{\partial G(\xi,\eta)}{\partial \xi^{(i)}} f(\eta) l_N(d\eta). \tag{7.2}$$

(b) *If in* (a) *there is an open set D on which f is continuous and satisfies a Hölder condition,*

$$|f(\xi) - f(\eta)| \le \text{const}\,|\xi-\eta|^p \qquad 0 < p \le 1,\ (\xi,\eta) \in D \times D,$$

then $u_{|D}$ is in class $\mathbb{C}^{(2)}(D)$, satisfies the Poisson equation $\Delta u = -\pi'_N f$, and

$$\frac{\partial^2 u(\xi)}{\partial \xi^{(i)}\partial \xi^{(j)}} = \int_{\mathbb{R}^N} \frac{\partial^2 G(\xi,\eta)}{\partial \xi^{(i)}\partial \xi^{(j)}} [f(\eta) - f(\xi)] l_N(d\eta) - \frac{\delta_{ij}\pi'_N}{N} f(\xi), \qquad \xi \in D. \tag{7.3}$$

Here δ_{ij} is the usual Kronecker symbol; that is, (δ_{ij}) is the identity matrix.

Proof of (a). Since $G(\xi,\cdot)$ is locally l_N integrable, the integral for $G\mu$ converges absolutely under the hypotheses of (a). Define $G^{(k)}$ as $G \wedge \log k$ when $N = 2$ and as $G \wedge k^{N-2}$ when $N > 2$. Then $G^{(k)}\mu$ is continuous and

$$|G\mu - G^{(k)}\mu| \le \begin{cases} \text{const}\,\dfrac{\log k}{k^2} & \text{if } N = 2, \\[2ex] \dfrac{\text{const}}{k^2} & \text{if } N > 2. \end{cases}$$

Hence $\lim_{k\to\infty} G^{(k)}\mu = G\mu$ uniformly on $\mathbb{R}^N$, and we conclude that $u = G\mu$ is continuous. In view of (7.1) the integral in (7.2) is absolutely convergent.

Round off the integrand in (7.2) when it exceeds k^{N-1} in absolute value to obtain a continuous finite-valued integrand and repeat the reasoning proving the continuity of $G\mu$ to show that the right side of (7.2) defines a continuous function of ξ. This function must be $\partial u/\partial\xi^{(i)}$. □

Proof of (b). In view of (7.1) and the Hölder condition satisfied by f on D the integrand in (7.3) for η in a neighborhood of ξ in D is majorized in absolute value by const $|\xi-\eta|^{-N+p}$; so the integral converges absolutely. Let β be a function from $\mathbb{R}^+$ into $\mathbb{R}^+$ satisfying the following conditions:

$$\beta\in\mathbb{C}^{(1)}(\mathbb{R}^+),\qquad 0\le\beta\le 1,$$
$$\beta(r)=\begin{cases}0 & \text{if } 0\le r\le \frac{1}{2},\\ 1 & \text{if } r\ge 1,\end{cases}$$

and define

$$\phi_i(\xi,\alpha)=\int_{\mathbb{R}^N}\frac{\partial G(\xi,\eta)}{\partial\xi^{(i)}}f(\eta)\beta\left(\frac{|\xi-\eta|}{\alpha}\right)l_N(d\eta)$$

for $\alpha>0$. Then

$$\left|\phi_i(\xi,\alpha)-\frac{\partial u(\xi)}{\partial\xi^{(i)}}\right|\le\sup|f|\int_{B(\xi,\alpha)}\left|\frac{\partial G(\xi,\eta)}{\partial\xi^{(i)}}\right|l_N(d\eta)\le\text{const}\,\alpha;$$

so $\lim_{\alpha\to0}\phi_i(\xi,\alpha)=\partial u(\xi)/\partial\xi^{(i)}$ uniformly on $\mathbb{R}^N$. The function $\phi_i(\cdot,\alpha)$ is in $\mathbb{C}^{(1)}(\mathbb{R}^N)$. To prove (b) it is sufficient to show that, on D, $\lim_{\alpha\to0}\partial\phi_i/\partial\xi^{(j)}$ is the right side of (7.3) and that the convergence is uniform on every compact subset of D. To show this, write $\partial\phi_i/\partial\xi^{(j)}$ in the form

$$\begin{aligned}\frac{\partial\phi_i(\xi,\alpha)}{\partial\xi^{(j)}}&=\int_{\mathbb{R}^N}\frac{\partial^2G(\xi,\eta)}{\partial\xi^{(i)}\partial\xi^{(j)}}[f(\eta)-f(\xi)]\beta\left(\frac{|\xi-\eta|}{\alpha}\right)l_N(d\eta)\\&\quad+f(\xi)\int_{\mathbb{R}^N}\frac{\partial^2G(\xi,\eta)}{\partial\xi^{(i)}\partial\xi^{(j)}}\beta\left(\frac{|\xi-\eta|}{\alpha}\right)l_N(d\eta)\\&\quad+\frac{1}{\alpha}\int_{\mathbb{R}^N}\frac{\partial G(\xi,\eta)}{\partial\xi^{(i)}}[f(\eta)-f(\xi)]\beta'\left(\frac{|\xi-\eta|}{\alpha}\right)\frac{\xi^{(j)}-\eta^{(j)}}{|\xi-\eta|}l_N(d\eta)\\&\quad+\frac{f(\xi)}{\alpha}\int_{\mathbb{R}^N}\frac{\partial G(\xi,\eta)}{\partial\xi^{(i)}}\beta'\left(\frac{|\xi-\eta|}{\alpha}\right)\frac{\xi^{(j)}-\eta^{(j)}}{|\xi-\eta|}l_N(d\eta)\end{aligned}$$

and denote the four terms on the right by I, II, III, IV, respectively. Observe that each integrand vanishes for η outside $B(\xi,\alpha)$.

The difference between I and the integral on the right side of (7.3) is at most const $\int_0^\alpha r^{p-1} l_1(dr)$ for $\alpha < |\xi - \partial D|$; so when $\alpha \to 0$, the term I has the integral on the right side of (7.3) as a limit uniformly on compact subsets of D.

If $i \neq j$ in II, the integral over $B(\xi, \alpha)$ vanishes because the integrand is odd in $\xi^{(i)} - \eta^{(i)}$. If $i = j$ in II, the integral becomes

$$\begin{aligned} &-(N-2)\int_{B(\xi,\alpha)} \frac{\beta(|\xi-\eta|/\alpha)}{|\xi-\eta|^N} l_N(d\eta) \\ &+ N(N-2)\int_{B(\xi,\alpha)} \frac{(\xi^{(i)}-\eta^{(i)})^2}{|\xi-\eta|^{N+2}} \beta\left(\frac{|\xi-\eta|}{\alpha}\right) l_N(d\eta) \end{aligned} \tag{7.4}$$

if $N > 2$ and $\alpha < |\xi - \partial D|$. The second integral does not depend on the choice of i, and so it has as value the average of its values for $i = 1, \ldots, N$. We conclude that the sum in (7.4) vanishes. The corresponding argument when $N = 2$ shows that the integral II also vanishes in this case.

If $\alpha < |\xi - \partial D|$,

$$|\text{III}| \leq \text{const}\frac{1}{\alpha}\int_0^\alpha r^p \beta'\left(\frac{r}{\alpha}\right) l_1(dr) \leq \text{const}\,\alpha^p;$$

so when $\alpha \to 0$, there is uniform convergence to 0 on compact subsets of D.

To evaluate IV observe that when $i \neq j$ this integral over $B(\xi, \alpha)$ vanishes because the integrand is odd in $\xi^{(i)} - \eta^{(i)}$. If $i = j$ and if $N > 2$ then

$$\text{IV} = -\frac{f(\xi)}{\alpha}(N-2)\int_{B(\xi,\alpha)} |\xi-\eta|^{-N-1}(\xi^{(i)}-\eta^{(i)})^2 \beta'\left(\frac{|\xi-\eta|}{\alpha}\right) l_N(d\eta).$$

The integral is the same for all i, and it is equal to the average of its values for $i = 1, \ldots, N$. Hence

$$\text{IV} = -\frac{\pi_N f(\xi)(N-2)}{\alpha N}\int_0^\alpha \beta'\left(\frac{r}{\alpha}\right) l_1(dr) = -\pi_N' \frac{f(\xi)}{N}.$$

If $N = 2$ the corresponding evaluation of IV yields $-\pi f(\xi) = -\pi_2' f(\xi)/2$, and the proof of the theorem is now complete. □

Extension. If ν is a signed measure on $\mathbb{R}^N$ for which the integral defining $G\nu$ converges absolutely and if the projection of ν on some bounded open set D is determined by a bounded density g relative to l_N, then $(G\nu)_{|D} \in \mathbb{C}^{(1)}(D)$ because if $f = g 1_D$ and $d\mu = f\,dl_N$, then Theorem 7 is applicable to $G\mu$ which differs on D from $G\nu$ by a harmonic function. The same argument shows that if g is continuous on D and satisfies a Hölder condition there, then $(G\nu)_{|D} (-\mathbb{C}^{(2)}(D)$ and $\Delta G\nu = -\pi_N' g$ there.

Extension of the Gauss Integral Theorem

Under the hypotheses of (b) the flux evaluation (1.3) yields, whenever B is a smooth domain with closure in D,

$$\mu(B) = -\frac{1}{\pi'_N}\int_B \Delta G\mu\, dl_N = -\frac{1}{\pi'_N}\int_{\partial B} \mathbf{D}_{\mathbf{n}} G\mu\, dl_{N-1}. \tag{7.5}$$

This evaluation generalizes the Gauss Integral Theorem (Section 6) in that the support of the measure μ is allowed to meet ∂B.

8. Harmonic Measure and the Riesz Decomposition

Let D be an open bounded subset of $\mathbb{R}^N$. Suppose that $\mu \in \mathbb{C}^{(2)}(D)$, and suppose that Δu is bounded on D. Define a signed measure μ on D by $d\mu = -\Delta u\, dl_N/\pi'_N$. Theorem 7 implies that $\Delta(u - G\mu) = 0$, so that $u = (u - G\mu) + G\mu$ is the sum of a harmonic function and the restriction to D of the potential $G\mu$. This fact will now be proved in a slightly different version and developed further.

If D is a smooth open subset of $\mathbb{R}^N$, if $\bar{B}(\xi, \delta) \subset D$, and if $u \in \mathbb{C}^{(2)}(\bar{D})$, an application of Green's identity to the pair of functions $[u, G(\xi, \cdot)]$ on the smooth open set $D - \bar{B}(\xi, \delta)$ yields

$$\begin{aligned}\int_{\partial B(\xi,\delta)} & G(\xi,\eta)\mathbf{D}_{\mathbf{n}_\eta} u\, l_{N-1}(d\eta) - \int_{\partial B(\xi,\delta)} u(\eta)\mathbf{D}_{\mathbf{n}_\eta} G(\xi,\cdot) l_{N-1}(d\eta)\\ &= \int_{\partial D} [G(\xi,\eta)\mathbf{D}_{\mathbf{n}_\eta} u - u(\eta)\mathbf{D}_{\mathbf{n}_\eta} G(\xi,\cdot)] l_{N-1}(d\eta) \\ &\quad - \int_{D-\bar{B}(\xi,\delta)} G(\xi,\eta)\,\Delta u(\eta) l_N(d\eta).\end{aligned} \tag{8.1}$$

Since $\mathbf{D}_{\mathbf{n}} u$ is bounded, the first integral on the left is majorized in absolute value by const $\delta|\log \delta|$ if $N = 2$ and by const δ^2 if $N > 2$; so this integral tends to 0 when $\delta \to 0$. The second integral on the left is equal to $-\pi'_N L(u, \xi, \delta)$ for $N \geq 2$ and therefore has limit $-\pi'_N u(\xi)$ when $\delta \to 0$. Thus when $\delta \to 0$ in (8.1), we find

$$\begin{aligned} u(\xi) = \frac{1}{\pi'_N}&\int_{\partial D} [G(\xi,\eta)\mathbf{D}_{\mathbf{n}_\eta} u - u(\eta)\mathbf{D}_{\mathbf{n}_\eta} G(\xi,\cdot)] l_{N-1}(d\eta)\\ &- \frac{1}{\pi'_N}\int_D G(\xi,\eta)\,\Delta u(\eta) l_N(d\eta)\end{aligned} \tag{8.2}$$

This representation of u is another version of that obtained above because the first term in (8.2) defines a function harmonic on D and the second term is the potential of the signed measure with density $-\Delta u/\pi'_N$. This representation of u remains valid if $G(\xi,\cdot)$ is replaced by $G(\xi,\cdot) - u(\xi,\cdot)$, where (i) $u(\xi,\cdot) \in \mathbb{C}^{(2)}(\bar{D})$ for each ξ in D and $u(\xi,\cdot)$ is harmonic on D. Suppose that $u(\xi,\cdot)$ can be chosen to satisfy (i) and also (ii) $u(\xi,\cdot) = G(\xi,\cdot)$ on ∂D. In this case the restriction to D of the difference $G_D(\xi,\cdot) = G(\xi,\cdot) - u(\xi,\cdot)$ is called the *Green function of D with pole* ξ, and G_D is called the *Green function of D*. The function $u(\xi,\cdot)$, if there is such a function, is uniquely determined because the difference between two such functions is harmonic on D with limit 0 at every boundary point and therefore vanishes identically (maximum-minimum theorem for harmonic functions). Recapitulating, $G_D(\xi,\cdot)$ is to satisfy the following conditions:

(i′) G_D is defined on $D \times D$ and $G_D(\xi,\xi) = +\infty$.

(ii′) $G_D(\xi,\cdot)$ is harmonic on $D - \{\xi\}$, and $G_D(\xi,\cdot) - G(\xi,\cdot)$ is harmonic on D if defined suitably at ξ.

(iii′) $G_D(\xi,\cdot)$ has limit 0 at every point of ∂D.

(iv′) $G_D(\xi,\cdot) - G(\xi,\cdot)$ if defined suitably at ξ and, if defined as $-G(\xi,\cdot)$ on ∂D, is in $\mathbb{C}^{(2)}(\bar{D})$.

An application of the harmonic function maximum-minimum theorem to $G_D(\xi,\cdot)$ on $D - \{\xi\}$ shows that $G_D \geq 0$. We shall show in Section II.1 that a ball has a Green function in this classical sense. The existence of a Green function satisfying conditions (i′)–(iv′) is a restriction on the smoothness of ∂D, but in Chapter VII we shall define a Green function much more loosely, keeping (i′) and (ii′), weakening (iii′), and dropping (iv′). It will be shown that every nonempty open subset of $\mathbb{R}^N$ for $N > 2$ and every not-too-large nonempty open subset of $\mathbb{R}^2$ (for example, every nonempty bounded open set) has a unique Green function in the looser sense. The present discussion shows what led to the more general definition and will suggest theorems to be proved.

Suppose then that D is smooth and that G_D exists satisfying (i′)–(iv′). If ν is a measure on D, the function $G_D\nu$ on D is called the *Green potential of* ν. Using G_D instead of G in (8.2) and defining $G_D(\xi,\cdot)$ as 0 on ∂D reduces (8.2) to

$$u(\xi) = -\frac{1}{\pi'_N}\int_{\partial D} u(\eta)\mathbf{D}_{\mathbf{n}_\eta} G_D(\xi,\cdot) l_{N-1}(d\eta) - \frac{1}{\pi'_N}\int_D G_D(\xi,\eta)\,\Delta u(\eta) l_N(d\eta). \tag{8.3}$$

In particular

$$1 = -\frac{1}{\pi'_N}\int_{\partial D} \mathbf{D}_{\mathbf{n}_\eta} G_D(\xi,\cdot) l_{N-1}(d\eta). \tag{8.4}$$

Define the measure $\mu_D(\xi, \cdot)$ of Borel subsets of ∂D by

$$\mu_D(\xi, A) = -\frac{1}{\pi'_N} \int_A \mathbf{D}_{\mathbf{n}_\eta} G_D(\xi, \cdot) l_{N-1}(d\eta). \tag{8.5}$$

Then $\mu_D(\xi, \cdot)$ is a positive measure because the normal derivative in the integrand is negative, and $\mu_D(\xi, \partial D) = 1$. The measure $\mu_D(\xi, \cdot)$ is called the *harmonic measure relative to* ξ. The representation (8.3) can now be written

$$u(\xi) = \int_{\partial D} u(\eta) \mu_D(\xi, d\eta) + G_D \nu(\xi), \qquad d\nu = -\Delta u \frac{dl_N}{\pi'_N}. \tag{8.6}$$

In our later general treatment we shall prove the following facts:

(a) G_D is symmetric and continuous on $D \times D$ ($= +\infty$ on the diagonal).
(b) The function $\mu_D(\cdot, A)$ is harmonic on D for every A, and more generally the function $\int_{\partial D} f(\eta) \mu_D(\cdot, d\eta)$ is harmonic on D for every bounded Borel measurable function f on ∂D.

These facts will be trivially verifiable when D is a ball (see Section II.1), the only case when they will be used explicitly before the general discussion. We therefore proceed without proving facts (a) and (b) in the present discussion. If u is harmonic on D the second term in (8.6) drops out; so

$$u(\xi) = \int_{\partial D} u(\eta) \mu_D(\xi, d\eta). \tag{8.7}$$

This representation of u makes $u(\xi)$ a weighted average of the values of u on ∂D. In particular, if D is a ball with center ξ, this averaging reduces to the defining average property of harmonic functions.

More generally, if $\Delta u \le 0$ and $u > 0$, (8.6) exhibits u as the sum of a positive harmonic function and the Green potential of a positive measure. One of the principal aims of the general theory is to generalize this result (the Riesz decomposition, Section IV.8) by dropping the smoothness conditions on u, D, and G_D.

Basic Properties of Harmonic, Subharmonic, and Superharmonic Functions

1. The Green Function of a Ball; The Poisson Integral

Let $B = B(\xi_0, \delta)$ and if $\xi \in B$ denote by ξ' the image of ξ under inversion in ∂B. That is, ξ' is on the ray from ξ_0 through ξ, and $|\xi - \xi_0|\,|\xi' - \xi_0| = \delta^2$. To simplify the notation take $\xi_0 = O$. Then G_B, as defined by

$$G_B(\xi, \eta) = \begin{cases} \log \dfrac{|\xi' - \eta|\,|\xi|}{\delta|\xi - \eta|} & \left(= \log \dfrac{\delta}{|\eta|} \text{ if } \xi = O\right) \\ & \text{for } N = 2 \\ |\xi - \eta|^{2-N} - \left(\dfrac{\delta}{|\xi|}\right)^{N-2} |\xi' - \eta|^{2-N} & (= |\eta|^{2-N} - \delta^{2-N} \text{ if } \xi = O) \\ & \text{for } N > 2 \end{cases} \tag{1.1}$$

with the understanding that $G_B(\xi, \xi) = +\infty$, satisfies items (i′)–(iv′) of Section I.8, so that harmonic measure for B is given by

$$\mu_B(\xi, d\eta) = -\frac{1}{\pi_N'} \mathbf{D}_{\mathbf{n}_\eta} G_B(\xi, \eta) l_{N-1}(d\eta) = \frac{K(\eta, \xi)}{\pi_N \delta^{N-1}} l_{N-1}(d\eta), \tag{1.2}$$

where l_{N-1} here refers to surface area on ∂B and

$$K(\eta, \xi) = \delta^{N-2} \frac{\delta^2 - |\xi|^2}{|\eta - \xi|^N}. \tag{1.3}$$

Hence, according to Section I.8, if u is harmonic on a neighthorbood of $\bar{B}$,

$$u(\xi) = \frac{1}{\pi_N \delta} \int_{\partial B} u(\eta) \frac{\delta^2 - |\xi|^2}{|\eta - \xi|^N} l_{N-1}(d\eta) = \frac{1}{\pi_N \delta^{N-1}} \int_{\partial B} u(\eta) K(\eta, \xi) l_{N-1}(d\eta). \tag{1.4}$$

The function $K(\eta, \cdot)$ is harmonic on $\mathbb{R}^N - \{\eta\}$ because $G_B(\cdot, \eta)$ is, and $K(\eta, \cdot)$ is normalized to be 1 at the origin. The function G_B is symmetric, positive ($= +\infty$ when the arguments are equal), and increases with δ. Moreover

$$\lim_{\delta\to\infty} G_B(\xi,\eta) = \begin{cases} +\infty & \text{if} \quad N=2, \\ G(\xi,\eta) & \text{if} \quad N>2. \end{cases} \tag{1.5}$$

Since $G_B(\cdot,\eta)$ is harmonic on $B - \{\eta\}$, all the partial derivatives of this function are also harmonic there so that in particular the harmonic measure density in (1.2) defines a harmonic function of ξ on B. It follows that if μ is a finite measure of Borel subsets of ∂B, the *Poisson integral* (PI)

$$\operatorname{PI}(B,\mu)(\xi) = \int_{\partial B} K(\eta,\xi)\mu(d\eta), \tag{1.6}$$

defines a harmonic function on B. It should cause no confusion if we write $\operatorname{PI}(B,f)$ instead of $\operatorname{PI}(B,\mu)$ when $\mu(d\eta) = f(\eta)l_{N-1}(d\eta)/(\pi_N\delta^{N-1})$. In later applications the ball may not have the origin as center, and the Poisson formula is modified accordingly. Moreover, if f is defined on a superset S of the boundary of a ball B, we shall write $\operatorname{PI}(B,f)$ instead of $\operatorname{PI}(B,f|_{\partial B})$.

Theorem. *If f is Lebesgue integrable on ∂B, $u = \operatorname{PI}(B,f)$ is harmonic on B and*

$$\limsup_{\substack{\xi\to\zeta\\ \xi\in B}} u(\xi) \le \limsup_{\substack{\xi\to\zeta\\ \xi\in\partial B}} f(\xi) \qquad (\zeta\in\partial B). \tag{1.7}$$

Combining (1.7) *with the corresponding inequality for inferior limits, it follows that u has limit $f(\zeta)$ at ζ if f is continuous at ζ.*

We have already noted that u is harmonic on B. Inequality (1.7) is a special case of Theorem 1 of Appendix VII with $X = \partial B$ and $n \to \infty$ replaced by $\xi \to \zeta$. The key fact is that for η outside a neighborhood of ζ and on ∂B the integrand $K(\eta,\xi)$ is at most $\operatorname{const}(\delta^2 - |\xi|^2)$, and this majorant tends to 0 when ξ in B tends to ζ.

Integral of K. The fact that $\int_B K(\eta,\xi)l_N(d\xi) = l_N(B)$ will be needed below. To derive this evaluation observe that the value $\alpha = \alpha(\eta)$ of this integral does not in fact depend on η; so

$$\alpha = L(\alpha, O, \delta) = \int_B \operatorname{PI}(B,1)l_N(d\eta) = l_N(B).$$

Solution of the Dirichlet Problem for a Ball

If f is a finite continuous function on the boundary of a ball B, the function $u = \operatorname{PI}(B,f)$ is harmonic on B and, according to Theorem 1 has limit $f(\zeta)$ at every boundary point ζ. There is only one harmonic function on B with this boundary limit property because the difference between two such functions is harmonic on B with limit 0 at every boundary point and therefore vanishes identically in view of the harmonic function maximum minimum theorem. The function u is thus the unique solution of the classical first boundary (Dirichlet) problem for harmonic functions on B. A generalized form of this problem for the relevant class of open subsets of $\mathbb{R}^N$ is treated in Chapter VIII.

If u is a finite-valued function defined and continuous on the closure of a ball B and harmonic on B, $u = \mathrm{PI}(B, u)$ on B because u is the unique solution of the Dirichlet problem for the boundary function $u|_{\partial B}$. This result weakens the conditions under which (1.3) was derived using the general theory in Section I.8.

If u is any Borel measurable function on an open subset D of $\mathbb{R}^N$, if B is a ball with closure in D, and if the restriction of u to ∂B is l_{N-1} integrable, we define

$$\tau_B u = \begin{cases} \mathrm{PI}(B, u) & \text{on} \quad B, \\ u & \text{on} \quad D - B. \end{cases} \tag{1.8}$$

If u is upper or lower semicontinuous, Theorem 1 implies that $\tau_B u$ has the same property.

2. Harnack's Inequality

Let A be a compact subset of the open connected subset D of $\mathbb{R}^N$. There is a function $(A, D) \mapsto c(A, D)$ such that if u is harmonic and strictly positive on D,

$$\frac{u(\xi)}{u(\eta)} \le c(A, D) \qquad [(\xi, \eta) \in A \times A]. \tag{2.1}$$

If $\bar{B}(\xi_0, \delta) \subset D$, the representation $u = \mathrm{PI}(B(\xi_0, \delta), u)$ yields

$$\frac{\delta^2 - \alpha^2}{(\delta + \alpha)^N} \delta^{N-2} \le \frac{u(\xi)}{u(\xi_0)} \le \frac{\delta^2 - \alpha^2}{(\delta - \alpha)^N} \delta^{N-2} \qquad (|\xi - \xi_0| \le \alpha < \delta). \tag{2.2}$$

Thus

$$\frac{u(\xi)}{u(\eta)} \le \left(\frac{\delta + \alpha}{\delta - \alpha}\right)^N \qquad [(\xi, \eta) \in B(\xi_0, \alpha) \times B(\xi_0, \alpha)]. \tag{2.3}$$

Harnack's inequality is therefore true if $A \subset B(\xi_0, \delta)$. More generally, it follows that Harnack's inequality is true if the compact subset A of D can be covered by finitely many balls $B_1, \ldots, B_k$, each with closure in D, in such a way that $B_{j+1} \cap B_j \neq \emptyset$ for $j = 1, \ldots, k-1$. Since D is connected, every compact subset A of D can be covered in this way.

Application to Lower-Bounded Harmonic Functions on $\mathbb{R}^N$

A lower-bounded harmonic function v on $\mathbb{R}^N$ must be identically constant because the function $u = v - \inf v + 1$ is a strictly positive harmonic function on $\mathbb{R}^N$ and (2.2) yields $u(\xi) = u(0)$ when $\delta \to +\infty$. (See Section 13 for the extension of this result, when $N = 2$, to lower-bounded superharmonic functions.) In particular, it follows that a bounded analytic function on the plane is identically constant (Liouville's theorem).

Application to Local Properties of Families of Harmonic Functions

Harnack's inequality implies that a family $u_{\bullet}$ of positive harmonic functions on a connected open set D is locally uniformly bounded if bounded at a single point. Moreover, an application of (2.3) with α near 0 shows that the family $u_{\bullet} + 1$ and therefore also the family $u_{\bullet}$ is equicontinuous at each point and thus uniformly continuous on each compact subset of D. Trivially, more generally, a locally uniformly bounded family Γ of harmonic functions on an open set D is uniformly continuous on each compact subset of D. This result can be strengthened as follows. Let Γ' be the family of partial derivatives of some specified order of the members of Γ. These derivatives are harmonic functions. Let B be a ball with closure in D. The representation $u = \mathrm{PI}(B, u)$ can be differentiated to show that Γ' is uniformly bounded in each smaller concentric ball, and therefore Γ' is locally uniformly bounded. This same reasoning shows that Γ is locally uniformly continuous, as already derived using Harnack's inequality.

3. Convergence of Directed Sets of Harmonic Functions

The following results are stated for directed sets (nets) of functions rather than sequences because in potential theory convergence problems commonly arise in the context of directed sets. In this section D is an open connected subset of $\mathbb{R}^N$ and $u_{\bullet}$ is a directed set of harmonic functions on D.

(a) **Harnack's Convergence Theorem.** *If $u_{\bullet}$ is directed upward, with limit u, there is locally uniform convergence on D to u, and u is either identically $+\infty$ or a harmonic function.*

To prove the theorem, observe that if A is a compact subset of D and if $\xi_0 \in A$, then by Harnack's inequality (2.1) if $\beta \geq \alpha$ so that $u_\beta \geq u_\alpha$, it follows that

$$c(A,D)^{-1}\sup_{\xi\in A}[u_\beta(\xi)-u_\alpha(\xi)]\le u_\beta(\xi_0)-u_\alpha(\xi_0)$$
$$\le c(A,D)\inf_{\xi\in A}[u_\beta(\xi)-u_\alpha(\xi)]. \tag{3.1}$$

If u is finite at a point, choose ξ_0 as that point. Then (3.1) implies that u is finite valued on D and that $u_.$ converges uniformly on A, so converges locally uniformly on D. The limit function u is harmonic because it is continuous and has the harmonic function average property. On the other hand, if u is infinite valued at some point, choose ξ_0 as that point, and then (3.1) implies that $u_.$ converges uniformly on A, so locally uniformly, to $+\infty$.

Observation. Whether u is identically $+\infty$ or harmonic in this convergence theorem, some increasing sequence in $u_.$ also has limit u according to Theorem 2 of Appendix VIII.

(b) If $u_.$ is locally bounded and converges to a function u, then the convergence is locally uniform, the limit function u is harmonic, and if $\mathbf{D}$ is any partial derivation operator, $\lim_\alpha \mathbf{D}u_\alpha = \mathbf{D}u$ locally uniformly.

The fact that the convergence is locally uniform follows from the local equicontinuity of $u_.$ (Section 2), and the last assertion follows on applying the operator $\mathbf{D}$ to the limit equation $\lim_\alpha \mathrm{PI}(B,u_\alpha)=\mathrm{PI}(B,u)$ with B a ball with closure in D. If $u_.$ is a locally bounded sequence of harmonic functions it is locally equicontinuous; so (by Ascoli's theorem) a subsequence converges locally uniformly.

4. Harmonic, Subharmonic, and Superharmonic Functions

A function u from an open subset D of $\mathbb{R}^N$ into $]-\infty,+\infty]$ is called *superharmonic* if

(a) u is lower semicontinuous.
(b) u is not identically $+\infty$ in any open connected component of D.
(c) $u(\xi)\ge L(u,\xi,\delta)$ if $\bar{B}(\xi,\delta)\subset D$.

Since u is necessarily locally lower bounded, the integral involved in (c) is well defined and not equal to $-\infty$. Applying (c) and I(2.2) we find that

(c′) $u(\xi)\ge A(u,\xi,\delta)>-\infty$ if $\bar{B}(\xi,\delta)\subset D$.

Hence finiteness of u at a point ξ implies finiteness l_N almost everywhere in $B(\xi,\delta)$ when $\bar{B}(\xi,\delta)\subset D$, and therefore a covering argument shows that u is finite l_N almost everywhere on D and that u is locally l_N integrable. In particular $A(u,\xi,\delta)$ is finite in (c′) even when $u(\xi)=+\infty$. We shall show in Section 6(a) that $L(u,\xi,\delta)$ in (c) is finite even when $u(\xi)=+\infty$, and we shall show in Section 6(c) that (c′) can replace (c) in the definition of superharmonicity.

If u_1 and u_2 are superharmonic functions on D and if c_1 and c_2 are positive constants, the functions c_1u_1, $c_1u_1+c_2u_2$, and $u_1\wedge u_2$ are superharmonic on D. Since superharmonic functions are lower semicontinuous, the limit of an upward-directed family of superharmonic functions on D is the limit of an increasing sequence of functions in the family (Theorem 2 of Appendix VIII). The limit function u is superharmonic if condition (b) is satisfied.

A *subharmonic* function is defined as the negative of a superharmonic function. A function is harmonic if and only if it is both subharmonic and superharmonic. It is unfortunate that it is natural in pure potential theory to consider superharmonic rather than subharmonic functions but that the applications of potential theory are likely to involve subharmonic functions rather than superharmonic functions.

Positive Integral Operations on Superharmonic Functions

Let D be a nonempty open subset of $\mathbb{R}^N$, and let $\mathscr{B}(D)$ be the class of Borel subsets of D. Let $\{u_\beta, \beta \in I\}$ be a family of superharmonic functions on D, indexed by a set I. If $(I, \mathscr{F}, \lambda)$ is a finite measure space and if the function $(\xi, \beta) \mapsto u_\beta(\xi)$ is measurable from $(D \times I, \mathscr{B}(D) \times \mathscr{F})$ into $(\mathbb{R}, \mathscr{B}(\bar{\mathbb{R}}))$, then the function $u' = \int_I u_\beta \lambda(d\beta)$ satisfies the superharmonic function average inequality,

$$L(u', \xi, \delta) = L\left(\int_I u_\beta \lambda(d\beta), \xi, \delta\right) = \int_I L(u_\beta, \xi, \delta)\lambda(d\beta) \leq u'(\xi) \qquad (|\xi - \partial D| > \delta),$$

if the double integral involved converges absolutely. Thus u' is superharmonic if the superharmonic function finiteness and lower semicontinuity conditions are satisfied.

Examples. Let δ be a strictly positive number so small that

$$D_\delta = \{\xi \in D : |\xi - \partial D| > \delta\}$$

is not empty. Then if u is a superharmonic function on D the function $u' = L(u, \cdot, \delta)$ on D_δ is a special case of the preceding integral operation, with $u_\beta(\xi) = u(\xi + \beta)$, $\beta \in \partial B(0, \delta)$ and with λ the normalized surface area of $B(0, \delta)$. In this case an application of Fatou's lemma shows that u' is lower semicontinuous. We shall show in Section 6(a) that $L(u, \cdot, \delta)$ is finite valued, and it follows that $L(u, \cdot, \delta)$ is superharmonic on D_δ. We leave to the reader the verification of the fact that if u is a superharmonic function on D, the functions $A(u, \cdot, \delta)$ and $A_\delta u$ are also special cases of the above integral operation. Since (from Section 2) $A(u, \cdot, \delta)$ is finite valued and continuous and $A_\delta u$ is infinitely differentiable, these two functions are superharmonic on D_δ.

For u superharmonic on D the three functions $L(u, \cdot, \delta)$, $A(u, \cdot, \delta)$, and $A_\delta u$ are all majorized by u, and it will be shown in Section 6(f) that these functions increase monotonely to u when $\delta \to 0$, in the sense that when $\delta_0 > 0$ and $\delta < \delta_0$, there is monotone convergence on D_{δ_0} to u when $\delta \to 0$.

5. Minimum Theorem for Superharmonic Functions

Theorem. *Let u be superharmonic on an open subset D of $\mathbb{R}^N$.*

(a) *If D is connected and if u attains its infimum at a point ξ, then u is identically constant.*

(b) *The infimum of u is the limit of u along a sequence of points approaching ∂D.*

(c) *If u has a lower semicontinuous extension to $D \cup \partial D$, the infimum of the extension is attained on ∂D.*

A typical implication of this theorem is the fact that a superharmonic function on D with a positive inferior limit at every boundary point must be positive. This theorem is the generalization for superharmonic functions of the harmonic function maximum minimum theorem (see Section I.4). The proof of Theorem 5(a) is precisely the same as that given of Theorem I.4(a) involving the infimum of u because that proof required only lower semicontinuity of u and the superharmonic function average inequality. Parts (b) and (c) follow easily from (a).

6. Application of the Operation τ_B

This operation, defined in Section 1, will be used to derive various important properties of superharmonic functions.

(a) *If u is superharmonic on D and if B is a ball with closure in D, then u is l_{N-1} integrable on ∂B and $\tau_B u \le u$.* In fact, since u is lower semicontinuous, it is bounded below on ∂B, and there is an increasing sequence $f_{\cdot}$ of finite-valued continuous functions on ∂B with limit $u|_{\partial B}$. Then $\mathrm{PI}(B, f_{\cdot})$ is an increasing sequence of harmonic functions on B with limit $\mathrm{PI}(B, u)$, and an application of the superharmonic function minimum theorem to $u - \mathrm{PI}(B, f_n)$ on B shows that the difference is positive there. Hence $(n \to \infty)$ u is l_{N-1} integrable on ∂B and $\tau_B u \le u$.

(b) *In* (a) *if $B = B(\xi, \delta)$ and if $u(\xi) = L(u, \xi, \delta)$, then u is harmonic on B.* In fact, under the stated conditions, $\tau_B u(\xi) = u(\xi)$ so that the restriction to B of $u - \tau_B u$ is positive superharmonic, vanishes at ξ, and therefore vanishes identically.

(c) *In the definition of superharmonicity in Section* 4, (c) *can be replaced by* (c′). We have already seen that superharmonic functions satisfy (c′). Conversely, if u satisfies Section 4(a), (b), (c′) the reasoning leading to the superharmonic function minimum theorem and thereby to (a) above remains valid, and then if $B = B(\xi, \delta)$, the inequality $\tau_B u(\xi) \le u(\xi)$ is condition Section 4(c).

(d) *In the definition of a superharmonic function in Section* 4 *condition* (c) [*or equivalently* (c′)] *there need be supposed true only for sufficiently small* δ, *depending on* ξ, because the reasoning leading to (a) of the present section used only this weakened condition, and (a) implies Section 4(c) when $B = B(\xi, \delta)$.

(e) *In* (a) *the function* $\tau_B u$ *is superharmonic on* D because according to Section 1 this function is lower semicontinuous, and it is trivial that for ξ in D the inequality $\tau_B u(\xi) \ge L(\tau_B u, \xi, \delta)$ is valid for sufficiently small δ. More generally, if B is a ball with closure in D, if u is superharmonic on D, and if v is a function defined and superharmonic on B, with $v \ge \tau_B u$ on B, then the function v' (equal on B to $v \wedge u$ and on $D - B$ to u) lies between u and $\tau_B u$ from which it follows easily that v' is superharmonic on D.

(f) If u is superharmonic on D and if $\xi \in D$, the functions $\delta \mapsto L(u, \xi, \delta)$, $\delta \mapsto A(u, \xi, \delta)$, and $\delta \mapsto A_\delta u(\xi)$ are monotone decreasing for $0 < \delta < |\xi - \partial D|$, with limit $u(\xi)$ when $\delta \to 0$. To prove the monotoneity of $L(u, \xi, \cdot)$ observe that if $B_i = B(\xi, \delta_i)$ with $\delta_1 < \delta_2 < |\xi - \partial D|$, then the inequality $\tau_{B_2}\tau_{B_1} u \le \tau_{B_1} u$ reduces at ξ to $L(u, \xi, \delta_2) \le L(u, \xi, \delta_1)$. Next apply the lower semicontinuity of u and Fatou's lemma to derive the inequality

$$u(\xi) \le \lim_{\delta \to 0} L(u, \xi, \delta) \le u(\xi)$$

and thereby complete the proof of (f) for $L(u, \xi, \cdot)$. The corresponding results for $A(u, \xi, \cdot)$ and $A_{.}u(\xi)$ follow from I(2.2) and I(2.4). The latter results imply that the relation $u \ge v$ or $u = v$, if satisifed l_N almost everywhere on their domain of definition D by superharmonic functions u and v, is satisfied everywhere on D.

The preceding results imply the truth of a slight strengthening of the lower semicontinuity property of a superharmonic function u, namely,

$$u(\xi) = \liminf_{\eta \to \xi} u(\eta) = \liminf_{\substack{\eta \to \xi \\ \eta \notin A}} u(\eta), \tag{6.1}$$

where A is an arbitrary l_N null set. We shall see in Section XI.1 that the natural class of sets A making (6.1) true is the class of sets thin at ξ in the sense of the fine topology.

Observation. If $\delta > 0$ and if ϕ_δ is a function on the interval $[0, \delta]$, Borel measurable and l_1 almost everywhere strictly positive, with $\int_0^\delta \phi_\delta(r)\,dr = 1$, then every superharmonic function u on an open set D obviously satisfies the inequality

$$u(\xi) \ge \int_0^\delta \phi_\delta(r) L(u, \xi, r)\,dr \tag{6.2}$$

when $\bar{B}(\xi, \delta) \subset D$. Now suppose that $\delta_0 > 0$ and that ϕ_δ is defined and satisfies the conditions imposed on ϕ_δ above, for $0 < \delta < \delta_0$. Then if u is a function on the open set D, satisfying the superharmonic function defining conditions (a) and (b) of Section 4, together with (6.2) for sufficiently small δ, depending on ξ, the function is superharmonic. This assertion has already been proved for $\phi_\delta(r) = N\delta^{-N}r^{N-1}$, in which case the right-hand side of (6.2) reduces to $A(u, \xi, \delta)$, and the proof needs no change in the general case.

EXAMPLE (The Fundamental Kernel and the Green Function of a Ball). If the fundamental kernel G is defined on $\mathbb{R}^N \times \mathbb{R}^N$, by $I(5.2)$ the function $G(\xi, \cdot)$ is for each point ξ harmonic on $\mathbb{R}^N - \{\xi\}$ and is superharmonic on $\mathbb{R}^N$. In fact we have noted in Section I.5 that this function is harmonic on $\mathbb{R}^N - \{\xi\}$. Moreover this function is continuous on $\mathbb{R}^N$, and in view of the local nature of superharmonicity proved in (d) above we need only observe, to prove that the function is superharmonic on $\mathbb{R}^N$, that the superharmonic function average inequality is trivially satisfied at ξ. Similarly, if B is a ball, the Green function G_B is defined on $B \times B$ by (1.1), and for each point ξ of B, $G_B(\xi, \cdot)$ is harmonic on $B - \{\xi\}$ and superharmonic on D.

7. Characterization of Superharmonic Functions in Terms of Harmonic Functions

It is important to characterize superharmonicity intrinsically, without the use of balls. Let u be a lower semicontinuous function from the open subset D of $\mathbb{R}^N$ into $]-\infty, +\infty]$, not identically $+\infty$ on any open connected component of D. Consider the following property of u: if D_0 is a relatively compact open subset of D and if v is a function defined and harmonic on a neighborhood of $\bar{D}_0$, with $u - v \geq 0$ on ∂D_0, then $u - v \geq 0$ on D_0. This condition is necessary and sufficient for u to be superharmonic and justifies the name "superharmonic." In fact a superharmonic function on D has this property according to the superharmonic function minimum theorem. Conversely, if u has this property, let $B = B(\xi, \delta)$ be a ball with closure in D. It is enough to prove that $u(\xi) \geq L(u, \xi, \delta)$ and even, since u is lower semicontinuous, to prove that $u(\xi) \geq L(f, \xi, \delta)$ for every finite-valued continuous function f defined on ∂B and majorized by u. For such a function f the function $v = \mathrm{PI}(B,f)$ is harmonic on B with a continuous extension to $\bar{B}$ obtained by setting $v = f$ on ∂B. The difference $u - v$ is lower semicontinuous on $\bar{B}$ and positive on ∂B, so that if $\varepsilon > 0$, then $u - v \geq -\varepsilon$ near ∂B (for example, on ∂D_0, for D_0 a slightly smaller concentric ball). The given condition implies that $u - v \geq -\varepsilon$ on D_0 and therefore $u - v \geq -\varepsilon$ on B, in particular at ξ; that is, $u(\xi) \geq L(f, \xi, \delta) - \varepsilon$. Since ε can be arbitrarily small, the desired inequality is true.

Whenever harmonic functions can be defined, for example, on a Riemann surface by Laplace's equation, superharmonic functions can be defined using the intrinsic condition of this section.

8. Differentiable Superharmonic Functions

Theorem. *If D is an open subset of $\mathbb{R}^N$ and if $u \in \mathbb{C}^{(2)}(D)$, then u is superharmonic if and only if $\Delta u \leq 0$.*

Since linear functions and products of two different coordinate functions are harmonic, the Taylor expansion of u about a point ξ of D yields

$$L(u, \xi, \delta) = u(\xi) + \frac{\delta^2}{2N}\Delta u(\xi) + 0(\delta^2), \tag{8.1}$$

so that $\Delta u \leq 0$ when u is superharmonic. Conversely if $\Delta u \leq 0$ define $v(\xi) = |\xi|^2$. Then $\Delta(u - \varepsilon v) < 0$ when $\varepsilon > 0$, so that by (8.1) with $u - \varepsilon v$ instead of u,

$$L(u - \varepsilon v, \xi, \delta) < u(\xi) - \varepsilon v(\xi), \tag{8.2}$$

for sufficiently small δ. Then $u - \varepsilon v$ is superharmonic; so (8.2) is valid whenever $|\xi - \partial D| > \delta$, and (8.2) becomes the superharmonic function inequality for u when $\varepsilon \to 0$.

Approximation of a Superharmonic Function by Differentiable Superharmonic Functions

If u is a superharmonic function on the open subset D of $\mathbb{R}^N$, the function $A_\delta u$ (see Section I.2) is defined and infinitely differentiable on the set $\{\xi \in D: |\xi - \partial D| > \delta\}$. According to Section 4 this function is superharmonic and majorized by u, and according to Section 6(f), for each point ξ of D the function $\delta \mapsto A_\delta u(\xi)$ is monotone decreasing for $\delta < |\xi - \partial D|$, and $\lim_{\delta \to 0} A_\delta u = u$. This approximation result will be improved in Section IV.10.

9. Application of Jensen's Inequality

This inequality implies that if ϕ is a convex function

$$\phi[L(u, \xi, \delta)] \leq L(\phi(u), \xi, \delta), \tag{9.1}$$

whenever u is Lebesgue measurable on $\partial B(\xi, \delta)$ and all integrals involved exist. This inequality has the following consequences:

(a) If u is harmonic and ϕ is convex, $\phi(u)$ is subharmonic.

(b) If u is subharmonic and ϕ is convex and monotone increasing, $\phi(u)$ is subharmonic.

Thus if u is harmonic or is positive and subharmonic, $|u|^p$ is subharmonic when $p \geq 1$. If u is positive and superharmonic, u^p is superharmonic when $0 < p \leq 1$.

10. Superharmonic Functions on an Annulus

Theorem. *Suppose that $0 \leq a < b$, let $D = \{\xi\colon a < |\xi| < b\}$ be an annulus in $\mathbb{R}^N$, and let u be a function from D into $\overline{\mathbb{R}}$.*

(a) *If u has the form $u(\xi) = f(|\xi|)$, then u is superharmonic if and only if u is a concave function of $G(\xi, 0)$ (that is, a concave function of $-\log|\xi|$ if $N = 2$ and of $|\xi|^{2-N}$ if $N > 2$). In particular, u is harmonic if and only if u is a linear function of $G(\xi, 0)$.*

(b) *If u is a superharmonic [harmonic] function on D, the function $\xi \mapsto L(u, 0, |\xi|)$ is also superharmonic [harmonic] on D, and therefore the function $\delta \mapsto L(u, 0, \delta)$ on $]a, b[$ is a concave [linear] function of $-\log \delta$ if $N = 2$ and of δ^{2-N} if $N > 2$.*

Observe that u is a concave function of $-\log|\xi|$ if and only if u is a concave function of $\log|\xi|$.

Proof of (a). Since we can replace u by $A_\alpha u$ and then let $\alpha \to 0$, we can suppose that f in (a) is infinitely differentiable. The following evaluation of Δu makes (a) trivial.

$$\Delta u(\xi) = f''(|\xi|) + \frac{N-1}{|\xi|} f'(|\xi|),$$

$$= \begin{cases} |\xi|^{-2} \dfrac{d^2 f(|\xi|)}{ds^2} & \text{if } N = 2 \text{ and } s = -\log|\xi|, \\[2ex] \dfrac{(N-2)^2}{|\xi|^{2N-2}} \dfrac{d^2 f(|\xi|)}{ds^2} & \text{if } N > 2 \text{ and } s = |\xi|^{2-N}. \end{cases} \tag{10.1}$$

□

Proof of (b). We prove (b) for u superharmonic; the proof for u harmonic is easier and is left to the reader. Since u is locally lower bounded we can assume in the proof of (b), at the expense of increasing a and decreasing b, that u is lower bounded on D. In addition suppose first that u is bounded on D. Since a space rotation around the origin preserves superharmonicity, integrating over all rotations (see the remark in Section 4 on positive integral operations on superharmonic functions) yields the fact that the function $\xi \mapsto L(u, 0, |\xi|)$ is superharmonic on D, as asserted in (b). If u is not bounded and $n \in \mathbb{Z}^+$, the function $u \wedge n$ is a bounded superharmonic function on D; so the function $\xi \mapsto L(u \wedge n, 0, |\xi|)$ is superharmonic on D and when $n \to \infty$, we find that the function $\xi \mapsto L(u, 0, |\xi|)$ is either superharmonic or identically $+\infty$. The latter case is excluded because u is locally l_N integrable so (by Fubini's theorem) $L(u, 0, \delta) < +\infty$ for l_1 almost every δ in $]a, b[$. □

The Minimum Function of a Superharmonic Function

Let D be as in Theorem 10, let u be a superharmonic function on D, and define

$$m(\delta) = \min\{u(\xi): |\xi| = \delta\}.$$

We now show that the function $\xi \mapsto m(|\xi|)$ is superharmonic on D; that is, the function $\delta \mapsto m(\delta)$ on $]a, b[$ is a concave function of $-\log\delta$ if $N = 2$ and of δ^{2-N} if $N > 2$. First observe that the function $\xi \mapsto m(|\xi|)$ is lower semicontinuous on D because u is. Next observe that if γ is an arbitrary rotation of D about the origin, the rotated function $\xi \mapsto u(\gamma\xi)$ is superharmonic on D, and the infimum of the class of all these rotated functions is the function $\xi \mapsto m(|\xi|)$. Since an elementary argument shows that the infimum of any locally lower bounded family of superharmonic functions on an open set is superharmonic if this infimum is lower semicontinuous, we conclude that the function $\xi \mapsto m(|\xi|)$ is superharmonic, as asserted.

11. Examples

(a) Suppose that $N = 2$ and let f be a not identically vanishing analytic function on the connected open set D. The real and imaginary parts of f are harmonic because they satisfy Laplace's equation, alternatively because the Cauchy integral formula applied to a ball yields the harmonic function average property for f. Taking absolute values in this average relation we conclude that $|f|$ is subharmonic and therefore (Section 9) that $|f|^p$ is subharmonic when $p \geq 1$. Actually $|f|^p$ is subharmonic when $p > 0$ because $|f|^p = |f^p|$ is the absolute value of an analytic function in a neighborhood of a nonzero of f, and it is trivial that the subharmonic function average property is satisfied at a zero of f. Similar reasoning shows that $\log|f| = \operatorname{Re}(\log f)$ is subharmonic when defined as $-\infty$ at a zero of f and is harmonic on the nonzero set of f. Since $|f|^p$ for $p > 0$ is a monotone increasing convex function of $\log|f|$, we have again that $|f|^p$ is subharmonic.

(b) If $N = 2$ and if f is analytic on $B(0, \delta)$ and does not vanish identically, and if $p > 0$, the function $|f|^p$ is subharmonic so

$$r \mapsto L(|f|^p, 0, r) = \frac{1}{2\pi}\int_0^{2\pi} |f(re^{i\theta})|^p \, d\theta \qquad (r < \delta) \tag{11.1}$$

is an increasing function which is a convex function of $\log r$. Define $M(r) = \max_{|z|=r}|f(z)|$ and let $\zeta_{.}$ be a sequence dense on the unit circle $\{|z| = 1\}$. Then if $u_n(z) = \log|f(\zeta_n z)|$ define $v_n = u_0 \vee \cdots \vee u_n$ to get an increasing sequence $v_{.}$ of subharmonic functions with limit the function $z \mapsto v(z) =$

$\log M(|z|)$. Since $r \mapsto L(v_n, 0, r)$ is a convex function of $\log r$, the same is true of the function $r \mapsto \log M(r)$. See a slightly different derivation of this same property of $M(\cdot)$ in the context of the minimum function of a superharmonic function in Section 10. Alternatively, the continuous function v is subharmonic because it has the subharmonic function average property and therefore the function

$$r \mapsto \log M(r) = L(v, 0, r)$$

is a convex function of $\log r$. This property of $M(\cdot)$ is known as the *Hadamard three circle theorem.*

12. The Kelvin Transformation ($N \geq 2$)

Inversion in a sphere $\partial B(\xi_0, \delta)$ is the transformation of the one-point compactification $\mathbb{R}^N \cup \{\infty\}$ onto itself which takes the point ξ into the point ξ' on the same ray through ξ_0, with $|\xi - \xi_0|\,|\xi' - \xi_0| = \delta^2$, under the convention that the points ξ_0 and ∞ are interchanged. This transformation is its own inverse. From now on to simplify the notation we take $\xi_0 = 0$. Let D be an open subset of $\mathbb{R}^N$, and let D' be the set of finite points of the image of D under inversion in $\partial B(0, \delta)$. If u is a function on D define u' on D' by $u'(\xi') = u(\xi)$. Then if $u \in \mathbb{C}^{(2)}(D)$ and if Δ' is the Laplacian in the variable ξ',

$$\Delta u = \frac{|\xi'|^{N+2}}{\delta^4} \Delta' \left(\frac{u'(\xi')}{|\xi'|^{N-2}} \right).$$

Thus the function

$$\xi' \mapsto v(\xi') = u'(\xi') \left(\frac{\delta}{|\xi'|} \right)^{N-2}$$

is harmonic on D' when u is harmonic on D. The characterization (Section 7) of superharmonicity in terms of harmonicity shows that v is superharmonic on D' if u is superharmonic on D. The function transformation from u into v is called the *Kelvin transformation* (relative to the inversion sphere). The Kelvin transformation is its own inverse.

If $N = 2$, if ϕ is an analytic not identically constant function on a connected open set D, and if $u \in \mathbb{C}^{(2)}(D)$, then $\Delta u(\phi) = |\phi'|^2 (\Delta u)(\phi)$. Thus $u(\phi)$ is harmonic on D if u is harmonic on $\phi(D)$, and as in the preceding discussion superharmonicity is also preserved. If ϕ is replaced by $\bar{\phi}$ the same argument is applicable, and this case includes inversion in a circle, already discussed.

13. Greenian Sets

It is useful to have a special name, "Greenian set," for an open subset of $\mathbb{R}^N$ which supports a positive nonconstant superharmonic function. If $N > 2$, every nonempty open subset of $\mathbb{R}^N$ is Greenian because $G(0, \cdot)$ is positive and superharmonic on $\mathbb{R}^N$. If $N = 2$ and D is not dense in $\mathbb{R}^2$, the set D is Greenian because if ξ is an inner point of $\mathbb{R}^2 - D$, the function $c - G(\xi, \cdot)$ is positive and harmonic on D for large c. In particular, if D is any open disconnected subset of $\mathbb{R}^2$, each open connected component of D is Greenian, as is D itself. On the other hand, $\mathbb{R}^2$ is not Greenian. To prove that a positive superharmonic function u on $\mathbb{R}^2$ must be identically constant, we can replace u by $A_\alpha u$ and thereby suppose that u is infinitely differentiable. Then if $D = B(0, \delta)$ in I(8.2) we find, with the help of I(1.3),

$$u(0) = L(u, 0, \delta) - \frac{1}{2\pi} \int_D \log \frac{\delta}{|\eta|} \Delta u(\eta) l_2(d\eta), \tag{13.1}$$

and $\Delta u \leq 0$. When $\delta \to +\infty$, this equation becomes impossible unless $\Delta u \equiv 0$; so u must be harmonic and, according to Section 2, must therefore be identically constant. [Alternatively, the fact that u must be harmonic follows from the fact that (Section 10) the function $\delta \mapsto L(u, \xi, \delta)$ is a positive concave decreasing function of $\log \delta$ for $0 < \delta < +\infty$ and so must be identically constant.]

Application. Liouville's classical theorem that a bounded holomorphic function f on the complex plane is identically constant is a special case of the fact that the plane is not Greenian. We need only observe that the real and imaginary parts of the function f are bounded harmonic functions and therefore can be made positive by addition of suitable constants.

Observe that the trace on $\mathbb{R}^N$ of the image of a Greenian subset of $\mathbb{R}^N$ under inversion in a sphere is Greenian.

14. The $\mathbf{L}^1(\mu_{B-})$ and $\mathbf{D}(\mu_{B-})$ Classes of Harmonic Functions on a Ball B; The Riesz–Herglotz Theorem

In a classification to be given in Chapter IX the harmonic functions characterized in part (a) of the following theorem will be the harmonic members of a class of functions denoted by $\mathbf{L}^1(\mu_{B-})$ and the harmonic functions characterized in part (b) will be the harmonic members of a class of functions denoted by $\mathbf{D}(\mu_{B-})$. The theorem will be a model that suggests corresponding characterizations in the context of relative harmonic and parabolic functions defined on wide classes of open sets and in the context of martingales. (See Chapter I of Part 3.)

Theorem. *Let u be a harmonic function on $B = B(0, \delta)$.*

(a) $\mathbf{L}^1(\mu_{B-})$ *harmonic functions. The following conditions on u are equivalent.*

(a1) $u = \mathrm{PI}(B, M_u)$ *for some signed measure M_u on ∂B.*

(a2) *u is the difference between two positive harmonic functions.*

(a3) *$|u|$ has a harmonic magorant.*

(a4) $\sup_{r<\delta} L(|u|, 0, r) < +\infty$.

Furthermore the map $u \mapsto M_u$ is a one-to-one, linear, order-preserving map from the class of $\mathbf{L}^1(\mu_{B-})$ harmonic functions onto the class of signed measures on ∂B, with

$$\lim_{r\to\delta} \int_{\partial B} f(\eta) u\left(\frac{r\eta}{\delta}\right) \frac{l_{N-1}(d\eta)}{\pi_N \delta^{N-1}} = \int_{\partial B} f\, dM_u \tag{14.1}$$

for every continuous function f on ∂B.

(b) $\mathbf{D}(\mu_{B-})$ *harmonic functions. The following more restrictive conditions on u are equivalent.*

(b1) $u = \mathrm{PI}(B, f_u)$ *for some l_{N-1} measurable and integrable function f_u on ∂B.*

(b2) *There is a uniform integrability test function (Appendix V) Φ for which $\Phi(|u|)$ has a harmonic majorant.*

(b3) *The family $\{u_{|\partial B(0,r)}, l_{N-1}/(\pi_N r^{N-1}), 0 < r < \delta\}$ of paired functions and measures is uniformly integrable.*

(b4) *(If $u > 0$) u is the limit of an increasing sequence of bounded positive harmonic functions.*

Furthermore the map $u \mapsto f_u$ is a one-to-one, linear, order-preserving map from the class of $\mathbf{D}(\mu_{B-})$ harmonic functions on B onto the class $L^1(\partial B, l_{N-1})$, and $dM_u = f_u\, dl_{N-1}/(\pi_N \delta^{N-1})$.

Observe that $|u|$ in (a3) and $\Phi(|u|)$ in (b2) are subharmonic functions; so $L(|u|, 0, \cdot)$ and $L(\Phi(|u|), 0, \cdot)$ are monotone increasing functions on $]0, \delta[$. In (a) the signed measure M_u is the zero measure if and only if $u = 0$, and $M_u \geq 0$ if and only if $u \geq 0$. In (b) the function f_u vanishes l_{N-1} almost everywhere if and only if $u = 0$, and $f_u \geq 0$ l_{N-1} almost everywhere if and only if $u \geq 0$. The function f_u will be identified in Section 15 as the nontangential boundary limit function of u.

Nomenclature. The fact that a positive harmonic function on a ball has a *Poisson–Stieltjes* representation (a1) with $M_u \geq 0$ is usually called the *Herglotz theorem*, although Herglotz himself referred to Riesz. To avoid confusion this theorem will be called the *Riesz–Herglotz theorem* in this book.

(a1) ⇒ (a2) If $M_u = \mu_1 - \mu_2$ is a Jordan decomposition of M_u into the difference between two (positive) measures, the equation

$$u = \mathrm{PI}(B, \mu_1) - \mathrm{PI}(B, \mu_2)$$

expresses u as the difference between two positive harmonic functions.

(a2) $\Rightarrow$ (a3) If $u = u_1 - u_2$ with u_i positive and harmonic, the function $|u|$ has the harmonic majorant $u_1 + u_2$.

(a3) $\Rightarrow$ (a4) If $|u|$ has the harmonic majorant v and if $0 < r < \delta$, then

$$L(|u|, 0, r) \leq L(v, 0, r) = v(0).$$

(a4) $\Rightarrow$ (a1) If $0 < r < \delta$, the function $\xi \mapsto u(r\xi/\delta)$ is harmonic on a neighborhood of $\bar{B}$ and so is given on B by a Poisson integral,

$$u = \mathrm{PI}(B, \mu_r), \qquad \mu_r(d\eta) = u\left(\frac{r\eta}{\delta}\right) \frac{l_{N-1}(d\eta)}{\pi_N \delta^{N-1}}. \tag{14.2}$$

The total variation of μ_r is

$$\int_{\partial B} \left| u\left(\frac{r\eta}{\delta}\right) \right| \frac{l_{N-1}(d\eta)}{\pi_N \delta^{N-1}} = L(|u|, 0, r),$$

and therefore, in view of the hypothesis of (a4), there is a strictly monotone increasing sequence $r_{\bullet}$ with limit δ for which the sequence $\mu_{r_{\bullet}}$ of signed measures has a vague limit; denote this limit by M_u. When $r \to \delta$ along this sequence in (14.2), the Poisson representation becomes $u = \mathrm{PI}(B, M_u)$. To show that this representation of u on B determines M_u uniquely, we show that any representation $u = \mathrm{PI}(B, \mu)$ with μ a signed measure on ∂B implies that vague $\lim_{r \to \delta} \mu_r = \mu$ when μ_r is defined by (14.2). That is, we show (14.1):

$$\lim_{r \to \delta} \int_{\partial B} f \, d\mu_r = \int_{\partial B} f \, d\mu$$

whenever f is continuous on ∂B. Using the fact that $|\eta - r\xi/\delta| = |\xi - r\eta/\delta|$ for ξ and η on ∂B, an interchange of orders of integration yields

$$\int_{\partial B} f \, d\mu_r = \int_{\partial B} \mathrm{PI}(B, f)\left(\frac{r\eta}{\delta}\right) \mu(d\eta). \tag{14.3}$$

By Theorem 1 the right-hand side has limit $\int_{\partial B} f \, d\mu$ when $r \to \delta$, as was to be proved.

The relation between $\mathbf{L}^1(\mu_{B-})$ harmonic functions on B and signed measures on ∂B is obviously linear and positivity preserving. If the space of signed measures on ∂B is ordered by setting $\mu_1 \leq \mu_2$ when the difference $\mu_1 - \mu_2$ is positive, the relation $u \Leftrightarrow M_u$ is order preserving. Since the space of signed measures on ∂B is a conditionally complete vector lattice (Appendix IV.7), the space $\mathbf{L}^1(\mu_{B-})$ of harmonic functions is also a conditionally complete vector lattice (in the order determined by pointwise inequality). The vector lattice isomorphism deduced here has an exact analog for harmonic functions on Greenian sets provided with their Martin boundaries (Section XII.9).

(b1) $\Rightarrow$ (b2) Choose a uniform integrability test function Φ for which $\int_{\partial B}\Phi(|f_u|)\,dl_{N-1} < +\infty$. Then by Jensen's inequality

$$\Phi(|u|) = \Phi[|\mathrm{PI}(B,f_u)|] \le \mathrm{PI}(B, \Phi(|f_u|)). \tag{14.4}$$

The last term is a harmonic majorant of $\Phi(|u|)$.

(b2) $\Rightarrow$ (b3) Let v be a harmonic majorant of $\Phi(|u|)$. Then

$$L(\Phi(|u|), 0, r) \le L(v, 0, r) = v(0);$$

so (b3) is true.

(b3) $\Rightarrow$ (b1) Since (b3) is stronger than (a4), $u = \mathrm{PI}(B, M_u)$, and we have proved above that the signed measure μ_r tends to M_u (vague convergence) when $r \to \delta$. Now the uniform integrability described in (b3) is equivalent to uniform integrability of the family $\{\eta \mapsto u(r\eta/\delta), \frac{1}{2} < r < \delta\}$ of functions on ∂B relative to the measure l_{N-1}. Hence (Appendix V) there is a strictly monotone increasing sequence $r_{\bullet}$ with limit δ for which the sequence

$$\eta \mapsto u(r_{\bullet}\eta/\delta)$$

has a weak limit function f on ∂B in the sense that

$$\lim_{n\to\infty} \int_{\partial B} u\left(\frac{r_n\eta}{\delta}\right) g(\eta) l_{N-1}(d\eta) = \int_{\partial B} f(\eta) g(\eta) l_{N-1}(d\eta) \tag{14.5}$$

for every bounded l_{N-1} measurable function g on ∂B. This limit relation implies (take g continuous) that the sequence $\mu_{r_{\bullet}}$, already known to be vaguely convergent to M_u, is vaguely convergent to the signed measure $f\,dl_{N-1}/(\pi_N\delta^{N-1})$. That is, M_u is absolutely continuous relative to l_{N-1} and $dM_u = f\,dl_{N-1}/(\pi_N\delta^{N-1})$. Thus (b1) is true and f_u is uniquely determined up to l_{N-1} null sets because M_u is uniquely determined. The map $u \mapsto f_u$ has the stated properties because the map $u \mapsto M_u$ has the corresponding properties.

(b1) $\Rightarrow$ (b4) Under (b1), if $u > 0$, $u = \mathrm{PI}(B,f_u)$ with $f_u \ge 0$, according to what we have proved, and $u = \lim_{n\to\infty}\mathrm{PI}(B,f_u \wedge n)$ represents u as the limit of an increasing sequence of positive bounded harmonic functions.

(b4) $\Rightarrow$ (b1) If u_n is bounded positive and harmonic on B, condition (b2) is satisfied, so (b1) is satisfied. Moreover, if $u_{\bullet}$ is an increasing sequence of positive bounded harmonic functions on B with limit u, the sequence $f_{u_{\bullet}}$

is an increasing sequence (up to l_{N-1} null sets) with limit some function f, and the equation $u_n = \mathrm{PI}(B, f_{u_n})$ becomes in the limit $u = \mathrm{PI}(B, f)$.

Observation. If $u \in \mathbf{D}(\mu_{D-})$ in the theorem, we have proved that $\lim_{r \to \delta} \mu_r = M_u$ (vague convergence of signed measures on ∂B), where μ_r is the signed measure defined by (14.2). This convergence, or alternatively, a slight variation of the convergence proof, shows that if v_r is the signed measure on $\bar{B}$ supported by $\partial B(0, r)$ and defined by

$$v_r(d\eta) = u(\eta) \frac{l_{N-1}(d\eta)}{\pi_N r^{N-1}} \qquad (r = |\eta|),$$

then $\lim_{r \to \delta} v_r = M_u$ (vague convergence of signed measures on $\bar{B}$).

15. The Fatou Boundary Limit Theorem

Let $B = B(0, \delta)$ in $\mathbb{R}^N$. A "Stolz domain" in B with vertex ζ on ∂B is defined as the intersection of B with an open cone of revolution with vertex ζ, axis of rotation the ray from ζ through the origin, half-angle $< \pi/2$. A "deleted nontangential neighborhood" of ζ is defined as a subset A of B with the property that if S is a Stolz domain with vertex ζ, then $S \cap B(\zeta, \varepsilon) \subset A$ for sufficiently small ε, depending on S. The class of deleted nontangential neighborhoods of ζ is closed under finite intersections. A function w on B is said to have radial limit q at ζ if $\lim_{r \to 1} w(r\zeta) = q$. The function is said to have nontangential limit q at ζ if $\lim_{\xi \to \zeta} w(\xi) = q$ whenever the approach to ζ is in a Stolz domain with vertex ζ, that is, if w has limit q at ζ along the filter of deleted nontangential neighborhoods of ζ.

A signed measure μ on the Borel subsets of ∂B can be considered as a signed measure of Borel subsets of $\mathbb{R}^N$, supported by ∂B. In particular, $(N-1)$-dimensional "area" on ∂B when so extended to $\mathbb{R}^N$ will be denoted by l'_N. If μ and v are signed measures on $\mathbb{R}^N$ supported by ∂B, the symmetric variational derivate $d\mu/dv$ (see Appendix IV.12) can thus be defined as the symmetric derivate for measures on $\mathbb{R}^N$ or equivalently in the obvious way for measures on ∂B.

Theorem. *If $u \in \mathbf{L}^1(\mu_{B-})$ and if h is a strictly positive harmonic function on B then if $\zeta \in \partial B$,*

$$(\text{nontang}) \lim_{\xi \to \zeta} \frac{u(\xi)}{h(\xi)} = \frac{dM_u}{dM_h}(\zeta) \qquad (\textit{symmetric variational derivate}) \tag{15.1}$$

whenever both the indicated derivate and the symmetric derivate $dl'_N/dM_h(\zeta)$ exist.

According to this theorem the limit in (15.1) exists at M_h almost every boundary point ζ.

Observation. In the classical version of this theorem, as proved by Fatou, $h \equiv 1$, so that M_h is a normalized Lebesgue measure $l_{N-1}/(\pi_N \delta^{N-1})$ on ∂B. In particular the theorem implies that if f is an l_{N-1} measurable and integrable function on ∂B then the harmonic function $\mathrm{PI}(B,f)$ has nontangential boundary limit $f(\zeta)$ at l_{N-1} almost every point ζ of ∂B.

Reduction of the Theorem to a Special Case

Suppose that it is known that u/h has radial limit 0 at a boundary point ζ whenever

(R1) $u \geq 0$,
(R2) The symmetric derivate dM_u/dM_h exists and is 0 at ζ.
(R3) The symmetric derivate dl'_{N-1}/dM_h exists and is finite at ζ.

It will now be shown that then the theorem follows. In the first place, when $u \geq 0$ and u/h has radial limit 0 at ζ then u/h also has nontangential limit 0 there. In fact let S be a Stolz domain with vertex ζ and let S_1 be a second Stolz domain with vertex ζ but with a larger vertex angle than S. Choose ρ with $0 < \rho < 1$ and let $B_\rho[B_{1\rho}]$ be the ball with center $\rho\zeta$, internally tangent to $\partial S[\partial S_1]$. According to the Harnack inequality for balls as applied to u and h in $B_{1\rho}$ there is a strictly positive constant c with the property that

$$\frac{u(\xi)}{h(\xi)} \leq c\frac{u(\rho\zeta)}{h(\rho\zeta)}, \qquad \xi \in B_\rho.$$

The constant c does not depend on ρ because the constant in Harnack's inequality is unaffected by a similarity transformation of $\mathbb{R}^N$. It follows that u/h has limit 0 on approach to ζ in S and therefore on nontangential approach. In the second place we need only remark that if the derivate value in (15.1) is α, to reduce the general case to that of positive u and $\alpha = 0$ we can replace M_u by $|M_u - \alpha M_h|$, that is, replace u by the corresponding vector lattice maximum $(u - \alpha h) \curlyvee (\alpha h - u)$ furnished by Theorem 14.

To finish the proof of the theorem, we prove that under (R1)–(R3) the function u/h has radial limit 0 at ζ. Let $\alpha_u(\theta)$ and $\alpha_h(\theta)$ be, respectively, the M_u and M_h measure of the closed spherical cap on ∂B cut off by the closed cone of revolution with vertex the origin, axis the radius to ζ, half-angle $\theta > 0$, and define $\alpha_u(0) = \alpha_h(0) = 0$. The functions α_u and α_h are monotone increasing on $[0, \pi]$ and right continuous except perhaps at 0, and $\alpha_h(\theta) > 0$ when $\theta > 0$ in view of (R3). To simplify the notation take $\delta = 1$. Define

$$K_s(\theta) = \frac{1}{\pi_N} \frac{1 - s^2}{[(1 - s)^2 + 4s\sin^2(\theta/2)]^{N/2}} \tag{15.2}$$

so that

$$u(s\zeta) = \int_0^\pi K_s(\theta)\, d\alpha_u(\theta), \qquad h(s\zeta) = \int_0^\pi K_s(\theta)\, d\alpha_h(\theta). \tag{15.3}$$

The function K_s is a decreasing function on $[0, \pi]$,

$$\int_0^a \liminf_{s\to 1} \frac{K_s(\theta)\theta^{N-2}}{K_s(a)} l_1(d\theta) = \int_0^a \frac{\sin^N(a/2)\theta^{N-2}}{\sin^N(\theta/2)} l_1(d\theta) = +\infty \qquad (a > 0), \tag{15.4}$$

and according to hypotheses (R2) and (R3),

$$\lim_{\theta\to 0} \frac{\alpha_u(\theta)}{\alpha_h(\theta)} = 0, \qquad \lim_{\theta\to 0} \frac{\theta^{N-1}}{\alpha_h(\theta)} < +\infty. \tag{15.5}$$

According to Appendix VII.3 with $f \equiv 1$, $p = N - 1$, and $n \to \infty$ replaced by $s \to 1$, relations (15.4) and (15.5) imply that $\lim_{s\to 1} u(s\zeta)/h(s\zeta) = 0$, as was to be proved.

16. Minimal Harmonic Functions

A harmonic function u on an open connected subset D of $\mathbb{R}^N$ is called *minimal* if u is positive and if every harmonic function v on D satisfying the inequality $0 \le v \le u$ is a constant multiple of u. If u is minimal, the positive multiples of u are also minimal. If $\xi_0 \in D$ and if Γ is the class of positive harmonic functions on D with value 1 at ξ_0, the class Γ is convex, and the minimal members of Γ are the extreme elements of this convex set. In fact, if u is minimal and in Γ and if u is not extremal, then u has a representation

$$u = pu_1 + qu_2, \qquad p + q = 1, \qquad 0 < p < 1, \tag{16.1}$$

with u_1 and u_2 distinct elements of Γ. But then pu_1 and qu_2 must be multiples of u with values p and q, respectively, at ξ_0, so $u = u_1 = u_2$, contrary to hypothesis. Conversely, if u is an extremal element of Γ and if v is harmonic on D with $0 \le v \le u$, then either (a) v vanishes at ξ_0 and so vanishes identically or (b) $u - v$ vanishes at ξ_0 and so vanishes identically or

(c) $$u = v(\xi_0)\left\{\frac{v}{v(\xi_0)}\right\} + [u(\xi_0) - v(\xi_0)]\left\{\frac{u - v}{u(\xi_0) - v(\xi_0)}\right\}.$$

The two functions in the braces are in Γ and so must be identical. Thus v is a constant multiple of u in all three cases.

EXAMPLE. Let $D = B(0, \delta)$ and let K be the normalized Poisson kernel density function,

$$K(\eta, \xi) = \delta^{N-2}\frac{\delta^2 - |\xi|^2}{|\eta - \xi|^N}, \qquad \eta \in \partial B(0, \delta), \qquad \xi \in B(0, \delta). \tag{16.2}$$

The class of minimal harmonic functions with value 1 at the origin is $\{K(\eta, \cdot), \eta \in \partial D\}$. In fact $K(\eta, \cdot)$ is minimal because the Riesz–Herglotz measure of this function is supported by the singleton $\{\eta\}$ and the correspondence between positive harmonic functions and their Riesz–Herglotz measures is order preserving; so any positive harmonic function majorized by $K(\eta, \cdot)$ also has Riesz–Herglotz measure supported by $\{\eta\}$. Conversely, suppose that u is minimal with Riesz–Herglotz measure μ. Unless μ is supported by a singleton $\{\eta\}$ in which case u is a multiple of $K(\eta, \cdot)$, μ is the sum of two nonidentically vanishing measures, say $\mu = \mu_1 + \mu_2$, supported by disjoint sets, so that μ_1 is not a multiple of μ, and μ_1 is then the Riesz–Herglotz measure of a positive harmonic function majorized by u but not a multiple of u. The point is that the minimal measures on $\partial B(0, \delta)$ in the obvious definition must correspond to the minimal harmonic functions, and the minimal measures are those supported by singletons.

Chapter III

Infima of Families of Superharmonic Functions

1. Least Superharmonic Majorant (LM) and Greatest Subharmonic Minorant (GM)

If Γ is a class of extended real-valued functions on a set D, a function v on D will be called a *majorant* [*minorant*] of Γ if $v \geq u$ [$v \leq u$] for every u in Γ. If D is an open subset of $\mathbb{R}^N$, the least superharmonic majorant [greatest subharmonic minorant] of Γ, if such a function exists, will be denoted by $\mathrm{LM}_D\Gamma$ [$\mathrm{GM}_D\Gamma$], or by $\mathrm{LM}_D u$ [$\mathrm{GM}_D u$] if $\Gamma = \{u\}$ is a singleton.

Theorem. *If a superharmonic function u on D has a subharmonic minorant, then* $\mathrm{GM}_D u$ *exists and is harmonic.*

Let $B_{\bullet}$ be a sequence of (not necessarily distinct) balls with closures in D and with the property that if ξ is in D, some neighborhood of ξ is contained in infinitely many of the balls. Define $u_n = \tau_{B_n} \cdots \tau_{B_0} u$. Let u_∞ be the limit of the decreasing sequence $u_{\bullet}$ of superharmonic functions on D. If v is a subharmonic minorant of u,

$$u_n \geq \tau_{B_n} \cdots \tau_{B_0} v \geq v,$$

and if ξ is in D, some subsequence of $u_{\bullet}$, in some ball with center ξ, is a decreasing sequence of harmonic functions. It follows (Harnack convergence theorem) that u_∞ is harmonic and majorizes v, and therefore $u_\infty = \mathrm{GM}_D u$.

Without the hypothesis that u has a subharmonic minorant the limit function u_∞ is either identically $-\infty$ or harmonic on each connected open component of D and, if harmonic on D, is $\mathrm{GM}_D u$.

Observation (a). If D is a ball or is $\mathbb{R}^N$, a simple choice of $B_{\bullet}$ is an increasing sequence of balls with closures in D and with union D. Choice of $B_{\bullet}$ as an increasing sequence of open relatively compact subsets of D with union D will be possible for every open set D once τ_B is defined (in Section VIII.11) whenever B is an open relatively compact subset of D.

Observation (b). As the proof of the theorem shows,

$$\mathrm{GM}_D(c_1u_1 + c_2u_2) = c_1\mathrm{GM}_D u_1 + c_2\mathrm{GM}_D u_2$$

for superharmonic functions u_1, u_2 and positive constants c_1, c_2.

EXAMPLE (a). *Let u be superharmonic on D and let $B = B(\xi, \delta)$ have closure in D. Then* $\mathrm{GM}_B(u_{|B}) = \mathrm{PI}(B, u)$. We already know that there is inequality ($\geq$) here because $\mathrm{PI}(B, u) = (\tau_B u)_{|B} \leq u$. If $B_n = B(\xi, \delta - 1/n)$, the minorant in question is $u_\infty = (\lim_{n\to\infty} \tau_{B_n} u)_{|B}$. The function u_∞ is harmonic on B and majorizes $\mathrm{PI}(B, u)$, and there is equality at ξ because

$$u_\infty(\xi) = \lim_{n\to\infty} L\left(u, \xi, \delta - \frac{1}{n}\right) = L(u, \xi, \delta) = \mathrm{PI}(B, u)(\xi). \tag{1.1}$$

Hence (harmonic function maximum-minimum theorem) there is equality on B, as asserted.

EXAMPLE (b). $\mathrm{GM}_D G_D(\xi, \cdot) = 0$. In fact, if D has a Green function G_D in the sense of Section I.8 then $G_D(\xi, \cdot)$ is a positive superharmonic function on D with limit 0 at every boundary point. The function $\mathrm{GM}_D G_D(\xi, \cdot)$ is positive and harmonic and also has limit 0 at every boundary point. It follows (harmonic function maximum-minimum theorem) that this minorant vanishes identically. Exactly the same proof, in which ∞ is the only boundary point, shows that $\mathrm{GM}_{\mathbb{R}^N} G(\xi, \cdot) = 0$ when $N > 2$. In Section VII.1 the Green function G_D will be defined for every Greenian set D. The property $\mathrm{GM}_D G_D(\xi, \cdot) = 0$ remains true and in fact is very nearly a defining property of G_D.

2. Generalization of Theorem 1

Theorem 1 is included in the following theorem but was proved separately because of the importance of its constructive proof.

Theorem. *If a class Γ of superharmonic functions on an open subset D of $\mathbb{R}^N$ has a subharmonic minorant, then* $\mathrm{GM}\Gamma$ *exists and is harmonic.*

Let Γ_0 be the class of subharmonic minorants of Γ. The class Γ_0 contains $u_1 \vee u_2$ with u_1, u_2 and is therefore directed upward in the order of pointwise inequality, with limit u', a function majorized by Γ. We prove the theorem by showing that u' is harmonic. Let B be a ball with closure in D and suppose that $v \in \Gamma_0$. For every function u in Γ, $v \leq \tau_B v \leq \tau_B u \leq u$. Thus $\tau_B v \in \Gamma_0$, and the supremum of Γ_0 is the same as that of $\{\tau_B v : v \in \Gamma_0\}$. This class on B is an upward-directed family of harmonic functions majorized by Γ on B; so u' is harmonic on B (Harnack convergence theorem) and therefore on D, as was to be proved.

3. Fundamental Convergence Theorem (Preliminary Version)

If u is a function from a topological space into $\overline{\mathbb{R}}$, we denote (Appendix VIII.1) by $\underset{+}{u}$ the lower semicontinuous smoothing of u, that is, the maximal lower semicontinuous minorant of u:

$$\underset{+}{u}(\xi) = u(\xi) \wedge \liminf_{\eta \to \xi} u(\eta). \tag{3.1}$$

In the following theorem the trivial inequality $\underset{+}{u} \le u$ is stated for completeness and to facilitate reference.

Theorem. *Let Γ: $\{u_\alpha, \alpha \in I\}$ be a family of superharmonic functions on an open subset of $\mathbb{R}^N$, locally uniformly bounded below, and define the lower envelope (infimum) u of Γ by $u(\xi) = \inf_{\alpha \in I} u_\alpha(\xi)$. Then $\underset{+}{u} \le u$,*

$$\underset{+}{u}(\xi) = \liminf_{\eta \to \xi} u(\eta) \tag{3.2}$$

and

(a) *$\underset{+}{u}$ is superharmonic.*
(b) *$\underset{+}{u} = u$ on each open set on which u is superharmonic.*
(c) *$\underset{+}{u} = u$ l_N almost everywhere.*
(d) *There is a countable subset of Γ whose lower envelope has the same lower semicontinuous smoothing $\underset{+}{u}$.*

Observation (1). The hypotheses of the theorem are not effectively weakened by the added assumption that $u_.$ is directed downward because if all finite minima $u_{\alpha_1} \wedge \cdots \wedge u_{\alpha_k}$ are adjoined to the family, the enlarged family of superharmonic functions is directed downward and has the same lower envelope u, and (d) is true for the original family if true for the enlarged family. The added assumption justifies the nomenclature "convergence theorem."

Observation (2). According to Choquet's topological lemma (Appendix VIII.3) Theorem 3(d) is true, and we can even assume that Γ is a decreasing sequence. Furthermore, since the theorem is local, it can be assumed that u is bounded below on its domain, say $u \ge c$, and replacing u_α by $u_\alpha - c$, it can be assumed that the members of Γ are positive.

In accordance with these observations the following proof assumes that $u_.$ is a decreasing sequence of positive superharmonic functions. Apply II(6.1) to find

$$u_n(\xi) = \liminf_{\eta \to \xi} u_n(\eta) \ge \liminf_{\eta \to \xi} u(\eta),$$

so that $u(\xi)$ is at least equal to the last term on the right, and therefore (3.1) implies (3.2). Since u_n is superharmonic,

$$A(u_n, \xi, \delta) \le u_n(\xi), \tag{3.3}$$

so that

$$A(\underset{+}{u}, \xi, \delta) \le A(u, \xi, \delta) \le u(\xi). \tag{3.4}$$

Moreover continuity of $A(u, \cdot, \delta)$ (Section I.2) implies that

$$A(\underset{+}{u}, \xi, \delta) \le A(u, \xi, \delta) \le \underset{+}{u}(\xi) \le u(\xi) \tag{3.5}$$

because $\underset{+}{u}$ is the largest lower semicontinuous function majorized by u. Inequality (3.5) implies that $\underset{+}{u}$ is superharmonic. Finally $\underset{+}{u} = u\, l_N$ almost everywhere on D because when $\delta \to 0$ inequality (3.5) becomes $\underset{+}{u} \le u \le \underset{+}{u} \le u\, l_N$ almost everywhere on D, by a standard derivation theorem.

Application to $\mathrm{GM}_D\Gamma$

Let Γ: $\{u_\alpha, \alpha \in I\}$ be a family of superharmonic functions on an open subset D of $\mathbb{R}^N$ and suppose that Γ has a subharmonic minorant. Then $\mathrm{GM}_D\Gamma$ exists; so Γ is locally lower bounded. Define $u = \inf u_\alpha$. According to Theorem 3, the function $\underset{+}{u}$ is superharmonic, and we now show that $\mathrm{GM}_D\Gamma = \mathrm{GM}_D\underset{+}{u}$. It is trivial that $\mathrm{GM}_D\Gamma \ge \mathrm{GM}_D\underset{+}{u}$ and in the other direction $\mathrm{GM}_D\Gamma$ is a continuous minorant of u and is therefore a minorant of $\underset{+}{u}$ by definition of $\underset{+}{u}$; so $\mathrm{GM}_D\Gamma \le \mathrm{GM}_D\underset{+}{u}$.

4. The Reduction Operation

Let D be an open subset of $\mathbb{R}^N$ coupled with a boundary ∂D provided by a metric compactification. Let A be a subset of $D \cup \partial D$, let v be a positive superharmonic function on D, and let Γ be the class of positive superharmonic functions u on D majorizing v on $A \cap D$ and near $A \cap \partial D$ in the sense that to each function u corresponds a neighborhood A_0 of $A \cap \partial D$ in the compact space $D \cup \partial D$ such that $\{u \ge v\} \supset (A \cup A_0) \cap D$. If u_1 and u_2 are in Γ, their minimum $u_1 \wedge u_2$ is also in Γ; so Γ is directed downward in the order of pointwise inequality. The reduction R_v^A of v on A is defined as the infimum, that is, the limit, of the directed set Γ. Since $v \in \Gamma$, this reduction is also the infimum and limit of the class of functions $v \wedge u$ with u in Γ, that is, of the class of positive superharmonic functions on D equal to v on $A \cap D$ and near $A \cap \partial D$. Thus $R_v^A \le v$, with equality on $A \cap D$. The set D is part of the context but is omitted from the notation. In most applications $A \subset D$. According to the Fundamental Convergence Theorem in the preceding section, the lower semicontinuous smoothed reduction $\underset{+v}{R}{}^A$ is majorized by R_v^A, is superharmonic, and coincides l_N almost everywhere on D with R_v^A. (According to the more refined version of this theorem in Section VI.1, the set $\{R_v^A > \underset{+v}{R}{}^A\}$ is not only l_N null but polar, a more restrictive characterization to be defined in Section V.1, and it will be proved in Section VI.4 that $R_v^A = \underset{+v}{R}{}^A$ on $D - A$.) The notation $\llbracket v \rrbracket^A$ will sometimes be used instead of $\underset{+v}{R}{}^A$.

Obviously R_v^A and $\underset{+v}{R}{}^A$ increase when A or v increases. Moreover the reduction operation is subadditive in the sense that

$$R_{u+v}^{A \cup B} \leq R_u^A + R_u^B + R_v^A + R_v^B, \tag{4.1}$$

and the corresponding inequality (4.1sm) for smoothed reductions is also true. To prove (4.1) we observe that if $u_A\,[v_A]$ majorizes $u\,[v]$ on $A \cap D$ and near $A \cap \partial D$ and if $u_B\,[v_B]$ majorizes $u\,[v]$ on $B \cap D$ and near $B \cap \partial D$, then $u_A + u_B + v_A + v_B$ majorizes $u + v$ on $(A \cup B) \cap D$ and near $(A \cup B) \cap \partial D$. To prove (4.1sm), observe that the two sides are superharmonic functions and that the inequality is true l_N almost everywhere on D and therefore everywhere on D.

If $D - \bar{A}$ is not empty and if B is a ball with closure in $D - \bar{A}$, the function $\tau_B u$ is in Γ whenever u is in Γ and $\tau_B u \leq u$; so R_v^A on B is the limit of the downward-directed family $\{(\tau_B u)_{|B}, u \in \Gamma\}$ of harmonic functions. Hence (by the Harnack convergence theorem) R_v^A is harmonic on B and so is harmonic on $D - \bar{A}$ and equal to $\underset{+v}{R}{}^A$ there. In particular, when $A \subset \partial D$, the reduction R_v^A is harmonic on D and $R_v^A = \underset{+v}{R}{}^A$.

Since $R_v^A = v$ on $A \cap D$, it follows that $R_v^A = \underset{+v}{R}{}^A = v$ on the interior of $A \cap D$. Hence, if A is open in $D \cup \partial D$, the smoothed reduction $\underset{+v}{R}{}^A$ is a positive superharmonic function majorizing v on $A \cap D$ and near $A \cap \partial D$ and majorized by R_v^A; so $R_v^A = \underset{+v}{R}{}^A$ on D. More generally we shall show in Section 5(d) that $R_v^A = \underset{+v}{R}{}^A$ whenever $A \cap D$ is open.

Example (a). *Let D be a ball and let A be a relatively compact subset of D. Then R_v^A (and therefore also $\underset{+v}{R}{}^A$) has limit 0 at every ball boundary point.* It is instructive to prove this result directly here, although the result also follows naturally at a later stage from the remark that when A is relatively compact in D, the function $\underset{+v}{R}{}^A$ is (Section IV.8) the potential $G_D\mu$ of a measure with compact support in the ball and as such (Section IV.1) has limit 0 at every ball boundary point. To prove the stated result, let ξ be the center of D, let A_1 be an open neighborhood of $\bar{A}$ and be relatively compact in D, and let B be an open neighborhood of $\bar{A}_1$ and be relatively compact in D. Then $R_v^{A_1} = v$ on A; so $R_v^A \leq R_v^{A_1}$, and $R_v^{A_1}$ is harmonic on a neighborhood of ∂B. Choose a constant c so large that $cG_D(\xi,\cdot) > R_v^{A_1}$ on ∂B and therefore on a neighborhood of ∂B. The function u equal to $R_v^{A_1}$ on B and equal to $R_v^{A_1} \wedge cG_D(\xi,\cdot)$ on $D - B$ is superharmonic on D, majorizes R_v^A, and is a minorant of $cG_D(\xi,\cdot)$ on $D - B$; so $R_v^A \le u \le cG_D(\xi,\cdot)$ near ∂D, and therefore R_v^A has limit 0 at every boundary point of D. A similar argument shows that when $N > 2$, $D = \mathbb{R}^N$, and A is a relatively compact subset of D, then R_v^A has limit 0 at the point ∞.

Example (b). Let D be a ball, let $A = \{\xi\} \subset D$ be a singleton, and let $v \equiv 1$. Then $R_v^A(\xi) = v(\xi) = 1$, but since $\varepsilon G_D(\xi,\cdot)$ majorizes v on A whenever $\varepsilon > 0$, it follows that $R_v^A = 0$ on $D - \{\xi\}$ and the smoothed reduction $\underset{+v}{R}^A$ vanishes identically.

Example (c). Let D be a ball with center the origin, and let v be a positive harmonic function on D with Riesz–Herglotz representation (Section II.14)

$$v = \int_{\partial D} K(\eta,\cdot)M_v(d\eta). \tag{4.2}$$

We shall now prove

$$R_v^A = \underset{+v}{R}^A = \int_A K(\eta,\cdot)M_v(d\eta) \tag{4.3}$$

for every Borel boundary subset A. The first equality in (4.3) is trivial because R_v^A is harmonic. Denote the integral on the right by v_A. If u is a positive superharmonic function on D that majorizes v near A and if A_0 is a closed subset of A, the difference $u - v_{A_0}$ is superharmonic on D and positive near A_0. Moreover the fact that $\lim_{\xi\to\zeta} K(\cdot,\xi) = 0$ uniformly on A_0 when $\zeta \in \partial D - A_0$ implies that v_{A_0} has boundary limit 0 on $\partial D - A_0$. Hence $u - v_{A_0}$ has a positive inferior limit at every boundary point of D, and the superharmonic function minimum theorem implies that $u - v_{A_0} \ge 0$; so $R_v^A \ge v_{A_0}$. If A_1 is an open boundary superset of A, then $\lim_{\xi\to\zeta} K(\cdot,\xi) = 0$ uniformly on $\partial D - A_1$ when $\zeta \in A$; so $v - v_{A_1}$ has boundary limit 0 at every point of A. Hence, if $\varepsilon > 0$ the function $v_{A_1} + \varepsilon$ majorizes v near A, and it follows that $v_{A_1} \ge R_v^A$. Thus $v_{A_0} \le R_v^A \le v_{A_1}$, and this inequality implies the desired equality (4.3).

Example (d). *Let D be a Greenian subset of $\mathbb{R}^N$, let B be an open relatively compact subset of D, and let v be a positive superharmonic function on D, finite and continuous at each point of ∂B. Then $R_v^{D-B} = \mathrm{GM}_B v$ on B.* In fact, on the one hand, R_v^{D-B} is harmonic on B and is majorized by v on B; so $R_v^{D-B} \le \mathrm{GM}_B v$ on B. On the other hand, if u is a positive superharmonic function on D, with $u \ge v$ on $D - B$, and if v_0 is a harmonic function on B and is majorized by v on B, then $u - v_0$ is superharmonic on B and has a positive inferior limit at every point of ∂B and so is positive on B, and it follows that $R_v^{D-B} \ge \mathrm{GM}_B v$ on B.

According to Section VIII.18, the hypothesis in Example (d) that v be finite valued and continuous at each point of ∂B is stronger than necessary. It is sufficient, for example, if v is finite valued on ∂B. More specifically, according to Theorem VIII.18(c), in which $\mu_D(\cdot, v1_B)$ can be identified with R_v^{D-B}, the evaluation $R_v^{D-B} = \mathrm{GM}_B v$ on B is true if and only if, in the terminology to be developed below, the set of irregular boundary points of B is a null set for the Riesz measure associated with v.

5. Reduction Properties

The following list of reduction properties will be useful, and expanded, in later chapters. The reductions are relative to a Greenian subset D of $\mathbb{R}^N$, and no restrictions not specifically stated are imposed on the positive superharmonic functions on D to be reduced or on the subsets of $D \cup \partial D$ on which these functions are reduced.

(a) *R_v^A in terms of reductions on subsets of D.*

$$\begin{aligned} R_v^A &= \inf\{R_v^{A\cup B}: A \cap \partial D \subset B,\ B \text{ open in } D \cup \partial D\} \\ &= \inf\{R_v^{(A\cup B)\cap D}: A \cap \partial D \subset B,\ B \text{ open in } D \cup \partial D\}. \end{aligned} \tag{5.1}$$

These evaluations follow from the fact that if u is a superharmonic function on D majorizing v on $A \cap D$ and near $A \cap \partial D$, then there is a neighborhood B of $A \cap \partial D$ so small that u also majorizes v near $B \cap \partial D$. The smoothed version of (5.1) is

$$\begin{aligned} \underset{+v}{R}{}^A &= \underset{+}{\inf}\{\underset{+v}{R}{}^{A\cup B}: A \cap \partial D \subset B,\ B \text{ open in } D \cup \partial D\} \\ &= \underset{+}{\inf}\{\underset{+v}{R}{}^{(A\cup B)\cap D}: A \cap \partial D \subset B,\ B \text{ open in } D \cup \partial D\}. \end{aligned} \tag{5.1sm}$$

To prove (5.1sm), observe that in view of (5.1) and the Fundamental Convergence Theorem the right sides are superharmonic functions majorized by R_v^A and therefore by $\underset{+v}{R}{}^A$, and the reverse inequalities are trivial.

If $A \cap \partial D$ is compact, the sets B in (5.1) and (5.1sm) need only run through a decreasing sequence of relatively compact open neighborhoods of $A \cap \partial D$ with that set as intersection.

(b) *If p is a positive superharmonic function on D with $\mathrm{GM}_D p = 0$, if $A \subset \partial D$, and if v is a positive superharmonic function on D, then $R_{v+p}^A = R_v^A$. In particular,* $R_p^A = 0$. This property follows from the inequality

$$R_v^A \le R_{v+p}^A \le R_v^A + R_p^A$$

because R_p^A, a positive harmonic function majorized by p, must vanish identically.

Observation. In Section IV.8 the positive superharmonic functions p with $\mathrm{GM}_D p = 0$ will be identified with the superharmonic potentials $G_D\mu$.

(c) *R_v^A in terms of separate reductions on subsets of D and of ∂D.* Recall (Section 4) that $R_v^{A\cap\partial D}$ is harmonic. We now prove that *if $v' = v - R_v^{A\cap\partial D}$ then*

$$R_v^A = R_v^{A\cap\partial D} + R_{v'}^{A\cap D} \tag{5.2}$$

and also prove the corresponding inequality (5.2sm) for smoothed reductions. In particular, if $\mathrm{GM}_D v = 0$ then $R_v^A = R_v^{A\cap D}$ and $\underset{+v}{R}{}^A = \underset{+v}{R}{}^{A\cap D}$. In fact, if v_1 is a positive superharmonic function on D, majorizing v near $A \cap \partial D$ and if v_2 is a positive superharmonic function on D, majorizing v' on $A \cap D$, then $v_1 + v_2 \geq R_v^A$, and therefore

$$R_v^{A\cap\partial D} + R_{v'}^{A\cap D} \geq R_v^A. \tag{5.3}$$

Conversely, if v_3 is a positive superharmonic function on D, majorizing v on $A \cap D$ and near $A \cap \partial D$, the positive superharmonic function $v_3 - R_v^{A\cap\partial D}$ majorizes v' on $A \cap D$; so

$$R_{v'}^{A\cap D} \leq R_v^A - R_v^{A\cap\partial D}, \tag{5.4}$$

and this inequality combines with (5.3) to yield (5.2). Equality (5.2) implies that the corresponding equality for smoothed reductions, an equality between two superharmonic functions, is true l_N almost everywhere on D and therefore is true everywhere on D.

(d) *$R_v^A = \underset{+v}{R}{}^A$ whenever $A \cap D$ is open.* If A is open, this property was pointed out in Section 4, and the more general property follows from (5.2) and (5.2sm).

(e) *If v is finite valued and continuous at each point of $A \cap D$, then*

$$R_v^A = \inf\{R_v^B\colon A \subset B,\ B \text{ open in } D \cup \partial D\}. \tag{5.5}$$

In fact, if u is a positive superharmonic function on D which majorizes v on $A \cap D$ and near $A \cap \partial D$, then for $n \geq 1$ the function $(1 + 1/n)u$ majorizes v on a neighborhood of $A \cap D$ and on the trace on D of a neighborhood of $A \cap \partial D$ and therefore majorizes R_v^B for some open superset B of A.

6. A Smallness Property of Reductions on Compact Sets

Theorem. *If D is a Greenian subset of $\mathbb{R}^N$, if A is a relatively compact subset of D, and if v is a positive superharmonic function on D, then* $\mathrm{GM}_D \underset{+v}{R}{}^A = 0$.

By hypothesis there is a nonconstant positive superharmonic function u on D. It can be supposed that D is connected and, replacing u by $u^{1/2}$ if necessary, that u is not harmonic. Replacing u by $u - \mathrm{GM}_D u$ if necessary, it can also be supposed that $\mathrm{GM}_D u = 0$. If A_1 is an open relatively compact subset of D containing $\bar{A}$, then $v_1 = R_v^{A_1} = \underset{+v}{R}^{A_1}$ is a positive superharmonic function on D, harmonic on $D - \bar{A}_1$, and is equal to v on A; so v_1 and v have the same reduction on A. Replacing v by v_1 if necessary, it can be supposed that v is harmonic on $D - \bar{A}_1$. Thus, if B is an open relatively compact subset of D containing $\bar{A}_1$, the function v is harmonic on a neighborhood of ∂B. Choose c so large that $cu > v$ on ∂B. This inequality will then hold on a neighborhood of ∂B, so that $(cu) \wedge v = v$ on this neighborhood. Define $u_1 = v$ on $\bar{B}$ and $u_1 = (cu) \wedge v$ on $D - \bar{B}$ to get a positive superharmonic function on D majorizing v on A and therefore majorizing R_v^A on D. Hence

$$\mathrm{GM}_D \underset{+v}{R}^A \le \underset{+v}{R}^A = R_v^A \le cu$$

on $D - \bar{B}$, and (by the subharmonic function maximum theorem) therefore $\mathrm{GM}_D \underset{+v}{R}^A \le cu$ on D. Since $\mathrm{GM}_D u = 0$, it follows that $\mathrm{GM}_D \underset{+v}{R}^A = 0$, as was to be proved.

7. The Natural (Pointwise) Order Decomposition for Positive Superharmonic Functions

Theorem. *If u, u_1, and u_2 are positive superharmonic functions on an open subset D of $\mathbb{R}^N$, with $u \le u_1 + u_2$, then there are positive superharmonic functions u_1', u_2' on D for which*

$$u_1' \le u_1, \qquad u_2' \le u_2, \qquad u = u_1' + u_2'. \tag{7.1}$$

Let u_1' be the smoothed infimum (given by the Fundamental Convergence Theorem) of the class of functions v, positive and superharmonic on D and satisfying the inequality $u \le v + u_2$. Then $u \le u_1' + u_2$ l_N almost everywhere on D and therefore everywhere on D. Let u_2' be the smoothed infimum of the class of functions v, positive and superharmonic on D and satisfying the inequality $u \le u_1' + v$. Then $u \le u_1' + u_2'$ on D, and if v is a positive superharmonic function on D for which $u \le u_1' + v$ $[u \le v + u_2']$, it follows that $v \ge u_2'$ $[v \ge u_1']$. In particular (choose $v = u$), it follows that $u \ge u_1'$ and $u \ge u_2'$. To prove the theorem, it will be shown that $u = u_1' + u_2'$. Define $(u - u_2')(\xi)$ as the indicated difference whenever $u_2'(\xi) < +\infty$ and define

$$(u - u_2')(\xi) = \liminf_{\eta \to \xi,\, u_2'(\eta) < +\infty} [u(\eta) - u_2'(\eta)]$$

whenever $u_2'(\xi) = +\infty$. The function $u - u_2'$ is thereby lower semicontinuous at each infinity of u_2', and we shall now prove that $u - u_2'$ is also lower semicontinuous at the other points of D. Let B be a ball with closure in D. Then

$$u = \tau_B u + u - \tau_B u \le u_1' + [\tau_B u_2' + u - \tau_B u] \tag{7.2}$$

on B, and the bracketed function v is superharmonic on B and majorizes $\tau_B u_2'$ there. Hence [Section II.6(e)] the function $(u_2' \wedge v)_{|B}$ extended to D by u_2' is a positive superharmonic function v' on D, and $u \le u_1' + v'$. It follows that $v' \ge u_2'$; that is, $\tau_B u_2' + u - \tau_B u \ge u_2'$ on B. Thus

$$\tau_B(u - u_2') \le (u - u_2') \tag{7.3}$$

on B. If B has center ξ, if $u_2'(\xi) < +\infty$, and if B shrinks to ξ, the value $\tau_B(u - u_2')(\xi) = \tau_B u(\xi) - \tau_B u_2'(\xi)$ tends to $(u - u_2')(\xi)$, and since the left side of (7.3) is continuous (harmonic) on B we conclude that $u - u_2'$ is lower semicontinuous at ξ, as desired, and therefore is lower semicontinuous at every point of D. Furthermore (7.3) implies that the function $u - u_2'$ satisfies the superharmonic function average inequality; so this function is superharmonic. The relations $(u - u_2') \le u_1'$ and $u = (u - u_2') + u_2'$ are valid except possibly at the l_N null set of infinities of u_2' and are therefore valid everywhere on D. It follows that $u - u_2' = u_1'$ and $u = u_1' + u_2'$, as was to be proved.

Chapter IV

Potentials on Special Open Sets

1. Special Open Sets, and Potentials on Them

Throughout this book a *special open* subset of $\mathbb{R}^N$ is either a ball in $\mathbb{R}^N$ or $\mathbb{R}^N$ itself, but the latter only when $N > 2$. The Green function G_D for D a ball was defined in Section II.1. The Green function G_D for $D = \mathbb{R}^N$ with $N > 2$ is defined as G. If μ is a measure on a special open set D, define the function $G_D\mu$ on D by

$$G_D\mu(\xi) = \int_D G_D(\xi, \eta)\mu(d\eta). \tag{1.1}$$

An application of Fatou's lemma shows that $G_D\mu$ is lower semicontinuous, and $G_D\mu$ satisfies the superharmonic function inequality because $G_D(\cdot, \eta)$ does for each η. Hence $G_D\mu$ is either superharmonic or identically $+\infty$ on D. The function $G_D\mu$ is called the *potential* or sometimes the *Green potential* of μ. In Chapter VII a Green function G_D will be defined for every Greenian open set and will be shown to enjoy many of the properties of the Green function of a special open set. For example, if μ is a measure on the Greenian set D, the function $G_D\mu$ will be shown to be either superharmonic or identically $+\infty$ on each open connected component of D.

In Chapters IV–VI many theorems will be stated for potentials on Greenian sets but proved only for potentials on special sets. This is done because the results for these special sets will be used to develop many of the general results of classical potential theory that are needed before a general Green function can be comfortably defined and its properties studied. It will be obvious that circular reasoning will not be invoked, and in Sections VII.5 and VII.8 it will be pointed out that the proofs in the preceding chapters for special sets are also applicable when the sets are Greenian.

In this book *potential* in the present classical context always means a function $G_D\mu$ with D Greenian or (rarely and then always pointed out explicitly) $G\mu$ when $N = 2$ and the open set involved is the plane. The existence of the latter potential is discussed later in this section.

Suppose again that D is special open and consider a potential $G_D\mu$. It is clear that $G_D\mu$ is superharmonic if μ has compact support in D because this potential is finite off that support. More generally $G_D\mu$ is superharmonic if $\mu(D) < +\infty$ because if μ_1 is the projection of μ on a ball B with closure in D and if μ_2 is the projection of μ on $D - B$, then $G_D\mu = G_D\mu_1 + G_D\mu_2$, $G_D\mu_1$ is superharmonic because μ_1 has compact support in D, and $G_D\mu_2$ is superharmonic because this potential is finite on B. Moreover, if $G_D\mu$ is superharmonic and if D_0 is an open μ null subset of D, the potential $G_D\mu$ is harmonic on D_0 because this potential is Borel measurable and satisfies the harmonic function average property on D_0.

If μ has compact support in the special open set D, the Lebesgue dominated convergence theorem when applied in (1.1) yields the fact that the potential $G_D\mu$ has limit 0 at every boundary point of D, because the function $G_D(\cdot, \eta)$ has this property and is uniformly bounded in a neighborhood of ∂D when η is restricted to a compact subset of D. More generally, if D is a Greenian subset of $\mathbb{R}^N$ and if μ is a measure on D with $G_D\mu$ superharmonic, it will be shown that in various senses the potential $G_D\mu$, in particular $G_D(\cdot, \eta)$ for fixed η in D, has boundary limit function 0. See, for example, Section VII.4, VIII.11, XII.19, XII.23, and 2.X.8.

Potentials $G\mu$ When $N = 2$

The study of potentials $G\mu$ when $N = 2$ is complicated by the fact that G is then not a positive function. Observe first that for an arbitrary measure μ on $\mathbb{R}^2$ and $m > 0$ the integrand in

$$u_m(\xi) = \int_{B(0,m)} G(\xi, \eta)\mu(d\eta) = \int_{B(0,m)} \log|\xi - \eta|^{-1}\mu(d\eta), \tag{1.2}$$

is lower bounded for each ξ; so $u_m(\xi)$ is uniquely defined ($\leq +\infty$). If we write $u(\xi)$ in the form

$$u_m(\xi) = \int_{B(0,m)} \log(n|\xi - \eta|^{-1})\mu(d\eta) - \mu(B(0,m))\log n, \qquad n > m > 0, \tag{1.3}$$

the integrand is positive for ξ in $B(0, n - m)$. The reasoning used above in discussing potentials of measures on balls and on $\mathbb{R}^N$ when $N > 2$ when applied to the function defined by the integral in (1.3) shows that u_m is a superharmonic function on $B(0, n - m)$ and is harmonic on each μ null open subset of $B(0, n - m)$. Since n can be chosen arbitrarily large, the function u_m is a superharmonic function on $\mathbb{R}^2$ and is harmonic on $\mathbb{R}^2 - B(0, m)$ and on each open μ null subset of $B(0, m)$. In particular, if μ has compact support and if m is so large that $B(0, m)$ contains this support, it follows that $G\mu$ is superharmonic and is harmonic on the complement of the closed support of μ. Going back to an arbitrary choice of μ, define $A = \bigcup_{m=1}^{\infty} \{u_m = +\infty\}$. Then A is l_N null since u_m is superharmonic, and in fact A is what will be described in Section V.1 as a polar set. Observe that for $k > 0$ and $m > k + 1$ the function $u_{m+1} - u_m$ is harmonic and negative on $B(0, k)$. In view of the Harnack Convergence Theorem (Section II.3) we conclude that either (a) $G\mu = -\infty$ on $B(0, k) - A$ and $G\mu$ is not uniquely defined on A or (b) $G\mu$ is uniquely defined and superharmonic on $B(0, k)$. Since k can be chosen arbitrarily large, either (a) $G\mu = -\infty$ on $\mathbb{R}^2 - A$ and $G\mu$ is not uniquely defined on A or (b) $G\mu$ is uniquely defined and superharmonic on $\mathbb{R}^2$. Thus $G\mu$ is superharmonic if and only if

$$\int_{\mathbb{R}^2 - B(0,1)} \log|\eta|\,\mu(d\eta) < +\infty; \tag{1.4}$$

the condition $\mu(\mathbb{R}^2) < +\infty$ is necessary but not sufficient for (1.4). See Section 9 for a further analysis of superharmonic functions on $\mathbb{R}^2$.

Dependence of G_D and $G_D\mu$ on D

If D_1 and D_2 are special open subsets of $\mathbb{R}^N$ with $D_1 \subset D_2$, it follows from the maximum-minimum theorem for harmonic functions that $G_{D_1} \le G_{D_2}$ on $D_1 \times D_1$ because if $\xi \in D_1$, the function $G_{D_2}(\xi, \cdot) - G_{D_1}(\xi, \cdot)$ if properly defined at ξ is harmonic on D_1 with positive limit at every boundary point. In this sense the function $D \mapsto G_D$ is monotone increasing. In particular, $G_{B(0,r)}$ increases as r increases, and (direct calculation) $\lim_{r\to\infty} G_{B(0,r)}$ is either identically $+\infty$ or G on $\mathbb{R}^N \times \mathbb{R}^N$ according as $N = 2$ or $N > 2$. If μ is a not identically vanishing measure on $\mathbb{R}^N$ and if μ_r is the projection of μ on $B(0, r)$, then $\lim_{r\to\infty} G_{B(0,r)}\mu_r$ is identically $+\infty$ if $N = 2$ and is $G\mu$ if $N > 2$. Whatever the dimensionality, if μ is a finite measure on the ball B, the potentials $G_B\mu$ and $G\mu$ are superharmonic and differ on B by a harmonic function.

2. Examples

Let μ be a unit mass distributed uniformly on $\partial B(0, \delta)$. Then

$$G\mu(\eta) = \begin{cases} \delta^{-N+2} & \text{on } B(0,\delta), \\ |\eta|^{-N+2} & \text{on } \mathbb{R}^N - B(0,\delta) \end{cases} \qquad (N > 2), \tag{2.1}$$
$$G_{B(0,a)}\mu = G\mu - a^{-N+2} \quad \text{on } B(0,a) \text{ if } a > \delta.$$

In fact the potential $G\mu$ must be a function of the distance $|\eta|$ from the origin, must be superharmonic, and must be harmonic on $\mathbb{R}^N - \partial B(0, \delta)$, and therefore (according to Section II.10) must be a concave function of $|\eta|^{-N+2}$, and must be a linear function of $|\eta|^{-N+2}$ on each open component of $\mathbb{R}^N - \partial B(0,\delta)$. Finally, this potential is δ^{-N+2} at the origin and has limit 0 at ∞. Hence the first two lines of (2.1) are correct. If $a > \delta$, the difference $G\mu - G_{B(0,a)}\mu$ is a function of $|\eta|$, defined and harmonic on $B(0,a)$, with limit $G\mu = a^{-N+2}$ at the boundary. Hence (maximum-minimum theorem for harmonic functions) this difference is identically a^{-N+2}; so the third line of (2.1) is correct.

Similarly

$$G\mu(\eta) = \begin{cases} -\log\delta & \text{on } B(0,\delta), \\ -\log|\eta| & \text{on } \mathbb{R}^2 - B(0,\delta) \end{cases} \qquad (N = 2), \tag{2.2}$$
$$G_{B(0,a)}\mu = G\mu + \log a \quad \text{on } B(0,a) \text{ if } a > \delta.$$

If $\mu_{\xi\delta}$ is a unit mass distributed uniformly on $\partial B(\xi,\delta)$, then

$$L(G(\eta,\cdot),\xi,\delta) = G\mu_{\xi\delta}(\eta), \tag{2.3}$$

and it follows from (2.1) and (2.2) that for $N \geq 2$ the function of (ξ,η) defined by (2.3) is continuous on $\mathbb{R}^N \times \mathbb{R}^N$. If $a > \delta$ and if $\xi \in B(0, a-\delta)$, the difference $G\mu_{\xi\delta} - G_{B(0,\delta)}\mu_{\xi\delta} = h$ is harmonic on $B(0,a)$; so

$$L(G_{B(0,a)}(\eta,\cdot),\xi,\delta) = G\mu_{\xi\delta}(\eta) - h(\xi). \tag{2.4}$$

It follows that the function of (ξ,η) defined by (2.4) is continuous on $B(0,a-\delta) \times B(0,a)$.

3. A Fundamental Smallness Property of Potentials

Theorem. *If D is a Greenian subset of $\mathbb{R}^N$ and if $u = G_D\mu$ is a superharmonic potential on D, then* $\mathrm{GM}_D u = 0$.

If D is special, choose a sequence $B_\bullet$ of balls with closures in D and with the property that if ξ is in D, some neighborhood of ξ is contained in infinitely many of the balls. Then (Section III.1) $\tau_{B_n} \cdots \tau_{B_1}\mu \downarrow \mathrm{GM}_D u$. Moreover

$$\tau_{B_n} \cdots \tau_{B_1} u(\xi) = \int_D (\tau_{B_n} \cdots \tau_{B_1}) G_D(\cdot,\eta)(\xi)\mu(d\eta),$$

and the integrand decreases to 0 when $n \to \infty$ because (Section III.1) $\mathrm{GM}_D G_D(\cdot,\eta) = 0$. Hence $\mathrm{GM}_D u = 0$, as was to be proved. Observe that when D is special, the simplest choice of $B_\bullet$ is an increasing sequence of concentric balls with union D. The proof as stated will be valid for arbitrary

Greenian D once G_D has been defined in Section VII.1. The converse of Theorem 3 is also true: it will be seen in Section 8 that for u positive and superharmonic on a Greenian set D the condition $\mathrm{GM}_D u = 0$ is necessary and sufficient for u to be a potential on D.

4. Increasing Sequences of Potentials

Theorem. *If D is a Greenian subset of $\mathbb{R}^N$, if $\mu_\cdot$ is a sequence of measures supported by a compact subset A of D, if* vague $\lim_{n\to\infty} \mu_n = \mu$, *and if $G_D\mu_1 \le G_D\mu_2 \le \cdots$, then* $\lim_{n\to\infty} G_D\mu_n = G_D\mu$.

We defer the proof for Greenian sets and suppose that D is special. Since $\sup_n \mu_n(A) < +\infty$, the function $u = \lim_{n\to\infty} G_D\mu_n$ is finite on $D - A$ and is therefore superharmonic on D. Apply Fubini's theorem and the fact that $(\xi, \eta) \mapsto L(G_D(\eta, \cdot), \xi, r)$ is continuous (Section 2) to deduce

$$L(u, \xi, r) = \lim_{n\to\infty} L(G_D\mu_n, \xi, r) = \int_A L(G_D(\eta, \cdot), \xi, r)\mu(d\eta) = L(G_D\mu, \xi, r). \quad (4.1)$$

When $r \to 0$, we find that $u = G_D\mu$, as was to be proved.

5. Smoothing of a Potential

Let μ be a measure on $\mathbb{R}^N$ with compact support A. An elementary calculation shows that $A_\alpha(G\mu) = G\mu_\alpha$, where μ_α is absolutely continuous relative to l_N with infinitely differentiable density

$$\frac{d\mu_\alpha}{dl_N}(\xi) = \alpha^{-N} \int_{\mathbb{R}^N} \gamma_N\left(\frac{|\xi - \eta|}{\alpha}\right)\mu(d\eta). \quad (5.1)$$

Moreover $\mu_\alpha(\mathbb{R}^N) = \mu(A)$, the measure μ_α is supported by $\{\xi : |\xi - A| < \alpha\}$, and $\lim_{\alpha\to 0} \mu_\alpha = \mu$ (vague convergence). Finally (from Section II.6) $G\mu_\alpha$ increases when α decreases and $\lim_{\alpha\to 0} G\mu_\alpha = G\mu$. According to Theorem I.7, the density (5.1) is equal to $-\Delta(G\mu_\alpha)/\pi'_N$.

If B is a ball containing A and if $\alpha < |A - \partial B|$, the function $A_\alpha(G_B\mu)$ is defined and superharmonic on $\{\xi \in B\colon |\xi - \partial B| > \alpha\}$ and is harmonic and equal to $G_B\mu$ on $\{\xi \in B\colon |\xi - A| > \alpha, |\xi - \partial B| > \alpha\}$. Define $A_\alpha(G_B\mu)$ on $\{\xi \in B\colon |\xi - \partial B| \le \alpha\}$ as $G_B\mu$ to obtain an infinitely differentiable superharmonic function on B. Since $G_B\mu$ differs from $G\mu$ on B by a harmonic function the results in the preceding paragraph yield the fact that $A_\alpha(G_B\mu) = G_B\mu_\alpha$, where μ_α, the same measure as above, has density $-\Delta(G_B\mu_\alpha)/\pi'_N = -\Delta(G\mu_\alpha)/\pi'_N$ relative to l_N, and $G_B\mu_\alpha$ increases to the limit $G_B\mu$ when $\alpha \downarrow 0$.

6. Uniqueness of the Measure Determining a Potential

Theorem. *Let D be a Greenian subset of $\mathbb{R}^N$ $(N \geq 2)$ or $D = \mathbb{R}^2$, and let μ and ν be measures on D for which $G_D\mu$ and $G_D\nu$ are superharmonic. Suppose that A is an open subset of D and that there is a function h defined and harmonic on A for which $G_D\mu = G_D\nu + h$ on A. Then the projections of μ and ν on A are identical.*

The proof will be given for $D = \mathbb{R}^N$ and requires only trivial changes for D a ball but the remarks validating the proof for a general Greenian set are deferred to Section VII.8. It is sufficient to prove the result for A relatively compact in $\mathbb{R}^N$, and therefore it can be supposed, replacing μ and ν by their projections on a ball containing $\bar{A}$ if necessary, that μ and ν have compact supports. According to Section 5, the functions $A_\alpha(G\mu)$, $A_\alpha(G\nu)$ are infinitely differentiable potentials of measures μ_α, ν_α with respective densities $-\Delta A_\alpha(G\mu)/\pi_N'$, $-\Delta A_\alpha(G\nu)/\pi_N'$ relative to l_N. Since these densities are equal on the set $\{\xi \in A : |\xi - \partial A| > \alpha\}$, the projections of μ_α and ν_α on this set are identical, and since μ_α and ν_α have vague limits μ and ν when $\alpha \to 0$, the projections of μ and ν on A are identical, as was to be proved.

Extension of Theorem 6 to Charges

In the following extension of Theorem 6 we consider charges μ on D (see Appendix IV.7) with the following property: μ has positive and negative variations μ^+ and $-\mu^-$, respectively, with the property that $G_D\mu^+$ and $G_D\mu^-$ are superharmonic. As usual μ^+ and μ^- are finite on compact subsets of D, but if $D \neq \mathbb{R}^2$ both $\mu^+(D)$ and $\mu^-(D)$ may be $+\infty$, so μ need not be a signed measure. The potential $G_D\mu$ is defined as $G_D\mu^+ - G_D\mu^-$ on the subset of D on which at least one of the potentials $G_D\mu^+$, $G_D\mu^-$ is finite. If ν is a second such charge on D, if A is an open subset of D and if h is a function defined and harmonic on A, suppose that $G_D\mu = G_D\nu + h$ on A in the sense that $G_D(\mu^+ + \nu^-) = G_D(\mu^- + \nu^+) + h$ on A. According to Theorem 6, if this condition is satisfied, it follows that $\mu^+ + \nu^- = \mu^- + \nu^+$ on subsets of A. Since the four measures involved here are finite valued on compact subsets of D, we conclude that $\mu = \nu$ on the compact subsets of A, so these measures have identical projections on A. Thus we have extended Theorem 6 to cover potentials of charges as restricted above.

In particular, if $G_D\mu$ is harmonic on the open subset A of D, so that we can take ν as the zero charge, we conclude that A is μ null. As a second particular case suppose that $G_D\mu^+$ and $G_D\mu^-$ are finite valued on an open subset A of D and that $(G_D\mu)_{|A}$ is in the class $\mathbb{C}^{(3)}(A)$. Let ν be the charge on A with density $-(\Delta G_D\mu)/\pi_N'$ relative to l_N. If A_0 is an open relatively compact subset of A and if ν_0 is the projection of ν on A_0 then (Section I.7) $(G_D\nu_0)_{|A}$ is in the class $\mathbb{C}^{(2)}(A)$, and $\Delta(G_D\mu - G_D\nu_0) = 0$ on A_0; so $G_D\mu$ differs from $G_D\nu_0$ on A_0 by a harmonic function and it follows that $\mu = \nu_0$ on subsets of A_0 and so on subsets of A; that is, $d\mu = -\Delta(G_D\mu)/\pi'_N\, dl_N$ on A. The hypothesis that $(G_D\mu)_{|A} \in \mathbb{C}^{(3)}(A)$ can be weakened, for example, to the condition $G_D\mu \in \mathbb{C}^{(2)}(A)$ by applying the result just obtained to a smoothing $A_\alpha(G_D\mu)$, but we shall not need this refinement.

7. Riesz Measure Associated with a Superharmonic Function

Theorem. *If u is a superharmonic function on an open subset D of $\mathbb{R}^N$ $(N \geq 2)$, there is a unique associated measure μ on D with the property that if A is an open relatively compact subset of D, and if μ_A is the projection of μ on A, there is a superharmonic function h_A on D, harmonic on A, with*

$$u = G\mu_A + h_A \tag{7.1}$$

on D.

The uniqueness of μ_A is obvious from Theorem 6, and incidentally this uniqueness implies that the measure determining a potential is the measure associated with the potential. If the theorem is true with h_A supposed only defined and harmonic on A and satisfying (7.1) on A, this function can be extended to a superharmonic function on D which satisfies (7.1) on D. In fact, if B is an open superset of A, relatively compact in D, and if v is the projection of μ_B on $B - A$, then $h_A = Gv + h_B$ on A. Thus h_A has a superharmonic extension $Gv + h_B$ to B, and when h_A is so extended, $u = G\mu_A + h_A$ on B. In this way h_A can be extended in steps to D. Finally, it is sufficient to prove that *if A is a ball with closure in D, there is a measure μ_A defined on A and a harmonic function h_A defined on A for which $u = G\mu_A + h_A$ on A.* In fact, if two such balls overlap the parts of the two associated measures on the intersection of the balls must be identical by Theorem 6, and the desired measure on D is obtained by piecing together the measures associated with the balls.

To prove the italicized statement we first replace D by a subset if necessary to make u bounded from below on D, and adding a constant if necessary, we can even suppose that u is positive on D. Since only the restriction of u to A is relevant to the desired result, we can replace u on D by R^A_{+u}, a positive superharmonic function on D, equal to u on A, harmonic on $D - \bar{A}$. The proof now proceeds as follows. Let B be a ball concentric with and larger than A, with closure in D. Choose α so small that α is strictly exceeded by the difference between the radii of A and B and by $|B - \partial D|$. The function $A_\alpha u$ is superharmonic and infinitely differentiable on a neighborhood of $\bar{B}$, equal to u, and harmonic on a neighborhood of ∂B, and $\lim_{\alpha\to 0} A_\alpha u = u$ on D. According to I(8.6),

$$A_\alpha u = G_B\mu_\alpha + \mathrm{PI}(B, u) \tag{7.2}$$

on B, where μ_α is the measure on B with density $-\Delta A_\alpha u/\pi'_N$ relative to l_N. The value of $\mu_\alpha(B)$ can be found by applying the Gauss integral theorem (integrate over ∂B), and since $A_\alpha u = u$ on B, $\mu_\alpha(B)$ does not depend on α. Thus as $\alpha \to 0$ along a suitable sequence μ_α has a vague limit measure μ_0 supported by $\bar{A}$. Applying Theorem 4, we find that if μ_A is the projection of μ_0 on A, then

$$u = \begin{cases} G_B\mu_0 + \mathrm{PI}(B,u) & \text{on } B, \\ G\mu_A + h_A & \text{on } A, \end{cases} \tag{7.3}$$

where h_A is harmonic on A, and we have used the fact that $G_B\mu_A$ and $G\mu_A$ differ on A by a harmonic function.

Observation. It follows from Theorem 6 that the map $u \mapsto \mu$ from superharmonic functions into their associated measures is additive ($u_1 \mapsto \mu_1$ and $u_2 \mapsto \mu_2$ imply that $u_1 + u_2 \mapsto \mu_1 + \mu_2$) and positive homogeneous ($u \mapsto \mu$ implies that $cu \mapsto c\mu$ for c a positive constant).

8. Riesz Decomposition Theorem

Theorem. *If D is a Greenian subset of $\mathbb{R}^N$ $(N \geq 2)$, if u is a superharmonic function on D with associated measure μ, and if u has a subharmonic minorant, then*

$$u = G_D\mu + \mathrm{GM}_D u. \tag{8.1}$$

In the most common application u is supposed positive, in which case the harmonic function $\mathrm{GM}_D u$ is also positive. We defer the general Greenian case to Section VII.8 and treat here only special sets D. If D is a ball, let B be a smaller concentric ball. In the notation of Theorem 7, $u = G\mu_B + h_B = G_D\mu_B + h'_B$, where h'_B is superharmonic on D and harmonic on B. Now according to Section III.1, the minorant $\mathrm{GM}_D v$ depends additively on the superharmonic function v, and according to Section 3, this minorant vanishes if v is a potential. Hence $\mathrm{GM}_D u = \mathrm{GM}_D h'_B \leq h'_B$. When B increases to D, we conclude that $G_D\mu$ is superharmonic, that h'_B decreases to a harmonic function h on D, and that $u = G_D\mu + h$. Then $h = \mathrm{GM}_D u$, and the proof is complete. The proof for $\mathbb{R}^N$ when $N > 2$ is essentially the same.

The following assertions about a positive superharmonic function u on D are easy consequences of the Riesz theorem.

(a) The function u is a potential $G_D\mu$ if and only if $\mathrm{GM}_D u = 0$, equivalently, when $D = B(\xi, \delta)$, if and only if $\lim_{r\to\delta} L(u,\xi,r) = \lim_{r\to\delta} \tau_{B(\xi,r)}u(\xi) = 0$ (or when $D = \mathbb{R}^N$ with $N > 2$, if and only if this limit is 0 when $r \to +\infty$).
(b) If u is majorized by a superharmonic potential u is itself a potential.
(c) Special case of (b). If u is a superharmonic potential, then $\underset{+u}{R^A}$ is also, for every choice of A; in particular, $\underset{+u}{R^{\partial D}} = 0$ for every choice of ∂D.

(d) The smoothed reduction $\underset{+u}{R}^A$ is a potential if A is a relatively compact subset of D (because $\mathrm{GM}_D \underset{+u}{R}^A = 0$ according to Theorem III.6).

Observe that a superharmonic function u on D is the sum of a potential and a harmonic function if and only if u has a subharmonic minorant.

9. Counterpart for Superharmonic Functions on $\mathbb{R}^2$ of the Riesz Decomposition

Let u be a superharmonic function on $\mathbb{R}^2$ and let δ be a strictly positive number. Then (Section II.4) the function $L(u, \cdot, \delta)$ is superharmonic on $\mathbb{R}^2$. For fixed ξ in $\mathbb{R}^2$ the function

$$\delta \mapsto \frac{L(u, \xi, \delta) - L(u, \xi, 1)}{-\log \delta} = m(u, \xi, \delta)$$

is a positive increasing function on $]1, +\infty[$ in view of the superharmonic function average inequality and the fact (Section II.10) that the function $L(u, \xi, \cdot)$ is a concave function of $\log \delta$. Hence (Section II.4) the function $m(u) = \lim_{\delta \to \infty} m(u, \cdot, \delta)$ is either identically $+\infty$ or a positive superharmonic function, and in the latter case it is identically constant because (Section II.13) the plane is not Greenian.

Theorem. (a) *If u is a superharmonic function on $\mathbb{R}^2$, then u is harmonic if and only if $m(u) = 0$.*

(b) *If μ is a measure on $\mathbb{R}^2$ and if $g(\delta) = \mu(\bar{B}(0, \delta))$, then $G\mu$ is superharmonic on $\mathbb{R}^2$ if and only if*

$$\int_1^\infty \log \delta \, dg(\delta) < +\infty. \tag{9.1}$$

(c) *If μ is a measure on $\mathbb{R}^2$ satisfying* (9.1) *and if $u = G\mu$, then*

$$\mu(\mathbb{R}^2) = m(u) < +\infty, \qquad \lim_{\delta \to \infty} [L(u, 0, \delta) + g(\delta) \log \delta] = 0, \tag{9.2}$$

and (9.2) *is also true if $g(\delta)$ is replaced by $\mu(\mathbb{R}^2)$.*

(d) *Let u be a superharmonic function on $\mathbb{R}^2$ with associated Riesz measure μ and define g as in* (b). *If*

$$\liminf_{\delta \to \infty} [L(u, 0, \delta) + g(\delta) \log \delta] < +\infty, \tag{9.3}$$

then $G\mu$ is superharmonic and $u = G\mu + h$, with h a harmonic function on $\mathbb{R}^2$ given by

$$h(\xi) = \lim_{\delta \to \infty} [L(u, \xi, \delta) + \mu(\mathbb{R}^2) \log \delta]. \tag{9.4}$$

Proof. (a) If u is harmonic on $\mathbb{R}^2$ the function $m(u, \cdot, \delta)$ vanishes identically; so $m(u) = 0$. Conversely, if u is superharmonic on $\mathbb{R}^2$ and $m(u) = 0$, then since $m(u, \xi, \cdot)$ is a positive increasing function on the interval $]1, +\infty[$, it follows that this function vanishes on the interval; that is, $L(u, \xi, \cdot)$ is a constant function there. Since $L(u, \xi, \cdot)$ is a decreasing concave function of $\log \delta$ for $\delta > 0$, it follows that $L(u, \xi, \cdot)$ is identically constant; that is, u satisfies the harmonic function average equality and so is harmonic.

(b) and (c) The condition (9.1) that $G\mu$ be superharmonic was derived in Section 1 [see (1.4)] in a trivially different form. If (9.1) is satisfied, the evaluation of $L(G(\xi, \cdot), \cdot, \cdot)$ in Section 2 or Section I.5 yields

$$\begin{aligned} L(G\mu, 0, \delta) &= -\mu(\bar{B}(0, \delta)) \log \delta - \int_{\mathbb{R}^2 - \bar{B}(0,\delta)} \log |\eta| \, \mu(d\eta) \\ &= -g(\delta) \log \delta - \int_{]\delta, +\infty[} \log s \, dg(s). \end{aligned} \tag{9.5}$$

The function g is positive, monotone increasing, and right continuous, with limit $\mu(\mathbb{R}^2)$ at $+\infty$. Thus (9.2) is true. Moreover (9.1) implies that

$$\lim_{\delta \to \infty} [\mu(\mathbb{R}^2) - g(\delta)] \log \delta = 0; \tag{9.6}$$

so $g(\delta)$ can be replaced in (9.2) by $\mu(\mathbb{R}^2)$.

(d) Under (d) let μ_δ be the projection of μ on $B(0, \delta)$. According to Theorem 7, there is a superharmonic function h_δ on $\mathbb{R}^2$, harmonic on $B(0, \delta)$, such that

$$u = G\mu_\delta + h_\delta. \tag{9.7}$$

Now (9.5) with μ replaced by μ_δ yields the equality $L(G\mu_\delta, 0, \delta) = -g(\delta-) \log \delta$. Furthermore the function $r \mapsto L(h_\delta, 0, r)$ is continuous and equal to $h_\delta(0)$ for $r < \delta$; so $L(h_\delta, 0, \delta) = h_\delta(0)$. Hence

$$L(u, 0, \delta) = -g(\delta-) \log \delta + h_\delta(0). \tag{9.8}$$

For fixed η the function $\delta \mapsto G\mu_\delta(\eta)$ is monotone decreasing on the interval $]1 + |\eta|, +\infty[$, so the function $\delta \mapsto h_\delta(\eta)$ is increasing on this interval. Hence (Harnack convergence theorem) either $\lim_{\delta \to \infty} h_\delta = +\infty$ on $\mathbb{R}^2$ or $\lim_{\delta \to \infty} h_\delta = h$ is a harmonic function on $\mathbb{R}^2$. In view of (9.8) the first case is excluded by (9.3). Hence in the limit (9.7) yields the fact that $G\mu$ is superharmonic and $u = G\mu + h$. Moreover from (9.8)

$$h(0) = \lim_{\delta \to \infty} [L(u, 0, \delta) + g(\delta) \log \delta],$$

and as noted in the proof of (c), the value $g(\delta)$ can be replaced by $\mu(\mathbb{R}^2)$ in this limit relation. Thus (9.4) is true for $\xi = 0$ and therefore for all ξ since ξ rather than the origin can be chosen as the reference point in this discussion. □

10. An Approximation Theorem

The following theorem is an example of the application of the reduction and related operations.

Theorem. *Every superharmonic function u on an open subset D of $\mathbb{R}^N$ is the limit of an increasing sequence $u_{\bullet}$ of superharmonic functions with the following properties:*

(a) *u_n is upper bounded.*
(b) $\inf_D u_n = \inf_D u$.
(c) *u_n is infinitely differentiable.*
(d) *[At the possible sacrifice of* (a)] *if u is harmonic outside a compact subset A of D, then $u_n = u$ outside a compact neighborhood A_n of A with $A_{n+1} \subset A_n$ and $\bigcap_0^\infty A_n = A$.*

If u is harmonic outside a compact subset A of D, we can choose $u_{\bullet}$ to satisfy (b)–(d) as follows. Choose a sequence $\alpha_{\bullet}$ in $\mathbb{R}^+$ satisfying

$$\tfrac{1}{3}|A - \partial D| > \alpha_0 > \alpha_1 > \cdots, \qquad \lim_{n \to \infty} \alpha_n = 0,$$

and define $u_n = A_{\alpha_n} u$ on the set $B_n = \{\xi : |\xi - A| < 2\alpha_n\}$. The function u_n is superharmonic and is equal to u near ∂B_n. Define $u_n = u$ on $D - B_n$ to obtain a sequence $u_{\bullet}$ satisfying (b)–(d), with $A_n = \bar{B}_n$.

To find a sequence $u_{\bullet}$ satisfying (a)–(c) for general superharmonic u suppose first that $\inf_D u = 0$ and let $D_{\bullet}$ be an increasing sequence of open relatively compact subsets of D with union D. Define $v_n = R^{D_n}_{u \wedge n}$, a positive superharmonic function on D, bounded by n, equal to $u \wedge n$ on D_n, and harmonic on $D - \bar{D}_n$. Now proceed as suggested by the method used in the preceding paragraph. Define $\alpha_n = |D_n - \partial D|/3$, except that $\alpha_n = 1/n$ if $D = \mathbb{R}^N$, and define $u_n = A_{\alpha_n}(v_n, \cdot)$ on the set $B_n = \{\xi : |\xi - D_n| < 2\alpha_n\}$. The function u_n is superharmonic, majorized by v_n, and equal to v_n near ∂B_n. Define $u_n = v_n$ on $D - B_n$ to obtain a sequence $u_{\bullet}$ satisfying (a)–(c).

If $\inf_D u = \beta > -\infty$, find a sequence $u_{\bullet}$ satisfying (a)–(c) by adding β to each member of a sequence satisfying these conditions for the function $u - \beta$.

If u is not lower bounded, modify the approximation procedure as follows. Choose D_n as above and also to satisfy $\bar{D}_n \subset D_{n+1}$, and define v_n by

$$v_n = \inf\left\{v\colon v \text{ superharmonic on } D,\ v \geq u \wedge n \text{ on } D_{2n} \cup \left(\bigcup_n^\infty \partial D_{2j}\right)\right\} \tag{10.1}$$

so that the lower semicontinuous smoothing $\underset{+}{v}_n$ is majorized by $u \wedge n$, coincides with $u \wedge n$ on D_{2n}, and (Fundamental Convergence Theorem) is superharmonic on D and harmonic on $\bigcup_n^\infty (D_{2j+2} - \bar{D}_{2j})$. Define $\alpha_.$ as a decreasing sequence of strictly positive numbers satisfying

$$\alpha_n < |\partial D_{2n+2} - D_{2n+1}| \wedge |\partial D_{2n+1} - D_{2n}|, \qquad \lim_{n\to\infty} \alpha_n = 0,$$

and define u_n by

$$u_n = \begin{cases} A_{\alpha_n} \underset{+}{v}_n & \text{on } \bar{D}_{2n+1}, \\ A_{\alpha_k} \underset{+}{v}_n & \text{on } D_{2k+1} - \bar{D}_{2k-1} \text{ for } k > n. \end{cases} \tag{10.2}$$

The sequence $u_.$ has the desired properties (a)–(c).

Specialization to Positive Superharmonic Functions and Potentials

If D in Theorem 10 is Greenian and if u is a positive superharmonic function on D, each approximation u_n can be chosen to satisfy (a)–(c) and also to be the potential of a measure with compact support. If u is itself the potential of a measure with compact support A, each approximation u_n can be chosen to satisfy (a)–(d) with u_n the potential of a measure supported by A_n [notation of (d)].

It is convenient to suppose first that u is the potential of a measure with compact support A. Then u is harmonic on $D - A$. Furthermore u is bounded outside each neighborhood of A because the function $(\xi, \eta) \mapsto G_D(\xi, \eta)$ is a bounded function for η restricted to A and ξ restricted to the complement of a neighborhood of A. (This property of the Green function is trivial for D special and will be proved in the general case.) The first paragraph of the proof of Theorem 10 furnishes a sequence $u_.$ with the properties (a)–(d); u_n is a potential because [Section 8(b)] u_n is majorized by the potential u; the Riesz measure associated with u_n is supported by A_n because u_n is harmonic on $D - A_n$. Thus the assertion for u when u is the potential of a measure with compact support in D is true. For u superharmonic and positive the proof of Theorem 10 for the case $\inf_D u = 0$ provides a sequence $u_.$ with the desired properties. In fact each function u_n is a potential because by compactness of $\bar{D}_n$ the function $v_n = R^{D_n}_{u \wedge n}$ is a potential [Section 8(d)], and the positive superharmonic function u_n is a potential because it is majorized by v_n.

Chapter V

Polar Sets and Their Applications

1. Definition

A *polar* subset of $\mathbb{R}^N$ is a set to each point of which corresponds an open neighborhood of the point that carries a superharmonic function equal to $+\infty$ at each point of the set in the neighborhood. An *inner polar* set is a set whose compact subsets are polar. It will be shown in Section VI.2 that an analytic inner polar set is polar. If a set is (inner) polar its Kelvin transforms are also.

In particular, the set of infinities of a superharmonic function is a polar subset of its domain. Conversely, it will be shown (Theorem 2) that a polar set is always a subset of the set of infinities of a single superharmonic function defined on $\mathbb{R}^N$.

The polar sets are the negligible sets of classical potential theory. An assumption about points of $\mathbb{R}^N$ true except for the points of an [inner] polar set is said to be true [inner] quasi everywhere. A subset of an [inner] polar set is [inner] polar. A singleton $\{\xi\}$ is polar because $G(\xi, \cdot)$ is superharmonic on $\mathbb{R}^N$ and equal to $+\infty$ at ξ. Although the point ∞ is not in $\mathbb{R}^N$, that point is considered a Euclidean boundary point of every unbounded set. In a context allowing ∞ in the domain of harmonic and superharmonic functions, this point is polar for $N = 2$ but not for $N > 2$.

Since a superharmonic function on an open subset of $\mathbb{R}^N$ is l_N integrable on every closed ball in its domain, and since every polar set A can be covered by a countable number of open sets, each carrying a positive superharmonic function with value $+\infty$ on the part of A in its domain, a polar set has l_N measure 0. It follows that an l_N measurable inner polar set also has l_N measure 0.

If u and v are superharmonic functions on an open subset of $\mathbb{R}^N$ and if $u = v$ inner quasi everywhere, or if $u \geq v$ inner quasi everywhere, then the same relation holds l_N almost everywhere and therefore [Section II.6(f)] everywhere.

2. Superharmonic Functions Associated with a Polar Set

Theorem. *If A is a polar subset of $\mathbb{R}^N$, there is a function superharmonic on $\mathbb{R}^N$ and identically $+\infty$ on A. This function can be chosen to be the potential $G\mu$ of a measure μ with $\mu(\mathbb{R}^N)$ finite and to be finite at any preassigned point of $\mathbb{R}^N - A$.*

To prove the theorem suppose that $\xi \in \mathbb{R}^N - A$ and apply the Lindelöf covering theorem to cover A by balls $B_0, B_1, \ldots$ so small that ξ is not in any ball closure and that to each ball B_k corresponds a function u_k defined and superharmonic on an open neighborhood of $\bar{B}_k$ and identically $+\infty$ on $B_k \cap A$. Let μ_k be the projection on B_k of the Riesz measure associated with u_k; choose a strictly positive constant c_k so small that $c_k\mu_k(B_k) < 2^{-k}$, that $c_k|G\mu_k(\xi)| < 2^{-k}$, and if $N = 2$, that $c_k \int_1^\infty \log|\eta|\,\mu_k(d\eta) < 2^{-k}$. The superharmonic potential $G\sum_0^\infty c_k\mu_k$ is $+\infty$ on A and finite at ξ.

Observation (a). Since the set of infinities of a superharmonic function v is the G_δ set $\bigcap_0^\infty \{v > n\}$, every polar set is a subset of a G_δ polar set.

Observation (b). Since a superharmonic function is l_{N-1} integrable on every ball boundary in its domain, a polar set meets a ball boundary in an l_{N-1} null set.

Observation (c). The complement of a closed polar subset A of $\mathbb{R}^N$ is connected. To see this, let B be an open connected component of $\mathbb{R}^N - A$, let u be a superharmonic function on $\mathbb{R}^N$, identically $+\infty$ on A, and define $v = +\infty$ on B and $v = u$ on $\mathbb{R}^N - B$. The function v satisfies the conditions for a function to be superharmonic on $\mathbb{R}^N$ except for the finiteness condition. Hence v is either identically $+\infty$ or superharmonic. Both alternatives are impossible unless $\mathbb{R}^N - A$ is connected.

Observation (d). If D is an open subset of $\mathbb{R}^N$, the set $\mathbb{R}^N - D$ is polar if and only if the finite part $\partial^\circ D = \mathbb{R}^N \cap \partial D$ of the boundary is polar, and then $\partial^\circ D = \mathbb{R}^N - D$. In fact, if $\partial^\circ D$ is polar, its complement is connected and everywhere dense and so is equal to D, and therefore $\mathbb{R}^N - D = \partial^\circ D$ is polar. Conversely, if $\mathbb{R}^N - D$ is polar, the set D is everywhere dense; so $\partial^\circ D = \mathbb{R}^N - D$ is polar.

Extension. If D is a Greenian subset of $\mathbb{R}^N$ and if A is a polar subset of D, then there is a positive function superharmonic on D and identically $+\infty$ on A. This function can be chosen to be the potential $G_D\mu$ of a measure μ with $\mu(D)$ finite and to be finite at any preassigned point of $D - A$.

If D is special, the proof of Theorem 2 for $N > 2$ with $\mathbb{R}^N$ replaced by D and G replaced by G_D is valid in the present context. This same proof will be valid in the case of general Greenian D once (Section VII.1) G_D has been defined.

3. Countable Unions of Polar Sets

Theorem. *A countable union of polar sets is polar. In particular, an inner polar F_σ set is polar.*

In fact, if $A_0, A_1, \ldots$ are polar, if $\xi \in \mathbb{R}^N - \bigcup_0^\infty A_k$, and if μ_k is a measure on $\mathbb{R}^N$ with $|G\mu_k(\xi)| + \mu_k(\mathbb{R}^N) < 2^{-k}$, with $G\mu_k = +\infty$ on A_k and (if $N = 2$) with $\int_1^\infty \log|\eta|\mu_k(d\eta) < 2^{-k}$, then the superharmonic potential $G(\Sigma_0^\infty \mu_k)$ is $+\infty$ on $\bigcup_0^\infty A_k$.

Since singletons are polar, this theorem implies that countable sets are polar. For example, suppose that A is a countable dense subset of $\mathbb{R}^N$, and let u be a superharmonic function on $\mathbb{R}^N$, equal to $+\infty$ on A. The set of infinities of u is a polar dense G_δ set and therefore is not countable.

It will be shown in Section VI.2 that an analytic inner polar set is polar.

According to Theorem 3, if u and v are superharmonic functions on an open subset of $\mathbb{R}^N$ and if $v \leq u$ inner quasi everywhere on an F_σ set A, then this inequality must hold quasi everywhere on A because the set

$$\{v > u\} \cap A = \bigcup \{v > r\} \cap \{r \geq u\} \cap A \qquad (r \text{ rational})$$

is an F_σ inner polar set and is therefore polar.

4. Properties of Polar Sets

Theorem. *The following conditions on a subset A of a connected Greenian set D are equivalent:*

(a) *A is polar.*
(b) *There is a superharmonic potential $u = G_D\mu$ with $\mu(D) < +\infty$ and $u = +\infty$ on A.*
(c) *If u is superharmonic and strictly positive on D, $\underset{+u}{R^A} \equiv 0$.*
(d) *There is a strictly positive superharmonic function u on D for which R_u^A has a zero.*

In the following argument D is special. We defer the general Greenian case to Section VII.8.

Proof. (a) $\Rightarrow$ (b) Has already been proved for $D = \mathbb{R}^N$. The proof for a ball is similar.

(b) $\Rightarrow$ (c) Let u be a positive superharmonic function on D, identically $+\infty$ on A. Then (for $n \geq 1$) $u/n \geq u$ on A; so $R_u^A \leq u/n$, and therefore $R_u^A = 0$ at every point where u is finite. Hence the positive superharmonic function $\underset{+u}{R}^A$ has a zero and accordingly must vanish identically (superharmonic function minimum theorem).

(c) ⇒ (d) The result follows because $\underset{+u}{R}^A = R_u^A$ l_N almost everywhere.

(d) ⇒ (a) The result follows because if $R_u^A(\xi) = 0$, there is a function v_n positive superharmonic on D, $\geq u$ on A, $\leq 2^{-n}$ at ξ, so that $\Sigma_0^\infty v_n$ is superharmonic on D, identically $+\infty$ on A. □

5. Extension of a Superharmonic Function

Theorem. *Let D be an open connected subset of $\mathbb{R}^N$, let A be a polar subset of D, and let u be an extended real-valued function on $D - A$ satisfying the following conditions: $-\infty < u \leq +\infty$; $u \not\equiv +\infty$; u is locally lower bounded; u is lower semicontinuous; if $\xi \in D - A$, then $u(\xi) \geq L(u, \xi, \delta)$ for sufficiently small δ, depending on ξ. Then u has a unique superharmonic extension u' to D, determined on A by*

$$u'(\xi) = \liminf_{\eta \to \xi} u(\eta). \tag{5.1}$$

In this theorem when $\bar{B}(\xi, \delta) \subset D$, the function u is defined l_{N-1} almost everywhere on $\partial B(\xi, \delta)$, and u is l_{N-1} measurable and lower bounded on this set; so $L(u, \xi, \delta)$ is well defined. In the simplest case A is relatively closed in D, so that the hypotheses make u superharmonic on $D - A$, but these hypotheses allow A to be dense in D. A superharmonic extension u' of u satisfies (5.1) because of the strengthening of the lower semicontinuity property of superharmonic functions in II(6.1). If the theorem is true locally, it is true as stated; so we can assume that D is a ball, that u is not identically $+\infty$, and (Section 4) that there is a positive superharmonic function v on D, identically $+\infty$ on A but finite at some point of $D - A$ at which u is finite. When $\varepsilon > 0$, the function $u + \varepsilon v$, if defined as $+\infty$ on A, is superharmonic on D, and according to the Fundamental Convergence Theorem (Section III.3), the function $u_0 = \lim_{\varepsilon \to 0}(u + \varepsilon v)$ has lower semicontinuous smoothing $\underset{+0}{u}$ superharmonic on D, equal l_N almost everywhere on D to u_0. If $\xi \in D - A$, the given inequality $u(\xi) \geq L(u, \xi, \delta)$ implies [see I(2.2)] that $u(\xi) \geq A(u, \xi, \delta)$ (for sufficiently small δ). Applying lower semicontinuity of u, we find that $\liminf_{\delta \to 0} A(u, \xi, \delta) \geq u(\xi)$. Hence

$$u(\xi) = \lim_{\delta \to 0} A(u, \xi, \delta) \qquad (\xi \in D - A). \tag{5.2}$$

Moreover $u = u_0 = \underset{+0}{u}$ l_N almost everywhere on D. Therefore

$$u(\xi) = \lim_{\delta \to 0} A(\underset{+0}{u}, \xi, \delta) = \underset{+0}{u}(\xi) \qquad (\xi \in D - A); \tag{5.3}$$

so $\underset{+0}{u}$ is the desired superharmonic extension of u.

Special Case

If in the theorem u is bounded and if $u(\xi) = L(u, \xi, \delta)$ for ξ in $D - A$ and sufficiently small δ depending on ξ, then u has a unique harmonic extension to D. In fact, under these stronger hypotheses if u' is the bounded superharmonic extension of u provided by Theorem 5 then when $\xi \in D - A$ and δ is sufficiently small the function $u' - \tau_{B(\xi,\delta)} u$ is positive and superharmonic on $B(\xi, \delta)$, vanishes at ξ, and therefore vanishes identically on $B(\xi, \delta)$. Thus u must be continuous on $D - A$, and Theorem 9 can now be applied to $-u$ to provide a subharmonic extension u'' of u to D. The function $u' - u''$ vanishes identically on D because this difference is superharmonic on D and coincides l_N almost everywhere with the harmonic function 0; so u' is the desired harmonic extension of u.

Application to Analytic Functions

If A is a polar subset of the open subset D of $\mathbb{R}^2$, if A is closed in D, and if f is a bounded analytic function on $D - A$, the real and imaginary parts of f have harmonic extensions to D according to the preceding paragraph. The extension to D of f obtained in this way is analytic because the Cauchy–Riemann equations are satisfied. If A is a singleton this result reduces to Cauchy's classical theorem on isolated singularities of bounded analytic functions.

Application to Isolated Singularities of Harmonic Functions

Let u be a function defined and harmonic on an open deleted neighborhood B of a point ξ and satisfying

$$\begin{aligned} &\liminf_{\eta \to \xi} \frac{u(\eta)}{-\log|\eta - \xi|} > -\infty \quad \text{if } N = 2, \\ &\liminf_{\eta \to \xi} u(\eta)|\eta - \xi|^{N-2} > -\infty \quad \text{if } N > 2. \end{aligned} \tag{5.4}$$

Then Theorem 5 implies that there is a constant c such that the function

$$\begin{aligned} &u + c \log|\cdot - \xi| \quad \text{if } N = 2, \\ &u - c|\cdot - \xi|^{2-N} \quad \text{if } N > 2 \end{aligned}$$

has a harmonic extension to $B \cup \{\xi\}$. We give the proof for $N = 2$. According to (5.4) there is a constant c_1 for which the function $u + c_1 \log|\cdot - \xi|$ is positive on a deleted neighborhood of ξ; so this harmonic function has a superharmonic extension to $B \cup \{\xi\}$. Since the Riesz measure associated with the extension must be supported by $\{\xi\}$, there is a constant c such that the function $u + c\log|\cdot - \xi|$ has a superharmonic extension u' to $B \cup \{\xi\}$ and that the Riesz measure associated with u' vanishes identically. Hence u' is harmonic. Observe that by way of inversion in a sphere of center ξ the result just obtained implies that if v is a function defined and harmonic on a deleted open neighborhood of the point ∞ of $\mathbb{R}^N$ and if

$$\begin{aligned} &\liminf_{\eta\to\infty} \frac{v(\eta)}{\log|\eta|} > -\infty \quad &\text{if } N = 2,\\ &\liminf_{\eta\to\infty} v(\eta) > -\infty \quad &\text{if } N > 2, \end{aligned} \tag{5.5}$$

then there is a constant c such that the function

$$\begin{aligned} &v - c\log|\cdot| \quad &\text{if } N = 2,\\ &(v - c)|\cdot|^{N-2} \quad &\text{if } N > 2 \end{aligned}$$

has a finite limit at the point ∞.

Application to a Generalized Liouville Theorem ($N = 2$)

According to Section II.2 a lower-bounded harmonic function on $\mathbb{R}^N$ for $N \geq 2$ must be identically constant. The preceding application of Theorem 5 implies when $N = 2$ that a harmonic function on $\mathbb{R}^2$ which satisfies (5.5) must be identically constant because if c is chosen so that $v - c\log|\cdot|$ has a finite limit at the point ∞, then v must have a limit α at the point ∞, finite or infinite according as $c = 0$ or $c \neq 0$. In view of the maximum-minimum theorem for harmonic functions α is finite and $v \equiv \alpha$.

Application to the Greatest Subharmonic Minorants of Superharmonic Functions

Let D be an open subset of $\mathbb{R}^N$ and suppose that u is a superharmonic function on D for which $\mathrm{GM}_D u$ exists. Let A be a subset of D, closed relative to D, and define $D_0 = D - A$. Then $\mathrm{GM}_{D_0}(u_{|D_0}) \geq \mathrm{GM}_D u$ on D_0. We now show that there is equality if A is polar and is null for the Riesz measure associated with u. We can assume, replacing u by $u - \mathrm{GM}_D u$ if necessary, that $\mathrm{GM}_D u = 0$, and it is then sufficient to prove that if h is a function defined and harmonic on D_0, with $0 \leq h \leq u$, it follows that $h = 0$. According to Theorem 5, such a function h has a positive superharmonic extension h_1 from D_0 to D and similarly $u - h$ has a positive superharmonic extension v from D_0 to D. Since $u = h_1 + v$ on D_0, it follows that $u = h_1 + v$ on D and therefore A must be null for the Riesz measure associated with h_1. Hence h_1 is positive and harmonic on D and is majorized there by u so $h_1 = 0$, and therefore $h = 0$, as was to be proved.

6. Greenian Sets in $\mathbb{R}^2$ as the Complements of Nonpolar Sets

Theorem. *A nonempty open subset D of $\mathbb{R}^2$ is Greenian if and only if $\mathbb{R}^2 - D$ is not polar, equivalently, if and only if $\mathbb{R}^2 \cap \partial D$ is not polar.*

If $\mathbb{R}^2 - D$ is polar (equivalently, according to Section 2, if $\mathbb{R}^2 \cap \partial D$ is polar) D is not Greenian because a positive superharmonic function on D can be extended to be a positive superharmonic function on $\mathbb{R}^2$ and therefore (Theorem II.13) is a constant function. Conversely, suppose that $\mathbb{R}^2 \cap \partial D$ is not polar. If D is bounded, it was noted in Section II.13 that D is Greenian. If D is unbounded, let A be the intersection of $\mathbb{R}^2 - D$ with a closed disk so large that A is not polar. Let B be a disk containing A and define v on $\mathbb{R}^2 - A$ by

$$v = \begin{cases} R_{+1}^A & \text{on } B - A \text{ (reduction relative to } B\text{)}, \\ 0 & \text{on } \mathbb{R}^2 - B. \end{cases}$$

In view of the fact that v is harmonic and positive on $B - A$ with limit 0 at every boundary point of B [Section III.4, Example (a)], the function v is subharmonic; so $1 - v$ is positive and superharmonic on D but not identically constant there because $v = 0$ on $D - B$ but (Theorem 4) $v > 0$ on $D \cap B$. Hence D is Greenian.

7. Superharmonic Function Minimum Theorem (Extension of Theorem II.5)

The obvious application of the following theorem to harmonic functions is left to the reader.

Theorem. *Let D be a Greenian subset of $\mathbb{R}^N$ and let u be a lower bounded superharmonic function on D. Suppose that there is a constant c such that $\liminf_{\eta \to \zeta} u(\eta) \geq c$ at quasi every finite point ζ of ∂D, and also at $\zeta = \infty$ if $N > 2$ and D is unbounded. Then $u \geq c$.*

Observation (a). We are only excluding trivia by the hypothesis that D be Greenian because if $N > 2$, all nonempty open subsets of $\mathbb{R}^N$ are Greenian and if $N = 2$, a lower-bounded superharmonic function on a non-Greenian open set is identically constant. (Moreover, according to Theorem 6, a condition on quasi every finite boundary point of a non-Greenian open subset of $\mathbb{R}^2$ is necessarily satisfied vacuously.)

Observation (b). Since the set A of finite boundary points of D at which u has inferior limit $< c$ is a countable union of compact sets,

$$A = \bigcup_{n=1}^{\infty} \left\{ \zeta \in \partial D : |\zeta| \le n,\ \liminf_{\eta \to \zeta} u(\eta) \le c - \frac{1}{n} \right\},$$

the set A is polar if it is inner polar. Thus Theorem 7 is true if "quasi every" is replaced by "inner quasi every."

To prove the theorem, observe that according to Theorem 2 if $N > 2$ there is a positive superharmonic function v on $\mathbb{R}^N$, identically $+\infty$ on A; if $N = 2$ and D is bounded, there is a positive superharmonic function v on a ball containing $\bar{D}$, identically $+\infty$ on A. In either case if $\varepsilon > 0$ the function $(u + \varepsilon v)_{|D}$ is superharmonic on D with inferior limit $\ge c$ at every point of ∂D, including ∞ if $N > 2$ and D is unbounded. Hence $u + \varepsilon v \ge c$ on D by the superharmonic function minimum theorem of Section II.5, and therefore $u \ge c$ quasi everywhere on D and so everywhere on D. If $N = 2$ and D is unbounded, we can suppose that $\liminf_{\eta \to \infty} u(\eta) \ge c$ because the plane can be inverted in a circle with center a finite boundary point of D not in A, so that the transformed superharmonic function on the image of D has this property. Thus if $c' < c$, there is a disk B so large that $u \ge c'$ on $D - B$. The part of the theorem already proved yields the inequality $u \ge c'$ on $D \cap B$ from which it follows that $u \ge c$ on D, as was to be proved.

Application to Analytic Function Theory

Let f be a bounded complex-valued function defined and analytic on a Greenian subset D of the plane; that is, the complement of D is nonpolar. The function $|f|$ is subharmonic; so it follows from Theorem 7 applied to $-|f|$ that if $\limsup_{z' \to z} |f(z')| \le c$ at quasi every finite boundary point of D, then $|f| \le c$ on D.

8. Evans–Vasilesco Theorem

Theorem. *Let D be either $\mathbb{R}^2$ or a Greenian subset of $\mathbb{R}^N$ $(N \ge 2)$, and let $G_D\mu$ be a superharmonic potential on D. Then if A is a closed (in D) support of μ, continuity of $(G_D\mu)_{|A}$ at a point ξ of A implies continuity of $G_D\mu$ at ξ.*

The triviality of the extension of this theorem from D special open to D general Greenian will be explained in Section VII.8. Accordingly we assume here that D is either $\mathbb{R}^2$ or a special open subset of $\mathbb{R}^N$ ($N \geq 2$). We can suppose that μ has compact support because the G_D potential of the projection of μ on $A - B(\xi, \delta)$ is harmonic and therefore continuous on $D \cap B(\xi, \delta)$ for every $\delta > 0$. Under the hypothesis of compactness of A, the restriction $(G\mu)_{|D}$ is superharmonic and is the sum of $G_D\mu$ and a function harmonic on D (Section IV.1); so it follows that under the hypotheses of Theorem 8, the function $(G\mu)_{|A}$ is continuous at ξ and is finite there, and it is sufficient to prove that $G\mu$ is continuous at ξ. Suppose first that $N > 2$. If $\mu^{\xi\delta}$ is the projection of μ on $B(\xi, \delta)$, finiteness of $G\mu(\xi)$ implies that $\mu(\{\xi\}) = 0$ and that $\lim_{\delta\to 0} G\mu^{\xi\delta}(\xi) = 0$. Furthermore, since $G(\mu - \mu^{\xi\delta})$ is continuous on $B(\xi, \delta)$, the function $(G\mu^{\xi\delta})_{|A}$ must be continuous at ξ. If $\varepsilon > 0$ and if δ is sufficiently small, depending on ε, $G\mu^{\xi\delta}(\xi) < \varepsilon$. For such a value of δ continuity of $(G\mu^{\xi\delta})_{|A}$ at ξ implies that $(G\mu^{\xi\delta})_{|A} < \varepsilon$ in some neighborhood of ξ, and hence this inequality is true for all smaller values of δ. It follows that

$$\lim_{\delta\to 0} \sup_{\zeta' \in A \cap B(\xi, \delta)} G\mu^{\xi\delta}(\zeta') = 0. \tag{8.1}$$

If ζ is a point of D, let ζ' be a point of A at minimum distance from ζ so that if $\eta \in A$,

$$|\zeta' - \eta| \leq |\zeta' - \zeta| + |\zeta - \eta| \leq 2|\zeta - \eta|. \tag{8.2}$$

The oscillation of $G\mu$ at ξ can be majorized as follows:

$$\begin{aligned} \operatorname{Osc} G\mu(\xi) = \operatorname{Osc} G\mu^{\xi\delta}(\xi) &\leq \sup_{\zeta \in B(\xi, \delta/2)} G\mu^{\xi\delta}(\zeta) \\ &\leq 2^{N-2} \sup_{\zeta \in B(\xi, \delta/2)} G\mu^{\xi\delta}(\zeta'). \end{aligned} \tag{8.3}$$

Now $\zeta' \in B(\xi, \delta) \cap A$ when $\zeta \in B(\xi, \delta/2)$ because $|\xi - \zeta| < \delta/2$; so the right side of (8.3) has limit 0 when $\delta \to 0$, by (8.1). That is, $G\mu$ is continuous at ξ, as was to be proved. When $N = 2$, the only change needed in the preceding argument is that in (8.3) it should be supposed that $\delta < \frac{1}{2}$ to ensure positivity of the potentials involved, and the last term in (8.3) should be replaced by

$$\sup_{\zeta \in B(\xi, \delta/2)} G\mu^{\xi\delta}(\zeta') + \mu(B(\xi, \delta)) \log 2,$$

which tends to 0 with δ.

Observation. If in the theorem it is supposed that $(G_D\mu)_{|A}$ is finite valued and continuous, it follows that $G_D\mu$ is finite valued and continuous on D. If D is special open and if μ has compact support in D, the function $G_D\mu$ is then necessarily bounded because this potential has limit 0 at every boundary point of D. For arbitrary Greenian D the boundedness of a continuous finite-valued potential $G_D\mu$ for μ of compact support in D follows from the boundedness properties of G_D to be proved in Section VII.5.

9. Approximation of a Potential by Continuous Potentials

Theorem. *Let D be either $\mathbb{R}^2$ or a Greenian subset of $\mathbb{R}^N$ $(N \geq 2)$ and let μ be a measure on D. Suppose that $G_D\mu$ is superharmonic and that $G_D\mu < +\infty$ on some support A of D. There is then a sequence $\mu_.$ of measures on D for which μ_n is supported by a compact subset of A, $G_D\mu_n$ is finite valued and continuous (bounded if $D \neq \mathbb{R}^2$), and $\mu = \sum_{n=0}^{\infty} \mu_n$; so $G_D\mu = \sum_{n=0}^{\infty} G_D\mu_n$.*

Recall our convention that "supported by A" means that A is μ measurable and that the complement of A is μ null. The generality of the statement of the theorem is convenient for reference, but there is no loss of generality in supposing that A is a Borel set because there is always a Borel support of μ that is a subset of A. Apply Lusin's theorem to find a sequence $A_.$ of disjoint compact subsets of A with the property that $\mu(A - \bigcup_{n=0}^{\infty} A_n) = 0$ and that $(G_D\mu)_{|A_n}$ is bounded and continuous. If μ_n is the projection of μ on A_n, the continuous function $(G_D\mu)_{|A_n}$ is the sum of the restrictions to A_n of the lower semicontinuous functions $G_D\mu_n$ and $G_D(\mu - \mu_n)$. Hence these restrictions are continuous, and therefore (by the observation in Section 8) $G_D\mu_n$ is bounded and continuous on D unless $D = \mathbb{R}^2$, in which case $G_D\mu_n = G\mu_n$ is at least finite valued and continuous.

Corollary. *If B is a compact (but see the extension below) nonpolar subset of $\mathbb{R}^N$, there is a measure ν on $\mathbb{R}^N$ supported by B for which if $N = 2$, the function $G\nu$ is finite valued, continuous, and not identically 0 and if $N \geq 2$, the function $G_D\nu$ is strictly positive, continuous, and bounded whenever D is a connected Greenian superset of B.*

If D is a Greenian superset of B, the smoothed reduction (relative to D) $\underset{+1}{R}^B$ is the potential of a measure μ supported by B [Section IV.8(d)], and according to Theorem 9, there is a measure ν supported by B with $G_D\nu$ as described. In particular, if $N = 2$, the potential $G\nu$ is then finite valued and continuous on $\mathbb{R}^2$.

Extension of the Corollary to Analytic Sets B

It will be proved in Section VI.2 that a nonpolar analytic subset of $\mathbb{R}^N$ has a nonpolar compact subset. This fact will imply that the above Corollary is true if B is an analytic nonpolar set.

10. The Domination Principle

The following theorem will be referred to as the *domination principle*, a designation sometimes used more restrictively.

Theorem. *Let D be a Greenian subset of $\mathbb{R}^N$, let μ be a measure on D, and let v be a positive superharmonic function of D. Then each of the following three conditions implies that $G_D\mu \le v$:*

(a) *$G_D\mu < +\infty$ μ almost everywhere and $G_D\mu \le v$ inner quasi everywhere on some Borel support of μ.*
(b) *The inequalities $G_D\mu < +\infty$ and $G_D\mu \le v$ are true μ almost everywhere.*
(c) *Polar sets are μ null and $G_D\mu \le v$ μ almost everywhere.*

It is not obvious that conditions (a)–(c) are equivalent, and we shall not need this equivalence in the proof of Theorem 10, but according to Theorem 11, both conditions (a) and (b) imply that polar sets are μ null and thereby that (a)–(c) are equivalent. It is a defect of Theorem 10 that polar sets must be μ null, but it will be shown in Section XI.23 that $G_D\mu \le v$ if

$$\limsup_{\substack{\eta\to\xi \\ G_D\mu(\eta) < +\infty}} \frac{G_D\mu(\eta)}{v(\eta)} \le 1, \tag{10.1}$$

for μ almost every ξ. In fact the condition in Theorem XI.23 is considerably weaker than (10.1). If μ is a probability measure supported by a singleton $\{\xi_0\}$, so that $G_D\mu = G_D(\xi_0, \cdot)$, and if $v = 2G_D(\xi_0, \cdot)$, the condition (10.1) but not Theorem 10 is applicable to show that $G_D\mu \le v$.

It is sufficient to show that $G_D\mu \le v$ when condition (a) is satisfied. In fact, if B is an arbitrary Borel support of μ, the set $B \cap \{G_D\mu < +\infty, G_D\mu \le v\}$ is, under either (b) or (c), a support of μ on which $G_D\mu < +\infty$ and $G_D\mu \le v$; so condition (a) is satisfied. To prove that condition (a) implies that $G_D\mu \le v$ we assume that D is special, deferring to Section VII.8 the explanation of why this specialization is trivial. Let A be a Borel support of μ on which $G_D\mu < +\infty$ and $G_D\mu \le v$ inner quasi everywhere. Since (by Theorem 9) the function $u = G_D\mu$ is the limit of an increasing sequence of bounded continuous potentials of measures supported by compact subsets of A, we can suppose that A is compact and that u is bounded and continuous. If $\xi \in A$ and if $u(\xi) \le v(\xi)$, then

$$\liminf_{\eta\to\xi} v(\eta) \ge v(\xi) \ge u(\xi) = \lim_{\eta\to\xi} u(\eta). \tag{10.2}$$

Thus the restriction to $D - A$ of $v - u$ is superharmonic, is lower bounded, and has positive inferior limit at inner quasi every point of $A \cap \partial(D - A)$.

Moreover every limiting value of this restriction at a point of ∂D is positive because (Section IV.1) u has limit 0 at every such point. (If D is not special, it will be seen in Section VII.5 that u has limit 0 at quasi every point of ∂D, including the point ∞ if D is unbounded and $N > 2$.) It follows (Theorem 7) that $v \geq u$ on $D - A$, so $v \geq u$ inner quasi everywhere on D and hence (Section 1) everywhere on D, as was to be proved.

Special Hypotheses on v

If v in Theorem 10 is a constant function, the theorem is sometimes called the *principle of the maximum*. If v is the sum of a constant function and a potential, the theorem is sometimes called the *complete principle of the maximum*.

The Domination Principle for Potentials $G\mu$ When $N = 2$

If $N = 2$, if $G\mu$ is the superharmonic potential of a measure μ on $\mathbb{R}^2$, and if there is a constant c such that $G\mu \leq c$ μ almost everywhere then the method of proof of Theorem 10 shows that $G\mu \leq c$ on $\mathbb{R}^2$. Observe that if c is replaced by a harmonic function h in this hypothesis then any further hypothesis on h leading to the conclusion that $G\mu \leq h$ on $\mathbb{R}^2$ would imply (according to Section IV.9) that $\liminf_{|\xi| \to \infty} h(\xi)/\log|\xi| > -\infty$ and that therefore (Section 5) h is identically constant.

11. The Infinity Set of a Potential and the Riesz Measure

Theorem. *If D is a Greenian subset of $\mathbb{R}^N$, if A is a polar subset of D, and if $u = G_D\mu$ is the potential of a measure on D, then $u = +\infty$ μ almost everywhere on A.*

(The phraseology of the theorem requires that μ be a completed measure.) We can assume that D is connected. If u is not superharmonic then $u \equiv +\infty$, so the theorem becomes trivial. If u is superharmonic the conclusion of the theorem can be restated in the following form: if u is finite valued on A then $\mu(A) = 0$. It is this assertion that will be proved. The triviality of the extension from special to Greenian sets will be explained in Section VII.8. Suppose then that D is special and that $u = G_D\mu$ is finite valued on the polar set A. Since the theorem is local it can be assumed that A is relatively compact in D. It can also be assumed that A is a Borel set because there is a polar G_δ superset of A and the intersection of this superset with the set of finiteness of u is a Borel set. If $A_n = \{\xi \in A : u \leq n\}$ and if μ_n is the projection of μ on A_n then an application of the domination principle to the pair $u_n = G_D\mu_n$, $v \equiv n$ shows that $u_n \leq n$ on D. Let $G_D\nu$ be a superharmonic potential identically $+\infty$ on A, with $\nu(D) < +\infty$. Apply Fubini's theorem and the symmetry of G_D to derive

$$n\nu(D) \geq \int_D G_D\mu_n \, d\nu = \int_D G_D\nu \, d\mu_n = +\infty \, \mu(A_n).$$

It follows that $\mu(A_n) = 0$; so $\mu(A) = 0$, as was to be proved.

Application to $G\mu$ When $N = 2$

If $N = 2$, if μ is a measure on $\mathbb{R}^2$, and if $u = G\mu$ is superharmonic, the conclusion of the theorem remains true. It is sufficient to prove that $u = +\infty$ μ almost everywhere on any bounded polar set A. Let D be a disk containing A and let μ_D be the projection of μ on D. According to Theorem 11, $G_D\mu_D = +\infty$ μ_D almost everywhere on A. Since $G_D\mu_D - G\mu$ is harmonic on D if defined suitably on the set of common infinities of $G_D\mu_D$ and $G\mu$, it follows that $G\mu = +\infty$ μ almost everywhere on A, as asserted.

Chapter VI

The Fundamental Convergence Theorem and the Reduction Operation

1. The Fundamental Convergence Theorem

Theorem. *Let Γ: $\{u_\alpha, \alpha \in I\}$ be a family of superharmonic functions defined on an open subset of $\mathbb{R}^N$, locally uniformly bounded below, and define the lower envelope u by $u(\xi) = \inf_{\alpha \in I} u_\alpha(\xi)$. Then $\underset{+}{u} \le u$,*

$$\underset{+}{u}(\xi) = \liminf_{\eta \to \xi} u(\eta), \tag{1.1}$$

and

(a) $\underset{+}{u}$ *is superharmonic.*
(b) $\underset{+}{u} = u$ *on each open set on which u is superharmonic.*
(c) $\underset{+}{u} = u$ *quasi everywhere.*
(d) *There is a countable subset of Γ whose lower envelope has the same lower semicontinuous smoothing $\underset{+}{u}$.*

Conversely, if A is a polar subset of a Greenian subset D of $\mathbb{R}^N$, there is a decreasing sequence $v_\cdot$ of positive superharmonic functions on D with limit v such that $v > \underset{+}{v}$ on A.

The direct part of the present theorem is identical with Theorem III.3 except that Theorem III.3(c) allows a larger exceptional set than Theorem 1(c). Thus there remains only the proof of Theorem 1(c) and of the converse part of Theorem 1. Since Theorem 1(c) is a local assertion it can be assumed in its proof that the functions are defined on a ball D, and in view of the discussion in Section III.3 it can be assumed that Γ is a decreasing sequence of positive superharmonic functions on D. The limit function need only be analyzed on a strictly smaller concentric ball B, with u_n replaced by $R^B_{u_n}$ (reduction relative to D). This reduction, equal to its lower semicontinuous smoothing because B is open, is a superharmonic potential $G_D\mu_n$ (Section IV.8) and is equal to u_n on B, so the replacement is legitimate. The measure μ_n is supported by $\overline{B}$. On $D - \overline{B}$ the sequence $u_\cdot$ is a locally uniformly convergent sequence of harmonic functions; so the sequences of partial derivatives are also locally uniformly convergent on $D - \overline{B}$ (Theorem II.3).

Evaluation of $\mu_n(D)$ by the Gauss Integral Theorem shows that $\mu_{\bullet}(D)$ is a convergent sequence. Going to a subsequence if necessary, it can be assumed that the sequence $\mu_{\bullet}$ is vaguely convergent to a measure μ supported by $\bar{B}$. If c is a strictly positive constant, the function $G_D \wedge c$ is continuous and

$$\lim_{n\to\infty} u_n = u \geq \lim_{n\to\infty} \int_D G_D(\cdot,\eta) \wedge c\, \mu_n(d\eta) = \int_D G_D(\cdot,\eta) \wedge c\, \mu(d\eta), \quad (1.2)$$

so that $u \geq G_D\mu$. Let ν be a measure having compact support in D, with $G_D\nu$ finite valued and continuous. Then

$$\int_D G_D\mu_n\, d\nu = \int_D G_D\nu\, d\mu_n < +\infty.$$

The equality

$$\int_D u\, d\nu = \lim_{n\to\infty} \int_D G_D\mu_n\, d\nu = \lim_{n\to\infty} \int_D G_D\nu\, d\mu_n = \int_D G_D\nu\, d\mu = \int_D G_D\mu\, d\nu, \quad (1.3)$$

combined with the inequality $u \geq G_D\mu$, implies that $u = G_D\mu$ at ν almost every point of D. Since $\underset{+}{u}$ is the maximal lower semicontinuous minorant of u, it follows that $u \geq \underset{+}{u} \geq G_D\mu$ with equality ν almost everywhere. Thus, if $A = \{u > \underset{+}{u}\}$, it has now been shown that $\nu(A) = 0$ whenever $G_D\nu$ is finite valued and continuous. In view of Corollary V.9 this fact means that every compact subset of A is polar; that is, the set A is a Borel inner polar set. Theorem 1(c) follows from the next theorem, whose proof uses the partial result just obtained.

Conversely, suppose that A is a polar subset of the Greenian subset D of $\mathbb{R}^N$, and let Γ be the class of positive superharmonic functions on D equal at least to 1 on A. The infimum of the class is (reduction relative to D) R_1^A, and (Theorem VI.4) $\underset{+1}{R^A} \equiv 0$. According to Theorem 1(d), there is a sequence in Γ whose lower semicontinuous smoothed infimum vanishes identically. If v_n is the minimum of the first n members of this sequence, $v_{\bullet}$ has the properties stated in the converse of Theorem 1.

2. Inner Polar versus Polar Sets

Theorem. *An analytic inner polar subset of* $\mathbb{R}^N$ *is polar.*

It is obviously sufficient to prove the theorem for bounded sets so we shall consider subsets of a Greenian set D, say a ball. Reductions below are relative to D. The proof will be in several steps. Some of the preliminary results proved below are more general than needed; this is to avoid later repetition when the results will be strengthened.

(a) *If A is a compact subset of D and if v is a finite-valued positive continuous superharmonic function on D, then* $\underline{R}_v^A = v$ *quasi everywhere on A*. In fact, according to the part of Theorem 1 actually proved in Section 1, the set

$$A \cap \{\underline{R}_v^A < v\} = \bigcup_1^\infty A \cap \{\underline{R}_v^A \leq (1 - 1/n)v\}$$

is inner polar. Since each set in the union is compact, the union is polar.

(b) *If v is a finite-valued positive superharmonic function on D and if* $\xi \in D$, *the set functions* $R_v^{\bullet}(\xi)$ *and* $\underline{R}_v^{\bullet}(\xi)$ *are strongly subadditive on the class of compact subsets of D*. To prove this, consider the strong subadditivity inequality

$$R_v^{A \cup B} + R_v^{A \cap B} \leq R_v^A + R_v^B \tag{2.1}$$

and the corresponding inequality (2.1sm) for smoothed reductions. Inequality (2.1) is trivially satisfied on $A \cap B$. On one of the sets but not on the other, say on $A - B$, (2.1) reduces to the inequality $v + R_v^{A \cap B} \leq v + R_v^B$. Thus (2.1) is satisfied on $A \cup B$, and hence (2.1sm), an inequality between superharmonic potentials, is satisfied inner quasi everywhere on $A \cup B$, a support of the measures associated with these potentials. It follows from the domination principle that (2.1sm) is true on D. Since (2.1sm) is true, (2.1) is true on the open set $D - (A \cup B)$ on which all the reductions are harmonic and equal to their smoothings. We have already verified (2.1) on $A \cup B$.

The foregoing proof depended on the domination principle, not available in a form strong enough for this proof in parabolic potential theory. A proof of reduction operator strong subadditivity not depending on the domination principle is given in Section 4 [proof of Section 3(j)].

(c) *If v is a finite-valued positive continuous superharmonic function on D and if* $A_{\bullet}$ *is an increasing sequence of compact subsets of D with compact union* A $(\subset D)$, *then*

$$\lim_{n\to\infty} R_v^{A_n} = R_v^A, \tag{2.2}$$

and the corresponding equation (2.2sm) for smoothed reductions is also true. Under these hypotheses $\lim_{n\to\infty} \underline{R}_v^{A_n} = v$ quasi everywhere on A because $\underline{R}_v^{A_n} = v$ quasi everywhere on A_n, and the two superharmonic potentials $\lim_{n\to\infty} \underline{R}_v^{A_n}$ and $\underline{R}_v^A$ are therefore equal quasi everywhere on the common support A of their associated Riesz measures. It follows from the domination principle that these potentials are identical. The functions $\lim_{n\to\infty} R_v^{A_n}$ and R_v^A are trivially equal on A and are equal on $D - A$ (on which they are harmonic and equal to their smoothings) because (2.2sm) is true.

(d) *If v is a finite-valued positive continuous superharmonic function on D* *and if $A_{\bullet}$ is a decreasing sequence of compact subsets of D with intersection A, then* (2.2) *is true*. In fact; if B is an open neighborhood of A the set B must also be a neighborhood of A_n for sufficiently large n, so $\lim_{n\to\infty} R_v^{A_n} \le R_v^B$, and since v is finite valued and continuous, it follows from III(5.5) that (2.2) is true with equality replaced by the inequality $\le$. Since the reverse inequality is trivial, (2.2) is true.

(e) *If v is a finite-valued positive continuous superharmonic function on D and if $\xi \in D$, the set function $R_v^{\bullet}$ is a Choquet capacity on D relative to the class of compact subsets of D.* According to (b)–(d) and the monotoneity of the reduction operator and under the stated hypotheses on v, the restriction of the set function $R_v^{\bullet}(\xi)$ to the class of compact subsets of D is a topological precapacity. This restriction therefore (Appendix II.8) has an extension to a Choquet capacity $I(\xi, \cdot)$ on D relative to the class of compact subsets of D, for which $I(\xi, F) = R_v^F(\xi)$ when F is compact and

$$I(\xi, B) = \sup\{R_v^F(\xi): F \subset B, F \text{ compact}\} \qquad (B \text{ open}), \tag{2.3}$$

$$I(\xi, A) = \inf\{I(\xi, B): A \subset B, B \text{ open}\} \qquad (A \text{ arbitrary}). \tag{2.4}$$

To prove (e) we show that $I(\xi, A) = R_v^A(\xi)$ for every subset A of D. Let $F_{\bullet}^{\circ}$ be an increasing sequence of relatively compact open subsets of the open subset B of D, with union B, and define $F_n = \overline{F_n^{\circ}}$. Then $R_v^{F_n^{\circ}} = \underset{+v}{R^{F_n^{\circ}}}$, and $I(\cdot, B) = \lim_{n\to\infty} \underset{+v}{R^{F_n^{\circ}}}$ is a positive superharmonic function, identically v on B. Hence $I(\cdot, B) \ge R_v^B$, and since the reverse inequality is trivial, it follows that $I(\cdot, B) = R_v^B$ when B is open. Since both $R_v^{\bullet}(\xi)$ and $I(\xi, \cdot)$ satisfy (2.4), the set functions $R_v^{\bullet}(\xi)$ and $I(\xi, \cdot)$ are identical.

(f) *Proof of the theorem.* Let v be any finite-valued strictly positive continuous superharmonic function on D, say $v \equiv 1$. Suppose that A is an inner polar analytic subset of D and choose ξ in $D - A$. The set A is capacitable for the Choquet capacity $R_v^{\bullet}(\xi)$; that is,

$$R_v^A(\xi) = \sup\{R_v^F(\xi): F \subset A, F \text{ compact}\}. \tag{2.5}$$

According to Theorem V.4, the right side of (2.5) is 0, and the consequent vanishing of the left side implies that A is polar. □

Characterization of a Nonpolar Set

According to Theorem 2 every analytic nonpolar subset of $\mathbb{R}^N$ has a compact nonpolar subset. Hence Corollary V.9, which asserts the existence of a not identically 0 continuous finite-valued potential supported by a nonpolar compact subset B of $\mathbb{R}^N$, remains valid if B is supposed only nonpolar analytic.

3. Properties of the Reduction Operation

In the following D is a Greenian subset of $\mathbb{R}^N$, provided with a boundary ∂D by a metric compactification. Reductions of positive superharmonic functions $u, v, \ldots$ on subsets of $D \cup \partial D$ are considered, and no further hypotheses not stated explicitly are imposed on these functions and sets. The following list of properties includes for the reader's convenience some properties already discussed in Sections III.4 and III.5. Reduction properties linked directly to notions of capacity are treated in Section 5.

Although it may seem logical and efficient to prove Theorem 3 and related theorems by methods also applicable to the parabolic potential theory treated in Chapters XV to XIX, it is unfortunately true that the most natural methods in the present context are not all applicable to parabolic potential theory. For this reason the methods used here will be those specially adapted to the present theory, and the parabolic counterparts of the reduction properties listed below and proved in the next section will be proved by different methods and in a different order, although those methods can also be used in the present context.

(a) $R_v^A = v$ on $A \cap D$; $R_v^A = \underset{+v}{R}{}^A = v$ on the interior of $A \cap D$; $R_v^A = \underset{+v}{R}{}^A$ when $A \cap D$ is open; R_v^A is harmonic on $D - \bar{A}$ and is equal to $\underset{+v}{R}{}^A$ there; $\underset{+v}{R}{}^A(\xi) = \liminf_{\eta \to \xi} R_v^A(\eta)$. [See also Chapter III(5.1), (5.1sm), (5.2).]

(b) $\underset{+v}{R}{}^A \le R_v^A$ on D, with equality on $D - A$ and quasi everywhere on $A \cap D$.

(c) If A_1 and A_2 differ by a polar subset of D, then $R_v^{A_1} = R_v^{A_2}$ on $D - (A_1 \cup A_2)$ and $\underset{+v}{R}{}^{A_1} = \underset{+v}{R}{}^{A_2}$ on D.

(d) $\underset{+v}{R}{}^A = \inf\{R_v^{A_0}: A - A_0 \text{ a polar subset of } D\} = \inf\{u: u \ge 0, u$ superharmonic on D, $u \ge v$ near $A \cap \partial D$ and quasi everywhere on $A \cap D\}$.

(e) If $A_\bullet$ is an increasing sequence of subsets of D with union A and if $v_\bullet$ is an increasing sequence of positive superharmonic functions on D with superharmonic limit v, then

$$\lim_{n \to \infty} R_{v_n}^{A_n} = R_v^A, \tag{3.1}$$

and the corresponding equation (3.1sm) for smoothed reductions is also true. If $v_n = v$ for all n, then (3.1) and (3.1sm) are true for $A_\bullet$ an increasing sequence of subsets of $D \cup \partial D$.

Observation. The following example shows that the last assertion of (e) is false without the hypothesis that $v_n = v$ for all n. Let D be a ball, let u be a minimal harmonic function on D corresponding to some boundary point ζ, that is, v is a strictly positive multiple of the Poisson kernel for the boundary point ζ, and define $v_n = u \wedge n$. An application of Section IV.8(a) shows that the function v_n is a potential, because this function is positive, bounded, and superharmonic and has limit 0 at each boundary point except ζ. Finally, $R_v^{\partial D} = \underset{+v}{R}^{\partial D} = v$, $R_{v_n}^{\partial D} = \underset{+v_n}{R}^{\partial D} = 0$, and $\lim_{n\to\infty} v_n = v$, contrary to (3.1).

(f) If $v = \sum_0^\infty v_n$ is a sum of positive superharmonic functions on D and is superharmonic, then

$$R_v^A = \sum_0^\infty R_{v_n}^A, \tag{3.2}$$

and the corresponding equation (3.2sm) for smoothed reductions is also true.

(g) If $A \subset \partial D$ and if $v_{\textstyle .}$ is a decreasing sequence of positive superharmonic functions on D with limit v, then $\lim_{n\to\infty} R_{v_n}^A = \underset{+}{R_v^A}$.

(h) If $A \subset B$,

$$[\![[\![v]\!]^A]\!]^B = [\![[\![v]\!]^B]\!]^A = [\![v]\!]^A. \tag{3.3}$$

(i) If $v' = \underset{+v}{R}^A \wedge \underset{+v}{R}^B$, then

$$R_v^{A\cup B} + R_{v'}^{A\cup B} = R_v^A + R_v^B, \tag{3.4}$$

and the corresponding equation (3.4sm) for smoothed reductions is also true.

(j) The set functions $R_v^{\cdot}$ and $\underset{+v}{R}^{\cdot}$ are countably strongly subadditive.

(k) If v is finite valued and if $\xi \in D$, the set function $R_v^{\cdot}(\xi)$ is a Choquet capacity on $D \cup \partial D$ relative to the class of compact subsets of $D \cup \partial D$.

(l) If A is analytic

$$R_v^A = \sup\{R_v^F : F \subset A, F \text{ compact}\}, \tag{3.5}$$

and the corresponding equation (3.5sm) for smoothed reductions is also true.

(m) The equality

$$R_v^A = \inf\{R_v^B : A \subset B, B \text{ open}\} \tag{3.6}$$

is true if either (i) v is finite valued and continuous at each point of $A \cap D$ (for example, if v is arbitrary and $A \subset \partial D$) or (ii) if v is finite valued and $A \cap D$ is analytic. Moreover (3.6) implies

$$\underset{+v}{R}^A = [\inf\{\underset{+v}{R}^B : A \subset B, B \text{ open}\}]_+. \tag{3.7}$$

(n) If u and v are bounded then

$$\sup_D |R_u^A - R_v^A| \le \sup_D |u - v|, \tag{3.8}$$

and the corresponding inequality (3.8sm) for smoothed reductions is also true.

(o) Denote $[\![u]\!]^{C_1}$, $[\![[\![u]\!]^{C_1}]\!]^{C_2}, \ldots$, respectively, by $u_{C_1}, u_{C_1C_2}, \ldots$.

(o_1) Let v, v', v'', h' be functions from D into $\mathbb{R}^+$ with $v'' = v' + h'$, and suppose that v, v'', and h' are positive superharmonic functions. Define $A = \{v \le v'\}$ and $B = \{v \ge v''\}$. Then

$$h'_A + h'_{ABA} + h'_{ABABA} + \cdots \le v'' \tag{3.9}$$

and

$$h'_{AB} + h'_{ABAB} + \cdots \le v''_B \le v \wedge v'' \tag{3.10}$$

(o_2) Let v and h be positive superharmonic functions on D, let a and b be numbers with $0 \le a < b$, and define $A = \{v \le ah\}$ and $B = \{v \ge bh\}$. Then

$$h_{AB} \le \frac{v \wedge (bh)}{b}, \qquad h_{ABAB} \le \left(\frac{a}{b}\right) h_{AB}, \qquad h_{ABABAB} \le \left(\frac{a}{b}\right)^2 h_{AB}, \ldots; \tag{3.11}$$

so

$$h_{AB} + h_{ABAB} + \cdots \le \frac{v \wedge (bh)}{b - a}. \tag{3.12}$$

Furthermore

$$h_{BA} \le \frac{v \wedge (ah)}{b}, \qquad h_{BABA} \le \left(\frac{a}{b}\right) h_{BA}, \qquad h_{BABABA} \le \left(\frac{a}{b}\right)^2 h_{BA}, \ldots; \tag{3.13}$$

so

$$h_{BA} + h_{BABA} + \cdots \le \frac{v \wedge (ah)}{b - a}. \tag{3.14}$$

Observation. Since $h'_B \le h'$, it is trivial that (3.9) is true with the left side replaced by $h'_{BA} + h'_{BABA} + \cdots$. We shall see in Section XI.4 that the inequalities under (o) yield limit properties of superharmonic functions and of ratios of superharmonic functions. See Sections 2.III.12 and 2.III.22 for the probability counterparts of these inequalities.

4. Proofs of the Reduction Properties

Proof of (a). See Sections III.4, III.5(d), VI.1. □

Proof of (b). According to the Fundamental Convergence Theorem in Section 1, $\underset{+v}{R}^A \leq R_v^A$, and there is equality quasi everywhere on D. If $\xi \in D - A$, let u be a positive superharmonic function on D, identically $+\infty$ on the polar set $A \cap \{\underset{+v}{R}^{A \cap D} < R_v^{A \cap D}\}$ but finite at ξ (see Theorem V.2). Then $v \leq \underset{+v}{R}^{A \cap D} + \varepsilon u$ on A when $\varepsilon > 0$ so $R_v^{A \cap D}(\xi) \leq \underset{+v}{R}^{A \cap D}(\xi) + \varepsilon u(\xi)$, and therefore $R_v^{A \cap D}(\xi) \leq \underset{+v}{R}^{A \cap D}(\xi)$. So there must be equality; that is, $R_v^A = \underset{+v}{R}^A$ on $D - A$ when $A \subset D$. If $A \cap \partial D$ is not empty the expressions III(5.2) and III(5.2sm) for reductions and smoothed reductions in terms of those on $A \cap D$ and $A \cap \partial D$ separately yield the equality of R_v^A and $\underset{+v}{R}^A$ on $D - A$. □

Proof of (c). Let ξ be in $D - (A_1 \cup A_2)$, and (Theorem V.2) let u_1 be a positive superharmonic function on D, identically $+\infty$ on the polar set $(A_1 - A_2) \cup (A_2 - A_1)$ but finite at ξ. If u is a positive superharmonic function on D, majorizing v on $A_1 \cap D$ and near $A_1 \cap \partial D$, then for every $\varepsilon > 0$ the function $u + \varepsilon u_1$ majorizes v on $A_2 \cap D$ and near $A_2 \cap \partial D$. Hence $u(\xi) + \varepsilon u_1(\xi) \geq R_v^{A_2}(\xi)$; so $R_v^{A_1}(\xi) \geq R_v^{A_2}(\xi)$, and by symmetry the reverse inequality is true so $R_v^{A_1} = R_v^{A_2}$ on $D - (A_1 \cup A_2)$. It then follows that the superharmonic functions $\underset{+v}{R}^{A_1}$ and $\underset{+v}{R}^{A_2}$ are equal quasi everywhere on D and therefore everywhere on D. □

Proof of (d). The first line of (d) and the equality of the two infima are immediate consequences of (b) and (c). □

Proof of (e). Under the hypotheses of (e) the reduction $R_{v_n}^{A_n}$ increases with n. Define $u = \lim_{n \to \infty} \underset{+v_n}{R}^{A_n}$, and observe that u is superharmonic and that $u = v$ quasi everywhere on A because $u_n = v_n$ quasi everywhere on A_n. It follows from (d) that $u \geq \underset{+v}{R}^A$, and since the reverse inequality is trivial, (3.1sm) is true. Equality (3.1) is trivial on A and is also true on $D - A$ because on that set $R_{v_n}^A = \underset{+v_n}{R}^A$ and $R_v^A = \underset{+v}{R}^A$ according to (b). In proving the second assertion of (e) we shall suppose that D is connected to avoid irr-evant notational complexity. Assume then that $v_n = v$ for all n, choose a point ξ in D, and let u_n be a positive superharmonic function on D, majorizing v near $A_n \cap \partial D$, with

$$u_n(\xi) \leq R_v^{A_n \cap \partial D}(\xi) + 2^{-n}.$$

The function

$$v_k' = u + \sum_{n=k}^{\infty} (u_n - R_v^{A_n \cap \partial D}), \qquad k \geq 0, \tag{4.1}$$

is positive and superharmonic on D because the sum is the limit of an increasing sequence of positive superharmonic functions and is finite at ξ. Moreover $v_k' \geq u \geq v$ quasi everywhere on $A \cap D$ because $\underset{\smile}{R}{}^{A_n}_{v} = v$ quasi everywhere on $A_n \cap D$, and

$$v_k' \geq R_v^{A_n \cap \partial D} + (u_n - R_v^{A_n \cap \partial D}) \geq u_n$$

when $n \geq k$; so v_k' majorizes v near $A \cap \partial D$, and therefore $v_k' \geq R_v^A$. When $k \to \infty$, we find that $u \geq \underset{\smile}{R}{}^{A}_{v}$ quasi everywhere on D and therefore everywhere on D. Since the reverse inequality is trivial, (3.1sm) is true, and as in the first case of (e), it follows that (3.1) is true. □

Proof of (f). Suppose first that there are only two summands in (3.2); so the equalities

$$R_{v_1+v_2}^A = R_{v_1}^A + R_{v_2}^A \tag{4.2}$$

and (4.2sm) are to be proved. By subadditivity (4.2sm) is true for "$\leq$"; so (natural order decomposition) there are positive superharmonic functions u_1, u_2 on D, satisfying

$$\underset{\smile}{R}{}^{A}_{v_1+v_2} = u_1 + u_2, \qquad u_i \leq \underset{\smile}{R}{}^{A}_{v_i}. \tag{4.3}$$

If A is an open subset of D, then $\underset{\smile}{R}{}^{A}_{v_i} = v_i$ on A and $\underset{\smile}{R}{}^{A}_{v_1+v_2} = v_1 + v_2$ on A so $u_i = v_i$ on A, and it follows that $u_i \geq \underset{\smile}{R}{}^{A}_{v_i}$. So (4.2sm) is true. If now A is still an open subset of D and if v_1 and v_2 are supposed finite valued and continuous, equation III(5.5) gives the value of a reduction on an arbitrary set in terms of reductions on open supersets, and it follows from (4.2sm) [the same as (4.2) for open sets] that (4.2) is true for an arbitrary subset A of D. Since a positive superharmonic function on D is the limit of an increasing sequence of finite-valued positive continuous superharmonic functions, (e) implies that (4.2) is true for an arbitrary subset A of D with no restriction on v_1, v_2, and then (4.2sm) must also be true with this generality because the two sides of (4.2sm) are superharmonic functions and are equal quasi everywhere on D. The evaluations in Section III.5 of reductions and their smoothings in terms of reductions and their smoothings on subsets of D show that (4.2) and (4.2sm) are true with no restrictions on the sets or functions. Thus (3.2) and (3.2sm) are true for two and therefore any finite number of summands. If there are infinitely many summands, write r_n for $\sum_{n+1}^{\infty} \underset{\smile}{R}{}^{A}_{v_k}$. Then

$$\underset{\smile}{R}{}^{A}_{v} = \sum_0^n \underset{\smile}{R}{}^{A}_{v_k} + \underset{\smile}{R}{}^{A}_{r_n},$$

and since $R^A_{\underset{+}{r_n}} \le \sum_{n+1}^{\infty} v_k$, it follows that (3.2sm) is true quasi everywhere and therefore everywhere on D. Equation (3.2) is then true on $D - A$, and this equation is trivial on $A \cap D$. □

Proof of (g). Define $h = \mathrm{GM}_D \underset{+}{v}$ and define $u = \lim_{n\to\infty} R^A_{v_n}$, a positive harmonic function. Obviously

$$0 \le \lim_{n\to\infty} R^A_{v_n - h} = u - R^A_h.$$

The difference $u - R^A_h$ is a positive harmonic minorant of $v_n - h$ for all n, and therefore $u - R^A_h \le \underset{+}{v} - h$; so $u = R^A_h$ by definition of h. Finally, $u = R^A_{\underset{+}{v}}$ because $R^A_{\underset{+}{v}-h}$ is a positive harmonic minorant of $\underset{+}{v} - h$ and so vanishes identically. □

Proof of (h). In view of the trivial fact that the smoothed successive reductions of v on A and B in either order lie between $\|\|v\|^A\|^A$ and $\|v\|^A$, it is sufficient to prove idempotence, that is, to prove that $\|\|v\|^A\|^A = \|v\|^A$. The proof will be carried through in several steps.

(h_1) If $A \subset D$, the desired idempotence is a consequence of (d) because the condition on u in (d) is unchanged if v is replaced by $\|v\|^A$.

(h_2) If $A \subset \partial D$, the following argument yields idempotence. Use Choquet's topological lemma to find a decreasing sequence $v_{\bullet}$ of positive superharmonic functions on D, each majorizing v near A, with sequence limit $\|v\|^A$. Replacing v_n by $v_n \wedge v$ if necessary, it can be supposed that $v_n = v$ near A so that $\|v_n\|^A = \|v\|^A$. It now follows from (g) that

$$\|v\|^A = \lim_{n\to\infty} \|v_n\|^A = \|\|v\|^A\|^A. \tag{4.4}$$

(h_3) For arbitrary A,

$$\|\|v\|^A\|^{A\cap D} = \|v\|^{A\cap D}, \qquad \|\|v\|^A\|^{A\cap\partial D} = \|v\|^{A\cap\partial D}. \tag{4.5}$$

These inequalities follow, respectively, from

$$\begin{aligned} \|v\|^{A\cap D} &= \|\|v\|^{A\cap D}\|^{A\cap D} \le \|\|v\|^A\|^{A\cap D} \le \|v\|^{A\cap D}, \\ \|v\|^{A\cap\partial D} &= \|\|v\|^{A\cap\partial D}\|^{A\cap\partial D} \le \|\|v\|^A\|^{A\cap\partial D} \le \|v\|^{A\cap\partial D}. \end{aligned} \tag{4.6}$$

Finally, to prove (h), apply III(5.2) and (f) to write $\|\|v\|^A\|^A$ in the form

$$\|\|v\|^A\|^A = \|\|v\|^A\|^{A\cap\partial D} + \|\|v\|^A\|^{A\cap D} - \|\|\|v\|^A\|^{A\cap\partial D}\|^{A\cap D},$$

and then apply (h_3) and III(5.2) to show that the right-hand side is $\|v\|^A$. □

Proof of (i). Observe that quasi everywhere on $(A \cup B) \cap D$,

$$v + v' = \|v\|^A + \|v\|^B$$

so that

$$\begin{aligned}\|v\|^{A\cup B} + \|v'\|^{A\cup B} = \|v + v'\|^{A\cup B} &= \big\| \|v\|^A \big\|^{A\cup B} + \big\| \|v\|^B \big\|^{A\cup B} \\ &= \|v\|^A + \|v\|^B.\end{aligned} \tag{4.7}$$

Thus (3.4sm) is true, so (3.4) is true off $A \cup B$, and the latter equation is trivial on $(A \cup B) \cap D$. □

Proof of (j). We have derived a weakened version of (*j*) in Section 2. Instead of going on from this result, we observe that (3.4sm) implies strong subadditivity of $\underset{+v}{R}^{\boldsymbol{\cdot}}$ because if v' is defined as in Section 3(i), it follows that $v' = v$ quasi everywhere on $A \cap B$; so

$$\underset{+v}{R}^{A\cup B} \geq \underset{+v'}{R}^{A\cap B} = \underset{+v}{R}^{A\cap B},$$

and therefore (3.4sm) implies

$$\underset{+v}{R}^{A\cup B} + \underset{+v}{R}^{A\cap B} \leq \underset{+v}{R}^{A} + \underset{+v}{R}^{B}, \tag{4.8}$$

the strong subadditivity inequality. This inequality is trivial on $A \cup B$ for the unsmoothed reductions and is true for these reductions elsewhere on D because it is true for the smoothed reductions. That is, both $R_v^{\boldsymbol{\cdot}}$ and $\underset{+v}{R}^{\boldsymbol{\cdot}}$ are strongly subadditive. Countable strong subadditivity follows from strong subadditivity by an application of (e). □

Proof of (k). (This proof involves the domination principle and thereby the Green function, so at this stage the arguments are relevant only for special open sets D, but the extension of the domination principle to all Greenian sets will be seen to be trivial once G_D has been defined for every Greenian set D in Chapter VII.) We shall not use in the following proof the partial result derived in the course of proving Theorem 2 that if v is continuous as well as finite valued, $R_v^{\boldsymbol{\cdot}}(\xi)$ is a Choquet capacity on D relative to the class of compact subsets of D. To prove (k) in the present context observe that all the capacity properties have been verified for $R_v^{\boldsymbol{\cdot}}(\xi)$ except the property that

$$\lim_{n\to\infty} R_v^{A_n} = R_v^A \tag{4.9}$$

whenever $A_{\bullet}$ is a decreasing sequence of compact subsets of $D \cup \partial D$ with intersection A. Define $u = \lim_{n\to\infty} \underset{+v}{R}^{A_n}$. The function $\underset{+}{u}$ is superharmonic, and $\underset{+}{u} = v$ quasi everywhere on $A \cap D$.

Proof of (4.9) when $A_0 \subset D$. In this case $\underset{+}{u}$ and $\underset{+v}{R}^A$ are finite-valued potentials on D (Section IV.8) whose associated Riesz measures are supported by A, and both potentials are equal to v quasi everywhere on A. According to the domination principle these potentials are therefore identical. Since $\underset{+v}{R}^{A_n}$ is harmonic on $D - A_n$, the function u is harmonic on $D - A$, so $u = \underset{+}{u}$ on that set. Since $R_v^{A_n} = \underset{+v}{R}^{A_n}$ on $D - A_n$ and $R_v^A = \underset{+v}{R}^A$ on $D - A$, equation (4.9) is true on $D - A$, and this equation is trivially satisfied on A.

Proof of (4.9) when v is a potential. If B is an open neighborhood of $A \cap \partial D$ the sequence $A_{\bullet} - B$ is a decreasing sequence of compact subsets of D with limit $A - B$; so by what we have just shown, $\lim_{n\to\infty} R_v^{A_n - B} = R_v^{A-B}$. By subadditivity of $R_v^{\bullet}$,

$$R_v^{A_m - B} + R_v^{A_k \cap B} \geq R_v^{A_m - B} + R_v^{A_m \cap B} \geq \lim_{n\to\infty} R_v^{A_n} \qquad (k \leq m). \tag{4.10}$$

Hence

$$R_v^{A-B} + R_v^B \geq \lim_{n\to\infty} R_v^{A_n}, \tag{4.11}$$

and as B shrinks to $A \cap \partial D$, the left side becomes $R_v^{A\cap D} + R_v^{A\cap\partial D}$ in view of III(5.1). The second reduction vanishes identically because v is a potential, so we find that

$$R_v^A \geq R_v^{A\cap D} \geq \lim_{n\to\infty} R_v^{A_n},$$

and the reverse inequality is trivial.

Proof of (4.9) in the general case. In view of the Riesz decomposition of a positive superharmonic function v and of the reduction additivity property Section 3(f), it is sufficient to prove (4.9) separately for v a potential and for v harmonic. The proof for v a potential was just given. The proof for v finite valued and continuous, and $A \subset D$, was given in Section 2 under part (d) of the proof of Theorem 2, and the proof is valid for A a compact subset of $D \cup \partial D$. Thus (4.9) is true for v harmonic. The proof of (4.9) and thereby that of (k) is now complete. □

Proof of (l). *Case 1.* If v is finite valued, (3.5) follows from (k) since analytic subsets of $D \cup \partial D$ are capacitable for the Choquet capacity $R_v^{\bullet}(\xi)$.

Case 2. If $A \subset D$ but if v is not necessarily finite valued, apply Section 3(e) to derive

$$R_v^A = \sup_{n \geq 0} R_{v \wedge n}^A = \sup_{n \geq 0} \{\sup R_{v \wedge n}^F : F \subset A,\ F \text{ compact}\}$$

$$= \sup \{R_{v \wedge n}^F : F \subset A,\ F \text{ compact},\ n \geq 0\} = \sup \{R_v^F : F \subset A,\ F \text{ compact}\};$$

so (3.5) is true.

Case 3. In the general case write v_F for $R_v^{F \cap \partial D}$ and observe that according to Section III.5(c),

$$R_v^F = v_F - R_{v - v_F}^{F \cap D} = v_F - R_v^{F \cap D} + R_{v_F}^{F \cap D}. \tag{4.12}$$

From now on F is to be a compact subset of A. The class of these sets ordered by inclusion is directed so each term on the right in (4.12) defines a directed set whose limit we now evaluate. Let h be the harmonic component of the Riesz decomposition of v. According to Section III.5, $v_F = R_h^{F \cap \partial D}$; so by Case 1 above

$$\lim_{F\uparrow} v_F = R_h^{A \cap \partial D} = R_v^{A \cap \partial D} = v_A.$$

Since the class of sets F includes the compact subsets of $A \cap D$, Case 2 above shows that $\lim_{F\uparrow} R_v^{F \cap D} = R_v^{A \cap D}$. The same argument shows that if F' is a compact subset of A, then

$$\lim_{F\uparrow} R_{v_F}^{F \cap D} \geq \lim_{F\uparrow} R_{v_{F'}}^{F \cap D} = R_{v_{F'}}^{A \cap D}. \tag{4.13}$$

Now $F \mapsto v_F$ is an upward-directed set of harmonic functions with limit v_A; so v_A is the limit of an increasing sequence $\{v_{F_n}, n \in \mathbb{Z}^+\}$, and therefore letting F' in (4.13) run through this sequence, we find from Section 3(e) that $\lim_{F\uparrow} R_{v_F}^{F \cap D} \geq R_{v_A}^{A \cap D}$. There must be equality because the reverse inequality is trivial, and (4.12) now yields

$$\lim_{F\uparrow} R_v^F = v_A - R_v^{A \cap D} + R_{v_A}^{A \cap D} = R_v^A;$$

so (3.5) is true. Equation (3.5sm) is true on $D - A$ because equations (3.5) and (3.5sm) are identical on $D - A$, and this equality on $D - A$ implies (3.5sm) on $A \cap D$ by the following argument. If $\xi \in A$, the left side of (3.5sm) is unchanged according to Section 3(c) when A is replaced by $A - \{\xi\}$. The right side is also unchanged in view of Section 3(c) and Section 3(e) [replace F in (3.5sm) by F less each of a sequence of balls of center ξ and radii tending to 0]. Thus (3.5sm) is reduced to (3.5). □

Proof of (m). Assertion (i) has already been proved in Section III(e). To prove (ii) observe that if v is finite valued on D, then in view of the properties (*j*) and (k) the restriction of the set function $R_v^{\cdot}(\xi)$ to the class of compact subsets of $D \cup \partial D$ is a topological precapacity. This topological precapacity generates a Choquet capacity $I(\xi,\cdot)$ on $D \cup \partial D$ relative to the paving of compact subsets of $D \cup \partial D$, as described in Appendix II.8. More specifically, $I(\xi,\cdot) = R_v^{\cdot}(\xi)$ on the class of compact sets, and when A is open

$$\begin{aligned} I(\xi, A) &= \sup\{I(\xi, F)\colon F \subset A,\, F \text{ compact}\} \\ &= \sup\{R_v^F(\xi)\colon F \subset A,\, F \text{ compact}\}. \end{aligned} \tag{4.14}$$

Hence $I(\xi, A) = R_v^A(\xi)$ when A is open, in view of Section 3(e) (since A is a countable union of compact sets) or Section 3(l). Finally, for arbitrary A,

$$\begin{aligned} I(\xi, A) &= \inf\{I(\xi, B\colon B \supset A,\, B \text{ open}\} \\ &= \inf\{R_v^B(\xi)\colon B \supset A,\, B \text{ open}\} = R_v^A(\xi)\}. \end{aligned}$$

If A is an $I(\xi,\cdot)$ capacitable set, in particular, if A is analytic, (4.14) is true and combined with (3.5) yields $I(\xi, A) = R_v^A(\xi)$ and thereby yields (3.6). According to property (a), each reduction on the right in (3.6) is equal to its smoothing; so an application of the Fundamental Convergence Theorem to (3.6) yields (3.7). □

Proof of (n). If $c = \sup_D|u - v|$, so that $v \le u + c$, we find that $R_v^A \le R_u^A + c$, and interchanging u and v yields the other half of (3.8). This argument applied to smoothed reductions yields (3.8sm). □

Proof of (o_1). Obviously $v_B'' \le v$ on D, and therefore $h' + v_B'' \le v''$ on A. Hence $h_A' + v_{BA}'' \le v''$; so $h_{ABA}' + v_{BABA}'' \le v_{BA}''$, ... It follows that (3.9) is true. To prove (3.10), observe first that $v_B'' \le v''$ and $v_B'' \le v$; so $v_B'' \le v \wedge v''$. Finally, take the smoothed reduction onto B of both sides of (3.9) to find that (3.10) is true. □

Proof of (o_2). To prove (3.11) and (3.12), observe first that $h_{AB} \le h_B \le v/b$ and that $v_A \le ah_A$. Hence

$$h_{ABA} \le h_{BA} \le \frac{v_A}{b} \le \left(\frac{a}{b}\right) h_A \tag{4.16}$$

from which it follows that $h_{ABAB} \le (a/b)h_{AB}$. Iterate and sum to derive (3.11) and (3.12) with v instead of $v \wedge (bh)$. Relations (3.11) and (3.12) are true as written because if v is replaced by $v \wedge (bh)$, the sets A and B are unchanged. To prove (3.13) and (3.14), observe that the inequality between the second and fourth terms in (4.16) implies

$$h_{BABA} \le \left(\frac{a}{b}\right) h_{ABA} \le \left(\frac{a}{b}\right) h_{BA}.$$

Iterate and sum to show that the left side of (3.14) is at most $bh_{BA}/(b-a)$. Since (4.16) implies that $bh_{BA} \le v$ and $bh_{BA} \le ah$, (3.13) and (3.14) are true. The reader is invited to derive (3.12) and (3.14) from (o_1). □

5. Reductions and Capacities

Let D be a Greenian subset of $\mathbb{R}^N$, provided with a boundary ∂D by a metric compactification, let ξ be a point of D, and let v be a positive finite-valued superharmonic function on D.

Relation between the Capacity $R_v^{\bullet}$ and Topological Precapacities

According to Section 3(k), the set function $R_v^{\bullet}(\xi)$ is a Choquet capacity on $D \cup \partial D$ relative to the class Γ of compact subsets of $D \cup \partial D$. This fact combined with Section 3(j) implies that the restriction of $R_v^{\bullet}(\xi)$ to Γ is a topological precapacity. Let $I(\xi, \cdot)$ be the Choquet capacity on $D \cup \partial D$ relative to Γ generated by this topological precapacity. We have seen in the course of proving Section 3(m) that $I(\xi, \cdot) \le R_v^{\bullet}(\xi)$, with equality on the class of $I(\xi, \cdot)$-capacitable sets and on the class of all boundary subsets, and that there is equality for all sets if v is continuous as well as finite valued.

Finiteness of v Is Necessary in Section 3(k)

Let D be a ball of radius 1, let ξ be the center of D, for $n > 1$, let A_n be a closed concentric ball of radius $1/n$, and define $v = G_D(\xi, \cdot)$. Then $A_{\bullet}$ is a decreasing sequence of compact subsets of D with intersection $A = \{\xi\}$, and $\lim_{n\to\infty} R_v^{A_n} = R_v^A$ is false because $R_v^{A_n} = v$ for all n even though R_v^A vanishes except at ξ. Thus the finiteness of v is necessary in Section 3(k). Observe that with D, A, v, ξ as just defined, (3.6) is false because the left-hand side of (3.6) is a function vanishing except at ξ whereas the right-hand side is v.

$\underset{+v}{R}^{\bullet}(\xi)$ Is Not a Choquet Capacity on D Relative to the Class Γ_0 of Compact Subsets of D

Choose v continuous and choose a compact subset A of D containing ξ in such a way that $\underset{+v}{R}^A(\xi) < v(\xi)$. If $A_{\bullet}$ is a decreasing sequence of compact neighborhoods of A with intersection A, then $\underset{+v}{R}^{A_n}(\xi) = v(\xi) > \underset{+v}{R}^A(\xi)$ for all n, and therefore even if v is finite valued and continuous, the set function $\underset{+v}{R}^{\bullet}(\xi)$ is not a topological precapacity on Γ_0 and $\underset{+v}{R}^{\bullet}(\xi)$ is not a Choquet capacity on D relative to Γ_0.

Chapter VII

Green Functions

1. Definition of the Green Function G_D

Let D be a nonempty open subset of $\mathbb{R}^N$. If $N > 2$ or if $N = 2$ and D is bounded, the function $G(\xi, \cdot)$ is lower bounded for each point ξ of D; so $\mathrm{GM}_D G(\xi, \cdot)$ exists (Section III.1). If $N = 2$, if D is unbounded, and if $G(\xi, \cdot)$ has a subharmonic minorant on D for some ξ in D, then the minorant $\mathrm{GM}_D G(\xi, \cdot)$ exists for every ξ in D. In fact $G(\xi', \cdot) - G(\xi, \cdot)$ is bounded below outside each neighborhood of ξ, and $G(\xi', \cdot)$ is bounded below on each compact neighborhood of ξ so that if $\mathrm{GM}_D G(\xi, \cdot)$ exists,

$$G(\xi', \cdot) \geq c + \mathrm{GM}_D G(\xi, \cdot), \qquad \mathrm{GM}_D G(\xi', \cdot) \geq c + \mathrm{GM}_D G(\xi, \cdot)$$

for some constant c depending on ξ' and ξ.

If $\mathrm{GM}_D G(\xi, \cdot)$ exists for ξ in D, define $G_D(\xi, \cdot)$ by

$$G_D(\xi, \cdot) = G(\xi, \cdot) - \mathrm{GM}_D G(\xi, \cdot). \tag{1.1}$$

The function G_D on $D \times D$ is called the *Green function of D*, and $G_D(\xi, \cdot)$ is called the *Green function of D with pole* ξ. The latter function is positive superharmonic, harmonic on $D - \{\xi\}$, with $\mathrm{GM}_D G_D(\xi, \cdot) = 0$.

It will be shown in Section 7 that D has a Green function if and only if D is Greenian. All we have proved so far is that D has a Green function if $N > 2$ or if $N = 2$ and if D is bounded. If $N = 2$ and if D is not everywhere dense in $\mathbb{R}^2$, the Green function G_D exists because if $\xi \in D$ and if ζ is an inner point of $\mathbb{R}^2 - D$, the restriction to D of $G(\xi, \cdot)$ has $G(\zeta, \cdot) - c$ as a harmonic minorant on D for sufficiently large c. If D is not connected, it follows that each open component of D has a Green function, and it is trivial to verify that G_D exists and that $G_D(\xi, \cdot)$ is defined for ξ in the open component D_0 of D by

$$G_D(\xi, \cdot) = \begin{cases} G_{D_0}(\xi, \cdot) & \text{on } D_0 \\ 0 & \text{on } D - D_0. \end{cases}$$

The existence of the Green function G_B of an open set B implies the existence of G_D whenever D is a nonempty open subset of B and also implies that $G_D \leq G_B$ on $D \times D$. Moreover G_D can be obtained from G_B just as G_D is obtained from G:

$$G_D(\xi, \cdot) = G_B(\xi, \cdot) - \mathrm{GM}_D G_B(\xi, \cdot). \tag{1.2}$$

To see this, note that if $\xi \in D$, then

$$\begin{aligned} G_D(\xi, \cdot) &= G(\xi, \cdot) - \mathrm{GM}_D G(\xi, \cdot) \\ &= G(\xi, \cdot) - \mathrm{GM}_D[G_B(\xi, \cdot) + \mathrm{GM}_B G(\xi, \cdot)], \end{aligned} \tag{1.3}$$

and by linearity of the GM operator (Section III.1)

$$G_D(\xi, \cdot) = G(\xi, \cdot) - \mathrm{GM}_D G_B(\xi, \cdot) - \mathrm{GM}_B G(\xi, \cdot), \tag{1.4}$$

which yields (1.2). In particular, if $B - D$ is polar then $G_D = G_B$ on $D \times D$ in view of the corresponding result for greatest subharmonic minorants in Section V.5. Conversely, if D is an open subset of $\mathbb{R}^N$ for which G_D exists and if B is an open superset of D with $B - D$ polar, then G_B also exists. In fact, if $\xi \in D$, the function $\mathrm{GM}_D G(\xi, \cdot)$ is locally upper bounded relative to B and so (Section V.5) has a subharmonic extension to B, and this extension is a subharmonic minorant of $G(\xi, \cdot)$ on B; so $\mathrm{GM}_B G(\xi, \cdot)$ exists.

Evaluation of G_D by Solving the First Boundary Value Problem

If D is an open subset of $\mathbb{R}^N$, bounded if $N = 2$, and if for ξ in D there is a harmonic function $u(\xi, \cdot)$ on D with limit $G(\xi, \eta)$ at each point η of ∂D (limit 0 at the point ∞ if $N > 2$ and D is unbounded), then an application of the superharmonic function minimum theorem and the harmonic function maximum-minimum theorem shows that $u(\xi, \cdot) = \mathrm{GM}_D G(\xi, \cdot)$, so that

$$G_D(\xi, \cdot) = G(\xi, \cdot) - u(\xi, \cdot). \tag{1.5}$$

Thus the Green function of a smooth open set as defined in Section I.8 and evaluated for a ball in Section II.1 is the Green function in the present sense, and $G_D = G$ when $N > 2$ and $D = \mathbb{R}^N$. We shall discuss the solution of the first boundary value problem for harmonic functions on Greenian subsets of $\mathbb{R}^N$ in Chapter VIII. Let $u(\xi, \cdot)$ be the solution in the generalized sense of Chapter VIII, that is, the "PWB" (Perron–Wiener–Brelot) solution, of this first boundary value problem on a Greenian set D for the boundary function

$f_\xi = G(\xi,\cdot)_{|D}$; if D is unbounded, define $f_\xi(\infty)$ as $-\infty$ or 0 according as $N=2$ or $N>2$. According to Theorem VIII.18, the evaluation (1.5) is correct with this interpretation of u if $N>2$ or if $N=2$ and D is bounded. If $N=2$ and D is unbounded, (1.5) is correct (Theorem VIII.19) if and only if D does not certain too much of a neighborhood of the point ∞ in a sense made precise in the statement of Theorem VIII.19.

2. Extremal Property of G_D

Theorem. *Let D be an open subset of $\mathbb{R}^N$ and let ξ be a point of D. If G_D exists, then*

$$\begin{aligned} G_D(\xi,\cdot) &= \inf\{u \geq 0\colon u = G(\xi,\cdot) + h \textit{ with } h \textit{ defined and harmonic on } D\} \\ &= \inf\{u \geq 0\colon u = G(\xi,\cdot) + h \textit{ with } h \textit{ defined and} \\ &\qquad \textit{superharmonic on } D\} \qquad (2.1) \\ &= \inf\{u \geq 0\colon u \textit{ superharmonic on } D,\ u \geq G(\xi,\cdot) \\ &\qquad \textit{on some neighborhood of } \xi\} \\ &= \inf\left\{u \geq 0\colon u \textit{ superharmonic on } D,\ \liminf_{\eta\to\xi}\frac{u(\eta)}{G(\xi,\eta)} \geq 1\right\}. \end{aligned}$$

In the following proof we denote the jth class on the right in (2.1) by Γ_j. These classes are not empty if G_D exists, but if G_D does not exist, these classes are empty because they all involve the existence of positive nonconstant superharmonic functions on D, and we shall see in Section 7 that G_D exists if and only if D is Greenian.

Proof for Γ_2. If G_D exists and if $u\in\Gamma_2$, then $-h \leq G(\xi,\cdot)$ on D; so $-h \leq \mathrm{GM}_D G(\xi,\cdot)$, and therefore $G_D(\xi,\cdot) \leq u$. On the other hand, $G_D(\xi,\cdot)\in\Gamma_2$. □

Proof that $\Gamma_1\cup\Gamma_3\cup\Gamma_4\subset\Gamma_2$. If G_D exists and if $u\in\Gamma_4$, let μ be the Riesz measure associated with u. If $c<1$, the function $u - cG(\xi,\cdot)$ is positive and superharmonic on an open deleted neighborhood B of ξ. Hence (Theorem V.5) u has a superharmonic extension to $B\cup\{\xi\}$; so $\mu\{\xi\}\geq c$, and therefore $\mu\{\xi\}\geq 1$; so $u\in\Gamma_2$. Thus $\Gamma_4\subset\Gamma_2$. The inclusions $\Gamma_1\subset\Gamma_2$ and $\Gamma_3\subset\Gamma_4$ are trivial. □

Proof for $\Gamma_1, \Gamma_3, \Gamma_4$. If $u\in\Gamma_1\cup\Gamma_3\cup\Gamma_4$, then we have just shown that $u\in\Gamma_2$, and we have already seen it follows that $u\geq G_D(\xi,\cdot)$. Conversely, $G_D(\xi,\cdot)\in\Gamma_1\cap\Gamma_2\cap\Gamma_4$ and $(1+\varepsilon)G_D(\xi,\cdot)\in\Gamma_3$ for every $\varepsilon>0$. □

3. Boundedness Properties of G_D

Theorem. *Let D be an open subset of $\mathbb{R}^N$ with a Green function G_D, and let ξ be a point of D.*

(a) *If B is an open neighborhood of ξ in D, relatively compact in D, and if v is a positive superharmonic function defined on an open superset of $D - B$, with $v \geq G_D(\xi, \cdot)$ on a neighborhood of ∂B, then $v \geq G_D(\xi, \cdot)$ on $D - B$.*

(b) $R^B_{G_D(\xi,\cdot)} = \hat{R}^B_{G_D(\xi,\cdot)} = G_D(\xi, \cdot)$ *(reductions relative to D) whenever B is a neighborhood of ξ.*

(c) *If v is a strictly positive superharmonic function on D, then outside each neighborhood of ξ, $G_D(\xi, \cdot) \leq \text{const}\, v$. In particular $(v \equiv 1)$, $G_D(\xi, \cdot)$ is bounded outside each neighborhood of ξ, and if ξ_1 is a point in the same open connected component of D as ξ, then outside each neighborhood of ξ, $G_D(\xi, \cdot) \leq \text{const}\, G_D(\xi_1, \cdot)$.*

(d) *Let B_1 be a compact subset of D and let B_2 be D less a neighborhood of B_1. Then G_D is bounded on $B_1 \times B_2$.*

Proof of (a). Define

$$v_1 = \begin{cases} G_D(\xi, \cdot) & \text{on } B, \\ G_D(\xi, \cdot) \wedge v & \text{on } D - B. \end{cases}$$

The function v_1 is superharmonic on D, and $G(\xi, \cdot) - v_1$, defined as 0 at ξ, is a subharmonic minorant of $G(\xi, \cdot)$ and is therefore majorized by $\text{GM}_D G(\xi, \cdot)$; that is, $v_1 \geq G_D(\xi, \cdot)$ on $D - B$. □

Proof of (b). The function $\hat{R}^B_{G_D(\xi,\cdot)}$ is a positive superharmonic minorant of $G_D(\xi, \cdot)$ on D, equal to $G_D(\xi, \cdot)$ on a neighborhood of ξ, and therefore by (a) must majorize $G_D(\xi, \cdot)$ outside this neighborhood. Hence these two functions are identical. Since the unsmoothed reduction lies between $G_D(\xi, \cdot)$ and the smoothed reduction, the three functions are identical on D. □

Proof of (c) *and* (d). In (c) we can suppose that the neighborhood in question is open and relatively compact in D. Then if v is a strictly positive superharmonic function on D, the function cv majorizes $G_D(\xi, \cdot)$ on a neighborhood of ∂B for sufficiently large c; so (a) can be applied to yield (c). The particular cases of (c) follow trivially, and (d) follows from the first of these cases by an application of the Heine–Borel theorem. □

The Set D_α: $\{\eta: G_D(\xi, \eta) > \alpha\}$

(*Note:* ξ is fixed throughout the discussion.) If $\alpha > 0$, the set D_α is an open subset of D containing ξ. No open connected component of D_α not containing

ξ is relatively compact in D, but D_α is itself connected and relatively compact in D if α is sufficiently large. In fact, if B is a relatively compact in D open connected component of D_α not containing ξ, then the function $G_D(\xi, \cdot)$ is harmonic on B, with boundary limit α at every boundary point of B, and so must be identically α on B, an impossibility. On the other hand, if β is the supremum of $G_D(\xi, \cdot)$ on the complement of a compact neighborhood of ξ in D and if $\alpha > \beta$, then the set D_α is relatively compact in the interior of this neighborhood and so also relatively compact in D and is connected by what was just proved.

Observation ($N = 2$). According to Theorem 3, the function $G_D(\xi, \cdot) = G(\xi, \cdot) - u(\xi, \cdot)$ is bounded outside each neighborhood of ξ. It will now be shown that if ξ_1 and ξ_2 are in D and if c is an arbitrary constant, the minorants

$$\mathrm{GM}_D[G(\xi_1, \cdot) \wedge G(\xi_2, \cdot)], \qquad \mathrm{GM}_D[G(\xi_1, \cdot) \wedge c] \tag{3.1}$$

exist and that the differences

$$\begin{aligned} &G(\xi_1, \cdot) \wedge G(\xi_2, \cdot) - \mathrm{GM}_D[G(\xi_1, \cdot) \wedge G(\xi_2, \cdot)], \\ &G(\xi_1, \cdot) \wedge c - \mathrm{GM}_D[G(\xi_1, \cdot) \wedge c] \end{aligned} \tag{3.2}$$

are bounded outside any neighborhood of ξ_1 and ξ_2, outside any neighborhood of ξ_1, respectively. These facts imply that if η is a finite boundary point of D, the minorants in (3.1) are bounded on the trace on D of a compact neighborhood of η in $\mathbb{R}^2$, a fact needed later.

Since $G(\xi_1, \cdot) - G(\xi_2, \cdot)$ is bounded outside any neighborhood of ξ_1 and ξ_2 and since $G_D(\xi_i, \cdot)$ is bounded outside any neighborhood of ξ_i, the difference $u(\xi_1, \cdot) - u(\xi_2, \cdot)$ is bounded outside any neighborhood A of ξ_1 and ξ_2. Choose a compact neighborhood A of these two points, so that $|u(\xi_1, \cdot) - u(\xi_2, \cdot)| \leq b$ on $D - A$, where b depends on A, ξ_1, and ξ_2. This inequality must hold on all of D in view of the maximum theorem for subharmonic functions. Then

$$G(\xi_1, \cdot) \wedge G(\xi_2, \cdot) \geq \mathrm{GM}_D[u(\xi_1, \cdot) \wedge u(\xi_2, \cdot)] \geq u(\xi_1, \cdot) - b; \tag{3.3}$$

so the first minorant in (3.1) exists. The harmonic function $u(\xi_1, \cdot)$ is bounded above by $c_1 = \sup_{\partial D} G(\xi_1, \cdot) < +\infty$; so

$$G(\xi_1, \cdot) \wedge c_1 \geq u(\xi_1, \cdot) \wedge c_1 = u(\xi_1, \cdot),$$

and it follows that for any constant c,

$$G(\xi_1, \cdot) \wedge c \geq G(\xi_1, \cdot) \wedge c_1 - |c - c_1| \geq u(\xi_1, \cdot) - |c - c_1|; \tag{3.4}$$

so the second minorant in (3.1) exists. Finally, the first difference in (3.2) is at most

$$G(\xi_1, \cdot) - u(\xi_1, \cdot) + b = G_D(\xi_1, \cdot) + b, \tag{3.5}$$

and similarly the second is at most $G_D(\xi_1, \cdot) + \text{const}$; so these differences are bounded outside each neighborhood of ξ_1, ξ_2 or each neighborhood of ξ_1, respectively.

4. Further Properties of G_D

As already remarked we shall prove in Section 7 that the open sets with Green functions are the Greenian sets. In the following theorem the boundary involved is the Euclidean boundary.

Theorem. *If D is an open subset of $\mathbb{R}^N$ for which there is a Green function, the Green function G_D has the following properties:*

(a) *G_D is continuous and symmetric on $D \times D$ ($= +\infty$ on the diagonal).*
(b) *For quasi every finite point ζ of ∂D and also for $\zeta = \infty$ if $N > 2$ and D is unbounded, $\lim_{\eta \to \zeta} G_D(\xi, \eta) = 0$ for every ξ in D.*
(c) *If $\xi \in D$, the function $G_D(\xi, \cdot)$ has a positive extension $G_{\bar{D}}^{=}(\xi, \cdot)$ to $\mathbb{R}^N$, uniquely characterized by the following properties:*
(c1) *$G_{\bar{D}}^{=}(\xi, \cdot) = 0$ on $\mathbb{R}^N - \bar{D}$ and at quasi every finite point of ∂D.*
(c2) *$G_{\bar{D}}^{=}(\xi, \cdot)$ is subharmonic on $\mathbb{R}^N - \{\xi\}$.*
(c3) *If ζ is a finite boundary point of D,*

$$G_{\bar{D}}^{=}(\xi, \zeta) = \limsup_{D \ni \eta \to \zeta} G(\xi, \eta); \tag{4.1}$$

if $N > 2$ and if D is unbounded, $\lim_{\eta \to \infty} G_{\bar{D}}^{=}(\xi, \eta) = 0$.

Observation (1). If D is connected, the finite points ζ of ∂D for which $G_{\bar{D}}^{=}(\cdot, \zeta) = 0$ on D, equivalently, by lower semicontinuity of subharmonic functions the boundary points ζ for which $\lim_{\eta \to \zeta} G_D(\xi, \eta) = 0$ for all ξ in D, will be seen in Section VIII.14 to be the finite regular boundary points of D relative to the Dirichlet problem for harmonic functions on D.

Observation (2). We shall prove (b) as a consequence of (c). Alternatively it is possible to prove (b) after the Dirichlet problem for harmonic functions on D has been treated, by showing that quasi every finite boundary point of D is regular and that the regular boundary points are the points at which $G_D(\xi, \cdot)$ has limit 0. The extension $G_{\bar{D}}^{=}(\xi, \cdot)$ can then be obtained by an application of the extension Theorem V.5 to $-G_D(\xi, \cdot)$ on $\mathbb{R}^N - \{\xi\}$.

Proof of (a). We have seen in Section III.1 that if G_D exists, then $u(\xi,\cdot) = \mathrm{GM}_D G(\xi,\cdot)$ can be obtained as follows. A sequence $B_{\bullet}$ of balls is chosen with closures in D and with the property that each point of D has a neighborhood which lies in B_n for infinitely many values of n. If τ_{B_j} is the operator II(1.8) but applied here to functions on $\mathbb{R}^N$, and if for $\xi \in D$ we define

$$u_n'(\xi,\cdot) = \tau_{B_n} \cdots \tau_{B_0} G(\xi,\cdot), \qquad u_n(\xi,\cdot) = u_n'(\xi,\cdot)_{|D},$$

then $\{u_n'(\xi,\cdot), n \in \mathbb{Z}^+\}$ and $\{u_n(\xi,\cdot), n \in \mathbb{Z}^+\}$ are decreasing sequences of superharmonic functions on $\mathbb{R}^N$ and D, respectively. The limit of the second sequence is $u(\xi,\cdot)$. We shall deal with the first sequence later. Since G is continuous on $D \times D$, the functions $u_0, u_1, \ldots$ are successively continuous, so u is upper semicontinuous. Let ξ_0 be any point of D and choose B_n in such a way that $\xi_0 \in B_0$ and that there is an open neighborhood $B \subset B_0$ of ξ_0 which is a subset either of B_n or of $D - \bar{B}_n$ for each n. A glance at the formula for $\tau_{B_0} G(\xi,\cdot)$ shows that for each point η of D the function $u_0(\cdot,\eta)$ is harmonic on B, and the same reasoning shows that $u_1(\cdot,\eta)$, $u_2(\cdot,\eta)$, ... are harmonic on B. It follows that $u(\cdot,\eta)$ is harmonic on B, and since u is independent of the choice of $B_{\bullet}$, the function $u(\cdot,\eta)$ is harmonic on D, a minorant of $G(\cdot,\eta)$. By definition of u it follows that $u(\cdot,\eta) \le u(\eta,\cdot)$, and reversing arguments in this inequality yields the symmetry of u on $D \times D$. Hence G_D is symmetric on $D \times D$.

To prove continuity of G_D, we prove that u is continuous. Let B' and B'' be balls with closures in D, and express u on $B' \times B''$ using the Poisson integral on the boundary of each ball. If the balls either are the same or have disjoint closures, this representation of u shows that u is continuous on $B' \times B''$. Thus u is continuous on $D \times D$. □

Proof of (b) and (c). The sequence $\{u_n'(\xi,\cdot), n \in \mathbb{Z}^+\}$ is a decreasing sequence of superharmonic functions on $\mathbb{R}^N$, all equal to $G(\xi,\cdot)$ on $\mathbb{R}^N - D$. Moreover the sequence is locally uniformly bounded below on $\mathbb{R}^N - \partial D$. In fact $u_n'(\xi,\cdot) \ge u(\xi,\cdot)$ on D, and $u_n'(\xi,\cdot) = G(\xi,\cdot)$ on $\mathbb{R}^N - \bar{D}$. We now show that this sequence is uniformly bounded below on a neighborhood of each finite boundary point of D. This fact is trivial if $N > 2$ because u_n' is then positive and is trivial if $N = 2$ and D is bounded because a lower bound of $G(\xi,\cdot)$ on a neighborhood of $\bar{D}$ is also a lower bound of $u_n(\xi,\cdot)$ on the neighborhood. If $N = 2$ and D is unbounded, the function $G_D(\xi,\cdot)$ is bounded on D in a neighborhood of ∂D (Theorem 3), say $G_D(\xi,\cdot) \ge c$ there; so

$$u_n'(\xi,\cdot) \ge u(\xi,\cdot) = G(\xi,\cdot) - G_D(\xi,\cdot) \ge G(\xi,\cdot) - c$$

on D near ∂D, and $u_n'(\xi,\cdot) = G(\xi,\cdot)$ on $\mathbb{R}^2 - D$; so the sequence $\{u_n'(\xi,\cdot), n \in \mathbb{Z}^+\}$ is locally lower bounded on a neighborhood of each finite point of ∂D, as asserted. Thus $\{u_n'(\xi,\cdot), n \in \mathbb{Z}^+\}$ is locally uniformly lower bounded on $\mathbb{R}^N$. It now follows from the Fundamental Convergence Theorem that the function $u'(\xi, \cdot) = \lim_{n\to\infty} u_n'(\xi, \cdot)$ is equal quasi everywhere on $\mathbb{R}^N$ to a superharmonic function $\underset{+}{u}'(\xi, \cdot)$. More specifically,

$$\underset{+}{u}'(\xi, \cdot) = \begin{cases} u(\xi, \cdot) = \mathrm{GM}_D G(\xi, \cdot) & \text{on } D, \\ G(\xi, \cdot) & \text{on } \mathbb{R}^N - \bar{D} \text{ and at quasi every finite} \\ & \text{point of } \partial D. \end{cases}$$

The function $G_{\bar{D}}^{=}(\xi, \cdot) = G(\xi, \cdot) - \underset{+}{u}'(\xi, \cdot)$ satisfies the conditions in (c) and is obviously uniquely determined by these conditions; that is, (c) is true. By the upper semicontinuity of subharmonic functions the positive function $G_{\bar{D}}^{=}(\xi, \cdot)$ is continuous at every zero; in particular, $G_D(\xi, \cdot)$ has limit 0 at quasi every finite point of ∂D. If for some point ξ_1 of D the function $G_D(\xi_1, \cdot)$ has limit 0 at a boundary point ζ, then $G_D(\xi, \cdot)$ has limit 0 at ζ for all ξ in the same open connected component of D as ξ_1 because (Theorem 3) outside each neighborhood of ξ the inequality $G_D(\xi, \cdot) \le \text{const}\, G_D(\xi_1, \cdot)$ is valid. Since the number of open connected components of D is countable, (b) is true. If $N > 2$ and D is unbounded, the function $G_{\bar{D}}^{=}(\xi, \cdot)$ has limit 0 at ∞ because $G_{\bar{D}}^{=}(\xi, \cdot) \le G(\xi, \cdot)$. If ζ is a finite boundary point of D and $N \ge 2$, then

$$G_{\bar{D}}^{=}(\xi, \zeta) = \limsup_{\eta \to \zeta} G_{\bar{D}}^{=}(\xi, \eta) \tag{4.2}$$

because $G_{\bar{D}}^{=}(\xi, \cdot)$ is subharmonic on $\mathbb{R}^N - \{\xi\}$. This limit relation remains correct if η tends to ζ along the nonzero set of $G_{\bar{D}}^{=}(\xi, \cdot)$ and also [see II(6.1)] if η avoids the polar set of points η' of ∂D for which $G(\cdot, \eta') \ne 0$; that is, (4.1) is true. □

Observation (3). The method of proof of (b) can be applied to show that the function differences in (3.2) also have limit 0 at quasi every finite point of ∂D.

5. The Potential $G_D\mu$ of a Measure μ

Let D be an open subset of $\mathbb{R}^N$ for which G_D exists. If μ is a measure on D, the potential $G_D\mu$ is defined by

$$G_D\mu(\xi) = \int_D G_D(\xi, \eta)\, \mu(d\eta). \tag{5.1}$$

The following facts have been derived for D special but are true in general, and the proofs for special D are applicable. A potential $G_D\mu$ is either superharmonic or identically $+\infty$ on each open connected component of D and, if superharmonic, is harmonic on every μ null open subset of D. Moreover $GM_D G_D\mu = 0$ if $G_D\mu$ is superharmonic. In particular, $G_D\mu$ is superharmonic if $\mu(D) < +\infty$.

If μ has compact support in D, Theorem 3(c) implies that $G_D\mu$ is bounded outside each neighborhood of this support. Moreover in this case $G_D\mu$ has limit 0 at every Euclidean boundary point ζ at which $\lim_{\eta\to\zeta} G_D(\xi,\eta) = 0$ for every ξ because one can integrate to the limit in (5.1) when μ has compact support. It follows (Section 4) that when μ has compact support in D, the potential $G_D\mu$ has limit 0 at quasi every finite Euclidean boundary point of D and also the point ∞ if D is unbounded and $N > 2$. Since a measure μ on D can be chosen, supported by a countable dense subset of D, making $G_D\mu$ a superharmonic function $+\infty$ on a dense subset of D, the hypothesis of compact support for μ was necessary in the preceding sentence.

If μ has compact support in D, the potential $G_D\mu$ differs from $G\mu$ on D by a harmonic function. In fact, if $u(\xi,\cdot) = GM_D G(\xi,\cdot)$,

$$G_D\mu(\xi) = \int_A G(\xi,\eta)\mu(d\eta) - \int_A u(\xi,\eta)\mu(d\eta). \tag{5.2}$$

The second integral defines a harmonic function on D because the function is Borel measurable, has the harmonic function average property, and is finite valued on a dense set. Observe that what we have proved implies that if A is a compact subset of $\mathbb{R}^N$, if μ is a measure on $\mathbb{R}^N$ supported by A, and if D_1 and D_2 are open supersets of A whose Green functions exist, then $G_{D_1}\mu$ and $G_{D_2}\mu$ differ on $D_1 \cap D_2$ by a harmonic function.

In order to complete the proof of the Riesz Decomposition Theorem (Section IV.8) by proving it for open sets D for which G_D exists—it will be seen in Section 7 that these sets are the Greenian sets—we need only follow the steps of the proof for D special. According to (5.2), the conclusion of Theorem IV.7 can be stated in the form: $u = G_D\mu_A + h_A$, where h_A is superharmonic on D and harmonic on A. This representation of u was the basis for the proof of the Riesz Decomposition Theorem for special open sets, and the continuation of that proof needs no change in the present more general context. Note again that according to that theorem, a positive superharmonic function u on D is a potential $G_D\mu$ if and only if $GM_D u = 0$.

In view of (5.2) the Gauss Integral Theorem (Section I.6) and also Theorem I.7 on the smoothness of the potentials $G\mu$ of measures given by densities are valid for the more general potentials $G_D\mu$.

Application to the Function $D \mapsto G_D$

Let D_2 be a Greenian subset of $\mathbb{R}^N$, let D_1 be a nonempty relatively compact open subset of D_2, and let μ be a measure on D_2 with compact support in

D_1. Then $G_{D_2}\mu = G_{D_1}\mu + \mathrm{GM}_{D_1}G_{D_2}\mu$ on D_1. The minorant is [Section III Example (d)] the restriction to D_1 of $\| G_{D_2} \|^{D_2 - D_1}$ (reduction relative to D_2). This smoothed reduction must be a potential $G_{D_2\mu_1}$ on D_2; so $G_{D_2}\mu = G_{D_1}\mu + G_{D_2\mu_1}$ on D_1, where μ_1 is a measure supported by ∂D_1.

6. Increasing Sequences of Open Sets and the Corresponding Green Function Sequences

Theorem. *Let $D_\bullet$ be a monotone sequence of open subsets of $\mathbb{R}^N$ with limit D, and suppose that G_{D_n} exists for all n.*

(a) *If $D_n \uparrow D$ then the increasing sequence $G_{D_\cdot}$ has limit G_D if G_D exists and has limit $+\infty$ otherwise.*

(b) *If $D_n \downarrow D$ and if D is open and not empty then the decreasing sequence $G_{D_\cdot}$ has limit G_D.*

Proof of (a). To avoid trivia suppose that D is connected. If ξ is a point of D and if $u_n(\xi, \cdot) = \mathrm{GM}_{D_n}G(\xi, \cdot)$, then $u_n(\xi, \cdot) \geq u_{n+1}(\xi, \cdot)$ on D_n, and therefore (Theorem II.3) the limit $u(\xi, \cdot)$ of the sequence $\{u_n(\xi, \cdot), n \geq 0\}$ on D is either identically $-\infty$ or harmonic, and it is easy to check that in the latter case $u(\xi, \cdot) = \mathrm{GM}_D G(\xi, \cdot)$. Conversely, if $\mathrm{GM}_D G(\xi, \cdot)$ exists, the sequence $\{u_n(\xi, \cdot), n \geq 0\}$ is locally bounded below; so the limit function $u(\xi, \cdot)$ is harmonic on D. Since $G_D(\xi, \cdot) = G(\xi, \cdot) - \mathrm{GM}_D G(\xi, \cdot)$, assertion (a) follows. □

Proof of (b). Define $G'_D = \lim_{n\to\infty} G_{D_n}$ on $D \times D$. For ξ in D the function $G'_D(\xi, \cdot)$ is harmonic on $D - \{\xi\}$. The inequality $G_D \leq G'_D \leq G_{D_n}$ implies that $G'_D(\xi, \cdot)$ is continuous with value $+\infty$ at ξ and that the difference $G'_D(\xi, \cdot) - G_D(\xi, \cdot)$ is a positive harmonic function on D if defined suitably at ξ. This difference is bounded [because $G_{D_n}(\xi, \cdot)$ is bounded outside a neighborhood of ξ] and has limit 0 at quasi every Euclidean boundary point of D, including the point ∞ if $N > 2$ and D is unbounded [because $G_{D_n}(\xi, \cdot)$ has this boundary limit property]; so this difference vanishes identically according to the extended maximum-minimum theorem for harmonic functions in Section V.7. □

7. The Existence of G_D versus the Greenian Character of D

Theorem. *An open subset D of $\mathbb{R}^N$ is Greenian if and only if G_D exists.*

Since all open nonempty subsets of $\mathbb{R}^N$ for $N > 2$ and all bounded nonempty open or not connected open subsets of $\mathbb{R}^2$ are both Greenian and have Green functions, only unbounded open connected subsets of $\mathbb{R}^2$ remain to be considered. If a set has a Green function, the set is trivially Greenian. Conversely, suppose that D is an open unbounded Greenian subset of $\mathbb{R}^2$, so that there is a positive nonconstant superharmonic function u on D, with associated nonnull Riesz measure μ. Let D_n be the part of D in $B(0, n)$ and let μ_n be the projection of μ on D_n. According to Theorem 6, unless D has a Green function, $\lim_{n\to\infty} G_{D_n} = +\infty$, but this limit relation is impossible because (Riesz decomposition)

$$u = G_{D_n}\mu_n + h_n \geq G_{D_n}\mu_n \qquad \text{on } D_n. \tag{7.1}$$

8. From Special to Greenian Sets

Many theorems have been stated for Greenian sets D but proved only for special sets. The justification for this awkward procedure is that the properties of G_D make the proofs already given for special D valid whenever D is Greenian, that is, whenever G_D exists. Some of these extensions to Greenian D have already been checked in Sections 4 and 5. The rest are equally easy to check.

9. Approximation Lemma

Lemma. *If D is a Greenian subset of $\mathbb{R}^N$, if ϕ is a finite-valued continuous function on D, with compact support B, and if B_1 is a compact neighborhood of B in D, there is a sequence $u_\cdot - v_\cdot$ of differences of finite-valued continuous potentials whose associated Riesz measures are supported by B_1, with $u_n = v_n$ on $D - B_1$ and* $\lim_{n\to\infty}(u_n - v_n) = \phi$ *uniformly on D.*

For n sufficiently large, say $n \geq n_0$, the function $A_{1/n}\phi$ is defined on a neighborhood of B_1 and vanishes off B_1. For n this large define $A_{1/n}\phi = 0$ on $D - B_1$ where this function is not already defined. Then $\{A_{1/n}\phi, n \geq n_0\}$ is a sequence of infinitely differentiable functions converging uniformly to ϕ on D. Define u_n $[v_n]$ as the potential of the measure with density $-((\Delta A_{1/n}\phi) \wedge 0)/\pi'_N$ $[((\Delta A_{1/n}\phi) \vee 0)/\pi'_N]$. Then $A_{1/n}\phi - (u_n - v_n) = h_n$ is a function harmonic on D, and $v_n - u_n = h_n$ on $D - B_1$. The inequalities $v_n \geq h_n$ and $u_n \geq -h_n$ are valid on $D - B_1$ and therefore (superharmonic function minimum theorem) also on D; so $h_n = 0$ since $\mathrm{GM}_D v_n = \mathrm{GM}_D u_n = 0$. That is, $A_{1/n}\phi = u_n - v_n$, and the lemma follows.

Application to the Ordering of Measures

Suppose that μ and ν are finite measures on a Greenian set D, with the property that whenever u and v are finite-valued positive continuous superharmonic functions on D, with $u \geq v$, then

$$\int_D (u - v)\, d\mu \le \int_D (u - v)\, d\nu. \tag{9.1}$$

We now show that it follows that $\mu \le \nu$. It is sufficient to show that if ϕ is a positive continuous function on D with compact support then

$$\int_D \phi\, d\mu \le \int_D \phi\, d\nu. \tag{9.2}$$

According to the lemma, the function ϕ can be approximated uniformly by a sequence $u_{\bullet} - v_{\bullet}$ of differences between continuous potentials, equal outside a neighborhood of the support of ϕ. By hypothesis

$$\int_D \{[u_n - \inf_D (u_n - v_n)] - v_n\}\, d\mu \le \int_D \{[u_n - \inf_D (u_n - v_n)] - v_n\}\, d\nu,$$

and this inequality yields (9.2) when $n \to \infty$.

10. The Function $G_D(\cdot, \zeta)_{|D-\{\zeta\}}$ as a Minimal Harmonic Function

Let D be a connected Greenian subset of $\mathbb{R}^N$. We now show that if $\zeta \in D$, the restriction of $G_D(\cdot, \zeta)$ to $D_0 = D - \{\zeta\}$ is a minimal harmonic function for D_0. In fact let u_0 be a positive harmonic function on D_0, majorized there by $G_D(\cdot, \zeta)$. We are to show that u_0 must be a constant multiple of $G_D(\cdot, \zeta)$ on D_0. The function u_0 has a superharmonic extension u to D (Theorem V.5) whose associated Riesz measure is supported by $\{\zeta\}$, $u = cG_D(\cdot, \zeta) + h$, where c is a positive constant and h is a positive harmonic function on D, majorized on D_0 by u and therefore majorized on D by $G_D(\cdot, \zeta)$. Hence $h \equiv 0$ and the proof is complete.

If v is a positive superharmonic function on D, with associated Riesz measure ν, then

$$\inf_{D_0} \frac{v}{G_D(\cdot, \zeta)} = \nu(\{\zeta\}). \tag{10.1}$$

To prove this, let c be the infimum in question. Then $v \ge cG_D(\cdot, \zeta)$ on D_0, so the difference $v - cG_D(\cdot, \zeta)$ is positive and superharmonic on D_0 and therefore has a superharmonic extension v_1 to D. Thus $v = cG_D(\cdot, \zeta) + v_1$ on D_0, and the equality is valid at ζ because two superharmonic functions equal on a deleted neighborhood of a point are equal at the point. But then $\nu(\{\zeta\}) \ge c$, and there must be equality by definition of c, so (10.1) is true. Furthermore

$$\liminf_{\eta \to \zeta} \frac{v(\eta)}{G_D(\eta, \zeta)} = \nu(\{\zeta\}). \tag{10.2}$$

To prove this, observe that according to (10.1) the limit inferior α in question is at least $\nu(\{\zeta\})$. If there is strict inequality and if $\alpha > \beta > \nu(\{\zeta\})$, then $v > \beta G_D(\cdot, \zeta)$ on a deleted neighborhood of ζ and therefore (Theorem 3) on D; so the infimum in (10.1) is $\geq \beta$. Hence $\nu(\{\zeta\}) \geq \beta$, contrary to hypothesis.

We shall prove (Theorem XI.4) that the limit inferior in (10.2) is a limit in the context of the fine topology discussed in Chapter XI.

Generalization. If ζ is a point of a compact polar subset A of D and if now D_0 is $D - A$, then a trivial modification of the preceding discussion shows that the restriction of $G_D(\cdot, \zeta)$ to $D - A$ is a minimal harmonic function for D_0 and that (10.1) and (10.2) are true, with the understanding that the infimum in (10.1) may be taken over either D_0 or $D - \{\zeta\}$ and that η in (10.2) may tend to ζ on either D_0 or D.

Chapter VIII

The Dirichlet Problem for Relative Harmonic Functions

1. Relative Harmonic, Superharmonic, and Subharmonic Functions

The class of relative harmonic functions is suggested by the following trivial remark. Let $(D, \mathscr{D})$ be a measurable space, and suppose that to each point ξ of D is assigned some set (perhaps empty) $\{\mu_\alpha(\xi, \cdot), \alpha \in I_\xi\}$ of probability measures on D. Call a function *generalized harmonic* if it satisfies specified smoothness conditions and if

$$v(\xi) = \int_D v(\eta)\mu_\alpha(\xi, d\eta) = \mu_\alpha(\xi, v) \tag{1.1}$$

for ξ in D and α in I_ξ. For example, if D is an open subset of $\mathbb{R}^N$, if for each ξ the index α represents a ball B of center ξ with closure in D, if I_ξ is the class of all such balls, and if $\mu_B(\xi, v)$ is the unweighted average of v on ∂B, then the class of continuous functions on D satisfying (1.1) is the class of harmonic functions on D. Going back to the general case, suppose that h is a strictly positive generalized harmonic function and define $\mu_\alpha^h(\xi, \cdot)$ by

$$\mu_\alpha^h(\xi, A) = \int_A h(\eta)\frac{\mu_\alpha(\xi, d\eta)}{h(\xi)}. \tag{1.2}$$

Then $\mu_\alpha^h(\xi, \cdot)$ is a probability measure, and if v is a generalized harmonic function, the function $u = v/h$ satisfies

$$u(\xi) = \int_D u(\eta)\mu_\alpha^h(\xi, d\eta). \tag{1.3}$$

Conversely, if u is a function satisfying (1.3), the function $v = uh$ is generalized harmonic. (We omit, here and in the following, possible side conditions on the functions.) The functions $u = v/h$ thus satisfy the same kind of averaging condition as the generalized harmonic functions, but with μ_α replaced by μ_α^h. Theorems on generalized harmonic functions v relative to the averaging $\{\mu_\alpha\}$

correspond to theorems on functions v/h relative to the averaging system $\{\mu_\alpha^h\}$. This remark suggests the following definition.

Let D be an open subset of $\mathbb{R}^N$ and let h be a strictly positive harmonic function on D. A function $u = v/h$ will be called *h-harmonic*, *h-superharmonic*, or *h-subharmonic* if v is harmonic, superharmonic, or subharmonic on D, respectively. This definition is of interest only if h is not identically constant, so that D is always supposed Greenian in discussing these matters. If $v = G_D\mu$ is a potential, the function v/h will be called an *h potential*. The functions of classical potential theory relativized by a strictly positive harmonic h play an essential role in the study of the case $h \equiv 1$. Many properties of the relativized functions follow trivially from the case $h \equiv 1$ or can be deduced using the proofs for that case. For example, the proof of the superharmonic function minimum theorem translates at once into a proof of the same result for h-superharmonic functions. The operator τ_B^h is defined by $\tau_B^h(v/h) = (\tau_B v/h$. The notations GM^h, LM^h, ${}^hR_u^A$, and ${}^h\|u\|^A$ (for u a positive h-superharmonic function) have the obvious interpretations.

Minimal harmonic functions were defined in Section II.16, and minimal h-harmonic functions are defined correspondingly. The positive h-harmonic function v/h on D is a minimal h-harmonic function if and only if v is a minimal harmonic function. In particular, if h is minimal harmonic, the positive constant functions are minimal h-harmonic, and conversely.

A function u is h harmonic if and only if (in the usual inner product notation) $\Delta u = -2(\operatorname{grad} u, \operatorname{grad} h)/h$. If u is a function in $\mathbb{C}^{(2)}(D)$ the function is h superharmonic if and only if $\Delta u \le -2(\operatorname{grad} u, \operatorname{grad} h)/h$.

Relative Superharmonic (Harmonic) Functions and the Kelvin Transformation

Let v $[h]$ be a superharmonic [harmonic] function on a Greenian subset D of $\mathbb{R}^N$, and let ϕ be an inversion of $\mathbb{R}^N$ in a sphere of center ξ_0. Define $v' = v(\phi)$ and $h' = h(\phi)$, and let v_1 $[h_1]$ be the Kelvin transform of v $[h]$. Then $v_1/h_1 = v'/h'$, and this function is an h_1-superharmonic function on $\phi(D - \{\xi_0\})$ if $h > 0$, h_1-harmonic if h is harmonic. According to the definition of the Kelvin transformation, the function v_1/h_1 is not the Kelvin transform of v/h unless $N = 2$, in which case $v_1 = v'$ and $h_1 = h'$.

2. The PWB Method

Let D be a Greenian subset of $\mathbb{R}^N$, provided with a boundary ∂D by a metric compactification, and let h be a strictly positive harmonic function on D. Let f be an extended real-valued function defined on ∂D. The traditional Dirichlet (first boundary value) problem in the context of h-harmonic functions is to find a function u which is h-harmonic on D and has limit $f(\zeta)$

at each boundary point ζ. (In the usual formulation of such a problem the boundary function f is supposed finite valued and continuous.) Since a function on ∂D that is the boundary limit of a function on D is continuous, this Dirichlet problem cannot have a solution unless f is continuous. On the other hand, the following example shows that boundary function finiteness and continuity may not be enough to ensure the existence of a solution.

Example (a). Let D be a ball less the center, let $h \equiv 1$, and let ∂D be the Euclidean boundary. Let $f = 0$ at the ball center and $f = 1$ on the rest of the boundary. Then f is continuous on ∂D. The Dirichlet problem on D for this boundary function has no solution because if u were a solution, then $0 \leq u \leq 1$, and (Section V.5) u would be harmonic on the ball if defined as 0 at the center. But then u would attain its infimum 0 at the center and therefore would vanish identically, contrary to the hypothesis of limit 1 at the ball boundary.

The following example illustrates the influence of the choice of the relativizing function h,

Example (b). Dirichlet problem for a ball. Let D be a ball, and let ∂D be the Euclidean boundary. According to Theorem II.1, the Poisson integral $\mathrm{PI}(D,f)$ solves the Dirichlet problem with $h \equiv 1$ for any finite-valued continuous boundary function f. If h is minimal harmonic however, that is (Section II.16), if h is a multiple of $K(\zeta, \cdot)$ for some boundary point ζ, the Dirichlet problem for h-harmonic functions with a specified bounded function f cannot have a solution unless f is identically constant. In fact a solution would be an h-harmonic function bounded by the bound of f and so would be identically constant because h is minimal harmonic.

The method of attacking the Dirichlet problem will be one devised by Perron, developed by Wiener, perfected by Brelot, now called the *PWB method*. In general the PWB method assigns a "PWB solution" to certain "resolutive" boundary functions. A finite-valued continuous boundary function f for which the traditional Dirichlet problem has a solution u_f is resolutive with PWB solution u_f, but the PWB method goes further. In fact the class of resolutive boundary functions is the class of integrable boundary functions relative to a certain measure, and one task of the theory is to show that the PWB solution for a resolutive boundary function has that boundary function as a boundary limit function in some reasonable sense. In the context of h-harmonic functions we shall introduce h into the notation and denote by H_f^h the PWB^h solution determined by an h-resolutive boundary function f. Although the method itself provides a not unreasonable justification for describing H_f^h as a solution corresponding to f, we shall show that H_f^h has f as a boundary limit function along certain paths.

The PWB method starts with a Greenian set D provided with a boundary ∂D by a metric compactification and assigns to each extended real-valued function f on ∂D an upper and a lower PWB^h class of function on D. A function v on ∂D is in the upper [lower] PWB^h class if on each open connected component of D this function is either identically $+\infty$ $[-\infty]$ or h-superharmonic [h-subharmonic], is bounded below [above], and satisfies the inequality

$$\liminf_{\eta\to\zeta} v(\eta) \geq f(\zeta) \qquad [\limsup_{\eta\to\zeta} v(\eta) \leq f(\zeta)]$$

for every boundary point ζ. The functions in the upper PWB^h class for f are the negatives of the functions in the lower PWB^h class for $-f$. An application of the minimum theorem for h-superharmonic functions yields the fact that every function in the upper PWB^h class for f is a majorant of every function in the lower class. If u and v are in the upper class, then $u \wedge v$ is also. Thus the upper class is directed downward, and dually the lower class is directed upward. If v is an h-superharmonic function in the upper class and if B is a ball with $\bar{B} \subset D$, then $\tau_B^h v$ is also in the upper class and is a minorant of v. Thus the infimum $\bar{H}_f^h$ of the upper class is, in each such ball B, the limit of a downward-directed family of h-harmonic functions unless each member of the upper class is identically $+\infty$ in B. In each open connected component of D the function $\bar{H}_f^h$ is therefore either the constant function $+\infty$, the constant function $-\infty$, or an h-harmonic function. The supremum $\underline{H}_f^h$ of the lower class must also have this property, and $\underline{H}_f^h \leq \bar{H}_f^h$. The function $\underline{H}_f^h$ $[\bar{H}_f^h]$ is called the lower [upper] PWB^h solution for f. It is immediate that $\bar{H}_{1_A}^h = R_h^A/h = {}^hR_1^A$ for A a subset of ∂D. If the upper and lower solution are identical and h-harmonic, they are denoted by H_f^h, f will be called *h-resolutive*, or simply *resolutive* when $h \equiv 1$, and H_f^h will be called the *PWB^h solution for f*. A boundary of a Greenian set D will be called *h-resolutive*, or simply *resolutive* when $h \equiv 1$, if every finite-valued continuous boundary function is h-resolutive, equivalently according to Section 6 below, if the bounded Borel measurable boundary functions are h-resolutive. If a boundary is h-resolutive for every h, the boundary will be called *universally resolutive*.

Disconnected versus Connected Greenian Sets

The set D has not been supposed connected in the preceding discussion. If D is not connected, let $D_0, D_1, \ldots$ be the open connected components of D, and let ∂D_j be the boundary of D_j relative to $D \cup \partial D$, where ∂D has been determined by a metric compactification of D. Then $\partial D = \bigcup_0^\infty \partial D_j$. If f is a function on ∂D let f_j be the restriction of f to ∂D_j. A function u on D is in the upper [lower] PWB^h class for the boundary function f if and only if for each j the restriction of u to D_j is in the upper [lower] PWB^h class on D_j for the boundary function f_j. Thus the restriction to D_j of $\bar{H}_f^h$ [$\underline{H}_f^h$] is the upper [lower] PWB^h solution on D_j for the boundary function f_j, and f is an h-resolutive boundary function for D if and only if for all j the boundary function f_j of D_j is an h-resolutive boundary function for D_j. Let $D_{a_1}, D_{a_2}, \ldots$ be the components of D on which $\bar{H}_f^h$ is not identically $+\infty$; for $j \geq 1$, choose a point ξ_{a_j} (to be held fast below) in D_{a_j}, and choose $\varepsilon > 0$. Then there is a function u on D, in the upper PWB^h class for f, such that for $j \geq 1$,

$$u(\xi_{a_j}) \leq \begin{cases} \bar{H}_f^h(\xi_{a_j}) + \varepsilon & \text{if } \bar{H}_f^h \text{ is harmonic on } D_{a_j}, \\ -\dfrac{1}{\varepsilon} & \text{if } \bar{H}_f^h \equiv -\infty \text{ on } D_{a_j}. \end{cases} \tag{2.1}$$

(A dual assertion is true for lower PWB^h solutions.) To see this, observe that by definition of upper PWB^h classes there is a function u_j in the upper PWB^h class on D for f such that (2.1) is satisfied by u_j. On each set D_{a_j} define $u = u_j$ and define $u = +\infty$ on each remaining component of D.

Internal Resolutivity

It is trivial that the one-point boundary of a Greenian set D is universally resolutive and that the PWB^h solutions are the constant functions. At the other extreme, a boundary of D will be called *internally h-resolutive* if every bounded h-harmonic function is the PWB^h solution of some boundary function, and a boundary will be called *universally internally resolutive* if the boundary is internally h-resolutive for every strictly positive harmonic h.

Observe that the characterization "h-resolutive" is not applied to any proper subset of a boundary.

EXAMPLE (c) (The Classical Solution). If f is a finite-valued continuous boundary function for which the traditional first boundary value problem has a solution, that is, there is an h-harmonic function u on D with boundary limit function f, then f is h-resolutive, and $H_f^h = u$ because u is in both upper and lower PWB^h classes; so $u = \underline{H}_f^h = \bar{H}_f^h$. Thus [see Example (b) above] the Euclidean boundary of a ball is resolutive. This ball boundary will be shown to be universally resolutive in Section 9 and will be shown to be universally internally resolutive in Section IX.12.

h-Regularity of Boundary Points

A boundary point ζ of D will be called *h-regular*, or *regular* when $h \equiv 1$, if whenever f is a finite-valued continuous boundary function

$$\lim_{\eta\to\zeta} \bar{H}_f^h(\eta) = \lim_{\eta\to\zeta} \underline{H}_f^h(\eta) = f(\zeta).$$

If every boundary point is h-regular, the boundary will be called *h-regular*, or *regular* when $h \equiv 1$. An h-regular boundary is h-resolutive because whenever f is a finite-valued continuous boundary function $\bar{H}_f^h = \underline{H}_f^h$ in view of the h-harmonic maximum-minimum theorem. For each such function f the PWB^h solution H_f^h for an h-regular boundary is the solution of the traditional Dirichlet problem. Conversely, if this problem has a solution for every finite-valued continuous f, the boundary is h-resolutive and h-regular.

Elementary Properties of Upper and Lower Solutions

It is trivial that $\bar{H}_{-f}^h = -\underline{H}_f^h$, that $\bar{H}_f^h$ and $\underline{H}_f^h$ increase with f, that $\bar{H}_{cf}^h = c\bar{H}_f^h$ and $\underline{H}_{cf}^h = c\underline{H}_f^h$ when c is a positive constant, and that $\bar{H}_{f+c}^h = \bar{H}_f^h + c$ and $\underline{H}_{f+c}^h = \underline{H}_f^h + c$ when c is an arbitrary constant. Moreover, if $\varepsilon > 0$ and if f and g are finite-valued boundary functions whose upper and lower solutions are finite-valued, and if $|f - g| \le \varepsilon$, then $|\bar{H}_f^h - \bar{H}_g^h| \le \varepsilon$ and $|\underline{H}_f^h - \underline{H}_g^h| \le \varepsilon$ because

$$\bar{H}_f^h - \varepsilon = \bar{H}_{f-\varepsilon}^h \le \bar{H}_g^h \le \bar{H}_{f+\varepsilon}^h = \bar{H}_f^h + \varepsilon$$

and the same inequalities are valid for the lower solutions. It follows that if in addition f is h-resolutive,

$$H_f^h - \varepsilon \le \underline{H}_g^h \le \bar{H}_g^h \le H_f^h + \varepsilon,$$

and we conclude that if $f_\cdot$ is a uniformly convergent sequence of finite-valued h-resolutive boundary functions with limit g, then g is h-resolutive, and the sequence $H_{f_\cdot}^h$ is uniformly convergent with limit H_g^h.

The Set of h-Irregular Boundary Points

This set is an F_σ set, that is, a countable union of compact boundary subsets. In fact for an arbitrary function ϕ from D into $\bar{\mathbb{R}}$ the boundary function $\zeta \mapsto \limsup_{\xi\to\zeta\in\partial D} \phi(\xi)$ is upper semicontinuous, so if f is a bounded boundary function, the boundary set

$$A(f, \varepsilon) = \left\{\zeta \in \partial D \colon \limsup_{\xi\to\zeta} \bar{H}_f^h(\xi) - \liminf_{\xi\to\zeta} \underline{H}_f^h(\xi) \ge \varepsilon\right\}$$

is compact. If $f_\cdot$ is a sequence of boundary functions dense in $\mathbb{C}(D)$, the set of irregular boundary points is $\bigcup_{m,n=0}^{\infty} A\{f_m, 2^{-n}\}$, an F_σ set.

The PWB Method and the Kelvin Transform

Let D be a Greenian subset of $\mathbb{R}^N$ provided with a boundary ∂D by a metric compactification of D, and let ϕ be an inversion of $\mathbb{R}^N$ in a sphere with center not in D. Then $D' = \phi(D)$ is Greenian. The $D \cup \partial D$ metric induces a metric on D', and the completion of D' in this metric is a metric homeomorph of $D \cup \partial D$ and thereby defines $\partial D'$. Let h be a strictly positive harmonic function on D with Kelvin transform h_1 on D'. A function f on ∂D induces a corresponding function f_1 on $\partial D'$. If v/h is in the upper PWB^h class on D for f and if v_1 is the Kelvin transform of v, then v_1/h_1 is in the upper PWB^{h_1} class on D for f_1, and the corresponding assertion is valid for the lower classes. Thus the PWB method on D is easily translated into that on D'. For example, f is h-resolutive on ∂D if and only if f_1 is h_1-resolutive on $\partial D'$, and a point of ∂D is h-regular if and only if its image on $\partial D'$ is h_1 regular. Observe that if ∂D is the Euclidean boundary, then $\partial D' = \phi(\partial D)$ is also the Euclidean boundary. It will be proved in Section 14 that in this Euclidean boundary context a point ξ of ∂D is not only regular (that is, h-regular with $h \equiv 1$) if and only if $\phi(\xi)$ is an h_1-regular point of $\partial D'$ (with h_1 the Kelvin transform of the constant function 1) but that ξ is regular if and only if $\phi(\xi)$ is regular, under the additional hypothesis that when $N > 2$ neither boundary point is the point ∞.

3. Examples

Example (a) (Euclidean Boundary). Let u be an h-superharmonic lower-bounded function on the Greenian subset D of $\mathbb{R}^N$, and suppose that u has a finite or infinite limit at every point of ∂D, thereby defining a continuous boundary function f. Then $\mathrm{GM}_D^h u$ exists, and if this minorant is bounded above, it follows that f is h-resolutive with $H_f^h = \mathrm{GM}_D^h u$. In fact under these hypotheses, u is in the upper and $\mathrm{GM}_D^h u$ is in the lower PWB^h class for f; so $\mathrm{GM}_D^h u \leq \underline{H}_f^h \leq \overline{H}_f^h \leq u$, and therefore by definition of $\mathrm{GM}_D^h u$ the first three terms of this inequality are equal.

Application: Green Functions and the Dirichlet Problem

Suppose that $h \equiv 1$ and that D is arbitrary if $N > 2$ but bounded if $N = 2$, choose ξ in D and define $u = G(\xi, \cdot)_{|D}, f = G(\xi, \cdot)_{|\partial D}$, with $f(\infty) = 0$ if $N > 2$ and D is unbounded. Then f is bounded so $\mathrm{GM}_D u$ is bounded (harmonic function maximum-minimum theorem). We conclude that f is resolutive with PWB solution $\mathrm{GM}_D G(\xi, \cdot)$ and that $G_D(\xi, \cdot) = G(\xi, \cdot) - H_f$. This evaluation of the Green function in terms of a Dirichlet solution is also correct if $N = 2$ and D is unbounded whenever D is sufficiently sparse near the point ∞. More precisely, as we shall show in Section 19, this evaluation is correct if and only if ∞ is a regular boundary point of D, equivalently (Section XI.12) if and only if D is not a deleted fine neighborhood of ∞.

EXAMPLE (b) (Euclidean Boundary, $h \equiv 1$). Let u_1 and u_2 be superharmonic functions on the Greenian subset D of $\mathbb{R}^N$, and suppose that $\mathrm{GM}_D u_i = u_i'$ exists. Furthermore suppose that $u = u_1 - u_2$ is well defined near ∂D and has a finite limit at every point of ∂D, determining a boundary function f. Then f is bounded and continuous and we now show (i) that f is resolutive with $H_f = u_1' - u_2'$. If, in addition, u_1 and u_2 have superharmonic extensions to a neighborhood of $\mathbb{R}^N \cap \bar{D}$, if these extensions have finite limits at every point of ∂D, and if $u_i - u_i'$ is bounded outside some compact subset of D, we shall show (ii) that $\lim_{\eta \to \zeta} H_f(\eta) = f(\zeta)$ at quasi every finite boundary point ζ of D and also at the point ∞ if $N > 2$ and D is unbounded. (These conditions are convenient to verify but are more stringent than necessary.)

The function $u_1 - u_2'$ is superharmonic with limit inferior $\geq f(\zeta)$ at every boundary point ζ of D. This difference is therefore lower bounded and is in the upper PWB class for f; so $u_1 \geq u_2' + \bar{H}_f$, and therefore $u_1' \geq u_2' + \bar{H}_f$. Interchanging u_1 with u_2 yields the inequality $u_2' \geq u_1' - \underline{H}_f$ and we conclude that $\underline{H}_f = \bar{H}_f = u_1' - u_2'$, that is, (i) is true. In Section VII.4 the Fundamental Convergence Theorem was used to show that the function $G_D(\xi, \cdot) = G(\xi, \cdot) - \mathrm{GM}_D G(\xi, \cdot)$ has limit 0 at quasi every finite point of ∂D. Under the hypotheses of (ii) the same reasoning shows that $u_i - u_i'$ has limit 0 at quasi every finite point of ∂D and therefore that H_f has limit $f(\zeta)$ at quasi every finite boundary point ζ. To prove that H_f has limit $f(\infty)$ at ∞ when $N > 2$ and D is unbounded, we apply a barrier argument, to be developed more generally in Section 14. Define $u(\xi) = |\xi|^{2-N}$ for ξ in D, choose $b > f(\infty)$, choose a neighborhood of ∞ so small that $f \leq b$ in the neighborhood, and then choose n so large that $b + nu \geq f$ on ∂D outside this neighborhood. The restriction to D of the function $b + nu$ is in the upper PWB class on D for f and has limit b at ∞. Hence $\limsup_{\xi \to \infty} H_f(\xi) \leq b$; so this limit superior is at most $f(\infty)$. Since the same reasoning is applicable to $-f$, the function H_f has limit $f(\infty)$ at ∞, as was to be proved.

Special case (b′). Let D be bounded and let f be the restriction to ∂D of a polynomial u. Then u can be written as the difference between two polynomials superharmonic on a heighborhood of $\bar{D}$:

$$u(\xi) = [u(\xi) - c|\xi|^2] - (-c|\xi|^2) \tag{3.1}$$

for sufficiently large c. The assertions (b)(i) and (b)(ii) are therefore applicable to polynomial boundary functions when D is bounded.

Special case (b″). According to Section VII.3, the hypotheses of (b)(i) and (b)(ii) are satisfied by $u_i = G(\xi_i, \cdot) \wedge c_{i|D}$ for ξ_i in D and $-\infty < c_i \leq +\infty$.

Special case (b‴). If $N > 2$ or if $N = 2$ and D is bounded, the hypotheses of (b)(i) and (b)(ii) are satisfied by $u_1 = G(\xi, \cdot)_{|D}$, $u_2 \equiv 0$, for each point ξ in D.

4. Continuous Boundary Functions on the Euclidean Boundary ($h \equiv 1$)

Theorem. *The Euclidean boundary of a Greenian set D is resolutive. Moreover quasi every finite point of this boundary is regular, and ∞ is a regular Euclidean boundary point whenever D is unbounded and $N > 2$.*

This theorem, together with the fact that the PWB method yields the classical solution whenever there is one, justifies this method. In the following, $\mathbb{C}(\partial D)$ denotes the class of finite-valued continuous functions on ∂D, metrized by the supremum norm. In view of the fact (Section 2) that the limit f of a uniformly convergent sequence $f_\cdot$ of finite-valued resolutive boundary functions is resolutive and that $H_{f_\cdot}$ is uniformly convergent to H_f on D, it is sufficient to prove that there is a countable dense subset Γ: $\{g_\alpha, \alpha \in I\}$ of $\mathbb{C}(\partial D)$ with the property that for each index value α, the function g_α is resolutive, and H_{g_α} has limit $g_\alpha(\zeta)$ at quasi every finite boundary point ζ, as well as at $\zeta = \infty$ when $N > 2$ and D is unbounded. In fact the Euclidean boundary is then resolutive, and if A_α is the exceptional polar boundary subset for g_α, the finite boundary points not in $\bigcup_{\alpha \in I} A_\alpha$ are regular.

In view of Section 3, Example (b′), when D is bounded Γ can be taken as any countable dense subset of $\mathbb{C}(\partial D)$ consisting of restrictions to ∂D of polynomials. The following proof is applicable to both unbounded and bounded sets D.

If $N > 2$, let Γ_1 be the class of positive superharmonic functions on $\mathbb{R}^N$ which are finite valued and continuous with finite limit at ∞. Define $u(\infty)$ for u in Γ_1 as the limit of u at ∞. Then Γ_1 contains the positive constant functions and $u \wedge v$, and $au + bv$ are in Γ_1 if u and v are and if a and b are positive constants. Let Γ_2 be the class of differences $u_1 - u_2$ with u_i in Γ_1, so that Γ_2 is a vector lattice in the order determined by pointwise inequality. The set Γ_3 of restrictions to ∂D of the members of Γ_2 is a vector lattice of finite continuous boundary functions, which contains the constant functions and separates ∂D because $c \wedge G(\xi, \cdot)_{|\partial D}$ (defined as 0 at ∞) is in Γ_3 for an arbitrary positive constant c and an arbitrary point ξ. The set Γ_3 is dense in $\mathbb{C}(\partial D)$ (Stone–Weierstrass theorem), and in view of Section 3, Example (b) any countable subset of Γ_3 dense in $\mathbb{C}(\partial D)$ can serve as the desired set Γ.

If $N = 2$, let ξ_0 be a point of $\mathbb{R}^2$ and define Γ_1 as the class of functions u satisfying the following two conditions:

(a) u is a finite-valued continuous superharmonic function on $\mathbb{R}^2$.

(b) There is a strictly positive constant α and a constant β, both depending on u, such that u is identically $\alpha G(\xi_0, \cdot) + \beta$ outside some bounded set which may depend on u.

When u is in Γ_1, the function cu is also in Γ_1 when c is a strictly positive constant, and $u + c$ is in Γ_1 when c is an arbitrary constant. When u_1 and u_2 are in Γ_1, the functions $u_1 + u_2$ and $u_1 \wedge u_2$ are in Γ_1. Let Γ_2 be the class of differences $u_1 - u_2$ for which u_1 and u_2 are in Γ_1 and have the same multiplier α in (b). Then $u_1 - u_2$ is constant outside some bounded set. Define $u_1 - u_2$ at ∞ as this constant value. The set Γ_2 is a lattice in the order determined by pointwise inequality. If u is in this lattice, cu and $u + c$ are also in the lattice, for every constant c. If Γ_3 is the class of restrictions to ∂D of the members of Γ_2, Γ_3 is a lattice of finite continuous boundary functions which contains cf and $f + c$ with f. Moreover Γ_3 separates ∂D. To see this, let ξ_1 and ξ_2 be distinct points of ∂D, and let B be a ball containing ξ_0 and either ξ_1 or ξ_2, say ξ_1, but not the other. The boundary function

$$f = [\tau_B G(\xi_0, \cdot) - 1 \wedge G(\xi_0, \cdot)]_{|\partial D}$$

is then in Γ_3. If f has the same value at ξ_1 as at ξ_2 decrease the radius of B to increase f at ξ_1 but not at ξ_2. Thus Γ_3 separates ∂D and therefore (Stone–Weierstrass theorem) is dense in $\mathbb{C}(\partial D)$. In view of Section 3 Example (b) any countable subset of Γ_3 dense in $\mathbb{C}(\partial D)$ will serve as the desired set Γ.

Partial Generalization to PWB^h Solutions

Suppose that D is Greenian, that h is a strictly positive harmonic function on D and that B is an open relatively compact subset of D. If f is a finite-valued continuous function on ∂B, a function u is in the upper [lower] PWB^h class on B for the boundary function f if and only if the function uh on B is in the upper [lower] PWB class on B for the boundary function $fh|_{\partial B}$. It follows that f is h-resolutive, with $H_f^h = H_{fh}/h$, and that a boundary point of B is h-regular if and only if the point is regular. Thus quasi every boundary point of B is h-regular.

This partial generalization of Theorem 4 will be completed in Section 8 where it will be shown that if D is given an h-resolutive boundary by a metric compactification, if B is an open subset of D, and if ∂B is the boundary of B relative to $D \cup \partial D$, then ∂B is h-resolutive. This fact can be deduced now when $h = 1$ and ∂D is the one-point boundary provided by the Alexandrov compactification of D. In fact, in this specialization if f is a finite continuous function on ∂B, let f' be the function on the Euclidean boundary $\partial' B$ of B defined as f on $\partial D \cap \partial' B$ and (if B is not relatively compact in D) as $f(\partial B \cap \partial D)$ on the rest of $\partial' B$. The function f' is finite and continuous and therefore resolutive on B for $\partial' B$. Moreover the upper and lower PWB classes on B for f' on $\partial' B$ are the same as those classes for f on ∂B. Hence f is resolutive, as was to be proved.

5. *h*-Harmonic Measure Null Sets

Let D be a Greenian subset of $\mathbb{R}^N$ provided with a boundary by a metric compactification, and let A be a boundary subset. If the upper solution $\bar{H}^h_{1_A}$ $(= {}^hR^A_1 = R^A_h/h = \underset{+}{R}^A_h/h)$ vanishes identically on D, the set A will be called an *h-harmonic measure null set*, or *harmonic measure null set* if $h \equiv 1$.

In the following discussion $D_1, D_2, \ldots$ are the open connected components of D and ξ_j is a point of D_j, held fast throughout. The proofs are given for infinitely many components and the simplifications to be made when there are only finitely many are left to the reader.

(a) A countable union $A = \bigcup_0^\infty A_n$ of h-harmonic measure null sets is h-harmonic measure null.

If $\varepsilon > 0$ and if (Section 2) u_n is an h-superharmonic function in the upper PWB^h class on D for the boundary function 1_{A_n}, with $u_n(\xi_j) < \varepsilon 2^{-n-1}$ for all j, then the function $v = \sum_0^\infty u_n$ is in the upper PWB^h class for the boundary function 1_A, and $\bar{H}^h_{1_A} = 0$ because $\bar{H}^h_{1_A}(\xi_j) \le v(\xi_j) < \varepsilon$ for all j.

(b) A boundary subset A is h-harmonic measure null if and only if there is a positive h-superharmonic function u on D with limit $+\infty$ at every point of A.

If A is h-harmonic measure null, define $A_n = A$ and observe that then the positive h-superharmonic function v in (a) has limit $+\infty$ at every point of A. Conversely, if a function u as described in (b) exists, then for every $\varepsilon > 0$ the function εu is in the upper PWB^h class for the boundary function 1_A; so $\bar{H}^h_{1_A} = 0$ except possibly on the polar set of infinities of u, and so $\bar{H}^h_{1_A} = 0$ on D.

(c) If f is a positive boundary function and if $\bar{H}^h_f = 0$, then the set $\{f > 0\}$ is h-harmonic measure null.

If $n \ge 1$ and if f_n is the indicator function on ∂D of the set $\{f > 1/n\}$, then

$$0 = \bar{H}^h_f \ge \frac{\bar{H}^h_{f_n}}{n} = \frac{\bar{H}^h_{f_n}}{n};$$

so the set $\{f > 1/n\}$ is h-harmonic measure null. Hence $\bigcup_{n=1}^\infty \{f > 1/n\} = \{f > 0\}$ is h-harmonic measure null.

(d) If f is a boundary function for which $\bar{H}^h_f < +\infty$, then the set $A = \{f = +\infty\}$ is h-harmonic measure null.

Choose (Section 2) an h-superharmonic function u on D in the upper PWB^h class for the boundary function f. The function u is positive and has limit $+\infty$ at every point of A so A is h-harmonic measure null according to (b).

(e) Euclidean boundary, $h \equiv 1$. A polar subset A of ∂D is a harmonic measure null set.

If $N > 2$, let u be a positive superharmonic function on $\mathbb{R}^N$, identically $+\infty$ on A. Then $v = u_{|D}$ has limit $+\infty$ at every point of A; so according to (b), the set A is h-harmonic measure null. If $N = 2$ and if ζ is a finite boundary point of D, choose a ball B of center ζ so small that $\mathbb{R}^2 - (D \cup B)$ is not polar, that is, so that $D' = D \cup B$ is Greenian. Then if $A \cap B$ is not empty, there is a positive superharmonic function u on D', identically $+\infty$ on $A \cap B$; so $u_{|D}$ has limit $+\infty$ at every point of $A \cap B$. Thus according to (b), A is locally harmonic measure null, and it follows that A is a countable union of harmonic measure null sets and is therefore itself harmonic measure null.

The converse of (e) is false: a harmonic measure null subset of a Euclidean boundary ∂D need not be polar. For example, if D is a ball in $\mathbb{R}^N$ it will be seen in Section 9 that the harmonic measure null subsets of the Euclidean boundary are the l_{N-1} null boundary subsets, and it is not difficult to find examples of l_{N-1} null subsets of a sphere which are not polar. The following is however a near converse to (e).

(f) Let D be an arbitrary nonempty open subset of $\mathbb{R}^N$, and let A be a nonpolar proper subset of D, closed relative to D. Provide D with a boundary by a metric compactification, and let $\partial(D - A)$ be the boundary of the Greenian set $D - A$ in this compactification. Then the set $A \cap \partial(D - A)$ is not harmonic measure null relative to $D - A$. In fact, if $A \cap \partial(D - A)$ is harmonic measure null, then by (b) there is a positive superharmonic function u on $D - A$ with limit $+\infty$ at each point of $A \cap \partial(D - A)$. If u is extended to D by defining $u = +\infty$ on A, the resulting function is superharmonic on D with value $+\infty$ on A; so A is polar, contrary to hypothesis.

h-Harmonic Measure Null Sets and the Kelvin Transformation

We use here the notation of the discussion in Section 2 of the PWB method and the Kelvin transformation. It is clear from that discussion that an h-harmonic measure null subset of ∂D is transformed under an inversion into an h_1-harmonic measure null subset of $\partial D'$. We shall use the following additional fact when $h \equiv 1$. If ∂D and $\partial D'$ are Euclidean boundaries and if A is a harmonic measure null subset of ∂D, then $\phi(A)$ is a harmonic measure null subset of $\partial D'$, under the additional hypothesis when $N > 2$ that all points of $\phi(A)$ are finite. This fact follows from the criterion for h-harmonic measure null sets in (b) because if v is a positive superharmonic function on D with limit $+\infty$ at every point of A, then the Kelvin transform of v is a positive superharmonic function on D' with limit $+\infty$ at every finite point of $\phi(A)$, every point of $\phi(A)$ if $N = 2$.

Example (a) (Euclidean Boundary, $h \equiv 1$). If $N \geq 2$ and if ζ is a finite boundary point of D, the singleton $\{\zeta\}$ is polar and therefore is harmonic measure null. If $n = 2$, the singleton $\{\infty\}$ is harmonic measure null for every unbounded Greenian set D because the point ∞ can be made finite by an inversion of the plane relative to a circle centered at another boundary point and we have just seen that such a transformation of the plane preserves the harmonic measure null property. If $N > 2$, the singleton $\{\infty\}$ may not be a harmonic measure null set for an unbounded Greenian set. For example, if $D = \mathbb{R}^N$ and if u is in the upper PWB class on $\mathbb{R}^N$ for the boundary function $1_{\{\infty\}}$, that is, if u is superharmonic and lower bounded on $\mathbb{R}^N$, with inferior limit ≥ 1 at ∞, then $u \geq 1$ by the superharmonic function minimum theorem.

Example (b) (Euclidean Boundary). A polar subset of ∂D need not be an h-harmonic measure null set for all h. For example, let $D = B(0, \delta)$, and let h be a minimal harmonic function on D corresponding to the boundary point ζ (Section II.16), say

$$h(\xi) = \frac{\delta^2 - |\xi|^2}{|\zeta - \xi|^N}.$$

Let g be the indicator function of $A = \partial D - \{\zeta\}$. Since the function ε/h is in the upper PWB^h class on D for g whenever $\varepsilon > 0$, the set A is h-harmonic measure null; so $\{\zeta\}$ is not, because ∂D is not an h-harmonic measure null set.

6. Properties of PWB^h Solutions

The following properties of PWB^h solution will be needed in addition to those derived in Section 2. Proofs are given in the next section. The functions f and g below are boundary functions.

(a) If $f = g$ up to an h-harmonic measure null set, then $\underline{H}_f^h = \underline{H}_g^h$ and $\bar{H}_f^h = \bar{H}_g^h$.

(b) $\bar{H}_f^h < +\infty$ if and only if $\bar{H}_{f \vee 0}^h < +\infty$.

(c) If $\bar{H}_f^h < +\infty$, if $\bar{H}_g^h < +\infty$, and if $f + g$ is defined arbitrarily on the h-harmonic measure null set, where f or g is $-\infty$ and the other is $+\infty$, then $\bar{H}_{f+g}^h \leq \bar{H}_f^h + \bar{H}_g^h$.

(d) The class of h-resolutive boundary functions contains the constant functions and is a vector lattice (that is, the class is linear and contains $f \vee g$ and $f \wedge g$ when it contains f and g). Moreover, if f and g are h-resolutive, then

$$H_{\alpha f + \beta g}^h = \alpha H_f^h + \beta H_g^h, \qquad H_{f \vee g}^h = \mathrm{LM}_D^h(H_f^h \vee H_g^h),$$
$$H_{f \wedge g}^h = \mathrm{GM}_D^h(H_f^h \wedge H_g^h).$$

(e) If $f_\cdot$ is a monotone increasing sequence of boundary functions with limit f and if $\bar{H}_{f_0}^h > -\infty$, then $\bar{H}_f^h = \lim_{n \to \infty} \bar{H}_{f_n}^h$. Moreover $\bar{H}_f^h = \underline{H}_f^h$ if each function f_n is h-resolutive.

(f) If f is an h-resolutive boundary function, there are Borel measurable h-resolutive boundary functions f_1 and f_2 such that $f_1 \leq f \leq f_2$, $H^h_{f_1} = H^h_f = H^h_{f_2}$, and $f_1 = f = f_2$ up to an h-harmonic measure null set. In particular, if A is a boundary subset with h-resolutive boundary indicator function $f = 1_A$, there are Borel boundary subsets A_1 and A_2 such that the functions $f_1 = 1_{A_1}$ and $f_2 = 1_{A_2}$ are h-resolutive, that $A_1 \subset A \subset A_2$, that $H^h_{f_1} = H^h_f = H^h_{f_2}$, and that $A_2 - A_1$ is h-harmonic measure null.

(g) If f is h-resolutive, if D_0 is an open subset of D, and if f_0 is defined on ∂D_0 as f on $\partial D_0 \cap \partial D$ and as H^h_f on $D \cap \partial D_0$, then f_0 is h-resolutive for D_0 and the PWB^h solution on D_0 for f_0 is the restriction of H^h_f to D_0.

(h) For every subset A of ∂D,

$$\begin{aligned} \bar{H}^h_{1_A} &= \inf\{\bar{H}^h_{1_B}: \partial D \supset B \supset A, B \text{ open in } \partial D\}, \\ \underline{H}^h_{1_A} &= \sup\{\underline{H}^h_{1_B}: B \subset A, B \text{ compact}\}. \end{aligned} \tag{6.1}$$

For each point ξ of D the set function $A \mapsto \bar{H}^h_{1_A}(\xi)$ is a Choquet capacity relative to the class of compact boundary subsets.

h-Resolutive Boundaries. Property (e) and its dual for decreasing sequences imply that if ∂D is h-resolutive, all bounded Borel measurable boundary functions are h-resolutive. This result will be extended in Section 8.

7. Proofs for Section 6

Since the proofs do not depend on the choice of h they will be given for $h \equiv 1$ to simplify notation. The point is that the assertions in Section 6 are valid for the PWB method in a very general context.

Proof of (a). Let A be the harmonic measure null set in question. It is enough to prove the second equality, and (by symmetry) it is even enough to prove that $\bar{H}_f \leq \bar{H}_g$. If v is in the upper PWB class for g and if v_1 is a positive superharmonic function on D with limit $+\infty$ at every point of A, then $v + \varepsilon v_1$ is in the upper PWB class for f whenever $\varepsilon > 0$; so $\bar{H}_f \leq \bar{H}_g$ quasi everywhere on D and therefore everywhere on D. □

Proof of (b). If $\bar{H}_f < +\infty$, there is a lower-bounded superharmonic function v on D, say $v \geq c$, in the upper PWB class for f. Then $v + |c|$ is in the upper PWB class for $f \vee 0$; so $\bar{H}_{f \vee 0} < +\infty$. Conversely, if $\bar{H}_{f \vee 0} < +\infty$, then $\bar{H}_f \leq \bar{H}_{f \vee 0} < +\infty$. □

Proof of (c). Redefine f and g to be finite on the harmonic measure null set on which either function has the value $+\infty$. Then $f + g$ is well defined, and if u $[v]$ is in the upper PWB class for f $[g]$, the function $u+v$ is in the upper PWB class for $f+g$; so $\bar{H}_{f+g} \le u+v$ and (c) follows. □

Proof of (d). The fact that the class of resolutive boundary functions contains the finite constant functions, that the class is linear, and that the map $f \mapsto H_f$ is linear on this class follows from the PWB properties listed in Section 2 together with (c) and its dual for lower solutions. We need not discuss the resolutivity of both $f \vee g$ and $f \wedge g$ because $f \wedge g = -[(-f) \vee (-g)]$. Moreover in treating $f \vee g$ we can choose $g = 0$ because

$$f \vee g = [(f-g) \vee 0] + g.$$

We assume therefore that f is resolutive and prove that $f \vee 0$ is resolutive, with $H_{f \vee 0} = \mathrm{LM}_D(H_f \vee 0)$. Choose a point ξ_j in each open connected component of D, and choose v_n in the upper PWB class for f with

$$v_n(\xi_j) \le H_f(\xi_j) + 2^{-n}$$

for all j. The series $\sum_k^\infty (v_n - H_f)$ of positive superharmonic functions has a superharmonic sum because the sum is finite at a point of each open connected component of D. The positive function

$$\mathrm{LM}_D(H_f \vee 0) + \sum_k^\infty (v_n - H_f)$$

is therefore superharmonic. Moreover this function majorizes v_k and therefore is in the upper PWB class for $f \vee 0$. When $k \to \infty$, it follows that $\bar{H}_{f \vee 0} \le \mathrm{LM}_D(H_f \vee 0)$ quasi everywhere on D and therefore everywhere on D. On the other hand, $\underline{H}_{f \vee 0}$ majorizes both H_f and 0; so $\underline{H}_{f \vee 0} \ge \mathrm{LM}_D(H_f \vee 0)$ and (d) follows. □

Proof of (e). Let $f_\cdot$ be a monotone increasing sequence of boundary functions with limit f. The first assertion in (e) is trivially true in any open connected component of D on which some $\bar{H}_{f_n}$ is identically $+\infty$. We therefore decrease D if necessary and assume that $\bar{H}_{f_n} < +\infty$ on D for all n. Choose a point ξ_k in each open connected component of D, and choose u_m in the upper PWB class for f_m with $u_m(\xi_k) < \bar{H}_{f_m}(\xi_k) + 2^{-m}$ for all k. The series $\Sigma_0^\infty (u_m - \bar{H}_{f_m})$ of positive superharmonic functions has a superharmonic sum because the sum is finite at a point in each open connected component of D. The superharmonic function v_n defined by

$$v_n = \lim_{m \to \infty} \bar{H}_{f_m} + \sum_{n+1}^\infty (u_m - \bar{H}_{f_m})$$

is a majorant of u_m for $m > n$, and v_n is therefore in the upper PWB class for f, with

$$\lim_{n\to\infty} \bar{H}_{f_n}(\xi_k) \leq \bar{H}_f(\xi_k) \leq \lim_{n\to\infty} v_n(\xi_k) = \lim_{n\to\infty} \bar{H}_{f_n}(\xi_k).$$

Thus the first assertion of (e) is true at each point ξ_k and therefore everywhere on D in view of the trivial inequality $\bar{H}_f \geq \lim_{n\to\infty} \bar{H}_{f_n}$. Under the hypotheses of the second assertion,

$$\bar{H}_f \geq \underline{H}_f \geq \lim_{n\to\infty} H_{f_n} = \bar{H}_f;$$

so there is equality throught, and the proof is complete. □

Proof of (f). Suppose first that f is an arbitrary boundary function. Choose a point ξ_k in each open connected component of D, and let $v_\cdot$ be a decreasing sequence of members of the upper PWB class for f with $\lim_{n\to\infty} v_n(\xi_k) = \bar{H}_f(\xi_k)$ for all k. There is such a sequence $v_\cdot$ because the upper PWB class for f is directed downward. Define $g_n(\zeta) = \liminf_{\eta\to\zeta} v_n(\eta)$ for $\zeta \in \partial D$. Then $g_\cdot$ is a decreasing sequence of lower semicontinuous boundary functions with Borel measurable limit function $f_2 \geq f$; so $\bar{H}_{f_2} \geq \bar{H}_f$, and there is equality on the set $\xi_\cdot$. Since an upper PWB solution for f is either harmonic or identically $+\infty$ or identically $-\infty$ on each connected open component of D, we conclude that $\bar{H}_{f_2} = \bar{H}_f$. A dual argument (or apply this result to $-f$) yields a Borel measureable boundary function f_1 with $f_1 \leq f$ and $\underline{H}_{f_1} = \underline{H}_f$. In particular, if f is PWB resolutive,

$$H_f = \underline{H}_{f_1} \leq \bar{H}_{f_1} \leq H_f \leq \underline{H}_{f_2} \leq \bar{H}_{f_2} = H_f;$$

so f_1 and f_2 are resolutive, and the equation $H_{f_2} - H_{f_1} = 0$ together with the inequality $f_1 \leq f \leq f_2$ implies (Section 5) that $f_1 = f = f_2$ up to a harmonic measure null set, as was to be proved. If $f = 1_A$ is the resolutive indicator function of a set A, we can replace f_2 [f_1] by the indicator function of the set $A_2 = \{f_2 = 1\}$ [$A_1 = \{f_1 = 1\}$] to obtain the second part of (f). □

Proof of (g). If v is in the upper PWB class on D for f, the restriction of v to D_0 is in the upper PWB class on D_0 for f_0. Hence the upper PWB solution ${}_0\bar{H}_{f_0}$ on D_0 for f_0 satisfies the inequality ${}_0\bar{H}_{f_0} \leq H_f$ on D_0. Similarly, ${}_0\underline{H}_{f_0} \geq H_f$ on D_0 so (g) is true. □

Proof of (h). Since $\underline{H}_{1_A} = 1 - \bar{H}_{1_{\partial D - A}}$, only the first equality in (6.1) need be proved. A function v on D is in the upper PWB class for 1_A if and only if on each open connected component of D the function v is either identically

$+\infty$ or positive superharmonic with $\liminf_{\eta\to\zeta} v(\eta) \geq 1$ when $\zeta \in A$. For such a function v, $0 < \alpha < 1$ implies that $\liminf_{\eta\to\zeta} v(\eta) > \alpha$ for ζ in some neighborhood B of A relative to ∂D, so that v/α is in the upper PWB class for 1_B, and $\bar{H}_{1_B} \leq v/\alpha$. Since α can be chosen arbitrarily close to 1, the infimum in (6.1) is at most $\bar{H}_{1_A}$. The inequality in the other direction is trivial. The set function $\xi \to \bar{H}_{1_A}(\xi) = R_1^A(\xi)$ is a Choquet capacity of boundary subsets relative to the class of compact boundary subsets according to Section VI.3(k). [In the present context we have already proved all the desired properties of this set function except that if $A_\bullet$ is a decreasing sequence of compact boundary subsets with intersection A, then $\lim_{n\to\infty} H_{1_{A_n}} = H_{1_A}$, and this fact is elementary in view of the first equation in (6.1).] □

8. h-Harmonic Measure

Let D be a Greenian subset of $\mathbb{R}^N$, coupled with a boundary ∂D provided by a metric compactification of D, and let h be a strictly positive harmonic function on D. In view of Sections 6(d) and 6(e) the class of boundary subsets whose indicator boundary functions are h-resolutive is a σ algebra, and if we define $\mu_D^h(\cdot, A) = H_{1_A}^h$ for each set A in this σ algebra, the set function $\mu_D^h(\xi, \cdot)$ is a probability measure for each point ξ of D. The sets of this σ algebra will be called the *μ_D^h measurable sets*, μ_D^h will be called *h-harmonic measure*, and $\mu_D^h(\xi, \cdot)$ will be called *h-harmonic measure relative to* ξ. The h will be omitted when $h \equiv 1$. Observe that a boundary subset A is an h-harmonic measure null set in the sense of Section 5 if and only if the set is μ_D^h measurable and $\mu_D^h(\cdot, A) \equiv 0$, equivalently, if and only if the function $\mu_D^h(\cdot, A)$ has a zero in each open connected component of D. According to Section 6(f), a boundary subset A is μ_D^h measurable if and only if there are Borel μ_D^h measurable boundary subsets A_1 and A_2 such that $A_1 \subset A \subset A_2$ and $A_2 - A_1$ is μ_D^h null. Thus for ξ in D, $\mu_D^h(\xi, \cdot)$ is the completion of the restriction of $\mu_D^h(\xi, \cdot)$ to the μ_D^h measurable Borel boundary subsets if D is connected. If f is a boundary function, we write $\mu_D^h(\cdot, f)$ for $\int_D f(\eta)\mu_D^h(\cdot, d\eta)$ when this integral is defined. Under this convention $\mu_D^h(\cdot, A)$ can also be written $\mu_D^h(\cdot, 1_A)$. If f is defined on a superset of ∂D, the notation $\mu_D^h(\cdot, f)$ is to be interpreted as $\mu_D^h(\cdot, f_{|\partial D})$. It is trivial that for f a linear combination of indicator functions of μ_D^h measurable sets, the function $\mu_D^h(\cdot, f)$ is h-harmonic. Since a bounded μ_D^h measurable boundary function is the limit of a uniformly convergent sequence of such linear combinations, the function $\mu_D^h(\cdot, f)$ is h-harmonic if f is bounded and μ_D^h measurable. If f is a μ_D^h measurable boundary function,

$$\mu_D^h(\cdot, |f|) = \lim_{n\to\infty} \mu_D^h(\cdot, |f| \wedge n) \leq +\infty,$$

and (Section II.3) on each open connected component of D the limit is either identically $+\infty$ or is h-harmonic. If this limit is h-harmonic on D, the function f will be called μ_D^h *integrable*; the class of such functions f will be denoted by $L^1(\mu_D^h)$. Then $f \in L^1(\mu_D^h)$ if and only if $f \vee 0$ and $f \wedge 0$ are in this class and, if so,

$$\mu_D^h(\cdot, f) = \mu_D^h(\cdot, f \vee 0) + \mu_D^h(\cdot, f \wedge 0).$$

Let $D_0, D_1, \ldots$ be the open connected components of D, and let ∂D_k be the boundary of D_k relative to $D \cup \partial D$. Let h be a strictly positive superharmonic function on D, and denote by h_k the restriction of h to D_k. Let f be a function on ∂D, and denote by f_k the restriction of f to ∂D_k. Then f is an h-resolutive boundary function for D if and only if f_k is h-resolutive for D_k for all k; f is μ_D^h measurable if and only if f_k is $\mu_{D_k}^{h_k}$ measurable for all k; $f \in L^1(\mu_D^h)$ if and only if $f_k \in L^1(\mu_{D_k}^{h_k})$ for all k. When A is a μ_D^h measurable subset of ∂D, the function $\mu_{D_k}^{h_k}(\cdot, A \cap \partial D_k)$ is the restriction of $\mu_D^h(\cdot, A)$ to D_k.

Theorem. *If D is a Greenian subset of $\mathbb{R}^N$ coupled with a boundary ∂D provided by a metric compactification of D and if h is a strictly positive harmonic function on D, then a boundary function f is h-resolutive if and only if $f \in L^1(\mu_D^h)$, equivalently, if and only if f is μ_D^h measurable with both $\bar{H}_f^h$ and $\underline{H}_f^h$ finite valued, and then*

$$H_f^h = \mu_D^h(\cdot, f). \tag{8.1}$$

Proof That h-Resolutive Boundary Functions Are μ_D^h Measurable

Let $\mathbb{C}_0$ be the class of continuous functions from $\mathbb{R}$ into $\mathbb{R}$ with limit 0 at $\pm\infty$ and let f be a finite-valued h-resolutive boundary function. The class of functions ϕ in $\mathbb{C}_0$ for which $\phi(f)$ is h-resolutive is a vector lattice which is closed under uniform convergence and which separates points of $\mathbb{R}$ because the class includes the function $x \mapsto (1 - |x - n|) \vee 0$ for $n \in \mathbb{Z}^+$. Hence (Weierstrass approximation theorem) the class is $\mathbb{C}_0$. Since the class of functions ψ from $\mathbb{R}$ into $\mathbb{R}$ for which $\psi(f)$ is h-resolutive is a class closed under bounded monotone convergence [Section 6(e)] and includes $\mathbb{C}_0$, this class includes every bounded Borel measurable function. When ψ is the indicator function of an interval in $\mathbb{R}$, we find that f is μ_D^h measurable. Finally, if f is an arbitrary h-resolutive boundary function, redefine f as 0 at its infinities to find an h-resolutive boundary function differing from f on a set of h-harmonic measure 0 and thereby to conclude that f is μ_D^h measurable.

Proof of Theorem 8. If f is the indicator function of a μ_D^h measurable boundary subset, then f is h-resolutive, and (8.1) becomes the definition of μ_D^h. More generally (8.1) is therefore true if f is a finite linear combination of indicator functions of μ_D^h measurable boundary subsets, that is, if f is an h-resolutive boundary function taking on only finitely many values, all finite. Denote this class of boundary functions by Γ. Apply the dominated convergence theorem and Section 6(e) to show that if f is a positive and h-resolutive [μ_D^h measurable and integrable] boundary function, then f, as the limit up to an h-harmonic measure null set of an increasing sequence of positive functions in Γ, is in $L^1(\mu_D^h)$ [is h-resolutive], and (8.1) is true. This result applied to $f \vee 0$ and $(-f) \vee 0$ shows that an arbitrary boundary function f is h-resolutive if and only if it is in $L^1(\mu_D^h)$ and that then (8.1) is true. Finally we prove that if the boundary function f is μ_D^h measurable, with $\bar{H}_f^h$ and $\underline{H}_f^h$ finite valued, then $f \in L^1(\mu_D^h)$. According to Section 6(b), $\bar{H}_{f\vee 0}^h < +\infty$; so in view of Section 6(e) and what we have just proved,

$$\mu_D^h(\cdot, f \vee 0) = \lim_{n\to\infty} \mu_D^h(\cdot, (f \vee 0) \wedge n) = \lim_{n\to\infty} H_{(f\vee 0)\wedge n}^h = H_{f\vee 0}^h.$$

Hence $f \vee 0 \in L^1(\mu_D^h)$. Similarly $f \wedge 0 \in L^1(\mu_D^h)$; so $f \in L^1(\mu_D^h)$, as was to be proved. □

h-Regularity in Terms of h-Harmonic Measure

If ∂D is h-resolutive, the condition for h-regularity of a boundary point ζ becomes

$$\lim_{\xi\to\zeta} \mu_D^h(\zeta, \cdot) = \delta_\zeta$$

(vague convergence of measures on ∂D), where δ_ζ is the probability measure supported by $\{\zeta\}$; equivalently, ζ is h-regular if and only if

$$\lim_{\xi\to\zeta} \mu_D^h(\xi, A) = 1$$

whenever A is the trace on ∂D of a neighborhood of ζ.

Example (a) (Relation between μ_B and μ_B^h for B Relatively Compact in D). If D is Greenian, if h is strictly positive and harmonic on D, and if B is an open relatively compact subset of D, it was shown in Section 4 that ∂B is h-resolutive with $H_f^h = H_{fh}/h$. It follows that

$$\mu_B^h(\xi, d\eta) = \frac{\mu_B(\xi, d\eta)h(\eta)}{h(\xi)} \qquad (\xi \in B). \tag{8.2}$$

Example (b) (Extension of Example (a) to Non-relatively Compact B). Let D be Greenian, provided with a boundary ∂D by a metric compactification, and let B be an open subset of D, with boundary ∂B relative to $D \cup \partial D$. Let h be a strictly positive harmonic function on D. Then the class of h-

resolutive boundary functions on ∂B includes the indicator functions of the Borel subsets of $D \cap \partial B$, and for such a subset A and point ξ in B the value $\mu_B^h(\xi, A)$ does not depend on the choice of ∂D. Furthermore (8.2) is true for $\eta \in D \cap \partial B$ when μ_B is defined using the Euclidean boundary of B. To prove these assertions, let A be a compact subset of $D \cap \partial B$, define f as the indicator function of A on ∂B, and define f' as the indicator function of A on the Euclidean boundary of B. A function u is in the upper PWB^h class on B for f on ∂B if and only if uh is in the upper PWB class on B for $f'h$ on the Euclidean boundary. In view of the fact that Euclidean boundaries are resolutive (Theorem 4), it follows that $\bar{H}_f^h = H_{fh}/h$, where the notation refers to Dirichlet solutions on B. If $\varepsilon > 0$ and if uh is in the lower PWB class on B for $f'h$ on the Euclidean boundary, then $u - \varepsilon/h$ is in the lower PWB^h class on B for f on ∂B. It follows that $\underline{H}_f^h \geq H_{fh}/h$. This conclusion implies the truth of the assertions made above.

Example (b) (Continued). An assertion made in Section 4 can now be strengthened and proved. Suppose in Example (b) that ∂D is h-resolutive. Then ∂B is h-resolutive for B, and if A is a Borel subset of ∂D,

$$\mu_D^h(\xi, A) = \mu_B^h(\xi, A \cap \partial B) + \int_{D \cap \partial B} \mu_D^h(\eta, A)\mu_B^h(\xi, d\eta) \qquad (\xi \in \mathrm{B}). \tag{8.3}$$

In view of Example (b) and Section 6(h) it is sufficient to prove that for A a compact subset of ∂D the function $1_{A \cap \partial B}$ on ∂B is an h-resolutive boundary function for B and that (8.3) is true. Observe that if v_2 $[u_2]$ is in the upper [lower] PWB^h class on D for the boundary function 1_A on ∂D and if v_1 $[u_1]$ is in the lower [upper] PWB^h class on B for the boundary function defined as $\mu_D^h(\cdot, A)$ on $D \cap \partial B$ and as 0 elsewhere on ∂B, the difference $v_2 - v_1$ $[u_2 - u_1]$ on B is in the upper [lower] PWB^h class for the boundary function $1_{A \cap \partial B}$ on ∂B. It follows that $1_{A \cap \partial B}$ is an h-resolutive boundary function on ∂B and that (8.3) is true.

See Section 3.II.3(d) for the simple (after the necessary foundations have been laid) probabilistic derivation of (8.3).

Application to the h-Superharmonic Function Inequality

Let u be an h-superharmonic function on D, and let B be an open relatively compact subset of D. Then the restriction of u to B is in the upper PWB^h class for the boundary function $u_{|\partial B}$; so $u \geq \mu_B^h(\cdot, u)$. A more delicate result is the following. Suppose that u is superharmonic and lower bounded on D, and define a lower semicontinuous boundary function f on ∂D by

$$f(\zeta) = \liminf_{\xi \to \zeta \in \partial D} u(\xi).$$

Then if ∂D is h-resolutive, we show that f is h-resolutive and that

$$u \geq \mu_D^h(\cdot, f). \tag{8.4}$$

In fact, for $n \in \mathbb{Z}^+$ the boundary function $f \wedge n$ is h-resolutive [Section 6(d)], and u is in the upper PWB^h class for this boundary function; so

$$u \geq H_{f \wedge n}^h = \mu_D^h(\cdot, f \wedge n)$$

and (8.4) follows. The function f is h-resolutive because it is in $L^1(\mu_D^h)$.

Example (c). Let D_1 be a Greenian subset of $\mathbb{R}^N$, let D be an open subset of D_1, and let ξ be a point of D. We consider PWB^h solutions on $D_0 = D - \{\xi\}$ with ∂D_0 the Euclidean boundary and $h = G_{D_1}(\xi, \cdot)_{|D_0}$. Let f be the function $1_{\{\xi\}}$ on ∂D_0. We show that f is a PWB^h resolutive boundary function and has the PWB^h solution

$$\mu_{D_0}^h(\cdot, \{\xi\}) = \frac{G_D(\xi, \cdot)}{G_{D_1}(\xi, \cdot)} \quad \text{on } D_0. \tag{8.5}$$

In particular, if $D = D_1$ it follows that $\mu_{D_0}^h(\cdot, \{\xi\}) = 1$ on D_0. Let u be the restriction to D_0 of the function on the right in (8.5). Then u is h-harmonic and is in the upper PWB^h class on D_0 for the boundary function f. The Green function $G_D(\xi, \cdot)$ has boundary limit 0 at quasi every finite point of ∂D and at the point ∞ if $N > 2$ and D is unbounded (Theorem VII.4), and therefore at quasi every finite point of $\partial D_0 = \partial D \cup \{\xi\}$ and at ∞ if $N > 2$ and D is unbounded. The polar exceptional set A is harmonic measure null (Section 5), so according to Section 5, there is a positive subharmonic function w on D_0 with limit $+\infty$ at every point of A. Hence if $\varepsilon > 0$, the h-subharmonic function

$$\frac{G_D(\xi, \cdot) - \varepsilon(1 + w)}{G_{D_1}(\xi, \cdot)}$$

on D_0 is bounded above because (Section 1) $G_D \leq G_{D_1}$, and this h-subharmonic function has limit superior $\leq f$ at every point of ∂D_0. This function is therefore in the lower PWB^h class for f; so ($\varepsilon \to 0$) the lower PWB^h solution for f is at least u. Hence u is the PWB^h solution for f, as stated in (8.5).

9. h-Resolutive Boundaries

Section 6 implies that a boundary is h-resolutive if and only if the Borel boundary subsets have h-resolutive indicator functions, equivalently, if and only if the compact boundary subsets have h-resolutive indicator functions.

The following theorem gives a useful criterion for this h-resolutivity condition.

Theorem. *A boundary of a Greenian set is h-resolutive if and only if the set function* $A \mapsto \bar{H}^h_{1_A} = R^A_h/h$ *is additive on the class of compact boundary subsets.*

The equality $\bar{H}^h_{1_A} = R^A_h/h$ was pointed out in Section 2. If ∂D is h-resolutive $A \mapsto \bar{H}^h_{1_A} = \mu^h_D(\cdot, A)$ for A compact. To complete the proof of the theorem, it will be shown that conversely if there is the stated additivity, then the boundary function 1_A is h-resolutive whenever A is compact. Let B be a compact subset of $\partial D - A$. By hypothesis

$$\bar{H}^h_{1_A} + \bar{H}^h_{1_B} = \bar{H}_{1_{A \cup B}},$$

and when B increases to $\partial D - A$ through a sequence of compact sets, this equation becomes, in view of Section 6(e),

$$\bar{H}^h_{1_A} + \bar{H}^h_{1_{\partial D - A}} = 1$$

so that $\bar{H}^h_{1_A} = \underline{H}^h_{1_A}$. That is, 1_A is h-resolutive, as was to be proved.

Observation. For an arbitrary boundary the set function $A \mapsto \bar{H}^h_{1_A}(\xi)$ (A compact) is subadditive; so there is additivity for all ξ in D if and only if there is additivity for a point ξ in each open connected component of D.

Application to Balls

The Euclidean boundary of a ball B is universally resolutive. In fact, if h is a strictly positive harmonic function on B and if M_h is the Riesz–Herglotz measure for h, then (Section III.4)

$$^hR^A_1 = \frac{R^A_h}{h} = \bar{H}^h_{1_A} = \int_A K(\zeta, \cdot)\frac{M_h(d\zeta)}{h} \tag{9.1}$$

for every Borel boundary subset A. Hence the set function in Theorem 9 is additive; so the Euclidean ball boundary is universally resolutive. Moreover equation (9.1) implies that

$$\mu^h_D(\xi, d\zeta) = K(\zeta, \xi)\frac{M_h(d\zeta)}{h(\xi)}, \tag{9.2}$$

so that $H^h_f = \mathrm{PI}(B, f\,dM_h)/h$. In particular ($h \equiv 1$), the class of harmonic measure null boundary subsets is the class of l_{N-1} null boundary subsets; the class of resolutive boundary functions, that is, the class $L^1(\mu_B)$, is the class of l_{N-1} measurable and integrable boundary functions; and $H_f =$

$\mathrm{PI}(B,f)$. The class of PWB solutions H_f is thus [Theorem II.14(b)] the class $\mathbf{D}(\mu_{B-})$ which includes all bounded harmonic functions on B. Hence the Euclidean ball boundary is internally resolutive. The generalization of Theorem II.14 to h-harmonic functions in Section IX.12 will make this reasoning applicable to show that the Euclidean ball boundary is universally internally resolutive. The following example exhibits h-resolutivity and internal h-resolutivity in an extreme case. If ζ is a Euclidean ball boundary point and if $h = K(\zeta, \cdot)$, then (9.2) implies that $\mu_B^h(\cdot, \{\zeta\}) \equiv 1$. In this case every boundary function f finite at ζ is h-resolutive, and $H_f^h = f(\zeta)$; the class of PWB^h solutions is the class of finite constant functions. Since h is minimal (Section II.16), a bounded h-harmonic function is necessarily a constant function; so we have proved that the Euclidean boundary is h-resolutive and internally h-resolutive for this special choice of h.

Application to Half-spaces

Denote by d_ξ the Nth coordinate of the point ξ of $\mathbb{R}^N$ and define $D = \{\xi: d_\xi > 0\}$. For η in D let η^* be the reflection of η in the boundary hyperplane of D. Then G_D is given by

$$G_D(\xi, \eta) = \begin{cases} \log \dfrac{|\xi - \eta^*|}{|\xi - \eta|} & \text{if } N = 2 \\ |\xi - \eta|^{2-N} - |\xi - \eta^*|^{2-N} & \text{if } N > 2 \end{cases} \tag{9.3}$$

because, as so defined, $G_D(\xi, \cdot)$ is harmonic on $D - \{\xi\}$ with the right singularity at ξ and has limit 0 at every boundary point of D, including ∞. In view of I(8.5) it is to be expected that $\mu_D(\xi, d\eta)$ (Euclidean boundary) is given by $-\mathbf{D}_{\mathbf{n}_\eta} G_D(\xi, \cdot) l_{N-1}(d\eta)/\pi_N'$, augmented possibly by a contribution from the singleton $\{\infty\}$. Evaluation of this normal derivative leads to the density function in (9.4). The l_{N-1} integral over $\mathbb{R}^{N-1}$ of this density is 1; so the natural conclusion is that $\mu_D(\xi, d\eta)$ is given by

$$\mu_D(\xi, d\eta) = \frac{2d_\xi}{\pi_N' |\xi - \eta|^N} l_{N-1}(d\eta), \qquad \mu_D(\xi, \{\infty\}) = 0. \tag{9.4}$$

In fact this evaluation of μ_D is correct because it is easily checked that if f is a finite continuous function on ∂D, the function $\mu_D(\cdot, f)$, as defined using (9.4), is harmonic on D with boundary limit function f. The function $\mu_D(\cdot, f)$ is called the *Poisson integral of f*, just as in the ball case. Just as in the ball case, to each boundary point η of D corresponds a minimal positive harmonic function $K(\eta, \cdot)$ on D, a constant multiple of the harmonic measure density in (9.4) for that value of η, with a special provision for $\eta = \infty$. More specifically, if ξ_0 is a point of D with $d_{\xi_0} = 1$ and if K is normalized to make $K(\cdot, \xi_0) \equiv 1$, then

$$K(\eta, \xi) = \begin{cases} \dfrac{d_\xi}{|\eta - \xi|^N} |\eta - \xi_0|^N & \text{if } \eta \neq \infty \\ d_\xi & \text{if } \eta = \infty. \end{cases} \tag{9.5}$$

It is shown that each function $K(\eta, \cdot)$ is minimal harmonic by showing that there is a Riesz–Herglotz-type representation of an arbitrary positive harmonic function on D by means of a unique measure M_u on ∂D:

$$u(\xi) = \int_{\partial D} K(\zeta, \xi) M_u(d\zeta). \tag{9.6}$$

The proof follows that in the ball case and is omitted. Just as in the ball case, it is shown that the Euclidean boundary is universally resolutive and that (9.2) is true in the present context.

Some of these results are easily reduced to the ball case by means of an inversion in a sphere taking D into a ball. It will be shown in Chapter XII that if D is an arbitrary Greenian subset of $\mathbb{R}^N$, there is a universally resolutive and universally internally resolutive boundary $\partial^M D$, the Martin boundary, and a function K on $\partial^M D \times D$, such that $K(\eta, \cdot)$ is a minimal positive harmonic function on D when η is in a certain subset $\partial_1^M D$ of $\partial^M D$, and that to each positive harmonic function u on D corresponds a unique measure on $\partial^M D$, supported by $\partial_1^M D$, for which the counterparts of the Riesz–Herglotz-type representation (9.6) (known as the Martin representation in this general context) and of (9.2) are valid. The Martin boundary reduces to the Euclidean boundary if D is a ball or half-space, in which cases $\partial_1^M D = \partial^M D$.

The PWB Method and the Fatou Boundary Limit Theorem

According to Theorem II.15, if h is a strictly positive harmonic function on a ball B, the Dirichlet problem solution $H_f^h = \mathrm{PI}(B, f\,dM_h)/h$ has nontangential limit $f(\zeta)$ at M_h almost every boundary point ζ. This fact is at least a partial justification of the PWB method. Furthermore, if D is an arbitrary Greenian subset of $\mathbb{R}^N$, provided with a boundary by a metric compactification, and if f is an h-resolutive boundary function, then (Theorem 3.II.2) H_f^h has f as a boundary limit function along h-Brownian paths from a point of D to the boundary in the sense that almost every such path tends to some boundary subset A depending on the path (A is a singleton if the boundary is h-resolutive), f is constant on each set A, and H_f^h has limit $f(A)$ along the path.

10. Relations between Reductions and Dirichlet Solutions

Let D be a Greenian set provided with a boundary ∂D by a metric compactification of D, and let B be an open subset of D with boundary ∂B relative to $D \cup \partial D$. Let h be a strictly positive harmonic function on D, and let u be an h-superharmonic function on D. Define a function f on ∂B by setting $f = u$ on $D \cap \partial B$ and (if B is not relatively compact in D) $f = 0$ on $\partial D \cap \partial B$. Observe that a function on B in the upper or lower PWB^h class for f will be in the same class for any other choice of ∂D and corresponding choices of ∂B and of f. The simplest choice for ∂D in this context is the Alexandrov compactification one-point boundary.

(a) If $u \geq 0$, the function f is an h-resolutive boundary function for B, with PWB^h solution the restriction to B of ${}^hR_u^{D-B}$ and

$$u(\xi) \geq \mu_B^h(\xi, u1_D) = \frac{\mu_B(\xi, uh1_D)}{h(\xi)} \qquad (\xi \in B). \tag{10.1}$$

If B_1 and B_2 are open subsets of D with $B_1 \subset B_2$, then

$$\mu_{B_1}^h(\xi, u1_D) \geq \mu_{B_2}^h(\xi, u1_D) \qquad (\xi \in B_1). \tag{10.2}$$

In proving these assertions we assume, as we can without loss of generality, that ∂D is the one-point boundary. Since this boundary of D is trivially h-resolutive, it follows [Section 8, continuation of Example (b)] that ∂B is h-resolutive. If u' is a positive h-superharmonic function on D and majorizes u on $D - B$, then the function $u'_{|B}$ is in the upper PWB^h class on B for f; so ${}^hR_v^{D-B} \geq \bar{H}_f^h$. Since $0 \leq \underline{H}_f^h \leq \bar{H}_f^h$, we conclude (Theorem 8) that f is h-resolutive with $H_f^h \leq {}^hR_u^{D-B}$. In the other direction, if u' is now a function on B in the upper PWB^h class on B for the boundary function f, then the function $u' \wedge (u_{|B})$ is also in this class and, when extended to D by u, is a positive h-superharmonic function on D majorizing u on $D - B$ and therefore majorizing ${}^hR_u^{D-B}$ on D. Hence $\bar{H}_f^h = H_f^h \geq {}^hR_u^{D-B}$ on B; so there is equality on B; that is, in terms of h-harmonic measure, $\mu_B^h(\cdot, u1_D) = {}^hR_u^{D-B}$ on B. The inequality in (10.1) and the inequality (10.2) are now both trivial. The equality in (10.1) follows from the relation between h-harmonic measure and harmonic measure on $D \cap \partial B$ established in Section 8, Example (b).

(b) The positivity hypothesis imposed on u in (a) was made only to allow the use of reductions. We now drop this hypothesis on u but suppose that B is an open relatively compact subset of D. Then a trivial modification of the proof in (a) shows that the restriction to B of the infimum of the class Γ of h-superharmonic functions on D majorizing v on $D - B$ is $\mu_B^h(\cdot, u)$, the PWB^h solution on B for the boundary function $f = u_{|\partial B}$. Finally, (10.2) remains true in the present context, and in this context $u1_D = u$ on B. If B_1 and B_2 are balls of center ξ and if $h \equiv 1$, the inequality (10.2) reduces to the fact that the function $r \mapsto L(u, \xi, r)$ is a decreasing function for $0 < r < |\xi - \partial D|$.

11. Generalization of the Operator τ_B^h and Application to GM^h

If u is an h-superharmonic function on an open subset D of $\mathbb{R}^N$ and if B is an open relatively compact subset of D, we define $\tau_B^h u$ as the smoothed infimum of the class of h-superharmonic functions on D, majorizing u on $D - B$, so that $\tau_B^h u$ is h-superharmonic on D, equal (Section 10) on B to the PWB^h solution $\mu_B^h(\cdot, u)$, and equal to u quasi everywhere on $D - B$, in particular, equal to u on the interior of $D - B$. When B is a ball, this definition agrees on $D - \partial B$ with the Section II.1 definition of $\tau_B u$ and the Section 1 definition of $\tau_B^h u$, and therefore the definitions agree everywhere on D because two h-superharmonic functions equal l_N almost everywhere are equal everywhere. Obviously, in all cases $\tau_{B_1}^h u \geq \tau_{B_2}^h u$ when $B_1 \subset B_2$. If u is h-subharmonic, $\tau_B^h u$ is defined as $-\tau_B^h(-u)$.

In view of Section III.1 [especially Observation (a)] as generalized trivially to allow arbitrary h and of the present extended definition of τ_B^h, if D is a Greenian set, if $B_{\bullet}$ is an increasing sequence of open relatively compact subsets of D with union D, and if u is an h-superharmonic function on D, the limit of the decreasing sequence $\tau_{B_{\bullet}}^h u$ of h-superharmonic functions is, on each open connected component of D, either identically $-\infty$ or h-harmonic and, if h-harmonic on D, is $\mathrm{GM}_D^h u$. That is,

$$\mathrm{GM}_D^h u = \lim_{n \to \infty} \tau_{B_n}^h u = \lim_{n \to \infty} \mu_{B_n}^h(\cdot, u) \tag{11.1}$$

if u has an h-subharmonic minorant. In particular, an h-potential u is characterized among the positive h-superharmonic functions on D by the condition $\lim_{n \to \infty} \mu_{B_n}^h(\cdot, u) \equiv 0$.

A related result is the following, in which u is supposed positive to make reduction notation possible. Let D, h, and $B_{\bullet}$ be as in the preceding paragraph, let u be a positive h-superharmonic function on D, define a bounary ∂D for D by a metric compactification, and write $\bar{D}$ for $D \cup \partial D$. Then if $A \subset \bar{D}$,

$$\mathrm{GM}_D^h {}^h\!R_u^A = \lim_{n \to \infty} {}^h\!R_{+u}^{A - B_n}. \tag{11.2}$$

We prove this for $h \equiv 1$ to avoid irrelevant notational complexities, and we use the alternative reduction notation because iterated reductions will be needed. Observe that

$$\big\| \|u\|^A \big\|^{D - B_n} = \big\| \|u\|^A \big\|^{\bar{D} - B_n} \geq \big\| \|u\|^A \big\|^{A - B_n} = \|u\|^{A - B_n},$$

and we have just proved that the limit ($n \to \infty$) on the left is $\mathrm{GM}_D \|u\|^A$; so in (11.2) the left side is at least equal to the right side. In the other direction,

$$\big\|\|u\|^A\big\|^{D-B_n} \le \big\|\|u\|^{A-B_m}\big\|^{D-B_n} + \big\|\|u\|^{A\cap B_m}\big\|^{D-B_n}$$
$$\le \|u\|^{A-B_m} + \big\|\|u\|^{A\cap B_m}\big\|^{D-B_n}.$$

When $n \to \infty$, this inequality yields

$$\mathrm{GM}_D\|u\|^A \le \lim_{m\to\infty} \|u\|^{A-B_m} + \mathrm{GM}_D\|u\|^{A\cap B_m},$$

and the last term vanishes because $A \cap B_m$ is relatively compact in D; so $\|u\|^{A\cap B_m}$ is a potential. Hence in (11.2) the left side is at most equal to the right side so there is equality.

A Local Property of τ^h

If u is an h-subharmonic or h-superharmonic function and if $B_\cdot$ is a decreasing sequence of open relatively compact subsets of D with intersection ξ, then $\lim_{n\to\infty} \tau^h_{B_n} u(\xi) = u(\xi)$. In fact the proof for $h \equiv 1$ and $B_\cdot$ a sequence of balls of center ξ [see Section II.6(f) for this result in a slightly different context] is applicable in the general case.

12. Barriers

Let D be a Greenian subset of $\mathbb{R}^N$, coupled with a boundary ∂D provided by a metric compactification, let h be a strictly positive harmonic function on D, and let ζ be a point of ∂D. Usable conditions that ζ be h-regular are most easily formulated in terms of "h-barriers" (see Section 13). A strictly positive h-superharmonic function u on D will be called an *h-barrier* for D at ζ if $\lim_{\eta\to\zeta} u(\eta) = 0$ and if $(*)$ $\inf_{D-B} u > 0$ whenever B is a neighborhood of ζ. If the condition $(*)$ is omitted, u will be called a *weak h-barrier* for D at ζ. By an easy application of the h-superharmonic function minimum theorem, $(*)$ is true if and only if $\liminf_{\eta\to\xi} u(\eta) > 0$ for ξ in $\partial D - \{\zeta\}$. As usual, h will be omitted from the notation and the nomenclature when $h \equiv 1$.

Local Nature of the Existence of an h-Barrier

Let D_0 be an open subset of D with boundary that relative to the topology of $D \cup \partial D$, and let h_0 be the restriction of h to D_0. Then if ζ is a common boundary point of D_0 and D, the restriction to D_0 of an h-barrier for D at ζ is an h_0-barrier for D_0 at ζ. Conversely, if some neighborhood of ζ has the same trace on D_0 as on D, the existence of an h_0 barrier u_0 for D_0 at ζ implies the existence of an h-barrier for D at ζ. In fact, let B be the trace on D_0 of some open neighborhood of ζ so small that $D \cap \partial B = D_0 \cap \partial B \neq \varnothing$, and set $\alpha = \inf_{D_0 - B} u_0$. Then $\alpha > 0$, and if u is defined as $\alpha \wedge u_0$ on $D \cap B$ and as α on $D - B$, the function u is an h-barrier for D at ζ. Thus the existence of an h-barrier for D at ζ is a local property of D near ζ, depending on h.

Lemma (Euclidean Boundary, $h \equiv 1$). *If there is a weak barrier at a boundary point ζ, there is a barrier at the point.*

If $N > 2$ every unbounded open subset D of $\mathbb{R}^N$ has a barrier at the point ∞, namely, the restriction of the function $G(0,\cdot)$ to D. An unbounded open subset D of $\mathbb{R}^2$ has a [weak] barrier at the point ∞ if and only if the image of D under an inversion in a circle has a [weak] barrier at the circle center. Hence we can assume in the following proof that the boundary point ζ in question is finite. In view of the fact that the existence of a barrier at ζ is a local property of D, it is sufficient to show that there is a barrier on the trace on D of an open neighborhood of ζ, so that D can be supposed bounded. Suppose then that D has diameter $\delta < +\infty$ and that u is a weak barrier for D at the boundary point ζ. Define $\phi(\eta) = |\eta - \zeta|$ and $f = \phi_{|\partial D}$. The function ϕ is subharmonic, and $\phi_{|D}$ is in the lower PWB class on D for f. Hence $H_f \geq \phi$ on D, and it will be shown that H_f is a barrier for D at ζ by showing that H_f has limit 0 at ζ. Fix $r > 0$, let $B = B(\zeta, r)$, let A be a compact subset of $D \cap \partial B$, and let ψ be the indicator function of $(\partial B - A) \cap D$ on ∂B, so that $\mathrm{PI}(B, \psi)$ is harmonic on B with limit 1 at every point of $\partial B - A$. If u_0 is in the lower PWB class on D for f, the function u_0 on $B \cap D$ is at most r if ∂B does not meet D and (by the maximum theorem for subharmonic functions on $B \cap D$) is at most $r + \delta u/\inf_A u + \delta \mathrm{PI}(B, \psi)$ if ∂B does meet D. Thus in both cases

$$H_f \leq r + \frac{\delta u}{\inf_A u} + \delta \mathrm{PI}(B, \psi) \tag{12.1}$$

on $B \cap D$. The sum on the right has limit $r + \delta \mathrm{PI}(B, \psi)(\zeta)$ at ζ, and this limit is at most $2r$ if A is sufficiently large. Since r is arbitrary, the function H_f has limit 0 at ζ, as was to be proved.

Example (a). If h is a strictly positive harmonic function on D with a finite continuous strictly positive extension to $\bar{D}$, then if u is a barrier for D at a boundary point ζ, the function u/h is an h-barrier for D at ζ.

Example (b) (Euclidean Boundary, $h \equiv 1$). If D is unbounded and $N > 2$, we have already noted that the function $G(0,\cdot)$ is a barrier for D at the point ∞.

Example (c) (Poincaré) (Euclidean Boundary, $h \equiv 1$). If ζ is a finite boundary point of D with the property that some closed ball meets $\bar{D}$ at ζ but at no other point the restriction to D of the function $G(\xi, \zeta) - G(\xi, \cdot)$, with ξ the ball center, is a barrier for D at ζ.

We shall show in Section 15 that it is sufficient for the existence of a barrier at ζ if in this Poincaré criterion the ball is replaced by a cone with vertex ζ.

EXAMPLE (d) ($N = 2$, Euclidean Boundary, $h \equiv 1$). If ζ is a finite boundary point of D and if there is a simple continuous arc with initial point ζ, in $\mathbb{R}^2 - D$ except for ζ, then there is a barrier for D at ζ. In fact, if B is a ball of center ζ so small that the arc hits ∂B, let B_0 be B less the part of the arc from ζ to the first hit of ∂B. Then B_0 is simply connected; so the function $\log(\cdot - \zeta)$ has a single-valued analytic branch ϕ in B_0. The restriction to $B \cap D$ of the real part of $1/\phi$ is a barrier for $B \cap D$ at ζ; so there is a barrier for D at ζ.

13. h-Barriers and Boundary Point h-Regularity

Theorem. *Let D be a Greenian subset of $\mathbb{R}^N$, coupled with a boundary ∂D provided by a metric compactification. If there is an h-barrier for D at the boundary point ζ and if f is an upper bounded boundary function, then*

$$\limsup_{\eta \to \zeta} \bar{H}_f^h(\eta) \leq \limsup_{\eta \to \zeta} f(\eta) \vee f(\zeta); \tag{13.1}$$

in particular, if f is bounded and is continuous at ζ, then

$$\lim_{\eta \to \zeta} \underline{H}_f^h(\eta) = \lim_{\eta \to \zeta} \bar{H}_f^h(\eta) = f(\zeta), \tag{13.2}$$

and therefore ζ is h-regular.

The second assertion follows from the first applied to f and $-f$. To prove the first assertion, let u be an h-barrier at ζ. Let b be any number strictly larger than the right side of (13.1), let B be a neighborhood of ζ so small that $f \leq b$ in the neighborhood, and let δ be the infimum of u outside $D \cap B$. Choose n so large that $b + n\delta$ exceeds the supremum of f. The function $b + nu$ is in the upper PWB^h class on D for f and has limit b at ζ. Hence the left side of (13.1) is at most b; so the theorem is true.

Extension. Since a change of f on a set of h-harmonic measure 0 does not change $\underline{H}_f^h$ or $\bar{H}_f^h$, the point η can tend to ζ on the right side of (13.1) on the complement of such a boundary set, and $f(\zeta)$ can be omitted on the right hand side if $\{\zeta\}$ is h-harmonic null, as is true when $h \equiv 1$ and ∂D is the Euclidean boundary, unless $N > 2$ and $\zeta = \infty$. This extension of Theorem 13 reduces to Theorem II.1 when D is a ball, $h \equiv 1$, and ∂D is the Euclidean boundary.

14. Barriers and Euclidean Boundary Point Regularity

Theorem. (Euclidean Boundary, $h \equiv 1$). *A boundary point is regular if and only if there is a barrier at the point.*

If there is a barrier at a boundary point, the point is regular by Theorem 13. Conversely, suppose that the point ζ is a regular boundary point of the Greenian set D. If $N > 2$ and if $\zeta = \infty$, there is a barrier, exhibited in Section 12, Example (b). If $N = 2$ and $\zeta = \infty$, make ζ finite by an inversion in a sphere with center a finite boundary point. Thus ζ can be supposed finite in proving the existence of a barrier there.

If D is bounded, define $f = |\cdot - \zeta|_{|\partial D}$ and $u = |\cdot - \zeta|_{|D}$. Then u is subharmonic and is in the lower PWB class on D for f; so $H_f \geq u$, and H_f is a barrier at ζ because (regularity of ζ) H_f has limit 0 at ζ. If D may not be bounded, the cases $N > 2$ and $N = 2$ will be treated separately. If $N > 2$, define f and u as the restrictions to ∂D and D, respectively, of $\Sigma_1^\infty n^{-3}\,[n - G(\zeta, \cdot) \wedge n]$, with $f(\infty) = \Sigma_1^\infty n^{-2}$ if $\infty \in \partial D$. The function f is a finite-valued continuous positive boundary function vanishing at ζ and only there, and the function u is a continuous positive subharmonic function in the lower PWB class for f; so $H_f \geq u$. Moreover H_f has limit 0 at ζ because ζ is regular. Hence H_f is a barrier at ζ. If $N = 2$, let B be a ball of center ζ so small that $D_1 = D \cup B$ is Greenian. The function $\Sigma_1^\infty n^{-3}\,[n - G_{D_1}(\zeta, \cdot) \wedge n]$ is a positive bounded subharmonic function on D_1. Define u as the restriction to D of this function, and define f at each point η of ∂D by $f(\eta) = \limsup_{\xi \to \zeta} u(\xi)$. Then f is positive, bounded, upper semicontinuous, and vanishes at ζ and only there. The function u is in the lower PWB class for f on D; so $H_f \geq u$. Moreover H_f has limit 0 at ζ because ζ is regular. Hence H_f is a barrier at ζ.

Application (a): Local Property of Regularity. (Euclidean Boundary, $h \equiv 1$.) Since the existence of a barrier at a boundary point is a local property of a Greenian set D near the point, regularity at a point of the Euclidean boundary is also a local property of D near the point.

Application (b): Regularity of a Boundary Point of a Disconnected Set. (Euclidean Boundary, $h \equiv 1$.) Let D be a Greenian subset of $\mathbb{R}^N$, let ζ be a boundary point of D, and let $D_1, D_2, \ldots$ be the open connected components of D with boundary point ζ. *Then ζ is a regular boundary point of D if and only if ζ is a regular boundary point of each set D_k.* In one direction, if ζ is a regular boundary point of D, the restriction to D_k of a barrier for D at ζ is a barrier for D_k. Conversely, if ζ is a regular boundary point of each set D_k and if u_k is a weak barrier for D_k at ζ, then the function u defined on D by setting $u = 2^{-k}(u_k \wedge 1)$ on D_k is a weak barrier for D at ζ.

Application (c): Regularity of a Boundary Point in Terms of the Green Function. (Euclidean Boundary, $h \equiv 1$.) A boundary point ζ of a Greenian set D is regular if and only if $\lim_{\eta \to \zeta} G_D(\xi, \eta) = 0$ for some (equivalently every) *point of each open connected component of D with boundary point ζ.* In view of application (b) we can assume in the proof that D is connected. If the condition on G_D is satisfied for a single point ξ, the function $G_D(\xi, \cdot)$ is a weak barrier at ζ; so ζ is regular. If $N > 2$ or if $N = 2$ and if D is bounded, the expression for $G_D(\xi, \cdot)$ in terms of a Dirichlet solution (Section 3) implies the truth of the converse, for all ξ. The following proof of the converse is valid in all cases. If ζ is regular and if u is a barrier for D at ζ, let ξ be a point of D, and let B_c be the set $\{G_D(\xi, \cdot) > c\}$, where c is a constant chosen so large that B_c is relatively compact in D. Such a choice is possible because $G_D(\xi, \cdot)$ is bounded in a neighborhood of ∂D. Choose n so large that $nu > c$ on B_c. Then (Section VII.3) $G_D(\xi, \cdot) \leq nu$ outside B_c; so $G_D(\xi, \cdot)$ has limit 0 at ζ.

Application (d): Relative Boundaries. If D is a Greenian subset of $\mathbb{R}^N$, provided with a boundary ∂D by a metric compactification, if h is a strictly positive harmonic function on D, and if B is an open subset of D with relative boundary ∂B in $D \cup \partial D$, then quasi every point of ∂B in D is h-regular because if u is a local barrier for B at a boundary point of B in D, then u/h is a local h-barrier for B at the point and (Section 12) can be extended to B to be an h-barrier there.

Application (e): The Kelvin Transformation and Regularity. (Euclidean Boundary, $h \equiv 1$.) We use here the notation of the discussion in Section 2 of the PWB method and the Kelvin transformation. If ζ is a regular finite boundary point of D, then its image under inversion in a sphere is a regular boundary point of D' because if u is a barrier for D at ζ, the Kelvin transform of u is a barrier for D' at $\phi(\zeta)$ except possibly when $N > 2$, and the inversion sphere has center ζ. In this case, however, the image of ζ is the point ∞, which is a regular boundary point of every unbounded Greenian set when $N > 2$ (Theorem 4). If $N > 2$ and if ∞ is a (necessarily regular) boundary point of D, the image of ∞ under an inversion may or may not be a regular boundary point of D'. If $N = 2$ and if ∞ is a regular boundary point of D, the image of ∞ under an inversion is a regular boundary point of D' because the Kelvin transform of a barrier for D at ∞ is a barrier for D' at the image of ∞.

15. The Geometrical Significance of Regularity (Euclidean Boundary, $h \equiv 1$)

If ζ is a boundary point of D with the property that some neighborhood of ζ meets ∂D in a harmonic measure null set, the point ζ is irregular because a PWB solution H_f is not affected by a change of f on such a set. This fact together with Poincaré's criterion [Section 12, Example (c)] for the existence of a barrier suggests that regularity of ζ, equivalently, the existence of a barrier at ζ, amounts to the requirement that D not fill up "too much" of a neighborhood of ζ. Thus an isolated finite boundary point ζ is irregular because $\{\zeta\}$ is polar and therefore harmonic measure null, and when $N = 2$, the point ∞ (if an isolated point of ∂D) is irregular because D can be mapped into a Greenian set by an inversion leaving harmonic measure invariant and taking ∞ into a finite boundary point. The point ∞ is exceptional in that when $N > 2$, the harmonic measure $\mu_D(\cdot, \{\infty\})$ may be strictly positive and, in fact, is identically 1 when $D = \mathbb{R}^N$. (Recall that when $N > 2$, the point ∞ is a regular boundary point of every unbounded Greenian set.) It will be shown in Section XI.12 that a finite boundary point ζ is regular if and only if ζ is a limit point of $\mathbb{R}^N - D$ in the *fine topology*, and this criterion will be stated probabilistically in Section 2.IX.15.

Poincaré–Zaremba Regularity Criterion

Poincaré's regularity criterion [Section 12, Example (c)] was improved to the following: if ζ is a finite boundary point of D with the property that some open solid cone of revolution with vertex ζ does not meet D in a neighborhood of ζ, there is a barrier at ζ (so ζ is a regular boundary point).

In view of the discussion of barriers in Section 12, it is sufficient to prove that if A is a closed cone of revolution with vertex the origin and if $D = B(0, 1) - A$, then D has a barrier at the origin. Define $\phi(\xi) = |\xi|$ and $f = \phi_{|\partial D}$. Since ϕ is subharmonic on D, $\phi_{|D}$ is in the lower PWB class on D for f; so $H_f \geq \phi_{|D}$, and we show that H_f is a barrier at the origin by showing that H_f has limit 0 there. Let D_0 be the part of D at distance $< \frac{1}{2}$ from the origin. Observe that $H_f \leq 1$ on D and that H_f has limit f at every point of $\partial D - \{0\}$ because (Poincaré criterion) every point of $\partial D - \{0\}$ is regular. By the harmonic function maximum theorem $\alpha = \sup_{D_0} H_f < 1$. The function $\xi \mapsto H_f(\xi) - (\alpha \vee \frac{1}{2})H_f(2\xi)$ is a bounded harmonic function on D_0 and has a negative limit at every point of $\partial D_0 - \{0\}$. Hence (Section V.7) by the extended harmonic function maximum theorem $H_f(\xi) - (\alpha \vee \frac{1}{2})H_f(2\xi) \leq 0$, and therefore

$$\limsup_{\xi \to 0} H_f(\xi) \leq (\alpha \vee \tfrac{1}{2}) \limsup_{\xi \to 0} H_f(\xi);$$

so the superior limit is 0, as was to be proved.

Strengthened Poincaré–Zaremba Criterion

In the preceding proof, if A_1 is an $(N-1)$-dimensional hyperplane containing the axis of the cone A, $A \cap A_1$ is a flattened cone, and the preceding proof with A replaced by $A \cap A_1$ is valid with trivial changes. Thus a finite boundary point of an open subset of $\mathbb{R}^N$ is regular if some flattened cone in this sense has vertex ζ and does not meet D in a neighborhood of ζ. The reader is invited to strengthed this criterion still further by weakening the conditions on A_1.

Regularity of Classical Boundaries: The Lebesgue Spine

Most of the open sets used in classical analysis have regular boundaries because there are Poincaré–Zaremba barriers at their boundary points. On the other hand, if the excluded cone is sharpened into a cusp that is sufficiently sharp, a barrier may no longer exist at the vertex. For example, suppose that $N = 3$, denote a point ξ of $\mathbb{R}^3$ by its coordinates $\xi^{(1)}$, $\xi^{(2)}$, $\xi^{(3)}$, let A be the closed line segment with endpoints the origin and the point $(1, 0, 0)$, and let μ be the measure supported by A and determined by $\mu(d\xi) = \xi^{(1)}l_1(d\xi^{(1)})$. The potential $G\mu$ has value 1 at the origin, value $+\infty$ elsewhere on A, and limit 1 at the origin along the negative $\xi^{(1)}$ axis. The set $D = \{\xi \neq 0: G\mu(\xi) < 2\}$ is a solid of revolution about the $\xi^{(1)}$ axis, includes the negative $\xi^{(1)}$ axis except for the origin, and has an exponential cusp at the origin; D is the *Lebesgue spine*. We now show that the origin is an irregular boundary point of D. If f is the boundary function equal to 2 at the finite boundary points of D and equal to 0 at ∞, then f is resolutive because the Euclidean boundary is always resolutive. Moreover H_f is the restriction of $G\mu$ to D because this restriction is a bounded harmonic function that has the prescribed limit at every boundary point with the exception of the origin and $\{0\}$ is a harmonic measure null set. The origin is not a regular boundary point of D because H_f has limit 1 at the origin along the negative $\xi^{(1)}$ axis.

16. Continuation of Section 13

Let D be a Greenian subset of $\mathbb{R}^N$, provided with a boundary ∂D by a metric compactification, and let h be a strictly positive harmonic function on D.

(a) If ∂D is h-resolutive and if ζ is an h-regular boundary point with $\mu_D^h(\cdot, \{\zeta\}) < 1$, then there is a weak h-barrier at ζ. In fact, if A_n is the part of the boundary at distance $\geq 2^{-n}$ from ζ, the function $\Sigma_0^\infty 2^{-n}\mu_D^h(\cdot, A_n)$ is a weak h-barrier at ζ.

(b) If ∂D is h-resolutive and if ζ is a boundary point with $\mu_D^h(\cdot, \{\zeta\}) \equiv 1$, then ζ is necessarily h-regular, but there may be no weak h-barrier at ζ. For example, if $h \equiv 1$ and if $\partial D = \{\zeta\}$ is the one-point boundary, then it is trivial that ∂D is resolutive and that $\mu_D(\cdot, \{\zeta\}) \equiv 1$. In this case there can in general be no weak barrier u for D at ζ because such a function would be a strictly positive superharmonic function on D with limit 0 at every Euclidean boundary point of D. Then u would be a weak barrier for D at every Euclidean boundary point; so the Euclidean boundary would be regular, and this regularity would be a restriction on D.

17. h-Harmonic Measure μ_D^h as a Function of D

Theorem. *Let D be a Greenian subset of $\mathbb{R}^N$ provided with a boundary ∂D by a metric compactification, let h be a strictly positive harmonic function on D, and suppose that ∂D is h-resolutive. Boundaries of subsets of D are to be relative to $D \cup \partial D$.*

(a) *If D_0 is an open subset of D, then ∂D_0 is h-resolutive and*

$$\mu_{D_0}^h(\cdot, A) \le \mu_D^h(\cdot, A) \tag{17.1}$$

on D_0 whenever A is a Borel subset of $\partial D_0 \cap \partial D$.

(b) *If $D_\cdot$ is an increasing sequence of open subsets of D with union D, then for ξ in D,*

$$\lim_{n\to\infty} \mu_{D_n}^h(\xi, \cdot) = \mu_D^h(\xi, \cdot) \tag{17.2}$$

(vague convergence on $D \cup \partial D$). Moreover, if $\xi \in D_m$ and if A is a Borel subset of $\partial D_m \cap \partial D$, then

$$\mu_{D_m}^h(\xi, A) \le \mu_{D_{m+1}}^h(\xi, A) \le \cdots \le \mu_D^h(\xi, A). \tag{17.3}$$

The h-resolutivity of ∂D_0 was proved in Section 8. Inequality (17.1) is a trivial consequence of (8.3), and (17.3) is simply a repeated application of (17.1). To prove (17.2), it will be shown that if f is a finite-valued continuous function on $D \cup \partial D$, then

$$\lim_{n\to\infty} \mu_{D_n}^h(\cdot, f) = \mu_D^h(\cdot, f). \tag{17.4}$$

If u is in the upper PWB^h class on D for $f_{|\partial D}$ and is bounded, extend u to u' on $D \cup \partial D$ by setting $u'(\zeta) = \liminf_{\eta\to\zeta} u(\eta)$ for $\zeta \in \partial D$. Then if $\varepsilon > 0$, the function $u' + \varepsilon - f$ is lower semicontinuous on $D \cup \partial D$, strictly positive on ∂D, and therefore also strictly positive on ∂D_n for sufficiently large n. Hence for sufficiently large n the restriction to D_n of $u + \varepsilon$ is in the upper PWB^h class on D_n for the boundary function $f_{|\partial D}$; so $\mu_{D_n}^h(\cdot, f) \le u + \varepsilon$ on D_n. It follows that $\limsup_{n\to\infty} \mu_{D_n}^h(\cdot, f) \le \mu_D^h(\cdot, f)$, and this inequality together with the corresponding inequality for $-f$ yields (17.2).

Application. In Theorem 17 suppose that $h \equiv 1$ and that ∂D is the Euclidean boundary. Let $D'_\cdot$ be an increasing sequence of subsets of $D \cup \partial D$, open relative to $D \cup \partial D$, define $D_n = D \cap D'_n$, and suppose that $\bigcup_0^\infty D_n = D$. Suppose that A_1 is a subset of $D \cup \partial D$, open relative to $D \cup \partial D$, and that $A_1 \subset D'_n$ for sufficiently large n. *Then if A is a Borel subset of $A_1 \cap \partial D$ and if $\xi \in D$,*

$$\lim_{n\to\infty} \mu_{D_n}(\xi, A) = \mu_D(\xi, A). \tag{17.5}$$

To see this, define $A_2 = A_1 \cap \partial D$, and observe that if n is so large that $\xi \in D_n$ and $A_1 \subset D'_n$, then $\mu_{D_n}(\xi, A_1) = \mu_{D_n}(\xi, A_2)$ and $\mu_D(\xi, A_1) = \mu_D(\xi, A_2)$. In view of the vague convergence in Theorem 17(b),

$$\liminf_{n\to\infty} \mu_{D_n}(\xi, A_2) = \liminf_{n\to\infty} \mu_{D_n}(\xi, A_1) \geq \mu_D(\xi, A_1) = \mu_D(\xi, A_2), \tag{17.6}$$

and in view of (17.3) with $A = A_2$, inequality (17.6) implies that (17.5) is true when $A = A_2$. Now by the monotoneity in (17.3),

$$\lim_{n\to\infty} \mu_{D_n}(\xi, A) \leq \mu_D(\xi, A), \qquad \lim_{n\to\infty} \mu_{D_n}(\xi, A_2 - A) \leq \mu_D(\xi, A_2 - A) \tag{17.7}$$

for $A \subset A_2$, and since the sum of these two inequalities yields an equality, there must be equality in each; that is, (17.5) is true.

18. The Extension $G_D^{=}$ of G_D and the Harmonic Average $\mu_D(\xi, G_B^{=}(\eta, \cdot))$ When $D \subset B$

In the following theorem D and B are Greenian subsets of $\mathbb{R}^N$ with $D \subset B$, and boundaries are Euclidean. Recall from Section VII.4 that for ξ in D the function $G_D^{=}(\xi, \cdot)$ is an extension of $G_D(\xi, \cdot)$ to $\mathbb{R}^N$, subharmonic on $\mathbb{R}^N - \{\xi\}$, vanishing on $\mathbb{R}^N - \bar{D}$ and quasi everywhere on $\mathbb{R}^N \cap \partial D$, with

$$G_D^{=}(\xi, \zeta) = \limsup_{D \ni \eta \to \zeta} G_D(\xi, \eta) \qquad (\zeta \in \mathbb{R}^N \cap \partial D). \tag{18.1}$$

In particular, if ζ is a finite regular boundary point of D, the function $G_D^{=}(\xi, \cdot)$ is continuous at ζ with value 0 there; if D is not bounded, this function has limit 0 at ∞ when $N > 2$ and also when $N = 2$ if ∞ is a regular boundary point of D.

In discussing $\mu_D(\xi, G_B^{=}(\eta, \cdot))$ when D is unbounded and $N = 2$, the value assigned to $G_B^{=}(\eta, \infty)$ is irrelevant because (Section 5) the singleton $\{\infty\}$ is μ_D null. When $N > 2$ and D is unbounded, the singleton $\{\infty\}$ may not be μ_D null, but we define $G_B^{=}(\eta, \infty) = 0$, corresponding to the fact that $G_B^{=}(\eta, \cdot)$ has limit 0 at ∞.

If u is a superharmonic function on a superset of D, we write $\mathrm{GM}_D u$ instead of $\mathrm{GM}_D(u_{|D})$.

Theorem. (a) *For each η in B the function $G_B^{=}(\eta, \cdot)_{|\partial D}$, defined as 0 at ∞ if D is unbounded, is a resolutive boundary function.*

(b) *If $\xi \in D$, then*

$$G_D^{=}(\xi, \eta) = G_B(\xi, \eta) - \mu_D(\xi, G_B^{=}(\eta, \cdot)) \qquad (\eta \in B); \tag{18.2}$$

that is,

$$G_D^{=}(\xi,\eta) = G_B(\xi,\eta) - \mu_D(\xi, G_B(\eta,\cdot)1_B) \qquad (\eta \in B). \qquad (18.2')$$

(c) *If $v = G_B\nu$ is a superharmonic potential on B and if ν^i is the projection of ν on the set of irregular boundary points of D in B, then on D,*

$$\mathrm{GM}_D v = \mu_D(\cdot, v1_B) + G_D^{=}\nu^i. \qquad (18.3)$$

If D is relatively compact in B, (18.3) is true whenever v is a superharmonic function on B and ν is its associated Riesz measure.

Observation (1). If $\eta \in D$, (18.2) becomes

$$G_D(\xi,\eta) = G_B(\xi,\eta) - \mu_D(\xi, G_B(\eta,\cdot)1_B) \qquad [(\xi,\eta)\in D \times D], \quad (18.4)$$

and we thereby have found the important symmetry relation

$$\mu_D(\xi, G_B(\eta,\cdot)1_B) = \mu_D(\eta, G_B(\xi,\cdot)1_B) \qquad [(\xi,\eta)\in D \times D]. \quad (18.5)$$

Note that we can take $B = \mathbb{R}^N$ here when $N > 2$. With this specialization (18.4) was derived in Section 3, application of Example (a). It will be shown in Section 19 that (18.4) and (18.5) are true when $N = 2$ and $B = \mathbb{R}^2$ if and only if $\mathbb{R}^2 - D$ is not too sparse near ∞, more precisely, if and only if D is bounded or if unbounded D has ∞ as a regular boundary point.

Observation (2). If we write $\mu_D(\xi,\cdot)_{|B}$ for the restriction of the measure $\mu_D(\xi,\cdot)$ to the class of Borel subsets of B, then according to Theorem 18,

$$\mu_D(\xi, G_B^{=}(\eta,\cdot)) = G_B[\mu_D(\xi,\cdot)_{|B}](\eta) < +\infty \qquad [(\xi,\eta)\in D \times B]; \quad (18.6)$$

so for fixed ξ in D the potential $G_B[\mu_D(\xi,\cdot)_{|B}]$ is a finite-valued superharmonic function on B. Thus Theorem 18(b) implies that the measure on $\mathbb{R}^N - \{\xi\}$ associated with the superharmonic function $-G_D^{=}(\xi,\cdot)$ on this set is the restriction of the harmonic measure $\mu_D(\xi,\cdot)$ to the class of Borel subsets of $\mathbb{R}^N - \{\xi\}$.

Proof of (a). For η in B define f_η on ∂D as $G_D(\eta,\cdot)$ on $B \cap \partial D$ and as 0 on $\partial B \cap \partial D$. Since $G_B^{=}(\eta,\cdot) = 0$ at quasi every point of ∂B, assertion (a) states that the function f_η is a resolutive boundary function. This fact was proved in Section 10 (but observe that the notation here reverses the roles of D and B in Section 10). □

Proof of (b). (When $\eta \in B - \partial D$.) If $\eta \in B$, the function $G_B(\eta,\cdot)_{|D}$ is in the upper PWB class on D for the boundary function f_η. If $\eta \in B - \partial D$, the function $\mathrm{GM}_D G_B(\eta,\cdot)$ is in the lower PWB class on D for f_η when f_η is increased to $+\infty$ on the polar set of irregular boundary points of D. Since a change of boundary function on a polar set does not change PWB solutions,

$$\mathrm{GM}_D G_B(\eta,\cdot) \le \mu_D(\cdot, G_B^{=}(\eta,\cdot)) \le G_B(\eta,\cdot)_{|D} \qquad (\eta \in B - \partial D), \quad (18.7)$$

and since the middle term in (18.7) is harmonic, the last term can be replaced by $\mathrm{GM}_D G_B(\eta,\cdot)$ to yield the equality

$$\mathrm{GM}_D G_B(\eta,\cdot) = \mu_D(\cdot, G_B^{=}(\eta,\cdot)) \qquad (\eta \in B - \partial D). \qquad (18.8)$$

Under this equality equation (18.2) with η in D reduces to the expression for G_D in terms of G_B derived in Section VII.1 [see VII(1.2)]. If $\eta \in B - \bar{D}$, (18.2) reduces to

$$0 = G_B(\xi,\eta) - \mu_D(\xi, G_B^{=}(\eta,\cdot)) \qquad (\xi \in D),$$

and since $G_B(\cdot,\eta)$ is harmonic on D, this equation is a trivial consequence of (18.8). Thus we have proved (18.2) for η in $B - \partial D$.

We now prove

$$\| G_B(\eta,\cdot) \|^{B-D}(\xi) = \| G_B(\xi,\cdot) \|^{B-D}(\eta) \qquad [(\xi,\eta) \in D \times B]. \qquad (18.9)$$

We shall need this symmetry relation which is a special case of a fundamental symmetry relation to be derived in Section X.3. For ξ and η in D equation (18.9) reduces to (18.5) [see Section 10(a)], and (18.5) is true because, as just proved, (b) is true when η is in D. Thus, if $(\xi,\eta) \in D \times (B - D)$, equation (18.9) is true if D is enlarged to $D \cup B(\eta, 1/n)$ with $|\eta - \partial D| < 1/n$. When $n \to \infty$, it follows [from Section VI.3(e)] that (18.9) is true if D is replaced by $D \cup \{\eta\}$ and therefore is true with no replacement because [by Section VI.3(c)] a smoothed reduction is unchanged if the target set is changed by a polar set.

(b) (When $\eta \in B \cap \partial D$.) Suppose first that η is a regular boundary point of D in B. With this choice of η, (18.2′) reduces to

$$\mu_D(\xi, G_B(\eta,\cdot) 1_B) = G_B(\xi,\eta), \qquad (18.10)$$

or equivalently [Section 10(a)],

$$\| G_B(\eta,\cdot) \|^{B-D}(\xi) = G_B(\xi,\eta). \qquad (18.11)$$

Now in view of the symmetry of the left side of (18.10), proved above, and of the special lower semicontinuity property II(6.1) of superharmonic functions,

$$\begin{aligned}\llbracket G_B(\eta,\cdot)\rrbracket^{B-D}(\xi) &= \llbracket G_B(\xi,\cdot)\rrbracket^{B-D}(\eta)\\ &= \liminf_{D\ni\zeta\to\eta}\llbracket G_B(\xi,\cdot)\rrbracket^{B-D}(\zeta) \wedge \liminf_{B-D\ni\zeta\to\eta}\llbracket G_B(\xi,\cdot)\rrbracket^{B-D}(\zeta),\end{aligned} \tag{18.12}$$

and if convenient, ζ can be allowed to tend to η on B less an arbitrary l_N null set. The first limit inferior in (18.12) can be written in the form $\liminf_{D\ni\zeta\to\eta}\mu_D(\zeta, G_B(\xi,\cdot)1_B)$, and since ζ is a regular boundary point of D, the limit inferior is actually a limit and is $G_B(\xi,\eta)$. Since $\llbracket G_B(\xi,\cdot)\rrbracket^{B-D} = G_B(\xi,\cdot)$ quasi everywhere on $B-D$, the second limit inferior in (18.12) is $G_B(\xi,\eta)$ unless some neighborhood of η meets $B-D$ in an l_N null set, in which case we can ignore this limit inferior. Hence (18.11) is true. We have now proved that (b) is true if η either is in $B-\partial D$ or is a regular boundary point of D in B; so (b) is true for quasi every point η of B. Since the two sides of (18.2) are equal when $\eta = \xi$ and define subharmonic functions of η on $B-\{\xi\}$, these two sides are equal for all η in B, and the proof of (b) is complete. □

Proof of (c). If $v = G_B\nu$, apply (18.2) to find that

$$G_D^{\equiv}\nu^i = G_D^{\equiv}\nu = v - \mu_D(\cdot, v1_B), \tag{18.13}$$

and then an application of the linear operation GM_D to (18.13) yields (18.3). If D is relatively compact in B and if v is superharmonic on B, it can be supposed in proving (18.3) that v is lower bounded on B, after decreasing B if necessary. In view of the Riesz decomposition it is then sufficient to prove (18.3) separately for v a potential, in which case the proof has just been given, and for v harmonic, in which case (18.3) is trivial. □

Application to the Vanishing of h-Potentials at the Boundaries of Their Domains

Let D be a Greenian subset of $\mathbb{R}^N$, let h be a strictly positive harmonic function on D, let $u = G_D\nu/h$ be an h-superharmonic h-potential on D, and define $D_c = \{u > c\}$, for $c > 0$. Then D_c is an open subset of D, and we now prove that if D_c is not empty, then (Euclidean boundaries)

$$\mu_{D_c}^h(\cdot, \partial D \cap \partial D_c) \equiv 0.$$

Define $v = G_D\nu$. We can assume that $u \le c+1$, that is, $v \le (c+1)h$, because D_c is unaltered if we replace u by $u \wedge (c+1)$. The measure ν vanishes on polar sets because v is finite valued; so (Theorem 18)

$$\mathrm{GM}_{D_c} v = \mu_{D_c}(\cdot, v1_D). \tag{18.14}$$

Now on the one hand $(\mathrm{GM}_{D_c}v)/h = {}^h\mathrm{GM}_{D_c}v$ on D_c and on the other hand (Section 10(a)) the evaluation $\mu^h_{D_c}(\cdot, u1_D) = \mu_{D_c}(\cdot, v1_D)h$ combined with (18.14) yields

$$ {}^h\mathrm{GM}_{D_c}u = \mu^h_{D_c}(\cdot, u1_D). \tag{18.15} $$

The left side of (18.15) is a majorant of the constant function c and $u \le c$ on $D \cap \partial D_c$ by lower semicontinuity of h-potentials, so (18.15) implies that $c \le c\mu^h_D(\cdot, D \cap \partial D_c)$ and we conclude that there is equality here and therefore that $\mu^h_{D_c}(\cdot, \partial D \cap \partial D_c) \equiv 0$, as asserted.

19. Modification of Section 18 for $D = \mathbb{R}^2$

In Section 18 the set B is Greenian and so cannot be chosen to be $\mathbb{R}^N$ unless $N > 2$. In this section $N = 2$, the set D is a Greenian subset of $\mathbb{R}^2$, and the work in Section 18 is adapted to the choice $B = \mathbb{R}^2$. As in Section 18 the topology defining boundaries is the Euclidean topology. We adopt the convention that $G(\xi, \infty) = -\infty$. The counterpart of the Section 18 set $\partial B \cap \partial D$ is the empty set if D is bounded and the singleton $\{\infty\}$ if D is unbounded. Recall from Section 5, Example (a), that when D is unbounded, the singleton $\{\infty\}$ is a μ_D null boundary set. According to the following theorem, an awkward new term appears in the counterpart of (18.2).

Theorem. (a) *For each η in $\mathbb{R}^2$ the function $G(\eta, \cdot)$ is a resolutive boundary function; that is, this function is in $L^1(\mu_D)$.*

(b) *If $\xi \in D$, then*

$$ G_D^{=}(\xi, \eta) = G(\xi, \eta) - \mu_D(\xi, G(\eta, \cdot)) + \phi_D(\xi) \qquad (\eta \in \mathbb{R}^2), \tag{19.1} $$

where ϕ_D is a positive harmonic function on D, defined in (19.10), *and*

- (b1) *ϕ_D has limit 0 at every finite regular boundary point of D.*
- (b2) *ϕ_D is bounded on bounded sets.*
- (b3) *$\phi_D = 0$ if D is bounded.*
- (b4) *if D is unbounded and connected, $\phi_D \equiv 0$ if and only if ∞ is a regular boundary point of D.*

(c) *If $v = Gv$ is a superharmonic potential on $\mathbb{R}^2$ and if v^i is the projection of v on the set of finite irregular boundary points of D, then on D,*

$$ \mathrm{GM}_D v = \mu_D(\cdot, v) + G_D^{=} v^i - v(\mathbb{R}^2)\phi_D. \tag{19.2} $$

If D is bounded, (19.2) *is true with $\phi_D \equiv 0$ whenever v is a superharmonic function on D and v is the associated measure.*

Observation (1). If $\eta \in D$, then (19.1) becomes

$$G_D(\xi, \eta) = G(\xi, \eta) - \mu_D(\xi, G(\eta, \cdot)) + \phi_D(\xi) \qquad [(\xi, \eta) \in D \times D]. \quad (19.3)$$

According to (b3) and (b4), the following two important implications of (19.3) are true if and only if $\phi_D = 0$, that is, if and only if D is bounded or is unbounded with regular boundary point ∞:

$$G_D(\xi, \eta) = G(\xi, \eta) - \mu_D(\xi, G(\eta, \cdot)) \qquad [(\xi, \eta) \in D \times D], \quad (19.4)$$

$$\mu_D(\xi, G(\eta, \cdot)) = \mu_D(\eta, G(\xi, \cdot)) \qquad [(\xi, \eta) \in D \times D]. \quad (19.5)$$

According to Section 18, Observation (1), both these relations are true when $N > 2$ with no restriction on the nonempty open subset D of $\mathbb{R}^N$. Equation (19.4) is a natural approach to finding the Green function G_D, and Theorem 19 exhibits the conditions under which it is valid, that is, under which $\mathrm{GM}_D G(\xi, \cdot) = \mu_D(\xi, G(\eta, \cdot))$ when $N = 2$. If there is no restriction on D, the difference between left and right sides of (19.5) is $\phi_D(\xi) - \phi_D(\eta)$.

Observation (2). If ζ is a finite regular boundary point of D or is an inner point of $\mathbb{R}^2 - D$, equation (19.1) yields

$$G(\xi, \zeta) = \mu_D(\xi, G(\zeta, \cdot)) - \phi_D(\xi). \quad (19.6)$$

Observation (3). According to Theorem 19, the integral

$$\mu_D(\xi, G(\eta, \cdot)) = G\mu_D(\xi, \cdot)(\eta) \qquad [(\xi, \eta) \in D \times \mathbb{R}^2] \quad (19.7)$$

is well defined and finite; so for fixed ξ in D the potential $G\mu_D(\xi, \cdot)$ is a finite-valued superharmonic function on $\mathbb{R}^2$. [There is a slight abuse of language here because when D is unbounded, $\mu_D(\xi, \cdot)$ is not a measure of subsets of $\mathbb{R}^2$, but we have already noted that $\mu_D(\cdot, \infty) = 0$ in the present context.] In view of (19.7) Theorem 19(b) implies that the Riesz measure on $\mathbb{R}^2 - \{\xi\}$ associated with the superharmonic function $-G_D^{=}(\xi, \cdot)$ on that set is $\mu_D(\xi, \cdot)$.

Proof of (a). For $\eta \in \mathbb{R}^2 - D$ and, if D is bounded, for $\eta \in \partial D$. If we define $f_\eta = G(\eta, \cdot)_{|\partial D}$, the function f_η is resolutive under the stated restrictions on η and D because Euclidean boundaries are resolutive and

$$-\infty < \mathrm{GM}_D G(\eta, \cdot) \le \underline{H}_{f_\eta} \le \bar{H}_{f_\eta} \le \begin{array}{ll} \sup f_\eta < +\infty & \text{if } \eta \in \mathbb{R}^2 - \partial D, \\ G(\eta, \cdot)_{|D} < +\infty & \text{if } \eta \in \partial D,\ D \text{ bounded}. \end{array} \quad (19.8)$$

The function $\eta \mapsto \mu_D(\xi, G(\eta, \cdot)) = \mu_D(\xi, f_\eta)$ is the potential for the kernel G of the measure $\mu_D(\xi, \cdot)$ and is superharmonic because it is finite on $\mathbb{R}^2 - \partial D$.

Proof of (19.1) when D is bounded. If D is bounded and if $B = B(0, \delta)$ contains $\bar{D}$, then

$$G_{\bar{D}}^{=}(\xi, \cdot) = G_B(\xi, \cdot) - \mu_D(\xi, G_B(\eta, \cdot)) \tag{19.9}$$

according to Theorem 18. Furthermore according to the formula for G_B in II(1.1),

$$G_B(\xi, \eta) = G(\xi, \eta) + h_\delta(\xi, \eta),$$

where h_δ is a symmetric function on $B \times B$ and $h_\delta(\xi, \cdot)$ has a harmonic extension to a neighborhood of $\bar{B}$. If this evaluation of G_B is substituted in (19.9), we obtain (19.1) with $\phi_D \equiv 0$.

Proof of (a) and (19.1) when D is unbounded. If D is unbounded, let $B'_{\cdot}$ be an increasing sequence of relatively compact open subsets of $\mathbb{R}^2$ with union $\mathbb{R}^2$ and define $B_n = B'_n \cap D$. If we write (19.1) for B_n, let $n \to \infty$, and take into account the fact that $\mu_D(\xi, f_\eta) > -\infty$, we find (19.1) first for quasi every η and then for all η, with

$$\phi_D(\xi) = -\lim_{n \to \infty} \int_{D \cap \partial B_n} G(\eta, \zeta) \mu_{B_n}(\xi, d\zeta); \tag{19.10}$$

for example, (19.10) can be particularized to

$$\phi_D(\xi) = \lim_{n \to \infty} [\mu_{B_n}(\xi, D \cap \partial B_n) \log n] \qquad [B_n = D \cap B(0, n)]. \tag{19.11}$$

Thus (a) and (19.1) are now completely proved. □

Proof of (b1)–(b3). If ζ is a finite regular boundary point of D, fix η in D and observe that since f_η is upper bounded, the PWB solution $\mu_D(\cdot, G(\eta, \cdot))$ has superior limit $\leq G(\eta, \zeta)$ at ζ (Theorem 13). Moreover $G_D(\eta, \cdot)$ has limit 0 at ζ; so (19.1) implies that

$$0 \geq G(\zeta, \eta) - \limsup_{\xi \to \zeta} \mu_D(\xi, G(\eta, \cdot)) + \limsup_{\xi \to \zeta} \phi_D(\xi),$$

and so (b1) is true. The function ϕ_D is bounded on bounded sets, that is, (b2) is true, because in (19.1) for fixed η in D the function $G(\eta, \cdot)$ is bounded above on ∂D and bounded below on bounded sets, and $G_D(\cdot, \eta)$ is bounded outside each neighborhood of η. Finally we have already proved (19.1) with $\phi_B = 0$ when D is bounded; that is, (b3) is true. □

Proof of (b4). Observe first that with no hypotheses on D,

$$\limsup_{D \ni \eta \to \infty} [G(\xi, \eta) - \mu_D(\xi, G(\eta, \cdot))] = \limsup_{D \ni \eta \to \infty} \int_{\partial D} \log \frac{|\zeta - \eta|}{|\xi - \eta|} \mu_D(\xi, d\zeta) \leq 0 \tag{19.12}$$

in view of Fatou's lemma, because the integrand has limit 0 when $\eta \to \infty$ and is at most

$$\log\left(1 + \frac{|\zeta - \xi|}{|\xi - \eta|}\right) \leq \log(1 + |\zeta - \xi|) \quad \text{if } |\xi - \eta| > 1. \tag{19.13}$$

Here the right side of (19.13) defines a function of ζ in $L^1(\mu_D)$, according to Theorem 19(a). If $\phi_D \equiv 0$, inequality (19.12) combined with (19.1) shows that $G_D(\xi, \cdot)$ has limit 0 at ∞; so (Section 14) ∞ is a regular boundary point of D. To prove the converse, apply in (19.1) the evaluation of $\lim_{\delta \to \infty} L(u, 0, \delta)$ with u a potential (kernel G) on $\mathbb{R}^2$ to find

$$\lim_{\delta \to \infty} L(G_{\bar{D}}^{=}(\xi, \cdot), 0, \delta) = \phi_D(\xi). \tag{19.14}$$

In particular, if ∞ is a regular boundary point of D, the function $G_{\bar{D}}^{=}(\xi, \cdot)$ has limit 0 at ∞; so the left side of (19.14) is 0, and $\phi_D \equiv 0$, as was to be proved. □

Proof of (c). See the proof of Theorem 18(c). □

20. Interpretation of ϕ_D as a Green Function with Pole ∞ ($N = 2$)

If D in Section 19 is a deleted neighborhood of the point ∞, this point is an irregular boundary point of D, and the limit equation (19.12) simplifies to

$$\lim_{\eta \to \infty} [G(\xi, \eta) - \mu_D(\xi, G(\eta, \cdot))] = 0. \tag{20.1}$$

Equation (19.1) now yields

$$\lim_{\eta \to \infty} G_D(\xi, \eta) = \phi_D(\xi). \tag{20.2}$$

It is natural to write the limit on the left as $G_D(\xi, \infty)$ and to think of $G_D(\cdot, \infty)$ as the Green function of D with pole ∞. See Section XIII.18 for further remarks on this Green function.

21. Variant of the Operator τ_B

If u is a superharmonic function on an open subset D of $\mathbb{R}^N$ ($N \geq 2$) and if B is an open relatively compact subset of D, we have defined $\tau_B u$ in Section 11 as the smoothed infimum of the class of superharmonic functions on D majorizing u on $D - B$. Under this definition $\tau_B u = \mu_B(\cdot, u)$ on B and $\tau_B u = u$ quasi everywhere on $D - B$. Define $\tau'_B u$ as the superharmonic function on D equal quasi everywhere on $D - B$ to u and equal on B to $\mathrm{GM}_B u$. The construction of $\mathrm{GM}_B u$ in Section III.1 in terms of a decreasing sequence of superharmonic functions shows that $\tau'_B u$ exists. Since $\tau_B u \leq u$ on B, it follows that $\tau_B u \leq \tau'_B u$. According to Theorem 18(c) (in which the roles of B and D are reversed), $\tau_B u = \tau'_B u$ if and only if the set of irregular Euclidean boundary points of B is a null set for the Riesz measure associated with u; so there is equality if ∂B is regular. In the general case both $\tau_B u$ and $\tau'_B u$ decrease when B increases. Moreover, if B_1 and B_2 are open relatively compact subsets of D, with $\overline{B}_1 \subset B_2$, then

$$\tau'_{B_1} u \geq \tau_{B_1} u \geq \tau'_{B_2} u \geq \tau_{B_2} u. \tag{21.1}$$

For many purposes we can use τ'_B and τ_B interchangeably. For example, if B varies through the open relatively compact subsets of D, then (21.1) shows that $\tau_B u$ and $\tau'_B u$ as B varies have a common infimum; the infimum is $\mathrm{GM}_D u$ if this harmonic minorant exists.

Chapter IX

Lattices and Related Classes of Functions

1. Introduction

In this chapter certain function classes that arise naturally in potential theory will be discussed. These classes, the corresponding identically named classes in parabolic potential theory (Section XVIII.19) and in stochastic process theory (Chapter V of Part 2), are discussed together in Chapter I of Part 3.

Throughout this chapter D is a Greenian subset of $\mathbb{R}^N$, $N \geq 2$, and h is a strictly positive harmonic function on D. Both D and h are held fast throughout the chapter.

2. $\mathrm{LM}_D^h u$ for an h-Subharmonic Function u

Suppose that D is connected and that u is an h-subharmonic function on D. According to Section VIII.11 (where the discussion is in the dual context, in which u is h-superharmonic), if A and B are open relatively compact subsets of D with $A \subset B$, then $u \leq \tau_A^h u \leq \tau_B^h u$, and if $B_\bullet$ is an increasing sequence of open relatively compact subsets of D with union D, then $\tau_{B_\bullet}^h u$ is an increasing sequence of h-subharmonic functions, and either $\mathrm{LM}_D^h u$ exists and is given by $\lim_{n\to\infty} \tau_{B_n}^h u$, an h-harmonic function, or this limit is identically $+\infty$. An equivalent formulation is: the set of functions $\tau_B^h u$ for B ranging through the class of open relatively compact subsets of D is directed upward and either $\mathrm{LM}_D^h u$ exists and is given by

$$\mathrm{LM}_D^h u = \sup\{\mu_B^h(\cdot, u) : B \text{ open relatively compact subset of } D\}, \tag{2.1}$$

or else the supremum in (2.1) is identically $+\infty$. [We adopt the convention here and in similar contexts below that the domain of $\mu_B^h(\cdot, u)$ is B and that the supremum of a set of functions at a point is the supremum at the point of the values of those functions defined there.] Observe that the supremum (directed limit) in question is unchanged if the sets B are all supposed to contain a specified compact subset of D; that is, the analysis relates to the properties of u near (any choice of) the boundary of D.

If u is positive, the existence of $\mathrm{LM}_D^h u$ is equivalent to the L^1 boundedness of the class

$$\{[u_{|\partial B}, \mu_B^h(\xi, \cdot)] : B \text{ open relatively compact subset of } D\} \tag{2.2}$$

of functions coupled with the indicated measures, for some, equivalently each, point ξ of D. Alternatively the set B can range through a nested sequence $B_\bullet$ with union D, as above. In particular, suppose that u is positive and that the class (2.2) is uniformly integrable for some point ξ in D, that is, that there is a uniform integrability test function Φ for which

$$\sup\{\mu_B^h(\xi, \Phi(u)) : B \text{ open relatively compact subset of } D\} < +\infty. \tag{2.3}$$

Then the class (2.2) is L^1 bounded; so $\mathrm{LM}_D^h u$ exists, and since $\Phi(u)$ is a positive h-subharmonic function, the supremum in (2.3) is $\mathrm{LM}_D^h\Phi(u)(\xi)$. Thus in this case $\mathrm{LM}_D^h\Phi(u)$ exists, and the supremum in (2.3) is finite for all ξ; that is, the class (2.2) is uniformly integrable for each point ξ in D. Furthermore the h-harmonic function $\mathrm{LM}_D^h u$ satisfies the same uniform integrability condition as u, with the same test function Φ; (2.3) is true with u replaced by $\mathrm{LM}_D^h u$. In fact we now show, under the hypothesis that (2.3) is true as written, that

$$\begin{aligned} \sup\{\mu_B^h(\cdot, \Phi(\mathrm{LM}_D^h u)) &: B \text{ open relatively compact subset of } D\} \\ &= \mathrm{LM}_D^h(\Phi[\mathrm{LM}_D^h u]) = \mathrm{LM}_D^h\Phi(u). \end{aligned} \tag{2.4}$$

To see this, let B and B' be open relatively compact subsets of D. Then as B' varies,

$$\begin{aligned} \mu_B^h(\xi, \Phi[\mathrm{LM}_D^h u]) &= \mu_B^h\left(\xi, \Phi\left[\sup_{B'} \mu_{B'}^h(\cdot, u)\right]\right) \le \mu_B^h\left(\xi, \sup_{B'} \mu_{B'}^h(\cdot, \Phi(u))\right) \\ &= \mu_B^h(\xi, \mathrm{LM}_D^h\Phi(u)) = \mathrm{LM}_D^h\Phi(u)(\xi) < +\infty. \end{aligned} \tag{2.5}$$

Take the supremum as B varies to find that the first two terms in (2.4) are majorized by the third. The reverse inequality is trivial.

3. The Class $\mathbf{D}(\mu_{D-}^h)$

(See the corresponding stochastic process class in Section 2.II.11. This class is the linear class of real-valued Borel measurable functions u on D for which if ξ is in D and if $B_\bullet$ is an increasing sequence of open relatively compact subsets of D with union D, then the sequence

$$\{[u_{|\partial B_n}, \mu_{B_n}^h(\xi, \cdot)], n \in \mathbb{Z}^+\} \tag{3.1}$$

of coupled functions and measures is uniformly integrable. It follows from the definition of uniform integrability that a Borel measurable function u on D is in $\mathbf{D}(\mu^h_{D-})$ if and only if, for each point ζ in B, $\mu^h_B(\xi, |u|) < +\infty$ whenever B is an open relatively compact subset of D and there is a compact subset $A = A_\xi$ of D for which the family (2.2) of coupled functions and measures, under the added restriction that $A \subset B$, is uniformly integrable. This condition is satisfied if and only if there is a uniform integrability test function $\Phi = \Phi_\xi$ such that

$$\sup\{\mu^h_B(\xi, \Phi(|u|)): A \subset B,\ B \text{ open relatively compact subset of } D\} < +\infty. \tag{3.2}$$

The notation $D-$ is to remind the reader that not h-harmonic measure on some boundary of D but h-harmonic measures on the Euclidean boundaries of open relatively compact subsets of D are involved. When D and h are specified and there is no danger of confusion, we shall sometimes write $\mathbf{D}$ instead of $\mathbf{D}(\mu^h_{D-})$. Since $\mu^h_B(\cdot, f) = \mu_B(\cdot, hf)/h$ a function u is in $\mathbf{D}(\mu^h_{D-})$ if and only if uh is in $\mathbf{D}(\mu_{D-})$. (See Section II.14 for the $\mathbf{D}$ class of harmonic functions on a ball.)

When the function u on D is h-subharmonic the situation is simpler because then the map $B \mapsto \mu^h_B(\cdot, u)$ is monotone increasing. In particular, if u is positive and h-subharmonic, the supremum in (3.2) is independent of the choice of A, and (Section 2) if (3.2) is true, the left side of (3.2) is $\mathrm{LM}^h_D\Phi(u)(\xi)$. It is thus natural that the case of most interest in potential theory is that in which $|u|$ is h-subharmonic, in particular, if u is h-harmonic.

Theorem. *If u is a Borel measurable real-valued function on the connected Greenian set D and if $|u|$ is an h-subharmonic function on D, the following conditions are equivalent:*

(a) $u \in \mathbf{D}(\mu^h_{D-})$.
(b) *The family* (2.2) *of coupled functions and measures is uniformly integrable for every point ξ in D.*
(c) *Condition* (b) *is satisfied for a single point ξ.*
(d) *There is a uniform integrability test function Φ such that the h-subharmonic function $\Phi(|u|)$ has an h-harmonic majorant.*
(e) *(If u is h-harmonic) $u = u_1 - u_2$, where u_i is a positive h-harmonic function in $\mathbf{D}(\mu^h_{D-})$.*

Moreover, if u satisfies (a)–(c) *and if Φ satisfies* (d), *then* $\mathrm{LM}^h_D|u| \in \mathbf{D}(\mu^h_{D-})$,

$$\mathrm{LM}^h_D\Phi(\mathrm{LM}^h_D|u|) = \mathrm{LM}^h_D\Phi(|u|), \tag{3.3}$$

and each function u_i in (e) *can be chosen so that the h-subharmonic function $\Phi(u_i)$ has an h-harmonic majorant.*

If $|u|$ here is identified with u in Section 2, this theorem follows at once from the discussion in Section 2, except for the assertion involving (e). To prove that an h-harmonic function u in $\mathbf{D}(\mu_{D-}^h)$ has the representation (e), observe that since $|u|$ is h-subharmonic and $\mathbf{LM}_D^h|u| \in \mathbf{D}(\mu_{D-}^h)$, it follows that

$$u = \mathbf{LM}_D^h|u| - (\mathbf{LM}_D^h|u| - u)$$

is the desired representation; with this choice of u_i, equality (3.3 shows that $\Phi(u_i)$ has $\mathrm{LM}^h\Phi(|u|)$ as an h-harmonic majorant. Conversely, if $u = u_1 - u_2$ with u_i positive h-harmonic and in $\mathbf{D}(\mu_{D-}^h)$, then $|u| \le u_1 + u_2$; so $u \in \mathbf{D}(\mu_{D-}^h)$.

It will be shown in Section 3.I.9 that an h-harmonic function on a Greenian subset D of $\mathbb{R}^N$ is in $\mathbf{D}(\mu_{D-}^h)$ if and only if the function is quasi bounded (a property defined in Section 9). More detailed results on the class of harmonic functions in $\mathbf{D}(\mu_{D-})$ for D a ball were obtained in Section II.14, and these results will be extended to h-harmonic functions on a ball in Section 12, to h-harmonic functions on a Greenian set in Section XII.9.

4. The Class $\mathbf{L}^p(\mu_{D-}^h)$ $(p \ge 1)$

This class is the linear class of extended real-valued Borel measurable functions u on D for which if ξ is in D and if $B_{\bullet}$ is an increasing sequence of open relatively compact subsets of D with union D, then $\sup_{n\ge 0} \mu_{B_n}^h(\xi, |u|^p) < +\infty$. It follows for $p > 1$ that $\mathbf{L}^p(\mu_{D-}^h) \subset \mathbf{D}(\mu_{D-}^h)$, because the function $s \mapsto s^p$ is a uniform integrability test function. A Borel measurable function u on D is in $\mathbf{L}^p(\mu_{D-}^h)$ if and only if, for each point ξ in D and each open relatively compact subset B of D with ξ in B, $\mu_B^h(\xi, |u|^p) < +\infty$ and there is a compact subset $A = A_\xi$ of D for which

$$\sup\{\mu_B^h(\xi, |u|^p)\colon A \subset B,\ B \text{ an open relatively compact subset of } D\} < +\infty \tag{4.1}$$

When D, h, and p are specified and there is no danger of confusion, we shall sometimes write $\mathbf{L}^p$ instead of $\mathbf{L}^p(\mu_{D-}^h)$. A function u is in $\mathbf{L}^p(\mu_{D-}^h)$ if and only if uh is in $\mathbf{L}^p(\mu_{D-})$. As in Section 3, the case of most interest in potential theory is that in which $|u|$ is h-subharmonic, in particular, if u is h-harmonic. The following theorem is in part a specialization of Theorem 3.

Theorem. *If u is a Borel measurable real-valued function on the connected Greenian set D and if $|u|$ is h-subharmonic, the following conditions are equivalent:*

(a) $u \in \mathbf{L}^p(\mu_{D-}^h)$.
(b) *The family* (2.2) *of functions coupled with the indicated measures is L^p bounded for every point ξ in D.*
(c) *Condition* (b) *is satisfied for a single point ξ.*

(d) *The function* $|u|^p$ *has an h-harmonic majorant.*
(e) *(If u is h-harmonic)* $u = u_1 - u_2$, *with* u_i *positive, h-harmonic, and in* $\mathbf{L}^p(\mu_{D-}^h)$.

If u satisfies (a)–(d), *then* $\mathrm{LM}_D^h|u| \in \mathbf{L}^p(\mu_{D-}^h)$, *and in fact*

$$\mathrm{LM}_D^h(\mathrm{LM}_D^h|u|)^p = \mathrm{LM}_D^h|u|^p. \tag{4.2}$$

The proof of this theorem is left to the reader because the theorem follows easily from the discussion in Section 2. Observe that part (e) of the present theorem is slightly stronger than Theorem 3(e). In fact, in Theorem 3(e) it is not asserted that if $u = u_1 - u_2$ and if $\Phi(u_1)$ and $\Phi(u_2)$ have h-harmonic majorants, then $\Phi(|u|)$ has an h harmonic majorant, although the counterpart of this assertion is contained in Theorem 4(e) with $\Phi(s) = s^p$. However, for $\Phi(s) = s^p$ and $p \geq 1$ this assertion is true because then $(u_1 + u_2)^p \leq 2^{p-1}(u_1^p + u_2^p)$.

In Theorem 4, as in Theorem 3, the condition (b) involves a set B in (2.2) increasing to D, but as in Theorem 3, it is sufficient if the set B runs through a nested sequence with union D as described in Section 2. If D is a ball, it is natural to choose B to increase through balls concentric with B so that if $D = B(0, \delta)$, the subharmonic function $|u|$ on D is in $\mathbf{L}^p(\mu_{D-}^h)$ if and only if $\sup_r L(|u|^p, 0, r) = \lim_{r \uparrow \delta} L(|u|^p, 0, r) < +\infty$. The class of harmonic functions on $B(0, \delta)$ in the class $\mathbf{L}^1(\mu_{B(0,\delta)-})$ was discussed in Section II.14.

Example. Let u be a harmonic function on the Greenian set D. Then the function u^2 is subharmonic with associated Riesz measure $d\lambda = (\Delta u^2/\pi_N')\, dl_N$, so that $\lambda(D)$ is a multiple of the Dirichlet integral of u,

$$\lambda(D) = 2 \int_D |\mathrm{grad}\, u|^2\, dl_N/\pi_N' \leq +\infty.$$

We now prove that $u \in \mathbf{L}^2(\mu_{D-})$ if and only if $G_D\lambda$ is superharmonic. In the one direction if $u \in \mathbf{L}^2(\mu_{D-})$, that is, if $\mathrm{LM}_D u^2 = v$ exists, then $\mathrm{GM}_D(v - u^2) = 0$; so $v - u^2$ is a potential and necessarily $v - u^2 = G_D\lambda$ by direct calculation of Δu^2. Conversely, if $G_D\lambda$ is superharmonic, the function $u^2 + G_D\lambda$ is a majorant of u^2 and is harmonic because $\Delta(u^2 + G_D\lambda) = 0$; so $u \in \mathbf{L}^2(\mu_{D-})$. In particular, if the Dirichlet integral of u is finite, that is, if $\lambda(D) < +\infty$, then $G_D\lambda$ is superharmonic; so $u \in \mathbf{L}^2(\mu_{D-})$.

5. The Lattices ($\mathbf{S}^{\pm}, \leq$) and ($\mathbf{S}^{+}, \leq$)

(See the corresponding martingale theory lattices in Section 2.V.5.)

The Lattice ($\mathbf{S}^{\pm}, \leq$). Denote in this way the class of h-superharmonic functions on D with positive h-superharmonic majorants, ordered by pointwise inequality in the notation $\leq$, $\geq$, $\wedge$, $\vee$. If $\Gamma \subset \mathbf{S}^\pm$ and if Γ has an h-superharmonic minorant, then according to the Fundamental Convergence Theorem (Section VI.1) as trivially adapted to h-superharmonic functions, the lower semicontinuous smoothing of the pointwise infimum of Γ is h-superharmonic, and this function is obviously the $(\mathbf{S}^\pm, \leq)$ infimum $\wedge\,\Gamma$. If $\Gamma \subset \mathbf{S}^\pm$ and if Γ has an h-superharmonic majorant, then $\vee\,\Gamma$ exists and is the $(\mathbf{S}^\pm, \leq)$ infimum of the class of h-superharmonic majorants of Γ. Thus $(\mathbf{S}^\pm, \leq)$ is a conditionally complete lattice. The pointwise order will sometimes be called the *essential order* to match the essential order of stochastic process theory defined in Section 2.I.8.

Let Γ be a subset of $\mathbf{S}^\pm$. According to the Fundamental Convergence Theorem, if Γ has an h-superharmonic minorant, some countable subset of Γ has the same $(\mathbf{S}^\pm, \leq)$ infimum as Γ. If Γ has an h-superharmonic majorant, some countable subset of Γ has the same $(\mathbf{S}^\pm, \leq)$ supremum as Γ. In fact, if Γ is directed upward, $\vee\,\Gamma$ is the pointwise supremum, and the assertion for pointwise suprema was proved in Section II.4; if Γ is not directed upward, apply the result in the directed case to the set of $(\mathbf{S}^\pm, \leq)$ suprema of finite subsets of Γ.

The Lattice $(\mathbf{S}^+, \leq)$. The sublattice $(\mathbf{S}^+, \leq)$ of $(\mathbf{S}^\pm, \leq)$ is the class of positive h-superharmonic functions on D in the pointwise order.

6. The Vector Lattice $(\mathbf{S}, \preceq)$

(See the corresponding martingale theory lattice in Section 2.V.6.) The set $\mathbf{S}^+$ has the unique subtraction property, that is, for u_1, u_2, u_2' in $\mathbf{S}^+$ the equality $u_1 + u_2 = u_1 + u_2'$ implies that $u_2 = u_2'$. This property, trivial for finite-valued functions, is true in the general case because $u_2 = u_2'$ on the set of finiteness of u_1; so $u_2 = u_2'$ quasi everywhere and therefore everywhere. Thus $\mathbf{S}^+$ is a cone as defined in Appendix III.3 and therefore determines a specific order on itself. The specific order symbols will be $\preceq$, $\succeq$, $\curlyvee$, $\curlywedge$, and $\mathbf{S}^+$ in the specific order will be denoted by $(\mathbf{S}^+, \preceq)$. Define $\mathbf{S} = \mathbf{S}^+ - \mathbf{S}^+$ so that each member of $\mathbf{S}$ can be identified with a function $u_1 - u_2$ with u_i in $\mathbf{S}^+$; this difference is well defined off the polar set of common infinities of u_1 and u_2. Order $\mathbf{S}$ by the specific order with positive cone $\mathbf{S}^+$ (Appendix III.4) to obtain a partially ordered vector space $(\mathbf{S}, \preceq)$.

Theorem. (a) *The space* $(\mathbf{S}, \preceq)$ *is a conditionally complete vector lattice.*
In (b)–(d) *let* Γ *be a subset of* $\mathbf{S}$ *with a specific order majorant.*

(b) $\curlyvee\Gamma$ *is the specific order supremum of a countable subset of* Γ.

(c) *If* Γ' *is the class of specific order majorants of* Γ, *then* $\vee\,\Gamma \preceq \Gamma'$.

(d) *If* Γ *is directed upward in the specific order, then* $\curlyvee\Gamma = \vee\,\Gamma$.

The duals of (b)–(d), involving specific order infima, are obtained by replacing Γ by $-\Gamma$. Since Γ' is directed downward (specific order) in (c), the dual of (d) implies that $\wedge\,\Gamma' = \curlywedge\Gamma' = \curlyvee\Gamma$.

It will be convenient in the following proof to use a generalized reduction: the function v in R_v^A will not be required to be superharmonic on the specified Greenian set D but may be an arbitrary extended positive-valued function; no other change is made in the reduction definition. In the applications to be made, $A = D$ and it will be obvious that v has a positive superharmonic majorant on D so that $\underset{+v}{R^D}$ will exist and be superharmonic on D. In the following proof Γ and Γ' are as defined in the statement of the theorem.

Proof that $\wedge\Gamma' \in \Gamma'$ if $\Gamma \subset \mathbf{S}^+$. If u is in Γ and u' is in Γ', there is a function v in $\mathbf{S}^+$ for which $u + v = u'$. Let u' run through a sequence in Γ' with essential order infimum $\wedge\Gamma'$ to find that $u \preceq \wedge\Gamma'$.

Proof that $\wedge\Gamma' = \curlywedge\Gamma' = \curlyvee\Gamma$ if $\Gamma \subset \mathbf{S}^+$. Let u' be in Γ', and define

$$\phi(\xi) = \begin{cases} +\infty & \text{if } u'(\xi) = +\infty, \\ (u' - \wedge\Gamma')(\xi) & \text{if } u'(\xi) < +\infty. \end{cases}$$

Then $u' \leq \wedge\Gamma' + \underset{+\phi}{R^D}$ quasi everywhere on D and therefore everywhere on D. According to the natural order decomposition theorem, there are functions v and v_1 in $\mathbf{S}^+$ such that $v \leq \wedge\Gamma'$, $v_1 \leq \underset{+\phi}{R^D}$, and $u' = v + v_1$. But then $v_1 \geq \phi$ quasi everywhere on D; so $v_1 \geq \underset{+\phi}{R^D}$, and there must be equality; so

$$u' = v + \underset{+\phi}{R^D}. \tag{6.1}$$

Now let u be a member of Γ. Since u' and $\wedge\Gamma'$ are in Γ', there are members v_2 and v_3 of $\mathbf{S}^+$ such that

$$u' = u + v_2, \qquad \wedge\Gamma' = u + v_3. \tag{6.2}$$

Then $v_2 \leq v_3 + \underset{+\phi}{R^D}$; so repeating an argument just used, there is a member w of $\mathbf{S}^+$ such that

$$v_2 = w + \underset{+\phi}{R^D}. \tag{6.3}$$

Then (6.2) implies that $u' = u + w + \underset{+\phi}{R^D}$; so $v = u + w$ from (6.1). If we fix u', the function v is also fixed; so $u \preceq v$ for all u, and it follows that $v \in \Gamma'$. Hence $v \geq \wedge\Gamma'$, and there must be equality because by its definition, v satisfies the reverse inequality. Thus $u' = \wedge\Gamma' + \underset{+\phi}{R^D}$ from (6.1), and since u' is an arbitrary member of Γ', it follows that $\wedge\Gamma' \preceq \Gamma'$. Since we have already proved that $\wedge\Gamma' \in \Gamma'$, we now deduce that $\wedge\Gamma' = \curlywedge\Gamma'$; so $\wedge\Gamma' = \curlyvee\Gamma$ by definition of Γ'.

Proof of (a). The fact that $\curlyvee\Gamma = \curlywedge\Gamma'$ exists shows that $(\mathbf{S}^+, \preceq)$ is a conditionally complete lattice and therefore that $(\mathbf{S}, \preceq)$ is a conditionally complete vector lattice. □

Proof of (b) *and* (d). In proving (b) and (d) we can assume that Γ is directed upward in the specific order, at the possible expense of replacing Γ by the set of specific order suprema of finite subsets of Γ. We can then also assume that $\Gamma \subset \mathbf{S}^+$, at the possible expense of choosing some member u_0 of Γ and then replacing Γ by $\{u - u_0 : u \in \Gamma, u \succeq u_0\}$. Under these hypotheses Γ is also directed upward and bounded in the essential order so that $\vee\Gamma = u'$ exists (Section 5) and is the pointwise supremum of Γ and in fact is the pointwise supremum of a sequence $u_.$ in Γ. Choose any member u of Γ and choose $v_0, v_1, \ldots$ successively in Γ to satisfy $v_0 = u$, $v_{n+1} \succeq (v_n \curlyvee u_n)$ for $u \geq 0$. The sequence $v_.$ is a specific order increasing sequence with pointwise limit u'. There is a member w_{n1} of $\mathbf{S}^+$ such that $u + w_{n1} = v_n$, and if $v' \in \Gamma'$, there is a member w_{n2} of $\mathbf{S}^+$ such that $v_{n1} + w_{n2} = v'$. Hence

$$u + \lim_{n\to\infty} w_{n1} = u', \qquad u' + \left(\lim_{n\to\infty} w_{n2}\right)_+ = v';$$

so $\Gamma \preceq u' \preceq \Gamma'$, and we conclude that $u' = \curlyvee\Gamma = \curlyvee_0^\infty u_n$. Thus Theorem 6(b) and (d) are true. □

Proof of (c). Let u' be in Γ' and define

$$\psi(\xi) = \begin{cases} +\infty & \text{if } u'(\xi) = +\infty, \\ (u' - \vee\Gamma)(\xi) & \text{if } u'(\xi) < +\infty. \end{cases}$$

Then $u' = \vee\Gamma + \psi \preceq \vee\Gamma + R^D_{+\psi}$ quasi everywhere on D, and therefore $u' \leq \vee\Gamma + R^D_{+\psi}$ everywhere on D; so repeating the reasoning already used twice above, we find that there is a function w in $\mathbf{S}^+$ such that

$$w \leq \vee\Gamma, \qquad u' = w + R^D_{+\psi}. \tag{6.4}$$

Furthermore by definition of Γ' if u is in Γ, there is a function w_1 in $\mathbf{S}^+$ such that $u' = u + w_1$. Then $w_1 \leq \psi$ quasi everywhere on D; so $w_1 \leq R^D_{+\psi}$ and

$$u + R^D_{+\psi} \leq u + w_1 = u' = w + R^D_{+\psi}.$$

It follows that $u \leq w$ for all u in Γ; so $\vee\Gamma \leq w$, and by (6.4) there must be equality. The second equality in (6.4) thus becomes $u' = \vee\Gamma + R^D_{+\psi}$; so $\vee\Gamma \preceq u'$, as was to be proved. □

7. The Vector Lattice $\mathbf{S}_m$

(See the corresponding martingale theory vector lattice in Section 2.V.7.) An h-superharmonic function specific order majorized by an h-harmonic function is itself h-harmonic and if Γ is a set of positive h-harmonic functions with $\curlyvee\Gamma = u$, then u is a specific order majorant of each member of Γ; so u is positive and h-superharmonic, and $\mathrm{GM}_D^h u$ is a specific order majorant of Γ. Hence $u = \mathrm{GM}_D^h u$ and is h-harmonic. It follows that if $\mathbf{S}_m^+$ is the cone of positive h-harmonic functions on D and if $\mathbf{S}_m = \mathbf{S}_m^+ - \mathbf{S}_m^+$, then $\mathbf{S}_m$ is a band in $\mathbf{S}$. The essential and specific orders coincide on $\mathbf{S}_m$. If $\Gamma \subset \mathbf{S}_m$,

$$\curlyvee\Gamma = \mathrm{LM}_D^h\Gamma, \qquad \curlywedge\Gamma = \mathrm{GM}_D^h\Gamma$$

in the sense that if one side of an equation exists, the other side exists and there is equality.

If D is a ball, it was shown in Section II.14 by means of the Riesz–Herglotz representation theorem that $\mathbf{S}_m$ is lattice isomorphic to the conditionally complete vector lattice of finite signed measures on ∂D. For Greenian D a corresponding result will be proved in Section XII.9 by means of the Martin boundary and the Martin representation theorem.

According to Section 4, an h-harmonic function is in $\mathbf{S}_m$ if and only if the function is in $\mathbf{L}^1(\mu_{D-}^h)$.

8. The Vector Lattice $\mathbf{S}_p$

(See the corresponding martingale theory vector lattice in Section 2.V.8.) An element of $\mathbf{S}^+$ specific order majorized by an h-superharmonic h-potential $G_D\mu/h$ of a measure is itself such a potential. If Γ is a set of such potentials with $\curlyvee\Gamma = u$, then u is a specific order majorant of each element of Γ; so u is a positive h-superharmonic function, and linearity of the GM_D^h operation implies that the h-potential u-$\mathrm{GM}_D^h u$ is also a specific order majorant of Γ and therefore must be u; that is, u is an h-potential. The class $\mathbf{S}_p^+$ is therefore a cone satisfying the conditions (Appendix III.8) implying that the set $\mathbf{S}_p = \mathbf{S}_p^+ - \mathbf{S}_p^+$ is a band in $\mathbf{S}$.

$\mathbf{S}_p$ *and the Corresponding Lattice of Charges.* Let $\mathbf{M}^+$ be the set of measures on D and let $\mathbf{M} = \mathbf{M}^+ - \mathbf{M}^+$, the set of charges on D. The set $\mathbf{M}$ ordered by the positive cone $\mathbf{M}^+$ is a conditionally complete vector lattice. Let $\mathbf{M}_p^+$ be the set of measures in $\mathbf{M}^+$ whose h-potentials are h-superharmonic and define $\mathbf{M}_p = \mathbf{M}_p^+ - \mathbf{M}_p^+$. Then $\mathbf{M}_p$ is a conditionally complete vector lattice, a sublattice of $\mathbf{M}$. The map $\mu \mapsto G_D\mu/h$ is a one-to-one order-preserving map from $\mathbf{M}_p^+$ onto $\mathbf{S}_p^+$ inducing a one-to-one linear order-preserving map from $\mathbf{M}_p$ onto $\mathbf{S}_p$.

$\mathbf{S}_p = \mathbf{S}_m^\perp$. In fact, on the one hand, if $u \in \mathbf{S}_p^+$ and $v \in \mathbf{S}_m^+$, then $u \curlywedge v = 0$ because this lattice infimum is h-harmonic as a specific order minorant of the h-harmonic function v and is a minorant in both specific and pointwise orders of the h-potential u. Hence $\mathbf{S}_p^+ \subset \mathbf{S}_m^\perp$; so $\mathbf{S}_p \subset \mathbf{S}_m^\perp$. Conversely, if $u \in (\mathbf{S}_m^\perp)^+$, then the two relations $\mathrm{GM}_D^h u \in \mathbf{S}_m^+$ and $\mathrm{GM}_D^h u \preceq u$ imply that

$$0 = u \curlywedge \mathrm{GM}_D^h u = \mathrm{GM}_D^h u;$$

that is, $u \in \mathbf{S}_p^+$. It follows that $\mathbf{S}_m^\perp \subset \mathbf{S}_p$, and so there is equality, as was to be proved.

9. The Vector Lattice $\mathbf{S}_{qb}$

(See the corresponding martingale theory vector lattice in Section 2.V.9.) The class $\mathbf{S}_{qb}^+$ is the class of elements u in $\mathbf{S}^+$ which satisfy the following equivalent conditions:

(a) The function u is the specific order supremum of a set of bounded elements of $\mathbf{S}^+$.
(b) The function u is the limit of a specific order increasing sequence of bounded elements of $\mathbf{S}^+$; that is, u is the sum of a series of bounded elements of $\mathbf{S}^+$.

Observe that if "specific order" in (a) were replaced by "essential order," the resulting class of elements u would be $\mathbf{S}^+$ itself. The class $\mathbf{S}_{qb}^+$ is a cone satisfying the conditions (Appendix III.8) implying that the set $\mathbf{S}_{qb} = \mathbf{S}_{qb}^+ - \mathbf{S}_{qb}^+$ is a band in $\mathbf{S}$; the members of this band are called *quasi bounded.*

The bands $\mathbf{S}_{mqb} = \mathbf{S}_m \cap \mathbf{S}_{qb}$ *and* $\mathbf{S}_{pqb} = \mathbf{S}_p \cap \mathbf{S}_{qb}$. In view of Section 8 these two bands are orthogonal and $\mathbf{S}_{qb} = \mathbf{S}_{mqb} + \mathbf{S}_{pqb}$. The band $\mathbf{S}_{mqb}$ is the band in $\mathbf{S}_m$ generated by the constant function 1. It will be shown in Section 3.I.9 that an h-superharmonic potential $G_D\mu/h$ is quasi bounded if and only if μ vanishes on polar sets, equivalently, if and only if $u \in \mathbf{D}(\mu_{D-}^h)$.

EXAMPLE (Quasi Bounded h-harmonic Functions and Dirichlet Solutions). If D is provided with a boundary ∂D by a metric compactification, the PWB^h solutions for this boundary are quasi bounded. To see this, observe that if u is a positive PWB^h solution, say $u = H_f^h$, then the relation

$$u = \lim_{n\to\infty} H_{f\wedge n}^h = \lim_{n\to\infty} \mu_D^h(\cdot, f \wedge n)$$

exhibits u as the limit of a specific order increasing sequence of positive bounded h-harmonic functions. Conversely, if the boundary is internally h-resolutive, that is (Section VIII.2), if every bounded h-harmonic function u is a PWB^h solution, $u = H_f^h = \mu_D^h(\cdot, f)$, then every quasi-bounded h-harmonic function on D is a PWB^h solution. It will be shown in Section 3.I.5 that $\mathbf{S}_{mqb} = \mathbf{S}_m \cap \mathbf{D}(\mu_{D-}^h)$. This class includes the PWB^h solutions whatever the choice of ∂D, and the inclusion becomes equality when the boundary is internally h-resolutive. We have already proved in Section VIII.9 that if D is a ball, its Euclidean boundary is universally resolutive and is internally resolutive, and we remarked there that the argument will be generalized in

Section 12 of the present chapter to show that this boundary is universally internally resolutive.

Roughly, a boundary is internally h-resolutive if it has enough points and is h-resolutive if it does not have too many. For example ($h \equiv 1$), if D is a disk less a radius, the Euclidean boundary is resolutive (Theorem VIII.4) but is not internally resolutive because a bounded harmonic function on D with a limit at every boundary point except that there are different limits at the two sides of the deleted radius is not a PWB solution. Adding boundary points by ramification to separate the two sides of the deleted radius makes the boundary internally resolutive. On the other hand, if h is the minimal harmonic function corresponding to a boundary point ζ of the disk and if the disk boundary is ramified so that ζ explodes into a set containing more than one point, the new boundary of the disk is h-internally resolutive because quasibounded h-harmonic functions are identically constant, but the new disk boundary may not be h-resolutive.

10. The Vector Lattice $\mathbf{S}_s$

(See the corresponding martingale theory vector lattice in Section 2.V.10.) Define $\mathbf{S}_s = \mathbf{S}_{qb}^{\perp}$, the class of *singular* elements of $\mathbf{S}$, and define $\mathbf{S}_{ms} = \mathbf{S}_m \cap \mathbf{S}_s$, $\mathbf{S}_{ps} = \mathbf{S}_p \cap \mathbf{S}_s$, $\mathbf{S}_s^+ = \mathbf{S}_s \cap \mathbf{S}^+$, $\mathbf{S}_{ms}^+ = \mathbf{S}_{ms} \cap \mathbf{S}^+$, $\mathbf{S}_{ps}^+ = \mathbf{S}_{ps} \cap \mathbf{S}^+$. Then $\mathbf{S}_s$, $\mathbf{S}_{ms}$, $\mathbf{S}_{ps}$ are bands, $\mathbf{S}_{ms} \perp \mathbf{S}_{ps}$, and $\mathbf{S}_s = \mathbf{S}_{ms} + \mathbf{S}_{ps}$. A function u in $\mathbf{S}^+$ is singular if and only if 0 is the only bounded function in $\mathbf{S}^+$ which is a specific order minorant of u.

$\mathbf{S}_{ms}$. It will be shown in Section XII.9 that an h-harmonic function $u = v/h$ is singular if and only if in the Martin representation (Section XII.9) of v and h, which generalizes the Riesz–Herglotz representation (Section II.14), the representing measure M_v is singular relative to the representing measure M_h. We prove now that for a function u in $\mathbf{S}_m^+$ to be singular, it is sufficient that $u \wedge c \in \mathbf{S}_p^+$ for some strictly positive constant c and necessary that $u \wedge c \in \mathbf{S}_p^+$ for every strictly positive constant c. If $u \in \mathbf{S}_{ms}^+$, then $\mathbf{GM}_D^h(u \wedge c) = u \curlywedge c = 0$ for every positive c because $u \curlywedge c$ is a bounded positive specific order minorant of u. Hence u is an h-potential. Conversely, if $u \in \mathbf{S}_m^+$, if $u \wedge c$ is an h-potential for some strictly positive c, and if v is a bounded member of $\mathbf{S}^+$ with $\sup_D v = a$, then

$$\left(\frac{a}{c}u\right) \curlywedge v = \mathbf{GM}_D^h\left[\left(\frac{a}{c}u\right) \wedge v\right] \le \frac{a}{c}\mathbf{GM}_D^h(u \wedge c) = 0;$$

so $(a/c)u \perp v$, and therefore $u \perp v$. Hence $u \in \mathbf{S}_{ms}^+$.

$\mathbf{S}_{ps}$. It will be shown in Section 3.I.10 that an h-superharmonic h-potential $G_D\mu/h$ is singular if and only if μ is supported by a polar set.

11. A Refinement of the Riesz Decomposition

The Riesz decomposition of a positive superharmonic function can be expressed in vector lattice language in the form $\mathbf{S}_m \perp \mathbf{S}_p$, $\mathbf{S} = \mathbf{S}_m + \mathbf{S}_p$. We have now found a refinement of this decomposition: the four bands $\mathbf{S}_{mqb}$, $\mathbf{S}_{ms}$, $\mathbf{S}_{pqb}$, $\mathbf{S}_{ps}$ are orthogonal and

$$\mathbf{S}_m = \mathbf{S}_{mqb} + \mathbf{S}_{mp}, \qquad \mathbf{S}_p = \mathbf{S}_{pqb} + \mathbf{S}_{ps}, \qquad \mathbf{S} = \mathbf{S}_{mqb} + \mathbf{S}_{ms} + \mathbf{S}_{pqb} + \mathbf{S}_{ps}.$$

Thus if $u \in \mathbf{S}$ and if $u_{mqb}, \ldots$ are the respective projections of u on the bands $\mathbf{S}_{mqb}, \ldots$, we have derived a unique decomposition

$$u = u_{mqb} + u_{ms} + u_{pqb} + u_{ps},$$

in which all functions on the right are positive in the specific order if u is a positive h-superharmonic function.

12. Lattices of h-Harmonic Functions on a Ball

Let B be a ball and let h be a strictly positive harmonic function on B. If $u = v/h$ is an h-harmonic function on B, then $u \in \mathbf{S}_m = \mathbf{S}_m \cap \mathbf{L}^1(\mu_{B-}^h)$ if and only if v is the difference between two positive harmonic functions, that is, if and only if there is a Riesz–Herglotz measure M_v on ∂B (Euclidean boundary) such that $v = \mathrm{PI}(B, M_v)$. It follows from Theorem II.14 that the map $u \mapsto \mathrm{PI}(B, M_v)/h$ is a linear one-to-one order-preserving map from $\mathbf{S}_m$ onto the vector lattice of signed measures on ∂B. The classes $\mathbf{S}_{mqb}$, $\mathbf{S}_{ms}$, $\mathbf{D}(\mu_{B-}^h)$, $\mathbf{L}^p(\mu_{B-}^h)$ of h-harmonic functions will now be described in terms of the corresponding classes of Riesz–Herglotz measures.

The Classes $\mathbf{S}_{mqb}$ *and* $\mathbf{S}_{ms}$. In the map $u = v/h \mapsto M_v$ the constant h-harmonic function 1 corresponds to M_h. The band $\mathbf{S}_{mqb}$ generated by the constant function 1 therefore corresponds to the band generated by M_h, which is (Appendix IV.8) the class of signed measures absolutely continuous relative to M_h. Thus $u \in \mathbf{S}_{mqb}$ if and only if $u = \mathrm{PI}(B, \phi dM_h)/h$ for some M_h measurable and integrable function ϕ $(= dM_v/dM_h)$, that is, if and only if (Section VIII.9) u is the PWB^h solution for some boundary function ϕ. According to Theorem II.15, the function ϕ is the nontangential boundary limit function of u. The band of singular h-harmonic functions corresponds to the band of signed measures on ∂B lattice orthogonal to M_h, that is (Appendix IV.8), the class of signed measures singular relative to M_h.

The Classes $\mathbf{L}^1(\mu_{B-}^h)$ *and* $\mathbf{D}(\mu_{B-}^h)$*: Theorem II.14 for General* h. As already noted, the $\mathbf{L}^1(\mu_{B-}^h)$ class of h-harmonic functions is the class $\mathbf{S}_m$. For $h \equiv 1$ the classes $\mathbf{L}^1(\mu_{B-}^h)$ and $\mathbf{D}(\mu_{B-}^h)$ of h-harmonic functions were characterized in Theorem II.14. For general h this theorem becomes the following.

Theorem. *Let $u = v/h$ be an h-harmonic function on the ball $B = B(0, \delta)$.*

(a) [$\mathbf{L}^1(\mu_{B-}^h)$ *h-harmonic functions.*] *The following conditions on u are equivalent:*

(a1) $u = \mathrm{PI}(B, M_v)/h$ *for some signed measure M_v on ∂B.*

(a2) *u is the difference between two positive h-harmonic functions.*

(a3) $|u|$ *has an h-harmonic majorant.*

(a4) $\sup_{r<\delta} \mu_{B(0,r)}^h(0, |u|) = \sup_{r<\delta} L(|u|h, 0, r)/h < +\infty$.

Furthermore the map $u \mapsto M_v$ is a one-to-one linear order-preserving map from the $\mathbf{L}^1(\mu_{B-}^h)$ class of h-harmonic functions onto the class of signed measures on ∂B.

(b) [$\mathbf{D}(\mu_{B-}^h)$ *h-harmonic functions.*] *The following more restrictive conditions on u are equivalent:*

(b1) $u = \mathrm{PI}(B, f_v dM_h)/h$ *for some M_h measurable and integrable function f_v on ∂B.*

(b2) *There is a uniform integrability test function Φ for which $\Phi(|u|)$ has an h-harmonic majorant.*

(b3) *The family $\{u_{|\partial B(0,r)}, \mu_{B(0,r)}^h(0, \cdot), 0 < r < \delta\}$ of paired functions and measures is uniformly integrable.*

(b4) $u \in \mathbf{S}_{mqb}$.

Furthermore the map $u \mapsto f_v$ is a one-to-one linear order-preserving map from the class of $\mathbf{D}(\mu_{B-}^h)$ h-harmonic functions on B onto the class $L^1(\partial B, M_h)$, and $dM_v = f_v dM_h$.

The proof follows that of Theorem II.14 and is omitted. Parts of the theorem merely repeat results already discussed at the beginning of this section as implications of Theorem II.14.

Universal Internal Resolutivity of ∂B. We have already shown in Section VIII.9 that ∂B is universally resolutive and that the class of PWB^h solutions is the class of h-harmonic functions given by (b1) above, which is the class of $\mathbf{D}(\mu_{B-}^h)$ h-harmonic functions. Since this class includes the bounded h-harmonic functions [criterion (b2) above], it follows that ∂B is universally internally resolutive.

The Class $\mathbf{L}^p(\mu_{B-}^h)$ for $p > 1$. We now show that $u \in \mathbf{S}_m \cap \mathbf{L}^p(\mu_{B-}^h)$ if and only if $u = \mathrm{PI}(B, \phi dM_h)/h$ for some function ϕ in $L^p(\partial B, M_h)$. Observe first that if u has this form, then

$$|u|^p \leq \frac{\mathrm{PI}(B, |\phi|^p dM_h)}{h}$$

by Jensen's inequality; so $|u|^p$ is majorized by an h-harmonic function and therefore is in $\mathbf{L}^p(\mu_{B-}^h)$. Conversely, if $u \in \mathbf{S}_m \cap \mathbf{L}^p(\mu_{B-}^h)$ for $p > 1$, it follows that $u \in \mathbf{D}(\mu_{B-}^h)$; so $u = v/h$ with $dM_v = f_v dM_h$, $u = \mathrm{PI}(B, f_v dM_h)/h$, and it will now be proved that $|f_v|^p$ is M_h integrable. It can be assumed that f_v

is Borel measurable. According to Theorem II.14, if f is a continuous function on ∂B, then

$$\lim_{r\uparrow\delta}\int_{\partial B} f(\eta)v\left(\frac{r\eta}{\delta}\right)\frac{l_{N-1}(d\eta)}{\pi_N\delta^{N-1}} = \int_{\partial B} f\,dM_v = \int_{\partial B} ff_v\,dM_h; \tag{12.1}$$

so if $1/p + 1/q = 1$, and if v_1/h is an h-harmonic majorant of $|u|^p$, the absolute value of the integral average on the left in (12.1) for fixed r is at most

$$\left[\int_{\partial B}|f(\eta)|^q h\left(\frac{r\eta}{\delta}\right)\frac{l_{N-1}(d\eta)}{\pi_N\delta^{N-1}}\right]^{1/q}\left[\int_{\partial B} v_1\left(\frac{r\eta}{\delta}\right)\frac{l_{N-1}(d\eta)}{\pi_N\delta^{N-1}}\right]^{1/p}, \tag{12.2}$$

and the second factor in (12.2) is $v_1(0)^{1/p}$; so

$$\left|\int_{\partial B} ff_v\,dM_h\right| \le v_1(0)^{1/p}\left[\int_{\partial B}|f|^q\,dM_h\right]^{1/q}. \tag{12.3}$$

Since this inequality is true for continuous f, it is true for f bounded and Borel measurable. Substitute in (12.3) the choice $f = |f_v|^{p-1}\operatorname{sgn} f_v$, where $|f_v| < n$ and $f = 0$ elsewhere to find

$$\int_{\{|f_v|<n\}}|f_v|^p\,dM_h \le v_1(0),$$

which imples that $|f_v|^p$ is M_h integrable, as was to be proved.

It is instructive to attempt an alternative proof of this integrability, based on h-harmonic measure. In the first place

$$\frac{\mu_{B(0,r)}(0,|u|^p h)}{h(0)} = \mu^h_{B(0,r)}(0,|u|^p) \le \mu^h_{B(0,r)}(0,u_1) = u_1(0) \tag{12.4}$$

if u_1 is an h-harmonic majorant of $|u|^p$. Next try to put this inequality in a context in which Fatou's lemma is applicable when $r \to \delta$. For example, if $h \equiv 1$, (12.4) implies

$$\int_{\partial B}\left|u\left(\frac{r\eta}{\delta}\right)\right|^p\frac{l_{N-1}(d\eta)}{\pi_N\delta^{N-1}} \le u_1(0), \tag{12.5}$$

and therefore Fatou's lemma yields the desired l_{N-1} integrability of the l_{N-1} almost everywhere limit $\lim_{r\to\delta}|u(r\eta/\delta)|^p$. This application of Fatou's lemma depends on the fact that when $h \equiv 1$, the h-harmonic average $\mu^h_{B(0,r)}(0,|u|^p)$ can be written as an integral average over a measure space that does not depend on r. It will be seen that such a representation of this harmonic average is possible in a probabilistic approach (Section 2.IX.13) even without the hypothesis that B is a ball.

Chapter X

The Sweeping Operation

1. Sweeping Context and Terminology

Throughout this chapter D is a Greenian subset of $\mathbb{R}^N$, coupled with a boundary ∂D provided by a metric compactification when a boundary is relevant, and the boundary of a subset of $D \cup \partial D$ is that relative to the compactification. In most of the discussion the nature of the boundary is irrelevant, and ∂D can be taken, for example, as the Euclidean boundary or one-point boundary of D.

If $A \subset D$ and if $u = G_D\mu$ is the superharmonic potential of a measure μ on D, the function $[\![u]\!]^A$ is a positive superharmonic function majorized by u and is therefore also the potential of a measure; this measure will be denoted by $[\![\mu]\!]^A$. The map $\mu \mapsto [\![\mu]\!]^A$ is called the *balayage*, that is, the *sweeping*, of μ onto A, and $[\![\mu]\!]^A$ is called the *swept measure* or, with an unfortunate grammatical ambiguity, the *sweeping* of μ. The reduction operation properties lead to sweeping operation properties, some of which will now be listed. The notation is as above: μ is a measure on D with a superharmonic potential, and all sets on which μ is swept are subsets of D.

(a) $G_D[\![\mu]\!]^A \le G_D\mu$ with equality quasi everywhere on A [cf. Section VI.3(b)].

(b) $[\![\mu]\!]^{A_1} = [\![\mu]\!]^{A_2}$ if A_1 differs from A_2 by a polar set [cf. Section VI.3(c)]. In particular, if A is polar, $[\![\mu]\!]^A \equiv 0$.

(c) If $A \subset B$,

$$[\![[\![\mu]\!]^A]\!]^B = [\![[\![\mu]\!]^B]\!]^A = [\![\mu]\!]^A \tag{1.1}$$

[cf. Section VI.3(h)]. In particular, the sweeping operation is idempotent.

(d) $[\![\mu]\!]^A$ is supported by $\bar{A} \cap D$ [cf. Section VI.3(a)].

(e) If A is a Borel support of μ and if $G_D\mu < +\infty$ on A, then $\mu = [\![\mu]\!]^A$, equivalently, $G_D\mu = G_D[\![\mu]\!]^A$. In fact, on the one hand, $G_D[\![\mu]\!]^A \le G_D\mu$ by (a), and on the other hand, $G_D\mu = G_D[\![\mu]\!]^A < +\infty$ on A, a support for μ; so (by the domination principle) $G_D\mu \le G_D[\![\mu]\!]^A$.

(f) If B is a μ null open subset of A, then $\|\mu\|^A(B) = 0$ [cf. Section VI.3(a)]. In particular, strengthening (d), if A is μ null the swept measure $\|\mu\|^A$ is supported by $D \cap \partial A$.
(g) If A is a Borel subset of D, the measure μ can be written as the sum $\mu_1 + \mu_2 + \mu_3$ of the projections of μ, respectively, on the sets $A \cap \{G_D\mu < +\infty\}$, $A \cap \{G_D\mu = +\infty\}$, $D - A$.

In view of the reduction additivity property of Section VI.3(f), $\|\mu\|^A = \|\mu_1\|^A + \|\mu_2\|^A + \|\mu_3\|^A$. According to (d)–(f), the sweeping operation of μ onto A leaves μ_1 invariant and sweeps μ_3 into a measure supported by $D \cap \partial A$. The measure $\|\mu_2\|^A$ is supported by $\bar{A} \cap \partial D$. More detailed information on $\|\mu\|^A$ will be furnished by Theorems 5 and XI.18. In the most common classical application of sweeping, $A = D - B$ with B a ball with closure in D. In this case the fact that the operation $G_D\mu \mapsto \|G_D\mu\|^{D-B} = \tau_B[G_D\mu]$ transfers the component of μ on B to a measure supported by the set $\partial B = D \cap \partial(D - B)$ is particularly clear.

Characterization of $\|\mu\|^A$ (Refined in Section XI.15)

If A is a subset of D, closed in D, and if μ is a measure on D with a finite-valued potential $G_D\mu$, the swept measure $\nu = \|\mu\|^A$ is uniquely determined by the following properties:

(S1) ν is supported by A.
(S2) $G_D\nu = G_D\mu$ quasi everywhere on A.

These properties of $\|\mu\|^A$ have already been noted. Conversely, if a measure ν has these properties, then $\nu = \|\mu\|^A$ by the domination principle because $G_D\nu = G_D\|\mu\|^A$ quasi everywhere on a common support A of ν and $\|\mu\|^A$, and both potentials are finite valued on A.

The Sweeping Kernel δ_D^A

If $\xi \in D$, the probability measure on D supported by $\{\xi\}$ will be denoted by $\delta_D(\xi, \cdot)$. The sweeping of this measure onto A, denoted by $\delta_D^A(\xi, \cdot)$, is determined by its defining equation

$$\|G_D(\xi, \cdot)\|^A(\eta) = \int_D G_D(\eta, \zeta)\delta_D^A(\xi, d\zeta) = \delta_D^A(\xi, G_D(\eta, \cdot)) \qquad (\eta \in D). \tag{1.2}$$

In the following we shall write $G_D^A(\xi, \eta)$ for $\|G_D(\xi, \cdot)\|^A(\eta)$. It will be shown in Section 3 that G_D^A is symmetric. The proof that $\delta_D^A(\cdot, B)$ is Borel measurable when B is a Borel set, so that δ_D^A is a kernel, and the proof that G_D^A is Borel measurable on $D \times D$ will be given in Section 4 and so cannot be used at the present stage. If B is open and relatively compact in D and if $v = \llbracket 1 \rrbracket^B$ then v is a potential, say $v = G_D \nu$, and integration with respect to $\nu(d\eta)$ in (1.2) yields $1 \geq \delta_D^A(\cdot, v) \geq \delta_D^A(\cdot, B)$. It follows that $\delta_D^A(\cdot, D) \leq 1$. It will be seen in Section 5 [see (5.1′)] that $\delta_D^A(\cdot, D) = \llbracket 1 \rrbracket^A$ for every set A. According to property (f) above, if ξ is not an interior point of A, the swept measure $\delta_D^A(\xi, \cdot)$ is supported by $D \cap \partial A$. On the other hand, if ξ is an interior point of A then $G_D^A(\xi, \cdot) = G_D(\xi, \cdot)$ according to Section VII.3(b), and therefore $\delta_D^A(\xi, \cdot) = \delta_D(\xi, \cdot)$; that is, $\delta_D^A(\xi, \{\xi\}) = 1$.

2. Relation between Harmonic Measure and the Sweeping Kernel

Theorem. *If D is a Greenian subset of $\mathbb{R}^N$, if A is a subset of D, closed relative to D, and if $\xi \in D - A$, then $\delta_D^A(\xi, \cdot) = \mu_{D-A}(\xi, \cdot)$ on the Borel subsets of $D \cap \partial A$.*

According to (1.2) and the relation between reductions and PWB solutions derived in Section VIII.10,

$$\mu_{D-A}(\eta, G_D(\xi, \cdot) 1_D) = \delta_D^A(\xi, G_D(\eta, \cdot)) \qquad [(\xi, \eta) \in D \times (D - A)]. \tag{2.1}$$

The harmonic measure symmetry property VIII(18.5), in which D and B correspond respectively to $D - A$ and D here, implies that (2.1) with both ξ and η in $D - A$ can be written in the form

$$\mu_{D-A}(\xi, G_D(\eta, \cdot) 1_D) = \delta_D^A(\xi, G_D(\eta, \cdot)) \qquad [(\xi, \eta) \in (D - A) \times (D - A)]. \tag{2.2}$$

Moreover, in the present context VIII(18.2′) becomes

$$G_D(\xi, \zeta) = \mu_{D-A}(\xi, G_D(\zeta, \cdot) 1_D) \qquad (\xi \in D) \tag{2.3}$$

whenever ζ is a regular boundary point of $D - A$ in D or is an inner point of A; so (2.3) is true for quasi every ζ in A. Now for fixed ξ in $D - A$ each side of (2.2) is the value at η of the G_D potential of a measure on D. These potentials are equal on $D - A$ according to (2.2) and are equal quasi everywhere on A according to (2.3). Hence these potentials are equal on D; so the measures $\mu_{D-A}(\xi, \cdot)$ (restricted to subsets of D) and $\delta_D^A(\xi, \cdot)$ generating the potentials are the same, as was to be proved.

3. Sweeping Symmetry Theorem

Theorem. *If D is a Greenian subset of $\mathbb{R}^N$ and if A is a subset of D, then*

$$\|G_D(\xi,\cdot)\|^A(\eta) = \|G_D(\eta,\cdot)\|^A(\xi) \qquad [(\xi,\eta)\in D\times D]; \tag{3.1}$$

that is G_D^A is symmetric on $D \times D$.

This fundamental theorem is a deep generalization of the theorem that the Green function G_D is symmetric, and reduces to the latter theorem when $A = D$. Suppose first that A is closed in D. For ξ and η in $D - A$ equation (3.1) is equivalent to the assertion that the function $(\xi,\eta)\mapsto \mu_{D-A}(\xi, G_D(\eta,\cdot))$ is symmetric on $(D - A) \times (D - A)$, a fact noted in Section VIII.18 and just applied in Section 2. If at least one of the points ξ, η is in the set A, then (3.1) is true for the pair (ξ,η) when A is replaced by

$$A_n = A - [B(\xi, 1/n) \cup B(\eta, 1/n)],$$

and therefore $(n\to\infty)$ equation (3.1) is true when A is replaced by $A - [\{\xi\} \cup \{\eta\}]$, in view of the reduction property Section VI.3(e). Finally, (3.1) is true as stated for A closed in D since [Section VI.3(c)] smoothed reductions on A are identical with smoothed reductions on A less two points. Apply Section VI.3(e) again to find that (3.1) is true if A is an F_σ set, in particular, if A is open. For an arbitrary set A it follows that (3.1) is true when A is replaced by an arbitrary open superset of A and therefore [Section VI.3(m)] that (3.1) is true for unsmoothed reductions if neither ξ nor η is in A. Since [Section VI.3(b)] reductions on A are equal on $D - A$ to their smoothings, (3.1) is true as stated if neither ξ nor η is in A. Finally, if at least one of these points is in A, apply (3.1) to A less these two points. The smoothed reductions are unchanged, and we deduce that (3.1) is true for the pair (ξ,η)

4. Kernel Property of δ_D^A

Lemma. *δ_D^A is a kernel.*

All that needs to be proved is that $\delta_D^A(\cdot, B)$ is Borel measurable whenever B is a Borel set. Consider the linear class Γ of functions of the form $f = u - v$, where u and v are bounded continuous potentials on D, the measures associated with u and v have compact support, and f has compact support. Define $f^A = \|u\|^A - \|v\|^A$, and observe that f^A is determined by f because if f has two representations, $f = u - v = \mu' - v'$, then $u + v' = u' + v$; so $\|u\|^A + \|v'\|^A = \|u'\|^A + \|v\|^A$, and finally $\|u\|^A - \|v\|^A = \|u'\|^A - \|v'\|^A$. If $\xi\in D$

the functional $f \mapsto L_\xi(f) = f^A(\xi)$ is linear on Γ. Let $\bar{\Gamma}$ be the class of uniform limits of sequences in Γ, so that (by Lemma VII.9) $\bar{\Gamma}$ includes the class of continuous functions on D with compact support. If $f \in \bar{\Gamma}$ with approximating sequence $f_\cdot$ in Γ, then [from Section VI.3(n)]

$$\sup_D |L_\xi(f_n - f_m)| \leq \sup_D |f_n - f_m|; \tag{4.1}$$

so the sequence $L_\xi(f_\cdot)$ is convergent. Since two approximating sequences for f can be combined into one, the limit of $L_\xi(f_\cdot)$ is independent of the approximating sequence and is $L_\xi(f)$ when $f \in \Gamma$. Define $L_\xi(f) = \lim_{n\to\infty} L_\xi(f_n)$ to get a linear functional on $\bar{\Gamma}$. The functional L_ξ is positive because if f in $\bar{\Gamma}$ is positive and if $f_\cdot = u_\cdot - v_\cdot$ is an approximating sequence in Γ for f, with $\inf_D(u_n - v_n) = -\varepsilon_n$, it follows that $u_n + \varepsilon_n \geq v_n$ and therefore that $\|u_n\|^A + \varepsilon_n \geq \|v_n\|^A$; so

$$L_\xi(f) = \lim_{n\to\infty} L_\xi(f_n) \geq -\lim_{n\to\infty} \varepsilon_n = 0. \tag{4.2}$$

Apply Section VI.3(n) again to obtain, for f in $\bar{\Gamma}$ with approximating sequence $f_\cdot = u_\cdot - v_\cdot$,

$$|L_\cdot(f)| = \lim_{n\to\infty} |L_\cdot(f_n)| = \lim_{n\to\infty} \big| \|u_n\|^A - \|v_n\|^A \big| \leq \lim_{n\to\infty} \sup_D |f_n| = \sup_D |f|; \tag{4.3}$$

that is, the linear functional L_ξ has bound 1. Thus (Riesz representation theorem) there is a measure $\delta_{DA}(\xi, \cdot)$ on D with $\delta_{DA}(\xi, D) \leq 1$ for which $L_\xi(f) = \delta_{DA}(\xi, f)$ whenever f is continuous on D and has compact support. Since this equation shows that $\delta_{DA}(\cdot, f)$ is Borel measurable, δ_{DA} is a kernel. Now suppose that $u = G_D\mu$ is the potential of a measure μ with compact support S, and let $D_\cdot$ be an increasing sequence of open relatively compact subsets of D with union D, for which $S \subset D_0$ and $\bar{D}_n \subset D_{n+1}$. According to Theorem IV.10, there is an increasing sequence $u_\cdot$ of continuous potentials with limit u for which $u_n = u$ on $D - D_0$, and there is an increasing sequence $u_{n\cdot}$ of continuous potentials with limit $\tau_{D_n} u$ for which $u_{nk} = \tau_{D_n} u = u$ on $D - D_{n+1}$. Then $u_n - u_{nn}$ is in Γ; so

$$\|u_n\|^A - \|u_{nn}\|^A = \delta_{DA}(\cdot, u_n - u_{nn}). \tag{4.4}$$

Moreover u is a potential; so (from Section VIII.11) $\lim_{n\to\infty} \tau_{D_n} u = 0$, and therefore

$$\limsup_{n\to\infty} \|u_{nn}\|^A \leq \limsup_{n\to\infty} u_{nn} \leq \lim_{n\to\infty} \tau_{D_n} u = 0. \tag{4.5}$$

Hence (4.4) yields

$$\|u\|^A = \delta_{DA}(\cdot, u). \tag{4.6}$$

In particular, when $u = G_D(\xi, \cdot)$, we find with the help of sweeping symmetry that δ_{DA} satisfies the defining equation for δ_D^A; so $\delta_D^A = \delta_{DA}$ and is a kernel.

5. Swept Measures and Functions

Theorem. *Let D be a Greenian subset of $\mathbb{R}^N$, and let A be a subset of D. Let $u = G_D\mu$ be a superharmonic potential, and let v be a positive superharmonic function on D with associated Riesz measure ν. Then*

$$\|u\|^A = G_D\|\mu\|^A = G_D^A\mu = \delta_D^A(\cdot, u), \tag{5.1}$$

$$\|v\|^A = \delta_D^A(\cdot, v), \tag{5,1'}$$

$$\|\mu\|^A = \int_D \delta_D^A(\zeta, \cdot)\mu(d\zeta), \tag{5.2}$$

$$\int_D \|v\|^A \, d\mu = \int_D v \, d\|\mu\|^A. \tag{5.3}$$

In particular, if v is a superharmonic potential, (5.3) *can be continued*:

$$\int_D \|v\|^A \, d\mu = \int_D v \, d\|\mu\|^A = \int_D u \, d\|\nu\|^A = \int_D \|u\|^A \, d\nu = \int_D \|u\|^A \, d\|\nu\|^A. \tag{5.3'}$$

In the course of proving Lemma 4 we proved (4.6); that is, the first and fourth terms of (5.1) are equal if μ has compact support. Since every positive superharmonic function is the limit of an increasing sequence of potentials of measures with compact supports (Section IV.10), the first and fourth terms of (5.1) are equal, and (5.1′) is true. The first term is (5.1) is equal to the second by definition of $\|\mu\|^A$. The third term is equal to the fourth because

$$\delta_D^A(\xi, u) = \int_D \delta_D^A(\xi, d\eta) \int_D G_D(\eta, \zeta)\mu(d\zeta),$$

and the iterated integral becomes $G_D^A\mu$ on reversal of the order of integration. Equation (5.2) follows immediately from (5.1); (5.3) and (5.3′) follow from (5.1), (5.1′), (5.2), and the fact that $\|\,\|v\|^A\|^A = \|v\|^A$.

Generalization of $\|\mu\|^A$

It is convenient to define $\|\mu\|^A$ by (5.2) when μ is a measure on D even when $G_D\mu$ is not superharmonic. Under this extended definition, (5.3) remains true.

Application to the Representation of a Reduction. Let D be a Greenian subset of $\mathbb{R}^N$, and let A be a subset of D. Then if v is a positive superharmonic function on D and if ν_A is the Riesz measure associated with $\llbracket v \rrbracket^A$ (reduction relative to D),

$$\llbracket v \rrbracket^A = G_D \nu_A + h, \qquad \nu_A = \llbracket \nu_A \rrbracket^A, \qquad h = \mathrm{GM}_D \llbracket v \rrbracket^A = \llbracket h \rrbracket^A. \quad (5.4)$$

The first equation in (5.4), with the stated identification of h, is simply the Riesz decomposition of $\llbracket v \rrbracket^A$. Since the smoothed reduction operation is idempotent, (5.4) leads to

$$\llbracket v \rrbracket^A = G_D \llbracket \nu_A \rrbracket^A + \llbracket h \rrbracket^A = G_D \lambda + h', \quad (5.5)$$

where λ is a measure on D and

$$h' = \mathrm{GM}_D \llbracket h \rrbracket^A \le \llbracket h \rrbracket^A \le h.$$

A comparison between (5.4) and (5.5) shows that $h' = h$ and $\lambda = \llbracket \nu_A \rrbracket^A = \nu_A$, so that the second and third equation in (5.4) are true.

Examples. If A is relatively compact in D, then $\llbracket v \rrbracket^A$ is a potential; so $h \equiv 0$. If A is sufficiently large, say $A = D$, then $\llbracket v \rrbracket^A = v$, and (5.4) becomes the Riesz decomposition of v.

6. Some Properties of δ_D^A

(a) Subadditivity of $\delta_D^{\boldsymbol{\cdot}}$. This set function is subadditive in the sense that

$$\delta_D^{A \cup B} \le \delta_D^A + \delta_D^B \quad (6.1)$$

for all subsets A and B of D. In fact [Section VI.3(i)]

$$\delta_D^{A \cup B}(\cdot, v) + \delta_D^{A \cup B}(\cdot, \llbracket v \rrbracket^A \wedge \llbracket v \rrbracket^B) = \delta_D^A(\cdot, v) + \delta_D^B(\cdot, v) \quad (6.2)$$

whenever v is positive and superharmonic on D; so if f is the difference between two finite-valued positive superharmonic functions on D, with $f \ge 0$, it follows that

$$\delta_D^{A \cup B}(\cdot, f) \le \delta_D^A(\cdot, f) + \delta_D^B(\cdot, f), \quad (6.3)$$

and this inequality implies (6.1) in view of the approximation application in Section VII.9.

(b) For each point ξ of D and each subset A of D the measure $\delta_D^A(\xi, \cdot)$ vanishes on polar sets not containing ξ. In fact, if B is a polar subset of D

not containing ξ, there is (by Theorem V.2) a positive superharmonic function on D, finite at ξ but identically $+\infty$ on B, and the inequality

$$+\infty > v(\xi) \geq \|v\|^A(\xi) = \delta_D^A(\xi, v)$$

implies that $\delta_D^A(\xi, B) = 0$.

(c) A zero-one law. Let ξ be a point of D, let A be a subset of D, let B be a neighborhood of ξ, and let v be a positive superharmonic function on D. Then either the equalities $\delta_D^{A\cap B}(\xi, \{\xi\}) = 1$ and $\|v\|^{A\cap B}(\xi) = v(\xi)$ are true for all B and all v or $\delta_D^{A\cap B}(\xi, \{\xi\}) = 0$ for all B, in which case whenever $v(\xi) < +\infty$, the function $\|v\|^{A\cap B}$ tends to 0 on D when B shrinks to ξ. The property that $\delta_D^A(\xi, \{\xi\}) = 1$ will be given a topological interpretation in Section XI.3, where incidentally it will also be shown that this property is independent of the choice of D containing ξ. Thus this property is a local property of A near ξ. To prove (c), we first observe that, using the notation of (c), $\delta_D^{A\cap B}(\xi, \{\xi\})$ decreases when B increases because when B_1 and B_2 are neighborhoods of ξ with $B_1 \subset B_2$ subadditivity of $\delta_D^{\bullet}$ implies

$$\delta_D^{B_2}(\xi, \{\xi\}) \leq \delta_D^{B_1}(\xi, \{\xi\}) + \delta_D^{B_2 - B_1}(\xi, \{\xi\}),$$

and the second term on the right is 0 because the measure $\delta_D^{B_2 - B_0}(\xi, \cdot)$ is supported by the closure of $B_2 - B_1$. Next observe that since $\|v\|^{A\ B} = \|\,\|v\|^A\,\|^{A\cap B}$, it follows that

$$\delta_D^{A\cap B}(\xi, \{\xi\}) = \int_D \delta_D^A(\eta, \{\xi\})\delta_D^{A\cap B}(\xi, d\eta), \tag{6.4}$$

and since the measure $\delta_D^A(\eta, \cdot)$ vanishes on polar sets not containing η, equation (6.4) reduces to

$$\delta_D^{A\cap B}(\xi, \{\xi\}) = \delta_D^A(\xi, \{\xi\})\delta_D^{A\cap B}(\xi, \{\xi\}). \tag{6.5}$$

When $B = D$, we conclude that $\delta_D^A(\xi, \{\xi\})$ must be either 0 or 1. In the first case (6.5) implies that $\delta_D^{A\cap B}(\xi, \{\xi\}) = 0$ for all B. On the other hand, if $\delta_D^A(\xi, \{\xi\}) = 1$, then monotoneity of $\delta_D^{A\cap B}(\xi, \{\xi\})$ as B varies implies that $\delta_D^{A\cap B}(\xi, \{\xi\}) = 1$ for all B.

If $\delta_D^{A\cap B}(\xi, \{\xi\}) = 1$, then $\|v\|^{A\cap B}(\xi) = \delta_D^{A\cap B}(\xi, v) = v(\xi)$. If $\delta_D^A(\xi, \{\xi\}) = 0$, let $B_{\bullet}$ be a sequence of balls of center ξ, relatively compact in D, with radii tending monotonely to 0. The sequence $\|v\|^{A\cap B_{\bullet}}$ is a decreasing sequence of potentials on D with limit a positive function v_0, harmonic on $D - \{\xi\}$. The restriction of v_0 to $D - \{\xi\}$ has a positive superharmonic extension v_1 to D, necessarily of the form const. $G_D(\xi, \cdot)$, and the constant is 0 if as we suppose from now on $v(\xi) < +\infty$, because $v_1(\xi) < v(\xi)$. Hence $v_0 = 0$ on $D - \{\xi\}$; so

$$\lim_{n\to\infty} \| v \|^{A\cap B_n}(\eta) = \lim_{n\to\infty} \| \| v \|^{A\cap B_n} \|^{A}(\eta) = \delta_D^A(\eta, v_0) = \delta_D^A(\eta, \{\xi\}) v_0(\xi).$$

The last term on the right is 0 by hypothesis when $\eta = \xi$ and vanishes at all other points η because according to (b) above, the measure $\delta_D^A(\eta, \cdot)$ vanishes on polar sets not containing η.

7. Poles of a Positive Harmonic Function

Let D be a connected Greenian subset of $\mathbb{R}^N$, coupled with a boundary provided by a metric compactification, and let A be a boundary subset. If h is a positive harmonic function on D, if $R_h^A = ch$ (as is true, for example, if h is minimal), and if $h \not\equiv 0$, then c is either 0 or 1 because the smoothed reduction operation is idempotent [Section VI.3(h)]. The boundary point ζ is said to be a pole of h if $R_h^{\{\zeta\}} = h$. The set of poles of h is obviously closed. If ζ is a pole of h, then ζ is also a pole of every positive harmonic minorant of h. In fact, if h_1 is such a minorant, the equality

$$\| h_1 \|^{\{\zeta\}} + \| h - h_1 \|^{\{\zeta\}} = h^{\{\zeta\}} = h$$

implies that the reduction operations performed on the left have not decreased h_1 or $h - h_1$, so $h_1 = \| h_1 \|^{\{\zeta\}}$.

Poles of a Minimal Harmonic Function

If h is minimal, if ζ is a pole of h, and if A is a boundary subset containing ζ, then

$$h \geq R_h^A \geq R_h^{\{\zeta\}} = h;$$

so $R_h^A = h$. On the other hand, if ζ is not a pole of h, there is a positive superharmonic function v on D majorizing h near ζ and strictly less than h at some point of D. It follows that $R_h^A \leq v < h$ at some point of D; so $R_h^A \equiv 0$ if A is a sufficiently small boundary neighborhood of ζ. Furthermore, if B is a boundary subset containing no poles of h, then $R_h^B \equiv 0$ because B can be covered by a countable collection $A_{\bullet}$ of boundary neighborhoods of nonpoles, with $R_h^{A_j} \equiv 0$; so $R_h^B \leq \Sigma_j R_h^{A_j} \equiv 0$. This result implies that there is at least one pole of a minimal harmonic function h because $R_h^{\partial D} = h$. It is left to the reader to check that the boundary is h-resolutive if and only if h has only one pole and that if ζ is this pole, the singleton $\{\zeta\}$ has h-harmonic measure 1. If h has more than one pole, the indicator function of the set of poles is h-resolutive, and the set of poles has h-harmonic measure 1.

8. Relative Harmonic Measure on a Polar Set

Let D be a connected Greenian subset of $\mathbb{R}^N$, let A be a compact polar subset of D, and let $h = G_D \nu_h$ be the potential of a not identically zero measure supported by A. Define $D_0 = D - A$ and ∂D_0 as the boundary of D_0 relative to the one-point compactification of D. Since A is nowhere dense, ∂D_0 consists of A and the Alexandrov boundary point α of D. Let h_0 be the restriction of h to D_0. Recall (from Section VII.1) that G_{D_0} is the restriction of G_D to $D_0 \times D_0$. We now prove that ∂D_0 is an h-resolutive boundary for D_0, that $\mu_{D_0}^{h_0}(\cdot, \{\alpha\}) \equiv 0$, and that for η in A and ξ in D_0

$$\mu_{D_0}^{h_0}(\xi, d\eta) = G_D(\xi, \eta)\frac{\nu_h(d\eta)}{h(\xi)}. \tag{8.1}$$

Reductions below not otherwise described are relative to D_0. If F is an open relatively compact subset of D containing A, the set $D_0 - F$ is a deleted neighborhood of α relative to $D_0 \cup \partial D_0$, $[\![h_0]\!]^{D_0 - F}$ is a minorant of h_0 and is harmonic and bounded on F, and $[\![h_0]\!]^{D_0 - F}/h_0$ is in the upper PWB^{h_0} class on D_0 for the boundary function $1_{\{\alpha\}}$. Moreover (by Theorem V.5) the function $[\![h_0]\!]^{D_0 - F}$ has an extension to D, harmonic on F and majorized by h. When F increases, tending to D, this extension decreases, tending to a harmonic minorant of the potential h, that is, tending to the zero function. Thus $\{\alpha\}$ is an h_0-harmonic measure null set. Next we prove that the evaluation

$$h(\xi) = \int_A G_D(\xi, \eta)\nu_h(d\eta)$$

implies that if C is a compact subset of A, then

$$[\![h]\!]^C = \int_C G_D(\cdot, \eta)\nu_h(d\eta). \tag{8.2}$$

If B_0 is the trace on D_0 of a neighborhood B of C,

$$\begin{aligned} [\![h_0]\!]^{B_0}(\xi) = \delta_{D_0}^{B_0}(\xi, h_0) &= \int_A \delta_{D_0}^{B_0}(\xi, G_D(\cdot, \eta))\nu_h(d\eta) \\ &= \int_A [\![G_{D_0}(\cdot, \eta)]\!]^{B_0}(\xi)\nu_h(d\eta). \end{aligned} \tag{8.3}$$

If smoothed reductions relative to D are denoted by $[\![\ ]\!]_D$, an easy argument involving the superharmonic function extension theorem (Theorem V.5) shows that for η in D

$$\|G_{D_0}(\cdot,\eta)\|^{B_0} = \|G_D(\cdot,\eta)\|_D^B$$

on B. Then for ξ in B_0

$$\|h_0\|^{B_0}(\xi) = \int_A \|G_D(\cdot,\eta)\|_D^B(\xi)\nu_h(d\eta). \tag{8.4}$$

Now [by Theorem VII.3(b)] when η is in C, the integrand is equal to $G_D(\xi,\eta)$, and on the other hand [Section VI.3(m)], when η is in A-C and B shrinks to C, the integrand tends to $\|G_D(\cdot,\eta)\|^C(\xi)$, which vanishes because C is polar. Thus (8.2) is true, and since $\|h_0\|^C/h_0$ is the upper PWB^{h_0} solution on D_0 for the boundary function 1_C, we have proved that this upper solution is an additive function of C. Since we can ignore the h_0-harmonic measure null singleton $\{\alpha\}$, it follows from Theorem VIII.9 that D_0 is h_0 resolutive. Finally (8.1) is now equivalent to (8.2).

Chapter XI

The Fine Topology

1. Definitions and Basic Properties

The topology of a topological space is the class of open subsets of the space. If T_1 and T_2 are topologies on a space, T_1 is said to be *finer* than T_2 (and then T_2 is said to be *coarser* than T_1) if $T_2 \subset T_1$. For any family of extended real-valued functions on a space there is a coarsest topology making every member of the family continuous, namely, the intersection of all the topologies doing this. The *fine* topology of classical potential theory is defined as the coarsest topology on $\mathbb{R}^N$ making continuous every superharmonic function on $\mathbb{R}^N$. It is easy to verify that the fine and Euclidean topologies coincide when $N = 1$ (see Chapter XIV for classical potential theory on $\mathbb{R}$), and we suppose from now on in this chapter that $N > 1$. Concepts relative to the fine topology will be distinguished by an "f," for example, f lim sup, $\partial^f A$. From now on any otherwise unqualified topological concept will refer to the Euclidean topology. Since the fine topology is defined intrinsically in terms of superharmonic functions, it is not surprising that this topology plays a fundamental role in classical potential theory.

If u is a superharmonic function, the function $u \wedge c$ is also, for every constant c, and it follows that the fine topology is the coarsest topology making continuous every upper-bounded superharmonic function on $\mathbb{R}^N$. Since $G(\xi, \cdot)$ is fine continuous and since $\{G(\xi, \cdot) > c\}$ is a ball with center ξ and an arbitrary radius depending on c, balls are fine open, and therefore the fine topology is at least as fine as the Euclidean topology. Since there are discontinuous superharmonic functions, the fine topology is strictly finer than the Euclidean topology.

Let D be a nonempty open subset of $\mathbb{R}^N$. Then the restriction to D of the fine topology is the coarsest topology making continuous every superharmonic function with domain D. To prove this, it is sufficient to prove that if u is superharmonic on D, then u is fine continuous, and it is even sufficient to prove that u is fine continuous on every open relatively compact subset D_0 of D. Now if μ is the measure associated with the restriction to D_0 of u, the potential $G\mu$ is superharmonic on $\mathbb{R}^N$, differs from u on D_0 by a continuous (harmonic) function, and is fine continuous on $\mathbb{R}^N$. Hence u is fine continuous on D_0, as was to be proved.

It will sometimes be useful to extend the fine topology to $\mathbb{R}^N \cup \{\infty\}$ by defining a set to be a deleted fine neighborhood of ∞ if after inversion in a sphere the image of the set is a deleted fine neighborhood of the sphere center (see Section 5). This criterion will be seen to be unaffected by the choice of inversion sphere, and in fact it will be seen that under an inversion the fine topology is invariant.

The fine derived set of a set A, that is, the (fine-closed) set of fine limit points of A, will be denoted by A^f. It will be shown (in Theorem 6) that $A^f = (A^f)^f$; that is, A^f is fine perfect.

It is immediate that one base for the class of fine neighborhoods of a finite point ξ is the class of sets of the form

$$\bigcap_1^n \{\eta \in B: u_j(\eta) < c_j\}, \tag{1.1}$$

where B is a ball containing ξ, u_j is an upper-bounded superharmonic function on B, vanishing at ξ, and each constant c_j is strictly positive. The neighborhood (1.1) is fine open. It follows that another base for the class of fine neighborhoods of ξ is the somewhat more convenient class of sets of the form

$$\bigcap_1^n \{\eta \in \bar{B}: u_j(\eta) \le c\}, \tag{1.2}$$

where B is a ball containing ξ, u_j is a bounded superharmonic function on a neighborhood of $\bar{B}$, vanishing at ξ, and c is a strictly positive constant. The neighborhood (1.2) is fine closed and is Euclidean topology compact. It follows that a point ξ is a fine limit point of a set A if and only if ξ is a fine limit point of every open superset of A.

A point ξ of $\mathbb{R}^N$ is a fine limit point of a set A if and only if ξ is a Euclidean limit point of A and if each superharmonic function u defined on an open neighborhood D of ξ has $u(\xi)$ as a cluster value at ξ along A. For example, according to II(6.1), the complement of an l_N null set is fine everywhere dense.

A subset A of $\mathbb{R}^N$ is said to be *thin* at a finite point ξ if ξ is not a fine limit point of A, that is, if ξ is not in A^f. According to the above remarks on fine neighborhoods, if A is thin at ξ, there is an open superset of A which is also thin at ξ. The corresponding remarks for $\xi = \infty$ are left to the reader.

The Baire Property of the Fine Topology

The intersection of a sequence of fine-open fine dense sets is fine dense. To prove this assertion, observe that the fine topology has a basis of Euclidean compact sets so that if $B_.$ is a sequence of fine-open fine dense sets and if B

is a nonempty fine-open set there is a sequence $B'_\cdot$ of compact sets with nonempty fine interiors such that $B'_0 \subset B_0 \cap B$ and that B'_{n+1} is a subset of the fine interior of $B'_n \cap B_n$ for all n. Then

$$\varnothing \neq \bigcap_0^\infty B'_n \subset \bigcap_0^\infty B_n \cap B$$

so $\bigcap_0^\infty B_n$ is fine dense.

2. A Thinness Criterion

Theorem. *If a set A has finite limit point ξ, then A is thin at ξ if and only if there is a superharmonic function u defined on an open neighborhood of ξ such that*

$$\liminf_{A \ni \eta \to \xi} u(\eta) > u(\xi). \tag{2.1}$$

If A is thin at ξ and if D is a Greenian set containing ξ, there is a potential $u = G_D\nu$ satisfying (2.1) *for which ν has compact support and the left side of* (2.1) *is* $+\infty$.

If ξ is a fine limit point of A and if u is superharmonic on an open neighborhood of ξ, u has fine limit $u(\xi)$ at ξ and is lower semicontinuous; so (2.1) is impossible. Conversely, if ξ is a limit point of A and if A is thin at ξ, some fine neighborhood (1.2) of ξ does not meet A except possibly at ξ. The functions $u = \Sigma_1^n u_j$ and $\min_{j \le n} u_j$ are superharmonic on B with $\min_{j \le n} u_j(\xi) = 0$, and

$$u > c + (n-1) \min_{j \le n} u_j$$

on $(A - \{\xi\}) \cap B$; so

$$\liminf_{A \ni \eta \to \xi} u(\eta) \ge c > u(\xi) = 0.$$

To prove the second assertion of the theorem, let u satisfy (2.1), and suppose first that $\bar{D}$ is bounded and contained in the domain of u. Then u is bounded below and can be made positive on D by addition of a suitable constant without affecting (2.1). Let B_r be the ball of center ξ and radius $r < |\xi - \partial D|$ and let μ_r be the projection on B_r of the measure associated with u. Then $G_D\mu_r(\xi) \le u(\xi) < +\infty$ and u differs from $G_D\mu_r$ by a function harmonic on B_r; so if δ ($\le +\infty$) is the strictly positive difference between the left and right sides of (2.1),

$$\liminf_{A \ni \eta \to \xi} G_D\mu_r(\eta) = G_D\mu_r(\xi) + \delta \geq \delta.$$

Choose $r = r_n$ so small that $G_D\mu_{r_n}(\xi) < 2^{-n}$. The superharmonic potential $G_D(\Sigma_0^\infty \mu_{r_n})$ has value ≤ 2 at ξ and has limit $+\infty$ at ξ along A as desired, and the associated measure has compact support in D. If $\bar{D}$ is not bounded and is in the domain of u, choose a ball B containing ξ, with $\bar{B}$ in the domain of u. Then the potential $G_B\nu$ of some measure ν with compact support in B is finite at ξ and has limit $+\infty$ at ξ along A. The potential $G_D\nu$ has the same property because $G_D\nu$ differs from $G_B\nu$ by a function harmonic on B.

Observation. If $N > 2$, the set D can be chosen to be $\mathbb{R}^N$ in the theorem. If $N = 2$, the proof yields a superharmonic potential $u = G\nu$ on $\mathbb{R}^N$ for which the left side of (2.1) is $+\infty$.

Application to Polar Sets and Reductions

A polar set A_0 has no fine limit points because according to Theorem V.2, if A_0 is polar and if $\xi \in A_0^f$, there is a superharmonic function defined on an open neighborhood of ξ, finite at ξ, and identically $+\infty$ on the part of A_0 in a deleted neighborhood of ξ; so (2.1) is satisfied, contradicting the hypothesis that $\xi \in A_0^f$. (In the fine topology of $\mathbb{R}^N \cup \{\infty\}$ as defined in Section 5 a polar set cannot have ∞ as a fine limit point; this assertion is reduced by an inversion in a sphere to the one just treated.) The converse result that $A_0^f = \emptyset$ implies that A_0 is polar will be proved in Section 6. The fact that a polar set has no fine limit point implies that if A is an arbitrary subset of a Greenian set D, then (reduction relative to D) $\llbracket v \rrbracket^A = v$ on $A^f \cap D$. In fact this equality is true quasi everywhere on A, that is, everywhere on $A - A_0$ for some polar set A_0; so the fine continuity of superharmonic functions implies that $v = \llbracket v \rrbracket^A$ on $(A - A_0^f) = A^f$.

3. Conditions That $\xi \in A^f$

Let D be a Greenian subset of $\mathbb{R}^N$, let A be a subset of D, let ξ be a point of D, and let v be a positive superharmonic function on D. All reductions will be relative to D. We shall write $\lim_{B \downarrow \xi}$ to mean the limit as B, a neighborhood of ξ, shrinks to ξ, that is, as its diameter tends to 0. Define $v_0 = \lim_{B \downarrow \xi} \llbracket v \rrbracket^{A \cap B}$, that is, v_0 is the infimum of the indicated class of smoothed reductions. The positive function v_0 is harmonic on $D - \{\xi\}$; so (by Theorem V.5) the restriction of v_0 to $D - \{\xi\}$ has a superharmonic extension to D, and (by the Riesz decomposition) this extension must have the form $cG_D(\xi, \cdot) + h$, where $c \geq 0$ and h is a positive harmonic function on D. Since this superharmonic function is majorized by a potential $\llbracket v \rrbracket^B$ for B relatively compact in D, the function h must vanish identically. The inequality $v_0 \leq v$

implies that $c = 0$ if $v(\xi) < +\infty$ and more generally $c = 0$ if $\|v\|^{A \cap B}(\xi) < +\infty$ for some B. According to the following theorem, if ν is the Riesz measure associated with v, then either $c = \nu(\{\xi\})$ or $c = 0$.

Theorem. *If $\xi \in A^f$, then*

(a) $\|v\|^{A \cap B}(\xi) = v(\xi)$ *for every neighborhood B of ξ.*
(b) $\delta_D^A(\xi, \{\xi\}) = 1$.
(c) $G_D^A(\xi, \cdot) = G_D(\xi, \cdot)$.

If $\xi \notin A^f$, then

(a′) $\lim_{B \downarrow \xi} \|v\|^{A \cap B} = 0$ *on $D - \{\xi\}$ and the limit is 0 at ξ if $v(\xi) < +\infty$.*
(b′) $\delta_D^A(\xi, \{\xi\}) = 0$.
(c′) $G_D^A(\xi, \cdot) \neq G_D(\xi, \cdot)$ *on $\{G_D(\xi, \cdot) > 0\}$, the open connected component of D containing ξ.*

Observation (1). The fact that when B is a neighborhood of ξ, the relations $\xi \in A^f$ and $\xi \in (A \cap B)^f$ are equivalent implies that properties (c) and (c′) are true under the stated conditions if A is replaced by $A \cap B$, and B need not appear in (a). Property (a) was phrased using B to contrast with (a′), in which B is essential.

Observation (2). We have already noted in Section X.6 the fact that the only possible values of $\delta_D^A(\xi, \{\xi\})$ are 0 and 1. We now see that the value is 0 if and only if A is thin at ξ.

Proof. (a)–(c) Suppose that $\xi \in A^f$. Then $\xi \in (A \cap B)^f$, and as pointed out in Section 2, it follows that $\|v\|^{A \cap B}(\xi) = v(\xi)$; so (a) is true, and in particular, $G_D^A(\cdot, \xi) = G_D(\cdot, \xi)$ on D; that is (sweeping symmetry), (c) is true. Finally, (c) $\Leftrightarrow$ (b) by definition of δ_D^A.

(a′)–(c′) Suppose that $\xi \in D - A^f$. We can suppose that ξ is a (Euclidean topology) limit point of A, adjoining a countable set with limit point ξ to A if necessary to achieve this. The adjunction does not change smoothed reductions on A and does not change δ_D^A. Then (by Theorem 2) there is a positive superharmonic function u on D, finite at ξ but with limit $+\infty$ at ξ along A. If $\varepsilon > 0$ and if B is so small that $u > 1/\varepsilon$ on $A \cap (B - \{\xi\})$, it follows that

$$\delta_D^{A \cap B}(\xi, \{\xi\}) \leq \delta_D^{A \cap B}(\xi, D) = \|1\|^{A \cap B}(\xi) \leq \varepsilon \|u\|^{A \cap B}(\xi) \leq \varepsilon u(\xi) \qquad (3.1)$$

Hence $\lim_{B \downarrow \xi} \delta_D^{A \cap B}(\xi, \{\xi\}) = 0$, and it follows from Section X.6(c) that $\delta_D^A(\xi, \{\xi\}) = 0$, as asserted in Theorem 3(b′). Theorem 3(a′) now also follows from Section X.6(c), and Theorem 3(c′) follows from the definition of δ_D^A.

□

Application. Let A be a Borel subset of $\mathbb{R}^N$, and let ξ be a point at which A is thin. Then if $A_r = A \cap \partial B(\xi, r)$,

$$\lim_{r\to 0} \frac{l_{N-1}(A_r)}{r^{N-1}} = 0. \tag{3.2}$$

In fact, in the first place according to Theorem 3 with reductions relative to $D = B(\xi, 1)$,

$$\lim_{r\to 0} \| 1 \|^{A_r}(\xi) \le \lim_{r\to 0} \| 1 \|^{A\cap B(0,2r)}(\xi) = 0.$$

In the second place the rth smoothed reduction on the left obviously majorizes the smoothed reduction at ξ relative to $B(\xi, r)$ of the function 1 onto the boundary set A_r; the value of the latter smoothed reduction is $l_{N-1}(A_r)/\pi_N r^{N-1}$ according to III(4.3).

This application of Theorem 3 implies that a solid cone of revolution in $\mathbb{R}^N$ is not thin at its vertex. According to the criterion of Theorem 12 for Dirichlet problem regularity of a boundary point in terms of the fine topology, the nonthinness of a solid cone of revolution at its vertex is equivalent to the Poincaré–Zaremba criterion for Dirichlet problem regularity of a boundary point (Section VIII.15).

Strengthening of (a′). The following slight strengthening of (a′) will be needed below: if $\xi \in A^{\mathrm{f}}$ and if $v(\xi) < +\infty$, then

$$\lim_{B\downarrow\xi} \| \| v \|^B \|^A(\xi) = 0. \tag{a''}$$

To see this, observe that this limit is $\lim_{B\downarrow\xi} \delta_D^A(\xi, \|v\|^B)$, that according to the remarks at the beginning of this section, $\lim_{B\downarrow\xi} \|v\|^B = 0$ on $D - \{\xi\}$, and that according to (b′), $\delta_D^A(\xi, \{\xi\}) = 0$.

4. An Internal Limit Theorem

If D is an open subset of $\mathbb{R}^N$, if v is a positive superharmonic function on D, and if h is a strictly positive superharmonic function on D, then $u = v/h$ is defined in the obvious way on the set of points at which at least one of the functions v, h is finite valued. As so defined, the function u is fine continuous. The following theorem provides information on the character of u near a point of the set of common infinities of v and h. Let ν_v and ν_h be, respectively, the Riesz measures associated with v and h. Recall our convention that $d\nu_v/d\nu_h$ is the Radon–Nikodym derivative of the absolutely continuous component of ν_v with respect to ν_h. The singular component of ν_v with respect to ν_h will be denoted by ν_v^s.

Theorem. (a) *The function u has a fine limit $u^*(\zeta)$ at quasi every and $\nu_v + \nu_h$ almost every point ζ of D. Moreover $u^* < +\infty$ quasi everywhere and ν_h almost everywhere on D, and $u^* = +\infty$ ν_v^s almost everywhere on $\{\xi: v(\xi) = +\infty\}$.*

(b) *At ν_h almost every point ζ of the set $\{\xi: h(\xi) = +\infty\}$, equivalently, at ν_h almost every point ζ of an arbitrary polar subset of D,*

$$u^*(\zeta) = \frac{d\nu_v}{d\nu_h}(\zeta). \tag{4.1}$$

(c) *In particular, if $\zeta \in D$,*

$$\operatorname{f\,lim}_{\eta \to \zeta} \frac{v(\eta)}{G_D(\eta, \zeta)} = \nu_v(\{\zeta\}) = \inf_{D-\zeta} \frac{v}{G_D(\cdot, \zeta)}. \tag{4.2}$$

(d) *Let ζ be a point of D, and let F be a subset of D with (Euclidean topology) limit point ζ. If F is thin at ζ, there is a superharmonic potential v on D for which*

$$\lim_{F \ni \eta \to \zeta} \frac{v(\eta)}{G_D(\eta, \zeta)} = +\infty. \tag{4.3}$$

Conversely, if there is a positive superharmonic function v on $D - \{\zeta\}$ satisfying (4.3), *then F is thin at ζ.*

Observation. In the first term of (4.2) and in (4.3) G_D can be replaced by G without altering the validity of these equations because $\lim_{\eta \to \zeta} G_D(\eta, \zeta)/G(\eta, \zeta) = 1$.

Since the theorem is local, we can suppose in the proof that D is connected.

Proof of (c). The special result (4.2), in which $h = G_D(\cdot, \zeta)$ and so ν_h is the unit measure supported by $\{\zeta\}$, is so much easier to prove than the more general case (4.1) that we prove (4.2) separately, as follows. We have already seen (Section VII.10) that the last two terms in (4.2) are equal. In proving (4.2) it can be supposed, replacing v by $v - \nu_v(\{\zeta\})G_D(\cdot, \zeta)$ if necessary, that these two terms vanish; under this hypothesis, unless (4.2) is true, there is a strictly positive number b such that the set $B = \{\eta: v(\eta) > bG_D(\eta, \zeta)\}$ is not thin at ζ. Apply Theorem 3(c) to find that

$$v \geq \|v\|^B \geq bG_D^B(\cdot, \zeta) = bG_D(\cdot, \zeta),$$

contrary to the hypothesis that $v/G_D(\cdot, \zeta)$ has infimum 0. Hence (4.2) is true. □

Proof of (d). If v is a positive superharmonic function on $D - \{\zeta\}$ and if (4.3) is true, then since v has a positive superharmonic extension to D (Theorem V.5), the set F must be thin at ζ, in view of (4.2). Thus the converse half of (d) is true. In the other direction, if F is thin at ζ, choose $\eta(\neq\zeta)$ in D. Then [by Theorem 3(a')] there is a decreasing sequence $B_.$ of neighborhoods of ζ, shrinking to ζ, with

$$G_D^{F\cap B_n}(\eta,\zeta) < 2^{-n}.$$

There is therefore a positive superharmonic function v_n on D, majorizing $G_D(\cdot,\zeta)$ on $F\cap B_n$ with $v_n(\eta) < 2^{-n}$. After replacing v_n by $v_n \wedge G_D(\cdot,\zeta)$ if necessary, we can suppose that v_n is a potential on D. The potential $v = \Sigma_0^\infty v_n$ is at least $(n+1)\, G_D(\cdot,\zeta)$ on $F\cap B_n$ and therefore satisfies (4.3). □

Proof of (a). If $0 \le a < b$, define

$$A = \{\xi: v(\xi) \le ah(\xi)\}, \qquad B = \{\xi: v(\xi) \ge bh(\xi)\}. \tag{4.4}$$

According to Section VI.3(o),

$$h_{AB} + h_{ABAB} + \cdots \le \frac{v \wedge bh}{b-a}. \tag{4.5}$$

Now

$$h_A = \|h\|^A \ge \|G_D \nu_h\|^A = G_D(\nu_h \delta_D^A),$$

and $\delta_D^A(\xi,\{\xi\}) = 1$ when $\xi \in A^f \cap D$; so if ν_{hA} is the projection of ν_h on $A^f\cap D$, it follows that $h_A \ge G_D \nu_{hA}$. Similarly

$$h_{AB} \ge \|G_D \nu_{hA}\|^B = G_D(\nu_{hA}\delta_D^B);$$

so if ν_{hC} is the projection of ν_h on $C = A^f \cap B^f \cap D$, that is, if ν_{hC} is the projection of ν_{hA} on $B^f\cap D$, it follows that $h_{AB} \ge G_D\nu_{hC}$. Similarly all the other summands in (4.5) majorize $G_D\nu_{hC}$, and we conclude that $\nu_{hC} \equiv 0$. Since every point of D at which u^* does not exist is in C for some rational pair a and b, we conclude that u^* exists ν_h almost everywhere. Apply this result to h/v to show that u^* exists ν_v almost everywhere. As noted at the beginning of this section, the function u is fine continuous at every point at which it is defined; that is, u^* exists and is equal to u quasi everywhere on D. Moreover $u^* < +\infty$ quasi everywhere on D because $u^* = u = +\infty$ can be true only on the polar infinity set of v. The rest of the last sentence of (a) will be proved in the course of proving (b). □

Proof of (b). We show first that if F is a Borel polar subset of D and if $\nu_v(F) = 0$, then $u^* = 0$ ν_h almost everywhere on F. It is sufficient to prove this for F compact. Under this hypothesis on F, we can assume that v and h are potentials because if v and h are replaced by their reductions on some open neighborhood of F, relatively compact in D, then v, h, ν_v, and ν_h are unchanged on this neighborhood, and v and h become potentials. Define B by (4.4). Then (reductions relative to D)

$$v \geq [\![v]\!]^B \geq b[\![h]\!]^B = bG_D(\nu_h \delta_D^B).$$

Since $\delta_D^B(\xi, \{\xi\}) = 1$ when $\xi \in B^f$, we can continue this inequality to find that if ν_h' is the projection of ν_h on $B^f \cap F$, then $v \geq bG_D\nu_h'$. Now G_{D-F} is the restriction to $(D-F)\times(D-F)$ of G_D, and the restriction to $D-F$ of $bG_D\nu_h'$ is harmonic and is majorized by a potential, the restriction to $D-F$ of v. Hence $G_D\nu_h' \equiv 0$ on $D-F$; so $G_D\nu_h' \equiv 0$ on D, and therefore $\nu_h(B^f \cap F) = 0$; that is, $\operatorname{f\,lim\,sup}_{\eta\to\zeta} u(\eta) \leq b$ for ν_h almost every point ξ of F for all $b>0$. Hence $u^* = 0$ ν_h almost everywhere on F, as asserted. If this result is applied to h/v with F the trace of a ν_h null support for ν_v^s on the set of infinities of v, we find that $u^* = +\infty$ ν_v^s almost everywhere on F, as asserted in (a).

There remains the proof of (4.1). Let ϕ be the Radon–Nikodym derivative in (4.1). It is sufficient to show that if F is a ν_v^s null Borel polar subset of D, then $u^* = \phi$ ν_h almost everywhere on F. By what we have just proved, if ν_h° is the projection of ν_h on F,

$$\operatorname*{f\,lim}_{\eta\to\zeta} \frac{h(\eta) - G_D\nu_h^\circ(\eta)}{h(\eta)} = 0$$

at ν_h almost every point ζ of F; so in discussing u^* on F we can assume that F supports ν_h and that $h = G_D\nu_h$. Define

$$F_\alpha = \{\xi\colon \phi(\xi) \leq \alpha\}, \qquad v_\alpha(\xi) = \int_{F_\alpha} G_D(\xi,\eta)\phi(\eta)\nu_h(d\eta).$$

Then $v_\alpha \leq \alpha h$; so $\operatorname{f\,lim}_{\eta\to\zeta} v_\alpha(\eta)/h(\eta) \leq \alpha$ at ν_h almost every point ζ of F. Moreover, by what we have proved above, $\operatorname{f\,lim}_{\eta\to\zeta}[v(\eta) - v_\alpha(\eta)]/h(\eta) = 0$ at ν_h almost every point ζ of $F_\alpha \cap F$. Hence $u^* \leq \alpha$ ν_h almost everywhere on F_α, that is, $u^* \leq \alpha$ ν_h almost everywhere on F where $\phi \leq \alpha$, and therefore $u^* \leq \phi$ ν_h almost everywhere on F. A similar argument gives the reverse inequality; so $u^* = \phi$ ν_h almost everywhere on F, as was to be proved. Incidentally it now follows that $u^* < +\infty$ ν_h almost everywhere on D, as asserted in (a), because $u^* = u < +\infty$ quasi everywhere on D, and we have just proved that on the remaining polar set, $u^* = d\nu_v/d\nu_h$, which is finite ν_h almost everywhere. □

5. Extension of the Fine Topology to $\mathbb{R}^N \cup \{\infty\}$

Theorem. *Let A be a subset of $\mathbb{R}^N$, and let ϕ be an inversion of $\mathbb{R}^N$ in a sphere.*

(a) *Either* (a1) *$\phi(A)$ is thin at $\phi(\infty)$ for every inversion ϕ, or* (a2) *$\phi(A)$ is thin at $\phi(\infty)$ for no inversion ϕ.*

(b) *An unbounded set A is under case* (a1) *if and only if there is a positive superharmonic function u on some deleted neighborhood of the point ∞ with the property that*

$$\lim_{A \ni \eta \to \infty} \frac{u(\eta)}{\log|\eta|} = +\infty \qquad (N = 2), \tag{5.1}$$

$$\lim_{A \ni \eta \to \infty} u(\eta) = +\infty \qquad (N > 2). \tag{5.2}$$

Moreover, under case (a1) *with $N > 2$, there is a positive superharmonic function on $\mathbb{R}^N$ satisfying* (5.2).

(c) *If ζ* and *$\phi(\zeta)$ are finite, $\phi(A)$ is thin at $\phi(\zeta)$ if and only if A is thin at ζ.*

Proof of (a) *and* (b). (For $N > 2$). To avoid trivialities we assume that A is unbounded. Let ϕ be an inversion in $\partial B(\zeta, \delta)$. Then $\phi(A)$ has limit point ζ, and we prove first that if $\phi(A)$ is thin at ζ, there is a positive superharmonic function u on $\mathbb{R}^N$ satisfying (5.2). In fact by Theorem 4(b) there is a positive superharmonic function v on $D = \mathbb{R}^N$ such that (4.1) is true, and therefore the function

$$\frac{v(\phi)}{G(\zeta, \phi)} = \delta^{2N-4} v(\phi) G(\zeta, \cdot)$$

has limit $+\infty$ at the point ∞ along A. The function on the right is a positive superharmonic function on $\mathbb{R}^N - \{\zeta\}$ and in fact is a multiple of the Kelvin transform of v. This function has a positive superharmonic extension u to $\mathbb{R}^N$ (Theorem V.5), and u is the desired function. Conversely, if there is a positive superharmonic function u on some open deleted neighborhood of the point ∞ satisfying (5.2) and if ϕ is an inversion in $\partial B(\zeta, \delta)$, we prove that $\phi(A)$ is thin at ζ. The Kelvin transform v of u is a multiple of $u(\phi)G(\zeta, \cdot)$, defined positive and superharmonic on a deleted open neighborhood B of ζ, and $v/G(\zeta, \cdot)$ has limit $+\infty$ at ζ along $\phi(A)$; so [by Theorem 4(b)] the set A is thin at ζ, as was to be proved. Thus (a) and (b) of the theorem are true for $N > 2$. □

Proof of (c). (For $N \geq 2$). Under the hypothesis of (c), the set A is thin at ζ if and only if (Theorem 2) there is a superharmonic function u defined on a neighborhood of ζ such that (2.1) is true. The Kelvin transform of u under an inversion ϕ is a positive superharmonic function defined on a neighborhood of $\phi(\zeta)$ and satisfies the version of condition (2.1) at $\phi(\zeta)$ if and only if u satisfies (2.1). Hence (c) is true. □

The proof of Theorem 5(a) and (b) for $N = 2$ follows the proof for $N > 2$ and is left to the reader.

Extension of the Fine Topology to $\mathbb{R}^N \cup \{\infty\}$. We make the definition that a subset A_1 of $\mathbb{R}^N \cup \{\infty\}$ does not have fine limit point ∞, and we describe A_1 as thin at ∞ if $A = A_1 \cap \mathbb{R}^N$ is subsumed under case (a1) of Theorem 5. The fine topology is thereby extended to $\mathbb{R}^N \cup \{\infty\}$, and according to Theorem 5, an inversion is a fine topology homeomorphism of $\mathbb{R}^N \cup \{\infty\}$ onto itself.

The Limit Relation (4.1) *at the Point* ∞. Let v be a positive superharmonic function defined on a deleted neighborhood of a finite point ζ. Then (by Theorem V.5) if $v(\zeta)$ is defined as $\liminf_{\eta\to\zeta} v(\eta)$, the extended function v is positive and superharmonic on a neighborhood of ζ, and v satisfies the fine limit relation (4.1). Trivially, v also satisfies the fine continuity relation

$$\operatorname*{f\,lim}_{\eta\to\zeta} v(\eta) = \liminf_{\eta\to\zeta} v(\eta).$$

If u is a positive superharmonic function defined on a deleted neighborhood of the point ∞, then inversion of $\mathbb{R}^N$ in a sphere yields a context to which (4.1) and (5.3) can be applied to yield the following.

If u is a positive superharmonic function on $\mathbb{R}^N - \bar{B}(0, \delta)$, *then*

$$\begin{aligned} \operatorname*{f\,lim}_{\eta\to\infty} u(\eta) &= \liminf_{\eta\to\infty} u(\eta)\ (\le +\infty) \quad \text{if } N = 2,\\ \operatorname*{f\,lim}_{\eta\to\infty} u(\eta)|\eta|^{N-2} &= \liminf_{\eta\to\infty} u(\eta)|\eta|^{N-2} \quad \text{if } N > 2, \end{aligned} \tag{5.4}$$

$$\begin{aligned} \operatorname*{f\,lim}_{\eta\to\infty} \frac{u(\eta)}{\log\delta|\eta|} &= \inf_{|\eta|>\delta} \frac{u(\eta)}{\log\delta|\eta|} \quad \text{if } N = 2,\\ \operatorname*{f\,lim}_{\eta\to\infty} u(\eta) &= \inf_{|\eta|>\delta} u(\eta)[1 - (\delta|\eta|)^{2-N}]^{-1} \quad \text{if } N > 2. \end{aligned} \tag{5.5}$$

In fact, if $\mathbb{R}^N$ is inverted in $B(0, 1)$, the Kelvin transform of u has domain of definition $B(0, 1/\delta) - \{0\}$ and is given there by $\eta \mapsto v(\eta) = u(\eta|\eta|^{-2})|\eta|^{2-N}$. The relation (5.3) with $\zeta = 0$ yields (5.4), and (4.1) with $\zeta = 0$ and $D = B(0, 1/\delta)$ yields (5.5). The Green function G_D is evaluated in Section II.1. If the domain of definition of u is an arbitrary Greenian set that is a deleted neighborhood of the point ∞, Example (c) of Section VIII.8 can be used to evalute the right-hand side of (5.5).

6. The Fine Topology Derived Set of a Subset of $\mathbb{R}^N$

Theorem. *A polar subset of $\mathbb{R}^N$ has no fine limit point. Conversely, a subset of $\mathbb{R}^N$ with no fine limit point is polar. If A is an arbitrary subset of $\mathbb{R}^N$, the set A^f is a fine perfect G_δ set including quasi every point of A.*

The converse is true whether or not a point ∞ is adjoined to $\mathbb{R}^N$.

It was pointed out in Section 2 that a polar set has no fine limit point. Since Theorem 6 is a local theorem, it can be assumed in proving the last assertion that A is a subset of some Greenian set D. (We can take $D = \mathbb{R}^N$ if $N > 2$.) Reductions below are relative to D. Let $B_\bullet$ be an enumeration of the balls with closures in D and with rational centers and rational radii. According to Theorem 3, the Euclidean F_σ set

$$\bigcup_{j=0}^{\infty} \{\xi \in B_j : \llbracket 1 \rrbracket^{A \cap B_j}(\xi) \leq \tfrac{1}{2}\}$$

is $D - A^f$. Since each set in this union meets A in a polar set, the set $D - A^f$ is an F_σ set meeting A in a polar set. In particular, A is polar if $A^f = \varnothing$. The set A^f is trivially fine closed, and is fine perfect because if ξ were a fine isolated point of A^f, that is, if some deleted fine neighborhood B of ξ contained no point of A^f, then the set $B \cap A$ would contain none of its fine limit points and thus would be polar, and ξ could not be in A^f, contrary to hypothesis.

The Fine Boundary

If A is fine perfect in $\mathbb{R}^N$, that is, if $A = A^f \cap \mathbb{R}^N$, its fine boundary $\partial^f A = A^f \cap (\mathbb{R}^N - A)^f$ is a Euclidean G_δ set, and therefore its fine interior $A - \partial^f A$ is a Borel set, the difference between two G_δ sets. If A is analytic, its fine closure in $\mathbb{R}^N$ is analytic, and its fine interior $A - (\mathbb{R}^N - A)^f$ is analytic; so its fine boundary is universally measurable. In particular, if A is a Borel set, its fine interior and fine boundary are also Borel sets according to this argument.

7. Application to the Fundamental Convergence Theorem and to Reductions

In the Fundamental Convergence Theorem the lower envelope u of the specified family of superharmonic functions is equal quasi everywhere to its lower semicontinuous superharmonic smoothing $\underset{+}{u}$, and $\underset{+}{u}(\xi) = \liminf_{\eta \to \xi} u(\eta)$. Hence u has a fine limit at every point,

$$\underset{+}{u}(\xi) = \liminf_{\eta \to \xi} u(\eta) = \operatorname{f}\lim_{\eta \to \xi} u(\eta).$$

If D is a Greenian subset of $\mathbb{R}^N$, if A is a subset of D, and if v is a positive superharmonic function on D,

$$\|v\|^A = \|v\|^{A\cup(A^f\cap D)} = \|v\|^{A\cap A^f} = \|v\|^{A^f\cap D}. \tag{7.1}$$

In fact the first two terms are equal because, on the one hand, $A \subset A \cup (A^f \cap D)$ and, on the other hand, a superharmonic function on D majorizing v on A necessarily majorizes v on $A^f \cap D$. The first term is equal to the third because A differs from $A \cap A^f$ by a polar set. The fourth term is equal to the others because $A \cap A^f \subset A^f \subset A \cup A^f$.

8. Fine Topology Limits and Euclidean Topology Limits

A function defined on a deleted Euclidean topology neighborhood of a point and with a fine topology limit at the point need not have a Euclidean topology limit there. For example, the indicator function of a countable dense subset of $\mathbb{R}^N$ has fine limit 0 at every point but has no Euclidean topology limit at any point. Nevertheless the following lemma makes trivial the proof of Theorem 9, which establishes a surprisingly close relation between the two kinds of limit.

Lemma. *Let ξ be a point of $\mathbb{R}^N \cup \{\infty\}$, and let $A_.$ be a decreasing sequence of subsets of $\mathbb{R}^N$ not containing ξ. Suppose that each set A_k is a deleted fine neighborhood of ξ [has fine limit point ξ]. Then there is a set A with the property that for each k the part of A in a sufficiently small neighborhood of ξ is in A_k and that A is a deleted fine neighborhood of ξ [has fine limit point ξ].*

It can be assumed that ξ is finite. Define $B(\delta) = B(\xi, \delta)$. Suppose that each set A_k is a deleted fine neighborhood of ξ, and define $F_k = B(1) - (A_k \cup \{\xi\})$. The set F_k is thin at ξ, and to show one part of the lemma, it is sufficient to show that there is a set F, thin at ξ, with the property that for each k the part of F_k in a sufficiently small neighborhood of ξ is a subset of F. If for all but a finite number of values of k the point ξ is not a limit point of F_k, we can take F as the union of those sets F_k for which ξ is a limit point of F_k. Otherwise, it is no restriction to assume that ξ is a limit point of F_k for all k, and we can apply Theorem 2 to find a positive superharmonic function u_k on $B(1)$ with $u_k(\xi) < 2^{-k}$ and with limit $+\infty$ at ξ along F_k. The function $u = \sum_0^\infty u_k$ is positive and superharmonic on $B(1)$ with $u(\xi) < 2$ and with limit $+\infty$ at ξ along each set F_k. Choose r_k so that $1 = r_0 > r_1 > \cdots$, $\lim_{k\to\infty} r_k = 0$, and so that $u \geq k$ on $F_k \cap B(r_k)$. If $F = \bigcup_0^\infty (F_k \cap B(r_k))$, the function u has limit $+\infty$ at ξ along F; so F is thin at ξ, and $F_k \cap B(r_k) \subset F$ as desired. To prove the second assertion of the lemma, observe that if each set A_k has ξ as fine limit point, then

$$\|1\|^{A_k \cap B(s)}(\xi) = 1 = \lim_{r\to 0} \|1\|^{A_k \cap (B(s) - B(r))}(\xi)$$

for $s > 0$, according to Theorem 3 and Section VI.3(e). Hence it is possible to choose $r_0 = 1 > r_1 > \ldots$ successively in such a way that $\lim_{n\to\infty} r_n = 0$ and that the smoothed reduction on the right with $s = r_k$ and $r = r_{k+1}$ is at least $\frac{1}{2}$. Define

$$A = \bigcap_0^\infty [A_n \cup (B(1) - B(r_{n+1}))]$$

so that $A_k \cap (B(r_k) - B(r_{k+1})) \subset A$. Then $\|1\|^{A \cap B(r_k)}(\xi) \geq \frac{1}{2}$. If follows (by Theorem 3) that A is not thin at ξ, and since $A \cap B(r_k) \subset A_k$, the proof is complete.

9. Fine Topology Limits and Euclidean Topology Limits (Continued)

If u is a function from a deleted Euclidean topology neighborhood of a point ξ of $\mathbb{R}^N$ into a topological space S, the fine cluster set of u at ξ is defined as $\bigcap u(B)^-$, where B ranges through the class of deleted fine neighborhoods of ξ. In particular, if S is metric α is a fine cluster value of u at ξ if and only if f lim $\inf_{\eta\to\xi}$ dist $(u(\eta), \alpha) = 0$.

Theorem. *A function u from a deleted Euclidean neighborhood of ξ into a metric space has fine limit [fine cluster value] α at ξ if and only if u has limit α at ξ along a subset of $\mathbb{R}^N$ that is a deleted fine neighborhood of ξ [that is not thin at ξ].*

These conditions are trivially sufficient. Conversely, if u has fine limit α at ξ, that is, if for $n > 0$ the set of points

$$A_n = \{\eta: \text{dist}\,(u(\eta), \alpha) < \frac{1}{n}\}$$

is a deleted fine neighborhood of ξ, the set A of Lemma 8 is a set along which u has limit α at ξ. The cluster value assertion is treated similarly.

If the domain of u merely has ξ as a fine limit point, the cluster value definition and Theorem 9 have obvious rephrasings.

Continuity of a Superharmonic Function

If v is a function defined and superharmonic on an open neighborhood of a point ξ and if v has associated measure ν, the function v is lower semicontinuous by definition. In addition the following properties have now been proved or are now trivial.

(a) The function v is fine continuous at ξ, and therefore (by Theorem 9) there is a fine neighborhood A of ξ such that $v_{|A}$ has (Euclidean topology) limit $v(\xi)$ at ξ.

(b) If $v(\xi) = +\infty$, then by lower semicontinuity v is continuous at ξ, but v cannot become infinite too fast. More precisely (Theorem 4), there is a fine neighborhood A of ξ with the property that $[v/G(\xi, \cdot)]_{|A}$ has a (Euclidean topology) finite limit $v(\{\xi\})$ at ξ. If D is a Greenian set containing ξ, G here can be replaced by G_D.

10. Identification of A^f in Terms of a Special Function u^*

Lemma. *If D is a Greenian subset of $\mathbb{R}^N$, there is a bounded continuous potential $u^* = G_D\mu$ with the property that, for every subset A of D,*

$$A^f \cap D = \{\xi: [\![u^*]\!]^A(\xi) = u^*(\xi)\}$$

(reduction relative to D).

Let $B_\bullet$ be a sequence of balls with closures in D, forming a basis for the topology of D, and define

$$u^* = \sum_0^\infty 2^{-n} [\![1]\!]^{B_n}.$$

The summands are continuous potentials because $[\![1]\!]^{B_n} = \mu_{D-B_n}(\cdot, \partial B_n)$ on $D - B_n$ according to Section VIII.10, and the points of ∂B_n are regular boundary points of $D - B_n$. The function u^* is therefore a bounded continuous potential on D and if $A \subset D$

$$[\![u^*]\!]^A(\xi) = \sum_0^\infty 2^{-n} [\![[\![1]\!]^{B_n}]\!]^A(\xi). \tag{10.2}$$

If $\xi \in A^f \cap D$, the left side of (10.2) is trivially $u^*(\xi)$ (Section 3). If $\xi \in D - A^f$ and if $\xi \in B_n$, then $[\![1]\!]^{B_n}(\xi) = 1$, but $[\![[\![1]\!]^{B_n}]\!]^A(\xi)$ is arbitrarily small for small B_n [Section 3(a″)]. Hence the right side of (10.2) is strictly less than $u^*(\xi)$, as was to be proved.

11. Quasi-Lindelöf Property

An arbitrary union of open subsets of a second countable Hausdorff space is equal to some countable subunion (Lindelöf property). The fine topology has a slightly weaker property (quasi-Lindelöf property) as follows.

Theorem. *An arbitrary union of fine-open subsets of $\mathbb{R}^N$ differs by a polar set from some countable subunion.*

We shall prove the corresponding complementary assertion: an arbitrary intersection $\bigcap_{\alpha\in I} A_\alpha = A$ of fine-closed subsets of $\mathbb{R}^N$ differs by a polar set from some countable subintersection. It is sufficient to prove this assertion for subsets of some Greenian set D. Reductions below are relative to D. Let u^* be a superharmonic potential on D with the properties described in Lemma 10. According to the Fundamental Convergence Theorem, the family $A_.$ has a countable subfamily $\{A_\alpha, \alpha\in J\}$ such that

$$\left(\inf_{\alpha\in J} \| u^* \|^{A_\alpha}\right)_+ = \left(\inf_{\alpha\in I} \| u^* \|^{A_\alpha}\right)_+ .$$

To prove the theorem, we define $B = \bigcap_{\alpha\in J} A_\alpha$ and prove that $B^f = A^f$. Observe that

$$\| u^* \|^{B} \le \left(\inf_{\alpha\in J} \| u^* \|^{A_\alpha}\right)_+ = \left(\inf_{\alpha\in I} \| u^* \|^{A_\alpha}\right)_+ \le \| u^* \|^{A_\alpha} \le u^* \tag{11.1}$$

for $\alpha\in I$, and we conclude that $B^f \subset A^f_\alpha \subset A_\alpha$ for $\alpha\in I$. Hence $B^f \subset A$, and therefore $(B^f)^f = B^f \subset A^f$. Since $A \subset B$, it follows that $B^f = A^f$; so $B - A$ is polar, as was to be proved.

Application to the Approximation of a Fine-Open Set by (Euclidean Topology) Compact Subsets

If A is a fine-open subset of $\mathbb{R}^N$ to each point of A corresponds (Section1), a Euclidean topology compact subset of A which is a fine neighborhood of the point. The union of these compact sets covers A, and therefore (quasi-Lindelöf property) some countable subunion covers A up to a polar set.

12. Regularity in Terms of the Fine Topology

Theorem. *A finite Euclidean boundary point ζ of a Greenian subset D of $\mathbb{R}^N$ (or the point ∞ if $N = 2$ and D is unbounded) is regular if and only if $\zeta\in(\mathbb{R}^N - D)^f$.*

Recall that if $N > 2$, the point ∞ is a regular boundary point of every unbounded open set. Since the case of an infinite boundary point when $N = 2$ can be reduced to that of a finite boundary point by an inversion, only finite boundary points will be considered in the following proof. Suppose then that ζ is a finite Euclidean boundary point of D, and let B be a ball of center ζ, so small if $N = 2$ that $B\cup D$ is still Greenian. According to Section VIII.10, the restriction to D of $\| 1 \|^{B-D}$ (reduction relative to $B\cup D$)

is $\mu_D(\cdot, B \cap \partial D)$. If ζ is a regular boundary point of D, it follows that this smoothed reduction has limit 1 at ζ on approach along D. Moreover $\llbracket 1 \rrbracket^{B-D} = 1$ quasi everywhere on $B - D$; so $\llbracket 1 \rrbracket^{B-D}$ has limit 1 on approach to ζ excluding a polar set. Hence (fine continuity of the smoothed reduction) $\llbracket 1 \rrbracket^{B-D}(\zeta) = 1$. This smoothed reduction is majorized by the smoothed reduction on $B - D$ relative to a Greenian superset D_1 of $B \cup D$; so the latter smoothed reduction is also 1 at ζ. Since this is true for arbitrarily small B, it follows (from Theorem 3) that $\zeta \in (\mathbb{R}^N - D)^f$. Conversely, if $\zeta \in (\mathbb{R}^N - D)^f$, then $\llbracket 1 \rrbracket^{B-D}(\zeta) = 1$ (reduction relative to $B \cup D$) by fine continuity; so the lower semicontinuity of superharmonic functions implies that this smoothed reduction is continuous at ζ. Hence $\mu_D(\cdot, B \cap \partial D)$ has limit 1 at ζ for every B, and this condition implies regularity of ζ (Section VIII.8).

Observation (a). Regularity of a finite Euclidean boundary point ζ requires only that ζ be in $(\mathbb{R}^N - D)^f$, but quasi every finite point of ∂D is not only regular and therefore in $(\mathbb{R}^N - D)^f$ but is even in $(\partial D)^f$ (Section 6).

Observation (b). Theorem 12 shows that every condition that a point be a fine limit point of a set is a condition for Euclidean boundary point regularity. For example (Theorem 3), a finite Euclidean boundary point ζ of D is regular if and only if whenever D' is a Greenian superset of D containing ζ, it follows that $G_{D'}^{D'-D}(\zeta, \cdot) = G_{D'}(\zeta, \cdot)$. In fact, since regularity is a local property, D' need not be a superset of D (but of course must contain ζ).

It is natural as a counterpart of Theorem 12 to investigate the set of points of ∂D in D^f. The basic result is the following.

13. The Euclidean Boundary Set of Thinness of a Greenian Set

Theorem. *A Greenian subset D of $\mathbb{R}^N$ is thin at μ_D almost no finite Euclidean boundary point of D.*

If $N > 2$ define $D_1 = \mathbb{R}^N$; if $N = 2$ define $D_1 = D \cup B$, where B is a ball with center a finite boundary point of D and is so small that $D \cup B$ is Greenian. It is sufficient to prove that D is thin at μ_D almost no Euclidean boundary point of D in D_1. Let u^* be the positive superharmonic function described in Lemma 10 but associated with D_1 rather than D, and define $u = \llbracket u^* \rrbracket^D$ (reduction relative to D_1). Also define $u = u^* = 0$ on $\partial D \cap \partial D_1$. To prove the theorem, it is sufficient to prove that $u = u^*$ at μ_D almost every point of $D_1 \cap \partial D$, equivalently, since $u \le u^*$, to prove that $\mu_D(\cdot, u) = \mu_D(\cdot, u^*)$ on D. Now (by Theorem V.11) the Riesz measures associated with u and u^* vanish on polar sets because u and u^* are bounded; so [by Theorem VIII.18(c)] $\mu_D(\cdot, u) = \mathrm{GM}_D u$ and $\mu_D(\cdot, u^*) = \mathrm{GM}_D u^*$ on D. These harmonic minorants are equal because $u = u^*$ on D; so the harmonic averages are also equal, as was to be proved.

14. The Support of a Swept Measure

Theorem. *Let D be a Greenian subset of $\mathbb{R}^N$, let A be a subset of D, and let ξ be a point of D.*

(a) *If $\xi \in A^f$, then $\delta_D^A(\xi, \{\xi\}) = 1$.*
(b) *If μ is a measure on D, then the measure $\|\mu\|^A$ [in particular, the measure $\delta_D^A(\xi, \cdot)$] is supported by $A^f \cap D$.*
(c) *If A is a Borel set, if A is fine dense in itself, and if μ is a measure supported by A, then $\mu = \|\mu\|^A$.*
(d) *If v is a positive superharmonic function on D, the Riesz measure associated with $\|v\|^A$ is supported by $A^f \cap D$.*

Observation. We shall sharpen (b) in Section 18 by proving that $\delta_D^A(\xi, \cdot)$ is supported by the fine boundary of A relative to D whenever $\xi \in D - A^f$.

Proof. (a) If $\xi \in A^f$, then $\delta_D^A(\xi, \{\xi\}) = 1$ according to Theorem 3(b).
(b) If $u^\#$ is a positive superharmonic function on D, then

$$\|u^\#\|^A = \|\,\|u^\#\|^A\|^A = \delta_D^A(\cdot, u^\#) = \delta_D^A(\cdot, \|u^\#\|^A); \tag{14.1}$$

so if $u^\#$ has the properties described in Lemma 10, the set

$$\{u^\# > \|u^\#\|^A\} = D - A^f$$

is $\delta_D^A(\xi, \cdot)$ null for every ξ; that is, $\delta_D^A(\xi, \cdot)$ is supported by $A^f \cap D$. The evaluation of $\|\mu\|^A$ in X(5.2) shows that this swept measure is supported by $A^f \cap D$.
(c) The evaluation of $\|\mu\|^A$ in X(5.2) yields (b).
(d) If ν_A is the Riesz measure associated with $\|v\|^A$, then $\nu_A = \|\nu_A\|^A$ according to Section X.5; so (d) follows from (b). □

15. Characterization of $\|\mu\|^A$

Theorem. *Let D be a Greenian subset of $\mathbb{R}^N$, let A be a subset of D, and let μ be a measure on D with superharmonic potential $G_D\mu$. Then the swept measure $\|\mu\|^A$ is characterized uniquely by the following properties:*

(a) *$\|\mu\|^A$ is supported by $A^f \cap D$.*
(b) *$G_D\|\mu\|^A = G_D\mu$ on $A^f \cap D$.*

We have already proved that $\|\mu\|^A$ has these two properties. Conversely, if a measure ν on D has these properties, then $\nu = \|\nu\|^A$ by Theorem 14; so since $\delta_D^A(\xi, \cdot)$ is supported by $A^f \cap D$,

$$G_D\nu = G_D\|\nu\|^A = \delta_D^A(\cdot, G_D\nu) = \delta_D^A(\cdot, G_D\mu) = G_D\|\mu\|^A,$$

and it follows that $\nu = \|\mu\|^A$.

Observation. This theorem extends and makes more precise the characterization of $\|\mu\|^A$ for A closed in D made in Section X.1. In comparing the two characterizations when A is closed in D, observe that in Theorem 15(b) the equation holds everywhere on $A^f \cap D$ if it is known to hold quasi everywhere on this set because A^f is fine perfect.

16. A Special Reduction

The following lemma is the first step in a delicate analysis of the support of a swept measure.

Lemma. *Let D be a Greenian subset of $\mathbb{R}^N$, let h be a strictly positive harmonic function on D, and let $u = G_D\mu/h$ be an h-superharmonic h-potential on D. Then for every positive constant c ($\leq +\infty$)*

$$^h\|u\|^{\{u\leq c\}} = {}^h\|u\|^{\{u<c\}} = u \wedge c. \tag{16.1}$$

Observe that if $v = G_D\nu$, a rephrasing of (16.1) is

$$\|v\|^{\{v\leq ch\}} = \|v\|^{\{v<ch\}} = v \wedge (ch), \tag{16.1'}$$

valid for every superharmonic potential v. To avoid trivialities, it will be assumed in the proof of (16.1) that $0 < c < +\infty$. It is sufficient to prove equality of the first and third terms in (16.1) because this equality with $c(1 - 1/n)$ instead of c yields equality of the second and third terms when $n \to +\infty$. Now

$$^h\|u\|^{\{u\leq c\}} = {}^h\|u \wedge c\|^{\{u\leq c\}} \leq u \wedge c. \tag{16.2}$$

Furthermore there is equality in (16.2) quasi everywhere on the closed in D set $\{u \leq c\}$, which is a support of the measures associated with the finite-valued h-potentials in (16.2). Hence (by the domination principle) there is equality in (16.2); so the lemma is true.

17. The Fine Interior of a Set of Constancy of a Superharmonic Function

Corollary. *Let D be an open subset of $\mathbb{R}^N$, let v be a superharmonic function on D with associated Riesz measure ν, and let c be a constant. Then the fine interior of the set $\{v = c\}$ is ν null.*

This corollary is a strengthening of the result that the Euclidean interior of the set $\{v = c\}$ is ν null because v is harmonic on this interior, and the corollary suggests that Euclidean open sets and fine-open sets play similar roles relative to superharmonic functions. This idea will be discussed further in Section 19.

Since the corollary is local, it is no restriction in its proof to assume that D is a ball on which v is lower bounded. Without loss of generality we can then assume that v is a potential, $v = G_D\nu$, because we can first add a constant to v to make v positive and then replace v by its smoothed reduction on a strictly smaller ball than D, concentric with D. The resulting function v is a potential on the ball D and on the smaller ball differs by a constant from the original function. To prove the corollary in this context, let A be the fine interior of the set $\{v = c\}$. Lemma 16 applied to the function $v \wedge c$ yields $[\![v \wedge c]\!]^{\{v<c\}} = v \wedge c$. Since (by Theorem 14) the Riesz measure ν_c associated with this smoothed reduction is supported by $D \cap \{v < c\}^f$, which contains no point of A, it follows that $\nu_c(A) = 0$. Next apply Lemma 16 to derive

$$[\![v]\!]^{\{v\leq c\}} = v \wedge c, \qquad [\![\nu]\!]^{\{v\leq c\}} = \nu_c, \tag{17.1}$$

so that

$$0 = \nu_c(A) = \int_D \delta_D^{\{v\leq c\}}(\xi, A)\nu(d\xi). \tag{17.2}$$

Now $A \subset \{v \leq c\}^f$, and therefore [by Theorem 3(b)] the integrand in (17.2) is 1 when $\xi \in A$; so $\nu(A) = 0$, as was to be proved.

Pseudogeneralization of the Corollary to h-Harmonic Functions

If h is a strictly positive harmonic function on D, the function $u = v/h$ is h-superharmonic, and the fine interior of the set $\{u = c\} = \{v - ch = 0\}$ is ν null according to the corollary because the superharmonic function $v - ch$ has the same associated measure as v. Thus the corollary is not effectively more general when stated for h-superharmonic functions.

18. The Support of a Swept Measure (Continuation of Section 14)

If A is a subset of $\mathbb{R}^N$, we denote the set of fine limit points of A, the fine interior of A, the fine boundary of A, and the Euclidean boundary of A by A^f, A^{fi}, $\partial^f A$, and ∂A, respectively, we leave to the reader the proof that $\partial^f A^f \subset \partial^f(A \cup A^f) \subset \partial^f A$. Recall from Section 6 that A^f, A^{fi}, and $\partial^f A^f$ are Borel sets.

According to Theorem 14, the measure $\delta_D^A(\xi,\cdot)$ is supported by $A^f \cap D$ for all ξ in D, and if v is a positive superharmonic function on D, the Riesz measure associated with the smoothed reduction $[\![v]\!]^A = \delta_D^A(\cdot, v)$ is also supported by $A^f \cap D$. The following theorem sharpens the more elementary result (given in Section X.1) that $\delta_D^A(\xi,\cdot)$ is supported by $D \cap \partial A$ when $\xi \in D - \bar{A}$ by putting the latter result into the context of the fine topology and thereby also sharpens Theorem 14.

Theorem. *Let D be a Greenian subset of $\mathbb{R}^N$, and let A be a subset of D.*

(a) *If $\xi \in D \cap A^f$, then $\delta_D^A(\xi, \{\xi\}) = 1$. If $\xi \in D - A^f$, the measure $\delta_D^A(\xi,\cdot)$ is supported by $D \cap \partial^f A^f$.*

(b) *If v is a positive superharmonic function on D, the Riesz measure ν_A associated with $[\![v]\!]^A$ is supported by $D \cap A^f$. In particular, if v is harmonic on a Euclidean open superset of A, the measure ν_A is supported by $D \cap \partial^f A^f$.*

Proof of (a). The first assertion of (a) is Theorem 3(b). To prove the second assertion, let ξ be a point of $D - A^f$, and observe that the restriction v to $D - \{\xi\}$ of the function $G_D^A(\xi,\cdot) - G_D(\xi,\cdot)$ is superharmonic with associated Riesz measure the restriction of $\delta_D^A(\xi,\cdot)$ to the class of $\delta_D^A(\xi,\cdot)$ measurable subsets of $D - \{\xi\}$. The set $\{v = 0\}$ is a Borel superset of the set $A^f \cap (D - \{\xi\})$, and it follows from Corollary 17 that the fine interior of this superset is $\delta_D^A(\xi,\cdot)$ null. Since it is already known (Theorem 14) that $\delta_D^A(\xi,\cdot)$ is supported by $D \cap A^f$, it follows that this measure is supported by $D \cap \partial^f A^f$, as was to be proved. □

Proof of (b). The first assertion of (b) is included in Theorem 14(c). To prove the second assertion, observe that if v is harmonic on the open superset B of A, then the set $\{[\![v]\!]^A = v\} \cap B$ is the set on which the function $[\![v]\!]^A - v$, superharmonic on B, vanishes. The fine interior of this set includes A^{fi} and is null for the Riesz measure associated with $[\![v]\!]^A - v$ on B. Hence ν_A is supported by $D \cap \partial^f A^f$. □

Extension of Corollary 17. *If u_1 and u_2 are superharmonic functions on an open subset of $\mathbb{R}^N$, if μ_i is the Riesz measure associated with u_i, and if $A = \{u_1 = u_2\}$, then $\mu_1 = \mu_2$ on the Borel subsets of the fine interior of A.*

Since this extension of Corollary 17 (like the corollary) is local, it is no restriction in the proof to assume that u_1 and u_2 are defined on a ball D on which u_1 and u_2 are lower bounded and even (after addition of a sufficiently large constant) positive. Thus we shall suppose that u_i is a positive superharmonic function on a ball D. Now $[\![u_i]\!]^{D-A}(\xi) = \delta_D^{D-A}(\xi, u_i) = u_i(\xi)$ for quasi every point ξ of $D - A$ and therefore (fine continuity of superharmonic functions) also for ξ in D and a fine boundary point of $D - A$, that is, a fine boundary point of A. For ξ a fine interior point of A the measure $\delta_D^{D-A}(\xi, \cdot)$ is supported by the trace in D of the fine boundary of A, on which $u_1 = u_2$; so $\|u_1\|^{D-A} + u_2 = u_1 + \|u_2\|^{D-A}$. Since (by Theorem 18) the measure associated with a reduction on $D - A$ vanishes on Borel subsets of the fine interior of A, we conclude that $\mu_1 = \mu_2$ on these subsets.

19. Superharmonic Functions on Fine-Open Sets

Let D be a Greenian subset of $\mathbb{R}^N$ and let B be an open relatively compact subset of D. If $\xi \in B$, Theorem 13 implies that the measure $\mu_B(\xi, \cdot)$ is supported by $\partial^f B$. If v is a superharmonic function on D then

$$v(\xi) \geq \mu_B(\xi, v) = \delta_D^{D-B}(\xi, v), \tag{19.1}$$

and there is equality if v is harmonic. Suppose now that B more generally is a fine-open relatively compact subset of D and that $\xi \in B$. Then $\mu_B(\xi, \cdot)$ is no longer defined, but (by Theorem X.5) if v is a positive superharmonic function on D,

$$v(\xi) \geq \|v\|^{D-B}(\xi) = \delta_D^{D-B}(\xi, v). \tag{19.2}$$

It follows easily that

$$v(\xi) \geq \delta_D^{D-B}(\xi, v) \tag{19.3}$$

whenever v is a superharmonic function on D. In particular, there is equality in (19.3) when v is harmonic because (19.3) in then true for both v and $-v$. Observe that the measure $\delta_D^{D-B}(\xi, \cdot)$ is supported by $D \cap \partial^f(D - B) = \partial^f B$, according to Theorem 18.

Thus the superharmonic function inequality and the harmonic function equality can be generalized to be valid on fine-open sets, and this suggests that much of classical potential theory can be extended by replacing Euclidean open sets by fine-open sets. The next step is to develop a *fine potential theory* based on generalizations of superharmonic and harmonic functions to *fine* superharmonic and harmonic functions, defined on fine-open sets. We omit further discussion in this direction.

20. A Generalized Reduction

If D is a Greenian subset of $\mathbb{R}^N$, if A is a subset of D, and if v is a positive superharmonic function on D, we have defined the reduction R_v^A relative to D as the infimum of the class of positive superharmonic functions on D

majorizing v on A, equivalently, as the infimum of the class of superharmonic functions on D majorizing $v1_A$ on D. In this section we replace $v1_A$ by a function on D that is arbitrary except for boundedness and smoothness conditions.

Let D be a Greenian subset $\mathbb{R}^N$, let f be a locally lower bounded function from D into $\overline{\mathbb{R}}$ majorized by some superharmonic function on D, and let $\hat{R}_f$ be the smoothed infimum of the class of superharmonic functions on D majorizing f. According to the Fundamental Convergence Theorem, $\hat{R}_f$ is superharmonic and $\hat{R}_f \geq f$ quasi everywhere on D; so $\hat{R}_f$ is the infimum of the class of superharmonic functions on D majorizing f quasi everywhere on D. Let μ_f be the Riesz measure associated with $\hat{R}_f$. Recall that when C is a subset of $\mathbb{R}^N$, the notation C^f refers to the set of fine limit points of C.

Theorem. *In the above context,*

(a) *For* $\varepsilon > 0$ *the measure* μ_f *is supported by the set*

$$\{\hat{R}_f = +\infty\} \cup \bigcap_{\varepsilon>0} \{f \leq \hat{R}_f \leq f + \varepsilon < +\infty\}^f.$$

(b) *If* f *is fine upper semicontinuous, the measure* μ_f *is supported by the set* $\{\hat{R}_f = f\}^f$.

This theorem generalizes Theorem 18(b). In fact, suppose that A is a subset of D, nonpolar in each open connected component of D to avoid trivialities, and suppose that v is a strictly positive superharmonic function on D. Define $f = 1_A v$; so $\hat{R}_f = \hat{R}_v^A$, and suppose first that A is fine closed in D. Then f is fine upper semicontinuous, and according to the present theorem, μ_f is supported by the set $\{\hat{R}_f = f\}^f$ and so is supported by $\{\hat{R}_f = f\}$ because this set is fine closed in D and so is supported by A^f because $\hat{R}_f > f = 0$ on $D - A$. If A is not fine closed in D, we can replace A here by the set $A^f \cap D$, fine closed in D, because by fine continuity of superharmonic functions any superharmonic function majorizing v on A majorizes v on $A^f \cap D$; so $\hat{R}_f$ and $\hat{R}_v^A$ are not changed by this set change. The result for fine-closed sets shows that μ_f is supported by $A^f \cap D$ as stated in Theorem 18(b).

Proof (a) To prove Theorem 20, suppose first that $f \geq 0$, define

$$A_{\varepsilon\delta} = \{f \leq \hat{R}_f \leq f + \varepsilon < +\infty\} \cup \left\{f \geq \frac{1}{\delta}\right\}$$

for $\varepsilon > 0$, $\delta > 0$, and observe that the function

$$\frac{1}{1+\varepsilon\delta}\hat{R}_f + \frac{\varepsilon\delta}{1+\varepsilon\delta}\|\hat{R}_f\|^{A_{\varepsilon\delta}}$$

is superharmonic on D, is majorized by $\underset{+f}{R}$, is equal to $\underset{+f}{R}$ quasi everywhere on $A_{\varepsilon\delta}$, and majorizes f quasi everywhere on $D - A_{\varepsilon\delta}$ because the first summand does. Hence this superharmonic function coincides with $\underset{+f}{R}$; so $\underset{+f}{R} = \|\underset{+f}{R}\|^{A_{\varepsilon\delta}}$. According to Theorem 18, it follows that $\underset{+f}{R}$ is supported by $A_{\varepsilon\delta}^f$ for all strictly positive ε and δ. Now

$$\left\{f \geq \frac{1}{\delta}\right\} \subset \left\{\underset{+f}{R} \geq \frac{1}{\delta}\right\}$$

up to a polar set; so

$$\left\{f \geq \frac{1}{\delta}\right\}^f \subset \left\{\underset{+f}{R} \geq \frac{1}{\delta}\right\}^f \subset \left\{\underset{+f}{R} \geq \frac{1}{\delta}\right\},$$

and when $\delta \to 0$, the set on the right decreases to $\{\underset{+f}{R} = +\infty\}$. Thus Theorem 20(a) has now been proved for $f \geq 0$.

To reduce the general case to the case of positive f, observe that if B is a compact subset of D and if D_0 is an open neighborhood of B, relatively compact in D, then the restriction of $\underset{+f}{R}$ to D_0 is the smoothed infimum of the class of superharmonic functions majorizing $\underset{+f}{R}$ on $D_0 - B$ and majorizing f on B. Without loss of generality in the discussion of μ_f we can suppose, adding a constant to f if necessary, that the restriction to D_0 of f is positive. It now follows from what we have already proved that for every B the projection of μ_f on B is supported by $A_{\varepsilon\delta}^f \cap D_0$ for all strictly positive ε and δ; so μ_f is supported by $A_{\varepsilon\delta}^f$ for all strictly positive ε and δ, and the rest of (a) follows as in the positive case.

(b) If f is fine upper semicontinuous, the set

$$A'_{\varepsilon\varepsilon} = \{\underset{+f}{R} \leq f + \varepsilon < +\infty\} \cup \left\{f \geq \frac{1}{\varepsilon}\right\}$$

differs from $A_{\varepsilon\varepsilon}$ by a polar set and is fine closed in D; so

$$A'^f_{\varepsilon\varepsilon} = A^f_{\varepsilon\varepsilon} \subset A'_{\varepsilon\varepsilon},$$

and part (a) of the theorem implies, when $\varepsilon \to 0$, that the measure μ_f is supported by the union of $\{\underset{+f}{R} = f\}$ and the polar set $\{\underset{+f}{R} < f\}$. Since (by Theorem V.11) $\underset{+f}{R} = +\infty$ at μ_f almost every point of a polar set, the set $\{\underset{+f}{R} < f\}$ must be μ_f null; so part (b) of the theorem is true. □

Special Positivity Case

Suppose in the theorem that f is positive (and as in the theorem is majorized on D by some superharmonic function), and define $A = \{f > 0\}$. Then $\underset{+f}{R}$

is a positive superharmonic function, and it is easy to check that $\underset{+f}{R}_f = \|\underset{+f}{R}_f\|^A$. We conclude from Theorem 14 that the measure associated with $\underset{+f}{R}_f$ is supported by $\partial^f A^f \cap D$.

21. Limits of Superharmonic Functions at Irregular Boundary Points of Their Domains

If u is a lower-bounded superharmonic function defined on a deleted open neighborhood of a point ζ of $\mathbb{R}^N$, then (by Theorem V.5) u can be extended to be superharmonic on the neighborhood; so u has a fine topology limit ($\leq +\infty$) at ζ. The following theorem shows that in this result the neighborhood of ζ need be merely a fine neighborhood. Recall (from Theorem 12) that a Euclidean boundary point of an open set is irregular if and only if the set is a deleted fine neighborhood of the point.

Theorem. *Let u be a lower-bounded superharmonic function on an open subset D of $\mathbb{R}^N$, and let ζ be an irregular Euclidean boundary point of D. Then* $\operatorname{f\,lim}_{\xi\to\zeta} u(\xi)$ *(denoted below by ${}^f u(\zeta)$) exists* ($\leq +\infty$).

Proof. If $N > 2$, the point ζ must be finite because (by Theorem VIII.4) ∞ is a regular boundary point of every unbounded open set. If $N = 2$ and if $\zeta = \infty$, an inversion of the plane in a finite boundary point reduces the theorem to one with ζ finite. Thus we can assume that ζ is finite and therefore also that D is bounded. If ζ is an isolated boundary point of D, u has a superharmonic extension to $D \cup \{\zeta\}$ and so has a fine limit at ζ, the value at ζ of the extension, namely, $\liminf_{\xi\to\zeta} u(\xi)$. If ζ is not an isolated boundary point of D, the fact that ζ is irregular implies that $\mathbb{R}^N - D$ is thin at ζ and therefore (by Theorem 2) that there is a positive superharmonic function v on an open neighborhood of $\bar{D}$, finite at ζ, with limit $+\infty$ at ζ along $\mathbb{R}^N - D$. Define $c = \liminf_{\xi\to\zeta}(u + v_{|D})(\xi)$. If $c = +\infty$, then u has fine limit $+\infty$ at ζ because v has the finite fine limit $v(\zeta)$ at ζ along D. If $c < +\infty$, the function $u_1 = (u + v_{|D}) \wedge (c + 1)$ is superharmonic on D, is majorized by $c + 1$, and has limit $c + 1$ at every boundary point of D in some open neighborhood B of ζ except at ζ itself. Extend u_1 to $D \cup (B - \{\zeta\})$ by setting $u_1 = c + 1$ on $B - (D \cup \{\zeta\})$. Then u_1 is superharmonic and lower bounded on a deleted neighborhood of ζ, and as noted above, such a function has a fine limit at ζ, necessarily the inferior limit of the function at ζ. Thus

$$\operatorname*{f\,lim}_{\xi\to\zeta} u_1(\xi) = \liminf_{\xi\to\zeta} u_1(\xi) = c.$$

Hence $u + v_{|D}$ has fine limit c at ζ; so u has fine limit $c - v(\zeta)$ at ζ. □

Observation. The preceding proof shows that if the fine limit of u in Theorem 21 is finite and if v is a superharmonic function on an open neighborhood of ζ, finite at ζ, with limit $+\infty$ at ζ along $\mathbb{R}^N - D$, then $u + v$ has a superharmonic extension to the union of D and an open neighborhood of ζ.

Application to Harmonic Measure

Let D be a Greenian subset of $\mathbb{R}^N$ with an irregular Euclidean boundary point ζ. If f is a finite-valued continuous function on the Euclidean boundary of D, then $\mu_D(\cdot,f) = H_f$ is a bounded harmonic function on D and thus has a finite fine limit at ζ. If $f_{\bullet}$ is a sequence of such continuous functions on D, dense in $\mathbb{C}(\partial D)$, then an application of Theorem 21 combined with Lemma 8 shows that there is a subset A of D, a deleted fine neighborhood of ζ, such that every function $\mu_D(\cdot,f_n)$ has a limit at ζ along A. It follows that $\lim_{A \ni \xi \to \zeta} \mu_D(\xi,\cdot)$ exists in the sense of vague convergence of measures on ∂D. Denote this limit by ${}^f\mu_D(\zeta,\cdot)$. In the next section we shall show that this limiting harmonic measure has many of the properties of ordinary harmonic measure.

Modifications of Theorem 21

If in Theorem 21 it is supposed only that u is lower bounded on a deleted fine neighborhood in D of ζ, say $u > \alpha$ on such a set, the theorem can be applied as originally stated to u on $\{u > \alpha\}$ to verify that the original conclusion remains valid. Moreover a variation of the argument of the theorem shows that if $u/G(\zeta,\cdot)$ is lower bounded on D or even merely lower bounded on a deleted fine neighborhood in D of ζ, then this ratio has a finite fine limit at ζ. The special case in which ζ is an isolated boundary point of D and u is positive on D is covered by Theorem 4(c).

22. The Limit Harmonic Measure ${}^f\mu_D$

In this section boundaries of subsets of $\mathbb{R}^N$ are relative to the Euclidean topology. Let D be a Greenian subset of $\mathbb{R}^N$, and let ζ be an irregular boundary point of D; that is, let D be a deleted fine neighborhood of ζ. Then (from Section 21) $\lim_{\xi \to \zeta} \mu_D(\xi,\cdot)$ exists in the sense of vague convergence of measures on ∂D when ξ tends to ζ along some subset of D that is a deleted fine neighborhood of ζ. In the following Γ is the class of open subsets of D that are deleted fine neighborhoods of ζ.

(a) If A is a Borel subset of ∂D and if $B \in \Gamma$, then

$$ {}^f\mu_D(\zeta,A) = {}^f\mu_B(\zeta,A \cap \partial B) + \int_{D \cap \partial B} \mu_D(\eta,A)\,{}^f\mu_B(\zeta,d\eta). \tag{22.1}$$

According to VIII(8.3), this equation is true when ζ is a point of B. Equation VIII(8.3) is equivalent to

$$\mu_D(\xi,f) = \mu_B(\xi,f_B) + \int_{D\cap\partial B} \mu_D(\eta,f)\mu_B(\xi,d\eta) \tag{22.2}$$

for all finite-valued positive continuous functions f on ∂D, with $f_B = f$ on $\partial B \cap \partial D$ and $f_B = 0$ elsewhere on ∂B. The function $\mu_B(\cdot,f_B)$ is harmonic on B and so (by Theorem 21) has a fine limit at ζ. The function equal to $\mu_D(\cdot,f)$ on $D \cap \partial B$ and equal to 0 elsewhere on ∂B is lower semicontinuous. Hence if we take fine limits in (22.2) when $\xi \to \zeta$, we find

$$^f\mu_D(\zeta,f) \geq \operatorname{f}\lim_{\xi\to\zeta} \mu_B(\xi,f_B) + \int_{D\cap\partial B} \mu_D(\eta,f)\,^f\mu_B(\zeta,d\eta), \tag{22.3}$$

and if this inequality is combined with the corresponding inequality for $-f + \max_D f$, we find that there is equality in (22.3). Since f_B is upper semicontinuous on ∂B, it follows that

$$^f\mu_D(\zeta,f) \leq {}^f\mu_B(\zeta,f_B) + \int_{D\cap\partial B} \mu_D(\eta,f)\,^f\mu_B(\zeta,d\eta). \tag{22.4}$$

Finally, this inequality combined with the corresponding inequality for $-f + \max_D f$ yields equality in (22.4); that is,

$$^f\mu_D(\zeta,f) = {}^f\mu_B(\zeta,f_B) + \int_{D\cap\partial B} \mu_D(\eta,f)\,^f\mu_B(\zeta,d\eta). \tag{22.5}$$

Since this equality is true for positive continuous f on ∂D, it is true for arbitrary one-side-bounded Borel measurable f on ∂B, and therefore (22.1) is true.

(b) If A is a harmonic measure null subset of ∂D, then $^f\mu_D(\zeta, A) = 0$.

We prove this in two steps, first proving it when ζ is not in A by proving it for A compact and not containing ζ. This result is implied by (22.1) if B is so small that $A \cap \partial B = \varnothing$. Second, we prove that $^f\mu_D(\zeta, \{\zeta\}) = 0$. To prove this, we use the fact that $\mathbb{R}^N - D$ is thin at ζ to find a positive superharmonic function v on the union of D with an open neighborhood of ζ, with $v(\zeta) < +\infty$ and with v having limit $+\infty$ at ζ along $\mathbb{R}^N - D$. Define the lower semicontinuous function v' on ∂D be setting $v'(\eta) = \liminf_{D\ni\xi\to\eta} v(\xi)$, and define v'' as v' on $\partial D - \{\zeta\}$ but $v''(\zeta) = +\infty$. Then v'' is also lower semicontinuous. If $\varepsilon > 0$,

$$\begin{aligned} ^f\mu_D(\zeta, \{\zeta\}) &\leq {}^f\mu_D(\zeta, \varepsilon v'') \leq \liminf_{\xi\to\zeta} \mu_D(\xi, \varepsilon v'') \\ &= \liminf_{\xi\to\zeta} \mu_D(\xi, \varepsilon v') \leq \varepsilon v(\zeta). \end{aligned} \tag{22.6}$$

Here we have used the fact that $v'' = v'$ off the harmonic measure null set $\{\zeta\}$ and that v is in the upper PWB class on D for the boundary function $\varepsilon v'$. Inequality (22.6) implies that ${}^f\mu_D(\zeta, \{\zeta\}) = 0$, as asserted.

From now on when u is a function in D, and $B \in \Gamma$, the averages $\mu_B(\xi, u)$ and ${}^f\mu_B(\zeta, u)$ will be used under the convention that in these averages $u(\eta)$ is to be taken as $\liminf_{B \ni \xi \to \eta} u(\xi)$ when $\eta \in \partial B \cap \partial D$. If u is superharmonic, this convention means that the function to be integrated on ∂B is lower semicontinuous.

(c) If u is a lower-bounded superharmonic function on D and if B_1 and B_2 are in Γ with $B_1 \subset B_2$, then

$$ {}^f u(\zeta) \geq {}^f\mu_{B_1}(\zeta, u) \geq {}^f\mu_{B_2}(\zeta, u), \tag{22.7} $$

and ${}^f u(\zeta) = \sup\{\mu_B(\zeta, u) \colon B \in \Gamma\}$.

Recall the convention just made for u on $\partial B_i \cap \partial D$, and observe that (from Section VIII.8) $u \geq \mu_{B_1}(\cdot, u)$ on B_1. (Here are below the convention on the values of u assigned at boundary points of the sets involved may not be the same as those in referenced inequalities, but the differences are in favorable directions.) On approach to ζ along a suitable deleted fine neighborhood of ζ this inequality yields the first inequality in (22.7). Similarly $u \geq \mu_D(\cdot, u)$ on D; so (22.5) with f replaced by u yields the inequality ${}^f\mu_D(\zeta, u) \leq {}^f\mu_B(\zeta, u)$ and thereby yields the second inequality in (22.7) if the pair B, D is replaced by the pair B_1, B_2. Finally, if $B = \{\xi \in D \colon u(\xi) > \alpha\}$ with $\alpha < {}^f u(\zeta)$, then ${}^f\mu_B(\zeta, u) \geq \alpha$; so the last assertion of (c) is true. It is trivial that the supremum in this assertion is not decreased by the additional condition on B that $\partial B \cap \partial D = \{\zeta\}$.

(d) The function $u \mapsto {}^f u(\zeta)$ is lower semicontinuous on the space of positive superharmonic functions on D in the topology of pointwise convergence.

In fact, if $u_\cdot$ is a convergent sequence in this space, with limit u, and if $\alpha < {}^f u(\zeta)$, choose B in Γ in such a way that $\partial B \cap \partial D = \{\zeta\}$ and that ${}^f\mu_B(\zeta, u) > \alpha$. Then apply Fatou's lemma to obtain

$$ \liminf_{n\to\infty} {}^f u_n(\zeta) \geq \liminf_{n\to\infty} {}^f\mu_B(\zeta, u_n) \geq {}^f\mu_B(\zeta, u) > \alpha. $$

This inequality implies the stated lower semicontinuity.

(e) By definition of regularity, if η is a regular boundary point of D, then $\lim_{\xi\to\eta} \mu_D(\xi, \cdot)$ in the sense of vague convergence is the unit measure supported by $\{\eta\}$, whereas if η is an irregular boundary point of D, then $\operatorname{f\,lim}_{\xi\to\eta} \mu_D(\xi, \cdot) = {}^f\mu_D(\eta, \cdot)$. That is, in terms of Dirichlet solutions if f is a finite-valued continuous function on ∂D, then $\lim_{\xi\to\eta} H_f(\xi) = f(\eta)$ if η is a regular boundary point, but $\operatorname{f\,lim}_{\xi\to\eta} H_f(\xi) = {}^f\mu_D(\eta, f)$ if η is an irregular boundary point.

(f) If ζ as above is an irregular boundary point of D and if $\eta \in D$, then according to Theorem 21, the function $G_D(\cdot, \eta)$ has a fine limit at ζ; we denote this fine limit by ${}^f G_D(\zeta, \eta)$. In view of the extension $G_D^{=}$ of G_D defined in Section VII.4, not only does this fine limit exist, but

$$^{f}G_D(\zeta,\eta) = \limsup_{\xi\to\zeta} G_D(\xi,\eta), \tag{22.8}$$

and VIII(18.2) implies that $^{f}G_D(\zeta,\cdot)$ is harmonic on D, as is also easily proved directly. When D is connected, the superior limit in (22.8) must be strictly positive; that is, $^{f}G_D(\zeta,\cdot)$ must be strictly positive, in view of the criterion [Section VIII.14, Application (c)] in terms of G_D that a boundary point of D be irregular.

23. Extension of the Domination Principle

Recall (from Section 4) that if h and v are strictly positive superharmonic functions on a Greenian subset D of $\mathbb{R}^N$ with respective associated measures ν_h and ν_v, if $u = v/h$ wherever either h or v is finite valued, and if $u^*(\zeta)$ is defined as the fine limit of u at ζ wherever this fine limit exists, then u^* is defined quasi everywhere and $\nu_h + \nu_v$ almost everywhere on D. The following extension of the domination principle (Theorem V.10), unlike Theorem V.10, does not suppose that polar sets are ν_h null.

Theorem. *Let $h = G_D\nu_h$ be a potential on a Greenian subset D of $\mathbb{R}^N$, and let v be a positive superharmonic function on D. Then $h \le v$ if (fine limit function notation) $(v/h)^* \ge 1$ ν_h almost everywhere.*

To prove this theorem, which obviously includes Theorem V.10, let B be a Borel support of ν_h on which $(v/h)^*$ is defined and ≥ 1. If $0 < c < 1$, the set $A = \{v \ge ch\}$ is a deleted fine neighborhood of every point of B; so $B \subset A^f$, and therefore $\delta_D^A(\xi,\{\xi\}) = 1$ when $\xi \in B$. Hence

$$v \ge \llbracket v \rrbracket^A \ge c\llbracket h \rrbracket^A = cG_D(\nu_h\delta_D^A) = ch;$$

so $v \ge h$.

Chapter XII

The Martin Boundary

1. Motivation

Let D be an open subset of $\mathbb{R}^N$. If D is a ball, its Euclidean boundary is so well adapted to it from a potential theoretic point of view that the following statements are true.

(a) The class of minimal harmonic functions on D is in a one-to-one correspondence with the Euclidean boundary points, $\eta \Leftrightarrow K(\eta, \cdot)$ (see Section II.16), and every positive harmonic function u on D is representable as an integral $\int_{\partial D} K(\eta, \cdot) M_u(d\eta)$ with M_u a uniquely defined measure on the ball boundary. The lattices of positive harmonic functions on D and of measures on ∂D are thereby in a one-to-one order-preserving correspondence, and the vector lattices of differences between positive harmonic functions on the ball and signed measures on the ball boundary are linearly and order isomorphic.

(b) The Euclidean boundary of the ball is universally resolutive and universally internally resolutive (Section IX.12).

(c) Every boundary point is regular, so that $G_D(\xi, \cdot)$ has limit 0 at every boundary point. Moreover (up to a multiplicative constant) the normal derivative of $G_D(\xi, \cdot)$ at the boundary is the density relative to l_{N-1} of harmonic measure.

(d) The Fatou theorem and its generalizations (Section II.15) assign to each positive h-harmonic function, and therefore to each h-harmonic function in $\mathbf{L}^1(\mu_{D-}^h)$ on the ball, a boundary function. In particular, the Fatou boundary function of a PWB^h solution H_f^h is equal M_h almost everywhere to f.

The Martin boundary of a Greenian subset D of $\mathbb{R}^N$ is defined in such a way that it has some of these properties. Observe, however, that other compactifications may be better adapted to other properties. For example, the Kuramochi boundary is specially well adapted to the second boundary value problem.

The only restriction to be imposed on D is that it be Greenian and connected. It will be necessary to generalize two concepts playing a central role in the ball case: special approach to the boundary (radial or nontangential) and the normal derivative of the Green function at the boundary. The fine topology will be extended to the Martin boundary, and approach to a Martin boundary point in this topology will replace both radial and nontangential approach to the boundary in the ball case. If $D = B(0, r)$ and if $\xi \in D$, then $G_D(\xi, \eta) \sim \text{const.}(r - |\eta|)$ when $|\eta| \to r$ so that if ν is a measure on D with compact support, $G_D\nu(\eta) \sim \text{const.}(r - |\eta|)$ near the boundary. This fact suggests that the normal derivative of $G_D(\xi, \cdot)$ at the boundary point ζ should be replaced by

$$\lim_{\eta \to \zeta} \frac{G_D(\xi, \eta)}{G_D\nu(\eta)}$$

when ζ is a Martin boundary point, if this limit exists, for some convenient measure ν. It was Martin's idea to define a boundary in such a way that this limit does exist and in fact to use the existence of this limit to define a boundary.

Topological Conventions

Recall that throughout this book in dealing with $\mathbb{R}^N$ and its subsets, if no topology is specified, the topology of $\mathbb{R}^N$ is to be understood as the usual topology, sometimes identified as the *Euclidean* topology, compactified by a point at infinity when boundaries of unbounded sets are in question.

2. The Martin Functions

Let D be a connected Greenian subset of $\mathbb{R}^N$, let ν be a not identically 0 measure on D with compact support $\sigma_\nu \subset D$, and define

$$K_\nu(\eta, \xi) = \frac{G_D(\eta, \xi)}{G_D\nu(\eta)} \tag{2.1}$$

for η and ξ in D except that $K_\nu(\xi, \xi)$ is left undefined when $G_D\nu(\xi) = +\infty$. The function K_ν will be called the *Martin function based on* ν, and K will sometimes be written for K_ν when the choice of ν is irrelevant. If ν is the probability measure supported by a singleton $\{\xi_0\}$, so that $K_\nu(\eta, \xi) = G_D(\eta, \xi)/G_D(\eta, \xi_0)$, the Martin function will be said to be based on ξ_0.

For η a point of finiteness of $G_D\nu$ the function $K_\nu(\eta, \cdot)$ is a strictly positive harmonic function on $D - \{\eta\}$, and for ξ in D the function $K_\nu(\cdot, \xi)$ is continuous on $D - \sigma_\nu$. In view of the properties of G_D listed in Section VII.3, if A is a compact subset of D and if A_1 is a neighborhood of $A \cup \sigma_\nu$, the functions K_ν and $1/K_\nu$ are bounded on $(D - A_1) \times A$. Moreover

$$\int_{\sigma_\nu} K_\nu(\eta, \xi)\nu(d\xi) = 1 \tag{2.2}$$

if $G_D\nu(\eta) < +\infty$.

3. The Martin Space

Theorem. *Let D be a connected Greenian subset of $\mathbb{R}^N$. There is a unique up to homeomorphisms metrizable compactification D^M of D with the following properties.*

(a) *Each Martin function K_ν on $(D - \sigma_\nu) \times D$ has a continuous extension (also denoted by K_ν) to $(D^M - \sigma_\nu) \times D$.*

(b) *$K_\nu(\eta_1, \cdot) = K_\nu(\eta_2, \cdot)$ if and only if $\eta_1 = \eta_2$.*

The boundary $\partial^M D = D^M - D$ obtained in this way will be called the Martin boundary, and any metric for D^M compatible with its topology will be called a Martin metric. The function K_ν is now defined on its original domain augmented by $\partial^M D \times D$.

Observation. According to the theorem, if $\eta_\cdot$ is a sequence of points of D with limit η in $\partial^M D$ and if K_ν is a Martin function, then $\lim_{n\to\infty} K_\nu(\eta_n, \cdot) = K_\nu(\eta, \cdot)$ is a strictly positive harmonic function satisfying (2.2) because in any open relatively compact subset of D the above limit is the limit of a uniformly bounded sequence of positive harmonic functions satisfying (2.2). Moreover, because of (b) and the convergence remarks in Section 2, a sequence $\eta_\cdot$ in D with no limit point in D converges to a boundary point if and only if $\lim_{n\to\infty} K_\nu(\eta_n, \cdot)$ exists; the limit function is then $K_\nu(\eta, \cdot)$ with $\eta = \lim_{n\to\infty} \eta_n$. Thus the existence assertion of the theorem implies that each point η of $D^M - \sigma_\nu$ is associated with a function $K_\nu(\eta, \cdot)$, harmonic on D if $\eta \in \partial^M D$, harmonic except for an infinity at η otherwise, and the class of associated functions under the topology of pointwise convergence is homeomorphic with $D^M - \sigma_\nu$. This homeomorphism implies the uniqueness of a compactification satisfying (a) and (b) above. Roughly, the Martin space D^M is the compactified space of the set of Green functions $\{G_D(\eta, \cdot), \eta \in D\}$ normalized to be conditionally compact under pointwise convergence. Observe that if K_ν and K_μ are Martin functions, then

$$K_\mu(\eta, \xi) = \frac{K_\nu(\eta, \xi)}{\displaystyle\int_{\sigma_\mu} K_\nu(\eta, \zeta)\mu(d\zeta)} \qquad [\eta \in D - (\sigma_\mu \cup \sigma_\nu)]. \tag{3.1}$$

Hence for η in $D - (\sigma_\mu \cup \sigma_\nu)$ the function $K_\mu(\eta, \cdot)$ associated with the point η and the Martin function K_μ is a constant multiple of the function $K_\nu(\eta, \cdot)$

associated with η and K_ν. The choice of reference measure for a Martin function is a matter of convenience, although the most common choice is one with σ_ν a singleton, that is, with K_ν based on a reference point.

Proof of theorem. Let K_ν be a Martin function, and suppose that ν has been chosen to make $G_D\nu$ finite valued and continuous. Let f be a strictly positive continuous function on D, l_N integrable over D, and define

$$d_\nu(\xi,\eta) = \int_D [1 \wedge |K_\nu(\xi,\zeta) - K_\nu(\eta,\zeta)|] f(\zeta) l_N(d\zeta). \tag{3.2}$$

The function d_ν is a metric for D, endowing D with the Euclidean topology: so if $\eta_{\bullet}$ is a Cauchy sequence for d_ν and if some subsequence $\eta_{a_{\bullet}}$ has Euclidean topology limit η in D, then $\lim_{n\to\infty} d_\nu(\eta,\eta_{a_n}) = 0$. It follows that $\eta_{\bullet}$ is convergent to η in the d_ν metric and therefore also in the Euclidean metric. If $\eta_{\bullet}$ is a sequence of points in D and if no subsequence of $\eta_{\bullet}$ has a Euclidean topology limit in D, then according to Section 2, if B is an open relatively compact subset of D, the sequence $n \mapsto K_\nu(\eta_n,\cdot)$ on B is a uniformly bounded sequence of harmonic functions if finitely many values of n are omitted. It follows from the convergence properties of harmonic function sequences (Section II.3) that the sequence $n \mapsto K_\nu(\eta_n,\cdot)$ has a subsequence $j \mapsto K_\nu(\eta_{a_j},\cdot)$ converging locally uniformly on D to a harmonic function h. This subsequence is a Cauchy sequence for d_ν. In particular, if $\eta_{\bullet}$ is a Cauchy sequence for d_ν and if the pair (ξ,η) in (3.2) is replaced by (η_m,η_{a_j}), we find when $j\to\infty$ that the integral

$$\int_D [1 \wedge |K_\nu(\eta_m,\zeta) - h(\zeta)|] f(\zeta) l_N(d\zeta)$$

tends to 0 when $m \to \infty$. We conclude that every convergent subsequence of the sequence $n \mapsto K_\nu(\eta_n,\cdot)$ has limit h, so that $\lim_{n\to\infty} K_\nu(\eta_n,\cdot) = h$ locally uniformly on D. □

This discussion shows that if D is completed in the d_ν metric, the resulting space D^M is a compact metric space and that each point ζ adjoined to D in the completion can be identified with a positive harmonic function, which we denote by $K_\nu(\zeta,\cdot)$, identified by the fact that $\eta\to\zeta$ in the D^M metric if and only if $K_\nu(\eta,\cdot) \to K_\nu(\zeta,\cdot)$ locally uniformly on D. Moreover, integrating to the limit shows that (2.2) is satisfied when $\eta = \zeta$. The conditions (a) and (b) of Theorem 3 are satisfied by K_ν and therefore also satisfied by an arbitrary Martin function K_μ in view of (3.1).

Example (Balls and Half-Spaces). Let $D = B(0,\delta)$, and let ν be a unit measure supported by the origin. Then $K_\nu(\eta, 0) = 1$ for $\eta \neq 0$ and

$$G_D(\eta, 0) \sim \begin{cases} \dfrac{\delta - |\eta|}{\delta} & \text{if } N = 2 \\ \dfrac{N-2}{\delta^{N-1}}(\delta - |\eta|) & \text{if } N > 2 \end{cases} \qquad (|\eta| \to \delta),$$

so that when η tends to ζ on the Euclidean boundary, the value $K_v(\eta, \xi)$ tends to a multiple $[\delta^{-1}$ if $N = 2$, $(N-2)\delta^{1-N}$ if $N > 2]$ of the inner normal derivative of $G_D(\cdot, \xi)$ at ζ, and the Martin function K_v becomes the Poisson kernel density denoted by K in Section II.1. Thus the Martin boundary of a ball is the Euclidean boundary. Similarly, if D is a half-space, the Martin function based on a suitably chosen reference point becomes the Poisson kernel density denoted by K in Section VIII.9, and again the Martin boundary is the Euclidean boundary. Alternatively the Martin boundary of a half-space can be derived from that of a ball since a ball can be mapped onto a half-space by an inversion. The same reasoning, based on the Riemann mapping theorem, shows that the Martin boundary of any plane Jordan domain is its Euclidean boundary and that more generally the Martin boundary of an arbitrary simply connected plane Greenian open set other than $\mathbb{R}^2$ is the Carathéodory prime end boundary.

4. Preliminary Representations of Positive Harmonic Functions and Their Reductions

Lemma. *Let D be a Greenian subset of $\mathbb{R}^N$ with Martin function K_v. If u is a positive harmonic function on D and if A is a subset of D with $(\partial A) \cap \sigma_v = \varnothing$, there is a measure λ_{uA} on D^M, supported by the boundary of A relative to D^M, for which* (*reduction relative to D*)

$$[\![u]\!]^A = \int_{D^M} K_v(\zeta, \cdot)\lambda_{uA}(d\zeta), \qquad \lambda_{uA}(D^M) = \int_D [\![u]\!]^A \, dv. \tag{4.1}$$

In particular, there is a measure λ_u $(= \lambda_{uD})$ supported by $\partial^M D$ for which

$$u = \int_{\partial^M D} K_v(\zeta, \cdot)\lambda_u(d\zeta), \qquad \lambda_u(\partial^M D) = \int_D u \, dv. \tag{4.2}$$

A measure λ_u on $\partial^M D$ satisfying (4.2) *must also satisfy*

$$[\![u]\!]^F = \int_{\partial^M D} [\![K_v(\zeta, \cdot)]\!]^F \lambda_u(d\zeta) \tag{4.3}$$

for every Borel subset F of $\partial^M D$.

Observation. In general the measure λ_u is not uniquely determined by u in (4.2), but it will be seen in Section 9 that λ_u can be chosen to be supported by the set of minimal Martin boundary points (to be defined in Section 5) and that if so chosen, λ_u is uniquely determined.

If A is relatively compact in D, the function $\|u\|^A$ is a potential, say $G_D\lambda'$, λ' is supported by ∂A, and (4.1) is true with $\lambda(d\zeta) = G_D\nu(\zeta)\lambda'(d\zeta)$. If A is not relatively compact in D, let $B_.$ be an increasing sequence of open subsets of D, relatively compact in D and with union D, so large that $\sigma_\nu \subset B_0$. According to what has just been proved, the potential $\|u\|^{B_n \cap A}$ can be represented in the form

$$\|u\|^{B_n \cap A} = \int_{D^M} K_\nu(\zeta, \cdot)\lambda_n(d\zeta), \tag{4.4}$$

where λ_n is a measure supported by $\partial(B_n \cap A)$ and

$$\int_{D^M} u\,d\nu \geq \int_{D^M} \|u\|^{B_n \cap A}\,d\nu = \lambda_n(D^M). \tag{4.5}$$

The sequence $\lambda_.$ is a bounded sequence of measures on the compact space D^M. If λ_u is the limit of a vaguely convergent subsequence, it is trivial that λ_u is supported by the boundary of A relative to D^M and that (4.4) yields (4.1).

We shall only need (4.3) when F is compact, but this equation has general interest and so we prove it as stated. In view of the properties of the reduction operation the class of sets F satisfying (4.3) includes limits of monotone sequences of its members; so it is sufficient to prove (4.3) for compact boundary subsets F. If F is such a set, let $F'_.$ be a decreasing sequence of compact neighborhoods of F in D^M with intersection F, and set $F_n = D \cap F'_n$. Then [from Section III.5(a)] $\lim_{n\to\infty} \|u\|^{F_n} = \|u\|^F$. Now

$$\delta_D^{F_n}(\cdot, u) = \int_{\partial^M D} \delta_D^{F_n}(\cdot, K_\nu(\zeta, \cdot))\lambda_u(d\zeta), \tag{4.6}$$

and (4.6) yields (4.3) when $n \to \infty$.

5. Minimal Harmonic Functions and Their Poles

Let D be a Greenian subset of $\mathbb{R}^N$ with Martin function $K\,(= K_\nu)$. A Martin boundary point ζ is called *minimal* if its associated function $K(\zeta, \cdot)$ is a minimal harmonic function for one and therefore for every choice of the Martin function K. The set of minimal boundary points is called the *minimal Martin boundary* and is denoted by $\partial_1^M D$. In the following theorem K is an arbitrary Martin function for D. Recall from Section X.7 that a Martin boundary point ζ is said to be a pole of a positive harmonic function u if

$\|u\|^{\{\zeta\}} = u$ and that then ζ is also a pole of every positive harmonic minorant of u.

Theorem. (a) *Every* ($\not\equiv 0$) *minimal harmonic function on D has a unique pole on $\partial^M D$. If a Martin boundary point ζ is the pole of a positive* ($\not\equiv 0$) *harmonic function u, then ζ is the only pole of u, $u = \text{const.}\, K(\zeta, \cdot)$, and ζ is a minimal boundary point. In particular, if η is a minimal Martin boundary point, the function $K(\eta, \cdot)$ has pole η.*

(b) *If ζ is a minimal Martin boundary point and if A is a set of minimal Martin boundary points, then $\|K(\zeta, \cdot)\|^A$ is either $K(\zeta, \cdot)$ or 0 according as $\zeta \in A$ or $\zeta \notin A$.*

Proof of (a). According to Section X.7, whatever the boundary assigned to D by a metric compactification, a minimal harmonic function on D has at least one pole. Now let u be a strictly positive harmonic function on D with pole ζ. We can assume in proving (a) that $K = K_v$ and that u has been normalized so that $\int_D u\, dv = 1$. Apply Lemma 4 with A the trace on D of a neighborhood of ζ to derive

$$u = \|u\|^A = \int_{D^M} K(\eta, \cdot)\lambda_{uA}(d\eta), \qquad \lambda_{uA}(D^M) = 1. \tag{5.1}$$

The measure λ_{uA} is supported by the boundary of A relative to D^M; so when A shrinks to ζ, the measure λ_{uA} tends (vague convergence) to the probability measure supported by the singleton $\{\zeta\}$, and therefore (5.1) becomes $u = K(\zeta, \cdot)$. Thus ζ is uniquely determined by u. Moreover, as already recalled at the beginning of this section, if v is a positive harmonic minorant of u, then v also has pole ζ. Hence $v = \text{const.}\, K(\zeta, \cdot)$, and therefore u is a minimal function; that is, ζ is a minimal boundary point. Finally, if η is a minimal Martin boundary point, then according to what we have just proved, $K(\eta, \cdot)$ has a pole ζ and $K(\eta, \cdot) = \text{const.}\, K(\zeta, \cdot)$. The constant must be 1; so $\eta = \zeta$. □

Proof of (b). If ζ is a minimal Martin boundary point, the function $K(\zeta, \cdot)$ has pole ζ according to (a); so surely $\|K(\zeta, \cdot)\|^A = K(\zeta, \cdot)$ if $\zeta \in A$. Furthermore the set $B = \partial^M D - \{\zeta\}$ is open in $\partial^M D$ and contains no pole of $K(\zeta, \cdot)$; so (from Section X.7) $\|K(\zeta, \cdot)\|^B \equiv 0$, and therefore $\|K(\zeta, \cdot)\|^A \equiv 0$ if $A \subset B$. □

6. Extension of Lemma 4

Lemma. (Context of Lemma 4). *If F is a compact subset of $\partial^M D$, there is a measure λ_{uF} supported by F such that*

$$\|u\|^F = \int_F K_v(\zeta, \cdot)\lambda_{uF}(d\zeta) = \int_F \|K_v(\zeta, \cdot)\|^F \lambda_{uF}(d\zeta). \tag{6.1}$$

Let $F'_\cdot$ be a decreasing sequence of compact neighborhoods of F in D^M with intersection F, and set $F_n = D \cap F'_n$. Then [Section III.5(a)] $\lim_{n\to\infty} \| u \|^{F_n} = \| u \|^F$, and according to Lemma 4, there is a measure μ_n supported by the boundary of F_n relative to D^M such that

$$\| u \|^{F_n} = \int_{D^M} K_\nu(\zeta, \cdot)\mu_n(d\zeta), \qquad \mu_n(D^M) \le \int_{D^M} u\, d\nu.$$

If λ_{uF} is the vague limit of a convergent subsequence of $\mu_\cdot$, the measure λ_{uF} is supported by F, and the first equation in (6.1) is true. Next apply the operator $\delta_D^{F_n}$ to the first equality to obtain

$$\begin{aligned} \| \| u \|^F \|^{F_n} = \delta_D^{F_n}(\cdot, \| u \|^F) &= \int_F \delta_D^{F_n}(\cdot, K_\nu(\zeta, \cdot))\lambda_{uF}(d\zeta) \\ &= \int_F \| K_\nu(\zeta, \cdot) \|^{F_n} \lambda_{uF}(d\zeta), \end{aligned}$$

which yields the second equation in (6.1) when $n \to \infty$ in view of the idempotency of the smoothed reduction operation.

Application. A Martin boundary point ζ is nonminimal if and only if $\| K_\nu(\zeta, \cdot) \|^{\{\zeta\}} \equiv 0$. In fact, according to Lemma 6 with $u = K_\nu(\zeta, \cdot)$ and $F = \{\zeta\}$,

$$\| K_\nu(\zeta, \cdot) \|^{\{\zeta\}} = cK_\nu(\zeta, \cdot), \qquad c = \lambda_{K_\nu(\zeta, \cdot)\{\zeta\}}(\{\zeta\}),$$

and c is either 1 or 0 because the smoothed reduction operation is idempotent. Since (by Theorem 5) the condition $c = 1$ characterizes minimality, the condition $c = 0$ must characterize nonminimality.

The following theorem strengthens this result.

7. The Set of Nonminimal Martin Boundary Points

Theorem. *The set of nonminimal Martin boundary points of a connected Greenian subset D of $\mathbb{R}^N$ is an F_σ set that is h-harmonic measure null for every strictly positive harmonic function h on D.*

Let h be a strictly positive harmonic function on D, let ξ_0 be a point of D, and let K be a Martin function for D. If B is an open subset of D^M, define

$$B' = \left\{ \zeta \in \partial^M D\colon \| K(\zeta, \cdot) \|^B(\xi_0) \le \frac{K(\zeta, \xi_0)}{2} \right\}.$$

Since a smoothed reduction on B is obtained by applying the operator $\delta_D^{B \cap D}$, an application of Fatou's lemma shows that B' is compact. Moreover, if A

is a compact subset of the F_σ set $B' \cap B$,

$$\|K(\zeta,\cdot)\|^A(\xi_0) \le \|K(\zeta,\cdot)\|^B(\xi_0) \le \frac{K(\zeta,\xi_0)}{2} \qquad (\zeta \in A). \tag{7.1}$$

An application of the representation (6.1) to $\|h\|^A$,

$$\begin{aligned}\|h\|^A(\xi_0) &= \int_A K(\zeta,\xi_0)\lambda_{hA}(d\zeta) = \int_A \|K(\zeta,\cdot)\|^A(\xi_0)\lambda_{hA}(d\zeta)\\ &\le \frac{\|h\|^A(\xi_0)}{2},\end{aligned} \tag{7.2}$$

shows that $\|h\|^A \equiv 0$; that is, A is h-harmonic measure null. The application in Section 6 implies that the points of $B' \cap B$ are not minimal and that when B runs through the open sets of a countable topological base of D^M, the class of F_σ h-harmonic measure null sets $B' \cap B$ covers the set of nonminimal Martin boundary points. The theorem follows.

8. Reductions on the Set of Minimal Martin Boundary Points

Lemma. *If h is a strictly positive harmonic function on the connected Greenian subset D of $\mathbb{R}^N$, then*

$$h = \|h\|^{\partial_1^M D} = \sup\{\|h\|^A : A \subset \partial_1^M D,\ A \text{ compact}\}. \tag{8.1}$$

It is sufficient to show that the first and third terms in (8.1) are equal because the second term lies between them. There is [from Section VIII.5(b)] a positive h-superharmonic function u on D that has limit $+\infty$ at every point of the h-harmonic measure null set of nonminimal Martin boundary points. Define

$$A_n = \left\{\zeta \in \partial^M D : \liminf_{\eta\to\zeta} u(\eta) \le n\right\}, \qquad A' = \bigcup_0^\infty A_n \qquad (n \in \mathbb{Z}^+).$$

The set A_n is a compact set of minimal boundary points, and since [from Section VI.3(e)] $\lim_{n\to\infty} \|h\|^{A_n} = \|h\|^{A'}$, it will suffice to prove that $\|h\|^{A'} = h$. Since u has limit $+\infty$ at every point of the set $B = \partial^M D - A'$, the set B is h-harmonic measure null [Section VIII.5(b)]; so $\|h\|^B = 0$, and by subadditivity of the set function $\|h\|^\cdot$,

$$h = \|h\|^{A'\cup B} \le \|h\|^{A'} + \|h\|^B = \|h\|^{A'} \le h;$$

so $\|h\|^{A'} = h$, as was to be proved.

9. The Martin Representation

Let D be a connected Greenian subset of $\mathbb{R}^N$. It will be convenient to expand the vector lattice notation in Chapter IX by introducing into the notation the relativizing strictly positive harmonic function: $\mathbf{S}, \mathbf{S}_m, \ldots$ will be written ${}^h\mathbf{S}, {}^h\mathbf{S}_m, \ldots$ when h is the relativizing function. For example, ${}^h\mathbf{S}_{mqb}$ is the class of quasi-bounded h-harmonic functions on D.

If K_v is a Martin function for D and if λ_u is a signed measure on $\partial^M D$, the function u defined by (4.2) is harmonic on D because u is continuous and has the harmonic function average property. In view of the Jordan decomposition of λ_u the function u is in the class ${}^1\mathbf{S}_m$. Conversely, according to Lemma 4, a positive harmonic function on D, and therefore also a harmonic function in ${}^1\mathbf{S}_m$, has a representation (4.2) in terms of a not necessarily uniquely defined signed measure on $\partial^M D$. The following Martin Representation Theorem details among other things the relation between harmonic functions in ${}^1\mathbf{S}_m$ and their unique Martin representing signed measures on $\partial_1^M D$.

Theorem. *Let D be a connected Greenian subset of $\mathbb{R}^N$, let K be a Martin function for D, and let h be a strictly positive harmonic function on D.*

(a) *To each function v in ${}^1\mathbf{S}_m$ corresponds a unique finite-valued signed measure M_v on $\partial^M D$, supported by the minimal Martin boundary $\partial_1^M D$, positive if v is, and satisfying*

$$v = \int_{\partial^M D} K(\zeta, \cdot) M_v(d\zeta). \tag{9.1}$$

(b) *For given K the correspondence $v/h \leftrightarrow M_v$ is an isomorphism between the vector lattice ${}^h\mathbf{S}_m$ and the vector lattice of signed measures on $\partial^M D$ supported by $\partial_1^M D$.*

(c) *A function v/h in ${}^h\mathbf{S}_m$ is in ${}^h\mathbf{S}_{mqb}$ [${}^h\mathbf{S}_{ms}$] if and only if M_v is absolutely continuous [singular] relative to M_h. In the quasi-bounded case*

$$v = \int_{\partial^M D} K(\zeta, \cdot) \frac{dM_v}{dM_h}(\zeta) M_h(d\zeta). \tag{9.2}$$

See Section 10 for the relation between Martin representing signed measures and harmonic measures.

Uniqueness proof. Suppose that v is a positive harmonic function on D and that there is a measure M_v supported by $\partial_1^M D$ for which (9.1) is satisfied. Then according to Lemma 4 and Theorem 5(b), if A is a Borel boundary subset,

$$\|v\|^A = \int_{\partial M_D} \|K(\zeta,\cdot)\|^A M_v(d\zeta) = \int_A K(\zeta,\cdot) M_v(d\zeta). \tag{9.3}$$

Hence $\int_D \|v\|^A \, dv = M_v(A)$; so the measure M_v is uniquely determined by v. If $v \in {}^1\mathbf{S}_m$ and if v has two representing signed measures, M_v and M_v', then the function 0 has the representing signed measure $M_v - M_v'$. The two positive measures whose difference is $M_v - M_v'$ (Jordan decomposition) are therefore representing measures for the same positive harmonic function on D and so must be identical according to what we have just proved; that is, $M_v = M_v'$. □

Proof of (a). Let v be a positive harmonic function on D, and let A be a compact subset of the minimal boundary, so that according to Lemma 6, there is a measure λ_{vA} supported by A and satisfying

$$\|v\|^A = \int_A K(\zeta,\cdot)\lambda_{vA}(d\zeta). \tag{9.4}$$

The measure λ_{vA} was just shown to be uniquely determined. If B is a compact subset of A, the equality $\big\| \|v\|^A \big\|^B = \|v\|^B$ implies by Lemma 4 and Theorem 5(b) that

$$\int_A \|K(\zeta,\cdot)\|^B \lambda_{vA}(d\zeta) = \int_B K(\zeta,\cdot)\lambda_{vA}(d\zeta) = \int_B K(\zeta,\cdot)\lambda_{vB}(d\zeta).$$

Hence $\lambda_{vB} = \lambda_{vA}$ on the Borel subsets of B. Let $A_\bullet$ be an increasing sequence of compact subsets of the minimal Martin boundary, chosen so that (Lemma 8) $\lim_{n\to\infty} \|v\|^{A_n} = v$ at some point of D, which implies (by the Harnack convergence theorem) that this limit relation is true locally uniformly on D. According to what we have just proved, $\lambda_{vA_n} = \lambda_{vA_{n+1}}$ on the Borel subsets of A_n. The increasing sequence $\lambda_{vA_\bullet}$ of measures has limit M_v, a measure (Appendix IV.4) of Borel sets supported by the minimal Martin boundary. If A in (9.4) runs through the sequence $A_\bullet$, this equation becomes (9.1) in the limit. □

Proof of (b). The uniqueness property has already been proved. The relation $v/h \leftrightarrow M_v$ is obviously linear and is specific order preserving because $v \ge 0$ if and only if $M_v \ge 0$. The vector lattices in question are therefore isomorphic. □

Proof of (c). The assertions of (c) follow from the vector lattice isomorphism just derived. On the one hand, ${}^h\mathbf{S}_{mqb}$ is the subband of ${}^h\mathbf{S}_m$ generated by the function 1, and ${}^h\mathbf{S}_{ms}$ is the orthogonal complement of ${}^h\mathbf{S}_{mqb}$ in ${}^h\mathbf{S}_m$; that is, ${}^h\mathbf{S}_{ms}$ is the subband of ${}^h\mathbf{S}_m$ orthogonal to the function 1. Equivalently, ${}^h\mathbf{S}_{mqb}$

is the subband of ${}^h\mathbf{S}_m$ consisting of the class of functions v/h with v in the subband of ${}^1\mathbf{S}_m$ generated by h, and ${}^h\mathbf{S}_{ms}$ is the subband of ${}^h\mathbf{S}_m$ consisting of the class of functions v/h with v in the subband of ${}^1\mathbf{S}_m$ orthogonal to h. On the other hand, it then follows from (b) that v/h is a quasi-bounded h-harmonic function if and only if M_v is in the band generated by M_h of signed measures (charges) on $\partial^M D$ supported by $\partial_1^M D$, that is, if and only if M_v is absolutely continuous relative to M_h. Furthermore v/h is a singular h-harmonic function if and only if $M_v \perp M_h$, that is, if and only if M_v is singular relative to M_h. □

The Martin Representation of a Minimal Harmonic Function

If u is a not identically 0 minimal harmonic function on D, then $u = cK(\eta, \cdot)$ for some uniquely determined minimal Martin boundary point η. In fact, more generally we now show that if u is a not identically vanishing minimal harmonic function on D and if

$$u = \int_{\partial^M D} K(\zeta, \cdot)\lambda_u(d\zeta)$$

for some measure λ_u on $\partial^M D$, then λ_u must be supported by a uniquely determined singleton $\{\eta\}$. To prove this, observe that if A is a Borel subset of $\partial^M D$, then $\int_A K(\zeta, \cdot)\lambda_u(d\zeta)$ is a positive harmonic function majorized by u and therefore proportional to u. If now ξ is in the compact support of λ_u and if A is the trace on $\partial^M D$ of an open Martin topology neighborhood of ξ and shrinks to ξ, it follows that

$$K(\xi, \cdot) = c(\xi)u, \qquad 1 = c(\xi)\int_D u\, dv.$$

Hence $c(\cdot)$ is a constant function on the support of λ_u, $K(\xi, \cdot)$ is the same function for all ξ in this support, and therefore λ_u is supported by a singleton $\{\eta\}$. The point η is minimal and therefore (Theorem 9) uniquely determined by u. In particular, if $u = K(\eta', \cdot)$ for some minimal boundary point η', it follows that the point η must be η'.

The Notation M_v

Let v be a positive superharmonic function on a Greenian subset D of $\mathbb{R}^N$, and let v_1 be the harmonic component of the Riesz decomposition of v. For a given Martin function K the function v_1 determines a unique measure M_{v_1} on $\partial_1^M D$, and we define $M_v = M_{v_1}$. A glance at (3.1) as extended to the Martin boundary shows that M_v and any other measure on the minimal Martin boundary induced by a different choice of K are mutually absolutely continuous. Slightly more generally, if $v \in {}^1\mathbf{S}$, we define M_v as the Martin representing signed measure on $\partial_1^M D$ of the ${}^1\mathbf{S}_m$ component of v.

10. Resolutivity of the Martin Boundary

Theorem. *The Martin boundary is universally internally resolutive and universally resolutive. If K is a Martin function and h is strictly positive and harmonic on D,* $\mu_D^h(\xi, d\zeta) = K(\zeta, \xi)M_h(d\zeta)/h(\xi)$. *An h-harmonic function* $u = v/h$ *is a* PWB^h *solution if and only if it is quasi bounded, equivalently, if and only if* M_v *is absolutely continuous relative to* M_h, *and then*

$$u = H_f^h = \int_{\partial^M D} f(\zeta)K(\zeta, \cdot)\frac{M_h(d\zeta)}{h(\xi)}, \qquad f = \frac{dM_v}{dM_h}. \tag{10.1}$$

Let h be a strictly positive harmonic function on D, and let A be a closed subset of $\partial^M D$. Apply (4.3) and Theorem 5 to derive

$$\bar{H}_{1_A}^h = \frac{[\![h]\!]^A}{h} = \int_{\partial^M D} [\![K(\zeta, \cdot)]\!]^A \frac{M_h(d\zeta)}{h} = \int_A K(\zeta, \cdot)\frac{M_h(d\zeta)}{h}. \tag{10.2}$$

The function $A \mapsto \bar{H}_{1_A}^h$ is therefore additive and (Section VIII.9) h-resolutivity of $\partial^M D$ follows, and also the evaluation $\mu_D^h(\cdot, d\zeta) = K(\zeta, \cdot)M_h(d\zeta)/h$. According to Section IX.9, every PWB^h solution is quasi bounded. Conversely, if $u = v/h$ is a quasi-bounded h-harmonic function, equivalently (Theorem 9), if M_v is absolutely continuous relative to M_h, and if $f = dM_v/dM_h$,

$$v = \int_{\partial^M D} K(\zeta, \cdot)f(\zeta)M_h(d\zeta) = \mu_D^h(\cdot, f)h, \tag{10.3}$$

so that u is the PWB^h solution for the boundary function f; that is, (10.1) is true. Thus the Martin boundary is universally internally resolutive as well as universally resolutive, and the proof of the theorem is complete.

Intrinsic Definition of h-Harmonic Measure. If $u = v/h$ is h-harmonic on D, if

$$0 \le u \le 1, \qquad u \curlywedge (1-u) = \mathrm{GM}_D^h[u \wedge (1-u)] = 0, \tag{10.4}$$

that is, if for $f = dM_v/dM_h$,

$$0 \le f \le 1, \qquad f \wedge (1-f) = 0 \quad M_h \text{ almost everywhere} \tag{10.4$'$}$$

then f coincides M_h almost everywhere on $\partial^M D$ with the indicator function of a Borel set A, and $u = \mu_D^h(\cdot, A)$. Conversely, if u is the h-harmonic measure of a Borel boundary subset A, the reverse argument shows that (10.4) is satisfied.

Relations between Martin Representing Measures and Harmonic Measures

According to Theorem 10, the measures $\mu_D^h(\xi,\cdot)$ and M_h are mutually absolutely continuous for all ξ in D; that is, a boundary subset is μ_D^h null if and only if it is M_h null. In fact the equality $\mu_D^h(\xi, d\zeta) = K(\zeta,\xi)M_h(d\zeta)/h(\xi)$ implies more: a boundary function is M_h measurable and integrable if and only if it is $\mu_D^h(\xi,\cdot)$ measurable and integrable for every (equivalently a single) point ξ of D.

Special Case: h Is Minimal. If $\zeta \in \partial_1^M D$ and if $h = K(\zeta,\cdot)$, then $\mu_D^h(\xi,\cdot) = M_h$ for every point ξ of D, and this measure is the unit measure supported by $\{\zeta\}$.

11. Minimal Thinness at a Martin Boundary Point

Theorem. *Let D be a Greenian subset of $\mathbb{R}^N$, let K be a Martin function for D, let A be a subset of D, and let ζ be a minimal Martin boundary point of D.*

(a) *The following conditions are equivalent:*
  (a1) $R^A_{K(\zeta,\cdot)} = K(\zeta,\cdot)$.
  (a2) $R^{A\cap B}_{K(\zeta,\cdot)} = K(\zeta,\cdot)$ *for every Martin topology neighborhood B of ζ.*

(b) *The following conditions are equivalent:*
  (b1) $R^A_{K(\zeta,\cdot)} \neq K(\zeta,\cdot)$.
  (b2) $\inf\{\underset{+}{R}{}^{A\cap B}_{K(\zeta,\cdot)} : B$ *is a Martin topology neighborhood of* $\zeta\} = 0$.
  (b3) $\underset{+}{R}{}^{A}_{K(\zeta,\cdot)}$ *is a potential.*

(c) *If B is a Martin topology neighborhood of ζ, then*
  (c1) $R^B_{K(\zeta,\cdot)} = K(\zeta,\cdot)$,
  (c2) $R^{D-B}_{K(\zeta,\cdot)} \neq K(\zeta,\cdot)$.

Each condition (a1), (a2), (b1), (c1), (c2) *is satisfied if and only if it is satisfied using the corresponding smoothed reduction.*

The set A is said to be *minimal thin* at ζ if the conditions (b) are satisfied. The last assertion of the theorem is trivial, and the proofs will be phrased accordingly. Observe that in view of the application in Section 6 if ζ is a nonminimal Martin boundary point, condition (b2) is satisfied because the indicated infimum is $R^{\{\zeta\}}_{K(\zeta,\cdot)}$.

The proof of the theorem will be carried through in several steps, numbered for convenience in reference.

Proof. Step 1. Proof that $\underset{+}{R}{}^{A}_{K(\zeta,\cdot)}$ is either $K(\zeta,\cdot)$ or a potential. Since $K(\zeta,\cdot)$ is minimal, the Riesz decomposition of $\underset{+}{R}{}^{A}_{K(\zeta,\cdot)}$ must have the form

$$R^A_{+K(\zeta,\cdot)} = v + cK(\zeta,\cdot), \tag{11.1}$$

where v is a potential and c is a positive constant. If we use the fact that the smoothed reduction operation is idempotent, we find that

$$R^A_{+K(\zeta,\cdot)} = R^A_{+v} + cR^A_{+K(\zeta,\cdot)}, \tag{11.2}$$

and since a function majorizes its smoothed reduction, it follows that the terms on the right in (11.1) and (11.2) are pairwise equal. Hence either $c = 0$ and $R^A_{+K(\zeta,\cdot)}$ is a potential or $c > 0$ and $R^A_{+K(\zeta,\cdot)} = K(\zeta,\cdot)$, in which case $c = 1$ and $v = 0$.

Step 2. Proof of (c2). Without loss of generality we can assume that B is so small that the compact support of the measure on which K is based does not meet $\bar{B}$. According to Lemma 4, there is a measure λ on D^M, supported by $\partial(D - B)$ (boundary relative to D^M), such that

$$R^{D-B}_{+K(\zeta,\cdot)} = \int_{D^M} K(\eta,\cdot)\lambda(d\eta).$$

If there were equality in (c2), the integral would define a harmonic function on D; so the measure λ would be supported by $\partial^M D$. However, according to Section 9, such an integral representation of a minimal harmonic function $K(\zeta,\cdot)$ is possible only if λ is supported by $\{\zeta\}$, contrary to the definition of B. Hence there cannot be equality in (c2).

Step 3. Proof that (a1)$\Leftrightarrow$(a2). The implication (a2)$\Rightarrow$(a1) is trivial. To prove the reverse implication observe that if (a1) is true and (a2) is false, then $R^{A\cap B}_{+K(\zeta,\cdot)}$ is a potential for sufficiently small B by Step 1, and $R^{A-B}_{+K(\zeta,\cdot)}$ is a potential according to Step 2, because this smoothed reduction is a positive superharmonic function majorized by the potential $R^{D-B}_{+K(\zeta,\cdot)}$. Hence by set subadditivity of reductions $R^A_{+K(\zeta,\cdot)}$ is a potential, contrary to hypothesis.

Step 4. Proof of (c1). Assertion (c1) is trivially true when $B = D$ and therefore true for arbitrary B by (a), which we have just proved. Alternatively (c1) is true because [by Theorem 5(a)] ζ is a pole of $K(\zeta,\cdot)$.

Step 5. Proof that (b1)$\Rightarrow$(b2). The function $R^{A\cap B}_{K(\zeta,\cdot)}$ is harmonic and positive on $D - \bar{B}$. Let B shrink to ζ, say along a sequence of balls of center ζ and radii tending to 0 (in terms of some Martin space metric). Then the limit of the corresponding sequence of reductions in (b2) is the indicated infimum and is a positive harmonic function, majorized by $R^A_{+K(\zeta,\cdot)}$. Since this smoothed reduction is a potential by Step 1, the harmonic function vanishes identically, as was to be proved.

Step 6. Proof that (b2)$\Rightarrow$(b3)$\Rightarrow$(b1). These implications follow trivially from Step 1 and the equivalence of (a1) and (a2). □

The proof of the theorem is now complete, and we turn to the definition of the minimal-fine topology of D^M.

12. The Minimal-Fine Topology

Let D be a connected Greenian subset of $\mathbb{R}^N$, let K be a Martin function for D, and let ζ be a minimal Martin boundary point. The class $\mathrm{MT}(\zeta)$ of subsets of D minimal thin at ζ has the following properties.

(P1) Every subset of a set in $\mathrm{MT}(\zeta)$ is itself in $\mathrm{MT}(\zeta)$ (trivial).

(P2) A finite union of sets in $\mathrm{MT}(\zeta)$ is in $\mathrm{MT}(\zeta)$, because the set function $A \mapsto R^A_{K(\zeta,\cdot)}$ is subadditive.

(P3) If $A \in \mathrm{MT}(\zeta)$, then the union A' of A and its set of fine limit points in D, as defined in Section XI.1, is in $\mathrm{MT}(\zeta)$, because according to Section XI.7,

$$R^A_{+K(\zeta,\cdot)} = R^{A'}_{+K(\zeta,\cdot)}.$$

(P4) Every set in $\mathrm{MT}(\zeta)$ has an open superset in $\mathrm{MT}(\zeta)$, because according to Section III.5(e),

$$R^A_{K(\zeta,\cdot)} = \inf\{R^B_{K(\zeta,\cdot)}: B \text{ open}, A \subset B \subset D\}.$$

The Minimal-Fine Topology

We define the *minimal-fine topology* of D^M by the following conventions:

A point ζ of D is a minimal-fine limit point of a set A if $A \cap D$ is not thin at ζ.

A point ζ of $\partial_1^M D$ is a minimal-fine limit point of a set A if $A \cap D$ is not minimal thin at ζ.

Each nonminimal Martin boundary point is a minimal-fine isolated point of D^M.

The minimal-fine topology of D^M has as relative topology on D the fine topology already defined on D in Section XI.1. According to Theorem 11(c), if B is a Martin topology neighborhood of the minimal Martin boundary point ζ, then $B \cap D$ is a deleted minimal-fine neighborhood of ζ, and $D - B$ does not have ζ as a minimal-fine limit point. Thus the minimal-fine topology of D^M is a (Hausdorff) topology finer than the Martin topology of D^M.

For some choices of D, for example, when D is a ball (Section 3), the Martin space D^M can be identified with the Euclidean closure of D. Observe that for such a choice of D if ζ is a boundary point and if A is a subset of

D, then thinness of A at ζ in the fine topology of $\mathbb{R}^N$ need not be equivalent to minimal thinness of A at ζ. For example, if $D = B(0, \delta)$, then every boundary point is a minimal Martin boundary point, and if B is a ball internally tangent to ∂D at ζ, then $D - B$ is minimal thin at ζ but is not thin at ζ. In fact, in this case $K(\zeta, \cdot)$ is a constant multiple of the function

$$\xi \mapsto \frac{\delta^2 - |\xi|^2}{|\zeta - \xi|^N}$$

(Section II.16), and $D - B$ is the locus of the inequality $K(\zeta, \cdot) \leq c$ for some strictly positive constant c. Since we shall prove [equation (12.3)] that the minimal-fine limit of $K(\zeta, \cdot)$ at ζ is $+\infty$, the set $D - B$ is minimal thin at ζ. The set $D - B$ is not thin at ζ because it contains the trace on a neighborhood of ζ of an open cone with vertex ζ (see Section XI.3). We shall use the notation "mf lim" for minimal-fine limits.

Example (a). Let D be a Greenian subset of $\mathbb{R}^N$, let ζ be a point of D, and define $D_0 = D - \{\zeta\}$. Then (from Section VII.1) G_{D_0} is the restriction of G_D to $D_0 \times D_0$, and (from Section VII.10) the restriction of $G_D(\zeta, \cdot)$ to D_0 is a minimal harmonic function on D_0. We conclude that the point ζ can be identified with a minimal Martin boundary point of D_0, the pole of the restriction of $G_D(\zeta, \cdot)$ to D_0. The Martin topology of D_0^M coincides on D with the Euclidean and Martin topologies of D. Thus the Martin space D_0^M can be identified with D^M. Finally, the minimal-thin topology of D_0^M on a Martin neighborhood of ζ is identical with the fine topology of D^M and of $\mathbb{R}^N$ on that set. Hence minimal-fine limit concepts on D_0 at ζ coincide with fine limit concepts on D at ζ. More generally, a trivial refinement of this reasoning shows that if A is a closed relative to D polar subset of D and if $D_0 = D - A$, the Martin space D_0^M can be identified with D^M by identifying each point ζ of A with a point of $\partial_1^M D_0$, the pole of the restriction to D_0 of $G_D(\zeta, \cdot)$; minimal-fine limits on D_0 at ζ coincide with fine limits on D at ζ. Finally, suppose that v is a positive superharmonic function on D whose Riesz measure ν is supported by A. The function v is harmonic on D_0 and thus has a Martin representation there in terms of a measure M_{0v} on $\partial_1^M D_0 = A \cup \partial_1^M D$. Choose a Martin function K° for D_0 based on a point ξ_0 in D_0. Then it is clear that for ζ in A,

$$K^\circ(\zeta, \xi) M_{0v}(d\zeta) = \frac{G_D(\xi, \zeta)}{G_D(\xi_0, \zeta)} M_{0v}(d\zeta) = G_D(\xi, \zeta)\nu(d\zeta),$$

so that $M_{0v}(d\zeta) = G_D(\xi_0, \zeta)\nu(d\zeta)$ for ζ in A. It is important that M_{0v} and ν are mutually absolutely continuous on A.

Minimal-Fine Limits at an Isolated Boundary Point. Example (a) implies that to each theorem on minimal-fine limits at a minimal Martin boundary point of a Greenian set D corresponds a theorem on fine limits at a point of D. For example, the fact that (Section XI.1) a set thin at a point of D has an open superset thin at the point corresponds to the fact (P4) that a set minimal thin at a minimal Martin boundary point of D has an open superset minimal thin at the point. The fact that [Theorem XI.4(a)] if v is a positive superharmonic function on D and if ζ is in D, then $v/G_D(\zeta,\cdot)$ has fine limit the value $\inf_{D-\{\zeta\}} v/G_D(\zeta,\cdot)$ at ζ corresponds to the fact that if ζ is a minimal Martin boundary point of D and if K is a Martin function for D, then $v/K(\zeta,\cdot)$ has fine limit the value $\inf_D v/K(\zeta,\cdot)$ at ζ. The latter result is proved in Section 13, and a dual result is proved in Section 14.

Example (b). Denote by d_ξ the Nth coordinate of the point ξ of $\mathbb{R}^N$, and define $D = \{\xi: d_\xi > 0\}$. Then (from Section 3) D^M is the closure of D in the one-point compactification of $\mathbb{R}_N$; so the Martin boundary is the Euclidean boundary. If u is a positive superharmonic function on D, if $c > 0$, if $u_c = u(\xi/c)$, if A is a subset of D, and if cA has the obvious meaning, then $\underset{+u}{R}^A(\xi) = \underset{+u_c}{R}^{cA}(c\xi)$. In particular, if $u(\xi) = |\xi|^{-N} d_\xi$, that is, if u is a minimal harmonic function on D with pole the origin, then $u_c = c^{N-1}u$ and

$$\underset{+u}{R}^{cA}(\xi) = c^{1-N}\underset{+u}{R}^A\left(\frac{\xi}{c}\right).$$

If A is relatively compact in D and not polar, then $\underset{+u}{R}^A$ is a nonzero potential, and the evaluation of G_D in Section VIII.9 shows that $\underset{+u}{R}^A(c\xi) \sim c^{N-1}\phi(\xi)$ when $c \to 0$, with ϕ a strictly positive finite-valued function. It follows that if $c_\cdot$ is an arbitrary sequence of strictly positive numbers with limit 0 and if B is the intersection of $\bigcup_0^\infty (c_n A) = A'$ with a Euclidean neighborhood of the origin, then

$$\underset{+u}{R}^B(\xi) \ge \underset{+u}{R}^{c_n A}(\xi) \ge \frac{\phi(\xi)}{2}$$

for sufficiently large n. Hence A' is not minimal thin at the origin.

A trivial example of the application of Example (b) shows that if $N = 2$, no initial segment of a ray from the origin into D is minimal thin at the origin. An analogous argument shows that if ζ' is a boundary point of a disk D', then no initial segment of a ray from ζ' into D' is minimal thin at ζ'.

In view of Example (b) if a function from the upper half-space of $\mathbb{R}^N$ into a Hausdorff space has both a minimal-fine and a nontangential limit at a boundary point, the two limits must be the same. An analogous argument leads to the same conclusion if D is a ball.

13. First Martin Boundary Counterpart of Theorem XI.4(c) and (d)

Theorem. *Let D be a connected Greenian subset of $\mathbb{R}^N$, let K be a Martin function for D, and let ζ be a minimal Martin boundary point of D.*

(a) *If v is a positive superharmonic function on D, then*

$$\operatorname{mf}\lim_{\eta\to\zeta}\frac{v(\eta)}{K(\zeta,\eta)}=\inf_{D}\frac{v}{K(\zeta,\cdot)}=M_v(\{\zeta\}). \tag{13.1}$$

(b) *Let A be a subset of D with Martin topology limit point ζ. If A is minimal thin at ζ, there is a positive superharmonic function v on D for which*

$$\lim_{A\ni\eta\to\zeta}\frac{v(\eta)}{K(\zeta,\eta)}=+\infty. \tag{13.2}$$

Conversely, if there is a positive superharmonic function v on D satisfying (13.2), *then A is minimal thin at ζ.*

We shall see (in Section 19) that Theorem 13(a) is a special case of the Fatou boundary limit theorem for a Martin space.

To prove Theorem 13, translate the proof of Theorem XI.4(c), (d) into the present context, replacing $G_D(\zeta,\cdot)$ in that proof by $K(\zeta,\cdot)$ and "thin" by "minimal thin." See Example (a) in Section 12 for a discussion of the relation between theorems on limits at a minimal Martin boundary point and limits at a point of D.

Application. If v is a potential in (a), we find that the minimal-fine limit is 0. If $v\equiv 1$ in (a), we find that

$$\operatorname{mf}\lim_{\eta\to\zeta}K(\zeta,\eta)=\sup_{D}K(\mu,\cdot). \tag{13.3}$$

14. Second Martin Boundary Counterpart of Theorem XI.4(c)

Theorem. *If D is a connected Greenian subset of $\mathbb{R}^N$, if v is a strictly positive superharmonic function on D, if $\zeta\in\partial_1^M D$, and if $\xi\in D$, then*

$$0<\operatorname{mf}\lim_{\eta\to\zeta}\frac{v(\eta)}{G_D(\xi,\eta)}=\liminf_{\eta\to\zeta}\frac{v(\eta)}{G_D(\xi,\eta)}\leq+\infty. \tag{14.1}$$

Observe that (14.1) is trivial if the indicated inferior limit is $+\infty$. We can therefore ignore this case and prove the equality in (14.1) by showing that whenever c is a finite number strictly larger than the inferior limit in (14.1), it follows that the set $A = \{\eta\colon v(\eta) \geq cG_D(\xi, \eta)\}$ is minimal thin at ζ. Let K be a Martin function for D based on the point ξ. By definition of the smoothed reduction on A the inequality $v \geq c\llbracket G_D(\xi, \cdot)\rrbracket^A$ is valid on D; so in view of sweeping symmetry, if $\eta \in D - \{\xi\}$,

$$c\llbracket K(\eta, \cdot)\rrbracket^A(\xi) = \frac{c\llbracket G_D(\eta, \cdot)\rrbracket^A(\xi)}{G_D(\xi, \eta)} = \frac{c\llbracket G_D(\xi, \cdot)\rrbracket^A(\eta)}{G_D(\xi, \eta)} \leq \frac{v(\eta)}{G_D(\xi, \eta)}, \tag{14.2}$$

and therefore since $\llbracket K(\eta, \cdot)\rrbracket^A(\xi) = \delta_D^A(\xi, K(\eta, \cdot))$, Fatou's lemma is applicable when $\eta \to \zeta$ in (14.2) and yields

$$c\llbracket K(\zeta, \cdot)\rrbracket^A(\xi) = c\delta_D^A(\xi, K(\zeta, \cdot)) \leq \liminf_{\eta \to \zeta} \frac{v(\eta)}{G_D(\xi, \eta)} < c. \tag{14.3}$$

Hence $\llbracket K(\zeta, \cdot)\rrbracket^A(\xi) < 1 = K(\zeta, \xi)$; so (from Section 11) the set A is minimal thin at ζ, as was to be shown in proving the equality in (14.1). To show that the minimal-fine limit in (14.1) is strictly positive, it can be assumed that v is a strictly positive potential $G_D\nu$, after replacing v if necessary by its reduction on a ball relatively compact in D. Under this hypothesis,

$$\liminf_{\eta \to \zeta} \frac{v(\eta)}{G_D(\xi, \eta)} \geq \int_D K(\zeta, \zeta')\nu(d\zeta') > 0,$$

as was to be proved.

Special case: $v \equiv 1$. If $G_D(\xi, \cdot)$ has minimal-fine limit 0 at ζ, as we shall prove (Section 18) is true at μ_D almost every minimal Martin boundary point ζ, it follows from Theorem 14 that $G_D(\xi, \cdot)$ has limit 0 at ζ on approach to ζ in the Martin topology. [Incidentally, this application of Theorem 14 to the function $1/G_D(\xi, \cdot)$ exhibits the fact that the minimal-fine limit $+\infty$ cannot be excluded in (14.1).] This vanishing of the Green function $G_D(\xi, \cdot)$ at the Martin boundary (to be extended by relativization in Section 18) is one indication that the Martin boundary is well adapted to classical potential theory.

Relation between Theorems 13, 14, and XI.4

The fact that the Laplacian is a self-adjoint differential operator leads to the symmetry of the Green function G_D, absent in the potential theory generated by a non-self-adjoint differential operator. In such a theory the counterpart of Theorem XI.4 splits into two theorems. See Section XVIII.14 for the versions of Theorem XI.4 in the potential theory corresponding to the heat equation and its adjoint. The self-dual character of classical potential theory is lost at the Martin boundary of a Greenian domain, however, and in fact Theorems 13 and 14 are dual to each other. This is suggested by the fact that the proof of Theorem 14, unlike the proof of Theorem 13, uses the sweeping symmetry of classical potential theory. In the probabilistic versions of these theorems (Theorem 3.III.5) Theorem 13 states that the function $v/K(\zeta, \cdot)$ has the indicated limit at ζ along almost every Brownian path conditioned to go from a point of D to ζ, whereas Theorem 14 states that $v/G_D(\xi, \cdot)$ has the indicated limit at ζ along almost every Brownian path conditioned to go from ζ to a point of D. If the two points of D here are taken to be the same, these conditional Brownian paths to ζ can be identified with those from ζ; so the same limit concept is involved, corresponding to minimal-fine limits at ζ. In particular, in classical potential theory if ζ is a point of a Greenian set D, then ζ can be identified with a minimal Martin boundary point of $D - \{\zeta\}$ [Section 12, Example (a)], the corresponding minimal harmonic function is a multiple of $G_D(\zeta, \cdot)$, and Theorems 13 and 14 coalesce to Theorem XI.4(c).

15. Minimal-Fine Topology Limits and Martin Topology Limits at a Minimal Martin Boundary Point

Let D be a connected Greenian subset of $\mathbb{R}^N$, let K be a Martin function for D, and let ζ be a minimal Martin boundary point. The following lemma is the analog of Lemma XI.8 in the present context.

Lemma. *Let $A_{\bullet}$ be a decreasing sequence of subsets of D, and suppose that each set A_k is a deleted minimal-fine neighborhood of ζ [has minimal-fine limit point ζ]. Then there is a subset A of D with the property that for each k the part of A in a sufficiently small Martin topology neighborhood of ζ is in A_k and that A is a deleted minimal-fine neighborhood of ζ [has minimal-fine limit point ζ].*

The proof is similar to that of Lemma XI.8; so only the unbracketed assertion will be proved. Suppose that each set A_k is a deleted minimal-fine neighborhood of ζ, and define $F_k = D - A_k$. The set F_k is minimal thin at ζ, and it is sufficient to show that there is a set F, minimal thin at ζ, with the property that for each k the part of F_k in a sufficiently small neighborhood of ζ is a subset of F. If for all but a finite number of values of k the point ζ is not a limit point of F_k, we can take F as the union of those sets F_k for which ζ is not a limit point of F_k. Otherwise, it is no restriction to assume that ζ is a limit point of F_k for all k. Let B_r be the intersection with D of a ball (in terms of some metric on D^M) of center ζ and radius r. Let ξ_0 be a point of D, and applying Theorem 13, let $u_k = v_k/K(\zeta, \cdot)$ for $k \in \mathbb{Z}^+$ be a positive $K(\zeta, \cdot)$-superharmonic function on D with $u(\xi_0) < 2^{-k}$ and with limit $+\infty$ at ζ on approach along F_k. The function $u = \Sigma_0^{\infty} u_k$ is positive and $K(\zeta, \cdot)$-

superharmonic on D with limit $+\infty$ at ζ on approach along each set F_k. Choose r_k so that $\lim_{k\to\infty} r_k = 0$ and that $u \geq k$ on $F_k \cap B_{r_k}$. If

$$F = \bigcup_0^\infty (F_k \cap B_{r_k}),$$

the function u has limit $+\infty$ at ζ along F; so [by Theorem 13(b)] F is minimal thin at ζ, and $F_k \cap B_{r_k} \subset F$, as desired.

16. Minimal-Fine Topology Limits and Martin Topology Limits at a Minimal Martin Boundary Point (Continued)

Theorem. *Let D be a connected Greenian subset of $\mathbb{R}^N$, and let ζ be a minimal Martin boundary point of D. A function u from the trace on D of a Martin topology neighborhood of ζ into a metric space has minimal-fine limit [minimal-fine cluster value] α at ζ if and only if u has limit α at ζ on approach along some subset of D that is a deleted minimal-fine neighborhood of ζ [is not minimal thin at ζ].*

The proof is that of Theorem XI.9 with trivial changes corresponding to the change of context. Observe that if the range space of u is $\overline{\mathbb{R}}$, then $\text{mf}\limsup_{\xi\to\zeta} u(\xi)$ is a minimal-fine topology cluster value of u at ζ, and therefore u has this value as a limit on approach to ζ along some subset of D that is not minimal thin at ζ; that is, this subset has ζ as a minimal-fine topology limit point.

17. Minimal-Fine Martin Boundary Limit Functions

Let D be a connected Greenian subset of $\mathbb{R}^N$. If A is a subset of D, denote by A^{mf} the set of minimal-fine limit points of A in D^M. Recall (Section 12) that $D^{mf} \cap \partial^M D = \partial_1^M D$.

Lemma. (a) *If A is a subset of D, the set A^{mf} is a Martin topology G_δ subset of D^M.*

(b) *If u is a function from D into $\overline{\mathbb{R}}$, the function $\zeta \mapsto mf\limsup_{\xi\to\zeta} u(\xi)$ on the minimal Martin boundary $\partial_1^M D$ is Borel measurable (Martin topology of D^M).*

(c) *If u is a function from D into a compact metric space, the set of Martin boundary points at which u has a minimal-fine limit is a Borel set, and the limit function on this set is Borel measurable (Martin topology of D^M).*

Proof of (a). According to Theorem XI.6, the set $A^{mf} \cap D$ is a Euclidean G_δ set, and this set is therefore a G_δ set in the Martin topology. To prove that $A^{mf} \cap \partial^M D$ is a G_δ set, let K be a Martin function for D, and let ξ_0 be a point of D. If B is a subset of D, let B' be the class of Martin boundary points ζ satisfying the inequality $\llbracket K(\zeta, \cdot) \rrbracket^B(\xi_0) \le K(\zeta, \xi_0)/2$. Since the smoothed reduction on the left is $\delta_D^B(\xi_0, K(\zeta, \cdot))$, Fatou's lemma implies that the set B' is compact. Let $B_\bullet$ be the sequence of traces on D of the sets of a countable topological base for D^M. If A is a subset of D, the set $\bigcup_0^\infty (A \cap B_n)'$ is an F_σ set, the set of Martin boundary points that are not minimal-fine limit points of A. □

Proof of (b). Assertion (b) follows from (a) because

$$\left\{\zeta \in \partial_1^M D\colon \operatorname{mf}\limsup_{\xi \to \zeta} u(\xi) \le c\right\} = \bigcap_{n=1}^{\infty} \left\{\xi \in D\colon u(\xi) \le c + \frac{1}{n}\right\}^{mf} \cap \partial^M D. \quad (17.1)$$

□

Proof of (c). According to (b), the boundary set on which u has a minimal-fine limit is a Borel set if u is extended real valued, because the set in question is the set on which the minimal-fine superior and inferior limits of u are equal. Moreover (b) implies that the limit function on this set is Borel measurable. If the range space S of u is compact metric and if $\phi_\bullet$ is a sequence of functions dense in $\mathbb{C}(S)$ in the metric of uniform convergence, then the map $\xi \mapsto \{\phi_n(\xi)/\sup_S |\phi_n|, n \in \mathbb{Z}^+\}$ is a one-to-one bicontinuous map of S onto a compact subset of the compact metric space $[0,1]^{\mathbb{Z}^+}$; so an application of (a) and (b) to each function $\phi_n(u)$ yields (c). □

EXAMPLE. *If h is a strictly positive harmonic function on the connected Greenian subset D of $\mathbb{R}^N$ and if A is a subset of D, then*

$$\mathrm{GM}_D^h({}^h R_{+1}^A) = \frac{\mathrm{GM}_D(R_{+h}^A)}{h} = \mu_D^h(\cdot, A^{mf} \cap \partial^M D). \quad (17.2)$$

Observe that the first equality is trivial and that the second equality asserts that the function $h\mu_D^h(\cdot, A^{mf} \cap \partial^M D)$ is the harmonic component of R_{+h}^A (Riesz decomposition). To prove this assertion, apply the kernel operator δ_D^A to the Martin representation of h to find

$$\begin{aligned} R_{+h}^A = \int_{\partial^M D} R_{+K(\zeta,\cdot)}^A M_h(d\zeta) &= \int_{A^{mf} \cap \partial^M D} K(\zeta, \cdot) M_h(d\zeta) \\ &\quad + \int_{\partial^M D - A^{mf}} R_{+K(\zeta,\cdot)}^A M_h(d\zeta). \end{aligned} \quad (17.3)$$

The first integral after the second equality sign is equal to $h\mu_D^h(\cdot, A^{mf} \cap \partial^M D)$ according to Section 10. According to Theorem 11, the function $\underset{+K(\zeta,\cdot)}{R^A}$ is a potential whenever ζ is not minimal or is not in A^{mf}, so the last integral is a potential. Thus (17.3) displays the Riesz decomposition of $\underset{+h}{R^A}$ and thereby yields the second equality in (17.2).

Observe that according to this example, the function $\|h\|^A$ is a potential if and only if $A^{mf} \cap \partial^M D$ is μ_D^h null.

Application. If D is a connected Greenian subset of $\mathbb{R}^N$, if h is a strictly positive harmonic function on D, if u is a positive h-superharmonic function on D, and if

$$B_c = \{\zeta \in \partial_1^M D \colon \operatorname{mf}\limsup_{\eta \to \zeta} u(\eta) \geq c\},$$

then $u \geq c\mu_D^h(\cdot, B_c)$. We can suppose in the proof that $c > 0$. If $A_\alpha = \{\xi \in D\colon u(\xi) \geq \alpha\}$, then according to the preceding example and the relation (Section VIII.2) between reductions and h-harmonic measure, if $0 < \alpha < c$,

$$u \geq {}^h\underset{+u}{R^{A_\alpha}} \geq \alpha\, {}^h\underset{+1}{R^{A_\alpha}} \geq \alpha\mu_D^h(\cdot, A_\alpha^{mf} \cap \partial^M D) \geq \alpha\mu_D^h(\cdot, B_c). \tag{17.4}$$

Let α tend to c to obtain the desired inequality.

18. The Fine Boundary Function of a Potential

Theorem. *If D is a connected Greenian subset of $\mathbb{R}^N$ a superharmonic h-potential, $u = G_D\mu/h$ has minimal-fine limit 0 at μ_D^h almost every (equivalently M_h almost every) point of $\partial^M D$.*

Define $A_\varepsilon = \{\zeta \in \partial_1^M D\colon \operatorname{mf}\limsup_{\eta \to \zeta} \geq \varepsilon\}$. Then according to Section 17 (Application), $u \geq \varepsilon\mu_D^h(\cdot, A_\varepsilon)$, and this is impossible unless A_ε is μ_D^h null, because $\mathrm{GM}_D^h u = 0$. Hence the theorem is true.

Application. According to Theorem 18, the function $G_D(\xi, \cdot)/h$ has minimal-fine limit 0 at μ_D^h almost every point of $\partial^M D$, and it then follows from Theorem 14 that

$$\lim_{\eta \to \zeta} \frac{G_D(\xi, \eta)}{h(\eta)} = 0 \qquad (\mu_D^h \text{ a.e. } \zeta \text{ in } \partial_1^M D). \tag{18.1}$$

In particular, if $\zeta \in \partial_1^M D$ and if K is a Martin function for D,

$$\lim_{\eta \to \zeta} \frac{G_D(\xi, \eta)}{K(\zeta, \eta)} = 0. \tag{18.1$'$}$$

19. The Fatou Boundary Limit Theorem for the Martin Space

Let D be a connected Greenian subset of $\mathbb{R}^N$. In this section the vector lattice notation of Chapter IX will be used, as further developed in Section 9 of the present chapter. For example, when h is a strictly positive harmonic function on D, we denote by ${}^h\mathbf{S}^+$ the cone of positive h-superharmonic functions on D. Let K be a Martin function for D. There is then (from Section 9) an isomorphism $v \leftrightarrow M_v$ between ${}^1\mathbf{S}_m$ and the vector lattice of signed measures on $\partial^M D$ supported by the minimal Martin boundary $\partial_1^M D$. When $v \in {}^1\mathbf{S}$, it will be convenient to denote by M_v the Martin representing signed measure of the component of v in ${}^1\mathbf{S}_m$; when $v \in {}^1\mathbf{S}^+$, this signed measure M_v is the Martin representing measure of the harmonic component of v in its Riesz decomposition. Recall that dM_v/dM_h under our conventions is the Radon–Nikodym derivative of the absolutely continuous component of M_v relative to M_h and that (Section 10) "M_h almost everywhere" for h in ${}^1\mathbf{S}_m^+$ is equivalent to "μ_D^h almost everywhere"; both M_h and μ_D^h are supported by $\partial_1^M D$.

Theorem. *If $u = v/h$ is in ${}^h\mathbf{S}$, then*

$$\operatorname{mf}\lim_{\eta \to \zeta} u(\eta) = \frac{dM_v}{dM_h}(\zeta) \qquad (M_h \text{ a.e. } \zeta \text{ on } \partial^M D). \tag{19.1}$$

In particular:

(a) *This boundary limit function vanishes M_h almost surely if $u \in {}^h\mathbf{S}_p$.*
(b) *If $u \in {}^h\mathbf{S}_m$, this boundary limit function vanishes M_h almost surely if and only if $u \in {}^h\mathbf{S}_{ms}$.*
(c) *If u is a* PWB^h *solution, that is, if $u = \mu_D^h(\cdot, f)$ for some h-resolutive boundary function f, then $f = dM_v/dM_h$ M_h almost everywhere on $\partial^M D$; so f is the minimal-fine boundary limit function of u up to an M_h null set.*

According to Section 10, the Martin boundary is universally resolutive and universally internally resolutive, and the class of PWB^h solutions is ${}^h\mathbf{S}_{mqb}$. In view of this fact and of the decomposition ${}^h\mathbf{S} = {}^h\mathbf{S}_p + {}^h\mathbf{S}_{ms} + {}^h\mathbf{S}_{mqb}$ together with the vector lattice isomorphism between ${}^h\mathbf{S}_m$ and the vector lattice of signed measures on $\partial_1^M D$ derived in Section 9 by means of the Martin representation theorem, it will be sufficient to prove that if $u \in {}^h\mathbf{S}_p \cup {}^h\mathbf{S}_{ms}$, then u has minimal-fine limit 0 M_h almost everywhere on $\partial^M D$ and that if $u = \mu_D^h(\cdot, f)$, then u has minimal-fine limit $f(\zeta)$ at M_h almost every point ζ of $\partial^M D$. Theorem 18 implies that a function in ${}^h\mathbf{S}_p$ has minimal-fine limit 0 M_h almost everywhere on $\partial^M D$ and therefore that the same is true for u in ${}^h\mathbf{S}_{ms}$ because according to Section IX.10, if $u \in {}^h\mathbf{S}_{ms}^+$, then $u \wedge 1 \in {}^h\mathbf{S}_p^+$.

Finally, if $u = {}^hH_f = \mu_D^h(\cdot, f)$, let u_1 and u_2 be in the lower and upper PWB^h classes, respectively, on D for f. Then (by the application in Section 17) when $\varepsilon > 0$,

$$u_2 - u_1 \geq \varepsilon\mu_D^h\left(\cdot, \left\{\zeta : \operatorname{mf}\limsup_{\eta\to\zeta}\left[u_2(\eta) - u_1(\eta)\right] \geq \varepsilon\right\}\right)$$

$$\geq \begin{cases} \varepsilon\mu_D^h\left(\cdot, \left\{\zeta : \operatorname{mf}\limsup_{\eta\to\zeta} u(\eta) - f(\zeta) \geq \varepsilon\right\}\right), \\ \varepsilon\mu_D^h\left(\cdot, \left\{\zeta : f(\zeta) - \operatorname{mf}\liminf_{\eta\to\zeta} u(\eta) \geq \varepsilon\right\}\right). \end{cases} \tag{19.2}$$

Since the left side of this inequality can be made arbitrarily small at any point of D, it follows that $\operatorname{mf}\lim_{\eta\to\zeta} u(\eta) = f(\zeta)$ for μ_D^h almost every boundary point ζ, equivalently, M_h almost every boundary point ζ, as was to be proved.

Special cases. If $h = K(\zeta, \cdot)$ is minimal, Theorem 19 states that $v/K(\zeta, \cdot)$ has minimal-fine limit $M_v(\{\zeta\})$ at ζ, as already proved in Section 13. If A is a Borel boundary subset, the function $\mu_D^h(\cdot, A)$ is the PWB^h solution for the boundary function 1_A and therefore has minimal-fine boundary limit function M_h almost everywhere equal to 1_A.

Application to the Internal Fine Limit Theorem of Section XI.4

Let v and h be positive superharmonic functions on D with respective Riesz measures ν_v and ν_h, and define $u = v/h$ at the points of D that are not in the polar set of infinities common to v and h. By definition of the fine topology the function u is fine continuous at every point at which it is defined. According to Theorem XI.4,

$$\operatorname{f}\lim_{\eta\to\zeta} u(\eta) = \frac{d\nu_v}{d\nu_h}(\zeta) \tag{19.3}$$

at ν_h almost every point ζ of the polar set of infinities of h, and we now show that this limit result is a consequence of Theorem 19. Let F be a compact subset of the set of infinities of h. To prove (19.3), it is sufficient to prove that u has fine limit $d\nu_v/d\nu_h$ at ν_h almost every point of F. In view of the Martin compactification of D less a compact polar set [Section 12, Example (a)], if ν_h' and ν_h'' are the projections of ν_h on F and $D - F$, respectively, Theorem 19 implies that

$$\operatorname{f}\lim_{\eta\to\zeta} \frac{v(\eta)}{G_D\nu_h'(\eta)} = \frac{d\nu_v}{d\nu_h'}(\zeta), \qquad \operatorname{f}\lim_{\eta\to\zeta} \frac{G_D\nu_h''(\eta)}{G_D\nu_h'(\eta)} = \frac{d\nu_h''}{d\nu_h'}(\zeta)$$

at ν_h' almost every point ζ of F. Since the second Radon–Nikodym derivative vanishes at ν_h' almost every point of F, equivalently, at ν_h almost every point of F, the first Radon–Nikodym derivative is equal to $d\nu_v/d\nu_h$ at ν_h almost every point of F, equation (19.3) follows when h is a potential; the general case then follows easily.

20. Classical versus Minimal-Fine Topology Boundary Limit Theorems for Relative Superharmonic Functions on a Ball in $\mathbb{R}^N$

Recall that the Martin boundary of a ball is the Euclidean boundary. We have proved two boundary limit theorems for a positive h-harmonic function u on a ball D. According to Theorem II.15 and Theorem 19, at μ_D^h almost every point of ∂D the function u has a finite limit both in the nontangential and minimal-fine topologies, necessarily the same limit (Section 12). The limit function is a certain Radon–Nikodym derivative. Observe, however, that Theorem 19 is more general in that Theorem II.15 is for h-harmonic functions whereas Theorem 19 is applicable to h-superharmonic functions. Furthermore Theorem 19 is applicable to functions on an arbitrary connected Greenian set D, but in this section we shall always assume that D is a ball.

Classical-type approach to a ball boundary is not always applicable in the general context of Theorem 19. In fact the following example shows that Theorem 19 is false for D a ball under nontangential rather than minimal-fine approach to the boundary.

Example (a). Let $v = G_D\mu$ be a superharmonic potential with μ supported by a countable dense subset of the ball D, so that the h-superharmonic function $u = v/h$ has nontangential limit superior $+\infty$ at every boundary point of D. Furthermore it is easy to choose μ in this example in such a way, depending on h, that u has nontangential (even radial) limit inferior 0 at every boundary point of D.

The following discussion shows that radial approach to a ball boundary allows a wider class of functions u in Fatou-type boundary limit theorems than nontangential approach but does not allow as wide a class as minimal-fine boundary approach. Let $v = G_D\mu$ be a superharmonic potential on the ball D. According to Theorem 19, the function v has minimal-fine limit 0 at μ_D almost every (equivalently, l_{N-1} almost every) boundary point. For $N = 2$ Littlewood showed that v has limit 0 along the radius to l_{N-1} almost every ball boundary point, and his result was later extended to $N > 2$. In view of the Riesz decomposition theorem it follows that for $h \equiv 1$ and D a ball

Theorem 19 is true for radial as well as minimal-fine approach to the boundary. The following example shows that this assertion is false for general h.

EXAMPLE (b). Let h be a strictly positive minimal harmonic function on a ball D, corresponding to a boundary point ζ. Then (from Section VIII.9) μ_D^h is supported by the singleton $\{\zeta\}$, and Theorem 19 states that a positive h-superharmonic function $u = v/h$ has a finite minimal-fine limit at ζ. This limit is $\inf_D u$ according to Theorem 13 and is 0 (by Theorem 19) if $v = G_D \nu$ is a potential. Since [Section 12, Example (b)] the radius to ζ is not minimal thin at ζ, the function u has radial limit inferior 0 at ζ. However, if ν is supported by a countable dense subset of this radius, the function u has radial limit superior $+\infty$ at ζ.

Thus the fine topology is better adapted to Fatou-type boundary limit theorems for a ball than nontangential and radial approach topologies. The difference is that the latter approach topologies are adapted to the ball rather than to the properties of the functions involved. To show the relation between Theorem 19 for a ball and the corresponding results for nontangential and radial approach to a ball boundary, it will be shown in the next sections that a positive h-harmonic function has a nontangential limit β at any ball boundary point at which there is a minimal-fine limit β and that a superharmonic potential has radial limit 0 at l_{N-1} almost every boundary point at which there is minimal-fine limit 0. Thus Theorem 19 for a ball implies the truth of the corresponding partial results described above for nontangential and radial boundary approach. The converse is false. More exactly, it will be convenient to prove the stated implications for a half-space rather than a ball. Since half-spaces and balls can be mapped onto each other by inversions, the assertions on nontangential convergence to boundaries correspond exactly in the two contexts. The assertion on radial convergence to a ball boundary corresponds very nearly to the corresponding assertion on convergence to a half-space boundary along normal lines.

21. Nontangential and Minimal-Fine Limits at a Half-space Boundary

Let d_ξ be the last coordinate of the point ξ of $\mathbb{R}^N$, and let D be the half-space $\{d_\xi > 0\}$.

Theorem. *If $u = v/h$ is a strictly positive h-harmonic function on the half-space D and if u has minimal-fine limit β at the boundary point ζ, then u has nontangential limit β at ζ.*

The converse theorem can be shown to be false. Theorem 21 implies Theorem II.15, that u has nontangential limit dM_v/dM_h at M_h almost every boundary point, in view of the minimal-fine topology Fatou theorem (Theorem 19) and the correspondence between ball and half-space given by an inversion.

We can assume in the following proof that the specified boundary point ζ of D: $\{d_\xi > 0\}$ is the origin. Let $\xi_\cdot$ be a sequence in D with nontangential limit the origin, and suppose that $\lim_{n\to\infty} u(\xi_n) = \beta'$. To prove Theorem 21, it will be shown that $\beta' = \beta$. Assume from now on that $0 < \beta < +\infty$. The modifications to be made in the argument if β is 0 or $+\infty$ will be obvious. It can be assumed that the ray from the origin through ξ_n tends to a limit ray L as $n \to \infty$. Let ξ_n' be the point of L at minimum distance from ξ_n, let A be a ball, closure in D, center ξ_0', let A_0 be a smaller concentric ball, and define $c_n = |\xi_n'|/|\xi_0'|$. The ball $c_n A$ has center ξ_n', and $\xi_n \in c_n A_n'$ for sufficiently large n. Finally (by the Harnack inequality), there is a strictly positive constant γ depending only on the ratio of the radii of A and A_0 such that $1/\gamma \le u(\eta)/u(\xi) \le \gamma$ for both ξ and η in $c_n A_0$. Since $\bigcup_0^\infty (c_n A_0)$ is not minimal-thin at the origin [Section 12, Example (b)], there is a sequence $\eta_\cdot$ along which u has limit β, with $\eta_n \in c_n A_0$. Hence $1/\gamma \le \beta/\beta' \le \gamma$. Since the Harnack constant γ can be made arbitrarily near 1 by choosing A_0 sufficiently small, it follows that $\beta = \beta'$, as was to be proved.

22. Normal Boundary Limits for a Half-space

Define the half-space D as in Section 21. A boundary point ζ will be called a *normal limit point* of a subset A of D if ζ is a limit point of the part of A on the normal to ∂D through ζ. A function on D will be said to have a normal limit at ζ if it has a limit at ζ along the normal.

Lemma. *If A is an (open) subset of the half-space D, l_{N-1} almost every normal boundary limit point of A is a minimal-fine limit point of A.*

Warning: "Open" was enclosed in parentheses in this statement because (Section 23) the lemma is true for an arbitrary subset A of D.

It is sufficient to prove the lemma for a bounded open set A. A careful application of the Vitali covering theorem shows that there is a subset A_k of A with the following properties: A_k is a countable union of closed $(N-1)$-dimensional intervals, each on a hyperplane $\{\xi: d_\xi = \text{const} < 1/k\}$, each with edges parallel to the coordinate axes; these intervals have disjoint projections on ∂D; the projection on ∂D of A_k covers l_{N-1} almost every point of the projection of $A \cap \{\xi: d_\xi < 1/k\}$. Define the potential v_k by

$$v_k(\xi) = \int_{A_k} \frac{G_D(\xi, \eta)}{d_\eta} l_{N-1}(d\eta). \tag{22.1}$$

The remainder of the proof assumes that $N > 2$. The argument is similar when $N = 2$. We first prove that there is a constant c_1 depending only on N

such that $v_k \leq c_1$ for all k and choices of A_k. To see this, let ξ be a point of D, and define

$$D_s = B(\xi, s), \qquad \psi(s) = \sup_{\eta \in D \cap \partial D_s} \frac{G_D(\xi, \eta)}{d_\eta}.$$

The evaluation (see Section VIII.9)

$$\frac{G_D(\xi, \eta)}{d_\eta} = \frac{s^{-(N-2)} - (s^2 + 4d_\xi d_\eta)^{-(N-2)/2}}{d_\eta} \qquad (|\eta - \xi| = s)$$

shows that the left side decreases when d_η increases, from which it follows that

$$\psi(s) = \begin{cases} 2(N-2)d_\xi s^{-N} & \text{if} \quad s > d_\xi, \\ \dfrac{s^{-(N-2)} - (2d_\xi - s)^{-(N-2)}}{d_\xi - s} & \text{if} \quad s \leq d_\xi, \end{cases} \tag{22.2}$$

so that $\psi(\cdot)$ is monotone decreasing. Thus

$$v_k(\xi) \leq \int_0^\infty \psi(s)\, d\phi(s) \quad \text{with} \quad \phi(s) = l_{N-1}(A \cap D_s). \tag{22.3}$$

Integration by parts, together with the fact that $\phi(s) \leq \pi_{N-1} s^{N-1}/(N-1)$, yields the majorant $\pi_{N-1} \int_0^\infty \psi(s) s^{N-2}\, ds$ of the right side of (22.3), and in view of (22.2) the integral $\int_0^\infty \psi(s) s^{N-2}\, ds$ is convergent with value c_1 independent of ξ, A, k, A_k. If v_k' is defined using only finitely many of the intervals in A_k, v_k'/c_1 is the potential of a measure supported by A_k and $v_k'/c_1 \leq 1$. Hence (domination principle) $v_k' \leq c_1 [\![1]\!]^{A}$, and therefore $v_k \leq c_1 [\![1]\!]^{A}$. To bound v_k from below, observe that when $k \to \infty$, the integrand in (22.1) tends to the normal derivative of $G_D(\xi, \cdot)$ at the boundary, that is [VIII(9.4)], to $\pi_N' \mu_D(\xi, d\eta)/l_{N-1}(d\eta)$, uniformly for ξ in any bounded set, so that (22.1) yields, if A'' is the set of normal limit points of A on ∂D,

$$\liminf_{k \to \infty} v_k \geq \pi_N' \mu_D(\cdot, A'').$$

Thus $\mu_D(\cdot, A'') \leq c_2 [\![1]\!]^{A}$ for some constant c_2. Denote by A' the set of minimal-fine limit points of A on ∂D. According to Section 17, $\mathrm{GM}_D [\![1]\!]^{A} = \mu_D(\cdot, A')$, so that $\mu_D(\cdot, A'') \leq c_2 \mu_D(\cdot, A')$. Since the harmonic measure of a Borel measurable boundary set B has minimal-fine boundary limit function 1_B up to an l_{N-1} null set, it follows that $A'' \subset A'$ up to an l_{N-1} null set, as was to be proved.

23. Boundary Limit Function (Minimal-Fine and Normal) of a Potential on a Half-space

Theorem. *Let D be a half-space of $\mathbb{R}^N$.*

(a) *A superharmonic potential on D has normal limit 0 at l_{N-1} almost every point of ∂D.*

(b) *If A is a subset of D, l_{N-1} almost every normal boundary limit point of A is a minimal-fine boundary limit point of A.*

Proof of (a). To prove (a), it is sufficient to show that for u a superharmonic potential on D and $\varepsilon > 0$ the open set $A_\varepsilon = \{u > \varepsilon\}$ has l_{N-1} almost no normal boundary limit point, and in view of Lemma 22 this follows from the fact (Theorem 18) that u has minimal-fine limit 0 at μ_D (equivalently, l_{N-1}) almost every boundary point, so that l_{N-1} almost no boundary point is a minimal-fine limit point of A_ε. □

Observation. We have now proved that a positive superharmonic function on a half-space of $\mathbb{R}^N$ has a limit at l_{N-1} almost every boundary point on both minimal-fine and normal approach to the boundary point and that the boundary limit functions in the two approaches are equal up to l_{N-1} null sets. In fact (Riesz decomposition) it was sufficient to prove the theorem separately for u harmonic and u a potential. The harmonic case was covered in Section 19, in which case nontangential boundary approach was admissible; the potential case, in which the limit vanishes l_{N-1} almost everywhere on ∂D, is covered in (a) of the present theorem.

Proof of (b). Let A'' $[A']$ be the set of normal [minimal-fine] boundary limit points of A. The function $[\![1]\!]^A$ is a positive superharmonic function on D, equal to 1 quasi everywhere on A. Let A_0 be the polar subset of A on which $[\![1]\!]^A < 1$, and let v_0 be a superharmonic potential on D, identically $+\infty$ on A_0. According to (a), the function v_0 has normal limit 0 at l_{N-1} almost every point of ∂D; so l_{N-1} almost no point of ∂D is a normal limit point of A_0, and we shall therefore assume from now on that A_0 is empty. In view of the Riesz decomposition theorem and (17.2),

$$[\![1]\!]^A = v + \mu_D(\cdot, A'), \tag{23.1}$$

where v is a potential on D. The smoothed reduction on the left side of (23.1) has normal boundary cluster value 1 at every point of A'' and therefore by (a) and the above observation has normal and minimal-fine limit 1 at l_{N-1} almost every point of A''. The function on the right side of (23.1) has minimal-fine boundary limit function $1_{A'}$ up to an l_{N-1} null set; so $A'' \subset A'$ up to an l_{N-1} null set, as was to be proved. □

Chapter XIII

Classical Energy and Capacity

1. Physical Context

Consider a distribution of positive and negative electric charges on $\mathbb{R}^3$ and the electrostatic potential induced by this charge. By definition of a conductor, if A is a connected conducting body in $\mathbb{R}^3$, the charge on A distributes itself in such a way that the net effect is that of an all-positive or all-negative charge, and the distribution on A is in equilibrium in the sense that the restriction to A of the potential of the charge distribution in $\mathbb{R}^3$ is a constant function.

Let D be an open subset of $\mathbb{R}^3$ with a conducting smooth boundary, and suppose that the boundary is grounded. The significance of grounding is that if a positive charge μ is imposed on D, an induced negative charge $-\mu^*$ appears on ∂D, and the potential $G(\mu - \mu^*)$ is identically 0 on ∂D. Thus, if μ is a unit positive charge at ξ in D, the restriction to D of the potential $G(\mu - \mu^*)$ is identified with the Green function $G_D(\xi,\cdot)$, and μ^* is identified with the sweeping of μ onto ∂D (relative to $\mathbb{R}^3$); that is, μ^* is identified with the harmonic measure $\mu_D(\xi,\cdot)$. It follows that for any μ the measure μ^* is identified with the sweeping of μ onto ∂D and that $G(\mu - \mu^*) = G_D\mu$ on D. In view of this physical context the existence of a mathematical version of the Green function of a reasonable set D was obvious long before there was a rigorous existence proof, and the sweeping of a measure was a natural concept to formalize.

Now suppose that a connected conducting body is introduced into D and given a positive charge μ. This charge necessarily distributes itself in such a way that $G(\mu - \mu^*) = G_D\mu$ is constant on A. Such a charge μ, that is, such a measure, is called an *equilibrium charge* (or *measure*, or *distribution*), and the corresponding potential $G_D\mu$ is called an *equilibrium potential* for A. Two equilibrium potentials for A are proportional, as are their potentials, and if the potential on A of an equilibrium measure has the constant value 1 the equilibrium measure [potential] is called the *capacitary measure* [*potential*]. In view of this physical context it was clear to Gauss that there must be a capacitary distribution in a suitable mathematical context for any reasonable pair A and D. The pair $(A, \partial D)$ is a condenser in the physical context, and the *capacity* of this condenser is defined as

$$\frac{\text{Total charge of an equilibrium distribution on } A}{\text{Value on } A \text{ of the corresponding potential}} \tag{1.1}$$
$$= \text{total charge of the capacitary distribution on } A.$$

The mathematical model of this physical context has already been discussed, at least in part. The set D is supposed Greenian. The Green function G_D has already been defined, and $G_D(\xi,\cdot)$ has been shown to have limit 0 at every regular boundary point of D. If μ is a measure on D, the function $G_D\mu$ is the mathematical version of the electrostatic potential generated by a distribution μ of electric charge. More generally μ will sometimes be allowed to be a suitably restricted signed measure or charge in the sense of Appendix IV.7. The energy of a charge and the mutual energy of a pair of charges will be defined, following mathematical tradition, as twice the values assigned by physicists. Equilibrium distributions will be derived, and the capacity of a subset A of D will be defined by (1.1) when A is analytic.

2. Measures and Their Energies

If D is a Greenian subset of $\mathbb{R}^N$, the mutual energy $[\mu, \nu]$ of a pair of measures on D is defined by

$$[\mu, \nu] = \int_D G_D\mu \, d\nu = \int_D \int_D G_D(\xi, \eta)\mu(d\xi)\nu(d\eta), \tag{2.1}$$

and the energy $\|\mu\|^2$ of a measure μ on D is defined as $[\mu, \mu]$. The form $[\cdot, \cdot]$ is symmetric,

$$[\mu, \nu] = [\nu, \mu], \tag{2.2}$$

and this symmetry is sometimes dignified by the name *reciprocity law*. The symmetry of the Green function G_D is a special case.

The Space $\mathscr{E}^+$

The space of measures on D of finite energy will be denoted by $\mathscr{E}^+$. It is trivial that a positive constant multiple of an element of $\mathscr{E}^+$ is in $\mathscr{E}^+$, but it is a much deeper fact that (Theorem 7) the sum of two elements in $\mathscr{E}^+$ is in $\mathscr{E}^+$, equivalently, that the mutual energy of two measures in $\mathscr{E}^+$ is finite.

Example (a). If μ is a measure on D and if A is a polar subset of D that is not μ null, then $\|\mu\| = +\infty$ because (Theorem V.11) $G_D\mu = +\infty$ μ almost everywhere on A.

Thus a measure of finite energy vanishes on polar sets. Conversely, if A is an analytic subset of a Greenian set D and is null for every measure on D

of finite energy supported by A, then A is polar in view of Corollary V.9 and the fact (Theorem VI.2) that an analytic nonpolar set has a compact nonpolar subset. It will be shown in Section 3.I.9 that a superharmonic potential $G_D\mu$ is quasi bounded if and only if μ vanishes on polar sets. It was noted in Section V.10 that the domination principle as applied to potentials of measures μ vanishing on polar sets has a simple form: if $G_D\mu$ is majorized μ almost everywhere on D by a positive superharmonic function v, then $G_D\mu \le v$ on D.

EXAMPLE (b). Let ξ be a point of D, choose $\alpha > 0$, and define $B = \{G_D(\xi,\cdot) > \alpha\}$, $A = D \cap \partial B$ (Euclidean boundary), and $\mu = \delta_D^A(\xi,\cdot)$. Since $G_D(\xi,\cdot)$ has limit 0 at quasi every point of ∂D, $\partial B - A$ is polar and thus is a $\mu_B(\xi,\cdot)$ null set; that is, $\mu_B(\xi, A) = 1$. (This fact is a special case of the application in Section VIII.18.) The measure μ is supported by A, and $G_D(\xi,\cdot) = \alpha$ on A; so

$$\|\mu\|^2 = \alpha\delta_B^A(\xi, A) = \alpha\mu_B(\xi, A) = \alpha. \tag{2.4}$$

In particular, if $D = B(\xi,\delta)$ and if $B = B(\xi,\beta)$ with $\beta < \delta$, then

$$B = \begin{cases} \left\{G(\xi,\cdot) > \log\dfrac{\delta}{\beta}\right\} & \text{if } N = 2,\\ \{G(\xi,\cdot) > \beta^{2-N} - \delta^{2-N}\} & \text{if } N > 2; \end{cases} \tag{2.5}$$

so the energy of the uniform distribution on ∂B of total value 1 is $\log(\delta/\beta)$ if $N = 2$ and is $\beta^{2-N} - \delta^{2-N}$ if $N > 2$. The latter energy decreases to β^{2-N}, the energy of μ relative to $\mathbb{R}^N$, when $\delta \to \infty$.

EXAMPLE (c). The general case of the preceding example is the following. Let D be a Greenian subset of $\mathbb{R}^N$, let B be an open subset of D, let ξ be a point of B, and define $A = D \cap \partial B$ (Euclidean boundary) and $\mu = \delta_D^A(\xi,\cdot)$. Then

$$\|\mu\|^2 = \delta_D^A(\xi, G_D(\xi,\cdot)) = G_D^A(\xi,\xi) = \mu_B(\xi, G_D(\xi,\cdot)1_D). \tag{2.6}$$

3. Charges and Their Energies

A charge (Radon measure) μ on the open subset D of $\mathbb{R}^N$ is (see Appendix IV.3) a pair (μ_1,μ_2) of measures on D under the identification $(\mu_1,\mu_2) = (\nu_1,\nu_2)$ when $\mu_1 + \nu_2 = \mu_2 + \nu_1$ and the operations

$$(\mu_1,\mu_2) + (\nu_1,\nu_2) = (\mu_1 + \nu_1, \mu_2 + \nu_2),$$

$$c(\mu_1,\mu_2) = \begin{cases} (c\mu_1, c\mu_2) & \text{if } c \ge 0,\\ (-c\mu_2, -c\mu_1) & \text{if } c < 0. \end{cases} \tag{3.1}$$

A charge has a minimal representation whose component measures have disjoint supports. A charge is called *positive* if its minimal representing pair has the form $(\mu_1, 0)$ and is then identified with the measure μ_1.

Let $\mathbf{N}$ be the space of pairs (f_1, f_2) of positive extended real-valued functions under the equivalence relation $(f_1, f_2) = (g_1, g_2)$ when $f_1 + g_2 = f_2 + g_1$ and the operations (3.1) interpreted in the present context. The potential $G_D\mu$ of a charge μ is defined as the point $(G_D\mu_1, G_D\mu_2)$ of $\mathbf{N}$ and is informally identified with the function $G_D\mu_1 - G_D\mu_2$ where this difference is defined. The only charges of interest are those charges μ the components of whose minimal representing pair (μ_1, μ_2) have superharmonic potentials, in which case $G_D\mu_1 - G_D\mu_2$ is defined quasi everywhere on D. The potential $G_D\mu$ of a positive charge $(\mu_1, 0)$ is identified with the function $G_D\mu_1$.

The space of charges that have representations with both components in $\mathscr{E}^+$ will be denoted by $\mathscr{E}$, and for elements of $\mathscr{E}$ only representations with components in $\mathscr{E}^+$ will be used. If μ: (μ_1, μ_2) and ν: (ν_1, ν_2) are charges in $\mathscr{E}$, their mutual energy is defined following the definition in the positive case,

$$[\mu, \nu] = [\mu_1, \nu_1] + [\mu_2, \nu_2] - [\mu_1, \nu_2] - [\mu_2, \nu_1], \tag{3.2}$$

and the energy of μ is defined as $[\mu, \mu]$. To justify this definition, it must be proved that the mutual energy of a pair of elements in $\mathscr{E}^+$ is finite (see Theorem 7), and until this fact is proved, the definition (3.2) will be used only when the finiteness of the summands is obvious from the context. The form $[\cdot, \cdot]$ is symmetric (reciprocity law). It will be proved that the energy of a charge in $\mathscr{E}$ is positive, and the energy will then be written $\|\cdot\|^2$. This notation will be used before positivity is proved whenever positivity is obvious from the context.

4. Inequalities between Potentials, and the Corresponding Energy Inequalities

Lemma. (a) *The inequalities $G_D\mu' \le G_D\nu'$ and $G_D\mu'' \le G_D\nu''$ for potentials of measures imply that $[\mu', \mu''] \le [\nu', \nu'']$.*

(b) *The inequality $G_D\mu \le G_D\nu$ for potentials of measures implies that $\|\mu\|^2 \le \|\nu\|^2$ and that $\mu = \nu$ if these energies are finite and equal.*

(c) *Let $G_D\mu_.$ and $G_D\nu_.$ be increasing [decreasing] sequences of potentials of measures, with respective [smoothed] limits $G_D\mu$ and $G_D\nu$. Then*

$$\lim_{n\to\infty} [\mu_n, \nu_n] = [\mu, \nu]$$

in the increasing case and, if $[\mu_0, \nu_0]$ is finite, also in the decreasing case.

Proof of (a). Apply the symmetry of G_D to obtain

$$\begin{aligned}[\mu', \mu''] &= \int_D G_D\mu' \, d\mu'' \le \int_D G_D\nu' \, d\mu'' = \int_D G_D\mu'' \, d\nu' \\ &\le \int_D G_D\nu'' \, d\nu' = [\nu', \nu'']. \quad \square\end{aligned} \tag{4.1}$$

Proof of (b). If $\mu' = \mu'' = \mu$ and $\nu' = \nu'' = \nu$ in (a), then $\|\mu\| \le \|\nu\|$, and with this specialization if the energies, that is, the integrals in (4.1), are finite, equality in (4.1) implies that $G_D\mu = G_D\nu$ both μ almost everywhere and ν almost everywhere, so that $G_D\mu = G_D\nu$ by the domination principle for potentials of measures of finite energy (Section 2), and therefore $\mu = \nu$. $\square$

Proof of (c). In the increasing case in view of (a) the sequence $[\mu_., \nu_.]$ is an increasing sequence with limit at most $[\mu, \nu]$. The limit is $[\mu, \nu]$ because for every m,

$$\begin{aligned}\lim_{n\to\infty} [\mu_n, \nu_n] &\ge \liminf_{n\to\infty} \int_D G_D\mu_m \, d\nu_n = \liminf_{n\to\infty} \int_D G_D\nu_n \, d\mu_m \\ &= \int_D G_D\nu \, d\mu_m = \int_D G_D\mu_m \, d\nu,\end{aligned} \tag{4.2}$$

and the last integral can be made arbitrarily near $[\mu, \nu]$ by choosing m large. The decreasing case is treated similarly, with the help of the fact that polar sets are λ null if λ is a measure of finite energy. $\square$

5. The Function $D \mapsto G_D\mu$

(a) Let $D_.$ be an increasing sequence of open subsets of $\mathbb{R}^N$ with union a Greenian set D. Since $G_{D_n} \le G_{D_{n+1}}$ on $D_n \times D_n$, it follows that if μ and ν are measures on D and if μ_n is the projection of μ on D_n, then

$$\lim_{n\to\infty} \int_D G_{D_n}\mu_n \, d\nu = \int_D G_D\mu \, d\nu; \tag{5.1}$$

that is, mutual energy relative to D_n increases to mutual energy relative to D. Observe that we have not supposed that the integrals in (5.1) are finite valued but that once we have proved (Section 7) that when the measures μ and ν are in $\mathscr{E}^+$ for D, their mutual energy for D is finite valued, it will follow trivially that (5.1) is valid for μ and ν in $\mathscr{E}$ for D, although the sequence $G_{D_.}\mu_.$ need then not be monotone.

(b) Let D_2 be a Greenian subset of $\mathbb{R}^N$, let D_1 be a relatively compact open subset of D_2, and let μ be a measure on D_2 with compact support in D_1. Energies, reductions, etc., relative to D_i will be distinguished by the subscript i: $[\cdot,\cdot]_i$, $\|\cdot\|_i^{\cdot}$, $\mathscr{E}_i$, ... According to Section VII.5,

$$G_{D_2}\mu = G_{D_1}\mu + G_{D_2}\mu_1 \tag{5.2}$$

on D_1, where $\mu_1 = \|\mu\|^{D_2-D_1}$ is supported by ∂D_1. Equation (5.2) is obviously also valid for a charge μ supported by a compact subset of D_1, with μ_1 a charge supported by ∂D_1. Since $G_{D_2}\mu_1$ is bounded on the compact support of μ_1, equation (5.2) implies that $\mu \in \mathscr{E}_2$ if and only if $\mu \in \mathscr{E}_1$. (We are abusing language slightly here in considering μ both as a measure on D_1 and as a measure on D_2.) If we assume that all the following mutual energies are well defined, we can combine (5.2) with X(5.3′) as applied to the measure components of the charges to find that if μ and ν are in $\mathscr{E}_2$ with compact supports in D_1, then

$$[\mu,\nu]_1 = [\mu - \mu_1, \nu - \nu_1]_2, \tag{5.3}$$

where ν_1 is defined in terms of ν in the way μ_1 was defined in terms of μ. Thus, if all the mutual energies involved are well defined (and according to Theorem 7, this is true if μ and ν are in $\mathscr{E}_2$), the map $\mu \mapsto \mu - \mu_1$ is a linear mutual-energy-preserving map from a subset of $\mathscr{E}_1$ into $\mathscr{E}_2$.

6. Classical Evaluation of Energy; Hilbert Space Methods

If D is an open subset of $\mathbb{R}^N$ and if u and v are functions from D to the extended reals, define

$$\mathscr{D}(u,v) = \int_D \langle \operatorname{grad} u, \operatorname{grad} v\rangle \, dl_N, \qquad \mathscr{D}(u) = \mathscr{D}(u,u) \tag{6.1}$$

whenever the integrals are meaningful. The value $\mathscr{D}(u)$ is the *Dirichlet integral* of u. If $u = G_D\mu$ and $v = G_D\nu$ are the potentials of charges in $\mathscr{E}$, on the Greenian set D, and are in class $\mathbb{C}^{(2)}(D)$, then $\pi_N'\, d\nu = -\Delta v \, dl_N$ (Section I.7), so that

$$[\mu,\nu] = -\frac{1}{\pi_N'} \int_D u \, \Delta v \, dl_N. \tag{6.2}$$

If D is bounded and smooth enough for the application of Green's identity and associated theorems and if the potentials u and v when defined as 0 on ∂D are in class $\mathbb{C}^{(2)}(\bar{D})$, (6.2) yields

$$[\mu,\nu] = \frac{\mathscr{D}(u,v)}{\pi_N'}, \qquad [\mu,\mu] = \frac{\mathscr{D}(u)}{\pi_N'}. \tag{6.3}$$

If $D = \mathbb{R}^N$ with $N \geq 3$ and if $u = G\mu$ and $v = G\nu$ are potentials in class $\mathbb{C}^{(2)}(\mathbb{R}^N)$ of charges with compact support, then if B is a ball containing the compact support of ν,

$$\begin{aligned}\int_{\mathbb{R}^N} G\mu\, d\nu &= \frac{-1}{\pi_N'} \int_{\mathbb{R}^N} u\, \Delta v\, dl_N \\ &= \frac{1}{\pi_N'} \int_B \langle \operatorname{grad} u, \operatorname{grad} v \rangle\, dl_N - \frac{1}{\pi_N'} \int_{\partial B} u \mathbf{D_n} v\, dl_{N-1},\end{aligned} \tag{6.4}$$

where $\mathbf{D_n} v$ is the directional derivative in the direction of the exterior normal. For large distances r from the origin, $|\operatorname{grad} u|$ and $|\operatorname{grad} v|$ are majorized by const r^{1-N}, and $|u|$ is majorized by const r^{2-N}. Hence the integral defining $\mathscr{D}(u, v)$ converges absolutely, and when B increases to $\mathbb{R}^N$, the last integral in (6.4) drops out; so (6.4) becomes (6.3). If $D = \mathbb{R}^2$, the preceding analysis is valid through (6.4) but must be modified thereafter. If λ is a charge on $\mathbb{R}^2$ with compact support A,

$$\lim_{|\xi| \to \infty} [G\lambda(\xi) + \lambda(A) \log|\xi|] = \lim_{|\xi| \to \infty} \int_A \log \frac{|\xi|}{|\xi - \eta|} \lambda(d\eta) = 0$$

and

$$\left| \operatorname{grad} G\lambda(\xi) + \frac{\operatorname{grad} |\xi|^2}{2|\xi|^2} \lambda(A) \right| \leq \text{const}\, |\xi|^{-2}$$

for large $|\xi|$. It follows that if μ and ν satisfy the additional condition $\mu(\mathbb{R}^2) = \nu(\mathbb{R}^2) = 0$, then the integral defining $\mathscr{D}(u, v)$ converges absolutely, and (6.4) becomes (6.3) when B increases to $\mathbb{R}^2$.

Returning to the hypotheses of the first paragraph, if $u = G_D\mu$ as described there but if v is harmonic and has an extension to $\bar{D}$ in class $\mathbb{C}^{(2)}(\bar{D})$, then Green's identity [Section I(1.2)] becomes

$$\int_D v\, \Delta u\, dl_N = \int_{\partial D} v \mathbf{D_n} u\, dl_{N-1}, \tag{6.5}$$

and integration by parts on ∂D yields $\mathscr{D}(u, v) = 0$.

This discussion suggests that Hilbert space methods are appropriate to the study of charges on D using the inner product $[\mu, \nu]$ and to functions on D using the inner product $\mathscr{D}(u, v)$. The following discussion sketches a basis for such a study. We suppose that D is connected.

(a) Charges. It will be shown in Section 7 that $\mathscr{E}$ is a linear space, that the energy of a charge in $\mathscr{E}$ is positive, and that this energy vanishes if and only if the charge is the zero charge (0, 0). Thus $\mathscr{E}$ is a space which, coupled with

the bilinear form $[\cdot,\cdot]$, is a pre-Hilbert space, that is, a space in which all the Hilbert space axioms except that of completeness are satisfied. It is known that the space $\mathscr{E}$ of charges is not complete but that its subset $\mathscr{E}^+$ of measures of finite energy is complete.

(b) Functions. Let $u_.$ be a sequence of infinitely differentiable functions on D for which $\mathscr{D}(u_n) < +\infty$ and $\lim_{m,n\to\infty} \mathscr{D}(u_n - u_m) = 0$. It is known that there is then a function u from D into $\overline{\mathbb{R}}$ with the following properties: grad u exists l_N almost everywhere on D, with $\mathscr{D}(u) < +\infty$; $\lim_{n\to\infty} \mathscr{D}(u - u_n) = 0$; u is the quasi everywhere limit of a subsequence of $u_.$; u is fine continuous quasi everywhere on D. A function u obtained in this way is called a "*BLD*" *function* (*Beppo–Levi–Deny*). The space of BLD functions is the natural space of functions on which to base the study of orthogonality and related topics involving the Dirichlet integral. The harmonic BLD functions are the harmonic functions with finite Dirichlet integrals. If the functions in the approximating sequence $u_.$ have compact supports, u is said to be of *potential type.* It has been proved that the class of BLD functions of potential type includes the potentials of charges in $\mathscr{E}$, in particular, the superharmonic potentials of measures of finite energy. If two BLD functions are identified when the restriction of their difference to the complement of a polar set is a constant function, the space of BLD functions coupled with the inner product $\mathscr{D}(\cdot,\cdot)$ is a Hilbert space in which the classes of harmonic BLD functions and of functions of potential type are orthogonal subspaces with direct sum the whole space. This fact generalizes our application of (6.5) according to which if u is a superharmonic potential on D and v is harmonic on D, if D is sufficiently smooth, and if u and v have sufficiently smooth extensions to $\overline{D}$, then $\mathscr{D}(u, v) = 0$.

In view of (6.3) as suitably generalized to the BLD context, the pre-Hilbert space $\mathscr{E}$ of charges can be immersed in the Hilbert space of BLD functions by identifying a charge μ in $\mathscr{E}$ with the function $(\pi'_N)^{-1/2} G_D\mu$. The Riesz decomposition of a positive superharmonic function into the sum of a positive harmonic function and the potential of a measure can be interpreted, if the given function is a BLD function, as the sum of its orthogonal projections on the subspaces of BLD harmonic functions and of functions of potential type.

The details of this Hilbert space approach will not be carried out in this book, and the stated results will not be needed with one exception. We shall need the fact that the energy of a charge in $\mathscr{E}$ is positive. This fact will be proved in the next section.

7. The Energy Functional (Relative to an Arbitrary Greenian Subset D of $\mathbb{R}^N$)

Theorem. (a) *The energy of a charge in $\mathscr{E}$ is positive and finite.*

(b) *If μ and ν are in $\mathscr{E}$, their mutual energy exists and*

$$|[\mu, \nu]| \leq \|\mu\| \, \|\nu\|. \tag{7.1}$$

(c) *A charge in $\mathscr{E}$ is the zero charge if and only if its energy vanishes.*
(d) *There is equality in* (7.1) *if and only if μ and ν are proportional.*

Observation. Theorem 7 implies that $\mathscr{E}$ is a linear space, a pre-Hilbert space when coupled with the inner product $[\cdot, \cdot]$. (See the discussion in Section 6.)
Assertion (b) for μ and ν in $\mathscr{E}^+$ will be denoted by (b^+), and we write (a) (sm), (b) (sm), and so on to refer to the smooth context in which the charges involved have component measures which have compact supports in D and have infinitely differentiable potentials. If μ and ν are in $\mathscr{E}^+$(sm) their mutual energy obviously exists and is finite. In the following proof of Theorem 7 all the assertions are relative to a specified Greenian set D unless otherwise qualified.

Proof. Step 1. Proof that (b^+) $\Rightarrow$ (a) $\Rightarrow$ (b) and that (b^+)(sm) $\Rightarrow$ (a)(sm) $\Rightarrow$ (b)(sm). The following argument without the smoothness condition is also applicable in the smooth context. Let $\mu : (\mu_1, \mu_2)$ and ν be in $\mathscr{E}$. Under (b^+)

$$[\mu, \mu] = \|\mu_1\|^2 - 2[\mu_1, \mu_2] + \|\mu_2\|^2 \geq (\|\mu_1\| - \|\mu_2\|)^2; \tag{7.2}$$

so (b^+) $\Rightarrow$ (a). Under (a) the equality in (7.2) implies that $[\mu_1, \mu_2] < +\infty$, which implies in turn that $\mathscr{E}$ is a linear space and that for $c \in \mathbb{R}$,

$$0 \leq \|\mu + c\nu\|^2 = \|\mu\|^2 + 2c[\mu, \nu] + c^2\|\nu\|^2. \tag{7.3}$$

If $\|\nu\| = 0$, inequality (7.3) is impossible unless $[\mu, \nu] = 0$; if $\|\nu\| > 0$, inequality (7.3) with $c = -[\mu, \nu] \, \|\nu\|^{-2}$ yields (7.1). Hence (a) $\Rightarrow$ (b).
Step 2. Proof that (b^+)(sm) $\Rightarrow$ (a) $\cap$ (b). According to Section IV.10 a superharmonic potential of a measure on D is the limit of an increasing sequence of infinitely differentiable potentials of measures with compact supports in D. Thus, if μ and ν are measures in $\mathscr{E}^+$, there are sequences $\mu_\cdot$ and $\nu_\cdot$ of measures in $\mathscr{E}^+$(sm) for which $G_D\mu_\cdot$ and $G_D\nu_\cdot$ are increasing sequences with respective limits $G_D\mu$ and $G_D\nu$. Moreover, if (b^+) (sm) is true,

$$[\mu_n, \nu_n] \leq \|\mu_n\| \, \|\nu_n\|,$$

and inequality (7.1) now follows from Lemma 4. Hence (b^+) is true and therefore (a) and (b) are true by Step 1.
Step 3. Proof that (a) and (b) are true when D is a ball. According to Step 2, it is sufficient to show that (b^+)(sm) is true when D is a ball. When μ is in $\mathscr{E}$(sm) relative to a ball D, the expression II(1.1) for G_D shows that $G_D\mu$, when defined as 0 on ∂D, is infinitely differentiable on $\bar{D}$. The context is therefore that of Section 6, and the evaluation (6.3) of $[\mu, \nu]$ shows that the inequality (b^+)(sm) follows from Schwarz's inequality.

Step 4. Proof of (a) and (b) when D is bounded. According to Steps 1 and 2, it is sufficient to prove that (a)(sm) is true when D is bounded. Let D_2 be a ball containing $\bar{D}$. If μ is a charge in $\mathscr{E}$(sm) relative to D, then μ is also in $\mathscr{E}$(sm) relative to D_2, and according to Section 5(b), the energy $[\mu, \mu]$ relative to D is the energy relative to D_2 of a certain charge supported by ∂D, and (by Step 3) this energy is positive.

Step 5. Proof of (a) and (b). According to Step 1, it is sufficient to prove (b$^+$). Let $D_\cdot$ be an increasing sequence of nonempty open bounded subsets of D with union D. If μ and ν are in $\mathscr{E}^+$ relative to D and if μ_n and ν_n are the projections of these measures on D_n, then (by Step 4) $[\mu_n, \nu_n] \le \|\mu_n\| \, \|\nu_n\|$ (energies relative to D_n), and (from Section 5) when $n \to \infty$, this inequality yields (7.1).

Step 6. Proof of (c) and (d). Suppose that $\mu\colon (\mu_1, \mu_2) \in \mathscr{E}$ and that $\|\mu\| = 0$. It follows from (7.1) that $[\mu, \nu] = 0$ whenever ν is in $\mathscr{E}$, in particular, when ν is a uniform distribution on the boundary of a ball with closure in D. With this choice of ν the vanishing of $[\mu, \nu]$ implies the equality of spherical averages of $G_D\mu_1$ and $G_D\mu_2$; that is,

$$L(G_D\mu_1, \xi, \delta) = L(G_D\mu_2, \xi, \delta)$$

when $\bar{B}(\xi, \delta) \subset D$. When $\delta \to 0$, it follows that $G_D\mu_1 = G_D\mu_2$; so $\mu_1 = \mu_2$, and μ must be the zero charge. If the charges in (7.1) are proportional, there is obviously equality in (7.1). Conversely, if there is equality in (7.1), the charges are trivially proportional if either has zero energy, that is, if either is the zero charge; otherwise, equality in (7.1) implies that

$$\|\mu - [\mu, \nu] \|\nu\|^{-2} \nu\|^2 = 0.$$

So $\mu = [\mu, \nu] \, \|\nu\|^{-2} \nu$, and the proof is complete. □

8. Alternative Proofs of Theorem 7(b$^+$)

The following proofs of the key inequality (7.1$^+$) are unnecessary but instructive.

(a) Heat equation potential theoretic-probabilistic proof of (7.1$^+$). Suppose that there is a positive Borel measurable function $(t, \xi, \eta) \mapsto b_D(t, \xi, \eta)$ from $]0, +\infty[\times D \times D$ into $\mathbb{R}^+$ with the property that $b_D(t, \cdot, \cdot)$ is symmetric, that

$$b_D(s + t, \xi, \eta) = \int_D b_D(s, \xi, \zeta) b_D(t, \zeta, \eta) l_N(d\zeta), \tag{8.1}$$

and that b_D is related to the Green function G_D by

$$G_D(\xi, \eta) = c \int_0^\infty b_D(t, \xi, \eta) l_1(dt) \tag{8.2}$$

for some positive constant c. If μ is a measure on D, denote $\int_D b_D(t, \xi, \eta)\mu(d\eta)$ by $b_D(t, \xi, \mu)$. Then

$$[\mu, \nu] = c \int_0^\infty l_1(dt) \int_D l_N(d\xi) b_D\left(\frac{t}{2}, \xi, \mu\right) b_D\left(\frac{t}{2}, \xi, \nu\right), \tag{8.3}$$

and Schwarz's inequality yields (7.1^+).

Every Greenian subset D of $\mathbb{R}^N$ has a function b_D with the stated properties. In fact we shall see that if $\dot{D} = D \times \mathbb{R}$ and if $\dot{G}$ is the heat equation Green function of $\dot{D}$, then b_D defined by

$$b_D(s - t, \xi, \eta) = \dot{G}_D((\xi, s), (\eta, t))$$

has the desired properties. This function b_D will also be identified with the transition density relative to l_N of Brownian motion in D (transition from ξ to η in time t). See Chapter XVII for a discussion of $\dot{G}_D$, Section 2.VII.9 for the Brownian motion transition density b_D, and Section 2.IX.17 for the identification of the potential theory b_D with the probability b_D.

This method of proving (7.1^+) can be applied without recourse to heat equation potential theory or probability as follows. Define a function b on $]0, +\infty[\times \mathbb{R}$ by

$$b(t, \xi) = (2\pi t)^{-N/2} \exp \frac{-|\xi|^2}{2t}.$$

Then if $N \geq 3$, the function $b_{\mathbb{R}^N}$ defined by $b_{\mathbb{R}^N}(t, \xi, \eta) = b(t, \xi - \eta)$ has the desired properties for $D = \mathbb{R}^N$. Hence (7.1^+) is true for $D = \mathbb{R}^N$, and with the help of Section 5 it follows that (7.1^+) is true for an arbitrary open subset of $\mathbb{R}^N$. If $N \geq 1$ and D is a half-space, denote by η' the reflection in ∂D of a point η in D. Then the function b_D defined by $b_D(t, \xi, \eta) = b(t, \xi - \eta) - b(t, \xi - \eta')$ has the desired properties, and with the help of Section 5 it follows that (7.1^+) is true first for an arbitrary nonempty open subset of a half-space and then for an arbitrary Greenian subset of $\mathbb{R}^N$.

(b) Proof of (7.1^+) when $N \geq 3$ by a splitting method. This method uses the fact that there is a constant c_N for which

$$\frac{1}{|\xi - \eta|^{N-2}} = c_N \int_{\mathbb{R}^N} \frac{l_N(d\zeta)}{|\xi - \zeta|^{N-1} |\eta - \zeta|^{N-1}}. \tag{8.4}$$

To verify (8.5), observe that the integral defines a function of $|\xi - \eta|$. Denote this function by ϕ, change the integration variable to $\zeta' = \zeta|\xi - \eta|^{-1}$, and thereby find that $\phi(|\xi - \eta|) = |\xi - \eta|^{-N+2}\phi(1)$. Hence (8.4) is true with $v_N = 1/\phi(1)$. Apply (8.4) to derive

$$[\mu, \nu] = c_N \int_{\mathbb{R}^N} l_N(d\xi)\left[\int_D |\xi - \eta|^{-N+1}\mu(d\eta)\right]\left[\int_D |\xi - \zeta|^{-N+1}\nu(d\zeta)\right], \tag{8.5}$$

and apply Schwarz's inequality to obtain (7.1$^+$) for $D = \mathbb{R}^N$. Proceed as in (a).

9. Sharpening of Lemma 4

The following theorem is needed in Section 10.

Theorem. *If $G_D\mu_\cdot$ is an increasing sequence of potentials of measures and if* $\sup_n\|\mu_n\| < +\infty$, *then* $\lim_{n\to\infty} G_D\mu_n$ *is the potential of a measure μ in $\mathscr{E}^+$, and*

$$\lim_{n\to\infty} \|\mu_n\| = \|\mu\|, \qquad \lim_{n\to\infty} \|\mu - \mu_n\| = 0. \tag{9.1}$$

(It is easy to see that the second limit result implies the first.) If u is the limit of the sequence of potentials and if ν is a measure on D,

$$\int_D u\, d\nu = \lim_{n\to\infty} [\mu_n, \nu] \le \sup_{n\ge 0} \|\mu_n\|\, \|\nu\|. \tag{9.2}$$

In particular, let ξ be a point of D, let B be an open relatively compact subset of D containing ξ, and choose $\nu = \mu_B(\xi, \cdot)$. Then [from Section 2, Example (c)] (9.2) becomes

$$\mu_B(\xi, u) \le \sup_{n\ge 0} \|\mu_n\| \mu_B(\xi, G_D(\xi, \cdot)), \tag{9.3}$$

and the right side has limit 0 when B increases to D. Hence (Section VIII.11) u is a potential, say $u = G_D\mu$, and (9.1) follows from Lemma 4.

10. The Classical Capacity Function

Let D be a Greenian subset of $\mathbb{R}^N$, let Γ_p be the class of those subsets A of D for which $[\![1]\!]^A$ (reduction relative to D) is a potential, and for A in Γ_p let λ_A be the measure associated with $[\![1]\!]^A$; that is, $[\![1]\!]^A = G_D\lambda_A$. The measure λ_A is called the *capacitary measure* of A, and $G_D\lambda_A$ is called the *capacitary potential* of A. If $c > 0$, the measure $c\lambda_A$ [potential $cG_D\lambda_A$] is called an *equilibrium measure* [*equilibrium potential*] of A. These concepts are relative to D. According to Section XII.17, $A \in \Gamma_p$ if and only if when D is provided with its Martin boundary, the set $A^{mf} \cap \partial^M D$ is μ_D null. It is unnecessary to use this characterization of Γ_p to prove the elementary facts that a subset of a set in Γ_p is in Γ_p, that every polar subset B of D is in Γ_p, with $\lambda_B = 0$, and that relatively compact subsets of D are in Γ_p. Since $G_D\lambda_A \le 1$, polar sets are λ_A null (Theorem V.11). If A and B are in Γ_p and if A differs from B by a polar set, then $\lambda_A = \lambda_B$. Finally, we show that if A is in Γ_p, some open superset of A is also in Γ_p. In fact, if $B_1 = \{[\![1]\!]^A > \frac{1}{2}\}$, an open set which covers quasi every point of A, let u be a superharmonic potential on D, identically $+\infty$ on the polar set $A - B_1$, and define $B = B_1 \cup \{u > 1\}$. Then B is an open superset of A and is in Γ_p because $[\![1]\!]^B$ is majorized by the potential $2[\![1]\!]^A + u$.

For A in Γ_p the measure λ_A is supported by $D \cap \partial A$ (Euclidean boundary) and even (Corollary XI.18) by $D \cap \partial^f A^f$. Recall (Section XI.6) that for an arbitrary subset A of $\mathbb{R}^N$ the sets A^f and $\partial^f A^f$ are Borel sets. Since $[\![1]\!]^A = 1$ on $D \cap \partial^f A^f$, we conclude that

$$\|\lambda_A\|^2 = \int_{D\cap\partial^f A^f} [\![1]\!]^A \, d\lambda_A = \lambda_A(D) \qquad (A \in \Gamma_p). \tag{10.1}$$

If A is a subset of D not in Γ_p, we do not define λ_A, but it is convenient to define $\|\lambda_A\|^2 = +\infty$ to obtain a set function $\|\lambda_\cdot\|^2$ defined on every subset of D.

Theorem. (a) *The set function $\|\lambda_\cdot\|^2$ is a countably strongly subadditive Choquet capacity on D relative to the paving of compact sets, and for A in Γ_p*

$$\|\lambda_A\|^2 = \inf\{\|\lambda_B\|^2 : A \subset B,\ B \text{ open}\} = \inf\{\lambda_B(D) : A \subset B,\ B \text{ open}\}. \tag{10.2}$$

(b) *If A is $\|\lambda_\cdot\|^2$ capacitable, then*

$$\|\lambda_A\|^2 = \sup\{\lambda(F) : F \subset A,\ F \text{ compact},\ \lambda \text{ supported by } F,\ G_D\lambda \le 1 \text{ on } D\}. \tag{10.3}$$

(c) *A subset A of D is polar if and only if $\|\lambda_A\|^2 = 0$.*

Observation. Assertions (a) and (c) together are equivalent to the assertion that the restriction of $\|\lambda_\cdot\|^2$ to the class of compact subsets of D is a topological precapacity and that $\|\lambda_\cdot\|^2$ is the Choquet capacity generated (Appendix II.8) by this topological precapacity.

Proof of (a). The set function $\|\lambda_\cdot\|^2$ is obviously an increasing set function, and if $A_\cdot$ is an increasing sequence of sets with union A ($\subset D$), then $\lim_{n\to\infty} \|\lambda_{A_n}\|^2 = \|\lambda_A\|^2$ because this limit equation is trivial if the limit is $+\infty$ and the limit equation follows from Theorem 9 otherwise. Next let B be an open relatively compact subset of D, and let A be a compact subset of B. Then λ_A is supported by A, $G_D\lambda_A = 1$ quasi everywhere on A, $G_D\lambda_B = 1$

on A^f, and [Section VI.3(b)] $G_D\lambda_A = R_1^A$ on $D - A$ and so surely on the compact support of λ_B. Since polar sets are λ_A null, it follows that

$$\|\lambda_A\|^2 = \int_D G_D\lambda_A\, d\lambda_A = \int_D G_D\lambda_B\, d\lambda_A = \int_D G_D\lambda_A\, d\lambda_B = \int_D R_1^A\, d\lambda_B. \tag{10.4}$$

Since (Theorem VI.5) the set function $R_1^{\cdot}(\xi)$ is a topological precapacity on the class of compact subsets of D, it follows that if $A_{\cdot}$ is a monotone sequence of compact subsets of B with limit the compact subset A of B, then $\lim_{n\to\infty}\|\lambda_{A_n}\|^2 = \|\lambda_A\|^2$. The proof that $\|\lambda_{\cdot}\|^2$ is a Choquet capacity relative to the paving of compact sets is now complete.

To prove (10.2), observe first that the second equality follows from (10.1) and that the first equality is trivial when $\|\lambda_A\|^2 = +\infty$. If $\|\lambda_A\|^2 < +\infty$, the set A must be in Γ_p, and we have already noted that then Γ_p contains open supersets of A. Furthermore, according to Section VI.3(m),

$$[\![1]\!]^A = [\inf\{[\![1]\!]^B : A \subset B,\, B \text{ open}\}]_+, \tag{10.5}$$

and therefore according to the Fundamental Convergence Theorem, there is a decreasing sequence $B_{\cdot}$ of open supersets of A such that

$$[\![1]\!]^A = \left[\lim_{n\to\infty} [\![1]\!]^{B_n}\right]_+. \tag{10.6}$$

The first equality in (10.2) now follows from Lemma 4(c). What we have proved implies that the restriction of the capacity $\|\lambda_{\cdot}\|^2$ to the class of compact subsets of D is a topological precapacity generating $\|\lambda_{\cdot}\|^2$; so this capacity is countably strongly subadditive. □

Proof of (b). If F is a compact subset of D, if λ is a measure supported by F, and if $G_D\lambda \le 1$ on D, then we can apply (10.1) and Lemma 4 to obtain

$$\lambda(F) \le \int_F G_D\lambda_F\, d\lambda \le \|\lambda_F\|^2 = \lambda_F(F). \tag{10.7}$$

Moreover, if A is $\|\lambda_{\cdot}\|^2$ capacitable, then by definition of capacitability

$$\begin{aligned}\|\lambda_A\|^2 &= \sup\{\|\lambda_F\|^2 : F \subset A, F \text{ compact}\}\\ &= \sup\{\lambda_F(F) : F \subset A, F \text{ compact}\}.\end{aligned} \tag{10.8}$$

The combination of (10.7) and (10.8) yields (b). □

Proof of (c). It is sufficient to prove (c) for relatively compact subsets A of an arbitrary relatively compact open subset B of D, and we apply (10.4).

If A is polar, then $R_1^A = 0$ quasi everywhere on D (Theorem V.4); so (10.4) implies that $\|\lambda_A\|^2 = 0$. Conversely, if $\|\lambda_A\|^2 = 0$, then equality of the first and third terms in (10.4) implies that $\lambda_A(D) = 0$; so $\underset{+1}{R^A}$ vanishes identically, and therefore (by Theorem V.4) A is polar.

Unique Characterization of λ_A

If $A \in \Gamma_p$, then a measure λ on D is λ_A if and only if λ is supported by A^f and $G_D\lambda \le 1$ with equality quasi everywhere on A. In fact, in view of Theorem XI.14 the measure λ_A satisfies these conditions. Conversely, if λ satisfies these conditions and if λ' is the projection of λ on a compact subset of A^f, then (domination principle) $G_D\lambda' \le G_D\lambda_A$; so $G_D\lambda \le G_D\lambda_A$. Interchanging λ and λ_A yields the reverse inequality; so $\lambda = \lambda_A$. Observe that this argument also shows that if A is closed relative to D in the Euclidean topology and if A is in Γ_p, then $\lambda = \lambda_A$ if and only if λ is supported by A and $G_D\lambda \le 1$ with equality quasi everywhere on A.

11. Inner and Outer Capacities (Notation of Section 10)

Let A be a subset of D, and define

$$C_*(A) = \sup\{\|\lambda_F\|^2 : F \subset A, F \text{ compact}\}, \qquad C^*(A) = \|\lambda_A\|^2. \tag{11.1}$$

Recall that with this definition $C^*(A) < +\infty$ implies that $A \in \Gamma_p$; so $\underset{+1}{R^A}$ is a potential, $G_D\lambda_A$, that λ_A is supported by $\partial^f A^f$ and that

$$C^*(A) = \|\lambda_A\|^2 = \lambda_A(D \cap \partial^f A^f).$$

On the other hand, there are sets A in Γ_p with $\lambda_A(D) = C^*(A) = +\infty$.

The values $C_*(A)$ and $C^*(A)$ are called, respectively, the *inner* and *outer* capacities of A (relative to D). Thus $C^*(A)$ is the infimum of the inner capacities of the open supersets of A. The set A is $\|\lambda_\cdot\|^2$ capacitable, that is, C^* capacitable, if and only if $C_*(A) = C^*(A)$, and then $C(A)$, called the *capacity* of A (relative to D), is defined as the common value of the inner and outer capacities of A.

If A is a subset of D with finite outer capacity and if $\varepsilon > 0$, there are an open superset A_ε'' of A and a compact subset A_ε' of A such that

$$C_*(A) < C(A_\varepsilon') + \varepsilon, \qquad C(A_\varepsilon'') < C^*(A) + \varepsilon,$$

and there exist a G_δ superset A'' of A and an F_σ subset A' of A such that

$$C(A') = C_*(A) \le C^*(A) = C(A'').$$

12. Extremal Property Characterizations of Equilibrium Potentials (Notation of Section 10)

The canonical equilibrium measure λ_A^*. We have used the capacitary measure λ_A as a canonical equilibrium measure. A second choice, which is sometimes convenient when A is known to be capacitable with $0 < C(A) < +\infty$, is λ_A^*, defined by $\lambda_A^* = \lambda_A/C(A)$. Then

$$\lambda_A^*(D) = 1, \qquad \|\lambda_A^*\|^2 = \frac{1}{C(A)}, \qquad G_D\lambda_A^* \le \frac{1}{C(A)} \quad \text{with equality quasi everywhere on } A.$$

A Special Class of Subsets of D

In discussing capacitary measures λ_A relative to D it will be convenient to restrict A somewhat. Observe (Theorem XI.6) that for an arbitrary subset A of D the set $A - A^f$ is polar and so has capacity 0 and is null for every measure that has finite energy or (Theorem V.11) whose potential is finite valued. Furthermore (Theorem XI.6) the set A^f is a Borel set and so is capacitable. Thus, if we restriqt A to be in Γ_p so that λ_A is defined and further restrict A to be fine closed in D, that is, $D \cap A^f \subset A$, then the representation $A = (A - A^f) \cup (D \cap A^f)$ shows that A is capacitable and is measurable for every measure λ on A of finite energy or with a finite-valued potential. This restriction on A does not restrict the class of capacitary measures because (Section XI.7) $[\![1]\!]^A = [\![1]\!]^{D\cap A^f}$. Furthermore the capacitary measure λ_A under this restriction on A is supported by A and in fact (by Theorem XI.18) is even supported by $D \cap \partial^f A^f$, and $C(A) = \lambda_A(A)$ according to (10.1). In particular, $C(A) < +\infty$ if A is also relatively compact (Euclidean topology) in D. We shall use the fact that if λ is a measure on D, vanishing on polar sets and supported by the fine closed in D set A, then

$$[\lambda, \lambda_A] = \int_A G_D\lambda_A \, d\lambda = \lambda(A). \tag{12.1}$$

In (a)–(c) below we characterize λ_A and λ_A^* by maximal and minimal properties of measures λ on D satisfying the stated side conditions under the hypothesis

$$A \in \Gamma_p, \qquad D \cap A^f \subset A, \qquad 0 < C(A) < +\infty.$$

(a) Side condition: $G_D\lambda \le 1$; λ is supported by A. Under this side condition

(a1) $\|\lambda\|^2 \le C(A)$.
(a2) $\lambda(A) \le C(A)$.

Equality in either inequality implies that $\lambda = \lambda_A$.

In fact (special domination principle in Section 2), $G_D\lambda \le G_D\lambda_A$; so (by Lemma 4) (a1) is true, and an application of Schwarz's inequality in (12.1) shows that

$$\lambda(A) \le \|\lambda_A\|^2 = C(A); \tag{12.2}$$

so (a2) is true. Equality in (a2) implies equality in (a1) according to (12.2), and [by Lemma 4(b)] equality in (a1) implies that $\lambda = \lambda_A$.

(b) Side condition: $\lambda(A) = 1$; λ is supported by A. Under this side condition,

(b1) $\sup_{\xi \in D} G_D\lambda(\xi) \ge 1/C(A)$.
(b2) $\|\lambda\|^2 \ge 1/C(A)$.

Equality in one of these inequalities implies that $\lambda = \lambda_A^*$.

Define $\lambda' = C(A)\lambda$. The measure λ' is supported by A, and $\lambda'(A) = C(A)$. If (b1) is false, then $G_D\lambda' < 1$ and $\lambda' = \lambda_A$ according to (a), a contradiction. If there is equality in (b1), this reasoning shows that $\lambda = \lambda_A^*$. Inequality (b2) follows from Schwarz's inequality applied to (12.1), and (by Theorem 7) there is equality in this application of Schwarz's inequality if and only if $\lambda = \text{const}\ \lambda_A$, that is, $\lambda = \lambda_A^*$.

(c) Side condition: $G_D\lambda \ge 1$ quasi everywhere on A. Under this side condition, $\|\lambda\|^2 \ge C(A)$, and equality implies that $\lambda = \lambda_A$.

According to the domination principle, $G_D\lambda \ge G_D\lambda_A$, and therefore (by Lemma 4) $\|\lambda\|^2 \ge \|\lambda_A\|^2 = C(A)$, and equality implies that $\lambda = \lambda(A)$.

Observation. Under the additional condition that λ is supported by A, the side condition (c) implies that $\lambda(A) \ge C(A)$ because

$$C(A) = \lambda_A(A) \le \int_A G_D\lambda\, d\lambda_A = \int_A G_D\lambda_A\, d\lambda = \lambda(A). \tag{12.3}$$

(d) Side condition: $\|\lambda\| < +\infty$; λ is supported by A. Under this side condition,

$$\int_A (G_D\lambda - 2)\, d\lambda \ge -C(A), \tag{12.4}$$

and equality implies that $\lambda = \lambda_A$. In fact (12.1) implies that $\lambda(A) < +\infty$ and that

$$\int_A (G_D\lambda - 2)\, d\lambda = \|\lambda\|^2 - 2\lambda(A) = \|\lambda - \lambda_A\|^2 - C(A). \tag{12.5}$$

13. Expressions for $C(A)$

A slight variation of the discussion in Section 12 yields the following theorem.

Theorem. *Let D be a Greenian subset of* $\mathbb{R}^N$, *and let A be an analytic subset of D of finite capacity. Then*

$$\begin{aligned}
C(A) &= \sup\left\{\int_A G_D\lambda\, d\lambda : \lambda \text{ supported by } A,\ G_D\lambda \le 1\right\}\\
&= \sup\{\lambda(A): \lambda \text{ supported by } A,\ G_D\lambda \le 1\}\\
&= \frac{1}{\inf\{\sup_{\xi\in D} G_D\lambda(\xi): \lambda \text{ supported by } A,\ \lambda(A) = 1\}}\\
&= \frac{1}{\inf\left\{\int_A G_D\lambda\, d\lambda : \lambda \text{ supported by } A,\ \lambda(A) = 1\right\}}\\
&= \inf\left\{\int_D G_D\lambda\, d\lambda : G_D\lambda \ge 1 \text{ quasi everywhere on } A\right\}\\
&= \inf\{\lambda(A): \lambda \text{ supported by } A,\ G_D\lambda \ge 1 \text{ quasi everywhere on } A\}
\end{aligned} \tag{13.1}$$

These expressions for $C(A)$ are so easily obtained that we shall prove only the first. If λ is supported by A and if $G_D\lambda \le 1$, then (domination principle) $G_D\lambda \le G_D\lambda_A$; so (by Lemma 4) $\|\lambda\|^2 \le \|\lambda_A\|^2 = C(A)$; that is, each integral on the right in the first line of (13.1) is at most $C(A)$. Furthermore $C(A)$ is the supremum of the capacities of the compact subsets of A; so if $\lambda = \lambda_F$ in (13.1), with F a compact subset of A, the integral can be made arbitrarily near $C(A)$.

Capacity of a Ball

If $N > 2$, if $D = \mathbb{R}^N$, and if A is a ball of radius δ, a uniform distribution on ∂A of total value 1 has potential identically δ^{-N+2} on $\bar{A}$ according to Section IV.2. Hence any uniform distribution on ∂A is an equilibrium distribution for ∂A and for $\bar{A}$. Both have capacity δ^{N-2}. Moreover, since $C(A)$ is the supremum of the capacities of the compact subsets of A, $C(A) = \delta^{N-2}$ also, and any uniform distribution on ∂A is an equilibrium measure for A.

If $N \ge 2$, if D is a ball of radius β, and if A is a concentric ball of radius $\alpha < \beta$, a uniform distribution on ∂A is an equilibrium distribution for A, $\bar{A}$, and ∂A, and

$$C(A) = C(\bar{A}) = C(\partial A) = \begin{cases} \left(\log\dfrac{\beta}{\alpha}\right)^{-1} & \text{if } N = 2,\\ (\alpha^{2-N} - \beta^{2-N})^{-1} & \text{if } N > 2. \end{cases} \tag{13.2}$$

14. The Gauss Minimum Problems and Their Relation to Reductions

Let D be a connected Greenian subset of D. Energies and reductions below are relative to D, and charges are on D. Let A be a nonpolar Borel subset of D, and let f be a Borel measurable function from A into $\mathbb{R}$, not vanishing quasi everywhere, with $|f|$ majorized quasi everywhere on A by the restriction to A of the potential $G_D\lambda$ of some measure λ of finite energy supported by A. Observe that if A is fine closed in D, the words "supported by A" are no restriction on f because if λ is a measure on D of finite energy and if $|f| \le G_D\lambda$ quasi everywhere on A, then $|f| \le \llbracket G_D\lambda \rrbracket^A = G_D\llbracket\lambda\rrbracket^A = G_D\lambda$ quasi everywhere on A, the measure $\llbracket\lambda\rrbracket^A$ is supported by A (Theorem XI.14), and $\|\llbracket\lambda\rrbracket^A\| \le \|\lambda\| < +\infty$ (Lemma 4).

Let Γ_A be the linear class of charges ν supported by A, of finite energy, with $\int_A |f|\, d|\nu| < +\infty$. This class contains λ. We consider two modifications of a problem studied in 1840 by Gauss (see-Historical Notes to this chapter) under hypotheses suited to his era.

(G1) Minimize $\int_A G_D\nu\, d\nu = \|\nu\|^2$ for $\nu \in \Gamma_A$ with $\int_A f\, d\nu = 1$.

(G2) Minimize $\int_A (G_D\nu - 2f)\, d\nu = \|\nu\|^2 - 2\int_A f\, d\nu$ for $\nu \in \Gamma_A$.

Let Γ_A^+ be the class of positive charges, that is, of measures, in Γ_A, and let problems (G1$^+$) and (G2$^+$) be, respectively, (G1) and (G2) with Γ_A replaced by Γ_A^+. We shall treat problems (G1$^+$) and (G2$^+$) only when $f \ge 0$. We shall write that a charge solves one of the above problems if the charge minimizes the relevant integral under the specified side conditions.

(a) *A charge μ solves problem* (G2) *if and only if*

$$[\mu, \nu] = \int_A f\, d\nu \tag{14.1}$$

for every ν in Γ_A, equivalently, if and only if $G_D\mu = f$ quasi everywhere on A. If so,

$$\int_A (G_D\mu - 2f)\, d\mu = -\|\mu\|^2 = -\int_A f\, d\mu < 0, \tag{14.2}$$

and for ν in Γ_A,

$$\int_A (G_D\nu - 2f)\, d\nu = -\|\mu\|^2 + \|\nu - \mu\|^2; \tag{14.3}$$

so μ is the only charge solving (G2). *Furthermore $\mu/\int_A f\, d\mu$ solves problem* (G1). In fact μ solves (G2) if and only if whenever $c \in \mathbb{R}$ and $\nu \in \Gamma_A$, the integral in

(G2) with ν replaced by $\mu + c\nu$ has its minimum value for fixed ν when $c = 0$. A condition necessary and sufficient for this is (14.1). Equation (14.1) implies that $\|\mu\|^2 = \int_A f\,d\mu$ and implies the evaluations (14.2) and (14.3) except for the strict inequality in (14.2), which is a consequence of the evaluation of $G_D\mu$ in terms of f to be made next. Equation (14.1) with ν replaced by its projection on an arbitrary Borel subset of A yields the inequality $G_D\mu = f$ ν almost everywhere on A for every ν, and therefore (Section 2) $G_D\mu = f$ quasi everywhere on A. Conversely, the latter condition implies (14.1) because (from Section 2) polar sets are $|\nu|$ null for charges ν of finite energy. Finally, if $\nu \in \Gamma_A$ and if $\int_A f\,d\nu = 1$, equation (14.1) implies that $\|\mu\|\,\|\nu\| \geq 1$; so

$$\|\nu\|^2 \geq \|\mu\|^{-2} = \left\| \frac{\mu}{\int_A f\,d\mu} \right\|^2,$$

and therefore $\mu/\int_A f\,d\mu$ solves problem (G1).

(b) *A charge μ solves problem* (G1) *if and only if $\|\mu\| > 0$ and $\mu\|\mu\|^{-2}$ solves problem* (G2), *equivalently, if and only if $\|\mu\| > 0$ and $G_D\mu = \|\mu\|^2 f$ quasi everywhere on A. There can be only one such charge.* In fact, if μ solves problem (G1), then the side condition implies that $\|\mu\| > 0$, and the minimizing property of μ implies that for ν in Γ_A,

$$\begin{aligned}\int_A (G_D\nu - 2f)\,d\nu &\geq \left(\int_A f\,d\nu\right)^2 \|\mu\|^2 - 2\int_A f\,d\nu \\ &= -\|\mu\|^{-2} + \left(\|\mu\| \int_A f\,d\nu - \|\mu\|^{-1}\right)^2\end{aligned}$$

with equality when $\nu = \mu\|\mu\|^{-2}$. Thus the latter charge solves problem (G2), and the rest of (b) is now trivial.

(a^+) *If $f \geq 0$, a measure μ solves problem* ($G2^+$) *if and only if*

$$[\mu, \nu] \geq \int_A f\,d\nu, \qquad \|\mu\|^2 = \int_A f\,d\mu \tag{14.1$^+$}$$

for every ν in Γ_A^+, equivalently, if and only if $G_D\mu \geq f$ quasi everywhere on A with equality μ almost everywhere on A. If so, (14.2) *is true, and for ν in Γ_A^+ the identity*

$$\int_A (G_D\nu - 2f)\,d\nu = -\|\mu\|^2 + \|\nu - \mu\|^2 + 2\left([\mu, \nu] - \int_A f\,d\nu\right) \tag{14.3$^+$}$$

shows that μ is the only measure solving ($G2^+$). *Furthermore $\mu/\int_A f\,d\mu$ solves problem* ($G1^+$). In fact, if μ solves problem ($G2^+$) and if $c \in \mathbb{R}^+$, the integral in (G2$^+$) with ν replaced by $c\mu$ has its minimum value when $c = 1$, and the integral in (G2$^+$) with ν replaced by $\mu + c\nu$ with c in $\mathbb{R}^+$ and ν in Γ_A^+ has its minimum value for fixed ν when $c = 0$. These two conditions are satisfied if and only if (14.1$^+$) is true. Conversely (14.1$^+$) implies that μ solves problem (G2$^+$) and is thereby uniquely determined because under (14.1$^+$) in the identity (14.3$^+$) the right-hand side $\geq -\|\mu\|^2$ with equality only when $\nu = \mu$. Equation (14.1$^+$) with ν replaced by its projection on an arbitrary Borel subset of A implies that $G_D\mu \geq f$ ν almost everywhere on A; so $G_D\mu \geq f$ quasi everywhere on A. There is equality μ almost everywhere on A in view of the equality in (14.1$^+$). Conversely, these two conditions on μ imply that (14.1$^+$) is true. Finally, if μ solves (G2$^+$), then $\mu/\int_A f\,d\mu$ solves (G1$^+$) by the same proof as that of the corresponding implication in (a).

(b$^+$) *If $f \geq 0$, a measure μ solves problem* (G1$^+$) *if and only if $\|\mu\| > 0$ and $\mu\|\mu\|^{-2}$ solves problem* (G2$^+$), *equivalently, if and only if $\|\mu\| > 0$ and $G_D\mu \geq \|\mu\|^2 f$ quasi everywhere on A with equality μ almost everywhere on A. There can be only one such measure.* The proof is left to the reader.

(c) Suppose that $f \geq 0$, and define f as 0 on $D - A$. Then if μ solves problem (G2$^+$), $G_D\mu = \underset{+f}{R}$. (See Section XI.20 for the reduction involved here.) *Conversely, if A is fine closed in D, then $\underset{+f}{R}$ is the potential of a measure μ solving problem* (G2$^+$). In fact, if μ solves problem (G2$^+$), then $G_D\mu = f < +\infty$ μ almost everywhere on A; so if v is a positive superharmonic function on D that majorizes f quasi everywhere on D, the domination principle states that $G_D\mu \leq v$. Since one choice of v is $G_D\mu$, it follows that $G_D\mu = \underset{+f}{R}$. Conversely, $\underset{+f}{R} \geq f$ quasi everywhere on A, and $\underset{+f}{R} \leq G_D\lambda$; so $\underset{+f}{R}$ is the potential of a measure μ of finite energy (Lemma 4). To prove that μ solves problem (G2$^+$), we need only verify that A supports μ, and according to Section XI.20 (special positivity case), μ is supported by $\partial^f A^f - D$, a subset of A since $A \subset A^f$ by hypothesis.

(d) Classical balayage. Suppose that A is a Borel subset of D fine closed in D and that f is equal quasi everywhere on A to the restriction to A of the potential $G_D\mu'$ of some measure μ' of finite energy not necessarily supported by A. We have seen at the beginning of this section that then f satisfies the conditions imposed throughout. In the present context, problem (G1) becomes the problem of minimizing $\|\nu\|^2$ for all charges ν of finite energy supported by A with $[\mu', \nu] = 1$, and problem (G2) becomes the problem of minimizing $\|\nu\|^2 - 2[\mu', \nu]$ for all charges ν of finite energy supported by A. The measure $[\![\mu']\!]^A$ solves (G2) because

$$G_D[\![\mu']\!]^A = [\![G_D\mu']\!]^A = f$$

quasi everywhere on A and (by Theorem XI.14) the measure $[\![\mu']\!]^A$ is supported by A.

The linearity properties of (G2) solutions noted in (a) imply that if f is equal quasi everywhere on the fine closed in D set A to the restriction to A of the potential of a charge μ' of finite energy, then problem (G2) has a solution: in fact, if $\mu' = \mu'^+ - \mu'^-$ (Hahn decomposition), then $\|\mu'^+\|^A - \|\mu'^-\|^A$ solves (G2).

In particular, if A is a fine-closed Borel relatively compact subset of D and if $f \equiv 1$ on A, f is the restriction to A of the potential of a measure of finite energy; for example, f is the restriction to A of the reduction of the function 1, reduced onto an open superset of A relatively compact in D. In this special case problem (G1) is the problem of minimizing $\|v\|^2$ for charges v of finite energy supported by A with $v(A) = 1$, and problem (G2) is the problem of minimizing $\|v\|^2 - 2v(A)$ for all charges of finite energy supported by A. The capacitary measure of A solves problem (G2) in this context.

The Frostman approach (see Historical Notes). If A is compact and if f is positive ($< +\infty$) and continuous, it is easy to see that problems (G1$^+$) and (G2$^+$) have solutions. In fact, if $\mu_.$ is a minimizing sequence of measures for (G1$^+$), then $\mu_.(A)$ is a bounded sequence, and any measure that is the vague limit of a subsequence of $\mu_.$ solves (G1$^+$).

15. Dependence of C^* on D

If D_1 and D_2 are Greenian subsets of $\mathbb{R}^N$ with $D_1 \subset D_2$, we have noted in Section VII.1 that $G_{D_1} \le G_{D_2}$. If outer capacity relative to D_i is denoted by C_i^*, it will now be shown that $C_1^* \ge C_2^*$ on subsets of D_1 and that to a compact subset B of D_1 corresponds a constant $\alpha = \alpha(B)$ such that $C_1^* \le \alpha C_2^*$ on subsets of B. In view of the relations between inner and outer capacities, the corresponding inequalities are true for inner capacities, and it is sufficient to prove these inequalities on compact sets, for which C_i^* reduces to the capacity function C_i. If A is a compact subset of D_1 with capacitary measure λ_{iA} relative to D_i, then $1 \ge G_{D_2}\lambda_{2A} \ge G_{D_1}\lambda_{2A}$ on D_1; so by (13.1)

$$C_1(A) = \sup\{\lambda(A): \lambda \text{ supported by } A,\, G_{D_1}\lambda \le 1\} \ge \lambda_{2A}(A) = C_2(A), \quad (15.1)$$

as asserted. To prove the second assertion, observe that if B is a compact subset of D_1 and $\alpha = \sup_{B \times B} G_{D_2}/G_{D_1}$ (the ratio is defined as 1 on the diagonal) and if A is a compact subset of B, then $G_{D_2}\lambda_{2A} \le \alpha G_{D_1}\lambda_{2A}$ on D_1; so $G_{D_1}(\alpha\lambda_{2A}) \ge 1$ quasi everywhere on A. Therefore by (13.1)

$$\begin{aligned} C_1(A) &= \inf\{\lambda(A): \lambda \text{ supported by } A,\, G_{D_1}\lambda \ge 1 \text{ quasi everywhere on } A\} \\ &\le \alpha\lambda_{2A}(A) = \alpha C_2(A), \end{aligned} \quad (15.2)$$

as was to be proved.

16. Energy Relative to $\mathbb{R}^2$

In this section we consider potentials $G\mu$ when $N = 2$. The Green function G is bounded below on compact subsets of $\mathbb{R}^2 \times \mathbb{R}^2$, and we have seen in Section IV.1 that $G\mu$ is superharmonic on $\mathbb{R}^2$ whenever μ is a measure on $\mathbb{R}^2$ with compact support. Hence, if μ and ν are measures on $\mathbb{R}^2$ with compact support, the integrals $\int_{\mathbb{R}^2} G\mu\, d\nu$ and $\int_{\mathbb{R}^2} G\nu\, d\mu$ are meaningful and equal. Their value, the *mutual energy* of the pair of measures, will be denoted by $[\mu, \nu]$.

Let $\mathscr{E}$ in the present context be the class of charges μ: (μ_1, μ_2) with the following properties (in which we suppose that μ_1 and μ_2 are minimal components):

E1. μ_1 and μ_2 have compact support.

E2. $\displaystyle\int_{\mathbb{R}^2} G\mu_i\, d\mu_i < +\infty,\ i = 1, 2.$

E3. $\mu(\mathbb{R}^2) = 0$.

Then the following two assertions are true.

(a) If μ and ν are measures in $\mathscr{E}$, then $[\mu, \nu] < +\infty$.

This assertion follows from the fact that if D is a ball containing the compact supports of μ and ν the potentials $G\mu$ $[G\nu]$ and $G_D\mu$ $[G_D\nu]$ differ on D by harmonic functions; so (a) can be reduced to the corresponding assertion for potentials relative to D, covered by Theorem 7.

In view of (a), if μ and ν are charges in $\mathscr{E}$, their mutual energy $[\mu, \nu]$ defined formally by (3.2) in terms of their minimal components is meaningful and finite valued. The evaluation (3.2) is independent of the choice of components of the charges as long as all the mutual energy integrals are finite valued. The class $\mathscr{E}$ is now seen to be a linear class.

We have not yet used the condition E3 in this discussion, but this condition is essential in the following. In view of the evaluation (6.3) of mutual energy in terms of the Dirichlet integral when the charges involved are in $\mathscr{E}$ and have $\mathbb{C}^{(2)}(\mathbb{R}^2)$ potentials, we can expect charges in $\mathscr{E}$ to have positive energy, and in fact we now prove the following statement.

(b) *Theorem 7 is true for $D = \mathbb{R}^2$ and charges in $\mathscr{E}$.*

To prove that $[\mu, \mu] \geq 0$ for μ in $\mathscr{E}$, let A be a compact support for μ, of diameter α, and let B be a ball containing A, with center in A, of radius $\beta > \alpha$. Define

$$u(\xi, \eta) = \operatorname*{GM}_{B} G(\xi, \cdot)(\eta) = \mu_B(\eta, G(\xi, \cdot)) \tag{16.1}$$

for ξ and η in B, so that $G - u = G_B$ on $B \times B$ and

$$-\log(\beta + \alpha) \leq u \leq -\log(\beta - \alpha). \tag{16.2}$$

In view of the positivity of energy relative to B (Theorem 7), if μ is a charge in $\mathscr{E}$ (for $\mathbb{R}^2$) with minimal representation (μ_1, μ_2),

$$\int_{\mathbb{R}^2} G\mu\, d\mu \geq \int_A \int_A u(\xi, \eta)\mu(d\xi)\mu(d\eta) \geq 2\mu_1(A)^2 \log\frac{\beta - \alpha}{\beta + \alpha}.$$

Since the last term has limit 0 when $\beta \to \infty$, it follows that the energy $[\mu, \mu]$ of μ relative to $\mathbb{R}^2$ is positive.

Schwarz's inequality (7.1) in the present context follows easily from positivity of energy [see (7.3)]. The rest of the proof of Theorem 7 in the present context follows that of Theorem 7 with one modification. To prove that $[\mu, \nu] = 0$ whenever ν is in $\mathscr{E}$ implies that μ is the zero charge, we cannot choose ν as a uniform distribution on a sphere since this distribution is not in $\mathscr{E}$. Instead choose $\delta_1 > \delta > 0$, and for ξ in $\mathbb{R}^2$ let ν be the charge supported by $\partial B(\xi, \delta) \cup \partial B(\xi, \delta_1)$, equal on the larger sphere to the uniform distribution of a unit mass and equal on the smaller sphere to the uniform distribution of a negative unit mass. Define $u = G_D\mu$. The equality $[\mu, \nu] = 0$ implies that $L(u, \xi, \delta) = L(u, \xi, \delta_1)$, and the condition E3 implies that $\lim_{\delta_1 \to \infty} L(u, \xi, \delta_1) = 0$. It follows that $L(u, \xi, \delta) = 0$ and therefore $(\delta \to 0)$ that $u \equiv 0$. Hence μ is the zero charge, as was to be proved.

17. The Wiener Thinness Criterion

Let D be a Greenian subset of $\mathbb{R}^N (N \geq 2)$, let A be a subset of D, and let ξ be a point of D. Let C^* be the outer capacity (Section 11) defined on subsets of D, relative to D. Let $\alpha \in]1, +\infty[$, $n \in \mathbb{Z}^+$, and define

$$B_n = \{\eta : \alpha^n \leq G(\xi, \eta) \leq \alpha^{n+1}\}, \qquad A_n = A \cap B_n, \qquad A^n = \bigcup_n^{\infty} A_m. \tag{17.1}$$

Let k be an integer so large that $\bigcup_k^\infty B_n \subset D$ and, if $N = 2$, so large that this union has diameter less than 1. In the following k is fixed, and it is to be understood that A_n and A^n are considered only for $n \geq k$. Nothing in the following theorem or its proof would have to be changed if one of the inequalities in (17.1) is changed to be a strict inequality. Observe that there is a constant c so large that

$$0 < \frac{G(\xi, \cdot)}{c} \leq G_D(\xi, \cdot) \leq cG(\xi, \cdot) \quad \text{on } A^k. \tag{17.2}$$

Theorem. *The set A is thin at ξ if and only if the following equivalent conditions (with reductions relative to D) are satisfied:*

(a) $\sum_0^\infty \alpha^n C^*(A_n) < +\infty.$

(b) $\sum_0^\infty R_1^{A_n}(\xi) < +\infty$.

Observe that since $R_1^{A_n}$ is harmonic on a neighborhood of ξ, it follows that $\underset{+1}{R}^{A_n}(\xi) = R_1^{A_n}(\xi)$. Since A_n is a relatively compact subset of D, the smoothed reduction $\underset{+1}{R}^{A_n}$ is a potential $G_D\lambda_{A_n}$, and [by (10.1)] $C^*(A_n) = \lambda_{A_n}(D)$. Since λ_{A_n} is supported by $\bar{A}_n$,

$$\alpha^n C^*(A_n) \le \underset{+1}{R}^{A_n}(\xi) = R_1^{A_n}(\xi) \le \alpha^{n+1} C^*(A_n). \tag{17.3}$$

Hence conditions (a) and (b) are equivalent. In view of Theorem X1.3 these conditions imply that A is thin at ξ because under (b)

$$\lim_{n\to\infty} R_1^{A^n}(\xi) \le \lim_{n\to\infty} \sum_n^\infty R_1^{A_m}(\xi) = 0, \tag{17.4}$$

where we have used the countable subadditivity of $R_1^{\cdot}(\xi)$ [Theorem VI.3(j)]. Conversely, if A is thin at ξ, we shall show that the sum in (b) is finite when the index n is even; the proof for n odd is the same. Define $A' = \bigcup_k^\infty A_{2m}$, a subset of A and therefore thin at ξ. If the part of A' in some neighborhood of ξ is polar, it is trivial that the series in (b) with n even converges. If A' meets every neighborhood of ξ in a nonpolar set, we use the fact that according to Theorem XI.2, there is a positive superharmonic function u_0 on D, finite at ξ, with limit $+\infty$ at ξ along A'. The function $\underset{+u_0}{R}^{A'}$ is a positive superharmonic function on D, finite at ξ, with limit $+\infty$ at ξ along A' less a polar set, and $\underset{+u_0}{R}^{A'}$ is a potential $G_D\mu$ because A' is relatively compact in D. The measure μ is supported by $\bar{A}'$, and $\mu(\{\xi\}) = 0$ because $u(\xi) < +\infty$. Let μ_{2n} and μ'_{2n} be the projections of μ on A_{2n} and $D - A_{2n}$, respectively. An elementary calculation shows that there is a constant c' such that for all $n \ge k$,

$$G(\zeta, \eta) \le c'G(\xi, \eta) \quad \text{for } \eta \in A' - A_{2n},\ \zeta \in A_{2n}. \tag{17.5}$$

It follows from the definition of G_D and the discussion in Section VII.3 that $\lim_{|\zeta-\eta|\to 0} [G_D(\zeta, \eta)/G(\zeta, \eta)] = 1$ when ζ and η are restricted to be in a compact subset of D. Hence for ζ and η in a compact subset B of D there is a constant $c'' = c''(B)$ such that $G_D \le c''G$ on $B \times B$. (If $N > 2$, the stronger relation $G_D \le G$ is valid on $D \times D$.) In view of this inequality for $B = \bar{A}_k$ and (17.2),

$$G_D(\zeta, \eta) \le c'c''cG_D(\xi, \eta) \quad \text{for } \eta \in A' - A_{2n},\ \zeta \in A_{2n},\ n \ge k.$$

Hence for $n \ge k$,

$$G_D\mu'_{2n} \le \text{const } G_D\mu'_{2n}(\xi) \le \text{const } u(\xi)$$

on A_{2n}, and the constant does not depend on n. It follows that $G_D\mu_{2n} \geq 1$ quasi everywhere on A_{2n} for sufficiently large n; so $\underset{+1}{R}^{A_{2n}} \leq G_D\mu_{2n}$ quasi everywhere on A_{2n} for sufficiently large n, and therefore this reduction inequality holds everywhere on D (domination principle) for sufficiently large n. Hence, if m is sufficiently large,

$$\sum_{n \geq m} R_1^{A_{2n}}(\xi) = \sum_{n \geq m} \underset{+1}{R}^{A_{2n}}(\xi) \leq \sum_{n \geq m} G_D\mu_{2n}(\xi) \leq G_D\mu(\xi) < +\infty. \quad (17.6)$$

Thus the sum in Theorem 17(b) over the terms with even n converges, as was to be proved.

Observation. In view of (17.2) the theorem is true, and the proof requires only trivial modification, if G in (17.1) is replaced by G_D.

18. The Robin Constant and Equilibrium Measures Relative to $\mathbb{R}^2$ ($N = 2$)

Since we shall not use the results of this section, detailed proofs will be omitted. The material is presented as an interesting and important application not readily available elsewhere.

Superharmonic and Harmonic Functions on Neighborhoods of ∞

A function defined on an open neighborhood of the point ∞ of $\mathbb{R}^2$ (that is, on an open set including this point) is said to be superharmonic [harmonic] there if the function is superharmonic [harmonic] on the deleted neighborhood and if the Kelvin transform of the function under an inversion is superharmonic [harmonic] on an open neighborhood of the image of ∞. We shall use obvious consequences of theorems on superharmonic and harmonic functions defined on open subsets of $\mathbb{R}^2$ even when ∞ is allowed in the domains of the functions. For example, if u is a positive superharmonic function defined on an open deleted neighborhood of ∞, then u has a superharmonic extension to the full neighborhood.

The Green Function of an Open Neighborhood of ∞ and the Robin Constant of Its Complement

Let A be a compact nonpolar subset of $\mathbb{R}^2$, and define $D = \mathbb{R}^2 - A$. The Green function G_D has a continuous extension to $(D \cup \{\infty\}) \times (D \cup \{\infty\})$, and we shall denote by $G_D(\infty, \cdot)$ the restriction to D of this extension with first argument fixed at ∞. The function $G_D(\infty, \cdot)$ is called the *Green function of D with pole* ∞. This function is positive and harmonic on D, is bounded outside each neighborhood of ∞, and has limit 0 at quasi every finite (Euclidean) boundary point of D, and the difference $G_D(\infty, \cdot) - \log|\cdot|$ has a finite limit $r(A)$ at ∞. [If this difference is defined as $r(A)$ at ∞, the difference becomes harmonic on a neighborhood of ∞.] The value $r(A)$ is called the *Robin constant* of A. Obviously $r(A)$ is invariant under rotation and translation of $\mathbb{R}^2$. The value $e^{-r(A)}$ is called the *logarithmic capacity* of A. Although the logarithmic capacity has the advantage of positivity, we shall see that the key set function in this context is $-r(\cdot)$.

EXAMPLE. If $A = \bar{B}(0, \delta)$, then $G_D(\infty, \cdot) = \log(|\cdot|/\delta)$, $r(A) = -\log\delta$, and the logarithmic capacity of A is therefore δ.

If $\xi_0 \in A$, then [all functions in (18.1) are defined on D]

$$\begin{aligned} G_D(\infty, \cdot) &= \inf\{u \geq 0\colon u = \log|\xi_0 - \cdot| + h,\ h \text{ superharmonic}\} \\ &= \inf\{u \geq 0\colon u = \log|\xi_0 - \cdot| + h,\ h \text{ harmonic}\} \\ &= \inf\{u \geq 0\colon u \text{ superharmonic}, \\ &\qquad u \geq \log|\cdot| \text{ on a deleted neighborhood of } \infty\}. \end{aligned} \tag{18.1}$$

If D is not connected the function $G_D(\infty, \cdot)$ vanishes identically on every open connected component of D except the component D_∞ containing a deleted neighborhood of ∞.

Canonical Equilibrium Potential of a Compact Set

The function $G_D(\infty, \cdot)$ can be extended to a positive subharmonic function $G_D^{=}(\infty, \cdot)$ on $\mathbb{R}^2$, vanishing quasi everywhere on A. The measure λ_A on $\mathbb{R}^2$ associated with the superharmonic function $-G^{=}(\infty, \cdot)$ is supported by A, in fact by the Euclidean boundary of D_∞, and even by $\partial^f D_\infty^f$. We can write $G_D^{=}(\infty, \cdot) = -G\lambda_A + h$, where h is a harmonic function on $\mathbb{R}^2$. In view of the form of G_D near ∞, it follows that $\lambda_A(A) = 1$ and [see IV(9.2)] that h has limit $r(A)$ at ∞. An application of the harmonic function maximum-minimum theorem to h shows that this function is identically $r(A)$; so

$$G_D^{=}(\infty, \cdot) = -G\lambda_A + r(A). \tag{18.2}$$

Conversely, suppose that α is a constant, that λ is a measure supported by A, that $\lambda(A) = 1$, and that $G\lambda \leq \alpha$ with equality quasi everywhere on A. Then we now show that $\alpha = r(A)$ and $\lambda = \lambda_A$. Observe first that λ and λ_A vanish on polar sets (Section V.11), and we can therefore ignore the subsets of A on which $G\lambda \neq \alpha$ or $G\lambda_A \neq r(A)$ in the following evaluations:

$$\alpha = \int_A G\lambda\, d\lambda_A = \int_A G\lambda_A\, d\lambda = r(A).$$

The domination principle now implies that $G\lambda = G\lambda_A$; so $\lambda = \lambda_A$. The measure λ_A is a canonical equilibrium measure for A in the present context.

Right and Left Continuity of the Function $-r(\cdot)$

The function $-r(\cdot)$ is an increasing function on the class of nonpolar compact subsets of $\mathbb{R}^2$. Now let $A_{\bullet}$ be a monotone sequence of compact nonpolar sets, with limit A. In view of Theorem VII.6, if $A_{\bullet}$ is an increasing sequence and if A is compact, then $\lim_{n\to\infty} r(A_n) = r(A)$; if $A_{\bullet}$ is a decreasing sequence, then $\lim_{n\to\infty} r(A_n) = r(A)$ if A is nonpolar and $\lim_{n\to\infty} r(A_n) = +\infty$ if A is polar. If A is compact and polar, we therefore define its Robin constant $r(A)$ to be $+\infty$ and its logarithmic capacity to be 0.

Strong Subadditivity of $-r(\cdot)$

Let B be a nonpolar compact subset of $\mathbb{R}^2$, and define $D = \mathbb{R}^2 - B$. If A is an arbitrary compact subset of D, let v_A be the reduction relative to D of $G_D(\infty, \cdot)$ on A. This reduction is harmonic and bounded on a deleted open neighborhood of the point ∞; so (Section V.5) v_A has an extension harmonic on the full neighborhood and therefore has a finite limit $v_A(\infty)$ at ∞. The restriction of the function $G_D(\infty, \cdot) - v_A$ to $D - A$ is $G_{D-A}(\infty, \cdot)$. Observe that the set function $A \mapsto v_A$ is strongly subadditive [Section VI.3(j)]. It follows that the set function $A \mapsto v_A(\infty) = -r(A \cup B) + r(B)$ and so also the set function $A \mapsto -r(A \cup B)$ are strongly subadditive on the class of compact subsets of $\mathbb{R}^2 - B$. When B shrinks to a point, we find that $-r(\cdot)$ is strongly subadditive on the class of compact (including polar compact) subsets of $\mathbb{R}^2$. Unfortunately the logarithmic capacity function is not strongly subadditive, in fact, not even subadditive.

Application of Section VIII.19

According to Section VIII.19, if D is a deleted open neighborhood of ∞,

$$G_D(\xi, \eta) = G(\xi, \eta) - \mu_D(\xi, G(\eta, \cdot)) + \phi_D(\xi), \tag{18.3}$$

where ϕ_D is a positive harmonic function on D, bounded on bounded subsets of D, with boundary limit 0 at quasi every Euclidean boundary point of D. Since the term on the left in (18.3) and the second term on the right both, for fixed η, define functions of ξ bounded near ∞, the function ϕ_D must have the form $\phi_D = \log|\cdot| + h$, where h is harmonic on D and bounded on a deleted neighborhood of ∞; equivalently, $\phi_D = G_D(\infty, \cdot) + h'$, where h' is harmonic on D, h' is bounded on a deleted neighborhood of ∞, and h' has limit 0 at quasi every Euclidean boundary point of D. Moreover h' is bounded on bounded subsets of D because $G_D(\infty,\cdot)$ is, and it follows that h' is bounded; so (Section V.7) h' vanishes identically. Thus $\phi_D = G_D(\infty,\cdot)$ and

$$G_D(\xi,\eta) = G(\xi,\eta) - \mu_D(\xi, G(\eta,\cdot)) + G_D(\infty,\xi), \tag{18.4}$$

or in view of the definition of G_D,

$$G_D(\infty,\xi) = \mu_D(\xi, G(\eta,\cdot)) - \mathrm{GM}_D G(\xi,\cdot)(\eta) \tag{18.5}$$

for every point η of D.

Extension of $r(\cdot)$ to the Class of Analytic Subsets of $\mathbb{R}^2$

The set function $-r(\cdot)$ very nearly satisfies the conditions defining a topological precapacity but is not positive. Let A_0 be a compact nonpolar subset of $\mathbb{R}^2$, and consider the set function $-r(\cdot) + r(A_0)$ defined on the class of compact supersets of A_0. This set function is a positive strongly subadditive set function, and it is easy to modify the topological precapacity extension theorem, using its methods to extend $-r(\cdot) + r(A_0)$, and therefore $-r(\cdot)$, to the class of analytic supersets of A_0. The value $-r(A)$ obtained in this way does not depend on the choice of the compact nonpolar subset A_0 of A. Define $-r(A) = -\infty$ if A is polar. Since (from Section VI.2) every analytic nonpolar set has a compact nonpolar subset, we have now defined $-r(\cdot)$ on the class of analytic subsets of $\mathbb{R}^2$. The set function $-r(\cdot)$ is now an increasing set function and is regular in the sense that

$$\begin{aligned} -r(A) &= \sup\{-r(B): B \subset A,\ B \text{ compact}\} \\ &= \inf\{-r(B): A \subset B,\ B \text{ open}\}. \end{aligned} \tag{18.6}$$

The logarithmic capacity set function $e^{-r(\cdot)}$ is then also regular in this sense but as already noted is not strongly subadditive, in fact, not even subadditive.

Equilibrium Measures of Analytic Subsets of $\mathbb{R}^2$

Let A be an analytic nonpolar subset of $\mathbb{R}^2$, and let A' be a compact nonpolar subset of A. Define $D' = \mathbb{R}^2 - A'$ and $D = \mathbb{R}^2 - A$. The positive function

$$G_{D'}(\infty,\cdot) - [\![G_{D'}(\infty,\cdot)]\!]^{A-A'}$$

(smoothed reduction relative to D') is subharmonic on D and can be extended to be a positive subharmonic function on $\mathbb{R}^2$, 0 at quasi every point of A. Denote this extension by $G_D^{=}(\infty,\cdot)$. Suppose first that A is bounded.

Then $G_{\bar{D}}^{=}(\infty,\cdot)$ is harmonic on a deleted open neighborhood of ∞, and if λ_A is the Riesz measure associated with $-G_{\bar{D}}^{=}(\infty,\cdot)$, this measure has compact support, $\lambda(\mathbb{R}^2) = 1$, and there is a harmonic function h on $\mathbb{R}^2$ such that $G_{\bar{D}}^{=}(\infty,\cdot) = -G\lambda_A + h$. It can be shown that h is identically $r(A)$. Thus (18.2) is true, and $G_D\lambda_A \le r(A)$ with equality quasi everywhere on A. If A is not bounded, this reasoning must be expanded. In the first place, since $G_{D'}(\infty,\cdot)$ is a minimal harmonic function on D' (Section VII.10), we must suppose (by Theorem XI.3) that A is thin at ∞ to ensure that $[\![G_{D'}(\infty,\cdot)]\!]^{A-A'}$ is not identically $G_{D'}(\infty,\cdot)$. This condition on A can be shown to be necessary and sufficient for A to have finite logarithmic capacity, and then if λ_A is the Riesz measure associated with $-G_{\bar{D}}^{=}(\infty,\cdot)$, it can be shown that $\lambda_A(\mathbb{R}^2) = 1$, that (18.2) is true, and that therefore again $G\lambda_A \le r(A)$ with equality quasi everywhere on A.

Chapter XIV

One-Dimensional Potential Theory

1. Introduction

The one-dimensional version of classical potential theory is so special that its discussion has been deferred to this chapter, and much of this theory is so elementary that it will be left to the reader to formulate and justify. A ball in $\mathbb{R}$ with center ξ is an open interval with midpoint ξ, and the averages $L(u, \xi, \delta)$, $A(u, \xi, \delta)$, and $A_\alpha u$ can play the same role when $N = 1$ as when $N > 1$, but more direct methods are sometimes clearer.

Since an open subset of $\mathbb{R}$ is a countable union of disjoint intervals, it is usually possible to consider functions defined on intervals.

2. Harmonic, Superharmonic, and Subharmonic Functions

A function u defined on a nonempty open subset D of $\mathbb{R}$ will be called *harmonic* (*superharmonic*, *subharmonic*) if on each component interval of D the function u is finite valued and linear (concave, convex, respectively). Such a function is necessarily continuous.

Every nonempty open subset of $\mathbb{R}$ except $\mathbb{R}$ itself supports a positive not identically constant superharmonic function. In the terminology of the multidimensional case $\mathbb{R}$ is the only non-Greenian open subset of $\mathbb{R}$ aside from the empty set. In agreement with the definition of a polar set when $N > 1$, only the empty set is defined as polar when $N = 1$. That is, the only negligible set in one-dimensional classical potential theory is the empty set.

3. Convergence Theorems

It is trivial that the supremum of an upper directed family of harmonic [superharmonic] functions on an open subset D of $\mathbb{R}$ is harmonic [superharmonic] if finite at a point of each component interval of D. It is just as trivial that the infimum of any family of superharmonic functions on D is superharmonic if finite valued. Thus the Fundamental Convergence Theorem of classical potential theory is both true and trivial when $N = 1$.

If u is superharmonic or subharmonic on D and if B is an open subinterval of D, relatively compact in D, the function $\tau_B u$ is defined as the function equal to u on $D - B$ and linear on $\bar{B}$. If u is superharmonic, the function τ_B is also superharmonic and is majorized by u.

If u is superharmonic on D and has a subharmonic minorant, the function $\mathrm{GM}_D u$ exists as in the case $N > 1$ and is harmonic. The proof for $N > 1$ is applicable to the case $N = 1$. A superharmonic function u on a finite interval D has a subharmonic minorant if and only if the limit of u at each endpoint of D is finite, and in that case $\mathrm{GM}_D u$ is the restriction to D of the linear function equal at each endpoint of D to the limit of u at that endpoint. A superharmonic function u on the infinite interval $]a, +\infty[$ with a finite has a subharmonic minorant if and only if the limit of u at a is finite and the limit α of u'_r at $+\infty$ is not $-\infty$. In that case $\mathrm{GM}_D u$ is the restriction to D of the linear function with value at a the limit of u there and with slope α.

4. Smoothness Properties of Superharmonic and Subharmonic Functions

We omit the elementary proofs of the following properties of a convex (that is, subharmonic) function u defined on an open interval D.

(a) u is continuous, is absolutely continuous on each compact subinterval of D, and has finite or infinite limits at the endpoints of D.
(b) u has a right [left] derivate u'_r $[u'_l]$ at each point of D.
(c) The function u'_r $[u'_l]$ is monotone increasing and right [left] continuous.
(d) $u'_l \leq u'_r$, and there is equality except at a countable subset of D. At a point of equality the derivative u' $(= u'_r = u'_l)$ exists.

In one dimension the Laplacian of a function u is the second derivate u''. A function u in $\mathbb{C}^{(2)}(D)$ is subharmonic if and only if $\Delta u \geq 0$.

5. The Dirichlet Problem (Euclidean Boundary)

If D is a finite interval $]a, b[$ and if f is a boundary function, it is trivial that the PWB method in the present context yields a solution if and only if f is finite valued and that for finite-valued f,

$$H_f = \mu_D(\cdot, f), \qquad \mu_D(\xi, \{a\}) = \frac{b - \xi}{b - a}, \qquad \mu_D(\xi, \{b\}) = \frac{\xi - a}{b - a}. \tag{5.1}$$

On the other hand, if the interval D has one finite and one infinite endpoint, the PWB method yields a solution if and only if the specified boundary function f is finite valued, say β, at the finite endpoint b of D. The PWB solution H_f is then identically β, that is, the harmonic measure of $\{b\}$ is identically 1. If D is not connected but is a proper subset of $\mathbb{R}$, the PWB method is applied separately to each interval component of D, or equivalently, to D itself, but the Dirichlet problem is not treated for $D = \mathbb{R}$ because $\mathbb{R}$ is not Greenian.

The Dirichlet problem for conditionally harmonic functions is left to the reader.

6. Green Functions

When $N = 1$ and D is a finite interval $]a, b[$, it is natural to define G_D as a function on $D \times D$ with the property that $G_D(\xi, \cdot)$ is harmonic on $D - \{\xi\}$ and superharmonic on D, with limit 0 at each endpoint of D. These conditions determine $G_D(\xi, \cdot)$ up to a multiplicative positive constant. Let d_1 $[d_2]$ be the derivative of $G_D(\xi, \cdot)$ to the left [right] of ξ. If now $u \in \mathbb{C}^{(2)}(\bar{D})$, integration by parts yields

$$\int_a^b G_D(\xi, \cdot) u''\, dl_1 = u(a)d_1 - u(b)d_2 - u(\xi)(d_1 - d_2). \tag{6.1}$$

To simplify this formula, we specify G_D completely by prescribing $d_1 - d_2 = 1$, so that

$$G_D(\xi, \eta) = \begin{cases} \dfrac{(b - \xi)(\eta - a)}{b - a} & \text{if } \eta \le \xi, \\[2ex] \dfrac{(b - \eta)(\xi - a)}{b - a} & \text{if } \eta \ge \xi, \end{cases} \tag{6.2}$$

and then (6.1) becomes

$$u(\xi) = \mu_D(\xi, u) - \int_a^b G_D(\xi, \cdot) u''\, dl_1. \tag{6.3}$$

The representation (6.3) should be compared with the corresponding representation I(8.6). Thus if $u \in \mathbb{C}^{(2)}(\bar{D})$ and if u is superharmonic and positive, u is the sum of $\mathrm{GM}_D u$ and the potential of the measure $-\Delta u\, dl_1$. This example of the Riesz decomposition will be extended to the general case in Section 9.

If D is the interval $]a, +\infty[$ with a finite, we let b tend to $+\infty$ in (6.2) and thereby are led to the definition

$$G_D(\xi, \eta) = (\xi - a) \wedge (\eta - a). \tag{6.4}$$

If $D =]-\infty, b[$ with b finite, we define G_D correspondingly. If D is a union of two or more disjoint nonempty open intervals, $G_D(\xi, \eta)$ is defined as

$G_{D_0}(\xi,\eta)$ for ξ, η in the same component interval D_0 of D and defined as 0 for ξ, η in different component intervals. Thus G_D is now defined for every Greenian set D.

7. Potentials of Measures

If D is Greenian and if μ is a measure of Borel subsets of D, then the potential $G_D\mu$ is superharmonic, that is, concave, on each component interval of D on which the potential is finite valued, because $G_D(\cdot,\eta)$ is concave for each η. We now show that $\mathrm{GM}_D u = 0$ when $u = G_D\mu$ is a superharmonic potential. It will follow from the Riesz decomposition (Section 9) that conversely a positive superharmonic function u with $\mathrm{GM}_D u = 0$ is the potential of a measure. We can assume in the following that D is an interval $]a,b[$. Let $B_.$ be an increasing sequence of subintervals of D with compact closures in D and union D. Then $\tau_{B_.}G_D\mu$ is a decreasing sequence of superharmonic functions. Moreover, if $\xi\in B_n$, then

$$\tau_{B_n}(G_D\mu)(\xi) = \int_{\partial B_n} \mu_{B_n}(\xi, G_D(\cdot,\eta))\mu(d\eta). \tag{7.1}$$

Here $n\mapsto\mu_{B_n}(\xi,G_D(\cdot,\eta))$ is a decreasing sequence, for n so large that $\xi\in B_n$, with limit 0. Hence $\lim_{n\to\infty}\tau_{B_n}G_D\mu = 0$. It follows that $u = G_D\mu$ has limit 0 at each finite endpoint of D and that if D has an infinite endpoint, the limit of right and left derivatives u_r' and u_l' at that endpoint is 0. Hence (see Section 4) $\mathrm{GM}_D u = 0$. Alternatively [see the N-dimensional context in Section III.1, Observation (a)], the sequence $\tau_{B_.}u$ can be proved directly to have limit $\mathrm{GM}_D u$.

8. Identification of the Measure Defining a Potential

Theorem. *If $u = G_D\mu$ is a superharmonic potential, then $d\mu = du_r'$ in the sense that*

$$u_r'(\beta) - u_r'(\alpha) = -\mu(]\alpha,\beta]) \quad \text{for }]\alpha,\beta]\subset D. \tag{8.1}$$

In proving the theorem we can suppose that D is an interval $]a,b[$ with at least one finite endpoint. If both endpoints are finite, define functions ϕ_a and ϕ_b by

$$\begin{aligned}\phi_a(\xi) &= \int_a^\xi (\eta-a)\mu(d\eta), \qquad a<\xi<b,\\ \phi_b(\xi) &= -\int_\xi^b (b-\eta)\mu(d\eta), \qquad a<\xi<b.\end{aligned} \tag{8.2}$$

Here $\int_\alpha^\beta$ means the integral over $]\alpha,\beta]$. The functions ϕ_a and ϕ_b are finite valued, monotone increasing, and right continuous and satisfy

$$\int_\alpha^\beta (b-\eta)\,d\phi_a(\eta) = \int_\alpha^\beta (\eta - a)\,d\phi_b(\eta), \qquad a<\alpha<\beta<b, \tag{8.3}$$

and

$$u(\xi) = \frac{(b-\xi)\phi_a(\xi) - (\xi - a)\phi_b(\xi)}{b-a}. \tag{8.4}$$

Apply (8.3) and (8.4) to find

$$u_r'(\xi) = -\frac{\phi_a(\xi)+\phi_b(\xi)}{b-a} \tag{8.5}$$

first when $\mu(\{\xi\}) = 0$ and then (right continuity) for all ξ. Then for $a < \alpha < \beta < b$

$$u_r'(\beta) - u_r'(\alpha) = -\frac{\int_\alpha^\beta (\eta-a)\mu(d\eta) + \int_\alpha^\beta (b-\eta)\mu(d\eta)}{b-a} = -\mu(]\alpha,\beta]), \tag{8.6}$$

as was to be proved. If one endpoint, say b, is not finite, define ϕ_a as in (8.2), but define $\phi_b(\xi) = -\mu(]\xi,+\infty[)$. Then (8.3), (8.4), (8.5) are replaced by

$$\phi_a(\beta) - \phi_a(\alpha) = \int_\alpha^\beta (\eta - a)\,d\phi_b(\eta), \tag{8.3$'$}$$

$$u(\xi) = \phi_a(\xi) - (\xi - a)\phi_b(\xi), \tag{8.4$'$}$$

$$u_r'(\xi) = -\phi_b(\xi); \tag{8.5$'$}$$

so in this case also (8.1) is true.

9. Riesz Decomposition

Theorem. *If u is a positive superharmonic function on an open proper subset D of $\mathbb{R}$, then*

$$u = \mathrm{GM}_D u + G_D\mu, \tag{9.1}$$

where $d\mu = du_r'$ in the sense of Theorem 8.

We can assume that D is an interval $]a, b[$ with at least one finite endpoint. In fact, however, we shall assume that both endpoints are finite, leaving the other case to the reader. Equation (8.6) is trivial, and when $\alpha \downarrow a$ and $\beta \uparrow b$, we find that ϕ_a and ϕ_b as defined by (8.2) are finite valued, monotone increasing, and right continuous and satisfy (8.3). Then (8.5) is true up to an additive constant; so (8.4) is true up to a linear term; that is, $u = G_D\mu + v$, where v is linear. Since the GM_D operation is linear, just as in the multidimensional case, and since this operation on a superharmonic potential yields 0, it follows that $v = \mathrm{GM}_D u$, and the proof of the theorem is complete.

10. The Martin Boundary

Martin boundary theory in one dimension is rather trivial. For example, if $D =]0, +\infty[$ and if $\xi_0 \in D$, we define the Martin function with reference point ξ_0 in the obvious way.

$$K(\eta, \xi) = \frac{\xi \wedge \eta}{\xi_0 \wedge \eta}.$$

We then find $(\eta \to 0, +\infty)$ that the Martin space can be identified with $\overline{\mathbb{R}}^+$; the Martin boundary point 0 is associated with the minimal harmonic function $K(0, \cdot) \equiv 1$; the Martin boundary point $+\infty$ is associated with the minimal harmonic function $\xi \mapsto K(+\infty, \xi) = \xi/\xi_0$.

Chapter XV

Parabolic Potential Theory: Basic Facts

1. Conventions

The potential theory based on the Laplace operator, developed in the preceding chapters, will be called *classical potential theory* below. The potential theory based on the heat operator $\dot{\Delta}$ and its adjoint $\overset{*}{\Delta}$, called *parabolic potential theory*, will be developed in Chapters XV to XIX. Concepts that are parabolic counterparts of classical concepts will be distinguished by dots or asterisks, depending on whether the concepts are related to $\dot{\Delta}$ or to $\overset{*}{\Delta}$. Just as the domains of classical potential theory are subsets of $\mathbb{R}^N$, the domains of parabolic potential theory are subsets of "space time" $\mathbb{R}^{N+1}$, which we denote in this context by $\dot{\mathbb{R}}^N$. Here $N \geq 1$, and the case $N = 1$ is not exceptional. A point $\dot{\xi} = (\xi, s)$ of $\dot{\mathbb{R}}^N$ has space coordinate ξ in $\mathbb{R}^N$ and time coordinate $s = \operatorname{ord} \dot{\xi}$ (the *ordinate* of $\dot{\xi}$), a point of $\mathbb{R}$. The point $\dot{\eta}:(\eta, t)$ will be said to be [strictly] below $\dot{\xi}:(\xi, s)$ if $t \leq s$ [$t < s$]. If $\dot{\xi}$ is a point of an open subset $\dot{D}$ of $\dot{\mathbb{R}}^N$, the set of points of $\dot{D}$ [*strictly*] *below* $\dot{\xi}$ *relative to* $\dot{D}$ is the set of points of $\dot{D}$ that are endpoints of continuous [strictly] downward-directed arcs from $\dot{\xi}$. That is, $\dot{\eta}$ is [strictly] below $\dot{\xi}$ relative to $\dot{D}$ if and only if there is a continuous function f from $[0, 1]$ into $\dot{D}$ for which $f(0) = \dot{\xi}$, $f(1) = \dot{\eta}$, and $\operatorname{ord} f$ is a [strictly] decreasing function. The *upper* [*lower*] *half-space* of $\dot{\mathbb{R}}^N$ is the set $\{\operatorname{ord} \dot{\xi} > 0\}$ [$\{\operatorname{ord} \dot{\xi} < 0\}$] and the *abscissa hyperplane* is the set $\{\operatorname{ord} \dot{\xi} = 0\}$. The boundary of a subset of $\dot{\mathbb{R}}^N$ relative to the one-point compactification of $\dot{\mathbb{R}}^N$ will be called the *Euclidean boundary*, and *boundary* will mean this boundary unless a different one is specified.

Let $\dot{D}$ be an interval in $\dot{\mathbb{R}}^N$, $\dot{D} =]a_1, b_1[\times \cdots \times]a_N, b_N[\times]s_1, s_2[$. The set of boundary points with ordinate value s_2 will be called the *upper boundary*, the set of boundary points with ordinate value s_1 will be called the *lower boundary*, and the closure of the rest of the boundary will be called the *lateral boundary*.

2. The Parabolic and Coparabolic Operators

In the following discussion the Laplace operator Δ acting on a function $\dot{\xi} = (\xi, s) \mapsto \dot{u}(\dot{\xi})$ defined on an open subset of $\dot{\mathbb{R}}^N$ is to act only on the space variable ξ. Choose a strictly positive number σ (fixed throughout the discussion). Define the parabolic operator $\dot{\Delta}$ and the coparabolic operator $\overset{*}{\Delta}$, operating on sufficiently smooth functions defined on open subsets of $\dot{\mathbb{R}}^N$, by

$$\begin{aligned}\dot{\Delta}\dot{u}(\xi, s) &= \frac{\sigma^2}{2}\Delta\dot{u}(\xi, s) - \frac{\partial\dot{u}(\xi, s)}{\partial s},\\ \overset{*}{\Delta}\dot{u}(\xi, s) &= \frac{\sigma^2}{2}\Delta\dot{u}(\xi, s) + \frac{\partial\dot{u}(\xi, s)}{\partial s}.\end{aligned} \tag{2.1}$$

Parabolic potential theory is based on the pair $\dot{\Delta}$, $\overset{*}{\Delta}$ and is similar in many respects to classical potential theory, but the fact that both $\dot{\Delta}$ and $\overset{*}{\Delta}$ are involved means that two theories dual to each other must be considered simultaneously.

A function $\dot{u}$ from an open subset of $\dot{\mathbb{R}}^N$ into $\mathbb{R}$, in class $\mathbb{C}^{(1)}$ there and also in class $\mathbb{C}^{(2)}$ relative to the space variable, and satisfying the heat equation $\dot{\Delta}\dot{u} = 0$ will be called *parabolic*; a solution (satisfying the same smoothness conditions) of the adjoint equation $\overset{*}{\Delta}\dot{u} = 0$ will be called *coparabolic*. A function $\dot{u}$ is coparabolic if and only if the function $(\xi, s) \mapsto \dot{u}(\xi, -s)$ is parabolic. If u is a function on an open subset D of $\mathbb{R}^N$, if $\dot{D} = D \times \mathbb{R}$, and if $\dot{u}(\dot{\xi}) = u(\xi)$ for $\dot{\xi} = (\xi, s)$ in $\dot{D}$, the function $\dot{u}$ is parabolic (equivalently, coparabolic) on $\dot{D}$ if and only if u is harmonic on D.

EXAMPLE (a). If γ is an N-dimensional vector, the function

$$\dot{u}: \dot{\xi} = (\xi, s) \mapsto \exp\left[\langle\gamma, \xi\rangle + \sigma^2|\gamma|^2\frac{s}{2}\right]$$

is a positive parabolic function on $\dot{\mathbb{R}}^N$. Recall that in contrast with this example there is no nonconstant positive harmonic function on $\mathbb{R}^N$.

Minimal Parabolic Functions

A positive parabolic function $\dot{u}$ on an open subset $\dot{D}$ of $\dot{\mathbb{R}}^N$ will be called *minimal* if every positive parabolic function on $\dot{D}$ majorized by $\dot{u}$ is a constant multiple of $\dot{u}$. If $\dot{D}$ is $\dot{\mathbb{R}}^N$ or is the lower half-space, it will be shown in Section XVI.8 that the restrictions to $\dot{D}$ of the functions in Example (a) are minimal and in fact are the only parabolic minimal functions up to constant multiples.

EXAMPLE (b). If $\dot{u}$ is parabolic on $\dot{\mathbb{R}}^N$ and is of the form $(\xi, s) \mapsto f(|\xi|)g(s)$, the functions f and g satisfy the equations $f''(r) + [(N-1)/r]f'(r) = cf(r)$ for $r \geq 0$ and $g' = c\sigma^2 g/2$ on $\mathbb{R}$, and we thereby find the parabolic function $\dot{u}$ defined on $\dot{\mathbb{R}}^N$ by

$$\begin{aligned} \dot{u}(\dot{\xi}) &= e^{\sigma^2 cs/2} \sum_{m=0}^{\infty} \frac{(c|\xi|^2)^m}{4^m m!\Gamma(m+N/2)} = \left(\frac{2}{i\sqrt{c}|\xi|}\right)^{(N-2)/2} J_{(N-2)/2}(i\sqrt{c}|\xi|)e^{\sigma^2 cs/2} \\ &= \left(\frac{2}{\sqrt{c}|\xi|}\right)^{(N-2)/2} I_{(N-2)/2}(\sqrt{c}|\xi|)e^{\sigma^2 cs/2}, \end{aligned} \tag{2.2}$$

where c is an arbitrary constant, J_k is the Bessel function of the first kind, and I_k is the modified Bessel function. The parabolic function $\dot{u}/\Gamma(N/2)$ can also be obtained by taking the parabolic function in Example (a) and averaging it over the values of γ with $|\gamma|^2 = c$. This representation of $\dot{u}/\Gamma(N/2)$ is an example of the fact that every positive parabolic function on $\dot{\mathbb{R}}^N$ is an integral over the set of minimal parabolic functions.

3. Coparabolic Polynomials

Define the Hermite polynomial $H_{m_1 \cdots m_N}$ on $\mathbb{R}^N$ by

$$H_{m_1 \cdots m_N}(\eta) = e^{|\eta|^2} \frac{\partial^{m_1 + \cdots + m_N}}{\partial \eta^{(1)m_1} \cdots \partial \eta^{(N)m_N}} e^{-|\eta|^2} \qquad (m_j \geq 0)$$

so that if γ is a vector in $\mathbb{R}^N$, the Taylor expansion

$$e^{-|\eta+\gamma|^2} = e^{-|\eta|^2} \sum_{n=0}^{\infty} \sum_{m_1 + \cdots + m_N = n} \frac{\gamma^{(1)m_1} \cdots \gamma^{(N)m_N}}{m_1! \cdots m_N!} H_{m_1 \cdots m_N}(\eta) \tag{3.2}$$

yields

$$e^{-2\langle\gamma,\eta\rangle - |\gamma|^2} = \sum_{n=0}^{\infty} \sum_{m_1 + \cdots + m_N = n} \frac{\gamma^{(1)m_1} \cdots \gamma^{(N)m_N}}{m_1! \cdots m_N!} H_{m_1 \cdots m_N}(\eta). \tag{3.3}$$

According to (3.1),

$$\begin{aligned} \frac{\partial H_{m_1 \cdots m_N}}{\partial \eta^{(j)}} &= e^{|\eta|^2} \frac{\partial^{m_1 + \cdots + m_N}}{\partial \eta^{(1)m_1} \cdots \partial \eta^{(N)m_N}} (-2\eta^{(j)} e^{-|\eta|^2}) \\ &= -2m_j H_{m_1' \cdots m_N'}, \qquad m_i' = m_i - \delta_{ij} \quad \text{if } m_j > 0, \end{aligned} \tag{3.4}$$

and this partial derivative vanishes if $m_j = 0$; so (3.4) is valid in all cases when interpreted reasonably. Repeated integration by parts leads to

$$\int_{\mathbb{R}^N} e^{-|\eta|^2} H_{m_1\cdots m_N}(\eta) H_{n_1\cdots n_N}(\eta) l_N(d\eta)$$
$$= \begin{cases} m_1! \cdots m_N! 2^{m_1+\cdots+m_N} \pi^{N/2} & \text{if } m_\cdot = n_\cdot, \\ 0 & \text{otherwise.} \end{cases}$$

Thus the sequence of Hermite polynomials is an orthogonal sequence relative to the measure $\exp(-|\eta|^2) l_N(d\eta)$, and (3.3) is the corresponding Fourier series of the left side.

The function $(\eta, t) \mapsto \exp(\langle \gamma, \eta \rangle - |\gamma|^2 \sigma^2 t/2)$ is coparabolic. Define the space-time Hermite polynomial $\dot{H}_{m_1\cdots m_N}$, homogeneous of degree $m_1 + \cdots + m_N$ in the variables $\eta^{(1)}, \ldots, \eta^{(N)}, t^{1/2}$, by the Taylor expansion

$$e^{\langle \gamma, \eta \rangle - |\gamma|^2 \sigma^2 t/2} = \sum_{n=0}^{\infty} \sum_{m_1+\cdots+m_N=n} \frac{\gamma^{(1)m_1} \cdots \gamma^{(N)m_N}}{m_1! \cdots m_N!} \dot{H}_{m_1\cdots m_N}(\eta, t).$$

The polynomial $\dot{H}_{m_1\cdots m_N}$ is coparabolic because it is obtained by repeated differentiation of the coparabolic function on the left side of (3.6). If γ and η in (3.3) are replaced by $(\sigma^2 t/2)^{1/2}\gamma$ and $-(2\sigma^2 t)^{-1/2}\eta$, respectively, the left sides of (3.3) and (3.6) become identical; so

$$\begin{aligned} \dot{H}_{m_1\cdots m_N}(\eta, t) &= \left(\frac{\sigma^2 t}{2}\right)^{n/2} H_{m_1\cdots m_N}(-(2\sigma^2 t)^{-1/2}\eta) \qquad \left(n = \sum_j m_j\right) \\ &= (-\sigma^2 t)^n \exp\frac{|\eta|^2}{2\sigma^2 t} \frac{\partial^n}{\partial \eta^{(1)m_1} \cdots \partial \eta^{(N)m_N}} \exp\frac{-|\eta|^2}{2\sigma^2 t} \\ &= \eta^{(1)m_1} \cdots \eta^{(N)m_N} + \cdots. \end{aligned} \tag{3.7}$$

The term written on the last line is the only term not involving t. The relation (3.4) becomes

$$\frac{\partial \dot{H}_{m_1\cdots m_N}}{\partial \eta^{(j)}} = m_j \dot{H}_{m'_1\cdots m'_N}, \qquad m'_i = \begin{cases} m_i - \delta_{ij} & \text{if } m_j > 0, \\ 0 & \text{if } m_j = 0, \end{cases} \tag{3.8}$$

and (3.5) becomes

$$\int_{\mathbb{R}^N} e^{-|\eta|^2/2\sigma^2 t} \dot{H}_{m_1\cdots m_N}(\eta, t) \dot{H}_{n_1\cdots n_N}(\eta, t) l_N d\eta)$$
$$= \begin{cases} (\sigma^2 t)^{n+N/2} (2\pi)^{N/2} m_1! \cdots m_N! & \text{if } m_\cdot = n_\cdot, \\ 0 & \text{otherwise.} \end{cases} \tag{3.9}$$

For each value of t the sequence of space-time Hermite polynomials is an orthgononal sequence in the space variables, and (3.6) is the corresponding Fourier expansion of the left side.

A coparabolic polynomial f ($\not\equiv 0$) must contain a term not involving t because

$$f(\eta, t) = \sum_{j=1}^{m} t^{n_j} f_j(\eta), \qquad 0 \leq n_1 < \cdots < n_m,$$

where f_j is a not identically 0 polynomial in the components of η, and f cannot satisfy the heat equation dual unless $n_1 = 0$. In view of (3.7) some linear combination $\dot{H}$ of space-time Hermite polynomials can be chosen with the same terms not involving t as f, and therefore $f - \dot{H}$ vanishes identically. Thus every coparabolic polynomial is a linear combination of the space-time Hermite polynomials.

EXAMPLE. If $N = 1$, the successive Hermite and space-time Hermite polynomials are

$$H_0 = 1, \quad H_1(\eta) = -2\eta, \quad H_2(\eta) = 4\eta^2 - 2, \quad H_3(\eta) = -8\eta^3 + 12\eta, \ldots,$$

$$\dot{H}_0 = 1, \quad \dot{H}_1(\eta, t) = \eta, \quad \dot{H}_2(\eta, t) = \eta^2 - \sigma^2 t, \quad \dot{H}_3(\eta, t) = \eta^3 - 3\eta\sigma^2 t, \ldots.$$

4. The Parabolic Green Function of $\dot{\mathbb{R}}^N$

Define the function ℓ on $\dot{\mathbb{R}}^N$ by

$$\ell(t, \eta) = \begin{cases} (2\pi\sigma^2 t)^{-N/2} \exp \dfrac{-|\eta|^2}{2\sigma^2 t} & \text{if } t > 0, \\ 0 & \text{if } t \leq 0. \end{cases} \tag{4.1}$$

Note that the time variable is placed first in this notation, as appropriate to the probability interpretation to be given later. The parabolic Green function $\dot{G}$ of $\dot{\mathbb{R}}^N$ is defined on $\dot{\mathbb{R}}^N \times \dot{\mathbb{R}}^N$ ($N \geq 1$) by

$$\dot{G}(\dot{\xi}, \dot{\eta}) = \dot{G}((\xi, s), (\eta, t)) = \ell(s - t, \xi - \eta). \tag{4.2}$$

In more detail,

(a) The function $\dot{G}(\cdot, \dot{\eta})$ is the Green function with pole $\dot{\eta}$ for the heat equation. This function is positive, parabolic on $\dot{\mathbb{R}}^N - \{\dot{\eta}\}$, and vanishes below $\dot{\eta}$ and in the limit at the point ∞.

(a*) The function $\dot{G}(\dot{\xi}, \cdot)$ is the Green function with pole $\dot{\xi}$ for the adjoint equation. This function is positive, coparabolic on $\dot{\mathbb{R}}^N - \{\dot{\xi}\}$, and vanishes above $\dot{\xi}$ and in the limit at the point ∞.

Observe that when $N > 2$, there is an intimate connection between the parabolic Green function of $\dot{\mathbb{R}}^N$ and the classical Green function of $\mathbb{R}^N$:

$$\begin{aligned}\int_{-\infty}^{\infty} \dot{G}((\xi, s), (\eta, t)) l_1(dt) &= \int_0^{\infty} \ell(s, \xi - \eta) l_1(ds) \\ &= a_N |\xi - \eta|^{2-N} = a_N G(\xi, \eta)\end{aligned} \tag{4.3}$$

with

$$a_N = \frac{\Gamma(N/2 - 1)}{2\sigma^2 \pi^{N/2}}.$$

This relation between G and $\dot{G}$ will be generalized to Greenian subsets of $\mathbb{R}^N$ in Section XVII.18.

The following inequality will be used below. Let $\mathbf{D}^{(k)}$ be a (possibly mixed) partial derivation operator of order k on space variables, and let β be a positive number. Then if $\dot{\xi} = (\xi, s)$ and $\dot{\eta} = (\eta, t)$,

$$|\mathbf{D}_\eta^{(k)} \dot{G}(\dot{\xi}, \dot{\eta})|\,|\xi - \eta|^\beta \le c(\beta, \sigma)(s - t)^{-(N+k-\beta)/2} \exp \frac{-|\xi - \eta|^2}{4\sigma^2(s - t)} \quad (s > t). \tag{4.4}$$

This inequality will be proved for $k = 1$. The proof in the general case involves more notation but no additional ideas. For $k = 1$ and $s > t$

$$\begin{aligned}\left|\frac{\partial \dot{G}(\dot{\xi}, \dot{\eta})}{\partial \eta^{(i)}}\right| |\xi - \eta|^\beta &\le (2\pi)^{-N/2} \sigma^{-N-2} \frac{|\xi - \eta|^{\beta+1}}{(s - t)^{(N+2)/2}} \exp \frac{-|\xi - \eta|^2}{2\sigma^2(s - t)} \\ &= (2\pi)^{-N/2} \sigma^{-N-2} (s - t)^{-(N+1-\beta)/2} \exp \frac{-|\xi - \eta|^2}{4\sigma^2(s - t)} \left[\left(\frac{|\xi - \eta|^2}{s - t}\right)^{(\beta+1)/2} \exp \frac{-|\xi - \eta|^2}{4\sigma^2(s - t)}\right],\end{aligned} \tag{4.5}$$

and the right side is majorized by the right side of (4.4) when $k = 1$.

If $\dot{\mu}$ is a measure on $\dot{\mathbb{R}}^N$, the functions $\dot{G}\dot{\mu}$ and $\dot{\mu}\dot{G}$ defined by

$$\dot{G}\dot{\mu}(\dot{\xi}) = \int_{\dot{\mathbb{R}}^N} \dot{G}(\dot{\xi}, \dot{\eta}) \dot{\mu}(d\dot{\eta}), \qquad \dot{\mu}\dot{G}(\dot{\eta}) = \int_{\dot{\mathbb{R}}^N} \dot{G}(\dot{\xi}, \dot{\eta}) \dot{\mu}(d\dot{\xi}) \tag{4.6}$$

will be called respectively the *potential* and *copotential* of $\dot{\mu}$ on $\dot{\mathbb{R}}^N$. These definitions will be generalized to potentials and copotentials of measures on an open subset $\dot{D}$ of $\dot{\mathbb{R}}^N$ with parabolic Green function $\dot{G}_{\dot{D}}$.

5. Maximum-Minimum Parabolic Function Theorem

This theorem will follow from the superparabolic function minimum theorem (Section 13), but the following direct proof is instructive. Since the function $\dot{G}(\cdot, \dot{\eta})$ is parabolic on $\dot{D} = \dot{\mathbb{R}}^N - \{\dot{\eta}\}$ and takes on its minimum value at every point of $\dot{D}$ below $\dot{\eta}$, the maximum-minimum parabolic function theorem is necessarily weaker than its harmonic function counterpart.

Theorem. *Let $\dot{\zeta}$ be a point of an open subset $\dot{D}$ of $\dot{\mathbb{R}}^N$, and let $\dot{D}(\dot{\zeta})$ be the set of points of $\dot{D}$ below $\dot{\zeta}$ relative to $\dot{D}$. If $\dot{u}$ is a parabolic function on $\dot{D}$ and if $\dot{u}(\dot{\zeta})$ is the supremum or infimum of the restriction of $\dot{u}$ to $\dot{D}(\dot{\zeta})$, then $\dot{u} = \dot{u}(\dot{\zeta})$ on $\dot{D}(\dot{\zeta})$.*

Observation. This theorem when applied to the part of $\dot{D}$ strictly below an arbitrary horizontal hyperplane implies that if m is the supremum of $\dot{u}$, there is a sequence $\dot{\eta}_{\cdot}$ of boundary points of $\dot{D}$ with ordinate values strictly less than $\sup_{\dot{D}} \operatorname{ord} \dot{\eta}$ such that

$$\lim_{n\to\infty} \limsup_{\dot{D}\ni\dot{\xi}\to\dot{\eta}_n} \dot{u}(\dot{\xi}) = m.$$

Thus if $\dot{D}$ is an interval, there is a sequence in $\dot{D}$ converging to a point of the union of lower and lateral boundaries of $\dot{D}$ such that $\dot{u}$ tends to m along this sequence. If $\dot{u}$ is parabolic on a neighborhood of the closure of the interval $\dot{D}$, the restriction of $\dot{u}$ to $\bar{\dot{D}}$ attains its maximum and minimum on the union of lower and lateral boundaries.

In proving the theorem we need consider only suprema, and it is sufficient to prove the following result. Let S be a closed line segment in $\dot{D}$, not orthogonal to the ordinate axis, and let m be the supremum of $\dot{u}$ on the set of points of $\dot{D}$ below the highest point of S. We prove that $\dot{u} < m$ on S if this inequality is true at the lowest point of S. Let $\dot{\xi}:(\xi, s)$ be the upper endpoint of S, and to simplify the notation, let the origin be the lower endpoint of S. If $\dot{u}(0) < m$, choose r so that $0 < r < 1$, so that the set

$$\dot{D}_0 = \left\{(\eta, t): \left|\eta - \frac{\xi t}{s}\right| < r, 0 < t < s\right\}$$

has closure in D, and so that $m_1 = \sup_{|\eta|<r} \dot{u}(\eta, 0) < m$. Define $\delta = \delta(\eta, t) = r^2 - |\eta - \xi t/s|^2$ and define $\dot{v}$ on $\dot{\mathbb{R}}^N$ by

$$\dot{v}(\eta, t) = m - (m - m_1)\delta^2 e^{-\alpha t}, \tag{5.1}$$

where α is a positive constant to be chosen below. Then if ξ and η have jth coordinates $\xi^{(j)}$ and $\eta^{(j)}$, respectively,

$$\begin{aligned}\dot{\Delta}\dot{v}(\eta, t) = (m - m_1)e^{-\alpha t}\Bigg\{-\alpha\delta^2 + 2\delta\Bigg[\sigma^2(N+2)\\ + 2\sum_{j=1}^{N}\left(\eta^{(j)} - \frac{\xi^{(j)}t}{s}\right)\left(\frac{\xi^{(j)}}{s}\right)\Bigg] - 4\sigma^2 r^2\Bigg\}.\end{aligned} \tag{5.2}$$

As a function of δ the quantity in braces has maximum value

$$-4\sigma^2 r^2 + \frac{\left[\sigma^2(N+2) + 2\sum_{j=1}^{N} (\eta^{(j)} - \xi^{(j)}t/s)(\xi^{(j)}/s)\right]^2}{\alpha}. \tag{5.3}$$

Choose α so that this maximum value is strictly negative on a neighborhood of $\bar{\dot{D}}_0$, on which therefore $\dot{\Delta}(\dot{v} - \dot{u}) < 0$. On the lower boundary of $\dot{D}_0$, that is, on its lower face, $\dot{v} - \dot{u} \geq m - (m - m_1) - m_1 = 0$, and on the lateral boundary of $\dot{D}_0$ we fine that $\dot{v} - \dot{u} \geq m - m = 0$. Now if the minimum value of $\dot{v} - \dot{u}$ on $\bar{\dot{D}}_0$ is attained at a point (η, t) either in the interior of $\dot{D}_0$ or in the interior of the upper face of $\dot{D}_0$, then $\partial(\dot{v} - \dot{u})/\partial t \leq 0$ at the point; so $\Delta(\dot{v} - \dot{u}) < 0$ (Laplacian applied to the space coordinates) there. Since $(\dot{v} - \dot{u})(\cdot, t)$ has a local minimum at η, the latter inequality is impossible, and it follows that the restriction of $\dot{v} - \dot{u}$ to $\bar{\dot{D}}_0$ attains its minimum value at a point of the union of the lateral boundary and lower face; so $\dot{v} - \dot{u} \geq 0$. In particular, this inequality on the segment S becomes

$$\dot{u}\left(\frac{t\xi}{s}, t\right) \leq m - (m - m_1)r^4 e^{-\alpha t} < m \qquad (0 \leq t \leq s), \tag{5.4}$$

and the proof is complete.

6. Application of Green's Theorem

Let $\dot{D}$ be an open subset of $\dot{\mathbb{R}}^N$. Suppose that for t in the interval $[t', t'']$ the set $\dot{D}(t) = \{\eta: (\eta, t) \in \dot{D}\}$ is a nonempty connected subset of $\dot{\mathbb{R}}^N$ smooth enough for the application of Green's theorem in N dimensions. Denote by $\partial\dot{D}(t)$ the boundary of $\dot{D}(t)$ relative to the hyperplane $\{\text{ord}\,\xi = t\}$, and denote by $\mathbf{D}_\mathrm{n}$ the directional derivative operator at a point of $\partial\dot{D}(t)$ in the direction of the outward normal to $\partial D(t)$. Define $\dot{D}(t_1, t_2) = \dot{D} \cap (\mathbb{R}^N \times\,]t_1, t_2[)$ for $t' \leq t_1 < t_2 \leq t''$. Denote by $\dot{\mathbf{D}}_\mathrm{n}$ the directional derivative operator at a point of $\partial\dot{D}(t_1, t_2)$ in the direction of the outward normal. The angle between this normal and the upward-directed ordinate axis will be denoted by γ. In the integrals below the differential element will always refer to Lebesgue measure of the indicated dimensionality on the indicated set. Thus $\int_{\dot{D}} - \,dl_{N+1}$ means integration over $\dot{D}$ with respect to $(N+1)$-dimensional measure and $\int_{\partial\dot{D}} - \,dl_N$ means integration over $\partial\dot{D}$ with respect to N-dimensional "surface" measure.

If $N > 1$ and if the functions $\dot{u}$ and $\dot{v}$ are defined on the closure of $\dot{D}$ and are $\mathbb{C}^{(2)}$ in the space variables and $\mathbb{C}^{(1)}$ in space time on $\bar{\dot{D}}$,

$$\int_{\dot{D}(s)} (\dot{u}\,\Delta\dot{v} - \dot{v}\,\Delta\dot{u})\,dl_N = \int_{\partial\dot{D}(s)} (\dot{u}\mathbf{D}_\mathrm{n}\dot{v} - \dot{v}\mathbf{D}_\mathrm{n}\dot{u})\,dl_{N-1}. \tag{6.1}$$

Equivalently,

$$\int_{\dot{D}(s)} (\dot{u}\,\overset{*}{\dot{\Delta}}\dot{v} - \dot{v}\,\dot{\Delta}\dot{u})\,dl_N = \int_{\dot{D}(s)} \frac{\partial(\dot{u}\dot{v})}{\partial s}\,dl_N + \frac{\sigma^2}{2}\int_{\partial\dot{D}(s)} (\dot{u}\mathbf{D}_{\mathbf{n}}\dot{v} - \dot{v}\mathbf{D}_{\mathbf{n}}\dot{u})\,dl_{N-1}. \tag{6.2}$$

Integrating with respect to s yields, if $\dot{D}$ is sufficiently smooth,

$$\begin{aligned}\int_{\dot{D}(t_1,t_2)} (\dot{u}\,\overset{*}{\dot{\Delta}}\dot{v} - \dot{v}\,\dot{\Delta}\dot{u})\,dl_{N+1} = {}& \int_{\partial\dot{D}(t_1,t_2)} \dot{u}\dot{v}\cos\gamma\,dl_N \\ &+ \frac{\sigma^2}{2}\int_{\partial\dot{D}(t_1,t_2)} (\dot{u}\mathbf{D}_{\mathbf{n}}\dot{v} - \dot{v}\mathbf{D}_{\mathbf{n}}\dot{u})\sin\gamma\,dl_N.\end{aligned} \tag{6.3}$$

If $\mathbf{D}_{\mathbf{n}}$ is replaced by $\dot{\mathbf{D}}_{\mathbf{n}}$ in the last integral, $\sin\gamma$ should be replaced by $\sin^2\gamma$. In particular, if $\dot{v} \equiv 1$, this equation reduces to

$$\int_{\dot{D}(t_1,t_2)} \dot{\Delta}\dot{u}\,dl_{N+1} = -\int_{\partial\dot{D}(t_1,t_2)} \dot{u}\cos\gamma\,dl_N + \frac{\sigma^2}{2}\int_{\partial\dot{D}(t_1,t_2)} \mathbf{D}_{\mathbf{n}}\dot{u}\sin\gamma\,dl_N. \tag{6.4}$$

The right side of (6.4) can be described as the heat flow of $\dot{u}$ out of $\dot{D}(t_1, t_2)$. This flow vanishes if $\dot{u}$ is parabolic.

Equation (6.4) is valid when $N = 1$, in which case dl_1 on the right is the differential of arc length.

If $\dot{\eta}_0$ is a point of $\dot{D}$, the heat flow of $\dot{G}(\cdot, \dot{\eta}_0)$ out of $\dot{D}(t_1, t_2)$, with t_1 and t_2 chosen so that $\dot{\eta}_0$ is in this set, is the same as that out of an interval containing $\dot{\eta}_0$ and relatively compact in $\dot{D}(t_1, t_2)$. The heat flow of $\dot{G}(\cdot, \dot{\eta}_0)$ out of an interval containing $\dot{\eta}_0$ is 1 by direct computation.

7. The Parabolic Green Function of a Smooth Domain; The Riesz Decomposition and Parabolic Measure (Formal Treatment)

Continuing the discussion in Section 6, choose $\alpha > 0$, $\dot{\xi} = (\xi, s)$ in $\dot{D}$, and apply (6.3) with $t_2 = s$ and $\dot{v} = \dot{G}((\xi, s+\alpha), \cdot)$ to obtain

$$\begin{aligned}\int_{\partial\dot{D}(t_1,s)} \dot{u}\dot{G}((\xi, s+\alpha),\cdot)\cos\gamma\,dl_N = {}& -\int_{\dot{D}(t_1,s)} \dot{G}((\xi, s+\alpha),\cdot)\,\dot{\Delta}\dot{u}\,dl_{N+1} \\ & - \frac{\sigma^2}{2}\int_{\partial\dot{D}(t_1,s)} [\dot{u}\dot{\mathbf{D}}_{\mathbf{n}}\dot{G}((\xi, s+\alpha),\cdot) - \dot{G}((\xi, s+\alpha),\cdot)\dot{\mathbf{D}}_{\mathbf{n}}\dot{u}]\sin^2\gamma\,dl_N.\end{aligned} \tag{7.1}$$

The dot replacing a variable refers to the integration variable, and the normal derivative is with respect to this variable. Apply Theorem 1 of Appendix VII to find when $\alpha \to 0$ that the part of the integral on the left over $\dot{D}(s)$ has limit $\dot{u}(\dot{\xi})$; so (7.1) yields

$$\begin{aligned} \dot{u}(\dot{\xi}) = &- \int_{\dot{D}(t_1,s)} \dot{G}(\dot{\xi},\cdot)\,\dot{\Delta}\dot{u}\,dl_{N+1} - \int_{\partial\dot{D}(t_1,s)} \dot{u}\dot{G}(\dot{\xi},\cdot)\cos\gamma\,dl_N \\ &- \frac{\sigma^2}{2}\int_{\partial\dot{D}(t_1,s)} [\dot{u}\dot{\mathbf{D}}_{\mathbf{n}}\dot{G}(\dot{\xi},\cdot) - \dot{G}(\dot{\xi},\cdot)\dot{\mathbf{D}}_{\mathbf{n}}\dot{u}]\sin^2\gamma\,dl_N. \end{aligned} \tag{7.2}$$

In particular, if $\dot{u}$ is parabolic, the first integral on the right vanishes, and the representation (7.2) then shows that a parabolic function (and therefore also a coparabolic function) is infinitely differentiable and is analytic in its space variables for each fixed ordinate value. In view of the corresponding development in classical potential theory it will be natural (see Section 12) to define superparabolic functions in such a way that a sufficiently smooth function $\dot{u}$ is superparabolic if and only if $\dot{\Delta}u \le 0$. Under such a definition, if $\dot{u}$ is smooth and superparabolic, the representation of $\dot{u}$ in (7.2) exhibits $\dot{u}$ as the sum of the potential of the positive measure with density $-\dot{\Delta}\dot{u}$ and a parabolic function.

Following the reasoning in the classical context, we next observe (cf. Section I.8) that the work leading to (7.2) can be carried through when $\dot{G}$ is replaced by a function $\dot{G}_{\dot{D}}$ defined on $\dot{D}\times\dot{D}$ and enjoying the following properties, stated for $\dot{D} = \dot{D}(t_1, t_2)$.

(a) For $\dot{\xi}$ in $\dot{D}$ the function $\phi(\dot{\xi},\cdot) = \dot{G}(\dot{\xi},\cdot) - \dot{G}_{\dot{D}}(\dot{\xi},\cdot)$ is coparabolic on $\dot{D}$; for $\dot{\eta}$ in $\dot{D}$ the function $\phi(\cdot,\dot{\eta})$ is parabolic on $\dot{D}$.
(b) $\dot{G}_{\dot{D}}(\dot{\xi},\cdot)$ has limit 0 at every point of $\partial\dot{D}$ with ordinate strictly between t_1 and ord $\dot{\xi}$ and has value 0 at all points of $\dot{D}$ above $\dot{\xi}$.
(c) The function $\phi(\dot{\xi},\cdot)$ can be extended to be of class $\mathbb{C}^{(1)}(\bar{\dot{D}})$ and of class $\mathbb{C}^{(2)}(\bar{\dot{D}})$ in the space variables.

If the argument leading to (7.2) is carried through with $\dot{G}_{\dot{D}}$ instead of $\dot{G}$, (7.2) becomes

$$\begin{aligned} \dot{u}(\dot{\xi}) = &- \int_{\dot{D}(t_1,s)} \dot{G}_{\dot{D}}(\dot{\xi},\cdot)\dot{\Delta}\dot{u}\,dl_{N+1} + \int_{\dot{D}(t_1)} \dot{u}\dot{G}_{\dot{D}}(\dot{\xi},\cdot)\,dl_N \\ &- \frac{\sigma^2}{2}\int_{\partial\dot{D}(t_1,s)} \dot{u}\dot{\mathbf{D}}_{\mathbf{n}}\dot{G}_{\dot{D}}(\dot{\xi},\cdot)\sin^2\gamma\,dl_N. \end{aligned} \tag{7.3}$$

The function $\dot{G}_{\dot{D}}$ is unique if it exists because if ψ defined on $\dot{D}(t_1,t_2)$ is the difference between two functions with the properties (a)–(c), for fixed $\dot{\xi}$ the function ψ is coparabolic on $\dot{D}(t_1,t_2)$ with boundary function 0 except possibly at the points of $\dot{D}(t_1)$. The maximum-minimum theorem for parabolic functions as dualized for coparabolic functions and applied to ψ implies that ψ vanishes identically.

The restriction of $\dot{G}_{\dot{D}}(\dot{\xi},\cdot)$ to the set of points of $\dot{D}$ strictly below $\dot{\xi}$ is coparabolic, and by the maximum-minimum theorem it follows that

$\dot{G}_{\dot{D}}(\dot{\xi},\cdot) \geq 0$, so that $\mathbf{\dot{D}_n}\dot{G}_{\dot{D}}(\dot{\xi},\cdot) \leq 0$ in (7.3). When $\dot{u} \equiv 1$ in (7.3), the equation reduces to

$$1 = \int_{\dot{D}(t_1)} \dot{G}_{\dot{D}}(\dot{\xi},\cdot)\,dl_N - \frac{\sigma^2}{2}\int_{\partial\dot{D}(t_1,t)} \mathbf{\dot{D}_n}\dot{G}_{\dot{D}}(\dot{\xi},\cdot)\sin^2\gamma\,dl_N. \tag{7.4}$$

Thus, if $\dot{u}$ is parabolic, (7.3) exhibits $\dot{u}$ on $\dot{D}(t_1, t_2)$ as a weighted average of its values on the boundary. This weighting is the analog in the parabolic context of harmonic measure and will therefore be called *parabolic measure*. Observe that the parabolic measure relative to $\dot{\xi}$ assigns value 0 to the part of the boundary above $\dot{\xi}$. This property will be proved in the general case (Section XVIII.2) when parabolic measure is defined on the boundary of an arbitrary open subset $\dot{D}$ of $\dot{\mathbb{R}}^N$.

8. The Green Function of an Interval

Suppose that $N = 1$ and that $\dot{B}$ is the infinite strip $]a,b[\, \times \mathbb{R}$, and define $c = b - a$. The Green function of $\dot{B}$ in the present context, like the classical Green function of a ball, is found by the use of reflections in the boundary, the *method of images*. Consider the series

$$\sum_{n=-\infty}^{\infty} \ell(s - t, 2nc - \xi + \eta), \qquad \dot{\xi} = (\xi, s), \qquad \dot{\eta} = (\eta, t). \tag{8.1}$$

If $|\xi - \eta| < 2kc$, if $|n| > k$, and if $s > t$,

$$\ell(s - t, 2nc - \xi + \eta) \leq [2\pi\sigma^2(s - t)]^{-1/2}\exp\frac{-(2|n| - k)^2c^2}{2\sigma^2(s - t)} \leq \frac{\sigma(s - t)^{1/2}}{c^2n^2}. \tag{8.2}$$

Thus after dropping a finite number of summands, the series (8.1) and similarly the series of partial derivatives of each order with respect to ξ, η, s, t converge uniformly on bounded subsets of $\dot{\mathbb{R}} \times \dot{\mathbb{R}}$. For fixed $\dot{\eta}$ $[\dot{\xi}]$ in $\dot{B}$ each term of the series (8.1) except $\ell(s - t, -\xi + \eta)$, the term with $n = 0$, defines a parabolic [coparabolic] function of $\dot{\xi}$ $[\dot{\eta}]$ on $\dot{B}$. Similarly, after dropping a finite number of summands, the series

$$\sum_{n=-\infty}^{\infty} \ell(s - t, 2nc + 2a - \xi - \eta) \tag{8.3}$$

and each derived series converge uniformly on bounded subsets of $\dot{\mathbb{R}} \times \dot{\mathbb{R}}$. The sum for fixed $\dot{\eta}$ $[\dot{\xi}]$ in $\dot{B}$ defines a parabolic [coparabolic] function of $\dot{\xi}$ $[\dot{\eta}]$ on $\dot{B}$. The function of $(\dot{\xi}, \dot{\eta})$ on $\dot{B} \times \dot{B}$ defined by

$$\phi(a,b)(\dot{\xi},\dot{\eta}) = \sum_{n=-\infty}^{\infty} [\ell(s-t, 2nc - \xi + \eta) - \ell(s-t, 2nc + 2a - \xi - \eta)] \tag{8.4}$$

is, for $\dot{\xi}$ fixed in $\dot{B}$, a function on $\dot{B}$ which differs from the function $\dot{\eta} \mapsto \ell(s-t, \xi-\eta)$ by a function coparabolic on $\dot{B}$, and $\phi(a,b)(\dot{\xi},\cdot)$ has limit 0 at every finite point of the boundary and vanishes when $t \geq s$. It is natural to accept $\phi(a,b)$ as $\dot{G}_{\dot{B}}$, the Green function of $\dot{B}$. Now choose $t_1 < t_2$ and define $\dot{D} =]a,b[\times]t_1,t_2[$. Let $\phi(a,b,t_1,t_2)$ be the restriction of $\phi(a,b)$ to $\dot{D} \times \dot{D}$. For $\dot{\xi}$ in $\dot{D}$ the function $\phi(a,b,t_1,t_2)(\dot{\xi},\cdot)$ differs from the function $\dot{\eta} \mapsto \ell(s-t, \eta-\xi)$ by a function coparabolic on $\dot{D}$, has limit 0 at every lateral boundary point of $\dot{D}$, and vanishes when $t \geq s$. Hence we accept $\phi(a,b,t_1,t_2)$ as the Green function $\dot{G}_{\dot{D}}$ of $\dot{D}$. These definitions of $\dot{G}_{\dot{B}}$ and $\dot{G}_{\dot{D}}$ are in agreement with the definition of the Green function of an arbitrary open subset of $\dot{\mathbb{R}}^N$ to be given in Section XVII.4.

If $N \geq 1$, if $\dot{D} =]a_1,b_1[\times \cdots \times]a_N,b_N[\times]t_1,t_2[$, and if $c_j = b_j - a_j$, the Green function $\dot{G}_{\dot{D}}$ is defined by

$$\dot{G}_{\dot{D}}(\dot{\xi},\dot{\eta}) = \prod_{j=1}^{N} \sum_{n=-\infty}^{\infty} [\ell(s-t, 2nc_j - \xi^{(j)} + \eta^{(j)}) - \ell(s-t, 2nc_j + 2a_j - \xi^{(j)} - \eta^{(j)})] \tag{8.5}$$

for $\dot{\xi} = (\xi^{(1)},\ldots,\xi^{(N)},s), \dot{\eta} = (\eta^{(1)},\ldots,\eta^{(N)},t)$, and $-\infty \leq t_1 < t_2 \leq +\infty$. Observe that the Green function $\dot{G}_{\dot{D}}$ is the restriction to $\dot{D} \times \dot{D}$ of the Green function of $]a_1,b_1[\times \cdots \times]a_N,b_N[\times]t_1',t_2'[$ for $-\infty \leq t_1' \leq t_1$ and $t_2 \leq t_2' \leq +\infty$. An application of the coparabolic function minimum theorem to $\dot{G}_{\dot{D}}(\dot{\xi},\cdot)$ on $\dot{D} - \{\dot{\xi}\}$ shows that $\dot{G}_{\dot{D}}(\dot{\xi},\dot{\eta})$ is strictly positive when $\operatorname{ord} \dot{\eta} < \operatorname{ord} \dot{\xi}$ and vanishes otherwise.

9. Parabolic Measure for an Interval

The Green function of an interval $\dot{D}$ is given by (8.5). Suppose for the rest of this section that the interval is finite, that is, that $-\infty < t_1 < t_2 < +\infty$. Since Section 7 is applicable to a finite interval, the parabolic measure $\dot{\mu}_{\dot{D}}$ can be written explicitly. In fact, for $\dot{\xi}$ in $\dot{D}$ the measure $\dot{\mu}_{\dot{D}}(\dot{\xi},\cdot)$ is supported by the part of $\partial\dot{D}$ strictly below $\dot{\xi}$, and this measure is absolutely continuous relative to l_N with finite continuous density $\dot{G}_{\dot{D}}(\dot{\xi},\cdot)$ on the lower boundary and finite continuous density

$$\dot{\eta} \mapsto -\frac{\sigma^2}{2}\frac{\partial}{\partial\eta^{(j)}}\dot{G}_{\dot{D}}(\dot{\xi},\dot{\eta})$$

for $\dot{\eta}$ on the part of the lateral boundary with jth coordinate b_j and with density the negative of this derivative for $\dot{\eta}$ on the part of the lateral boundary with jth coordinate a_j. This statement is to be understood to mean that on the l_N null set of points common to the lateral and lower boundaries where the statement gives more than one possible value for the density, either value can be used. For example, if $N = 1$ and if $\dot{D} =]a, b[\times]t_1, t_2[$, $c = b - a$, $\dot{\xi} = (\xi, s)$, and $\dot{\eta} = (\eta, t)$, the density is given by

$$\dot{G}_{\dot{D}}(\dot{\xi}, (\eta, t_1)) = \sum_{n=-\infty}^{\infty} [b(s - t_1, 2nc + \eta - \xi) - b(s - t_1, 2nc + 2a - \xi - \eta)], \tag{9.1}$$

$$-\frac{\sigma^2}{2}\mathbf{D}_{\mathbf{n}_{\dot{\xi}}}\dot{G}_{\dot{D}}(\dot{\xi}, \dot{\eta}) = \begin{cases} \dfrac{1}{s - t}\displaystyle\sum_{n=-\infty}^{\infty} (2nc + b - \xi)\mathscr{E}(s - t, 2nc + b - \xi) & \text{if } \eta = b, \\ \dfrac{1}{s - t}\displaystyle\sum_{n=-\infty}^{\infty} (2nc + a - \xi)\mathscr{E}(s - t, 2nc + a - \xi) & \text{if } \eta = a. \end{cases}$$

According to Section 7, if $\dot{u}$ is a parabolic function on a neighborhood of $\bar{\dot{D}}$, then $\dot{u} = \dot{\mu}_{\dot{D}}(\cdot, \dot{u})$ on $\dot{D}$. Furthermore, if f is an l_N measurable and integrable function on the union of the lateral and lower boundaries of $\dot{D}$, the function $\dot{\mu}_{\dot{D}}(\cdot, f)$ is parabolic on $\dot{D}$, by direct differentiation. The function $\dot{\mu}_{\dot{D}}(\cdot, f)$ will sometimes be called the *parabolic Poisson integral* of f and will accordingly sometimes be denoted by $\dot{\mathrm{PI}}(\dot{D}, f)$. If $\dot{D}_1$ is an interval obtained from $\dot{D}$ by raising the upper boundary, that is, by increasing t_2, and if f is extended to the lateral and lower boundaries of $\dot{D}_1$ by defining f as 0 at the added lateral boundary points, then $\dot{\mathrm{PI}}(\dot{D}, f) = \dot{\mathrm{PI}}(\dot{D}_1, f)$ on $\dot{D}$; so $\dot{\mathrm{PI}}(\dot{D}, f)$ has a parabolic extension to $\dot{D}_1$. It follows that $\dot{\mathrm{PI}}(\dot{D}, f)$ has a limit at every inner point of the upper boundary of $\dot{D}$. For fixed $\dot{\eta}$ on the lateral or lower boundary of $\dot{D}$ the parabolic measure density has limit 0 when $\dot{\xi}$ tends to a point of the lateral or lower boundary of $\dot{D}$ other than $\dot{\eta}$. Apply Theorem 1 of Appendix VII to see that if $\dot{\eta}$ is a lateral or lower boundary point of $\dot{D}$, then

$$\limsup_{\dot{\xi} \to \dot{\eta}} \dot{\mu}_{\dot{D}}(\dot{\xi}, f) \le \limsup_{\dot{D} \ni \dot{\xi} \to \dot{\eta}} f(\dot{\xi})$$

and thereby to see that if f is finite valued and continuous, the function $\dot{\mu}_{\dot{D}}(\cdot, f)$ has boundary limit $f(\dot{\eta})$ at every point $\dot{\eta}$ of the lateral or lower boundary. The inner points of the upper boundary of $\dot{D}$ are to be considered as irregular boundary points for the first boundary value problem for parabolic functions. (See Chapter XVIII for a discussion of the first boundary value problem in the parabolic context.)

The Operation $\dot{\tau}_{\dot{B}}$. If $\dot{u}$ is a Borel measurable function on an open subset $\dot{D}$ of $\dot{\mathbb{R}}^N$, if $\dot{B}$ is an interval with closure in $\dot{D}$, and if the restriction of $\dot{u}$ to the lower and lateral boundaries of $\dot{B}$ is l_N integrable, define

$$\dot{\tau}_{\dot{B}}\dot{u} = \begin{cases} \dot{\mathrm{PI}}(\dot{B}, \dot{u}) & \text{on } \dot{B} \\ \dot{u} & \text{on } \dot{D} - \dot{B} \end{cases}$$

except that if $\dot{\eta}$ is an inner point of the upper boundary of $\dot{B}$, define

$$\dot{\tau}_{\dot{B}}\dot{u}(\dot{\eta}) = \lim_{\substack{\dot{\xi}\to\dot{\eta} \\ \operatorname{ord}\dot{\xi}<\operatorname{ord}\dot{\eta}}} \dot{\mathrm{P}}\mathrm{I}(\dot{B},\dot{u})(\dot{\xi}).$$

If $\dot{u}$ is upper or lower semicontinuous, the above remarks imply that $\dot{\tau}_{\dot{B}}\dot{u}$ has the same property except possibly at the inner points of the upper boundary of $\dot{B}$ and that $\dot{\tau}_{\dot{B}}\dot{u}$ is parabolic on $\dot{B}$ with a finite limit from below at each inner point of the upper boundary of $\dot{B}$.

10. Parabolic Averages

Let $\dot{\xi}:(\xi,s)$ be a point of $\dot{\mathbb{R}}^N$, and when $\delta>0$, let $\dot{B}(\dot{\xi},\delta)$ be the interval

$$]\xi^{(1)}-\delta,\xi^{(1)}+\delta[\times\cdots\times]\xi^{(N)}-\delta,\xi^{(N)}+\delta[\times]s-\delta^2,s[$$

in $\dot{\mathbb{R}}^N$. The interval $\dot{B}(\dot{\xi},\delta)$ will play the same role in the study of parabolic potential theory as the ball $B(\xi,\delta)$ in the classical theory. In the following we take $N=1$. The added complications in the general case are merely notational, and the results will be valid, and applied, for all N. Define the function f_1 on the one-dimensional interval $[-1,1]$ by

$$f_1(\xi) = \sum_{-\infty}^{\infty}[\mathscr{E}(1,4n+\xi)-\mathscr{E}(1,4n-2-\xi)], \tag{10.1}$$

and define f_2 on $[-1,0]$ by

$$f_2(s) = -\frac{1}{s}\sum_{-\infty}^{\infty}(4n+1)\mathscr{E}(-s,4n+1). \tag{10.2}$$

According to (9.1), the functions $\xi\mapsto f_1(\xi/\delta)/\delta$ and $s\mapsto f_2(s/\delta^2)/\delta^2$ are respectively the densities relative to l_1 of $\dot{\mu}_{\dot{B}(0,\delta)}(0,\cdot)$ on the lower and lateral boundaries of $\dot{B}(0,\delta)$. If $\dot{u}$ is a Borel measurable function on $\partial\dot{B}(\dot{\eta},\delta)$ for which the following integrals exist, we define $L(\dot{u},\dot{\eta},\delta)$ by

$$\begin{aligned}\dot{L}(\dot{u},\dot{\eta},\delta) = \dot{\mu}_{\dot{B}(0,\delta)}(\dot{\eta},\dot{u}) &= \int_{-\delta}^{\delta}\dot{u}(\eta+\xi,t-\delta^2)f_1\left(\frac{\xi}{\delta}\right)\frac{l_1(d\xi)}{\delta} \\ &+\int_{t-\delta^2}^{0}[\dot{u}(\eta-\delta,s)+\dot{u}(\eta+\delta,s)]f_2\left(\frac{s}{\delta^2}\right)\frac{l_1(ds)}{\delta^2} \qquad [\dot{\eta}=(\eta,t)].\end{aligned} \tag{10.3}$$

In order to treat the parabolic analogs of the volume averages in Section I.2, let ϕ be a positive Borel measurable function on $\mathbb{R}^+$ vanishing on $]1,+\infty[$, and consider the integral

$$\int_0^\infty \phi(r)L(\dot{u}, 0, r\delta)l_1(dr) = \frac{1}{\delta}\int_{\dot{B}(0,\delta)} \dot{u}(\dot{\xi})f(\dot{\xi})l_2(d\dot{\xi}), \tag{10.4}$$

where

$$f(\dot{\xi}) = \begin{cases} -\dfrac{1}{2\delta s} f_1(\xi(-s)^{-1/2})\phi\left(\dfrac{(-s)^{1/2}}{\delta}\right) & \text{if } s < -\xi^2, \\[2ex] f_2\left(\dfrac{s}{\xi^2}\right)\dfrac{\phi(|\xi|/\delta)}{\xi^2} & \text{if } s > -\xi^2. \end{cases}$$

When $\phi = 1$ on $[0, 1]$, we denote the value in (10.4) by $\dot{A}(\dot{u}, 0, \delta)$ and define $\dot{A}(\dot{u}, \dot{\eta}, \delta) = \dot{A}(\dot{u}(\dot{\eta} + \cdot), 0, \delta)$. When $\phi(r) = c\exp[r^{-2}(1 - r^2)^{-1}]$ for $0 < r < 1$ and $\phi(0) = \phi(1) = 0$ and c is chosen so that $\int_0^1 \phi(r)l_1(dr) = 1$, we denote the value in (10.4) by $\dot{A}_\delta(\dot{u}, 0)$ and define $\dot{A}_\delta(\dot{u}, \dot{\eta}) = \dot{A}_\delta(\dot{u}(\dot{\eta} + \cdot), 0)$. If $\dot{u}$ is a Borel measurable function defined on an open subset $\dot{D}$ of $\dot{\mathbb{R}}$, the values $\dot{A}(\dot{u}, \dot{\eta}, \delta)$ and $\dot{A}_\delta(\dot{u}, \dot{\eta})$ are defined whenever $\overline{\dot{B}}(\dot{\eta}, \delta) \subset \dot{D}$, that is, whenever $\delta^2 + \delta^4 < |\dot{\eta} - \partial\dot{D}|^2$, if $\dot{u}$ is locally l_2 integrable on $\dot{D}$. Under the latter condition, $\dot{A}(\dot{u}, \cdot, \delta)$ is continuous, and $\dot{A}_\delta(\dot{u}, \cdot)$ is infinitely differentiable.

Application. It is trivial from the definition that if $\dot{u}$ is parabolic on $\dot{D}$, then

$$\dot{u}(\dot{\eta}) = \dot{L}(\dot{u}, \dot{\eta}, \delta) = \dot{A}(\dot{u}, \dot{\eta}, \delta) = \dot{A}_\delta(\dot{u}, \dot{\eta}) \tag{10.5}$$

for δ so small that $\overline{\dot{B}}(\dot{\eta}, \delta) \subset \dot{D}$. Conversely, if $\dot{u}$ is Borel measurable and locally l_2 integrable and if $\dot{u}(\dot{\eta})$ is equal to the third (or fourth) term in (10.5) whenever $\overline{\dot{B}}(\dot{\eta}, \delta) \subset \dot{D}$, then $\dot{u}$ is parabolic. In fact then $\dot{u}$ is finite valued and continuous, and if $\dot{B}$ is an interval with closure in $\dot{D}$, the difference $\dot{v} = \dot{u} - \dot{\mathrm{P}}\mathrm{I}(\dot{B}, \dot{u})$ has the same average property as $\dot{u}$ in $\dot{B}$, and a trivial argument shows that therefore $\dot{v}$ satisfies the maximum-minimum parabolic function Theorem 5. Hence $\dot{v} = 0$ since $\dot{v}$ has limit 0 at every lateral or lower boundary point of $\dot{B}$. Finally, if $\dot{u}$ is supposed Borel measurable on $\dot{D}$, locally l_1 integrable on lines parallel to a coordinate axis, and either bounded on one side or locally l_2 integrable and also if $\dot{u}(\dot{\eta}) = \dot{L}(\dot{u}, \dot{\eta}, \delta)$ whenever $\overline{\dot{B}}(\dot{\eta}, \delta) \subset \dot{D}$, then $\dot{u}$ is parabolic because the hypotheses imply that $\dot{u}(\dot{\eta})$ is equal to the third and fourth terms in (10.5) as well as the first. These criteria for parabolicity will be weakened to be local in Section 14.

11. Harnack's Theorems in the Parabolic Context

(a) Convergence Theorem. *If $\dot{u}_\cdot$ is an upward-directed family of parabolic functions on $\dot{D}$ and if the limit function $\dot{u}$ is finite at a point $\dot{\eta}$, then $\dot{u}$ is parabolic on the open set $\dot{D}_0$ of points of $\dot{D}$ strictly below $\dot{\eta}$ relative to $\dot{D}$, and the convergence is locally uniform on $\dot{D}_0$.*

In fact we can suppose that $\dot{u}_{\cdot}$ is an increasing sequence (Theorem 2 of Appendix VIII). If $\bar{\dot{B}}(\dot{\xi}, \delta) \subset \dot{D}$, integration to the limit yields the equality $\dot{u}(\dot{\xi}) = \dot{A}(\dot{u}, \dot{\xi}, \delta)$; so taking $\dot{\xi} = \dot{\eta}$, we find that $\dot{u}$ is l_{N+1} integrable on $\dot{B}(\dot{\eta}, \delta)$. A covering argument shows that $\dot{u}$ is locally l_{N+1} integrable on $\dot{D}_0$, and therefore according to the application in Section 10, the function $\dot{u}$ is parabolic on $\dot{D}_0$. Dini's theorem implies that the convergence is locally uniform on $\dot{D}_0$.

(b) Inequality Theorem. *Let $\dot{D}$ be an open subset of $\dot{\mathbb{R}}^N$, and let $\dot{\lambda}$ be a measure on $\dot{D}$ with minimal closed in $\dot{D}$ support $\dot{S}$. Let $\dot{A}$ be a compact subset of $\dot{D}$ for which to each point of $\dot{A}$ there corresponds a point in $\dot{S}$ over the first point relative to $\dot{D}$. Then there is a constant c depending only on $\dot{D}$, $\dot{S}$, $\dot{A}$, $\dot{\lambda}$ such that if $\dot{u}$ is a positive parabolic function on $\dot{D}$,*

$$\max_{\dot{A}} \dot{u} \leq c \int_{\dot{S}} \dot{u}\, d\dot{\lambda}. \tag{11.1}$$

In fact, if there is no such number c, then for some choice of $\dot{D}$, $\dot{S}$, $\dot{A}$, $\dot{\lambda}$ there corresponds to each positive integer n a positive parabolic function $\dot{u}_n$ on $\dot{D}$ such that $\max_{\dot{A}} \dot{u}_n \geq 1$ but $\int_{\dot{S}} \dot{u}_n\, d\dot{\lambda} \leq 2^{-n}$. The series $\Sigma_0^\infty \dot{u}_n$ converges $\dot{\lambda}$ almost everywhere on $\dot{S}$, and the set of points of convergence is therefore dense in $\dot{S}$. It follows from the above convergence theorem that $\Sigma_0^\infty \dot{u}_n$ converges uniformly on $\dot{A}$; so the sequence $\dot{u}_{\cdot}$ converges uniformly to 0 on $\dot{A}$, contrary to hypothesis.

Special Case. If $\dot{\lambda}$ is supported by a singleton $\{\dot{\xi}\}$, Harnack's inequality states that to each compact subset $\dot{A}$ of $\dot{D}$, all of whose points are strictly below $\dot{\xi}$ relative to $\dot{D}$, corresponds a constant c, depending only on $\dot{D}$, $\dot{\xi}$, $\dot{A}$, such that if $\dot{u}$ is a positive parabolic function on $\dot{D}$, then

$$\max_{\dot{A}} \dot{u} \leq c\dot{u}(\dot{\xi}). \tag{11.2}$$

12. Superparabolic Functions

A function $\dot{u}$ from an open subset $\dot{D}$ of $\dot{\mathbb{R}}^N$ into $]-\infty, +\infty]$ is called *superparabolic* if

(a) $\dot{u}$ is lower semicontinuous.
(b) $\dot{u}$ is finite on a dense subset of $\dot{D}$.
(c) $\dot{u}(\dot{\xi}) \geq \dot{L}(\dot{u}, \dot{\xi}, \delta)$ if $\bar{\dot{B}}(\dot{\xi}, \delta) \subset \dot{D}$.

Just as in the classical context (Section II.4), $\dot{u}$ is locally bounded below and $\dot{u}(\dot{\xi}) \geq \dot{A}(\dot{u}, \dot{\xi}, \delta)$ when $\bar{\dot{B}}(\dot{\xi}, \delta) \subset \dot{D}$, and it follows that a superparabolic function is locally l_{N+1} integrable and therefore is finite l_{N+1} almost everywhere on its domain. If u_1 and u_2 are superparabolic functions and if c_1 and c_2 are positive constants, then $c_1u_1 + c_2u_2$ is superparabolic.

A *subparabolic* function is defined as the negative of a superparabolic function, and it follows that a function is parabolic if and only if it is both superparabolic and subparabolic. A *cosuperparabolic* function is defined as a function on an open set $\dot{D}$ for which the function $(\xi, s) \mapsto \dot{u}(\xi, -s)$ is superparabolic on the reflection of $\dot{D}$ in the abscissa hyperplane, and a *cosubparabolic* function is defined as the negative of a cosuperparabolic function.

Smooth Superparabolic Functions

If $\dot{u}$ is a $\mathbb{C}^{(1)}$ function on an open subset of $\dot{\mathbb{R}}^N$ and is $\mathbb{C}^{(2)}$ in the space variables, then $\dot{u}$ is superparabolic if and only if $\dot{\Delta}\dot{u} \leq 0$, in view of (7.3) with $\dot{D}$ an interval.

Application of Jensen's Inequality

The application of Jensen's inequality in the classical context (Section II.9) is carried through in exactly the same way in the present context. For example, a convex function of a parabolic function is subparabolic.

EXAMPLE (a). The function $\dot{u}$: $(\xi, s) \mapsto c|\xi|^2 + s$ is superparabolic on $\dot{\mathbb{R}}^N$ if $c \leq 1(\sigma^2 N)$, subparabolic if $c \geq 1/(\sigma^2 N)$, and parabolic if $c = 1/(\sigma^2 N)$, because $\dot{\Delta}\dot{u} = \sigma^2 cN - 1$.

EXAMPLE (b). If f is a monotone increasing left continuous function on $\mathbb{R}$ and if $\dot{u}(\eta, t) = f(t)$ for (η, t) in $\dot{\mathbb{R}}^N$, the function $\dot{u}$ is superparabolic on $\dot{\mathbb{R}}^N$. If in addition $\lim_{t \to \infty} f(t) = \alpha > -\infty$, then the function $\dot{u}$ is the sum of the constant parabolic function α and of a superparabolic potential on $\dot{\mathbb{R}}^N$, the potential $\dot{G}\dot{\mu}$ of the product measure $\dot{\mu} = l_N \times \nu$, with $d\nu = df$.

Convergence of Families of Superparabolic and Parabolic Functions

In view of the parabolic Poisson integral for an interval it is clear that if $\dot{D}$ is an open subset of $\dot{\mathbb{R}}^N$, if $k \in \mathbb{Z}^+$, and if $\dot{u}_{\bullet}$ is a locally uniformly bounded family of parabolic functions on $\dot{D}$, the family of partial derivatives of these functions of order $\leq k$ is also a locally uniformly bounded family of parabolic functions. It follows that $\dot{u}_{\bullet}$ is locally uniformly equicontinuous. Furthermore, if a sequence of parabolic functions on $\dot{D}$ converges locally uniformly, the corresponding sequence of partial derivatives of any prescribed order also converges locally uniformly to the corresponding partial derivative of the limit function. Thus the limit function is also parabolic. An application of Ascoli's theorem shows that a locally uniformly bounded sequence of parabolic functions has a locally uniformly convergent subsequence.

Let $\dot{u}_\bullet$ be an upward-directed family of superparabolic functions on $\dot{D}$ with limit $\dot{u}$. Since $\dot{u}_\bullet$ is an upward-directed family of lower semicontinuous functions, there is (by Theorem 2 of Appendix VIII) an increasing sequence in the family with limit $\dot{u}$. The function $\dot{u}$ has the superparabolic average property and is therefore superparabolic if it is finite on a dense subset of $\dot{D}$; the latter condition is satisfied if every point of $\dot{D}$ is below, relative to $\dot{D}$, a point of finiteness of $\dot{u}$.

Positive Integral Operations on Superparabolic Functions. Such operations yield superparabolic functions and yield parabolic functions if the given functions are parabolic. (See Section II.4 for the argument in the classical context.) For example, if $\dot{u}$ is superparabolic on an open set $\dot{D}$, the functions $\dot{L}(\dot{u}, \cdot, \delta)$, $\dot{A}(\dot{u}, \cdot, \delta)$, and $\dot{A}_\delta \dot{u}$ are superparabolic on their domains of definition.

13. Superparabolic Function Minimum Theorem

Theorem. *Let $\dot{u}$ be a superparabolic function on an open subset $\dot{D}$ of $\dot{\mathbb{R}}^N$.*

(a) *If $\dot{u}$ attains its infimum at a point $\dot{\xi}$ of $\dot{D}$, then $\dot{u}$ is identically that infimum on the set of points below $\dot{\xi}$ relative to $\dot{D}$.*
(b) *The infimum of $\dot{u}$ is the limit of $\dot{u}$ along some sequence of points tending to $\partial\dot{D}$.*
(c) *Any lower semicontinuous extension of $\dot{u}$ to $\bar{\dot{D}}$ attains its infimum on the boundary.*

Assertion (a) is an easy consequence of the fact that $\dot{u}$ is lower semicontinuous and that $\dot{u}(\dot{\xi}) \geq \dot{A}(\dot{u}, \dot{\xi}, \delta)$ whenever $\bar{\dot{B}}(\dot{\xi}, \delta) \subset \dot{D}$. Assertions (b) and (c) follow from (a). The dual version of this theorem for subparabolic functions will be called the *subparabolic maximum theorem*. Observe that Theorem 13 includes Theorem 5.

Application to Functions on Slabs

A nonempty open subset of $\dot{\mathbb{R}}^N$ that is either a half-space bounded by a horizontal hyperplane or is the intersection of two such half-spaces will be called a *slab*. If a slab $\dot{D}$ has a lower hyperplane boundary and if $\dot{u}$ is a superparabolic function on $\dot{D}$, then the infimum of $\dot{u}$ is the limit of $\dot{u}$ along some sequence tending either to the point ∞ or to a point of the lower hyperplane boundary. To show this we show that if $\dot{D}(\dot{\xi})$ is the part of $\dot{D}$ strictly below $\dot{\xi}$ and if $m(\dot{\xi})$ is the infimum of $\dot{u}$ on $\dot{D}(\dot{\xi})$, then there is a sequence $\dot{\eta}_\bullet$ in $\dot{D}(\dot{\xi})$ tending to the point ∞ or to a point of the lower boundary of $\dot{D}$ along which $\dot{u}$ tends to $m(\dot{\xi})$. According to the superparabolic minimum theorem, there is a sequence $\dot{\eta}_\bullet$ in $\dot{D}(\dot{\xi})$ tending to a point $\dot{\eta}$ of $\partial\dot{D}(\dot{\xi})$ along which u tends to $m(\dot{\xi})$. Unless $\dot{\eta}$ is on the upper boundary of $\dot{D}(\dot{\xi})$, we are done. If $\dot{\eta}$ is on the upper boundary, then (by the lower semicontinuity of $\dot{u}$) $u(\dot{\eta}) \leq m(\dot{\xi})$, and a trivial variation of the proof of the superparabolic minimum theorem shows that $\dot{u}$ is identically $m(\dot{\xi})$ on $\dot{D}(\dot{\xi})$. Hence there is a sequence $\dot{\eta}_{\bullet}$ with the desired properties.

14. The Operation $\dot{\tau}_{\dot{B}}$ and the Defining Average Properties of Superparabolic Functions

We follow the reasoning of Section II.6 to derive the corresponding results in the present context.

(a) If $\dot{u}$ is superparabolic on $\dot{D}$ and if $\dot{B}$ is an interval with closure in $\dot{D}$, then $\dot{u}$ is locally l_N integrable on the intersection with $\dot{D}$ of any hyperplane parallel to a coordinate hyperplane, $\dot{\tau}_{\dot{B}}\dot{u}$ is parabolic on $\dot{B}$, and $\dot{\tau}_{\dot{B}}\dot{u} \leq \dot{u}$. In fact, by lower semicontinuity and local lower boundedness of $\dot{u}$ there is an increasing sequence $f_{\bullet}$ of finite continuous functions of $\partial\dot{D}$ with limit $\dot{u}$ there. Then $\dot{\text{P}}\text{I}(\dot{B}, f_{\bullet})$ is an increasing sequence of parabolic functions on $\dot{B}$ with limit $\dot{\text{P}}\text{I}(\dot{B}, \dot{u})$, and an application of the superparabolic minimum theorem to $\dot{u}$-$\dot{\text{P}}\text{I}(\dot{B}, f_n)$ on $\dot{B}$ shows that this difference is positive on $\dot{B}$. Hence (from Section 11) $\dot{\tau}_{\dot{B}}\dot{u}$ is parabolic on $\dot{B}$ and $\dot{\tau}_{\dot{B}}\dot{u} \leq \dot{u}$. Let $\dot{\xi}$ be a point of $\dot{B}$. The foregoing proof shows that $\dot{u}$ is l_N integrable on any Borel subset of $\partial\dot{B}$ on which the derivative of $\dot{\mu}_{\dot{B}}(\dot{\xi}, \cdot)$ with respect to l_N has a strictly positive infimum. Trivial adjustments of $\dot{\xi}$ and $\dot{B}$ now show that $\dot{u}$ is l_N integrable on $\partial\dot{B}$; in fact it is locally l_N integrable on the intersection with $\dot{D}$ of any hyperplane parallel to a coordinate hyperplane. Since rotations around a vertical axis preserve parabolicity, $\dot{u}$ is locally l_N integrable on the intersection with $\dot{D}$ of any hyperplane parallel to the ordinate axis.

(b) In (a) if $\dot{\tau}_{\dot{B}}\dot{u} = \dot{u}$ at a point $\dot{\xi}$ of $\dot{B}$, then there is equality at the points of $\dot{B}$ below $\dot{\xi}$ because $\dot{u} - \dot{\tau}_{\dot{B}}\dot{u}$ is a positive superparabolic function on $\dot{B}$.

(c) In the definition of superparabolic function in Section 12, condition (c) can be replaced by

(c′) $$\dot{u}(\dot{\xi}) \geq \dot{A}(\dot{u}, \dot{\xi}, \delta)$$

or

(c″) $$\dot{u}(\dot{\xi}) \geq \dot{A}_{\delta}\dot{u}(\dot{\xi})$$

[in both cases for δ so small that $\overline{\dot{B}}(\dot{\xi}, \delta) \subset \dot{D}$].

In fact it is trivial that superparabolic functions satisfy (c′) and (c″), and the proof of the converse follows the corresponding proof in the classical context [Section II.6(c)].

We can now proceed precisely as in Section II.6 (so proofs will be omitted) to obtain the following results.

(d) In the definition of superparabolic functions in Section 12, condition (c) [or (c′) or (c″)] need only be supposed true locally, that is, for sufficiently small δ, depending on $\dot{\xi}$.

(e) In (a) the function $\dot{\tau}_{\dot{B}}\dot{u}$ is superparabolic on $\dot{D}$.

(f) If $\dot{u}$ is superparabolic on $\dot{D}$ and if $\dot{\xi} \in \dot{D}$, the functions $\delta \mapsto \dot{L}(\dot{u}, \dot{\xi}, \delta)$, $\delta \mapsto \dot{A}(\dot{u}, \dot{\xi}, \delta)$, $\delta \mapsto \dot{A}_{\delta}\dot{u}(\dot{\xi})$ are monotone decreasing, with limit $\dot{u}(\dot{\xi})$ when $\delta \to 0$.

It follows that the relation $\dot{u} \geq \dot{v}$ or $\dot{u} = \dot{v}$, if satisfied l_{N+1} almost everywhere on their domain of definition $\dot{D}$ by superparabolic functions $\dot{u}$ and $\dot{v}$, is satisfied everywhere on $\dot{D}$. Moreover, if $\dot{u}$ is superparabolic, then

$$\dot{u}(\dot{\xi}) = \liminf_{\dot{\eta} \to \dot{\xi}} \dot{u}(\dot{\eta}) = \liminf_{\substack{\dot{\eta} \to \dot{\xi} \\ \dot{\eta} \notin \dot{B}}} \dot{u}(\dot{\eta}),$$

where $\dot{B}$ is an arbitrary l_{N+1} null set.

(g) If $\dot{u}$ is a lower semicontinuous function from an open subset $\dot{D}$ of $\dot{\mathbb{R}}^N$ into $]-\infty, +\infty]$, finite on a dense subset of $\dot{D}$, then $\dot{u}$ is superparabolic if and only if whenever $\dot{D}_0$ is an open relatively compact subset of $\dot{D}$ and $\dot{v}$ is parabolic on an open neighborhood of $\dot{D}_0$, with $\dot{u} - \dot{v} > 0$ on $\partial\dot{D}_0$, then $\dot{u} - \dot{v} \geq 0$ on $\dot{D}_0$.

Approximation of a Superparabolic Function by Infinitely Differentiable Superparabolic Functions

According to (f), if $\dot{u}$ is superparabolic on an open set $\dot{D}$, this function is the limit on each open relatively compact subset of $\dot{D}$ of an increasing sequence $\{\dot{A}_{1/n}\dot{u}, n \geq 1\}$ of infinitely differentiable superparabolic functions. See the strengthening of this result in Section XVII.7(e).

Example (The Green Function $\dot{G}$ of $\dot{\mathbb{R}}^N$). For fixed $\dot{\eta}$ in $\dot{\mathbb{R}}^N$ the function $\dot{G}(\cdot, \dot{\eta})$ is parabolic on $\dot{\mathbb{R}}^N - \{\dot{\eta}\}$ and therefore satisfies the parabolic function average equality there. Since $\dot{G}(\cdot, \dot{\eta})$ is lower semicontinuous on $\dot{\mathbb{R}}^N$ and satisfies the superparabolic function average inequality at $\dot{\eta}$, this function is superparabolic on $\dot{\mathbb{R}}^N$. Dually, for fixed $\dot{\xi}$ in $\dot{\mathbb{R}}^N$ the function $\dot{G}(\dot{\xi}, \cdot)$ is coparabolic on $\dot{\mathbb{R}}^N - \{\dot{\xi}\}$ and cosuperparabolic on $\dot{\mathbb{R}}^N$.

15. Superparabolic and Parabolic Functions on a Cylinder

Let D be an open nonempty subset of $\mathbb{R}^N (N \geq 1)$, and define $\dot{D} = D \times \mathbb{R}$. The following results (a) and (b) are rather trivial but key results in the relations between [super] harmonic and [super] parabolic functions.

(a) Let u be a function on D, and define $\dot{u}$ on $\dot{D}$ by setting $\dot{u}(\xi, s) = u(\xi)$. Then $\dot{u}$ is [super] parabolic if and only if u is [super] harmonic. If u is of class $\mathbb{C}^{(2)}$, the assertion is trivial because $\dot{\Delta}\dot{u} = \Delta u$. In the general case we need only discuss superharmonic functions u and superparabolic functions $\dot{u}$. If u is superharmonic, u is (Section IV.10) the limit of an increasing sequence of infinitely differentiable superharmonic functions; so $\dot{u}$ is the limit of an increasing sequence of superparabolic functions and therefore is superparabolic. Conversely, if $\dot{u}$ is superparabolic, let D_0 be an open relatively compact subset of D. For sufficiently small δ the function $\dot{A}_\delta\dot{u}$ is defined on $D_0 \times \mathbb{R}$, is infinitely differentiable and superparabolic, and depends only on the space coordinate, $\dot{A}_\delta\dot{u}(\xi, s) = u_\delta(\xi)$. Hence u_δ is superharmonic on D_0, and $u = \lim_{\delta\to 0} u_\delta$ is also superharmonic on D_0 and therefore is superharmonic on D.

(b) Suppose that D is connected, let $\dot{v}$ be a positive superparabolic function on $\dot{D}$, and define v on D by the following positive integral operation on the family of time translates of $\dot{v}$:

$$v(\xi) = \int_{-\infty}^{+\infty} \dot{v}(\xi, t) l_1(dt) = \int_{-\infty}^{+\infty} \dot{v}(\xi, s+t) l_1(dt). \tag{15.1}$$

The function v is lower semicontinuous (Fatou's lemma); so considered as a function on $\dot{D}$, v is superparabolic if finite at points with arbitrarily large ordinate values, as is true unless $v \equiv +\infty$. According to (a) above it follows that v is either identically $+\infty$ or is superharmonic on D. In particular, if $\dot{v}$ is parabolic, the function v is either identically $+\infty$ or harmonic on D. For example, we shall show that the Green function $\dot{G}_{\dot{D}}$ of $\dot{D}$ (parabolic context) has the form $\dot{G}_{\dot{D}}((\xi, s), (\eta, t)) = \mathscr{E}_D(s - t, \xi, \eta)$, where for fixed η in D the function $(\xi, s) \mapsto \mathscr{E}_D(s, \xi, \eta)$ is a positive parabolic function on $\dot{D} - \{(\eta, 0)\}$, vanishing if and only if $s \le 0$, and is superparabolic on $\dot{D}$; according to XVII(18.2), integration of $\mathscr{E}_D$ yields G_D, the Green function of D in the classical context,

$$\int_{-\infty}^{+\infty} \mathscr{E}_D(t, \xi, \eta) l_1(dt) = a_N G_D(\xi, \eta).$$

Here a_N is a positive constant.

16. The Appell Transformation

If α is a nonzero constant, the map

$$\dot{\eta}: (\eta, t) \mapsto T\dot{\eta} = \left(\frac{\alpha\eta}{t}, \frac{-\alpha^2}{t}\right)$$

takes the upper [lower] half-space of $\dot{\mathbb{R}}^N$ in a one-to-one way onto the lower [upper] half-space, and $T^{-1}\dot{\eta} = (-\alpha\eta/t, -\alpha^2/t)$. Define

$$b_0(\dot\eta) = (2\pi\sigma^2|t|)^{-N/2}\exp\left(\frac{-|\eta|^2}{2\sigma^2 t}\right)$$

for $\dot\eta$ in $\dot{\mathbb{R}}^N$ less the abscissa hyperplane. If $\dot u$ is a function with domain a subset $\dot B$ of $\dot{\mathbb{R}}^N$ not meeting the abscissa hyperplane, define $T\dot u$ on $T\dot B$ by $T\dot u(\dot\eta) = b_0(\dot\eta)\dot u(T^{-1}\dot\eta)$, so that formally

$$\dot\Delta T\dot u(\dot\eta) = \frac{\alpha^2}{t^2} b_0(\dot\eta)(\dot\Delta\dot u)(T^{-1}\dot\eta) = \frac{\alpha^2}{t^2} T(\dot\Delta\dot u)(\dot\eta)$$

on $T\dot B$. We conclude that if $\dot u$ is parabolic on $\dot B$, then $T\dot u$ is parabolic on $T\dot B$, and if $\dot u$ is superparabolic on $\dot B$ and in class $\mathbb{C}^{(2)}(\dot B)$, then $T\dot u$ is superparabolic on $T\dot B$. A trivial approximation argument then shows that the $\mathbb{C}^{(2)}(\dot B)$ hypothesis is unnecessary. The transformation T is known as the *Appell transformation*.

EXAMPLE. If $\gamma \in \mathbb{R}^N$ and $\alpha = -1/\sigma^2$, the Appell transformation of the restriction of the function $\dot\eta \mapsto \ell(t, \eta - \gamma)$ to the upper half-space of $\dot{\mathbb{R}}^N$ is the function

$$\dot\eta \mapsto (2\pi)^{-N}\exp\left(<\gamma, \eta> + \sigma^2|\gamma|^2\frac{t}{2}\right)$$

on the lower half-space. This parabolic function [without the factor $(2\pi)^{-N}$] was noted in Section 2 and will be seen in Section XVI.8 to play the same role in the lower half-space for Poisson-integral-type representations and minimal parabolic functions that $\ell(\cdot, \cdot - \gamma)$ plays in the upper half-space.

17. Extensions of a Parabolic Function Defined on a Cylinder

Lemma. *Let D be an open nonempty subset of $\mathbb{R}^N$ ($N \geq 1$), let $\dot D = D \times]a, b[$ with $-\infty \leq a < b \leq +\infty$, and let $\dot v$ be a positive parabolic function on $\dot D$. Then if $a < b' < b$ and if $\dot v'$ is a positive parabolic function on $\dot D' = D \times]a, b'[$, majorized there by $\dot v$, the function $\dot v'$ has a positive parabolic extension $\dot v''$ to $\dot D$, majorized there by $\dot v$.*

Observation. The lemma does not assert that the extension is unique. In the following proof we shall find the maximum extension.

Let Γ be the class of subparabolic minorants of $\dot v$ on $\dot D$ which are majorized on $\dot D'$ by $\dot v'$ and define $\dot v'' = \sup\{\dot u : \dot u \in \Gamma\}$. If $a < b'' < b'$ and if $\dot u$ is defined on $\dot D$ by

$$\dot u = \begin{cases} \dot v' & \text{on } D \times]a, b''], \\ 0 & \text{on } D \times]b'', b[, \end{cases}$$

then $\dot{u} \in \Gamma$. Hence $\dot{v}''$ is an extension of $\dot{v}'$. The class Γ contains $\dot{u}_1 \vee \dot{u}_2$ if it contains $\dot{u}_1$ and $\dot{u}_2$, and Γ contains the majorant $\dot{\tau}_{\dot{B}}\dot{v}$ of $\dot{v}$ if $\dot{v} \in \Gamma$ and if $\dot{B}$ is an interval relatively compact in $\dot{D}$. Thus $\dot{v}''_{|\dot{B}}$ is the limit of an upward-directed set of parabolic minorants of $\dot{v}_{|\dot{B}}$; so $\dot{v}''$ is parabolic and is the desired extension of $\dot{v}'$.

Application to Minimal Parabolic Functions

Let $\dot{D}$ and $\dot{D}'$ be as in the lemma, and let $\dot{v}$ be a minimal parabolic function on $\dot{D}$. Then $\dot{v}_{|\dot{D}'}$ is minimal on $\dot{D}'$. In fact, if $\dot{v}'$ is a positive parabolic minorant of $\dot{v}_{|\dot{D}'}$, the extension of $\dot{v}'$ to $\dot{D}$ provided by the lemma must be proportional to $\dot{v}$; so $\dot{v}'$ is proportional to $\dot{v}_{|\dot{D}'}$.

Special Case: Extension of a Bounded Parabolic Function. Let $\dot{D}$ and $\dot{D}'$ be as in the lemma. Then an arbitrary bounded parabolic function $\dot{v}'$ defined on $\dot{D}'$ has a parabolic extension to $\dot{D}$ with the same infimum and supremum there as $\dot{v}$ has on $\dot{D}'$. In proving this we can assume that $\dot{v}'$ has infimum 0. Let $\gamma = \sup_{\dot{D}} \dot{v}'$. Then the function $\dot{v} \equiv \gamma$ on $\dot{D}$ is a parabolic majorant of $\dot{v}'$ on $\dot{D}'$; so $\dot{v}'$ has a positive parabolic extension to $\dot{D}$, majorized there by γ.

If D is a finite interval in $\mathbb{R}^N$, the preceding result follows easily from our discussion of the parabolic context Poisson integral.

Chapter XVI

Subparabolic, Superparabolic, and Parabolic Functions on a Slab

1. The Parabolic Poisson Integral for a Slab

If $\dot{D}$ is the slab $\mathbb{R}^N \times]0, \delta[$, with $0 < \delta \leq +\infty$, the restriction to $\dot{D} \times \dot{D}$ of $\dot{G}$ satisfies the rather vague description of the Green function $\dot{G}_{\dot{D}}$ given in Section XV.7 for smooth regions. It is therefore to be expected from XV(7.3) that the upper boundary of $\dot{D}$ if $\delta < +\infty$ is a parabolic measure null set and that parabolic measure on the lower boundary is given by

$$\dot{\mu}_{\dot{D}}(\dot{\xi}, d\eta) = \ell(s, \xi - \eta) l_N(d\eta) = \dot{G}(\dot{\xi}, (\eta, 0)) l_N(d\eta) \qquad [\dot{\xi} = (\xi, s)],$$

so that if $\dot{u}$ is parabolic on $\dot{D}$ with boundary function f in some suitable sense on the lower boundary and if $\dot{u}$ is appropriately restricted, then

$$\dot{u}(\dot{\xi}) = \int_{\mathbb{R}^N} \ell(s, \xi - \eta) f(\eta) l_N(d\eta) \qquad [\dot{\xi} = (\xi, s)]. \tag{1.1}$$

Such representations will be derived below. Moreover we shall see that if μ is a suitably restricted charge on $\mathbb{R}^N$, the Poisson integral

$$\dot{P}I(\dot{D}, \mu)(\dot{\xi}) = \int_{\mathbb{R}^N} \ell(s, \xi - \eta) \mu(d\eta) \qquad [\dot{\xi} = (\xi, s)]$$

defines a parabolic function on $\dot{D}$. The integral in (1.1) will be denoted by $\dot{P}I(\dot{D}, f)$ $(\dot{\xi})$, and the Poisson integral will be modified in the obvious way when the lower boundary of $\dot{D}$ is not the abscissa hyperplane.

Theorem. *If f is a Lebesgue measurable function on $\mathbb{R}^N$ and if for some δ in* $]0, +\infty]$

$$\int_{\mathbb{R}^N} |f(\eta)| \exp \frac{-|\eta|^2}{2\sigma^2 \delta'} l_N(d\eta) < +\infty \tag{1.2}$$

whenever $\delta' < \delta$, then the function $\dot{u}$ defined on $\dot{D} = \mathbb{R}^N \times]0, \delta[$ by (1.1) *is parabolic. Moreover,*

$$\limsup_{\dot{D} \ni \dot{\xi} \to (\zeta, 0)} \dot{u}(\dot{\xi}) \leq \limsup_{\mathbb{R}^N \ni \eta \to \zeta} f(\eta) \qquad (\zeta \in \mathbb{R}^N) \tag{1.3}$$

and

$$\limsup_{\dot{D}\ni\dot{\xi}\to\infty} \dot{u}(\dot{\xi}) \le \limsup_{\eta\in\mathbb{R}^N,\,|\eta|\to+\infty} f(\eta). \tag{1.4}$$

On combining (1.3) with the corresponding inequality for inferior limits we find that $\dot{u}$ has limit $f(\zeta)$ at $(\zeta, 0)$ if f is continuous at ζ, and the corresponding specialization of (1.4) is valid. The fact that $\dot{u}$ is defined and parabolic on $\dot{D}$ is trivial. Inequality (1.3) is a special case of Theorem 1 of Appendix VII because if $0 < \delta' < \delta$ and $\dot{\xi} = (\xi, s)$,

$$\lim_{\dot{\xi}\to(\zeta,0)} \frac{\mathscr{E}(s, \xi - \eta)}{\mathscr{E}(\delta', \eta)} = 0$$

uniformly for η outside an arbitrary open neighborhood A of ζ in $\mathbb{R}^N$; so if $\varepsilon > 0$ and if $\dot{\xi}$ is sufficiently near $(\zeta, 0)$,

$$\int_{\mathbb{R}^N - A} \mathscr{E}(s, \xi - \eta)|f(\eta)|l_N(d\eta) \le \varepsilon \int_{\mathbb{R}^N} \mathscr{E}(\delta', \eta)|f(\eta)|l_N(d\eta). \tag{1.5}$$

Thus the left side of (1.5) tends to 0 when $\dot{\xi} \to (\zeta, 0)$, and the same limit relation holds when f is replaced by the constant function 1, as required for the application of Theorem 1 of Appendix VII. Inequality (1.4) is also a special case of Theorem 1 of Appendix VII because

$$t^{-N/2}e^{-\alpha/t} \le t^{-N/2}\left[\left(\frac{N}{2e}\right)^{N/2} \alpha^{-N/2}\right] \qquad (t > 0, \alpha > 0);$$

so if A is an open neighborhood of the point ∞ of $\mathbb{R}^N$,

$$\int_{\mathbb{R}^N - A} \mathscr{E}(s, \xi - \eta)|f(\eta)|l_N(d\eta) \le \int_{\mathbb{R}^N - A} s^{-N/2}\left[\left(\frac{N}{e}\right)^{N/2} \left|\xi - \eta\right|^{-N}\right] |f(\eta)|l_N(d\eta). \tag{1.6}$$

Hence the integral on the left and the same integral with f replaced by the constant function 1 tend to 0 when $\dot{\xi} \to \infty$ in $\dot{D}$, as required for the application of Theorem 1 of Appendix VII.

Extension. According to the theorem as applied to $-f$, if $\lim_{|\eta|\to\infty} f(\eta) = +\infty$, then $\lim_{\dot{D}\ni\dot{\xi}\to\infty} \dot{u}(\dot{\xi}) = +\infty$ also. It will be useful to sharpen this result, as stated in the following extension of the theorem. *If (1.2) is true whenever $\delta' < \delta$ and if, for some $\beta > 0$,*

$$\liminf_{|\eta|\to\infty} f(\eta)\exp(-\beta|\eta|^2) > 0,$$

then for $0 < \alpha < \beta$, $\delta_1 < (2\beta\sigma^2)^{-1}$, and $\dot{D}_1 = \mathbb{R}^N \times]0, \delta_1[$ it follows that

$$\lim_{\dot{D}_1 \ni \dot{\xi} \to \infty} \dot{u}(\dot{\xi}) \exp(-\alpha|\xi|^2) = +\infty.$$

Note that $\beta \le (2\sigma^2\delta)^{-1}$ because of (1.2). To prove this extension of Theorem 1, observe that by hypothesis there is a constant $c > 0$ such that $f(\eta) \ge c \exp(\beta|\eta|^2)$ for sufficiently large $|\eta|$, say for $|\eta| \ge r$. Then

$$\liminf_{\dot{D}_1 \ni \dot{\xi} \to \infty} \dot{u}(\dot{\xi}) \exp(-\alpha|\xi|^2) \ge \liminf_{\dot{D}_1 \ni \dot{\xi} \to \infty} c \int_{\mathbb{R}^N} \ell(s, \xi - \eta) \exp(\beta|\eta|^2 - \alpha|\xi|^2) l_N(d\eta) \tag{1.7}$$

because the part of the integral over $\{|\eta| \le r\}$ has limit 0 when $|\xi| \to +\infty$. The value of this integral is

$$(1 - 2\beta\sigma^2 s)^{-N/2} \exp \frac{|\xi|^2(\beta - \alpha + 2\alpha\beta\sigma^2 s)}{1 - 2\beta\sigma^2 s},$$

which tends to $+\infty$ when $|\xi| \to +\infty$ with $\dot{\xi}$ in $\dot{D}_1$. The proof of the extension is complete.

2. A Generalized Superparabolic Function Inequality

Lemma. *Let $\dot{u}$ be a positive superparabolic function on the slab $\dot{D} = \mathbb{R}^N \times$ $]0, \delta[$, and define $f(\zeta) = \liminf_{s \to 0} \dot{u}(\zeta, s)$ for ζ in $\mathbb{R}^N$. Then $\dot{u} \ge \dot{\mathrm{P}}\mathrm{I}(\dot{D}, f)$.*

Let $\dot{D}_k$ be the slab $\mathbb{R}^N \times]1/k, \delta[$ for k so large that $k > 1/\delta$. Let $f_{k\cdot}$ be an increasing sequence of positive functions on $\mathbb{R}^N$, finite valued and continuous with compact support and with limit the lower semicontinuous function $\dot{u}(\cdot, 1/k)$. According to Theorem 1, the function $\dot{\mathrm{P}}\mathrm{I}(\dot{D}_k, f_{kn})$ is a positive parabolic function on $\dot{D}_k$ with limit $f_{kn}(\zeta)$ at each point $(\zeta, 1/k)$ of the lower boundary of $\dot{D}_k$ and limit 0 at the point ∞; so $\dot{u} - \dot{\mathrm{P}}\mathrm{I}(\dot{D}_k, f_{kn}) \ge 0$ on $\dot{D}_k$ by the minimum theorem for superparabolic functions on a slab (Section XV.13). The lemma follows when $n \to \infty$ and then $k \to \infty$.

Application (a). It follows trivially from Lemma 2 that

$$\dot{u}(\xi', s') \ge \int_{\mathbb{R}^N} \ell(s' - s, \xi' - \xi) \dot{u}(\xi, s) l_N(d\xi), \qquad 0 < s < s' < \delta, \tag{2.1}$$

and therefore if $s' < t < \delta$,

$$\int_{\mathbb{R}^N} \ell(t - s', \eta - \xi') \dot{u}(\xi', s') l_N(d\xi') \ge \int_{\mathbb{R}^N} \ell(t - s, \eta - \xi) \dot{u}(\xi, s) l_N(d\xi). \tag{2.2}$$

Thus the parabolic average relative to (η, t) of the positive superparabolic function $\dot{u}$ on the hyperplane of constant ordinate value s' $(<t)$ defines an

increasing function of s' on $]0, t[$. This is an example of the fact that roughly (see Section VIII.10 for the classical counterpart) the parabolic average of a superparabolic function over a set boundary decreases as the set increases. According to Theorem 5 below, when $\dot{u}$ in Lemma 2 is parabolic and positive, the parabolic average on the left side of (2.2) is constant, equal to $\dot{u}(\eta, t)$, as s' varies.

Application (b). If $\dot{u}$ is bounded and parabolic on the above slab $\dot{D}$ and if $\dot{u}$ has normal limit $f(\zeta) = \lim_{s\to 0} \dot{u}(\zeta, s)$ at l_N almost every lower boundary point $(\zeta, 0)$, then an application of Lemma 2 to the function $\dot{u} + \sup_{\dot{D}} \dot{u}$ and $-\dot{u} + \sup_{\dot{D}} \dot{u}$ shows that $\dot{u} = \dot{\mathrm{P}}\mathrm{I}(\dot{D}, f)$. According to Section 5, the boundedness of $\dot{u}$ can be replaced here by positivity if $f(\zeta) = \lim_{\dot{\xi}\to(\zeta,0)} \dot{u}(\dot{\xi})$ for all ζ in $\mathbb{R}^N$. Theorem 6(b) gives necessary and sufficient conditions that a parabolic function on a slab be representable as the Poisson integral of a function.

3. A Criterion of a Subparabolic Function Supremum

Lemma. *If $0 < \delta \le +\infty$, if $\alpha > 0$, if $\dot{u}$ is subparabolic on the slab $\dot{D} = \mathbb{R}^N \times]0, \delta[$, and if*

$$\dot{u}(\dot{\xi}) \le \exp[\alpha(|\xi|^2 + 1)], \qquad \limsup_{\dot{\eta}\to(\zeta,0)} \dot{u}(\dot{\eta}) \le 0 \qquad [\dot{\xi} = (\xi, s)] \quad (3.1)$$

for all $\dot{\xi}$ in $\dot{D}$ and ζ in $\mathbb{R}^N$, then $\dot{u} \le 0$ on $\dot{D}$.

In fact, if $\dot{D}_1 = \mathbb{R}^N \times]0, (8\alpha\sigma^2)^{-1}[$, if $f(\eta) = \exp[2\alpha(|\eta|^2 + 1)]$, and if $\dot{v} = \dot{\mathrm{P}}\mathrm{I}(\dot{D}_1, f)$, apply the extension in Section 1 with $\beta = 2\alpha$ and $\delta_1 = (8\alpha\sigma^2)^{-1}$ to find that

$$\lim_{\dot{D}_1 \ni \dot{\xi}\to\infty} \dot{v}(\dot{\xi}) \exp[-\alpha(|\xi|^2 + 1)] = +\infty.$$

If $\varepsilon > 0$, it now follows from the subparabolic maximum theorem applied to $\dot{u} - \varepsilon\dot{v}$ in $\dot{D} \cap \dot{D}_1 = \mathbb{R}^N \times]0, \delta \wedge \delta_1[$ that $\dot{u} \le \varepsilon\dot{v}$ on $\dot{D} \cap \dot{D}_1$. Hence $\dot{u}(\dot{\xi}) \le 0$ when $\operatorname{ord} \dot{\xi} < \delta \wedge \delta_1$. If $\delta \le \delta_1$, there is nothing more to prove. If $\delta > \delta_1$, iteration of the preceding reasoning shows that $\dot{u}(\dot{\xi}) \le 0$ when $\operatorname{ord} \dot{\xi} < \delta \wedge (2\delta_1), \ldots$; so $\dot{u} \le 0$ on $\dot{D}$.

4. A Boundary Limit Criterion for the Identically Vanishing of a Positive Parabolic Function

Lemma. *Let $\dot{u}$ be a positive parabolic function on the slab $\mathbb{R}^N \times]0, \delta[$, and suppose that $\lim_{\dot{\xi}\to(\zeta,0)} \dot{u}(\dot{\xi}) = 0$ for every ζ in $\mathbb{R}^N$. Then $\dot{u} \equiv 0$.*

It can be assumed that (a) $\dot{u}(\xi, s) = \dot{u}(\mathbf{R}\xi, s)$ whenever $\mathbf{R}$ is a rotation of $\mathbb{R}^N$ about the origin and that (b) the function $\dot{u}(\xi, \cdot)$ is an increasing function.

To see that (a) can be assumed, observe that for any choice of rotation **R** the function $(\xi, s) \mapsto \dot{u}(\mathbf{R}\xi, s)$ is parabolic; so if $\dot{u}(\xi, s)$ is replaced by the average of $u(\cdot, s)$ over the sphere in $\mathbb{R}^N$ of radius $|\xi|$ and center the origin, the new function will be parabolic and positive and have limit 0 at every point of the abscissa hyperplane. It is sufficient to prove the lemma for this new function. To see that (b) can be assumed, we show that the function

$$\dot{v}: (\xi, s) \mapsto \int_0^s \dot{u}(\dot{\xi}, r) l_1(dr)$$

is parabolic on the given slab. This function satisfies (b), and it is sufficient to prove the lemma for this function, which satisfies (a) if $\dot{u}$ does. If $\alpha > 0$, define $\dot{v}_\alpha$ by

$$\dot{v}_\alpha(\xi, s) = \int_\alpha^s \dot{u}(\xi, r) l_1(dr), \qquad 0 < \alpha < s < \delta.$$

Then $\dot{\Delta}\dot{v}_\alpha(\xi, s) = -\dot{u}(\xi, \alpha)$; so $\dot{v}$ is the locally uniform limit of the increasing sequence $\{\dot{v}_{1/n}, n \geq 1\}$, with $\lim_{n\to\infty} \dot{\Delta}\dot{v}_{1/n} = 0$ locally uniformly. It follows that $\dot{v}$ is parabolic [for example, apply XV(7.2) to $\dot{v}_{1/n}$ and go to the limit, $n \to \infty$].

Thus we now have $\dot{u}(\xi, s) = f(|\xi|, s)$ with $f(r, \cdot)$ monotone increasing for every value of $r \geq 0$. In view of this monotoneity and of the parabolic function maximum-minimum theorem the maximum of $\dot{u}$ on the cylinder $\{0 < \operatorname{ord} \dot{\xi} < t, |\xi| \leq r\}$ must be attained on the top of the lateral boundary, that is, at a point $\dot{\xi}$ with $|\xi| = r$, $\operatorname{ord} \dot{\xi} = t$, and it follows that $f(\cdot, s)$ is an increasing function for each value of $s > 0$. According to Lemma 2,

$$[2\pi\sigma^2(s-t)]^{-N/2}\pi_N \int_0^\infty r^{N-1} f(r, t) \exp\frac{-r^2}{2\sigma^2(s-t)} l_1(dr) \leq f(0, s), \tag{4.1}$$
$$0 < t < s < \delta,$$

and therefore if $r > 1$,

$$[2\pi\sigma^2(s-t)]^{-N/2}\pi_N f(r, t) \int_r^\infty \exp\frac{-\alpha^2}{2\sigma^2(s-t)} l_1(d\alpha) \leq f(0, s), \tag{4.2}$$

so that if s is fixed,

$$\dot{u}(\xi, t) = f(r, t) \leq \text{const } (1 + r) \exp\frac{r^2}{\sigma^2 \varepsilon s}, \qquad r = |\xi|, \qquad t \leq (1-\varepsilon)s. \tag{4.3}$$

It now follows from Lemma 3 that $\dot{u} \leq 0$ on the slab $\mathbb{R}^N \times]0, (1-\varepsilon)s[$; so $\dot{u} = 0$ on this slab and so on $\mathbb{R}^N \times]0, \delta[$.

5. A Condition that a Positive Parabolic Function Be Representable by a Poisson Integral

Theorem. *If* $0 < \delta \le +\infty$, *if* $\dot{u}$ *is a positive parabolic function on* $\dot{D} = \mathbb{R}^N \times]0, \delta[$, *and if* $\lim_{\dot{\xi} \to (\zeta, 0)} \dot{u}(\dot{\xi}) = f(\zeta) < +\infty$ *exists for all* ζ *in* $\mathbb{R}^N$, *then* $\dot{u} = \dot{\mathrm{P}}\mathrm{I}(\dot{D}, f)$.

The boundary limit function f is necessarily continuous since there is a limit at every point of the abscissa hyperplane. Furthermore $\dot{u} - \dot{\mathrm{P}}\mathrm{I}(\dot{D}, f) \ge 0$ by Lemma 2, and the difference has limit 0 at every point of the abscissa hyperplane according to Theorem 1. Hence (Lemma 4) the difference vanishes identically.

6. The $\mathbf{L}^1(\dot{\mu}_{\dot{B}-})$ and $\mathbf{D}(\dot{\mu}_{\dot{B}-})$ Classes of Parabolic Functions on a Slab

Theorem II.14 for harmonic functions on a ball has the following analog for parabolic functions on a slab. Recall that if $\dot{\mu}$ is a charge with minimal Jordan decomposition $\dot{\mu}^+ - \dot{\mu}^-$, we denote the absolute variation measure $\dot{\mu}^+ + \dot{\mu}^-$ by $|\dot{\mu}|$. In the following theorem if $0 < \delta \le +\infty$, we denote by $\dot{B}(0, \delta)$ the slab $\mathbb{R}^N \times]0, \delta[$.

Theorem. *Let* $\dot{u}$ *be a parabolic function on the slab* $\dot{B} = \dot{B}(0, \delta)$.

(a) $\mathbf{L}^1(\dot{\mu}_{\dot{B}-})$ *parabolic functions. The following conditions on* $\dot{u}$ *are equivalent*:

(a1) $\dot{u} = \dot{\mathrm{P}}\mathrm{I}(\dot{B}, \dot{N}_{\dot{u}})$ *for some charge* $\dot{N}_{\dot{u}}$ *on* $\mathbb{R}^N$ *for which*

$$\int_{\mathbb{R}^N} \exp \frac{-|\eta|^2}{2\sigma^2 s} |\dot{N}_{\dot{u}}|(d\eta) < +\infty, \qquad 0 < s < \delta. \tag{6.1}$$

(a2) $\dot{u}$ *is the difference between two positive parabolic functions.*

(a3) $|\dot{u}|$ *has a parabolic majorant.*

(a4) $\sup_{0<t<s} \int_{-\infty}^{+\infty} \ell(s - t, \xi - \eta)|\dot{u}(\eta, t)| l_N(d\eta) < +\infty$ *for every point* (ξ, s) *of* $\dot{B}$.

The map $\dot{u} \mapsto \dot{N}_{\dot{u}}$ *is a one-to-one linear order-preserving map from the class of parabolic functions satisfying these conditions onto the vector lattice of charges on* $\mathbb{R}^N$ *satisfying* (6.1).

(b) $\mathbf{D}(\dot{\mu}_{\dot{B}-})$ *parabolic functions. The following more restrictive conditions on* $\dot{u}$ *are equivalent:*

(b1) $\dot{u} = \dot{\mathrm{P}}\mathrm{I}(\dot{B}, f_{\dot{u}})$ *for some* l_N *measurable function* $f_{\dot{u}}$ *on* $\mathbb{R}^N$ *satisfying*

$$\int_{\mathbb{R}^N} \exp \frac{-|\eta|^2}{2\sigma^2 s} |f_{\dot{u}}(\eta)| l_N(d\eta) < +\infty, \qquad 0 < s < \delta. \tag{6.2}$$

(b2) *If* $0 < s < \delta$, *there is a uniform integrability test function* Φ_s *for which the restriction to* $\dot B(0,s)$ *of* $\Phi_s(|\dot u|)$ *has a parabolic majorant.*

(b3) *For every point* (ξ, s) *in* $\dot B$ *the family*

$$\{t\mapsto[\dot u(\eta,t),\mathscr{E}(s-t,\xi-\eta)l_N(d\eta)],0<t<s\}$$

of paired functions and measures on $\mathbb{R}^N$ *is uniformly integrable.*

(b4) *If* $\dot u > 0$, $\dot u$ *is the limit of an increasing sequence of bounded positive parabolic functions.*

The map $\dot u\mapsto f_{\dot u}$ *is a one-to-one linear order-preserving map from the class of parabolic functions satisfying these conditions onto the vector lattice* $L^1(\mathbb{R}^N, l_N)$, *and* $d\dot N_{\dot u} = f_{\dot u} dl_N$.

As will be noted in Section XVIII.19, the notation introduced in Chapter IX for various classes of functions in classical potential theory is readily adapted to the context of parabolic potential theory. These classes together with their martingale theory counterparts are discussed in Chapter I of Part 3. In this spirit the parabolic functions in Theorem 6(a) are the $\mathbf{L}^1(\dot\mu_{\dot B-})$ parabolic functions, those in Theorem 6(b) are the $\mathbf{D}(\dot\mu_{\dot B-})$ parabolic functions, and Theorem 6(b4) asserts that the positive $\mathbf{D}(\dot\mu_{\dot B-})$ parabolic functions are quasi bounded. A uniform notation will be used in Chapter I of Part 3 both for classes of functions in the classical and parabolic potential theory contexts and for classes of stochastic processes in the martingale theory context. The assertion of Theorem 6(b4) in all three contexts is covered by Theorem 3.I.5.

To prove Theorem 6, define $\dot{\mathrm{P}}\mathrm{I}(\dot B(s,t),f)$ on $\dot B(s,t) = \mathbb{R}^N \times]s,t[$ by

$$\dot{\mathrm{P}}\mathrm{I}(\dot B(s,t),f)(\xi',s') = \int_{\mathbb{R}^N} \mathscr{E}(s'-t,\xi'-\eta)f(\eta)l_N(d\eta).$$

Observe that $|\dot u|$ in (a3) and $\Phi_s(|\dot u|)$ in (b2) are positive subparabolic functions with parabolic majorants. It follows that for fixed $s > 0$ and fixed (ξ', s') in $\dot B(0,s)$ the parabolic averages

$$\dot{\mathrm{P}}\mathrm{I}(\dot B(s,t),|\dot u(\cdot,t)|)(\xi',s'),\qquad \dot{\mathrm{P}}\mathrm{I}(\dot B(s,t),\Phi_s[\dot u(\cdot,t)])(\xi',s')$$

define decreasing functions of t on the interval $]0,s'[$. To see this, for example, for the first parabolic average, note that if $\dot v$ is a parabolic majorant of $|\dot u|$ on $\dot B$, then (by Theorem 5) $\dot{\mathrm{P}}\mathrm{I}(\dot B(s,t),\dot v(\cdot,t)) = \dot v$ on $\dot B(s,t)$; so (from Section 2) the function

$$t\mapsto \dot{\mathrm{P}}\mathrm{I}(\dot B(s,t),\dot v(\cdot,t)-|\dot u(\cdot,t)|)(\xi',s') = \dot v(\xi',s') - \dot{\mathrm{P}}\mathrm{I}(\dot B(s,t),|\dot u(\cdot,t)|)(\xi',s')$$

is an increasing function on $]0,s'[$. With the help of this monotoneity result the proof of Theorem 6 becomes so close to that of Theorem II.14 in its ideas that the details will be omitted. Choose δ' with $0<\delta'<\delta$, and for $0<r<\delta'$ define charges $\dot\mu_r$ and $\dot\mu_r'$ on $\mathbb{R}^N$ by

$$\dot\mu_r(d\eta)=\dot u(\eta,r)l_N(d\eta),\qquad \dot\mu_r'(d\eta)=\dot u(\eta,r)\exp\frac{-|\eta|^2}{2\sigma^2\delta'}l_N(d\eta).$$

It is convenient to consider $\dot\mu_r'$ as a measure on the space $\mathbb{R}^N$ compactified by a point at infinity, with the infinite singleton $\dot\mu_r'$ null. Instead of showing directly that $\dot\mu_r$ tends to a limit measure $\dot N_{\dot u}$ when $r\to 0$, it is easier to show that under (a3) the total variation of $\dot\mu_r'$ is bounded independently of r, that (vague)$\lim_{r\to 0}\dot\mu_r'=\dot\mu'$ exists, and that (a1) is true with

$$\dot N_{\dot u}(d\eta)=\exp\frac{|\eta|^2}{2\sigma^2\delta'}\dot\mu'(d\eta).$$

Finally observe that the supremum in (a4) defines a parabolic majorant of $|\dot u|$ on $\dot B$. Further details of the proof of Theorem 6 are left to the reader.

7. The Parabolic Boundary Limit Theorem

A function f on a slab $\mathbb{R}^N\times\,]0,\delta[$ is said to have normal limit q at the boundary point $\dot\zeta$: $(\zeta,0)$ if $\lim_{s\to 0}f(\zeta,s)=q$. The function is said to have parabolic limit q at $(\zeta,0)$ if $\lim_{\dot\xi\to\dot\zeta}f(\dot\xi)=q$ whenever $\dot\xi$: $(\xi,s)\to\dot\zeta$ in a paraboloid of revolution with vertex $\dot\zeta$ and opening upward, that is, whenever $\dot\xi\to\dot\zeta$ with $\liminf_{\dot\xi\to\dot\zeta}s|\xi-\zeta|^{-2}>0$. In more sophisticated language, if a subset $\dot A$ of the upper half-space is called a *deleted coparabolic neighborhood* of $\dot\zeta$ whenever for some $a>0$ the set $\dot A$ contains the intersection of the paraboloid $\{s>a|\xi-\zeta|^2\}$ with some Euclidean neighborhood of $\dot\zeta$, then f has parabolic limit q at $\dot\zeta$ if f has limit q along the filter of deleted coparabolic neighborhoods.

Theorem. *Let $\dot B=\mathbb{R}^N\times\,]0,\delta[$ be a slab in $\dot{\mathbb{R}}^N$, let $\dot u$ be a parabolic function in $\mathbf{L}^1(\dot\mu_{\dot B-})$, and let $\dot h$ be a strictly positive parabolic function on $\dot B$. Then if ζ is a point of $\mathbb{R}^N$ at which the convex variational derivates $dl_N/d\dot N_{\dot h}$ and $d\dot N_{\dot u}/d\dot N_{\dot h}$ both exist, the function $\dot u/\dot h$ has parabolic limit $(d\dot N_{\dot u}/d\dot N_{\dot h})(\zeta)$ at $(\zeta,0)$.*

Observe that the stated conditions are satisfied at $\dot N_{\dot h}$ almost every point ζ of $\mathbb{R}^N$. In the more common version of this theorem $\dot h\equiv 1$; so $\dot N_{\dot h}=l_N$. In particular, if $\dot u\in\mathbf{D}(\dot\mu_{\dot B-})$ so that $\dot u=\dot{\mathrm{P}}\mathrm{I}(\dot B,f)$ for some function f on $\mathbb{R}^N$, the theorem states that $\dot u$ has parabolic limit $f(\zeta)$ at l_N almost every slab boundary point $(\zeta,0)$.

Theorem 3 of Appendix VII can be applied to prove a modified version of Theorem 7, namely, that under the hypothesis that $dl_N/d\dot N_{\dot h}$ and $d\dot N_{\dot u}/d\dot N_{\dot h}$

both exist as symmetric derivates at ζ, the function $\dot{u}/\dot{h}$ has normal limit $(d\dot{N}_{\dot{u}}/d\dot{N}_{\dot{h}})(\zeta)$ at $(\zeta, 0)$. The proof of this modification follows the proof of the corresponding result (Theorem II.15) in the harmonic function context. In the latter context the Harnack inequality made it possible to go from normal approach to nontangential approach, but this step does not seem possible for general $\dot{h}$ in the parabolic context. We therefore prove Theorem 7 as an application of Theorem 4 of Appendix VII. In the latter theorem if we set

$$K(s,r) = \exp\frac{-r^2}{2\sigma^2 s}, \qquad \delta(s) = \text{const}\, s^{1/2}, \qquad s_0(a) = \frac{\delta}{2},$$

we obtain Theorem 7.

Application. We have proved in Section 2 that if $\dot{u}$ is a bounded parabolic function on the slab $\dot{B}$ and if $\dot{u}$ has normal limit $f(\zeta)$ at l_N almost every point $(\zeta, 0)$, then $\dot{u} = \dot{\text{P}}\text{I}(\dot{B}, f)$. According to Theorems 6 and 7, the weaker hypothesis that the parabolic function $\dot{u}$ on $\dot{B}$ is in $\mathbf{D}(\dot{\mu}_{\dot{B}-})$ implies that the parabolic limit, say $f(\zeta)$, exists at l_N almost every boundary point $(\zeta, 0)$ and that $\dot{u} = \dot{\text{P}}\text{I}(\dot{B}, f)$.

8. Minimal Parabolic Functions on a Slab

(a) $\dot{D} = \mathbb{R}^N \times]0, \delta[,\ 0 < \delta \leq +\infty$. In view of the Riesz–Herglotz-type representation (Theorem 6) of a positive parabolic function on $\dot{D}$, such a function is minimal if and only if it is a positive multiple of the function $(\xi, s) \mapsto \ell(s, \xi - \zeta)$ on $\dot{D}$ for some point ζ of $\mathbb{R}^N$.

(b) $\dot{D} = \mathbb{R}^N \times]-\infty, 0[$, the lower half-space. In view of the Appell transformation which takes the upper half-space of $\dot{\mathbb{R}}^N$ into the lower half-space, a parabolic function $\dot{u}$ on the lower half-space is minimal if and only if it is the Appell transform of a minimal parabolic function for the upper half-space, that is (Section XV.16), if and only if $\dot{u}$ is a positive multiple of the function

$$(\xi, s) \mapsto \exp\left[\langle \gamma, \xi \rangle + \sigma^2 |\gamma|^2 \frac{s}{2}\right] \tag{8.1}$$

on the lower half-space, for some point γ of $\mathbb{R}^N$. In particular ($\gamma = 0$), the positive constant functions are minimal on the lower half-space. It follows (Liouville-type theorem) that a bounded parabolic function on the lower half-space, and therefore surely a bounded parabolic function on $\dot{\mathbb{R}}^N$, is a constant function. Furthermore the representation Theorem 6 transferred to the lower half-space by an Appell transformation yields a representation for a positive parabolic function $\dot{u}$ on the lower half-space:

$$\dot{u}(\xi, s) = \int_{\mathbb{R}^N} \exp\left[\langle \gamma, \xi \rangle + \sigma^2 |\gamma|^2 \frac{s}{2}\right] \dot{N}_{\dot{u}}(d\gamma), \tag{8.2}$$

where $\dot{N}_{\dot{u}}$ is a measure on $\mathbb{R}^N$ for which

$$\int_{\mathbb{R}^N} \exp(-\alpha|\gamma|^2) \dot{N}_{\dot{u}}(d\gamma) < +\infty \tag{8.3}$$

for all $\alpha > 0$. We conclude that every positive parabolic function on the lower half-space is either strictly positive or identically 0 and is monotone increasing in the ordinate variable. We leave to the reader the full formulations of Theorems 6 and 7 in the lower half-space context.

(c) $\dot{D} = \dot{\mathbb{R}}^N$. It follows easily from (b) that the minimal positive parabolic functions on $\dot{\mathbb{R}}^N$ are the positive multiples of the functions (8.1), now considered on $\dot{\mathbb{R}}^N$, and that every positive parabolic function $\dot{u}$ on $\dot{\mathbb{R}}^N$ has a representation of the form (8.2) with (8.3) true for all real α. Every positive parabolic function on $\dot{\mathbb{R}}^N$ is either strictly positive or identically 0 and is monotone increasing in the ordinate variable.

Chapter XVII

Parabolic Potential Theory (Continued)

1. Greatest Minorants and Least Majorants

If $\dot{D}$ is a nonempty open subset of $\dot{\mathbb{R}}^N$ and if Γ is a class of functions on $\dot{D}$, the greatest subparabolic minorant [least superparabolic majorant] of Γ, if there is one, is denoted by $\dot{\mathrm{G}}\mathrm{M}_{\dot{D}}\Gamma$ [$\dot{\mathrm{L}}\mathrm{M}_{\dot{D}}\Gamma$]. For example, if Γ is a class of superparabolic functions and if Γ has a subparabolic minorant then $\dot{\mathrm{G}}\mathrm{M}_{\dot{D}}\Gamma$ exists and is parabolic. The proof is a translation of that of Theorem III.2. The corresponding notation in the coparabolic context is $\overset{*}{\mathrm{G}}\mathrm{M}_{\dot{D}}\Gamma$ and $\overset{*}{\mathrm{L}}\mathrm{M}_{\dot{D}}\Gamma$.

Example. Let $\dot{D}$ be either $\dot{\mathbb{R}}^N$ ($N \geq 1$) or an interval in $\dot{\mathbb{R}}^N$. Then $\dot{\mathrm{G}}\mathrm{M}_{\dot{D}}\dot{G}_{\dot{D}}(\cdot, \dot{\xi}) = \overset{*}{\mathrm{G}}\mathrm{M}_{\dot{D}}\dot{G}_{\dot{D}}(\dot{\xi}, \cdot) = 0$ for every point $\dot{\xi}$ in $\dot{D}$. In fact, say for $\dot{\mathrm{G}}\mathrm{M}_{\dot{D}}\dot{G}_{\dot{D}}(\cdot, \dot{\xi})$ when $\dot{D}$ is an interval, the parabolic minorant in question is positive, is majorized by $\dot{G}_{\dot{D}}(\cdot, \dot{\xi})$, and so has limit 0 at every lateral and lower boundary point of $\dot{D}$. This minorant therefore vanishes identically, according to the parabolic function maximum-minimum theorem. More generally it will follow from the Riesz decomposition of a positive superparabolic function on a nonempty open subset $\dot{D}$ of $\dot{\mathbb{R}}^N$ that the parabolic potential of a measure on $\dot{D}$ if finite on a dense subset of $\dot{D}$ is superparabolic on $\dot{D}$ and has greatest subparabolic minorant 0.

2. The Parabolic Fundamental Convergence Theorem (Preliminary Version) and the Reduction Operation

The proof of the following counterpart of the first version of the Fundamental Convergence Theorem (Theorem III.3) in the classical context follows the proof of Theorem III.3 and is therefore omitted.

Theorem. *Let Γ: $\{\dot{u}_\alpha, \alpha \in I\}$ be a family of superparabolic functions on an open subset of $\dot{\mathbb{R}}^N$, locally uniformly bounded below, and define $\dot{u}(\dot{\xi}) = \inf_{\alpha \in I} \dot{u}_\alpha(\dot{\xi})$. Then $\underset{+}{\dot{u}} \leq \dot{u}$,*

$$\underset{+}{\dot{u}}(\dot{\xi}) = \liminf_{\dot{\eta} \to \dot{\xi}} \dot{u}(\dot{\eta}), \tag{2.1}$$

and

(a) $\underset{+}{\dot u}$ *is superparabolic.*
(b) $\underset{+}{\dot u} = \dot u$ *on each open set on which* $\dot u$ *is superparabolic.*
(c) $\underset{+}{\dot u} = \dot u \quad l_{N+1}$ *almost everywhere.*
(d) *There is a countable subfamily of* Γ *whose infimum has smoothing* $\underset{+}{\dot u}$.

Application: The Natural Order Decomposition Theorem

As application of this simple version of the Fundamental Convergence Theorem in the parabolic context we remark that the classical context Natural Order Decomposition Theorem (Theorem III.7) translates directly into the parabolic context: If $\dot u$, $\dot u_1$, $\dot u_2$ are positive superparabolic functions on $\dot D$ with $\dot u \le \dot u_1 + \dot u_2$, then there are positive superparabolic functions $\dot u_1'$, $\dot u_2'$ on $\dot D$ for which $\dot u_1' \le \dot u_1$, $\dot u_2' \le \dot u_2$, $\dot u = \dot u_1' + \dot u_2'$. The classical context proof requires only trivial changes. Observe that this decomposition and its proof are also valid for relative superharmonic and superparabolic functions. Alternatively the decomposition theorem for superharmonic and superparabolic functions implies trivially the decomposition theorem in the relative contexts.

3. The Parabolic Context Reduction Operations

If $\dot D$ is a nonempty open subset of $\dot{\mathbb{R}}^N$ coupled with a boundary $\partial\dot D$ provided by a metric compactification, if $\dot A \subset \dot D \cup \partial \dot D$, and if $\dot v$ is a positive superparabolic [cosuperparabolic] function on $\dot D$, the superparabolic [cosuperparabolic] reduction of $\dot v$ on $\dot A$, denoted by $\dot R_{\dot v}^{\dot A}$ [$\overset{*}{\dot R}{}_{\dot v}^{\dot A}$], is the infimum of the class of positive superparabolic [cosuperparabolic] functions on $\dot D$ which majorize $\dot v$ on $\dot A \cap \dot D$ and near $\dot A \cap \partial \dot D$, in the sense that each function in the class is to majorize $\dot v$ both on $\dot A \cap \dot D$ and on the trace on $\dot D$ of some neighborhood of $\dot A \cap \partial\dot D$. The smoothed reduction $\underset{+\dot v}{\dot R}{}^{\dot A}$ [$\underset{+\dot v}{\overset{*}{\dot R}}{}^{\dot A}$] is superparabolic [cosuperparabolic] according to Theorem 2 and will sometimes be denoted by $\dot{\|}\dot v\|^{\dot A}$ [$\overset{*}{\|}\dot v\|^{\dot A}$]. As in the classical context, it is trivial that $\dot R_{\dot v}^{\dot A}$ is the infimum of the class of positive superparabolic functions on $\dot D$ which are equal to $\dot v$ on $\dot A \cap \dot D$ and near $\dot A \cap \partial \dot D$. Thus $\dot R_{\dot v}^{\dot A} = \dot v$ on $\dot A \cap \dot D$, and obviously $\underset{+\dot v}{\dot R}{}^{\dot A} \le \dot R_{\dot v}^{\dot A} \le \dot v$ on $\dot D$.

Let ζ be a point of $\dot D$, let $\dot D_\zeta$ be the set of points of $\dot D$ strictly below ζ relative to $\dot D$, let $\dot v$ be a positive superparabolic function on $\dot D$, and let $\dot v_\zeta$ be the restriction of $\dot v$ to $\dot D_\zeta$. Then if $\dot A \subset \dot D$, we now prove that $\dot R_{\dot v}^{\dot A}$ (reduction relative to $\dot D$) is equal on $\dot D_\zeta$ to the reduction relative to $\dot D_\zeta$ of $\dot v_\zeta$ on $\dot A \cap \dot D_\zeta$. (This fact implies the truth of the corresponding statement for smoothed reductions.) The point is that roughly the reduction of $\dot v$ below a point depends only on $\dot v$ below that point and on the part of the target set below that point. To prove the assertion, we need only remark that on the one hand if $\dot u$ is a positive superparabolic function on $\dot D$ which majorizes $\dot v$ on $\dot A$, then the restriction of $\dot u$ to $\dot D_\xi$ majorizes $\dot v_\xi$ on $\dot A \cap \dot D_\xi$ and on the other hand if $\dot u'$ is a positive superparabolic function on $\dot D_\xi$ which majorizes $\dot v_\xi$ on $\dot A \cap \dot D_\xi$ and if $\dot u$ is a positive superparabolic function on $\dot D$ which majorizes $\dot v$ on $\dot A$, then the function $\dot u''$ equal to $\dot u$ on $\dot D - \dot D_\xi$ and to $\dot u \wedge \dot u'$ on $\dot D_\xi$ is a positive superparabolic function on $\dot D$ majorizing $\dot v$ on $\dot A$ with $\dot u'' \le \dot u$ on $\dot D$ and $\dot u'' \le \dot u'$ on $\dot D_\xi$.

The fact that a smoothed reduction $\underset{+v}{R}{}^{A}$ in the classical context is equal quasi everywhere on $A \cap D$ to v and that this smoothed reduction is unchanged when $A \cap D$ is changed by a polar set is considerably weakened in the parabolic context. In fact it will be shown that a smoothed reduction $\underset{+\dot v}{\dot R}{}^{\dot A}$ is in general equal to $\dot v$ on $\dot A$ only up to a parabolic-semipolar set (to be defined in Section 10) and may be changed if $\dot A$ is changed by a parabolic-semipolar set. This weakening entails that some of the proofs of properties in the classical context cannot be used to prove the corresponding properties in the parabolic context and that it is necessary to change the order of the derivation of the properties common to the two contexts.

Just as in the classical context (Section III.4), $\dot R^{\dot A}_{\dot v}$ and $\underset{+\dot v}{\dot R}{}^{\dot A}$ increase when $\dot A$ or $\dot v$ increases, the reduction and smoothed reduction operations are subadditive, the function $\dot R^{\dot A}_{\dot v}$ is parabolic on $\dot D - \overline{\dot A}$ and equal to $\underset{+\dot v}{\dot R}{}^{\dot A}$ there, and $\dot R^{\dot A}_{\dot v} = \underset{+\dot v}{\dot R}{}^{\dot A}$ when $\dot A$ is open. Furthermore (see Section III.5)

$$\begin{aligned}\dot R^{\dot A}_{\dot v} &= \inf\{\dot R^{\dot A\cup\dot B}_{\dot v}: \dot A \cap \partial\dot D \subset \dot B,\ \dot B \text{ open in } \dot D \cup \partial\dot D\}\\ &= \inf\{\dot R^{(\dot A\cup\dot B)\cap\dot D}_{\dot v}: \dot A \cap \partial\dot D \subset \dot B,\ \dot B \text{ open in } \dot D \cup \partial\dot D\}\end{aligned}\tag{3.1}$$

and

$$\begin{aligned}\underset{+\dot v}{\dot R}{}^{\dot A} &= \underset{+}{\inf}\{\underset{+\dot v}{\dot R}{}^{\dot A\cup\dot B}: \dot A \cap \partial\dot D \subset \dot B,\ \dot B \text{ open in } \dot D \cup \partial\dot D\}\\ &= \underset{+}{\inf}\{\underset{+\dot v}{\dot R}{}^{(\dot A\cup\dot B)\cap\dot D}: \dot A \cap \partial\dot D \subset \dot B,\ \dot B \text{ open in } \dot D \cup \partial\dot D\}.\end{aligned}\tag{3.1sm}$$

The counterparts of the other properties listed in Section III.5 will be listed below in Section 16.

Example (a). Let $\dot A$ be the open upper half-space of $\dot{\mathbb{R}}^N$, let $\dot A_0$ be an arbitrary subset of the abscissa hyperplane, and let $\dot v$ be a positive superparabolic function on $\dot{\mathbb{R}}^N$. The parabolic reduction (relative to $\dot{\mathbb{R}}^N$) and smoothed reduction of $\dot v$ on $\dot A \cup \dot A_0$ are trivially $\dot v$ on $\dot A \cup \dot A_0$ and 0 on the lower half-space. Hence (by lower semicontinuity of superparabolic functions) the smoothed reduction is $\dot v$ on the upper half-space and 0 otherwise. The reduction is 0 on $\dot{\mathbb{R}}^N - (\dot A \cup \dot A_0)$ because according to Section 16(f) the reduction and smoothed reduction are identical off the reduction set.

Example (b). If $\dot A$ is a horizontal plane in $\dot{\mathbb{R}}^N$, any positive superparabolic function $\dot u$ on $\dot{\mathbb{R}}^N$ with $\dot u \ge 1$ on $\dot A$ satisfies the same inequality above $\dot A$

(Lemma XVI.2), and it follows that, for reductions relative to $\dot{\mathbb{R}}^N$, $\dot{[\![}1\dot{]\!]}^{\dot{A}}$ is equal to 1 strictly above $\dot{A}$ and equal to 0 elsewhere. Thus $\dot{[\![}\dot{[\![}1\dot{]\!]}^{\dot{A}}\dot{]\!]}^{\dot{A}} = 0$, and so, unlike the situation in the classical context as given in Section VI.3(h), the smoothed parabolic reduction operation is not always idempotent. [However, according to Section 16(i) this operation is idempotent if $\dot{A} \cap \dot{D}$ is parabolic-fine open.]

4. The Parabolic Green Function

Let $\dot{D}$ be a nonempty open subset of $\dot{\mathbb{R}}^N (N \geq 1)$, and let $\dot{\xi}$, $\dot{\eta}$ be points of $\dot{D}$. The parabolic [coparabolic] Green function with pole $\dot{\eta}$ [$\dot{\xi}$] is defined on $\dot{D}$ by

$$\dot{G}_{\dot{D}}(\cdot, \dot{\eta}) = \dot{G}(\cdot, \dot{\eta}) - \dot{\mathrm{G}}\mathrm{M}_{\dot{D}}\dot{G}(\cdot, \dot{\eta}) \qquad [\overset{*}{G}_{\dot{D}}(\dot{\xi}, \cdot) = \dot{G}(\dot{\xi}, \cdot) - \overset{*}{\mathrm{G}}\mathrm{M}_{\dot{D}}\dot{G}(\dot{\xi}, \cdot)]. \tag{4.1}$$

It will be shown in this section that, corresponding to the symmetry of the Green function in the classical context, $\dot{G}_{\dot{D}} = \overset{*}{G}_{\dot{D}}$. Thus the notation $\overset{*}{G}_{\dot{D}}$ is unnecessary and will not be used in later sections. The function $\dot{G}_{\dot{D}}$ will be called the *parabolic Green function*.

As defined by (4.1), the function $\dot{G}_{\dot{D}}(\cdot, \dot{\eta})$ is positive and superparabolic on $\dot{D}$, is parabolic on $\dot{D} - \{\dot{\eta}\}$, and differs from $\dot{G}(\cdot, \dot{\eta})$ by a continuous function, and $\dot{\mathrm{G}}\mathrm{M}_{\dot{D}}\dot{G}_{\dot{D}}(\cdot, \dot{\eta}) = 0$. Conversely, these conditions uniquely determine $\dot{G}_{\dot{D}}(\cdot, \dot{\eta})$. The corresponding dual remarks for $\overset{*}{G}_{\dot{D}}(\dot{\xi}, \cdot)$ are omitted. The definition of the parabolic Green function $\dot{G}_{\dot{D}}$ is the counterpart of the definition of the classical Green function G_D, but no side condition on the domain, depending on the dimensionality, is necessary in the present context because $\dot{G}$ is positive for $N \geq 1$. Note that (Section 1) the present definition of $\dot{G}_{\dot{D}}$ agrees with that given in Sections XV.4, XV.8, and XVI.1 when $\dot{D}$ is $\dot{\mathbb{R}}^N$, an interval or a slab. In particular, $\dot{G}_{\dot{D}} = \dot{G}$ when $\dot{D} = \dot{\mathbb{R}}^N$. The properties assinged to $\dot{G}_{\dot{D}}(\dot{\xi}, \cdot)$ in the smooth region context discussed in Section XV.7 make $\dot{G}_{\dot{D}}(\dot{\xi}, \cdot)$ there the coparabolic Green function with pole $\dot{\xi}$, by an easy application of the coparabolic maximum theorem to $\phi(\dot{\xi}, \cdot)$ (notation of Section XV.7). This is as it should be because, as we shall now show, $\dot{G}_{\dot{D}} = \overset{*}{G}_{\dot{D}}$.

Proof that $\dot{G}_{\dot{D}} = \overset{*}{G}_{\dot{D}}$. Define $\dot{u}(\cdot, \dot{\eta}) = \dot{\mathrm{G}}\mathrm{M}_{\dot{D}}\dot{G}(\cdot, \dot{\eta})$, and let $\dot{B}_{\bullet}$ be a sequence of intervals with closures in $\dot{D}$ and with the property that each point of $\dot{D}$ has a neighborhood which lies in $\dot{B}_n$ for infinitely many values of n. Define

$$\dot{u}_n(\cdot, \dot{\eta}) = \dot{\tau}_{\dot{B}_n} \cdots \dot{\tau}_{\dot{B}_0}\dot{G}(\cdot, \dot{\eta})_{|\dot{D}}$$

to obtain a decreasing sequence $\dot{u}_{\bullet}(\cdot, \dot{\eta})$ of superparabolic functions on $\dot{D}$ with limit $\dot{u}(\cdot, \dot{\eta})$ (cf. the corresponding discussion for the classical context in Section VII.4). Let $\dot{\eta}_0$ be a point of $\dot{D}$. The sequence $\dot{B}_{\bullet}$ can be chosen in such a way that $\dot{\eta}_0 \in \dot{B}_0$ and that there is a neighborhood of $\dot{\eta}_0$ which is either a subset of $\dot{B}_n$ or of $\dot{D} - \dot{B}_n$ for each n. For $\dot{\xi}$ in $\dot{D} - \{\dot{\eta}_0\}$ the function $\dot{G}(\dot{\xi}, \cdot)$ is coparabolic on a neighborhood of $\dot{\xi}_0$, so the functions $\dot{u}_0(\dot{\xi}, \cdot), \ldots$ are also. In fact each of these functions after the first is an integral average of its predecessor, averaged over values of the first argument. Thus $\dot{u}(\dot{\xi}, \cdot)$ is coparabolic on $\dot{D}$ and therefore $\dot{u}(\dot{\xi}, \cdot) \leq \mathring{\mathrm{G}}\mathrm{M}_{\dot{D}}\dot{G}(\dot{\xi}, \cdot)$. Define $\mathring{u}(\dot{\xi}, \cdot)$ as the right side of this inequality so that $\dot{u} \leq \mathring{u}$. If we had begun with $\mathring{u}$ instead of $\dot{u}$ and carried through the dual argument, we would have obtained the reverse inequality, and it follows that $\dot{u} = \mathring{u}$; that is,

$$\dot{\mathrm{G}}\mathrm{M}_{\dot{D}}\dot{G}(\cdot, \dot{\eta})(\dot{\xi}) = \mathring{\mathrm{G}}\mathrm{M}_{\dot{D}}\dot{G}(\dot{\xi}, \cdot)(\dot{\eta}), \qquad \dot{G}_{\dot{D}} = \mathring{G}_{\dot{D}}. \tag{4.2}$$

The counterpart of the proof in Section VII.4 that u is continuous on $D \times D$ proves that $\dot{u}$ is continuous on $\dot{D} \times \dot{D}$. We shall use the fact (cf. Theorem VII.3) that $\dot{G}_{\dot{D}}(\dot{\xi}, \cdot)$ and $\dot{G}(\cdot, \dot{\eta})$ are bounded outside neighborhoods of their poles.

We leave to the reader the easy translation of the extremal properties of Green functions in the classical context (Theorem VII.2) into the parabolic context.

Relativization of the Green Function

The reasoning used in the discussion of the Green function can be relativized with no change in detail, just as in the classical context (Section VII.1) to use an arbitrary nonempty open subset of $\dot{\mathbb{R}}^N$ as a reference set instead of $\dot{\mathbb{R}}^N$. That is, if $\dot{D}$ and $\dot{B}$ are nonempty open subsets of $\dot{\mathbb{R}}^N$ with $\dot{D} \subset \dot{B}$, then

$$\dot{G}_{\dot{D}}(\cdot, \dot{\eta}) = \dot{G}_{\dot{B}}(\cdot, \dot{\eta}) - \dot{\mathrm{G}}\mathrm{M}_{\dot{D}}\dot{G}_{\dot{B}}(\cdot, \dot{\eta}), \qquad \dot{G}_{\dot{D}}(\dot{\xi}, \cdot) = \dot{G}_{\dot{B}}(\dot{\xi}, \cdot) - \mathring{\mathrm{G}}\mathrm{M}_{\dot{D}}\dot{G}_{\dot{B}}(\dot{\xi}, \cdot) \tag{4.1$'$}$$

on $\dot{D}$ and

$$\dot{\mathrm{G}}\mathrm{M}_{\dot{D}}\dot{G}_{\dot{B}}(\cdot, \dot{\eta})(\dot{\xi}) = \mathring{\mathrm{G}}\mathrm{M}_{\dot{D}}\dot{G}_{\dot{B}}(\dot{\xi}, \cdot)(\dot{\eta}). \tag{4.2$'$}$$

Example. Suppose that D is a half-space of $\mathbb{R}^N$, define $\dot{D} = D \times \mathbb{R}$, and if $\dot{\eta} \in \dot{D}$, denote by $\dot{\eta}'$ the reflection of $\dot{\eta}$ in $\partial\dot{D}$. Then

$$\dot{G}_{\dot{D}}(\dot{\xi}, \dot{\eta}) = \dot{G}(\dot{\xi}, \dot{\eta}) - \dot{G}(\dot{\xi}, \dot{\eta}') \tag{4.3}$$

because an application of the superparabolic minimum theorem shows that $\dot{G}(\cdot, \dot{\eta}') = \dot{\mathrm{G}}\mathrm{M}_{\dot{D}}\dot{G}(\cdot, \dot{\eta})$ on $\dot{D}$. More generally, if $j = 1, \ldots, N$ successively and if for each j the set D_j is either $\mathbb{R}$ or a half-space of $\mathbb{R}$, define

$$\dot{D} = \left(\prod_{j=1}^{N} D_j\right) \times \mathbb{R}, \qquad \dot{D}_j = D_j \times \mathbb{R}.$$

Then the function

$$(\dot{\xi},\dot{\eta}) = ((\xi^{(1)},\ldots,\xi^{(N)},s),(\eta^{(1)},\ldots,\eta^{(N)},t)) \mapsto \dot{G}_{\dot{D}}(\dot{\xi},\dot{\eta})$$

is the product of the Green functions for $N = 1$ of $\dot{D}_1,\ldots,\dot{D}_N$ written in the respective variables $((\xi^{(1)},s),(\eta^{(1)},t)),\ldots,((\xi^{(N)},s),(\eta^{(N)},t))$.

Strict Positivity Set of $\dot{G}_{\dot{D}}$

$\dot{G}_{\dot{D}}(\dot{\xi},\dot{\eta}) > 0$ if and only if $\dot{\eta}$ is in the set $\dot{D}_{\dot{\xi}}$ of points of $\dot{D}$ strictly below $\dot{\xi}$ relative to $\dot{D}$. To see this, observe first that the inequality $0 \le \dot{G}_{\dot{D}} \le \dot{G}$ implies that $\dot{G}_{\dot{D}}(\dot{\xi},\dot{\eta}) = 0$ if $\operatorname{ord}\dot{\eta} \ge \operatorname{ord}\dot{\xi}$. Next observe that (coparabolic maximum-minimum theorem) if $\dot{G}_{\dot{D}}(\dot{\xi},\dot{\eta}) = 0$ and if $\dot{\eta}'$ is strictly above $\dot{\eta}$ relative to $\dot{D}$, then $\dot{G}_{\dot{D}}(\dot{\xi},\dot{\eta}') = 0$. Since $\dot{G}_{\dot{D}}(\dot{\xi},\cdot) > 0$ at points arbitrarily close to $\dot{\xi}$ and strictly below $\dot{\xi}$, it follows that $\dot{G}_{\dot{D}}(\dot{\xi},\dot{\eta}) > 0$ whenever $\dot{\eta}\in\dot{D}_{\dot{\xi}}$. If now we fix $\dot{\xi}$ and define $\dot{u}(\dot{\eta}) = \dot{G}_{\dot{D}}(\dot{\xi},\dot{\eta})$ when $\dot{\eta}\in\dot{D}_{\dot{\xi}}$ and define $\dot{u}(\dot{\eta}) = 0$ when $\dot{\eta}\in\dot{D} - \dot{D}_{\dot{\xi}}$, then $\dot{u}$ is cosuperparabolic on $\dot{D}$ and $\dot{G}_{\dot{D}}(\dot{\xi},\cdot) - \dot{u}$ is a positive coparabolic function on $\dot{D}$. It follows that this difference vanishes identically; so $\dot{D}_{\dot{\xi}}$ is the strict positivity set of $\dot{G}_{\dot{D}}(\dot{\xi},\cdot)$, as asserted.

Extensions and Contractions of Green Functions

If $\dot{\zeta}$ is a point of $\dot{D}$, then $\dot{G}_{\dot{D}_{\dot{\zeta}}}$ is the restriction of $\dot{G}_{\dot{D}}$ to $\dot{D}_{\dot{\zeta}} \times \dot{D}_{\dot{\zeta}}$. In fact this restriction has the required properties except perhaps the property that $\mathring{\text{G}}\text{M}_{\dot{D}_{\dot{\zeta}}}\dot{G}_{\dot{D}}(\dot{\xi},\cdot) \equiv 0$ when $\dot{\xi}\in\dot{D}_{\dot{\zeta}}$ (or the equivalent dual property). Now if $\dot{h}$ is a positive coparabolic function on $\dot{D}_{\dot{\zeta}}$ and is a minorant of $\dot{G}_{\dot{D}}(\dot{\xi},\cdot)$, then $\dot{h} = 0$ at the points of $\dot{D}_{\dot{\zeta}}$ with ordinate values $\ge \operatorname{ord}\dot{\xi}$; so if $\dot{h}$ is extended to $\dot{D}$ by defining $\dot{h} = 0$ on $\dot{D} - \dot{D}_{\dot{\zeta}}$, the extended function is a positive coparabolic minorant of $\dot{G}_{\dot{D}}(\dot{\xi},\cdot)$ on $\dot{D}$ and so vanishes identically. Hence $\dot{h} \equiv 0$ on $\dot{D}$; so $\mathring{\text{G}}\text{M}_{\dot{D}_{\dot{\zeta}}}\dot{G}_{\dot{D}}(\dot{\xi},\cdot) \equiv 0$, as required. A similar dual argument shows that the Green function of the set $\dot{D}_0$ of points of $\dot{D}$ with ordinate values strictly greater than $\operatorname{ord}\dot{\zeta}$ is the restriction of $\dot{G}_{\dot{D}}$ to $\dot{D}_0 \times \dot{D}_0$.

An observation in the reverse direction will be useful. Suppose that $\dot{\xi}$ is a finite Euclidean boundary point of $\dot{D}$, that every point of $\dot{D}$ is strictly below $\dot{\xi}$, and that the part of some open neighborhood $\dot{B}$ of $\dot{\xi}$ strictly below $\dot{\xi}$ is in $\dot{D}$. Then for $\dot{\eta}$ in $\dot{D}$ the function $\dot{G}_{\dot{D}}(\cdot,\dot{\eta})$ has the extension $\dot{G}_{\dot{D}\cup\dot{B}}(\cdot,\dot{\eta})$, parabolic on $\dot{D}\cup\dot{B} - \{\dot{\eta}\}$.

5. Potentials

If $\dot{D}$ is a nonempty open subset of $\dot{\mathbb{R}}^N$ and if $\dot{\mu}$ is a measure on $\dot{D}$, the functions

$$\dot{G}_{\dot{D}}\dot{\mu} = \int_{\dot{D}} \dot{G}_{\dot{D}}(\cdot,\dot{\eta})\dot{\mu}(d\dot{\eta}), \qquad \dot{\mu}\dot{G}_{\dot{D}} = \int_{\dot{D}} \dot{G}_{\dot{D}}(\dot{\xi},\cdot)\dot{\mu}(d\dot{\xi})$$

are, respectively, the (Green) potential and copotential of $\dot{\mu}$. Since the properties of copotentials follow trivially from those of potentials, we shall consider only the latter unless the interplay between the two is involved. The potential $\dot{G}_{\dot{D}}\dot{\mu}$ is lower semicontinuous (Fatou's lemma) and has the superparabolic function average property (Fubini's theorem) so $\dot{G}_{\dot{D}}\dot{\mu}$ is superparabolic on $\dot{D}$ if to each point $\dot{\eta}$ of $\dot{D}$ corresponds a point of finiteness of $\dot{G}_{\dot{D}}\dot{\mu}$ above $\dot{\eta}$ relative to $\dot{D}$. If superparabolic, the potential $\dot{G}_{\dot{D}}\dot{\mu}$ is parabolic off the closed support of $\dot{\mu}$. If $\dot{\mu}(\dot{D}) < +\infty$, the counterpart of an argument in Section IV.1 shows that $\dot{G}_{\dot{D}}\dot{\mu}$ is superparabolic on $\dot{D}$. In particular suppose that $\dot{\mu}$ has compact support $\dot{A}$ in $\dot{D}$, and let $\dot{B}$ be a neighborhood of $\dot{A}$. Since $\dot{G}_{\dot{D}}$ is bounded on the set $(\dot{D} - \dot{B}) \times \dot{A}$ it follows that the superparabolic potential $\dot{G}_{\dot{D}}\dot{\mu}$ is bounded on $\dot{D} - \dot{B}$ and has limit 0 at every boundary point $\dot{\zeta}$ of $\dot{D}$ for which $\lim_{\dot{\xi}\to\dot{\zeta}} \dot{G}_{\dot{D}}(\dot{\xi}, \dot{\eta}) = 0$ when $\dot{\eta} \in \dot{A}$.

If $\dot{G}_{\dot{D}}\mu$ is a superparabolic potential then $\dot{G}\mathrm{M}_{\dot{D}}\dot{G}_{\dot{D}}\dot{\mu} \equiv 0$ by the counterpart of the proof of the corresponding classical context result (Section IV.3).

Example (a). Let $\dot{u}$ be the indicator function of the upper half-space $\dot{A}$ of $\dot{\mathbb{R}}^N$. Then $\dot{u}$ is a superparabolic function on $\dot{D} = \dot{\mathbb{R}}^N$ and is the potential of the measure l_N on the abscissa hyperplane. This potential is discontinuous and vanishes on the support of its measure. Thus the domination principle (Theorem V.10) is false in the parabolic context. A trivial variation of this example in which the measure has compact support shows that the Evans–Vasilesco theorem (Section V.8) is also false in the parabolic context. Versions of the domination principle adapted to the parabolic context will be proved in Section XVIII.16.

Example (b). Define $\dot{D} = \mathbb{R}^N \times]0, \delta[$ and $\dot{D}' = \mathbb{R}^N \times]-\infty, \delta[$ with $0 < \delta \leq +\infty$. Recall (Section 4) that $\dot{G}_{\dot{D}}$ $[\dot{G}_{\dot{D}'}]$ is the restriction of $\dot{G}$ to $\dot{D} \times \dot{D}$ $[\dot{D}' \times \dot{D}']$. Let $\dot{v}$ be a positive superparabolic function on $\dot{D}$ with associated Riesz measure $\dot{\nu}$ and extend $\dot{v}$ to $\dot{v}'$ on $\dot{D}'$ by setting $\dot{v}' = 0$ on $\dot{D}' - \dot{D}$. Then $\dot{v}'$ is superparabolic and $\dot{\nu}$ is the projection on $\dot{D}$ of the Riesz measure $\dot{\nu}'$ associated with $\dot{v}'$. For example, if $\delta = +\infty$ and $\dot{v} \equiv 1$, the function $\dot{v}'$ is the potential $\dot{u}$ on $\dot{\mathbb{R}}^N$ discussed in Example (a). Now suppose that $\dot{v}$ is a potential, $\dot{v} = \dot{G}_{\dot{D}}\dot{\nu} = \dot{G}_{\dot{D}'}\dot{\nu}$ on $\dot{D}$. Since $\dot{G}_{\dot{D}'}\dot{\nu} = 0$ on $\dot{D}' - \dot{D}$, it follows that $\dot{v}' = \dot{G}_{\dot{D}'}\dot{\nu}$, and this representation shows that $\dot{D}' - \dot{D}$ is $\dot{\nu}'$ null. On the other hand, if $\dot{v}$ is parabolic there is (by Theorem XVI.6) a measure $\dot{N}_{\dot{v}}$ on the abscissa hyperplane such that $\dot{v}' = \dot{G}_{\dot{D}'}\dot{N}_{\dot{v}}$. Thus in this case $\dot{N}_{\dot{v}}$ is the projection of $\dot{\nu}'$ on the abscissa hyperplane. We have now proved that whatever the choice of $\dot{v}$ on $\dot{D}$, the extension $\dot{v}'$ to $\dot{D}'$ is a potential. One way of phrasing the fact (Theorem XVI.6) that a positive parabolic function on the slab $\dot{D}$ is given by a Poisson–Stieltjes integral is to state that every such function when extended by 0 to $\dot{D}'$ becomes a potential. This fact suggests that one way of deriving the Poisson–Stieltjes representation is to prove directly that the extended function $\dot{v}'$ is a potential; it is trivial that the corresponding Riesz measure must be supported by the abscissa hyperplane.

EXAMPLE (c). Consider the following potential $\dot{G}\dot{\mu}$ in $\dot{\mathbb{R}}^N$, for which $\dot{\mu}$ is supported by the surface $\{(\eta, t): t = -|\eta|^4, |\eta| < 1\}$,

$$\dot{v}(\xi, s) = \begin{cases} \displaystyle\int_{\{-s<|\eta|^4<1\}} \frac{\exp \dfrac{-|\xi-\eta|^2}{2\sigma^2(s+|\eta|^4)}}{(s+|\eta|^4)^{N/2}} \left(|\eta|^{N+1/2} \exp \frac{1}{2\sigma^2|\eta|^2}\right) l_N(d\eta) & \text{if } s > -1, \\ 0 & \text{if } s \leq -1. \end{cases} \tag{5.1}$$

Obviously $\dot{v}(\xi, s) = +\infty$ when $s > 0$; so $\dot{v}$ is not superparabolic. On the other hand, if β is any constant the function $\beta \wedge \dot{v}$ is superparabolic. The evaluation

$$\dot{v}(\xi, 0) = \int_{\{|\eta|<1\}} \left(\exp \frac{-|\xi|^2 + 2\langle\xi, \eta\rangle}{2\sigma^2|\eta|^4}\right) |\eta|^{-N+1|2} l_N(d\eta) \tag{5.2}$$

shows that $\dot{v}(\cdot, 0)$ is a finite-valued function of $|\xi|$. We now show that $\lim_{\xi\to 0} \dot{v}(\xi, 0) = +\infty$, for which it is sufficient to show that $\dot{v}(\cdot, 0)$ has limit $+\infty$ along a ray in $\mathbb{R}^N$ to the origin. Let ξ be on this ray from now on, and let θ be the angle between the rays from the origin to ξ and η. Then for $|\xi| < \frac{1}{2}$,

$$\dot{v}(\xi, 0) \geq \int_{\substack{\{|\xi|<|\eta|<2|\xi|\}\\ \{|\theta|<\pi/6\}}} \left(\exp \frac{\alpha|\xi|^2}{|\eta|^4}\right) |\eta|^{-N+1/2} l_N(d\eta), \qquad \alpha = \frac{3^{1/2}-1}{2\sigma^2}. \tag{5.3}$$

Since the integrand for fixed $|\xi|$ is a function of $|\eta|$, there is a constant c such that

$$\dot{v}(\xi, 0) \geq c \int_{|\xi|}^{2|\xi|} r^{-1/2} \exp \frac{\alpha|\xi|^2}{r^4} l_1(dr) \geq c 2^{-1/2} |\xi|^{1/2} \exp \frac{\alpha}{16|\xi|^2} \tag{5.4}$$

and therefore $\dot{v}(\cdot, 0)$ has limit $+\infty$ at the origin, as stated.

Thinness

Recall that a superparabolic function has limit inferior at a point equal to its value at that point. In Example (a) the value $\dot{u}(0)$ is strictly less than the limit of $\dot{u}$ at 0 along the upper half-space. In this sense the upper half-space of $\dot{\mathbb{R}}^N$ is thin at the origin. The abscissa hyperplane is thin at the origin in this same sense because in Example (c) the superparabolic function $\beta \wedge \dot{v}$ with $\beta > \dot{v}(0, 0)$ is strictly less at the origin than its limit along the abscissa hyperplane. This concept of thinness, the counterpart of the classical concept, will be given a topological interpretation (the parabolic-fine topology) in Section 9.

Dependence of $\dot{G}_{\dot{D}}\dot{\mu}$ on $\dot{D}$

(Cf. Sections IV.1 and VII.5.) If $\dot{D}_1 \subset \dot{D}_2$, the difference $\dot{G}_{\dot{D}_2} - \dot{G}_{\dot{D}_1}$ on $\dot{D}_1 \times \dot{D}_1$ is a positive function, parabolic in the first argument and coparabolic in the second. If $\dot{\mu}$ is a measure on $\dot{D}_1$ whose potential $\dot{G}_{\dot{D}_2}\dot{\mu}$ is superparabolic, then the difference $\dot{G}_{\dot{D}_2}\dot{\mu} - \dot{G}_{\dot{D}_1}\dot{\mu}$ if defined suitably at the common infinities of the two potentials is positive and parabolic on $\dot{D}_1$.

6. The Smoothness of Potentials

The following theorem is the parabolic context analog of Theorem I.7. In view of the dependence of $\dot{G}_{\dot{D}}\dot{\mu}$ on $\dot{D}$ described at the end of Section 5 the smoothness conclusions of the theorem are applicable in their obvious adaptations to potentials $\dot{G}_{\dot{D}}\dot{\mu}$ as well as $\dot{G}\dot{\mu}$.

Theorem. *Suppose that $d\dot{\mu} = \dot{f}\,dl_{N+1}$, where $\dot{f}$ is l_{N+1} measurable and is supported by a slab $\mathbb{R}^N \times [a, +\infty[$ with $a > -\infty$.*

(a) *If $\dot{f}$ is bounded, the potential $\dot{u} = \dot{G}\dot{\mu}$ is finite and continuous on $\dot{\mathbb{R}}^N$ and has continuous first partial derivatives with respect to the space variables, given by formal derivation of the integral defining $\dot{u}$.*

(b) *If in* (a) *$\dot{f}$ is continuous and satisfies a uniform Lipschitz condition of exponent β, $0 < \beta \leq 1$, in the space variables,*

$$|\dot{f}(\eta, s) - \dot{f}(\xi, s)| \leq \text{const}\,|\eta - \xi|^\beta, \tag{6.1}$$

then $\dot{u}$ has continuous second partial derivatives in the space variables and a continuous first partial derivative in the ordinate variable, given by

$$\frac{\partial^2 \dot{u}(\dot{\xi})}{\partial \xi^{(i)} \partial \xi^{(j)}} = \int_{\dot{\mathbb{R}}^N} \frac{\partial^2 \dot{G}(\dot{\xi}, \dot{\eta})}{\partial \xi^{(i)} \partial \xi^{(j)}} [\dot{f}(\eta, t) - \dot{f}(\xi, t)] l_{N+1}(d\dot{\eta})$$
$$\dot{\xi} = (\xi, s) = (\xi^{(1)}, \ldots, \xi^{(N)}, s) \tag{6.2}$$
$$\dot{\eta} = (\eta, t) = (\eta^{(1)}, \ldots, \eta^{(N)}, t)$$

$$\frac{\partial \dot{u}(\dot{\xi})}{\partial s} = \int_{\dot{\mathbb{R}}^N} \frac{\partial}{\partial s} \dot{G}(\dot{\xi}, \dot{\eta}) [\dot{f}(\eta, t) - \dot{f}(\xi, t)] l_{N+1}(d\dot{\eta}) + \dot{f}(\dot{\xi}). \tag{6.3}$$

Hence $\dot{\Delta}\dot{u} = -\dot{f}$.

Proof of (a). Since $\int_{\mathbb{R}^N} \dot{G}(\dot{\xi}, (\eta, t)) l_N(d\eta) \le 1$, the potential $\dot{u}$ is bounded under the hypotheses of (a). To prove continuity, observe that if $\dot{u}_\delta$ is defined by

$$\dot{u}_\delta(\dot{\xi}) = \int_{-\infty}^{s-\delta} dt \int_{\mathbb{R}^N} \dot{G}(\dot{\xi}, (\eta, t)) \dot{f}(\dot{\eta}) l_N(d\eta), \tag{6.4}$$

the function $\dot{u}_\delta$ is continuous (dominated convergence theorem), and

$$|\dot{u} - \dot{u}_\delta| \le \sup |\dot{f}| \int_{s-\delta}^{s} dt = \delta \sup |\dot{f}|; \tag{6.5}$$

so $\lim_{\delta \to 0} \dot{u}_\delta = \dot{u}$ uniformly on $\dot{\mathbb{R}}^N$, and $\dot{u}$ is continuous. Formal differentiation of $\dot{u}$ yields

$$\frac{\partial \dot{u}(\dot{\xi})}{\partial \xi^{(i)}} = \int_{\dot{\mathbb{R}}^N} \frac{\partial \dot{G}(\dot{\xi}, \dot{\eta})}{\partial \xi^{(i)}} \dot{f}(\dot{\eta}) l_{N+1}(d\dot{\eta}) = -\frac{1}{\sigma^2} \int_{\dot{\mathbb{R}}^N} \dot{G}(\dot{\xi}, \dot{\eta}) \frac{\xi^{(i)} - \eta^{(i)}}{s - t} \dot{f}(\dot{\eta}) l_{N+1}(d\dot{\eta}). \tag{6.6}$$

The last integral in (6.6) converges absolutely because if $t < s$,

$$\int_{\mathbb{R}^N} \dot{G}(\dot{\xi}, \dot{\eta}) |\xi^{(i)} - \eta^{(i)}| (s - t)^{-1} l_N(d\eta) = \left(\frac{2}{\pi}\right)^{1/2} \sigma (s - t)^{-1/2}. \tag{6.7}$$

The last integral in (6.6) defines a continuous function of $\dot{\xi}$ because if the slab $\mathbb{R}^N \times [s - \delta, s[$ is excluded from the domain of integration, continuity becomes trivial, and the error in excluding this slab is [by (6.7)] at most $\sigma 2 \delta^{1/2} \sup |\dot{f}|$. Hence (6.6) is true. □

Proof of (b). Under the hypotheses of (b) the integral on the right in (6.2) is absolutely convergent, in view of the majorant of the integrand provided by XV(4.4) with $k = 2$. Moreover the integral on the right in (6.2) defines a continuous function of $\dot{\xi}$ by an argument following that used in proving continuity of the two integrals in the proof of (a). Since the function $\eta \mapsto \partial \dot{G}(\dot{\xi}, \dot{\eta}) / \partial \xi^{(i)}$ is odd about ξ, the first equality in (6.6) can be written in the form

$$\frac{\partial \dot{u}(\dot{\xi})}{\partial \xi^{(i)}} = \int_{\dot{\mathbb{R}}^N} \frac{\partial \dot{G}(\dot{\xi}, \dot{\eta})}{\partial \xi^{(i)}} [\dot{f}(\eta, t) - \dot{f}(\xi, t)] l_{N+1}(d\dot{\eta}). \tag{6.8}$$

If the integral on the right in (6.2) is integrated in $\xi^{(j)}$ over an interval but evaluated by first integrating in η, then $\xi^{(j)}$, then t, the result is the difference between the values of $\partial \dot{u}(\dot{\xi}) / \partial \xi^{(i)}$ at the endpoints of the interval. It follows that (6.2) is correct. The absolute convergence of the integral in (6.3) and the continuity of the function of $\dot{\xi}$ thereby defined are proved just as the corresponding assertions for the other integrals were proved. Moreover

$$\dot u(\dot\xi) = \int_{\dot{\mathbb{R}}^N} \dot G(\dot\xi, \dot\eta)[f(\eta, t) - f(\xi, t)]\, l_{N+1}(d\dot\eta) + \int_{-\infty}^{s} f(\xi, t)\, dt, \tag{6.9}$$

and differentiation in (6.9) yields (6.3). □

7. Riesz Decomposition Theorem

Theorem. *If $\dot D$ is an arbitrary nonempty open subset of $\dot{\mathbb{R}}^N$ and if $\dot u$ is a superparabolic function on $\dot D$, there is a unique measure $\dot\mu$ on $\dot D$ with the following properties.*

(a) *If $\dot B$ is an open nonempty relatively compact subset of $\dot D$ and if $\dot\mu_{\dot B}$ is the projection of $\dot\mu$ on $\dot B$, there is a superparabolic function $\dot h_{\dot B}$ on $\dot D$, parabolic on $\dot B$, with $\dot u = \dot G\dot\mu_{\dot B} + \dot h_{\dot B}$.*

(b) *If in addition $\dot u$ has a subparabolic minorant, then $\dot u$ has the representation $\dot u = \dot G_{\dot D}\dot\mu + \dot{\mathrm{G}}\mathrm{M}_{\dot D}\dot u$.*

The proof is a translation into the present context of the corresponding classical Riesz theorem for superharmonic functions (Theorems IV.7 and IV.8) and will therefore be omitted.

The Riesz decomposition leads at once to the following facts for a positive superparabolic function $\dot u$ on $\dot D$ (cf. their counterparts for the classical context in Sections IV.8 and IV.10).

(a) The function $\dot u$ is a potential $\dot G_{\dot D}\dot\mu$ if and only if $\dot{\mathrm{G}}\mathrm{M}_{\dot D}\dot u = 0$.

(b) If $\dot u$ is majorized by a superparabolic potential, then $\dot u$ is itself a potential.

(c) Special case of (b). If $\dot u$ is a superparabolic potential, $\underset{+\dot u}{\dot R^{\dot A}}$ is also, for every $\dot A$; in particular, $\underset{+\dot u}{\dot R^{\partial\dot D}} \equiv 0$ for every choice of $\partial\dot D$.

(d) The smoothed reduction $\underset{+\dot u}{\dot R^{\dot A}}$ is a potential if $\dot A$ is a relatively compact subset of $\dot D$.

(e) The function $\dot u$ is the limit of an increasing sequence of bounded infinitely differentiable potentials of measures with compact supports.

The formulation of the full counterpart of Theorem IV.10 is left to the reader; this counterpart is a slight extension of (e).

8. Parabolic-Polar Sets

A *parabolic-polar subset* of $\dot{\mathbb{R}}^N$ ($N \geq 1$) is defined as a subset $\dot A$ satisfying the following equivalent conditions:

(a) To each point of $\dot{A}$ corresponds an open neighborhood of the point which carries a superparabolic function identically $+\infty$ on the part of $\dot{A}$ in that neighborhood.

(b) If $\dot{D}$ is an open superset of $\dot{A}$, there is a function superparabolic on $\dot{D}$ and identically $+\infty$ on $\dot{A}$. This function can be chosen to be the potential $\dot{G}_{\dot{D}}\dot{\mu}$ of a finite measure.

Observe that in the discussion of parabolic-polar sets the case $N = 1$ is not exceptional. The equivalence of (a) and (b) is proved just as in the classical context (Sections V.1 and V.2). As in the classical context, a polar set is a subset of a G_δ polar set. A set all of whose compact subsets are parabolic polar will be called *inner parabolic-polar*.

Counterparts of Theorems V.3 to V.5.

A countable union of parabolic-polar sets is parabolic polar. If $\dot{A}$ is parabolic polar, the smoothed reduction of a positive superparabolic function on $\dot{A}$ vanishes identically; conversely, if the smoothed reduction of some strictly positive superparabolic function on a set $\dot{A}$ vanishes identically, then $\dot{A}$ is parabolic polar. The qualification "strictly positive" is necessary in this converse because, for example, if $\dot{A}$ is the abscissa hyperplane and $\dot{v}(\dot{\xi}) = 1_{\{\mathrm{ord}\,\cdot > 0\}}(\dot{\xi})$, then (reduction relative to $\dot{\mathbb{R}}^N$) $\hat{\dot{v}}^{\dot{A}} = 0$ even though $\dot{A}$ is not parabolic polar. If $\dot{A}$ is parabolic polar and $\dot{\xi}$ is a point not in $\dot{A}$, there is a positive superparabolic function finite at $\dot{\xi}$ but identically $+\infty$ on $\dot{A}$. This result implies, as in the classical context [Section VI.4(c)], that if $\dot{v}$ is a positive superparabolic function on an open subset $\dot{D}$ of $\dot{\mathbb{R}}^N$ and if $\dot{A}_1$ and $\dot{A}_2$ are subsets of $\dot{D}$ differing by a parabolic-polar set, then the reductions of $\dot{v}$ on $\dot{A}_1$ and $\dot{A}_2$ coincide off the symmetric difference of these sets, and the smoothed reductions of $\dot{v}$ on $\dot{A}_1$ and $\dot{A}_2$ coincide on $\dot{D}$. The classical context extension Theorem V.5 translates directly into the parabolic context along with its applications. For example, if $\dot{u}$ is subparabolic on $\dot{D}$ and if $\dot{A}$ is a closed in $\dot{D}$ parabolic-polar set, null for the measure associated with $\dot{u}$, then $\dot{\mathrm{L}}\mathrm{M}\,\dot{u}$ relative to $\dot{D}$ and $\dot{\mathrm{L}}\mathrm{M}\,\dot{u}$ relative to $\dot{D} - \dot{A}$ are equal on $\dot{D} - \dot{A}$. It follows that $\dot{G}_{\dot{D}} = \dot{G}_{\dot{D}-\dot{A}}$ on $(\dot{D} - \dot{A}) \times (\dot{D} - \dot{A})$. The extended superharmonic function minimum theorem (Theorem V.7) becomes (for all N): If $\dot{u}$ is a lower-bounded superparabolic function on an open subset $\dot{D}$ of $\dot{\mathbb{R}}^N$ and if at parabolic quasi every finite point $\dot{\zeta}$ of $\partial\dot{D}$ and also at $\dot{\zeta} = \infty$ if $\dot{D}$ is unbounded, $\liminf_{\dot{\eta}\to\dot{\zeta}} \dot{u}(\dot{\eta}) \geq c$, then $\dot{u} \geq c$ on $\dot{D}$.

Minimality of $\dot{G}_{\dot{D}}(\cdot, \dot{\zeta})$

In the classical context it was proved in Section VII.10 as an application of an extension theorem that the restriction of $G_D(\zeta, \cdot)$ to $D - \{\zeta\}$ is minimal harmonic, and it then followed easily that for v positive and superharmonic on D, with associated Riesz measure ν,

$$\inf_{D-\{\zeta\}} \frac{v}{G_D(\zeta,\cdot)} = \nu(\{\zeta\}). \tag{8.1}$$

The corresponding argument in the present context shows that the restriction of $\dot{G}_{\dot{D}}(\cdot,\dot{\zeta})$ to $\dot{D} - \{\dot{\zeta}\}$ is minimal parabolic and that if $\dot{v}$ is positive and superparabolic on $\dot{D}$, with associated Riesz measure $\dot{\nu}$,

$$\inf_{\{\dot{\eta}:\dot{G}_{\dot{D}}(\dot{\eta},\dot{\zeta})>0\}} \frac{\dot{v}(\dot{\eta})}{\dot{G}_{\dot{D}}(\dot{\eta},\dot{\zeta})} = \dot{\nu}(\{\dot{\zeta}\}). \tag{8.2}$$

Similarly the restriction of $\dot{G}_{\dot{D}}(\dot{\zeta},\cdot)$ to $\dot{D} - \{\dot{\zeta}\}$ is minimal coparabolic, and the dual of (8.2) has the obvious formulation.

Example (a) (Classical-polar versus parabolic-polar sets). Let $\dot{A}$ be a subset of $\dot{\mathbb{R}}^N$ with projection A on the abscissa hyperplane. If A is classical polar, then $\dot{A}$ is parabolic polar, and in fact $A \times \mathbb{R}$ is parabolic polar because a superharmonic function on $\mathbb{R}^N$ identically $+\infty$ on A can be considered as a superparabolic function on $\dot{\mathbb{R}}^N$ (Section XV.15) and as such is identically $+\infty$ on $A \times \mathbb{R}$. In particular, when $N > 1$ a line in $\dot{\mathbb{R}}^N$ parallel to the ordinate axis is parabolic polar. Conversely, it will be proved in Section XVIII.11 that A is classical polar whenever $A \times \mathbb{R}$ is parabolic polar.

Example (b). If $N \geq 1$, a singleton is parabolic polar. In fact, if $N > 1$, this follows from Example (a). To prove the assertion for $N \geq 1$, suppose that the singleton is $\{0\}$, and let $\dot{\mu}$ be the measure supported by the set of points $\{(0, -2^{-k}), k \geq 1\}$ with $\dot{\mu}\{(0, -2^{-k})\} = k^{-2}$. The superparabolic potential $\dot{G}\dot{\mu}$ is $+\infty$ at the origin; so $\{0\}$ is parabolic polar. It follows that countable subsets of $\dot{\mathbb{R}}^N$ are parabolic polar.

Example (c). Let A be an l_N measurable subset of $\mathbb{R}^N$, with $l_N(A) > 0$. The set $A \times \{0\}$ is not parabolic polar in $\dot{\mathbb{R}}^N$ because a positive superparabolic function on $\dot{\mathbb{R}}^N$, identically $+\infty$ on $A \times \{0\}$, would be identically $+\infty$ on the upper half-space (Lemma XVI.2).

Coparabolic-Polar Sets

A *coparabolic-polar subset* of $\dot{\mathbb{R}}^N$ is defined as a set satisfying conditions (a) and (b) at the beginning of this section with "superparabolic" replaced by "cosuperparabolic." That is, a set $\dot{A}$ is coparabolic polar if and only if its reflection in the abscissa hyperplane is parabolic polar. The application of this section to coparabolic-polar sets is obvious. It will be proved in Section XVIII.11 that a set is parabolic polar if and only if it is coparabolic polar.

9. The Parabolic-Fine Topology

It will be convenient to introduce the fine topology in the parabolic context at a much earlier stage than in the classical context. In the following discussion otherwise unspecified topological concepts refer to the Euclidean topology. In the present context there are two fine topologies, the parabolic and the coparabolic. All concepts relative to the parabolic-fine topology are trivially translatable into the corresponding concepts relative to the coparabolic-fine topology by a reflection of $\dot{\mathbb{R}}^N$ in the abscissa hyperplane; so no separate discussion of the coparabolic-fine topology is given.

The *parabolic-fine topology* of $\dot{\mathbb{R}}^N$ $(N \geq 1)$ is defined as the coarsest topology of $\dot{\mathbb{R}}^N$ making every superparabolic function (equivalently, every upper-bounded superparabolic function) on $\dot{\mathbb{R}}^N$ continuous. Since linear functions of the space coordinate functions are parabolic and since the ordinate function is superparabolic, it follows that every Euclidean open interval in $\dot{\mathbb{R}}^N$ is parabolic-fine open. Hence the parabolic-fine topology is at least as fine as the Euclidean topology, and in fact it is finer because there are superparabolic functions discontinuous in the Euclidean topology; for example, the indicator function of the (open) upper half-space is superparabolic and thus is parabolic-fine continuous. The parabolic-fine continuity of this function implies that no point of the abscissa hyperplane is a parabolic-fine limit point of the upper half-space; that is, the closure of the lower half-space is a parabolic-fine neighborhood of each point of the abscissa hyperplane. This result can be strengthened as follows. Let $\dot{v}$ be the potential on $\dot{\mathbb{R}}^N$ defined in Section 5, Example (c), for which $\dot{v} = +\infty$ on the upper half-space but $\dot{v} < +\infty$ otherwise and $\dot{v}$ has limit $+\infty$ at the origin along the abscissa hyperplane. If $\beta > \dot{v}(0)$, the function $\dot{v} \wedge \beta$ is superparabolic and at the origin is strictly less than its limit there along the closure of the upper half-space. Since $\dot{v} \wedge \beta$ is parabolic-fine continuous, it follows that the origin is not a parabolic-fine limit point of the closure of the upper half-space; that is, the lower half-space is a deleted parabolic-fine neighborhood of the origin and so also of every point of the abscissa hyperplane. This fact implies that the abscissa hyperplane, in fact each hyperplane parallel to it, has no parabolic-fine limit point even though the hyperplane is not parabolic-polar. Recall however that according to Theorem XI.6, every nonpolar set in the classical context has a fine limit point. A more elegant example showing that the lower half-space is a deleted parabolic-fine neighborhood of each point of the abscissa hyperplane is furnished by the Green function $\dot{G}$. According to the dual of Theorem XVIII.14(f) (whose proof depends on an analysis of reductions not available at this stage), for fixed $\dot{\xi}$ in $\dot{\mathbb{R}}^N$,

$$\operatorname{pf}\lim_{\dot{\eta} \to \dot{\xi}} \dot{G}(\dot{\xi}, \dot{\eta}) = +\infty.$$

Thus the set $\{\dot{\eta}\colon \dot{G}(\dot{\xi}, \dot{\eta}) > c\}$ is a parabolic-fine deleted neighborhood of $\dot{\xi}$ for every constant c.

A set will be said to be "*parabolic thin*" at a point if the point is not a parabolic-fine limit point of the set. Limit concepts relative to the parabolic-fine topology will be distinguished by the prefix pf, for example, pf lim sup. The set of parabolic-fine limit points of a set $\dot{A}$ will be denoted by $\dot{A}^{pf}$. An asterisk will be added for the corresponding coparabolic concept: p*f lim sup, $\dot{A}^{p*f}$.

The following properties of the parabolic-fine topology are derived in essentially the same way as the corresponding properties of the classical fine topology and the proofs are therefore omitted.

(a) The restriction of the parabolic-fine topology to an open subset $\dot{D}$ of $\dot{\mathbb{R}}^N$ is the coarsest topology making continuous every superparabolic function with domain $\dot{D}$.
(b) The parabolic-fine topology has a basis of (Euclidean) compact sets.
(c) The parabolic-fine topology has the Baire property that the intersection of a sequence of parabolic-fine open parabolic-fine dense sets is parabolic-fine dense.
(d) If $\dot{A}$ has limit point $\dot{\xi}$ then $\dot{A}$ is parabolic thin at $\dot{\xi}$ if and only if there is a superparabolic function $\dot{u}$ defined on a neighborhood of $\dot{\xi}$ such that

$$\liminf_{\dot{A} \ni \dot{\eta} \to \dot{\xi}} \dot{u}(\dot{\eta}) > \dot{u}(\dot{\xi}). \tag{9.1}$$

A point $\dot{\xi}$ is a parabolic-fine limit point of a set $\dot{A}$ if and only if $\dot{\xi}$ is a Euclidean limit point of $\dot{A}$ and if each superparabolic function $\dot{u}$ defined on an open neighborhood of $\dot{\xi}$ has $\dot{u}(\dot{\xi})$ as a cluster value at $\dot{\xi}$ along $\dot{A}$. For example, according to XV(14.1) and the example in this section, if $\dot{A}$ is the complement of an l_{N+1} null set, the part of A strictly below a point of $\dot{\mathbb{R}}^N$ has that point as parabolic-fine limit point. Hence the space $\dot{\mathbb{R}}^N$ has no parabolic-fine isolated point, and it follows that a nonempty parabolic-fine open subset $\dot{A}$ of $\dot{\mathbb{R}}^N$ is parabolic-fine dense in itself, $\dot{A} \subset \dot{A}^{pf}$. If $\dot{A}$ is a set parabolic thin at $\dot{\xi}$ and if $\dot{D}$ is an arbitrary open neighborhood of $\dot{\xi}$, there is a potential $\dot{u} = \dot{G}_{\dot{D}}\dot{\mu}$ satisfying (9.1) for which $\dot{\mu}$ has compact support in $\dot{D}$ and the left side of (9.1) is $+\infty$. Thus a parabolic-polar set has no parabolic-fine limit point. As already noted, the converse is false.

10. Semipolar Sets

It was proved in Section XI.6 that in the classical context a set is polar if and only if it has no fine limit point, but it was pointed out in Section 9 that although a parabolic-polar set has no parabolic-fine limit point, the abscissa hyperplane of $\dot{\mathbb{R}}^N$ is not parabolic-polar even though it has no parabolic-fine limit point. The following definition is therefore natural. A subset $\dot{A}$ of $\dot{\mathbb{R}}^N$ will be called *parabolic-semipolar* [*coparabolic-semipolar*] if

$\dot{A}$ is a countable union of sets each of which has no parabolic-[coparabolic-] fine limit point. If "parabolic" ["coparabolic"] is omitted here, that is, if the definition is applied in the classical context, a semipolar set is polar so the terminology "semipolar" is not used in the classical context. It will be shown in Section XVIII.12 that a parabolic-semipolar set is necessarily coparabolic semipolar, and conversely, and that such a set is l_{N+1} null. Trivially a countable union of parabolic-semipolar sets is parabolic semipolar. Thus a countable union of horizontal hyperplanes is parabolic semipolar. Such a union may have parabolic-fine limit points however.

A parabolic-semipolar set $\dot{A}$ is the union of countably many parabolic-fine nowhere dense parabolic-fine closed sets; that is, the complement of $\dot{A}$ is the intersection of a sequence of parabolic-fine open parabolic-fine dense sets, and (Baire property) the complement of $\dot{A}$ is therefore parabolic-fine everywhere dense. It follows that if $\dot{u}$ and $\dot{v}$ are superparabolic functions on some open set $\dot{D}$ and if $\dot{u} = \dot{v}$ (or $\dot{u} \leq \dot{v}$) up to a parabolic-semipolar set, then the relation is true everywhere on $\dot{D}$.

11. Preliminary List of Reduction Properties

Let $\dot{D}$ be an open subset of $\dot{\mathbb{R}}^N$, coupled with a boundary $\partial\dot{D}$ provided by a metric compactification of $\dot{D}$, and let $\dot{A}$ be a subset of $\dot{D} \cup \partial\dot{D}$. In this section we prove certain basic reduction properties, some under restrictions to be removed or weakened later (see Section 16). We shall use repeatedly the fact that in view of the preliminary version of the Fundamental Convergence Theorem in Section 2 an equality or inequality between unsmoothed reductions relative to $\dot{D}$ is true l_{N+1} almost everywhere on $\dot{D}$ for the smoothed reductions and is therefore true everywhere on $\dot{D}$ for the latter.

(a) If $\dot{v}$ is a positive superparabolic function on $\dot{D}$, finite valued and continuous at each point of $\dot{A} \cap \dot{D}$, then

$$\dot{R}_{\dot{v}}^{\dot{A}} = \inf\{\dot{R}_{\dot{v}}^{\dot{B}} : \dot{A} \subset \dot{B},\ \dot{B} \text{ open in } \dot{D} \cup \partial\dot{D}\}. \tag{11.1}$$

See the proof of the corresponding fact in the classical context in Section III.5(e).

(b) If $\dot{A}$ and $\dot{B}$ are open subsets of $\dot{D}$ with $\dot{A} \subset \dot{B}$ and if $\dot{v}$ is a positive superparabolic function on $\dot{D}$, then

$$\dot{\llbracket}\dot{\llbracket}\dot{v}\rrbracket^{\dot{A}}\rrbracket^{\dot{B}} = \dot{\llbracket}\dot{\llbracket}\dot{v}\rrbracket^{\dot{B}}\rrbracket^{\dot{A}} = \dot{\llbracket}\dot{v}\rrbracket^{\dot{A}} \tag{11.2}$$

Just as in the classical context (Section VI.4), in view of the fact that the smoothed successive reductions of $\dot{v}$ on $\dot{A}$ and $\dot{B}$ in either order lie between $\dot{\llbracket}\dot{\llbracket}\dot{v}\rrbracket^{\dot{A}}\rrbracket^{\dot{A}}$ and $\dot{\llbracket}\dot{v}\rrbracket^{\dot{A}}$, it is sufficient to prove that (11.2) is true when $\dot{A} = \dot{B}$. To prove this, we need only observe that since $\dot{\llbracket}\dot{v}\rrbracket^{\dot{A}} = \dot{v}$ on $\dot{A}$ a parabolic function $\dot{u}$ on $\dot{D}$ majorizes $\dot{v}$ on $\dot{A}$ if and only if $\dot{u}$ majorizes $\dot{\llbracket}\dot{v}\rrbracket^{\dot{A}}$ on $\dot{A}$.

(c) If $\dot{u}$ and $\dot{v}$ are positive superparabolic functions on $\dot{D}$ and if $\dot{A}$ is an open subset of $\dot{D}$, then

$$\dot{R}^{\dot{A}}_{\dot{v}_1+\dot{v}_2} = \dot{R}^{\dot{A}}_{\dot{v}_1} + \dot{R}^{\dot{A}}_{\dot{v}_2}. \tag{11.3}$$

See the proof of the corresponding fact in the classical context in Section VI.4. Property (c) will be extended in Section 16(g) to cover countable sums of positive superparabolic functions and arbitrary subsets $\dot{A}$ of $\dot{D} \cup \partial\dot{D}$.

(d) If $\dot{v}$ is a finite-valued continuous positive superparabolic function on $\dot{D}$, the set functions $\dot{A} \mapsto \dot{R}^{\dot{A}}_{\dot{v}}$ and $\dot{A} \mapsto \underset{+\dot{v}}{\dot{R}^{\dot{A}}}$ are strongly subadditive on the class of subsets of $\dot{D} \cup \partial\dot{D}$.

Property (d) will be extended in Section 16(k) where the restriction that $\dot{v}$ be finite valued and continuous will be dropped. To prove (d), we first choose open subsets $\dot{A}$ and $\dot{B}$ of $\dot{D}$, set $\dot{v}' = \underset{+\dot{v}}{\dot{R}^{\dot{A}}} \wedge \underset{+\dot{v}}{\dot{R}^{\dot{B}}}$, and prove

$$\dot{R}^{\dot{A}\cup\dot{B}}_{\dot{v}} + \dot{R}^{\dot{A}\cup\dot{B}}_{\dot{v}'} = \dot{R}^{\dot{A}}_{\dot{v}} + \dot{R}^{\dot{B}}_{\dot{v}}. \tag{11.4}$$

The classical context proof of this reduction equality in Section VI.4 needs no change except simplification: "quasi everywhere" is to be replaced by "everywhere." Moreover, when $\dot{A}$ and $\dot{B}$ are open, $\dot{R}^{\dot{A}\cup\dot{B}}_{\dot{v}'} = \dot{v}$ on $\dot{A} \cap \dot{B}$, so $\dot{R}^{\dot{A}\cup\dot{B}}_{\dot{v}'} \geq \dot{R}^{\dot{A}\cap\dot{B}}_{\dot{v}}$. Thus (11.4) implies the validity of the strong subadditivity inequality

$$\dot{R}^{\dot{A}\cup\dot{B}}_{\dot{v}} + \dot{R}^{\dot{A}\cap\dot{B}}_{\dot{v}} \leq \dot{R}^{\dot{A}}_{\dot{v}} + \dot{R}^{\dot{B}}_{\dot{v}} \tag{11.5}$$

when $\dot{A}$ and $\dot{B}$ are open subsets of $\dot{D}$. In view of (a) this inequality is valid for arbitrary subsets of $\dot{D} \cup \partial\dot{D}$, and the corresponding inequality for smoothed reductions, true l_{N+1} almost everywhere on $\dot{D}$, must be true everywhere on $\dot{D}$.

(e) If $\dot{A}_{\bullet}$ is an increasing sequence of subsets of $\dot{D}$ with union $\dot{A}$ and if $\dot{v}_{\bullet}$ is an increasing sequence of finite-valued positive continuous superparabolic functions on $\dot{D}$ with finite-valued continuous limit $\dot{v}$, then

$$\lim_{n\to\infty} \dot{R}^{\dot{A}_n}_{\dot{v}_n} = \dot{R}^{\dot{A}}_{\dot{v}}, \tag{11.6}$$

and the corresponding equation (11.6sm) for smoothed reductions is also true. [See Section 16(e) for a stronger property.]

It is sufficient to prove (11.6). This limit relation is trivial on any open connected component of $\dot{D}$ on which $\dot{v}$ vanishes identically; so we suppose from now on that $\dot{v}$ is strictly positive. First suppose that each set $\dot{A}_n$ is open. Then (11.6) coincides with (11.6sm) and the limit function $\dot{v}' = \lim_{n\to\infty} \dot{R}^{\dot{A}_n}_{\dot{v}_n}$ is superparabolic on $\dot{D}$, equal to $\dot{v}$ on $\dot{A}$. It follows that $\dot{v}' \geq \dot{R}^{\dot{A}}_{\dot{v}}$. The reverse inequality is trivial so (11.6) is true for open sets. In the general case choose $\dot{\xi}$ in $\dot{D}$, $\varepsilon > 0$, $\alpha > 1$, define $\dot{A}^{\alpha}_n = \dot{A}_n \cap \{\dot{v} < \alpha\dot{v}_n\}$ and apply (a) to obtain an open superset $\dot{B}_j$ of $\dot{A}^{\alpha}_j$ satisfying

$$\dot{R}_{\dot{v}}^{\dot{B}_j}(\dot{\xi}) < \dot{R}_{\dot{v}}^{\dot{A}_j^\alpha}(\dot{\xi}) + 2^{-j}\varepsilon.$$

The sequence $\dot{A}_{\bullet}^{\alpha}$ is an increasing sequence of sets with union $\dot{A}$, and we define $\dot{B} = \bigcup_0^\infty \dot{B}_j \supset \dot{A}$. Apply the strong subadditivity property of reductions to deduce

$$\dot{R}_{\dot{v}}^{\bigcup_0^n \dot{B}_j}(\dot{\xi}) \leq \dot{R}_{\dot{v}}^{\dot{A}_n^\alpha}(\dot{\xi}) + \sum_0^n \dot{R}_{\dot{v}}^{\dot{B}_j}(\dot{\xi}) - \sum_0^n \dot{R}_{\dot{v}}^{\dot{A}_j^\alpha}(\dot{\xi}) \leq \dot{R}_{\dot{v}}^{\dot{A}_n^\alpha}(\dot{\xi}) + 2\varepsilon. \tag{11.7}$$

Since (11.6) is true when the sets involved are open, (11.7) yields $(n \to \infty)$

$$\dot{R}_{\dot{v}}^{\dot{A}}(\dot{\xi}) \leq \dot{R}_{\dot{v}}^{\dot{B}}(\dot{\xi}) \leq \lim_{n\to\infty} \dot{R}_{\dot{v}}^{\dot{A}_n^\alpha}(\dot{\xi}) + 2\varepsilon \leq \alpha \lim_{n\to\infty} \dot{R}_{\dot{v}_n}^{\dot{A}_n}(\dot{\xi}) + 2\varepsilon.$$

Hence

$$\dot{R}_{\dot{v}}^{\dot{A}}(\dot{\xi}) \leq \lim_{n\to\infty} \dot{R}_{\dot{v}}^{\dot{A}_n}(\dot{\xi}), \tag{11.8}$$

and the reverse inequality is trivial.

(f) $\dot{R}_{\dot{v}}^{\dot{A}} = \dot{v}$ on $\dot{A}^{pf} \cap \dot{D}$; if $\dot{v}$ is finite valued and continuous, then $\dot{R}_{\dot{v}}^{\dot{A}} = \underset{+\dot{v}}{\dot{R}}{}^{\dot{A}}$ on $\dot{D} - \dot{A}$ and $\dot{R}_{\dot{v}}^{\dot{A}} = \underset{+\dot{v}}{\dot{R}}{}^{\dot{A}} = \dot{v}$ on $\dot{A}^{pf} \cap \dot{D}$.

If $\dot{u}$ is a positive superparabolic function on $\dot{D}$, majorizing $\dot{v}$ on $\dot{A} \cap \dot{D}$ and near $\dot{A} \cap \partial\dot{D}$, then (by the parabolic-fine continuity of $\dot{u}$ and $\dot{v}$) $\dot{u}$ also majorizes $\dot{v}$ on $\dot{A}^{pf}$; so $\dot{R}_{\dot{v}}^{\dot{A}} = \dot{v}$ on $\dot{A}^{pf} \cap \dot{D}$. If $\dot{\xi} \in \dot{D} - \dot{A}$, let $\dot{B}_{\bullet}$ be a decreasing sequence of open neighborhoods of $\dot{\xi}$ with intersection $\{\dot{\xi}\}$, and define $\dot{A}_n = \dot{D} - \dot{B}_n$. Then $\dot{R}_{\dot{v}}^{\dot{A}_n}$ is parabolic on $\dot{B}_n$; so $\dot{R}_{\dot{v}}^{\dot{A}_n}(\dot{\xi}) = \underset{+\dot{v}}{\dot{R}}{}^{\dot{A}_n}(\dot{\xi})$, and $(n \to \infty)$ property (e) yields, under the stated conditions on $\dot{v}$,

$$\dot{R}_{\dot{v}}^{\dot{A}}(\dot{\xi}) = \underset{+\dot{v}}{\dot{R}}{}^{\dot{A}}(\dot{\xi}). \tag{11.9}$$

There remains the proof that (11.9) is true when $\dot{\xi} \in \dot{A}^{pf} \cap \dot{A}$. To see this, observe that for such a point $\dot{\xi}$ we have now proved at least that $\dot{R}_{\dot{v}}^{\dot{A}-\{\dot{\xi}\}}(\dot{\xi}) = \underset{+\dot{v}}{\dot{R}}{}^{\dot{A}-\{\dot{\xi}\}}(\dot{\xi})$. Moreover the reduction value on the left is equal to $\dot{v}(\dot{\xi})$ because $\dot{\xi} \in (\dot{A} - \{\dot{\xi}\})^{pf}$, and the value on the right is equal to $\underset{+\dot{v}}{\dot{R}}{}^{\dot{A}}(\dot{\xi})$ because (Section 8) a smoothed reduction is unaffected by a parabolic-polar change of the target set. Thus (11.9) is true for the present choice of $\dot{\xi}$.

The proof of (f) is based on (e) and therefore all but the first assertion of (f) requires that $\dot{v}$ be finite valued and continuous. It will be proved [statement in Section 16(e), proof in Section 17] that the conclusion of (e) remains true for arbitrary positive superparabolic $\dot{v}$, and it follows that the conclusions of (f), restated in Section 16, remain true for arbitrary positive superparabolic $\dot{v}$.

12. A Criterion of Parabolic Thinness

The following lemma will be needed in the proof (Section 13) of the Fundamental Convergence Theorem in the parabolic context.

Lemma. *Let $\dot{D}$ be a nonempty open subset of $\dot{\mathbb{R}}^N$, let $\dot{v}$ be a positive continuous finite-valued superparabolic function on $\dot{D}$, let $\dot{\xi}$ be a point of $\dot{D}$, and let $\dot{A}$ be a subset of $\dot{D}$ that is parabolic thin at $\dot{\xi}$. Then*

$$\lim_{\dot{B}\downarrow\dot{\xi}} \mathring{\llbracket}\dot{v}\rrbracket^{\dot{A}\cap\dot{B}}(\dot{\xi}) = 0 \qquad (\dot{B} \text{ a neighborhood of } \dot{\xi}). \tag{12.1}$$

Recall from Section 11(f) that the reduction in (12.1) is $\dot{v}(\dot{\xi})$ for every choice of $\dot{B}$ if $\dot{A}$ is not parabolic thin at $\dot{\xi}$.

We can assume in the proof of the lemma that $\dot{\xi} \notin \dot{A}$. In fact, if $\dot{A}$ contains $\dot{\xi}$, then replacing $\dot{A}$ by $\dot{A} - \{\dot{\xi}\}$ does not affect the hypothesis that $\dot{A}$ is parabolic thin at $\dot{\xi}$ and does not affect (12.1) because (Section 8) a smoothed reduction on a set is unchanged if the set is decreased by a parabolic-polar set. The lemma is trivial if $\dot{v}(\dot{\xi}) = 0$; so we assume strict positivity below. Since $\dot{A}$ is parabolic thin at $\dot{\xi}$, there is (from Section 9) a positive superparabolic function $\dot{u}$ on $\dot{D}$, majorized by $\dot{v}$ at $\dot{\xi}$ and with limit $+\infty$ at $\dot{\xi}$ along $\dot{A}$. If $\dot{B}$ is so small that $\dot{v} < 2\dot{v}(\dot{\xi})$ on $\dot{B}$, then

$$\begin{aligned} +\infty > \dot{u}(\dot{\xi}) \geq \mathring{\llbracket}\dot{u}\rrbracket^{\dot{A}\cap\dot{B}}(\dot{\xi}) &\geq \inf_{\dot{A}\cap\dot{B}} \dot{u} \; \frac{\mathring{\llbracket}\dot{v}\rrbracket^{\dot{A}\cap\dot{B}}(\dot{\xi})}{\sup_{\dot{A}\cap\dot{B}} \dot{v}} \\ &\geq \left(\frac{1}{2\dot{v}(\dot{\xi})} \inf_{\dot{A}\cap\dot{B}} \dot{u}\right) \mathring{\llbracket}\dot{v}\rrbracket^{\dot{A}\cap\dot{B}}(\dot{\xi}), \end{aligned}$$

from which inequality the lemma follows.

Just as in the classical context [Section XI.3(a″)], we shall need a slight variation of (12.1): if $\dot{A}$ is parabolic thin at $\dot{\xi}$, then

$$\lim_{\dot{B}\downarrow\dot{\xi}} \mathring{\llbracket}\mathring{\llbracket}\dot{v}\rrbracket^{\dot{B}}\rrbracket^{\dot{A}}(\dot{\xi}) = 0. \tag{12.1′}$$

Since sweeping has not yet been treated in the present context, the method of proof in the classical context is not applicable. To prove (12.1′), choose $\varepsilon > 0$, let $\dot{u}_1$ be a positive superparabolic function on $\dot{D}$ with $\dot{u}_1(\dot{\xi}) = +\infty$, and let $\dot{B}_0$ be a neighborhood of $\dot{\xi}$. Then

$$\begin{aligned} \mathring{\llbracket}\mathring{\llbracket}\dot{v}\rrbracket^{\dot{B}}\rrbracket^{\dot{A}}(\dot{\xi}) &\leq \mathring{\llbracket}\dot{v}\rrbracket^{\dot{A}\cap\dot{B}_0}(\dot{\xi}) + \mathring{\llbracket}\mathring{\llbracket}\dot{v}\rrbracket^{\dot{B}}\rrbracket^{\dot{A}-\dot{B}_0}(\dot{\xi}) \\ &\leq \mathring{\llbracket}\dot{v}\rrbracket^{\dot{A}\cap\dot{B}_0}(\dot{\xi}) + \frac{\mathring{\llbracket}\dot{u}_1\rrbracket^{\dot{A}-\dot{B}_0}(\dot{\xi})}{\inf_{\dot{B}} \dot{u}_1} \sup_{\dot{B}} \dot{v}. \end{aligned} \tag{12.2}$$

Apply (12.1) to find $\dot{B}_0$ so small that the first term in the second inequality is at most $\varepsilon/2$, and with this choice of $\dot{B}_0$ observe that the second term is at most $\varepsilon/2$ if $\dot{B}$ is sufficiently small, since $\dot{u}_1$ is continuous at $\dot{\xi}$.

Parabolic-Fine Limits and Cluster Values

A function $\dot{u}$ from a deleted neighborhood of a point $\dot{\xi}$ of $\dot{\mathbb{R}}^N$ into a metric space has parabolic-fine limit [parabolic-fine cluster value] α at $\dot{\xi}$ if and only if $\dot{u}$ has limit α at $\dot{\xi}$ along a subset of $\dot{\mathbb{R}}^N$ which is a deleted parabolic-fine neighborhood of $\dot{\xi}$ [is not parabolic thin at $\dot{\xi}$]. The proof is a translation into the parabolic context of that of Theorem XI.9, using the obvious parabolic context version of Lemma XI.8.

13. The Parabolic Fundamental Convergence Theorem

Theorem. *Let* Γ: $\{\dot{u}_\alpha, \alpha \in I\}$ *be a family of superparabolic functions on an open subset* $\dot{D}$ *of* $\mathbb{R}^N$, *locally uniformly bounded below, and define* $\dot{u} = \inf_{\alpha \in I} \dot{u}_\alpha$. *Then* $\underset{+}{\dot{u}} \le \dot{u}$,

$$\underset{+}{\dot{u}}(\dot{\xi}) = \liminf_{\dot{\eta} \to \dot{\xi}} \dot{u}(\dot{\eta}) = \operatorname{pf}\lim_{\dot{\eta} \to \dot{\xi}} \dot{u}(\dot{\eta}), \tag{13.1}$$

and

(a) $\underset{+}{\dot{u}}$ *is superparabolic,*
(b) $\underset{+}{\dot{u}} = \dot{u}$ *on each open set on which* $\dot{u}$ *is superparabolic,*
(c) $\underset{+}{\dot{u}} = \dot{u}$ *except on a parabolic-semipolar set,*
(d) *there is a countable subfamily of* Γ *whose infimum has smoothing* $\underset{+}{\dot{u}}$.

Conversely, if $\dot{A}$ *is a parabolic-semipolar subset of* $\dot{D}$, *there is a decreasing sequence* $\dot{v}_\bullet$ *of positive superparabolic functions on* $\dot{D}$ *with limit* $\dot{v}$ *such that* $\dot{v} > \underset{+}{\dot{v}}$ *on* $\dot{A}$.

Assertions (a), (b), (d) and the first equation in (13.1) are contained in Theorem 2. In view of (d) it will be assumed in the proof of the direct half of the theorem that $\dot{u}_\bullet$ is a sequence of parabolic functions, and it can even be assumed, as in the classical context (Section III.3), that these functions are positive. We now choose r_1 and r_2 with $r_1 < r_2$, define

$$\dot{A}_{r_1 r_2} = \{\dot{u} > r_2, r_1 > \underset{+}{\dot{u}}\},$$

choose $\dot{\zeta}$ in $\dot{D}$, and show that the set $\dot{A} = \dot{A}_{r_1 r_2}$ is parabolic thin at $\dot{\zeta}$. If $\dot{\zeta}$ is not a limit point of $\dot{A}$, the assertion is trivial. If $\dot{\zeta}$ is a limit point of $\dot{A}$, then on a sufficiently small neighborhood $\dot{B}$ of $\dot{\zeta}$ we have $\underset{+}{\dot{u}} > \underset{+}{\dot{u}}(\dot{\zeta}) - (r_2 - r_1)/2$ by lower semicontinuity of $\underset{+}{\dot{u}}$; so

$$\dot{u}_n \geq \dot{u} > \underset{+}{\dot{u}}(\dot{\zeta}) + \frac{r_2 - r_1}{2} = c \tag{13.2}$$

on $\dot{A} \cap \dot{B}$. Hence (reduction relative to $\dot{D}$) $\dot{u}_n \geq c\dot{R}_1^{\dot{A} \cap \dot{B}}$ and $\underset{+}{\dot{u}} \geq c\underset{+1}{\dot{R}}^{\dot{A} \cap \dot{B}}$. Since $\underset{+}{\dot{u}}(\dot{\zeta}) < c$, we conclude that $1 > \underset{+1}{\dot{R}}^{\dot{A} \cap \dot{B}}(\dot{\zeta})$, and therefore in view of Section 11(f) we conclude that $\dot{A}$ is parabolic thin at every point $\dot{\zeta}$ of $\dot{D}$. Since each function $\dot{u}_n$ is parabolic-fine continuous, the function $\dot{u}$ is parabolic-fine upper semicontinuous; so

$$\dot{u}(\dot{\zeta}) \geq \operatorname{pf}\limsup_{\dot{\eta} \to \dot{\zeta}} \dot{u}(\dot{\eta}) \geq \operatorname{pf}\liminf_{\dot{\eta} \to \dot{\zeta}} \dot{u}(\dot{\eta}) \geq \underset{+}{\dot{u}}(\dot{\zeta}). \tag{13.3}$$

If there were a point $\dot{\zeta}$ in $\dot{D}$ at which the second and fourth terms in (13.3) were unequal, there would be two numbers, r_1 and r_2, strictly between the values of these terms at $\dot{\zeta}$, with $r_1 < r_2$. Since $\underset{+}{\dot{u}}$ is parabolic-fine continuous, the set $\dot{A}_{r_1 r_2}$ would not be parabolic thin at $\dot{\zeta}$, and consequently there can be no such point $\dot{\zeta}$. Thus (13.1) is true. Finally (c) is true because

$$\{\dot{u} > \underset{+}{\dot{u}}\} = \bigcup_{r_1, r_2} \dot{A}_{r_1 r_2} \qquad (r_1, r_2 \text{ rational}). \tag{13.4}$$

Conversely, suppose that $\dot{A}$ is a parabolic-semipolar subset of $\dot{D}$. To prove the converse half of the theorem, write $\dot{A} = \bigcup_0^\infty \dot{A}_k$ with $\dot{A}_k^{pf} \cap \dot{D} = \varnothing$, and let $\dot{B}_{\bullet}$ be a sequence of open subsets of $\dot{D}$ forming a basis for the Euclidean topology of $\dot{D}$. Define

$$\dot{A}_{km} = \dot{A}_k \cap \dot{B}_m \cap \{\underset{+1}{\dot{R}}^{\dot{A}_k \cap \dot{B}_m} < 1\}$$

so that (Lemma 12)

$$\dot{A}_{km} \subset \{\underset{+1}{\dot{R}}^{\dot{A}_{km}} < \dot{R}_1^{\dot{A}_{km}}\}, \qquad \bigcup_{m=0}^{\infty} \dot{A}_{km} = \dot{A}_k.$$

Apply (d) to find a decreasing sequence $\dot{v}_{km\bullet}$ of positive superparabolic functions, each at most 1 and equal to 1 on $\dot{A}_{km}$, with limit $\dot{v}_{km\infty}$ for which $\underset{+km\infty}{\dot{v}} = \underset{+1}{\dot{R}}^{\dot{A}_{km}}$. The sequence $\dot{v}_{\bullet}$ defined by

$$\dot{v}_n = \sum_{k,m} \dot{v}_{kmn} 2^{-k-m}$$

has the properties described in the converse half of Theorem 14. In fact $\dot{v}_{\bullet}$ is a monotone decreasing sequence of positive superparabolic functions with

$$\dot{v} = \lim_{n\to\infty} \dot{v}_n = \sum_{k,m} \dot{v}_{+km\infty} 2^{-k-m};$$

so $\underset{+}{\dot{v}} = \sum_{k,m} \dot{v}_{+km\infty} 2^{-k-m}$ up to a parabolic-semipolar set and therefore everywhere on $\dot{D}$. Hence $\dot{v} > \underset{+}{\dot{v}}$ on $\dot{A}$.

14. Applications of the Fundamental Convergence Theorem to Reductions and to Green Functions

Application to Reductions

According to the parabolic Fundamental Convergence Theorem, a parabolic reduction $\dot{R}_{\dot{v}}^{\dot{A}}$ satisfies the relation $\dot{R}_{\dot{v}}^{\dot{A}} = \dot{R}_{+\dot{v}}^{\dot{A}}$ up to a parabolic semipolar set. In view of the parabolic-fine continuity of superparabolic functions, $\dot{R}_{\dot{v}}^{\dot{A}} = \dot{v}$ on $(\dot{A} \cup \dot{A}^{pf}) \cap \dot{D}$, and this inequality combined with (13.1) implies that $\dot{R}_{+\dot{v}}^{\dot{A}} = \dot{R}_{\dot{v}}^{\dot{A}} = \dot{v}$ on $\dot{A}^{pf} \cap \dot{D}$; the latter fact was proved under stronger hypotheses on $\dot{v}$ in Section 11(f).

Example. If $\dot{D} = \dot{\mathbb{R}}^N$ and if $\dot{A} = \{\dot{\xi}: \text{ord}\,\dot{\xi} \geq 0\}$, then

$$\dot{R}_{\dot{v}}^{\dot{A}} = \begin{cases} \dot{v} & \text{on } \dot{A}, \\ 0 & \text{elsewhere,} \end{cases} \qquad \dot{R}_{+\dot{v}}^{\dot{A}} = \begin{cases} \dot{v} & \text{on the upper half-space,} \\ 0 & \text{elsewhere.} \end{cases}$$

Thus in this example if $\dot{v}$ is strictly positive, the set $\{\dot{R}_{\dot{v}}^{\dot{A}} > \dot{R}_{+\dot{v}}^{\dot{A}}\}$ is the abscissa hyperplane, which is parabolic semipolar but is not parabolic polar.

Application to Green Functions

The counterpart of the argument in Section VII.4 showing by way of the classical context Fundamental Convergence Theorem that the classical context Green function with a given pole has limit 0 at quasi every finite Euclidean boundary point shows in the parabolic context that $\dot{G}_{\dot{D}}(\cdot, \dot{\eta})$ $[\dot{G}_{\dot{D}}(\dot{\xi}, \cdot)]$ has limit 0 at every finite Euclidean boundary point except possibly for those boundary points in some parabolic-semipolar [coparabolic-semipolar] set. (It will be shown in Section XVIII.12 that a subset of $\dot{\mathbb{R}}^N$ is parabolic semipolar if and only if it is coparabolic semipolar.) This boundary limit result may be vacuous, however. For example, if $\dot{D}$ is the upper half-space, the finite part of the Euclidean boundary is the abscissa hyperplane which is both parabolic-semipolar and coparabolic-semipolar. Actually in this case since $\dot{G}_{\dot{D}}(\dot{\xi}, \dot{\eta}) = 0$ when ord $\dot{\xi} \leq$ ord $\dot{\eta}$, it is trivial that $\dot{G}_{\dot{D}}(\cdot, \dot{\eta})$ has limit 0 at every point of the bounding hyperplane.

15. Applications of the Fundamental Convergence Theorem to the Parabolic-Fine Topology

Application to the Smoothness of a Parabolic-Semipolar Set

A parabolic-semipolar set is a subset of a Borel semipolar set which is a countable union of Borel sets each of which has no parabolic-fine limit point. In fact on the one hand according to Theorem 13 (converse assertion), a parabolic-semipolar set is a subset of the exceptional set $\{\dot{u} > \underset{+}{\dot{u}}\}$ in some application of the theorem, and on the other hand according to the proof of Theorem 13, the exceptional set $\{\dot{u} > \underset{+}{\dot{u}}\}$ in each application is a subset of a countable union (13.4) of Borel sets each of which has no parabolic-fine limit point.

Character of $\dot{A}^{pf}$

The counterpart of the argument (Section XI.6) that in the classical context the set A^f is a Euclidean G_δ set and that $A - A^f$ is polar yields in the parabolic context that $\dot{A}^{pf}$ is a Euclidean G_δ set and that $\dot{A} - \dot{A}^{pf}$ is parabolic semipolar. Although the set A^f is fine perfect, the set $\dot{A}^{pf}$ need not be parabolic-fine perfect. For example, if $\dot{A} = \bigcup_1^\infty \{\operatorname{ord} \dot{\xi} = -1/n\}$, the set $\dot{A}^{pf}$ is the abscissa hyperplane, which has no parabolic-fine limit point.

Borel Measurability of the Parabolic-Fine Limit Superior Function

If $\dot{u}$ is a function from an open subset of $\dot{\mathbb{R}}^N$ into $\overline{\mathbb{R}}$, the function $\dot{\xi} \mapsto \operatorname{pf\,lim\,sup}_{\dot{\eta} \to \dot{\xi}} \dot{u}(\dot{\eta})$ is Borel measurable (and therefore the corresponding inferior limit function is also Borel measurable) because if $\alpha \in \mathbb{R}$, the set

$$\{\dot{\xi}: \operatorname{pf}\limsup_{\dot{\eta}\to\dot{\xi}} \dot{u}(\dot{\eta}) > \alpha\} = \bigcup_{n=1}^{\infty} \left\{\dot{\xi}: \dot{u}(\dot{\xi}) \geq \alpha + \frac{1}{n}\right\}^{pf}$$

is a countable union of G_δ sets.

16. Parabolic-Reduction Properties

The list of reduction properties in this section includes for completeness some already discussed. The reductions are relative to an open subset $\dot{D}$ of $\dot{\mathbb{R}}^N$, provided with a boundary $\partial\dot{D}$ by a metric compactification. The sets on which positive superparabolic functions are reduced are subsets of $\dot{D} \cup \partial\dot{D}$, and no further hypotheses not stated explicitly are imposed on either functions or sets. Proofs are given in Section 17 and consist merely of the reference to the proof in the classical context when the latter proof requires only translation into the present context. We stress that every property in the following list is a property of reductions in the classical context also, in which "(super)parabolic" is to be interpreted as "(super)harmonic" and "semipolar" as "polar." Some of the proofs given in the present context are unnecessarily indirect for the classical context but have the advantage that they are applicable in many general contexts.

(a) If $\dot v' = \dot v - \dot R_{\dot v}^{\dot A \cap \partial \dot D}$, then

$$\dot R_{\dot v}^{\dot A} = \dot R_{\dot v}^{\dot A \cap \partial \dot D} + \dot R_{\dot v'}^{\dot A \cap \dot D}, \tag{16.1}$$

and the corresponding equation (16.1sm) for smoothed reductions is also true.

(b) $\underset{+\dot v}{\dot R}{}^{\dot A} \le \dot R_{\dot v}^{\dot A} \le \dot v$ on $\dot D$,
$\dot R_{\dot v}^{\dot A} = \dot v$ on $(\dot A \cup \dot A^{pf}) \cap \dot D$,
$\dot R_{\dot v}^{\dot A} = \underset{+\dot v}{\dot R}{}^{\dot A} = \dot v$ on $\dot A^{pf} \cap \dot D$, in particular on the parabolic-fine interior of $\dot A \cap \dot D$.
$\dot R_{\dot v}^{\dot A} = \underset{+\dot v}{\dot R}{}^{\dot A}$ on $\dot D$ when $\dot A \cap \dot D$ is parabolic-fine open.
$\dot R_{\dot v}^{\dot A}$ is parabolic on $\dot D - \bar{\dot A}$ and equal to $\underset{+\dot v}{\dot R}$ there.
$\underset{+\dot v}{\dot R}{}^{\dot A}(\dot \xi) = \liminf_{\dot \eta \to \dot \xi} \dot R_{\dot v}^{\dot A}(\dot \eta) = \text{pf}\lim_{\dot \eta \to \dot \xi} \dot R_{\dot v}^{\dot A}(\dot \eta)$.

(c) [See also (3.1) and (3.1sm).] If $\dot v$ is finite valued on $\dot A \cap \dot D$, then

$$\dot R_{\dot v}^{\dot A} = \inf\{\dot R_{\dot v}^{\dot B} : \dot A \subset \dot B,\ \dot B \cap \dot D \text{ is parabolic-fine open, } \dot B \text{ contains a neighborhood of } \dot A \cap \partial \dot D\}. \tag{16.2}$$

If in addition $\dot v$ is continuous at each point of $\dot A \cap \dot D$, the set $\dot B$ in (16.2) can be restricted to be open in $\dot D \cup \partial \dot D$.

(d) If $\dot A_1$ and $\dot A_2$ differ by a parabolic-polar subset $\dot B$ of $\dot D$, then $\dot R_{\dot v}^{\dot A_1} = \dot R_{\dot v}^{\dot A_2}$ on $\dot D - \dot B$ and $\underset{+\dot v}{\dot R}{}^{\dot A_1} = \underset{+\dot v}{\dot R}{}^{\dot A_2}$ on $\dot D$.

(e) If $\dot A_\bullet$ is an increasing sequence of subsets of $\dot D$ with union $\dot A$ and if $\dot v_\bullet$ is an increasing sequence of positive superparabolic functions on $\dot D$ with superparabolic limit $\dot v$, then

$$\lim_{n\to\infty} \dot R_{\dot v_n}^{\dot A_n} = \dot R_{\dot v}^{\dot A}, \tag{16.3}$$

and the corresponding equation (16.3sm) for smoothed reductions is also true. If $\dot v_n = \dot v$ for all n, then (16.3) and (16.3sm) are true for $\dot A_\bullet$ an increasing sequence of subsets of $\dot D \cup \partial \dot D$.

(f) $\underset{+\dot v}{\dot R}{}^{\dot A} \le \dot R_{\dot v}^{\dot A}$ on $\dot D$ with equality on $\dot D - \dot A$, and also equality on $\dot A \cap \dot D$ up to a parabolic-semipolar set.

(g) If $\dot v = \Sigma_0^\infty \dot v_n$ is a superparabolic sum of positive superparabolic functions on $\dot D$, then

$$\dot{R}_{\dot{v}}^{\dot{A}} = \sum_0^\infty \dot{R}_{\dot{v}_n}^{\dot{A}}, \tag{16.4}$$

and the corresponding equation (16.4sm) for unsmoothed reductions is also true.

(h) If $\dot{A} \subset \partial\dot{D}$ and if $\dot{v}_\bullet$ is a decreasing sequence of positive superparabolic functions on $\dot{D}$ with limit $\dot{v}$, then $\lim_{n\to\infty} \dot{R}_{\dot{v}_n}^{\dot{A}} = \dot{R}_{\dot{v}}^{\dot{A}}$.

(i) If $\dot{A} \subset \dot{B} \subset \dot{D} \cup \partial\dot{D}$ and if $\dot{A} \cap \dot{D}$ is parabolic-fine open, then $\dot{[\![} [\![\dot{v}]\!]^{\dot{A}}]\!]^{\dot{B}} = [\![\dot{v}]\!]^{\dot{A}}$. If in addition the set $\dot{B} \cap \dot{D}$ is parabolic-fine open, then $\dot{[\![} [\![\dot{v}]\!]^{\dot{B}}]\!]^{\dot{A}} = [\![\dot{v}]\!]^{\dot{A}}$.

(j) If $\dot{A} \cap \dot{D}$ and $\dot{B} \cap \dot{D}$ are parabolic-fine open and if $\dot{v}' = \dot{R}_{\dot{v}}^{\dot{A}} \wedge \dot{R}_{\dot{v}}^{\dot{B}}$, then

$$\dot{R}_{\dot{v}}^{\dot{A}\cup\dot{B}} + \dot{R}_{\dot{v}'}^{\dot{A}\cup\dot{B}} = \dot{R}_{\dot{v}}^{\dot{A}} + \dot{R}_{\dot{v}}^{\dot{B}}. \tag{16.5}$$

(k) The set functions $\dot{R}_{\dot{v}}^{\bullet}$ and $\underset{+\dot{v}}{\dot{R}^{\bullet}}$ are countably strongly subadditive on the class of subsets of $\dot{D} \cup \partial\dot{D}$.

(l) (Strengthening of the assertion in (f) that $\underset{+\dot{v}}{\dot{R}^{\dot{A}}} = \dot{R}_{\dot{v}}^{\dot{A}}$ on $\dot{D} - \dot{A}$.)

If $\varepsilon > 0$ and if $\dot{C}$ is a compact subset of $\dot{D} - \dot{A}$, there is a positive superparabolic function $\dot{u}$ on $\dot{D}$, equal to $\dot{v}$ on $\dot{A} \cap \dot{D}$ and near $\dot{A} \cap \partial\dot{D}$ and satisfying the inequality $\dot{u} \le \underset{+\dot{v}}{\dot{R}^{\dot{A}}} + \varepsilon$ on $\dot{C}$.

(m) If $\dot{A}$ is a relatively compact subset of $\dot{D}$, then $\dot{G}\mathrm{M}_{\dot{D}} \underset{+\dot{v}}{\dot{R}^{\dot{A}}} = 0$; that is, $\underset{+\dot{v}}{\dot{R}^{\dot{A}}}$ is a potential.

(n) If $\dot{v}$ is a potential on $\dot{D}$, then $\dot{R}_{\dot{v}}^{\dot{A}} = \dot{R}_{\dot{v}}^{\dot{A}\cap\dot{D}}$ and $\underset{+\dot{v}}{\dot{R}^{\dot{A}}} = \underset{+\dot{v}}{\dot{R}^{\dot{A}\cap\dot{D}}}$.

(o) If $\dot{v}$ is finite valued and continuous, then for every point $\dot{\xi}$ of $\dot{D}$ the set function $\dot{R}_{\dot{v}}^{\bullet}(\dot{\xi})$ is a Choquet capacity on $\dot{D} \cup \partial\dot{D}$ relative to the class of compact subsets of $\dot{D} \cup \partial\dot{D}$.

Observation. In the classical context (Section VI.5) the set function $R_v^{\bullet}(\xi)$ was shown to be a Choquet capacity on $D \cup \partial D$ relative to the class of compact subsets of $D \cup \partial D$ when v is finite valued. In the present context finite valuedness of $\dot{v}$ is not sufficient for the validity of (o) according to the following example. Let A be the closure of a ball in $\mathbb{R}^N$, and let $\dot{A}_n$ be the subset $A \times [0, 1/n]$ of $\dot{\mathbb{R}}^N$. Then $\dot{A}_\bullet$ is a decreasing sequence of compact subsets of $\dot{\mathbb{R}}^N$ with intersection $\dot{A} = A \times \{0\}$. Let $\dot{v}$ be the indicator function of the upper half-space. Then (reductions relative to $\dot{D} = \dot{\mathbb{R}}^N$) the reduction $\dot{R}_{\dot{v}}^{\dot{A}}$ vanishes identically, but $\lim_{n\to\infty} \dot{R}_{\dot{v}}^{\dot{A}_n} > 0$ on the upper half-space because according to Lemma XVI.2, the smoothed reduction $\dot{R}_{\dot{v}}^{\dot{A}_n}$ is at least equal to the parabolic Poisson integral on the upper half-space with boundary function the indicator function of A.

(p) If $\dot{A}$ is an analytic subset of $\dot{D} \cup \partial\dot{D}$, then

$$\dot{R}_{\dot{v}}^{\dot{A}} = \sup\{\dot{R}_{\dot{v}}^{\dot{F}} : \dot{F} \subset \dot{A}, \dot{F} \text{ compact}\}, \tag{16.6}$$

and the corresponding equation (16.6sm) for smoothed reductions is also true.

(q) If $\dot{u}$ and $\dot{v}$ are bounded, then

$$\sup_{\dot{D}} |\dot{R}_{\dot{v}}^{\dot{A}} - \dot{R}_{\dot{u}}^{\dot{A}}| \leq \sup_{\dot{D}} |\dot{u} - \dot{v}|, \tag{16.7}$$

and the corresponding inequality (16.7sm) for smoothed reductions is also true.

(r) There is a bounded continuous superparabolic potential $\dot{u}^*$ on $\dot{D}$ for which, for each subset $\dot{A}$ of $\dot{D}$,

$$\dot{A}^{pf} \cap \dot{D} = \{\underset{+\dot{u}^*}{\dot{R}}{}^{\dot{A}} = \dot{u}^*\}. \tag{16.8}$$

(s) Parabolic context counterpart of Section VI.3(o). To avoid repetition of complicated inequalities, we refrain from writing this property explicitly. It is the set of reduction inequalities obtained by translating Section VI.3(o) into the parabolic context; that is, "superharmonic" is replaced by "superparabolic," and dots are inserted over set and reduction symbols as required by the parabolic context.

17. Proofs of the Reduction Properties in Section 16

Proof of (a). See the proof of the corresponding property in the classical context in Section III.5(c). □

Proof of (b). The properties listed under (b) have already been proved except for the identification of $\dot{R}_{\dot{v}}^{\dot{A}}$ with $\underset{+\dot{v}}{\dot{R}}{}^{\dot{A}}$ when $\dot{A} \cap \dot{D}$ is parabolic-fine open. If $\dot{A}$ is a parabolic-fine open subset of $\dot{D}$, the function $\underset{+\dot{v}}{\dot{R}}{}^{\dot{A}}$ is a positive superparabolic function equal to $\dot{v}$ on $\dot{A}$ according to the third line of (b); so $\underset{+\dot{v}}{\dot{R}}{}^{\dot{A}} \geq \dot{R}_{\dot{v}}^{\dot{A}}$, and the reverse inequality is listed on the first line of (b). For general $\dot{A}$ with $\dot{A} \cap \dot{D}$ parabolic-fine open, combine property (a) with the fifth assertion in (b) and the special case just considered to obtain the stated identification. □

Proof of (c). The second assertion was proved in Section 11(a) by referral back to the proof in the classical context. Since superparabolic functions are parabolic-fine continuous, the same proof is applicable to prove the first assertion. □

Proof of (d). See Section 8. □

Proof of (e). If equation (16.3) is true, then (16.3sm) is true up to a parabolic semipolar subset of $\dot{D}$ and therefore is true everywhere on $\dot{D}$ because both sides of (16.3sm) are superparabolic functions. In the following we therefore consider only (16.3). The first assertion of (e) was proved in Section 11(e) for $\dot{v}_n$ and $\dot{v}$ finite valued and continuous, using Section 11(a), that is, using the second assertion of (c) of the present section. This proof is applicable as long as $\dot{v}$ is finite valued on $\dot{A}$ if we use the first assertion of (c) and replace the open sets in the Section 11(e) proof by parabolic-fine open sets. For arbitrary $\dot{v}_n$, $\dot{v}$, $\dot{A}_n$, $\dot{A}$ with $\dot{A} \subset \dot{D}$ it follows that

$$\lim_{n\to\infty} \dot{R}_{\dot{v}_n}^{\dot{A}_n \cap \{\dot{v} < +\infty\}} = \dot{R}_{\dot{v}}^{\dot{A} \cap \{\dot{v} < +\infty\}}, \tag{17.1}$$

and therefore (16.3) is true, in view of (d), except possibly on the parabolic-polar set $\dot{A} \cap \{\dot{v} = +\infty\}$, on which, however, (16.3) is trivial. To prove the second assertion of (e) suppose first that $\dot{A}_n \subset \partial\dot{D}$. Let $\dot{\xi}_{\bullet}$ be a sequence dense in $\dot{D}$, and choose a positive superparabolic function $\dot{u}_n$ on $\dot{D}$, majorizing $\dot{v}$ near $\dot{A}_n \cap \partial\dot{D}$ and satisfying

$$\dot{u}_n(\dot{\xi}_j) \le \dot{R}_{\dot{v}}^{\dot{A}_n}(\dot{\xi}_j) + 2^{-n}, \qquad j \le n.$$

The function

$$\dot{u}_k' = \lim_{n\to\infty} \dot{R}_{\dot{v}}^{\dot{A}_n} + \sum_{n=k}^{\infty} (\dot{u}_n - \dot{R}_{\dot{v}}^{\dot{A}_n})$$

is positive and is superparabolic on $\dot{D}$ because the indicated limit is parabolic on $\dot{D}$, and the sum is a sum of positive superparabolic functions and is finite at each point $\dot{\xi}_j$. The function $\lim_{n\to\infty} \dot{R}_{\dot{v}}^{\dot{A}_n}$ majorizes $\dot{R}_{\dot{v}}^{\dot{A}_n}$ for all n, so $\dot{u}_k' \ge \dot{u}_n$ for $n \ge k$. It follows that $\dot{u}_k' \ge \dot{R}_{\dot{v}}^{\dot{A}}$, and when $k \to \infty$, we find that $\lim_{n\to\infty} \dot{R}_{\dot{v}}^{\dot{A}_n} \ge \dot{R}_{\dot{v}}^{\dot{A}}$ up to a parabolic-semipolar subset of $\dot{D}$ and therefore everywhere on $\dot{D}$ since both sides of this equation are parabolic functions on $\dot{D}$. Since the reverse inequality is trivial, (16.3) is true when $\dot{A} \subset \partial\dot{D}$ and $\dot{v}_n = \dot{v}$ for all n. According to the first assertion of (e), equation (16.3) is true when $\dot{A} \subset \dot{D}$ with no restriction on $\dot{v}_{\bullet}$. Now apply (a) to find

$$\dot{R}_{\dot{v}}^{\dot{A}_n} = \dot{R}_{\dot{v}}^{\dot{A}_n \cap \partial\dot{D}} + \dot{R}_{\dot{v}_n'}^{\dot{A}_n \cap \dot{D}} \ge \dot{R}_{\dot{v}}^{\dot{A}_n \cap \partial\dot{D}} + \dot{R}_{\dot{v}'}^{\dot{A}_n \cap \dot{D}},$$

where $\dot{v}_n' = \dot{v} - \dot{R}_{\dot{v}}^{\dot{A}_n \cap \partial\dot{D}}$ and $\dot{v}' = \dot{v} - \dot{R}_{\dot{v}}^{\dot{A} \cap \partial\dot{D}}$.

Hence

$$\lim_{n\to\infty} \dot{R}_{\dot{v}}^{\dot{A}_n} \ge \dot{R}_{\dot{v}}^{\dot{A} \cap \partial\dot{D}} + \dot{R}_{\dot{v}'}^{\dot{A} \cap \dot{D}} = \dot{R}_{\dot{v}}^{\dot{A}}.$$

Since the reverse inequality is trivial, (16.3) is true, as was to be proved. □

Proof of (f). We have already applied the Fundamental Convergence Theorem to find that $\dot{R}_{\dot{v}}^{\dot{A}} = \dot{R}_{+\dot{v}}^{\dot{A}}$ up to a parabolic-semipolar subset of $\dot{D}$.

Equality on $\dot{D} - \dot{A}$ was proved in Section 11(f) for $\dot{v}$ finite valued and continuous but it was pointed out there (in an observation following the proof) that the present property (e) implies equality on $\dot{D} - \dot{A}$ without this restriction. □

Proof of (g). To prove (g) for two summands, that is, to prove

$$\dot{R}^{\dot{A}}_{\dot{v}_1+\dot{v}_2} = \dot{R}^{\dot{A}}_{\dot{v}_1} + \dot{R}^{\dot{A}}_{\dot{v}_2} \tag{17.2}$$

and the corresponding equation for smoothed reductions, observe that (17.2) appears as (11.3) in Section 11(c) for $\dot{A}$ an open subset of $\dot{D}$ but that the proof of this special case was there referred back to the proof of the corresponding special case in the classical context. That proof [Section VI.4(f)] translates trivially into the present context for $\dot{A}$ a parabolic-fine open subset of $\dot{D}$, with no restriction on $\dot{v}_1$, $\dot{v}_2$. In view of the evaluation in (3.1) of a reduction in terms of reductions on subsets of $\dot{D}$, (17.2) is true whenever $\dot{A} \cap \dot{D}$ is parabolic-fine open. It then follows from (c) that (17.2) is true with no restriction on $\dot{A}$ if $\dot{v}_1$ and $\dot{v}_2$ are finite valued on $\dot{A} \cap \dot{D}$. Hence with no restriction on $\dot{v}_1$ and $\dot{v}_2$ and with $\dot{A}_0 = \dot{A} \cap \{\dot{v}_1 + \dot{v}_2 < +\infty\}$,

$$\dot{R}^{\dot{A}_0}_{\dot{v}_1+\dot{v}_2} = \dot{R}^{\dot{A}_0}_{\dot{v}_1} + \dot{R}^{\dot{A}_0}_{\dot{v}_2}.$$

According to (d), these reductions are equal to the corresponding reductions on $\dot{A}$, that is, to the reductions in (17.2), except possibly on the parabolic-polar set $\dot{A} - \dot{A}_0$ on which in fact (17.2) is trivial. Hence (17.2) is true; that is, (16.4) is true for two summands and therefore for finitely many summands. Hence for finitely many summands (16.4sm) is true up to a parabolic-semipolar subset of $\dot{D}$ and is therefore true everywhere on $\dot{D}$ because each side of (16.4) is a superparabolic function. The extension to infinitely many summands is effected as in the classical context [Section VI.4(f)]. □

Proof of (h). The proof follows that of the corresponding classical property in Section VI.4(g). □

Proof of (i). The proof follows that of the corresponding classical property in Section VI.4(h) but observe that in the argument there for (h_1) no extra hypothesis on the smoothness of A was used, whereas the hypothesis that $\dot{A} \cap \dot{D}$ is parabolic-fine open plays an essential role in proving the parabolic counterpart of (h_1). □

Proof of (j). This property was proved under added restrictions (or rather referred back to its classical counterpart) in Section 11(d). The method of proof referred to is applicable in the present context whenever the sets $\dot{A} \cap \dot{D}$ and $\dot{B} \cap \dot{D}$ are parabolic-fine open. □

Proof of (k). Equation (16.5) implies the strong subadditivity of the set function $\dot{A} \mapsto \dot{R}_{\dot{v}}^{\dot{A}}$ on the class of sets $\dot{A}$ with $\dot{A} \cap \dot{D}$ parabolic-fine open [the argument in Section 11(d) for $\dot{A} \cap D$ open is applicable when $\dot{A} \cap \dot{D}$ is merely fine open]. An application of (c) then shows that the strong subadditivity inequality (11.5) is true whenever $\dot{v}$ is finite valued on $(\dot{A} \cup \dot{B}) \cap \dot{D}$; so (11.5) is true with no restriction if $\dot{A}$ and $\dot{B}$ are replaced by $\dot{A} \cap \{\dot{v} < +\infty\}$ and $\dot{B} \cap \{\dot{v} < +\infty\}$, respectively. Hence by (d) the inequality (11.5) is true as written except possibly on the set $(\dot{A} \cup \dot{B}) \cap \{\dot{v} = +\infty\}$, and (11.5) is therefore true on $\dot{D}$ because the inequality is trivially true on $(\dot{A} \cup \dot{B}) \cap \dot{D}$. The strong subadditivity inequality for smoothed reductions is true on $\dot{D}$ because the inequality is true up to a parabolic-semipolar subset of $\dot{D}$ and the two sides of this inequality are superparabolic functions. Properties (e) and (g) imply that there must be countable strong subadditivity when there is strong subadditivity. □

Proof of (l). Denote by $\dot{A}_k$ the set of points of $\dot{D} \cap \dot{A} \cap \{\dot{v} < +\infty\}$ at distance $\geq 1/k$ from $\dot{C}$. The function $\dot{R}_{\dot{v}}^{\dot{A}_k \cup (\dot{A} \cap \partial \dot{D})}$ is continuous, in fact parabolic, on the neighborhood $\{\dot{\xi} \in \dot{D}\colon |\dot{\xi} - \dot{C}| < 1/k\}$ of $\dot{C}$. The downward-directed family

$$\begin{aligned}&\{\dot{R}_{\dot{v}}^{\dot{B}}\colon \dot{A}_k \cup (\dot{A} \cap \partial \dot{D}) \subset \dot{B},\ \dot{B} \cap \dot{D} \text{ parabolic-fine open},\\ &\dot{B} \text{ a neighborhood of } \dot{A} \cap \dot{D}\}_{|\dot{C}}\end{aligned} \tag{17.3}$$

of continuous functions on $\dot{C}$, with limit $\dot{R}_{\dot{v}}^{\dot{A}_k \cup (\dot{A} \cap \partial \dot{D})}$ according to (c), is uniformly convergent (Dini's theorem). Thus if $\varepsilon > 0$ and $k \geq 1$, there is a choice $\dot{B}_k$ of $\dot{B}$ in (17.3) so small that

$$\dot{R}_{\dot{v}}^{\dot{B}_k} \leq \dot{R}_{\dot{v}}^{\dot{A}_k \cup (\dot{A} \cap \partial \dot{D})} + 2^{-k-1}\varepsilon$$

on $\dot{C}$. For $k \geq 1$ let $\dot{u}_k$ be a positive superparabolic function on $\dot{D}$, identically $+\infty$ on the set $\{\dot{\xi}\colon |\dot{\xi} - \dot{C}| \geq 1/k,\ \dot{v}(\dot{\xi}) = +\infty\}$ and at most $2^{-k-1}\varepsilon$ on $\dot{C}$. (For example, if $\dot{u}_k'$ is a positive superparabolic function on $\dot{D}$, identically $+\infty$ on the set $\{\dot{v} = +\infty\}$, let $\dot{u}_k''$ be the reduction of $\dot{u}_k'$ on the set $\{|\dot{\xi} - \dot{C}| > 1/k\}$. Then $\dot{u}_k''$ is continuous on the compact set $\dot{C}$, and we define $\dot{u}_k = 2^{-k-1}\varepsilon \dot{u}_k'' / \sup_{\dot{C}} \dot{u}_k''$.) By countable strong subadditivity (k)

$$\dot{R}_{\dot{v}}^{\bigcup_1^\infty \dot{B}_k} \leq \dot{R}_{\dot{v}}^{\bigcup_1^\infty [\dot{A}_k \cup (\dot{A} \cap \partial \dot{D})]} + \frac{\varepsilon}{2} \leq \dot{R}_{\dot{v}}^{\dot{A}} + \frac{\varepsilon}{2}$$

on $\dot{C}$. The function

$$\dot{v} \wedge \left[\dot{R}_{\dot{v}}^{\bigcup_1^\infty \dot{B}_k} + \sum_1^\infty \dot{u}_k\right]$$

is superparabolic on $\dot{D}$, equal to $\dot{v}$ on $\dot{A} \cap \dot{D}$ and near $\dot{A} \cap \partial\dot{D}$, and is at most $\dot{R}_{\dot{v}}^{\dot{A}} + \varepsilon$ on $\dot{C}$, as desired. □

Proofs of (m) *and* (n). The proofs follow those of the corresponding classical properties in Sections III.6 and III.5, respectively. Note however that the counterpart $\dot{u}$ of the function u in III.6 should be superparabolic, but not parabolic on any open subset of $\dot{D}$; choose for example the potential of a measure supported by a countable dense subset of $\dot{D}$. □

Proof of (o). In view of (e) all that remains to be proved is that $\lim_{n\to\infty} \dot{R}_{\dot{v}}^{\dot{A}_n} = \dot{R}_{\dot{v}}^{\dot{A}}$ whenever $\dot{v}$ is finite valued and continuous and $\dot{A}_{\cdot}$ is a decreasing sequence of compact subsets of $\dot{D} \cup \partial\dot{D}$ with intersection $\dot{A}$. If $\dot{A} \subset \dot{B}$ and if $\dot{B}$ is open, then $\dot{A}_n \subset \dot{B}$ for sufficiently large n so that $\dot{R}_{\dot{v}}^{\dot{B}} \geq \lim_{n\to\infty} \dot{R}_{\dot{v}}^{\dot{A}_n}$, and therefore in view of (c) $\dot{R}_{\dot{v}}^{\dot{A}} \geq \lim_{n\to\infty} \dot{R}_{\dot{v}}^{\dot{A}_n}$. The latter inequality is actually an equality because the reverse inequality is obviously true. □

Proof of (p). Suppose first that $\dot{v}$ is a finite-valued continuous positive superparabolic function on $\dot{D}$. In this case (16.6) is true because according to (o) for each point $\dot{\xi}$ of $\dot{D}$ the set function $\dot{R}_{\dot{v}}^{\bullet}(\dot{\xi})$ is a Choquet capacity relative to the paving of compact subsets of $\dot{D} \cup \partial\dot{D}$. For this choice of $\dot{v}$ the $\dot{R}_{\dot{v}}^{\bullet}(\dot{\xi})$ capacitability of the analytic set $\dot{A}$ implies (16.6). Equation (16.6sm) follows on $\dot{D} - \dot{A}$ because by (f) the equations (16.1) and (16.6sm) are identical on $\dot{D} - \dot{A}$; equation (16.6sm) on $\dot{A}$ is deduced by the following argument. According to (d), when $\dot{\xi} \in \dot{A}$ the left side of (16.6sm) is unchanged when $\dot{A}$ is replaced by $\dot{A} - \{\dot{\xi}\}$; the right side is also unchanged in view of (d) and (e) [if $\dot{\xi} \in \dot{F}$ in (16.6sm), replace $\dot{F}$ in (16.6sm) by $\dot{F}$ less each member of a sequence of balls of center $\dot{\xi}$ and radii tending to 0]. Thus (16.6sm) is reduced to (16.6), and we have proved (p) when $\dot{v}$ is finite valued and continuous, in particular, when $\dot{v}$ is parabolic. Next we prove (p) when $\dot{v}$ is a potential. Let $\dot{v}_{\cdot}$ be an increasing sequence of finite-valued continuous potentials on $\dot{D}$ with limit $\dot{v}$. According to (n) $\dot{R}_{\dot{v}}^{\dot{A}} = \dot{R}_{\dot{v}}^{\dot{A}\cap\dot{D}}$ and $\dot{R}_{+\dot{v}}^{\dot{A}} = \dot{R}_{+\dot{v}}^{\dot{A}\cap\dot{D}}$; so we can apply (e) and the fact that (p) has been shown to be true for $\dot{v}_n$ to deduce

$$\begin{aligned}
\dot{R}_{\dot{v}}^{\dot{A}} = \dot{R}_{\dot{v}}^{\dot{A}\cap\dot{D}} &= \sup\{\dot{R}_{\dot{v}_n}^{\dot{A}\cap\dot{D}} : n \in \mathbb{Z}^+\} \\
&= \sup\{\dot{R}_{\dot{v}_n}^{\dot{F}} : n \in \mathbb{Z}^+, \dot{F} \subset \dot{A} \cap \dot{D}, \dot{F}\ \text{compact}\} \\
&= \sup\{\dot{R}_{\dot{v}}^{\dot{F}} : \dot{F} \subset \dot{A} \cap \dot{D}, \dot{F}\ \text{compact}\}.
\end{aligned}$$

Since the last supremum is at most the supremum in (16.6) which is itself majorized by the left side of (16.6), equation (16.6) is true. The same argument involving smoothed reductions yields (16.6 sm) when $\dot{v}$ is a potential. We have now proved (p) when $\dot{v}$ is either parabolic or a potential so (p) is true as stated in view of (g) and the Riesz decomposition of a positive superparabolic function. □

Proof of (q). The proof follows that of the corresponding classical property in Section VI.4(n). □

Proof of (r). The proof must be more than a translation of the proof of Lemma XI.10 into the parabolic context because the parabolic Dirichlet problem has not yet been treated. The basic method of the proof of Lemma XI.10 will be used however. Let $\dot{D}$ be a nonempty open subset of $\dot{\mathbb{R}}^N$, let $\dot{I}$ be an interval with closure in $\dot{D}$, and let a [b] be the ordinate value on the lower [upper] face of $\dot{I}$. Define $\dot{u}(\dot{\xi}) = \dot{u}(\xi, s)$ on $\dot{D}$ by

$$\dot{u}(\xi, s) = \begin{cases} 1 & \text{if } s > b, \\ \dfrac{s-a}{b-a} & \text{if } a \le s \le b, \\ 0 & \text{if } s < a. \end{cases}$$

Then $\dot{u}$ is superparabolic, and we now show that (reduction relative to $\dot{D}$) the function $\dot{R}_{\dot{u}}^{\dot{I}}$ is continuous. The function is equal to the smoothed reduction according to the fourth property in (b). The function is lower semicontinuous, and $\dot{R}_{\dot{u}}^{\dot{I}} \le \dot{u}$; so $\dot{R}_{\dot{u}}^{\dot{I}}$ is continuous at a point if this inequality is actually an equality at that point. Since $\dot{R}_{\dot{u}}^{\dot{I}}$ is parabolic and therefore continuous on $\dot{D} - \partial\dot{I}$, we need only prove that $\dot{R}_{\dot{u}}^{\dot{I}} = \dot{u}$ on $\partial\dot{I}$. On the lower face of $\dot{I}$, $0 \le \dot{R}_{\dot{u}}^{\dot{I}} \le \dot{u} = 0$ so there is equality on this face. If $\dot{\xi}$ is a boundary point of $\dot{I}$ not on this face, it will be proved in the next paragraph that $\dot{I}$ is not parabolic thin at $\dot{\xi}$; so since $\dot{R}_{\dot{u}}^{\dot{I}}$ is parabolic-fine continuous and equal to $\dot{u}$ on $\dot{I}$, there is equality at $\dot{\xi}$. (Or, apply the second assertion of (b).) Thus $\dot{R}_{\dot{u}}^{\dot{I}}$ is continuous. Let $\dot{I}_{\bullet}$ be a sequence of intervals with closure in $\dot{D}$ forming a basic for the topology of $\dot{D}$, let $\dot{u}_j$ be defined for $\dot{I}_j$ as $\dot{u}$ was for $\dot{I}$, and set

$$\dot{u}^{\#} = \sum_0^{\infty} 2^{-j} \dot{R}_{\dot{u}_j}^{\dot{I}_j}. \tag{17.4}$$

The function $\dot{u}^{\#}$ satisfies the conditions demanded in (r) by a translation into the present context of the corresponding proof in the classical case in Section XI.10.

There remains the proof that each point $\dot{\xi}$ not on the lower face of $\dot{I}$ is in $\dot{I}^{pf}$. Since (from Section 9) each point of the abscissa hyperplane is a parabolic-fine limit point of the lower half-space, it follows that $\dot{I}^{pf}$ contains the interior points of the upper face. Finally suppose that $\dot{\xi}$ is a point of a lateral face, not on the lower face. If $N = 1$, let $\dot{I}_1$ be the reflection of $\dot{I}$ in the bounding side of $\dot{I}$ containing $\dot{\xi}$. A trivial symmetry argument shows that $\dot{I}_1$ is parabolic thin at $\dot{\xi}$ if and only if $\dot{I}$ is. If both are parabolic thin at $\dot{\xi}$, their union is also. However, the set $\dot{I} \cup \dot{I}_1$ contains all but an l_2 null set of the part of some Euclidean neighborhood of $\dot{\xi}$ strictly below $\dot{\xi}$, and so [Section 9(d)] $\dot{I} \cup \dot{I}_1$ cannot be parabolic thin at $\dot{\xi}$. Hence $\dot{\xi} \in \dot{I}^{pf}$. The similar treatment of the case $N > 1$ is left to the reader. □

Proof of (s). The proof follows that of the corresponding inequalities in the classical context in Section VI.4(o). □

18. The Classical Context Green Function in Terms of the Parabolic Context Green Function ($N \geq 1$)

Let D be a nonempty open subset of $\mathbb{R}^N$, and define $\dot{D} = D \times \mathbb{R}$. The parabolic context Green function $\dot{G}_{\dot{D}}$ is then invariant under translations of the ordinate axis,

$$\dot{G}_{\dot{D}}((\xi, s), (\eta, t)) = \dot{G}_{\dot{D}}((\xi, 0), (\eta, t - s)).$$

It follows that for fixed ξ and t the function $(\eta, s) \mapsto \dot{G}_{\dot{D}}((\xi, s), (\eta, t))$ is superparabolic on $\dot{D}$ and is parabolic on $\dot{D} - \{(\xi, t)\}$ with the canonical isolated singularity at (ξ, t); so $\dot{G}_{\dot{D}}((\xi, s), (\eta, t)) \geq \dot{G}_{\dot{D}}((\eta, s), (\xi, t))$. Repeat this interchange of space variables to derive equality here, that is, to find that the function $(\xi, \eta) \mapsto \dot{G}_{\dot{D}}((\xi, s), (\eta, t))$ is symmetric. Define the function ℓ_D on $\mathbb{R} \times D \times D$ by

$$\ell_D(t, \xi, \eta) = \dot{G}_{\dot{D}}((\xi, 0), (\eta, -t)). \tag{18.1}$$

If $D = \mathbb{R}^N$ or more generally if D is $\mathbb{R}^N$ less a closed classical context polar set, then (from Section 8, Example (a)] $\dot{D}$ is $\dot{\mathbb{R}}^N$ less a closed parabolic-polar set; so (from Section 8) $\dot{G}_{\dot{D}} = \dot{G}$ on $\dot{D}$ and $\ell_D(t, \xi, \eta) = \ell(t, \xi - \eta)$. Whatever the choice of D, the function $\ell_D(t, \cdot, \cdot)$ is symmetric on $D \times D$, and the function $\ell_D(\cdot, \cdot, \eta)$ is superparabolic on $\dot{D}$, parabolic on $\dot{D} - \{(\eta, 0)\}$. Moreover $\ell_D(t, \xi, \eta) \leq \ell(t, \xi - \eta)$. The function ℓ_D will be identified in Section 2.IX.17 with the transition density function of Brownian motion in D.

Theorem. *If D is a Greenian subset of $\mathbb{R}^N$,*

$$\int_{-\infty}^{\infty} \dot{G}_{\dot{D}}((\xi, s), (\eta, t)) l_1(dt) = \int_0^{\infty} \ell_D(t, \xi, \eta) l_1(dt) = a_N G_D(\xi, \eta);$$
$$a_N = \frac{\Gamma(N/2 - 1)}{2\pi^{N/2}\sigma^2} \quad \text{if } N > 2, \qquad a_2 = \frac{1}{2\pi\sigma^2}, \qquad a_1 = \frac{2}{\sigma^2}. \tag{18.2}$$

We first verify (18.2) in three special cases. If $N > 2$ and $D = \mathbb{R}^N$, then $\dot{D} = \dot{\mathbb{R}}^N$, $\dot{G}_{\dot{D}} = \dot{G}$, and (18.2) follows by direct evaluation. Incidentally it also follows that for any choice of D the integrals in (18.2) converge when $N > 2$ and $\xi \neq \eta$. If $N = 2$ and if $D = D^+$ is a half-plane, let η^* be the reflection of η in ∂D^+. Then (from Section VIII.9)

$$G_{D^+}(\xi, \eta) = \log \frac{|\xi - \eta^*|}{|\xi - \eta|}. \tag{18.3}$$

On the other hand, according to (4.3), the function ℓ_{D^+} is given by

$$\ell_{D^+}(t,\xi,\eta) = \begin{cases} (2\pi\sigma^2 t)^{-1}\left(\exp\dfrac{-|\xi-\eta|^2}{2\sigma^2 t} - \exp\dfrac{-|\xi-\eta^*|^2}{2\sigma^2 t}\right) & \text{if } t>0 \\ 0 & \text{if } t \le 0. \end{cases} \tag{18.4}$$

Integration yields (18.2) in this case also. If $N = 1$, let $D = D^+$ be the half-line $]0, +\infty[$. In this case $G_D(\xi,\eta) = \xi \wedge \eta$ according to Section XIV.6, and according to (4.3), the function ℓ_D is given by (18.4) except that the first factor on the right has exponent $-\frac{1}{2}$ instead of -1. In this case $\eta = -\eta^*$. Again (18.2) can be verified by direct integration.

To prove (18.2) in the general case for $N > 2$, denote by $g_D(\xi,\eta)$ the value of the equal integrals on the left. For fixed η the function $(\xi,s) \mapsto g_D(\xi,\eta)$ is superparabolic on $\dot{D}$ because the superparabolic function inequality is satisfied; so $g_D(\cdot,\eta)$ is superharmonic on D. The corresponding argument shows that $g_D(\cdot,\eta)$ is harmonic on $D - \{\eta\}$. Furthermore the first integral in (18.2) is the value at (ξ,s) of the superparabolic potential of the measure l_1 on the line parallel to the ordinate axis through $(\eta,0)$. Hence the function $g_D(\cdot,\eta)$ has no positive harmonic minorant other than 0. That is, $g_D(\cdot,\eta)$ is the superharmonic potential of a measure supported by $\{\eta\}$, and it follows that $g_D(\cdot,\eta) = cG_D(\cdot,\eta)$ for some positive constant c. Now

$$\dot{G}_{\dot{D}}(\dot{\xi},\dot{\eta}) = \dot{G}(\dot{\xi},\dot{\eta}) - \dot{u}(\dot{\xi},\dot{\eta}), \qquad \dot{\xi} = (\xi,s),\ \dot{\eta} = (\eta,t), \tag{18.5}$$

with $\dot{u}(\cdot,\dot{\eta})$ a positive parabolic function on $\dot{D}$, and integrating this equality with respect to l_1 over the line through $(\eta,0)$ parallel to the ordinate axis yields

$$cG_D(\xi,\eta) = a_N G(\xi,\eta) - \int_{-\infty}^{\infty} \dot{u}(\dot{\xi},(\eta,t))\,l_1(dt). \tag{18.6}$$

The integral on the right defines a harmonic function of ξ on D, and it follows that $c = a_N$, as was to be proved.

If $N = 2$, it is enough to prove (18.2) for a bounded set D because when $D_.$ is an increasing sequence of bounded open subsets of D with union D, it follows that $\lim_{n\to\infty} G_{D_n} = G_D$ and $\lim_{n\to\infty} \dot{G}_{\dot{D}_n} = \dot{G}_{\dot{D}}$ according to Theorem VII.6 and its parabolic counterpart. Let D^+ be a half-plane containing D, and define $\dot{D}^+ = D^+ \times \mathbb{R}$. Although (18.5) is valid when $N = 2$, we cannot integrate as in the case $N > 2$ because

$$\int_{-\infty}^{\infty} \dot{G}((\xi,s),(\eta,t))\,l_1(dt) = +\infty$$

when $N < 3$. We therefore replace (18.5) by a relative equation (see Section XVII.4)

$$\dot{G}_{\dot{D}}(\dot{\xi}, \dot{\eta}) = \dot{G}_{\dot{D}^+}(\dot{\xi}, \dot{\eta}) - \dot{v}(\dot{\xi}, \dot{\eta}), \tag{18.7}$$

where $\dot{v}(\cdot, \dot{\eta})$ is a positive parabolic function. The proof continues as in the case $N > 2$, using the truth of the theorem when $D = D^+$. When $N = 1$, there remains only the proof of (18.2) for D a finite interval, and this is carried through just as in the case of bounded open sets when $N = 2$.

Observation. According to Theorem 18, if D is Greenian, if μ is a measure on D, and if $\dot{\mu}$ is the product measure $\mu \times l_1$ on $\dot{D} = D \times \mathbb{R}$, then $\dot{G}_{\dot{D}}\dot{\mu}(\xi, s) = a_N G_D \mu(\xi)$.

19. The Quasi-Lindelöf Property

Theorem. *An arbitrary union of parabolic-fine open sets differs by a parabolic-semipolar set from some countable subunion.*

The proof follows that of the corresponding property in the classical context (Theorem XI.11).

Chapter XVIII

The Parabolic Dirichlet Problem, Sweeping, and Exceptional Sets

1. Relativization of the Parabolic Context; The PWB Method in this Context

Let $\dot{D}$ be a nonempty open subset of $\dot{\mathbb{R}}^N$, and let $\dot{h}$ be a strictly positive parabolic function on $\dot{D}$. A function $\dot{v}/\dot{h}$ on $\dot{D}$ will be called *$\dot{h}$-parabolic*, *$\dot{h}$-superparabolic*, or *$\dot{h}$-subparabolic* if $\dot{v}$ is parabolic, superparabolic, or subparabolic, respectively. The notation will be parallel to that in the classical context, with $\dot{h}$ omitted when $\dot{h} \equiv 1$. Thus $\dot{\mathrm{G}}\mathrm{M}_{\dot{D}}^{\dot{h}}$, ${}^{\dot{h}}\dot{R}_{\dot{v}}^{\dot{A}}$, $\dot{\tau}_{\dot{B}}^{\dot{h}}$, $\dot{H}_f^{\dot{h}}, \ldots$ need no further identification. In the dual context in which $\dot{h}$ is coparabolic we write $\overset{*}{\mathrm{G}}\mathrm{M}_{\dot{D}}^{\dot{h}}$, ${}^{\dot{h}}\overset{*}{R}{}_{\dot{v}}^{\dot{A}}$, $\overset{*}{\tau}{}_{\dot{B}}^{\dot{h}}$, $\overset{*}{H}{}_f^{\dot{h}}, \ldots$.

The PWB method for solving the first boundary value (Dirichlet) problem for relative harmonic functions translates directly into the parabolic context. As in the classical context, the boundary is that obtained by a metric compactification of the given open subset $\dot{D}$ of $\dot{\mathbb{R}}^N$; the boundary of $\dot{D}$ relative to the one-point Alexandrov compactification of $\dot{\mathbb{R}}^N$ will be called the *Euclidean boundary*. Upper and lower $\dot{\mathrm{P}}\mathrm{WB}^{\dot{h}}$ solutions $\bar{\dot{H}}{}_f^{\dot{h}}$ and $\underline{\dot{H}}{}_f^{\dot{h}}$ are defined on $\dot{D}$ corresponding to a specified boundary function f, and if these solutions are equal and parabolic, the function f is parabolic $\dot{h}$-resolutive with $\dot{\mathrm{P}}\mathrm{WB}^{\dot{h}}$ solution $\dot{H}_f^{\dot{h}} = \underline{\dot{H}}{}_f^{\dot{h}}$. The definitions of $\dot{h}$-resolutive boundaries, $\dot{h}$-regular boundary points, $\dot{h}$-parabolic measure boundary subsets, and so on are translations of the corresponding definitions in the classical context. The properties of h-harmonic measure null sets and of PWB^h solutions derived in Sections VIII.5 to VIII.7 go over into the parabolic context with trivial changes in the derivations. The relations between PWB^h solutions and reductions (Section VIII.2) are preserved in the present context:

$$\bar{\dot{H}}{}_{1_{\dot{A}}}^{\dot{h}} = \frac{\dot{R}_{\dot{h}}^{\dot{A}}}{\dot{h}} = {}^{\dot{h}}\dot{R}_1^{\dot{A}}, \qquad \bar{\overset{*}{H}}{}_{1_{\dot{A}}}^{\dot{h}} = \frac{\overset{*}{R}{}_{\dot{h}}^{\dot{A}}}{\dot{h}} = {}^{\dot{h}}\overset{*}{R}{}_1^{\dot{A}}$$

when $\dot{A}$ is a boundary subset.

Example (a) (Euclidean boundary). If $\dot{\xi} \in \dot{D}$ and if $\dot{A} = \{\dot{\eta} \in \partial\dot{D}\colon \operatorname{ord}\dot{\eta} \geq \operatorname{ord}\dot{\xi}\}$, then $\bar{\dot{H}}{}_{1_{\dot{A}}}^{\dot{h}}(\dot{\xi}) = 0$. In fact, if $\dot{\xi} = (\xi, s)$, if $\dot{\xi}_n = (\xi, s - 1/n)$, and if $\dot{u}_n$ is the indicator function of the set $\{\dot\eta \in \dot D\colon \operatorname{ord}\dot\eta > s - 1/n\}$, the function $\dot u_n$ is in the upper $\dot{\mathrm{P}}\mathrm{WB}^{\dot h}$ class on $\dot D$ for the boundary function $1_{\dot A}$; so $\bar{\dot H}{}^{\dot h}_{1_{\dot A}}(\dot\xi_n) = 0$ when n is so large that $\dot\xi_n \in \dot D$. Hence $\bar{\dot H}{}^{\dot h}_{1_{\dot A}}(\dot\xi) = 0$.

EXAMPLE (b). Let D be a Greenian subset of $\mathbb{R}^N$ coupled with a boundary ∂D provided by a metric compactification of D, define $\dot D = D \times \mathbb{R}$, and define $\partial\dot D$ as the boundary of $\dot D$ provided by the compactification $(D \cup \partial D) \times \bar{\mathbb{R}}$, where $\bar{\mathbb{R}}$ is the two-point compactification of $\mathbb{R}$. A trivial adaptation of Example (a) shows that the upper boundary $(D \cup \partial D) \times \{+\infty\}$ is an $\dot h$-parabolic measure zero subset of $\partial\dot D$ for every choice of strictly positive parabolic function $\dot h$ on $\dot D$. Alternatively this follows from the fact that the function $\dot u\colon (\xi, s) \mapsto e^s$ is a positive $\dot h$-superparabolic function on $\dot D$ for every positive superparabolic function $\dot h$ on $\dot D$ and $\dot u$ has limit $+\infty$ at the upper boundary of $\dot D$. The role of the lower boundary of $\dot D$ is more delicate. We show that the lower boundary is parabolic measure null if D is on one side of a hyperplane of $\mathbb{R}^N$. It is sufficient to exhibit a sequence $\dot u_{\bullet}$ of positive parabolic functions on $\dot D$ with limit 0 and with $\dot u_n(\dot\eta) = 0$ when $\operatorname{ord}\dot\eta \le -n$, and it is therefore sufficient to exhibit such a sequence when D is the half-space on one side of a hyperplane of $\mathbb{R}^N$; we can even assume that the hyperplane is a coordinate hyperplane and that D is the set on which the corresponding coordinate function is strictly positive. Finally we need only consider the case $N = 1$ because a sequence $\dot u_{\bullet}$ in this case induces one in the general case. The sequence $\dot u_{\bullet}$ on $\dot D = \{(\xi, s)\colon \xi > 0\}$ $(N = 1)$ defined by

$$\dot u_n(\xi, s) = \begin{cases} 1 - \dfrac{(2/\pi)^{1/2}}{(s+n)^{1/2}} \displaystyle\int_\xi^\infty \exp\frac{-\alpha^2}{2(s+n)}\, l_1(d\alpha) & \text{if } s > -n \\ 1 & \text{if } s \le -n \end{cases}$$

has the stated properties. (We have simplified the notation by taking $\sigma = 1$.) It is not difficult to show, although not necessary for present purposes, that $\dot u_n(\xi, s)$ is the parabolic measure of $\{(0, t)\colon t \le -n\}$, the Euclidean boundary subset relative to (ξ, s).

EXAMPLE (c) (Euclidean boundary, $\dot h \equiv 1$). Let $\dot D$ be an interval in $\dot{\mathbb{R}}^N$. According to Example (a), the upper boundary of $\dot D$ is a parabolic measure null set. We show that $\partial\dot D$ is parabolic resolutive by showing that every finite-valued continuous boundary function f is parabolic resolutive with $\dot H_f = \dot{\mathrm{P}}\mathrm{I}(\dot D, f)$. Define f_1 on $\partial\dot D$ as f except that $f_1 \equiv \inf_D f$ on the upper boundary. Then $\bar{\dot H}_f = \bar{\dot H}_{f_1}$ because $f = f_1$ up to a parabolic measure null set. Since $\dot{\mathrm{P}}\mathrm{I}(\dot D, f)$ is in the upper $\dot{\mathrm{P}}\mathrm{WB}$ class on $\dot D$ for f_1, it follows that $\dot{\mathrm{P}}\mathrm{I}(\dot D, f) \ge \bar{\dot H}_{f_1} = \bar{\dot H}_f$ and similarly $\dot{\mathrm{P}}\mathrm{I}(\dot D, f) \le \underline{\dot H}_f$; so $\dot{\mathrm{P}}\mathrm{I}(\dot D, f) = \dot H_f$, as asserted.

EXAMPLE (d) (Euclidean boundary, $\dot h \equiv 1$). Let $\dot D$ be a slab $\mathbb{R}^N \times]0, \delta[$ with $0 < \delta \le +\infty$. According to Example (a), the finite part of the upper slab boundary is a parabolic measure null set. We show that the boundary is parabolic resolutive with $\dot{H}_f = \dot{\mathrm{P}}\mathrm{I}(\dot{D}, f_0)$ when f is a finite-valued continuous boundary function and $f_0(\xi) = f((\xi, 0))$. To prove these assertions, observe first that the function f_0 has limit $f(\infty)$ at ∞ and (from Section XVI.1) if $\dot{\zeta}$ is either ∞ or a point of the lower slab boundary the function $\dot{\mathrm{P}}\mathrm{I}(\dot{D}, f_0)$ has limit $f(\dot{\zeta})$ at $\dot{\zeta}$. When $\delta = +\infty$, the function $\dot{\mathrm{P}}\mathrm{I}(\dot{D}, f_0)$ is in both the lower and upper $\dot{\mathrm{P}}$WB classes on $\dot{D}$ for f and so can be identified with $\dot{H}_f$; when $\delta < +\infty$, the method used in Example (b) shows that $\dot{H}_f = \dot{\mathrm{P}}\mathrm{I}(\dot{D}, f_0)$.

Example (e) (Euclidean boundary). (Cf. Section VIII.3.) Let $\dot{u}$ be $\dot{h}$-superparabolic on $\dot{D}$, and suppose that $\dot{u}$ has a finite or infinite limit at every boundary point, thereby defining a boundary function f. Suppose also that $\dot{u}$ is lower bounded and that $\dot{\mathrm{G}}\mathrm{M}_{\dot{D}}^{\dot{h}}\dot{u}$ is upper bounded. Then f is parabolic $\dot{h}$-resolutive with $\dot{H}_f^{\dot{h}} = \dot{\mathrm{G}}\mathrm{M}_{\dot{D}}^{\dot{h}}\dot{u}$ because $\dot{u}$ is in the upper and $\dot{\mathrm{G}}\mathrm{M}_{\dot{D}}^{\dot{h}}\dot{u}$ is in the lower $\dot{\mathrm{P}}\mathrm{WB}^{\dot{h}}$ class on $\dot{D}$ for f. Moreover if $\dot{h} \equiv 1$, if $\dot{u}$ has a superparabolic extension to a neighborhood of $\dot{\mathbb{R}}^N \cap \overline{\dot{D}}$, and if this extension has a finite limit at every point of $\partial\dot{D}$, then $\lim_{\dot{\eta}\to\dot{\zeta}} \dot{H}_f(\dot{\eta}) = f(\dot{\zeta})$ whenever the boundary point $\dot{\zeta}$ is not in some at most parabolic-semipolar set. In fact $\dot{\mathrm{G}}\mathrm{M}_{\dot{D}}\dot{u}$ can be obtained as the restriction to $\dot{D}$ of the limit of a decreasing sequence of locally lower bounded superparabolic functions on a neighborhood of $\dot{\mathbb{R}}^N - \overline{\dot{D}}$, each equal to $\dot{u}$ outside a compact subset of $\dot{D}$, depending on the function. [See the corresponding argument in Section XVII.4 as applied to analyze $\dot{\mathrm{G}}\mathrm{M}_{\dot{D}}\dot{G}(\cdot, \dot{\eta})$, or see the general discussion of GM_D in the classical context in Section III.1.] According to the parabolic Fundamental Convergence Theorem, the limit function $\dot{u}'$ majorizes its smoothing, which is parabolic on $\dot{D}$, and the two functions are equal up to a parabolic-semipolar set. At a nonexceptional boundary point $\dot{\zeta}$ of $\dot{D}$ the function $\dot{\mathrm{G}}\mathrm{M}_{\dot{D}}\dot{u}$ has limit $f(\dot{\zeta}) = \underset{+}{\dot{u}'}(\dot{\zeta})$ on approach from within $\dot{D}$ because $\underset{+}{\dot{u}'}$ is lower semicontinuous, $\underset{+}{\dot{u}'} \le \dot{u}$, and $\dot{u}$ has this limit.

Application to the Parabolic Resolutivity of the Euclidean Boundary

The Euclidean boundary of an open subset of $\dot{\mathbb{R}}^N$ is parabolic resolutive, and the set of its finite parabolic irregular boundary points is parabolic-semipolar. In fact this assertion follows from a trivial modification of the proof for $N > 2$ of Theorem VIII.4, the counterpart in the classical context of this assertion. Observe that the set of finite parabolic irregular boundary points is parabolic-semipolar rather than parabolic-polar.

Application to a Representation of Parabolic Green Functions (Euclidean Boundary, $\dot{h} \equiv 1$)

If $\dot{\eta}$ is a point of $\dot{D}$, the restriction to $\partial\dot{D}$ of $\dot{G}(\cdot, \dot{\eta})$ (defined as 0 at ∞) can be taken as f in Example (e), and it follows that $\dot{G}_{\dot{D}}(\cdot, \dot{\eta}) = \dot{G}(\cdot, \dot{\eta}) - \dot{H}_f$.

More generally, if $\dot{B}$ is an open subset of $\dot{D}$, if $\dot{\eta} \in \dot{B}$, and if f is the restriction to $\partial\dot{B}$ of $\dot{G}_{\dot{D}}(\cdot, \dot{\eta})$ (defined as 0 on $\partial\dot{B} \cap \partial\dot{D}$), then $\dot{G}_{\dot{B}}(\cdot, \dot{\eta})$ is $\dot{G}_{\dot{D}}(\cdot, \dot{\eta})$ less the $\dot{\mathrm{P}}$WB solution for $\dot{B}$ with the boundary function f.

2. $\dot{h}$-Parabolic Measure

Let $\dot{D}$ be a nonempty open subset of $\dot{\mathbb{R}}^N$ coupled with a boundary provided by a metric compactification, and let $\dot{h}$ be a strictly positive parabolic function on $\dot{D}$. The development of $\dot{h}$-parabolic measure follows that of h-harmonic measure; so the details will be omitted. If $\dot{A}$ is a subset of $\partial\dot{D}$ with an $\dot{h}$-parabolic-resolutive indicator function, we define $\dot{\mu}_{\dot{D}}^{\dot{h}}(\cdot, \dot{A}) = \dot{H}_{1_{\dot{A}}}^{\dot{h}}$, call $\dot{A}$ a $\dot{\mu}_{\dot{D}}^{\dot{h}}$ *measurable set*, and call $\dot{\mu}_{\dot{D}}^{\dot{h}}(\xi, \dot{A})$ the *$\dot{h}$-parabolic measure of $\dot{A}$ relative to ξ*. The set function $\dot{\mu}_{\dot{D}}^{\dot{h}}(\xi, \cdot)$ is a probability measure, and the completion of the restriction of this measure to the class of $\dot{\mu}_{\dot{D}}^{\dot{h}}$ measurable Borel sets is an extension of this measure. The $\dot{h}$-parabolic measure null sets defined above are the sets $\dot{A}$ for which $\dot{\mu}_{\dot{D}}^{\dot{h}}(\cdot, \dot{A})$ vanishes identically. A boundary function measurable with respect to the σ algebra of $\dot{\mu}_{\dot{D}}^{\dot{h}}$ measurable sets will be called $\dot{\mu}_{\dot{D}}^{\dot{h}}$ measurable, and $L^p(\dot{\mu}_{\dot{D}}^{\dot{h}})$ is defined as the class of $\dot{\mu}_{\dot{D}}^{\dot{h}}$ measurable functions from $\partial\dot{D}$ into $\overline{\mathbb{R}}$, with two functions identified when they are equal $\dot{\mu}_{\dot{D}}^{\dot{h}}$ almost everywhere (that is, up to an $\dot{h}$-parabolic measure null set) for which $|f|^p$ is $\dot{\mu}_{\dot{D}}^{\dot{h}}(\xi, \cdot)$ integrable for every point ξ of $\dot{D}$. A function f from $\partial\dot{D}$ into $\overline{\mathbb{R}}$ is $\dot{h}$-parabolic resolutive if and only if f is in $L^1(\dot{\mu}_{\dot{D}}^{\dot{h}})$, and if so, then $\dot{H}_f^{\dot{h}} = \dot{\mu}_{\dot{D}}^{\dot{h}}(\cdot, f)$.

The work in Sections VIII.8 to VIII.10 goes through in the present context except that even if $\dot{D}$ is connected, it is not true that a positive parabolic function on $\dot{D}$ which vanishes at a point must vanish identically; so the L^1 class for $\dot{\mu}_{\dot{D}}^{\dot{h}}(\xi, \cdot)$ depends on ξ; the class of $\dot{\mu}_{\dot{D}}^{\dot{h}}$ measurable sets and the class $L^1(\dot{\mu}_{\dot{D}}^{\dot{h}})$ were defined taking this fact into account.

In the present context the notation τ_B^h of Section VIII.11 becomes $\dot{\tau}_{\dot{B}}^{\dot{h}}$. Thus if $\dot{v}$ is an $\dot{h}$-superparabolic function on $\dot{D}$ and if $\dot{B}$ is an open relatively compact subset of $\dot{D}$, the function $\dot{\tau}_{\dot{B}}^{\dot{h}}\dot{v}$ is $\dot{h}$-superparabolic on $\dot{D}$ and is $\dot{h}$-parabolic on $\dot{B}$, and $\dot{\tau}_{\dot{B}}^{\dot{h}}\dot{v} \le \dot{v}$, with equality on $\dot{D} - \dot{B}$ up to a parabolic-semipolar subset of $\partial\dot{B}$ (Euclidean boundary). In particular, if $\dot{v} \ge 0$, then $\dot{\tau}_{\dot{B}}^{\dot{h}}\dot{v} = {}^{\dot{h}}\dot{R}_{+\dot{v}}^{\dot{D}-\dot{B}}$.

Example (a). If $\dot{D}$ is an interval, we have seen in Section 1 that the Euclidean boundary is parabolic resolutive, that the upper boundary is a parabolic measure null set, and that $\dot{\mu}_{\dot{D}}(\cdot, f) = \dot{\mathrm{P}}\mathrm{I}(\dot{D}, f)$ so that parabolic measure is given by the densities evaluated in Section XV.9. The inner points of the upper boundary are parabolic irregular, but all other boundary points are parabolic regular according to Section XV.9.

Example (b). If $\dot{D} = \mathbb{R}^N \times]0, \delta[$ is a slab we have seen in Section 1 that the Euclidean boundary is parabolic resolutive, that the abscissa hyperplane part of the boundary has parabolic measure 1, and that

$$\dot\mu_{\dot D}(\dot\xi, d\eta) = \ell(s,\xi,\eta)l_N(d\eta) \qquad [\dot\xi = (\xi, s)]$$

for $(\eta, 0)$ on the abscissa hyperplane. The points of the abscissa hyperplane and the point ∞ are parabolic regular, but all other boundary points are parabolic irregular. It is easy to see (cf. the classical context for D a ball in Section VIII.9) that the Euclidean boundary is parabolic universally resolutive, that the abscissa hyperplane has $\dot h$-parabolic measure 1 for every choice of $\dot h$, and if (Riesz–Herglotz type representation, Theorem XVI.6), $\dot h = \dot{\mathrm{P}}\mathrm{I}(\dot D, \dot N_h)$, then

$$\dot\mu_{\dot D}^{\dot h}(\dot\xi, d\eta) = \ell(s,\xi,\eta)\frac{\dot N_h(d\eta)}{\dot h(\dot\xi)} \qquad [\dot\xi = (\xi, s)]$$

for $(\eta, 0)$ on the abscissa hyperplane.

The Support of $\dot\mu_{\dot D}$ (Euclidean Boundary)

If $\dot\xi \in \dot D$, define $\dot D_{\dot\xi} = \{\dot\eta \in \dot D : \operatorname{ord}\dot\eta < \operatorname{ord}\dot\xi\}$. We have seen in Section 1, Example (a), that $\dot\mu_{\dot D}(\dot\xi, \cdot)$ is supported by the part of $\partial\dot D$ strictly below $\dot\xi$. Moreover, if $\dot A$ is a Borel subset of $\partial\dot D$ strictly below $\dot\xi$, then the function $\dot\mu_{\dot D_{\dot\xi}}(\cdot, \dot A)$ is the restriction to $\dot D_{\dot\xi}$ of $\dot\mu_{\dot D}(\cdot, \dot A)$ because if $\dot v$ $[\dot u]$ is in the upper [lower] $\dot{\mathrm{P}}$WB class on $\dot D$ for the boundary function $1_{\dot A}$ on $\partial\dot D$, the restrictions of these functions to $\dot D_{\dot\xi}$ are in the corresponding $\dot{\mathrm{P}}$WB classes on $\dot D_{\dot\xi}$ for the restriction of $1_{\dot A}$ to $\partial\dot D_{\dot\xi}$. This fact has the following useful implication, in which the roles of $\dot D_{\dot\xi}$ and $\dot D$ are reversed. Let now $\dot D$ be an open subset of $\dot{\mathbb{R}}^N$, and let $\dot\xi$ be a boundary point of $\dot D$. Suppose that every point of $\dot D$ is strictly below $\dot\xi$ and that the part of some open neighborhood $\dot B$ of $\dot\xi$ strictly below $\dot\xi$ lies in $\dot D$. Then if $\dot A$ is a Borel subset of $\partial\dot D$, the function $\dot\mu_{\dot D}(\cdot, \dot A)$ is defined and parabolic on $\dot D$ and has the parabolic extension $\dot\mu_{\dot D\cup\dot B}(\cdot, \dot A)$ to $\dot D \cup \dot B$.

3. Parabolic Barriers

The classical context definitions of weak h-barrier and h-barrier translate directly into the parabolic context, as does the proof that if there is an h-barrier locally at a boundary point, then there is an h-barrier defined on the whole open set in question. The classical context Bouligand theorem (Section VIII.12) is also true in the present context: If there is a weak parabolic barrier at a Euclidean boundary point $\dot\zeta$ of an open subset $\dot D$ of $\dot{\mathbb{R}}^N$, then there is a parabolic barrier there. To see this, adapt the classical context proof as follows. In view of the fact that the existence of a parabolic barrier at $\dot{\zeta}$ is a local property it is sufficient to show that there is a parabolic barrier on the trace on $\dot{D}$ of an open neighborhood of $\dot{\zeta}$; so $\dot{D}$ can be supposed bounded. Suppose then that $\dot{D}$ has diameter $\delta < +\infty$ and that $\dot{u}$ is a weak parabolic barrier for $\dot{D}$ at $\dot{\zeta}$. For notational simplicity assume that $\dot{\zeta}$ is the origin. Define $\dot{\phi}: \dot{\eta} = (\eta, t) \mapsto c|\eta|^2 + t^2$ on $\dot{D}$, choosing c so large that $\dot{\phi}$ is subparabolic. The function $\dot{\phi}$ is in the lower $\dot{\text{P}}$WB class on $\dot{D}$ for the boundary function $f: \dot{\eta} = (\eta, t) \mapsto c|\eta|^2 + t^2$ on $\dot{D}$. Hence $\dot{H}_f \geq \dot{\phi}$, and it will be shown that $\dot{H}_f$ is a parabolic barrier for $\dot{D}$ at the origin by showing that $\dot{H}_f$ has limit 0 there. Let $\dot{B}$ be an interval containing the origin, of diameter r, let $\dot{A}$ be a compact subset of $\dot{D} \cap \partial\dot{B}$, and let $\dot{\psi}$ be the indicator function of $(\partial\dot{B} - \dot{A}) \cap \dot{D}$ on $\partial\dot{B}$, so that $\dot{\mu}_{\dot{B}}(\cdot, \dot{\psi})$ is parabolic on $\dot{B}$ with limit 1 at every point of $\partial\dot{B} - \dot{A}$ not on the upper boundary. Since the upper boundary of $\dot{B}$ has parabolic measure 0, there is a positive superparabolic function $\dot{v}$ on $\dot{B}$ with $\dot{v}(0) < +\infty$ and with limit $+\infty$ at every point of the upper boundary. If $\dot{u}_0$ is in the lower $\dot{\text{P}}$WB class on $\dot{D}$ for f, apply the maximum theorem for subparabolic functions to find that $\dot{u}_0$ on $\dot{B} \cap \dot{D}$ is at most $(c + 1)r^2$ if $\partial\dot{B}$ does not meet $\dot{D}$ and is at most

$$(c + 1)r^2 + (c + 1)\delta^2 \left[\frac{\dot{u}}{\inf_{\dot{A}} \dot{u}} + \dot{\mu}_{\dot{B}}(\cdot, \dot{\psi})\right] + r\dot{v} \tag{3.1}$$

if $\partial\dot{B}$ does meet $\dot{D}$. Thus in both cases $\dot{H}_f$ is majorized on $\dot{B} \cap \dot{D}$ by the sum (3.1). This sum has limit

$$(c + 1)r^2 + (c + 1)\delta^2 \dot{\mu}_{\dot{B}}(0, \dot{\psi}) + r\dot{v}(0)$$

at the origin, and this limit is at most $2(c + 1)r^2 + r\dot{v}(0)$ if $\dot{A}$ is sufficiently large. Since r can be made arbitrarily small, the function $\dot{H}_f$ has limit 0 at the origin, as was to be proved.

Theorems VIII.13 and VIII.14 on the relations between regularity and barriers translate directly into the parabolic context.

4. Relations between the Classical Dirichlet Problem and the Parabolic Context Dirichlet Problem

Let D be a Greenian subset of $\mathbb{R}^N$ coupled with a boundary ∂D by a metric compactification of D. Define $\dot{D} = D \times \mathbb{R}$, and define $\partial\dot{D}$ as the boundary provided by the compactification $(D \cup \partial D) \times \overline{\mathbb{R}}$, where $\overline{\mathbb{R}}$ is the two-point compactification of $\mathbb{R}$. Let h be a strictly positive harmonic function on D, and define the parabolic function $\dot{h}$ on $\dot{D}$ by setting $\dot{h}(\xi, s) = h(\xi)$. We now consider the Dirichlet problem for h-harmonic functions on D and for $\dot{h}$-parabolic functions on $\dot{D}$. Recall [Section 1, Example (b)] that the upper boundary $(D \cup \partial D) \times \{+\infty\}$ of $\dot{D}$ is an $\dot{h}$-parabolic measure null set. We suppose that the lower boundary of $\dot{D}$ is also $\dot{h}$-parabolic measure null. This is true if D is not too large, for example [Section 1, Example (b)], if D is on one side of a hyperplane in $\mathbb{R}^N$ and $\dot{h} \equiv 1$. Let f be an extended real-valued function on ∂D, and define $\dot{f}$ on $\partial\dot{D}$ by setting $\dot{f}(\xi, s) = f(\xi)$, except that no restriction is imposed on $\dot{f}$ on the upper and lower boundaries of $\dot{D}$. This looseness is convenient, and the actual definition of $\dot{f}$ on the upper and lower boundaries does not affect the $\dot{\text{P}}\text{WB}^{\dot{h}}$ solutions for $\dot{f}$. If u is an h-superharmonic function in the upper PWB^h class on D for f, the function $\dot{u}$: $(\xi, s) \mapsto u(\xi)$ is in the upper $\dot{\text{P}}\text{WB}^{\dot{h}}$ class on $\dot{D}$ for $\dot{f}$ if $\dot{f}$ is defined as $-\infty$ on the upper and lower boundaries of $\dot{D}$. Hence $\bar{\dot{H}}{}^{\dot{h}}_{\dot{f}}(\xi, s) \le \bar{H}^h_f(\xi)$, and similarly $\underline{\dot{H}}{}^{\dot{h}}_{\dot{f}}(\xi, s) \ge \underline{H}^h_f(\xi)$. We conclude that $\dot{f}$ is $\dot{h}$-parabolic resolutive on $\dot{D}$ whenever f is h-harmonic resolutive on D, and then $\dot{H}^{\dot{h}}_{\dot{f}}(\xi, s) = H^h_f(\xi)$. In particular, if A is a μ^h_D measurable subset of ∂D, the product set $\dot{A} = A \times \mathbb{R}$ is $\dot{\mu}^{\dot{h}}_{\dot{D}}$ measurable and $\dot{\mu}^{\dot{h}}_{\dot{D}}((\xi, s), \dot{A}) = \mu^h_D(\xi, A)$. Conversely, if $\dot{f}$ is Borel measurable and $\dot{\text{P}}\text{WB}^{\dot{h}}$ resolutive and if ∂D is h-resolutive, then f is Borel measurable, and we show it is PWB^h resolutive. In fact there is nothing to prove if f is bounded, and in the general case

$$\mu^h_D(\cdot, |f|) = \lim_{n\to\infty} \mu^h_D(\cdot, |f| \wedge n) = \lim_{n\to\infty} \dot{\mu}^{\dot{h}}_{\dot{D}}(\cdot, |\dot{f}| \wedge n) = \dot{\mu}^{\dot{h}}_{\dot{D}}(\cdot, |\dot{f}|) < +\infty.$$

Suppose in this paragraph that ∂D is PWB^h resolutive and that $\partial\dot{D}$ is $\dot{\text{P}}\text{WB}^{\dot{h}}$ resolutive. This hypothesis is satisfied if $h \equiv 1$ and ∂D is the Euclidean boundary. (Note that $\partial\dot{D}$ is not then the Euclidean boundary, but the $\dot{\text{P}}\text{WB}$ resolutivity of $\partial\dot{D}$ follows from that of the Euclidean boundary of $\dot{D}$ and the fact that the upper and lower boundaries of $\dot{D}$ are parabolic measure null.) If (ξ, s) is a finite $\dot{h}$-regular point of $\partial\dot{D}$, the point ξ is an h-regular point of ∂D. In fact a bounded h-resolutive function f on ∂D, continuous at ξ, defines a bounded function $\dot{f}$ on $\partial\dot{D}$ as above (setting $\dot{f} \equiv 0$ on the upper and lower boundaries). The function $\dot{f}$ is $\dot{h}$-resolutive, bounded, and continuous at (ξ, s), so $\dot{H}^{\dot{h}}_{\dot{f}}$ has limit $f(\xi)$ at (ξ, s); that is, H^h_f has limit $f(\xi)$ at ξ. If $h \equiv 1$ and if ξ is a finite regular point of ∂D, then each finite point (ξ, s) of $\partial\dot{D}$ is regular because a weak barrier for D at ξ when considered as a function on $\dot{D}$ is a weak barrier for $\dot{D}$ at (ξ, s).

5. Classical Reductions in the Parabolic Context

Let D be a Greenian subset of $\mathbb{R}^N$, coupled with a boundary provided by a metric compactification, define $\dot{D} \times \mathbb{R}$, and compactify $\dot{D}$ as in Section 4 to obtain $\partial\dot{D}$. Suppose that the lower boundary of $\dot{D}$ is parabolic measure null; we have seen in Section 1 that this condition is satisfied if D is on one side of a hyperplane of $\mathbb{R}^N$. Let v be a positive superharmonic function on D, and define the superharmonic function $\dot{v}$ on $\dot{D}$ by setting $\dot{v}(\xi, s) = v(\xi)$. Let A be a subset of D, and define $\dot{A} = A \times \mathbb{R}$. We now show that if A is analytic, then

$$\dot{R}_{\dot{v}}^{\dot{A}}(\xi, s) = R_v^A(\xi), \tag{5.1}$$

$$\dot{\underset{+\dot{v}}{R}}^{\dot{A}}(\xi, s) = \underset{+v}{R}^A(\xi). \tag{5.1sm}$$

If A is a compact subset of D, then (from Section VIII.10) R_v^A on $D - A$ is the PWB solution for the set $D - A$ with boundary function equal to 0 on $\partial D \cap \partial(D - A)$ and equal to v on $A \cap \partial(D - A)$. Similarly $\dot{R}_{\dot{v}}^{\dot{A}}$ on $\dot{D} - \dot{A}$ is the $\dot{\text{P}}$WB solution for the set $\dot{D} - \dot{A}$ with boundary function equal to 0 on $\partial\dot{D} \cap \partial(\dot{D} - \dot{A})$ and equal to $\dot{v}$ on $\dot{A} \cap \partial(\dot{D} - \dot{A})$. In view of the relation (Section 4) between the Dirichlet problems for harmonic and for parabolic functions it follows that (5.1) is true for ξ in $D - A$; this equation is trivial for ξ in A. If A is an analytic subset of D, then $\dot{A}$ is an analytic subset of $\dot{D}$; so [from Section XVII.16(p)] the value $\dot{R}_{\dot{v}}^{\dot{A}}(\xi, s)$ can be approximated arbitrarily closely from below by $\dot{R}_{\dot{v}}^{\dot{F}}(\xi, s)$ with $\dot{F}$ a compact subset of $\dot{A}$; if F is the projection of $\dot{F}$ on the abscissa hyperplane, the approximation is improved if $\dot{F}$ is replaced by $F \times \mathbb{R}$. Since F is a compact subset of D and since [from Section VI.3(l)] $R_v^A(\xi)$ can be approximated arbitrarily closely by $R_v^B(\xi)$ with B a compact subset of A, it follows that (5.1) is true. Equations (5.1) and (5.1sm) are identical when $\xi \in D - A$ and the left [right] sides of these equations are equal up to a parabolic-semipolar subset of $\dot{D}$ [parabolic-polar subset of D]. Since a classical-polar subset of D is the projection of a parabolic-polar subset of $\dot{D}$ (because a positive superharmonic function on D identically $+\infty$ on a set B can be identified with a positive superparabolic function on $\dot{D}$ identically $+\infty$ on $B \times \mathbb{R}$), equation (5.1sm) is true up to a parabolic-semipolar subset of $\dot{D}$ and therefore is true on $\dot{D}$.

Observation. There are three possible approaches to the properties of reductions in the classical context.

(1) These properties can be proved using the specific classical context. Such proofs were used in the preceding chapters. Unfortunately some of the most natural proofs cannot be applied in the parabolic context because of the weaker version of the Fundamental Convergence Theorem in the parabolic context and (a related fact) because the domination principle is false in the parabolic context.

(2) The reduction properties in the parabolic context can be proved, and then either it can be noted that these proofs are valid in the classical context or

(3) It can be noted that in view of (5.1) and (5.1sm) the properties of reductions in the classical context can usually be read off from those in the parabolic context, or at least be deduced from them.

The choice (1) adopted in this book is inefficient and repetitious but was made because the proofs in the classical context are thereby clearer and more natural than the more generally usable proofs in choices (2) and (3).

6. Parabolic Regularity of Boundary Points

In the following examples $\dot{D}$ is a nonempty open subset of $\dot{\mathbb{R}}^N$ provided with its Euclidean boundary. In this section *barrier* means "parabolic context barrier." Since the existence of a weak barrier for $\dot{D}$ at a boundary point $\dot{\xi}_0$ is a property of $\dot{D}$ in a neighborhood of $\dot{\xi}_0$, one can prove that a barrier for $\dot{D}$ at $\dot{\xi}_0$ exists and thereby prove that $\dot{\xi}_0$ is parabolic regular by exhibiting a weak barrier for the part of $\dot{D}$ in some neighborhood of $\dot{\xi}_0$.

Example (a). If every point of $\dot{D}$ in some neighborhood of the finite boundary point $\dot{\xi}_0$ is either above the horizontal hyperplane through $\dot{\xi}_0$ or on one side of some other hyperplane through $\dot{\xi}_0$, then $\dot{\xi}_0$ is a parabolic-regular boundary point. In fact, to construct a local weak barrier for $\dot{D}$ at $\dot{\xi}$, define $\dot{\phi}$ by

$$\dot{\xi} = (\xi, s) \mapsto \dot{\phi}(\xi, s) = as + \phi(\xi),$$

where $\dot{\phi} = 0$ is the equation of a hyperplane through $\dot{\xi}_0$. If the hyperplane is not parallel to the ordinate axis, it can be supposed that $a = 1$. Under this hypothesis the half-space strictly above the hyperplane is the set $\{\dot{\phi} > 0\}$, and the restriction of $\dot{\phi}$ to the part of $\dot{D}$ in an open neighborhood of $\dot{\xi}_0$ is a local weak barrier at $\dot{\xi}_0$ if every point of $\dot{D}$ in that neighborhood is above the hyperplane. If every point of $\dot{D}$ in some neighborhood of $\dot{\xi}_0$ is below the hyperplane and if α is a strictly positive number, the function $\dot{u} = 1 - \exp \alpha\dot{\phi}$ is strictly positive strictly below the hyperplane and for sufficiently large α is superparabolic there; so the restriction of $\dot{u}$ to the part of $\dot{D}$ in some open neighborhood of $\dot{\xi}_0$ is a local barrier at $\dot{\xi}_0$. Finally, if the hyperplane is parallel to the ordinate axis, then $a = 0$. If the part of $\dot{D}$ in some open neighborhood of $\dot{\xi}_0$ is on one side of this hyperplane, then the restriction of either $\dot{\phi}$ or $-\dot{\phi}$ to this part of $\dot{D}$ is a local barrier at $\dot{\xi}_0$.

Observe that if the complement of $\dot{D}$ includes the part in a neighborhood of $\dot{\xi}_0$ of some not horizontal hyperplane through $\dot{\xi}_0$, then if $\dot{D}$ contains points arbitrarily near $\dot{\xi}_0$ on both sides of this hyperplane there is a local weak barrier for the part of $\dot{D}$ on each side of the hyperplane and therefore a local barrier for $\dot{D}$ itself at $\dot{\xi}_0$. Such combinations will not be mentioned further.

Example (a) implies that the lower and lateral boundary points of an interval are parabolic regular. Conversely, our discussion of the parabolic Poisson integral for an interval can be used to derive the conclusions of Example (a) involving hyperplanes parallel or perpendicular to the ordinate axis.

Example (b). If $\dot{\xi}_0 \colon (\xi_0, s_0)$ is a boundary point of $\dot{D}$ with the property that the boundary in some neighborhood of $\dot{\xi}_0$ consists of points with ordinate values $\geq s_0$ and if $\dot{D}$ contains points arbitrarily near $\dot{\xi}_0$ with ordinate values $< s_0$, then $\dot{\xi}_0$ is a parabolic-irregular boundary point of $\dot{D}$. In fact (Section 1) some boundary neighborhood of $\dot{\xi}_0$ has parabolic measure 0 relative to points of $\dot{D}$ near but below $\dot{\xi}_0$; therefore the boundary function values assigned near $\dot{\xi}_0$ are irrelevant to the Dirichlet solution at points of $\dot{D}$ near but below $\dot{\xi}_0$. For example, if $N = 1$ and if $\dot{D}$ is the interior of the simple polygon with vertices $(-1, 0)$, $(1, 0)$, $(1, 1)$, $(0, \frac{1}{2})$, and $(-1, 1)$, the boundary point $(0, \frac{1}{2})$ is parabolic irregular. Observe that for this choice of $\dot{D}$ there is no increasing sequence of open sets with parabolic-regular boundaries and with union $\dot{D}$, but this lack of parabolic-regular approximation is only a minor inconvenience. Theorem VIII.17 on the approximation of harmonic measure by harmonic measures of boundaries of subsets translates directly into the parabolic context.

Example (c). [This example strengthens (a) except when the boundary hyperplane in (a) is horizontal.] Suppose that $\dot{\xi}_0$: (ξ_0, s_0) is a boundary point of $\dot{D}$, and suppose that there is a ball of center $\dot{\xi}$: (ξ, s) with $\xi \neq \xi_0$ such that the ball closure meets $\bar{\dot{D}}$ in $\dot{\xi}_0$ but in no other point. Then $\dot{\xi}_0$ is a parabolic-regular boundary point of $\dot{D}$. To see this, let δ be the ball radius, and define $\dot{u}$ on $\dot{D}$ by

$$\dot{u}(\dot{\eta}) = \delta^{-p} - |\dot{\eta} - \dot{\xi}|^{-p}, \tag{6.1}$$

where p is a strictly positive number to be chosen below. Then $\dot{u} > 0$, $\dot{u}$ has limit 0 at $\dot{\xi}_0$, $\dot{u}$ has a strictly positive lower bound outside each neighborhood of $\dot{\xi}_0$, and if $\dot{\eta} = (\eta, t)$,

$$\begin{aligned} \dot{\Delta}\dot{u}(\dot{\eta}) = p|\dot{\eta} - \dot{\xi}|^{-p-4}\Big\{&\frac{N\sigma^2}{2}|\dot{\eta} - \dot{\xi}|^2 \\ &- \frac{\sigma^2}{2}(p+2)|\eta - \xi|^2 - |\dot{\eta} - \dot{\xi}|^2(t - s)\Big\}. \end{aligned} \tag{6.2}$$

Then $\dot{\Delta}\dot{u} < 0$ sufficiently near $\dot{\xi}_0$ if p is large so there is a local barrier at $\dot{\xi}_0$.

Example (d) (Iterated logarithm criterion for parabolic regularity). [This example strengthens (a) when the boundary hyperplane in (a) is horizontal.] Let $\dot{D}$ be the set below the abscissa hyperplane defined by the inequalities

$$|\xi|^2 < 2\sigma^2|s|\log\big|\log|s|\big|, \qquad -1 < s < 0. \tag{6.3}$$

We show that the origin is a parabolic-regular boundary point of $\dot{D}$ by showing that the function $\dot{v}$ defined by $\dot{D}$ by

$$\begin{aligned} \dot{v}(\xi, s) &= \big|\log|s|\big|^{-1}\big[1 - |s|^{(N+2)/2}\dot{u}(\xi, s)\big], \\ \dot{u}(\xi, s) &= |s|^{-N/2}\exp\frac{-|\xi|^2}{2\sigma^2 s}, \end{aligned} \tag{6.4}$$

is a barrier for $\dot{D}$ at the origin. The inequality (6.3) can be written in the form

$$|s|^{(N+2)/2}\dot{u}(\xi, s) < |s|\,|\log|s||,$$

which makes it clear that $\dot{v} > 0$ on $\dot{D}$ and that $\dot{v}$ has limit 0 at the origin. Furthermore, using the fact that $\dot{u}$ is parabolic, we find that

$$\dot{\Delta}\dot{v}(\xi, s) = -\frac{N+2}{2}|s|^{N/2}|\log|s||^{-1}\dot{u}(\xi, s) - |s\log|s||^{-1}\dot{v}(\xi, s) < 0; \tag{6.5}$$

so $\dot{v}$ is a barrier.

Application. This example shows that the top point of a ball in $\dot{\mathbb{R}}^N$ is parabolic regular. The other boundary points of a ball are parabolic regular according to Example (a).

Example (e). (Iterated logarithm criterion for parabolic regularity, continued). Let $\dot{D}_c$ be the set defined by the inequalities

$$|\xi|^2 < 2\sigma^2|s|c\log|\log|s||, \qquad 0 > s > s', \qquad s' > -e^{-1}. \tag{6.6}$$

The value of s' will be chosen so near 0 that certain inequalities below will be true. It will now be shown, in contrast with Example (d) in which $c = 1$, that the origin is a parabolic-irregular boundary point of $\dot{D}_c$ when $c > 1$. Define the function $\dot{v}$ on $\dot{\mathbb{R}}^N \times]s', 0[$ by

$$\dot{v}(\xi, s) = |\log|s||^{-1-\varepsilon}\exp\frac{k|\xi|^2}{2\sigma^2|s|} - (\log|\log|s||)^{-1}, \tag{6.7}$$

where $0 < k < 1$ and $\varepsilon > 0$. Then

$$\dot{v}(\xi, s) > |\log|s||^{-1-\varepsilon} - (\log|\log|s||)^{-1} > 0$$

if s' is sufficiently near 0. Furthermore

$$\begin{aligned}\dot{\Delta}\dot{v}(\xi, s) = -|s|^{-1}|\log|s||^{-1-\varepsilon}\exp\frac{k|\xi|^2}{2\sigma^2|s|}\bigg\{&-\frac{Nk}{2} - \frac{1+\varepsilon}{|\log|s||}\\ &+\frac{(k-k^2)|\xi|^2}{2\sigma^2|s|} + \frac{|\log|s||^{\varepsilon}}{(\log|\log|s||)^2}\exp\frac{-k|\xi|^2}{2\sigma^2|s|}\bigg\}.\end{aligned} \tag{6.8}$$

Choose s' so near 0 that $(1+\varepsilon)|\log|s'||^{-1} < Nk/2$. The function $\dot{v}$ is then superparabolic if

$$Nk < \frac{(k-k^2)|\xi|^2}{2\sigma^2|s|} + \frac{\left|\log|s|\right|^{\varepsilon}}{(\log\left|\log|s|\right|)^2}\exp\frac{-k|\xi|^2}{2\sigma^2|s|}. \tag{6.9}$$

This inequality will be satisfied if

$$Nk < \frac{(k-k^2)|\xi|^2}{2\sigma^2|s|}. \tag{6.10}$$

On the other hand, if (6.10) is not satisfied at (ξ, s), the second term on the right in (6.9) is at least

$$\frac{\left|\log|s|\right|^{\varepsilon}}{(\log\left|\log|s|\right|)^2}\exp\frac{-Nk}{1-k},$$

which is at least Nk on $\dot{D}_c$ if s' is sufficiently near 0, depending on ε and k. Thus for any choice of ε and k the function $\dot{v}$ is superparabolic and positive on $\dot{D}_c$ when s' is sufficiently small. Now choose $c = (1+\varepsilon)/k$, and define the continuous function f on $\partial\dot{D}_c$ as 1 at the origin and as $\dot{v}$ at the other boundary points. The $\dot{\text{P}}$WB solution $\dot{H}_f$ on $\dot{D}_c$ for f is the same as the $\dot{\text{P}}$WB solution for a boundary function differing from f only at a single point; therefore $\dot{H}_f = \dot{H}_{f_1}$ when $f = f_1$ except at the origin, where f_1 is defined as 0. Since $\dot{v}$ is in the upper $\dot{\text{P}}$WB class on $\dot{D}_c$ for f_1, we conclude that $\dot{H}_f \le \dot{v}$ and therefore

$$\lim_{s\to 0} \dot{H}_f(0, s) \le \lim_{s\to 0} \dot{v}(0, s) = 0.$$

Since f is continuous with value 1 at the origin, the origin is a parabolic-irregular boundary point for $\dot{D}_c$. Moreover c can be made arbitrarily near 1 by choosing ε near 0 and k near 1. It follows that for any choice of $c > 1$ and any choice of s' the origin is a parabolic-irregular boundary point of $\dot{D}_c$.

Parabolic Balls

A *parabolic ball* in $\dot{\mathbb{R}}^N$ of radius δ and center (ξ_0, s_0) is defined as the set

$$\{(\xi, s)\colon \ell(s_0 - s, \xi_0 - \xi) > (2\pi\sigma^2\delta)^{-N/2}\}. \tag{6.11}$$

Observe that the center of the parabolic ball is a boundary point. The set (6.11) lies strictly below the center and is a solid of revolution with vertical axis through the center. The highest and lowest points of the boundary are, respectively, the center and $(\xi_0, s_0 - \delta)$. Example (a) shows that every boundary point of a parabolic ball except possibly the center is parabolic regular, and Example (c) shows that every boundary point of a parabolic ball except possibly the center and the lowest boundary point is parabolic regular. The center is a parabolic-irregular boundary point according to Example (e) because the set $\dot{D}_c$ of that example is, for an arbitrary value of $c > 1$ and an arbitrary $\delta > 0$, included in the parabolic ball of center the origin and radius δ when s' is sufficiently near 0.

7. Parabolic Regularity in Terms of the Fine Topology

Theorem. *A finite Euclidean boundary point $\dot{\zeta}$ of an open subset $\dot{D}$ of $\dot{\mathbb{R}}^N$ is parabolic regular if and only if $\dot{\zeta} \in (\dot{\mathbb{R}}^N - \dot{D})^{pf}$.*

This theorem is the parabolic context version of Theorem XI.12, and the proof is omitted because the only change needed in the proof of that theorem to make it applicable in the present context is to change "polar" to "parabolic semipolar."

Example. If $\dot{D}$ is a parabolic ball (Section 6), the center is a parabolic-irregular boundary point; so $\dot{D}$ is a deleted parabolic-fine neighborhood of the center. Similarly the other examples of parabolic-regular and parabolic-irregular boundary points in Section 6 have simple interpretations in the parabolic-fine topology.

8. Sweeping in the Parabolic Context

The sweeping operation defined in the classical context in Section X.1 becomes a pair of linked operations in the present context. If $\dot{D}$ is a non-empty open subset of $\dot{\mathbb{R}}^N$, $\dot{A} \subset \dot{D}$, and if $\dot{G}_{\dot{D}}\dot{\mu}$ $[\dot{\mu}\dot{G}_{\dot{D}}]$ is a superparabolic [cosuperparabolic] potential, the smoothed reduction $\dot{\llbracket}\dot{G}_{\dot{D}}\dot{\mu}\rrbracket^{\dot{A}}$ $[\overset{*}{\llbracket}\dot{\mu}\dot{G}_{\dot{D}}\rrbracket^{\dot{A}}]$ is also a potential, namely that of a "swept measure" $\overset{*}{\llbracket}\dot{\mu}\rrbracket^{\dot{A}}$ $[\dot{\llbracket}\dot{\mu}\rrbracket^{\dot{A}}]$, ·

$$\dot{\llbracket}\dot{G}_{\dot{D}}\dot{\mu}\rrbracket^{\dot{A}} = \dot{G}_{\dot{D}}\overset{*}{\llbracket}\dot{\mu}\rrbracket^{\dot{A}}, \qquad \overset{*}{\llbracket}\dot{\mu}\dot{G}_{\dot{D}}\rrbracket^{\dot{A}} = \dot{\llbracket}\dot{\mu}\rrbracket^{\dot{A}}\dot{G}_{\dot{D}}. \tag{8.1}$$

The swept measures are supported by $\overline{\dot{A}} \cap \dot{D}$, and it will be shown in Section 13 that if $\dot{A}$ is a Borel set, then $\llbracket\dot{\mu}\rrbracket^{\dot{A}}$ $[\overset{*}{\llbracket}\dot{\mu}\rrbracket^{\dot{A}}]$ is supported by the in general smaller parabolic-fine [coparabolic-fine] closure of $\dot{A}$ in $\dot{D}$. Since $\llbracket\dot{G}_{\dot{D}}\dot{\mu}\rrbracket^{\dot{A}}$ is unaffected when $\dot{A}$ is changed by a parabolic-polar set, $\overset{*}{\llbracket}\dot{\mu}\rrbracket^{\dot{A}}$ is also unaffected by such a change in $\dot{A}$. Dually $\llbracket\dot{\mu}\rrbracket^{\dot{A}}$ is unaffected when $\dot{A}$ is changed by a coparabolic-polar set. It will be shown in Section 11 that a set is parabolic polar if and only if it is coparabolic polar.

The notation of Section X.1 is adapted to the present context by writing $\dot{\delta}_{\dot{D}}(\dot{\xi}, \cdot)$ for the probability measure on $\dot{D}$ supported by $\{\dot{\xi}\}$ and by defining $\dot{\delta}_{\dot{D}}^{\dot{A}}$ and $\overset{*}{\delta}{}_{\dot{D}}^{\dot{A}}$ by

$$\dot\delta_{\dot D}^{\dot A}(\dot\xi,\cdot) = \overset{\cdot}{[\![}\dot\delta_{\dot D}(\dot\xi,\cdot)]\!]^{\dot A}, \qquad \overset{*}{\delta}{}_{\dot D}^{\dot A}(\dot\xi,\cdot) = \overset{*}{[\![}\dot\delta_{\dot D}(\dot\xi,\cdot)]\!]^{\dot A}. \tag{8.2}$$

These swept measures are determined uniquely by

$$\overset{\cdot}{[\![}\dot G_{\dot D}(\cdot,\dot\eta)]\!]^{\dot A}(\dot\xi) = \overset{*}{\delta}{}_{\dot D}^{\dot A}(\dot\eta, \dot G_{\dot D}(\dot\xi,\cdot)), \qquad \overset{*}{[\![}\dot G_{\dot D}(\dot\xi,\cdot)]\!]^{\dot A}(\dot\eta) = \dot\delta_{\dot D}^{\dot A}(\dot\xi, \dot G_{\dot D}(\cdot,\dot\eta)) \tag{8.3}$$

and the reasoning in the classical case (Section X.1) shows that $\dot\delta_{\dot D}^{\dot A} \le 1$ and $\overset{*}{\delta}{}_{\dot D}^{\dot A} \le 1$.

The discussion proceeds as in the classical context, very slightly complicated by the existence of two kernels instead of one, and the details are left to the reader. It is first shown that if $\dot B$ is an open subset of $\dot D$ and if $\dot\xi$ is in $\dot B$, then

$$\dot\delta_{\dot D}^{\dot D-\dot B}(\dot\xi,\cdot) = \dot\mu_{\dot B}(\dot\xi,\cdot), \qquad \overset{*}{\delta}{}_{\dot D}^{\dot D-\dot B}(\dot\xi,\cdot) = \overset{*}{\mu}_{\dot B}(\dot\xi,\cdot) \tag{8.4}$$

on the Borel subsets of $\dot D \cap \partial \dot B$. Next the classical context symmetry argument is adapted, yielding here for an arbitrary subset $\dot A$ of $\dot D$,

$$\overset{*}{[\![}\dot G_{\dot D}(\dot\xi,\cdot)]\!]^{\dot A}(\dot\eta) = \overset{\cdot}{[\![}\dot G_{\dot D}(\cdot,\dot\eta)]\!]^{\dot A}(\dot\xi); \tag{8.5}$$

that is,

$$\dot\delta_{\dot D}^{\dot A}(\dot\xi, \dot G_{\dot D}(\cdot,\dot\eta)) = \overset{*}{\delta}{}_{\dot D}^{\dot A}(\dot\eta, \dot G_{\dot D}(\dot\xi,\cdot)). \tag{8.6}$$

The common value in (8.5) will be denoted by $\dot G_{\dot D}^{\dot A}(\dot\xi,\dot\eta)$. It is then shown that $\dot\delta_{\dot D}^{\dot A}$ and $\overset{*}{\delta}{}_{\dot D}^{\dot A}$ are kernels and that if $\dot v$ [$\overset{*}{v}$] is a positive superparabolic [cosuperparabolic] function,

$$\overset{\cdot}{[\![}\dot v]\!]^{\dot A} = \dot\delta_{\dot D}^{\dot A}(\cdot,\dot v), \qquad \overset{*}{[\![}\overset{*}{v}]\!]^{\dot A} = \overset{*}{\delta}{}_{\dot D}^{\dot A}(\cdot,\overset{*}{v}). \tag{8.7}$$

In particular, if $\dot\mu$ is a measure on $\dot D$,

$$\begin{aligned} \overset{\cdot}{[\![}\dot G_{\dot D}\dot\mu]\!]^{\dot A} &= \dot\delta_{\dot D}^{\dot A}(\cdot, \dot G_{\dot D}\dot\mu) = \dot G_{\dot D}\overset{*}{[\![}\dot\mu]\!]^{\dot A} = \dot G_{\dot D}^{\dot A}\dot\mu, \\ \overset{*}{[\![}\dot\mu\dot G_{\dot D}]\!]^{\dot A} &= \overset{*}{\delta}{}_{\dot D}^{\dot A}(\cdot, \dot\mu\dot G_{\dot D}) = \overset{\cdot}{[\![}\dot\mu]\!]^{\dot A}\dot G_{\dot D} = \dot\mu\dot G_{\dot D}^{\dot A}. \end{aligned} \tag{8.8}$$

Moreover

$$\overset{\cdot}{[\![}\mu]\!]^{\dot A} = \dot\mu\dot\delta_{\dot D}^{\dot A}, \qquad \overset{*}{[\![}\dot\mu]\!]^{\dot A} = \dot\mu\overset{*}{\delta}{}_{\dot D}^{\dot A}. \tag{8.9}$$

More precisely the equations in (8.9) are correct according to our definitions if the potentials involved are superparabolic or cosuperparabolic as the case may be; if then $\overset{\cdot}{[\![}\dot\mu]\!]^{\dot A}$ and $\overset{*}{[\![}\dot\mu]\!]^{\dot A}$ are defined by (8.9) for measures $\dot\mu$ on $\dot D$ for which the swept measures are not already defined, every equation in (8.8) is true. Finally a trivial integration yields

$$\int_{\dot{D}} \|\dot{\mu}\|^{\dot{A}} \, d\dot{v} = \int_{\dot{D}} \dot{u} \, d\|\dot{v}\|^{\dot{A}} \leq +\infty \tag{8.10}$$

first for a measure $\dot{v}$ and a parabolic potential $\dot{G}_{\dot{D}}\dot{\mu}$ and then by a limit procedure for $\dot{v}$ and an arbitrary positive superparabolic function $\dot{u}$ with associated Riesz measure $\dot{\mu}$. The formulation of the dual of (8.10) is left to the reader.

Subadditivity of $\dot{A} \mapsto \dot{\delta}_{\dot{D}}^{\dot{A}}$

This subadditivity, and the dual subadditivity, are shown by a slight refinement of the proof in the classical context (SectionX.6).

9. The Extension $\dot{G}_{\dot{D}}^{=}$ of $\dot{G}_{\dot{D}}$ and the Parabolic Average $\dot{\mu}_{\dot{D}}(\dot{\xi}, \dot{G}_{\dot{B}}^{=}(\cdot, \dot{\eta}))$ when $\dot{D} \subset \dot{B}$

Boundaries in this section are Euclidean. Let $\dot{D}$ be a nonempty open subset of $\dot{\mathbb{R}}^N$, and let $\dot{\eta}$ be a point of $\dot{D}$. We extend $\dot{G}_{\dot{D}}(\cdot, \dot{\eta})$ to $\dot{\mathbb{R}}^N$ to obtain a function $\dot{G}_{\dot{D}}^{=}$ with the following properties:

(a) $\dot{G}_{\dot{D}}^{=}(\cdot, \dot{\eta})$ is a positive subparabolic function on $\dot{\mathbb{R}}^N - \{\dot{\eta}\}$.
(b) $\dot{G}_{\dot{D}}^{=}(\cdot, \dot{\eta}) = 0$ on $\dot{\mathbb{R}}^N - \dot{\bar{D}}$, and $\dot{G}_{\dot{D}}^{=}(\cdot, \dot{\eta}) = 0$ at a finite point of $\partial\dot{D}$ if the point is parabolic regular.

Observe that such an extension must be unique because two such extensions would be parabolic-fine continuous and equal up to a parabolic-fine nowhere dense set and therefore would be identical. To define $\dot{G}_{\dot{D}}^{=}(\cdot, \dot{\eta})$, recall (from Section XVII.4) that the function $\mathrm{GM}_{\dot{D}}\dot{G}(\cdot, \dot{\eta})$ is the restriction to $\dot{D}$ of the limit $\dot{u}(\cdot, \dot{\eta})$ of a decreasing sequence $n \mapsto \dot{\tau}_{\dot{B}_n} \cdots \dot{\tau}_{\dot{B}_0}\dot{G}(\cdot, \dot{\eta})$ of superparabolic functions on $\dot{\mathbb{R}}^N$; each set $\dot{B}_k$ is an interval with closure in $\dot{D}$. Then $\dot{u}(\cdot, \eta) = \dot{G}(\cdot, \dot{\eta})$ on $\dot{\mathbb{R}}^N - \dot{D}$, $\underset{+}{\dot{u}}(\cdot, \dot{\eta}) = \dot{u}(\cdot, \dot{\eta})$ up to a parabolic-semipolar set, $\underset{+}{\dot{u}}(\cdot, \dot{\eta})$ is superparabolic on $\dot{\mathbb{R}}^N$, and we define $\dot{G}_{\dot{D}}^{=}(\cdot, \dot{\eta})$ on $\dot{\mathbb{R}}^N$ as $\dot{G}(\cdot, \dot{\eta}) - \underset{+}{\dot{u}}(\cdot, \dot{\eta})$. According to the parabolic context Fundamental Convergence Theorem, the function $\dot{G}_{\dot{D}}(\cdot, \dot{\eta})$ has the stated properties (a) and (b) except possibly the vanishing at the set of parabolic-regular boundary points of D. If $\dot{\zeta}$ is a parabolic-regular boundary point of $\dot{D}$, the function $\dot{G}_{\dot{D}}(\cdot, \dot{\eta})$ has limit 0 at $\dot{\zeta}$ on approach along $\dot{D}$ in view of the evaluation of a Green function in terms of a Dirichlet solution. The function $\dot{G}_{\dot{D}}^{=}(\cdot, \dot{\eta})$ has limit 0 at $\dot{\zeta}$ on approach along $\dot{\mathbb{R}}^N - \dot{D}$ if a parabolic-semipolar set is excluded. Thus the parabolic-fine continuous function $\dot{G}_{\dot{D}}^{=}(\cdot, \dot{\eta})$ has limit 0 at $\dot{\zeta}$ on approach along a parabolic-fine dense set, so $\dot{G}_{\dot{D}}^{=}(\dot{\zeta}, \dot{\eta}) = 0$. Dually, for $\dot{\xi}$ in $\dot{D}$ there is an extension $\dot{G}_{\dot{D}}^{=}(\dot{\xi}, \cdot)$ of $\dot{G}_{\dot{D}}(\dot{\xi}, \cdot)$, positive and cosubparabolic on $\dot{\mathbb{R}}^N - \{\dot{\xi}\}$, vanishing on $\dot{\mathbb{R}}^N - \dot{\bar{D}}$ and vanishing at a finite boundary point of $\dot{D}$ if the point is coparabolic regular. Thus we have extended $\dot{G}_{\dot{D}}$ to $(\dot{D} \times \dot{\mathbb{R}}^N) \cup (\dot{\mathbb{R}}^N \times \dot{D})$.

We can now state the parabolic context version of Theorem VIII.18. Observe that there is no exceptional value of N in this context because $\dot{G}_{\dot{D}}$ is positive for all values of N.

Theorem. *Let $\dot{D}$ and $\dot{B}$ be nonempty open subsets of $\dot{\mathbb{R}}^N$ with $\dot{D} \subset \dot{B}$.*

(a) *For $\dot{\eta} \in \dot{B}$ $[\dot{\xi} \in \dot{B}]$ the function $\dot{G}_{\dot{B}}^{=}(\cdot, \dot{\eta})_{|\partial\dot{D}}$ $[\dot{G}_{\dot{B}}^{=}(\dot{\xi}, \cdot)_{|\partial\dot{D}}]$, defined as 0 on $\partial\dot{B} \cap \partial\dot{D}$, is a [co] parabolic-resolutive boundary function.*

(b) *The function $\dot{G}_{\dot{D}}^{=}$ has the representations*

$$\dot{G}_{\dot{D}}^{=}(\dot{\xi}, \dot{\eta}) = \dot{G}_{\dot{B}}(\dot{\xi}, \dot{\eta}) - \dot{\mu}_{\dot{D}}(\dot{\xi}, \dot{G}_{\dot{B}}^{=}(\cdot, \dot{\eta})) \qquad [(\dot{\xi}, \dot{\eta}) \in \dot{D} \times \dot{B}], \tag{9.1}$$

$$\dot{G}_{\dot{D}}^{=}(\dot{\xi}, \dot{\eta}) = \dot{G}_{\dot{B}}(\dot{\xi}, \dot{\eta}) - \mathring{\mu}_{\dot{D}}(\dot{\eta}, \dot{G}_{\dot{B}}^{=}(\dot{\xi}, \cdot)) \qquad [(\dot{\xi}, \dot{\eta}) \in \dot{B} \times \dot{D}]. \tag{9.1*}$$

(c) *If $\dot{v} = \dot{G}_{\dot{B}}\dot{\nu}$ is a superparabolic potential on $\dot{B}$ and if $\dot{\nu}^i$ is the projection of $\dot{\nu}$ on the set of parabolic-irregular boundary points of $\dot{D}$ in $\dot{B}$, then*

$$\dot{\mathrm{G}}\mathrm{M}_{\dot{D}}\dot{v} = \dot{\mu}_{\dot{D}}(\cdot, \dot{v}1_{\dot{B}}) + \dot{G}_{\dot{D}}^{=}\dot{\nu}^i. \tag{9.2}$$

If $\dot{D}$ is relatively compact in $\dot{B}$, then (9.2) is true whenever $\dot{v}$ is a superparabolic function on $\dot{B}$ and $\dot{\nu}$ is its associated Riesz measure.

We leave to the reader the dual statement of (c), for cosuperparabolic functions.

Observation (1). If $\dot{\xi}$ and $\dot{\eta}$ are points of $\dot{D}$, Theorem 9(b) yields the important symmetry relation

$$\dot{\mu}_{\dot{D}}(\dot{\xi}, \dot{G}_{\dot{B}}(\cdot, \dot{\eta})1_{\dot{B}}) = \mathring{\mu}_{\dot{D}}(\eta, \dot{G}_{\dot{B}}(\dot{\xi}, \cdot)1_{\dot{B}}) \tag{9.3}$$

in which we can take $\dot{B} = \dot{\mathbb{R}}^N$ if desired.

Observation (2). If the write $\dot{\mu}_{\dot{D}}(\dot{\xi}, \cdot)_{|\dot{B}}$ and $\mathring{\mu}_{\dot{D}}(\dot{\eta}, \cdot)_{|\dot{B}}$ for the respective restrictions of the indicated measures to the class of Borel subsets of $\dot{B}$, then

$$\dot{\mu}_{\dot{D}}(\dot{\xi}, \dot{G}_{\dot{B}}^{=}(\cdot, \dot{\eta})) = [\dot{\mu}_{\dot{D}}(\dot{\xi}, \cdot)_{|\dot{B}}\dot{G}_{\dot{B}}](\dot{\eta}) < +\infty \qquad [(\dot{\xi}, \dot{\eta}) \in \dot{D} \times \dot{B}], \tag{9.4}$$

$$\mathring{\mu}_{\dot{D}}(\dot{\eta}, \dot{G}_{\dot{B}}^{=}(\dot{\xi}, \cdot)) = [\dot{G}_{\dot{B}}\mathring{\mu}_{\dot{D}}(\dot{\eta}, \cdot)_{|\dot{B}}](\dot{\xi}) < +\infty \qquad [(\dot{\xi}, \dot{\eta}) \in \dot{B} \times \dot{D}]; \tag{9.4*}$$

so the respective cosuperparabolic and superparabolic potentials on the right are finite valued. According to Theorem 9(b), the measure on $\dot{\mathbb{R}} - \{\dot{\xi}\}$ $[\dot{\mathbb{R}} - \{\dot{\eta}\}]$ associated with the cosuperparabolic [superparabolic] function $-\dot{G}_{\dot{D}}^{=}(\dot{\xi}, \cdot)$ $[-\dot{G}_{\dot{D}}^{=}(\cdot, \dot{\eta})]$ on this set is the restriction of the Borel measure $\dot{\mu}_{\dot{D}}(\xi, \cdot)$ $[\mathring{\mu}_{\dot{D}}(\eta, \cdot)]$ to the class of Borel subsets of $\dot{\mathbb{R}}^N - \{\dot{\xi}\}$ $[\dot{\mathbb{R}}^N - \{\dot{\eta}\}]$.

Proof of (a). Since Euclidean boundaries are parabolic resolutive, the parabolic resolutivity of the boundary function $f_{\dot{\eta}}$, defined as $\dot{G}_{\dot{B}}^{=}(\cdot, \dot{\eta})$ on $\dot{B} \cap \partial\dot{D}$

and as 0 elsewhere on $\partial\dot{D}$, reduces to the $\dot{\mu}_{\dot{D}}$ integrability of this boundary function. If $\dot{\eta} \in \dot{B} - \partial\dot{D}$, the function $f_{\dot{\eta}}$ is bounded and so is $\dot{\mu}_{\dot{D}}$ integrable. If $\dot{\eta} \in \dot{B} \cap \partial\dot{D}$, then $\dot{G}_{\dot{B}}(\cdot, \dot{\eta})_{|\dot{D}}$ is in the upper $\dot{\text{P}}$WB class on $\dot{D}$ for $f_{\dot{\eta}}$; so again ((see Section VIII.10) for the justification in the classical context) $f_{\dot{\eta}}$ is $\dot{\mu}_{\dot{D}}$ integrable. The coparabolic resolutivity of the other boundary function in (a) is derived by a dual argument. □

Proof of (b). Let $\dot{\xi}$ be in $\dot{D}$. If $\dot{\eta}$ is also in $\dot{D}$, equation (9.1) reduces to the evaluation of $\dot{G}_{\dot{D}}$ in terms of $\dot{G}_{\dot{B}}$ already derived in Section 1. If $\dot{\eta} \in \dot{B} - \bar{\dot{D}}$, the left side of (9.1) vanishes, and $f_{\dot{\eta}}$ is a bounded function on $\partial\dot{D}$ with $\dot{\text{P}}$WB solution $\dot{G}_{\dot{B}}(\cdot, \dot{\eta})_{|\dot{D}}$. Hence (9.1) is true for $\dot{\eta}$ in $\dot{B} - \bar{\dot{D}}$ and therefore for $\dot{\eta}$ in $\dot{B} - \partial\dot{D}$. Next suppose that $\dot{\eta}$ is a coparabolic-regular boundary point of $\dot{D}$ in $\dot{B}$, so that (9.1) reduces to

$$\dot{\mu}_{\dot{D}}(\dot{\xi}, \dot{G}_{\dot{B}}(\cdot, \dot{\eta})1_{\dot{B}}) = \dot{G}_{\dot{B}}(\dot{\xi}, \dot{\eta}) \tag{9.5}$$

or equivalently (see Section VIII.10 for the expression of a Dirichlet solution in terms of a smoothed reduction in the classical context; no change is needed in the present context) in terms of reductions relative to $\dot{B}$,

$$\dot{G}_{\dot{B}}^{\dot{B}-\dot{D}}(\dot{\xi}, \dot{\eta}) = \dot{G}_{\dot{B}}(\dot{\xi}, \dot{\eta}). \tag{9.6}$$

Furthermore $\dot{G}_{\dot{B}}^{\dot{B}-\dot{D}}(\dot{\xi}, \cdot) = \dot{G}_{\dot{B}}(\dot{\xi}, \cdot)$ on $\dot{B} - \dot{D}$ up to a coparabolic-semipolar set. Since $\dot{G}_{\dot{B}}^{\dot{B}-\dot{D}}(\dot{\xi}, \cdot)$ is coparabolic-fine continuous and since a coparabolic-semipolar set is coparabolic-fine nowhere dense, we conclude that (9.6) is true if $\dot{\eta}$ is a coparabolic-fine limit point of $\dot{B} - \dot{D}$, as $\dot{\eta}$ is because $\dot{\eta}$ is coparabolic regular. We have now proved that for fixed $\dot{\xi}$ equation (9.1) is true except possibly when $\dot{\eta}$ is in the coparabolic-fine nowhere dense set of coparabolic-irregular boundary points of $\dot{D}$ in $\dot{B}$. Since both sides of this equation define coparabolic-fine continuous functions of $\dot{\eta}$, equation (9.1) is true for all $\dot{\eta}$. Equation (9.1*) can be proved by the dual argument or can be reduced to (9.1) by a reflection of $\dot{\mathbb{R}}^N$ in the abscissa hyperplane. □

Proof of (c). See the proof of Theorem VIII.18(c). □

The application of Theorem VIII.18 to the vanishing of *h*-potentials at the boundaries of their domains, as detailed in Section VIII.18, goes over into the present context with no change.

10. Conditions that $\dot{\xi} \in \dot{A}^{pf}$

The following is the parabolic context counterpart of Theorem XI.3. Let $\dot{D}$ be a nonempty open subset of $\dot{\mathbb{R}}^N$, let $\dot{\xi}$ be a point of $\dot{D}$, and let $\dot{v}$ be a positive superparabolic function on $\dot{D}$. All reductions will be relative to $\dot{D}$. We shall write $\lim_{\dot{B}\downarrow\dot{\xi}}$ to mean the limit as $\dot{B}$, a neighborhood of $\dot{\xi}$, shrinks to $\dot{\xi}$.

Theorem. *If* $\dot{\xi} \in \dot{A}^{pf}$, *then*

(a) $\mathring{\|}\dot{v}\|^{\dot{A} \cup \dot{B}}(\dot{\xi}) = \dot{v}(\dot{\xi})$ *for every neighborhood* $\dot{B}$ *of* $\dot{\xi}$,

(b) $\dot{\delta}_{\dot{D}}^{\dot{A}}(\dot{\xi}, \{\dot{\xi}\}) = 1$,

(c) $\dot{G}_{\dot{D}}^{\dot{A}}(\dot{\xi}, \cdot) = \dot{G}_{\dot{D}}(\dot{\xi}, \cdot)$.

If $\dot{\xi} \notin \dot{A}^{pf}$, *then*

(a′) $\lim_{\dot{B} \downarrow \dot{\xi}} \mathring{\|}\dot{v}\|^{\dot{A} \cap \dot{B}} = 0$ *on* $\dot{D} - \{\dot{\xi}\}$, *and the limit is* 0 *at* $\dot{\xi}$ *if* $\dot{v}(\dot{\xi}) < +\infty$,

(b′) $\dot{\delta}_{\dot{D}}^{\dot{A}}(\dot{\xi}, \{\dot{\xi}\}) = 0$,

(c′) $\dot{G}_{\dot{D}}^{\dot{A}}(\dot{\xi}, \cdot) \neq \dot{G}_{\dot{D}}(\dot{\xi}, \cdot)$ *on the set* $\{\dot{G}_{\dot{D}}(\dot{\xi}, \cdot) > 0\}$.

Theorem 10(a) is already known (Section XVII.16(b)) and implies the truth of Theorem 10(b) and (c). In proving assertions (a′) and (c′) we can [Section XVII.9(b)] enlarge $\dot{A}$ to be open but still parabolic-thin at $\dot{\xi}$; so it is sufficient to prove these assertions for $\dot{A}$ open. Under this condition on $\dot{A}$ the proof of Theorem XI.3(a′)–(c′), the classical counterpart of Theorem 10(a′)–(c′), gives the latter when translated into the parabolic context. Thus there remains only the proof of Theorem 10(b′) without the hypothesis that $\dot{A}$ is open. If we take $\dot{v} \equiv 1$ and apply Theorem 10(a′), we find

$$0 = \lim_{\dot{B} \downarrow \dot{\xi}} \mathring{\|}1\|^{\dot{A} \cap \dot{B}}(\dot{\xi}) = \lim_{\dot{B} \downarrow \dot{\xi}} \dot{\delta}_{\dot{D}}^{\dot{A} \cap \dot{B}}(\dot{\xi}, \dot{D}) \geq \lim_{\dot{B} \downarrow \dot{\xi}} \dot{\delta}_{\dot{D}}^{\dot{A} \cap \dot{B}}(\dot{\xi}, \{\dot{\xi}\}). \tag{10.1}$$

Now in the classical context we have proved [Section X.6(a)] that the set function $\delta_{\dot{D}}$ is subadditive on the class of subsets of D, and this led to a proof that $\delta_D^{A \cap B}(\xi, \{\xi\})$ increases when the neighborhood B of ξ decreases. This reasoning will now be refined to the parabolic context. In the first place recall that [Section XVII.16(j)] if $\dot{A}$ and $\dot{B}$ are open subsets of $\dot{D}$ and if $\dot{v}_i$ is a finite-valued continuous superparabolic function on $\dot{D}$, then

$$\mathring{\|}\dot{v}_i\|^{\dot{A} \cup \dot{B}} + \mathring{\|}\,\mathring{\|}\dot{v}_i\|^{\dot{A}} \wedge \mathring{\|}\dot{v}_i\|^{\dot{B}}\,\|^{\dot{A} \cup \dot{B}} = \mathring{\|}\dot{v}_i\|^{\dot{A}} + \mathring{\|}\dot{v}_i\|^{\dot{B}}; \tag{10.2}$$

so if $\dot{v}_1 \leq \dot{v}_2$,

$$\mathring{\|}\dot{v}_2\|^{\dot{A} \cup \dot{B}} - \mathring{\|}\dot{v}_1\|^{\dot{A} \cup \dot{B}} \leq \mathring{\|}\dot{v}_2\|^{\dot{A}} - \mathring{\|}\dot{v}_1\|^{\dot{A}} + \mathring{\|}\dot{v}_2\|^{\dot{B}} - \mathring{\|}\dot{v}_1\|^{\dot{B}}. \tag{10.3}$$

In view of approximation of smoothed reductions on sets by reductions on open supersets (10.3) must be valid for arbitrary subsets $\dot{A}$ and $\dot{B}$ of $\dot{D}$, equivalently,

$$\dot{\delta}_{\dot{D}}^{\dot{A} \cup \dot{B}}(\cdot, \dot{v}_2 - \dot{v}_1) \leq \dot{\delta}_{\dot{D}}^{\dot{A}}(\cdot, \dot{v}_2 - \dot{v}_1) + \dot{\delta}_{\dot{D}}^{\dot{B}}(\cdot, \dot{v}_2 - \dot{v}_1).$$

Finally the counterpart of the classical context reasoning in Section X.6(a) shows that this inequality implies the subadditivity inequality $\dot{\delta}_{\dot{D}}^{\dot{A} \cup \dot{B}} \leq \dot{\delta}_{\dot{D}}^{\dot{A}} + \dot{\delta}_{\dot{D}}^{\dot{B}}$. If the pair $\dot{A}$, $\dot{B}$ is replaced here by the pair $\dot{A} \cap \dot{B}$,

$\dot{A} - \dot{B}$ and if $\dot{B}$ is a neighborhood of $\dot{\xi}$, we find that

$$\delta_{\dot{D}}^{\dot{A}}(\dot{\xi}, \{\dot{\xi}\}) \le \delta_{\dot{D}}^{\dot{A}\cap\dot{B}}(\dot{\xi}, \{\dot{\xi}\}) + \delta_{\dot{D}}^{\dot{A}-\dot{B}}(\dot{\xi}, \{\dot{\xi}\}) = \delta_{\dot{D}}^{\dot{A}\cap\dot{B}}(\dot{\xi}, \{\dot{\xi}\}),$$

where we have used the fact that the measure $\delta_{\dot{D}}^{\dot{A}-\dot{B}}(\dot{\xi}, \cdot)$ is supported by the closure of $\dot{A} - \dot{B}$. Thus $\delta_{\dot{D}}^{\dot{A}\cap\dot{B}}(\dot{\xi}, \{\dot{\xi}\})$ in (10.1) increases as $\dot{B}$ decreases; so $\delta_{\dot{D}}^{\dot{A}}(\dot{\xi}, \{\dot{\xi}\}) = 0$, and the proof of Theorem 10(b′) is now complete.

11. Parabolic- and Coparabolic-Polar Sets

Theorem. *The following six conditions on a subset $\dot{A}$ of a nonempty open subset $\dot{D}$ of $\dot{\mathbb{R}}^N$ are equivalent:*

(a) *$\dot{A}$ is parabolic polar.* (a*) *$\dot{A}$ is coparabolic polar.*
(b) *$\delta_{\dot{D}}^{\dot{A}}$ vanishes identically.* (b*) *$\overset{*}{\delta}{}_{\dot{D}}^{\dot{A}}$ vanishes identically.*
(c) [(c*)] *(If $\dot{A}$ is Borel) the only measure $\dot{\nu}$ on $\dot{D}$ supported by $\dot{A}$ and with $\dot{G}_{\dot{D}}\dot{\nu}$ [$\dot{\nu}\dot{G}_{\dot{D}}$] bounded is the null measure.*

Observation. In connection with (c) and (c*) recall (from Section XVII.8) that a parabolic-polar set has a parabolic-polar Borel superset; the dual assertion for coparabolic-polar sets follows trivially. In view of this observation it is no restriction on generality in the following proof to assume that $\dot{A}$ is a Borel set. Furthermore, if $\dot{\eta}$ is a point of $\dot{\mathbb{R}}^N$, the superparabolic potential $\dot{G}(\cdot, \dot{\eta})$ is the finite-valued potential on $\dot{\mathbb{R}}^N$ of a nonnull measure supported by the polar singleton $\{\dot{\eta}\}$. Hence "bounded" in (c) and (c*) cannot be replaced by "finite-valued."

Proof. (a)$\Leftrightarrow$(a*) If $\dot{A}$ is parabolic polar, then (from Section XVII.8) $\dot{G}_{\dot{D}}^{\dot{A}}(\cdot, \dot{\eta}) \equiv 0$ for every point $\dot{\eta}$ in $\dot{D}$; so $\dot{G}_{\dot{D}}^{\dot{A}}(\dot{\xi}, \cdot) \equiv 0$ for every point $\dot{\xi}$ in $\dot{D}$, and therefore $\dot{A}$ is coparabolic polar. The dual argument yields the reverse implication.

(a), (a*)$\Leftrightarrow$(b), (b*) The evaluation (8.3) of $\delta_{\dot{D}}^{\dot{A}}\dot{G}_{\dot{D}}$ and $\overset{*}{\delta}{}_{\dot{D}}^{\dot{A}}\dot{G}_{\dot{D}}$ shows that (b) and (b*) are equivalent to each other and to the pair (a), (a*).

(a), (a*)$\Leftrightarrow$(c), (c*) If $\dot{A}$ is a parabolic-polar and so also a coparabolic-polar Borel set, if $\dot{\nu}$ is a measure on $\dot{D}$ supported by $\dot{A}$, if $\dot{G}_{\dot{D}}\dot{\nu}$ is a bounded potential, and if $\dot{\lambda}\dot{G}_{\dot{D}}$ is the cosuperparabolic potential of a finite measure and is identically $+\infty$ on $\dot{A}$, then

$$+\infty > \int_{\dot{D}} \dot{G}_{\dot{D}}\dot{\nu}\, d\dot{\lambda} = \int_{\dot{D}} \dot{\lambda}\dot{G}_{\dot{D}}\, d\dot{\nu} = +\infty$$

unless $\dot{\nu} = 0$. Hence (a)$\Rightarrow$(c). On the other hand, if $\dot{A}$ is not parabolic polar, the superparabolic function $\dot{[\![} 1]\!]^{\dot{A}}$ does not vanish identically; so [from Section XVII.16(p)] the set $\dot{A}$ has a compact subset $\dot{B}$ for which the function

$\dot{\|}1\dot{\|}^{\dot{B}}$ does not vanish identically. This function is a bounded superparabolic potential with associated Riesz measure supported by $\dot{B}$. Since (c) implies that this potential vanishes identically, (c) ⇒ (a). Thus (a) ⇔ (c); so dually (a*) ⇔ (c*), and since (a) ⇔ (a*), the proof of Theorem 11 is complete. □

Application to the Projection of a Parabolic-Polar Set

According to Section XVII.8, Example (a), if A is a classical-polar subset of $\mathbb{R}^N$, then $A \times \mathbb{R}$ is a parabolic-polar subset of $\dot{\mathbb{R}}^N$. Conversely, we now prove that if $\dot{A} = A \times \mathbb{R}$ is a parabolic-polar subset of $\dot{\mathbb{R}}^N$, then A is a classical-polar subset of $\mathbb{R}^N$. It is sufficient to prove this assertion for A bounded and $\dot{A}$ a G^δ set, in which case A is also a G_δ set. Hence (from the application in Section VI.2) if A is not polar and if D is a Greenian superset of A, there is a measure μ on D supported by A and making $G_D\mu$ strictly positive bounded and continuous. However (from Section XVII.18), $G_D\mu$ considered as a function on $\dot{D} = D \times \mathbb{R}$ is a bounded superparabolic potential, $G_D\mu = \dot{G}_{\dot{D}}\dot{\mu}$ with $\dot{\mu} = (\mu \times l_1)/a_N$ supported by $\dot{A}$. Hence $\dot{A}$ is not parabolic polar, contrary to hypothesis.

12. Parabolic- and Coparabolic-Semipolar Sets

Theorem. *The following four conditions on a subset $\dot{A}$ of a nonempty open subset $\dot{D}$ of $\dot{\mathbb{R}}^N$ are equivalent:*

(a) *$\dot{A}$ is parabolic semipolar.* (a*) *$\dot{A}$ is coparabolic semipolar.*
(b) [(b*)] *(If $\dot{A}$ is Borel) the only measure $\dot{\nu}$ on $\dot{D}$ supported by $\dot{A}$ and with $\dot{G}_{\dot{D}}\dot{\nu}$ [$\dot{\nu}\dot{G}_{\dot{D}}$] finite valued and continuous is the null measure.*

Observation (a). In connection with (b) and (b*) recall (from Section XVII.15) that a parabolic-semipolar set has a parabolic-semipolar Borel superset; the dual assertion for coparabolic-semipolar sets follows trivially. In view of this observation it is no restriction on generality in the following proof to assume that $\dot{A}$ is a Borel set.

Observation (b). As the following proof shows, "finite-valued" in (b) and (b*) can be replaced by "bounded."

Observation (c). Condition (b) of the theorem implies that $\dot{A}$ is l_{N+1} null because if $\dot{A}$ is a bounded Borel l_{N+1} nonnull subset of $\dot{\mathbb{R}}^N$ and if $\dot{\nu}$ is the projection of l_{N+1} on $\dot{A}$, then (by Theorem XVII.6) the potential $\dot{G}_{\dot{D}}\dot{\nu}$ is bounded and continuous.

Proof. (a) ⇒ (b*) Let $\dot{\nu}$ be a measure on $\dot{D}$ supported by a compact parabolic-semipolar subset $\dot{A}$ of $\dot{D}$, and suppose that $\dot{\nu}\dot{G}_{\dot{D}}$ is finite valued and continuous. Recall that a set is exceptional for the parabolic context Fundamental Convergence Theorem in the sense of the converse statement of Theorem XVII.13 if and only if the set is parabolic semipolar. Now the proof of the classical Fundamental Convergence Theorem (Theorem VI.1) when translated into the present context shows that a measure $\dot{\nu}$ supported by a compact subset $\dot{A}$ of $\dot{D}$ is necessarily the null measure if (1) $\dot{A}$ is exceptional for the parabolic context Fundamental Convergence Theorem and if (2) $\dot{\nu}\dot{G}_{\dot{D}}$ is finite valued and continuous. Hence $\dot{\nu}$ is null in the present context. More generally suppose that $\dot{\nu}$ is a measure supported by a Borel parabolic-semipolar subset $\dot{A}$ of $\dot{D}$, with $\dot{\nu}\dot{G}_{\dot{D}}$ finite valued and continuous, and let $\dot{A}_0$ be a compact subset of $\dot{A}$. If $\dot{\nu}_0$ is the projection of $\dot{\nu}$ on $\dot{A}_0$, then $\dot{\nu}_0\dot{G}_{\dot{D}}$ and $(\dot{\nu} - \dot{\nu}_0)\dot{G}_{\dot{D}}$ are finite valued and lower semicontinuous with a continuous sum; so $\dot{\nu}_0\dot{G}_{\dot{D}}$ is finite valued and continuous, and we have just proved that therefore $\dot{\nu}_0$ is null. Since this is true for all $\dot{A}_0$, we conclude that $\dot{\nu}$ is null, as was to be proved.

(b*) ⇒ (a*) It is sufficient to prove that a Borel subset $\dot{A}$ of $\dot{D}$ is coparabolic semipolar if every finite-valued continuous potential $\dot{\nu}\dot{G}_{\dot{D}}$ of a measure $\dot{\nu}$ supported by $\dot{A}$ vanishes identically. Actually we shall only use this implication when $\dot{\nu}\dot{G}_{\dot{D}}$ is bounded and continuous. Let ${}^{*}\dot{u} = {}^{*}\dot{\mu}\dot{G}_{\dot{D}}$ be a bounded continuous cosuperparabolic potential satisfying the dual property of that satisfied by $\dot{u}^{*}$ in Section XVII.16(r). The cosuperparabolic potential $\mathring{[\!]}\,{}^{*}\dot{u}[\!]^{\dot{A}} = [\!]\,{}^{*}\dot{\mu}[\!]^{\dot{A}}\dot{G}_{\dot{D}}$ is lower semicontinuous, $\leq {}^{*}\dot{u}$, and continuous at every point where there is equality, in particular, at every point of $\dot{A}^{p*f} \cap \dot{D}$. If $\dot{\mu}_1$ is the projection of $[\!]{}^{*}\dot{\mu}[\!]^{\dot{A}}$ on a compact subset $\dot{A}_1$ of $\dot{A}^{p*f} \cap \dot{A}$, the potentials $\dot{\mu}_1\dot{G}_{\dot{D}}$ and $([\!]{}^{*}\dot{\mu}[\!]^{\dot{A}} - \dot{\mu}_1)\dot{G}_{\dot{D}}$ are lower semicontinuous with continuous sum at each point of $\dot{A}_1$, and so both are continuous at such a point. Since $\dot{\mu}_1\dot{G}_{\dot{D}}$ is coparabolic and therefore continuous on $\dot{D} - \dot{A}_1$, it follows that $\dot{\mu}_1\dot{G}_{\dot{D}}$ is bounded and continuous on $\dot{D}$; so (b*) implies that $\dot{\mu}_1 = 0$. Thus $[\!]{}^{*}\dot{\mu}[\!]^{\dot{A}}$ vanishes on compact subsets of $\dot{A}^{p*f} \cap \dot{A}$ and therefore vanishes on this set; that is,

$$\int_{\dot{D}} \dot{\delta}_{\dot{D}}^{\dot{A}}(\xi, \dot{A}^{p*f} \cap \dot{A})\, {}^{*}\dot{\mu}(d\xi) = 0.$$

Thus the positive superparabolic function $\dot{\delta}_{\dot{D}}^{\dot{A}}(\cdot, \dot{A}^{p*f} \cap \dot{A})$ vanishes ${}^{*}\dot{\mu}$ almost everywhere on $\dot{D}$, certainly on a dense subset of $\dot{D}$, and therefore vanishes identically on $\dot{D}$. It follows (by Theorem 11) that the set $\dot{A}^{p*f} \cap \dot{A}$ is parabolic polar and so coparabolic polar, and therefore the union $\dot{A}$ of the coparabolic-polar set $\dot{A}^{p*f} \cap \dot{A}$ and the coparabolic-semipolar set $\dot{A} - \dot{A}^{p*f}$ is coparabolic semipolar, as was to be proved.

(a*) ⇒ (b) ⇒ (a) These implications are dual to the already proved implications (a) ⇒ (b*) ⇒ (a*) and are therefore true. The set of these implications implies the truth of Theorem 12. □

13. The Support of a Swept Measure

Let $\dot{D} = \dot{\mathbb{R}}^N$, and let $\dot{A}$ be the abscissa hyperplane. Then $\dot{A}^{pf} = \dot{A}^{p*f} = \varnothing$, and the measures $\dot{\delta}_{\dot{D}}^{\dot{A}}(\dot{\xi}, \cdot)$ and $\overset{*}{\delta}{}_{\dot{D}}^{\dot{A}}(\dot{\xi}, \cdot)$ are supported by $\dot{A}$. The situation in the classical context (Section XI.14 and XI.18) is quite different in that the sweeping operation is self-dual and that $\delta_D^A(\xi, \cdot)$ is supported by A^{pf}. Furthermore, although $\dot{A}^{pf}$ and $\dot{A}^{p*f}$ are Borel sets, the sets $\dot{A} \cup \dot{A}^{pf}$ and $\dot{A} \cup \dot{A}^{p*f}$, which are involved in Theorem 13 below and its dual, are not necessarily Borel sets. For example, if $\dot{A}$ is an arbitrary subset of the abscissa hyperplane of $\dot{\mathbb{R}}^N$, then $\dot{A} \cup \dot{A}^{pf} = \dot{A} \cup \dot{A}^{p*f} = \dot{A}$. For this reason in the following counterpart of Theorem XI.14 the set $\dot{A}$ is supposed Borel. Recall that if $\dot{u} = \dot{G}_{\dot{D}}\dot{\mu}$, the measure associated with $\dot{\delta}_{\dot{D}}^{\dot{A}}(\cdot, \dot{u}) = \|\dot{u}\|^{\dot{A}}$ is not $\|\dot{\mu}\|^{\dot{A}}$ but $\overset{*}{\|}\dot{\mu}\|^{\dot{A}}$.

Theorem. *Let $\dot{D}$ be a nonempty open subset of $\dot{\mathbb{R}}^N$, let $\dot{A}$ be a Borel subset of $\dot{D}$, and let $\dot{\xi}$ be a point of $\dot{D}$.*

(a) *If $\dot{\xi} \in \dot{A}^{pf}$, then $\dot{\delta}_{\dot{D}}^{\dot{A}}(\dot{\xi}, \{\dot{\xi}\}) = 1$.*

(b) *If $\dot{\mu}$ is a measure on $\dot{D}$, then the measure $\|\dot{\mu}\|^{\dot{A}}$ [in particular, the measure $\dot{\delta}_{\dot{D}}^{\dot{A}}(\dot{\xi}, \cdot)$] is supported by $(\dot{A} \cup \dot{A}^{pf}) \cap \dot{D}$.*

(c) *If $\dot{A}$ is parabolic-fine dense in itself and if $\dot{\mu}$ is supported by $\dot{A}$, then $\dot{\mu} = \|\dot{\mu}\|^{\dot{A}}$.*

Proof. (a) If $\dot{\xi} \in \dot{A}^{pf}$, then $\dot{\delta}_{\dot{D}}^{\dot{A}}(\dot{\xi}, \{\dot{\xi}\}) = 1$ according to Theorem 10(b).

(b) Suppose that $\dot{v}$ and $\dot{v}'$ are positive superparabolic functions on $\dot{D}$ and that $\dot{v}' \leq \dot{v}$, with equality on $\dot{A}$. Then

$$\|\dot{v}'\|^{\dot{A}} = \|\dot{v}\|^{\dot{A}} = \dot{\delta}_{\dot{D}}^{\dot{A}}(\cdot, \dot{v}) = \dot{\delta}_{\dot{D}}^{\dot{A}}(\cdot, \dot{v}'); \tag{13.1}$$

so for every point $\dot{\xi}$ in $\dot{D}$ the measure $\dot{\delta}_{\dot{D}}^{\dot{A}}(\dot{\xi}, \cdot)$ is supported by the set $\{\dot{v} = \dot{v}'\}$. Now let $\dot{v}_{\bullet}$ be a decreasing sequence of positive superparabolic functions on $\dot{D}$ chosen (Choquet topological lemma) so that $\lim_{n\to\infty} \dot{v}_n = \dot{v}_\infty$ has smoothing $\|\dot{v}\|^{\dot{A}}$. After replacing $\dot{v}_n$ by $\dot{v}_n \wedge \dot{v}$ if necessary, we can suppose that $\dot{v}_n \leq \dot{v}$ with equality on $\dot{A}$. It follows that the measure $\dot{\delta}_{\dot{D}}^{\dot{A}}(\dot{\xi}, \cdot)$ is supported by $\{\dot{v}_n = \dot{v}\}$ for all n and therefore is supported by the intersection $\{\dot{v}_\infty = \dot{v}\}$ of these sets. Thus $\dot{\delta}_{\dot{D}}^{\dot{A}}$ is supported by the set $\dot{A} \cup [\{\dot{v}_\infty = \dot{v}\} \cap (\dot{D} - \dot{A})]$. Suppose that $\dot{C}$ is a compact subset of $\dot{D} - \dot{A}$. According to Section XVII.16(l), the sequence $\dot{v}_{\bullet}$ can be chosen in such a way that $\dot{v}_\infty = \|\dot{v}\|^{\dot{A}}$ on $\dot{C}$. Hence, if $\dot{\xi}$ is fixed, the projection on $\dot{D} - \dot{A}$ of the measure $\dot{\delta}_{\dot{D}}^{\dot{A}}(\dot{\xi}, \cdot)$ is supported by the set $\{\|\dot{v}\|^{\dot{A}} = \dot{v}\} \cap (\dot{D} - \dot{A})$. Finally, if $\dot{v}$ is chosen as the function $\dot{u}^{\#}$ defined in Section XVII.16(r), it follows that $\dot{\delta}_{\dot{D}}^{\dot{A}}(\dot{\xi}, \cdot)$ is supported by $(\dot{A} \cup \dot{A}^{pf}) \cap \dot{D}$. The evaluation of $\|\dot{\mu}\|^{\dot{A}}$ in (8.9) shows that this swept measure is also supported by $(\dot{A} \cup \dot{A}^{pf}) \cap \dot{D}$.

(c) Follows trivially from Theorem 13(b). □

14. An Internal Limit Theorem; The Coparabolic-Fine Topology Smoothness of Superparabolic Functions

According to Sections 11 and 12 a subset of $\dot{\mathbb{R}}^N$ is, respectively, coparabolic polar or coparabolic semipolar if and only if it is parabolic polar or parabolic semipolar; we shall write "parabolic" in both cases from now on. This self-duality is less deep than the relations between parabolic and coparabolic concepts displayed in the following parabolic context counterpart of Theorem XI.4. If $\dot{D}$ is an open subset of $\dot{\mathbb{R}}^N$ and if $\dot{v}$ and $\dot{h}$ are positive superparabolic functions on $\dot{D}$, the function $\dot{u} = \dot{v}/\dot{h}$ is defined in the obvious way when the ratio is neither $0/0$ nor $+\infty/+\infty$. The function $\dot{u}$ is thereby defined parabolic quasi everywhere on the strict positivity set of $\dot{h}$ and is also defined on the $\dot{v}$ strict positivity subset of the $\dot{h}$ zero set. We shall see that it is important to allow $\dot{h}$ to vanish. Let $\dot{\nu}_{\dot{v}}$ and $\dot{\nu}_{\dot{h}}$ be respectively the Riesz measures associated with $\dot{v}$ and $\dot{h}$. The singular component of $\dot{\nu}_{\dot{v}}$ relative to $\dot{\nu}_{\dot{h}}$ will be denoted by $\dot{\nu}_{\dot{v}}^s$; in particular, $\dot{\nu}_{\dot{v}}^s = \dot{\nu}_{\dot{v}}$ if $\dot{h}$ is parabolic. If ϕ is a function defined on a subset of $\dot{D}$ and if $\dot{\zeta} \in \dot{D}$, define $\phi^*(\dot{\zeta}) = \text{p*f}\lim_{\dot{\eta}\to\dot{\zeta}} \phi(\dot{\eta})$ if this limit exists.

The Zero Set of $\dot{h}$

If $\dot{\zeta}$ is in the zero set $\dot{Z}$ of $\dot{h}$, then every point of $\dot{D}$ below $\dot{\zeta}$ relative to $\dot{D}$ is also in $\dot{Z}$, and in particular, if π is the horizontal hyperplane containing $\dot{\zeta}$, the set $\dot{Z}$ contains the open in π connected component of $\pi \cap \dot{D}$ containing $\dot{\zeta}$. It follows that the trace in $\dot{D}$ of the Euclidean boundary of $\dot{Z}$ consists of countably many such components and that if $\dot{\zeta}$ is a point of such a boundary component, $\dot{h} > 0$ on the part strictly above $\dot{\zeta}$ of a sufficiently small Euclidean neighborhood of $\dot{\zeta}$.

Theorem. (a) *The function $\dot{u}^*$ is defined*

(a1) $\dot{\nu}_{\dot{v}} + \dot{\nu}_{\dot{h}}$ *almost everywhere on $\dot{D}$ and*
(a2) *Parabolic quasi everywhere on the strict positivity set of $\dot{v} + \dot{h}$.*

(b) $\dot{u}^* < +\infty$ $\dot{\nu}_{\dot{h}}$ *almost everywhere on $\dot{D}$ and also parabolic quasi everywhere on the strict positivity set of $\dot{h}$.*

(c) *Let $\dot{\nu}_{\dot{v}\dot{Z}}$ and $\dot{\nu}_{\dot{h}\dot{Z}}$ be the respective projections of $\dot{\nu}_{\dot{v}}$ and $\dot{\nu}_{\dot{h}}$ on the zero set $\dot{Z}$ of $\dot{h}$. Then at $\dot{\nu}_{\dot{h}}$ almost every point of $\dot{Z}$,*

$$\dot{u}^* = \frac{d\dot{\nu}_{\dot{v}}}{d\dot{\nu}_{\dot{h}}} = \frac{d\dot{\nu}_{\dot{v}\dot{Z}}}{d\dot{\nu}_{\dot{h}\dot{Z}}}. \tag{14.1}$$

(d) *Let $\dot{F}$ be a parabolic-polar subset of $\dot{D}$. Then*

(d1) (14.1) *is true $\dot{\nu}_{\dot{h}}$ almost everywhere on $\dot{F}$.*
(d2) $\dot{h}^* = +\infty$ $\dot{\nu}_{\dot{h}}$ *almost everywhere on $\dot{F}$.*
(d3) $\dot{u}^* = +\infty$ $\dot{\nu}_{\dot{v}}^s$ *almost everywhere on $\dot{F}$.*

(e) *The function* $\dot{u}^*$ *is coparabolic-fine continuous, and the set where* $\dot{u}^* \neq \dot{u}$ *is parabolic semipolar.*
(f) *In particular, if* $\dot{\zeta} \in \dot{D}$,

$$\mathrm{p^*f}\lim_{\dot{\eta}\to\dot{\zeta}} \frac{\dot{v}(\dot{\eta})}{\dot{G}_{\dot{D}}(\dot{\eta},\dot{\zeta})} = \dot{\nu}_{\dot{v}}(\{\dot{\zeta}\}) = \inf_{\{\dot{\eta}:\dot{G}_{\dot{D}}(\dot{\eta},\dot{\zeta})>0\}} \frac{\dot{v}(\dot{\eta})}{\dot{G}_{\dot{D}}(\dot{\eta},\dot{\zeta})}. \tag{14.2}$$

(g) *Let* $\dot{\zeta}$ *be a point of* $\dot{D}$, *let* $\dot{F}$ *be a subset of* $\dot{D}$ *with Euclidean topology limit point* $\dot{\zeta}$, *and suppose that* $\operatorname{ord}\dot{\eta} > \operatorname{ord}\dot{\zeta}$ *when* $\dot{\eta} \in \dot{F}$. *Then if* $\dot{F}$ *is coparabolic thin at* $\dot{\zeta}$, *there is a superparabolic potential* $\dot{v}$ *on* $\dot{D}$, *strictly positive on a Euclidean neighborhood of* $\dot{\zeta}$, *for which*

$$\lim_{\dot{F}\ni\dot{\eta}\to\dot{\zeta}} \frac{\dot{v}(\dot{\eta})}{\dot{G}_{\dot{D}}(\dot{\eta},\dot{\zeta})} = +\infty. \tag{14.3}$$

Conversely, if there is a positive superparabolic function $\dot{v}$ *on* $\dot{D}$, *satisfying* (14.3), *the set* $\dot{F}$ *is coparabolic thin at* $\dot{\zeta}$.

Observation. Since $\dot{v}$ and $\dot{h}$ are positive superparabolic functions but are otherwise unrestricted, (d2) implies that $\dot{v}^* = +\infty$ $\dot{\nu}_{\dot{v}}$ almost everywhere on $\dot{F}$. Other obvious implications of Theorem 14 can be obtained by trivial manipulations. For example, (b) implies, after an interchange of $\dot{v}$ and $\dot{h}$, that $\dot{u}^* > 0$ $\dot{\nu}_{\dot{v}}$ almost everywhere on $\dot{D}$. Note that (d2) is the parabolic context counterpart of Theorem V.11.

Proof of (f). Assertion (f) is a special case both of (c) and (d1). Its direct proof is a translation of that of Theorem XI.4(c), but we give the translation because a direct proof is so much easier than the proof of the general case and because the direct proof illustrates the adaptation of the Theorem XI.4(c) proof technique to the present context. The equality of the first and second terms on the right side of (14.2) was pointed out in Section XVII.8. It can be supposed, replacing $\dot{v}$ by $\dot{v} - \dot{\nu}_{\dot{v}}(\{\dot{\zeta}\})\dot{G}_{\dot{D}}(\cdot,\dot{\zeta})$, that $\dot{\nu}_{\dot{v}}(\{\dot{\zeta}\}) = 0$. Under this condition, unless (14.2) is true, there is a strictly positive number b such that the set $\dot{B} = \{\dot{\eta} : \dot{v}(\dot{\eta}) > b\dot{G}_{\dot{D}}(\dot{\eta},\dot{\zeta})\}$ is not coparabolic thin at $\dot{\zeta}$. Apply the dual version of Theorem 10 to obtain

$$\dot{v}(\dot{\eta}) \geq \mathring{[\![}\dot{v}]\!]^{\dot{B}}(\dot{\eta}) \geq b\mathring{[\![}\dot{G}_{\dot{D}}(\cdot,\dot{\zeta})]\!]^{\dot{B}}(\dot{\eta}) = b\,\mathring{\dagger}[\![\dot{G}_{\dot{D}}(\dot{\eta},\cdot)]\!]^{\dot{B}}(\dot{\zeta}) = b\dot{G}_{\dot{D}}(\dot{\eta},\dot{\zeta}),$$

contrary to the hypothesis that $\dot{v}/\dot{G}_{\dot{D}}(\cdot,\dot{\zeta})$ has infimum 0. □

Proof of (g). The proof is the counterpart of that of Theorem XI.4(d) and is omitted. □

Proof of (a1). If $0 \leq a < b$, define

$$\dot{A} = \{\dot{v} \leq a\dot{h}\}, \qquad \dot{B} = \{\dot{v} \geq b\dot{h}\}, \qquad \dot{C} = \dot{A}^{p^*f} \cap \dot{B}^{p^*f} \cap \dot{D}, \tag{14.4}$$

and denote by $\dot{v}_{\dot{h}\dot{C}}$ the projection of $\dot{v}_{\dot{h}}$ on $\dot{C}$. Following the proof of Theorem XI.4(a), we find that in the reduction notation of Section XVII.16(s) as translated from the classical context [see Section VI.3(o)] the smoothed reductions $\dot{h}_{\dot{A}\dot{B}}$, $\dot{h}_{\dot{A}\dot{B}\dot{A}\dot{B}}, \ldots$ all majorize $\dot{G}_{\dot{D}}\dot{v}_{\dot{h}\dot{C}}$. According to the parabolic context version of VI(3.12), the sum of these smoothed reductions is at most $[\dot{v} \wedge (b\dot{h})]/(b-a)$; so $\dot{v}_{\dot{h}\dot{C}} \equiv 0$, that is, $\dot{v}_{\dot{h}}(\dot{C}) = 0$. Now according to the analysis of the zero set $\dot{Z}$ of $\dot{h}$ given at the beginning of this section, a point of $\dot{D}$ must be either a Euclidean interior point of $\dot{Z}$ or a coparabolic-fine interior point of $\dot{D} - \dot{Z}$. The Euclidean interior of $\dot{Z}$ is $\dot{v}_{\dot{h}}$ null and every coparabolic-fine interior point of $\dot{D} - \dot{Z}$ at which $\dot{u}^*$ does not exist is in $\dot{C}$ for some rational pair a, b. It follows that $\dot{u}^*$ exists $\dot{v}_{\dot{h}}$ almost everywhere on $\dot{D}$. Apply this result to $\dot{h}/\dot{v}$ to find that $\dot{u}^*$ also exists $\dot{v}_{\dot{v}}$ almost everywhere on $\dot{D}$. □

Proof of (a2). In the context of Theorem XI.4 the fact that u^* exists quasi everywhere on D is a triviality, and in fact $u^* = u$ quasi everywhere on D, but in the present context assertion (a2) that $\dot{u}^*$ exists parabolic quasi everywhere on the strict positivity set of $\dot{h} + \dot{v}$ is by no means trivial. Since (a2) is a local assertion, we can assume that $\dot{h}$ is bounded below on $\dot{D}$ by a strictly positive number β. Let $\dot{h}_0 = \dot{G}_{\dot{D}}\dot{v}_{\dot{h}_0}$ be a potential majorized by β, with measure $\dot{v}_{\dot{h}_0}$ supported by the set $\dot{C}$ defined in (14.4). Then

$$\dot{h}_{0\dot{A}\dot{B}} + \dot{h}_{0\dot{A}\dot{B}\dot{A}\dot{B}} + \cdots \leq \dot{h}_{\dot{A}\dot{B}} + \dot{h}_{\dot{A}\dot{B}\dot{A}\dot{B}} + \cdots \leq \frac{\dot{v} \wedge (b\dot{h})}{b-a},$$

and the reasoning showing that each summand of the second series majorizes $\dot{G}_{\dot{D}}\dot{v}_{\dot{h}\dot{C}}$ can be applied to $\dot{h}_0$ and shows that each summand of the first series majorizes $\dot{G}_{\dot{D}}\dot{v}_{\dot{h}_0\dot{C}}$, where $\dot{v}_{\dot{h}_0\dot{C}}$ is the projection of $\dot{v}_{\dot{h}_0}$ on $\dot{C}$. Hence $\dot{v}_{\dot{h}_0}(\dot{C}) = 0$; that is, the zero measure is the only measure supported by $\dot{C}$ whose potential is bounded. It follows (by Theorem 11) that the set $\dot{C}$ is parabolic polar for all pairs a, b with $0 < a < b$, and this implies that u^* exists parabolic quasi everywhere on the strict positivity set of $\dot{h}$. Apply this result to $\dot{h}/\dot{v}$ to complete the proof of (a2). □

Proof of (b). Write $\dot{B}_b$ for $\dot{B}$ defined by (14.4) and observe that the infinity set of $\dot{u}^*$ is included in $\dot{B}_\infty = \bigcap_{b=0}^{\infty} \dot{B}_b^{p*f}$. If $\dot{v}_{\dot{h}\dot{B}_\infty}$ is the projection of $\dot{v}_{\dot{h}}$ on $\dot{B}_\infty$, then

$$\dot{v} \geq \mathring{[}\dot{v}\mathring{]}^{\dot{B}_b} \geq b\mathring{[}\dot{h}\mathring{]}^{\dot{B}_b} = b\dot{G}_{\dot{D}}(\dot{v}_{\dot{h}}\mathring{\delta}_{\dot{D}}^{\dot{B}_b}) \geq b\dot{G}_{\dot{D}}\dot{v}_{\dot{h}\dot{B}_\infty} \tag{14.5}$$

because $\mathring{\delta}_{\dot{D}}^{\dot{B}_b}(\dot{\xi}, \{\dot{\xi}\}) = 1$ when $\dot{\xi} \in \dot{B}_b^{p*f}$. Hence $\dot{v}_{\dot{h}}(\dot{B}_\infty) = 0$; so $\dot{u}^* < +\infty$ $\dot{v}_{\dot{h}}$ almost everywhere on $\dot{D}$. To show that $u^* < +\infty$ parabolic quasi everywhere on the strict positivity set of $\dot{h}$, we can suppose, localizing the context, that $\dot{h}$ has a strictly positive lower bound β on $\dot{D}$. Let $\dot{h}_0 = \dot{G}_{\dot{D}}\dot{v}_{\dot{h}_0}$ be a potential on $\dot{D}$ majorized by β with $\dot{v}_{\dot{h}_0}$ supported by $\dot{B}_\infty$. Then (14.5) with $\dot{h}$ replaced by $\dot{h}_0$ shows that $\dot{\nu}_{\dot{h}_0}(\dot{B}_\infty) = 0$; so the zero measure is the only measure supported by $\dot{B}_\infty$ whose potential is bounded, and therefore $\dot{B}_\infty$ is parabolic polar. Thus $\dot{u}^* < +\infty$ parabolic quasi everywhere on the strict positivity set of $\dot{h}$. □

Proof of (c). Since the Euclidean interior of $\dot{Z}$ is $\dot{\nu}_{\dot{h}}$ null, we need only consider the Euclidean boundary of $\dot{Z}$ in proving (c). In view of the properties of this boundary, discussed at the beginning of this section, it is sufficient to prove that (14.1) is true $\dot{\nu}_{\dot{h}}$ almost everywhere on an arbitrary compact subset $\dot{F}$ of this boundary lying on a horizontal hyperplane. We can suppose that $\dot{v}$ and $\dot{h}$ are potentials, after replacing these functions by their reductions on an open neighborhood of $\dot{F}$, relatively compact in $\dot{D}$. We first prove that if $\dot{\nu}_{\dot{v}}(\dot{F}) = 0$, then $\dot{u}^* = 0$ $\dot{\nu}_{\dot{h}}$ almost everywhere on $\dot{F}$. Define $\dot{B}$ by (14.4). Then

$$\dot{v} \geq \,\rrbracket\dot{v}\rrbracket^{\dot{B}} \geq b\rrbracket\dot{h}\rrbracket^{\dot{B}} = b\dot{G}_{\dot{D}}(\dot{\nu}_{\dot{h}}\overset{*}{\delta}{}_{\dot{D}}^{\dot{B}}) \geq b\dot{G}_{\dot{D}}\dot{\nu}'_{\dot{h}},$$

where $\dot{\nu}'_{\dot{h}}$ is the projection of $\dot{\nu}_{\dot{h}}$ on $\dot{B}^{p^*f} \cap \dot{F}$. Consider the subset $\dot{D}_0$ of $\dot{D}$ strictly above the hyperplane of $\dot{F}$. The Green function $\dot{G}_{\dot{D}_0}$ is the restriction to $\dot{D}_0 \times \dot{D}_0$ of $\dot{G}_{\dot{D}}$ (Section XVII.4); therefore the restriction of $\dot{v}$ to $\dot{D}_0$ is a potential, and this potential majorizes the restriction to $\dot{D}_0$ of $b\dot{G}_{\dot{D}}\dot{\nu}'_{\dot{h}}$. This restriction is a parabolic function and so $\dot{\nu}'_{\dot{h}} \equiv 0$, that is, $\dot{\nu}_{\dot{h}}(\dot{B}^{p^*f} \cap \dot{F}) = 0$. So $\mathrm{p^*f}\limsup_{\dot{\eta}\to\dot{\zeta}} \dot{u}(\dot{\eta}) \leq b$ for $\dot{\nu}_{\dot{h}}$ almost every point $\dot{\zeta}$ of $\dot{F}$, and therefore $\dot{u}^* = 0$ $\dot{\nu}_{\dot{h}}$ almost everywhere on $\dot{F}$, as asserted. It follows from this result that $(\dot{h} - \dot{G}_{\dot{D}}\dot{\nu}_{\dot{h}\dot{Z}})/\dot{h}$ has coparabolic-fine limit 0 $\dot{\nu}_{\dot{h}}$ almost everywhere on $\dot{F}$ and a corresponding result holds for $\dot{v}$. Only the projections of the measures on $\dot{Z}$ are relevant. The proof of the first equality in (14.1) now follows that of XI(4.1) and is therefore omitted. □

Proof of (d). Following the proof of Theorem XI.4(b), it is proved first that if $\dot{F}$ is a Borel parabolic-polar subset of $\dot{D}$ and if $\dot{\nu}_{\dot{v}}(\dot{F}) = 0$, then $\dot{u}^* = 0$ $\dot{\nu}_{\dot{h}}$ almost everywhere on $\dot{F}$. Since the proof follows closely that of Theorem XI.4(b), it is omitted. Apply this result to $1/\dot{h}$ to find that (d2) is true, and apply the same result to $\dot{v}/\dot{h}$ on the trace of a $\dot{\nu}_{\dot{h}}$ null support of the projection of $\dot{\nu}^s_{\dot{v}}$ on $\dot{F}$ to find that (d3) is true. The proof of (d1) follows that of Theorem XI.4(b) and so is omitted. □

Proof of (e). It is trivial from the definition of $\dot{u}^*$ that this function is coparabolic-fine continuous. To prove that $\dot{u}^* = \dot{u}$ up to a parabolic-semipolar set, it is sufficient to prove that if $\varepsilon > 0$, the set

$$\dot{F} = \{\dot{\eta} : \dot{u}^* \text{ is defined at } \dot{\eta}, |\arctan \dot{u}(\dot{\eta}) - \arctan \dot{u}^*(\dot{\eta})| > \varepsilon\}$$

is parabolic semipolar. Since (from Section XVII.15) $\dot{F} - \dot{F}^{p^*f}$ is parabolic semipolar, it is sufficient to show that $\dot{F}^{p^*f}$ is parabolic-semipolar. Actually, if $\dot{u}^*$ is defined at $\dot{\zeta}$, then

$$\mathrm{p^*f}\lim_{\dot\eta\to\dot\zeta}\left|\arctan \dot u(\dot\eta) - \arctan \dot u^*(\dot\eta)\right| = 0$$

so $\dot\zeta$ cannot be in $\dot F^{p^*f}$. Hence $\dot F^{p^*f}$ is even parabolic polar so $\dot F$ is parabolic semipolar. □

Application to the Fundamental Convergence Theorem, Reductions, and the Fine Topologies

(a) Application to the Fundamental Convergence Theorem. In that theorem (Section XVII.13) a locally lower bounded family Γ of superparabolic functions on an open subset $\dot D$ of $\dot{\mathbb{R}}^N$ is given, with pointwise infimum $\dot u$. According to that theorem, $\dot u$ has a parabolic-fine limit at every point of $\dot D$,

$$\underset{+}{\dot u}(\dot\xi) = \liminf_{\dot\eta\to\dot\xi} \dot u(\dot\eta) = \mathrm{pf}\lim_{\dot\eta\to\dot\xi} \dot u(\dot\eta), \tag{14.6}$$

and $\underset{+}{\dot u} = \dot u$ up to a parabolic-semipolar set. We shall now add to this conclusion by proving that, in the notation of this section, $\dot u^*$ is defined and equal to $\underset{+}{\dot u}{}^*$ parabolic quasi everywhere on $\dot D$ and that $\underset{+}{\dot u} = \dot u^* = \dot u$ up to a parabolic-semipolar set. Suppose first that Γ is a decreasing sequence $\dot u_\bullet$. According to Theorem 14 there is a parabolic-semipolar set $\dot A$ such that the coparabolic limit function $\dot u_n^*$ exists and is equal to $\dot u_n$ for all n at every point of $\dot D - \dot A$. The function $\dot u$ is therefore coparabolic-fine upper semicontinuous at each point of $\dot D - \dot A$, so for $\dot\xi$ in $\dot D$ less some parabolic-semipolar set

$$\underset{+}{\dot u}(\dot\xi) \le \mathrm{p^*f}\liminf_{\dot\eta\to\dot\xi} \dot u(\dot\eta) \le \mathrm{p^*f}\limsup_{\dot\eta\to\dot\xi} \dot u(\dot\eta) \le \dot u(\dot\xi) = \underset{+}{\dot u}(\dot\xi).$$

Thus up to a coparabolic-semipolar subset of $\dot D$ the function $\dot u^*$ exists and is equal to $\underset{+}{\dot u}$. To prove that $\dot u^*$ exists parabolic quasi everywhere on $\dot D$, choose a, a', and b with $0 \le a < a' < b$, and define

$$A = \{\dot u \le a\}, \qquad A_n' = \{\dot u_n \le a'\}, \qquad B_n = \{\dot u_n \ge b\}, \qquad B = \{\dot u \ge b\}.$$

Since the problem is local and the functions are locally lower bounded, we can suppose as usual that they are positive. Apply the reduction property in Section XVII.16(s) to find

$$1_{A_n'B} + 1_{A_n'BA_n'B} + \cdots \le 1_{A_n'B_n} + 1_{A_n'B_nA_n'B_n} + \cdots \le \frac{\dot u_n \wedge b}{b - a'}. \tag{14.7}$$

Now $A_\bullet'$ is an increasing sequence of sets with union a superset A'' of A. It follows from repeated application of the reduction property in Section XVII.16(e) that (14.7) yields $(n \to \infty)$

$$1_{AB} + 1_{ABAB} + \cdots \le 1_{A''B} + 1_{A''BA''B} + \cdots \le \frac{\dot u \wedge b}{b-a}. \tag{14.8}$$

The parabolic quasi everywhere existence of u^* now follows as in the proof of Theorem 14(a2). Further $\underset{+}{\dot u} \le \dot u$; so $\underset{+}{\dot u}{}^* \le \dot u^*$ where both functions are defined, that is, parabolic quasi everywhere on $\dot D$. Since $\underset{+}{\dot u}{}^* = \underset{+}{\dot u} = \dot u^*$ up to a parabolic-semipolar set, the functions $\underset{+}{\dot u}{}^*$ and $\dot u^*$, which are continuous functions in the coparabolic-fine topology, are equal on a dense-in-this-topology subset of their domains and so are equal where both are defined, that is, are equal parabolic quasi everywhere. Thus the proof is complete when Γ is a decreasing sequence. In the general case according to the Fundamental Convergence Theorem, more specifically according to Choquet's topological lemma (Appendix VIII.3), there is a sequence $\dot u_{\bullet}$ in Γ with pointwise infimum $\dot u_\infty$ such that $\underset{+}{\dot u}{}_\infty = \underset{+}{\dot u}$. We can suppose that $\dot u_{\bullet}$ is a decreasing sequence, after replacing $\dot u_n$ by $\dot u_0 \wedge \cdots \wedge \dot u_n$ if necessary. Apply the result for Γ a decreasing sequence to $\dot u_{\bullet}$ to find that the inequality $\underset{+}{\dot u} \le \dot u \le \dot u_\infty$ implies

$$\underset{+}{\dot u}{}^*(\dot\xi) \le \mathrm{p^*f}\liminf_{\dot\eta \to \dot\xi} \dot u(\dot\eta) \le \mathrm{p^*f}\limsup_{\dot\eta\to\dot\xi} \dot u(\dot\eta) \le \dot u_\infty^*(\dot\xi) = \underset{+}{\dot u}{}^*(\dot\xi)$$

for parabolic quasi every point $\dot\xi$ of $\dot D$; so $\dot u^*$ is defined and equal to $\underset{+}{\dot u}{}^*$ parabolic quasi everywhere on $\dot D$. Hence $\underset{+}{\dot u} = \dot u^* = \dot u$ up to a parabolic-semipolar set, and the proof of the application of Theorem 14 to the Fundamental Convergence Theorem is complete.

(b) Application to Reductions. Apply the preceding result to find that if $\dot v$ is a positive superparabolic function on $\dot D$ and $\dot A$ is an arbitrary subset of $\dot D$, then $\underset{+\dot v}{\dot R}{}^{\dot A}(\dot\xi) = \mathrm{p^*f}\lim_{\dot\eta\to\dot\xi} \dot R_{\dot v}^{\dot A}(\dot\eta)$ if $\dot\xi$ is not in some parabolic-semipolar set. Since $\dot R_{\dot v}^{\dot A} = \dot v$ on $\dot A$, we conclude that $\underset{+\dot v}{\dot R}{}^{\dot A} = \dot v^*$ on $\dot A^{p^*f}$ less a parabolic-semipolar set, and since $\dot v^* = \dot v$ up to a parabolic-semipolar set, we find that $\underset{+\dot v}{\dot R}{}^{\dot A} = \dot v$ on $\dot A^{p^*f}$ less a parabolic-semipolar set. In contrast recall [from Section XVII.16(b)] that this equality is true everywhere on $\dot A^{pf}$.

(c) Application to the Fine Topologies. In particular, in application (b) if $\dot D = \dot{\mathbb R}^N$ and if $\dot v$ is the function $\dot u^*$ with the properties listed in Section XVII.16(r), we conclude that $\dot A^{p^*f} \subset \dot A^{pf}$ up to a parabolic-semipolar set, and since the dual result reverses the inclusion, it follows that $\dot A^{p^*f} = \dot A^{pf}$ up to a parabolic-semipolar set. Equivalently, the parabolic-fine and coparabolic-fine closures of $\dot A$ differ by a parabolic-semipolar set.

(d) Application to Parabolic-Fine Lower Semicontinuous Functions. Let f be an arbitrary parabolic-fine lower semicontinuous funtion from an open subset $\dot D$ of $\dot{\mathbb R}^N$ into $\overline{\mathbb R}$, and let f_0 be the coparabolic-fine lower semicontinuous smoothing of f, $f_0(\dot\xi) = f(\dot\xi) \wedge \mathrm{p^*f}\liminf_{\dot\eta\to\dot\xi} f(\dot\eta)$. We show that the set

$\{f > f_0\}$ is parabolic semipolar. It is sufficient to show that for arbitrary α in $\mathbb{R}$ the set $\{f_0 < \alpha < f\}$ is parabolic semipolar. By definition of f_0 each point of $\{f_0 < \alpha\}$ is in the complement of the coparabolic-fine interior $\dot{B}$ of the parabolic-fine open set $\{f > \alpha\}$. Hence $\{f_0 < \alpha < f\} \subset \{f > \alpha\} - \dot{B}$. The set difference is parabolic semipolar according to application (c). In particular, if f is a parabolic-fine continuous function, apply the preceding result to both f and $-f$ to show that the set of coparabolic-fine discontinuities of a parabolic-fine continuous function is parabolic semipolar.

15. Application to a Version of the Parabolic Context Fatou Boundary Limit Theorem on a Slab

Suppose that $\dot{D} = \mathbb{R}^N \times]0, \delta[$ with $0 < \delta \leq +\infty$, and recall that $\dot{G}_{\dot{D}}$ is the restriction of $\dot{G}$ to $\dot{D} \times \dot{D}$. Let $\dot{v}$ and $\dot{h}$ be positive superparabolic functions on $\dot{D}$ with respective Riesz measures $\dot{v}_{\dot{v}}$ and $\dot{v}_{\dot{h}}$. If we extend $\dot{v}$ and $\dot{h}$ to functions $\dot{v}'$ and $\dot{h}'$, respectively, by defining $\dot{v}' = \dot{h}' = 0$ on the closed lower half-space of $\dot{\mathbb{R}}^N$, the extended functions are positive superparabolic functions of $\dot{D}' = \mathbb{R}^N \times]-\infty, \delta[$. The measure $\dot{v}_{\dot{v}}$ $[\dot{v}_{\dot{h}}]$ is the projection on $\dot{D}$ of the Riesz measure $\dot{v}_{\dot{v}'}$ $[\dot{v}_{\dot{h}'}]$ associated with $\dot{v}'$ $[\dot{h}']$. We have seen (Section XVII.5 Example (b)) that $\dot{v}'$ and $\dot{h}'$ are potentials on $\dot{D}'$ and that if $\dot{v}$ $[\dot{h}]$ is a potential on $\dot{D}$, then the abscissa hyperplane is $\dot{v}_{\dot{v}'}$ $[\dot{v}_{\dot{h}'}]$ null. Denote by $\dot{N}_{\dot{v}}$ $[\dot{N}_{\dot{h}}]$ the projection of $\dot{v}_{\dot{v}'}$ $[\dot{v}_{\dot{h}'}]$ on the abscissa hyperplane so that the parabolic component of $\dot{v}$ $[\dot{h}]$ in its Riesz decomposition is the restriction of $\dot{G}\dot{N}_{\dot{v}}$ $[\dot{G}\dot{N}_{\dot{h}}]$ to $\dot{D}$.

Theorem. *With the preceding definitions*

$$\mathrm{p^*f}\lim_{\dot{\eta}\to\dot{\zeta}} \frac{\dot{v}(\dot{\eta})}{\dot{h}(\dot{\eta})} = \frac{d\dot{N}_{\dot{v}}}{d\dot{N}_{\dot{h}}}(\dot{\zeta}) \tag{15.1}$$

at $\dot{N}_{\dot{h}}$ almost every point $\dot{\zeta}$ of the abscissa hyperplane.

Observation. If $\dot{h}$ is a potential, we have already remarked that $\dot{N}_{\dot{h}}$ is the zero measure so the theorem is vacuous in this case. If $\dot{v}$ is a potential, $\dot{N}_{\dot{v}}$ is the zero measure so the limit in (15.1) is 0 at $\dot{N}_{\dot{h}}$ almost every point of the abscissa hyperplane, as would be expected. If both $\dot{v}$ and $\dot{h}$ are parabolic, Theorem XVI.7 states that (15.1) is true for approach in the parabolic sense as defined in Section XVI.7. Such a restatement of (15.1) in terms of a geometrically simple approach to $\dot{\zeta}$ is not possible in the general case, however. For example, if $N = 1$ and $\dot{h} \equiv 1$, a potential $\dot{v}$ can be defined for which $\dot{v}/\dot{h}$ does not have a limit at any point $\dot{\zeta}$ of the abscissa axis on approach along the line normal to the axis.

To prove the theorem we need only apply Theorem 14(c) to the ratio $\dot{v}'/\dot{h}'$ on $\dot{D}'$.

16. The Parabolic Context Domination Principle

As noted in Section XVII.5, the classical context domination principle is not valid in the parabolic context. The following theorem, stated in the fine limit notation of Theorem 14, is the parabolic version of Theorem XI.23, the extended classical domination principle.

Theorem. *Let $\dot{h} = \dot{G}_{\dot{D}}\dot{\nu}_{\dot{h}}$ be a potential on an open subset $\dot{D}$ of $\dot{\mathbb{R}}^N$, and let $\dot{v}$ be a positive superparabolic function on $\dot{D}$. Then $\dot{h} \leq \dot{v}$ on $\dot{D}$ if $(\dot{v}/\dot{h})^* \geq 1$ $\dot{\nu}_{\dot{h}}$ almost everywhere.*

For the convenience of the reader we also state a weaker result, the parabolic context version of Theorem V.10, the classical context domination principle: $\dot{h} \leq \dot{v}$ under each of the following three conditions.

(a) *$\dot{h}^* < +\infty$ $\dot{\nu}_{\dot{h}}$ almost everywhere, and $\dot{h}^* \leq \dot{v}^*$ parabolic quasi everywhere on some Borel support of $\dot{\nu}_{\dot{h}}$.*
(b) *The inequalities $\dot{h}^* < +\infty$ and $\dot{h}^* \leq \dot{v}^*$ are true $\dot{\nu}_{\dot{h}}$ almost everywhere.*
(c) *Parabolic-polar sets are $\dot{\nu}_{\dot{h}}$ null, and $\dot{h}^* \leq \dot{v}^*$ $\dot{\nu}_{\dot{h}}$ almost everywhere.*

Observation. Each of these three conditions implies the validity of the condition of Theorem 16, namely, that $(\dot{v}/\dot{h})^* \geq 1$ $\dot{\nu}_{\dot{h}}$ almost everywhere. Since [by Theorem 14(d2)] $\dot{h}^* = +\infty$ $\dot{\nu}_{\dot{h}}$ almost everywhere on each parabolic-polar subset of $\dot{D}$, conditions (a) and (b) as well as (c) require (but Theorem 16 does not) that parabolic-polar sets are $\dot{\nu}_{\dot{h}}$ null. A useful although stronger condition than (c) is

(c′) *Parabolic-semipolar sets are $\dot{\nu}_{\dot{h}}$ null, and $\dot{h} \leq \dot{v}$ $\dot{\nu}_{\dot{h}}$ almost everywhere.*

This condition is stronger than (c) because $\dot{u} = \dot{u}^*$ and $\dot{v} = \dot{v}^*$ up to parabolic-semipolar sets.

The proof of Theorem 16 is merely a translation into the parabolic context of the proof of Theorem XI.23. That is, if $\dot{B}$ is a Borel support of $\dot{\nu}_{\dot{h}}$ on which $(\dot{v}/\dot{h})^*$ is defined and ≥ 1, and if $0 < c < 1$, the set $\dot{A} = \{\dot{v} \geq c\dot{h}\}$ is a deleted coparabolic-fine neighborhood of every point of $\dot{B}$; so $\dot{B} \subset \dot{A}^{p^*f}$, and therefor $\overset{*}{\delta}{}^{\dot{A}}_{\dot{D}}(\dot{\xi}, \{\dot{\xi}\}) = 1$ when $\dot{\xi} \in \dot{B}$. Hence

$$\dot{v} \geq \llbracket \dot{v} \rrbracket^{\dot{A}} \geq c \llbracket \dot{h} \rrbracket^{\dot{A}} = c\dot{G}_{\dot{D}}(\dot{\nu}_{\dot{h}}\overset{*}{\delta}{}^{\dot{A}}_{\dot{D}}) = c\dot{h};$$

so $\dot{v} \geq \dot{h}$.

17. Limits of Superparabolic Functions at Parabolic-Irregular Boundary Points of Their Domains

We sketch here the parabolic context versions of the classical context results in Sections XI.21 and XI.22. The boundaries of subsets of $\dot{\mathbb{R}}^N$ are relative to the Euclidean topology. The proofs are direct translations of the classical context proofs and so are omitted.

Let $\dot{D}$ be an open subset of $\dot{\mathbb{R}}^N$, let $\dot{\zeta}$ be a finite parabolic-irregular boundary point of $\dot{D}$, that is, $\dot{D}$ is a parabolic-fine deleted neighborhood of $\dot{\zeta}$, and let $\dot{\Gamma}$ be the class of open subsets of $\dot{D}$ which are parabolic-fine deleted neighborhoods of $\dot{\zeta}$. The class $\dot{\Gamma}$ is ordered by inclusion. The basic result for present purposes is that each lower-bounded superparabolic function $\dot{u}$ on $\dot{D}$ has a parabolic-fine limit ($\leq +\infty$) at $\dot{\zeta}$. This limit will be denoted by ${}^{pf}\dot{u}(\dot{\zeta})$. It then follows that $\dot{\mu}_{\dot{D}}(\dot{\xi},\cdot)$ has a limit measure ${}^{pf}\dot{\mu}_{\dot{D}}(\dot{\zeta},\cdot)$ in the sense of vague convergence of measures on $\partial\dot{D}$ when $\dot{\xi}$ tends to $\dot{\zeta}$ along some deleted parabolic-fine neighborhood of $\dot{\zeta}$. The relation XI(22.1) between ${}^{f}\mu_D$ and ${}^{f}\mu_B$ for B in Γ translates directly into the parabolic context. Furthermore, if $\dot{A}$ is a parabolic measure null subset of $\partial\dot{D}$, then $\dot{A}$ is also ${}^{pf}\dot{\mu}_{\dot{D}}(\dot{\zeta},\cdot)$ null. If $\dot{u}$ is a lower-bounded superparabolic function on $\dot{D}$, the function $\dot{B}\mapsto {}^{pf}\dot{\mu}_{\dot{B}}(\dot{\zeta},\dot{u})$ from $\dot{\Gamma}$ into $\mathbb{R}^+$ is monotone decreasing with supremum ${}^{pf}\dot{u}(\dot{\zeta})$. Here the obvious convention corresponding to that in the classical context (Section XI.22) is made for the values assigned to $\dot{u}$ on $\partial\dot{B}\cap\partial\dot{D}$. The function $\dot{u}\mapsto {}^{pf}\dot{u}(\dot{\zeta})$ is lower semicontinuous on the space of positive superparabolic functions on $\dot{D}$ in the topology of pointwise convergence.

In particular, if $\dot{\eta}\in\dot{D}$, the positive superparabolic function $\dot{G}_{\dot{D}}(\cdot,\dot{\eta})$ has a parabolic-fine limit at $\dot{\zeta}$; we denote this parabolic-fine limit by ${}^{pf}\dot{G}_{\dot{D}}(\dot{\zeta},\dot{\eta})$. Then

$$ {}^{pf}\dot{G}_{\dot{D}}(\dot{\zeta},\dot{\eta}) = \limsup_{\dot{\xi}\to\dot{\zeta}} \dot{G}_{\dot{D}}(\dot{\xi},\dot{\eta}); $$

the function ${}^{pf}\dot{G}_{\dot{D}}(\dot{\zeta},\cdot)$ is coparabolic on $\dot{D}$.

Martin Point Set Pairs

It will be useful to formulate a strengthened connectedness hypothesis adapted to the present context. In view of the application to be made to parabolic context Martin boundaries we shall call a pair $(\dot{\zeta},\dot{D})$ consisting of a point $\dot{\zeta}$ of $\dot{\mathbb{R}}^N$ and an open subset $\dot{D}$ of $\dot{\mathbb{R}}^N$ a *Martin point set pair* if the following two conditions are satisfied:

MPS(1) The point $\dot{\zeta}$ is a parabolic-irregular boundary point of $\dot{D}$.

MPS(2) If $\dot{\eta}\in\dot{D}$, there is a deleted parabolic-fine neighborhood of $\dot{\zeta}$ whose points are all strictly above $\dot{\eta}$ relative to $\dot{D}$.

Observe that MPS(2) implies that $\dot{D}$ is connected and that every point of $\dot{D}$ is strictly below $\dot{\zeta}$. We now show under MPS(1) that MPS(2) is equivalent to

MPS(2′) ${}^{pf}\dot{G}_{\dot{D}}(\dot{\zeta},\cdot) > 0$ on $\dot{D}$.

In fact under MPS(2) if $\dot{\eta}\in\dot{D}$ the set $\dot{B}_1$ of strict positivity of $\dot{G}_{\dot{D}}(\cdot,\dot{\eta})$ is a deleted parabolic fine neighborhood of $\dot{\zeta}$. If ${}^{pf}\dot{G}_{\dot{D}}(\dot{\zeta},\dot{\eta}) = 0$ there is a smaller deleted parabolic fine neighborhood $\dot{B}$ of $\dot{\zeta}$ along which $\dot{G}_{\dot{D}}(\cdot,\dot{\eta})$ has limit 0, and $\dot{B}$ can be chosen to be open because $\dot{G}_{\dot{D}}(\cdot,\dot{\eta})$ is continuous. Then $\dot{\zeta}$ is a parabolic regular boundary point of $\dot{B}$ because $\dot{G}_{\dot{D}}(\cdot,\dot{\eta})_{|\dot{B}}$ is a weak parabolic barrier for $\dot{B}$ at $\dot{\zeta}$. In view of the parabolic fine topology characterization of parabolic regularity it follows that $\dot{\zeta} \in (\dot{\mathbb{R}}^N - \dot{B})^{pf}$ and therefore that $\dot{\zeta} \in (\dot{\mathbb{R}}^N - \dot{D})^{pf}$. Hence $\dot{\zeta}$ is a parabolic regular boundary point of $\dot{D}$, contrary to hypothesis. Thus MPS(2) $\Rightarrow$ MPS(2′). Conversely MPS(2′) implies that when $\dot{\eta} \in \dot{D}$ the function $\dot{G}_{\dot{D}}(\cdot, \dot{\eta})$ is strictly positive on a deleted parabolic fine neighborhood of $\dot{\zeta}$ and therefore (by the discussion in Section XVII.4 of the strict positivity set of $\dot{G}_{\dot{D}}$) implies MPS(2).

Application to the Zero Set of a Positive Superparabolic Function [Martin Point Set Pair $(\dot{\zeta}, \dot{D})$]

If $\dot{u}$ is a positive superparabolic function on $\dot{D}$ and if ${}^{pf}\dot{u}(\dot{\zeta}) = 0$, then $\dot{u} \equiv 0$. In fact the zero set $\dot{A}$ of $\dot{u}$ is closed in $\dot{D}$, and $\dot{\xi} \in \dot{A}$ implies by the superparabolic minimum theorem that $\dot{D}_{\dot{\xi}} \subset \dot{A}$. Furthermore, if $\dot{v}$ is defined on $\dot{D}$ as $1_{\dot{D}-\dot{A}}$, the function $\dot{v}$ is superparabolic. Hence $\dot{v}$ has a parabolic-fine limit at $\dot{\zeta}$. Since

$$0 = {}^{pf}\dot{\mu}_{\dot{B}}(\dot{\zeta}, \dot{u}) = {}^{pf}\dot{\mu}_{\dot{B}}(\dot{\zeta}, \dot{v}) \quad \text{for } \dot{B} \text{ in } \dot{\Gamma},$$

it follows that ${}^{pf}\dot{v}(\dot{\zeta}) = 0$, and therefore $\dot{A}$ is a deleted parabolic-fine neighborhood of $\dot{\zeta}$; so by MPS(2) $\dot{A} = \dot{D}$, as was to be proved. This result is a generalization of the superparabolic minimum theorem.

Application to the Harnack Inequality [Martin Point Set Pair $(\dot{\zeta}, \dot{D})$]

To each compact subset $\dot{A}$ of $\dot{D}$ corresponds a constant c depending only on $\dot{\zeta}$, $\dot{D}$, $\dot{A}$ such that if $\dot{u}$ is a positive parabolic function on $\dot{D}$, then

$$\max_{\dot{A}} \dot{u} \leq c\,{}^{pf}\dot{u}(\dot{\zeta}). \tag{17.1}$$

In fact, if there is no such constant c, then there corresponds to each positive integer n a positive parabolic function $\dot{u}_n$ on $\dot{D}$ such that $\max_{\dot{A}} \dot{u}_n \geq 1$, although ${}^{pf}\dot{u}_n(\dot{\zeta}) \leq 2^{-n}$. Define $\dot{u} = \Sigma_0^\infty \dot{u}_n$. Then if $\alpha > 0$, the function $\dot{u} \wedge \alpha = \lim_{n\to\infty} (\Sigma_0^n \dot{u}_m) \wedge \alpha$ is superparabolic on $\dot{D}$ with

$${}^{pf}(\dot{u} \wedge \alpha) = \operatorname{pf}\lim_{\dot{\xi}\to\dot{\zeta}} (\dot{u} \wedge \alpha)(\dot{\xi}) \leq 1;$$

so $\operatorname{pf}\lim_{\dot{\xi}\to\dot{\zeta}} \dot{u}(\dot{\xi})$ exists and is at most 1. The finiteness set of $\dot{u}$ is therefore a set including a deleted parabolic-fine neighborhood of $\dot{\zeta}$, and (Harnack inequality applied to each function $\dot{u}_n$) if $\dot{u}(\dot{\xi}) < +\infty$, then $\dot{u} < +\infty$ on $\dot{D}_{\dot{\xi}}$. Hence the condition MPS(2) implies that $\dot{u}$ is finite valued on $\dot{D}$ and so is parabolic there. The convergence is locally uniform (Dini's theorem) so the sequence $\dot{u}_{\bullet}$ converges uniformly to 0 on $\dot{A}$, contrary to hypothesis. The proof of this generalization of Harnack's inequality is now complete.

18. Martin Flat Point Set Pairs

Let $\dot{D}$ be a nonempty open subset of $\dot{\mathbb{R}}^N$, and let $\dot{\zeta}$ be a point of $\dot{\mathbb{R}}^N$ linked to $\dot{D}$ by the following conditions:

MFPS(1) The part strictly below $\dot{\zeta}$ of some neighborhood of $\dot{\zeta}$ is in $\dot{D}$.
MFPS(2) If $\dot{\xi} \in \dot{D}$, there is a continuous function ϕ from the interval $[0, 1]$ into $\dot{D} \cup \{\dot{\xi}\}$ such that ord ϕ is a strictly monotone decreasing function with $\phi(0) = \dot{\zeta}$ and $\phi(1) = \dot{\xi}$.

According to MFPS(2), every point of $\dot{D}$ is strictly below $\dot{\zeta}$. Let $\dot{B}$ be an interval containing $\dot{\zeta}$ so small that every point of $\dot{B}$ strictly below $\dot{\zeta}$ is in $\dot{D}$, and define $\dot{D}_1 = \dot{D} \cup \dot{B}$. According to Section XVII.9, every point of $\dot{B}$ with the same ordinate value as $\dot{\zeta}$ is a parabolic-irregular boundary point of $\dot{D}$. It is now clear that $(\dot{\zeta}, \dot{D})$ is a Martin point set pair; we shall characterize the pair as *flat* in view of the geometry of $\dot{D}$ near $\dot{\zeta}$. The following analysis shows that the flatness trivializes the study of ${}^{pf}\dot{G}_{\dot{D}}(\dot{\zeta}, \cdot)$ and ${}^{pf}\dot{\mu}_{\dot{D}}(\dot{\zeta}, \cdot)$.

Define $\dot{B}$ and $\dot{D}_1$ as in the preceding paragraph. According to Section XV.17, a bounded parabolic function defined on $\dot{B} \cap \dot{D}$ has a parabolic extension to $\dot{B}$. Thus, if $\dot{\eta} \in \dot{D}$, the superparabolic function $\dot{G}_{\dot{D}}(\cdot, \dot{\eta})$, which is bounded and parabolic on $\dot{D}$ outside an arbitrary neighborhood of $\dot{\eta}$, has a limit in the Euclidean topology at $\dot{\zeta}$, a limit already denoted by ${}^{pf}\dot{G}_{\dot{D}}(\dot{\zeta}, \dot{\eta})$. In the present context even this identification of the parabolic-fine limit ${}^{pf}\dot{G}_{\dot{D}}(\dot{\zeta}, \dot{\eta})$ as a Euclidean topology limit can be simplified and made more concrete because (from Section XVII.4) $\dot{G}_{\dot{D}}$ is the restriction to $\dot{D} \times \dot{D}$ of $\dot{G}_{\dot{D}_1}$; so $\dot{G}_{\dot{D}_1}(\cdot, \dot{\eta})$ is an extension of $\dot{G}_{\dot{D}}(\cdot, \dot{\eta})$ to $\dot{D}_1$. Moreover (Section XVII.4), when $\dot{\xi} \in \dot{D}_1$, the function $\dot{G}_{\dot{D}_1}(\dot{\xi}, \cdot)$ is strictly positive at the points of $\dot{D}_1$ strictly below $\dot{\xi}$ relative to $\dot{D}_1$; so ${}^{pf}\dot{G}_{\dot{D}}(\dot{\zeta}, \cdot) = \dot{G}_{\dot{D}_1}(\dot{\zeta}, \cdot) > 0$ on $\dot{D}$. If f is a bounded continuous function on $\partial\dot{D}$, the bounded parabolic function $\dot{\mu}_{\dot{D}}(\cdot, f)$ on $\dot{D}$ has a parabolic extension to $\dot{D}_1$ according to Section XV.17; so the measure $\dot{\mu}_{\dot{D}}(\dot{\xi}, \cdot)$ tends to a limit measure ${}^{pf}\dot{\mu}_{\dot{D}}(\dot{\zeta}, \cdot)$ in the sense of vague convergence when $\dot{\xi} \to \dot{\zeta}$ in the Euclidean topology. More specifically, according to Section 2, if $\dot{A}$ is a Borel subset of $\partial\dot{D}$, the function $\dot{\mu}_{\dot{D}}(\cdot, \dot{A})$ is the restriction to $\dot{D}$ of $\dot{\mu}_{\dot{D}_1}(\cdot, \dot{A})$; so ${}^{pf}\dot{\mu}_{\dot{D}}(\dot{\zeta}, \dot{A}) = \dot{\mu}_{\dot{D}_1}(\dot{\zeta}, \dot{A})$.

19. Lattices and Related Classes of Functions in the Parabolic Context

The classes of functions defined in the classical context in Chapter IX have parabolic potential theory counterparts. In the latter context we fix an arbitrary nonempty open subset $\dot{D}$ of $\dot{\mathbb{R}}^N$ $(N \geq 1)$ and an arbitrary strictly positive parabolic function $\dot{h}$ on $\dot{D}$ and go on very nearly as in Chapter IX, replacing "h-harmonic measure" by "$\dot{h}$-parabolic measure" and so on. The change involves a slight complication left to the reader to take into account as needed, due to the fact that $\dot{h}$-parabolic measure relative to a point of $\dot{D}$ vanishes on sets above the point.

The change needed to apply these concepts to the coparabolic context will always be obvious.

Example. The class $\mathbf{L}^p(\dot{\mu}_{\dot{D}-}^{\dot{h}})$ is the class of extended real-valued Borel measurable functions on $\dot{D}$ for which, if $\dot{\xi}$ is in $\dot{D}$ and if $\dot{B}_{\bullet}$ is an increasing sequence of open relatively compact subsets of $\dot{D}$ with union $\dot{D}$ then $\sup_{n\geq 0}\dot{\mu}_{\dot{B}_n}^{\dot{h}}(\dot{\xi}, |\dot{u}|^p) < +\infty$. This is the exact counterpart of the classical context definition in Section IX.4, but observe that, in contrast to that context, even if $\dot{D}$ is connected and $|\dot{u}|^p$ is $\dot{h}$-subparabolic, the fact that this condition is satisfied for one point $\dot{\xi}$ of $\dot{D}$ does not imply that the condition is satisfied for every point $\dot{\xi}$ of $\dot{D}$.

Chapter XIX

The Martin Boundary in the Parabolic Context

1. Introduction

In discussing the parabolic context Martin boundary of an open subset $\dot{D}$ of $\dot{\mathbb{R}}^N$ for $N \geq 1$ we first make the obvious remark that there are necessarily two boundaries, one adapted to the operator $\dot{\Delta}$ and superparabolic potentials, the other adapted to the operator $\overset{*}{\dot{\Delta}}$ and cosuperparabolic potentials. The first is called the *exit* boundary; the second is called the *entrance* boundary. These dual contexts are interchanged by a reflection of $\dot{\mathbb{R}}^N$ in the abscissa hyperplane. We shall treat the exit boundary but shall omit the word "exit" unless both boundaries are involved. The following remarks are offered to orient the reader to the new features that arise in parabolic context Martin boundary theory.

Let $\dot{D}$ be a nonempty open subset of $\dot{\mathbb{R}}^N$. Throughout this chapter $\dot{D}_{\dot{\xi}}$ will denote the set of points of $\dot{D}$ strictly below $\dot{\xi}$ relative to $\dot{D}$. Suppose that a Martin function $\dot{K}$ based on a point $\dot{\zeta}$ of $\dot{D}$ is defined in the natural way,

$$\dot{K}(\dot{\eta}, \dot{\xi}) = \frac{\dot{G}_{\dot{D}}(\dot{\xi}, \dot{\eta})}{\dot{G}_{\dot{D}}(\dot{\zeta}, \dot{\eta})}.$$

Since (Section XVII.4) $\dot{G}_{\dot{D}}(\dot{\zeta}, \dot{\eta}) > 0$ if and only if $\dot{\eta} \in \dot{D}_{\dot{\zeta}}$ the function $\dot{K}$ has domain $\dot{D} \times \dot{D}_{\dot{\zeta}}$. For $\dot{\eta}$ in $\dot{D}_{\dot{\zeta}}$ the function $\dot{K}(\dot{\eta}, \cdot)$ is superparabolic on $\dot{D}$, parabolic on $\dot{D} - \{\dot{\eta}\}$, with value 1 at $\dot{\zeta}$. The classical Martin boundary treatment suggests that the Martin boundary points to be introduced should correspond to limit functions of the family $\{\dot{K}(\dot{\eta}, \cdot), \dot{\eta} \in \dot{D}_{\dot{\zeta}}\}$ when $\dot{\eta}$ in $\dot{D}_{\dot{\zeta}}$ tends to the Euclidean boundary $\partial\dot{D}$ of $\dot{D}$. This procedure will only yield Martin boundary points of $\dot{D}$ below $\dot{\zeta}$; more precisely this procedure will yield a boundary for $\dot{D}_{\dot{\zeta}}$. A second new complication is the one-sided nature of Harnack's parabolic context inequality, which is relied on to bound $\dot{K}(\dot{\eta}, \cdot)$ as $\dot{\eta}$ tends to $\partial\dot{D}$. In fact this inequality, based on the normalization $\dot{K}(\dot{\eta}, \dot{\zeta}) = 1$, only bounds $\dot{K}(\dot{\eta}, \cdot)$ strictly below $\dot{\zeta}$. Thus a limit function of $\dot{K}(\dot{\eta}, \cdot)$ as $\dot{\eta} \to \partial\dot{D}$ cannot be defined at $\dot{\zeta}$ so the normalization of such a function at $\dot{\zeta}$ requires a new formulation. This problem does not arise in the classical context.

These difficulties will be at least partially surmounted in two ways, without renouncing the Martin approach. One way is based on a Martin point set pair, as defined in Section XVIII.17; the other way is based on a Martin measure set pair, to be defined in the next section.

2. The Martin Functions of Martin Point Set and Measure Set Pairs

The Martin Function of a Martin Point Set Pair

The Martin function $\dot{K}$ of a Martin point set pair $(\dot{\zeta}, \dot{D})$ is defined on $\dot{D} \times \dot{D}$ by

$$\dot{K}(\dot{\eta}, \dot{\xi}) = \frac{\dot{G}_{\dot{D}}(\dot{\xi}, \dot{\eta})}{{}^{pf}\dot{G}_{\dot{D}}(\dot{\zeta}, \dot{\eta})}. \tag{2.1}$$

For fixed $\dot{\eta}$ the function $\dot{K}(\dot{\eta}, \cdot)$ is superparabolic on $\dot{D}$, parabolic on $\dot{D} - \{\dot{\eta}\}$, and has the normalization ${}^{pf}\dot{K}(\dot{\eta}, \dot{\zeta}) = 1$, that is, $\dot{K}(\dot{\eta}, \cdot)$ has parabolic-fine limit 1 at $\dot{\zeta}$. This normalization displays a limitation of the approach to Martin boundaries by way of Martin point set pairs: every positive superparabolic function $\dot{u}$ to be considered on $\dot{D}$ will have to satisfy the condition ${}^{pf}\dot{u}(\dot{\zeta}) < +\infty$. In particular, the minimal parabolic functions on $\dot{D}$ and the positive parabolic functions on $\dot{D}$ represented by the parabolic context Martin representation will have to satisfy this condition.

If $\dot{D}'$ is an arbitrary nonempty open subset of $\dot{\mathbb{R}}^N$ and if $\dot{\xi} \in \dot{D}'$, the pair $(\dot{\xi}, \dot{D}'_{\dot{\xi}})$ is a flat Martin point set pair to which Martin boundary theory based on Martin point set pairs can be applied. In some cases we shall see that we can then vary $\dot{\xi}$ to obtain a Martin-type boundary for $\dot{D}$ itself and a Martin-type representation of an arbitrary positive parabolic function on $\dot{D}'$.

Martin Measure Set Pairs and Their Martin Functions

Let $\dot{D}$ be a nonempty open subset of $\dot{\mathbb{R}}^N$, and let $\dot{\nu}$ be a measure on $\dot{D}$. Then the pair $(\dot{\nu}, \dot{D})$ will be called a *Martin measure set pair* if the following two conditions are satisfied:

MMS(1) The copotential $\dot{\nu}\dot{G}_{\dot{D}}$ is finite valued and continuous on the complement of a compact subset $\dot{\sigma}_{\dot{\nu}}$ of $\dot{D}$.

MMS(2) If $\dot{\xi} \in \dot{D}$, there is a point $\dot{\xi}'$ strictly above $\dot{\xi}$ relative to $\dot{D}$ such that no neighborhood of $\dot{\xi}'$ is $\dot{\nu}$ null; that is, $\dot{\xi}'$ is in the minimal closed in $\dot{D}$ support of $\dot{\nu}$.

Observe that condition MMS(2) implies that $\dot{\nu}$ cannot have compact support in $\dot{D}$. It is trivial that to every nonempty open subset $\dot{D}$ of $\dot{\mathbb{R}}^N$ corresponds a measure $\dot{\nu}$ such that $(\dot{\nu}, \dot{D})$ is a Martin measure set pair, and in fact measures can be chosen for which $\dot{v}\dot{G}_{\dot{D}}$ is finite valued and continuous on $\dot{D}$ (so $\dot{\sigma}_{\dot{v}} = \emptyset$) and $\dot{D}$ is itself the smallest closed in $\dot{D}$ support of $\dot{v}$. Since $\dot{G}_{\dot{D}}(\dot{\xi}, \cdot) > 0$ on $\dot{D}_{\dot{\xi}}$, condition MMS(2) implies that $\dot{v}\dot{G}_{\dot{D}} > 0$ on $\dot{D}$. The Martin function $\dot{K}$ for the pair $(\dot{v}, \dot{D})$ is defined on $\dot{D} \times \dot{D}$ by

$$\dot{K}(\dot{\eta}, \dot{\xi}) = \frac{\dot{G}_{\dot{D}}(\dot{\xi}, \dot{\eta})}{\dot{v}\dot{G}_{\dot{D}}(\dot{\eta})} \tag{2.2}$$

($= 0$ at the infinities of $\dot{v}\dot{G}_{\dot{D}}$). For fixed $\dot{\eta}$ with $\dot{v}\dot{G}_{\dot{D}}(\dot{\eta}) < +\infty$ the function $\dot{K}(\dot{\eta}, \cdot)$ is superparabolic on $\dot{D}$, parabolic on $\dot{D} - \{\dot{\eta}\}$, and has the normalization $\int_{\dot{D}} \dot{K}(\dot{\eta}, \cdot)\, d\dot{v} = 1$. This normalization displays a limitation of the approach to Martin boundaries by way of measure set pairs: every positive superparabolic function $\dot{u}$ to be considered on $\dot{D}$ will have to satisfy the condition $\int_{\dot{D}} \dot{u}\, d\dot{v} < +\infty$. In particular, the minimal parabolic functions on $\dot{D}$ and the positive parabolic functions on $\dot{D}$ represented by the parabolic context Martin representation will have to satisfy this condition. Every positive parabolic function on $\dot{D}$ satisfies this condition, however, for a suitable choice of $\dot{v}$, depending on the specified function. Observe that for a Martin measure set pair $(\dot{v}, \dot{D})$ the function $\dot{u} \mapsto \int_{\dot{D}} \dot{u}\, d\dot{v}$ from the space of positive superparabolic functions on $\dot{D}$, with the topology of pointwise convergence, into $\dot{\mathbb{R}}^+$ is lower semicontinuous (Fatou's lemma), as is (Section XVIII.17) the function $\dot{u} \mapsto {}^{pf}\dot{u}(\dot{\zeta})$ for $(\dot{\zeta}, \dot{D})$ a Martin point set pair. The fact that there is lower semicontinuity here rather than continuity as in the corresponding classical context is a complicating feature of the parabolic context Martin boundary constructions.

Although the approaches to the parabolic context Martin boundary by way either of Martin point pairs or set pairs involve restrictions on the classes of positive parabolic functions representable by the Martin representation, in certain cases (for example, if the domain $\dot{D}$ involved is a slab), we shall see that the Martin representation leads to a representation of every positive parabolic function on $\dot{D}$.

Generalization of Martin Point and Measure Set Pairs

It is clear that if $\dot{D}$ is an arbitrary nonempty open subset of $\dot{\mathbb{R}}^N$ and if we assign a measure $\dot{v}$ on the union of $\dot{D}$ with the set of parabolic-irregular boundary points of $\dot{D}$, and if we suppose that $\dot{v}$ has suitable properties, we can combine the Martin boundary developments based on Martin point set and measure set pairs. We shall not need this refinement, however.

Admissible Superparabolic Functions

If $(\dot{\zeta}, \dot{D})$ $[(\dot{v}, \dot{D})]$ is a Martin point [measure] set pair we shall call a positive superparabolic function $\dot{u}$ on $\dot{D}$ *admissible* if ${}^{pf}\dot{u}(\dot{\zeta}) < +\infty$ $[\int_{\dot{D}} \dot{u}\, d\dot{v} < +\infty]$.

If $\dot{h}$ is an admissible strictly positive parabolic function on $\dot{D}$, a function $\dot{v}/\dot{h}$ on $\dot{D}$ will be called *admissible* if $\dot{v}$ is the difference between two admissible positive superparabolic functions on $\dot{D}$.

3. The Martin Space $\dot{D}^M$

Theorem. *If $(\dot{\zeta}', \dot{D})$ $[(\dot{\nu}, \dot{D})]$ is a Martin Point [measure] set pair there is a unique up to homeomorphisms metrizable compactification $\dot{D}^M$ of $\dot{D}$ with the following properties:*

(a) *The Martin function $\dot{K}$ has an extension (also denoted by $\dot{K}$) to $\dot{D}^M \times \dot{D}[(\dot{D}^M - \dot{\sigma}_{\dot{\nu}}) \times \dot{D}]$ for which, when $\dot{\xi} \in \dot{D}$, the function $\dot{K}(\cdot, \dot{\xi})$ is finite valued and continuous on $\dot{D}^M - \{\dot{\xi}\}$ $[\dot{D} - \dot{\sigma}_{\dot{\nu}}]$.*

(b) *$\dot{K}(\dot{\eta}_1, \cdot) = \dot{K}(\dot{\eta}_2, \cdot)$ if and only if $\dot{\eta}_1 = \dot{\eta}_2$.*

The set $\partial^M \dot{D} = \dot{D}^M - \dot{D}$ will be called the *Martin boundary* of $(\dot{\zeta}', \dot{D})$ or $(\dot{\nu}, \dot{D})$ as the case may be. This theorem implies (see the discussion in the corresponding classical context of Theorem XII.3) that $\dot{K}(\dot{\eta}, \cdot)$ is a positive parabolic function on $\dot{D}$ when $\dot{\eta}$ is a Martin boundary point. The proof follows that of Theorem XII.3, but observe that $\dot{D}^M$ is defined in terms of a specified pair $(\dot{\zeta}', \dot{D})$ or $(\dot{\nu}, \dot{D})$ so $\dot{K}$ is uniquely determined. Thus $\dot{D}$ itself has not been assigned a Martin boundary. Let f be a strictly positive continuous function on $\dot{D}$, l_{N+1} integrable over $\dot{D}$, and for $\dot{\eta}_1$ and $\dot{\eta}_2$ in $\dot{D}$ define

$$d(\dot{\eta}_1, \dot{\eta}_2) = \int_{\dot{D}} [1 \wedge |\dot{K}(\dot{\eta}_1, \dot{\xi}) - \dot{K}(\dot{\eta}_2, \dot{\xi})|] f(\dot{\xi}) l_{N+1}(d\dot{\xi})$$

to obtain a metric on $\dot{D}$ compatible with the Euclidean topology of $\dot{D}$. The proof of the theorem then follows that of Theorem XII.3 [with a minor variation for $(\dot{\nu}, \dot{D})$ when $\dot{\sigma}_{\dot{\nu}} \neq \varnothing$] and will be omitted. The appropriate versions of the Harnack inequality have been proved for $(\dot{\zeta}', \dot{D})$ in Section XVIII.17 and for $(\dot{\nu}, \dot{D})$ in Section XV.11, and in view of these versions $\dot{\zeta} \in \partial^M \dot{D}$ implies that $\lim_{\dot{\eta} \to \dot{\zeta}} \dot{K}(\dot{\eta}, \cdot) = \dot{K}(\dot{\zeta}, \cdot)$ locally uniformly on $\dot{D}$ and that a sequence $\dot{\eta}_{\bullet}$ of points of $\dot{D}$ converges to a Martin boundary point $\dot{\zeta}$ if and only if $\lim_{n \to \infty} \dot{K}(\dot{\eta}_n, \cdot) = \dot{K}(\dot{\zeta}, \cdot)$ locally uniformly on $\dot{D}$. The lower semicontinuity property of our normalizations (Section 2) implies that for $\dot{\zeta}$ in $\partial^M \dot{D}$

$$(\dot{\zeta}', \dot{D}): \qquad {}^{pf}\dot{K}(\dot{\zeta}, \dot{\zeta}') \leq 1;$$

$$(\dot{\nu}, \dot{D}): \qquad \int_{\dot{D}} \dot{K}(\dot{\zeta}, \cdot)\, d\dot{\nu} \leq 1. \tag{3.1}$$

We shall need more general inequalities. Suppose that $\dot{\lambda}$ is a finite measure on $\dot{D}^M$, supported by $\dot{D}^M - \dot{\sigma}_{\dot{\nu}}$ in the $(\dot{\nu}, \dot{D})$ case. Then the function

$\int_{\dot{D}^M} \dot{K}(\dot{\zeta},\cdot)\dot{\lambda}(d\dot{\zeta})$ is superparabolic on $\dot{D}$, parabolic if $\dot{\lambda}$ is supported by $\partial^M \dot{D}$, and we now show that

$$
\begin{aligned}
&(\dot{\zeta}', \dot{D}): \quad {}^{pf}\!\left(\int_{\dot{D}^M} \dot{K}(\dot{\zeta},\cdot)\dot{\lambda}(d\dot{\zeta})\right)(\dot{\zeta}') = \int_{\dot{D}^M} {}^{pf}\dot{K}(\dot{\zeta},\dot{\zeta}')\dot{\lambda}(d\dot{\zeta}) \le \dot{\lambda}(\dot{D}^M)\\
&(\dot{v}, \dot{D}): \quad \int_{\dot{D}} \dot{v}(d\dot{\xi}) \int_{\dot{D}^M} \dot{K}(\dot{\zeta},\dot{\xi})\dot{\lambda}(d\dot{\zeta}) \le \dot{\lambda}(\dot{D}^M),
\end{aligned}
\tag{3.2}
$$

with equality if and only if $\dot{\lambda}$ is supported by the set of points $\dot{\zeta}$ for which there is equality in the corresponding inequality under (3.1). Inequality (3.2) for $(\dot{v}, \dot{D})$, with equality under the stated condition, follows trivially from (3.1) for $(\dot{v}, \dot{D})$. To analyze (3.2) for $(\dot{\zeta}', \dot{D})$ observe (see Section XVIII.17, but note that $\dot{\zeta}$ there is replaced here by $\dot{\zeta}'$) that if $\dot{B}$ is a Euclidean open subset of $\dot{D}$ with every Euclidean boundary point except $\dot{\zeta}'$ in $\dot{D}$ and if $\dot{B}$ is a parabolic-fine deleted neighborhood of $\dot{\zeta}'$, then

$$
{}^{pf}\dot{\mu}_{\dot{B}}\left(\dot{\zeta}', \int_{\dot{D}^M} \dot{K}(\dot{\zeta},\cdot)\dot{\lambda}(d\dot{\zeta})\right) = \int_{\dot{D}^M} {}^{pf}\dot{\mu}_{\dot{B}}(\dot{\zeta}', \dot{K}(\dot{\zeta},\cdot))\dot{\lambda}(d\dot{\zeta}) \le \dot{\lambda}(\dot{D}^M). \tag{3.3}
$$

If we order the class of sets $\dot{B}$ by inclusion, then (from Section XVIII.17) when $\dot{B}$ shrinks to $\dot{\zeta}$, the first term of (3.3) increases to the first term of (3.2) for $(\dot{\zeta}', \dot{D})$ and the integrand in the second term of (3.3) increases to ${}^{pf}\dot{K}(\dot{\zeta}, \dot{\zeta}')$. Thus (3.2) for $(\dot{\zeta}', \dot{D})$ is true if integration of the second term of (3.3) to the limit when $\dot{B}$ shrinks to $\dot{\zeta}$ can be justified; on the latter point it need only be pointed out that (Fatou's lemma) the function $\dot{\zeta} \mapsto {}^{pf}\dot{\mu}_{\dot{B}}(\dot{\zeta}', \dot{K}(\dot{\zeta},\cdot))$ is lower semicontinuous. Finally, it is trivial that there is equality in (3.2) for $(\dot{\zeta}', \dot{D})$ if $\dot{\lambda}$ is supported by the set of points $\dot{\zeta}$ for which there is equality in (3.1) for $(\dot{\zeta}', \dot{D})$.

The Zero Set of $\dot{K}(\dot{\zeta},\cdot)$

Since (from Section XVII.4) $\dot{G}_{\dot{D}}(\dot{\xi}, \dot{\eta}) > 0$ if and only if $\dot{\eta}$ is strictly below $\dot{\xi}$ relative to $\dot{D}$, it follows that if $\dot{\eta}_{\cdot}$ is a sequence of points in $\dot{D}$ with limit a Martin boundary point $\dot{\zeta}$ and if the numerical sequence $\operatorname{ord} \dot{\eta}_{\cdot}$ has limit α, then $\dot{K}(\dot{\zeta}, \dot{\xi}) = 0$ when $\operatorname{ord} \dot{\xi} \le \alpha$.

4. Preparatory Material for the Parabolic Context Martin Representation Theorem

In this section we shall derive the parabolic context counterpart of Lemma XII.4. The latter lemma was the key lemma in the derivation of the Martin representation theorem, and it will be clear from the following proof that the reasoning in the derivation of that theorem can be translated without difficulty into the parabolic context. We shall therefore omit the proofs of the counterparts of the other results in Sections XII.4 to XII.8 leading to the Martin representation theorem, but we shall state the basic results in the parabolic context for ease in later reference.

Lemma (Parabolic Counterpart of Lemma XII.4). *Let* $(\dot\zeta', \dot D)$ $[(\dot\nu, \dot D)]$ *be a Martin point [measure] set pair with Martin function* $\dot K$. *If* $\dot u$ *is a positive admissible parabolic function on* $\dot D$ *and if* $\dot A$ *is a subset of* $\dot D$ [*with* $\dot\sigma_{\dot\nu} \cap \partial \dot A = \varnothing$] *there is a measure* $\dot\lambda_{\dot u \dot A}$ *on* $\dot D^M$, *supported by the boundary of* $\dot A$ *relative to* $\dot D^M$, *for which (reduction relative to* $\dot D$*)*

$$\llbracket \dot u \rrbracket^{\dot A} = \int_{\dot D^M} \dot K(\dot\zeta, \cdot)\dot\lambda_{\dot u \dot A}(d\dot\zeta) \tag{4.1}$$

and

$$\begin{aligned} (\dot\zeta', \dot D)&: \qquad \dot\lambda_{\dot u \dot A}(\dot D^M) = {}^{pf}\llbracket \dot u \rrbracket^{\dot A}(\dot\zeta'); \\ (\dot\nu, \dot D)&: \qquad \dot\lambda_{\dot u \dot A}(\dot D^M) = \int_{\dot D} \llbracket \dot u \rrbracket^{\dot A}\, d\dot\nu. \end{aligned} \tag{4.2}$$

In particular, there is a measure $\dot\lambda_{\dot u}(= \dot\lambda_{\dot u \dot D})$ *supported by* $\partial^M \dot D$ *for which*

$$\dot u = \int_{\partial^M \dot D} \dot K(\dot\zeta, \cdot)\dot\lambda_{\dot u}(d\dot\zeta) \tag{4.3}$$

and

$$\begin{aligned} (\dot\zeta', \dot D)&: \qquad \dot\lambda_{\dot u}(\partial^M \dot D) = {}^{pf}\dot u(\dot\zeta'); \\ (\dot\nu, \dot D)&: \qquad \dot\lambda_{\dot u}(\partial^M \dot D) = \int_{\dot D} \dot u\, d\dot\nu. \end{aligned} \tag{4.4}$$

A measure $\dot\lambda_{\dot u}$ *on* $\partial^M \dot D$ *satisfying* (4.3) *must also satisfy*

$$\llbracket \dot u \rrbracket^{\dot F} = \int_{\partial^M \dot D} \llbracket \dot K(\dot\zeta, \cdot) \rrbracket^{\dot F} \dot\lambda_{\dot u}(d\dot\zeta) \tag{4.5}$$

for every Borel subset $\dot F$ *of* $\partial^M \dot D$.

We treat first the $(\dot\zeta', \dot D)$ case. If $\dot A$ is relatively compact in $\dot D$, then $\llbracket \dot u \rrbracket^{\dot A}$ is the potential $\dot G_{\dot D}\dot\lambda'$ of a measure $\dot\lambda'$ supported by $\partial \dot A$; so

$$\llbracket \dot u \rrbracket^{\dot A} = \dot G_{\dot D}\dot\lambda' = \int_{\partial \dot A} \dot K(\dot\zeta, \cdot)\dot\lambda(d\dot\zeta), \qquad \dot\lambda(d\dot\zeta) = {}^{pf}\dot G_{\dot D}(\dot\zeta', \dot\zeta)\dot\lambda'(d\dot\zeta), \tag{4.6}$$

and by (3.2)

$$^{pf}\mathring{|}\dot{u}\mathring{|}^{\dot{A}}(\dot{\zeta}') = \int_{\partial\dot{A}} {}^{pf}\dot{K}(\dot{\zeta},\cdot)\dot{\lambda}(d\dot{\zeta}) = \dot{\lambda}(\partial\dot{A}).$$

If $\dot{A}$ is not relatively compact in $\dot{D}$, let $\dot{A}_n$ be the intersection of $\dot{A}$ with the nth member of an increasing sequence of relatively compact open subsets of $\dot{D}$ with union $\dot{D}$. According to what has just been proved, the potential $\mathring{|}\dot{u}\mathring{|}^{\dot{A}_n}$ can be represented in the form

$$\mathring{|}\dot{u}\mathring{|}^{\dot{A}_n} = \int_{\dot{D}^M} \dot{K}(\dot{\zeta},\cdot)\dot{\lambda}_n(d\dot{\zeta}), \tag{4.7}$$

where $\dot{\lambda}_n$ is a measure supported by $\partial\dot{A}_n$ and

$$\dot{\lambda}_n(\dot{D}^M) = {}^{pf}\mathring{|}\dot{u}\mathring{|}^{\dot{A}_n}(\dot{\zeta}') \le {}^{pf}\mathring{|}\dot{u}\mathring{|}(\dot{\zeta}'). \tag{4.8}$$

The sequence $\dot{\lambda}_.$ is a bounded sequence of measures on the compact space $\dot{D}^M$. If $\dot{\lambda}$ is the limit of a vaguely convergent subsequence, then $\dot{\lambda}$ is supported by the boundary of $\dot{A}$ relative to $\dot{D}^M$, and (4.7) yields (4.1); equality (4.2) for $(\dot{\zeta}', \dot{D})$ follows from (4.8) and the lower semicontinuity of the operator $\dot{v} \mapsto {}^{pf}\dot{v}(\dot{\zeta}')$, operating in this case on an increasing sequence of positive superparabolic functions. The proof that (4.3) implies (4.4) is carried through by a direct translation of the corresponding implication proof in Theorem XII.4.

In the $(\dot{\nu}, \dot{D})$ case $\dot{\lambda}(d\dot{\zeta})$ in (4.6) is $\dot{\nu}\dot{G}_{\dot{D}}(\dot{\zeta})\dot{\lambda}'(d\dot{\zeta})$, and we go on as in the $(\dot{\zeta}', \dot{D})$ case.

5. Minimal Parabolic Functions and Their Poles

Poles of Parabolic Functions

If $\dot{D}$ is a nonempty open subset of $\dot{\mathbb{R}}^N$ provided with a boundary by a metric compactification, a boundary point $\dot{\zeta}$ is said to be a *pole* of a positive parabolic function $\dot{h}$ on $\dot{D}$ if (reduction relative to $\dot{D}$) $\mathring{|}\dot{h}\mathring{|}^{\{\dot{\zeta}\}} = \dot{h}$. If $\dot{h}$ is minimal parabolic on $\dot{D}$ and is not identically 0, either $\dot{\zeta}$ is a pole of $\dot{h}$ or $\mathring{|}\dot{h}\mathring{|}^{\{\dot{\zeta}\}} \equiv 0$ because the parabolic reduction operation on a boundary subset is idempotent. The elementary properties of poles in the classical context (Section X.7) are valid in the present context with no change in derivation.

Minimal Martin Boundary Points

A Martin boundary point $\dot{\zeta}$ of $(\dot{\zeta}', \dot{D})$ $[(\dot{\nu}, \dot{D})]$ will be called *minimal* if its associated parabolic function $\dot{K}(\dot{\zeta},\cdot)$ is minimal with

$$ {}^{pf}\dot{K}(\dot{\zeta}, \dot{\zeta}') = 1 \qquad \left[\int_{\dot{D}} \dot{K}(\dot{\zeta}, \cdot)\, d\dot{v} = 1\right] \tag{5.1}$$

and the set of minimal boundary points will be denoted by $\partial_1^M \dot{D}$. Observe that the counterpart of (5.1) in the classical context is trivially true for each function associated with a Martin boundary point but that in the present context one Martin boundary point, say ζ_0, associated with the identically vanishing function, is associated with a minimal function but is not a minimal boundary point. According to the following counterpart of Theorem XII.5, however, every minimal parabolic function which is admissible and not identically 0 is a constant multiple of $\dot{K}(\dot{\zeta}, \cdot)$ for some uniquely determined minimal Martin boundary point $\dot{\zeta}$. The proof of this theorem follows that of Theorem XII.5 and is omitted.

Theorem (Parabolic Counterpart of Theorem XII.5). *Let $(\dot{\zeta}', \dot{D})$ $[(\dot{v}, \dot{D})]$ be a Martin point [measure] set pair with Martin function $\dot{K}$ determining a Martin boundary $\partial^M \dot{D}$.*

(a) *Every minimal parabolic function on $\dot{D}$ has a pole on $\partial^M \dot{D}$. If a Martin boundary point $\dot{\zeta}$ is the pole of a positive $\not\equiv 0$ admissible parabolic function $\dot{u}$, then $\dot{\zeta}$ is the only pole of $\dot{u}$, $\dot{u} = \mathrm{const}\, \dot{K}(\dot{\zeta}, \cdot)$, and $\dot{\zeta}$ is a minimal Martin boundary point. In particular, if $\dot{\eta}$ is a minimal Martin boundary point, the function $\dot{K}(\dot{\eta}, \cdot)$ has pole $\dot{\eta}$.*

(b) *If $\dot{\zeta}$ is a minimal Martin boundary point and if $\dot{A}$ is a set of minimal Martin boundary points, then $[\![\dot{K}(\dot{\zeta}, \cdot)]\!]^{\dot{A}}$ is either $\dot{K}(\dot{\zeta}, \cdot)$ or 0 according as $\dot{\zeta} \in \dot{A}$ or $\dot{\zeta} \notin \dot{A}$.*

6. The Set of Nonminimal Martin Boundary Points

We leave to the reader the easy formulation and proof of the parabolic counterpart of Lemma XII.6. In the parabolic context Theorem XII.7 becomes the following.

Theorem. *If $\dot{D}$ is a nonempty open subset of $\dot{\mathbb{R}}^N$, then in the context of either Martin point or Martin measure set pairs for $\dot{D}$ the set $\dot{A}$ of nonminimal Martin boundary points is an F_σ set, and for every positive admissible parabolic function $\dot{h}$ on $\dot{D}$ (reduction relative to $\dot{D}$), $[\![\dot{h}]\!]^{\dot{A}} \equiv 0$. In particular, if $\dot{h}$ is strictly positive the set $\dot{A}$ is $\dot{h}$-parabolic measure null.*

The proof follows that of Theorem XII.7.

We also leave to the reader the application of Theorem 6 to prove the parabolic context counterpart of Lemma XII.8. The embarrassing possibility in this context that $\dot{h}$ may not be strictly positive can be treated by proving the desired result first for $\dot{h} + \varepsilon$ with $\varepsilon > 0$ and then letting ε tend to 0.

7. The Martin Representation in the Parabolic Context

Let $\dot{D}$ be a nonempty open subset of $\dot{\mathbb{R}}^N$. We shall use the notation corresponding to that of the classical context Martin representation theorem (Theorem XII.9). Thus ${}^{\dot{h}}\dot{\mathbf{S}}_m$ refers to the class of differences $\dot{v}/\dot{h} = (\dot{v}_1 - \dot{v}_2)/\dot{h}$, where $\dot{v}_i$ is a positive parabolic function on $\dot{D}$ and $\dot{h}$ is a strictly positive parabolic function on $\dot{D}$, omitted from the notation if $\dot{h} \equiv 1$. The following is a parabolic version of the classical Martin representation theorem.

Theorem. *Let $(\dot{\zeta}', \dot{D})$ $[(\dot{\nu}, \dot{D})]$ be a Martin point [measure] set pair determining a Martin function $\dot{K}$ and Martin boundary $\partial^M \dot{D}$.*

(a) *To each admissible function $\dot{v}$ in $\dot{\mathbf{S}}_m$ corresponds a unique finite-valued signed measure $\dot{M}_{\dot{v}}$ on $\partial^M \dot{D}$, supported by the minimal Martin boundary $\partial_1^M \dot{D}$, positive if $\dot{v}$ is, and satisfying*

$$\dot{v} = \int_{\partial^M \dot{D}} \dot{K}(\dot{\zeta}, \cdot) \dot{M}_{\dot{v}}(d\dot{\zeta}), \qquad \dot{M}_{\dot{v}}(\partial^M \dot{D}) = {}^{PJ}\dot{v}(\dot{\zeta}') \qquad \left[= \int_{\dot{D}} \dot{v}\, d\dot{\nu}\right]. \tag{7.1}$$

(b) *If $\dot{h}$ is a strictly positive admissible parabolic function on $\dot{D}$, the correspondence $\dot{v}/\dot{h} \leftrightarrow \dot{M}_{\dot{v}}$ is an isomorphism between the vector lattice of admissible functions in ${}^{\dot{h}}\dot{\mathbf{S}}_m$ and the vector lattice of finite signed measures on $\partial^M \dot{D}$ supported by $\partial_1^M \dot{D}$.*

(c) *An admissible function $\dot{v}/\dot{h}$ in ${}^{\dot{h}}\dot{\mathbf{S}}_m$ is quasi bounded or singular if and only if $\dot{M}_{\dot{v}}$ is, respectively, absolutely continuous or singular relative to $\dot{M}_{\dot{h}}$. In the quasi-bounded case*

$$\dot{v} = \int_{\partial^M \dot{D}} \dot{K}(\dot{\zeta}, \cdot) \frac{d\dot{M}_{\dot{v}}}{d\dot{M}_{\dot{h}}} \dot{M}_{\dot{h}}(d\dot{\zeta}). \tag{7.2}$$

The proof follows that of Theorem XII.9 with slight variations.

8. Martin Boundary of a Slab $\dot{D} = \mathbb{R}^N \times]0, \delta[$ with $0 < \delta \le +\infty$

Boundary for a Martin Point Set Pair

Suppose first that $\delta < +\infty$. Let ξ' be a point of $\mathbb{R}^N$, and define $\dot{\zeta}' = (\xi', \delta)$. Then $(\dot{\zeta}', \dot{D})$ is a Martin point set pair, and if $\dot{K}$ is the Martin function for the pair,

$$\dot{G}_{\dot{D}}(\dot{\xi}, \dot{\eta}) = \ell(s - t, \xi - \eta), \qquad \dot{K}(\dot{\eta}, \dot{\xi}) = \frac{\ell(s - t, \xi - \eta)}{\ell(\delta - t, \xi' - \eta)},$$
$$\dot{\xi} = (\xi, s), \qquad \dot{\eta} = (\eta, t). \tag{8.1}$$

The limit functions of $\dot{K}$ are as follows:

(a) If $\dot{\zeta} = (\zeta, 0)$ is a point of the abscissa hyperplane,

$$\lim_{\dot{\eta}\to\dot{\zeta}} \dot{K}(\dot{\eta}, \dot{\xi}) = \frac{\ell(s, \xi - \zeta)}{\ell(\delta, \xi' - \zeta)}.$$

(b) If $\dot{\eta} = (\eta, t) \in \dot{D}$ and $t \to \delta$ with no restriction on the varying of η or if $|\eta| \to +\infty$ with no restriction on the varying of t, then $\lim_{\dot{\eta}\to} \dot{K}(\dot{\eta}, \cdot) \equiv 0$.

The Martin space of $(\dot{\zeta}', \dot{D})$ is therefore $\dot{D}$ compactified by adjunction of the abscissa hyperplane and a single point $\mathbf{0}$: $\dot{K}((\zeta, 0), \dot{\xi}) = \ell(s, \xi - \zeta/\ell(\delta, \xi' - \zeta)$, $\dot{K}(\mathbf{0}, \cdot) \equiv 0$; $\dot{\eta}$ in $\dot{D}$ tends to $(\zeta, 0)$ in the Martin topology if and only if $\dot{\eta} \to (\zeta, 0)$ in the Euclidean topology; $\dot{\eta}$ in $\dot{D}$ tends to $\mathbf{0}$ in the Martin topology under the conditions in (b). The point $\mathbf{0}$ is a nonminimal Martin boundary point. It is trivial that every point or no point of the abscissa hyperplane is a minimal Martin boundary point so every point of the abscissa hyperplane is minimal. In the present context the Martin representation theorem states that if $\dot{u}$ is a positive parabolic function on $\dot{D}$ with ${}^{pf}\dot{u}(\xi', \delta) < +\infty$, then there is a unique finite measure $\dot{M}_{\dot{u}}$ on the abscissa hyperplane such that

$$\dot{u}(\dot{\xi}) = \int_{\mathbb{R}^N} \dot{K}((\zeta, 0), \dot{\xi}) \dot{M}_{\dot{u}}(d\zeta) \tag{8.2}$$

and

$${}^{pf}\dot{u}(\xi', \delta) = \dot{M}_{\dot{u}}(\mathbb{R}^N); \tag{8.3}$$

that is [define $\dot{N}_{\dot{u}}(d\zeta) = \dot{M}_{\dot{u}}(d\zeta)/\ell(\delta, \xi' - \zeta)$], there is a unique measure $\dot{N}_{\dot{u}}$ on $\mathbb{R}^N$ such that

$$\dot{u}(\dot{\xi}) = \int_{\mathbb{R}^N} \ell(s, \xi - \zeta) \dot{N}_{\dot{u}}(d\zeta) \qquad [\dot{\xi} = (\xi, s)] \tag{8.4}$$

and

$${}^{pf}\dot{u}(\xi', \delta) = \int_{\mathbb{R}^N} \ell(\delta, \xi' - \zeta) \dot{N}_{\dot{u}}(d\zeta). \tag{8.5}$$

Conversely, the Martin representation theorem states that if $\dot{M}_{\dot{u}}$ is a finite measure on $\mathbb{R}^N$ [if $\dot{N}_{\dot{u}}$ is a measure on $\mathbb{R}^N$ making the integral in (8.5) finite], then (8.2) [(8.4)] defines a positive parabolic function on $\dot{D}$ with ${}^{pf}\dot{u}(\xi', \delta)$ given by (8.3) [(8.5)].

Observe that for $\dot{u}$ determined in this way if ξ' is replaced by another point

ξ'', with ${}^{pf}\dot{u}(\xi'', \delta) < +\infty$, then $\dot{K}$ is changed but the representing measure $\dot{N}_{\dot{u}}$ in (8.4) is unchanged. Observe also that if $\dot{N}_{\dot{u}}$ is an arbitrary measure on $\mathbb{R}^N$ making $\dot{u}$ as defined by (8.4) finite at points of $\dot{D}$ with ordinate values arbitrarily near δ, then $\dot{u}$ is necessarily parabolic on $\dot{D}$. Moreover we have seen that (8.5) is then true if either side is known to be finite so (8.5) is true in general.

Finally we show that if $0 < \delta \le +\infty$ and if $\dot{u}$ is an arbitrary positive parabolic function on $\dot{D} = \mathbb{R}^N \times]0, \delta[$, then there is a unique measure $\dot{N}_{\dot{u}}$ on $\mathbb{R}^N$ for which (8.4) is true, and (8.5) is true for every point ξ' of $\mathbb{R}^N$ when $\delta < +\infty$, and that conversely, if $\dot{N}_{\dot{u}}$ is a measure on $\mathbb{R}^N$ for which $\dot{u}$ as defined by (8.4) is finite valued on $\dot{D}$, then $\dot{u}$ is a positive parabolic function on $\dot{D}$ and, if $\delta < +\infty$, (8.5) is true for every point ξ' of $\mathbb{R}^N$. All that remains to be proved is the existence and uniqueness of $\dot{N}_{\dot{u}}$ satisfying (8.4) when $\dot{u}$ is specified. To prove this, choose $\delta' < \delta$ and apply the preceding work to the Martin point set pair $((\xi', \delta'), \mathbb{R} \times]0, \delta'[)$ with arbitrary ξ'. The restriction of $\dot{u}$ to $\mathbb{R}^N \times]0, \delta'[$ determines a unique measure $\dot{N}_{\dot{u}}$ for which (8.4) is true for $s < \delta'$, and the uniqueness of $\dot{N}_{\dot{u}}$ shows that this measure depends neither on ξ' nor on δ'. Hence (8.4) is valid for $s < \delta$ as desired. Observe that we have now found a new proof of the representation part of Theorem XVI.6(a) and in addition have derived (8.5) in the positive case when $\delta < +\infty$. As already remarked in Section XVI.8, the representation (8.4) shows that a positive parabolic function on $\dot{D} = \mathbb{R}^N \times]0, \delta[$ is minimal if and only if the function is a positive multiple of the function $(\xi, s) \mapsto \ell(s, \xi - \zeta)$ on $\dot{D}$ for some point ζ of $\mathbb{R}^N$.

Boundary for a Martin Measure Set Pair

If $0 < \delta \le +\infty$ and if $\dot{D} = \mathbb{R}^N \times]0, \delta[$, choose a measure $\dot{v}$ for a Martin measure set pair $(\dot{v}, \dot{D})$ with $\dot{v}$ supported by a set whose intersection with each slab $\mathbb{R}^N \times]0, \delta'[$ with $\delta' < \delta$ is compact. We can also suppose that $\dot{v}\dot{G}$ is cosuperparabolic on $\dot{\mathbb{R}}^N$ and therefore coparabolic on a neighborhood of the abscissa hyperplane. Since $\dot{v}\dot{G}_{\dot{D}}$ is the restriction to $\dot{D}$ of $\dot{v}\dot{G}$, the function $\dot{v}\dot{G}_{\dot{D}}$ has the strictly positive limit $\dot{v}\dot{G}(\dot{\zeta})$ at every point $\dot{\zeta} = (\zeta, 0)$ of the abscissa hyperplane. Hence the Martin function $\dot{K}$ for $(\dot{v}, \dot{D})$ has a limit at $\dot{\zeta}$:

$$\lim_{\dot{\eta} \to \dot{\zeta}} \dot{K}(\dot{\eta}, \dot{\xi}) = \frac{\ell(s, \xi - \zeta)}{\dot{v}\dot{G}(\dot{\zeta})}, \qquad \dot{\xi} = (\xi, s).$$

Furthermore the function $\dot{K}(\dot{\eta}, \cdot)$ tends to 0 when $\dot{\eta}$ tends to any Euclidean boundary point of $\dot{D}$ not on the abscissa hyperplane. Thus the Martin boundary for $(\dot{v}, \dot{D})$ is the abscissa hyperplane and a point $\mathbf{0}$, as in the Martin point set context. The class of positive parabolic functions on $\dot{D}$ given by the Martin representation is now limited by the side condition $\int_{\dot{D}} \dot{u}\, d\dot{v} < +\infty$. Just as in the Martin point set pair case, this form of the Martin representation can be applied to derive a representation of an arbitrary positive parabolic function on $\dot{D}$.

9. Martin Boundaries for the Lower Half-space of $\dot{\mathbb{R}}^N$ and for $\dot{\mathbb{R}}^N$

Martin Boundaries for the Lower Half-space of $\dot{\mathbb{R}}^N$

Let $\dot{D}$ be this lower half-space, and let $\dot{\xi}' = (\xi', 0)$ be a point of the abscissa hyperplane. Then $(\dot{\xi}', \dot{D})$ is a Martin point set pair. According to Section XVII.4, the Green function $\dot{G}_{\dot{D}}$ is the restriction of $\dot{G}$ to $\dot{D} \times \dot{D}$. The possible limit functions of $\dot{K}(\dot{\eta}, \cdot)$ when $\dot{\eta} \to \partial D$ are the following [with $\dot{\xi} = (\xi, s)$, $\dot{\eta} = (\eta, t)$]:

(a) If $t \to -\infty$ and $\eta/t \to -\sigma^2 \gamma$, then

$$\lim_{\dot{\eta}\to} K(\dot{\eta}, \dot{\xi}) = c \exp\left(\langle \gamma, \xi \rangle + \sigma^2 |\gamma|^2 \frac{s}{2}\right), \qquad c = \exp\langle -\gamma, \xi' \rangle. \quad (9.1)$$

(b) If $|\eta|(1 - t)^{-1} \to +\infty$, then $\dot{K}(\dot{\eta}, \cdot) \to 0$.

The Martin boundary therefore consists of a point $\boldsymbol{\gamma}$ corresponding to each point γ of $\mathbb{R}^N$, with $\dot{K}(\boldsymbol{\gamma}, \dot{\xi})$ given by the right side of (9.1), and a point $\mathbf{0}$ with $\dot{K}(\mathbf{0}, \cdot) \equiv 0$. Since the context here reduces to that in Section 8 under an Appell transformation taking the lower into the upper half-plane, we go no further except to remark that $\dot{D}$ has many parabolic irregular boundary points but that the upper half-space has a quite different configuration, which suggests the interest of a more careful analysis than we have given of regularity at ∞ in the parabolic context.

Martin Boundaries for $\dot{\mathbb{R}}^N$

Martin point set pairs cannot assign a boundary to $\dot{\mathbb{R}}^N$, but if a measure $\dot{\nu}$ is chosen on $\dot{\mathbb{R}}^N$ to make $(\dot{\nu}, \dot{\mathbb{R}}^N)$ a Martin measure set pair (for example, if $\dot{\nu}$ is chosen in such a way that $\dot{\nu}\dot{G}$ is cosuperparabolic on $\dot{\mathbb{R}}^N$, that $\dot{\nu}$ is supported by a set whose intersection with the slab $\mathbb{R}^N \times]-\infty, \alpha[$ is bounded for each α, and that the smallest closed support of $\dot{\nu}$ has points with arbitrarily large ordinate values), then we obtain a Martin boundary for $(\dot{\nu}, \dot{\mathbb{R}}^N)$. This boundary consists of a point $\mathbf{0}$ corresponding to the identically vanishing parabolic function and of points $\boldsymbol{\gamma}$ obtained like the corresponding Martin boundary points of the lower half-space obtained above.

It was shown in Section XVI.8 as an application of the Appell transformation to the known properties of parabolic functions on a slab that a positive parabolic function on $\dot{\mathbb{R}}^N$ is minimal if and only if it is a positive multiple of a function $(\xi, s) \mapsto \exp(\langle \gamma, \xi \rangle + \sigma^2 |\gamma|^2 s/2)$ and that a positive parabolic function on $\dot{\mathbb{R}}^N$ can be represented in a unique way as an integral XVI(8.2) over the class of minimal parabolic functions. The reader is invited to derive these results as applications of the Martin boundary theory developed in this section.

10. The Martin Boundary of $\dot{D} =]0, +\infty[\times]-\infty, \delta[$

This Martin boundary has features not present in those discussed in Sections 8 and 9. To simplify the notation, we have taken $N = 1$, and we leave to the reader the formulations for $N > 1$. Suppose first that $\delta < +\infty$. Let ξ' be a strictly positive number, and define $\dot{\zeta}' = (\xi', \delta)$. Then $(\dot{\zeta}', \dot{D})$ is a Martin point set pair, and if $\dot{K}$ is the Martin function for this pair,

$$\dot{G}_{\dot{D}}(\dot{\xi}, \dot{\eta}) = \ell(s-t, \xi-\eta) - \ell(s-t, \xi+\eta)$$
$$\dot{K}(\dot{\eta}, \dot{\xi}) = \frac{\ell(s-t, \xi-\eta) - \ell(s-t, \xi+\eta)}{\ell(\delta-t, \xi'-\eta) - \ell(\delta-t, \xi'+\eta)} \qquad \dot{\xi} = (\xi, s), \qquad \dot{\eta} = (\eta, t). \tag{10.1}$$

Corresponding to each point τ of $\mathbb{R}$ we introduce the symbol τ' and for $\dot{\xi}$ in the right half-plane $]0, +\infty[\times \mathbb{R}$ define

$$\hat{K}(\tau', \dot{\xi}) = \begin{cases} \dfrac{\xi}{\sigma(2\pi)^{1/2}(s-\tau)^{3/2}} \exp \dfrac{-\xi^2}{2\sigma^2(s-\tau)} & \text{if } s > \tau \\ 0 & \text{if } s \le \tau. \end{cases} \tag{10.2}$$

Corresponding to each point γ of $-\mathbb{R}^+$ we introduce the symbol γ'' and for $\dot{\xi}$ in the right half-plane define

$$\hat{K}(\gamma'', \dot{\xi}) = \begin{cases} (e^{-\gamma\xi} - e^{\gamma\xi}) \exp \dfrac{\gamma^2\sigma^2 s}{2} & \text{if } \gamma < 0, \\ \xi & \text{if } \gamma = 0. \end{cases} \tag{10.3}$$

The functions $\hat{K}(\tau', \cdot)$ and $\hat{K}(\gamma'', \cdot)$ are positive and parabolic on the right half-plane. The limit functions of $\dot{K}$ are the following:

(a) When $\dot{\eta} \to \dot{\tau} = (0, \tau)$ with $\tau < \delta$,

$$\lim_{\dot{\eta} \to \dot{\tau}} \dot{K}(\dot{\eta}, \cdot) = \frac{\hat{K}(\tau', \cdot)}{\hat{K}(\tau', \dot{\zeta}')} \quad \text{on } \dot{D}. \tag{10.4}$$

(b) When $t \to -\infty$ and $\eta/t \to \sigma^2\gamma \le 0$,

$$\lim_{\dot{\eta} \to} \dot{K}(\dot{\eta}, \cdot) = \frac{\hat{K}(\gamma'', \cdot)}{\hat{K}(\gamma'', \dot{\zeta}')} \quad \text{on } \dot{D}. \tag{10.5}$$

Observe that

$$\lim_{\tau\to-\infty}\frac{\hat{K}(\tau',\dot{\xi})}{\hat{K}(\tau',\dot{\zeta}')}=\frac{\xi}{\xi'}=\frac{\hat{K}(0'',\dot{\xi})}{\hat{K}(0'',\dot{\zeta}')}\quad\text{for } \dot{\xi}\in\dot{D}.$$

(c) If $\dot{\eta}=(\eta,t)\in\dot{D}$ and either (c1) $\eta\to+\infty$ and $\eta/(1+|t|)\to+\infty$ or (c2) $t\to\delta$ with no restriction on the varying of η, then $\lim K(\dot{\eta},\cdot)\equiv 0$.

The Martin space of $(\dot{\zeta}',\dot{D})$ is therefore $\dot{D}$ compactified by adjunction of the set $\{0\}\times]-\infty,\delta[$ identified with the set $\{\tau':-\infty<\tau<\delta\}$, the set $\{\gamma'':\gamma\in-\mathbb{R}^+\}$, and a single point $\mathbf{0}$: $\dot{K}(\tau',\cdot)$ and $\dot{K}(\gamma'',\cdot)$ are defined respectively by the right sides of (10.4) and (10.5); $\dot{K}(\mathbf{0},\cdot)\equiv 0$; $\dot{\eta}$ in $\dot{D}$ tends to τ' or γ'' or $\mathbf{0}$ in the Martin topology if and only if the respective conditions (a) or (b) or (c) are satisfied. Next we show that every Martin boundary point τ' is minimal. According to Theorem 5, it is sufficient to show that each function $\dot{K}(\tau',\cdot)$ has pole τ'. Let $\dot{A}$ be the set of all Martin boundary points except the point τ'. If $\varepsilon>0$, the function identically equal to ε on $\dot{D}$ is positive and superparabolic and majorizes $\dot{K}(\tau',\cdot)$ on a deleted Martin neighborhood of each point of $\dot{A}$. Hence (reductions relative to the Martin space) $\mathring{[\![}\dot{K}(\tau',\cdot)]\!]^{\dot{A}}\equiv 0$; so

$$\begin{aligned}\dot{K}(\tau',\cdot)&=\mathring{[\![}\dot{K}(\tau',\cdot)]\!]^{\partial^M\dot{D}}\le\mathring{[\![}\dot{K}(\tau',\cdot)]\!]^{\{\tau'\}}+\mathring{[\![}\dot{K}(\tau',\cdot)]\!]^{\dot{A}}\\&=\mathring{[\![}\dot{K}(\tau',\cdot)]\!]^{\{\tau'\}}\le\dot{K}(\tau',\cdot),\end{aligned}$$

and therefore there is equality here; so $\dot{K}(\tau',\cdot)$ has pole τ'. Thus τ' is minimal, and similarly every boundary point γ'' is minimal. Hence $\mathbf{0}$ is the only nonminimal Martin boundary point.

In view of the corresponding discussion of the Martin representation for positive parabolic functions on a slab in Section 8 we omit details in the following. In the present context the Martin representation theorem states that if $\dot{u}$ is a positive parabolic function on $\dot{D}$ with ${}^{pf}\dot{u}(\xi',\delta)<+\infty$, then there is a unique measure $\dot{N}_{\dot{u}}'$ on $]-\infty,\delta[$ and a unique measure $\dot{N}_{\dot{u}}''$ on $-\mathbb{R}^+$ such that

$$\dot{u}(\dot{\xi})=\int_{-\infty}^{\delta}\hat{K}(\tau',\dot{\xi})\dot{N}_{\dot{u}}'(d\tau)+\int_{-\infty}^{0}\hat{K}(\gamma'',\dot{\xi})\dot{N}_{\dot{u}}''(d\gamma)\tag{10.6}$$

and [recall that $\dot{\zeta}'=(\xi',\delta)$]

$${}^{pf}\dot{u}(\dot{\zeta}')=\int_{-\infty}^{\delta}\hat{K}(\tau',\dot{\zeta}')\dot{N}_{\dot{u}}'(d\tau)+\int_{-\infty}^{0}\hat{K}(\gamma'',\dot{\zeta}')\dot{N}_{\dot{u}}''(d\gamma).\tag{10.7}$$

Conversely, measures $\dot{N}_{\dot{u}}'$ and $\dot{N}_{\dot{u}}''$ making the integrals on the right in (10.7) finite define a parabolic function $\dot{u}$ by (10.6), and (10.7) is then true. For example, if $\dot{N}'_{\dot{u}} = l_1$ on $]-\infty, \delta[$ and $\dot{N}''_{\dot{u}} \equiv 0$, then $\dot{u} \equiv 1$. More generally, if measures $\dot{N}'_{\dot{u}}$ and $\dot{N}''_{\dot{u}}$ define a finite-valued function $\dot{u}$ by way of (10.6), then $\dot{u}$ is positive and parabolic, and (10.7) is true, but the two sides may be $+\infty$. Finally, if $0 < \delta \leq +\infty$ and if $\dot{u}$ is a positive parabolic function on $\dot{D} =]0, +\infty[\times]-\infty, \delta[$, then (see the corresponding discussion in Section 8) there is a unique measure $\dot{N}'_{\dot{u}}$ on $]-\infty, \delta[$ and a unique measure $\dot{N}''_{\dot{u}}$ on $-\mathbb{R}^+$ such that (10.6) is true and (10.7) is true when $\delta < +\infty$. A positive parabolic function on $\dot{D}$ is minimal if and only if it is a positive multiple of a function $\hat{K}(\tau', \cdot)$ or $\hat{K}(\gamma'', \cdot)$.

The discussion of Martin boundaries for the sets in this section in the context of Martin measure set pairs is left to the reader.

11. $\dot{\text{P}}\text{WB}^h$ Solutions on $\dot{D}^M$

Let $(\dot{\zeta}', \dot{D})$ $[(\dot{v}, \dot{D})]$ be a Martin point [measure] set pair determining a Martin function $\dot{K}$ and Martin boundary $\partial^M \dot{D}$. In this context we define *universal resolutivity* to mean that $\partial^M \dot{D}$ is $\dot{h}$-resolutive for every strictly positive admissible parabolic function $\dot{h}$, and *universal internal resolutivity* is to mean that for every such choice of $\dot{h}$ every bounded $\dot{h}$-parabolic function is a $\dot{\text{P}}\text{WB}^h$ solution. Under these conventions the classical Martin boundary resolutivity theorem (Theorem XII.10) goes over directly into the parabolic context. The parabolic version is stated for the record, but the proof follows that of Theorem XII.10 and is omitted.

Theorem. *The Martin boundary of a Martin point or Martin measure set pair is universally internally resolutive and universally resolutive, with*

$$\mu_{\dot{D}}^{\dot{h}}(\dot{\xi}, d\dot{\zeta}) = \dot{K}(\dot{\zeta}, \dot{\xi}) \frac{\dot{M}_{\dot{h}}(d\dot{\zeta})}{\dot{h}(\dot{\xi})} \tag{11.1}$$

(for $\dot{h}$ strictly positive and admissible). An admissible $\dot{h}$-parabolic function $\dot{u} = \dot{v}/\dot{h}$ is a $\dot{\text{P}}\text{WB}^h$ solution if and only if it is quasi bounded, equivalently, if and only if $\dot{M}_{\dot{v}}$ is absolutely continuous relative to $\dot{M}_{\dot{h}}$, and then

$$\dot{u} = \dot{H}_f^{\dot{h}} = \int_{\partial^M \dot{D}} f(\dot{\zeta}) \dot{K}(\dot{\zeta}, \cdot) \frac{\dot{M}_{\dot{h}}(d\dot{\zeta})}{\dot{h}(\dot{\xi})}, \qquad f = \frac{d\dot{M}_{\dot{v}}}{d\dot{M}_{\dot{h}}}. \tag{11.2}$$

12. The Minimal-Fine Topology in the Parabolic Context

Let $(\dot{\zeta}', \dot{D})$ $[(\dot{v}, \dot{D})]$ be a Martin point [measure] set pair defining a Martin function $\dot{K}$ and Martin boundary $\partial^M \dot{D}$.

Parabolic Minimal Thinness

Theorem XII.11, which led to the definition of minimal thinness in the classical context, is true when translated into the parabolic context, and the proof of Theorem XII.11 goes over without change except for the proof of Step 1 which involves idempotency of the smoothed reduction operator, a property not valid in general in the parabolic context. We therefore now prove the parabolic counterpart of this step; that is, we prove that if $\dot{\zeta}$ is a minimal Martin boundary point and if $\dot{A}$ is a subset of $\dot{D}$, then $\,\rVert \dot{K}(\dot{\zeta},\cdot)\rVert^{\dot{A}}$ is either $\dot{K}(\dot{\zeta},\cdot)$ or a potential. If $\dot{A}$ is open, the parabolic reduction operation on $\dot{A}$ is idempotent; so the proof of step 1 in Section XII.11 is applicable. In the general case we need only prove that if there is a point $\dot{\xi}$ of $\dot{D}$ such that $\,\rVert \dot{K}(\dot{\zeta},\cdot)\rVert^{\dot{A}}(\dot{\xi}) < \dot{K}(\dot{\zeta},\dot{\xi})$, then $\,\rVert \dot{K}(\dot{\zeta},\cdot)\rVert^{\dot{A}}$ is a potential. If there is such a point $\dot{\xi}$, apply Section XVII.11(a) on parabolic reductions to find a decreasing sequence $\dot{B}_{\cdot}$ of open subsets of $\dot{D}$ for which

$$\lim_{n\to\infty} \,\rVert \dot{K}(\dot{\zeta},\cdot)\rVert^{\dot{B}_n}(\dot{\xi}) = \,\rVert \dot{K}(\dot{\zeta},\cdot)\rVert^{\dot{A}}(\dot{\xi}) < \dot{K}(\dot{\zeta},\dot{\xi}).$$

Since the assertion to be proved is true when $\dot{A}$ is open, the smoothed reduction $\,\rVert \dot{K}(\dot{\zeta},\cdot)\rVert^{\dot{B}_n}$ must be a potential for sufficiently large n. Hence $\,\rVert \dot{K}(\dot{\zeta},\cdot)\rVert^{\dot{A}}$ is majorized by a potential and so is itself a potential, as was to be proved.

The subset $\dot{A}$ of $\dot{D}$ is said to be *coparabolic minimal thin* at the minimal Martin boundary point $\dot{\zeta}$ if the conditions of the parabolic context counterpart of Theorem XII.11(b) are satisfied, that is, if the following equivalent conditions are satisfied:

$\dot{R}^{\dot{A}}_{\dot{K}(\dot{\zeta},\cdot)} \neq \dot{K}(\dot{\zeta},\cdot)$ on the set $\{\dot{K}(\dot{\zeta},\cdot) > 0\}$.
$\inf\{\dot{R}^{\dot{A}\cap\dot{B}}_{+\dot{K}(\dot{\zeta},\cdot)} : \dot{B}$ is a Martin topology neighborhood of $\dot{\zeta}\} = 0$.
$\dot{R}^{\dot{A}}_{+\dot{K}(\dot{\zeta},\cdot)}$ is a potential.

The first condition is satisfied if and only if it is satisfied using the corresponding smoothed reduction. In particular [cf. Section XII.12, Example (a)], if $\dot{D}$ is a deleted Euclidean neighborhood of a finite point $\dot{\zeta}$, the point $\dot{\zeta}$ can be identified with a minimal Martin boundary point of $\dot{D} - \{\dot{\zeta}\}$, with associated parabolic function a multiple of $\dot{G}_{\dot{D}}(\cdot,\dot{\zeta})$, and according to the dual of Theorem XVIII.10, a subset $\dot{A}$ of $\dot{D} - \{\dot{\zeta}\}$ is coparabolic minimal thin at $\dot{\zeta}$ if and only if $\dot{A}$ is coparabolic thin at $\dot{\zeta}$.

Coparabolic Minimal-Fine Limits at a Minimal Martin Boundary Point

We shall not need and it may be confusing to define a coparabolic minimal-fine topology on a Martin space, but the coparabolic minimal-thinness definition makes it possible to define limit concepts at a minimal Martin boundary point. In fact it follows easily from the above criteria that the union of two subsets of $\dot{D}$ coparabolic minimal thin at a minimal Martin boundary point is also coparabolic minimal thin at the point. Hence, if we call a subset $\dot{A}$ of $\dot{D}$ a deleted coparabolic minimal-fine neighborhood of the point when $\dot{D} - \dot{A}$ is coparabolic minimal thin at the point, the class (filter) of these deleted neighborhoods defines a limit theory. Limit concepts for this filter will be distinguished by the prefix p*mf. A subset of $\dot{D}$ which is a deleted coparabolic minimal-fine neighborhood of a minimal Martin boundary point has a closed-in-$\dot{D}$ subset with the same property. Lemma XII.15 goes over into the present context and thereby leads to the parabolic context counterpart of Theorem XII.16. Thus, if $\dot{\zeta}$ is a minimal Martin boundary point, a function $\dot{u}$ from the trace on $\dot{D}$ of a Martin topology neighborhood of $\dot{\zeta}$ into a metric space has coparabolic minimal-fine limit [coparabolic minimal-fine cluster value] α at $\dot{\zeta}$ if and only if $\dot{u}$ has limit α at $\dot{\zeta}$ on approach along some subset of $\dot{D}$ which is a deleted coparabolic minimal-fine neighborhood of $\dot{\zeta}$ [is not coparabolic minimal thin at $\dot{\zeta}$].

Coparabolic Minimal Thinness at Martin Boundary Points of a Slab $\dot{D} = \mathbb{R}^N \times]0, \delta[$ with $0 < \delta < +\infty$

According to Section 8, if $\xi' \in \mathbb{R}^N$, the Martin boundary for the point set pair $((\xi', \delta), \dot{D})$ includes the abscissa hyperplane. If $\dot{K}$ is the Martin function and if $\dot{\zeta} = (\zeta, 0)$ is a point of the abscissa hyperplane, the function $\dot{K}(\dot{\zeta}, \cdot)$ is a multiple of the function $(\xi, s) \mapsto \mathscr{E}(s, \xi - \zeta) = \dot{G}((\xi, s), (\zeta, 0))$. It follows from the criteria of Theorem XVIII.10 for parabolic-fine limit points of a set, as dualized for the coparabolic-fine topology, that a subset of $\dot{D}$ is coparabolic thin at $\dot{\zeta}$ if and only if the subset is coparabolic minimal thin relative to $\dot{D}$ at $\dot{\zeta}$.

Coparabolic Minimal Thinness at Martin Boundary Points of $\dot{D} =]0, +\infty[\times]-\infty, \delta[$ with $\delta < +\infty$

According to Section 10, if ξ' is a strictly positive number, the Martin boundary for the point set pair $((\xi', \delta), \dot{D})$ includes the left boundary $\{0\} \times]-\infty, \delta[$. If $\dot{K}$ is the Martin function and if $\dot{\tau} = (0, \tau)$ with $\tau < \delta$, the corresponding Martin boundary point is denoted by τ', and $\dot{K}(\tau', \cdot)$ is a multiple of the function $\hat{K}(\tau', \cdot)$ defined by (10.2). It is trivial that the set $]0, +\infty[\times]-\infty, \tau]$ is coparabolic minimal thin relative to $\dot{D}$ at τ'.

13. Boundary Counterpart of Theorem XVIII.14(f)

Theorem. *Let $(\dot{\zeta}', \dot{D})$ $[(\dot{\nu}, \dot{D})]$ be a Martin point [measure] set pair determining a Martin function $\dot{K}$ and Martin boundary $\partial^M \dot{D}$, and let $\dot{\zeta}$ be a minimal Martin boundary point.*

(a) *If $\dot{v}$ is a positive superparabolic function on $\dot{D}$, then*

$$\mathrm{p^*mf}\lim_{\substack{\dot{\eta}\to\zeta\\ \dot{K}(\zeta,\dot{\eta})>0}} \frac{\dot{v}(\dot{\eta})}{\dot{K}(\zeta,\dot{\eta})} = \inf_{\substack{\dot{D}\\ \dot{K}(\zeta,\dot{\eta})>0}} \frac{\dot{v}(\dot{\eta})}{\dot{K}(\zeta,\dot{\eta})} = \dot{M}_{\dot{v}}(\{\zeta\}). \tag{13.1}$$

(b) *If $\dot{v}$ is a positive cosuperparabolic function on $\dot{D}$ and if $\dot{\xi}\in\dot{D}$ with $\dot{K}(\zeta,\dot{\xi})>0$, then $\dot{G}_{\dot{D}}(\dot{\xi},\cdot)>0$ on a deleted Martin topology neighborhood of ζ and*

$$\mathrm{p^*mf}\lim_{\dot{\eta}\to\zeta} \frac{\dot{v}(\dot{\eta})}{\dot{G}_{\dot{D}}(\dot{\xi},\dot{\eta})} = \liminf_{\dot{\eta}\to\zeta} \frac{\dot{v}(\dot{\eta})}{\dot{G}_{\dot{D}}(\dot{\xi},\dot{\eta})} \le +\infty. \tag{13.2}$$

Theorem 13(a) is the parabolic counterpart of Theorem XII.13 and is the boundary counterpart of Theorem XVIII.14(f). The proof follows that of Theorem XVIII.14(f) and so will be omitted. Theorem 13(b) is the parabolic counterpart of Theorem XII.14. It is stated for a cosuperparabolic function because the dual theorem for a superparabolic function involves the Martin entrance boundary, that is, the Martin boundary dual to the one we have studied. The proof of Theorem 13(b) is slightly more complicated than that of Theorem XII.14 and so will be given. We can assume that the inferior limit in (13.2) is finite, and we prove that if c is a finite number strictly larger than this inferior limit, then the set

$$\dot{A} = \{\dot{\eta}\in\dot{D}\colon \dot{v}(\dot{\eta}) \ge c\dot{G}_{\dot{D}}(\dot{\xi},\dot{\eta})\}$$

is coparabolic minimal thin at ζ. Under the hypotheses of (b) the function $\dot{G}_{\dot{D}}(\dot{\xi},\cdot)>0$ on a deleted Martin topology neighborhood of ζ, and

$$c\dot{\delta}_{\dot{D}}^{\dot{A}}(\dot{\xi},\dot{K}(\dot{\eta},\cdot)) = c\,\dot{\|}\dot{K}(\dot{\eta},\cdot)\|^{\dot{A}}(\dot{\xi}) = \frac{c\,\dot{\|}\dot{G}_{\dot{D}}(\cdot,\dot{\eta})\|^{\dot{A}}(\dot{\xi})\dot{K}(\dot{\eta},\dot{\xi})}{\dot{G}_{\dot{D}}(\dot{\xi},\dot{\eta})}$$
$$= \frac{c\,\overset{*}{\|}\dot{G}_{\dot{D}}(\dot{\xi},\cdot)\|^{\dot{A}}(\dot{\eta})\dot{K}(\dot{\eta},\dot{\xi})}{\dot{G}_{\dot{D}}(\dot{\xi},\dot{\eta})} \le \frac{\dot{v}(\dot{\eta})\dot{K}(\dot{\eta},\dot{\xi})}{\dot{G}_{\dot{D}}(\dot{\xi},\dot{\eta})}.$$

When $\dot{\eta}\to\zeta$, an application of Fatou's lemma yields $c\,\dot{\|}\dot{K}(\zeta,\cdot)\|^{\dot{A}}(\dot{\xi}) < c\dot{K}(\zeta,\dot{\xi})$, and therefore $\dot{A}$ is coparabolic minimal thin at ζ. Thus the coparabolic minimal-thin limit in (13.2) exists and has the indicated value.

Observation. The limit in (13.2) can be 0 even when $\dot{v}$ does not vanish identically, but if $\dot{v}$ is strictly positive, this limit is not 0, by the counterpart of the argument in the proof of Theorem XII.14 making the corresponding limit strictly positive.

14. The Vanishing of Potentials on $\partial^M \dot{D}$

Using the notation of the preceding sections, Lemma XII.17 on the character of the set of (classical context) minimal-fine limit points of a subset of D and the set of minimal-fine limit values of a function on D goes over into the parabolic context without difficulty.

Theorem (Counterpart of Theorem XII.18). *If $\dot{h}$ is a strictly positive parabolic function on $\dot{D}$, then each superparabolic $\dot{h}$-potential $\dot{u} = \dot{G}_{\dot{D}}\dot{\mu}/\dot{h}$ has coparabolic minimal-fine limit* 0 *at $\dot{\mu}_{\dot{D}}^{\dot{h}}$ almost every (equivalently, $\dot{M}_{\dot{h}}$ almost every) point of $\partial^M \dot{D}$.*

The proof follows that of Theorem XII.18.

15. The Parabolic Context Fatou Boundary Limit Theorem on Martin Spaces

(We use the notation of Section 7.) The following theorem is stated for ease of reference, but the proof is omitted because it follows that in the classical context (Theorem XII.19.)

Theorem. *If $\dot{h}$ is a strictly positive admissible parabolic function on $\dot{D}$ and if $\dot{u} = \dot{v}/\dot{h}$ is in ${}^{\dot{h}}\dot{\mathbf{S}}$ and is admissible, then*

$$\text{p*mf}\lim_{\dot{\eta}\to\dot{\zeta}} \dot{u}(\dot{\eta}) = \frac{d\dot{M}_{\dot{v}}}{d\dot{M}_{\dot{h}}}(\dot{\zeta}) \tag{15.1}$$

for $\dot{M}_{\dot{h}}$ almost every point $\dot{\zeta}$ of $\partial^M \dot{D}$. In particular,

(a) *This limit function vanishes $\dot{M}_{\dot{h}}$ almost surely if $\dot{u} \in {}^{\dot{h}}\mathbf{S}_{\dot{p}}$;*

(b) *If $\dot{u} \in {}^{\dot{h}}\dot{\mathbf{S}}_m$, the boundary function vanishes $\dot{M}_{\dot{h}}$ almost surely if and only if $\dot{u} \in {}^{\dot{h}}\dot{\mathbf{S}}_{ms}$;*

(c) *If $\dot{u}$ is a $\dot{\mathbf{P}}\mathbf{WB}^{\dot{h}}$ solution, that is, if $\dot{u} = \dot{\mu}_{\dot{D}}^{\dot{h}}(\cdot, f)$ for some $\dot{M}_{\dot{h}}$ integrable* boundary function f, then $f = d\dot{M}_{\dot{v}}/d\dot{M}_{\dot{h}}$ *$\dot{M}_{\dot{h}}$ almost everywhere on $\partial^M \dot{D}$; therefore f is the coparabolic minimal-fine boundary limit function of $\dot{u}$ up to an $\dot{M}_{\dot{h}}$ null set.*

Extension to Nonstrictly Positive $\dot{h}$

If $\dot{h}$ is not strictly positive, apply Theorem 15 to the ratio $1/(\dot{h}+1)$ to find that this ratio has a finite coparabolic minimal-fine limit $\dot{M}_{\dot{h}} + \dot{M}_1$ almost everywhere on the Martin boundary and so certainly $\dot{M}_{\dot{h}}$ almost everywhere there. Hence $\dot{h}$ has a strictly positive ($\leq +\infty$) coparabolic minimal-fine limit

$\dot{M}_{\dot{h}}$ almost everywhere on the Martin boundary. If Theorem 15 is applied to $\dot{v}/(\dot{h}+1)$ and to $\dot{h}/(\dot{h}+1)$, it follows that $\dot{v}/\dot{h}$ has a finite coparabolic minimal-fine limit $\dot{M}_{\dot{h}}$ almost everywhere on the Martin boundary, in fact the limit $d\dot{M}_{\dot{v}}/d\dot{M}_{\dot{h}}$. The key point in understanding the situation is a rephrasing of part of one fact just derived: the zero set of $\dot{h}$ is coparabolic minimal thin at $\dot{M}_{\dot{h}}$ almost every Martin boundary point. Thus we can ignore this zero set in the present discussion. The necessary rephrasing of Theorem 15(a)–(c) when $\dot{h}$ may vanish is left to the reader.

Application to Functions on a Slab $\mathbb{R}^N \times]0,\delta[$ with $0 < \delta \le +\infty$

Let $\dot{h}$ be a strictly positive parabolic function on the slab determined using (8.4) by a measure $\dot{N}_{\dot{h}}$ on $\mathbb{R}^N$. Let $\dot{v}$ be a positive superparabolic function on the slab with parabolic component determined by the measure $\dot{N}_{\dot{v}}$. Let $\dot{\xi}' = (\xi', s')$ be a point of finiteness of $\dot{v}$, and set $\dot{D} = \mathbb{R}^N \times]0,s'[$. Then $(\dot{\xi}', \dot{D})$ is a Martin point set pair to which Theorem 15 can be applied. In view of the relation (Section 12) between coparabolic thinness and coparabolic minimal thinness at a point $(\zeta, 0)$ and in view of the relation between the measures $\dot{N}_{\dot{h}}$, $\dot{N}_{\dot{v}}$ and the corresponding measures in the Martin representations of $\dot{h}$ and the parabolic component of $\dot{v}$, we conclude from Theorem 15 that $\dot{u} = \dot{v}/\dot{h}$ has coparabolic-fine limit $d\dot{N}_{\dot{v}}/d\dot{N}_{\dot{h}}$ $\dot{N}_{\dot{h}}$ almost everywhere on the abscissa hyperplane. Thus we have derived Theorem XVIII.15 as a special case of the Fatou boundary limit theorem in the parabolic Martin boundary context. Observe that if $\dot{v}$ is parabolic, we have proved (using Theorem XVI.7) that $\dot{v}/\dot{h}$ has parabolic limit $d\dot{N}_{\dot{v}}/d\dot{N}_{\dot{h}}$ $\dot{N}_{\dot{h}}$ almost everywhere on the abscissa hyperplane. On considering the corresponding situation in the classical context (Section XII.20) it is natural to conjecture that if $\dot{v}$ is a superparabolic potential on the slab, then $\dot{v}/\dot{h}$ has limit 0 $\dot{N}_{\dot{h}}$ almost everywhere on the abscissa hyperplane on approach along normal lines. This conjecture is false (see Historical Notes) even when $\dot{h} \equiv 1$, in which case $\dot{N}_{\dot{h}}$ reduces to l_N. Thus coparabolic minimal-fine boundary approach is the most appropriate boundary approach in treating Fatou boundary limit theorems, just as the minimal fine topology approach was the most appropriate in the classical context.

Application to Functions on the Set $]0,+\infty[\times]-\infty,\delta[$ with $\delta \le +\infty$

Let $\dot{h}$ be a strictly positive parabolic function on this set, determined using (10.6) by a measure $\dot{N}'_{\dot{h}}$ on $]-\infty,\delta[$ and a measure $\dot{N}''_{\dot{h}}$ on $-\mathbb{R}^+$. Let $\dot{v}$ be a positive superparabolic function on this set with positive parabolic component in the Riesz decomposition determined by measures $\dot{N}'_{\dot{v}}$ and $\dot{N}''_{\dot{v}}$. Let $\dot{\xi}' = (\xi', s')$ be a point of finiteness of $\dot{v}$ and set $\dot{D} =]0,+\infty[\times]-\infty,s'[$. Then $(\dot{\xi}', \dot{D})$ is a Martin point set pair to which Theorem 15 can be applied. The measures $\dot{N}''_{\dot{h}}$ and $\dot{N}''_{\dot{v}}$ do not depend on the choice of $\dot{\xi}'$, and if s' is increased to $s'' < \delta$, the measures $\dot{N}'_{\dot{h}}$ and $\dot{N}'_{\dot{v}}$ do not change on $]-\infty,s'[$.

Theorem 15 in this context, applied only to the left boundary $\{0\} \times]-\infty, \delta[$, asserts that $\dot{v}/\dot{h}$ has coparabolic minimal-fine limit $d\dot{N}'_{\dot{v}}/d\dot{N}'_{\dot{h}}$ at $\dot{N}'_{\dot{h}}$ almost every point of the left boundary. See Historical Notes for the details on this result for other than coparabolic minimal-fine approach to the left boundary. Observe that coparabolic minimal-fine approach to a point $(0, \tau)$ is significant only for approach by way of points with ordinate values strictly greater than τ because (from Section 12) the set $]0, +\infty[\times]-\infty, \tau]$ is coparabolic minimal thin at $(0, \tau)$. If $\dot{h}$ is not strictly positive, there is a number τ_0 such that $\dot{h}(\xi, s) > 0$ if and only if $\tau > \tau_0$. The boundary set $\{\tau' : \tau \leq \tau_0\}$ is then $\dot{N}'_{\dot{h}}$ null, and the discussion needs only trivial changes.

Part 2

Probabilistic Counterpart of Part 1

Chapter I

Fundamental Concepts of Probability

1. Adapted Families of Functions on Measurable Spaces

(a) Filtrations of a Measurable Space. Let $(\Omega, \mathscr{F})$ be a measurable space, and let $(I, \leq)$ be a linearly ordered set. A *filtration* of $(\Omega, \mathscr{F})$ is a map $t \mapsto \mathscr{F}(t)$ from I into the class of sub σ algebras of $\mathscr{F}$, increasing in the sense that $s \leq t$ implies that $\mathscr{F}(s) \subset \mathscr{F}(t)$. The triple $(\Omega, \mathscr{F}, \mathscr{F}(\cdot))$ is called a *filtered measurable space*.

If $(\Omega, \mathscr{F}, \mathscr{F}(\cdot))$ is a filtered measurable space and if the index set I is an interval in $\mathbb{R}$ ordered by $\leq$, define $\mathscr{F}^+(t) = \bigcap_{s>t} \mathscr{F}(s)$ for t in I to get a filtration $\mathscr{F}^+(\cdot)$ with $\mathscr{F}(t) \subset \mathscr{F}^+(t)$ for t in I. The filtration $\mathscr{F}(\cdot)$ is said to be *right continuous* if $\mathscr{F}(\cdot) = \mathscr{F}^+(\cdot)$. In particular, $\mathscr{F}^+(\cdot)$ is necessarily right continuous.

(b) Families of Functions. Let $\{x(t), t \in I\}$ be a family, indexed by a parameter set I, of functions from a space Ω into a "state space" Ω'. The value of the function $x(t)$ at the point ω of Ω will be denoted by $x(t, \omega)$; the function $x(t)$ will also be denoted by $x(t, \cdot)$. The function $x(\cdot, \omega)$: $t \mapsto x(t, \omega)$ is called the *sample function determined by* ω. The family $x(\cdot)$ of functions on Ω is described as continuous, or right continuous, or ... if the sample functions are all continuous, or right continuous, or ..., for a state space and parameter set structured to make the description meaningful.

(c) Adapted Families of Functions. If $\{x(t), t \in I\}$ is a family of measurable functions from a filtered measure space $(\Omega, \mathscr{F}, \mathscr{F}(\cdot))$ into a measurable space $(\Omega', \mathscr{F}')$ the family is said to be *adapted to* $\mathscr{F}(\cdot)$ if for each t in the parameter set I the function $x(t)$ is measurable from the measurable space $(\Omega, \mathscr{F}(t))$ into the measurable space $(\Omega', \mathscr{F}')$. The notation $\{x(\cdot), \mathscr{F}(\cdot)\}$ will always imply adaptedness unless the contrary is stated. If $\{x(t), t \in I\}$ is a family of measurable functions from a measurable space $(\Omega, \mathscr{F})$ into a measurable space $(\Omega', \mathscr{F}')$, and if I is linearly ordered, define $\mathscr{F}_0(t) = \mathscr{F}\{x(s), s \leq t\}$ for t in I (see Appendix IV.1 for the notation used here) to obtain a filtration $\mathscr{F}_0(\cdot)$ making $\{x(\cdot), \mathscr{F}_0(\cdot)\}$ an adapted family. The filtration $\mathscr{F}_0(\cdot)$ is minimal for adaptedness in the sense that if $\{\mathscr{F}(t), t \in I\}$ is a filtration with respect to which $\{x(\cdot), \mathscr{F}(\cdot)\}$ is adapted, then $\mathscr{F}_0(t) \subset \mathscr{F}(t)$ for t in I.

Filtered measurable spaces and their adapted families of functions provide a mathematical formalism modeling certain physical ideas. A measurable space $(\Omega, \mathscr{F})$ is a mathematical model of the set of possible events in some physical context, together with a distinguished class of compound events. If I is a subset of $\mathbb{R}$, a filtration $\{\mathscr{F}(t), t \in I\}$ of $(\Omega, \mathscr{F})$ is a mathematical model for the flow of events in time. Each pair (t, ω) represents a possible outcome of an experiment at time t, and $\mathscr{F}(t)$ represents the class of compound events observable before or at time t. The value $x(t, \omega)$ of a function $x(t, \cdot)$ at ω models the value of some observable at the outcome (t, ω), and the function $x(t, \cdot)$ itself is therefore incorporated in $\mathscr{F}(t)$ in the sense that this function is supposed $\mathscr{F}(t)$ measurable; that is, $\{x(\cdot), \mathscr{F}(\cdot)\}$ is an adapted process.

The Measurable Sets of a Topological Measurable Space

If a measurable space is given as a topological space, the σ algebra of measurable sets will always be the σ algebra of Borel subsets of the space unless some other σ algebra is specified. In particular, the *state space* $\mathbb{R}$ means the measurable space $(\mathbb{R}, \mathscr{B}(\mathbb{R}))$.

2. Progressive Measurability

If I is a subinterval or singleton of $\mathbb{R}$, we denote by $\mathscr{B}(I)$ the class of Borel subsets of I. If I is a singleton, $\mathscr{B}(I)$ consists of that singleton and the empty set. A family $\{x(t), t \in I\}$ of functions from a measurable space $(\Omega, \mathscr{F})$ into a measurable space $(\Omega', \mathscr{F}')$ is called *measurable* if the function $(t, \omega) \mapsto x(t, \omega)$ is measurable from the measurable space $(I \times \Omega, \mathscr{B}(I) \times \mathscr{F})$ into the measurable space $(\Omega', \mathscr{F}')$. This definition will frequently not be strong enough, however, when we deal with an adapted family $\{x(t), \mathscr{F}(t), t \in \mathbb{R}^+\}$. In fact in dealing with an adapted family one frequently wishes to deal with a subfamily of the form $\{x(t), \mathscr{F}(t), t \leq c\}$ and wishes the analysis of this subfamily, in particular, the measurability of the subfamily, to depend only on properties involving parameter values in the interval $[0, c]$. Unfortunately even if $\mathscr{F} = \bigvee_{t \in \mathbb{R}^+} \mathscr{F}(t)$, the measurability of the original family does not imply that the subfamilies of the stated form are measurable. The following definition is formulated precisely to provide this measurability.

Suppose then that $\{x(t), \mathscr{F}(t), t \in \mathbb{R}^+\}$ is an adapted family of functions. A subset A of $\mathbb{R}^+ \times \Omega$ will be called *progressively measurable* if $A \cap ([0, c] \times \Omega) \in \mathscr{B}([0, c]) \times \mathscr{F}(c)$ when $c \geq 0$. The class of progressively measurable sets is a σ algebra. The adapted family is called *progressively measurable* if the function $x(\cdot, \cdot)$ is measurable from the space $\mathbb{R}^+ \times \Omega$ coupled with the

σ algebra of progressively measurable sets into $(\Omega', \mathscr{F}')$, equivalently, if for $c \geq 0$ the restriction of the function $x(\cdot, \cdot)$ to $[0, c] \times \Omega$ is measurable from $([0, c] \times \Omega, \mathscr{B}([0, c]) \times \mathscr{F}(c))$ into $(\Omega', \mathscr{F}')$. (In this phraseology the condition that a set A be progressively measurable is equivalent to the condition that the adapted family $\{1_A(t, \cdot), \mathscr{F}(t), t \in \mathbb{R}^+\}$ with state space $(\mathbb{R}^+, \mathscr{B}(\mathbb{R}^+))$ be progressively measurable. Observe that even without the prior hypothesis of adaptedness the progressive measurability condition implies adaptedness. We leave to the reader the easy proof that if $\{x(t), \mathscr{F}(t), t \in \mathbb{R}^+\}$ is progressively measurable, then for $b > 0$ the family $\{x(b + t), \mathscr{F}(b + t), t \in \mathbb{R}^+\}$ is also progressively measurable. It is immediate that if $\{x(\cdot), \mathscr{F}(\cdot)\}$ is progressively measurable and if f is a measurable function from the state space $(\Omega', \mathscr{F}')$ into the measurable space $(\Omega'', \mathscr{F}'')$, then $\{f[x(\cdot)], \mathscr{F}(\cdot)\}$ is also progressively measurable, with state space $(\Omega'', \mathscr{F}'')$.

Alternative Characterization of Progressive Measurability

In the preceding context define a new filtration by setting $\mathscr{F}^-(0) = \mathscr{F}(0)$ and $\mathscr{F}^-(t) = \bigvee_{s<t} \mathscr{F}(s)$ for $t > 0$, and consider the following conditions on the family $\{x(t), t \in \mathbb{R}^+\}$ of functions from $(\Omega, \mathscr{F})$ into $(\Omega', \mathscr{F}')$:

(a) The family is adapted to $\mathscr{F}(\cdot)$.
(a′) $\mathscr{F}(\cdot)$ is right continuous.
(b) For $c > 0$ the restriction of the family $x(\cdot)$ to the set $[0, c[$ is $\mathscr{B}([0, c[) \times \mathscr{F}^-(c)$ measurable.

Observe that the pair (a′), (b) implies (a). We now show that $\{x(\cdot), \mathscr{F}(\cdot)\}$ is progressively measurable if and only if (a) and (b) are satisfied. To prove that (a) and (b) together imply progressive measurability, we need only observe that the subsets $[0, c[\times \Omega$ and $\{c\} \times \Omega$ of $[0, c] \times \Omega$ have the latter set as union and are in $\mathscr{B}([0, c]) \times \mathscr{F}(c)$ and that under (a) and (b) the function $x(\cdot, \cdot)$ is measurable on both subsets and so also on their union, relative to $\mathscr{B}([0, c]) \times \mathscr{F}(c)$. Conversely, progressive measurability implies (a), and another piecing together of measurable functions shows that progressive measurability also implies (b).

EXAMPLE. Let $\{x(t), \mathscr{F}(t), t \in \mathbb{R}^+\}$ be an adapted family whose state space is Polish. Then the family is progressively measurable if the sample functions are all right continuous or all left continuous. For example, in the right continuous case define

$$\begin{aligned} x_n(t, \omega) &= x\left(\frac{jc}{n}, \omega\right) \quad \text{if } (j-1)\frac{c}{n} < t \leq \frac{jc}{n},\ 1 \leq j \leq n, \\ x_n(0, \omega) &= x(0, \omega) \end{aligned} \tag{2.1}$$

when $c > 0$. Then the restriction of $x_n(\cdot,\cdot)$ to $[0, c] \times \Omega$ is $\mathscr{B}([0, c]) \times \mathscr{F}(c)$ measurable and $\lim_{n\to\infty} x_n(\cdot,\cdot) = x(\cdot,\cdot)$, so the progressive measurability condition is satisfied when $c > 0$. The condition when $c = 0$ is satisfied because the family is adapted. The left continuous case is treated similarly.

Observe that if whenever $\varepsilon > 0$ the restriction of the family $x(\cdot)$ to the set $[0, c]$ is $\mathscr{B}[0, c] \times \mathscr{F}(c + \varepsilon)$ measurable, then (b) is satisfied. In fact this condition implies that for $0 < \delta < c$ the restriction of this family to $[0, c - \delta]$ is

$$\mathscr{B}([0, c - \delta]) \times \mathscr{F}\left(c - \frac{\delta}{2}\right) \subset \mathscr{B}([0, c[) \times \mathscr{F}^-(c)$$

measurable, and this yields (b). In other words the adapted process $\{x(\cdot), \mathscr{F}(\cdot)\}$ is progressively measurable if the process $\{x(t), \mathscr{F}(t + \varepsilon), t \in \mathbb{R}^+\}$ is progressively measurable for every $\varepsilon > 0$.

3. Random Variables

A probability space is a triple $(\Omega, \mathscr{F}, P)$ consisting of a measurable space $(\Omega, \mathscr{F})$ and a measure P on $\mathscr{F}$ with $P\{\Omega\} = 1$. It will frequently be unnecessary to specify Ω or $\mathscr{F}$, but it will always be supposed even though not mentioned that P is complete, that is, that subsets of P null sets in $\mathscr{F}$ are also in $\mathscr{F}$.

A *random variable* is a measurable function from a probability space into a measurable "state" space. If the state space is topological, its measurable sets unless otherwise specified will be the Borel sets. The *distribution* of a random variable x is defined as the measure μ on the class of measurable sets of the state space induced by x: $\mu(A) = P\{x \in A\}$. The integral of a real-valued random variable x will be denoted by $E\{x\}$, and the integral of x over a set Λ will sometimes be denoted by $E\{x; \Lambda\}$.

Sometimes the context is expanded to allow measures P other than probability measures, even to allow $P\{\Omega\} = +\infty$, but if so, the expansion is mentioned explicitly.

We follow the common abuse of notation in measure theory, in which the same notation is used for equivalence classes of measurable functions under the equivalence relation of almost everywhere (a.e.) equality, or "almost sure" (a.s.) equality in the slang of probability, as for their members. The proper interpretation will be clear from the context. Thus the essential infimum $\operatorname{ess\,inf}_\alpha z_\alpha$ of a family of random variables (Appendix IV.9) is determined only up to a null set and properly speaking is an equivalence class, but the language and the notation will refer either to the equivalence class or to one of its members, as indicated by the context. The members of such an equivalence class will sometimes be called *versions* of the class.

4. Conditional Expectations

If x is a real integrable random variable on a probability space or a real random variable bounded on one side, and if $\mathscr{F}$ is a σ algebra of measurable sets, the conditional expectation $E\{x|\mathscr{F}\}$ is the uniquely determined up to null sets random variable which is $\mathscr{F}$ measurable and satisfies

$$E\{E\{x|\mathscr{F}\};\Lambda\} = E\{x;\Lambda\} \quad \text{for } \Lambda\in\mathscr{F}. \tag{4.1}$$

The notation $E\{x|\mathscr{F}\}$ always refers to any one of the possible choices of the conditional expectation rather than to the equivalence class. Unless the contrary is stated, it will always be assumed in discussing this conditional expectation that x is integrable. The Radon–Nikodym theorem ensures the existence and uniqueness up to null sets of conditional expectations, and the following six properties follow easily from the defining properties.

(a) $E\{y|\mathscr{F}\} = y$ a.s. if y is $\mathscr{F}$ measurable, in particular, if y is a constant function.

(b) If y is bounded and $\mathscr{F}$ measurable, $E\{xy|\mathscr{F}\} = yE\{x|\mathscr{F}\}$ a.s.

(c) $E\{x+y|\mathscr{F}\} = E\{x|\mathscr{F}\} + E\{y|\mathscr{F}\}$ a.s.

(d) $x \le y$ a.s. implies that $E\{x|\mathscr{F}\} \le E\{y|\mathscr{F}\}$ a.s. and almost sure equality in the first inequality implies almost sure equality in the second.

These properties imply that the map $x \mapsto E\{x|\mathscr{F}\}$ is a positive linear idempotent transformation from L^1 into L^1, of norm 1.

(e1) If $\Lambda\in\mathscr{F}$ and if $x = y$ a.s. on Λ, then $E\{x|\mathscr{F}\} = E\{y|\mathscr{F}\}$ a.s. on Λ.

(e2) If $\Lambda\in\mathscr{F}\cap\mathscr{G}$ and if the σ algebras $\mathscr{F}$ and $\mathscr{G}$ have the same trace on Λ (that is, if a subset of Λ is in $\mathscr{F}$ if and only if the subset is in $\mathscr{G}$), then $E\{x|\mathscr{F}\} = E\{x|\mathscr{G}\}$ a.s. on Λ.

(f) If $\mathscr{F}\subset\mathscr{G}$, then

$$E\{E\{x|\mathscr{F}\}|\mathscr{G}\} = E\{E\{x|\mathscr{G}\}|\mathscr{F}\} = E\{x|\mathscr{F}\} \quad \text{a.s.} \tag{4.2}$$

The next properties are less immediate, and their proofs will be sketched.

(g) If x is independent of $\mathscr{F}$ (that is, if x is independent of every $\mathscr{F}$ measurable random variable), $E\{x|\mathscr{F}\} = E\{x\}$ a.s.

In fact, if y is the indicator function of a set Λ in $\mathscr{F}$, the random variables x, y are independent and

$$E\{x;\Lambda\} = E\{xy\} = E\{x\}E\{y\} = E\{E\{x\};\Lambda\} \quad \text{a.s.} \tag{4.3}$$

(h) Jensen's inequality for conditional expectations. If f is a convex function and if x and $f(x)$ are integrable, then

$$f[E\{x|\mathscr{F}\}] \le E\{f(x)|\mathscr{F}\} \quad \text{a.s.} \tag{4.4}$$

In fact, if $f(\xi) \ge a\xi + b$ for all ξ,

$$E\{f(x)|\mathscr{F}\} \geq aE\{x|\mathscr{F}\} + b \quad \text{a.s.,} \tag{4.5}$$

and since a convex function is the supremum of all the linear functions it majorizes, and even the supremum of a properly chosen countable subset of those linear functions, (4.5) implies (4.4). If $\mathscr{F}$ contains only the space and the empty set, (4.4) reduces to Jensen's inequality.

According to (h), if $p \geq 1$, the transformation $x \mapsto E\{x|\mathscr{F}\}$ takes L^p into itself. In particular, let $\mathfrak{M}$ be the space of square integrable $\mathscr{F}$ measurable random variables. The transformation maps L^2 onto $\mathfrak{M}$, and it is easy to verify that $x - E\{x|\mathscr{F}\}$ is orthogonal to $\mathfrak{M}$, so that the transformation acting on L^2 is the orthogonal projection onto $\mathfrak{M}$.

(i) If x is integrable, the family of all conditional expectations of x is uniformly integrable.

In proving this it can be assumed that x is positive (or replace x by $|x|$), and then if we write $x_{\mathscr{F}}$ for $E\{x|\mathscr{F}\}$ and if c is a strictly positive constant,

$$E\{x_{\mathscr{F}}; x_{\mathscr{F}} > c\} = E\{x; x_{\mathscr{F}} > c\}, \qquad P\{x_{\mathscr{F}} > c\} \leq \frac{E\{x_{\mathscr{F}}\}}{c} = \frac{E\{x\}}{c}. \tag{4.6}$$

The second relation ensures that, for sufficiently large c, $P\{x_{\mathscr{F}} > c\}$ is uniformly small as $\mathscr{F}$ varies and the left side of the first relation must then also be uniformly small for large c as $\mathscr{F}$ varies, and this property is precisely the uniform integrability property.

Alternative Proof

The following less direct proof displays a technique which has many applications. Since x is integrable, this function constitutes a uniformly integrable class; so there is a uniform integrability test function Φ (Appendix V) for which $E\{\Phi(|x|)\} < +\infty$. Apply Jensen's inequality (h) above for conditional expectations to obtain

$$E\{\Phi(E\{|x||\mathscr{F}\})\} \leq E\{E\{\Phi(|x|)|\mathscr{F}\} = E\{\Phi(|x|)\}.$$

Since the left side is uniformly bounded as $\mathscr{F}$ varies, the family of conditional expectations of x is uniformly integrable.

More generally a trivial extension of either proof shows that the class of conditional expectations of the random variables of a uniformly integrable family is uniformly integrable.

(j) If $\{x(t), t \in I\}$ is a family of integrable random variables and is directed upward neglecting null sets, then (whether or not $\operatorname{ess\,sup}_{t \in I} x(t)$ is integrable)

$$\operatorname*{ess\,sup}_{t\in I} E\{x(t)|\mathscr{F}\} = E\left\{\operatorname*{ess\,sup}_{t\in I} x(t)|\mathscr{F}\right\} \quad \text{a.s.} \tag{4.7}$$

This fact for integrals, that is, in the present context for expectations, is proved in Appendix IV.9. Since the family $\{E\{x(t)|\mathscr{F}\}, t\in I\}$ is directed upward neglecting null sets, the result for integrals yields, when $\Lambda\in\mathscr{F}$,

$$E\left\{\operatorname*{ess\,sup}_{t\in I} E\{x(t)|\mathscr{F}\};\Lambda\right\} = \sup_{t\in I} E\{x(t);\Lambda\} = E\left\{\operatorname*{ess\,sup}_{t\in I} x(t);\Lambda\right\}, \tag{4.8}$$

and this equation is equivalent to (4.7) if we choose the essential supremum on the left, as we can, to be $\mathscr{F}$ measurable.

The corresponding result for downward-directed families is deduced by replacing $x(t)$ by $-x(t)$.

Specialization. If x is an integrable random variable and if $x(\cdot)$ is a family of random variables, $E\{x|x(\cdot)\}$ is defined as $E\{x|\mathscr{F}\{x(\cdot)\}\}$. In particular, if $x(\cdot)$ is a finite set of random variables, say $x(1), \ldots, x(n)$, it follows (Appendix IV.3) that there is a Borel measurable function ϕ on $\mathbb{R}^n$ such that

$$E\{x|x(1),\ldots,x(n)\} = \phi[x(1),\ldots,x(n)] \quad \text{a.s.} \tag{4.9}$$

5. Conditional Expectation Continuity Theorem

Theorem. *Let $\{\mathscr{F}(n), n\in\mathbb{Z}^+\}$ $[\{\mathscr{F}(n), n\in -\mathbb{Z}^+\}]$ be a filtration of a probability space $(\Omega,\mathscr{F},P)$, and define $\mathscr{F}(+\infty) = \bigvee_0^\infty \mathscr{F}(n)$ $[\mathscr{F}(-\infty) = \bigcap_{-\infty}^0 \mathscr{F}(n)]$. Then if x is an integrable random variable on Ω, it follows that in both the almost everywhere and L^1 senses*

$$\lim_{n\to\infty} E\{x|\mathscr{F}(n)\} = E\{x|\mathscr{F}(+\infty)\} \tag{5.1}$$

in the forward case and

$$\lim_{n\to-\infty} E\{x|\mathscr{F}(n)\} = E\{x|\mathscr{F}(-\infty)\} \tag{5.2}$$

in the backward case.

This theorem has applications in surprisingly many apparently unrelated contexts. It is a martingale convergence theorem and as such will be derived again (Sections III.14 and III.17), but an independent proof will be given here to avert suspicion of circular reasoning. To simplify the notation, define $x(n) = E\{x|\mathscr{F}(n)\}$ for n in $\overline{\mathbb{Z}}^+$ or in $-\overline{\mathbb{Z}}^+$ according to the context.

To prove (5.1), choose integers $j < k < l < m$ and real numbers $a < b$, and define

$$\begin{aligned} v &= \inf\{n \geq l : x(n) \geq b\}, \\ [j,k,l;r] &= \left\{\min_{j \leq n \leq k} x(n) \leq a; v = r\right\} \in \mathscr{F}(r), \\ [j,k;l,m] &= \left\{\min_{j \leq n \leq k} x(n) \leq a; \max_{l \leq n \leq m} x(n) \geq b\right\} = \bigcup_{r=l}^{m} [j,k,l;r], \\ [a;b] &= \left\{\liminf_{n\to\infty} x(n) \leq a, \limsup_{n\to\infty} x(n) \geq b\right\}. \end{aligned} \tag{5.3}$$

Then $x(r) \geq b$ on $[j,k,l;r]$; so

$$\begin{aligned} \int_{[j,k;l,m]} x\,dP &= \sum_{r=l}^{m} \int_{[j,k,l;r]} x(r)\,dP \geq \sum_{r=l}^{m} \int_{[j,k,l;r]} b\,dP \\ &= bP\{[j,k;l,m]\}. \end{aligned} \tag{5.4}$$

Observe that if "$\leq a$" had been "$< a$" or "$\geq b$" had been "$> b$" in the above definitions, the change would carry over throughout with no change in the reasoning, and a trivial continuity argument for varying a and b shows that here and in further reasoning below the inequalities allowing equality imply those prescribing strict inequality and conversely. In view of this fact (5.4) yields, when successively $m \to +\infty$, $l \to +\infty$, $k \to +\infty$, $j \to +\infty$,

$$\int_{[a;b]} x\,dP \geq bP\{[a;b]\}. \tag{5.5}$$

If the triple x, a, b is replaced by $-x, -b, -a$ in (5.5), this inequality becomes

$$\int_{[a;b]} x\,dP \leq aP\{[a;b]\}, \tag{5.6}$$

and we conclude that $P\{[a;b]\} = 0$. The equality

$$\{x(\cdot) \text{ diverges}\} = \bigcup_{\substack{a<b \\ a,b \text{ rational}}} [a;b]$$

implies that there is almost everywhere convergence, say $\lim_{n\to\infty} x(n) = x'(+\infty)$, and since according to Section 4(i) the sequence $x(\cdot)$ is uniformly integrable, the limit function $x'(+\infty)$ is integrable and there is L^1 convergence (Appendix V). Finally, $x'(+\infty)$ can be chosen $\mathscr{F}(+\infty)$ measurable, and if Λ is in the algebra $\bigcup_0^\infty \mathscr{F}(n)$ which generates $\mathscr{F}(+\infty)$, then $\int_\Lambda x\,dP = \int_\Lambda x(n)\,dP$ for sufficiently large n so in view of the L^1 convergence of $x(\cdot)$

$$\int_\Lambda x\,dP = \int_\Lambda x'(+\infty)\,dP.$$

The equality must also be true for Λ in $\mathscr{F}(+\infty)$, so $x'(+\infty) = x(+\infty)$ almost surely, as was to be proved.

The proof of (5.2) is similar to that of (5.1) but is easier. If $a < b$ and $m < n$, the same argument as that in the preceding proof shows that if v is defined as in (5.3), if

$$[l, m] = \{x(-\infty) \le a, \max_{l \le n \le m} x(n) \ge b\},$$

and if $[a; b]$ is defined as in (5.3) except that $n \to -\infty$, then

$$\int_{[l,m]} x\,dP \ge bP\{[l, m]\}, \tag{5.4'}$$

so that if successively $l \to -\infty$, $m \to -\infty$,

$$\int_{[a;b]} x\,dP \ge bP\{[a; b]\}. \tag{5.5'}$$

Replacing x, a, b by $-x$, $-b$, $-a$ as in the preceding proof, we find that $P\{[a; b]\} = 0$, which implies that the uniformly integrable sequence $x(\cdot)$ is almost surely and L^1 convergent to some random variable $x'(-\infty)$. This random variable can be chosen $\mathscr{F}(-\infty)$ measurable. Finally, if $\Lambda \in \mathscr{F}(-\infty)$

$$\int_\Lambda x(-\infty)\,dP = \int_\Lambda x\,dP = \int_\Lambda x(n)\,dP$$

for all finite n and when $n \to -\infty$, the L^1 convergence of $x(\cdot)$ implies that these three integrals are also equal to $\int x'(-\infty)\,dP$; so $x'(-\infty) = x(-\infty)$ almost surely, as was to be proved.

Generalization of Theorem 5 to Linearly Ordered Index Sets

Let $(I, \le)$ be a linearly ordered set and let $\{\mathscr{G}(t), t \in I\}$ be a monotone family of σ algebras of measurable subsets of a probability space. Suppose that I does not have a last element and denote by $\mathscr{G}(\infty)$ either $\bigvee_{t \in I} \mathscr{G}(t)$ or $\bigcap_{t \in I} \mathscr{G}(t)$ according as $\mathscr{G}(\cdot)$ is monotone increasing or decreasing. If x is an integrable random variable, we shall now show, generalizing Theorem 5, that

$$\lim_{t\uparrow} E\{x|\mathscr{G}(t)\} = E\{x|\mathscr{G}(\infty)\} \tag{5.7}$$

both as an almost everywhere essential limit and an L^1 limit. We shall use the notation $x(t) = E\{x|\mathscr{G}(t)\}$ for $t \in I \cup \{\infty\}$.

Case (a): I is countable. In this case there is according to Appendix Theorem IV.10 an increasing sequence $t_{\bullet}$ in I for which the essential limits inferior and superior of $x(\cdot)$ along I are, respectively, the same as along $t_{\bullet}$. The generalization of Theorem 5 for countable I thus follows from Theorem 5. Moreover the limit in (5.7) can be taken as an ordinary almost everywhere pointwise limit because I is countable.

Case (b): General case. If I has a countable cofinal subset, there is even (Appendix III.13) a countable cofinal subset J of I such that $x(\cdot)$ has the same essential limits inferior and superior, respectively, along J as along I. According to case (a), there is almost everywhere convergence along J, and it follows that there is essential order convergence along I. The identification of the limit and the L^1 convergence are left to the reader. If I has no countable cofinal subset, we prove a stronger result than (5.7): there is a point t_0 of I such that for $t > t_0$

$$P\{x(t) = x(t_0)\} = 1. \tag{5.8}$$

To see this when $G(\cdot)$ is increasing, observe that (Appendix IV.2) there is a countable subset J of I such that $x(\infty)$ is $\bigvee_{t \in J} G(t)$ measurable, and we leave to the reader the proof that (5.8) is true with t_0 any upper order bound of J. The following argument is more instructive and shows that t_0 exists satisfying (5.8) whether $G(\cdot)$ is increasing or decreasing. An order-preserving map ϕ will be exhibited in Section III.4, involving only elementary inequalities, taking I into a subset I' of $\mathbb{R}$, with the property that $\phi(s) = \phi(t)$ if and only if $P\{x(s) = x(t)\} = 1$. The set I' must have a last element t' because I' has a cofinal sequence $s'_{\bullet}$, and if $s_{\bullet}$ is a sequence in I with $\phi(s_{\bullet}) = s'_{\bullet}$ and if t is an order upper bound of $s_{\bullet}$, then $\phi(t) = t'$ is the last element of I'. Each element t_0 of $\phi^{-1}(t')$ satisfies (5.8).

6. Fatou's Lemma for Conditional Expectations

Lemma. *Define I, $\mathscr{G}(\cdot)$ as in the generalization of Theorem 5, and let $\{x(t), t \in I\}$ be a family of positive extended real-valued random variables. Then*

$$\operatorname{ess}\liminf_{t\uparrow} E\{x(t)|\mathscr{G}(t)\} \geq E\{\operatorname{ess}\liminf_{t\uparrow} x(t)|\mathscr{G}(\infty)\} \quad \text{a.s.} \tag{6.1}$$

In particular, if $I = \mathbb{Z}^+$,

$$\liminf_{n\to\infty} E\{x(n)|\mathscr{G}(n)\} \geq E\{\liminf_{n\to\infty} x(n)|\mathscr{G}(\infty)\} \quad \text{a.s.} \tag{6.1'}$$

For every k in $\mathbb{Z}^+$ and s in I the left side of (6.1) is almost surely at least

$$\operatorname{ess}\liminf_{t\uparrow} E\{k \wedge \operatorname{ess}\inf_{r\geq s} x(r)|\mathscr{G}(t)\} = E\{k \wedge \operatorname{ess}\inf_{r\geq s} x(r)|\mathscr{G}(\infty)\} \quad \text{a.s.} \tag{6.2}$$

Inequality (6.1) follows from the fact that the right side of (6.2) increases as k and s increase and has the essential limit $E\{\operatorname{ess}\liminf_{t\uparrow} x(t)|\mathscr{G}(\infty)\}$ according to (4.7).

7. Dominated Convergence Theorem for Conditional Expectations

Theorem. *Let $\{\mathscr{G}(n),\ n\in\mathbb{Z}^+\}$ be a monotone sequence of σ algebras of measurable subsets of a probability space, and denote by $\mathscr{G}(\infty)$ either $\bigvee_{n\in\mathbb{Z}^+}\mathscr{G}(n)$ or $\bigcap_{n\in\mathbb{Z}^+}\mathscr{G}(n)$ according as $\mathscr{G}(\cdot)$ is monotone increasing or decreasing. Let $\{x(n),\ n\in\mathbb{Z}^+\}$ be a sequence of random variables on the space, with almost sure limit $x(\infty)$ and $E\{\sup_{n\in\mathbb{Z}^+}|x(n)|\} < +\infty$. Then*

$$\lim_{n\to\infty} E\{x(n)|\mathscr{G}(n)\} = E\{x(\infty)|\mathscr{G}(\infty)\} \quad \text{a.s.} \tag{7.1}$$

To prove this theorem, apply Fatou's lemma for conditional expectations to the sequences

$$\{\sup_{m\in\mathbb{Z}^+}|x(m)| + x(n),\ n\in\mathbb{Z}^+\}, \qquad \{\sup_{m\in\mathbb{Z}^+}|x(m)| - x(n),\ n\in\mathbb{Z}^+\}.$$

Application. Let $\mathscr{F}$ be a σ algebra of measurable sets of a probability space. Let x and y be random variables on the space with x independent of $\mathscr{F}$, y measurable with respect to $\mathscr{F}$. Let f be a bounded Borel measurable function on $\mathbb{R}^2$, and define f_1 on $\mathbb{R}$ by $f_1(\eta) = E\{f(x,\eta)\}$. Then f_1 is Borel measurable, and $E\{f(x,y)|\mathscr{F}\} = f_1(y)$ almost surely. To prove this assertion, let Γ be the class of bounded Borel measurable functions f for which this assertion is true. The class Γ includes the functions of the form $(\xi,\eta)\mapsto g(\xi)h(\eta)$ with g and h bounded Borel measurable functions on $\mathbb{R}$ because in this case $f_1(\eta) = E\{g(x)\}h(\eta)$, and by Section 4(b) and (g)

$$E\{g(x)h(y)|\mathscr{F}\} = E\{g(x)|\mathscr{F}\}h(y) = f_1(y) \quad \text{a.e.} \tag{7.2}$$

The class Γ is linear and is closed under bounded pointwise convergence in view of the dominated convergence theorem for conditional expectations. Hence Γ is the class of bounded Borel measurable functions on $\mathbb{R}^2$, as was to be proved.

8. Stochastic Processes, "Evanescent," "Indistinguishable," "Standard Modification," "Nearly"

Let $(\Omega, \mathscr{F}, P)$ be a probability space, and let I be an arbitrary set (to be used as the parameter set of a family of random variables). A subset $\dot{A}$ of $I \times \Omega$ is called *evanescent* if there is a set A in $\mathscr{F}$ for which $P\{A\} = 0$ and $\dot{A} \subset I \times A$. A subset of an evanescent set is evanescent, and a countable union of evanescent sets is evanescent. For many purposes evanescent subsets of $I \times \Omega$ are counterparts of the polar subsets of the state space in classical potential theory, and we shall accordingly describe a relation true on $I \times \Omega$ up to an evanescent set as true quasi everywhere.

A stochastic process on $(\Omega, \mathscr{F}, P, I)$ is a family of random variables $\{x(t), t \in I\}$ with a common state space, defined on $(\Omega, \mathscr{F}, P)$, with parameter set I. The random variable $x(t)$ of this process has the value $x(t, \omega)$ at the point ω of Ω, and the process thus defines a function from $I \times \Omega$ into the state space of the process, that is, the state space of the random variables, with the property that each function $x(t) = x(t, \cdot)$ is measurable. Two processes $x(\cdot)$ and $y(\cdot)$ with common probability space, parameter set, and state space are called *indistinguishable* if $x(\cdot, \cdot) = y(\cdot, \cdot)$ quasi everywhere on $I \times \Omega$, that is, if

$$P\{\omega: x(t, \omega) = y(t, \omega) \text{ for all } t\} = 1. \tag{8.1}$$

Two subsets of $I \times \Omega$ are called *indistinguishable* if their indicator functions determine indistinguishable processes, that is, if the sets differ by an evanescent set. Two processes $x(\cdot)$ and $y(\cdot)$ with common probability space, parameter set, and state space are said to be *standard modifications* of each other if

$$P\{\omega: x(t, \omega) = y(t, \omega)\} = 1 \tag{8.2}$$

for all t. The adjective "standard" is sometimes omitted.

Sums of processes, functions of a process, and so on are defined in the obvious way in terms of the random variables of the processes.

The concept of an adapted family $\{x(t), \mathscr{F}(t), t \in I\}$ of functions on a measurable space was defined in Section 1. If the measurable space is a probability space, the family becomes an adapted stochastic process, and the above notation always implies adaptedness unless the contrary is stated.

A stochastic process indistinguishable from one having a specified character will sometimes be said to have *nearly* that character, and the corresponding description will be used for subsets of $I \times \Omega$. Thus a nearly progressively measurable process is a process indistinguishable from a progressively measurable process.

Normalizing Hypotheses on Filtrations

Let $\{\mathscr{F}(t), t \in I\}$ be a filtration of a measure space, and suppose that I has a first point, denoted by 0. If the given measure is complete, as is our convention for probability measure spaces, the hypothesis that $\mathscr{F}(0)$ contains the null sets implies that the restriction of the given measure to each σ algebra $\mathscr{F}(t)$ is complete, but the converse is false. In practically every investigation involving a filtration the parameter set has a first element, and it is possible without loss of generality to suppose that the first σ algebra contains the null sets, at the cost of enlarging each σ algebra $\mathscr{F}(t)$ to the σ algebra generated by $\mathscr{F}(t)$ and the null sets. If the parameter set is $\mathbb{R}^+$, it is usually also possible without loss of generality to replace $\mathscr{F}(\cdot)$ by the right continuous filtration $\mathscr{F}^+(\cdot)$.

Progressively Measurable Stochastic Processes

If $\{x(t), \mathscr{F}(t), t \in \mathbb{R}^+\}$ is a progressively measurable stochastic process every sample function $x(\cdot, \omega)$ is necessarily Borel measurable. If this process is only nearly progressively measurable, it need not be progressively measurable, but almost every sample function is Borel measurable. For example, define a real-valued process with $x(\cdot, \omega)$ arbitrary for ω in some null set of a probability space, define $x(\cdot, \omega) = 0$ otherwise, and define $\mathscr{F}(t)$ for all t as the σ algebra consisting of the null sets and their complements. The process $\{x(\cdot), \mathscr{F}(\cdot)\}$ is then indistinguishable from the identically 0 progressively measurable process but is progressively measurable only in trivial cases. It is nevertheless true that most theorems involving progressively measurable processes are valid with unimportant modifications for nearly progressively measurable processes.

Almost Surely [Right] [[Left]] Continuous Processes and Progressive Measurability

If $\{x(t), t \in \mathbb{R}^+\}$ is a stochastic process with a topological state space the process will be called [almost surely] continuous or right continuous or left continuous if [almost every] sample function has the stated property. If $\{x(t), \mathscr{F}(t), t \in \mathbb{R}^+\}$ is almost surely [right] [[left]] continuous and adapted and if $\mathscr{F}(0)$ contains the null sets the sample function $x(\cdot, \omega)$ can be redefined, say by setting $x(t, \omega) = x(0, \omega)$ for all t when ω is in the exceptional null set, to obtain a [right] [[left]] continuous process adapted to $\mathscr{F}(\cdot)$ and indistinguishable from the given process. If in addition the state space is Polish, this modified process is progressively measurable according to Section 2.

It would be consistent with our conventions to call an almost surely right or left continuous process *nearly* right or left continuous, but the *almost surely* terminology is the customary one.

Stochastic Processes in the Essential Order

The class of extended real-valued processes with a common probability space and parameter set is ordered by the *essential* order in which $y(\cdot)$ is a majorant of $x(\cdot)$, denoted by $x(\cdot) \le y(\cdot)$ or $y(\cdot) \ge x(\cdot)$, if $P\{x(t) \le y(t)\} = 1$ for all t. This order is thus an ordering of equivalence classes in which two processes are identified if each is a standard modification of the other. The usual inexact language in which a "process" may actually be an equivalence class will be used when there is no danger of confusion.

Essential Order Infima and Suprema

If $\Gamma: \{x_\alpha(\cdot), \alpha \in J\}$ is a family of stochastic processes with a common probability space, common parameter set, and extended real-valued random variables, the essential order infimum of the family, denoted by $\operatorname{ess\,inf}_{\alpha \in J} x_\alpha(\cdot)$ or $\operatorname{ess\,inf}\Gamma$, is the process whose tth random variable is $\operatorname{ess\,inf}_{\alpha \in J} x_\alpha(t)$ [chosen to be $\mathscr{F}(t)$ measurable if Γ is specified as a family of processes adapted to a common filtration $\mathscr{F}(\cdot)$]. The process $\operatorname{ess\,sup}\Gamma$ is defined and denoted dually. These processes are uniquely determined up to standard modifications and therefore are unique in the essential order.

L^p Bounded and Uniformly Integrable Processes

A stochastic process is said to be *L^p bounded* if the family of its random variables is L^p bounded and is said to be *uniformly integrable* if the family of its random variables is uniformly integrable.

Sample Function Integrals

If $\{x(t), \mathscr{F}(t), t \in \mathbb{R}^+\}$ is a progressively measurable stochastic process with state space the extended reals, the process sample functions are Borel measurable. If the process random variables are positive, the sample function integrals are well defined, that is, meaningful, and if $y(t) = \int_0^t x(s) l_1(ds)$, that is, if for each point ω of the probability space on which the process $x(\cdot)$ is defined $y(t, \omega) = \int_0^t x(s, \omega) l_1(ds)$, then the process $\{y(\cdot), \mathscr{F}(\cdot)\}$ is progressively measurable (Fubini's theorem). This integral process is also well defined and progressively measurable if instead of supposing that the random variables are positive, it is supposed that the $x(\cdot)$ process sample functions are l_1 absolutely integrable over finite intervals. If the process $\{x(\cdot), \mathscr{F}(\cdot)\}$ is not supposed progressively measurable but merely indistinguishable from a progressively measurable process, the above remarks are modified in the obvious way.

9. The Hitting of Sets and Progressive Measurability

Theorem. *Let $(\Omega, \mathscr{F}, P; \mathscr{F}(t), t \in \mathbb{R}^+)$ be a filtered probability space, suppose that the restriction of P to $\mathscr{F}(t)$ is complete for all t, and let c be a strictly positive constant.*

(a) *If A'' is a subset of $\mathbb{R}^+ \times \Omega$, analytic over the class of progressively measurable sets, then the projection of $A'' \cap ([0, c] \times \Omega)$ on Ω is in $\mathscr{F}(c)$.*

Let $\{x(\cdot), \mathscr{F}(\cdot)\}$ be a progressively measurable stochastic process on $(\Omega, P, \mathscr{F})$, with an arbitrary state space $(X, \mathscr{X})$.

(b) *If A' is a subset of $\mathbb{R}^+ \times X$ analytic over $\mathscr{B}(\mathbb{R}^+) \times \mathscr{X}$, then*

$$\{\omega: (t, x(t, \omega)) \in A' \text{ for some } t \leq c\} \in \mathscr{F}(c). \tag{9.1}$$

(c) *If A is a subset of X analytic over $\mathscr{X}$, then*

$$\{\omega; x(t, \omega) \in A \text{ for some } t \leq c\} \in \mathscr{F}(c). \tag{9.2}$$

Assertion (c) is a special case of (b) because in (b) the set A' can be chosen as $[0, c] \times A$. Assertion (b) is a special case of (a) by the following argument. The map $(t, \omega) \mapsto (t, x(t, \omega))$ from the measurable space $\mathbb{R}^+ \times \Omega$ coupled with the class of progressively measurable sets into the measurable space ($\mathbb{R}^+ \times X$, $\mathscr{B}(\mathbb{R}^+) \times \mathscr{X}$) is measurable because the inverse image of any product set $[0, b] \times X_0$ with X_0 in $\mathscr{X}$ is a progressively measurable set. Hence (Appendix Theorem I.8) the inverse image of any set A' analytic over $\mathscr{B}(\mathbb{R}^+) \times \mathscr{X}$ is a set A'' analytic over the class of progressively measurable sets. Finally the projection of $A'' \cap ([0, c] \times \Omega)$ is then the set in (9.1); so (a) implies (b). To prove (a), observe that by Theorem 7 of Appendix I the projection of A'' on Ω is analytic over $\mathscr{F}(c)$ and therefore is in $\mathscr{F}(c)$ by Lusin's theorem (Theorem 4 of Appendix II because the restriction of P to $\mathscr{F}(c)$ is a complete measure.

Variant of the Theorem

In (a) the projection of $A'' \cap ([0, c[\times \Omega)$ on Ω and in (b) and (c) the sets in (9.1) and (9.2) with "$t \leq c$" replaced by "$t < c$" are all in $\mathscr{F}^-(c) = \bigvee_{t<c} \mathscr{F}(t)$. The method of proof of the theorem yields this variant, or the variant can be deduced by an application of the theorem with c replaced by $c - 1/n$, $n \geq 1$.

Generalization

If in (b) and (c) the process is supposed merely indistinguishable from a progressively measurable process, the conclusion remains valid if $\mathscr{F}(0)$ contains the P null sets. This added condition implies the validity of the condition already imposed, that the restriction of P to $\mathscr{F}(t)$ be complete, because P as always is itself supposed complete.

Change of Origin

The analysis of the hitting of a set in the parameter interval $[0, c]$ or $[0, c[$ is valid with trivial changes for hitting in an interval $[b, c]$ or $[b, c[$ with $0 < b < c$.

10. Canonical Processes and Finite-Dimensional Distributions

Let $(X, \mathscr{X})$ be a measurable space, let I be a space and let $\mathring{\Omega}$ be a subset of the space X^I of functions $\mathring{\omega}$ from I into X. Denote $\mathring{\omega}(t)$ by $\mathring{x}(t, \mathring{\omega})$, and define $\mathring{\mathscr{F}} = \mathscr{F}\{\mathring{x}(t), t \in I\}$; that is, $\mathring{\mathscr{F}}$ is the σ algebra generated by the algebra $\bigcup \mathscr{F}\{\mathring{x}(t_1), \ldots, \mathring{x}(t_n)\}$, where the union is over all finite subsets $t_.$ of I. If $\mathring{P}$ is a probability measure on $\mathring{\mathscr{F}}$, extended by completion to a complete probability measure on $\mathring{\mathscr{F}}^* \supset \mathring{\mathscr{F}}$, $\{\mathring{x}(t), t \in I\}$ is a stochastic process on $(\mathring{\Omega}, \mathring{\mathscr{F}}^*, \mathring{P})$. A process defined in this way, by means of a measure on a subset of the function space X^I, will be called *canonical.*

If $\{x(t), t \in I\}$ is a stochastic process with state space $(X, \mathscr{X})$ on a probability space $(\Omega, \mathscr{F}, P)$, a *finite-dimensional distribution* of the process is the distribution of a finite set of the process random variables,

$$A \mapsto p(t_1, \ldots, t_n; A) = P\{[x(t_1), \ldots, x(t_n)] \in A\} \qquad (A \in \mathscr{X}^n). \quad (10.1)$$

An associated canonical process with the same parameter set, state space, and finite-dimensional distributions can be defined as follows. Set $\mathring{\Omega} = X^I$. The map $\phi\colon \omega \mapsto x(\cdot, \omega)$ is measurable from $(\Omega, \mathscr{F})$ into $(\mathring{\Omega}, \mathring{\mathscr{F}})$ because

$$\phi^{-1}\{[\mathring{x}(t_1), \ldots, \mathring{x}(t_n)] \in A\} = \{[x(t_1), \ldots, x(t_n)] \in A\} \in \mathscr{F} \qquad (A \in \mathscr{X}^n). \tag{10.2}$$

Define $\mathring{P}$ on $\mathring{\mathscr{F}}$ as the measure induced by P; that is, $\mathring{P}\{\mathring{\Lambda}\} = P\{\phi^{-1}(\mathring{\Lambda})\}$. If as usual P is supposed complete, it follows that $\phi^{-1}(\mathring{\mathscr{F}}^*) \subset \mathscr{F}$; so ϕ is a measurable map from $(\Omega, \mathscr{F})$ into $(\mathring{\Omega}, \mathring{\mathscr{F}}^*)$. Note that $\mathscr{F}$ may be strictly larger than $\phi^{-1}(\mathring{\mathscr{F}}^*)$ so that there may be sample function properties defining measurable subsets of Ω but not of $\mathring{\Omega}$. Sample function properties involving uncountably many parameter values are obvious candidates.

If I is a topological space and if $(X, \mathscr{X})$ is a topological space with $\mathscr{X}$ the σ algebra generated by the open subsets of X, suppose that $x(\cdot)$ in the preceding paragraph is a continuous process. Then we can choose an associated continuous canonical process by setting $\mathring{\Omega} = \mathring{\Omega}_c$, the space of continuous functions from I into X, a subset of X^I. The map ϕ above is now from Ω into $\mathring{\Omega}_c$, and P induces a measure on this function space. The family $\mathring{x}(\cdot)$ of coordinate variables on $\mathring{\Omega}_c$ is the canonical continuous process associated with $x(\cdot)$. If $x(\cdot)$ is an almost surely continuous process, that is, if there is a measurable subset Ω_c of Ω with $P\{\Omega_c\} = 1$ whose corresponding sample functions are continuous, the preceding procedure applied to Ω_c instead of Ω yields a canonical continuous process associated with $x(\cdot)$. If I is ordered, with a first point α, and if the distribution of $x(\alpha)$ is supported by some measurable set A, it is sometimes convenient to restrict the associated canonical process further by defining $\mathring{\Omega}_c$ as the class of continuous functions from I into X with initial value in A. If I is an interval of $\mathbb{R}$ and if $x(\cdot)$ is an almost surely right continuous process, the preceding discussion for continuous processes has the obvious modification in which $\mathring{\Omega}_c$ is replaced by the space of right continuous functions from I into X.

The class of finite-dimensional distributions of a stochastic process $x(\cdot)$ satisfies two consistency conditions: (a) if $A \in \mathscr{X}^n$

$$p(t_1, \ldots, t_{n+1}; A \times X) = p(t_1, \ldots, t_n; A) \tag{10.3}$$

and (b) if $t_1', \ldots, t_n'$ is a permutation of $t_1, \ldots, t_n$, the measure $p(t_1, \ldots, t_n; \cdot)$ becomes the measure $p(t_1', \ldots, t_n'; \cdot)$ when the coordinates of X^n undergo the same permutation.

Kolmogorov's Construction of Canonical Processes (see the Historical Notes to this Section)

Let $(X, \mathscr{X})$ be a Polish space coupled with its σ algebra of Borel sets. Then $(X^n, \mathscr{X}^n)$ is for $n \geq 1$ also a Polish space coupled with its σ algebra of Borel sets, and therefore (Appendix IV.11) every measure on $\mathscr{X}^n$ is inner regular. Kolmogorov showed that for an arbitrary parameter set I, state space $(X, \mathscr{X})$, and arbitrary finite-dimensional distributions for this choice of parameter set and state space, satisfying (a) and (b) above, there is a canonical stochastic process with $\mathring{\Omega} = X^I$ and the given finite-dimensional distributions. (Kolmogorov had $X = \mathbb{R}$, but the above inner regularity property was all his proof needed.) His method of proof was to define $\mathring{P}$ on the algebra $\bigcup \mathscr{F}\{\mathring{x}(t_1), \ldots, \mathring{x}(t_n)\}$ by

$$\mathring{P}\{[\mathring{x}(t_1), \ldots, \mathring{x}(t_n)] \in A\} = p(t_1, \ldots, t_n; A) \qquad (A \in \mathscr{X}^n) \tag{10.4}$$

and then to show that $\mathring{P}$ was countably additive. The extension of $\mathring{P}$ to $\mathring{\mathscr{F}}$ followed by completion makes $\mathring{P}$ the measure on function space inducing the desired process $\mathring{x}(\cdot)$.

Specification of a Stochastic Process and Analysis of Its Sample Functions

A stochastic process is usually specified (i) by its state space, parameter set, and finite-dimensional distributions, and sometimes in addition (ii) by such further properties as continuity, progressive measurability, and so on. In discussing process sample functions all probabilities should be derived from these specifications, and if there is a choice of probability measure space, process sample function probabilities should not depend on the choice. In analyzing the meaning of this admonition suppose first that there are no specifications under (ii) above. Let $(\Omega, \mathscr{F}, P)$ be a probability space, and let $\{x(t), t \in I\}$ be a process on this space with arbitrary state space $(X, \mathscr{X})$ and specified finite-dimensional distributions $p(\cdot;\cdot)$. If $t_1, \ldots, t_n$ are parameter values, if $A \in \mathscr{X}^n$, and if $\Lambda = \{[x(t_1), \ldots, x(t_n)] \in A\}$, then $P\{\Lambda\}$ is determined by the finite-dimensional distributions; so these distributions determine P on the algebra of all such sets Λ and so determine P on the σ algebra $\mathscr{F}' = \mathscr{F}\{x(t), t \in I\}$ generated by this algebra. Finally the finite-dimensional distributions determine P on the class of $\mathscr{F}$ sets obtained by completing the restriction of P to $\mathscr{F}'$. That is to say in the notation of the beginning of this section with $\mathring{\Omega} = X^I$, the finite-dimensional distributions determine P on $\phi^{-1}(\mathring{\mathscr{F}}^*)$. A set Λ not in $\phi^{-1}(\mathring{\mathscr{F}}^*)$ but defined by conditions on $x(\cdot)$ sample functions, that is, $\Lambda = \phi^{-1}(\mathring{\Lambda})$ with $\mathring{\Lambda}$ not in $\mathring{\mathscr{F}}^*$, may be measurable for some choices of $(\Omega, \mathscr{F}, P)$ and $x(\cdot)$, with the specified finite-dimensional distributions, but if so, $P\{\Lambda\}$ may depend on the choice. (See Section 11 for an example of this possibility.)

This discussion is modified in the obvious way if there are specifications prescribed under (ii) above. For example, suppose that I is an interval of $\mathbb{R}$, that $(X, \mathscr{X}) = (\mathbb{R}, \mathscr{B}(\mathbb{R}))$, and that the process $x(\cdot)$ is almost surely continuous. Define the associated continuous canonical process, setting $\mathring{\Omega} = \mathring{\Omega}_c$. The σ algebra $\mathring{\mathscr{F}}^*$ is now a σ algebra of subsets of $\mathring{\Omega}_c$, and new sample function properties can be analyzed. For example, it is now true that if I' is a countable dense subset of I and if

$$\mathring{\Lambda} = \left\{\sup_{t \in I} \mathring{x}(t) \le 1\right\}, \text{ then } \mathring{\Lambda} = \left\{\sup_{t \in I'} \mathring{x}(t) \le 1\right\} \in \mathring{\mathscr{F}} \text{ and, up to a null set,}$$

$$\phi^{-1}(\mathring{\Lambda}) = \left\{\sup_{t \in I} x(t) \le 1\right\} = \left\{\sup_{t \in I'} x(t) \le 1\right\} \in \phi^{-1}(\mathring{\mathscr{F}});$$

so $P\{\phi^{-1}(\mathring{\Lambda})\}$ is defined and equal to $\mathring{P}\{\mathring{\Lambda}\}$ no matter what the choice of $(\Omega, \mathscr{F}, P)$ and the process $x(\cdot)$ as long as the process has the given finite-dimensional distributions and is almost surely continuous.

11. Choice of the Basic Probability Space

In each example $x(\cdot)$ is a stochastic process with parameter set I and state space $(X, \mathscr{X})$ on the probability space $(\Omega, \mathscr{F}, P)$.

EXAMPLE (a). Let $I = \mathbb{Z}^+$, $(X, \mathscr{X}) = (\mathbb{R}, \mathscr{B}(\mathbb{R}))$, and let $x(0)$, $x(1)$, ... be mutually independent random variables. The theorem that the probability of convergence of $\Sigma_0^\infty x(n)$ is either 0 or 1 does not depend for its validity on the choice of probability space, and the value of this probability, 0 or 1 as the case may be, depends only on the distributions of the random variables, that is, on the finite-dimensional distributions of the process $x(\cdot)$.

Example (b). Let $I = \Omega = X = \mathbb{R}^+$, let $X = \mathscr{B}(\mathbb{R}^+)$, let $\mathscr{F}$ be the σ algebra of subsets of Ω measurable for the completion P of a probability measure on $\mathscr{B}(\mathbb{R}^+)$, and suppose that all singletons are P null. Define $\mathscr{F}(t) = \mathscr{F}$ for $t \in \mathbb{R}^+$, and define

$$x(t, \omega) = \begin{cases} 1 & \text{if } t = \omega, \\ 0 & \text{if } t \neq \omega, \end{cases} \qquad x_1(\cdot, \cdot) = 0 \text{ on } \mathbb{R}^+ \times \mathbb{R}^+.$$

The processes $\{x(\cdot), \mathscr{F}(\cdot)\}$ and $\{x_1(\cdot), \mathscr{F}(\cdot)\}$ are progressively measurable. Moreover these two processes have the same finite-dimensional distributions, and in fact $x_1(\cdot)$ is a standard modification of $x(\cdot)$. Since no $x(\cdot)$ process sample function is continuous and every $x_1(\cdot)$ process sample function is continuous, an assertion involving the value of the probability of a sample function continuity property for a progressively measurable process must involve specification of the process beyond its finite-dimensional distributions. The hypotheses of Theorem 9(b) and (c) are satisfied by both $\{x(\cdot), \mathscr{F}(\cdot)\}$ and $\{x_1(\cdot), \mathscr{F}(\cdot)\}$; so for each process that theorem, whose conclusion is trivial for these processes, asserts that if A is an analytic subset of $X = \mathbb{R}^+$, the probability that a sample function ever takes on a value in A, that is, the probability that a sample path hits A, is well defined. If A is the set $\mathbb{R} - \{0\}$, this hitting probability is 1 for the $x(\cdot)$ process and 0 for the $x_1(\cdot)$ process. Thus in Theorem 9(b) and (c) if the finite-dimensional distributions of $x(\cdot)$ are specified, the probability of hitting A depends on which progressively measurable process with those finite-dimensional distributions is chosen.

12. The Hitting of Sets by a Right Continuous Process

Example (a). If $\{x(t), t \in \mathbb{R}^+\}$ is a right continuous process with a topological state space and if $c > 0$, the probability that an $x(\cdot)$ path hits an open set B at some time $\leq c$ is the probability of the set

$$\bigcup_{t \leq c} \{x(t) \in B\} = \bigcup_{\substack{0 \leq r \leq 1 \\ r \text{ rational}}} \{x(rc) \in B\}. \tag{12.1}$$

Since only countably many parameter values are involved on the right-hand side the finite-dimensional distributions of the process determine the desired hitting probability. Thus for a right continuous (or almost surely right continuous) process the probability of hitting the open set B by time c

does not depend on the choice of almost surely right continuous process with the given finite-dimensional distributions.

EXAMPLE (b) (The hitting of an F_σ set by a continuous process). If $\{x(t), t\in\mathbb{R}^+\}$ is a continuous process with a metric state space, if A is a closed subset of the state space, and if B_n for $n \geq 1$ is the open set of points at distance $< 1/n$ from A, the probability of hitting A by time c is the probability of hitting every B_n by time c, and this probability is that of the set

$$\bigcup_{t\leq c}\{x(t)\in A\} = \bigcap_{n=1}^{\infty}\bigcup_{\substack{0\leq r\leq 1\\ r \text{ rational}}}\{x(rc)\in B_n\}. \tag{12.2}$$

Again only countably many parameter values are involved on the right-hand side. If A is a countable union of closed sets, the set A is hit by time c if and only if some summand is. Thus for a continuous (or almost surely continuous) process with a metric state space the probability of hitting an F_σ set by time c does not depend on the choice of almost surely continuous process with the given finite-dimensional distributions.

These examples should be compared with the Section 11 examples which show that two progressively measurable processes with state space the line and identical finite-dimensional distributions may have quite different probabilities for hitting an open set. The following discussion shows that the hypothesis of right continuity of a process prevents such a circumstance for the hitting of analytic subsets of a Polish state space.

Let $(X,\mathscr{X})$ be a Polish space together with its Borel subsets, let $\{x(t), \mathscr{F}(t), t\in\mathbb{R}^+\}$ be a right continuous adapted process with $(X,\mathscr{X})$ as state process, on the probability measure space $(\Omega,\mathscr{F},P)$, and let $\mathring{x}(\cdot)$ be the associated right continuous canonical process with the same state space and finite-dimensional distributions, on the probability measure space $(\mathring{\Omega},\mathring{\mathscr{F}},\mathring{P})$. As usual it is supposed that all process measures are complete. Suppose that each σ algebra has been enlarged if necessary to make $\mathscr{F}(0)$ contain the null sets. Define $\mathring{\mathscr{F}}(t)$ as the σ algebra generated by $\mathscr{F}\{\mathring{x}(s), s\leq t\}$ and the $\mathring{P}$ null sets. The processes $\{x(\cdot),\mathscr{F}(\cdot)\}$ and $\{\mathring{x}(\cdot),\mathring{\mathscr{F}}(\cdot)\}$ are progressively measurable according to Section 2; thus (by Theorem 9) if A' is a subset of $\mathbb{R}^+\times X$ analytic over $\mathscr{B}(\mathbb{R}^+)\times\mathscr{X}$, if $c>0$, if

$$\Lambda = \{\omega\colon (t, x(t,\omega))\in A' \text{ for some } t\leq c\}, \tag{12.3}$$

and if $\mathring{\Lambda}$ is defined similarly for $\mathring{x}(\cdot)$, then $\Lambda\in\mathscr{F}(c)$ and $\mathring{\Lambda}\in\mathring{\mathscr{F}}(c)$. Moreover since $\Lambda = \phi^{-1}(\mathring{\Lambda})$ (notation of Section 10), it follows that $P\{\Lambda\} = \mathring{P}\{\mathring{\Lambda}\}$. Thus $P\{\Lambda\}$ depends on the finite-dimensional distributions and the right continuity hypothesis but not otherwise on the choice of $(\Omega,\mathscr{F},P)$ and $x(\cdot)$. In particular, if A is an analytic subset of X and $A' = [0,c]\times A$, it follows that the probability of the set

$$\{\omega: x(t,\omega) \in A \text{ for some } t \leq c\} \tag{12.4}$$

depends only on the finite-dimensional distributions and the right continuity hypothesis.

If the process $x(\cdot)$ is supposed only almost surely right continuous, we can omit the P null subset of Ω corresponding to non-right continuous sample functions to get the same results as in the right continuous case.

If in (12.3) and (12.4) the restriction on t is $t < c$ instead of $t \leq c$, all of the above results remain valid, in fact can be slightly sharpened since the sets in (12.3) and (12.4) are then in $\mathscr{F}^-(c)$.

13. Measurability versus Progressive Measurablity of Stochastic Processes

Let $(\Omega, \mathscr{F}, P)$ be a probability space. We shall need a definition of convergence in measure for random variables from this space into $\overline{\mathbb{R}}$ when the random variables may have the values $\pm\infty$ with strictly positive probability. In this context we shall say that a sequence $x_{\bullet}$ of random variables converges in measure to x if the sequence $\arctan x_{\bullet}$ converges in measure to x in the usual sense, equivalently, if

$$\lim_{n\to\infty} E\{|\arctan x - \arctan x_n|\} = 0.$$

The metric

$$d(x,y) = E\{|\arctan x - \arctan y|\} \tag{13.1}$$

on the space S of extended real-valued random variables has the property that a sequence converges in the metric if and only if the sequence converges in measure in the above sense. This metric, together with the identification of two random variables whenever they are equal almost everywhere, makes S a complete metric space. Observe that if a sequence in S converges in the metric (13.1), then some subsequence of the corresponding random variables converges almost everywhere, in the topology of $\overline{\mathbb{R}}$. There is therefore a sequence $\varepsilon_{\bullet}$ of strictly positive numbers such that if $d(x, x_n) \leq \varepsilon_n$, then $\lim_{n\to\infty} x_n = x$ almost everywhere. We use here the standard abuse of language, leaving it to the reader to decide when a symbol represents a function on Ω and when it represents a point of S, that is, an equivalence class of functions. Observe that if we wished to treat only random variables bounded in modulus by 1, the L^1 metric could be used instead of the metric (13.1). In fact one way to carry out the work of this section is to replace each random variable x by $(2/\pi)\arctan x$ and thereby make all random variables bounded in modulus by 1.

In the present context a process $\{x(t), t \in \mathbb{R}^+\}$ with state space $\overline{\mathbb{R}}$ defines a function $\beta_{x(\cdot)}: t \mapsto x(t)$ from $\mathbb{R}^+$ into S. This map taking a process into a function is one to one and onto if two processes are identified when they are standard modifications of each other.

Theorem. *Let $(\Omega, \mathscr{F}, P)$ be a probability space, and let $\{x(t), t \in \mathbb{R}^+\}$ be a process on the space, with state space $\overline{\mathbb{R}}$.*

(a) *The process has a measurable standard modification if and only if the function $\beta_{x(\cdot)}$ is Borel measurable.*

Suppose that the probability space is provided with a right continuous filtration $\{\mathscr{F}(t), t \in \mathbb{R}^+\}$ for which $\mathscr{F}(0)$ contains the null sets.

(b) *If the process $\{x(\cdot), \mathscr{F}(\cdot)\}$ is adapted and measurable, it has a progressively measurable standard modification.*

(c) *If $x(\cdot, \cdot)$ is measurable on $\mathbb{R}^+ \times \Omega$ relative to the completion of the measure $l_1 \times P$, there is a progressively measurable process $x'(\cdot)$ such that*

$$P\{x(t) = x'(t)\} = 1 \quad \textit{for } l_1 \textit{ almost every } t. \tag{13.2}$$

Proof of (a). (i) *If $x(\cdot)$ is a measurable process, then $\beta_{x(\cdot)}$ is Borel measurable.* Since each measurable process $x(\cdot)$ is determined by a function $(t, \omega) \mapsto x(t, \omega)$ and since pointwise convergence of a sequence $\{x_n(\cdot, \cdot), n \in \mathbb{Z}^+\}$ implies pointwise convergence of the corresponding sequence $\{\beta_{x_n(\cdot)}, n \in \mathbb{Z}^+\}$, it is sufficient to show that if $x(\cdot, \cdot)$ has the form

$$x(t, \omega) = \sum_1^n c_j 1_{A_j}(t) 1_{\Lambda_j}(\omega)$$

with $A_1, \ldots, A_n$ disjoint Borel subsets of $\mathbb{R}^+$ and $\Lambda_j \in \mathscr{F}$, then $\beta_{x(\cdot)}$ is Borel measurable. This Borel measurability is trivial.

(ii) *If $\beta_{x(\cdot)}$ is Borel measurable, then $x(\cdot)$ has a measurable standard modification.* To show this, it is sufficient to show that the class of functions from $\mathbb{R}^+$ into S corresponding to processes which have measurable standard modifications includes the continuous functions and is closed under convergence. Suppose first that $\beta_{x(\cdot)}$ is continuous. Define

$$\phi_k(t) = j2^{-k} \quad \text{for } (j-1)2^{-k} \le t < j2^{-k} \quad (j = 1, 2, \ldots).$$

Since $\beta_{x(.)}$ is uniformly continuous on compact intervals, there are positive integers $k_1 < k_2 < \ldots$ such that (see the beginning of this section) $d[x(s), x(t)] < \varepsilon_n$ if $|s - t| \le 2^{-k_n}$ and $s \vee t \le n$; so for each fixed t in $\mathbb{R}^+$

$$\lim_{n \to \infty} x[\phi_{k_n}(t)] = x(t) \text{ a.s.} \tag{13.3}$$

If we define $x'(t) = \limsup_{n\to\infty} x[\phi_{k_n}(t)]$ the process $x'(\cdot)$ is a measurable standard modification of $x(\cdot)$. Next suppose that $\{x_n(\cdot), n \in \mathbb{Z}^+\}$ is a sequence of processes on $(\Omega, \mathscr{F}, P)$ with the property that each function $\beta_{x_n(\cdot)}$ is Borel measurable, that $x_n(\cdot)$ has a measurable standard modification, and that $\lim_{n\to\infty} \beta_{x_n(\cdot)} = \beta_{x(\cdot)}$ exists in the S metric on $\mathbb{R}^+$; that is, $\lim_{n\to\infty} x_n(t) = x(t)$ exists in measure for all t. We can suppose that $x_n(\cdot)$ has been chosen to be measurable. Define $m_n(t)$ by

$$m_n(t) = \min\{k: d[x(t), x_k(t)] < \varepsilon_n\}.$$

The set $A_{nj} = \{t: m_n(t) = j\}$ is a Borel subset of $\mathbb{R}^+$, and for each t

$$d[x(t), x_{m_n}(t)] < \varepsilon_n;$$

so

$$\lim_{n\to\infty} x_{m_n(t)}(t) = x(t) \text{ a.s.} \tag{13.4}$$

and $\lim_{n\to\infty} \beta_{x_n(\cdot)} = \beta_{x(\cdot)}$. The process defined by

$$(t, \omega) \mapsto x_{m_n(t)}(t, \omega) 1_{A_{nj}}(t) = x_j(t, \omega) 1_{A_{nj}}(t)$$

is measurable because it is the product of two measurable processes, so summing over j we find that the process $x_{m_n(\cdot)}(\cdot)$ is measurable. The process $\limsup_{n\to\infty} x_{m_n(\cdot)}(\cdot)$ is therefore a measurable standard modification of $x(\cdot)$, and the proof of (a) is complete. □

Proof of (b). There is no loss of generality in proving (b) if we suppose, after replacing $x(\cdot)$ by $(2/\pi)\arctan x(\cdot)$, that $|x(\cdot)| \le 1$, and we shall assume that all processes are bounded in this way throughout the proof of (b). To each process $x(\cdot)$ then corresponds a process $x^\circ(\cdot)$ defined by

$$x^\circ(t) = E\{x(t) | \mathscr{F}(t)\}.$$

Recall our convention that $x^\circ(t)$ is a version of the indicated conditional expectation and that $x^\circ(\cdot)$ is therefore determined only up to a standard modification. The process $x^\circ(\cdot)$ is adapted and is a standard modification of $x(\cdot)$ if the latter process is adapted. To prove (b), it is sufficient to show that whenever $x(\cdot)$ is measurable, in particular, whenever $\beta_{x(\cdot)}$ is a Borel measurable function, the process $x^\circ(\cdot)$ has a progressively measurable standard modification. It is therefore sufficient to show (i) that such a modification exists for $x^\circ(\cdot)$ if it exists when $\beta_{x(\cdot)}$ is continuous and (ii) that such a modification exists for $x^\circ(\cdot)$ if it exists for each process of the sequence $\{x_n^\circ(\cdot), n \in \mathbb{Z}^+\}$ and $\lim_{n\to\infty} \beta_{x_n(\cdot)} = \beta_{x(\cdot)}$.

Proof of (b)(i). If $\beta_{x(\cdot)}$ is continuous, the conditional expectation continuity theorem can be applied to (13.3) to obtain, in view of the right continuity of $\mathscr{F}(\cdot)$,

$$\lim_{n\to\infty} E\{x[\phi_{k_n}(t)]|\mathscr{F}[\phi_{k_n}(t)]\} = E\{x(t)|\mathscr{F}(t)\} = x^\circ(t). \quad \text{a.s.} \tag{13.5}$$

When $2^{-k_n} < \varepsilon$, the nth term on the left defines a process adapted to and progressively measurable relative to the filtration $\mathscr{F}(\cdot + \varepsilon)$. Hence if the limit in (13.5) is replaced by the limit superior, the resulting process is progressively measurable relative to $\mathscr{F}(\cdot + \varepsilon)$ for all strictly positive ε and so (Section I.2) is progressively measurable relative to $\mathscr{F}(\cdot)$, and this process is a standard modification of $x^\circ(\cdot)$.

Proof of (b)(ii). Apply the conditional expectation continuity theorem to (13.4) to obtain

$$\lim_{n\to\infty} E\{x_{m_n(t)}(t)|\mathscr{F}[\phi_n(t)]\} = x^\circ(t) \quad \text{a.s.,} \tag{13.6}$$

and repeat the reasoning just used, thereby showing that the process defined when the limit in (13.6) is replaced by the limit superior is a progressively measurable standard modification of $x^\circ(\cdot)$. The proof of (b) is now complete.

Proof of (c). If $x(\cdot,\cdot)$ is $l_1 \times P$ (completed) measurable, there is a function $(t,\omega)\mapsto x_1(t,\omega)$ on $\mathbb{R}^+ \times \Omega$ which is measurable relative to $\mathscr{B}(\mathbb{R}^+) \times \mathscr{F}$, that is, for which $x_1(\cdot)$ is a measurable process, such that $x(\cdot,\cdot) = x_1(\cdot,\cdot)$ $l_1 \times P$ almost everywhere. Hence (Fubini)

$$P\{x(t) = x_1(t)\} = 1 \quad (\text{for } l_1 \text{ a.e. } t).$$

According to (b), the process $x_1(\cdot)$ has a progressively measurable standard modification $x'(\cdot)$, and this is the process demanded by (c).

14. Predictable Families of Functions

Let $(\Omega, \mathscr{F}; \mathscr{F}(t), t\in I\}$ be a filtered measurable space. The concept of a predictable function family strengthens that of an adapted family in a way suggested by the name.

Discrete Parameter Case, $I = \mathbb{Z}^+$

A sequence $\{x(n); \mathscr{F}(n), n\in\mathbb{Z}^+\}$ of functions from Ω into a measurable space is called *predictable* (relative to $\mathscr{F}(\cdot)$) if $x(0)$ is $\mathscr{F}(0)$ measurable and if $x(n)$ is $\mathscr{F}(n-1)$ measurable for $n > 0$.

Continuous Parameter Case, $I = \mathbb{R}^+$

Consider the class of adapted families $\{x(t); \mathscr{F}(t), t \in \mathbb{R}^+\}$ of functions from Ω into $\mathbb{R}$ for which every sample function $x(\cdot, \omega)$ is left continuous, that is, for which the family $x(\cdot)$ is left continuous. (Left continuity at 0 is vacuously satisfied by every function on $\mathbb{R}^+$.) The smallest σ algebra of subsets of $\mathbb{R}^+ \times \Omega$ making the function $(t, \omega) \mapsto x(t, \omega)$ measurable for every such left continuous family $x(\cdot)$ is called the *predictable* σ algebra, and the sets in this σ algebra are called *predictable*. A family $\{y(t), t \in \mathbb{R}^+\}$ of functions from Ω into an arbitrary measurable space is called *predictable* if the function $(t, \omega) \mapsto y(t, \omega)$ on the space $\mathbb{R}^+ \times \Omega$ coupled with its predictable σ algebra is measurable. A predictable family is necessarily adapted to the given filtration. Since left continuity of a function family with state space $(\mathbb{R}, \mathscr{B}(\mathbb{R}))$ implies progressive measurability (Section 2), the predictable sets are progressively measurable, and a predictable function family is progressively measurable. The predictability definition implies that if $x(\cdot)$ is predictable, if $\alpha > 0$, and if $x_\alpha(t) = x((t - \alpha) \vee 0)$, then the family $x_\alpha(\cdot)$ is also predictable.

Example (Continuous parameter case). By definition of predictability a left continuous adapted family with state space $(\mathbb{R}, \mathscr{B}(\mathbb{R}))$ is predictable. More generally a left continuous adapted family with state space a metric space coupled with its Borel sets is predictable. In fact, whenever f is a continuous function from the state space into $\mathbb{R}$, the family $\{f[x(\cdot)], \mathscr{F}(\cdot)\}$ is predictable, and therefore the set $\{(t, \omega): x(t, \omega) \in A\}$ is a predictable set whenever A is a Borel subset of the state space; so $\{x(\cdot), \mathscr{F}(\cdot)\}$ is predictable.

Generators of the Predictable σ Algebra

If $0 \le a < b \le +\infty$ and if $\Lambda \in \mathscr{F}(a)$, the subset $]a, b] \times \Lambda$ of $\mathbb{R}^+ \times \Omega$ if $b < +\infty$ or $]a, +\infty[\times \Lambda$ if $b = +\infty$ is a predictable set, as is the product set $\{0\} \times \Lambda_0$ when $\Lambda_0 \in \mathscr{F}(0)$. The class of finite unions of sets in the class Γ of product sets of these two types is an algebra, and we now show that the σ algebra $\mathscr{G}$ generated by Γ is the predictable σ algebra. To see this, observe first that if $x(t) = 1_{\{0\}}(t)\phi$ with ϕ an $\mathscr{F}(0)$ measurable function from Ω into $\mathbb{R}$, or if $0 \le a < b$ and if $x(t) = 1_{]a,b]}(t)\phi$ with ϕ an $\mathscr{F}(a)$ measurable function from Ω into $\mathbb{R}$, then the family $x(\cdot)$ is $\mathscr{G}$ measurable. Second, observe that if $\{y(\cdot), \mathscr{F}(\cdot)\}$ is a left continuous adapted family of functions with state space $(\mathbb{R}, \mathscr{B}(\mathbb{R}))$ and if

$$y_n(t) = y(0)1_{\{0\}}(t) + \sum_{j=1}^{4^n} y((j-1)2^{-n})1_{](j-1)2^{-n}, j2^{-n}]}(t),$$

then the family $y_n(\cdot)$ is $\mathscr{G}$ measurable because it has just been shown that each summand family is. Finally $y(\cdot) = \lim_{n\to\infty} y_n(\cdot)$; so $y(\cdot)$ is $\mathscr{G}$ measurable, and therefore $\mathscr{G}$ is the predictable σ algebra, as asserted. The class of sets analytic over the paving Γ therefore (Theorem 4 of Appendix I) includes the predictable σ algebra (so is the same as the class of analytic sets over the predictable σ algebra).

For some purposes it is convenient to replace the generating family Γ by a generating family Γ' involving compact parameter intervals. Define Γ' as the class of product sets $[a,b] \times \Lambda$ with $0 \leq a < b$, $\Lambda \in \mathscr{F}(0)$ if $a = 0$ and $\Lambda \in \bigcup_{t<a} \mathscr{F}(t)$ if $a > 0$. The σ algebra generated by Γ' includes the predictable σ algebra because

$$\begin{aligned}]a,b] \times \Lambda &= \bigcup_{n=1}^{\infty} \left[a + \frac{1}{n}, b\right] \times \Lambda \qquad [a < b, \Lambda \in \mathscr{F}(a)], \\ \{0\} \times \Lambda &= \bigcap_{n=1}^{\infty} \left[0, \frac{1}{n}\right] \times \Lambda \qquad [\Lambda \in \mathscr{F}(0)]. \end{aligned} \tag{14.1}$$

Here the sets on the left are in Γ and those on the right are in the σ algebra generated by Γ'. Conversely, just as trivial a representation of Γ' sets in terms of predictable sets shows that the σ algebra of predictable sets includes Γ'. Thus the σ algebra generated by Γ' is the predictable σ algebra. The representation (14.1) shows that the class of analytic sets over Γ' includes the predictable sets. We leave to the reader the similar trivial argument showing that if $0 \leq a < b \leq +\infty$, the predictable σ algebra is the σ algebra generated by the class Γ'' of product sets $[a,b[\times \Lambda$ with $\Lambda \in \mathscr{F}(0)$ if $a = 0$ and $\Lambda \in \bigcup_{t<a} \mathscr{F}(t)$ if $a > 0$. The class of finite unions of these product sets is an algebra. The class of analytic sets over Γ'' includes the predictable sets.

The predictable σ algebra is the smallest σ algebra making the function $(t,\omega) \mapsto x(t,\omega)$ measurable whenever $\{x(\cdot), \mathscr{F}(\cdot)\}$ is an adapted continuous family with state space $(\mathbb{R}, \mathscr{B}(\mathbb{R}))$. To prove this, we need show only that if G'' is a σ algebra of subsets of $\mathbb{R}^+ \times \Omega$ making every such family measurable, then G'' contains Γ'. This inclusion follows from the fact that the indicator function of a compact interval of $\mathbb{R}^+$ is the limit of a decreasing sequence of continuous functions on $\mathbb{R}^+$ vanishing outside an arbitrarily small neighborhood of the interval.

Chapter II

Optional Times and Associated Concepts

1. The Context of Optional Times

Let $(\Omega, \mathscr{F}; \mathscr{F}(t), t \in I)$ be a filtered measurable space. If I does not have a last element, extend I to $I \cup \{+\infty\}$, where $+\infty$ is a new element ordered after every element of I, and define $\mathscr{F}(+\infty)$ as any sub σ algebra of $\mathscr{F}$ containing $\bigvee_{t<+\infty} \mathscr{F}(t)$. The choice of $\mathscr{F}(+\infty)$ within these limits is usually irrelevant. If I has a last element, that element will be denoted by $+\infty$ in this section. The most common choices of I are the set $\mathbb{Z}^+$ (discrete parameter case) and the set $\mathbb{R}^+$ (continuous parameter case). The index set I is thought of as representing a set of time points, and $\mathscr{F}(t)$ then represents the class of events up to time t. The problems of defining what is meant by a random time T corresponding to the arrival time of an event whose arrival is determined by preceding events and of defining the class $\mathscr{F}(T)$ of preceding events are solved by the following definitions.

Definition. An *optional time* (also called a *stopping time* or a *Markov time*) T is a function from Ω into $I \cup \{+\infty\}$ satisfying

$$\{T \leq c\} \in \mathscr{F}(c) \tag{1.1}$$

for $c < +\infty$. This relation is trivially true for $c = +\infty$.

Definition. If T is optional, denote by $\mathscr{F}(T)$ the σ algebra of subsets Λ of Ω for which

$$\Lambda \cap \{T \leq c\} \in \mathscr{F}(c) \tag{1.2}$$

for $c \leq +\infty$. Equivalently, $\mathscr{F}(T)$ is the σ algebra of sets Λ in $\mathscr{F}(+\infty)$ for which (1.2) is true when $c < +\infty$.

A function T from Ω into a countable subset of $I \cup \{+\infty\}$ is an optional time if and only if

$$\{T = c\} \in \mathscr{F}(c) \tag{1.1d}$$

for c in the range of T, and if T is optional, $\Lambda \in \mathscr{F}(T)$ if and only if

$$\Lambda \cap \{T = c\} \in \mathscr{F}(c) \tag{1.2d}$$

for c in the range of T. In particular, an identically constant function $T \equiv t$ on Ω with value in $I \cup \{+\infty\}$ is an optional time and in this case $\mathscr{F}(T)$, as defined above, reduces to the specified σ algebra $\mathscr{F}(t)$.

$\mathscr{F}(T)$ in the Continuous Parameter Case

If $I = \mathbb{R}^+$, the conditions

$$\{T < c\} \in \mathscr{F}(c) \tag{1.1'}$$

$$\Lambda \cap \{T < c\} \in \mathscr{F}(c) \tag{1.2'}$$

for all finite c are equivalent to $\{T \leq c\} \in \mathscr{F}(c+)$ and $\Lambda \cap \{T \leq c\} \in \mathscr{F}(c+)$, respectively, for all finite c. That is, (1.1′) is the condition that T is optional for the family $\mathscr{F}^+(\cdot)$, and (1.2′) together with $\Lambda \in \mathscr{F}(+\infty)$ is the condition that $\Lambda \in \mathscr{F}^+(T)$. If T is optional, then $T + \varepsilon$ is optional for positive ε and $\mathscr{F}(T) \subset \mathscr{F}(T + \varepsilon)$. Define $\mathscr{F}(T+) = \bigcap_{\varepsilon>0} \mathscr{F}(T + \varepsilon)$. Then $\Lambda \in \mathscr{F}(T+)$ if and only if $\Lambda \cap \{T \leq c\} \in \mathscr{F}^+(c)$. It follows that $\mathscr{F}(T+) = \mathscr{F}^+(T)$.

Stochastic Intervals (Parameter Set $\mathbb{R}^+$)

If S and T are optional with $S \leq T$, the *stochastic interval* $[\![S, T]\!]$, a subset of $\mathbb{R}^+ \times \Omega$, is defined by

$$[\![S, T]\!] = \{(t, \omega): S(\omega) \leq t \leq T(\omega), t \in \mathbb{R}^+\}.$$

In particular, we write $[\![T]\!]$ for the interval $[\![T, T]\!]$, the graph of T (but observe that this graph is a subset of $\mathbb{R}^+ \times \Omega$; if $T \equiv +\infty$, this graph is the empty set). The open stochastic interval $]\!]S, T[\![$ and the half-open ones are defined in the obvious way.

A stochastic interval of the form $]\!]S, T]\!]$ is a predictable set because each function $1_{]\!]S,T]\!]}(\cdot, \omega)$ is left continuous. The classes Γ and Γ' of generators of the predictable σ algebra, as defined in Section I.14, can be described in terms of stochastic intervals. For example, Γ consists of the graphs $[\![S]\!]$ with $S = 0$ on an $\mathscr{F}(0)$ set and $S = +\infty$ elsewhere and of the stochastic intervals $]\!]S, T]\!]$ with $S = a$ on an $\mathscr{F}(a)$ set Λ and $S = +\infty$ elsewhere, $T = b$ on Λ and $T = +\infty$ elsewhere.

Optional Times on a Filtered Probability Space

Suppose that the given measurable space is a probability space, and that each σ algebra $\mathscr{F}(t)$ of the filtrations contains the null sets. Then if T is optional, $\mathscr{F}(T)$ also contains the null sets, and if T_1 and T_2 are functions from Ω into $I \cup \{+\infty\}$ which differ only on a null set, T_1 is optional if and only if T_2 is. Moreover if both are optional, then $\mathscr{F}(T_1) = \mathscr{F}(T_2)$. Thus under this hypothesis on $\mathscr{F}(\cdot)$ we can ignore null sets in discussing simple relations between optional times and the σ algebras they determine.

2. Optional Time Properties (Continuous Parameter Context)

In this section the commonly used properties of optional times and their associated σ algebras on a filtered measurable space $(\Omega, \mathscr{F}; \mathscr{F}(t), t \in \mathbb{R}^+)$ are listed, and proofs of those properties not immediate consequences of the definitions are given. The corresponding properties in the discrete parameter context of the parameter set $\mathbb{Z}^+$ will be obvious.

(a) If S is optional, S is $\mathscr{F}(S)$ measurable.

(b) If S is optional and if T is a random variable for which $T \geq S$ and T is $\mathscr{F}(S)$ measurable, then T is optional.

Example (b1). If S is optional and α is a positive constant, then $S + \alpha$ is optional.

Example (b2). If S is optional define $[S]_n$ by

$$[S]_n = \begin{cases} j2^{-n} & \text{when } (j-1)2^{-n} \leq S < j2^{-n}, \\ +\infty & \text{when } S = +\infty. \end{cases}$$

Then $[S]_n \geq S$, with strict inequality where S is finite valued, and the sequence $[S]_.$ is a monotone decreasing sequence of optional times with limit S.

(c) If S and T are optional, then $S \vee T$ and $S \wedge T$ are optional.

(d) If S and T are optional and $S \leq T$, then $\mathscr{F}(S) \subset \mathscr{F}(T)$; if $S < T$, then $\mathscr{F}(S+) \subset \mathscr{F}(T)$.

In fact, if $\Lambda \in \mathscr{F}(S)$ and $S \leq T$, then

$$\Lambda \cap \{T \leq c\} = [\Lambda \cap \{S \leq c\}] \cap \{T \leq c\} \in \mathscr{F}(c); \tag{2.1}$$

so $\Lambda \in \mathscr{F}(T)$. Thus the first assertion of (d) is true. To prove the second assertion, replace $\{S \leq c\}$ by $\{S < c\}$ in (2.1).

(e) If S and T are optional $\mathscr{F}(S) \cap \mathscr{F}(T) = \mathscr{F}(S \wedge T)$. In particular, $\{S \leq c\} \in \mathscr{F}(S \wedge c)$.

(f) If S and T are optional, the sets $\{S < T\}$, $\{S \leq T\}$, $\{S = T\}$ are in $\mathscr{F}(S \wedge T)$.

In fact, since

$$\{S < T\} \cap \{S \wedge T \leq c\} = \bigcup_{\substack{0 < r \leq 1 \\ r \text{ rational}}} \{S \leq rc\} \cap \{T > rc\} \tag{2.2}$$

and $\{T > rc\} = \Omega - \{T \le rc\} \in \mathscr{F}(c)$, it follows that $\{S < T\} \in \mathscr{F}(S \wedge T)$. Interchanging S and T, we conclude that the set $\{T < S\}$ and therefore also the complementary set $\{S \ge T\}$ are in $\mathscr{F}(S \wedge T)$. Finally

$$\{S = T\} = \{S \le T\} \cap \{T \le S\} \in \mathscr{F}(S \wedge T).$$

(g) If S and T are optional and if $\Lambda \in \mathscr{F}(S)$, then

$$\Lambda \cap \{S \le T\} \in \mathscr{F}(S \wedge T).$$

In fact $\Lambda \cap \{S \le T\} \cap \{S \wedge T \le c\} \in \mathscr{F}(c)$ because this intersection is the same as $\Lambda \cap \{S \le T\} \cap \{S \le c\}$ in which expression both Λ and $\{S \le T\}$ are in $\mathscr{F}(S)$. Property (g) generalizes the defining relation of $\mathscr{F}(S)$. The same proof shows that $\{S \le T\}$ in (g) can be replaced by $\{S < T\}$ or $\{S = T\}$.

(h) Let S be optional and finite valued, and let T be a positive ($\le +\infty$) random variable. Then optionality of $S + T$ implies optionality of T relative to $\mathscr{F}(S + \cdot)$. Conversely, optionality of T relative to $\mathscr{F}(S + \cdot)$ implies optionality of $S + T$ for $\mathscr{F}(\cdot)$ if $\mathscr{F}(\cdot)$ is right continuous.

In fact, if $S + T$ is optional for $\mathscr{F}(\cdot)$,

$$\{T \le c\} = \{S + T \le S + c\} \in \mathscr{F}(S + c) \tag{2.3}$$

by (f); therefore T is optional for $\mathscr{F}(S + \cdot)$. Conversely, if T is optional for $\mathscr{F}(S + \cdot)$,

$$\{S + T < c\} = \bigcup_{\substack{a+b<c \\ a,b \text{ rational}}} \{S < a, T < b\}. \tag{2.4}$$

For each pair (a, b) the brace set on the right is in $\mathscr{F}(c)$ because by hypothesis $\{T < b\} \in \mathscr{F}(S + b\}$, that is,

$$\{T < b\} \cap \{S + b < a + b\} \in \mathscr{F}(a + b) \subset \mathscr{F}(c). \tag{2.5}$$

Hence $S + T$ is optional for $\mathscr{F}(\cdot)$ if this family is right continuous.

According to property (h) if $\mathscr{F}(\cdot)$ is right continuous and S is finite valued and optional and $\mathscr{F}_S(t)$ is defined as $\mathscr{F}(S + t)$, then optionality of T for $\mathscr{F}_S(\cdot)$ implies that $S + T$ is optional for $\mathscr{F}(\cdot)$ and $\mathscr{F}(S + T) = \mathscr{F}_S(T)$.

(i) Let $\{T_n, n \in \mathbb{Z}^+\}$ be a sequence of optional times.

(i1) $\sup_{\mathbb{Z}^+} T_\cdot$ is optional.

(i2) $\inf_{\mathbb{Z}^+} T_\cdot$ is optional for $\mathscr{F}^+(\cdot)$.

(i3) $\liminf_{n\to\infty} T_n$ and $\limsup_{n\to\infty} T_n$ are optional for $\mathscr{F}^+(\cdot)$.

(i4) If $\lim_{n\to\infty} T_n = T = \inf_{\mathbb{Z}^+} T_\cdot$, then $\mathscr{F}^+(T) = \bigcap_0^\infty \mathscr{F}^+(T_n)$.

(i5) If $T_n \uparrow T$ $[T_n \downarrow T]$ and $T_n = T$ for sufficiently large $n = n(\omega)$, then T is optional and $\mathscr{F}(T) = \bigvee_0^\infty \mathscr{F}(T_n)$ $[\mathscr{F}(T) = \bigcap_0^\infty \mathscr{F}(T_n)]$.

Only (i4) and (i5) need comment. In (i4) we need only prove that $\bigcap_0^\infty \mathscr{F}^+(T_n) \subset \mathscr{F}^+(T)$, and this inclusion follows from the fact that if $\Lambda \in \bigcap_0^\infty \mathscr{F}^+(T_n)$, then

$$\Lambda \cap \{T < c\} = \bigcup_0^\infty \Lambda \cap \{T_n < c\} \in \mathscr{F}(c).$$

To prove (i5), observe that if $T_n \uparrow T$, then $\bigvee_0^\infty \mathscr{F}(T_n) \subset \mathscr{F}(T)$ trivially, and the inclusion is true in the other direction because if $\Lambda \in \mathscr{F}(T)$, then $\Lambda \cap \{T = T_n\} \in \mathscr{F}(T_n)$ by (g) and summing over n yields $\Lambda \in \bigvee_0^\infty \mathscr{F}(T_n)$. In (i5) if $T_n \downarrow T$, the limit is optional because $\{T \le c\} = \bigcup_0^\infty \{T_n \le c\} \in \mathscr{F}(c)$. Moreover in this case $\bigcap_0^\infty \mathscr{F}(T_n) \supset \mathscr{F}(T)$, and the inclusion is true in the other direction because if $\Lambda \in \bigcap_0^\infty \mathscr{F}(T_n)$, then

$$\Lambda \cap \{T \le c\} = \bigcup_0^\infty [\Lambda \cap \{T_n \le c\}] \in \mathscr{F}(c)$$

so that $\Lambda \in \mathscr{F}(T)$.

(j) Suppose that $\{\mathscr{F}_0(t), t \in \mathbb{R}^+\}$ is a filtration of a probability space $(\Omega, \mathscr{F}, P)$, that $\mathscr{F}_0(+\infty)$ is a σ algebra satisfying $\bigvee_{t \in \mathbb{R}^+} \mathscr{F}_0(t) \subset \mathscr{F}(+\infty) \subset \mathscr{F}$, that $\mathscr{F}(t)$ is the σ algebra generated by $\mathscr{F}_0(t)$ and the null sets, that T is $\mathscr{F}(\cdot)$ optional, and that $\Lambda \in \mathscr{F}(T)$. Then there are an $\mathscr{F}_0^+(\cdot)$ optional time T_0 and a set $\Lambda_0 \in \mathscr{F}_0^+(T_0)$ such that $T = T_0$ almost surely and that Λ differs from Λ_0 by a null set.

For $j = 0, 1, \ldots, +\infty$ define M_{nj} as an $\mathscr{F}_0(j2^{-n})$ set differing from the $\mathscr{F}(j2^{-n})$ set $\{[T]_n = j2^{-n}\}$ by a null set, and define $S_n = j2^{-n}$ on $M_{nj} - \bigcup_{k<j} M_{nk}$. Then S_n is an $\mathscr{F}_0(\cdot)$ optional time, $S_.$ is a decreasing sequence, and the limit T_0 is an $\mathscr{F}_0^+(\cdot)$ optional time equal almost surely to T. If $\Lambda \in \mathscr{F}(T)$ and if $T = +\infty$ everywhere on Λ, let Λ_0 be a set in $\mathscr{F}_0(+\infty)$ differing from Λ by a null set. The set $\{T_0 = +\infty\} \cap \Lambda_0$ is then in $\mathscr{F}_0^+(T_0)$ and differs from Λ by a null set. If $\Lambda \in \mathscr{F}(T)$ and if $T < +\infty$ everywhere on Λ, define the $\mathscr{F}(\cdot)$ optional time T' by setting $T' = T$ on Λ and $T' = +\infty$ elsewhere. We have shown above that there is an $\mathscr{F}_0^+(\cdot)$ optional time T_0' equal almost surely to T'. The set $\{T_0' = T_0 < +\infty\}$ is in $\mathscr{F}_0^+(T_0)$ and differs from Λ by a null set. If Λ is an arbitrary member of $\mathscr{F}(T)$, apply the two results just obtained to the sets $\Lambda \cap \{T = +\infty\}$ and $\Lambda \cap \{T < +\infty\}$ to complete the proof of (j).

3. Process Functions at Optional Times

If $\{x(t), t \in I\}$ is a family indexed by I of functions on a space Ω and if T is a function from Ω into a set including I, we write $x(T)$ for the function $\omega \mapsto x(T(\omega), \omega)$, defined on the set $\{\omega\colon T(\omega) \in I\}$.

Theorem. *If* $\{x(t), \mathscr{F}(t), t \in \mathbb{R}^+\}$ *is a progressively measurable family of functions and if T is optional, then* $x(T)$ *is* $\mathscr{F}(T)$ *measurable.*

According to our conventions, the function $x(T)$ is defined on the set $\{T < +\infty\}$ which [Section 2(a)] is $\mathscr{F}(T)$ measurable. To prove the theorem, let $(\Omega', \mathscr{F}')$ be the state space of the family $x(\cdot)$, and suppose that $\Lambda' \in \mathscr{F}'$. The theorem asserts that $\{T < +\infty, x(T) \in \Lambda'\} \in \mathscr{F}(T)$, that is, for $c \geq 0$,

$$\{T \leq c, x(T) \in \Lambda'\} \in \mathscr{F}(c). \tag{3.1}$$

Let $\mathscr{F}_c$ be the class of intersections with $\{T \leq c\}$ of $\mathscr{F}(c)$ sets, and consider the three measurable spaces

$$(\{T \leq c\}, \mathscr{F}_c), \qquad ([0, c] \times \Omega, \mathscr{B}([0, c]) \times \mathscr{F}(c)), \qquad (\Omega', \mathscr{F}').$$

The condition (3.1) is that the restriction to $\{T \leq c\}$ of $x(T)$ be measurable as a function from the first to the third space. This restriction is the composition of the two functions $\omega \mapsto (T(\omega), \omega)$ and $(t, \omega) \mapsto x(t, \omega)$, respectively, from the first space to the second and the second space to the third. The first function is measurable because by optionality of T the inverse image of a product set $[a, b] \times \Lambda$ is in $\mathscr{F}(c)$ when $a < b \leq c$ and $\Lambda \in \mathscr{F}(c)$. The second function is measurable by definition of progressive measurability. Hence the composed function is measurable.

Special Case. If $\{x(t), t \in \mathbb{R}^+\}$ is a measurable family of functions on the measurable space $(\Omega, \mathscr{F})$ and if T is a measurable function from Ω into $\overline{\mathbb{R}}^+$, then Theorem 3 with $\mathscr{F}(t) = \mathscr{F}$ for all t implies that $x(T)$ is measurable.

Observation. If the process $\{x(t), \mathscr{F}(t), t \in \mathbb{R}^+\}$ on a probability space is nearly progressively measurable and if $\mathscr{F}(0)$ contains the null sets, then Theorem 3 implies that $x(T)$ is measurable whenever T is an optional time.

Predictability of a Family of Functions at a Predicted Time (Continuous Parameter Context, Arbitrary State Space)

If $\{x(\cdot), \mathscr{F}(\cdot)\}$ is a predictable function family and if $S_\cdot$ is an increasing sequence of optional times with limit S, then $x(S)|_{\{S < +\infty\}}$ is a $\bigvee_0^\infty \mathscr{F}(S_n)$ measurable function. To prove this assertion, Let I be a compact interval of $\mathbb{R}$, and suppose first that the state space is $(I, \mathscr{B}(I))$. The assertion is trivially true for this state space if $x(\cdot)$ is a left continuous family. Since the property to be proved is invariant under convergence of function families, the assertion is true in general for this state space. The assertion is therefore true for an arbitrary state space because whenever ϕ is a measurable function from the state space into $(I, \mathscr{B}(I))$, the function $\phi[x(S)]|_{\{S < +\infty\}}$ is $\bigvee_0^\infty \mathscr{F}(S_n)$ measurable, and so $x(S)|_{\{S < +\infty\}}$ is.

4. Hitting and Entry Times

Let $(\Omega, \mathscr{F}; \mathscr{F}(t), t \in I)$ be a filtered measurable space, with I either $\mathbb{Z}^+$ or $\mathbb{R}^+$. If A'' is a subset of $I \times \Omega$, the *hitting time* of A'' (by $\{(t, \omega), t \in I\}$) is defined as

$$\inf\{t > 0: (t, \omega) \in A''\} \qquad (t \in I). \tag{4.1}$$

Here and below the infimum of the empty set is defined as $+\infty$. If $\{x(t), t \in I\}$ is a family of functions on Ω with arbitrary state space $(X, \mathscr{X})$ and if A' is a subset of $I \times X$, the hitting time of A' (by $\{[t, x(t)], t \in I\}$) is defined as

$$\inf\{t > 0: [t, x(t)] \in A'\} \qquad (t \in I). \tag{4.2}$$

Finally, if A is a subset of X, the hitting time of A (by $\{x(t), t \in I\}$) is defined as

$$\inf\{t > 0: x(t, \omega) \in A\} \qquad (t \in I). \tag{4.3}$$

In each case the *entry time* of the set in question is defined in the same way as the hitting time except that "$t > 0$" is replaced by "$t \geq 0$." Observe that If $A' = I \times A$, the hitting and entry times of A' reduce to those of A and that if $A'' = \{(t, \omega): [t, x(t, \omega)] \in A'\}$, the hitting and entry times of A'' reduce to those of A'.

The analysis of measurability of hitting and entry times is trivial when $I = \mathbb{Z}^+$. For example, if $A \in \mathscr{X}$, it is clear that the entry and hitting times of A by $x(\cdot)$ are measurable and in fact are optional for $\mathscr{F}(\cdot)$. From now on we shall therefore assume that $I = \mathbb{R}^+$. Observe that if T' $[T'']$ is the entry [hitting] time of A'' and if $c > 0$, then

$$\begin{gathered}\{T' < c\} = \{\omega: (t, \omega) \in A'' \text{ for some } t < c\},\\ \{T'' < c\} = \bigcup_{n=1}^{\infty} \left\{\omega: (t, \omega) \in A'' \text{ for some } t \in \left[\frac{1}{n}, c\right]\right\},\end{gathered} \tag{4.4}$$

and the corresponding observation is valid for the entry and hitting times of A' and A. The evaluations in (4.4) and the corresponding ones for A' and A make it easy to deduce optionality of entry and hitting times in a suitable context, as follows.

Theorem. *Let $(\Omega, \mathscr{F}, P; \mathscr{F}(t), t \in \mathbb{R}^+)$ be a filtered probability space, and suppose that for each t the restriction of P to $\mathscr{F}(t)$ is a complete measure.*

(a) *If A'' is a subset of $\mathbb{R}^+ \times \Omega$, analytic over the class of progressively measurable sets, the entry and hitting times of A'' are optional for $\mathscr{F}^+(\cdot)$.*

(b) *Let $\{x(\cdot), \mathscr{F}(\cdot)\}$ be a progressively measurable process on the probability space, with an arbitrary state space $(X, \mathscr{X})$, let A' be a subset of $\mathbb{R}^+ \times X$, analytic over $\mathscr{B}(\mathbb{R}^+) \times \mathscr{X}$, and let A be a subset of X analytic over $\mathscr{X}$. Then the entry and hitting times of A' and A by $\{[t, x(t)], t \in \mathbb{R}^+\}$ and $\{x(t), t \in \mathbb{R}^+\}$, respectively, are optional for $\mathscr{F}^+(\cdot)$. In particular, if the state space is Polish and if the process is right continuous, then the distributions of these times depend on A', A, and the finite-dimensional distributions of $x(\cdot)$ together with the assumption of right continuity but do not depend on the choice of the probability space.*

To prove the theorem, recall that the sets on the right in (4.4) and the corresponding ones for A' and A were shown in Section I.9 to be in $\mathscr{F}(c)$; so the entry and hitting times are optional for $\mathscr{F}^+(\cdot)$. The last assertion of (b) is trivial because the probability of hitting and entry in the stated contexts, by time c, does not depend on the choice of probability space according to Section I.9.

Generalization

For $n \le +\infty$ let T_n be the nth entry time of A''; that is,

$$T_n(\omega) = \inf\{t \ge 0: (t, \omega) \in A'' \text{ for at least } n \text{ values of } t\}.$$

Then T_1 is the entry time of A'' and under the hypotheses of (a) is optional for $\mathscr{F}^+(\cdot)$. If $n < +\infty$ and if T_n is optional, the stochastic interval $]\!]T_n, +\infty[\![$ is progressively measurable. Hence T_{n+1} is the entry time of the progressively measurable set $A'' \cap]\!]T_n, +\infty[\![$ and so is optional. It follows that $T_1, T_2, \ldots$ and therefore also $T_\infty = \lim_{n\to\infty} T_n$ are optional. The corresponding assertion is true for A' and A. Further, if $\mathscr{F}(0)$ contains the null sets, part (a) of the theorem as just generalized is valid if A'' differs by an evanescent set from one as described in (a), and part (b) of the theorem as just generalized is valid if $\{x(\cdot), \mathscr{F}(\cdot)\}$ is indistinguishable from a progressively measurable process.

Last Hitting Times

We use the notation introduced at the beginning of this section. The *last hitting time* of A'' by $\{(t, \omega), t \in I\}$ is defined as

$$S'' = \sup\{t > 0: (t, \omega) \in A''\} \tag{4.5}$$

under the convention that this supremum is 0 of the set in question is empty. The last hitting times of A' and A, in the respective contexts of (4.2) and

(4.3), are defined in the obvious way. If $c > 0$, let $S(c)$ be the hitting time of A'' by $\{(c+t,\omega), t \in \mathbb{R}^+\}$. The function S'' is not optional except in trivial cases, but $\{S'' > c\} = \{S(c) < +\infty\}$ so that measurability problems for last hitting times can be reduced to dual problems for hitting times.

5. Application to Continuity Properties of Sample Functions

Let $\{x(t), \mathscr{F}(t), t \in \mathbb{R}^+\}$ be a process on $(\Omega, \mathscr{F}, P)$ with a Polish state space. Let $d(\cdot,\cdot)$ be a metric for the state space, define the strict right oscillation $O_r^*(t,\omega)$ of the sample function $x(\cdot,\omega)$ at t by

$$O_r^*(t,\omega) = \lim_{\delta \to 0} \sup_{t < r_1, r_2 < t+\delta} d[x(r_1,\omega), x(r_2,\omega)],$$

and define the right oscillation by

$$O_r^{**}(t,\omega) = \lim_{\delta \to 0} \sup_{t < r < t+\delta} d[x(t,\omega), x(r,\omega)].$$

If $x(\cdot)$ is extended real valued, define

$$\bar{x}_r(t,\omega) = \limsup_{r \downarrow t} x(r,\omega),$$

and define $\underline{x}_r(t,\omega)$ as the corresponding limit inferior. (Recall that according to our conventions $r > t$ in these superior and inferior limits.) We shall not need an analysis of the left limit properties of sample functions.

Theorem. *Let $\{x(t), \mathscr{F}(t), t \in \mathbb{R}^+\}$ be a progressively measurable process with a Polish state space. Suppose that $\mathscr{F}(\cdot)$ is right continuous and that the restriction of P to each σ algebra $\mathscr{F}(t)$ is complete.*

(a) *The processes $\{O_r^*(\cdot), \mathscr{F}(\cdot)\}$ and $\{O_r^{**}(\cdot), \mathscr{F}(\cdot)\}$ are progressively measurable.*
(b) *If the state space is $(\bar{\mathbb{R}}, \mathscr{B}(\bar{\mathbb{R}}))$, the processes $\{\bar{x}_r(\cdot), \mathscr{F}(\cdot)\}$ and $\{\underline{x}_r(\cdot), \mathscr{F}(\cdot)\}$ are progressively measurable.*

We prove the first assertion of (a) but omit the similar proofs of the other parts of the theorem.

Fix $c > 0$, $\alpha > 0$, and define $\mathscr{F}(c-) = \bigvee_{b<c} \mathscr{F}(b)$, $\mathscr{G} = \mathscr{B}([0,c[) \times \mathscr{F}(c-)$, $T(t,\omega) = \inf\{s \in [t,c[: O_r^*(s,\omega) \ge \alpha\}$, for $t \in [0,c[$. Then

$$\{(s,\omega): O_r^*(s,\omega) \ge \alpha, 0 \le s < c\} = \{(s,\omega): T(s,\omega) = s\}. \tag{5.1}$$

To prove that $\{O_r^*(\cdot), \mathscr{F}(\cdot)\}$ is progressively measurable it is sufficient to prove that the function $T(\cdot,\cdot)$ is $\mathscr{G}$ measurable, because then the set in (5.1) is in $\mathscr{G}$. Define

$$\Lambda(\alpha, m) = \{(s, r_1, r_2, \omega) \colon d[x(r_1, \omega), x(r_2, \omega)] \geq \alpha,$$
$$0 \leq s < r_1, r_2 < (s + 1/m) \wedge c\}.$$

Then $\Lambda(\alpha, m) \in \mathscr{B}([0, c[)^3 \times \mathscr{F}(c-)$. According to Theorem 5 of Appendix I, the projection $\Lambda'(\alpha, m)$ of $\Lambda(\alpha, m)$ on (s, ω) space is analytic over $\mathscr{G}$, as is therefore also the set $\bigcap_{k=1}^{\infty} \bigcap_{m=1}^{\infty} \Lambda'(\alpha - 1/k, m)$, which is the set in (5.1). Project the latter set onto ω space to see that for each value of s the function $T(s, \cdot)$ is $\mathscr{F}(c-)$ measurable. The desired measurability of $T(\cdot, \cdot)$ follows from the fact that $T(\cdot, \omega)$ is monotone increasing and left continuous, so that

$$T(t, \omega) = \lim_{n \to \infty} \sum_{j=1}^{n} T((j-1)c/n, \omega) 1_{[(j-1)c/n, jc/n[}(t),$$

and each sum on the right is $\mathscr{G}$ measurable.

Observation. One can derive from Theorem 5 the measurability of the Ω subsets corresponding to the sets of sample functions which have right limits on an interval, or are right continuous there, or have oscillation at most some special value there, etc. Observe, however, that in this discussion there has been no assurance that another progressively measurable process with the same finite-dimensional distributions will assign the same probabilities as the given process to the subsets of its probability space discussed above. For example, there has been no assurance that the finite-dimensional distributions of the process $\bar{x}_r(\cdot)$ are independent of the choice of $(\Omega, \mathscr{F}, P)$ with the given finite-dimensional $x(\cdot)$ distributions. In fact examples in Section I.11 show that such an assurance would be false. If however, $x(\cdot)$ is given as a right continuous process and if either the family $\mathscr{F}(\cdot)$ is already assigned and has the desired properties or $\mathscr{F}(t)$ is defined as the σ algebra generated by the null sets and $\mathscr{F}\{x(s), s \leq t\}$, then all the work in this section is applicable in this stricter context, and (Section I.12) the probabilities mentioned above will be the same for any other choice of $x(\cdot)$ with the given finite-dimensional distributions as long as the process is right continuous, or even almost surely right continuous.

6. Continuation of Section 5

Theorem. *Let $\{x(t), \mathscr{F}(t), t \in \mathbb{R}^+\}$ be a progressively measurable process on $(\Omega, \mathscr{F}, P)$ with a Polish state space. Suppose that $\mathscr{F}(\cdot)$ is right continuous and that the restriction of P to each σ algebra $\mathscr{F}(t)$ is complete.*

(a) *If whenever T is a bounded optional time almost every $x(\cdot)$ sample function has a right limit at T, then almost every $x(\cdot)$ sample function has a right limit at every parameter value.*

(b) *If whenever T is a bounded optional time almost every $x(\cdot)$ sample function is right continuous at T, then the process $x(\cdot)$ is almost surely right continuous.*

We use the notation of Theorem 5. If $\varepsilon > 0$, let T_ε be the entry time of the set

$$\{t, \omega): O_r^*(t, \omega) \geq \varepsilon\}.$$

Then T_ε is optional and $(T_\varepsilon(\omega), \omega)$ is in this set when $T_\varepsilon(\omega) < +\infty$; so the hypothesis of (a) implies that $T_\varepsilon \wedge n = n$ almost surely for every $n > 0$, and (a) follows. The proof of (b) is similar.

7. Predictable Optional Times

If $(\Omega, \mathscr{F}; \mathscr{F}(t), t \in \mathbb{R}^+)$ is a filtered measurable space, an optional time T will be called *predictable* if there is an increasing sequence $T_\bullet$ of optional times such that

$$T_n \leq T, \qquad T_n < T \quad \text{on } \{T > 0\}, \qquad \lim_{n \to \infty} T_n = T. \tag{7.1}$$

The sequence $T_\bullet$ is said to *announce* T. If there is a complete probability measure P on $\mathscr{F}$ and if $\mathscr{F}(0)$ contains the P null sets then if each condition in (7.1) is true merely P almost surely, it follows that T is predictable according to the following argument. Let Λ be the null set where at least one of the conditions in (7.1) fails to hold. Then $\Lambda \in \mathscr{F}(0)$, and if we redefine T_n on Λ as $(T - 2^{-n}) \vee 0$ or $T \wedge n$, according as T is finite or not, the modified optional times are still optional, and (7.1) is now satisfied. Thus under the stated hypotheses on P and $\mathscr{F}(0)$, an optional time equal almost surely to a predictable time is itself predictable.

We leave to the reader the proof that the maximum and minimum of a finite number of predictable times is predictable. The following properties are slightly deeper. Let $T_\bullet$ be a sequence of predictable times, and let $T_{n\bullet}$ be a sequence announcing T_n.

(a) The optional time $\sup_{n \in \mathbb{Z}} T_n$ is predictable because it is announced by the sequence

$$\left\{\max_{j \leq n} T_{jn}, n \in \mathbb{Z}^+\right\}.$$

(b) The optional time $T = \inf_{n \in \mathbb{Z}} T_n$ is predictable if $\bigcup_0^\infty \{T_n = T\} = \Omega$ up to a null set because if $\phi_n(\omega) = \inf\{k: T_{kn}(\omega) < T(\omega)\}$, the sequence $S_\bullet$ defined by

$$S_n = \sum_{m=0}^{\infty} T_{mn} 1_{\{\phi_n = m\}}$$

announces T up to a null set.

The σ Algebra $\mathscr{F}(T-)$

If T is a predictable optional time and if $S_.$ and $T_.$ are sequences of optional times announcing T, then

$$\lim_{n\to\infty} S_n \wedge T_m = T_m, \qquad \bigcup_{n=0}^{\infty} \{S_n \wedge T_m = T_m\} = \Omega;$$

so

$$\bigvee_{n=0}^{\infty} \mathscr{F}(S_n) \supset \bigvee_{n=0}^{\infty} \mathscr{F}(S_n \wedge T_m) = \mathscr{F}(T_m)$$

by Section 2(i5). Hence $\bigvee_{n=0}^{\infty} \mathscr{F}(S_n) \supset \bigvee_{m=0}^{\infty} \mathscr{F}(T_m)$, and there must be equality because $S_.$ and $T_.$ can be interchanged. Thus $\bigvee_{n=0}^{\infty} \mathscr{F}(T_n)$ does not depend on the choice of the announcing sequence $T_.$, and we denote this σ algebra by $\mathscr{F}(T-)$. According to Section 3, if S is a predictable optional time and if $x(\cdot)$ is a predictable process, then $S|_{\{S<+\infty\}}$ is $\mathscr{F}(S-)$ measurable.

Predictable Filtrations

A filtration $\{\mathscr{F}(t), t \in \mathbb{R}^+\}$ of a probability space will be said to be predictable if whenever T is a predictable time, $\mathscr{F}(T) = \mathscr{F}(T-)$. Observe that this condition implies that $\mathscr{F}(+\infty)$ must have been defined as $\bigvee_{t\in\mathbb{R}^+} \mathscr{F}(t)$.

Example: A predictable process defined by a predictable optional time. If T is an optional time, the process $A^T(\cdot)$ defined by setting $A^T(t) = 1_{[T,+\infty[}(t)$ ($=0$ when $T = +\infty$) is a right continuous adapted process. In particular, if T is predictable and if $T_.$ is an increasing sequence of optional times announcing T, the process $A^{T_n}(\cdot -) = 1_{]T_n,+\infty[}(\cdot)$ is adapted and left continuous, therefore predictable, and $\lim_{n\to\infty} A^{T_n}(t) = A^T(t)$; so $A^T(\cdot)$ is predictable.

8. Section Theorems

Let $(\Omega, \mathscr{F}, P)$ be a probability space, and let $\dot{A}$ be a subset of $\mathbb{R}^+ \times \Omega$. A *section theorem* is a theorem stating that under certain conditions on $\dot{A}$ there is a function T with useful properties, from Ω into $\bar{\mathbb{R}}^+$, such that a significant part of the graph of T lies in $\dot{A}$. In applications to probability it is desirable that T be optional relative to a given filtration of the probability space.

The following context will be useful. Define an outer measure P^* on Ω by

$$P^*\{\Lambda\} = \inf P\{\mathrm{M}: \mathrm{M} \supset \Lambda, \mathrm{M} \in \mathscr{F}\}.$$

Let $\hat{\mathscr{F}}$ be the class of finite unions of subsets of $\mathbb{R}^+ \times \Omega$ of the form $C \times \Lambda$ with C a compact subset of $\mathbb{R}^+$ and Λ in $\mathscr{F}$. If $\dot{A} \subset \mathbb{R}^+ \times \Omega$, let $\pi(\dot{A})$ be the projection of $\dot{A}$ on Ω, and define $I(\dot{A}) = P^*\{\pi(\dot{A})\}$. According to Appendix II.5, the set function I is a Choquet capacity on $\mathbb{R}^+ \times \Omega$ relative to $\hat{\mathscr{F}}$, and

$$\begin{aligned} &\mathscr{A}(\hat{\mathscr{F}}) \supset \mathscr{B}(\mathbb{R}^+) \times \mathscr{F}, \qquad \pi(\mathscr{A}(\hat{\mathscr{F}})) = \mathscr{F}; \\ &I(\dot{A}) = P\{\pi(\dot{A})\} \quad \text{if } \dot{A} \in \mathscr{A}(\hat{\mathscr{F}}). \end{aligned} \tag{8.1}$$

Theorem. *Let $\{\Omega, \mathscr{F}, P; \mathscr{F}(t), t \in \mathbb{R}^+\}$ be a filtered probability space. Suppose that $\mathscr{F}(\cdot)$ is right continuous and that $\mathscr{F}(0)$ contains the null sets.*

(a) *If $\dot{A} \in \mathscr{B}(\mathbb{R}^+) \times \mathscr{F}$, there is an $\mathscr{F}$ measurable function T from Ω into $\bar{\mathbb{R}}^+$ for which*

$$[\![T]\!] \subset \dot{A}, \qquad P\{T < +\infty\} = P\{\pi(\dot{A})\}. \tag{8.2}$$

(b) *If $\dot{A}$ is predictable and if $\varepsilon > 0$, there is a predictable time T for which*

$$[\![T]\!] \subset \dot{A}, \qquad P\{T < +\infty\} + \varepsilon \geq P\{\pi(\dot{A})\}. \tag{8.3}$$

Proof of (a). According to the Choquet capacity theorem, the set $\dot{A}$ is I capacitable; so there is an $\hat{\mathscr{F}}_\delta$ set $\dot{B}_1$ (that is, $\dot{B}_1$ is a countable intersection of $\hat{\mathscr{F}}$ sets) for which $\dot{B}_1 \subset \dot{A}$ and $I(\dot{B}_1) > I(\dot{A}) - 2^{-1}$; equivalently, $P\{\pi(\dot{B}_1)\} > P\{\pi(\dot{A})\} - 2^{-1}$. Similarly since the set $\dot{A}_2 = \dot{A} - (\mathbb{R}^+ \times \pi(\dot{B}_1))$ is in $\mathscr{B}(\mathbb{R}^+) \times \mathscr{F}$, there is an $\hat{\mathscr{F}}_\delta$ and $\dot{B}_2$ for which $\dot{B}_2 \subset \dot{A}_2$ and $P\{\pi(\dot{B}_2)\} > P\{\pi(\dot{A})\} - 2^{-2}$. Continuing in this way we find sequences $\dot{A}_1 = \dot{A}, \dot{A}_2, \ldots$ and $\dot{B}_1, \dot{B}_2, \ldots$ of subsets of $\dot{A}$, in $\mathscr{B}(\mathbb{R}^+) \times \mathscr{F}$ such that for $n \geq 1$,

$$\dot{B}_n \in \hat{\mathscr{F}}_\delta, \qquad \dot{A}_{n+1} = \dot{A}_n - (\mathbb{R}^+ \times \pi(\dot{B}_n)), \qquad P\{\pi(\dot{B}_n)\} > P\{\pi(\dot{A}_n)\} - 2^{-n}.$$

The set $\dot{B} = \bigcup_1^\infty \dot{B}_n$ is a subset of $\dot{A}$, is in $\mathscr{B}(\mathbb{R}^+) \times \mathscr{F}$, $P\{\pi(\dot{B})\} = P\{\pi(\dot{A})\}$, and by definition of $\hat{\mathscr{F}}$ the set $\{t: (t, \omega) \in \dot{B}\}$ is compact for every ω. The entry time T of $\dot{B}$ is measurable because

$$\{T \leq c\} = \pi\{\dot{A} \cap ([0, c] \times \Omega)\} \in \mathscr{A}(\mathscr{F}),$$

and T obviously satisfies the other conditions prescribed in (a). □

Observation. If $\dot{A}$ is supposed merely in $\mathscr{A}(\hat{\mathscr{F}})$, only the first step in the above proof remains valid, but since the number 2^{-1} in the first step can be replaced by an arbitrary strictly positive ε, there is an $\mathscr{F}$ measurable function T on Ω, the entry time of $\dot{B}_1$, for which (8.3) is true.

Proof of (b). Observe first that if $\dot{A}$ is a set in the algebra Γ''' generated by Γ'' (defined in Section I.14), then (b) is trivial because the entry time T of $\dot{A}$ is predictable and satisfies (8.3) with $\varepsilon = 0$. Second, suppose that $\dot{A} \in \Gamma'''_\delta$, that is, $\dot{A} = \bigcap_0^\infty \dot{A}_n$ is a countable intersection of Γ''' sets. Then the entry time of $\dot{A}$ is predictable because it is the supremum of the sequence of entry times of $\dot{A}_1, \dot{A}_2, \ldots$, and again (8.3) is satisfied with $\varepsilon = 0$. The general case is reduced to this special case as follows. According to (a), if $\dot{A}$ is predictable, there is an $\mathscr{F}$ measurable function S from Ω into $\bar{\mathbb{R}}^+$ for which $[\![S]\!] \subset \dot{A}$ and $P\{S < +\infty\} = P\{\pi(\dot{A})\}$. Define

$$\lambda(\dot{B}) = P\{\omega\colon (S(\omega), \omega) \in \dot{B}\} \tag{8.4}$$

for $\dot{B} \in \Gamma'''$. The set function λ is a measure on the algebra Γ''' and therefore (Hahn–Kolmogorov theorem) has a unique measure extension to the σ algebra generated by Γ''', that is, to the predictable σ algebra, and (8.4) is valid for the extension. The extended measure λ is supported by $\dot{A}$, with $\lambda(\dot{A}) = P\{\pi(\dot{A})\}$. According to the full statement of the Hahn–Kolmogorov theorem, there is a set $\dot{B}_\varepsilon$ in Γ'''_δ such that $\dot{B}_\varepsilon \subset \dot{A}$ and $\lambda(\dot{B}_\varepsilon) \geq P\{\pi(\dot{A})\} - \varepsilon$. Let T be the entry time of $\dot{B}_\varepsilon$. According to the result in the special case of (b) already treated, $[\![T]\!] \subset \dot{B}_\varepsilon$ and $P\{T < +\infty\} = P\{\pi(\dot{B}_\varepsilon)\}$; so $P\{T < +\infty\} \geq \lambda(\dot{B}_\varepsilon) \geq P\{\pi(\dot{A})\} - \varepsilon$, and T therefore satisfies (8.3). □

9. The Graph of a Predictable Time and the Entry Time of a Predictable Set

Theorem. *Let $(\Omega, \mathscr{F}, P; \mathscr{F}(t), t \in \mathbb{R}^+)$ be a filtered probability space, Suppose that $\mathscr{F}(\cdot)$ is right continuous and that $\mathscr{F}(0)$ contains the null sets.*

(a) *An optional time is predictable if and only if its graph is predictable.*
(b) *The intersection of a predictable set with the graph of its entry time is a predictable set. In particular, the entry time of a predictable set is predictable if the entry time graph is included in the set.*

Proof of (a). If T is an optional time announced by $T_\bullet$, then

$$[\![T]\!] = \left(\bigcap_{n=0}^{\infty}]\!]T_n, +\infty[\![-]\!]T, +\infty[\![\right) \cup (\{0\} \times \{T = 0\}).$$

The sets $]\!]T_n, +\infty[\![$ and $]\!]T, +\infty[\![$ are predictable because their indicator functions define adapted left continuous processes. Hence $[\![T]\!]$ is a predictable set. Conversely, suppose that T is an optional time whose graph is a predictable set. Then [Theorem 8(b)] for $n \geq 1$ there is a predictable time T_n such that $P\{T_n \neq T\} < 1/n$ and that $T_n = +\infty$ where $T_n \neq T$. Hence $\bigcup_1^\infty \{T_n = T\} = \Omega$ up to a null set so [from Section 7, property (b)] it follows that T is predictable. □

Proof of (b). If a set A is predictable and if T is its entry time, the stochastic interval $]\!]T, +\infty[\![$ is predictable because its indicator function defines an adapted left continuous process. Hence the set $A -]\!]T, +\infty[\![= A \cap [\![T]\!]$ is predictable.

Generalization

Let Ω, $\mathscr{F}$, P, $\mathscr{F}(\cdot)$ be as in Theorem 9. Let A'' be a nearly predictable subset of $\mathbb{R}^+ \times \Omega$, and let T_n be the nth entry time of A'',

$$T_n(\omega) = \inf\{t \in \mathbb{R}^+ : (t, \omega) \in A'' \text{ for at least } n \text{ values of } t\},$$

and suppose that the graph of T_n is up to an evanescent set included in A''. Then we show that $T_1, T_2, \ldots, T_\infty = \lim_{n\to\infty} T_n$ are predictable optional times. We can suppose that A'' is predictable and therefore progressively measurable and apply Section 4 (Generalization) to find that these functions are optional times. We need only show in addition that each time T_n with finite n is predictable. According to Theorem 9 the optional time T_1 is predictable. For $n > 1$ the stochastic interval $]\!]T_{n-1}, +\infty[\![$ is predictable because its indicator function defines a left continuous adapted process; so T_n is the entry time of the predictable set $A'' \cap]\!]T_{n-1}, +\infty[\![$, and the hypotheses of Theorem 9 are satisfied so T_n is predictable.

10. Semipolar Subsets of $\mathbb{R}^+ \times \Omega$

Let $(\Omega, \mathscr{F}, P; \mathscr{F}(t), t \in \mathbb{R}^+)$ be a filtered probability space. Suppose that $\mathscr{F}(\cdot)$ is right continuous and that $\mathscr{F}(0)$ contains the null sets.

In discussing potential theory on $\mathbb{R}^+ \times \Omega$ we shall need appropriate definitions of small sets, the counterparts of the polar and semipolar sets of classical and parabolic potential theory. The counterpart of a polar set that we adopt in the present context is an evanescent set, and we have already (Section I.8) defined "quasi everywhere on $\mathbb{R}^+ \times \Omega$" accordingly. The following remark suggests the definition to be given of a semipolar subset of $\mathbb{R}^+ \times \Omega$.

It is trivial that if T is optional and if $[\![T]\!]$ is evanescent, then $P\{T < +\infty\} = 0$. Conversely, consider the class of subsets $\dot{A}$ of $\mathbb{R}^+ \times \Omega$ with the property that $P\{T < +\infty\} = 0$ whenever T is optional and $[\![T]\!] \subset \dot{A}$. If $\dot{A}$ satisfies this restriction and is sufficiently smooth, for example (by Theorem 9), if $\dot{A}$ is predictable, then $\dot{A}$ is evanescent. The negation of this property of $\dot{A}$ leads to the following definition.

A subset $\dot{A}$ of $\mathbb{R}^+ \times \Omega$ will be called *semipolar* if there is a sequence $T_\cdot$ of optional times such that $\dot{A} \subset \bigcup_0^\infty [\![T_n]\!]$ up to an evanescent set. Subsets of semipolar sets are semipolar, and countable unions of semipolar sets are semipolar. If $\dot{A}$ is semipolar, then the set of values of t with (t, ω) in $\dot{A}$ is countable for almost every ω. Conversely, if $\dot{A}$ is sufficiently smooth, for example if $\dot{A}$ is predictable, it has been shown that $\dot{A}$ is semipolar when this countability condition is satisfied, but we shall not need this result.

11. The Classes D and L^p of Stochastic Processes

In this section the concepts are relative to a specified filtered probability space $(\Omega, \mathscr{F}, P; \mathscr{F}(t), t \in I)$; I is an arbitrary linearly ordered parameter set. The processes involved are defined on this space, adapted, and have state space $\mathbb{R}$.

The Class D

This class of processes is the class of processes $x(\cdot)$ for which the family

$$\{x(T)\colon T \text{ optional, countably valued with values in } I\} \tag{11.1}$$

of random variables is uniformly integrable. A class **D** process is uniformly integrable but a uniformly integrable process need not be in the class **D**. The family (11.1) is uniformly integrable if and only if there is a uniform integrability test function Φ and a constant c such that

$$E\{\Phi[|x(T)|]\} \leq c \tag{11.2}$$

for all T in (11.1). Observe that if a point $+\infty$ and a σ algebra $\mathscr{F}(+\infty)$ are introduced as in Section 1 and if $x(+\infty)$ is an arbitrary $\mathscr{F}(+\infty)$ measurable and integrable random variable, then if the original process was in **D** relative to the original filtration the augmented process is in **D** relative to the augmented filtration. In fact the family (11.1) augmented by the random variable $x(+\infty)$ is uniformly integrable so there is a uniform integrability test function Φ and a constant c such that (11.2) is true for T as in (11.1) and also so that $E\{\Phi[|x(+\infty)|]\} \leq c$. The augmented process, in which T is allowed to have the value $+\infty$, then satisfies (11.2) with c replaced by $2c$.

In particular, if $I = \mathbb{R}^+$, if $x(\cdot)$ is an almost surely right continuous class **D** process, and if $\mathscr{F}(0)$ contains the null sets, then the family (11.1) is uniformly integrable even if T is allowed to be an arbitrary finite-valued optional time S. To see this observe that if (11.2) is true for T as in (11.1), then (11.2) is true for $[S]_n$ [see Section 2, Example (b2), for the definition of this discretized optional time]. When $n \to \infty$, Fatou's lemma implies that (11.2) is true for $T = S$. Finally, if $x(+\infty)$ is an $\mathscr{F}(+\infty)$ measurable and integrable random variable, we find as above that the augmented process is in **D** if the original process was, and if so, the augmented family (11.1) with T now an arbitrary optional time is uniformly integrable.

The Class $\mathbf{L}^p$ ($p \geq 1$)

This class of processes is the class of processes $x(\cdot)$ for which

$$\sup\{E\{|x(T)|^p\} : T \text{ optional, countably valued with values in } I\} < +\infty \tag{11.3}$$

Observe that $\mathbf{L}^p \subset \mathbf{D}$ when $p > 1$. Adjoin a pair $\{x(+\infty), \mathscr{F}(+\infty)\}$ to each process $x(\cdot)$ as in the discussion of the class **D**, but under the additional condition of finiteness of $E\{|x(+\infty)|^p\}$. Follow that discussion to show that (11.3) is true for the augmented process if true for the original process and that then in the continuous parameter context (11.3) is true with T an arbitrary ($\leq +\infty$) optional time.

12. Decomposition of Optional Times; Accessible and Totally Inaccessible Optional Times

In this section the reference space $(\Omega, \mathscr{F}, P; \mathscr{F}(t), t \in \mathbb{R}^+)$ is a probability space provided with a right continuous filtration for which $\mathscr{F}(0)$ contains the null sets.

Totally Inaccessible Optional Times

(The reader is warned that the now accepted definition of these optional times differs slightly from the original version.) An optional time T is said to be *totally inaccessible* if $P\{S = T < +\infty\} = 0$ whenever S is a predictable optional time. In particular, the distribution function of T on $\mathbb{R}^+$ is continuous; that is, $P\{T = c\} = 0$ for every finite constant c. Observe that this definition only involves the properties of T on $\{T < +\infty\}$; total inaccessibility is a property of T where T is finite valued. A striking token of this fact is that an optional time T is both predictable and totally inaccessible if and only if $T = +\infty$ almost surely. Observe also that predictability of an optional time is defined relative to a measurable space provided with a filtration, but total inaccessibility involves a measure space.

A Characterization of Total Inaccessibility

An optional time T is totally inaccessible if and only if $T > 0$ almost surely and if whenever $S_.$ is an increasing sequence of optional times with limit S, the increasing sequence $\{S_. \geq T, T < +\infty\}$ of sets has union $\{S \geq T, T < +\infty\}$. In fact for an arbitrary optional time the latter set includes the union, and the difference between the two is the $\mathscr{F}(S)$ set

$$\Lambda = \{S_n < S = T < +\infty \text{ for all } n\}.$$

If $S_n' = S_n$ on the $\mathscr{F}(S_n)$ set $\{S_n < S < +\infty\}$ and $S_n' = +\infty$ elsewhere on Ω, then the sequence $n \mapsto S_n' \wedge n$ is an increasing sequence of optional times whose limit, announced by this sequence, is predictable and equal to T on Λ. Hence Λ is null if T is totally inaccessible. Conversely, if Λ is null for all sequences $S_.$, it follows that $P\{S = T < +\infty\} = 0$ whenever S is a predictable optional time not almost surely 0; so T (by hypothesis almost surely strictly positive) is totally inaccessible.

Decomposition of an Optional Time

Let T be an optional time, and choose a sequence $S_.$ of predictable optional times maximizing $P\{\bigcup_0^\infty \{T = S_n < +\infty\}\}$ for all such sequences. Let α be the maximum probability.

Special Case: $\alpha = 0$. In this case and only in this case T is totally inaccessible.

Special Case: $\alpha = 1$. In this case T is called *accessible* and necessarily

$$[\![T]\!] = \bigcup_0^\infty [\![S_n]\!]$$

up to an evanescent set. Here the union can be replaced by a union $\bigcup_0^\infty [\![S_n']\!]$ of disjoint graphs of predictable optional times by the usual procedure: define

$$\Lambda_0 = \{S_0 < +\infty\},\ \Lambda_n = \bigcap_{k=1}^{n-1} \{S_n \neq S_k\} \cap \{S_n < +\infty\} \quad \text{if } n > 0,$$

$$S_n' = \begin{cases} S_n & \text{on } \Lambda_n \\ +\infty & \text{on } \Omega - \Lambda_n. \end{cases}$$

Thus to an accessible optional time T corresponds a sequence $S'_.$ of predictable optional times and a sequence $\Lambda_.$ of disjoint subsets of Ω such that for all n

$$T = S'_n < +\infty \quad \text{on } \Lambda_n, \qquad S'_n = +\infty \quad \text{on } \Omega - \Lambda_n, \quad \text{and} \quad P\left\{\bigcup_0^\infty \Lambda_m\right\} = 1.$$

Special Case: $0 < \alpha < 1$. In this case define the accessible optional time T' and the totally inaccessible optional time T'' by

$$T'(\omega) = \begin{cases} T(\omega) & \text{if } T(\omega) = S_n(\omega) < +\infty, \\ +\infty & \text{otherwise}; \end{cases}$$

$$T'' = \begin{cases} T(\omega) & \text{if } T'(\omega) = +\infty, \\ +\infty & \text{otherwise}. \end{cases}$$

These optional times are uniquely determined up to null sets, and there is a subset Λ of Ω such that $T' = T$ on Λ and $T'' = T$ on $\Omega - \Lambda$. Thus $[\![T]\!] = [\![T']\!] \cup [\![T'']\!]$ up to an evanescent set. The optional times T' and T'' are called, respectively, the accessible and totally inaccessible components of T.

EXAMPLE (Predictable process defined by an accessible optional time). It was pointed out in Section 7 that if T is a predictable optional time, the right continuous process $A^T(\cdot) = 1_{[T, +\infty[}(\cdot)$ is predictable. In view of the decomposition of an accessible optional time noted above, $A^T(\cdot)$ is nearly predictable if T is accessible. In fact, if up to an evanescent set $[\![T]\!] = \bigcup_0^\infty [\![T_n]\!]$ (disjoint union) with T_n predictable, then $A^T(\cdot)$ is indistinguishable from $\Sigma_0^\infty A^{T_n}(\cdot)$.

Chapter III

Elements of Martingale Theory

1. Definitions

Let $(\Omega, \mathscr{F}, P; \mathscr{F}(t), t \in I\}$ be a filtered probability space, and let $\{x(\cdot), \mathscr{F}(\cdot)\}$ be a process on this space, with state space $(\bar{\mathbb{R}}, \mathscr{B}(\bar{\mathbb{R}}))$. The process is called a *supermartingale* if the process random variables are integrable and if the *supermartingale inequality*

$$x(s) \geq E\{x(t)|\mathscr{F}(s)\} \quad \text{a.s. if } s < t \tag{1.1}$$

is satisfied. The exceptional null set may depend on s and t. If I is a set of consecutive integers, inequality (1.1) for $t = s + 1$ implies (1.1) for all pairs s, t with $s < t$. If the inequality is reversed the process is called a *submartingale*, and if there is equality in (1.1) the process is called a *martingale*. The martingale definition is sometimes also applied to complex-valued or vector-valued random variables, but in this book the state space will always be $(\bar{\mathbb{R}}, \mathscr{B}(\bar{\mathbb{R}}))$ unless some other state space is specified. *Martingale theory* refers to the mathematics of supermartingales and submartingales as well as martingales.

A supermartingale is a mathematical model for the fortune of a player of a game in which the player has fortune $x(s)$ at time s and given the past up to and including s the player expects his fortune at the later time t to be at most $x(s)$. Thus a supermartingale represents an unfavorable game; in this context a martingale represents a fair game, and a submartingale represents a favorable game.

The negative of a supermartingale is a submartingale, and an adapted process is a martingale if and only if it is both a supermartingale and a submartingale. In any one of the three cases the process is said to be *left* [*right*] *closed* if I has a first [last] element and is said to be *left* [*right*] *closable* if a first [last] element can be adjoined to I, together with an appropriate σ algebra, to get an enlarged process which is left [right] closed and of the same type as the given one. If a process is right closed, say with last parameter element ∞, the σ algebra $\mathscr{F}(\infty)$ is not involved in the appropriate version of (1.1) and is restricted only by the condition that the process is adapted. A martingale is always left closable by the pair

$[c, (\varnothing, \Omega)]$, where c is the common expected value of the process random variables.

EXAMPLE. Let $I = \mathbb{Z}$ in increasing order, let $\mathscr{F}(n)$ consist, for each n, of the empty set and the whole probability space, and consider the process all of whose sample functions are identically 1. This process is a martingale and becomes a right closed martingale if the constant function 1 is adjoined at the end or a right closed supermartingale if the constant function 0 is adjoined at the end.

Choice of Filtration

Let $\{x(\cdot), \mathscr{F}(\cdot)\}$ be a supermartingale. The following remarks are made for the supermartingale case but are valid with the obvious changes in the other two cases. Define $\mathscr{F}_0(t) = \mathscr{F}\{x(s), s \leq t\}$. Then $\mathscr{F}_0(t) \subset \mathscr{F}(t)$ and $\{x(\cdot), \mathscr{F}_0(\cdot)\}$ is a supermartingale because if $s < t$,

$$x(s) = E\{x(s)|\mathscr{F}_0(s)\} \geq E\{E\{x(t)|\mathscr{F}(s)\}|\mathscr{F}_0(s)\} = E\{x(t)|\mathscr{F}_0(s)\} \quad \text{a.s.} \tag{1.2}$$

Thus $\mathscr{F}_0(\cdot)$ is the minimal filtration $\mathscr{F}(\cdot)$ for which $\{x(\cdot), \mathscr{F}(\cdot)\}$ is a supermartingale. The larger the σ algebras of the filtration, the more one knows about the process, more precisely, the more (1.1) implies. It is sometimes convenient to suppose that each σ algebra $\mathscr{F}(t)$ contains all the null sets. If this inclusion is not already true, $\mathscr{F}(t)$ can be replaced by $\mathscr{F}_1(t)$, the smallest σ algebra containing $\mathscr{F}(t)$ and the null sets, to obtain a filtration $\mathscr{F}_1(\cdot)$ for which $\{x(\cdot), \mathscr{F}_1(\cdot)\}$ is a supermartingale and $\mathscr{F}_1(\cdot)$ contains the null sets.

Omission of the Filtration in Martingale Theory Notation

If a process $x(\cdot)$ is described as a supermartingale, martingale, or submartingale without specification of the reference filtration, it is to be understood that the reference filtration is $\mathscr{F}_0(\cdot)$ or $\mathscr{F}_1(\cdot)$ as defined in the preceding paragraph.

2. Examples

EXAMPLE (a). If $\mathscr{F}(\cdot)$ is a filtration (arbitrary linearly ordered parameter set I) of a probability space, if x is an integrable random variable, and if $x(t) = E\{x|\mathscr{F}(t)\}$, then the process $\{x(\cdot), \mathscr{F}(\cdot)\}$ is a martingale. Every right closable martingale is of this type. Suppose in this example that T is a countably valued optional time with values in I. Then we shall show that

$$x(T) = E\{x|\mathscr{F}(T)\} \quad \text{a.s.,} \tag{2.1}$$

and if S is a second such optional time, we shall show that

$$E\{x(S)|\mathscr{F}(T)\} = E\{x(T)|\mathscr{F}(S)\} = x(S \wedge T) \quad \text{a.s.} \tag{2.2}$$

In the most common application of these results the parameter set I is $0, 1, \ldots, +\infty$ in the indicated order, and then optional times are necessarily countably valued. A continuous parameter version of (2.1) and (2.2) will be derived in Section IV.2.

Since it is sufficient to prove (2.1) for $x \vee 0$ and $(-x) \vee 0$, we can assume in proving (2.1) that x is positive. If $\Lambda \in \mathscr{F}(T)$ and if $\Lambda_r = \Lambda \cap \{T = r\}$, then $\Lambda_r \in \mathscr{F}(r)$; so $x(T)$ is $\mathscr{F}(T)$ measurable, and

$$\int_{\Lambda_r} x\,dP = \int_{\Lambda_r} x(r)\,dP = \int_{\Lambda_r} x(T)\,dP. \tag{2.3}$$

Sum in (2.3) over the countably many values of r taken on by T to find that $x(T)$ is integrable. The equality (2.3) for all r is an integrated version of (2.1). In the proof of (2.2) we shall use freely the properties in Section II.2 of optional times and the σ algebras they determine. The translation of these properties into the present context of countably valued optional times is trivial. If $S \geq T$ $[S \leq T]$, then $\mathscr{F}(T) \subset \mathscr{F}(S)$ $[\mathscr{F}(S) \subset \mathscr{F}(T)]$; so (2.2) reduces to the special case I(4.2). We now reduce the general case to this special one. Observe that the set $\{S \leq T\}$ and its complement are in $\mathscr{F}(S \wedge T)$ and that $x(S \wedge T)$ is $\mathscr{F}(S \wedge T)$ $(\subset \mathscr{F}(T))$ measurable; so

$$\begin{aligned} E\{x(S)|\mathscr{F}(T)\} &= E\{1_{\{S \leq T\}} x(S \wedge T)|\mathscr{F}(T)\} + E\{1_{\{S > T\}} x(S \vee T)|\mathscr{F}(T)\} \\ &= 1_{\{S \leq T\}} x(S \wedge T) + 1_{\{S > T\}} E\{E\{x|\mathscr{F}(S \vee T)\}|\mathscr{F}(T)\} \\ &= 1_{\{S \leq T\}} x(S \wedge T) + 1_{\{S > T\}} x(T) = x(S \wedge T) \quad \text{a.s.} \end{aligned}$$

Interchange S and T to complete the proof of (2.2).

Example (b). Let $y(1), y(2), \ldots$ be mutually independent integrable random variables, and let $x(n) = \Sigma_1^n y(j)$, $\mathscr{F}(n) = \mathscr{F}\{y(j), j \leq n\}$. Then $\{x(\cdot), \mathscr{F}(\cdot)\}$ is a submartingale if $E\{y(j)\} \geq 0$ for $j > 1$ and is a martingale if these expectations vanish.

Example (c) [Continuous parameter version of Example (b)]. Let $\{x(t), t \in \mathbb{R}^+\}$ be a stochastic process with state space $\mathbb{R}$ and independent increments; that is, $0 \leq t_1 < \cdots < t_k$ implies that the increments $x(t_2) - x(t_1), \ldots, x(t_k) - x(t_{k-1})$ are mutually independent. Define $\mathscr{F}(t) = \mathscr{F}\{x(s) - x(0), s \leq t\}$. Then if every increment $x(t) - x(s)$ with $t > s$ is integrable and has a positive expectation, the process $\{x(\cdot) - x(0), \mathscr{F}(\cdot)\}$ is a submartingale.

If these expectations all vanish, the process is a martingale. We shall see that Brownian motion is a special case of this example. To clarify the ideas involved here and in Example (b), we prove the submartingale assertion. Using the fact that $x(t) - x(s)$ is independent of $\mathscr{F}(s)$ when $t > s$, we find

$$\begin{aligned} E\{x(t) - x(0)|\mathscr{F}(s)\} &= E\{x(s) - x(0)|\mathscr{F}(s)\} + E\{x(t) - x(s)|\mathscr{F}(s)\} \\ &= x(s) - x(0) + E\{x(t) - x(s)\} \geq x(s) - x(0) \quad \text{a.s.}, \end{aligned}$$

as was to be proved.

Example (d). Let Ω be the interval $[0, 1]$, let $\mathscr{F}$ be the class of Borel subsets of Ω, and let P be Lebesgue measure on $\mathscr{F}$. Define a stochastic process $\{x(\cdot), \mathscr{F}(\cdot)\}$ by

$$x(n) = \begin{cases} 2^n & \text{on } [0, 2^{-n}], \\ 0 & \text{on }]2^{-n}, 1], \end{cases} \qquad \mathscr{F}(n) = \mathscr{F}\{x(1), \ldots, x(n)\}, \qquad n > 0.$$

The process $\{x(\cdot), \mathscr{F}(\cdot)\}$ is a martingale and

$$E\{x(n)\} = 1, \qquad \lim_{n\to\infty} x(n) = 0 \quad \text{a.s.}$$

This process cannot be uniformly integrable because the sequence $x(\cdot)$ is not L^1 convergent to 0, although it is almost surely convergent to 0. The process is a right closable supermartingale, right closable by the constant random variable 0, for example. The process is not a right closable submartingale and hence not a right closable martingale because if x right closes this submartingale, then $E\{x|\mathscr{F}(n)\} \geq x(n)$ almost surely, and this inequality implies uniform integrability of $x(\cdot)$ because [Section I.4(i)] the family of conditional expectations of an integrable random variable is uniformly integrable.

3. Elementary Properties (Arbitrary Simply Ordered Parameter Set)

Proofs are omitted if trivial.

(a) If $\{x(\cdot), \mathscr{F}(\cdot)\}$ is a supermartingale, the function $t \mapsto E\{x(t)\}$ is decreasing, and if $s \leq t_1 \leq t_2$,

$$E\{x(t_2)|\mathscr{F}(s)\} \leq E\{x(t_1)|\mathscr{F}(s)\} \quad \text{a.s.} \tag{3.1}$$

The proof of the second assertion is simply a manipulation of conditional expectations:

$$E\{x(t_2)|\mathscr{F}(s)\} = E\{E\{x(t_2)|\mathscr{F}(t_1)\}|\mathscr{F}(s)\} \leq E\{x(t_1)|\mathscr{F}(s)\} \quad \text{a.s.} \tag{3.2}$$

(b) If for each $n \in \mathbb{Z}^+$ the process $\{x_n(\cdot), \mathscr{F}(\cdot)\}$ is a positive supermartingale and if $s < t$, then

$$\liminf_{n\to\infty} x_n(s) \geq \liminf_{n\to\infty} E\{x_n(t)|\mathscr{F}(s)\} \geq E\left\{\liminf_{n\to\infty} x_n(t)|\mathscr{F}(s)\right\} \quad \text{a.s.}, \tag{3.3}$$

so that the process $\{\liminf_{n\to\infty} x_n(\cdot), \mathscr{F}(\cdot)\}$ is a supermartingale if its random variables are integrable.

(c) If $\{x(\cdot), \mathscr{F}(\cdot)\}$ is a submartingale and if f is an increasing convex function with $E\{|f[x(t)]|\} < +\infty$ for all t, the process $\{f[x(\cdot)], \mathscr{F}(\cdot)\}$ is a submartingale. If the first process is a martingale, f need not be increasing for the conclusion to hold. The analogs of these properties in classical potential theory are in Section 1.II.9. To prove the first one, for example, apply Jensen's inequality for conditional expectations to deduce, when $s < t$,

$$f[x(s)] \leq f[E\{x(t)|\mathscr{F}(s)\}] \leq E\{f[x(t)]|\mathscr{F}(s)\} \quad \text{a.e.}, \tag{3.4}$$

and this is the desired submartingale inequality.

(d) If $\{x(t), \mathscr{F}(t); t \in I\}$ is a supermartingale, define $c = \sup_{t\in I} E\{x(t)\}$. If the supermartingale is left closed with smallest parameter value α, then $E\{x(\alpha)\} = c$; so left closability of the supermartingale implies finiteness of c. Conversely, if c is finite, the pair $[c, (\varnothing, \Omega)]$ left closes the supermartingale.

(e) If $\{x(\cdot), \mathscr{F}(\cdot)\}$ is a right closed positive submartingale with largest parameter value β, then this process is uniformly integrable [Section I.4(i)] because each random variable of the process is majorized by a conditional expectation of $x(\beta)$. Positivity is a necessary hypothesis here. In particular, a right closed martingale $\{y(\cdot), \mathscr{F}(\cdot)\}$ is uniformly integrable because $\{|y(\cdot)|, \mathscr{F}(\cdot)\}$ is a positive right closed submartingale. In the other direction, if $\{x(t), \mathscr{F}(t); t \in I\}$ is a uniformly integrable submartingale, it is right closable according to the following argument. If $\Lambda \in \mathscr{F}(s)$, the function $t \mapsto E\{x(t); \Lambda\}$ is an increasing function for $t \geq s$. Define $\phi(\Lambda)$ as the supremum (limit as t increases) of this function. The set function ϕ as so defined on the set algebra $\bigcup_{s\in I} \mathscr{F}(s)$ is additive, and in view of the uniform integrability hypothesis

$$\lim_{P\{\Lambda\}\to 0} \phi(\Lambda) = 0. \tag{3.5}$$

Hence ϕ is countably additive and thus has a countably additive extension to the σ algebra $\bigvee_{s\in I} \mathscr{F}(s)$, and this extension is absolutely continuous relative to ϕ. If $x = d\phi/dP$, the pair $[x, \bigvee_{s\in I}\mathscr{F}(s)]$ right closes the given submartingale. A trivial variation of this argument shows that a uniformly integrable martingale is right closable; the converse martingale result was noted above.

(f) A supermartingale which majorizes a uniformly integrable submartingale is right closable. In fact, if $\{y(\cdot), \mathscr{F}(\cdot)\}$ is a supermartingale which majorizes the uniformly integrable submartingale $\{x(\cdot), \mathscr{F}(\cdot)\}$, we use the notation of (e) and show that the pair just obtained which closes $x(\cdot)$ also closes $y(\cdot)$. For $\Lambda \in \mathscr{F}(s)$ and $t \geq s$,

$$E\{y(s);\Lambda\} \geq E\{y(t);\Lambda\} \geq E\{x(t);\Lambda\} \to \phi(\Lambda) \qquad (t\uparrow),$$

and the inequality here between first and last terms is the integrated form of the desired almost sure supermartingale inequality $y(s) \geq E\{x|\mathscr{F}(s)\}$.

4. The Parameter Set in Martingale Theory

If $\{x(t), \mathscr{F}(t), t\in I\}$ is a submartingale and b is a constant, the process $[x(\cdot) - b] \vee 0 = y_b(\cdot)$ is a submartingale [Section 3(c)]; so $s < t$ implies that $E\{y_b(t)\} \geq E\{y_b(s)\}$. If there is equality here for some triple s, t, b, then the pair $[y_b(s), \mathscr{F}(s)]$, $[y_b(t), \mathscr{F}(t)]$ is a martingale; so

$$\int_{\{x(s)\leq b\}} [x(t) - b] \vee 0\, dP = \int_{\{x(s)\leq b\}} [x(s) - b] \vee 0\, dP = 0, \tag{4.1}$$

and therefore $x(t) \leq b$ almost everywhere where $x(s) \leq b$. If $b_{\bullet}$ is a sequence dense in $\mathbb{R}$ and if $E\{y_{b_j}(t)\} = E\{y_{b_j}(s)\}$ for all j, it follows that $x(t) \leq x(s)$ almost surely, and therefore there is almost sure equality because $E\{x(t)\} \geq E\{x(s)\}$. Choose $c_j > 0$ so small that $\Sigma_0^\infty c_j(1 + |b_j|) < +\infty$. The process $y(\cdot) = \Sigma_0^\infty c_j y_{b_j}(\cdot)$ is a submartingale with the property that $t \mapsto E\{y(t)\}$ is an increasing function and that the equality $E\{y(s)\} = E\{y(t)\}$ implies that $x(s) = x(t)$ almost surely. If the parameter set I is mapped into $\mathbb{R}$ by ϕ: $t \mapsto E\{y(t)\}$, then ϕ is monotone increasing and $\phi(s) = \phi(t)$ if and only if $x(s) = x(t)$ almost surely. Thus it is not an essential restriction to suppose that the ordered parameter set in martingale theory is a subset of $\mathbb{R}$ ordered by inequality.

5. Convergence of Supermartingale Families

If Γ is a family of supermartingales adapted to $\mathscr{F}(\cdot)$, then ess inf Γ (defined in Section I.8) is also a supermartingale if its random variables have finite expectations because if s and t are parameter values with $s < t$,

$$E\{(\text{ess inf}\,\Gamma)(t)|\mathscr{F}(s)\} \leq x(s) \quad \text{a.s.} \tag{5.1}$$

for each process $x(\cdot)$ in Γ, and therefore the right side can be replaced by $(\operatorname{ess\,inf}\Gamma)(s)$ to obtain the supermartingale inequality. In the other direction, if Γ is a family of supermartingales which is directed upward, the process ess sup Γ is a supermartingale if its random variables have finite expectations in view of the monotone convergence theorem for conditional expectations. If the processes in this upward-directed set Γ are martingales, this reasoning shows that the process ess sup Γ is a martingale.

If Γ is an upward-directed family of supermartingales (essential order of processes) with essential limit a supermartingale, there is an increasing sequence $\{x_n(\cdot), n \in \mathbb{Z}^+\}$ in Γ with the property that $\operatorname{ess\,sup}\Gamma = \sup_{n \in \mathbb{Z}^+} x_n(\cdot)$ up to a standard modification. To see this, it can be supposed (Section 4) that the parameter set of the process $x'(\cdot) = \operatorname{ess\,sup}\Gamma$ is a subset of $\mathbb{R}$ ordered by inequality. The monotone decreasing function $t \mapsto E\{x'(t)\}$ is the supremum of the upward-directed family of decreasing functions $t \mapsto E\{x(t)\}$ for $x(\cdot)$ in Γ. Hence there is a sequence $\{x_n(\cdot), n \in \mathbb{Z}^+\}$, increasing in the essential order, for which

$$\lim_{n\to\infty} E\{x_n(t)\} = E\{x'(t)\}$$

for every parameter value t, and it follows that, for every parameter value t, $\sup_{n \in \mathbb{Z}^+} x_n(t) = x'(t)$ almost surely, as was to be proved. If the hypothesis that the essential limit of Γ is a supermartingale, equivalently, that the random variables of this process are integrable, is not satisfied, the conclusion still holds, as can be seen by applying the result just obtained to the family $\{x(\cdot) \wedge k : x(\cdot) \in \Gamma\}$ for each positive integer k.

Similarly, but somewhat more easily, if Γ is a family of supermartingales and if ess inf Γ is a supermartingale, that is, if the random variables of this process are integrable, there is a countable subset Γ_1 of Γ for which ess inf Γ $= \inf \Gamma_1$ up to a standard modification. See the Fundamental Convergence Theorem for supermartingales (Theorem IV.5) for a more complete study of the essential infimum of a family of right continuous supermartingales.

6. Optional Sampling Theorem (Bounded Optional Times)

The idea that if a game is unfavorable to a player, it looks unfavorable to him at random times chosen without foreknowledge suggests the following integral parameter theorem.

Theorem. *Let $\{x(n), \mathscr{F}(n), n \in \mathbb{Z}^+\}$ be a supermartingale. If T is a bounded optional time, then $x(T)$ is integrable. If S and T are bounded optional times with $S \le T$, then*

$$x(S) \ge E\{x(T) \mid \mathscr{F}(S)\} \quad \text{a.s.,} \tag{6.1}$$

and there is almost sure equality if the process is a martingale.

If T is bounded by k, then

$$E\{|x(T)|\} \leq \sum_0^k E\{|x(j)|\} < +\infty.$$

If $S \leq T$, if $\Lambda \in \mathscr{F}(S)$, and if $\Lambda_j = \Lambda \cap \{S = j\}$, then $\Lambda_j \in \mathscr{F}(j)$; if $\Lambda_{ji} = \Lambda_j \cap \{T > i\}$, then $\Lambda_{ji} \in \mathscr{F}(i)$ when $i \geq j$. Hence the supermartingale inequality yields

$$\begin{aligned} \int_\Lambda [x(T) - x(S)]\,dP &= \sum_{j=0}^k \int_{\Lambda_j} \sum_{i=j}^{T-1} [x(i+1) - x(i)]\,dP \\ &= \sum_{j=0}^k \sum_{i=j}^k \int_{\Lambda_{ji}} [x(i+1) - x(i)]\,dP \leq 0, \end{aligned} \tag{6.2}$$

and there is equality if the given process is a martingale. Thus the integrated version of the desired supermartingale inequality or martingale equality is true.

Observation. According to this theorem if $T_\bullet$ is an increasing sequence of bounded optional times the process $\{x(T_\bullet), \mathscr{F}(T_\bullet)\}$ is a supermartingale, or a martingale if the original process is a martingale.

Application to Right Closable Submartingales and the Class D

Let $\{x(t), \mathscr{F}(t), t \in I\}$ be a positive right closable, equivalently, uniformly integrable according to Section 3(e), submartingale, right closable by a positive random variable x. In view of Theorem 6 as applied to $-x(\cdot)$, $-x$,

$$x(T) \leq E\{x \mid \mathscr{F}(T)\} \quad \text{a.s.} \tag{6.6}$$

if T is optional, taking on finitely many values, and therefore (6.6) is true if T is optional and countably valued. Hence $x(\cdot)$ is a class **D** process because [Section I.4(i)] the family of conditional expectations of x is uniformly integrable. Thus there is a uniform integrability test function Φ and a constant c such that

$$E\{\Phi[x(T)]\} \leq c \tag{6.7}$$

whenever T is a countably valued optional time. In particular, if $I = \mathbb{R}^+$, if $x(\cdot)$ is almost surely right continuous, and if S is an arbitrary optional time, (6.7) can be applied to $[S]_n$ as defined in Section II, Example (b2), and $(n \to \infty)$ (6.7) is therefore true for $T = S$. Thus this submartingale is in the class **D** in the strong sense that the family $\{x(T)\colon T \text{ optional}\}$ is uniformly integrable. In particular, a right closable, equivalently, uniformly integrable, martingale $\{y(t), \mathscr{F}(t), t \in I\}$ is a class **D** process [set $x(t) = |y(t)|$]; if $I = \mathbb{R}^+$ and if the martingale is almost surely right continuous, the family $\{y(T)\colon T \text{ optional}\}$ is uniformly integrable.

7. Optional Sampling Theorem for Right Closed Processes

If the process in Theorem 6 is right closed, the optional times need not be bounded according to the following theorem.

Theorem. *If $\{x(n), \mathscr{F}(n), n \in \overline{\mathbb{Z}}^+\}$ } is a supermartingale and if T ($\leq +\infty$) is optional, then $x(T)$ is integrable. If S ($\leq +\infty$) is also optional with $S \leq T$, then* (6.1) *is true, and there is equality if the process is a martingale.*

If the process is a martingale, $x(T)$ is integrable and $x(T) = E\{x(+\infty)|\mathscr{F}(T)\}$ according to Section 2; so (6.1) with equality reduces to

$$x(S) = E\{E\{x(+\infty)|\mathscr{F}(T)\}|\mathscr{F}(S)\} \quad \text{a.s.} \tag{7.1}$$

which is true because $\mathscr{F}(S) \subset \mathscr{F}(T)$. If the process is a supermartingale,

$$x(n) = [x(n) - E\{x(+\infty)|\mathscr{F}(n)\}] + E\{x(+\infty)|\mathscr{F}(n)\} \quad \text{a.s.} \tag{7.2}$$

so that $x(\cdot)$ is the sûm of a positive supermartingale with last element 0 and a martingale, both relative to $\mathscr{F}(\cdot)$. Thus it is sufficient to prove the theorem for a positive supermartingale with last element 0. Under this hypothesis define $T_n = T \wedge n$, $S_n = S \wedge n$, and observe that according to Theorem 6 $E\{x(0)\} \geq E\{x(T_n)\}$. It follows ($n \to \infty$) that $E\{x(0)\} \geq E\{x(T)\}$. Thus $x(T)$ is integrable. Again according to Theorem 6, if $m \leq n$,

$$x(S_m) \geq E\{x(T_n)|\mathscr{F}(S_m)\} \quad \text{a.s.,} \tag{7.3}$$

and when $n \to \infty$ this inequality becomes

$$x(S_m) \geq E\{x(T)|\mathscr{F}(S_m)\} \quad \text{a.s.} \tag{7.4}$$

by Fatou's lemma for conditional expectations. Now $\Lambda \in \mathscr{F}(S)$ implies that $\Lambda_m = \Lambda \cap \{S \leq m\} \in \mathscr{F}(S_m)$; so (7.4) implies that

$$\int_{\Lambda_m} x(S)\, dP \geq \int_{\Lambda_m} x(T)\, dP. \tag{7.5}$$

This inequality when $m \to \infty$ yields the integrated version of (6.1).

EXAMPLE. If $\{x(n), \mathscr{F}(n), n \in \mathbb{Z}^+\}$ is a positive supermartingale, the theorem is applicable because the process can be closed on the right by defining $x(+\infty) = 0$. This is the most used example. More generally the theorem is applicable according to Section 3(f) to a supermartingale $\{x(n), \mathscr{F}(n), n \in \mathbb{Z}^+\}$ which majorizes a uniformly integrable submartingale.

Generalization. Let $\{x(t), \mathscr{F}(t), t \in I\}$ be a supermartingale with an arbitrary linearly ordered parameter set I for which there is a first and last element. If T is a countably valued optional time (with values in I), then we shall now prove that $x(T)$ is integrable. Moreover, if S is also a countably valued optional time with $S \leq T$, then we shall prove that (6.1) is true and that there is almost sure equality if the process is a martingale. The proof follows that of Theorem 7. Denote by 0 the first element of I and by ∞ the last element of I. Then just as in the proof of Theorem 7 the martingale case is covered by Section 2, and only the case of a positive supermartingale with last random variable 0 needs further examination. Let $r_{\bullet}$ be the sequence of parameter values taken on by S and T, define

$$T_n = \begin{cases} T & \text{on } \bigcup_{j=0}^{n} \{T = r_j\}, \\ \infty & \text{on } \bigcup_{j=n+1}^{+\infty} \{T = r_j\}, \end{cases}$$

and define S_n similarly in terms of S. Then S_n and T_n are optional, take on only finitely many values, and $S_n \leq T_n$. Theorem 6 is applicable to S_n and T_n because only the parameter values $r_0, \ldots, r_n, \infty$ are involved. An application of Theorem 6 to the pair $(0, T_n)$ of optional times shows that $E\{x(T_n)\} \leq E\{x(0)\}$; so when $n \to +\infty$, we find that $x(T)$ is integrable. An application of Theorem 6 to the pair (S_n, T_n) shows that

$$\int_{\{S=r_j\}} x(T_n)\, dP = \int_{\{S_n=r_j\}} x(T_n)\, dP \leq \int_{\{S_n=r_j\}} x(S_n)\, dP = \int_{\{S=r_j\}} x(S)\, dP \qquad (j \leq n);$$

so $(n \to +\infty)$

$$\int_{\{S=r_j\}} x(T)\, dP \leq \int_{\{S=r_j\}} x(S)\, dP,$$

and this inequality is an integrated version of (6.1).

8. Optional Stopping

Let $\{\mathscr{F}(t), t\in I\}$ be a filtration of a probability space, with I either $\mathbb{Z}^+$ or $\mathbb{R}^+$, and let T be an optional time. If $\{x(\cdot), \mathscr{F}(\cdot)\}$ is a stochastic process on the probability space, the process $\{x(T\wedge t), t\in I\}$ is described as the process stopped at T.

Suppose that $I=\mathbb{Z}^+$ and that $\{x(\cdot), \mathscr{F}(\cdot)\}$ is a supermartingale [martingale]. According to Theorem 6, the stopped process $\{x(T\wedge n), \mathscr{F}(T\wedge n), n\in\mathbb{Z}^+\}$ is also a supermartingale [martingale]. It is sometimes important that even the process $\{x(T\wedge n), \mathscr{F}(n), n\in\mathbb{Z}^+\}$ is a supermartingale [martingale]. To prove this result, suppose that $\Lambda\in\mathscr{F}(n)$. Then

$$E\{x(T\wedge n);\ \Lambda\cap\{T\le n\}\} = E\{x(T\wedge(n+1));\ \Lambda\cap\{T\le n\}\} \tag{8.1}$$

because the integrands are the same on the integration domain, and

$$E\{x(T\wedge n);\ \Lambda\cap\{T>n\}\} \ge E\{x(T\wedge(n+1));\ \Lambda\cap\{T>n\}\}, \tag{8.2}$$

with equality in the martingale case, because the integrands are $x(n)$ and $x(n+1)$, respectively, on the integration domain; so the supermartingale inequality [martingale equality] is applicable. Adding (8.1) and (8.2) we obtain an integrated version of the desired supermartingale inequality, or of the martingale equality in the martingale case.

9. Maximal Inequalities

Theorem. (a) *If $x(0), \ldots, x(n)$ is a submartingale and α is an arbitrary real number,*

$$\alpha P\{\max_{j\le n} x(j)\ge\alpha\} \le E\{x(n);\ \max_{j\le n} x(j)\ge\alpha\} \le E\{x(n)\vee 0\}, \tag{9.1}$$

$$\begin{aligned}\alpha P\{\min_{j\le n} x(j)\le\alpha\} &\ge E\{x(0)\} - E\{x(n);\ \min_{j\le n} x(j)>\alpha\}\\ &\ge E\{x(0)\} - E\{x(n)\vee 0\}.\end{aligned}$$

(b) *If $x(0), \ldots, x(n)$ is a positive supermartingale and $\alpha>0$,*

$$\alpha P\{\max_{j\le n} x(j)\ge\alpha\} \le E\{x(0)\}. \tag{9.2}$$

To prove the first inequality in (9.1), define $T=n\wedge\min\{j: x(j)\ge\alpha\}$ Then T is optional for the submartingale; so (Theorem 6) the ordered pair $[x(T), x(n)]$ is a submartingale. The submartingale inequality $E\{x(T)\}\le E\{x(n)\}$ yields

$$E\{x(n)\} \geq \alpha P\{\max_{j \leq n} x(j) \geq \alpha\} + E\{x(n); \max_{j \leq n} x(j) < \alpha\}, \tag{9.3}$$

which implies the first inequality in (9.1). To prove the second inequality, define $S = n \wedge \min\{j: x(j) \leq \alpha\}$. Then S is optional; so the ordered pair $[x(0), x(S)]$ is a submartingale, and the submartingale inequality $E\{x(0)\} \leq E\{x(S)\}$ yields

$$E\{x(0)\} \leq \alpha P\{\min_{j \leq n} x(j) \leq \alpha\} + E\{x(n); \min_{j \leq n} x(j) > \alpha\}, \tag{9.4}$$

which implies the second line in (9.1). To prove (9.2), apply (9.4) to $-x(\cdot)$.

Observe that in (9.1) and (9.2) if "$\geq\alpha$" and "$\leq\alpha$" are replaced by the corresponding strict inequalities, the resulting versions of (9.1) and (9.2) are apparently weaker than the old versions but actually imply them by a trivial continuity argument.

If the parameter set of the submartingale or positive supermartingale in question is any linearly ordered countable set A, the inequalities (9.1) or (9.2) as the case may be are valid for every finite set $x(t_0), \ldots, x(t_n)$, where $t_\bullet$ is a finite ordered subset of A. It follows that

$$\begin{aligned} &\alpha P\{\sup_{t \in A} x(t) \geq \alpha\} \leq \sup_{t \in A} E\{x(t) \vee 0\} \\ &\alpha P\{\inf_{t \in A} x(t) \leq \alpha\} \geq \inf_{t \in A} E\{x(t)\} - \sup_{t \in A} E\{x(t) \vee 0\} \end{aligned} \qquad [x(\cdot) \text{ a submartingale}] \tag{9.1'}$$

$$\alpha P\{\sup_{t \in A} x(t) \geq \alpha\} \leq \sup_{t \in A} E\{x(t)\} \qquad [x(\cdot) \text{ a positive supermartingale}]. \tag{9.2'}$$

If the process is left closed, by $x(a)$,

$$\inf_{t \in A} E\{x(t)\} = E\{x(a)\} \qquad [x(\cdot) \text{ a submartingale}]$$

$$\sup_{t \in A} E\{x(t)\} = E\{x(a)\} \qquad [x(\cdot) \text{ a supermartingale}].$$

If the process is right closed, by $x(b)$,

$$\sup_{t \in A} E\{x(t) \vee 0\} = E\{x(b) \vee 0\} \qquad [x(\cdot) \text{ a submartingale}].$$

Moreover in this right closed case the derivation of the first inequality in (9.1′) yields the in general smaller right-hand side $E\{x(b); \sup_{t \in A} x(t) \geq \alpha\}$.

If the parameter set of the submartingale or supermartingale is uncountable, (9.1′) and (9.2′) remain true if "sup" and "inf" on the left are replaced by "ess sup" and "ess inf" respectively, because these essential bounds are equal almost everywhere to ordinary bounds over suitably chosen countable parameter subsets. On the other hand, if the process parameter set A is an interval and if the process is almost surely right continuous, then (9.1′) and (9.2′) are correct as written since replacing A in these inequalities by a countable dense subset including the right-hand endpoint of A if A is right closed changes neither the left nor the right side.

10. Conditional Maximal Inequalities

The inequalities in Section 9 can be made more precise by conversion to conditional inequalities. For example, let $\{x(j), \mathscr{F}(j), 0 \le j \le n\}$ be a submartingale, and suppose that $\Lambda \in \mathscr{F}(0)$. An application of the first inequality in (9.1) to the submartingale $\{1_\Lambda x(j), \mathscr{F}(j), 0 \le j \le n\}$ leads to the integrated version of

$$\alpha P\{\max_{j \le n} x(j) \ge \alpha | \mathscr{F}(0)\} \le E\{x(n)z | \mathscr{F}(0)\} \le E\{x(n) \vee 0 | \mathscr{F}(0)\} \quad \text{a.s.,} \tag{10.1}$$

where z is the indicator function of the set $\{\max_{j \le n} x(j) \ge \alpha\}$. The other inequalities in Section 9 can be extended similarly. Such extensions to conditional inequalities will be omitted from now on.

11. An L^p Inequality for Submartingale Suprema

Theorem. *If $\{x(t), t \in I\}$ is a positive right closed submartingale, closed by the random variable x, if A is a countable subset of I, if $p > 1$, and if $1/p + 1/q = 1$, then*

$$E\{\sup_{t \in A} x(t)^p\} \le q^p E\{x^p\}. \tag{11.1}$$

According to Section 9,

$$\alpha P\{\sup_{t \in A} x(t) \ge \alpha\} \le E\{x; \sup_{t \in A} x(t) \ge \alpha\}, \tag{11.2}$$

and it is therefore sufficient to drop the martingale theory context and to show that if x and y are positive random variables satisfying

$$\alpha P\{y \ge \alpha\} \le E\{x; y \ge \alpha\}, \tag{11.3}$$

then

$$E\{y^p\} \le q^p E\{x^p\}. \tag{11.4}$$

Now

$$\begin{aligned} E\{y^p\} &= p\int_\Omega dP \int_0^{y(\omega)} \alpha^{p-1}\,d\alpha = p\int_0^{+\infty} P\{y \ge \alpha\}\alpha^{p-1}\,d\alpha \\ &\le p\int_0^{+\infty} \alpha^{p-2}\,d\alpha \int_{\{y\ge\alpha\}} x\,dP = p\int_\Omega x\,dP \int_0^{y(\omega)} \alpha^{p-2}\,d\alpha \quad (11.5) \\ &= qE\{xy^{p-1}\} \le qE^{1/p}\{x^p\}E^{1/q}\{y^p\}. \end{aligned}$$

If $E\{y^p\} < +\infty$, this inequality yields (11.4). Otherwise, replace y by $y \wedge k$, for which (11.3) is still true, thereby deduce (11.4) for $y \wedge k$, and let $k \to +\infty$.

A common application is to processes with parameter set $0, 1, \ldots, +\infty$ ordered as indicated. In this case one chooses $A = I$. If I is uncountable, (11.1) is true with A replaced by I and "sup" replaced by "ess sup" because there is a countable subset I_0 of I with the property that $\operatorname{ess\,sup}_{t\in I} x(t) = \sup_{t\in I_0} x(t)$ almost everywhere. If I is a right closed interval of $\mathbb{R}$ and if the given process is supposed almost surely right continuous, then (11.1) is true with $A = I$ because if A is a dense subset of I including the right-hand endpoint, then $\sup_{t\in I} x(t) = \sup_{t\in A} x(t)$ almost everywhere.

Modification of Theorem 11 if the Submartingale Is Not Right Closed

Observe that in Theorem 11 the process $\{x(\cdot)^p, \mathscr{F}(\cdot)\}$ is a submartingale if $x(t)^p$ is integrable for all t; so the function $t \mapsto E\{x(t)^p\}$ is monotone increasing. If $\sup_{t\in I} E\{x(t)^p\} = \lim_{t\uparrow} E\{x(t)^p\} < +\infty$, the process $x(\cdot)$ is uniformly integrable because the pth power is a uniform integrability test function when $p > 1$. Hence [Section 3(e)] the submartingale $\{x(\cdot), \mathscr{F}(\cdot)\}$ is right closable. The closability argument in Section 3(e) can be interpreted to imply that $x(t)$ has a weak limit x in $L^1(\Omega, \bigvee_{t\in I}\mathscr{F}(t), P)$ when t increases and that x closes the submartingale $x(\cdot)$. We shall see in Theorem 15 that if $I = \mathbb{Z}^+$, it is even true that $\lim_{n\to\infty} x(n) = x$ exists almost surely and in L^1. Then by Fatou's lemma $E\{x^p\} < +\infty$; so by (11.1) the submartingale $\{x(\cdot)^p, \mathscr{F}(\cdot)\}$ is also uniformly integrable and therefore is right closed by x^p; moreover (dominated convergence theorem) $\lim_{n\to\infty} x(n) = x$ in the L^p sense. We leave to the reader the easy extension of these results to other parameter sets.

12. Crossings

Let f, g, h be functions from a linearly ordered set I into $\overline{\mathbb{R}}$, with $g \le h$. The number $\mathrm{Dn}[f; g, h]$ of downcrossings of $[g, h]$ by f is defined as the supremum of the values of the positive integer k for which there exist $t_1 < \cdots$

$< t_{2k}$ in I satisfying $f(t_j) \geq h(t_j)$ when j is odd and $f(t_j) \leq g(t_j)$ when j is even. Upcrossings are defined in the obvious dual way. Observe that $\lim_{i\to\infty} f_i = f$ implies that

$$\mathrm{Dn}[f; g, h] \leq \inf_{m \in \mathbb{Z}^+} \liminf_{i\to\infty} \mathrm{Dn}[f_i; g + 2^{-m}, h - 2^{-m}] \tag{12.1}$$

when g and h are finite valued.

Theorem. *Let $n \in \mathbb{Z}^+$, and let $x(\cdot)$, $x'(\cdot)$, $x''(\cdot)$ be adapted processes on a filtered probability space $(\Omega, \mathscr{F}, P; \mathscr{F}(j), 0 \leq j \leq n)$. Suppose that $x'(\cdot) \geq 0$ and that $x(\cdot)$, $x''(\cdot)$ and $x''(\cdot) - x'(\cdot)$ are positive supermartingales relative to $\mathscr{F}(\cdot)$. Then*

$$E\{[x''(n) - x'(n)]\mathrm{Dn}[x(\cdot); x'(\cdot), x''(\cdot)]\} \leq E\{x(0) \wedge x''(0)\}. \tag{12.2}$$

In particular, if $x(\cdot)$ and $y(\cdot)$ are positive supermartingales relative to $\mathscr{F}(\cdot)$ and if a, b are numbers with $0 \leq a < b$,

$$E\{y(n)\mathrm{Dn}[x(\cdot); ay(\cdot), by(\cdot)]\} \leq \frac{E\{x(0) \wedge [by(0)]\}}{b - a} \tag{12.3}$$

and in fact with the convention that $0^0 = 1$

$$E\{y(n); \mathrm{Dn}[x(\cdot); ay(\cdot), by(\cdot)] \geq k\} \leq \left(\frac{a}{b}\right)^{k-1} \frac{E\{x(0) \wedge [by(0)]\}}{b} \quad (k \geq 1). \tag{12.4}$$

If $x(\cdot)$ is a (not necessarily positive) supermartingale and if $-\infty < a < b < +\infty$,

$$E\{\mathrm{Dn}[x(\cdot); a, b]\} \leq \frac{E\{x(0) \wedge b\} - E\{x(n) \wedge b\}}{b - a}. \tag{12.5}$$

Observation. The inequalities in Section 1.VI.3(o) limiting the oscillation of a superharmonic function by means of iterated reductions correspond to supermartingale downcrossing inequalities. In fact it will be shown in Section 23 that the exact counterpart of the reduction procedure in Section 1.VI.3(o) leads to (12.2) and (12.3). Observe also that if $y(\cdot) > 0$, the left side of (12.3) is $E\{y(n)\ \mathrm{Dn}[x(\cdot)/y(\cdot); a, b]\}$; the corresponding remark is of course applicable to (12.4). Finally observe that the right sides of (12.2–5) are majorized, respectively, by $E\{x''(0)\}$, $bE\{y(0)\}/(b-a)$, $(a/b)^{k-1}E\{y(0)\}$, $b/(b-a)$, which do not involve $x(\cdot)$.

To prove the theorem, define $S_\cdot$ and $T_\cdot$ by setting $S_0 = T_0 = 0$ and

$$S_1 = \min\{j \geq 0 \colon x(j) \geq x''(j)\},$$

$$S_k = \begin{cases} \min\{j > T_{k-1} : x(j) \geq x''(j)\} & (k \text{ odd}, \geq 3), \\ \min\{j > T_{k-1} : x(j) \leq x'(j)\} & (k \text{ even}, \geq 2), \end{cases} \tag{12.6}$$

$$T_k = \quad S_k \wedge n.$$

As usual the minimum of the empty set of numbers is defined as $+\infty$. The sequence $T_{\bullet}$ is an increasing sequence of optional times for $\mathscr{F}(\cdot)$.

Proof of (12.2). In this proof we abbreviate $\mathrm{Dn}[x(\cdot); x'(\cdot), x''(\cdot)]$ to Dn. In the first place if j is odd,

$$\begin{aligned} E\{x(T_j) - x(T_{j+1})\} &= E\{x(T_j) - x(T_{j+1});\ T_j < n\} \\ &\geq E\{x''(T_{j+1}) - x(T_{j+1});\ T_j < n\} \\ &\geq E\left\{x''(T_{j+1}); \mathrm{Dn} \geq \frac{j+1}{2}\right\} - E\left\{x'(T_{j+1}); \mathrm{Dn} \geq \frac{j+1}{2}\right\} \\ &\quad - E\left\{x(n); T_j < n, \mathrm{Dn} < \frac{j+1}{2}\right\} \\ &\geq E\left\{x''(n) - x'(n); \mathrm{Dn} \geq \frac{j+1}{2}\right\} \\ &\quad - E\left\{x(n); \mathrm{Dn} = \frac{j-1}{2}\right\}. \end{aligned} \tag{12.7}$$

In the second place (supermartingale inequality) $E\{x(T_k) - x(T_{k+1})\} \geq 0$ for all k. Hence we can drop the summands with even j in the first sum below to obtain

$$\begin{aligned} E\{x(0)\} &= \sum_{j=0}^{n} E\{x(T_j) - x(T_{j+1})\} + E\{x(n)\} \\ &\geq \sum_{k=1}^{n} E\{x''(n) - x'(n); \mathrm{Dn} \geq k\} \\ &\geq \sum_{k=1}^{n} E\{k[x''(n) - x'(n)]; \mathrm{Dn} = k\} = E\{[x''(n) - x'(n)]\mathrm{Dn}\}. \end{aligned} \tag{12.8}$$

If $x(\cdot)$ is replaced in this inequality by $x(\cdot) \wedge x''(\cdot)$, the number of downcrossings is unchanged, and (12.8) yields (12.2).

Proof of (12.3) *and* (12.4). Inequality (12.3) is a special case of (12.2). Alternatively (12.3) can be obtained by summing (12.4). To prove (12.4), define $S_{\bullet}$ and $T_{\bullet}$ by (12.6) with $x''(\cdot) = by(\cdot)$ and $x'(\cdot) = ay(\cdot)$. Observe that for $k \geq 1$,

$$
\begin{aligned}
E\{y(S_{2k+2}); S_{2k+2} \le n\} &\le E\{y(T_{2k+2}); S_{2k+1} \le n\} \\
&\le E\{y(S_{2k+1}); S_{2k+1} \le n\} \le \frac{E\{x(S_{2k+1}); S_{2k+1} \le n\}}{b} \\
&\le \frac{E\{x(T_{2k+1}); S_{2k} \le n\}}{b} \le \frac{E\{x(S_{2k}); S_{2k} \le n\}}{b} \qquad (12.9) \\
&\le \frac{a}{b} E\{y(S_{2k}); S_{2k} \le n\}
\end{aligned}
$$

and so for $k \ge 1$, under the convention $0^0 = 1$, needed when $a = 0$,

$$
\begin{aligned}
E\{y(S_{2k}); S_{2k} \le n\} &\le \left(\frac{a}{b}\right)^{k-1} E\{y(S_2); S_2 \le n\} \\
&\le \left(\frac{a}{b}\right)^{k-1} E\{y(T_2); S_1 \le n\} \le \left(\frac{a}{b}\right)^{k-1} E\{y(S_1); S_1 \le n\} \\
&\le \left(\frac{a}{b}\right)^{k-1} \frac{E\{x(S_1); S_1 \le n\}}{b} \le \left(\frac{a}{b}\right)^{k-1} \frac{E\{x(0)\}}{b}. \qquad (12.10)
\end{aligned}
$$

Furthermore for $k \ge 1$,

$$
\begin{aligned}
E\{y(n); \mathrm{Dn}[x(\cdot); ay(\cdot), by(\cdot)] \ge k\} &= E\{y(n); S_{2k} \le n\} \\
&\le E\{y(S_{2k}); S_{2k} \le n\} \le \left(\frac{a}{b}\right)^{k-1} \frac{E\{x(0)\}}{b}.
\end{aligned}
\qquad (12.11)
$$

If $x(\cdot)$ is replaced by $x(\cdot) \wedge [by(\cdot)]$, the number of downcrossings is unchanged and (12.11) becomes (12.4).

Proof of (12.5). To prove (12.5), define $S_.$ and $T_.$ by (12.6) with $x''(\cdot) \equiv b$ and $x'(\cdot) \equiv a$. Since $T_n \equiv n$,

$$
x(0) - x(n) = \sum_{j=0}^{n} [x(T_j) - x(T_{j+1})], \qquad (12.12)
$$

and (supermartingale inequality) each bracket has a positive expectation. We shall minorize each bracket with odd j. On the set $\Lambda_k = \{\mathrm{Dn}[x(\cdot); a, b] = k\}$ each of the first k brackets in (12.12) with odd j is $\ge b - a$, and of the later brackets with odd j only $[x(T_{2k+1}) - x(T_{2k+2})]$ can be nonzero, and if so,

$$
x(T_{2k+1}) \ge b,\ T_{2k+2} = n,\ x(n) > a.
$$

Thus on Λ_k,

$$x(T_{2k+1}) - x(T_{2k+2}) = x(T_{2k+1}) - x(n) \geq b - x(n) \tag{12.13}$$

and therefore

$$E\{x(0) - x(n)\} \geq (b-a)E\{\mathrm{Dn}[x(\cdot);a,b]\} + E\{[b - x(n)] \wedge 0\}. \tag{12.14}$$

If $x(\cdot)$ is replaced by $x(\cdot) \wedge b$, the number of downcrossings is unchanged and (12.14) yields (12.5).

Strict Downcrossings

If downcrossings had been defined using strict inequalities, $f > h$ and $f < g$, the resulting apparently weaker versions of the inequalities of Theorem 12 would imply the present versions by a trivial continuity argument.

Adaptation to Infinite Parameter Sets

If the parameter set of the theorem is replaced by a countably infinite linearly ordered set I, the inequalities obtained have obvious extensions. For example, (12.5) becomes

$$E\{\mathrm{Dn}[x(\cdot);a,b]\} \leq \frac{\sup_{t\in I} E\{x(t) \wedge b\} - \inf_{t\in I} E\{x(t) \wedge b\}}{b-a}, \tag{12.15}$$

and (12.15) is also valid if the parameter set I is an interval of $\mathbb{R}$ and the process is almost surely right continuous. In fact under this hypothesis the two sides of (12.15) are unchanged, for strict downcrossings and therefore for downcrossings, if I is replaced by a countable dense subset including each endpoint of I in I. In the context of (12.15) if the supermartingale $x(\cdot)$ is left closed, by a random variable $x(\alpha)$, the supremum in (12.15) is $E\{x(\alpha) \wedge b\}$. If this supermartingale is right closed, by a random variable $x(\beta)$, the infimum in (12.15) is $E\{x(\beta) \wedge b\}$.

The Role of Downcrossing Inequalities

The importance of the downcrossing concept lies in the following three facts. Let f be either (i) a sequence of numbers, or (ii) a function from an interval I into $\mathbb{R}$, or (iii) a function from a dense subset of an interval I into $\mathbb{R}$. Then the number of downcrossings $\mathrm{Dn}[f;a,b]$ by f of an interval $[a,b]$ is finite for every interval if and only if, respectively, (i) f is convergent to a (possibly infinite) limit; (ii) f has right and left (possibly infinite) limits at every point of I and at the endpoints of I not in I; (iii) f has an extension to I which has the property stated in (ii). Moreover in all three contexts it is sufficient if $\mathrm{Dn}[f;a,b] < +\infty$ for all intervals $[a,b]$ with endpoints in a countable dense subset of $\mathbb{R}$. The relative downcrossing function $\mathrm{Dn}[x(\cdot);ay(\cdot),by(\cdot)] = \mathrm{Dn}[x(\cdot)/y(\cdot);a,b]$ of a pair of positive supermartingales has not yet proved useful in probability theory, but its potential theory counterpart was useful in Part 1 of this book. In fact we shall see in Section 23 that Theorem 12 can be obtained by a reduction method, and it will then be clear that Theorem 12 is the exact counterpart of a set of inequalities [Sections 1.VI.3(o) and 1.XVII.16(s)] for iterated reductions. The latter inequalities were essential in proving Theorems 1.XI.4 and 1.XVIII.14 on the fine topology limit properties of the ratio of two positive superharmonic, respectively, superparabolic functions.

13. Forward Convergence in the L^1 Bounded Case

The fundamental martingale theory convergence theorems are the following forward one for the parameter set 0, 1, ... and backward one (Theorem 17) for the parameter set ..., −1, 0.

Theorem. *Let $\{x(n), \mathscr{F}(n), n \in \mathbb{Z}^+\}$ be an L^1 bounded supermartingale martingale or submartingale. Then $\lim_{n\to\infty} x(n)$ exists (finite) almost surely.*

It is sufficient to consider the supermartingale case. In that case if $r_1 < r_2$, Theorem 12 implies that $E\{\mathrm{Dn}[x(\cdot);r_1,r_2]\} < +\infty$ so that almost no sample sequence has infinitely many downcrossings over the interval $[r_1,r_2]$. Thus the summands on the right in the relation

$$\begin{aligned}&\left\{\limsup_{n\to\infty} x(n) > \liminf_{n\to\infty} x(n)\right\}\\ &\subset \bigcup_{\substack{r_1<r_2\\ r_i \text{ rational}}} \left\{\limsup_{n\to\infty} x(n) > r_2 > r_1 > \liminf_{n\to\infty} x(n)\right\}\end{aligned} \tag{13.1}$$

are null sets so $\lim_{n\to\infty} x(n)$ exists almost surely, and the limit is almost surely finite because

$$E\left\{\left|\lim_{n\to\infty} x(n)\right|\right\} \le \liminf_{n\to\infty} E\{|x(n)|\} < +\infty. \tag{13.2}$$

The hypotheses of the theorem are satisfied if the process is a positive supermartingale (the most natural case for potential theory) or more generally if the process is a right closable supermartingale because if $x(+\infty)$ right closes the supermartingale

$$E\{|x(n)|\} = E\{x(n)\} - 2E\{x(n) \wedge 0\} \le E\{x(0)\} - 2E\{x(+\infty) \wedge 0\}. \tag{13.3}$$

This inequality shows that $x(\cdot)$ is L^1 bounded if and only if $\inf_{n\ge 0} E\{x(n) \wedge 0\} > -\infty$. See Section 15 for further discussion of the convergence of a right closable supermartingale.

Generalization to Arbitrary Parameter Sets

Let $\{x(t), t \in I\}$ be an L^1 bounded supermartingale martingale or submartingale with an arbitrary linearly ordered parameter set I except that I has no last element. If I is countable, the method of proof of Theorem 13 is applicable to yield the fact that $\lim_{t\uparrow} x(t)$ exists and is integrable. If I is not countable, ess $\lim_{t\uparrow} x(t)$ exists and is integrable because we can assume (Section 4) that I is a subset of $\mathbb{R}$ ordered by $\le$, in which case I has a cofinal sequence and (Appendix Theorem IV.10) the stated result follows from the convergence result for countable I.

Continuous Parameter Case

Suppose that the process $x(\cdot)$ is a supermartingale martingale or submartingale with parameter set a subinterval I of $\mathbb{R}$ whose right-hand endpoint α ($\le +\infty$) is not in I. If the process is L^1 bounded, we have just seen that $\lim_{t\uparrow\alpha} x(t)$ exists almost surely when t tends to α along the rationals, and this fact implies, if the process is known to be almost surely right continuous, that this limit exists almost surely when t tends to α unrestrictedly. This is the typical application of the L^1 bounded martingale theory convergence theorem in the continuous parameter case, and it will be shown in Chapter IV that the hypothesis of almost sure right continuity is not very restrictive.

14. Convergence of a Uniformly Integrable Martingale

Let $(\Omega, \mathscr{F}, P; \mathscr{F}(n), n \in \mathbb{Z}^+)$ be a filtered probability space, and define $\mathscr{F}(+\infty) = \bigvee_0^\infty \mathscr{F}(n)$. Recall from Section 3 that a martingale is uniformly integrable if and only if it is right closable, so that the most general uniformly integrable martingale relative to the above unextended filtration is a sequence $\{x(n), \mathscr{F}(n), n \in \mathbb{Z}^+\}$ with $x(n) = E\{x|\mathscr{F}(n)\}$, where x closes the martingale. In this context define $x(+\infty) = E\{x|\mathscr{F}(+\infty)\}$. If $p \ge 1$, then (submartingale inequality if the expectations are finite) $E\{|x(0)|^p\} \le E\{|x(1)|^p\} \le \cdots \le E\{|x|^p\}$, so that the martingale is L^1 bounded (as is implied by the given uniform integrability without invoking martingale theory), and in general if the last expectation is finite, the martingale is L^p bounded. The following martingale theorem, in which $p \ge 1$ and the preceding notation is used, contains the forward conditional expectation continuity theorem proved directly in Section I.5.

Theorem. *If* $\{x(n), \mathscr{F}(n), n \in \mathbb{Z}^+\}$ *is a uniformly integrable martingale right closable by* x, *then* $\lim_{n\to\infty} x(n) = x(+\infty)$ *almost surely, and also in the* L^p *metric whenever* $x \in L^p$ [*equivalently, for* $p > 1$ *whenever* $\sup_{n\in\mathbb{Z}^+} E\{|x(n)|^p\} < +\infty$]. *The process*

$$[x(0), \mathscr{F}(0)], \ldots, [x(+\infty), \mathscr{F}(+\infty)], [x, \mathscr{F}] \tag{14.1}$$

is a martingale.

Since $x(\cdot)$ is L^1 bounded, Theorem 13 is applicable, and under the given uniform integrability hypothesis the almost everywhere convergence assured by Theorem 13 implies L^1 convergence. Furthermore, if $\Lambda \in \mathscr{F}(n)$, then (martingale equality)

$$E\{x(m); \Lambda\} = E\{x(n); \Lambda\} = E\{x; \Lambda\} \quad \text{if } m \geq n, \tag{14.2}$$

and when $m \to \infty$, this equation yields, in view of the L^1 convergence of $x(\cdot)$,

$$E\left\{\lim_{m\to\infty} x(m); \Lambda\right\} = E\{x; \Lambda\}. \tag{14.3}$$

This equality holds for Λ in the algebra $\bigcup_0^\infty \mathscr{F}(n)$ and therefore holds in the generated σ algebra $\mathscr{F}(+\infty)$; so this limit random variable is $x(+\infty)$. It is trivial that the process (14.1) is a martingale. If $p > 1$ and if $\sup_{n\in\mathbb{Z}^+} E\{|x(n)|^p\} < +\infty$, then (Fatou's lemma) $E\{|x(+\infty)|^p\} < +\infty$. If $p > 1$ and if $E\{|x(+\infty)|^p\} < +\infty$, then (Theorem 11 applied to $|x(\cdot)|$) $E\{\sup_{n\in\mathbb{Z}^+} |x(n)|^p\} < +\infty$. So $E\{\sup_{n\in\mathbb{Z}^+} |x(n) - x(+\infty)|^p\} < +\infty$, and therefore (dominated convergence) the sequence $x(\cdot)$ converges to $x(+\infty)$ in the L^p metric.

Extension to Other Parameter Sets

Suppose that $x(\cdot)$ is a uniformly integrable martingale, right closable by x, for which the parameter set I is either (a) a countable dense subset of an open interval $]0, \alpha[$ or (b) the interval $]0, \alpha[$. Theorem 14 (and its proof with trivial modifications) remains valid in both these contexts when suitably rephrased. More specifically, in (a) $\lim_{t\uparrow\alpha} x(t)$ exists almost surely and in L^1; in (b) $\operatorname{ess\,lim}_{t\uparrow\alpha} x(t)$ exists, and the limit also exists in L^1. Moreover in (b) if the process is almost surely right continuous, the essential limit is also an almost sure limit. In both cases the limit is almost surely $E\{x|\bigvee_{t\in I} \mathscr{F}(t)\}$. The extension when $x \in L^p$ is obvious.

Application to Approximation Theory

Let $\{x(t), t \in I\}$ be an arbitrary infinite collection of random variables on some probability space, and if $J \subset I$, define $\mathscr{F}(J) = \mathscr{F}\{x(t), t \in J\}$. Then if x is an $\mathscr{F}(I)$ measurable and integrable random variable, we shall now show that there is an increasing sequence $J_{\cdot}$ of finite subsets of I such that

$$\lim_{n\to\infty} E\{x|\mathscr{F}(J_n)\} = E\left\{x|\mathscr{F}\left(\bigcup_0^\infty J_n\right)\right\} = x \quad \text{a.s. and in } L^1. \qquad (14.4)$$

Since (Section I.4) $E\{x|\mathscr{F}(J_n)\}$ is equal almost surely to a Borel measurable function of the finitely many random variables with indices in J_n, equation (14.4) exhibits a canonical way to approximate x by functions of finitely many of the random variables. In fact let x be as stated, and (Appendix IV.2) choose J countable and so large that x is $\mathscr{F}(J)$ measurable. If J_n in (14.4) is chosen to be the set of first n members of J in some enumeration, (14.4) reduces to a special case of the conditional expectation continuity theorem, that is, a special case of Theorem 14. Observe that if x is an arbitrary integrable random variable on the given probability space, (14.4) should be replaced by

$$\lim_{n\to\infty} E\{x|\mathscr{F}(J_n)\} = E\left\{x|\mathscr{F}\left(\bigcup_0^\infty J_n\right)\right\} = E\{x|\mathscr{F}(I)\} \quad \text{a.s. and in } L^1. \qquad (14.5)$$

15. Forward Convergence of a Right Closable Supermartingale

Theorem. *If $\{x(n), \mathscr{F}(n), n \in \mathbb{Z}^+\}$ is a right closable supermartingale, $\lim_{n\to\infty} x(n) = x(+\infty)$ exists (finite) almost surely, and if x right closes the supermartingale the process,* (14.1) *is a supermartingale.*

It was pointed out in Section 13 that Theorem 13 is applicable to right closable supermartingales. Hence only the second assertion of the theorem requires proof. Since the process $x(\cdot)$ is the sum of a positive supermartingale and a right closable martingale according to

$$x(n) = [x(n) - E\{x|\mathscr{F}(n)\}] + E\{x|\mathscr{F}(n)\}, \qquad (15.1)$$

and since the corresponding right closure result for martingales was proved in Section 14, it can be supposed from now on that the given supermartingale is positive. Apply Fatou's lemma for conditional expectations to get

$$E\{x(+\infty)|\mathscr{F}(n)\} \leq \liminf_{m\to\infty} E\{x(m)|\mathscr{F}(n)\} \leq x(n) \quad \text{a.s.}, \qquad (15.2)$$

which shows that $x(+\infty)$ coupled with $\mathscr{F}(+\infty) = \bigvee_0^\infty \mathscr{F}(n)$ right closes the given supermartingale. Finally

$$E\{x|\mathscr{F}(+\infty)\} = \lim_{n\to\infty} E\{x|\mathscr{F}(n)\} \le \lim_{n\to\infty} x(n) = x(+\infty) \quad \text{a.s.,} \tag{15.3}$$

and therefore (14.1) is a supermartingale in the present context.

As the negative of Example (d) in Section 2 shows, an L^1 bounded supermartingale need not be right closable.

Extension to Other Parameter Sets

The rephrasing of Theorem 15 for a process with a parameter set which is either an interval $]0, \alpha[$ or a countable dense subset of this interval is similar to the corresponding rephrasing in Sections 13 and 14.

16. Backward Convergence of a Martingale

The fact that the parameter set $\mathbb{Z}^-$: ..., -1, 0 has a last element makes backward martingale theory convergence theorems stronger than forward ones. It will be convenient to treat the martingale case first.

Theorem (Backward Half of the Conditional Expectation Continuity Theorem Already Proved Directly in Section I.5). *Suppose that* $\{x(n), \mathscr{F}(n), n \in \mathbb{Z}^-\}$ *is a martingale, that is,* $x(n) = E\{x(0)|\mathscr{F}(n)\}$ a.s. *Then* $\lim_{n\to-\infty} x(n) = E\{x(0)|\bigcap_{-\infty}^0 \mathscr{F}(n)\}$ a.s. *and in* L^1 *norm.*

The process is uniformly integrable because it consists of conditional expectations of a random variable $x(0)$ [Section I.4(i)]. Just as in the proof of Theorem 13 the downcrossing inequality, easier here because the process is right closed, together with Fatou's lemma implies that there is an almost sure limit which is integrable. To identify this limit, which we denote by $x(-\infty)$, choose Λ in $\mathscr{F}(-\infty) = \bigcap_{-\infty}^0 \mathscr{F}(n)$. The martingale property and the uniform integrability of the process yield

$$\int_\Lambda x(-\infty)\, dP = \lim_{n\to-\infty} \int_\Lambda x(n)\, dP = \int_\Lambda x(0)\, dP, \tag{16.1}$$

and this equality identifies $x(-\infty)$ as the stated conditional expectation. The convergence is L^1 convergence because the process is uniformly integrable.

The rephrasing of Theorem 16 for other parameter sets follows that of the forward convergence theorems and is left to the reader.

17. Backward Convergence of a Supermartingale

Theorem. *Suppose that $\{x(n), \mathscr{F}(n), n \in \mathbb{Z}^-\}$ is a supermartingale. Then*

(a) $E\{x(0)\} \leq E\{x(-1)\} \leq \cdots \to l \leq +\infty$.

(b) $\lim_{n\to-\infty} x(n) = x(-\infty)$ *exists almost surely*; $x(-\infty) \wedge 0$ *is integrable*; $-\infty < x(-\infty) \leq +\infty$ *almost surely.*

(c) *If $l < +\infty$, then $x(-\infty)$ is almost surely finite and left closes the supermartingale, the supermartingale is uniformly integrable, and the convergence in* (b) *is in L^1 as well as almost sure.*

The example $x(n) \equiv -n$ shows that the limit in (b) may be identically $+\infty$. Part (a) is trivial and is stated only for orientation. Just as in the proof of Theorem 13 but more easily here because the process is right closed, the downcrossing inequality implies that there is an almost sure backward limit $x(-\infty)$. Apply Fatou's lemma and the supermartingale inequality to obtain

$$E\{x(-\infty) \wedge 0\} \geq \lim_{n\to-\infty} E\{x(n) \wedge 0\} \geq E\{x(0) \wedge 0\}, \tag{17.1}$$

which implies that $x(-\infty) \wedge 0$ is integrable. In view of Theorem 16 and the fact that the process can be written as the sum of a positive supermartingale and a uniformly integrable martingale according to

$$x(n) = [x(n) - E\{x(0)|\mathscr{F}(n)\}] + E\{x(0)|\mathscr{F}(n)\}, \tag{17.2}$$

it is sufficient to prove the present theorem for a positive supermartingale, and positivity will be assumed from now on. If $l < +\infty$, the process is L^1 bounded; so $x(-\infty)$ is integrable. Finally, according to Fatou's lemma, $E\{x(-\infty)\} \leq l$, whereas for every c the process $x(\cdot) \wedge c$ is a bounded supermartingale; so for all n,

$$E\{x(-\infty) \wedge c\} = \lim_{m\to-\infty} E\{x(m) \wedge c\} \geq E\{x(n) \wedge c\}. \tag{17.3}$$

This inequality yields $E\{x(-\infty)\} \geq E\{x(n)\}$ and thereby $E\{x(-\infty)\} = l$. Thus there is equality in Fatou's lemma which implies uniform integrability of the process.

The rephrasing of Theorem 17 for other parameter sets is left to the reader (see Section 14).

18. The τ Operator

This probabilistic operator is the counterpart of the τ operator in classical potential theory. Let $\{x(\cdot), \mathscr{F}(\cdot)\}$ be a stochastic process with state space $(\overline{\mathbb{R}}, \mathscr{B}(\overline{\mathbb{R}}))$ and with an arbitrary linearly ordered parameter set I. It is supposed that the process random variables are integrable. Let S and T be countably valued optional times with values in I and with $S \le T$. Suppose that $x(T)$ is integrable, and define the process $\tau_{ST}x(\cdot)$, writing $\tau_{ST}x(t)$ instead of $\tau_{ST}x(\cdot)(t)$, by

$$\tau_{ST}x(t) = \begin{cases} x(t) & \text{on } \{S > t\} \cup \{T < t\}, \\ E\{x(T)|\mathscr{F}(t)\} & \text{on } \{S \le t \le T\}. \end{cases} \tag{18.1}$$

Observe that

$$\tau_{ST}x(t) = E\{x(T \vee t)|\mathscr{F}(t)\} \quad \text{a.s. on } \{S \le t\}. \tag{18.2}$$

The process $\tau_{ST}x(\cdot)$ is adapted to $\mathscr{F}(\cdot)$ and $\tau_{ST}x(T) = x(T)$ almost surely by Section 2, Example (a). Furthermore the process $\tau_{ST}x(\cdot)$ is a martingale between times S and T in the sense that the process

$$\{\tau_{ST}x((S \vee t) \wedge T), \mathscr{F}((S \vee t) \wedge T), t \in I\}$$

is a martingale; in fact

$$\tau_{ST}x((S \vee t) \wedge T) = E\{x(T)|\mathscr{F}((S \vee t) \wedge T)\} \quad \text{a.s.}$$

In the most important applications $\{x(\cdot), \mathscr{F}(\cdot)\}$ is a supermartingale or submartingale. Suppose, for example, that this process is a supermartingale. If T is order bounded above and below, it follows from the supermartingale optional sampling theorem that $x(T)$ is integrable and that $P\{\tau_{ST}x(t) \le x(t)\} = 1$ for all t, that is, $\tau_{ST}x(\cdot) \le x(\cdot)$ in the essential order. Furthermore the process $\{\tau_{ST}x(\cdot), \mathscr{F}(\cdot)\}$ is then also a supermartingale (and a martingale between times S and T in the above sense). In fact, if $s < t$, then

$$E\{\tau_{ST}x(t)|\mathscr{F}(s)\} \le E\{x(t)|\mathscr{F}(s)\} \le x(s) = \tau_{ST}x(s) \quad \text{a.e. on } \{S > s\}$$

and, almost everywhere on $\{S \le s\}$,

$$\begin{aligned} E\{\tau_{ST}x(t)|\mathscr{F}(s)\} &= E\{1_{\{S \le s\}}\tau_{ST}x(t)|\mathscr{F}(s)\} \\ &= E\{1_{\{S \le s\}}E\{x(T \vee t)|\mathscr{F}(t)\}|\mathscr{F}(s)\} \\ &= E\{x(T \vee t)|\mathscr{F}(s)\} \\ &= E\{E\{x(T \vee t)|\mathscr{F}(T \vee s)\}|\mathscr{F}(s)\} \le E\{x(T \vee s)|\mathscr{F}(s)\} \\ &= \tau_{ST}x(s). \end{aligned}$$

Define τ_T by (18.1) with all references to S deleted. If T is order bounded above and below, and if $\{x(\cdot), \mathscr{F}(\cdot)\}$ is a supermartingale [submartingale], then $\tau_T x(\cdot)$ is also one, is an essential order minorant [majorant] of $x(\cdot)$, and is a martingale up to time T in the sense that the process

$$\{\tau_T x(T \wedge t), \mathscr{F}(T \wedge t), t \in I\}$$

is a martingale. Observe that $\tau_T x(\cdot) = \tau_{aT} x(\cdot)$ if I has a first element a.

The operation $x(\cdot) \mapsto \tau_{ST} x(\cdot)$ on processes is the martingale theory counterpart of the classical potential theory operation $u \mapsto \tau_B u$ on functions, as defined in Section 1.II.1 for B a ball and in Section 1.VIII.11 for B an open set (in both cases relatively compact in the domain of u). In most probabilistic applications $I = \mathbb{R}^+$, the processes involved are almost surely right continuous, and the optional times S and T need not be countably valued. The adaptation to this context will be given in Section IV.14.

19. The Natural Order Decomposition Theorem for Supermartingales

In the following theorem and its proof all the supermartingales have the same (arbitrary) parameter set and filtration.

Theorem. *Let $x(\cdot)$, $x_1(\cdot)$, $x_2(\cdot)$ be positive supermartingales, and suppose that (essential order) $x(\cdot) \le x_1(\cdot) + x_2(\cdot)$, that is, for each parameter value the indicated inequality is true almost surely. Then there are positive supermartingales $x_1'(\cdot)$, $x_2'(\cdot)$ for which (essential order) $x_i'(\cdot) \le x_i(\cdot)$ and $x(\cdot) = x_1'(\cdot) + x_2'(\cdot)$.*

The proof is a translation of that of Theorem 1.III.7, a particularly easy example of the translation of potential theoretic reasoning into the corresponding probabilistic context. Choose $x_1'(\cdot)$ as the essential order infimum of the class of positive supermartingales $y(\cdot)$ for which $x(\cdot) \le y(\cdot) + x_2(\cdot)$, and then choose $x_2'(\cdot)$ as the essential order infimum of the class of positive supermartingales $y(\cdot)$ for which $x(\cdot) \le x_1'(\cdot) + y(\cdot)$. Then $x(\cdot) \le x_1'(\cdot) + x_2'(\cdot)$, $x(\cdot) \ge x_i'(\cdot)$, and for any parameter value t,

$$x(\cdot) = \tau_t x(\cdot) + [x(\cdot) - \tau_t x(\cdot)] \le x_1'(\cdot) + [\tau_t x_2'(\cdot) + x(\cdot) - \tau_t x(\cdot)].$$

It is easy to check that the bracketed process is the same as $x_2'(\cdot)$ for parameter values $\ge t$ and is a positive supermartingale. Hence this process majorizes $x_2'(\cdot)$, and it follows that

$$x(\cdot) - x_2'(\cdot) \ge \tau_t[x(\cdot) - x_2'(\cdot)].$$

If this inequality is evaluated at a parameter value $s < t$, we find that $x(\cdot) - x_2'(\cdot)$ is a positive supermartingale, majorized (essential order) by $x_1'(\cdot)$; impossible unless there is equality. The proof is complete.

See Section V.5 for Theorem 19 as modified for right continuous processes in the continuous parameter case.

20. The Operators LM and GM

Let Γ: $\{x_\alpha(\cdot), \mathscr{F}(\cdot), \alpha \in J\}$ be a class of stochastic processes with a common linearly ordered parameter set, on a common probability space, and adapted to a common filtration $\mathscr{F}(\cdot)$. If in the essential order of processes adapted to $\mathscr{F}(\cdot)$ there is a least supermartingale majorant [greatest submartingale minorant] of Γ, this majorant [minorant] will be denoted by LMΓ [GMΓ]. Strictly speaking LMΓ, for example, is an equivalence class, and we shall sometimes describe its members as its *versions*. The notation LMΓ can refer either to the equivalence class or to one of its versions, but the intended meaning will always be clear from the context.

Theorem. *If* Γ *is a class of supermartingales and has an essential order submartingale minorant, then* GMΓ *exists and is a martingale.*

The proof of this theorem is formally identical with that of its classical counterpart Theorem 1.III.2. The class Γ_0 of submartingale essential order minorants of Γ contains $x_1(\cdot) \vee x_2(\cdot)$ with $x_1(\cdot)$, $x_2(\cdot)$ and is therefore directed upward in the essential order with essential order supremum (limit), say $\{x'(\cdot), \mathscr{F}(\cdot)\}$, an essential order minorant of Γ. We prove the theorem by showing that $\{x'(\cdot), \mathscr{F}(\cdot)\}$ is a martingale. For every parameter value t, every $x(\cdot)$ in Γ, every $y(\cdot)$ in Γ_0,

$$y(\cdot) \le \tau_t y(\cdot) \le \tau_t x(\cdot) \le x(\cdot)$$

in the essential order. Thus $\tau_t y(\cdot) \in \Gamma_0$, and the essential order supremum of Γ_0 is the same as that of $\{\tau_t y(\cdot): y(\cdot) \in \Gamma_0\}$. For every t the latter class on the parameter set $\le t$ is an upward-directed set of martingales; so $\{x'(\cdot), \mathscr{F}(\cdot)\}$ is a martingale.

Special Case (Counterpart of Theorem 1.III.1). If $\{x(\cdot), \mathscr{F}(\cdot)\}$ is a supermartingale with a submartingale essential order minorant, then the greatest submartingale minorant of $x(\cdot)$ will be denoted by GM$x(\cdot)$, the tth random variable of this process will be denoted by GM$x(t)$, and GM$x(\cdot)$ can be obtained as follows. For each parameter value t the process $\tau_t x(\cdot)$ is a supermartingale essential order minorant of $x(\cdot)$ and is a martingale up to the parameter value t. As t increases, the supermartingale decreases, and the essential order infimum and limit of this decreasing family is GM$x(\cdot)$. The proof is formally identical with that of Theorem 1.III.1 and is left to the reader. According to Section 5, there is an increasing sequence $t_.$ of parameter values such that $\lim_{n\to\infty} \tau_{t_n} x(\cdot) = \text{GM}x(\cdot)$ up to a standard modification.

21. Supermartingale Potentials and the Riesz Decomposition

The special case in the preceding section yields the following result. *If* $\{x(\cdot), \mathscr{F}(\cdot)\}$ *is a supermartingale (arbitrary linearly ordered parameter set I) and if this process has an essential order submartingale minorant, then* $\{\mathrm{GM}\,x(\cdot), \mathscr{F}(\cdot)\}$ *is a martingale and* $\mathrm{GM}\,x(t) = \operatorname{ess\,inf}_{s\in I} \tau_s x(t)$. *Moreover*

$$x(\cdot) = \mathrm{GM}\,x(\cdot) + y(\cdot), \tag{21.1}$$

where $\{y(\cdot), \mathscr{F}(\cdot)\}$ *is a positive supermartingale with the following properties*:

(a) $\mathrm{GM}y(t) = 0$ *almost surely, for each t.*
(b) $\inf_{t\in I} E\{y(t)\} = 0$.
(c) $\operatorname{ess\,lim}_{s\uparrow} \tau_s y(t) = 0$ *almost surely, for each t.*

Since the function $t \mapsto E\{y(t)\}$ is monotone decreasing, the infimum in (b) is actually the limit as t increases. Each of these three properties of a positive supermartingale implies the other two, and a positive supermartingale with these properties will be called a *supermartingale potential.* Observe that in the present context if the parameter set has a last element β, a positive supermartingale $\{x(\cdot), \mathscr{F}(\cdot)\}$ is a potential if and only if $x(\beta) = 0$ almost surely.

Recall that in classical potential theory a positive superharmonic function u on a connected Greenian set D is a potential if and only if $\mathrm{GM}_D u = 0$, equivalently (for ξ a point of D), if and only if $\inf_B \mu_B(\xi, u) = 0$ when B ranges through the open relatively compact subsets of D containing ξ, equivalently, if and only if $\lim_{B\uparrow D} \tau_B u = \lim_{B\uparrow D} \mu_B(\cdot, u) = 0$ when B increases through the open relatively compact subsets of D. These three classical context conditions are respective counterparts of conditions (a)–(c) above. The decomposition (21.1) is the counterpart of the Riesz decomposition in classical potential theory and will accordingly also be called the *Riesz decomposition.*

22. Potential Theory Reductions in a Discrete Parameter Probability Context

(See Section 2.IV.17 for the corresponding continuous parameter theory.)

Let $(\Omega, \mathscr{F}, P; \mathscr{F}(j), j\in \mathbb{Z}^+)$ be a filtered probability space. Let $\{z(\cdot), \mathscr{F}(\cdot)\}$ be a positive process, and denote by Γ the class of supermartingales $\{z'(\cdot), \mathscr{F}(\cdot)\}$ majorizing $z(\cdot)$ in the essential order. It is supposed that Γ is not empty. The minimum of two supermartingales in Γ is in Γ; so Γ is directed downward. Denote the positive supermartingale ess inf Γ by $R_{z(\cdot)}(\cdot)$, and define $R_{z(\cdot)}(+\infty) = z(+\infty) = 0$. In reduction theory we are dealing with equivalence classes of adapted stochastic processes under standard modification; so $R_{z(\cdot)}(\cdot)$ is to mean either the appropriate equivalence class or one of its members, as indicated by the context.

Theorem. *Under the stated conditions*

$$R_{z(\cdot)}(j) = \operatorname{ess\,sup}\{E\{z(T)|\mathscr{F}(j)\}: T \text{ optional}, \geq j\}. \tag{22.1}$$

To prove the theorem, define $y(j)$ as the right side of (22.1), and observe (set $T \equiv j$) that $y(j) \geq z(j)$ almost surely. Next observe that for fixed j the class of conditional expectations in (22.1) is directed upward. In fact if S and T are optional and $\geq j$, define $U = T$ or $U = S$ according as the inequality

$$E\{z(T)|\mathscr{F}(j)\} \geq E\{z(S)|\mathscr{F}(j)\} \tag{22.2}$$

is true or false. Then U is optional and $E\{z(U)|\mathscr{F}(j)\}$ is almost surely the maximum of the conditional expectations in (22.2). Next observe that $\{y(\cdot), \mathscr{F}(\cdot)\}$ is a supermartingale; so $y(\cdot) \in \Gamma$. In fact, if $j > 0$ and if the optional time sequence $S_\bullet$ is chosen with $S_k \geq j$ for all k and is also chosen to make the sequence $E\{z(S_\bullet)|\mathscr{F}(j)\}$ monotone increasing with limit $y(j)$, then

$$\begin{aligned} E\{y(j)|\mathscr{F}(j-1)\} &= \lim_{k\to\infty} E\{E\{z(S_k)|\mathscr{F}(j)\}|\mathscr{F}(j-1)\} \\ &= \lim_{k\to\infty} E\{z(S_k)|\mathscr{F}(j-1)\} \leq y(j-1) \quad \text{a.s.} \end{aligned}$$

Finally, if $z'(\cdot) \in \Gamma$ and if T is optional and $\geq j$, then

$$z'(j) \geq E\{z'(T)|\mathscr{F}(j)\} \geq E\{z(T)|\mathscr{F}(j)\} \quad \text{a.s.};$$

so $z'(j) \geq y(j)$ almost surely; that is, $y(\cdot) = \operatorname{ess\,inf}\Gamma$, as asserted.

We shall not pursue the general case far enough to obtain a probability counterpart to Theorem 1.XI.20 but proceed at once to the probability counterpart of classical potential theory reductions.

Application to Reductions

Let $\dot{\Lambda}$ be a subset of $\mathbb{Z}^+ \times \Omega$ for which the process $j \to 1_{\dot{\Lambda}}(j, \cdot)$ is adapted to $\mathscr{F}(\cdot)$, and let $\{z(\cdot), \mathscr{F}(\cdot)\}$ be a positive supermartingale. Define $R_{z(\cdot)}^{\dot{\Lambda}}$, also to be denoted by $\llbracket z(\cdot) \rrbracket^{\dot{\Lambda}}$, as $R_{z(\cdot)1_{\dot{\Lambda}}(\cdot)}$. Then according to Theorem 22, if we define $1_{\dot{\Lambda}}(+\infty) = 0$,

$$R_{z(\cdot)}^{\dot{\Lambda}}(j) = \operatorname{ess\,sup}\{E\{z(T)1_{\dot{\Lambda}}(T)|\mathscr{F}(j)\}, T \text{ optional}, \geq j\}, \quad \text{a.s.} \tag{22.3}$$

and we show further that if T_j is the entry time $\geq j$ of $\dot{\Lambda}$, that is,

$$T_j(\omega) = \min\{k \geq j: (k, \omega) \in \dot{\Lambda}\},$$

then

$$R^{\dot{A}}_{z(\cdot)}(j) = E\{z(T_j)|\mathscr{F}(j)\} \quad \text{a.s.} \tag{22.4}$$

To prove (22.4), observe that if T is optional and $\geq j$, then

$$1_{\dot{A}}(T) \leq 1_{\dot{A}}(T \wedge T_j) \leq 1_{\dot{A}}(T_j),$$

and observe that $1_{\dot{A}}(T \wedge T_j) = 0$ when $T < T_j$; so

$$\begin{aligned} E\{z(T)1_{\dot{A}}(T)|\mathscr{F}(j)\} &\leq E\{E\{z(T)|\mathscr{F}(T \wedge T_j)\}1_{\dot{A}}(T \wedge T_j)|\mathscr{F}(j)\} \\ &\leq E\{z(T \wedge T_j)1_{\dot{A}}(T \wedge T_j)|\mathscr{F}(j)\} \leq E\{z(T_j)|\mathscr{F}(j)\} \quad \text{a.s.} \end{aligned} \tag{22.5}$$

Thus the supremum in (22.4) is attained when $T = T_j$, as asserted.

Specialization to the Parameter Set $\mathbb{Z}_n^+$

If the parameter set is $\mathbb{Z}_n^+$ instead of $\mathbb{Z}^+$, the corresponding analysis is easily carried through, or we can define $\mathscr{F}(j) = \mathscr{F}(n)$ and $z(j) = z(n)$ for $j > n$ to reduce the context to that of the parameter set $\mathbb{Z}^+$. Equations (22.1), (22.4), and (22.5) remain true under the convention that all processes are extended to vanish at the parameter value $+\infty$.

23. Application to the Crossing Inequalities

We suppose that we are in the context of the preceding section with parameter set $\mathbb{Z}_n^+$, and we shall use the notation $z_{\dot{C}_1}(\cdot)$, $z_{\dot{C}_1\dot{C}_2}(\cdot)$, ..., respectively, for $⟦z(\cdot)⟧^{\dot{C}_1}$, $⟦⟦z(\cdot)⟧^{\dot{C}_1}⟧^{\dot{C}_2}$, ... Let $x(\cdot)$, $x'(\cdot)$, $x''(\cdot)$ be adapted processes on the given filtered probability space. Suppose that $x'(\cdot) \geq 0$ and that $x(\cdot)$, $x''(\cdot)$, and $x''(\cdot) - x'(\cdot) = y'(\cdot)$ are positive supermartingales; that is, we are in the context of Theorem 12. In Section 12 we defined $\text{Dn}[x(\cdot); x'(\cdot), x''(\cdot)]$ as the number of times an $x(\cdot)$ sample sequence proceeds from above $x''(\cdot)$ to below $x'(\cdot)$. Define

$$\begin{aligned} \dot{A} &= \{(j,\omega): j \in \mathbb{Z}_n^+, x(j,\omega) \leq x'(j,\omega)\}, \\ \dot{B} &= \{(j,\omega): j \in \mathbb{Z}_n^+, x(j,\omega) \geq x''(j,\omega)\}. \end{aligned}$$

It is easy to check using (22.1) that (almost surely)

$$\begin{aligned} y'_{\dot{A}\dot{B}}(0) &\geq E\{y'(n); \text{Dn}[x(\cdot); x'(\cdot), x''(\cdot)] \geq 1|\mathscr{F}(0)\}; \\ y'_{\dot{A}\dot{B}\dot{A}\dot{B}}(0) &\geq E\{y'(n); \text{Dn}[x(\cdot); x'(\cdot), x''(\cdot)] \geq 2|\mathscr{F}(0)\}; \end{aligned} \tag{23.1}$$

etc., with equality if $y'(\cdot)$ is a martingale. Now the iterated reduction inequalities in Section 1.VI.3(o) were proved in the context of classical potential theory using elementary reduction properties valid in the present context. Hence (almost surely)

$$y'_{\dot{A}}(\cdot) + y'_{\dot{A}\dot{B}\dot{A}}(\cdot) + y'_{\dot{A}\dot{B}\dot{A}\dot{B}\dot{A}}(\cdot) + \cdots \le x''(\cdot) \tag{23.2}$$

and

$$y'_{\dot{A}\dot{B}}(\cdot) + y'_{\dot{A}\dot{B}\dot{A}\dot{B}}(\cdot) + \cdots \le x''_{\dot{B}}(\cdot) \le x(\cdot) \wedge x''(\cdot). \tag{23.3}$$

In view of (23.1) inequality (23.3) implies

$$E\{y'(n)\,\mathrm{Dn}[x(\cdot); x'(\cdot), x''(\cdot)] | \mathscr{F}(0)\} \le x(0) \wedge x''(0) \quad \text{a.s.,} \tag{23.4}$$

which yields (12.2) on integration. (See the remarks in Section 10 on conditional inequalities.) In particular let a and b be positive constants with $a < b$, let $\{y(\cdot), \mathscr{F}(\cdot)\}$ be a positive supermartingale, and set $x'(\cdot) = ay(\cdot)$, $x''(\cdot) = by(\cdot)$, in (23.3) to obtain the almost sure inequality

$$y_{\dot{A}\dot{B}}(\cdot) + y_{\dot{A}\dot{B}\dot{A}\dot{B}}(\cdot) + \cdots \le \frac{x(\cdot) \wedge by(\cdot)}{b-a}. \tag{23.5}$$

The corresponding particularization of (23.4) can be obtained by the same substitution or by summation and integration in (23.5). The martingale theory translation of 1.VI(3.11) is the sequence

$$y_{\dot{A}\dot{B}}(\cdot) \le \frac{x(\cdot) \wedge by(\cdot)}{b}, \qquad y_{\dot{A}\dot{B}\dot{A}\dot{B}}(\cdot) \le \left(\frac{a}{b}\right) y_{\dot{A}\dot{B}}(\cdot), \ \ldots \tag{23.6}$$

of almost sure inequalities. On integration these yield (12.4). The point is that the classical context iterated reduction inequalities in 1.VI.6(o) remain valid when translated into the martingale theory context, and that the translated inequalities yield, on integration, the crossing inequalities proved directly in Section 12.

Chapter IV

Basic Properties of Continuous Parameter Supermartingales

1. Continuity Properties

The continuity properties of continuous parameter supermartingales derived in this section are of course also valid for continuous parameter submartingales and martingales. Recall that a process is said to be [*almost surely*] *right continuous* if [almost] every sample function is right continuous and the process is said to be [*almost surely*] *right continuous with left limits* if [almost] every sample function is right continuous and has a left limit at every point.

The usual smoothness hypothesis to be imposed on a continuous parameter supermartingale is almost sure right continuity. Moreover the following remark shows that it is possible to change the reference filtration of an almost surely right continuous supermartingale to obtain a new one to which the supermartingale is adapted, which is right continuous, and for which each σ algebra contains the null sets. Let $\{x(t), \mathscr{F}(t), t \in \mathbb{R}^+\}$ be an almost surely right continuous supermartingale, and define $\mathscr{F}_1(t)$ as the σ algebra generated by $\mathscr{F}(t)$ and the null sets. Then $\{x(\cdot), \mathscr{F}_1(\cdot)\}$ is a supermartingale. Moreover the filtration $\mathscr{F}_1^+(\cdot)$ is right continuous, $\mathscr{F}_1^+(0)$ contains the null sets, and $\{x(\cdot), \mathscr{F}_1^+(\cdot)\}$ is a supermartingale because if $s' < t$,

$$E\{x(t) | \mathscr{F}_1(s')\} \leq x(s') \quad \text{a.s.},$$

and when $s' \downarrow s$ sequentially, we obtain the supermartingale inequality for the process $\{x(\cdot), \mathscr{F}_1^+(\cdot)\}$ in view of the conditional expectation continuity theorem.

In the following theorem convergence in measure has the usual definition, not the special definition for $\overline{\mathbb{R}}$ valued random variables needed in Section I.13.

Theorem. *Let $\{x(t), \mathscr{F}(t), t \in \mathbb{R}^+\}$ be a supermartingale, and suppose that $\mathscr{F}(0)$ contains the null sets. Let B $[B^+]$ be the set of values of t in $\mathbb{R}^+$ for which it is false that $\lim_{s \to t} x(s) = x(t)$ $[\lim_{s \downarrow t} x(s) = x(t)]$ in the sense of convergence in measure, and let A be a countable dense subset of $\mathbb{R}^+$.*

(a) *The set B is countable.*

(b) *The restriction to A of almost every $x(\cdot)$ sample function is bounded on the trace on A of every compact interval, has left and right limits at every point of $\mathbb{R}^+$, and for t in $\mathbb{R}^+ - B^+$,*

$$P\left\{\lim_{A \ni s \downarrow t} x(s) = x(t)\right\} = 1.$$

(c) *If $t \in B^+$, define $y(t) = x(t)$; if $t \in \mathbb{R}^+ - B^+$, define $y(t) = \lim_{A \ni s \downarrow t} x(s)$ where this limit exists, and define $y(t)$ arbitrarily where this limit does not exist. For all t define $\underset{+}{x}(t) = \lim_{A \ni s \downarrow t} x(s)$ where this limit exists, and define $\underset{+}{x}(t)$ arbitrarily where this limit does not exist. Then*

(c1) *The processes $\{y(\cdot), \mathscr{F}(\cdot)\}$, $\{\underset{+}{x}(\cdot), \mathscr{F}^+(\cdot)\}$ are supermartingales, martingales if $\{x(\cdot), \mathscr{F}(\cdot)\}$ is a martingale.*

(c2) *Almost every $y(\cdot)$ sample function and almost every $\underset{+}{x}(\cdot)$ sample function are bounded on compact intervals of $\mathbb{R}^+$.*

(c3) *Almost every $y(\cdot)$ sample function has right and left limits at every point of $\mathbb{R}^+$ and is right continuous at every point of $\mathbb{R}^+ - B^+$. Almost every $\underset{+}{x}(\cdot)$ sample function is right continuous and has a left limit at every point of $\mathbb{R}^+$. The processes $y(\cdot)$, $y(\cdot-)$, $y(\cdot+)$, $\underset{+}{x}(\cdot)$, $\underset{+}{x}(\cdot-)$ are L^1 bounded on compact intervals of $\mathbb{R}^+$.*

(c4) *The process $y(\cdot)$ is a standard modification of $x(\cdot)$, indistinguishable from $\underset{+}{x}(\cdot)$ if and only if B^+ is empty, as is true if $\{x(\cdot), \mathscr{F}(\cdot)\}$ is a martingale and $\mathscr{F}(\cdot) = \mathscr{F}^+(\cdot)$.*

Proof of (a) *and* (b). If $t_0 > 0$, the restriction to $A \cap [0, t_0]$ of almost every $x(\cdot)$ process sample function is bounded, according to the inequalities III(9.1′) applied to $-x(\cdot)$. If such a restriction does not have a left and right limit at every point of $[0, t_0[$, there must be a pair of rational numbers r_1, r_2 with $r_1 < r_2$ such that the sample function restriction to $A \cap [0, t_0[$ has infinitely many downcrossings of $[r_1, r_2]$. Since the downcrossing inequality III(12.5) yields

$$E\{\mathrm{Dn}[x(\cdot)_{|A}; r_1, r_2]\} \le \frac{E\{x(0) \wedge r_2 - x(t_0) \wedge r_2\}}{r_2 - r_1}, \tag{1.1}$$

the number of downcrossings is almost surely finite simultaneously for all rational pairs r_1, r_2 and values of t_0. Therefore the restriction to A of almost every sample function has one-sided limits $x(t-)$ and $x(t+)$ for every t. Moreover, since

$$E\{|x(t)|\} = E\{x(t)\} - 2E\{x(t) \wedge 0\} \le E\{x(0)\} - 2E\{x(t_0) \wedge 0\} \quad (t \le t_0), \tag{1.2}$$

an application of Fatou's lemma shows that these one-sided limits are integrable and that the processes $x(\cdot)$, $x(\cdot+)$, $x(\cdot-)$ are L^1 bounded on compact intervals of $\mathbb{R}^+$. Now consider the space of almost surely finite-valued random variables, identifying two random variables if they are equal almost surely, and define the distance between the random variables x and y as $E\{|x-y| \wedge 1\}$. Convergence of a sequence in this metric is equivalent to convergence in measure of the corresponding sequence of random variables. Let $\phi(t)$ be the point of this metric space corresponding to the random variable $x(t)$. Observe that if A_1 is a second countable dense subset of $\mathbb{R}^+$, the results proved above for A when applied to $A \cup A_1$ show that the left- and right-hand sample function limits are almost surely the same for A as for A_1. Among other things, this fact implies that when s tends sequentially and strictly monotonely to t, the random variable $x(s)$ tends almost surely to $x(t-)$ or $x(t+)$ depending on the direction of approach; so the function ϕ on $\mathbb{R}^+$ has left and right limits at all points. If $\varepsilon > 0$, the set of points t at which the oscillation of ϕ is at least ε is closed and can have no finite limit point; so this set is finite in each compact interval, and it follows that the set of discontinuities of ϕ, which is the set B of the theorem, is countable. Moreover the definitions of B and B^+ can now be strengthened: $t \in \mathbb{R}^+ - B$ [$t \in \mathbb{R}^+ - B^+$] if and only if whenever A is a countable subset of $\mathbb{R}^+$ with [left] limit point t, it follows that

$$\lim_{A \ni s \to t} x(s) = x(t) \qquad \Big[\lim_{A \ni s \downarrow t} x(s) = x(t)\Big] \quad \text{a.s.} \qquad \square$$

Proof of (c1)–(c4). If $\{x(\cdot), \mathscr{F}(\cdot)\}$ is a martingale, if $\mathscr{F}(\cdot)$ is right continuous, and if $0 \le s' < s < t$, the almost sure equality $x(s') = E\{x(t)|\mathscr{F}(s')\}$ combined with the conditional expectation continuity theorem yields $x(s+) = E\{x(t)|\mathscr{F}(s)\} = x(s)$ almost surely when s' tends sequentially and monotonely to s. Thus in this case B^+ is empty. To prove the rest of (c1)–(c4), observe that the process $y(\cdot)$ is a standard modification of $x(\cdot)$, and is adapted to $\mathscr{F}(\cdot)$ because $\mathscr{F}(0)$ contains the null sets. Hence $\{y(\cdot), \mathscr{F}(\cdot)\}$ is a supermartingale, a martingale if $\{x(\cdot), \mathscr{F}(\cdot)\}$ is a martingale. If we choose A to be a superset of B, an application of (a) and (b) shows that (c1)–(c4) are true for $y(\cdot)$ and that $\{\underset{+}{x}(\cdot), \mathscr{F}^+(\cdot)\}$ is an almost surely right continuous adapted process with left limits. If B^+ is empty, $\underset{+}{x}(\cdot)$ is adapted to $\mathscr{F}(\cdot)$. Furthermore $P\{\underset{+}{x}(t) = x(t)\} = 1$ if and only if $t \in \mathbb{R}^+ - B^+$. To prove that $\{\underset{+}{x}(\cdot), \mathscr{F}^+(\cdot)\}$ is a supermartingale, which implies that this process is a martingale if $\{x(\cdot), \mathscr{F}(\cdot)\}$ is [apply the first result to $-x(\cdot)$], suppose that $0 \le s < s' < t < t' < t_0$, and observe that

$$E\{x(t') - E\{x(t_0)|\mathscr{F}(t')\}|\mathscr{F}(s')\} + E\{x(t_0)|\mathscr{F}(s')\} \le x(s') \quad \text{a.s.} \tag{1.3}$$

When $t' \downarrow t$ sequentially and $s' \downarrow s$ sequentially, apply Fatou's lemma for conditional expectations together with the conditional expectation continuity theorem to derive

$$E\{\underset{+}{x}(t) - E\{x(t_0)|\mathscr{F}^+(t)\}|\mathscr{F}^+(s)\} + E\{x(t_0)|\mathscr{F}^+(s)\} \le \underset{+}{x}(s) \quad \text{a.s.}$$

Hence $E\{\underset{+}{x}(t)|\mathscr{F}^+(s)\} \le \underset{+}{x}(s)$ almost surely; so the process $\{\underset{+}{x}(\cdot), \mathscr{F}^+(\cdot)\}$ is a supermartingale, and the proof of (c1)–(c4) for $\underset{+}{x}(\cdot)$ is now complete. □

The Right Continuous Case

In most applications the process $\{x(\cdot), \mathscr{F}(\cdot)\}$ in Theorem 1 is given as an almost surely right continuous supermartingale. In this context Theorem 1 implies that the process almost surely has finite left limits, that the processes $x(\cdot)$ and $x(\cdot-)$ are L^1 bounded on compact intervals of $\mathbb{R}^+$, and that there is an at most countable subset B of $\mathbb{R}^+$ such that $P\{x(t) = x(t-)\} = 1$ if and only if $t \in \mathbb{R}^+ - B$. We stress that even if B is empty, the process is not necessarily almost surely continuous.

Semipolar Sets and Supermartingale Smoothing

Recall that in the Fundamental Convergence Theorem (Theorem 1.VI.1) of classical potential theory a basic operation was the lower semicontinuous smoothing of the infimum of a locally lower bounded family of superharmonic functions. This smoothing yielded a superharmonic function equal quasi everywhere to the infimum function. The corresponding smoothing in the parabolic context (Section 1.XVII.13) yielded a superparabolic function equal to the infimum function up to a semipolar set. In these contexts the infimum function satisfies the average property of the function class considered, the class of superharmonic or superparabolic functions as the case may be, but does not satisfy the lower semicontinuity property of the functions in the class except in trivial cases. The lower semicontinuous smoothing replaces the infimum function by a function in the class equal to the infimum function off a small set. The analogous smoothing operation in the probability context is the smoothing of $x(\cdot)$ into $\underset{+}{x}(\cdot)$ in Theorem 1. This will be seen in the context of the probabilistic Fundamental Convergence Theorem (Theorem 5 below). The following application of Theorem 1 is an illustration of this smoothing.

The Set of Discontinuities of a Supermartingale

Let $(\Omega, \mathscr{F}, P; \mathscr{F}(t), t \in \mathbb{R}^+)$ be a filtered probability space with $\mathscr{F}(\cdot)$ right continuous and $\mathscr{F}(0)$ containing the null sets. Let $\{x(\cdot), \mathscr{F}(\cdot)\}$ be a nearly progressively measurable supermartingale on the space, and suppose that almost no sample function has an oscillatory discontinuity; that is, the process almost surely has right and left limits. Define $\underset{+}{x}(t,\omega) = x(t+,\omega)$ $[\underline{x}(t,\omega) = x(t-,\omega)]$ if this right [left] limit exists, and define $\underset{+}{x}(t,\omega)$ $[\underline{x}(t,\omega)]$ arbitrarily otherwise. Then we show that $\underset{+}{x}(\cdot,\cdot) = x(\cdot,\cdot) = \underline{x}(\cdot,\cdot)$ up to a semipolar subset of $\mathbb{R}^+ \times \Omega$. Since an evanescent set is semipolar, we lose no generality if we adjust the process if necessary on a null set of Ω to make the process progressively measurable with no sample function having an oscillatory discontinuity, and we shall suppose that this has been done. Define

$$\dot{A}_k = \left\{(t,\omega): \left|x(t,\omega) - \underset{+}{x}(t,\omega)\right| \geq \frac{1}{k}\right\}.$$

The process $\{\underset{+}{x}(\cdot), \mathscr{F}(\cdot)\}$ is progressively measurable because it is right continuous (Section I.2); so the process $\{|x(\cdot) - \underset{+}{x}(\cdot)|, \mathscr{F}(\cdot)\}$ is progressively measurable, and therefore $\dot{A}_k$ is a progressively measurable subset of $\mathbb{R}^+ \times \Omega$. Since the sample function $x(\cdot,\omega)$ has no oscillatory discontinuity, the set of values of t with $(t,\omega) \in \dot{A}_k$ has no finite limit point; if k and n are strictly positive integers, let T_{kn} be the nth value of t in this set (with $T_{kn} = +\infty$ if there is no nth value). Then (Section II.4) T_{kn} is the nth entry time of the progressively measurable set $\dot{A}_k$; so T_{kn} is optional. Furthermore

$$\{\underset{+}{x}(\cdot,\cdot) \neq x(\cdot,\cdot)\} = \bigcup_{k=1}^{\infty} \dot{A}_k = \bigcup_{k,n=1}^{\infty} [\![T_{kn}]\!];$$

so the set on the left is semipolar as asserted, and similarly the set $\{\underline{x}(\cdot,\cdot) \neq x(\cdot,\cdot)\}$ is semipolar.

Returning to the original hypotheses of this application of Theorem 1, observe that one choice of $\underset{+}{x}(\cdot,\cdot)$, in fact the choice we shall always use below whenever $x(\cdot)$ almost surely has right limits, is $\underset{+}{x}(t) = \liminf_{s\downarrow t} x(s)$. This choice makes $\{\underset{+}{x}(\cdot), \mathscr{F}(\cdot)\}$ progressively measurable if $\mathscr{F}(\cdot)$ is right continuous even if $\{x(\cdot), \mathscr{F}(\cdot)\}$ is not progressively measurable.

Application of Theorem 1 to a Conditional Expectation Process $x(t) = E\{z|\mathscr{F}(t)\}$

Let $(\Omega, \mathscr{F}, P; \mathscr{F}(t), t \in \mathbb{R}^+)$ be a probability measure space with a right continuous filtration, let z be an integrable random variable on the space, and define a martingale $\{x(\cdot), \mathscr{F}(\cdot)\}$ by setting $x(t) = E\{z|\mathscr{F}(t)\}$. Right continuity of the filtration implies that the set B^+ of Theorem 1 is empty. According to Theorem 1, the process $\underset{+}{x}(\cdot)$ is a standard modification of $x(\cdot)$ and is almost surely right continuous. In other words the conditional expectation defining $x(t)$ can be chosen to make $x(\cdot)$ an almost surely right continuous process, or even right continuous if $\mathscr{F}(0)$ contains the null sets.

Application to the Decomposition of a Right Closable Supermartingale

Let $\{x(t), \mathscr{F}(t); t \in \mathbb{R}^+\}$ be a right closable almost surely right continuous supermartingale, and suppose that $\mathscr{F}(\cdot)$ is right continuous. If x right closes the supermartingale and if $E\{x|\mathscr{F}(t)\}$ is chosen properly, the equation

$$x(t) = [x(t) - E\{x|\mathscr{F}(t)\}] + E\{x|\mathscr{F}(t)\} \tag{1.4}$$

exhibits $x(\cdot)$ as the sum of a positive almost surely right continuous supermartingale and an almost surely right continuous uniformly integrable martingale. The discrete parameter version of this decomposition was used in Section III.15. Going farther, the representation

$$E\{x|\mathscr{F}(t)\} = E\{x \vee 0|\mathscr{F}(t)\} - E\{(-x) \vee 0|\mathscr{F}(t)\}, \tag{1.5}$$

with suitable choices of the conditional expectations, exhibits the martingale component in (1.5) as the difference between two positive almost surely right continuous uniformly integrable martingales.

2. Optional Sampling of Uniformly Integrable Continuous Parameter Martingales

Recall [Section III.3(e)] that a martingale is uniformly integrable if and only if it is right closable. The following theorem is a continuous parameter version of the results obtained in Section III.2, Example (a), and is proved using those results.

Theorem. *Let $(\Omega, \mathscr{F}, P; \mathscr{F}(t), t \in \mathbb{R}^+)$ be a filtered probability space. If $\{x(\cdot), \mathscr{F}(\cdot)\}$ is an almost surely right continuous uniformly integrable martingale, closable, say by $\{x(+\infty), \mathscr{F}(+\infty)\}$, and if S and T are optional ($\leq +\infty$), then*

$$x(T) = E\{x(+\infty)|\mathscr{F}(T)\} \quad \text{a.s.} \tag{2.1}$$

and

$$E\{x(S)|\mathscr{F}(T)\} = E\{x(T)|\mathscr{F}(S)\} = x(S \wedge T) \quad \text{a.s.} \tag{2.2}$$

Observe that we have not supposed that $\mathscr{F}(\cdot)$ is right continuous. To avoid trivia, enlarge the filtration if necessary to make $\mathscr{F}(0)$ contain the null sets. Define $[T]_n$, an optional time for $\mathscr{F}(\cdot)$ and also for the discrete filtration $\{\mathscr{F}(j2^{-n}), j = 0, 1, \ldots, +\infty\}$, as in Section II.2, and define $[S]_n$ correspondingly. According to Section III.2, Example (a),

$$x([S]_n) = E\{x(+\infty)|\mathscr{F}([S]_n)\} \quad \text{a.s.} \tag{2.3}$$

and

$$E\{x([S]_n)|\mathscr{F}([T]_n)\} = x([S]_n \wedge [T]_n) \quad \text{a.s.} \tag{2.4}$$

The integrated version of (2.4) is

$$\int_\Lambda x([S]_n)\,dP = \int_\Lambda x([S]_n \wedge [T]_n)\,dP \tag{2.5}$$

for $\Lambda \in \mathscr{F}([T]_n)$. Hence (2.5) is true for Λ in the smaller σ algebra $\mathscr{F}(T)$. When $n \to \infty$, we can integrate to the limit on the left side of (2.5) because the sequence of integrands, a sequence of conditional expectations of $x(+\infty)$ according to (2.3), is uniformly integrable [Section I.4(i)]. Since the same argument is applicable to the right side of (2.5), equation (2.5) implies that $x(S)$ and $x(S \wedge T)$ are integrable and that

$$\int_\Lambda x(S)\,dP = \int_\Lambda x(S \wedge T)\,dP, \qquad \Lambda \in \mathscr{F}(T). \tag{2.6}$$

Furthermore $x(S \wedge T)$ is $\mathscr{F}(S \wedge T)$ measurable (Section II.3) and so $\mathscr{F}(T)$ measurable. Equation (2.6) is the integrated version of the almost everywhere equality between the first and third terms of (2.2). Interchange S and T to derive the almost everywhere equality between the second and third terms. Set $S \equiv +\infty$ in (2.2) to obtain (2.1).

Application of Theorem 2 to Totally Inaccessible Optional Times

Let T be a totally inaccessible optional time, and define

$$a(t) = 1_{[T,+\infty[}(t),$$
$$a_n(t) = E\{a((k+1)2^{-n})|\mathscr{F}(t)\} \quad \text{on } I_{nk} = [k2^{-n}, (k+1)2^{-n}[, \quad k \in \mathbb{Z}^+.$$

Here we use the versions of the conditional expectations making $a_n(\cdot)$ an almost surely right continuous process. Observe that $1 \geq a_n(t) \geq a_{n+1}(t) \geq a(t)$ almost surely, simultaneously for all t. We now show that *almost surely* $\lim_{n\to\infty} a_n(t) = a(t)$ *uniformly on* $\mathbb{R}^+$. To see this, choose $\varepsilon > 0$, and define

$$T_n = \inf\{t: a_n(t) - a(t) > \varepsilon\}.$$

Then $T_\cdot$ is an increasing sequence of optional times. Let $T_\infty = \lim_{n\to\infty} T_n$. Since $a_n(t) = a(t) = 1$ for sufficiently large t almost everywhere on $\{T < +\infty\}$, and for t in I_{nk} the evaluation

$$a_n(t) = P\{T \le (k+1)2^{-n} | \mathscr{F}(t)\}$$

implies that $\lim_{t\to\infty} a_n(t) = \lim_{t\to\infty} a(t) = 0$ almost everywhere on $\{T = +\infty\}$ (by the conditional expectation continuity theorem), it is sufficient to prove that $T_\infty = +\infty$ almost surely. Now

$$\begin{aligned} E\{a_n(T_n)1_{\{T_n<+\infty\}}\} &= \sum_{k=0}^{\infty} \int_{\{T_n \in I_{nk}\}} E\{a((k+1)2^{-n}) | \mathscr{F}(T_n)\}\, dP \\ &= \sum_{k=0}^{\infty} \int_{\{T_n \in I_{nk}\}} a((k+1)2^{-n})\, dP \\ &\le \sum_{k=0}^{\infty} \int_{\{T_n \in I_{nk}\}} a(T_n + 2^{-n})\, dP \\ &= E\{a(T_n + 2^{-n})1_{\{T_n<+\infty\}}\} = P\{T \le T_n + 2^{-n} < +\infty\} \\ &\le P\{T \le T_\infty + 2^{-n}, T < +\infty\}. \end{aligned} \tag{2.7}$$

Moreover by definition of $a(\cdot)$

$$\lim_{n\to\infty} E\{a(T_n)1_{\{T_n<+\infty\}}\} = \lim_{n\to\infty} P\{T \le T_n, T < +\infty\}, \tag{2.8}$$

and by definition of T_n

$$E\{[a_n(T_n) - a(T_n)]1_{\{T_n<+\infty\}}\} \ge \varepsilon P\{T_n < +\infty\} \ge \varepsilon P\{T_\infty < +\infty\}. \tag{2.9}$$

Thus (2.7)–(2.9) lead to

$$P\{T \le T_\infty, T < +\infty\} - \lim_{n\to\infty} P\{T \le T_n, T < +\infty\} \ge \varepsilon P\{T_\infty < +\infty\}. \tag{2.10}$$

Finally the total inaccessibility of T implies that the limit in (2.10) is $P\{T \le T_\infty, T < +\infty\}$ and thereby implies that $P\{T_\infty < +\infty\} = 0$, as was to be proved.

3. Optional Sampling and Convergence of Continuous Parameter Supermartingales

Theorem. *Let $\{x(t), \mathscr{F}(t), t \in \mathbb{R}^+\}$ be a right continuous right closable supermartingale. Then*

(a) $\lim_{t\to\infty} x(t) = x(+\infty)$ *exists* a.s. *and right closes the supermartingale.*

(b) *If $T(\le +\infty)$ is optional, it follows that $x(T)$ is measurable and integrable.*

(c) *If S and T are optional with $S \le T \le +\infty$, it follows that*

$$E\{x(T)|\mathscr{F}(S)\} \le x(S) \quad \text{a.s.} \tag{3.1}$$

Observation (a). Almost sure right continuity of $x(\cdot)$, instead of right continuity, is sufficient for the proof of (a), but under this weakened hypothesis $x(T)$ may not be measurable, and so (b) and (c) may fail. If $\mathscr{F}(0)$ contains the null sets, however, the theorem is true with no change in proof even if $x(\cdot)$ is only almost surely right continuous because (Section I.8) $x(\cdot)$ is indistinguishable from a right continuous process adapted to $\mathscr{F}(\cdot)$. If the supermartingale is almost surely right continuous and if $\mathscr{F}(0)$ does not contain the null sets, each σ algebra $\mathscr{F}(t)$ can be replaced by the σ algebra generated by $\mathscr{F}(t)$, and the null sets to get an enlarged family of σ algebras for which the preceding argument becomes applicable.

Observation (b). If the process in the theorem is not right closable but if S and T are bounded, say $T \le c$, the process restricted to the parameter interval $[0, c[$ is right closable, and the theorem is applicable with a trivial reformulation to the process on this parameter interval. In particular, and we state the following for martingales as well as for supermartingales since the supermartingale result can be applied to both $x(\cdot)$ and $-x(\cdot)$ in the martingale case, if $\{x(t), \mathscr{F}(t), t \in \mathbb{R}^+\}$ is a right continuous supermartingale [martingale] and if T is optional, the process $\{x(T \wedge t), \mathscr{F}(T \wedge t), t \in \mathbb{R}^+\}$, the process $x(\cdot)$ stopped at T, is a supermartingale [martingale]. A slight refinement of the proof below (see Section III.8 for the discrete parameter case) yields the stronger result that $\{x(T \wedge t), \mathscr{F}(t), t \in \mathbb{R}^+\}$ is a supermartingale [martingale].

Since the given supermartingale is right closable, the almost sure limit $x(+\infty)$ exists (Section III.15), and the reasoning used in the discrete parameter case in that section shows that $x(+\infty)$ right closes the supermartingale. Since the process $x(\cdot)$ is right continuous and therefore progressively measurable, the function $x(T)$ is measurable. The process $\{x(j2^{-n}), \mathscr{F}(j2^{-n}), j = 0, 1, \ldots, \infty\}$ is a supermartingale and $[T]_n$ as defined in Section II.2 is optional for the indicated discrete σ algebra family. Hence (Theorem III.7)

$$\begin{aligned} x([S]_n) &\ge E\{x([T]_n)|\mathscr{F}([S]_n)\} \\ &= E\{x([T]_n) - E\{x(+\infty)|\mathscr{F}([T]_n)\}|\mathscr{F}([S]_n)\} \\ &\quad + E\{x(+\infty)|\mathscr{F}([S]_n)\} \quad \text{a.s.} \end{aligned} \tag{3.2}$$

In view of Fatou's lemma for conditional expectations this inequality yields ($n \to \infty$)

$$x(S) \ge E\{x(T) - E\{x(+\infty)|\mathscr{F}^+(T)\}|\mathscr{F}^+(S)\} + E\{x(+\infty)|\mathscr{F}^+(S)\} \quad \text{a.s.} \tag{3.3}$$

If S vanishes identically (3.3) implies that $x(T)$ is integrable. For general S inequality (3.3) yields

$$x(S) \geq E\{x(T) | \mathscr{F}^+(S)\} \quad \text{a.s.},$$

and the operation $E\{-|\mathscr{F}(S)\}$ performed on both sides of this inequality yields (c).

Extension of Theorem 3 Involving Left Limits

Under the hypotheses of Theorem 3 we have seen that the process $x(\cdot)$ has almost sure left limits. Suppose that S and T are as described in that theorem and that $T_\cdot$ is an increasing sequence of optional times with limit T and with $T_n = T$ on $\{S = T\}$. Define $x^-(T) = \lim_{n\to\infty} x(T_n)$. We shall now prove that the ordered triple

$$[x(S), \mathscr{F}(S)], \left[x^-(T), \bigvee_0^\infty \mathscr{F}(T_n)\right], [x(T), \mathscr{F}(T)]$$

is a supermartingale, In particular, if $T_n = T$ for all n, this result is a rewriting of part of Theorem 3. At the other extreme, if $P\{T_n = T > 0\} = 0$ for all n, which implies that T is a predictable optional time announced by $T_\cdot$, the assertion is that the ordered triple

$$[x(S), \mathscr{F}(S)], [x(T-), \mathscr{F}(T-)], [x(T), \mathscr{F}(T)]$$

is a supermartingale. [See the definition of $\mathscr{F}(T-)$ in Section II.7.] Moreover if the given process is a right closable martingale, these triples are martingales.

It must be proved that $x^-(T)$ is integrable and

$$\begin{aligned} &\text{(a)} \quad E\{x^-(T) | \mathscr{F}(S)\} \leq x(S), \\ &\text{(b)} \quad E\left\{x(T) \,\middle|\, \bigvee_0^\infty \mathscr{F}(T_n)\right\} \leq x^-(T) \quad \text{a.s.} \end{aligned} \tag{3.4}$$

According to Theorem 3, $E\{x(T)|\mathscr{F}(T_n)\} \leq x(T_n)$ almost surely, and when $n \to \infty$ this inequality becomes (3.4b) by virtue of the conditional expectation continuity theorem. In view of the decomposition (1.4) of $x(\cdot)$ into the sum of a positive supermartingale and a uniformly integrable martingale it is sufficient to prove that (3.4a) is true for a positive supermartingale and also for a uniformly integrable martingale. If $x(\cdot) \geq 0$ and if c is a positive constant, the following supermartingale inequalities for $x(\cdot)$

$$E\{x(T_n)\} \leq E\{x(0)\}, \qquad E\{x(T_n \vee S) | \mathscr{F}(S)\} \leq x(S) \quad \text{a.s.}$$

yield when $n \to \infty$ the fact that $x^-(T)$ is integrable and (3.4a) is true. If $x(\cdot)$ is a uniformly integrable martingale, it is a class **D** martingale (Section III.6); so the martingale equality

$$\int_\Lambda x(T_n \vee S)\,dP = \int_\Lambda x(S)\,dP \qquad [\Lambda \in \mathscr{F}(S)]$$

from Theorem 3 yields the integrated form of (3.4a) when $n \to \infty$.

4. Increasing Sequences of Supermartingales

Theorem. *Let $\{x_\cdot(t), \mathscr{F}(t), t \in \mathbb{R}^+\}$ be an increasing sequence of almost surely right continuous supermartingales, with limit process $\{x(\cdot), \mathscr{F}(\cdot)\}$.*

(a) *If $E\{x(0)\} < +\infty$, the process $\{x(\cdot), \mathscr{F}(\cdot)\}$ is a supermartingale.*
(b) *The $x(\cdot)$ process is almost surely right continuous with left limits, and if T is the hitting time by (t, ω) of the set $\{(t, \omega): x(t, \omega) < +\infty\}$, then for $t > T$ almost every $x(\cdot)$ sample function is finite valued and has finite left limits.*

Assertion (a) is trivial. In proving (b) it can be assumed (Section 1), enlarging each σ algebra $\mathscr{F}(t)$ if necessary, that $\mathscr{F}(\cdot)$ is right continuous, that $\mathscr{F}(0)$ contains the null sets, and that each process $x_n(\cdot)$ is right continuous. It is sufficient to prove the theorem for $x_n(\cdot)$ right closable because it is sufficient to derive the conclusions for each parameter interval $[0, b[$. We can nevertheless continue to use the parameter interval $\mathbb{R}^+$. Furthermore it can be assumed that each process $x_n(\cdot)$ is positive because (Section 1) $x_0(\cdot)$ is the sum of a right continuous martingale $y_0(\cdot)$ and a positive right continuous supermartingale, and we can replace $x_n(\cdot)$ by $x_n(\cdot) - y_0(\cdot)$ to get an increasing sequence of positive right continuous supermartingales. The truth of (b) for this sequence implies its truth for the given sequence.

Proof of almost sure right continuity of $x(\cdot)$. It is sufficient to prove almost sure right continuity for $x(\cdot)$ a bounded process since this result can be applied to the sequence $x_\cdot(\cdot) \wedge c$. To prove almost sure right continuity in this case, under the simplifying assumptions developed above, observe first that $\{x(\cdot), \mathscr{F}(\cdot)\}$ is progressively measurable since this process is the limit of a sequence of progressively measurable processes. Hence (Section II.6) it is sufficient to prove that whenever T is a finite-valued optional time, almost every $x(\cdot)$ sample function is right continuous at T. In fact in the present context it is sufficient to prove that almost every sample function is right continuous at 0 because this result can be applied to the increasing sequence $\{x_\cdot(T + t), \mathscr{F}(T + t), t \in \mathbb{R}^+\}$ of right continuous supermartingales to get $x(\cdot)$ almost surely right continuous at T. To prove almost sure right continuity at 0, observe that since $x_n(\cdot)$ is right continuous, the $x(\cdot)$ sample functions are right lower semicontinuous. Furthermore the restriction of the supermartingale $x(\cdot)$ to the parameter set of strictly positive rationals almost surely has a right limit y at the origin according to Section III.17, and

$$x(0) \leq \liminf_{t\to 0} x(t) \leq y \quad \text{a.s.} \tag{4.1}$$

On the other hand, the supermartingale inequality $E\{x(0)\} \geq E\{x(t)\}$ implies

$$E\{x(0)\} \geq E\{y\}. \tag{4.2}$$

It follows that

$$x(0) = \liminf_{t\to 0} x(t) = y \quad \text{a.s.}$$

and the right lower semicontinuity of $x(\cdot)$ then implies that

$$\limsup_{t\to 0} x(t) = \limsup_{r\to 0} x(r) = y \quad \text{a.s.} \qquad (r \text{ rational}).$$

Thus $x(\cdot)$ is almost surely right continuous at 0, and hence as pointed out above, $x(\cdot)$ is almost surely right continuous. Almost sure right continuity implies (Section 1) almost sure left limits at all points.

Proof of (b). To finish the proof of (b), it is sufficient to prove (b′): If α and β are strictly positive numbers, then for almost every ω in the set $\{x(\alpha) < \beta\}$ the values $x(t, \omega)$ and $x(t+, \omega)$ are finite for $t > \alpha$. To reduce this fact to what we have already proved, define $z_n(t)$ for $t \geq 0$ as $x_n(\alpha + t)$ on $\{x(\alpha) \leq \beta\}$ and as 0 elsewhere. Apply (a) to the increasing sequence $\{z_{\cdot}(t), \mathscr{F}(\alpha + t), t \in \mathbb{R}^+\}$ of supermartingales to find that $\lim_{n\to\infty} z_n(\cdot)$ according to what we have proved already is an almost surely right continuous supermartingale and that therefore almost every sample function is bounded on compact intervals. The remaining part of (b) follows.

Elementary Proof of Almost Sure Right Continuity of $x(\cdot)$

The proof given above of almost sure right continuity of $x(\cdot)$ is natural, but it is of interest to exhibit an elementary proof that does not involve the relatively deep theorems on hitting probabilities. The following proof does not even need the supermartingale convergence theorems. We can assume as in the proof already given that $\mathscr{F}(\cdot)$ is right continuous, that $\mathscr{F}(0)$ contains the null sets, that each process $x_n(\cdot)$ is positive and right continuous, and that $x(\cdot)$ is bounded. Define $x(+\infty) = x_n(+\infty) = 0$. Observe first that if $S_\cdot$ is a sequence of optional times and if $\Lambda = \{\lim_{n\to\infty} S_n = 0\}$, then $\Lambda \in \mathscr{F}(0)$, and an application of the right lower semicontinuity of $x(\cdot)$ sample functions and of the supermartingale inequality at optional times [valid for $x(\cdot)$ because it is valid for $x_n(\cdot)$] yields

$$E\{x(0);\Lambda\} \le E\left\{\liminf_{n\to\infty} x(S_n);\Lambda\right\} \le \liminf_{n\to\infty} E\{x(S_n);\Lambda\} \le E\{x(0);\Lambda\}$$

from which it follows in view of sample function right lower semicontinuity that

$$\liminf_{n\to\infty} x(S_n) = x(0) \tag{4.3}$$

almost everywhere on Λ. Next choose $\varepsilon > 0$ and define

$$T_n = \inf\left\{t: \left|x_n(t) - x(0)\right| > \frac{\varepsilon}{2}\right\}, \quad T = \limsup_{n\to\infty} T_n.$$

The time T_n is optional because up to a null set

$$\{T_n \ge c\} = \sup_{r<c} |x_n(r) - x(0)| \le \frac{\varepsilon}{2} \qquad (r \text{ rational}, c > 0)$$

and therefore T is also optional. Furthermore T is almost surely strictly positive because otherwise, there is a set of strictly positive probability on which $\lim_{n\to\infty} T_n = 0$ and

$$|x_n(T_n) - x(0)| \ge \frac{\varepsilon}{2} \tag{4.4}$$

for sufficiently large n. Now under (4.4) either $x(T_n) - x(0) \ge x_n(T_n) - x(0) \ge \varepsilon/2$ for sufficiently large n, which leads to $\liminf_{n\to\infty} x(T_n) - x(0) \ge \varepsilon/2$ and thereby contradicts (4.3) or under (4.4), for every k, $x_k(T_n) - x(0) \le x_n(T_n) - x(0) \le -\varepsilon/2$ for infinitely many values of n, in which case $x_k(0) - x(0) \le -\varepsilon/2$, which is impossible for large k. Thus T is almost surely strictly positive. Furthermore $|x_n(t) - x(0)| \le \varepsilon/2$ on $[0, T_n[$ and so almost surely $|x(t) - x(0)| \le \varepsilon/2$ on $[0, T[$, and if we set

$$r(t) = \lim_{\delta\to 0} \sup_{t<s<t+\delta} |x(s) - x(t)|,$$

then almost surely $r(t) \le \varepsilon$ on $[0, T[$. The class Γ_ε of almost surely strictly positive optional times T for which almost surely $r(t) \le \varepsilon$ on $[0, T[$ is not empty according to what we have just proved, is directed upward, and is closed under countable suprema; so Γ_ε contains its own essential supremum T'. This is impossible unless $T' = +\infty$ almost surely because the above proof that Γ_ε is not empty when applied to the process $x(T' \wedge k + \cdot)$ shows that there is a member of Γ_ε almost surely strictly larger than $T' \wedge k$ for every k. Thus for every ε the right oscillation of almost every $x(\cdot)$ sample function is at most ε at all points, and the proof is complete.

Generalization to Upward-Directed Supermartingale Families

Observe that Theorem 4 can be applied to an upward-directed family of supermartingales (essential order) since according to Section III.5 the essential order supremum of the family is a sequential supremum.

5. Probability Version of the Fundamental Convergence Theorem of Potential Theory

The Right Continuous Smoothing of a Supermartingale

Let $\{x(t), \mathscr{F}(t), t \in \mathbb{R}^+\}$ be a supermartingale with almost sure right limits; that is, we suppose that almost every $x(\cdot)$ sample function has a right limit at every parameter value. According to Section 1, the process also has almost sure left limits. From now on in this context we define

$$\underset{+}{x}(t) = \liminf_{s \downarrow t} x(s), \tag{5.1}$$

thus making the definition of $\underset{+}{x}(\cdot)$ slightly more specific than that in Section 1. According to Section 1, if $\mathscr{F}(0)$ contains the null sets and if $\mathscr{F}(\cdot)$ is right continuous, the process $\{\underset{+}{x}(\cdot), \mathscr{F}(\cdot)\}$ is an almost surely right continuous supermartingale and $x(\cdot,\cdot) = \underset{+}{x}(\cdot,\cdot)$ up to a semipolar subset of $\mathbb{R}^+ \times \Omega$; in particular, $P\{x(t) = \underset{+}{x}(t)\} = 1$ up to a countable set of values of t. The (almost surely) right continuous smoothing $\underset{+}{x}(\cdot)$ plays the same role for $x(\cdot)$ as the lower semicontinuous smoothing of a function plays in classical and parabolic potential theory, according to the theorem of this section.

The martingale theory concept corresponding to that of a locally lower bounded family of superharmonic functions is a locally lower bounded family of supermartingales. Although local lower boundedness is not difficult to formulate in the martingale theory context, we shall restrict ourselves to families of positive supermartingales to simplify the discussion.

Theorem. *Let $\{x_\alpha(\cdot), \mathscr{F}(\cdot), \alpha \in J\}$ be a family of positive almost surely right continuous supermartingales on a filtered probability space*

$$(\Omega, \mathscr{F}, P; \mathscr{F}(t), t \in \mathbb{R}^+).$$

It is supposed that $\mathscr{F}(\cdot)$ is right continuous and that $\mathscr{F}(0)$ contains the null sets.

(a) *There is then a process $\{x(\cdot), \mathscr{F}(\cdot)\}$ determined up to indistinguishability by the following conditions.*
   (a1) *$\{x(\cdot), \mathscr{F}(\cdot)\}$ is nearly progressively measurable.*
   (a2) *If $\alpha \in J$, then $x(\cdot,\cdot) \le x_\alpha(\cdot,\cdot)$ quasi everywhere on $\mathbb{R}^+ \times \Omega$.*
   (a3) *If $\{y(\cdot), \mathscr{F}(\cdot)\}$ is a process satisfying* (a1) *and* (a2), *then $y(\cdot,\cdot) \le x(\cdot,\cdot)$ quasi everywhere on $\mathbb{R}^+ \times \Omega$.*

(b) *The process $\{x(\cdot), \mathscr{F}(\cdot)\}$ in* (a) *is a supermartingale with almost sure right and left limits. If $\underset{+}{x}(\cdot)$ is the (almost surely) right continuous smoothing of $x(\cdot)$, then up to an evanescent subset of $\mathbb{R}^+ \times \Omega$*

$$\underset{+}{x}(t) = x(t+), \qquad \underset{+}{x}(t-) = x(t-), \qquad \underset{+}{x}(t) \le x(t), \tag{5.2}$$

and there is equality in the inequality up to a semipolar subset of $\mathbb{R}^+ \times \Omega$.

(c) *For some countable subset J_0 of J ($J_0 = J$ if J is countable) the process $x(\cdot) = \inf_{\alpha \in J_0} x_\alpha(\cdot)$ satisfies the conditions in* (a).

(d) *Conversely, if $\dot{A}$ is a semipolar subset of $\mathbb{R}^+ \times \Omega$, there is a decreasing sequence $\{x_n(\cdot), \mathscr{F}(\cdot), n \in \mathbb{Z}^+\}$ of right continuous positive supermartingales such that if $x(\cdot) = \inf_{n \in \mathbb{Z}^+} x_n(\cdot)$, then $\dot{A} \subset \{\underset{+}{x}(\cdot,\cdot) < x(\cdot,\cdot)\}$ up to an evanescent set.*

The notation ess* $\inf_{\alpha \in J} x_\alpha(\cdot)$. If $\{x(\cdot), \mathscr{F}(\cdot)\}$ satisfies the conditions under (a), then every process indistinguishable from $x(\cdot)$ also satisfies these conditions. The equivalence class of these processes under indistinguishability and in the usual notational abuse the individual members of this class will be denoted by ess* $\inf_{\alpha \in J} x_\alpha(\cdot)$. The equivalence class of right continuous smoothings $\underset{+}{x}(\cdot)$ and its individual members will be denoted by $\bigwedge_{\alpha \in J} x_\alpha(\cdot)$ when lattice concepts in martingale theory are discussed systematically in Chapter V. A right continuous smoothing $\underset{+}{x}(\cdot)$ is up to indistinguishability the maximum almost surely right continuous supermartingale majorized in the essential order by every supermartingale $\{x_\alpha(\cdot), \mathscr{F}(\cdot)\}$.

Theorem 5 as a Convergence Theorem

There is no loss of generality in supposing that the given supermartingale family is directed downward (pointwise order) because if the given family is enlarged to the family of minima of finite subsets, the enlarged family is directed downward and this enlargement does not change the scope of the theorem's statement. The process $x(\cdot)$ in the theorem is a version of the essential order limit (and infimum) of the downward-directed family. In particular, if J is countable, the enlarged supermartingale family is also countable and without loss of generality can be replaced by a decreasing sequence of its members. (If the given family has a smallest member, the decreasing sequence has all but a finite number of its members this smallest member of the original family.)

Deletion of "Almost Surely"

Without loss of generality we can suppose that every process $x_\alpha(\cdot)$ is right continuous because $x_\alpha(\cdot)$ can be made so by a change on a null set (depending on α) and this change does not affect the theorem. We shall suppose throughout the proof that this change has been made.

Proof of (a)–(c) *for J countable.* As observed above there is no loss of generality in this case if we suppose that the family of supermartingales is a decreasing sequence $\{x_n(\cdot), \mathscr{F}(\cdot), n \in \mathbb{Z}^+\}$ of positive right continuous supermartingales. The process $x(\cdot) = \inf_{n \geq 0} x_n(\cdot)$ is a supermartingale (Section III.5), progressively measurable because each process $\{x_n(\cdot), \mathscr{F}(\cdot)\}$ is right continuous and therefore progressively measurable. The other conditions under (a) are trivially satisfied. Observe that $E\{x_n(0)\} \leq E\{x_0(0)\}$ and thereby conclude that the expected number of downcrossings of an interval by $x(\cdot)$ is bounded because this expected number for $x_n(\cdot)$ is bounded by a number independent of n. Hence $x(\cdot)$ has almost sure left and right limits. To prove (5.2) observe that the first equation is the definition of $\underset{+}{x}(t)$, that the second equation follows from the first and that the inequality follows from the right continuity of $x_n(\cdot)$, which implies right upper semicontinuity of $x(\cdot)$. Finally there is equality up to a semipolar set in this inequality according to Section 1. □

Proof of (a)–(c) *for J uncountable.* Choose any countable subset J_0 of J, and define $x(\cdot) = \inf_{\alpha \in J_0} x_\alpha(\cdot)$. According to the preceding discussion with countable J, the process $\{x(\cdot), \mathscr{F}(\cdot)\}$ is a progressively measurable supermartingale and (b) is true. To prove (a)–(c), it suffices to show that J_0 can be chosen so large that (a2) is true. First choose J_0 so large (Section III.5) that $x(\cdot)$ is a version of $\operatorname{ess\,inf}_{\alpha \in J} x_\alpha(\cdot)$; that is, for each t the random variable $x(t)$ is a version of $\operatorname{ess\,inf}_{\alpha \in J} x_\alpha(t)$. Then for α in J it follows that $x(\cdot) \leq x_\alpha(\cdot)$ almost surely on the set of positive rationals; so $\underset{+}{x}(\cdot) \leq x_\alpha(\cdot)$ almost surely on $\mathbb{R}^+$ in view of sample function right continuity of $x_\alpha(\cdot)$. The exceptional null set depends on α. With this preliminary choice of J_0 define

$$\begin{aligned}\dot{M} = \{(t,\omega)\colon & x(\cdot,\omega) \text{ and } \underset{+}{x}(\cdot,\omega) \text{ have left and right limits at } t \text{ and}\\ & x(t+,\omega) = \underset{+}{x}(t,\omega) = \underset{+}{x}(t+,\omega),\ x(t-,\omega) = \underset{+}{x}(t-,\omega)\},\end{aligned}$$

$$\dot{B}_c = \dot{M} \cap \{(t,\omega)\colon x(\cdot,\omega) \text{ is continuous at } t\},$$

$$\dot{B}_d = \dot{M} \cap \{(t, \omega): x(\cdot, \omega) \text{ is discontinuous at } t\}.$$

The set $(\mathbb{R}^+ \times \Omega) - M$ is evanescent. When $(t, \omega) \in \dot{B}_c$, the sample functions $x(\cdot, \omega)$ and $\underset{+}{x}(\cdot, \omega)$ are continuous at t and equal there; so for each α in J, $x(\cdot, \cdot) = \underset{+}{x}(\cdot, \cdot) \le x_\alpha(\cdot, \cdot)$ quasi everywhere on $\dot{B}_c$. The problem is to enlarge J_0 to ensure that also $x(\cdot, \cdot) \le x_\alpha(\cdot, \cdot)$ quasi everywhere on $\dot{B}_d$. If J' is a countable superset of J_0 and if $x'(\cdot) = \inf_{\alpha \in J'} x_\alpha(\cdot)$, then $x'(\cdot) = x(\cdot)$ almost surely on the positive rationals; so the almost surely right continuous processes $\underset{+}{x}(\cdot)$ and $\underset{+}{x}'(\cdot)$ are indistinguishable. Moreover $\dot{M}$, $\dot{B}_c$, and $\dot{B}_d$ are changed by at most evanescent sets when J_0 is increased to J'. According to the discussion of the discontinuities of a supermartingale at the end of Section 1, there is a sequence $T_\cdot$ of optional times the union of whose graphs covers $\dot{B}_d$. Choose J' so large that $\inf_{\alpha \in J'} x_\alpha(T_n)$ is a version of $\operatorname{ess\,inf}_{\alpha \in J} x_\alpha(T_n)$ for all n. Then for each α in J, $x'(\cdot, \cdot) \le x_\alpha(\cdot, \cdot)$ quasi everywhere on $\mathbb{R}^+ \times \Omega$; the exceptional evanescent set depends on α. The countable set J' is the final choice of J_0 and has the properties described in (c). □

Proof of (d). We can suppose that $\dot{A} \subset \bigcup_0^\infty [\![S_k]\!]$ for some sequence of optional times. For k and n in $\mathbb{Z}^+$ define

$$x_{kn}(\cdot, \cdot) = \begin{cases} 2^{-k} & \text{on } [\![0, S_k + 2^{-n}[\![, \\ 0 & \text{on } [\![S_k + 2^{-n}, +\infty[\![, \end{cases} \qquad x_n(t) = \sum_{k=0}^{\infty} x_{kn}(t).$$

The process $\{x_n(\cdot), \mathscr{F}(\cdot)\}$ is a positive right continuous supermartingale because each summand process is. The sequence $\{x_n(\cdot), \mathscr{F}(\cdot), n \in \mathbb{Z}^+\}$ is a decreasing sequence with limit process $\{x(\cdot), \mathscr{F}(\cdot)\}$ for which

$$\dot{A} \subset \bigcup_0^\infty [\![S_k]\!] \subset \{\underset{+}{x}(\cdot, \cdot) < x(\cdot, \cdot)\};$$

so (d) is true. □

A Special Property of $\operatorname{ess}^* \inf_{\alpha \in I} x_\alpha$

Let $y(\cdot)$ be a version of this infimum process. Since $y(\cdot)$ is equal quasi everywhere to the infimum of a countable set of almost surely right continuous positive supermartingales, equivalently, equal quasi everywhere to the limit of a decreasing sequence of almost surely right continuous positive supermartingales, the fact that the optional sampling Theorem 5 is applicable to an almost surely right continuous positive supermartingale implies that this theorem is applicable to $y(\cdot)$.

6. Quasi-Bounded Positive Supermartingales; Generation of Supermartingale Potentials by Increasing Processes

Let I be a linearly ordered set, and let $\{\mathscr{F}(t), t \in I\}$ be a filtration of a probability space.

Quasi-Bounded Positive Supermartingales

A positive supermartingale $\{x(\cdot), \mathscr{F}(\cdot)\}$ is called *quasi-bounded* if it is the sum of a series of bounded (adapted to $\mathscr{F}(\cdot)$) positive supermartingales.

Increasing Processes

From now on in this section we shall suppose that I has a first point, denoted by 0, but no last point. A process $\{A(t), t \in I\}$, not necessarily adapted to $\mathscr{F}(\cdot)$, with state space $(\mathbb{R}^+, \mathscr{B}(\mathbb{R}^+))$, will be called an *increasing process* if the following conditions are satisfied.

(a) $A(0) = 0$, $P\{A(s) \le A(t)\} = 1$ when $s \le t$;
(b) $E\{A(t)\} < +\infty$ for $t \in I$.

Let "$+\infty$" denote a point not in I, and define $A(+\infty) = \operatorname{ess\,sup}_{t \in I} A(t)$. If $A(\cdot)$ is L^1 bounded, that is, if $E\{A(+\infty)\} < +\infty$, define (uniquely up to a standard modification) a supermartingale $\{x(t), \mathscr{F}(t), t \in I\}$ by

$$x(t) = E\{A(+\infty) - A(t) | \mathscr{F}(t)\}. \tag{6.1}$$

This supermartingale is a supermartingale potential because

$$\lim_{t\uparrow} E\{x(t)\} = \lim_{t\uparrow} E\{A(+\infty) - A(t)\} = 0,$$

and this supermartingale potential is said to be generated by $A(\cdot)$. Observe that if T is a countably valued optional time with values in I, an application of Section III.2, Example (a), shows that

$$x(T) \le E\{A(+\infty) | \mathscr{F}(T)\} \quad \text{a.s;}$$

so [Section I.4(i)] the family

$$\{x(T)\colon T \text{ optional, countably valued, with values in } I\}$$

is uniformly integrable; that is, the process $\{x(\cdot), \mathscr{F}(\cdot)\}$ is in the class **D** defined in Section II.11. Moreover, if k is a positive integer, the process $A_k(\cdot)$ defined by

$$A_k(t) = A(t) \wedge (k+1) - A(t) \wedge k, \qquad t \in I,$$

is also an L^1 bounded increasing process, generating a supermartingale potential $\{x_k(\cdot), \mathscr{F}(\cdot)\}$ which can be chosen to be bounded by 1, and $x(\cdot) = \Sigma_0^\infty x_k(\cdot)$ up to a standard modification. Thus a supermartingale potential generated by an L^1 bounded increasing process is quasi-bounded and in the class **D**. Conversely, we shall prove in Section 10 that every class **D** or quasi-bounded supermartingale potential can be generated by an L^1 bounded increasing process. The corresponding results for the discrete and continuous parameter cases are proved, respectively, in Sections 8 and 11.

Adapted Increasing Processes

If $A(\cdot)$ is adapted to $\mathscr{F}(\cdot)$, (6.1) can be written in the form

$$x(t) = E\{A(+\infty)|\mathscr{F}(t)\} - A(t) \quad \text{a.s.} \tag{6.1'}$$

Observe that if $A(\cdot)$ and $B(\cdot)$ are L^1 bounded increasing processes adapted to $\mathscr{F}(\cdot)$, they generate the same supermartingale potential up to a standard modification if and only if for each parameter value t,

$$E\{A(+\infty) - A(t)|\mathscr{F}(t)\} = E\{B(+\infty) - B(t)|\mathscr{F}(t)\} \quad \text{a.s.,}$$

that is, if and only if the difference process $\{A(\cdot) - B(\cdot), \mathscr{F}(\cdot)\}$ is a martingale. In particular, suppose that I is either $\mathbb{Z}^+$ or $\mathbb{R}^+$ and that $\{A(\cdot), \mathscr{F}(\cdot)\}$ is an adapted L^1 bounded increasing process generating a supermartingale $x(\cdot)$. If $I = \mathbb{R}^+$, it is supposed that $\mathscr{F}(\cdot)$ is right continuous and that $A(\cdot)$ is almost surely right continuous; so we can suppose that $x(\cdot)$ is also almost surely right continuous. Under these hypotheses if $x(+\infty)$ is defined as 0, equality (6.1′) can be strengthened.

In fact, if $T\,(\leq +\infty)$ is an arbitrary optional time,

$$x(T) = E\{A(+\infty)|\mathscr{F}(t)\} - A(T) \quad \text{a.s.} \tag{6.1''}$$

in view of Section III.2, Example (a), when $I = \mathbb{Z}^+$ and Theorem 2 when $I = \mathbb{R}^+$.

The Support of the Measure Defined by an Increasing Process

If $\{A(\cdot), \mathscr{F}(\cdot)\}$ is an increasing adapted process with I either $\mathbb{Z}^+$ or $\mathbb{R}^+$ and under the hypotheses of the end of the preceding paragraph, define the (closed if $I = \mathbb{R}^+$) support of the measure $A(dt)$, more precisely the support of the measure $d_t A(t, \omega)$, by

$$I = \mathbb{Z}^+ : \dot{A} = \{(n, \omega): n > 0,\ A(n, \omega) - A(n-1, \omega) > 0\},$$

$$I = \mathbb{R}^+ : \dot{A} = \bigcap_{n=1}^{\infty} \left\{(t, \omega): t \geq 0;\ A\left(\frac{t+1}{n}, \omega\right) - A\left(\frac{t-1}{n}, \omega\right) > 0\right\},$$

where we set $A(t) = 0$ when $t < 0$. In the continuous parameter case, the nth set on the right is nearly progressively measurable relative to the filtration $\mathscr{F}(\cdot + 1/n)$ so $\dot{A}$ is nearly progressively measurable relative to this filtration for all n, and therefore (Section I.2) $\dot{A}$ is nearly progressively measurable relative to $\mathscr{F}(\cdot)$.

Natural Increasing Processes with $I = \mathbb{Z}^+$ or $\mathbb{R}^+$

We now suppose that $(\Omega, \mathscr{F}, P; \mathscr{F}(t), t \in I)$ is a filtered probability space with I either $\mathbb{Z}^+$ or $\mathbb{R}^+$, that $\mathscr{F}(0)$ contains the null sets, and if $I = \mathbb{R}^+$, that $\mathscr{F}(\cdot)$ is right continuous. Observe that if $I = \mathbb{R}^+$ and if $A(\cdot)$ is an increasing almost surely right continuous process, then almost all its sample functions are monotone increasing. In this context if $\{f(t), t \in \mathbb{R}^+\}$ is a stochastic process with state space $\bar{\mathbb{R}}$ and if $H \subset \mathbb{R}^+$, the integral $\int_H f(t)\, dA(t)$ is defined as the integral $\int_H f(t, \omega)\, d_t A(t, \omega)$ at each point ω for which $A(\cdot, \omega)$ is monotone increasing and $f(\cdot, \omega)$ is integrable in the usual sense over H with respect to the measure $A(dt, \omega)$. When I is either $\mathbb{Z}^+$ or $\mathbb{R}^+$, the problem of finding a unique increasing process $A(\cdot)$ generating a specified supermartingale potential leads to a condition on $A(\cdot)$ which we shall now formulate. Let $\{y(\cdot), \mathscr{F}(\cdot)\}$ be a bounded martingale, almost surely right continuous if $I = \mathbb{R}^+$, and define $y(+\infty) = \lim_{t \to \infty} y(t)$; this limit exists almost surely. An adapted increasing L^1 bounded process $\{A(\cdot), \mathscr{F}(\cdot)\}$, almost surely right continuous if $I = \mathbb{R}^+$, is called *natural* if

$$I = \mathbb{Z}^+: \qquad E\left\{\sum_{k=0}^{\infty} [y(k+1) - y(k)][A(k+1) - A(k)]\right\} = 0$$

$$I = \mathbb{R}^+: \qquad E\left\{\int_0^{\infty} [y(t) - y(t-)]\, dA(t)\right\} = 0 \tag{6.2}$$

for every choice of $y(\cdot)$. It will be shown in Section 7 that this somewhat awkward condition on $A(\cdot)$ is satisfied if and only if $A(\cdot)$ is nearly predictable.

In the following examples $I = \mathbb{R}^+$, and $A(\cdot)$ is adapted, L^1 bounded, almost surely right continuous.

EXAMPLE (a). The process $A(\cdot)$ is natural if almost surely continuous because $(6.2, I = \mathbb{R}^+)$ is then trivial because almost every $y(\cdot)$ sample function in this equation has at most countably many discontinuities.

EXAMPLE (b). If T is a predictable optional time and if $A^T(t) = 1_{[T, +\infty[}(t)$, then $A^T(\cdot)$ is natural. In fact condition $(6.2, I = \mathbb{R}^+)$ for $A^T(\cdot)$ reduces to

$$E\{[y(T) - y(T-)]1_{\{T<+\infty\}}\} = 0, \tag{6.3}$$

and if $T_{\bullet}$ is an increasing sequence of optional times announcing T, the martingale equality $E\{y(T \wedge n) - y(T_n \wedge n)\} = 0$ leads to (6.3) when $n \to \infty$.

7. Natural versus Predictable Increasing Processes ($I = \mathbb{Z}^+$ or $\mathbb{R}^+$)

In this section $(\Omega, \mathscr{F}, P; \mathscr{F}(t), t \in I)$ is the reference space, a filtered probability space with I either $\mathbb{Z}^+$ or $\mathbb{R}^+$, $\mathscr{F}(0)$ contains the null sets, and, if $I = \mathbb{R}^+$, $\mathscr{F}(\cdot)$ is right continuous. In the following theorem it is supposed without explicit mention that the martingales and increasing processes involved are almost surely right continuous when $I = \mathbb{R}^+$. If $A(\cdot)$ is an increasing process $A(+\infty)$ is defined as $\lim_{t\to\infty} A(t)$. If z is an integrable random variable the martingale $E\{z|\mathscr{F}(\cdot)\}$ can be chosen to be right continuous [by Section 1], and we define $E\{z|\mathscr{F}(\cdot\,-)\}$ as the left limit process of this martingale. This left limit process is an almost surely left continuous martingale and is therefore nearly predictable.

Theorem. *Let $\{A(\cdot), \mathscr{F}(\cdot)\}$ be an adapted L^1-bounded increasing process.*

(a1) *If $\{y(\cdot), \mathscr{F}(\cdot)\}$ is a bounded martingale and if $y(+\infty)$ is defined as $\lim_{t\to\infty} y(t)$, then*

$$I = \mathbb{Z}^+: \qquad E\left\{\sum_{k=0}^{\infty} y(k+1)[A(k+1) - A(k)]\right\} = E\{y(+\infty)A(+\infty)\}, \tag{7.1}$$

$$I = \mathbb{R}^+: \qquad E\left\{\int_0^{\infty} y(t)\, dA(t)\right\} = E\{y(+\infty)A(+\infty)\}.$$

(a2) *The process $A(\cdot)$ is natural if and only if whenever α is a constant in I and z is a bounded $\mathscr{F}(\alpha)$ measurable random variable*

$$I = \mathbb{Z}^+: \qquad E\left\{\sum_{k=0}^{\alpha} E\{z|\mathscr{F}(k+1)\}[A(k+1) - A(k)]\right\} = E\{zA(\alpha+1)\}, \tag{7.2}$$

$$I = \mathbb{R}^+: \qquad E\left\{\int_{[0,\alpha]} E\{z|\mathscr{F}(t-)\}\, dA(t)\right\} = E\{zA(\alpha)\}.$$

(a3) *The process $A(\cdot)$ is natural if and only if it is nearly predictable.*

Let $\{A(\cdot), \mathscr{F}(\cdot)\}$ and $\{B(\cdot), \mathscr{F}(\cdot)\}$ be adapted L^1 bounded increasing processes for which $\{A(\cdot) - B(\cdot), \mathscr{F}(\cdot)\}$ is a martingale.

(b1) *If* $\{y(\cdot), \mathscr{F}(\cdot)\}$ *is a bounded predictable process then*

$$I = \mathbb{Z}^+: \qquad E\left\{\sum_{k=0}^{\infty} y(k+1)[A(k+1) - A(k)]\right\} = E\left\{\sum_{k=0}^{\infty} y(k+1)[B(k+1) - B(k)]\right\} \tag{7.3}$$

$$I = \mathbb{R}^+: \qquad E\left\{\int_0^{\infty} y(t)\,dA(t)\right\} = E\left\{\int_0^{\infty} y(t)\,dB(t)\right\}.$$

(b2) *If* $A(\cdot)$ *and* $B(\cdot)$ *are predictable then these processes are indistinguishable.*

Proof of (a1) *when* $I = \mathbb{Z}^+$. Since $y(\cdot)$ is a martingale in (a1),

$$E\{[y(k+1) - y(k)]A(k)\} = E\{A(k)E\{y(k+1) - y(k) | \mathscr{F}(k)\}\} = 0;$$

so

$$\begin{aligned} &E\left\{\sum_{k=0}^{\infty} y(k+1)[A(k+1) - A(k)]\right\} \\ &= E\left\{\sum_{k=0}^{\infty} [y(k+1)A(k+1) - y(k)A(k)]\right\} \\ &= E\{y(+\infty)A(+\infty)\}. \end{aligned}$$

□

Proof of (a2) *when* $I = \mathbb{Z}^+$. See the exactly parallel proof below for the case $I = \mathbb{R}^+$. □

Proof of (a3) *when* $I = \mathbb{Z}^+$. If $A(\cdot)$ is predictable apply $(7.1, I = \mathbb{Z}^+)$ to both $A(\cdot)$ and the increasing process $\{A(k+1) - A(1); \mathscr{F}(k), k \in \mathbb{Z}^+\}$ to derive $(6.2,\ I = \mathbb{Z}^+)$. Conversely, suppose that $\{A(\cdot), \mathscr{F}(\cdot)\}$ is an adapted L^1 bounded increasing process and that $(6.2, I = \mathbb{Z}^+)$ is satisfied for every choice of $y(\cdot)$. To show that then $A(\cdot)$ is nearly predictable, it is sufficient to observe that $A(0)$ is $\mathscr{F}(0)$ measurable and to prove that $A(1)$ is also $\mathscr{F}(0)$ measurable, because this result can then be applied to the increasing process $\{A(k+1) - A(1), \mathscr{F}(k+1), k \in \mathbb{Z}^+\}$ which satisfies the counterpart of $(6.2, I = \mathbb{Z}^+)$ for the filtration $\{\mathscr{F}(k+1), k \in \mathbb{Z}^+\}$ to show that $A(2)$ is $\mathscr{F}(1)$ measurable, and so on. To prove that $A(1)$ is $\mathscr{F}(0)$ measurable, let y be a bounded $\mathscr{F}(1)$ measurable random variable and define $y(\cdot)$ as the martingale $E\{y | \mathscr{F}(0)\}, y, y, \ldots$ Then $(6.2, I = \mathbb{Z}^+)$ yields

$$E\{[y - E\{y | \mathscr{F}(0)\}]A(1)\} = 0. \tag{7.4}$$

Hence

$$E\{yA(1)\} = E\{E\{y|\mathscr{F}(0)\}A(1)\} = E\big\{E\{E\{y|\mathscr{F}(0)\}A(1)|\mathscr{F}(0)\}\big\}$$
$$= E\{yE\{A(1)|\mathscr{F}(0)\}\}.$$

Since y can be the indicator function of an arbitrary $\mathscr{F}(1)$ set, it follows that

$$A(1) = E\{A(1)|\mathscr{F}(1)\} = E\{A(1)|\mathscr{F}(0)\} \quad \text{a.s.};$$

so $A(1)$ is $\mathscr{F}(0)$ measurable, as was to be proved.

Proof of (a1) *for* $I = \mathbb{R}^+$. If $\{y(t), t \in \mathbb{R}^+\}$ is a bounded almost surely right continuous martingale and if $y_n(\cdot)$ is the process defined by

$$y_n(0) = y(0), \qquad y_n(t) = \sum_{k=0}^{\infty} y((k+1)2^{-n})1_{]k2^{-n},(k+1)2^{-n}]}(t) \quad \text{if } t > 0,$$

then $(7.1, I = \mathbb{Z}^+)$ as applied to the martingale $\{y(k2^{-n}), \mathscr{F}(k2^{-n}), k \in \mathbb{Z}^+\}$ and the increasing process $\{A(k2^{-n}), k \in \mathbb{Z}^+\}$ yields

$$E\left\{\sum_{k=0}^{\infty} y((k+1)2^{-n})[A((k+1)2^{-n}) - A(k2^{-n})]\right\} = E\left\{\int_0^{\infty} y_n(t)\, dA(t)\right\}$$
$$= E\{y(+\infty)A(+\infty)\}. \tag{7.5}$$

Since $\lim_{n\to\infty} y_n(t) = y(t)$ almost surely, simultaneously for all t, (7.5) yields $(7.1, I = \mathbb{R}^+)$ by the Lebesgue dominated convergence theorem. □

Proof of (a2) *for* $I = \mathbb{R}^+$. If $A(\cdot)$ is natural apply both $(6.2, I = \mathbb{R}^+)$ and $(7.1, I = \mathbb{R}^+)$ with $y(t) = E\{z|\mathscr{F}(t)\}$ to derive

$$E\left\{\int_0^{\infty} E\{z|\mathscr{F}(t-)\}\, dA(t)\right\} = E\left\{\int_0^{\infty} E\{z|\mathscr{F}(t)\}\, dA(t)\right\} = E\{xA(+\infty)\}$$

and $(7.2, I = \mathbb{R}^+)$ now follows from the fact that $E\{z|\mathscr{F}(t-)\} = z$ almost surely when $t > \alpha$. Conversely, if $(7.2, I = \mathbb{R}^+)$ is true and if $y(\cdot)$ is a bounded almost surely right continuous martingale, apply $(7.2, I = \mathbb{R}^+)$ with $z = y(\alpha)$ and then let $\alpha \to +\infty$ to derive the condition $(6.2, I = \mathbb{R}^+)$ that $A(\cdot)$ be natural.

Proof of (a3) *for* $I = \mathbb{R}^+$. According to Section 6, if either $A(\cdot)$ is almost surely continuous or if there is a predictable optional time T such that $A(t) = c1_{[T,+\infty[}(t)$, then $A(\cdot)$ is natural. Thus to show that $A(\cdot)$ is natural if nearly predictable, it is sufficient to show that (dropping evanescent sets) if $A(\cdot)$ is predictable, right continuous with left limits, and if all its sample functions are increasing, then $A(\cdot)$ can be written as a countable sum of processes of these two types. Choose $\beta > 0$ and define $T_0 = 0$ and

$$T_{n+1} = \inf\{t > T_n : A(t) - A(t-) \geq \beta\}, \qquad n > 0.$$

The process $A(\cdot\, -)$ is predictable by left continuity; so the process $A(\cdot) - A(\cdot\, -)$ is predictable, and therefore the set $\{(t,\omega): A(t,\omega) - A(t-,\omega) \geq \beta\}$ is predictable. For $n > 1$ the nth entry time of this set is T_n. Hence (Section II.9) T_n is predictable; so (Section II.7) the process $A^{T_n}(\cdot) = 1_{[T_n, +\infty[}(\cdot)$ is predictable. The process

$$A(\cdot) - \beta \sum_{n=0}^{\infty} A^{T_n}(\cdot)$$

is a predictable increasing process for which the sample function discontinuities of magnitude $\geq \beta$ are decreased by β. First choose $\beta = 1$, thereby decreasing jumps ≥ 1 by 1, then apply the same procedure, but with $\beta = \frac{1}{2}$, to the above difference process, then with $\beta = \frac{1}{3}$ to the new difference process, and so on, to obtain an evaluation of $A(\cdot)$ as a sum of the desired type.

Conversely, suppose that $A(\cdot)$ is natural. We show first that $A(\cdot)$ does not charge any totally inaccessible optional time T, that is, $A(T) = A(T-)$ almost everywhere where $T < +\infty$. To see this, define $I'_{nk} =]k2^{-n}, (k+1)2^{-n}]$, and observe that

$$E\{A(T) - A(T-); T < +\infty\} = \lim_{n\to\infty} \sum_{k=0}^{\infty} \int_{\{T \in I'_{nk}\}} [A((k+1)2^{-n}) - A(k2^{-n})]\, dP; \tag{7.6}$$

so it is sufficient to show that the right side vanishes. The sum in (7.6) is equal to

$$\sum_{k=0}^{\infty} E\{1_{\{T\leq (k+1)2^{-n}\}}[A((k+1)2^{-n}) - A(k2^{-n})] - 1_{\{T \leq k2^{-n}\}}[A((k+1)2^{-n}) - A(k2^{-n})]\},$$

which by $(7.2, I = \mathbb{R}^+)$ is equal to

$$\sum_{k=0}^{\infty} E\left\{\int_{I'_{nk}} P\{T \leq (k+1)2^{-n} | \mathscr{F}(t-)\}\, dA(t)\right\} - \sum_{k=0}^{\infty} E\left\{\int_{I'_{nk}} 1_{\{T \leq k2^{-n}\}}\, dA(t)\right\}. \tag{7.7}$$

If we define the interval I_{nk} and the processes $a(\cdot)$ and $a_n(\cdot)$ as in Section 2, the first sum in (7.7) can be written in the form $E\{\int_0^\infty a_n(t-)\, dA(t)\}$, and this expectation, by the uniform convergence of the sequence $a_{\bullet}(\cdot)$ proved in Section 2, has limit ($n \to \infty$)

$$E\left\{\int_0^\infty a(t-)\,dA(t)\right\} = E\left\{\int_0^\infty 1_{]T,+\infty[}(t)\,dA(t)\right\}. \tag{7.8}$$

If now T_n is defined by

$$T_n = \begin{cases} (k+1)2^{-n} & \text{when } T \in I'_{nk}, \\ +\infty & \text{when } T = +\infty, \end{cases}$$

then $T_{\bullet}$ is a decreasing sequence of optional times with limit T and $T_n > T$ almost everywhere where T is finite valued and not equal to a dyadic rational number. Since the distribution function on $\mathbb{R}^+$ of a totally inaccessible optional time is continuous, $T_n > T$ almost surely where T is finite. The second sum in (7.7) is equal to

$$E\left\{\int_0^\infty 1_{[T_n,+\infty[}(t)\,dA(t)\right\},$$

which has limit ($n \to \infty$) the right side of (7.8); so $A(\cdot)$ does not charge T, as asserted. □

In the proof of (a3) it was shown that $A(\cdot)$ if nearly predictable can be expressed as a countable sum of increasing processes of which one is almost surely continuous and the others have the form $cA^T(\cdot)$ with T and $A^T(\cdot)$ predictable. That reasoning in the present context, in which $A(\cdot)$ is to be proved nearly predictable and in which we have just proved that $A(\cdot)$ charges no totally inaccessible optional time, leads to the same expression for $A(\cdot)$ except that as first defined each optional time T is accessible. In fact we need only point out that the discussion in the proof of (a3) leads to optional times T charged by $A(\cdot)$ and therefore without totally inaccessible components. Since $A^T(\cdot)$ is nearly predictable if T is accessible (Section II.12), $A(\cdot)$ is nearly predictable if natural.

Proof. (b1) If $x(\cdot) = A(\cdot) - B(\cdot)$ and $I = \mathbb{Z}^+$, then

$$E\{y(k+1)[x(k+1) - x(k)]\} =$$

$$E\{y(k+1)E\{x(k+1) - x(k)|\mathscr{F}(k)\}\} = 0$$

because $y(k+1)$ is $\mathscr{F}(k)$ measurable and $\{x(\cdot), \mathscr{F}(\cdot)\}$ is a martingale. Hence (b1) is true when $I = \mathbb{Z}^+$. If $I = \mathbb{R}^+$ and if $\{y(\cdot), \mathscr{F}(\cdot)\}$ is a bounded left continuous process, define

$$y_n(0) = y(0),\ y_n(t) = \sum_{k=0}^{\infty} y(k2^{-n}) 1_{]k2^{-n},\,(k+1)2^{-n}]}(t) \quad \text{if} \quad t > 0,$$

so that the process $\{y_n(k2^{-n}), \mathscr{F}(k2^{-n}), k \in \mathbb{Z}^+\}$ is predictable and therefore by what we have just proved

$$E\left\{\int_0^\infty y_n(t)\,dA(t)\right\} = E\left\{\int_0^\infty y_n(t)\,dB(t)\right\}. \tag{7.9}$$

When $n \to \infty$ this equation yields (7.3, $I = \mathbb{R}^+$). Thus the latter equation is true if $y(\cdot)$ is bounded and left continuous, and since the class of bounded predictable processes on $\mathbb{R}^+$ is the smallest class of bounded processes closed under bounded monotone convergence and containing the left continuous adapted processes, (b1) is true.

(b2) We give the proof for $I = \mathbb{R}^+$; the proof for $I = \mathbb{Z}^+$ is similar but less deep. Suppose first that $A(+\infty)$ and $B(+\infty)$ are bounded, and combine (7.1, $I = \mathbb{R}^+$) with (7.3, $I = \mathbb{R}^+$) to find that $E\{y(+\infty)[A(+\infty) - B(+\infty)]\} = 0$ whenever $\{y(\cdot), \mathscr{F}(\cdot)\}$ is a bounded almost surely right continuous predictable martingale. If now we choose $y(\cdot) = A(\cdot) - B(\cdot)$, we find that $A(+\infty) = B(+\infty)$ almost surely. Hence (martingale equality) for each value of t it follows that $A(t) = B(t)$ almost surely; so by the almost sure right continuity of $A(\cdot)$ and $B(\cdot)$ these processes must be indistinguishable. In the general case we need only find an increasing sequence $S_\bullet$ of optional times for which $\lim_{n\to\infty} S_n = +\infty$ almost surely and for which $A(S_n)$ and $B(S_n)$ are almost surely bounded for each n. In fact, if the result for bounded processes is applied to the pair $A(S_n \wedge \cdot)$, $B(S_n \wedge \cdot)$, it follows that these processes are indistinguishable so $A(\cdot)$ and $B(\cdot)$ are. To find a sequence $S_\bullet$, observe that for $n > 0$ the entry time T_n of the predictable set

$$\{(t, \omega): A(t, \omega) + B(t, \omega) \geq n\}$$

is predictable (Theorem II.9), and if $T_{n.}$ is a sequence of optional times announcing T_n, then $A(T_{nj}) + B(T_{nj}) < n$; so we can choose

$$S_n = T_{1n} \vee \cdots \vee T_{nn}.$$

8. Generation of Supermartingale Potentials by Increasing Processes in the Discrete Parameter Case

Theorem. *Every supermartingale potential $\{x(n), \mathscr{F}(n), n \in \mathbb{Z}^+\}$ is a quasi-bounded class* **D** *supermartingale and is generated by a unique up to standard modification, that is, unique up to indistinguishability, increasing predictable L^1 bounded process.*

In view of Section 6 it is necessary only to show that there is a unique generating process as described. Given a supermartingale potential $\{x(n), \mathscr{F}(n), n \in \mathbb{Z}^+\}$ define an increasing predictable process by

$$A(0) = 0, \qquad A(n) = \sum_{m=0}^{n-1} [x(m) - E\{x(m+1)|\mathscr{F}(m)\}] \quad \text{if } n > 0, \tag{8.1}$$

and define $A(+\infty) = \sup_{n \in \mathbb{Z}^+} A(n)$. Then $E\{A(+\infty)\} = E\{x(0)\}$ and

$$\begin{aligned} E\{A(+\infty) - A(n)|\mathscr{F}(n)\} &= \sum_{m=n}^{\infty} E\{[x(m) - E\{x(m+1)|\mathscr{F}(m)\}]|\mathscr{F}(n)\} \\ &= x(n) \quad \text{a.s.}; \end{aligned} \tag{8.2}$$

so $\{x(\cdot), \mathscr{F}(\cdot)\}$ is generated by $A(\cdot)$. Moreover, if $\{x(\cdot), \mathscr{F}(\cdot)\}$ is any supermartingale potential generated by an increasing predictable process $\{A(\cdot), \mathscr{F}(\cdot)\}$, it is immediate that

$$x(m) - E\{x(m+1)|\mathscr{F}(m)\} = A(m+1) - A(m) \quad \text{a.s.} \tag{8.3}$$

from which (8.1) follows as an almost sure equality; so $A(\cdot)$ is uniquely determined up to a standard modification. [Uniqueness here also follows from Theorem 7(b).]

Decomposition of a Submartingale

If $\{A(n), \mathscr{F}(n), n \in \mathbb{Z}^+\}$ is a predictable increasing process and if $\{y(n), \mathscr{F}(n), n \in \mathbb{Z}^+\}$ is a martingale, the process $\{y(\cdot) + A(\cdot), \mathscr{F}(\cdot)\}$ is a submartingale. Conversely, if $\{x(n), \mathscr{F}(n), n \in \mathbb{Z}^+\}$ is a submartingale and if $A(n)$ is defined by (8.1) with $x(\cdot)$ replaced by $-x(\cdot)$, then $\{A(\cdot), \mathscr{F}(\cdot)\}$ is a predictable increasing process and $x(\cdot) = y(\cdot) + A(\cdot)$, where $\{y(\cdot), \mathscr{F}(\cdot)\}$ is a martingale. It is left to the reader to check that $x(\cdot)$ is L^1 bounded if and only if $y(\cdot)$ and $A(\cdot)$ are. This decomposition of a submartingale makes it possible to deduce the almost sure convergence of an L^1 bounded discrete parameter submartingale from the almost sure convergence of an L^1 bounded discrete parameter martingale.

9. An Inequality for Predictable Increasing Processes

Lemma. *Let $\{x(n), \mathscr{F}(n), n \in \mathbb{Z}^+\}$ be a supermartingale potential generated by the predictable increasing process $A(\cdot)$, and define $A(+\infty) = \sup_{n \in \mathbb{Z}^+} A(n)$. Let c be a strictly positive constant, and define the optional time T_c by*

$$T_c = \begin{cases} +\infty & \text{if } A(+\infty) \le c \\ \inf\{j: A(j+1) > c\} & \text{if } A(+\infty) > c. \end{cases}$$

Then

$$\int_{\{A(+\infty)>c\}} A(+\infty)\,dP \le 3\int_{\{A(+\infty)>c/2\}} x(T_{c/2})\,dP. \tag{9.1}$$

By definition of T_c, $A(T_c) \le c$ and $\{A(+\infty) > c\} = \{T_c < +\infty\} \in \mathscr{F}(T_c)$ so that

$$\begin{aligned} cP\{A(+\infty) > c\} &\le 2\int_{\{A(+\infty)>c\}} [A(+\infty) - A(T_{c/2})]\,dP \\ &\le 2\int_{\{A(+\infty)>c/2\}} x(T_{c/2})\,dP, \end{aligned} \tag{9.2}$$

and this inequality combines with the supermartingale inequality to yield

$$\begin{aligned} \int_{\{A(+\infty)>c\}} A(+\infty)\,dP &\le \int_{\{A(+\infty)>c\}} x(T_c)\,dP + cP\{A(+\infty) > c\} \\ &\le 3\int_{\{A(+\infty)>c/2\}} x(T_{c/2})\,dP, \end{aligned} \tag{9.3}$$

as was to be proved.

10. Generation of Supermartingale Potentials by Increasing Processes for Arbitrary Parameter Sets

Let $(\Omega, \mathscr{F}, P; \mathscr{F}(t), t \in I)$ be a filtered probability space. It is supposed that the linearly ordered parameter set I has a first but not a last point.

Theorem. *The following conditions on a supermartingale potential $\{x(\cdot), \mathscr{F}(\cdot)\}$ are equivalent:*

(a) *The process is generated by an increasing L^1 bounded adapted process $\{A(\cdot), \mathscr{F}(\cdot)\}$; for each t in I,*

$$\begin{aligned} x(t) &= E\{A(+\infty) - A(t) \mid \mathscr{F}(t)\} \\ &= E\{A(+\infty) \mid \mathscr{F}(t)\} - A(t) \quad \text{a.s.} \end{aligned} \tag{10.1}$$

(b) *The process is in the class* **D**.
(c) *The process is quasi-bounded.*

Observe that when $I = \mathbb{Z}^+$, this theorem does not contain Theorem 8 because in that theorem the increasing process is predictable; in fact this predictability determines the increasing process in Theorem 8 up to a standard modification. Uniqueness cannot be asserted in the present theorem. Moreover, according to Theorem 8, a supermartingale potential for the parameter set $\mathbb{Z}^+$ is necessarily quasi-bounded and in the class **D**, whereas there is an example in Section IX.6 below for the parameter set $\mathbb{R}^+$ of a continuous supermartingale potential which has neither property.

In the following proof we denote $\bigvee_{t\in I}\mathscr{F}(t)$ by $\mathscr{F}(+\infty)$.

Proof. (a) $\Rightarrow$ (b) and (a) $\Rightarrow$ (c) See Section 6 where these implications are derived without the adaptedness hypothesis on $A(\cdot)$.

(b) $\Rightarrow$ (a) Assume first that I is a bounded subset of $\mathbb{R}^+$ and includes 0. The random variable $x(t)$ can be identified with a point of $L^1(\Omega, \mathscr{F}(+\infty), P)$. This function has a left [right] limit at each point of the closure $\bar{I}$ of I which is a limit point of I from the left [right] according to the supermartingale convergence theorems and the condition that the given process $x(\cdot)$ is in **D**, so is uniformly integrable. According to an elementary theorem on functions from a subset of $\mathbb{R}$ into a metric space, the set I_0 of points t of $\bar{I}$ for which two of the members $x(t-), x(t), x(t+)$ of $L^1(\Omega, \mathscr{F}(+\infty), P)$ exist but are not the same is countable. Let J be a countable subset of I, dense in I, including 0, I_0, and every point of I which is not a bilateral limit point of I, and let J_n be a finite subset of J, including 0, with $J_0 \subset J_1 \subset \cdots$, $\bigcup_0^\infty J_n = J$. The restriction of $x(\cdot)$ to J_n is a positive supermartingale to which Theorem 8 as trivialized for finite parameter sets can be applied. (Theorem 8 is stated for the parameter set $\mathbb{Z}^+$ but can be applied to a finite parameter set, say $0, \ldots, k$, by defining the process random variables to vanish identically at parameter values strictly greater than k.) According to Theorem 8, there is a process $\{A_n(t), t\in J_n\}$ for which $A_n(0) = 0$, $A_n(\cdot)$ is increasing and is predictable, and if we define $A_n(+\infty)$ as the value of $A_n(\cdot)$ at the maximum parameter value in J_n,

$$x(t) = E\{A_n(+\infty)|\mathscr{F}(t)\} - A_n(t) \quad \text{a.s.}, t\in J_n. \tag{10.2}$$

Moreover for $c > 0$,

$$cP\left\{A_n(+\infty) > \frac{c}{2}\right\} \le 2E\{A_n(+\infty)\} = 2E\{x(0)\}, \tag{10.3}$$

and according to Lemma 9, there is an optional time T (depending on n and c) with values in J_n, for which

$$\int_{\{A_n(+\infty)>c\}} A_n(+\infty)\,dP \le 3\int_{\{A_n(+\infty)>c/2\}} x(T)\,dP. \tag{10.4}$$

In view of (10.3), $\lim_{c\to\infty} P\{A_n(+\infty) > c/2\} = 0$ uniformly as n varies; so by the class **D** property of $x(\cdot)$ the right side of (10.4) has limit 0 when $c \to \infty$, uniformly as n varies, and it follows that the sequence $A_{\bullet}(+\infty)$ is uniformly integrable. Hence some subsequence $A_{\alpha}(+\infty)$ converges weakly in $L^1(\Omega, \mathscr{F}(+\infty), P)$ to a limit random variable $A(+\infty)$, and consequently for each parameter value t in I the sequence $E\{A_{\alpha}(+\infty)|\mathscr{F}(t)\}$ converges weakly in $L^1(\Omega, \mathscr{F}(t), P)$ to $E\{A(+\infty)|\mathscr{F}(t)\}$. From (10.2) if $t \in J$, the sequence $A_{\alpha}(t)$ also converges weakly in $L^1(\Omega, \mathscr{F}(t), P)$ to a positive $\mathscr{F}(t)$ measurable random variable $A(t)$, with $A(0) = 0$. Furthermore (10.1) is true for t in J, and $A(\cdot)$ is an increasing process on J. By definition of J each point t in $I - J$ is a bilateral limit point of J, and $x(t) = x(t+) = x(t-)$ almost surely if the one-sided limits are defined either as L^1 limits or as sequential almost sure limits. Hence for t in $I - J$,

$$\begin{aligned} x(t) &= E\{A(+\infty)|\mathscr{F}(t+)\} - A(t+) \\ &= E\left\{A(+\infty)\middle|\bigvee_{s<t}\mathscr{F}(s)\right\} - A(t-) \quad \text{a.s.} \end{aligned} \tag{10.5}$$

On taking expectations here it follows that $A(t+) = A(t-)$ almost surely when t is in $I - J$; so if $A(t)$ is defined when t is in $I - J$ as $\sup_{J \ni s<t} A(s)$, the process $\{A(t), \mathscr{F}(t), t \in I\}$ is an adapted increasing process, and for t in I

$$x(t) = E\{A(+\infty)|\mathscr{F}(t+)\} - A(t) \quad \text{a.s.}$$

Taking conditional expectations with respect to $\mathscr{F}(t)$ yields (10.1). Moreover, when t tends to the supremum of I along J, we find that

$$0 = E\{A(+\infty)|\mathscr{F}(+\infty)\} - \lim_{t\uparrow} A(t) = A(+\infty) - \operatorname{ess\,sup}_{t \in I} A(t) \quad \text{a.s.}$$

(b) $\Rightarrow$ (a) (in the general case) According to Section III.4, there is an order-preserving map $t \mapsto \phi(t)$ from I into $\mathbb{R}$ such that $\phi(s) = \phi(t)$ if and only if $x(s) = x(t)$ almost surely. We can choose ϕ to be bounded and to take the initial point of I into 0. Define $\hat{I} = \phi(I)$. If $\hat{t} \in \hat{I}$, define $\hat{\mathscr{F}}(\hat{t}) = \bigvee_{t \in \phi^{-1}(\hat{t})} \mathscr{F}(t)$, and define $\hat{x}(\hat{t})$ as $x(t)$ for some t in $\phi^{-1}(\hat{t})$; the choice of t does not matter. If $\hat{s}$ and $\hat{t}$ are in $\hat{I}$ and if $\hat{s} < \hat{t}$, then when $\phi(t) = \hat{t}$,

$$E\{\hat{x}(\hat{t})|\hat{\mathscr{F}}(\hat{s})\} = E\left\{x(t)\middle|\bigvee_{s \in \phi^{-1}(\hat{s})} \mathscr{F}(s)\right\} \quad \text{a.s.}$$

Here $s \leq t$; so by the conditional expectation continuity theorem the right hand side is almost surely at most $\hat{x}(\hat{s})$. Thus the process $\{\hat{x}(\hat{t}), \hat{\mathscr{F}}(\hat{t}), \hat{t} \in \hat{I}\}$ is a positive supermartingale. It follows from the special case of the assertion (b) $\Rightarrow$ (a) already proved that there is an increasing adapted process $\{\hat{A}(\hat{t}), \hat{\mathscr{F}}(\hat{t}), \hat{t} \in \hat{I}\}$ for which, defining $\hat{A}(+\infty)$ as in Section 6,

$$\hat{x}(\hat{t}) = E\{\hat{A}(+\infty)|\hat{\mathscr{F}}(\hat{t})\} - \hat{A}(\hat{t}) \quad \text{a.s., } \hat{t} \in \hat{I}.$$

Define $A(+\infty) = \hat{A}(+\infty)$, $A'(t) = \hat{A}(\hat{t})$ for $t \in \phi^{-1}(\hat{t})$. Then if $t \in \phi^{-1}(\hat{t})$,

$$x(t) = E\{A(+\infty)|\hat{\mathscr{F}}(\hat{t})\} - A'(t) \quad \text{a.s.}$$

Apply $E\{-|\mathscr{F}(t)\}$ to both sides to obtain

$$x(t) = E\{A(+\infty)|\mathscr{F}(t)\} - E\{A'(t)|\mathscr{F}(t)\} \quad \text{a.s.}$$

If $A(t) = E\{A'(t)|\mathscr{F}(t)\}$, the increasing adapted process $\{A(t), \mathscr{F}(t), t \in I\}$ has the desired properties.

(c) ⇒ (a) Under (c), $x(\cdot) = \Sigma_0^\infty y_k(\cdot)$ with $\{y_k(\cdot), \mathscr{F}(\cdot)\}$ a positive supermartingale which is bounded and therefore in the class **D**. Hence $y_k(\cdot)$ is generated by some increasing adapted process $\{B_k(\cdot), \mathscr{F}(\cdot)\}$; so $\Sigma_0^\infty B_k(\cdot)$ generates $x(\cdot)$; that is, (a) is true.

11. Generation of Supermartingale Potentials by Increasing Processes in the Continuous Parameter Case: The Meyer Decomposition

Let $(\Omega, \mathscr{F}, P; \mathscr{F}(t), t \in \mathbb{R}^+)$ be a filtered probability space for which $\mathscr{F}(\cdot)$ is right continuous and $\mathscr{F}(0)$ contains the null sets.

Theorem. *The following conditions on the almost surely right continuous supermartingale potential $\{x(\cdot), \mathscr{F}(\cdot)\}$ are equivalent:*

(a) *The process is generated by an increasing adapted L^1 bounded process $\{A(\cdot), \mathscr{F}(\cdot)\}$: for $t \in \mathbb{R}^+$ and $A(+\infty) = \sup_{t \in \mathbb{R}^+} A(t)$,*

$$x(t) = E\{A(+\infty)|\mathscr{F}(t)\} - A(t) \quad \text{a.s.}$$

(b) *The process is in the class* **D**.

(c) *The process is quasi-bounded.*

If these conditions are satisfied the process $\{A(\cdot), \mathscr{F}(\cdot)\}$ can be chosen to be almost surely right continuous and predictable, and is uniquely determined up to indistinguishability by these conditions.

If $A(\cdot)$ and $B(\cdot)$ are increasing processes adapted to $\mathscr{F}(\cdot)$ and generate the same supermartingale, then the process $\{A(\cdot) - B(\cdot), \mathscr{F}(\cdot)\}$ is a martingale (Section 6). Thus in view of Sections 6 and 7 to prove the theorem, all we need prove is that in the present context there is an almost surely right continuous predictable L^1 bounded increasing process that generates $x(\cdot)$. Define $\mathscr{F}(+\infty) = \bigvee_{t \in \mathbb{R}^+} \mathscr{F}(t)$, and define $\mathscr{F}(t) = \mathscr{F}(0)$ for $t < 0$. Let J be the set of dyadic points of $\mathbb{R}^+$, and let J_n be the set $\{m2^{-n}, m \in \mathbb{Z}^+\}$. Then according to Theorem 8, there is a process $\{A_n(t), \mathscr{F}(t), t \in J_n\}$ for which $A_n(\cdot)$

is increasing, L^1 bounded, and predictable, and (10.2) is true with $A_n(+\infty) = \sup_{t\in J_n} A_n(t)$. It follows as in the proof of Theorem 10 that the sequence $A_{\bullet}(+\infty)$ is uniformly integrable and that therefore some subsequence $A_{\alpha_\bullet}(+\infty)$ converges weakly in $L^1(\Omega, \mathscr{F}(+\infty), P)$ to a limit random variable $A(+\infty)$. When $t\in J$, the sequence $E\{A_{\alpha_\bullet}(+\infty)|\mathscr{F}(t)\}$ therefore converges weakly in $L^1(\Omega, \mathscr{F}(t), P)$ to $E\{A(+\infty)|\mathscr{F}(t)\}$, and so (10.2) implies that the sequence $A_{\alpha_\bullet}(t)$ converges weakly in $L^1(\Omega, \mathscr{F}(t), P)$ to an $\mathscr{F}(t)$ measurable random variable $A(t)$. In particular, the sequence $A_{\alpha_\bullet}(0)$ converges trivially to $A(0) = 0$. Equation (11.1) is true for $t\in J$, and $(t\uparrow +\infty)$

$$0 = E\{A(+\infty)|\mathscr{F}(+\infty)\} - \sup_{t\in J} A(t) = A(+\infty) - \sup_{t\in J} A(t) \quad \text{a.s.} \tag{11.2}$$

Observe that in view of the right continuity of $\mathscr{F}(\cdot)$ and the conditional expectation continuity theorem the process $A(\cdot)$ as we have defined it on J is almost surely right continuous. If now we apply Theorem 2 to choose a version of $E\{A(+\infty)|\mathscr{F}(t)\}$ for each t to obtain an almost surely right continuous process and define $A(t)$ for $t\in \mathbb{R}^+ - J$ by (11.1), the process $A(\cdot)$ is an almost surely right continuous increasing L^1 bounded process generating $x(\cdot)$. Finally we prove that $\{A(\cdot), \mathscr{F}(\cdot)\}$ is a natural increasing process. According to Theorem 7, it will follow that $\{A(\cdot), \mathscr{F}(\cdot)\}$ is nearly predictable; this process can be trivially adjusted to be predictable if desired. Let $\{y(\cdot), \mathscr{F}(\cdot)\}$ be a bounded almost surely right continuous martingale, and define $y(+\infty) = \lim_{t\to\infty} y(t)$. Then $\{A(\cdot), \mathscr{F}(\cdot)\}$ is a natural increasing process because

$$\begin{aligned}
&E\left\{\int_0^\infty y(s-)\,dA(s)\right\}\\
&\quad = \lim_{n\to\infty} \sum_{k=0}^{\infty} E\{y(k2^{-n})[A((k+1)2^{-n}) - A(k2^{-n})]\}\\
&\quad = \lim_{n\to\infty} \sum_{k=0}^{\infty} E\{y(k2^{-n})E\{A((k+1)2^{-n}) - A(k2^{-n})|\mathscr{F}(k2^{-n})\}\}\\
&\quad = \lim_{n\to\infty} \sum_{k=0}^{\infty} E\{y(k2^{-n})E\{x(k2^{-n}) - x((k+1)2^{-n})|\mathscr{F}(k2^{-n})\}\}\\
&\quad = \lim_{n\to\infty} \sum_{k=0}^{\infty} E\{y(k2^{-n})E\{A_n((k+1)2^{-n}) - A_n(k2^{-n})|\mathscr{F}(k2^{-n})\}\}\\
&\quad = \lim_{n\to\infty} \sum_{k=0}^{\infty} E\{y(k2^{-n})[A_n((k+1)2^{-n}) - A_n(k2^{-n})]\}\\
&\quad = \lim_{n\to\infty} E\{y(+\infty)A_n(+\infty)\} = E\{y(+\infty)A(+\infty)\},
\end{aligned} \tag{11.3}$$

where we have used the discrete parameter case of Theorem 7 to derive the last line, and in the last line $n\to\infty$ along the sequence $\alpha_\bullet$. Incidentally this proof shows that there is convergence in (11.3) when $n\to\infty$ unrestrictedly.

It is easily deduced that the original sequence $A_{\bullet}(+\infty)$ converges weakly to $A(+\infty)$, but we shall not need this fact.

Uniqueness Observation

If the supermartingale $\{x(\cdot), \mathscr{F}(\cdot)\}$ satisfies the conditions of Theorem 11 and if there are an almost surely right continuous martingale $\{x'(\cdot), \mathscr{F}(\cdot)\}$ and an almost surely right continuous predictable increasing L^1 bounded process $\{A'(\cdot), \mathscr{F}(\cdot)\}$ for which $x(\cdot) = x'(\cdot) - A'(\cdot)$, then $A'(\cdot)$ is the unique up to indistinguishability generator $A(\cdot)$ of $x(\cdot)$. In fact the process $\{A(\cdot) - A'(\cdot), \mathscr{F}(\cdot)\}$ is a martingale; so (by Theorem 7) the processes $A(\cdot)$ and $A'(\cdot)$ are indistinguishable.

Predictable Supermartingale Potentials

If the supermartingale $\{x(\cdot), \mathscr{F}(\cdot)\}$ in Theorem 11 is predictable then the martingale component of the Meyer decomposition must also be predictable because the increasing component $\{A(\cdot), \mathscr{F}(\cdot)\}$ is. We shall show (Section 23) that an almost surely right continuous predictable martingale is almost surely continuous, and it follows that the supermartingale sample function jumps are balanced out by the increasing process sample function jumps; that is, for P almost every ω,

$$x(t-, \omega) - x(t, \omega) = A(t, \omega) - A(t-, \omega)$$

simultaneously for all t. Trivially more generally this assertion is true if $\{x(\cdot), \mathscr{F}(\cdot)\}$ is indistinguishable from a predictable process. In particular if the given supermartingale potential is almost surely continuous it follows that the increasing process component $A(\cdot)$ is also almost surely continuous.

Meyer Decomposition for an Arbitrary Parameter Interval $[0, c[$.

If $c > 0$, the adaptation of the Meyer decomposition theorem to the parameter interval $[0, c[$ is accomplished trivially by the map $t' = (2c/\pi)\arctan t$ taking $\mathbb{R}^+$ into $[0, c[$.

12. Meyer Decomposition of a Submartingale

Let $(\Omega, \mathscr{F}, P; \mathscr{F}(t), t \in \mathbb{R}^+)$ be a filtered probability space for which $\mathscr{F}(\cdot)$ is right continuous and $\mathscr{F}(0)$ contains the null sets. Let $\{A(\cdot), \mathscr{F}(\cdot)\}$ be an almost surely right continuous adapted increasing process. The restriction of this process to a parameter interval $[0, c[$ is in the class **D** because if T is optional and if $T < c$, then $A(T)$ is majorized by the integrable random variable $A(c)$. Let $\{y(\cdot), \mathscr{F}(\cdot)\}$ be an almost surely right continuous martingale. The restriction of this process to $[0, c[$ is also in the class **D** according to Section III.6. The sum process $\{y(\cdot) + A(\cdot), \mathscr{F}(\cdot)\}$ is a submartingale whose restriction to any parameter interval $[0, c[$ is in the class **D**. Conversely, we shall now show that an almost surely right continuous submartingale $\{z(\cdot), \mathscr{F}(\cdot)\}$ whose restriction to each parameter interval $[0, c[$ is in the class **D** can be written as a sum $\{y(\cdot) + A(\cdot), \mathscr{F}(\cdot)\}$ with $\{y(\cdot), \mathscr{F}(\cdot)\}$ an almost surely right continuous martingale and $\{A(\cdot), \mathscr{F}(\cdot)\}$ an almost surely right continuous predictable increasing process; moreover both components are uniquely determined up to indistinguishability. Let $z_c(\cdot)$ be the restriction of $z(\cdot)$ to the parameter interval $[0, c[$. This submartingale is in the class **D** by hypothesis, and the right closable martingale

$$\{E\{z(c)|\mathscr{F}(t)\}, \mathscr{F}(t), 0 \le t < c\}$$

is also in the class **D** (Section III.6). The version of the conditional expectation in the process

$$\{E\{z(c)|\mathscr{F}(t)\} - z_c(t), \mathscr{F}(t), 0 \le t < c\}$$

can be chosen to make the difference process a positive almost surely right continuous class **D** supermartingale. The supermartingale potential component of the Riesz decomposition of the difference process is then also in the class **D** and so is generated by an almost surely right continuous predictable increasing L^1 bounded process $\{A_c(t), \mathscr{F}(t), 0 \le t < c\}$. That is, we can write $z(\cdot) = y_c(\cdot) + A_c(\cdot)$ on the interval $[0, c[$, where $\{y_c(\cdot), \mathscr{F}(\cdot)\}$ is an almost surely right continuous martingale on the parameter interval $[0, c[$. In view of the uniqueness property of the Meyer decomposition we can choose an almost surely right continuous predictable increasing process $\{A(\cdot), \mathscr{F}(\cdot)\}$ whose restriction to the parameter interval $[0, c[$ is indistinguishable from $A_c(\cdot)$, and therefore $\{z(\cdot) - A(\cdot), \mathscr{F}(\cdot)\}$ is a martingale on the parameter interval $\mathbb{R}^+$.

EXAMPLE. If $\{B(t), \mathscr{F}(t), t \in \mathbb{R}^+\}$ is an almost surely right continuous adapted increasing process, this process is a submartingale so there is an almost surely right continuous predictable increasing process $\{B'(t), \mathscr{F}(t), t \in \mathbb{R}^+\}$ such that $\{B(\cdot) - B'(\cdot), \mathscr{F}(\cdot)\}$ is a martingale; $B'(\cdot)$ is determined uniquely up to indistinguishability.

13. Role of the Measure Associated with a Supermartingale; The Supermartingale Domination Principle

Let $(\Omega, \mathscr{F}, P; \mathscr{F}(t), t \in \mathbb{R}^+)$ be a filtered probability space for which $\mathscr{F}(\cdot)$ is right continuous and $\mathscr{F}(0)$ contains the null sets. Let $\{x(\cdot), \mathscr{F}(\cdot)\}$ be a class **D** positive almost surely right continuous supermartingale with associated

(Meyer decomposition) nearly predictable almost surely right continuous L^1 bounded increasing process $A(\cdot)$, so that $x(\cdot)$ is the sum of an almost surely right continuous martingale and a class **D** almost surely right continuous potential,

$$x(t) = E\{x(+\infty-)|\mathscr{F}(t)\} + E\{A(+\infty)|\mathscr{F}(t)\} - A(t) \quad \text{a.s.} \qquad (13.1)$$

In this context $x(\cdot)$ is the martingale theory counterpart of a positive superharmonic function on an open subset of $\mathbb{R}^N$, and $x(\cdot)$ is also the martingale theory counterpart of a positive superparabolic function on an open subset of $\dot{\mathbb{R}}^N$. The counterpart of the Riesz measure associated with such a superharmonic or superparabolic function is the measure $A(dt)$. As a trivial illustration of this fact recall that a superharmonic [superparabolic] function is harmonic [parabolic] off the closed support of the associated Riesz measure. One probabilistic counterpart of this property is the following. Let S and T be finite optional times with $S \le T$, and suppose that $A(T) = A(S)$ almost surely. Then the process $x(\cdot)$ is a martingale between S and T; more precisely, the process

$$\{x((S+t)\wedge T); \mathscr{F}((S+t)\wedge T), t\in\mathbb{R}^+\}$$

is a martingale. In fact

$$\begin{aligned} x((S+t)\wedge T) = E\{x(+\infty-)|\mathscr{F}((S+t)\wedge T)\} \\ + E\{A(+\infty)|\mathscr{F}((S+t)\wedge T)\} - A(S) \quad \text{a.s.} \end{aligned}$$

Next consider the domination principle in classical and parabolic potential theory. First consider the following example. Let Ω be a singleton $\{\omega\}$, let $\mathscr{F}$ be the σ algebra consisting of Ω and $\varnothing$, let $\mathscr{F}(t) = \mathscr{F}$ for $t\in\mathbb{R}^+$, define $P\{\Omega\} = 1 = 1 - P\{\varnothing\}$, let $y(\cdot,\omega)$ be the indicator function of the interval $[0, 1[$, and define $y_0(\cdot,\omega) \equiv 0$. Then $y(\cdot)$ and $y_0(\cdot)$ are class **D** right continuous supermartingale potentials. One version of the predictable increasing process $B(\cdot)$ associated with $y(\cdot)$ is the indicator function of the interval $[1, +\infty[$; so $y(\cdot) = y_0(\cdot)$ on the support $\{(1,\omega)\}$ of the measure $B(dt)$, although the processes $y(\cdot)$ and $y_0(\cdot)$ are not indistinguishable. Thus the obvious counterpart of the classical domination principle (Theorem 1.V.10) is invalid. On the other hand, the parabolic domination principle (Theorem 1.XVIII.16) suggests that not $y(t)$ and $y_0(t)$ but $y(t-)$ and $y_0(t-)$ should be compared on the support of $B(dt)$, and in fact $y_0(t-) < y(t-)$ on this support. The statement of the supermartingale domination principle suggested by this example and by Theorem 1.XVIII.16 is the following, in which we return to the context and notation of the beginning of this section.

Theorem. *Let $\{x(\cdot), \mathscr{F}(\cdot)\}$ be a class* **D** *almost surely right continuous supermartingale potential with associated measure $A(dt)$ of closed support $\dot{A}$. Let*

$\{y(\cdot), \mathscr{F}(\cdot)\}$ be a positive almost surely right continuous supermartingale. Suppose that $x(t-,\omega) \le y(t-,\omega)$ quasi everywhere on

$$\{(t,\omega)\in \dot{A}\colon A(t,\omega) > A(t-,\omega)\}$$

and that $x(t,\omega) \le y(t,\omega)$ quasi everywhere on

$$\{(t,\omega)\in \dot{A}\colon A(t,\omega) = A(t-,\omega)\}.$$

Then $x(\cdot,\cdot) \le y(\cdot,\cdot)$ quasi everywhere on $\mathbb{R}^+ \times \Omega$; that is, $x(\cdot) \le y(\cdot)$.

(We have ignored the evanescent set on which one of the processes involved is not right continuous with left limits.)

Set $x(+\infty) = 0$, $y(+\infty) = y(+\infty-)$. For $t\in\mathbb{R}^+$ let $T(t)$ be the first entry time $\ge t$ of $\dot{A}$ by $x(\cdot)$. Since $\dot{A}$ is progressively measurable, $T(t)$ is optional. Choose t a point of almost sure continuity of $A(\cdot)$, that is, a point of continuity of the monotone function $s\mapsto E\{A(s)\}$. This choice of t excludes an at most countable set, and in view of the almost sure right continuity of $x(\cdot)$ and $y(\cdot)$ it is therefore sufficient to prove that $x(t) \le y(t)$ almost surely. Define, for the chosen value t,

$$T' = \begin{cases} T(t) & \text{on } \{A(T(t)) > A(T(t)-)\},\\ +\infty & \text{elsewhere,}\end{cases}$$

$$T'' = \begin{cases} T(t) & \text{on } \{A(T(t)) = A(T(t)-)\},\\ +\infty & \text{elsewhere.}\end{cases}$$

Then T' and T'' are optional, $t < T' \le +\infty$, $t \le T'' \le +\infty$, and

$$\{T' = T'' = +\infty\} = \{T(t) = +\infty\}.$$

(Null sets are ignored throughout.) The optional time T' can have no totally inaccessible component because [see the proof of Theorem 7(a3)] $A(\cdot)$ does not charge any totally inaccessible optional time. Hence T' is accessible. Let S be a predictable time whose graph is a subset of that of T', and let $S_\cdot$ be a sequence of optional times announcing S. It is no restriction to suppose that $S_0 \ge t$. The almost sure supermartingale inequality

$$y(t) \ge E\{y(S_n \wedge T'')|\mathscr{F}(t)\}$$

yields ($n \to \infty$)

$$\begin{aligned} y(t) &\ge E\{y(S-)1_{\{S<+\infty\}} + y(T'')1_{\{T''<+\infty\}}|\mathscr{F}(t)\} \\ &\ge E\{x(T'-)1_{\{S<+\infty\}} + x(T'')1_{\{T''<+\infty\}}|\mathscr{F}(t)\} \quad \text{a.s.} \end{aligned} \tag{13.2}$$

Now

$$\begin{aligned}
&E\{x(T'-)1_{\{S<+\infty\}}|\mathscr{F}(t)\}\\
&= E\left\{\lim_{n\to\infty} x(S_n)\,1_{\{S<+\infty\}}|\mathscr{F}(t)\right\}\\
&= E\left\{\lim_{n\to\infty} E\{A(+\infty)-A(S_n)|\mathscr{F}(S_n)\}\,1_{\{S<+\infty\}}|\mathscr{F}(t)\right\}\\
&= E\left\{\left[E\{A(+\infty)|\bigvee_0^\infty \mathscr{F}(S_n)\} - A(T'-)\right]1_{\{S<+\infty\}}|\mathscr{F}(t)\right\} \quad \text{a.s.}
\end{aligned} \tag{13.3}$$

Each optional time S_k is $\bigvee_0^\infty \mathscr{F}(S_n)$ measurable; so S is also, and therefore the random variable $1_{\{S<+\infty\}}$ can be put inside the inner conditional expectation operation in the last line of (13.3). Hence

$$E\{x(T'-)1_{\{S<+\infty\}}|\mathscr{F}(t)\} = E\{[A(+\infty)-A(T'-)]1_{\{S<+\infty\}}|\mathscr{F}(t)\} \quad \text{a.s.} \tag{13.4}$$

Now $A(T-)=A(t)$ when $T'<+\infty$; so

$$E\{x(T'-)1_{\{S<+\infty\}}|\mathscr{F}(t)\} = E\{[A(+\infty)-A(t)]1_{\{S<+\infty\}}|\mathscr{F}(t)\} \quad \text{a.s.} \tag{13.5}$$

A similar but easier argument shows that

$$E\{x(T'')1_{\{T''<+\infty\}}|\mathscr{F}(t)\} = E\{[A(+\infty)-A(t)]1_{\{T''<+\infty\}}|\mathscr{F}(t)\} \quad \text{a.s.} \tag{13.6}$$

Furthermore $\{S<+\infty\}\subset\{T'<+\infty\}$, and S can be chosen to make the probability of the difference arbitrarily small. Using this fact, we can replace S by T' in (13.2) and (13.5) so (13.2) yields

$$y(t) \geq E\{[A(+\infty)-A(t)]1_{\{T'\wedge T''<+\infty\}}|\mathscr{F}(t)\} \quad \text{a.s.} \tag{13.7}$$

Since $A(+\infty)=A(t)$ on the set $\{T'=T''=+\infty\}$, the right side of (13.7) is almost surely $x(t)$, and the proof is now complete.

Special Case: Almost Sure Continuity of $A(\cdot)$. The parabolic context domination principle (Theorem 1.XVIII.16) is the exact translation of the classical domination principle (Theorem 1.V.10) into the parabolic context if the Riesz measure corresponding to the given superparabolic potential vanishes on parabolic semipolar sets; that is, under this condition coparabolic-fine limits are not involved. Corresponding to this fact, in the present Theorem 13 if $A(dt)$ almost surely vanishes on semipolar subsets of $\mathbb{R}^+\times\Omega$, that is, if

$$P\left\{\int_{\dot B} d_tA(t,\omega)=0\right\}=1$$

whenever $\dot{B}$ is semipolar, then $A(\cdot)$ is almost surely continuous [apply Section 6 on the discontinuity set of a supermartingale to $-A(\cdot)$]; so the inequality hypothesis of Theorem 13 reduces to the inequality $x(\cdot) \leq y(\cdot)$ quasi everywhere on $\dot{A}$; that is, left limits are not involved.

14. The Operators τ, LM, and GM in the Continuous Parameter Context

Let $(\Omega, \mathscr{F}, P; \mathscr{F}(t), t \in \mathbb{R}^+)$ be a filtered probability space for which $\mathscr{F}(\cdot)$ is right continuous and $\mathscr{F}(0)$ contains the null sets.

The Operator τ. Let $\{x(\cdot), \mathscr{F}(\cdot)\}$ be an almost surely right continuous process, let S and T be optional, with $S \leq T < +\infty$, and suppose that $x(T)$ is integrable. For example, according to Section 3, if the process is a supermartingale, then $x(T)$ is integrable if T is bounded, or if T is not bounded and $x(\cdot)$ is right closable. Following Section III.18 in which, however, the parameter set was an arbitrary linearly ordered set and the optional times were countably valued, define

$$\tau_{ST}x(t) = \begin{cases} x(t) & \text{on } \{S > t\} \cup \{T < t\}, \\ E\{x(T) \mid \mathscr{F}(t)\} & \text{on } \{S \leq t \leq T\}. \end{cases} \tag{14.1}$$

Here the conditional expectations are to be chosen to make $\tau_{ST}x(\cdot)$ almost surely right continuous (Section 2). Obviously $\tau_{ST}x(T) = x(T)$ almost surely. Suppose now that $\{x(\cdot), \mathscr{F}(\cdot)\}$ is a supermartingale. It follows from Theorem 3 that $x(T)$ is integrable and that $\tau_{ST}x(\cdot) \leq x(\cdot)$ (essential order); so by almost sure right continuity this inequality is true quasi everywhere on $\mathbb{R}^+ \times \Omega$. The argument in Section III.18 translated into the present context shows that $\{\tau_{ST}x(\cdot), \mathscr{F}(\cdot)\}$ is a supermartingale, a martingale on $[\![S, T]\!]$, in the sense that the process

$$\{\tau_{ST}x((S \vee t) \wedge T), \mathscr{F}((S \vee t) \wedge T), t \in \mathbb{R}^+\}$$

is a martingale. The dual results for $\{x(\cdot), \mathscr{F}(\cdot)\}$ a submartingale are clear.

Observe that if $\{x(\cdot), \mathscr{F}(\cdot)\}$ is a positive supermartingale and if we define $x(+\infty) = 0$, then we need not restrict S and T to be finite valued in the preceding discussion.

The Operators LM and GM

It will be sufficient to discuss GM. Suppose that $\{x(\cdot), \mathscr{F}(\cdot)\}$ is an almost surely right continuous supermartingale which has an essential order submartingale minorant. Then (from Section III.20) one version of the martingale $\mathrm{GM}x(\cdot)$ is $\lim_{n\to\infty} \tau_{0n}x(\cdot)$. For each positive integer k the sequence $n \to \{\tau_{0n}x(t), \mathscr{F}(t), t \le k\}$ is, for $n \ge k$, a decreasing sequence of almost surely right continuous martingales. An application of Theorem 4 shows that one version of $\mathrm{GM}x(\cdot)$ is an almost surely right continuous martingale, and in the present continuous parameter context we shall accept only such versions. More generally, if Γ is a class of almost surely right continuous supermartingales adapted to $\mathscr{F}(\cdot)$, for which $\mathrm{GM}\Gamma$ exists, we now show that $\mathrm{GM}\Gamma$ has an almost surely right continuous version. Such a version is the only one we shall accept; it can be modified trivially to yield a right continuous version, if desired. To prove the assertion, observe first that we can assume that Γ is directed downward, at the price of adjoining to Γ the finite minima of its members; the enlargement of Γ does not change $\mathrm{GM}\Gamma$. The class $\{\mathrm{GM}x(\cdot)\colon x(\cdot) \in \Gamma\}$ is a downward-directed class of almost surely right continuous martingales with essential order infimum and limit $\mathrm{GM}\Gamma$, and according to Section III.5, there is a decreasing sequence of these martingales whose limit is a version of $\mathrm{GM}\Gamma$. An application of Theorem 4 shows that this version is almost surely right continuous.

15. Potential Theory on $\mathbb{R}^+ \times \Omega$

Let $(\Omega, \mathscr{F}, P; \mathscr{F}(t), t \in \mathbb{R}^+)$ be a filtered probability space for which $\mathscr{F}(\cdot)$ is right continuous and $\mathscr{F}(0)$ contains the null sets. Many of the results we have obtained suggest that one can construct a theory on $\mathbb{R}^+ \times \Omega$ which is formally parallel in many ways to parabolic potential theory and therefore also to classical potential theory, in which almost surely right continuous supermartingales play the role of superparabolic functions. The nomenclature, for example, Fundamental Convergence Theorem, GM and τ operators, has been chosen with this in mind. In later chapters when we compose superparabolic functions with space-time Brownian motion to get supermartingales, it will become clear that the relation between a suitably constructed potential theory on $\mathbb{R}^+ \times \Omega$ and parabolic potential theory is more than just a formal resemblance.

Probability theory and potential theory developed independently of each other aside from interrelations between random walks and Brownian motion with solutions of parabolic and elliptic differential equations, until on the one hand axiomatics of generalized potential theory were devised and on the other hand the theory of Markov stochastic processes led to a generalized potential theory in which potential theoretic ideas were defined in terms of probabilistic ones. In both theories generalized versions of superparabolic functions lay at the base of the subject, and under suitable conditions it was shown that the axiomatic potential theory could be generated by Markov process theory. It then became possible to prove theorems in axiomatic potential theory by probabilistic methods. We shall discuss in later chapters the Brownian motion process which generates classical potential theory and the space-time Brownian motion process which generates parabolic potential theory.

Although probabilistic methods have frequently been used to derive potential theoretic results, derivations in the reverse direction have been rare. To illustrate the possibilities and to give insight into the whole subject, we shall use the ideas and methods of potential theory to analyze reductions in the context of potential theory on $\mathbb{R}^+ \times \Omega$. We shall need a few properties of the fine topology on this space.

16. The Fine Topology of $\mathbb{R}^+ \times \Omega$

Let $(\Omega, \mathscr{F}, P; \mathscr{F}(t), t \in \mathbb{R}^+)$ be a filtered probability space for which $\mathscr{F}(\cdot)$ is right continuous and $\mathscr{F}(0)$ contains the null sets. The *fine* topology of $\mathbb{R}^+ \times \Omega$ is by definition the smallest topology on $\mathbb{R}^+ \times \Omega$ making every right continuous supermartingale $\{x(\cdot), \mathscr{F}(\cdot)\}$ into a (fine topology) continuous process in the sense that the function $(t, \omega) \to x(t, \omega)$ is to be continuous in the fine topology. The set of fine limit points of a set $\dot{A}$ will be denoted by $\dot{A}^f$.

Example (a). If A is a P null subset of Ω and if $x(\cdot, \cdot)$ is the indicator function of $\dot{A} = \mathbb{R}^+ \times A$, the process $\{x(\cdot), \mathscr{F}(\cdot)\}$ is a right continuous martingale; so $\dot{A}$ is both fine open and fine closed.

Example (b). Let T be optional and define $x(\cdot, \cdot)$ as the indicator function of the stochastic interval $[\![0, T[\![$. Then $\{x(\cdot), \mathscr{F}(\cdot)\}$ is a right continuous supermartingale; so both stochastic intervals $[\![0, T[\![$ and $[\![T, +\infty[\![$ are fine open and fine closed. If S and T are optional with $S \leq T$, we conclude that the stochastic interval $[\![S, T[\![= [\![0, T[\![\cap [\![S, +\infty[\![$ is fine open and fine closed. If we choose $T_n = T + 1/n$, we find that the stochastic interval $[\![S, T]\!] = \bigcap_1^\infty [\![S, T_n[\![$ is fine closed. In particular, the graph $[\![S]\!]$ is fine closed. If S is optional, if $A \in \mathscr{F}(S)$, and if $S_A = S$ on A but $S = +\infty$ otherwise, the stochastic interval $[\![S_A, T_A[\![= (\mathbb{R}^+ \times A) \cap [\![S, T[\![$ is fine open and fine closed. Observe that the intersection of this stochastic interval with another one $(\mathbb{R}^+ \times A') \cap [\![S', T'[\![$ of the same type is

$$(\mathbb{R}^+ \times A \cap A') \cap [\![S \vee S', T \wedge T'[\![,$$

also of this type, and that $S \vee S'$ is identically constant if S and S' are.

A Base for the Fine Topology. The class of sets (stochastic intervals) $(\mathbb{R}^+ \times A) \cap [\![\alpha, T[\![$ with α a positive constant and $A \in \mathscr{F}(\alpha)$ is a base for the fine topology because if $\{x(\cdot), \mathscr{F}(\cdot)\}$ is a right continuous process, not necessarily a supermartingale, and if T_t is the entry time of the set

$$\{(s, \omega): s > t, x(s, \omega) \leq c\},$$

then

$$\{(t, \omega): x(t, \omega) > c\} = \bigcup_{t \in \mathbb{R}^+} (\mathbb{R}^+ \times \{\omega: x(t, \omega) > c\}) \cap [\![t, T_t[\![.$$

If $\{\omega_0\}$ is a P null singleton and if $t_0 \in \mathbb{R}^+$, the set

$$(\mathbb{R}^+ \times \{\omega_0\}) \cap [\![t_0, t_0 + \varepsilon[\![$$

is a fine neighborhood of (t_0, ω_0) for every $\varepsilon > 0$, and the set of these fine neighborhoods for all $\varepsilon > 0$ is a base for the fine neighborhoods of (t_0, ω_0). Hence a subset $\dot{A}$ of $\mathbb{R}^+ \times \Omega$ has (t_0, ω_0) as a fine limit point if and only if $\dot{A}$ contains points (t_n, ω_0) with $t_n > t_0$, $\lim_{n \to \infty} t_n = t_0$. Thus if all Ω singletons are P null the fine topology is the direct product of the discrete topology on Ω and the one sided Euclidean topology on $\mathbb{R}^+$ in which a base for the topology is the class of left closed intervals.

Evanescent Sets and the Fine Topology

The fine closure of an evanescent set is evanescent because an evanescent set is a subset of an evanescent set of the form $\mathbb{R}^+ \times A$ with $P\{A\} = 0$ and [from Example (a) above] this product set is fine closed.

Almost Thinness at an Optional Time

If T is optional, call a subset $\dot{A}$ of $\mathbb{R}^+ \times \Omega$ *almost thin* at T if the set $[\![T]\!] \cap \dot{A}^f$ is evanescent. Then an evanescent set $\dot{A}$ is almost thin at every optional time. A progressively measurable set $\dot{A}$ almost thin at every optional time T is semipolar if all Ω singletons are null, according to the following argument. If T_1 is the entry time of $\dot{A}$, then $[\![T_1]\!] \subset \dot{A}$ up to an evanescent set, and the hitting time T_2 of $\dot{A}$ after T_1, that is, the second entry time of $\dot{A}$, is almost surely strictly larger than T_1 on $\{T_1 < +\infty\}$. Continuing in this way, we find an increasing sequence $T_.$ of optional times such that $T_n < T_{n+1}$ almost everywhere on $\{T_n < +\infty\}$, and $[\![T_n]\!] \subset \dot{A}$ up to an evanescent set. Going on by transfinite induction, the next optional time is the first entry time $\geq \lim_{n \to \infty} T_n, \ldots$. If α is a countable ordinal for which T_α has been defined, either $T_{\alpha+1} = +\infty$ almost surely, in which case $\dot{A}$ is semipolar because $\dot{A} \subset \bigcup_{n \leq \alpha} [\![T_n]\!]$ up to an evanescent set, or $P\{T_{\alpha+1} > T_\alpha\} > 0$; so

$$E\{\arctan T_{\alpha+1}\} > E\{\arctan T_\alpha\}.$$

It follows that the second possibility can occur only countably many times, that is, there is some countable ordinal α with $T_{\alpha+1} = +\infty$ almost surely, and we have just seen that then $\dot{A}$ must be semipolar.

17. Potential Theory Reductions in a Continuous Parameter Probability Context

(See Section III.22 for the corresponding discrete parameter theory.)

Let $(\Omega, \mathscr{F}, P; \mathscr{F}(t), t \in \mathbb{R}^+)$ be a filtered probability space for which $\mathscr{F}(\cdot)$ is right continuous and $\mathscr{F}(0)$ contains the null sets. All processes below are adapted processes on this space. Let $z(\cdot)$ be a positive process, and denote by Γ the class of almost surely right continuous supermartingale pointwise majorants of $z(\cdot)$. It is supposed that Γ is not empty. The class Γ is directed downward, and we define $R_{z(\cdot)} = \text{ess}^{*}\inf \Gamma$ (= either an equivalence class under identification of indistinguishable processes or a member of the class, as indicated by the context). Observe that the equivalence class $R_{z(\cdot)}$ is unchanged if Γ is defined as the class of almost surely right continuous supermartingales which are quasi everywhere majorants of $z(\cdot)$ or if "almost surely" is omitted here. Define $R_{z(\cdot)}(+\infty) = z(+\infty) = 0$. According to the Fundamental Convergence Theorem (Theorem 5), almost every sample function of the supermartingale $R_{z(\cdot)}$ has right and left limits at all points, and if $\underset{+z(\cdot)}{R}$ is defined as the right smoothing of $R_{z(\cdot)}$, then $\underset{+z(\cdot)}{R} \leq R_{z(\cdot)}$ quasi everywhere. For $t \in \mathbb{R}^+$ define

$$y(t) = \text{ess}\sup\{E\{z(T)|\mathscr{F}(t)\}: T \text{ optional}, \geq t\}; \tag{17.1}$$

this random variable is determined only up to a null set. The method of proof of Theorem III.22 shows that $\{y(\cdot), \mathscr{F}(\cdot)\}$ is a positive supermartingale which in the essential order is both a majorant of $z(\cdot)$ and a minorant of Γ. Thus no matter how $y(t)$ is chosen, $R_{z(\cdot)} \geq y(\cdot) \geq z(\cdot)$ in the essential order; that is, $P\{R_{z(\cdot)}(t) \geq y(t) \geq z(t)\} = 1$ for all t. According to Section 1 there is a countable parameter set B such that $y(t)$ can be chosen for each t in such a way that almost every $y(\cdot)$ sample function has right and left limits at all points, and is right continuous except possibly at points of B. Choose $y(t)$ in this way. With this choice $R_{z(\cdot)}(\cdot) \geq \underset{+z(\cdot)}{R}(\cdot) \geq y(\cdot+)$ quasi everywhere. If $y_+(t)$ is defined by the right side of (17.1) except that the optional time T is supposed strictly greater than t, then $y_+(\cdot)$ is a supermartingale, $y_+(t) \leq y(t)$ almost surely, and $t' \downarrow t$ sequentially implies that

$$y_+(t') \to y_+(t) = y(t+) \quad \text{a.s.}$$

We choose $y_+(t) = y(t+)$ and thereby have $y_+(\cdot) = y(\cdot+) \leq y(\cdot)$ quasi everywhere. In particular, if $z(\cdot)$ is almost surely right lower semicontinuous, $y_+(\cdot) \geq z(\cdot)$ quasi everywhere; so $y_+(\cdot) \in \Gamma$, and therefore $R_{z(\cdot)}(\cdot) = y_+(\cdot)$ quasi everywhere. We conclude that $R_{z(\cdot)}(\cdot)$ and $y(\cdot)$ are indistinguishable almost surely right continuous supermartingales. Unfortunately we shall not be able to make this lower semicontinuity hypothesis in the following application to reductions, in which the process corresponding to $z(\cdot)$ above is the product $z(\cdot)1_{\dot{\Lambda}}(\cdot)$, where $z(\cdot)$ is now an almost surely right continuous supermartingale and $\dot{\Lambda}$ is a subset of $\mathbb{R}^+ \times \Omega$.

Application to Reductions (on the Above Filtered Probability Space)

Let $\dot{\Lambda}$ be a subset of $\mathbb{R}^+ \times \Omega$ for which $t \to 1_{\dot{\Lambda}}(t)$ is a process adapted to $\mathscr{F}(\cdot)$, and let $z(\cdot)$ be a positive almost surely right continuous supermartingale. Define the reduction $R^{\dot{\Lambda}}_{z(\cdot)}$ as the supermartingale $R_{z(\cdot)1_{\dot{\Lambda}}(\cdot)}$. The right smoothing of the reduction will be denoted by $\underset{+z(\cdot)}{R}{}^{\dot{\Lambda}}$ or by $\|z(\cdot)\|^{\dot{\Lambda}}$. According to Theorem 5, the process $R^{\dot{\Lambda}}_{z(\cdot)}$ is determined up to indistinguishability by the following conditions:

The reduction is a progressively measurable process.

Whenever $y(\cdot)$ is in the class Γ of almost surely right continuous positive supermartingales $\geq z(\cdot)$ quasi everywhere on $\dot{\Lambda}$, the inequality $y(\cdot) \geq R^{\dot{\Lambda}}_{z(\cdot)}(\cdot)$ is true quasi everywhere on $\mathbb{R}^+ \times \Omega$.

If $x(\cdot)$ is a process satisfying the preceding conditions, then $x(\cdot) \geq R^{\dot{\Lambda}}_{z(\cdot)}(\cdot)$ quasi everywhere on $\mathbb{R}^+ \times \Omega$.

Furthermore a reduced process $R^{\dot{\Lambda}}_{z(\cdot)}(\cdot)$ is a supermartingale with almost sure right and left limits, coinciding quasi everywhere with the respective right and left limits of its right smoothing $\underset{+z(\cdot)}{R}(\cdot)$, and $\underset{+z(\cdot)}{R}(\cdot) \leq R_{z(\cdot)}(\cdot)$ quasi everywhere, with equality up to a semipolar set. Finally, a suitably chosen decreasing sequence of members of Γ has as limit a version of the reduction.

We shall see that many reduction properties in the present context are close counterparts of reduction properties in the context of classical and parabolic potential theory. In one respect the present context is simpler than those earlier ones. The fact that superharmonic and parabolic functions can have infinities complicates their theory, whereas almost every sample function of an almost surely right continuous supermartingale is finite valued with finite left limits.

18. Reduction Properties

The context is that of Section 17. We consider reductions of a positive almost surely right continuous supermartingale $\{z(\cdot), \mathscr{F}(\cdot)\}$ on subsets of $\mathbb{R}^+ \times \Omega$. Reduction properties in this context are counterparts of reduction properties in parabolic potential theory as listed in Sections 1.XVII.11 and 1.XVII.16 except that for simplicity we have not introduced counterparts of reductions on boundary sets. The following properties are listed as far as possible in the order of their parabolic counterparts, and their counterparts are always referred to. Proofs are given or referred to in Section 19. We leave to the reader the translation of the listed properties to the context of a countable parameter set.

(a) [See Section 1.XVII.16(b).]

$$\underset{+z(\cdot)}{R}^{\dot A}(t+) = R^{\dot A}_{z(\cdot)}(t+),\ \underset{+z(\cdot)}{R}^{\dot A}(t-) = R^{\dot A}_{z(\cdot)}(t-) \quad \text{q.e.};$$

$$z(t) \geq R^{\dot A}_{z(\cdot)}(t) \geq R^{\dot A}_{z(\cdot)}(t+) \quad \text{q.e.};$$

$$\underset{+z(\cdot)}{R}^{\dot A} = R^{\dot A}_{z(\cdot)} \quad \text{up to a semipolar set};$$

$$R^{\dot A}_{z(\cdot)} = R^{\dot B}_{z(\cdot)} \quad \text{q.e. if } \dot A \text{ differs from } \dot B \text{ by an evanescent set.}$$

(b) [See Section 1.XVII.16(b).]

$$R^{\dot A}_{z(\cdot)} = z(\cdot) \quad \text{q.e. on } \dot A \cap \dot A^f;$$

$$R^{\dot A}_{z(\cdot)} = \underset{+z(\cdot)}{R}^{\dot A} = z(\cdot) \quad \text{q.e. on } \dot A^f, \text{ in particular, q.e. on the fine interior of } \dot A;$$

$$\underset{+z(\cdot)}{R}^{\dot A} = R^{\dot A}_{z(\cdot)} \quad \text{q.e. when } \dot A \text{ is fine open.}$$

(c) [See Section 1.XVII.16(b).] If S and T are optional times with $S \leq T$ and if $\llbracket S, T\llbracket \cap \dot A$ is evanescent, then $R^{\dot A}_{z(\cdot)} = \underset{+z(\cdot)}{R}^{\dot A}$ quasi everywhere on $\llbracket S, T\llbracket$, and $\underset{+z(\cdot)}{R}^{\dot A}$ is an almost surely right continuous martingale on $\llbracket S, T\rrbracket$ in the sense that the process

$$\{R^{\dot A}_{z(\cdot)}((S \vee t) \wedge T), \mathscr{F}((S \vee t) \wedge T), t \in \mathbb{R}^+\} \tag{18.1}$$

is an almost surely right continuous martingale.

(d) [See Sections 1.XVII.11(a) and 1.XVII.16(c).]

$$R^{\dot A}_{z(\cdot)} = \text{ess}^*\text{inf}\,\{\underset{+z(\cdot)}{R}^{\dot B} : \dot A \subset \dot B,\ \dot B \text{ progressively measurable and fine open}\} \text{ q.e.}; \tag{18.2}$$

one version of the right-hand side is the infimum when $\dot B$ runs through some decreasing sequence of fine-open supersets of $\dot A$. In particular, if all almost surely right continuous supermartingales on the given space are almost surely lower semicontinuous, then whenever $z(\cdot)$ is almost surely continuous, the set $\dot B$ in (18.2) can be further restricted to have open t cross sections for fixed ω.

(e) [See Sections 1.XVII.11(b) and 1.XVII.16(i).] If $\dot A \subset \dot B$ and if $\dot A$ is fine open, then the supermartingales $\llbracket \llbracket z(\cdot)\rrbracket^{\dot A} \rrbracket^{\dot B}$ and $\llbracket z(\cdot)\rrbracket^{\dot A}$ are indistinguishable. If $\dot B$ is also fine open, these supermartingales are indistinguishable from $\llbracket \llbracket z(\cdot)\rrbracket^{\dot B} \rrbracket^{\dot A}$.

(f) [See Section 1.XVII.16(f).] If $\dot{A}$ is indistinguishable from a progressively measurable set, then $R^{\dot{A}}_{+z(\cdot)} = R^{\dot{A}}_{z(\cdot)}$ quasi everywhere on $(\mathbb{R}^+ \times \Omega) - \dot{A}$.

(g) [See Sections 1.XVII.11(d) and 1.XVII.16(k).] The set functions $\dot{A} \mapsto R^{\dot{A}}_{z(\cdot)}$ and $\dot{A} \mapsto R^{\dot{A}}_{+z(\cdot)}$ are strongly subadditive in the sense that quasi everywhere on $\mathbb{R}^+ \times \Omega$ both

$$R^{\dot{A}\cup\dot{B}}_{z(\cdot)} + R^{\dot{A}\cap\dot{B}}_{z(\cdot)} \leq R^{\dot{A}}_{z(\cdot)} + R^{\dot{B}}_{z(\cdot)} \tag{18.3}$$

and the corresponding inequality (18.sm) for smoothed reductions are true. The exceptional evanescent set depends on $z(\cdot)$, $\dot{A}$, $\dot{B}$. [See (i) for countable strong subadditivity.]

(h) [See Sections 1.XVII.11(e) and 1.XVII.16(e).] If $\dot{A}_{\bullet}$ is an increasing sequence of subsets of $\mathbb{R}^+ \times \Omega$ with union $\dot{A}$ and if $\{z_n(\cdot), n \in \mathbb{Z}^+\}$ is an increasing sequence of almost surely right continuous positive supermartingales with limit $z(\cdot)$, then quasi everywhere on $\mathbb{R}^+ \times \Omega$ both

$$\lim_{n\to\infty} R^{\dot{A}_n}_{z_n(\cdot)} = R^{\dot{A}}_{z(\cdot)} \tag{18.4}$$

and the corresponding equation (18.4sm) for smoothed reductions are true.

(i) [See Section 1.XVII.16(k).] The set functions $\dot{A} \mapsto R^{\dot{A}}_{z(\cdot)}$ and $\dot{A} \mapsto R^{\dot{A}}_{+z(\cdot)}$ are countably strongly subadditive in the sense that if $\dot{A}_{\bullet}$ and $\dot{B}_{\bullet}$ are sequences of subsets of $\mathbb{R}^+ \times \Omega$ with $\dot{A}_n \subset \dot{B}_n$, then quasi everywhere on $\mathbb{R}^+ \times \Omega$ both

$$R^{\bigcup_0^\infty \dot{B}_n}_{z(\cdot)} + \sum_0^\infty R^{\dot{A}_n}_{z(\cdot)} \leq R^{\bigcup_0^\infty \dot{A}_n}_{z(\cdot)} + \sum_0^\infty R^{\dot{B}_n}_{z(\cdot)} \tag{18.5}$$

and the corresponding inequality (18.5sm) for smoothed reductions are true.

(j) [See Section 1.XVII.16(g).] If $\{z_n(\cdot), n \in \mathbb{Z}^+\}$ is a sequence of almost surely right continuous positive supermartingales, if $z(\cdot) = \Sigma_0^\infty z_n(\cdot)$ (almost surely right continuous by Theorem 4) is a supermartingale, that is, if $\Sigma_0^\infty E\{z_n(0)\} < +\infty$, then quasi everywhere on $\mathbb{R}^+ \times \Omega$ both

$$R^{\dot{A}}_{z(\cdot)} = \sum_0^\infty R^{\dot{A}}_{z_n(\cdot)} \tag{18.6}$$

and the corresponding equation (18.6sm) for smoothed reductions are true.

(k) [See Section 1.XVII.16(m).] If S is a finite optional time and if $\dot{A} = [\![0, S]\!]$, then $\mathrm{GM}\, R^{\dot{A}}_{+z(\cdot)} = 0$ quasi everywhere (equivalently, $R^{\dot{A}}_{+z(\cdot)}$ is a supermartingale potential) if either S is bounded or S is unrestricted and $z(\cdot)$ is uniformly integrable.

(l) [See Section 1.XVII.16(o).] Suppose that $(\Omega, \mathscr{F}, \mathscr{F}(\cdot), P)$ has the property that every almost surely right continuous supermartingale is almost surely lower semicontinuous, and suppose that $\{z(\cdot), \mathscr{F}(\cdot)\}$ is an almost surely continuous class **D** positive supermartingale. Then the set function $\dot{A} \mapsto R^{\dot{A}}_{z(\cdot)}$

is a Choquet capacity on $\mathbb{R}^+ \times \Omega$ relative to the class Γ_c of progressively measurable sets whose t cross sections for fixed ω are compact, in the sense that

$$\dot{A} \subset \dot{B} \Rightarrow 0 \leq R_{z(\cdot)}^{\dot{A}} \leq R_{z(\cdot)}^{\dot{B}} \quad \text{q.e.} \tag{18.7$'$}$$

$$\dot{A}_n \uparrow \dot{A} \Rightarrow R_{z(\cdot)}^{\dot{A}_n} \uparrow R_{z(\cdot)}^{\dot{A}} \quad \text{q.e.} \tag{18.7$''$}$$

$$\dot{A}_n \in \Gamma_c, \qquad \dot{A}_n \downarrow \dot{A} \Rightarrow R_{z(\cdot)}^{\dot{A}_n} \downarrow R_{z(\cdot)}^{\dot{A}} \quad \text{q.e.} \tag{18.7$'''$}$$

(m) [See Section 1.XVII.16(p).] Suppose that $(\Omega, \mathscr{F}, \mathscr{F}(\cdot), P)$ has the property that every almost surely right continuous supermartingale is almost surely lower semicontinuous, and suppose that $z(\cdot)$ is a positive almost surely continuous supermartingale. Let Γ' be the class of predictable product sets defined in Section I.14, and let Γ_p be the class of finite unions of Γ' sets. Then if $\dot{A}$ is analytic over the class of predictable sets, there is an increasing sequence $\dot{F}_\bullet$ of $\Gamma_{p\delta}$ subsets of $\dot{A}$ for which quasi everywhere on $\mathbb{R}^+ \times \Omega$ both

$$\lim_{n\to\infty} R_{z(\cdot)}^{\dot{F}_n} = R_{z(\cdot)}^{\dot{A}} \tag{18.8}$$

and the corresponding equation (18.8.sm) for smoothed reductions are true.

(n) [See Section 1.XVII.16(q).] If $y(\cdot)$ and $z(\cdot)$ are bounded positive almost surely right continuous supermartingales, then

$$\sup{}' \left| R_{y(\cdot)}^{\dot{A}} - R_{z(\cdot)}^{\dot{A}} \right| \leq \sup{}' \left| y(\cdot) - z(\cdot) \right|, \tag{18.9}$$

where sup$'$ means the supremum up to evanescent sets, and the corresponding inequality (18.9sm) is true for smoothed reductions.

(o) [See Section 1.XVII.16(s).] Let $\{x(\cdot), \mathscr{F}(\cdot)\}$ and $\{y(\cdot), \mathscr{F}(\cdot)\}$ be almost surely right continuous positive supermartingales, let a and b be positive numbers with $0 \leq a < b$, and define

$$\dot{A} = \{(t, \omega)\colon x(t, \omega) \leq ay(t, \omega)\}, \qquad \dot{B} = \{(t, \omega)\colon x(t, \omega) \geq by(t, \omega)\}.$$

Denote $[\![y(\cdot)]\!]^{\dot{A}}$, $[\![[\![y(\cdot)]\!]^{\dot{A}}]\!]^{\dot{B}}$, ... by $y_{\dot{A}}(\cdot), y_{\dot{A}\dot{B}}(\cdot)$, ... Then the iterated reduction and downcrossing inequalities III(23.2)–III(23.6) are true in the present context, as are also the other martingale theory translations of the classical context reduction inequalities in Section 1.VI.3(o). Moreover these inequalities are true for the parameter restricted to an interval. In particular if $\mathrm{Dn}_t[x(\cdot), y(\cdot); a, b]$ refers to downcrossings on the parameter interval $[0, t]$, then (almost surely) for $t \in \mathbb{R}^+$,

$$E\{y(t)\mathrm{Dn}_t[x(\cdot), y(\cdot); a, b] | \mathscr{F}(0)\} \leq \frac{x(0) \wedge [by(0)]}{b - a}. \tag{18.10}$$

19. Proofs of the Reduction Properties in Section 18

(a) The properties listed in Section 18(a) were all derived in Section 17.
(b) The first two lines in Section 18(b) follow easily from two facts:

(i) There is a decreasing sequence of right continuous supermartingales equal to $z(\cdot)$ on $\dot{A}$ and so also on $\dot{A}^f$, and tending quasi everywhere to $R^{\dot{A}}_{z(\cdot)}$.
(ii) Each fine neighborhood of a point (t_0, ω_0) contains a set of the form $[t_0, t_0 + \varepsilon] \times \{\omega_0\}$ with $\varepsilon > 0$.

To prove the third line, observe that $\underset{+z(\cdot)}{R^{\dot{A}}} = R^{\dot{A}}_{z(\cdot)}$ quasi everywhere on $\dot{A}$; so $\underset{+z(\cdot)}{R^{\dot{A}}}$ is a version of the unsmoothed reduction.

(c) To prove Section 18(c), observe (see Section 14) that if T is a finite-valued optional time and if $\{x_n(\cdot), n \in \mathbb{Z}^+\}$ is a decreasing sequence of almost surely right continuous positive supermartingales with quasi everywhere limit $R^{\dot{A}}_{z(\cdot)}$, then $\tau_{ST}x_n(\cdot)$ is an almost surely right continuous positive martingale on $[\![S, T]\!]$, $\tau_{ST}x_n(\cdot)$ majorizes $z(\cdot)$ quasi everywhere on $\dot{A}$, and $\tau_{ST}x_n(\cdot) \leq x_n(\cdot)$ quasi everywhere. Hence $\lim_{n\to\infty} \tau_{ST}x_n(\cdot) = \tau_{ST}R^{\dot{A}}_{z(\cdot)}$ is indistinguishable from $R^{\dot{A}}_{z(\cdot)}$ and is (by Theorem 4) an almost surely right continuous martingale on $[\![S, T]\!]$ and so is equal quasi everywhere on $[\![S, T]\!]$ to its right smoothing. If T is not finite valued, the same argument is applicable without any change if we define $x_n(+\infty) = R^{\dot{A}}_{z(\cdot)}(+\infty) = 0$ and allow T to be infinite valued in the definition of τ_{ST}. The needed properties of the operator τ_{ST} remain valid under this extension.

(d) The first assertion of Section 18(d), in which $\underset{+z(\cdot)}{R^{\dot{B}}}$ is indistinguishable from $R^{\dot{B}}_{z(\cdot)}$, follows from the fact that, assuming as we can in this proof that $z(\cdot)$ is right continuous, if $y(\cdot)$ is a right continuous positive supermartingale essential order majorant of $z(\cdot)$ on $\dot{A}$ and if $\varepsilon > 0$, then $y(\cdot) + \varepsilon$ is an essential order majorant of $z(\cdot)$ on a progressively measurable fine-open neighborhood of $\dot{A}$. If every almost surely right continuous supermartingale on the probability space is almost surely lower semicontinuous "fine open" in the preceding argument can be replaced by "open t cross sections for fixed ω." The right-hand side of (18.2) is a countable infimum according to Theorem 5.

(e) See the proof of Section 18(e) in the parabolic context [Section 1.XVII.11(b)].

(f) To prove the equality of reduction and smoothed reduction quasi everywhere on $\dot{B} = (\mathbb{R}^+ \times \Omega) - \dot{A}$, observe first that the strict inequality set is not only semipolar but (see the proof of the Fundamental Convergence Theorem) is, up to an evanescent set, a countable union $\bigcup_0^\infty [\![S_n]\!]$ of optional time graphs. It is to be shown that, for each value of n, the set $\Lambda = \{\omega\colon [S_n(\omega), \omega] \in \dot{B}\}$ is null. This set, the projection on Ω of the set $[\![S_n]\!] \cap \dot{B}$, is in $\mathscr{F}(S_n)$ because, for $c \geq 0$, the set $\Lambda \cap \{S_n \leq c\}$, the projection on Ω of $[\![S_n]\!] \cap \dot{B} \cap ([0, c] \times \Omega)$, is in $\mathscr{F}(c)$. Define an optional time S_n' by setting $S_n' = S_n$ on Λ and $S_n' = +\infty$ elsewhere. If T_n is the time of

(t,ω) entry into the nearly progressively measurable set $[\![S_n']\!] \cap \dot{A}$, then $\{T_n = S_n' < +\infty\} \subset \dot{A}^f$ up to a null set; it then follows from (b) that $T_n > S_n'$ almost everywhere on Λ. Since the set $[\![S_n', T_n[\![\cap \dot{A}$ is evanescent it follows from (c) that $R^{\dot{A}}_{+z(\cdot)}(S_n') = R^{\dot{A}}_{z(\cdot)}(S_n')$ almost everywhere on Λ; this equality implies that Λ is null.

(g) To prove strong subadditivity of the reduction operator and its smoothing, suppose first that $\dot{A}$ and $\dot{B}$ are fine open, in which case (18.3sm) reduces to (18.3), and define $z'(\cdot) = R^{\dot{A}}_{z(\cdot)} \wedge R^{\dot{B}}_{z(\cdot)}$. Then [see the proof of the corresponding potential theory equality 1.VI(3.4) and 1.XVII(11.4)]

$$R^{\dot{A}\cup\dot{B}}_{z(\cdot)} + R^{\dot{A}\cup\dot{B}}_{z'(\cdot)} = R^{\dot{A}}_{z(\cdot)} + R^{\dot{B}}_{z(\cdot)} \quad \text{q.e.} \tag{19.1}$$

This equality implies inequality (18.3) for fine-open $\dot{A}$ and $\dot{B}$ and in view of (18.2) thereby implies (18.3) for all $\dot{A}$ and $\dot{B}$. Inequality (18.3sm) follows trivially from (18.3).

(h) The proof of (18.4) follows the proof of the parabolic potential theory equation 1.XVII(11.6). If each set $\dot{A}_n$ is fine open, (18.4sm) reduces to (18.4) and follows from the fact that on the one hand (quasi everywhere) the limit on the left side of (18.4) exists and is majorized by the right side and on the other hand the limit process is almost surely right continuous (Theorem 4) and (quasi everywhere) majorizes $z(\cdot)$ on $\dot{A}$ and so majorizes the right side on $\mathbb{R}^+ \times \Omega$. If the sets involved are not fine open, let T be a finite valued positive $\bigvee_{t\geq 0}\mathscr{F}(t)$ measurable function, and let $x_k(\cdot,\cdot)$ be the indicator function of the set $\{z(\cdot,\cdot) \leq k\}$. Then

$$E\{R^{\dot{A}_n}_{z_n(\cdot,\cdot)}(T)x_k(T)\} \leq E\{R^{\dot{A}}_{z(\cdot)}(T)x_k(T)\} \leq k.$$

Choose $\varepsilon > 0$, $\alpha > 1$, define

$$\dot{A}^\alpha_n = \dot{A}_n \cap \{z(\cdot,\cdot) < \alpha z_n(\cdot,\cdot)\},$$

and apply (18.2) to find a fine-open superset $\dot{B}_j$ of $\dot{A}^\alpha_j$ satisfying

$$E\{R^{\dot{B}_j}_{z(\cdot)}(T)x_k(T)\} < E\{R^{\dot{A}_j}_{z(\cdot)}(T)x_k(T)\} + 2^{-j}\varepsilon. \tag{19.2}$$

The sequence $\dot{A}^\alpha_\cdot$ is an increasing sequence with union $\dot{A}$, and we define $\dot{B} = \bigcup_0^\infty \dot{B}_j \supset \dot{A}$. By strong subadditivity of reductions

$$\begin{aligned} E\{R^{\bigcup_0^n \dot{B}_j}_{z(\cdot)}(T)x_k(T)\} &\leq E\{R^{\dot{A}^\alpha_n}_{z(\cdot)}(T)x_k(T)\} + E\left\{\sum_0^n R^{\dot{B}_j}_{z(\cdot)}(T)x_k(T)\right\} \\ &\quad - E\left\{\sum_0^n R^{\dot{A}^\alpha_j}_{z(\cdot)}(T)x_k(T)\right\} \\ &\leq E\{R^{\dot{A}^\alpha_n}_{z(\cdot)}(T)x_k(T)\} + 2\varepsilon. \end{aligned} \tag{19.3}$$

Since (18.4) is true when the sets involved are fine open, (19.3) yields (when $n \to \infty$)

$$E\{R^{\dot{A}}_{z(\cdot)}(T)x_k(T)\} \le E\{R^{\dot{B}}_{z(\cdot)}(T)x_k(T)\} \le \lim_{n\to\infty} E\{R^{\dot{A}^{\alpha}_n}_{z(\cdot)}(T)x_k(T)\} + 2\varepsilon$$
$$\le \alpha \lim_{n\to\infty} E\{R^{\dot{A}_n}_{z_n(\cdot)}(T)x_k(T)\} + 2\varepsilon. \tag{19.4}$$

Hence

$$E\{R^{\dot{A}}_{z(\cdot)}(T)x_k(T)\} \le \lim_{n\to\infty} E\{R^{\dot{A}_n}_{z_n(\cdot)}(T)x_k(T)\}, \tag{19.5}$$

and the reverse inequality is trivial. Since there is monotone convergence of integrands in (19.5), we conclude that for every k

$$\lim_{n\to\infty} R^{\dot{A}_n}_{z_n(\cdot)}(T)x_k(T) = R^{\dot{A}}_{z(\cdot)}(T)x_k(T) \qquad \text{a.s.}$$

The set

$$\dot{C}_k\colon \{R^{\dot{A}}_{z(\cdot)}(\cdot,\cdot)x_k(\cdot,\cdot) > \lim_{n\to\infty} R^{\dot{A}_n}_{z_n(\cdot)}(\cdot,\cdot)x_k(\cdot,\cdot)\}$$

has the property that its intersection with the graph of T has a null projection on Ω, whatever the choice of T. It follows (from Theorem II.8) that $\dot{C}_k$ is evanescent and that therefore (18.4) is true. Then (18.4sm) is true up to a semipolar set and since both sides of this equation are almost surely right continuous processes the two sides must be indistinguishable.

(i) Inequality (18.5) and (18.5sm) follow from strong subadditivity combined with (18.4) and (18.4sm).

(j) To prove (18.6) for two summands, it is sufficient to give the proof for $\dot{A}$ fine open, in which case the proof of the classical context counterpart 1.VI(4.2) translates easily into one valid in the present context. It then follows from (18.2) that (18.6) is true for arbitrary $\dot{A}$ and finitely many summands and therefore by (18.4) for countably many summands. Finally (18.6sm) has now been shown to be true up to a semipolar set, and since both sides of (18.6sm) are almost surely right continuous processes, the two sides must be indistinguishable.

(k) If $\dot{A} = [\![0, S]\!]$ and if $\dot{B} = [\![S, +\infty[\![$, it is trivial that $R^{\dot{A}}_{+z(\cdot)} = 0$ quasi everywhere on $\dot{B}$. Hence the martingale GM$R^{\dot{A}}_{+z(\cdot)}$ vanishes quasi everywhere if S is bounded. If S is not bounded

$$\lim_{t\to\infty} E\{R^{\dot{A}}_{+z(\cdot)}(t)\} = \lim_{t\to\infty} \int_{\{S>t\}} z(t)\,dP, \tag{19.6}$$

and this limit is 0 if $z(\cdot)$ is uniformly integrable; so $R^{\dot{A}}_{+z(\cdot)}$ is a supermartingale potential.

(l) Assertion (18.7′) is trivial, and (18.7″) is included in Section 18(h). To prove (18.7‴), suppose first that $\dot{A} = \varnothing$. Then if T_n is the entry time of $\dot{A}_n$, the sequence $T_{\bullet}$ is increasing, $\bigcup_0^\infty \{T_n = +\infty\} = \Omega$ almost surely, and by Section 18(c) the process $R_{z(\cdot)}^{\dot{A}}$ is an almost surely right continuous process on $[\![0, T_n]\!]$. Furthermore, if we define $z(+\infty) = 0$,

$$R_{z(\cdot)}^{\dot{A}_n} \leq \tau_{T_n} z(\cdot) = E\{z(T_n)|\mathscr{F}(\cdot)\}$$

quasi everywhere on this set. Hence the submartingale maximal inequality applied to the martingale on the right, which can be supposed right continuous, yields

$$\alpha P\{\sup_{t \leq T_n} R_{z(\cdot)}^{\dot{A}_n}(t) \geq \alpha\} \leq E\{z(T_n)\}. \tag{19.7}$$

Since $T_n = +\infty$ for sufficiently large n, almost surely, the class **D** property of $z(\cdot)$ implies that the expectation in (19.7) has limit 0 as $n \to \infty$. Hence

$$\lim_{n\to\infty} \sup_{t \leq T_n} R_{z(\cdot)}^{\dot{A}_n}(t) = 0 \quad \text{a.s.},$$

and therefore (18.7‴) is true when $\dot{A} = \varnothing$. In the general case let $\dot{B}$ be a progressively measurable superset of $\dot{A}$ with open cross sections for each fixed ω. Then by subadditivity of the reduction and by (18.7′)

$$R_{z(\cdot)}^{\dot{A}_n} \leq R_{z(\cdot)}^{\dot{B}} + R_{z(\cdot)}^{\dot{A}_n - \dot{B}} \quad \text{q.e.},$$

and $\dot{A}_{\bullet} - \dot{B}$ is a decreasing sequence to which the special case just treated is applicable. We conclude that $\lim_{n\to\infty} R_{z(\cdot)}^{\dot{A}_n} \leq R_{z(\cdot)}^{\dot{B}}$ quasi everywhere, and Section 18.(l) follows, in view of Section 18(d).

(m) The class Γ_p is closed under finite unions and intersections, and the sets in Γ_p have compact cross sections for fixed ω. Furthermore (Section I.14) the class of analytic sets over Γ_p includes the predictable sets. Suppose first that $z(\cdot)$ is bounded. It follows from Section 18(l) that the set function $\dot{A} \mapsto R_{z(\cdot)}^{\dot{A}}$ is a Choquet capacity on $\mathbb{R}^+ \times \Omega$ relative to Γ_p in the sense stated. Hence, if T is an optional time and if we define $R_{z(\cdot)}^{\dot{A}}(+\infty) = 0$, the set function $\dot{A} \mapsto E\{R_{z(\cdot)}^{\dot{A}}(T)\}$ is a Choquet capacity on $\mathbb{R}^+ \times \Omega$ relative to Γ_p. Now choose $\dot{A}$ analytic over Γ_p and therefore capacitable. Choose an increasing sequence $\dot{F}_{\bullet}$ of countable intersections of Γ_p sets satisfying

$$\dot{F}_n \subset \dot{A}, \qquad \lim_{n\to\infty} E\{R_{z(\cdot)}^{\dot{F}_n}(T)\} = E\{R_{z(\cdot)}^{\dot{A}}(T)\}. \tag{19.8}$$

Define $\dot{F} = \bigcup_0^\infty \dot{F}_n$. In view of the monotoneity of the integrand sequence in (19.8) and in view of Section 18(h),

$$\lim_{n\to\infty} R_{z(\cdot)}^{\dot{F}_n}(T) = R_{z(\cdot)}^{\dot{F}}(T) = R_{z(\cdot)}^{\dot{A}}(T) \quad \text{a.s.} \tag{19.9}$$

An easy diagonal procedure shows that $\dot{F}_\bullet$ can be chosen to make (19.9) valid almost surely simultaneously for an arbitrary prescribed countable set of optional times T, and we choose $\dot{F}_\bullet$ for the following countable set of optional times:

T is identically a positive rational number, for each such number.

$T = S_n$, where $S_\bullet$ is chosen so that the semipolar set $\{R^{\dot{A}}_{+z(\cdot)} < R^{\dot{A}}_{z(\cdot)}\}$ is a subset of $\bigcup_0^\infty [\![S_n]\!]$, up to an evanescent set.

With $\dot{F}_\bullet$ so chosen the processes $R^{\dot{F}}_{+z(\cdot)}$ and $R^{\dot{A}}_{+z(\cdot)}$ are indistinguishable, since they are almost surely right continuous and are equal almost surely at the rational parameter values. Thus (18.8sm) is true. Moreover quasi everywhere on $\mathbb{R}^+ \times \Omega$,

$$R^{\dot{A}}_{+z(\cdot)} = R^{\dot{F}}_{+z(\cdot)} \leq R^{\dot{F}}_{z(\cdot)} \leq R^{\dot{A}}_{z(\cdot)},$$

and there is equality throughout except possibly on $\bigcup_0^\infty [\![S_n]\!]$. Since $\dot{F}_\bullet$ was defined to make $R^{\dot{F}}_{z(\cdot)} = R^{\dot{A}}_{z(\cdot)}$ quasi everywhere on this union of graphs, we have proved that $R^{\dot{F}}_{z(\cdot)} = R^{\dot{A}}_{z(\cdot)}$ quasi everywhere; that is, (18.8) is true. Without the boundedness condition on $z(\cdot)$ that we have imposed in this proof (18.8) and (18.8sm) are true with $z(\cdot)$ replaced by $z(\cdot) \wedge n$ with $n \in \mathbb{Z}^+$, and the desired equations then follow when $n \to \infty$, using Section 18(h).

(n) The proof of (18.9) is omitted because this proof is the exact counterpart of that of the corresponding classical inequality 1.VI(3.8).

(o) The proof is obtained by a translation into martingale theory of the corresponding classical context reduction inequalities in Section 1.VI.3(o). The discrete parameter argument is in Section III.23.

20. Evaluation of Reductions

Example (a). Let $\dot{A}$ be a fine-open (Section 16) stochastic interval $[\![S, T[\![$ with $S < T$. Then (Section 18(b)) $R^{\dot{A}}_{+z(\cdot)} = R^{\dot{A}}_{z(\cdot)}$ quasi everywhere. If $y(\cdot)$ is a positive almost surely right continuous supermartingale majorizing $z(\cdot)$ on $\dot{A}$, then under the convention that all processes are 0 at the parameter value $+\infty$,

$$y(t) \geq E\{y(S)|\mathcal{F}(t)\} \geq E\{z(S)|\mathcal{F}(t)\} \quad \text{a.s. on } \{S > t\}. \tag{20.1}$$

The process $z_0(\cdot)$ defined by

$$z_0(t) = \begin{cases} E\{z(S)|\mathscr{F}(t)\} & \text{if } t < S \\ z(t) & \text{if } S \le t < T \\ 0 & \text{if } +\infty > t \ge T \end{cases}$$

is a positive right continuous supermartingale if the conditional expectations are defined suitably (Theorem 2), is equal to $z(\cdot)$ on $\dot{A}$, and is a quasi everywhere minorant of $y(\cdot)$. Hence $z_0(\cdot)$ is a version of $R^{\dot{A}}_{z(\cdot)}$.

EXAMPLE (b). If $\dot{A}$ is as in Example (a) and $\dot{B} = [\![S, T]\!]$, then any positive right continuous supermartingale $y(\cdot)$ majorizing $z(\cdot)$ on $\dot{B}$ majorizes $R^{\dot{A}}_{z(\cdot)}$ quasi everywhere and satisfies (20.1). Moreover for $\varepsilon > 0$ the process $y(\cdot)1_{[0,T+\varepsilon[}(\cdot)$ is also a positive right continuous supermartingale majorizing $z(\cdot)$ on $\dot{B}$. It follows that quasi everywhere

$$R^{\dot{B}}_{z(\cdot)}(t) = \begin{cases} E\{z(S)|\mathscr{F}(t)\} & \text{if } t < S, \\ z(t) & \text{if } S \le t \le T,\ t < +\infty, \\ 0 & \text{if } +\infty > t > T, \end{cases}$$

$R^{\dot{B}}_{+z(\cdot)} = R^{\dot{A}}_{+z(\cdot)}$ quasi everywhere, and $R^{\dot{B}}_{+z(\cdot)} < R^{\dot{B}}_{z(\cdot)}$ on $[\![T]\!]$, a semipolar set.

EXAMPLE (c). If $0 \le a < b$ and if $\Lambda \in \mathscr{F}(0)$ when $a = 0$ but $\Lambda \in \bigcup_{t<a} \mathscr{F}(t)$ when $a > 0$, then $[a, b] \times \Lambda = [\![S, T]\!]$ with $S = a$ on Λ and $S = +\infty$ otherwise, $T = b$ on Λ and $T = +\infty$ otherwise; so the reduction evaluation in Example (b) is applicable.

Theorem. *Suppose that $(\Omega, \mathscr{F}, \mathscr{F}(\cdot), P)$ has the property that every almost surely right continuous supermartingale is almost surely lower semicontinuous, and suppose that $z(\cdot)$ is an almost surely continuous positive supermartingale. Let $\dot{A}$ be predictable, and let T'_t $[T_t]$ be the entry [hitting] time of $\dot{A} \cap [\![t, +\infty]\!]$. Then for each $t \ge 0$*

$$R^{\dot{A}}_{z(\cdot)}(t) = E\{z(T'_t)|\mathscr{F}(t)\} \quad \text{a.s.} \tag{20.2}$$

$$R^{\dot{A}}_{+z(\cdot)}(t) = E\{z(T_t)|\mathscr{F}(t)\} \quad \text{a.s.} \tag{20.2sm}$$

It is sufficient to prove this theorem for $z(\cdot)$ bounded since (20.2) and (20.2sm) are true for $z(\cdot)$ if true for $z(\cdot) \wedge n$ when $n \in \mathbb{Z}^+$. Moreover the truth of (20.2) implies that of (20.2sm) by the following argument. Since $R^{\dot{A}}_{+z(\cdot)}$ is the smoothing of $R^{\dot{A}}_{z(\cdot)}$, there is a countable subset of $\mathbb{R}^+$ such that for t not in this set $R^{\dot{A}}_{+z(\cdot)}(t) = R^{\dot{A}}_{z(\cdot)}(t)$ up to a null set which depends on t. Now the process $T_\bullet$ is the smoothing of $T'_\bullet$, $T_t = \lim_{s \downarrow t} T'_s$; so there is a countable subset of $\mathbb{R}^+$ such that for t not in this set $T_t = T'_t$ up to a null set which depends on t. Thus, if (20.2) is true, it follows that (20.2sm) is also true for t

not in some countable set. Finally if $s \geq 0$ and if t decreases sequentially to s, not hitting the exceptional countable set, (20.2) yields (20.2sm) for s, in view of the almost sure right continuity of the left side and the dominated convergence theorem for conditional expectations as applied on the right.

We proceed to prove (20.2) for $z(\cdot)$ bounded. Let $\dot{A}_{\bullet}$ be a sequence of sets for which (20.2) is true. To prove (20.2) for arbitrary predictable $\dot{A}$, it is sufficient to prove the following two assertions.

(i) If $\dot{A}_0 \subset \dot{A}_1 \subset \cdots$, and $\bigcup_0^\infty \dot{A}_n = \dot{A}$, then (20.2) is true.
(ii) If $\dot{A}_0 \supset \dot{A}_1 \supset \cdots$, and $\bigcap_0^\infty \dot{A}_n = \dot{A}$, and if each set $\dot{A}_n$ is progressively measurable and has compact t cross sections for fixed ω, then (20.2) is true.

In fact, if Γ_p is the class of finite unions of the sets of the type considered in Example (c), it follows from (ii) that (20.2) is true for countable intersections of Γ_p sets, and therefore by (i) combined with Section 18(m) it follows that (20.2) is true for every predictable set $\dot{A}$. To prove (i) and (ii), let S_{nt} be the entry time of $\dot{A}_n \cap [\![T, +\infty[\![$. In (i) $S_{0t} \geq S_{1t} \geq \cdots$, and $\lim_{n\to\infty} S_{nt} = T_t'$. On the one hand, $\lim_{n\to\infty} R_{z(\cdot)}^{\dot{A}_n} = R_{z(\cdot)}^{\dot{A}}$ quasi everywhere by Section 18(h). On the other hand (dominated convergence theorem for conditional expectations),

$$\lim_{n\to\infty} E\{z(S_{nt})|\mathscr{F}(t)\} = E\{z(T_t')|\mathscr{F}(t)\} \quad \text{a.s.} \tag{20.3}$$

so (20.2) for $\dot{A}_n$ becomes (20.2) for $\dot{A}$ when $n \to \infty$. In (ii) $S_{0t} \leq S_{1t} \leq \cdots$, and $\lim_{n\to\infty} S_{nt} = T_t'$. On the one hand, $\lim_{n\to\infty} R_{z(\cdot)}^{\dot{A}_n} = R_{z(\cdot)}^{\dot{A}}$ by (18.7‴). On the other hand, (20.3) is true; so again (20.2) for $\dot{A}_n$ becomes (20.2) for $\dot{A}$ when $n \to \infty$.

Observation. The reduction $R_{z(\cdot)}^{\dot{A}}$ is identified in (20.2) only up to a standard modification. In fact, however, we have defined $R_{z(\cdot)}^{\dot{A}}$ as a very special standard modification, unique up to an evanescent set.

21. The Energy of a Supermartingale Potential

Let $\{A(\cdot), \mathscr{F}(\cdot)\}$ be an almost surely right continuous increasing predictable L^1 bounded process generating the almost surely right continuous supermartingale potential $\{x(\cdot), \mathscr{F}(\cdot)\}$. In view of the classical definition (Section 1.XVII.2) of the energy of a measure it is natural to define the energy of $A(\cdot)$ by $\int_0^\infty x(t)\,dA(t)$, but the standard more useful definition is

$$\|A(\cdot)\|^2 = \tfrac{1}{2}\int_0^\infty [x(t) + x(t-)]\,dA(t), \tag{21.1}$$

and we shall now show that if $E\{A(+\infty)^2\} < +\infty$, then according to this definition,

$$\|A(\cdot)\|^2 = \tfrac{1}{2}E\{A(+\infty)^2\}. \tag{21.2}$$

Suppose first that $A(\cdot)$ is bounded. Choose the conditional expectation $y(t) = E\{A(+\infty)|\mathscr{F}(t)\}$ in such a way that $\{y(\cdot), \mathscr{F}(\cdot)\}$ is a right continuous martingale. Then

$$\int_0^\infty x(t\pm)\,dA(t) = \int_0^\infty y(t\pm)\,dA(t) - \int_0^\infty A(t\pm)\,dA(t). \tag{21.3}$$

Since $A(\cdot)$ is bounded, the processes $x(\cdot)$ and $y(\cdot)$ are also bounded; so (by Theorem 7)

$$E\left\{\int_0^\infty y(t\pm)\,dA(t)\right\} = E\{y(+\infty)A(+\infty)\} = E\{A(+\infty)^2\}. \tag{21.4}$$

An application of Fubini's theorem to the product measure $d_tA(t,\omega) \times d_sA(s,\omega)$ yields

$$E\left\{\int_0^\infty A(t\pm)\,dA(t)\right\} = \tfrac{1}{2}E\left\{A(+\infty)^2 + \tfrac{1}{2}\int_0^\infty [A(t\pm) - A(t\mp)]\,dA(t)\right\}$$

and thereby yields (21.2). If $A(+\infty)$ is not bounded, let (Section 7) $S_\cdot$ be an increasing sequence of finite optional times for which $\lim_{n\to\infty} S_n = +\infty$ and for which $A(S_n)$ is bounded for each n. Then (21.2) for the increasing process $A(S_n \wedge \cdot)$ yields (21.2) for $A(\cdot)$ when $n \to \infty$.

22. The Subtraction of a Supermartingale Discontinuity

Let $\{x(t), \mathscr{F}(t), t \in \mathbb{R}^+\}$ be a right closable almost surely right continuous supermartingale, suppose that $\mathscr{F}(0)$ contains the null sets, and define $\mathscr{F}(+\infty) = \bigvee_{t\in\mathbb{R}^+} \mathscr{F}(t)$. According to Section III.15 the random variable $x(+\infty) = \lim_{t\to\infty} x(t)$ right closes the supermartingale. In the following discussion $\mathscr{F}(\cdot)$ will always be the reference filtration but will not always be mentioned explicitly. Almost every $x(\cdot)$ sample function necessarily has a left limit at every strictly positive parameter value. The left limit process, defined as $x(0)$ at the parameter value 0 and defined arbitrarily at strictly positive parameter values when the left limit does not exist, will be denoted by $x(\cdot -)$. Let T be a predictable optional time and define

$$J = \begin{cases} x(T-) - x(T) & \text{on } \{T < +\infty\}, \\ 0 & \text{on } \{T = +\infty\}, \end{cases}$$

and for $t \in \mathbb{R}^+$ define $y(t) = J 1_{\{T \le t\}}$ ($=0$ on the set $\{T = +\infty\}$). Let $T_{\bullet}$ be a sequence of optional times predicting T. Then

(a) *the process* $x(\cdot) + y(\cdot)$ *is an almost surely right continuous supermartingale, almost surely continuous at* $T|_{\{T<+\infty\}}$, *right closable by* $x(+\infty) + J$, *and* $E\{J\} \le E\{x(0) - x(+\infty)\}$.

In fact according to Section 3 (extension of Theorem 3), $x(T-)$ is integrable, and the following supermartingale inequalities are true:

$$E\{x(+\infty)\} \le E\{x(T)\} \le E\{x(T-)\} \le E\{x(0)\}. \tag{22.1}$$

Hence J is integrable, and $E\{J\}$ satisfies the stated inequality. To prove that $x(\cdot) + y(\cdot)$ is a supermartingale observe that $y(\cdot)$ is adapted to $\mathscr{F}(\cdot)$ and that $x(\cdot) + y(\cdot)$ satisfies the supermartingale inequality if and only if $0 \le s < t$ implies that

$$E\{x(t) + J 1_{\{s<T\le t\}} | \mathscr{F}(s)\} \le x(s) \quad \text{a.s.} \tag{22.2}$$

The sequence $(T_{\bullet} \vee s) \wedge t$ is an increasing sequence of optional times with limit $(T \vee s) \wedge t$. Define $\hat{x} = \lim_{n\to\infty} x((T_n \vee s) \wedge t)$. According to Section 3 (extension of Theorem 3), the ordered triple

$$[x(s), \mathscr{F}(s)], \left[\hat{x}, \bigvee_0^\infty \mathscr{F}((T_n \vee s) \wedge t)\right], [x((T \vee s) \wedge t), \mathscr{F}((T \vee s) \wedge t)]$$

is a supermartingale, and it follows that

$$x(s) \ge E\{\hat{x}|\mathscr{F}(s)\} \ge E\{x((T \vee s) \wedge t)|\mathscr{F}(s)\} \ge E\{x(t)|\mathscr{F}(s)\} \quad \text{a.s.} \tag{22.3}$$

Hence

$$E\{\hat{x} - x((T \vee s) \wedge t)|\mathscr{F}(s)\} \le x(s) - E\{x(t)|\mathscr{F}(s)\} \quad \text{a.s.} \tag{22.4}$$

and this inequality leads to (22.2) because $\hat{x} - x((T \vee s) \wedge t) = J 1_{\{s<T\le t\}}$ almost surely. Thus $x(\cdot) + y(\cdot)$ is a supermartingale. An elementary calculation shows that this supermartingale is right closed by the random variable $x(+\infty) - |J|$ and therefore is right closed by the supermartingale limit $x(+\infty) + J$.

Finally we prove

(b) *If* $x(T)$ *is* $\bigvee_0^\infty \mathscr{F}(T_n)$ *measurable, for example (Section II.3), if* $x(\cdot)$ *is nearly predictable, it follows that* $J \ge 0$ *almost surely and that the process* $y(\cdot)$ *is nearly predictable.*

The almost sure inequality $E\{x(T)|\mathscr{F}(T_n)\} \le x(T_n)$ yields $(n \to \infty)$ $x(T) \le x(T-)$ almost surely; that is, $J \ge 0$ almost surely. We prove that $y(\cdot)$ is nearly predictable by proving that each of the two processes with respective tth random variables

$$x(T-)1_{\{T\le t\}}, \qquad x(T)1_{\{T\le t\}} \tag{22.5}$$

(defined as 0 on $\{T = +\infty\}$) is nearly predictable. The processes with respective tth random variables

$$x(T_n)1_{\{T_n<t\}}, \qquad E\{x(T)|\mathscr{F}(T_n)\}1_{\{T_n<t\}} \tag{22.6}$$

are nearly predictable because they are adapted and left continuous; so their almost sure limit processes ($n \to \infty$) are nearly predictable. The first almost sure limit process is that with tth random variable the first in (22.5). The second almost sure limit process is that with tth random variable the second in (22.5). Thus $y(\cdot)$ is nearly predictable, as asserted.

23. Supermartingale Decompositions and Discontinuities

All processes in this section have parameter set $\mathbb{R}^+$, and as in Section 22 the left limit process of an almost surely right continuous supermartingale $x(\cdot)$ will be denoted by $x(\cdot-)$. Recall that almost sure lower semicontinuity of $x(\cdot)$ means that $x(t-) \ge x(t)$ almost surely, simultaneously for all t.

Theorem. *Let $\mathscr{F}(\cdot)$ be a right continuous filtration of a probability space, and suppose that $\mathscr{F}(0)$ contains the null sets.*

(a) *A nearly predictable almost surely right continuous supermartingale $\{x(\cdot), \mathscr{F}(\cdot)\}$ is almost surely lower semicontinuous, almost surely continuous if the process is a martingale.*

(b) *If $x(\cdot)$ in* (a) *is right closable it can be written as a sum $x''(\cdot) - x'(\cdot)$ of processes adapted to $\mathscr{F}(\cdot)$, nearly predictable, for which*

(b1) *Almost every $x'(\cdot)$ sample function is monotone increasing, constant except for the $-x(\cdot)$ sample function jumps, and $x'(0) = 0$.*

(b2) *$x''(\cdot)$ is an almost surely continuous supermartingale and is right closable.*

(c) *If every $\mathscr{F}(\cdot)$ optional time is predictable and if $\mathscr{F}(\cdot)$ is predictable, then every almost surely right continuous supermartingale $\{x(\cdot), \mathscr{F}(\cdot)\}$ is nearly predictable and therefore is almost surely lower semicontinuous, almost surely continuous if the process is a martingale.*

Proof of (a). Since it is sufficient to prove (a) for $x(\cdot \wedge n)$ for all $n > 0$, we can assume that $x(\cdot)$ is a right closable supermartingale, or a right closable martingale if the process is a martingale. The process $\{x(\cdot-), \mathscr{F}(\cdot)\}$ is almost surely left continuous and therefore nearly predictable so the process $\{x(\cdot) - x(\cdot-), \mathscr{F}(\cdot)\}$ is nearly predictable, and it follows that when $c > 0$, the set

$$\dot{H}_c = \{(t, \omega)\colon x(t, \omega) - x(t-, \omega) \ge c\}$$

is a nearly predictable set. Furthermore the graph of the entry time T of $\dot{H}_c$ is in $\dot{H}_c$ up to an evanescent set; so (by Theorem II.9) T is nearly predictable. Hence $x(T-) \geq x(T)$ almost everywhere where $T < +\infty$ according to Section 22(b), and this is impossible unless $T = +\infty$ almost surely. We conclude that the supermartingale assertion of (a) is true. If $x(\cdot)$ is a (right closable) martingale, this result applied to $x(\cdot)$ and $-x(\cdot)$ shows that the martingale assertion of (a) is true.

It will be convenient to prove (c) before (b). □

Proof of (c). Suppose first that $x(\cdot)$ is right closable as a supermartingale. If $n \in \mathbb{Z}^+$ and $k \in \mathbb{Z}$, define $T_{0k} = 0$, and if $T_{n-1\,k}$ is already defined, define $T_{nk} = +\infty$ if $T_{n-1\,k} = +\infty$ and

$$T_{nk} = \inf\{t > T_{n-1\,k}: 2^k < |x(t) - x(t-)| \leq 2^{k+1}\} \quad \text{otherwise.}$$

Then T_{nk} is optional, therefore predictable by hypothesis, and therefore by Section 22(b) $x(T_{nk}-) > x(T_{nk})$ when $n > 0$ and $T_{nk} < +\infty$. Define $J_{nk} = x(T_{nk}-) - x(T_{nk})$ when $T_{nk} < +\infty$ and $J_{nk} = 0$ otherwise, and define a right continuous process $x_{nk}(\cdot)$ by

$$x_{nk}(t) = J_{nk} 1_{\{T_{nk} \leq t\}} \qquad (=0 \text{ when } T_{nk} = +\infty) \text{ for } t \in \mathbb{R}^+.$$

Since $x(\cdot)$ is right closable it is right closable by $\lim_{n\to\infty} x(t)$ which we denote by $x(+\infty)$. According to Section 22, $J_{nk} \geq 0$ almost surely, the process $x_{nk}(\cdot)$ is nearly predictable [the reference filtration is $\mathscr{F}(\cdot)$ here and below], the process $x(\cdot) + x_{nk}(\cdot)$ is a right closable (by $x(+\infty) + J_{nk}$) almost surely right continuous supermartingale, and $E\{J_{nk}\} \leq E\{x(0) - x(+\infty)\}$. If the jump processes are added successively to $x(\cdot)$ and if we define (summing over all n and k) $J = \Sigma J_{nk}$ and $x'(t) = \Sigma x_{nk}$, then $E\{J\} \leq E\{x(0) - x(+\infty)\}$. It follows that J and $x'(t)$ are almost surely finite. The process $x'(\cdot)$ is nearly predictable, and its sample functions are almost all right continuous except for positive jumps J_{nk} at T_{nk}. The process $x''(\cdot) = x(\cdot) + x'(\cdot)$ is almost surely continuous and therefore is nearly predictable, and we conclude that $x(\cdot)$ (supposed right closable) is nearly predictable. The supermartingale $x''(\cdot)$ is right closable by $x(+\infty) + J$. If $x(\cdot)$ is not right closable, this result implies that $x(\cdot \wedge n)$ is nearly predictable for all strictly positive n; so $x(\cdot)$ is nearly predictable. □

Proof of (b). In the proof of (c) a decomposition of $x(\cdot)$ was obtained satisfying (b1) and (b2), under the hypothesis that $x(\cdot)$ was a right closable almost surely continuous supermartingale satisfying certain conditions on the reference filtration $\mathscr{F}(\cdot)$. A glance at the proof shows that wherever these conditions on $\mathscr{F}(\cdot)$ are used, near predictability of $x(\cdot)$ suffices. □

Chapter V

Lattices and Related Classes of Stochastic Processes

1. Conventions; The Essential Order

In this chapter certain stochastic process classes which arise naturally in martingale theory will be discussed. These classes and their relations with the identically named classes in Chapter IX of Part 1 will be studied in later chapters. See Appendix III for the lattice theory to be used.

All stochastic process concepts in this chapter are relative to a specified filtered probability space $(\Omega, \mathscr{F}, P; \mathscr{F}(t), t \in I)$, where I is a linearly ordered set, arbitrary unless specifically limited. As always the probability measure is supposed complete; in addition it is supposed that each σ algebra $\mathscr{F}(t)$ contains the null sets. The stochastic processes considered are adapted to $\mathscr{F}(\cdot)$ and have state space $(\overline{\mathbb{R}}, \mathscr{B}(\overline{\mathbb{R}}))$.

Recall from Section I.8 that in the essential order stochastic processes are grouped into equivalence classes by identifying processes which are standard modifications of each other and a process $y(\cdot)$ is an essential order majorant of $x(\cdot)$ if

$$P\{y(t) \geq x(t)\} = 1, \qquad t \in I.$$

Recall further that the essential order infimum $\operatorname{ess\,inf} \Gamma$ of a set Γ of stochastic processes, that is, the essential order infimum of their equivalence classes, can be obtained as follows. If we write $x(\cdot) \in \Gamma$ to mean that the equivalence class containing $x(\cdot)$ is in Γ, in other words that $x(\cdot)$ is a version of an equivalence class in Γ (see the remark on the abuse of notation in Section I.1), then $\operatorname{ess\,inf} \Gamma$ is the equivalence class consisting of all the versions of

$$\{\operatorname*{ess\,inf}_{x(\cdot) \in \Gamma} x(t), t \in I\}.$$

Recall that the essential infimum of a family of random variables is determined only up to a null set. The class of processes in the essential order is a complete lattice for I a singleton and therefore for arbitrary I.

The Continuous Parameter Context

This is the context in which $I = \mathbb{R}^+$, $\mathscr{F}(\cdot)$ is right continuous, $\mathscr{F}(0)$ contains the null sets, and the processes to be classified will be almost surely right continuous. Recall that two almost surely right continuous processes on the parameter set $\mathbb{R}^+$ which are standard modifications of each other are indistinguishable, that is, equal quasi everywhere on $\Omega \times \mathbb{R}^+$.

2. LM $x(\cdot)$ when $\{x(\cdot), \mathscr{F}(\cdot)\}$ Is a Submartingale

(See Section 1.IX.2 for the potential theory counterpart of this section.)

Let $\{x(\cdot), \mathscr{F}(\cdot)\}$ be a submartingale with an arbitrary linearly ordered parameter set I having a first element. If S and T are countably valued optional times, upper bounded in I, with $S \le T$, then $x(\cdot) \le \tau_S x(\cdot) \le \tau_T x(\cdot)$ in the essential order. According to the dual of the *Special case* in Section III.20, either $x(\cdot)$ has no essential order supermartingale majorant and $\lim_{s\uparrow} E\{x(s)\} = +\infty$, or $x(\cdot)$ has an essential order supermartingale majorant, LM $x(\cdot)$ exists, the process $\{\text{LM}\, x(\cdot), \mathscr{F}(\cdot)\}$ is a martingale,

$$\text{LM}\, x(\cdot) = \operatorname*{ess\,lim}_{s\uparrow} \tau_s x(\cdot) = \operatorname*{ess\,lim}_{s\uparrow} E\{x(s) \mid \mathscr{F}(\cdot)\} \tag{2.1}$$

up to a standard modification, and

$$E\{\text{LM}\, x(t)\} = \lim_{s\uparrow} E\{x(s)\} = \sup_{s \in I} E\{x(s)\}. \tag{2.1'}$$

(As usual we write LM $x(t)$ for $[\text{LM}\, x(\cdot)](t)$.) In view of one of the forms of the submartingale sampling theorem (Theorem III.7) the supremum and directed limit

$$\sup\{E\{x(S)\}: S \text{ optional, countably valued, bounded}\} \tag{2.2}$$

is equal to the supremum in (2.1′). Thus LM $x(\cdot)$ exists if and only if the set of expectations of the random variable class

$$\{x(S): S \text{ optional, countably valued, bounded}\} \tag{2.3}$$

is bounded.

If $\{x(\cdot), \mathscr{F}(\cdot)\}$ is a positive submartingale, the existence of LM $x(\cdot)$ is equivalent to the L^1 boundedness of the random variable class (2.3) and in this case a trivial argument shows that S in (2.3) need not be bounded in I. Moreover the hypothesis that I has a first element can be dropped because in any case the process is left closable by the random variable 0 coupled with the trivial σ algebra $(\varnothing, \Omega)$. In particular, suppose that the positive submartingale is uniformly integrable; that is, suppose that there is a uniform integrability test function Φ for which

$$\sup_{s\in I} E\{\Phi[x(s)]\} < +\infty. \tag{2.4}$$

In this case LM $x(\cdot)$ exists, and, since $\{\Phi[x(\cdot)], \mathscr{F}(\cdot)\}$ is a positive submartingale, (2.4) is true if and only if LM $\Phi[x(\cdot)]$ exists. We now show that then LMΦ[LM $x(\cdot)$] also exists and

$$\text{LM}\,\Phi[\text{LM}\,x(\cdot)] = \text{LM}\,\Phi[x(\cdot)] \tag{2.5}$$

up to a standard modification. To see this observe that up to standard modifications

$$\begin{aligned} E\{\Phi[\text{LM}\,x(s)]|\mathscr{F}(\cdot)\} &= E\{\Phi[\operatorname*{ess\,lim}_{s'\uparrow} E\{x(s')|\mathscr{F}(s)\}]|\mathscr{F}(\cdot)\} \\ &\le E\{\operatorname*{ess\,lim}_{s'\uparrow} E\{\Phi[x(s')]|\mathscr{F}(s)\} \,|\, \mathscr{F}(\cdot)\} \\ &= E\{\text{LM}\,\Phi[x(s)]|\mathscr{F}(\cdot)\}. \end{aligned}$$

Take the essential limit as s increases to find that the left-hand side of (2.5) is an essential order minorant of the right-hand side. The reverse order relation is trivial.

Continuous Parameter Context

Recall from Section IV.14 that in the continuous parameter context (defined in Section 1) if $\{x(\cdot), \mathscr{F}(\cdot)\}$ is an almost surely right continuous submartingale for which LM $x(\cdot)$ exists, this martingale majorant can be chosen to be right continuous; the notation LM $x(\cdot)$ will always refer to an almost surely right continuous version. Moreover, if $\{x(\cdot), \mathscr{F}(\cdot)\}$ is a positive almost surely right continuous submartingale, not only is the existence of LM $x(\cdot)$ equivalent to the finiteness of

$$\sup\{E\{x(S)\}: S \text{ optional } (< +\infty) \text{ countably valued}\}, \tag{2.6}$$

as already stated in the general context, but the restriction that S be countably valued is unnecessary. In fact if the supremum in (2.6) is c, let T be an arbitrary optional time except that $T < +\infty$. Then if $[T]_n$ is defined as in Section II.2, Example (b2), it follows from Fatou's lemma that

$$E\{x(T)\} \le \liminf_{n\to\infty} E\{x([T]_n)\} \le c.$$

If an integrable random variable $x(+\infty)$ can be adjoined to right close the submartingale, as is possible if the submartingale is uniformly integrable

[Section III.3(e)], then this argument is valid for an arbitrary optional time $T \leq +\infty$; so in this case S in (2.6) can be an entirely unrestricted optional time.

3. Uniformly Integrable Positive Submartingales

In Section 1.IX.3 a class $\mathbf{D}(\mu_{D-}^h)$ of functions on a Greenian subset of $\mathbb{R}^N$ was defined, and Theorem 1.IX.3 treated functions u for which $u \in \mathbf{D}(\mu_{D-}^h)$. The applications in view were to positive h-subharmonic functions and to h-harmonic functions. The probabilistic counterparts of class $\mathbf{D}(\mu_{D-}^h)$ functions on a specified set are the class $\mathbf{D}$ stochastic processes on a probability space, relative to a given parameter set and filtration, as defined in Section II.11. The following theorem is the probabilistic counterpart of Theorem 1.IX.3, and the applications in view are to positive submartingales and to martingales. All the processes in the theorem are defined on the same filtered probability space.

Theorem. *Let $\{x(\cdot), \mathscr{F}(\cdot)\}$ be a stochastic process with state space $(\overline{\mathbb{R}}, \mathscr{B}(\overline{\mathbb{R}}))$ and an arbitrary linearly ordered parameter set. If $\{|x(\cdot)|, \mathscr{F}(\cdot)\}$ is a submartingale, the following conditions are equivalent:*

(a) $x(\cdot) \in \mathbf{D}$.
(b) *$x(\cdot)$ is uniformly integrable.*
(c) *There is a uniform integrability test function Φ for which the submartingale $\{\Phi[|x(\cdot)|], \mathscr{F}(\cdot)\}$ has a martingale essential order majorant.*
(d) *(If $\{x(\cdot), \mathscr{F}(\cdot)\}$ is a martingale) $x(\cdot) = x_1(\cdot) - x_2(\cdot)$, where $\{x_i(\cdot), \mathscr{F}(\cdot)\}$ is a positive martingale, in $\mathbf{D}$, and therefore uniformly integrable.*
(e) *The submartingale $\{|x(\cdot)|, \mathscr{F}(\cdot)\}$ is right closable.*

Moreover, if $x(\cdot)$ satisfies (a) *and* (b), *and if Φ satisfies* (c), *then the martingale $\{\mathrm{LM}\,|x(\cdot)|, \mathscr{F}(\cdot)\}$ is uniformly integrable,*

$$\mathrm{LM}\,\Phi[\mathrm{LM}\,|x(\cdot)|] = \mathrm{LM}\,\Phi[|x(\cdot)|] \tag{3.1}$$

up to a standard modification, and, if $\{x(\cdot), \mathscr{F}(\cdot)\}$ is a martingale, each process $x_i(\cdot)$ in (d) *can be chosen so that the submartingale $\{\Phi[x_i(\cdot)], \mathscr{F}(\cdot)\}$ has a martingale essential order majorant.*

If $|x(\cdot)|$ here is identified with $x(\cdot)$ in Section III.2 and Section 2, the present theorem follows; if $\{x(\cdot), \mathscr{F}(\cdot)\}$ is a martingale, one representation of $x(\cdot)$ with the properties stated in (e) and the final assertion of the theorem is

$$x(\cdot) = \mathrm{LM}\,|x(\cdot)| - [\mathrm{LM}\,|x(\cdot)| - x(\cdot)]. \tag{3.2}$$

(See the corresponding discussion in the proof of Theorem 1.IX.3.)

Theorem 3 versus Theorem 1.IX.3. The parallelism between these two theorems is obvious except that no potential theory counterpart of Theorem 3(e) has been suggested. To find such a counterpart, suppose that $|u|$ is a class $\mathbf{D}(\mu_{D-}^h)$ h-subharmonic function on a connected Greenian subset D of $\mathbb{R}^N$, as in Theorem 1.IX.3, and define $v = \mathrm{LM}_D^h|u|$, a class $\mathbf{D}(\mu_{D-}^h)$ h-harmonic function according to that theorem. The function $v - |u|$ is an h-potential because

$$\mathrm{GM}_D^h(v - |u|) = v - \mathrm{LM}_D^h|u| = 0;$$

so (Theorem 1.XII.18) the function $v - |u|$ has μ_D^h almost everywhere minimal-fine boundary limit 0 on the Martin boundary $\partial^M D$. According to Theorem 1.XII.19, a quasi-bounded h-harmonic function on D is the PWB^h solution H_f^h for some h-resolutive boundary function f on $\partial^M D$, and H_f^h has f as its μ_D^h almost everywhere minimal-fine Martin boundary limit function. Moreover we shall show (Theorem 3.I.5) that an h-harmonic function on D is quasi bounded if and only if it is in $\mathbf{D}(\mu_{D-}^h)$. Hence there is an h-resolutive function f on $\partial^M D$ which is the μ_D^h almost everywhere minimal-fine boundary limit function of both v and $|u|$, and $|u| \le H_f^h = v$. The μ_D^h measurable and integral Martin boundary functions f_1, that is, the h-resolutive Martin boundary functions f_1, for which $f_1 \ge f$ are potential theory counterparts of the random variables which close the submartingale $\{|x(\cdot)|, \mathscr{F}(\cdot)\}$ in Theorem 3(e).

Continuous Parameter Context

In this context, as already noted in Section 2, when $x(\cdot)$ in Theorem 3 is supposed almost surely right continuous, all the least majorants in the theorem can be supposed almost surely right continuous and in view of (3.2) each process $x_i(\cdot)$ in Theorem 3(e) can be chosen to be almost surely right continuous and to satisfy the last assertion of the theorem.

4. L^p Bounded Stochastic Processes ($p \ge 1$)

The following theorem is in the context of Theorem 3 and when $p > 1$ is merely a specialization of that theorem. Recall that a process $\{x(t), t \in I\}$ is called L^p bounded if $\sup_{t \in I} E\{|x(t)|^p\} < +\infty$.

Theorem. *Let $\{x(\cdot), \mathscr{F}(\cdot)\}$ be a stochastic process with state space $(\bar{\mathbb{R}}, \mathscr{B}(\bar{\mathbb{R}}))$ and an arbitrary linearly ordered parameter set, and suppose that $p \ge 1$. If $\{|x(\cdot)|, \mathscr{F}(\cdot)\}$ is a submartingale, the following conditions are equivalent:*

(a) *$x(\cdot)$ is L^p bounded.*
(b) *The submartingale $\{|x(\cdot)|^p, \mathscr{F}(\cdot)\}$ has a martingale essential order majorant.*

(c) *(If* $\{x(\cdot), \mathscr{F}(\cdot)\}$ *is a martingale)*; $x(\cdot) = x_1(\cdot) - x_2(\cdot)$, *where* $\{x_i(\cdot), \mathscr{F}(\cdot)\}$ *is a positive* L^p *bounded martingale.*

(d) *(If* $p > 1$*); the submartingale* $\{|x(\cdot)|, \mathscr{F}(\cdot)\}$ *is right closable by a random variable in* L^p.

Moreover, if $p \geq 1$ *and if* $x(\cdot)$ *satisfies* (a) *and* (b), *then the martingale* $\{\mathrm{LM}|x(\cdot)|, \mathscr{F}(\cdot)\}$ *is* L^p *bounded,*

$$\mathrm{LM}[(\mathrm{LM}|x(\cdot)|)^p] = \mathrm{LM}(|x(\cdot)|^p) \tag{4.1}$$

up to a stochastic modification, and if $\{x(\cdot), \mathscr{F}(\cdot)\}$ *is a martingale, each process* $x_i(\cdot)$ *in* (c) *can be chosen so that the submartingale* $\{x_i^p(\cdot), \mathscr{F}(\cdot)\}$ *has a martingale essential order majorant.*

The proof is left to the reader because whatever is not already covered by Theorem 3 with $\Phi(s) = s^p$ follows easily from the discussion in Section 2. Observe that assertion (c) of the present theorem with $p > 1$ is slightly stronger than Theorem 3(d) with $\Phi(s) = s^p$. In fact in Theorem 3(d) it is not asserted that if $x(\cdot) = x_1(\cdot) - x_2(\cdot)$ and if $\Phi[x_i(\cdot)]$ has a martingale essential order majorant, then $\Phi[|x(\cdot)|]$ has a martingale essential order majorant. However, for $\Phi(s) = s^p$ and $p \geq 1$ this assertion is true because then

$$|x(\cdot)|^p \leq [x_1(\cdot) + x_2(\cdot)]^p \leq 2^{p-1}[x_1^p(\cdot) + x_2^p(\cdot)].$$

Continuous Parameter Context

In this context (see the corresponding remarks in Section 3) if $x(\cdot)$ is almost surely right continuous, all the other processes involved in Theorem 4 can also be supposed almost surely right continuous.

Observation (Recall the definition of $\mathbf{L}^p$ *in Section* II.11.). It is trivial that a process in $\mathbf{L}^p$ is L^p bounded. Conversely (context of Theorem 4) if $x(\cdot)$ is L^p bounded with $p > 1$ and if $\{|x(\cdot)|, \mathscr{F}(\cdot)\}$ is a submartingale, it follows that $x(\cdot) \in \mathbf{L}^p$. In fact it is sufficient to prove that if T is an optional time with finitely many values then $E\{|x(T)|^p\} \leq \sup_{t \in I} E\{|x(t)|^p\}$, and this inequality is true because if s is the maximum value of T in the parameter set order then $E\{|x(T)|^p\} \leq E\{|x(s)|^p\}$.

5. The Lattices ('$\mathbf{S}^{\pm}$, ≤), ('$\mathbf{S}^{+}$, ≤), ($\mathbf{S}^{\pm}$, ≤), ($\mathbf{S}^{+}$, ≤)

(See the corresponding potential theory lattices in Section 1.IX.5.)

The Lattice ('$\mathbf{S}^{\pm}$, ≤)

We assume, as described in Section 1, a specified filtered probability space with an arbitrary linearly ordered parameter set. Denote by ('$\mathbf{S}^{\pm}$, ≤) the lattice, in the essential order, of those stochastic process equivalence classes under standard modification which contain supermartingales having positive supermartingale essential order majorants. Recall the essential order notation $\leq, \geq, =, \vee, \wedge$. Let Γ be a subset of $'\mathbf{S}^{\pm}$. If the set Γ_0 of supermartingales in the Γ equivalence classes has an essential order minorant in $'\mathbf{S}^{\pm}$, then (from Section III.5) the equivalence class containing the versions of $\operatorname{ess\,inf}\Gamma_0$ is $\bigwedge\Gamma$. It follows that $('\mathbf{S}^{\pm}, \leq)$ is a conditionally complete lattice, but observe that if Γ is a subset of $'\mathbf{S}^{\pm}$ with an essential order majorant in $'\mathbf{S}^{\pm}$, then $\bigvee\Gamma$ is not necessarily the equivalence class containing the essential suprema of the class of supermartingales in the equivalence classes of Γ but is the essential order infimum of the class of essential order $'\mathbf{S}^{\pm}$ majorants of Γ.

The careful language used above is correct but inconvenient, and we shall frequently follow the usual incorrect but convenient abuse of language in which, for example, Γ_0 above is identified with Γ and $\bigwedge\Gamma$ is described as the essential infimum of Γ even though Γ is not a set of processes but a set of equivalence classes and $\operatorname{ess\,inf}\Gamma_0$ is not an equivalence class but a process.

If $\Gamma \subset {'\mathbf{S}^{\pm}}$ and if Γ has an essential order minorant in $'\mathbf{S}^{\pm}$, then (Section III.5) $\bigwedge\Gamma = \bigwedge\Gamma_1$ for some countable subset Γ_1 of Γ. If $\Gamma \subset {'\mathbf{S}^{\pm}}$ and if Γ has an essential order majorant in $'\mathbf{S}^{\pm}$, then $\bigvee\Gamma = \bigvee\Gamma_1$ for some countable subset Γ_1 of Γ. In fact, if Γ is directed upward in the essential order, then $\bigvee\Gamma = \operatorname{ess\,sup}\Gamma$, and the assertion was proved in Section III.5 for essential suprema. If Γ is not directed upward, apply this result in the directed case to the set of $'\mathbf{S}^{\pm}$ suprema of finite subsets of Γ.

In the following, we sometimes prime $\bigvee$ and $\bigwedge$ when referring to $'\mathbf{S}^{\pm}$.

The Lattice $('\mathbf{S}^{+}, \leq)$

The sublattice $('\mathbf{S}^{+}, \leq)$ of $('\mathbf{S}^{\pm}, \leq)$ consists of the $'\mathbf{S}^{\pm}$ equivalence classes containing positive supermartingales.

The Continuous Parameter Context: $(\mathbf{S}^{\pm}, \leq)$, $(\mathbf{S}^{+}, \leq)$

Observe that if $I = \mathbb{R}^{+}$ an equivalence class in $'\mathbf{S}^{\pm}$ contains a right continuous supermartingale if the class contains an almost surely right continuous supermartingale and that two almost surely right continuous supermartingales in the same equivalence class are indistinguishable, that is, are equal quasi everywhere on $\mathbb{R}^{+} \times \Omega$, or in our other terminology are equal up to an evanescent subset of this product space. If $x(\cdot) \in {'\mathbf{S}^{\pm}}$, define

$$\underset{+}{x}(t) = \liminf_{r \downarrow t} x(t) \qquad (r \text{ rational}). \tag{5.1}$$

Then (Section IV.1) the process $\underset{+}{x}(\cdot)$ is an almost surely right continuous supermartingale, and

$$P\{x(t) = \underset{+}{x}(t)\} = 1$$

except possibly for a countable set of values of t. Moreover the limit inferior in (5.1) is an almost sure limit for each t. Let $\mathbf{S}^{\pm}$ be the set of equivalence classes of almost surely right continuous supermartingales, under the relation of indistinguishability. Then $\mathbf{S}^{\pm}$ can be imbedded in $'\mathbf{S}^{\pm}$ (in the present continuous parameter context) in an obvious way. If Γ is a subset of $\mathbf{S}^{\pm}$ and if we denote by $'\Gamma$ the subset of $'\mathbf{S}^{\pm}$ consisting of the equivalence classes of the latter set which contain those of Γ, we have noted above that if $\bigwedge '\Gamma$ exists there is a sequence $\{x_n(\cdot), n \in \mathbb{Z}^+\}$ in Γ such that the process $x(\cdot) = \inf_{n \geq 0} x_n(\cdot)$ determines $\bigwedge '\Gamma$. The process $x(\cdot)$ is almost surely right upper semicontinuous so $\underset{+}{x}(\cdot) \leq x(\cdot)$ in the essential order, and there may be strict inequality. The process $\underset{+}{x}(\cdot)$ is in the equivalence class of the maximal essential order $\mathbf{S}^{\pm}$ minorant of Γ. Thus, if we denote by $(\mathbf{S}^{\pm}, \leq)$ the set $\mathbf{S}^{\pm}$ in the essential order, this set becomes a conditionally complete lattice, for which we shall use the order symbols $\leq$, $\geq$, $\vee$, $\wedge$, but the $\mathbf{S}^{\pm}$ order infima and suprema are not inherited from the natural imbedding in $('\mathbf{S}^{\pm}, \leq)$. The argument just given together with the analysis of $('\mathbf{S}^{\pm}, \leq)$ shows that if $\Gamma \subset \mathbf{S}^{\pm}$ then the $(\mathbf{S}^{\pm}, \leq)$ infimum [supremum] of Γ, if it exists, is the infimum [supremum] of a countable subset of Γ. If Γ is countable and if Γ_0 is a set of supermartingales consisting of one member from each equivalence class in Γ, then the equivalence class $\bigwedge\Gamma$ has as one member the process $\underset{+}{x}(\cdot)$ for $x(\cdot)$ the pointwise infimum of Γ_0, and (Theorem IV.4) if Γ is directed upward and is bounded above in $(\mathbf{S}^{\pm}, \leq)$, the equivalence class $\bigvee\Gamma$ has as one member the pointwise supremum of Γ_0.

The Lattice $(\mathbf{S}^{+}, \leq)$

In the continuous parameter context the sublattice $(\mathbf{S}^{+}, \leq)$ of $(\mathbf{S}^{\pm}, \leq)$ consists of the $\mathbf{S}^{\pm}$ equivalence classes containing positive right continuous supermartingales.

The Natural Decomposition in the Continuous Parameter Context

The natural decomposition theorem (Section III.19) is valid in the continuous parameter context, in which all the supermartingales in the theorem are almost surely right continuous. In fact, if $x(\cdot)$, $x_1(\cdot)$, $x_2(\cdot)$ are almost surely right continuous positive supermartingales and if $x(\cdot) \leq x_1(\cdot) + x_2(\cdot)$ up to an evanescent set, then according to the natural decomposition theorem of Section III.19 there are positive supermartingales $x_1'(\cdot)$, $x_2'(\cdot)$ such that

$$x_i'(\cdot) \leq x_i(\cdot), \qquad x(\cdot) = x_1'(\cdot) + x_2'(\cdot)$$

up to a standard modification, and it follows that

$$\underset{+i}{x}'(\cdot) \leq x_i(\cdot), \qquad x(\cdot) = \underset{+1}{x}'(\cdot) + \underset{+2}{x}'(\cdot) \quad \text{q.e.}$$

6. The Vector Lattices ('**S**, $\preceq$) and (**S**, $\preceq$)

(See the corresponding potential theory vector lattice in Section 1.IX.6.)

The Vector Lattice ('**S**, $\preceq$)

The set '$\mathbf{S}^+$ is a cone as defined in Appendix III.3 and therefore defines a specific order on itself for which we use the order symbols $\preceq$, $\succeq$, $\curlyvee$, $\curlywedge$, and '$\mathbf{S}^+$ in the specific order will be denoted by ('$\mathbf{S}^+$, $\preceq$). Define '$\mathbf{S}$ = '$\mathbf{S}^+$ − '$\mathbf{S}^+$, so that each member of '**S** can be identified with an equivalence class of differences $x_1(\cdot) - x_2(\cdot)$ between two positive supermartingales. Each random variable $x_1(t) - x_2(t)$ is well defined off the probability null set of common infinities of $x_1(t)$ and $x_2(t)$. If '**S** is ordered by the specific order with positive cone '$\mathbf{S}^+$, we obtain a partially ordered vector space ('**S**, $\preceq$).

Theorem. (a) *The space ('**S**, $\preceq$) is a conditionally complete vector lattice.* [*In* (b), (c), (d) *let* Γ *be a subset of* '**S** *with a specific order majorant.*]

(b) $\curlyvee\Gamma$ *is the specific order supremum of a countable subset of* Γ.

(c) *If* Γ' *is the class of specific order majorants of* Γ, *then* $\vee\Gamma \preceq \Gamma'$.

(d) *If* Γ *is directed upward in the specific order, then* $\curlyvee\Gamma = \vee\Gamma$.

The duals of (b), (c), (d), involving infima, are obtained by replacing Γ by $-\Gamma$. Since Γ' is directed downward in the specific order in (c), the dual of (d) implies that $\wedge\Gamma' = \curlywedge\Gamma' = \curlyvee\Gamma$.

The reader will observe that this theorem has precisely the same statement as Theorem 1.IX.6, although **S** in that theorem does not have the same meaning as '**S** here. The point is that the contexts of the two theorems are quite different but the order properties in the two contexts are identical. The proof of Theorem 6 is simply a translation of that of Theorem 1.IX.6 into the present context. For example, to prove that $\wedge\Gamma' = \curlywedge\Gamma' = \curlyvee\Gamma$ if $\Gamma \subset$ '$\mathbf{S}^+$, we can follow the proof of this assertion in Section 1.IX.6. That is, we now interpret the members u, v, ϕ, ... of $\mathbf{S}^+$, Γ, Γ' in that section as positive supermartingales and interpret $\underset{+\phi}{R}^D$ there as a generalized probabilistic reduction, namely, as the equivalence class of essential infima of the set of positive supermartingale essential majorants of $\phi(\cdot)$. The details of the translated proof are left to the reader.

Continuous Parameter Context: The Vector Lattice (**S**, $\preceq$)

The set $\mathbf{S}^+$ is a cone and therefore defines a specific order on itself, for which we use the order symbols $\preceq$, $\succeq$, $\curlyvee$, $\curlywedge$, and $\mathbf{S}^+$ in the specific order will be denoted by $(\mathbf{S}^+, \preceq)$. If $x(\cdot)$ and $y(\cdot)$ are positive almost surely right continuous supermartingales, we describe $x(\cdot)$ as a specific minorant of $y(\cdot)$ and write $x(\cdot) \preceq y(\cdot)$ if this relation holds between the equivalence classes determined by the processes. The corresponding significance is given to $x(\cdot) \curlyvee y(\cdot)$ and other abuses of notation involving processes and equivalence classes.

Define $\mathbf{S} = \mathbf{S}^+ - \mathbf{S}^+$, and denote by $(\mathbf{S}, \preceq)$ the vector space $\mathbf{S}$ ordered by the specific order with positive cone $\mathbf{S}^+$. Then $\mathbf{S}$ can be identified with a subset of $'\mathbf{S}$. Moreover the following relations between $('\mathbf{S}, \preceq)$ and $(\mathbf{S}^+, \preceq)$ are true.

(r1) If $x(\cdot)$ and $y(\cdot)$ are in $\mathbf{S}^+$, then $x(\cdot) \preceq y(\cdot)$ in $('\mathbf{S}, \preceq)$ implies that $\underset{+}{x}(\cdot) \preceq \underset{+}{y}(\cdot)$ in $(\mathbf{S}, \preceq)$. In fact by hypothesis there is a $z(\cdot)$ in $'\mathbf{S}^+$ such that $x(\cdot) + z(\cdot)$ is a stochastic modification of $y(\cdot)$, and therefore $\underset{+}{x}(\cdot) + \underset{+}{z}(\cdot)$ is a stochastic modification of and in fact is indistinguishable from $\underset{+}{y}(\cdot)$; so $\underset{+}{x}(\cdot) \preceq \underset{+}{y}(\cdot)$ in $(\mathbf{S}, \preceq)$.

(r2) If $x(\cdot)$ and $y(\cdot)$ are in $\mathbf{S}^+$, then $x(\cdot) \preceq y(\cdot)$ in $(\mathbf{S}, \preceq)$ if and only if $x(\cdot) \preceq y(\cdot)$ in $('\mathbf{S}, \preceq)$. In fact "if" is trivial, and "only if" follows from (r1).

(r3) If $x(\cdot) \in {}'\mathbf{S}$ and if $x(\cdot)$ is almost surely right continuous, then $x(\cdot) \in \mathbf{S}$ because (up to a standard modification) by hypothesis $x(\cdot) = x_1(\cdot) - x_2(\cdot)$ with $x_i(\cdot) \in {}'\mathbf{S}^+$; so $x(\cdot) = \underset{+1}{x}(\cdot) - \underset{+2}{x}(\cdot)$ with $\underset{+i}{x}(\cdot) \in \mathbf{S}^+$.

If $\Gamma \subset \mathbf{S}^+$ we can identify Γ with a subset $'\Gamma$ of $'\mathbf{S}^+$. If $x(\cdot)$ is in the equivalence class $\bigwedge {}'\Gamma$ then $\underset{+}{x}(\cdot) \preceq {}'\Gamma$ in $('\mathbf{S}, \preceq)$ by (r1). Moreover, if $x_1(\cdot)$ is in $\mathbf{S}^+$ and if $x_1(\cdot) \preceq \Gamma$ in $(\mathbf{S}, \preceq)$, then $x_1(\cdot) \preceq \Gamma$ in $('\mathbf{S}, \preceq)$; so $x_1(\cdot) \preceq x(\cdot)$ in $('\mathbf{S}, \preceq)$, and therefore $x_1(\cdot) \preceq \underset{+}{x}(\cdot)$ in $(\mathbf{S}, \preceq)$. So $\wedge\,\Gamma$ exists and is the equivalence class in $\mathbf{S}^+$ containing $\underset{+}{x}(\cdot)$. Hence $(\mathbf{S}^+, \preceq)$ is a conditionally complete vector lattice. We leave to the reader the verification that Theorem 6(b)–(d) holds for $(\mathbf{S}, \preceq)$.

Intrinsic Definition of S

It will be shown in Section 13 that the processes in $\mathbf{S}$ can be characterized without involving supermartingales.

7. The Vector Lattices ($'\mathbf{S}_m, \preceq$) and ($\mathbf{S}_m, \preceq$)

(See the corresponding potential theory vector lattice in Section 1.IX.7.)

The Vector Lattice ($'\mathbf{S}_m, \preceq$)

A supermartingale specific order majorized by a martingale is itself a martingale. If Γ is a set of positive martingales with $\bigvee \Gamma = x(\cdot)$ then $x(\cdot)$ is a specific order majorant of each member of Γ; so $x(\cdot)$ is a positive supermartingale, and since $\mathrm{GM}x(\cdot)$ is also a specific order majorant of Γ, it follows that

$x(\cdot) = \mathrm{GM}x(\cdot)$, and so this process is a martingale. Hence (Appendix III.8) if $'\mathbf{S}_m^+$ is the cone in $'\mathbf{S}$ whose equivalence classes contain the positive martingales, the set $'\mathbf{S}_m = {}'\mathbf{S}_m^+ - {}'\mathbf{S}_m^+$ is a band in $('\mathbf{S}, \preceq)$. The restrictions to $'\mathbf{S}_m^+$ of the essential and specific orders coincide. If $\Gamma \subset {}'\mathbf{S}_m$, then, with some abuse of the notation LM and GM,

$$\bigvee\Gamma = \mathrm{LM}\Gamma, \qquad \bigwedge\Gamma = \mathrm{GM}\Gamma$$

in the sense that if one side of an equation exists, the other side also exists and there is equality. According to Section 4, a martingale is in an $'\mathbf{S}_m$ equivalence class if and only if the martingale is L^1 bounded.

Continuous Parameter Context: The Vector Lattice $(\mathbf{S}_m, \preceq)$

In the context of $(\mathbf{S}, \preceq)$ an almost surely right continuous supermartingale specific order majorized by an almost surely right continuous martingale is itself a martingale. If Γ is a set of positive almost surely right continuous martingales with $\bigvee\Gamma = x(\cdot)$, then as above it follows that $x(\cdot)$ is a martingale. Thus if $\mathbf{S}_m^+$ is the cone in $\mathbf{S}$ whose equivalence classes contain the positive martingales, the set $\mathbf{S}_m = \mathbf{S}_m^+ - \mathbf{S}_m^+$ is a band in $(\mathbf{S}, \preceq)$ and as such is a conditionally complete sublattice of $(\mathbf{S}, \preceq)$.

Observe that in the continuous parameter context if $x(\cdot)$ is an L^1 bounded almost surely right continuous martingale, then $x(\cdot) \in \mathbf{S}_m$ because up to a standard modification $x(\cdot) = x_1(\cdot) - x_2(\cdot)$ with $x_i(\cdot) \in \mathbf{S}_m^+$; so up to an evanescent set $x(\cdot) = \underset{+1}{x}(\cdot) - \underset{+2}{x}(\cdot)$, with $\underset{+i}{x}(\cdot) \in \mathbf{S}_m^+$.

8. The Vector Lattices $('\mathbf{S}_p, \preceq)$ and $(\mathbf{S}_p, \preceq)$

(See the corresponding potential theory vector lattice in Section 1.IX.8.)

The Vector Lattice $('\mathbf{S}_p, \preceq)$

Recall that in Section III.21 we defined a supermartingale potential as a positive supermartingale $\{x(\cdot), \mathcal{F}(\cdot)\}$ with $\inf_{t \in I} E\{x(t)\} = 0$, equivalently with $\mathrm{GM}x(\cdot) = 0$ up to a standard modification. A positive supermartingale specific order majorized by a supermartingale potential is itself one and if Γ is a set of supermartingale potentials with $\bigvee\Gamma = x(\cdot)$ then $x(\cdot)$ must be a positive supermartingale for which $x(\cdot) - \mathrm{GM}x(\cdot)$ is also a specific order majorant of Γ and therefore must also be a version of $\bigvee\Gamma$ so $\mathrm{GM}x(\cdot) = 0$ and $x(\cdot)$ is a potential. It follows that if $'\mathbf{S}_p^+$ is the cone in $'\mathbf{S}$ whose equivalence classes contain the supermartingale potentials then the set $'\mathbf{S}_p = {}'\mathbf{S}_p^+ - {}'\mathbf{S}_p^+$ is a band in $('\mathbf{S}, \preceq)$ and as such is a conditionally complete vector sublattice of $('\mathbf{S}, \preceq)$.

Continuous Parameter Context: The Vector Lattice ($\mathbf{S}_p, \preceq$)

In this context observe that if $x(\cdot)$ is a supermartingale potential, then $\underset{+}{x}(\cdot)$ is also a supermartingale potential and conversely. The reasoning used above with trivial adaptation to the context of $\mathbf{S}$ shows that if $\mathbf{S}_p^+$ is the cone in $\mathbf{S}$ whose equivalence classes contain the right continuous supermartingale potentials, then the set $\mathbf{S}_p = \mathbf{S}_p^+ - \mathbf{S}_p^+$ is a band in ($\mathbf{S}, \preceq$) and as such is a conditionally complete vector sublattice of ($\mathbf{S}, \preceq$).

Theorem. $'\mathbf{S}_p = {'\mathbf{S}}_m^\perp$ *and* $\mathbf{S}_p = \mathbf{S}_m^\perp$.

The proof of these relations in the potential theory context (Section 1.IX.8) is applicable in the present context.

9. The Vector Lattices ($'\mathbf{S}_{qb}, \preceq$) and ($\mathbf{S}_{qb}, \preceq$)

(See the corresponding potential theory vector lattice in Section 1.IX.9.)

The Vector Lattice ($'\mathbf{S}_{qb}, \preceq$)

The class $'\mathbf{S}_{qb}^+$ is defined as the subset of $'\mathbf{S}^+$ whose equivalence classes contain the quasi-bounded positive supermartingales, that is (Section IV.6), contain the supermartingales $x(\cdot)$ which satisfy the following equivalent conditions, in which all processes are adapted to the specified filtration.

(a) The process $x(\cdot)$ is the specific order essential supremum of a set of bounded positive supermartingales.
(b) The process $x(\cdot)$ is the limit of a specific order increasing sequence of bounded positive supermartingales; that is, $x(\cdot)$ is the sum of a series of bounded positive supermartingales.

The class $'\mathbf{S}_{qb}^+$ is a cone which satisfies the conditions (Appendix III.8) implying that the set $'\mathbf{S}_{qb} = {'\mathbf{S}}_{qb}^+ - {'\mathbf{S}}_{qb}^+$ is a band in ($'\mathbf{S}, \preceq$) and as such is a conditionally complete vector lattice. The equivalence classes in this band, and also their stochastic process members, are called *quasi bounded*. Observe that if $x(\cdot)$ in (a) and (b) above is a martingale, then the bounded positive supermartingales in (a) and (b) must also be martingales because they are specific order minorants of $x(\cdot)$.

The Bands $'\mathbf{S}_{mqb} = {'\mathbf{S}}_m \cap {'\mathbf{S}}_{qb}$ and $'\mathbf{S}_{pqb} = {'\mathbf{S}}_p \cap {'\mathbf{S}}_{qb}$

In view of Section 8 these two bands are mutually orthogonal, and $'\mathbf{S}_{qb} = {'\mathbf{S}}_{mqb} + {'\mathbf{S}}_{pqb}$. The band $'\mathbf{S}_{mqb}$ is the band in $'\mathbf{S}$ generated by the equivalence class of the process all of whose random variables are identically 1. According to Theorem IV.10, $'\mathbf{S}_{pqb} = '\mathbf{S}_p \cap \mathbf{D}$ if the parameter set has a first point but not a last point.

Continuous Parameter Context: The Vector Lattice $(\mathbf{S}_{qb}, \preceq)$

In this context a quasi-bounded positive supermartingale $x(\cdot)$ which is almost surely right continuous and which is the sum $\Sigma_0^\infty x_n(\cdot)$ of a series of bounded positive supermartingales is the sum of a series of bounded positive almost surely right continuous supermartingales. In fact the sum $\Sigma_0^\infty \underset{+}{x}_n(\cdot)$ of bounded positive almost surely right continuous supermartingales is almost surely right continuous (Theorem IV.4), and except for a countable parameter set, $P\{x(t) = \Sigma_0^\infty x_n(t)\} = 1$. Hence by almost sure right continuity $\underset{+}{x}(\cdot) = x(\cdot) = \Sigma_0^\infty \underset{+}{x}_n(\cdot)$ up to an evanescent set. With the help of this result the reasoning at the beginning of this section, with trivial adaptation to the context of $\mathbf{S}$, shows that if $\mathbf{S}_{qb}^+$ is the cone in $\mathbf{S}$ whose equivalence classes contain the supermartingales which satisfy the equivalent conditions (a) and (b) above with all the positive bounded supermartingales involved supposed almost surely right continuous, then the set $\mathbf{S}_{qb} = \mathbf{S}_{qb}^+ - \mathbf{S}_{qb}^+$ is a band in $\mathbf{S}$ and as such is a conditionally complete vector sublattice of $\mathbf{S}$. Define $\mathbf{S}_{mqb} = \mathbf{S}_m \cap \mathbf{S}_{qb}$ and $\mathbf{S}_{pqb} = \mathbf{S}_p \cap \mathbf{S}_{qb}$ to obtain orthogonal bands with vector sum $\mathbf{S}_{qb}$. According to Theorem IV.11, $\mathbf{S}_{pqb} = \mathbf{S}_p \cap \mathbf{D}$.

In the following theorem **UI** denotes the class of uniformly integrable processes in the given context or, with the usual abuse of language, the set of equivalence classes in $'\mathbf{S}$ or $\mathbf{S}$ whose members are uniformly integrable.

Theorem. $'\mathbf{S}_{mqb} = '\mathbf{S}_m \cap \mathbf{UI}$ *and* $\mathbf{S}_{mqb} = \mathbf{S}_m \cap \mathbf{UI}$.

It is sufficient to prove the first equality, and [by Theorem 3(d)] even sufficient to prove that $'\mathbf{S}_{mqb}^+ = '\mathbf{S}_m^+ \cap \mathbf{UI}$. In this context if $x(\cdot)$ is a uniformly integrable positive martingale, it is right closable [Section III.3(e)] by some random variable x; so $x(t) = E\{x|\mathscr{F}(t)\}$ almost surely. If $x_n(\cdot)$ is the bounded martingale defined by $x_n(t) = E\{x \wedge n|\mathscr{F}(t)\}$, then $\lim_{n\to\infty} x_n(t)$, $t \subset \mathbb{R}^+$, defines a martingale in the equivalence class of $x(\cdot)$; so $x(\cdot)$ is quasi bounded. Conversely, if $x(\cdot)$ is a quasi-bounded martingale, so that there is a sequence of bounded martingales, $\{y_n(\cdot), n \in \mathbb{Z}^+\}$, such that $x(\cdot) = \Sigma_0^\infty y_n(\cdot)$, then $y_n(\cdot)$ is right closable, say by a random variable y_n, and $x(\cdot)$ must be right closable by $\Sigma_0^\infty y_n$; so $x(\cdot)$ is uniformly integrable.

10. The Vector Lattices $('\mathbf{S}_s, \preceq)$ and $(\mathbf{S}_s, \preceq)$

(See the corresponding potential theory vector lattice in Section 1.IX.10.) Define $'\mathbf{S}_s = '\mathbf{S}_{qb}^\perp$ relative to $('\mathbf{S}, \preceq)$, and define $'\mathbf{S}_s^+ = '\mathbf{S}_s \cap '\mathbf{S}^+$. Define $\mathbf{S}_s = \mathbf{S}_{qb}^\perp$ relative to $(\mathbf{S}, \preceq)$, and define $\mathbf{S}_s^+ = \mathbf{S}_s \cap \mathbf{S}^+$. The equivalence classes in the bands $'\mathbf{S}_s$ and $\mathbf{S}_s$ and the processes they contain are called *singular*. A process in $'\mathbf{S}^+$ $[\mathbf{S}^+]$ is singular if and only if every bounded $'\mathbf{S}^+$ $[\mathbf{S}^+]$ specific order minorant of the process is a standard modification of [indistinguishable from] the identically zero process. We leave to the reader the verification of the fact that in the continuous parameter context if $x(\cdot)$ is an almost surely right continuous supermartingale, then $x(\cdot) \in {'\mathbf{S}_s^+}$ if and only if $x(\cdot) \in \mathbf{S}_s^+$ and that $x(\cdot) \in {'\mathbf{S}_s}$ if and only if $x(\cdot) \in \mathbf{S}_s$. Thus in the continuous parameter context the equivalence classes in $\mathbf{S}_s^+$ $[\mathbf{S}_s]$ can be identified with those in $'\mathbf{S}_s^+$ $['\mathbf{S}]$ which contain almost surely right continuous processes.

We shall denote by $'\mathbf{S}_{ms}$ and $\mathbf{S}_{ms}$, respectively, the bands $'\mathbf{S}_m \cap {'\mathbf{S}_s}$ and $\mathbf{S}_m \cap \mathbf{S}_s$ of singular martingales in their respective vector lattices. The bands $'\mathbf{S}_{ps}$ and $\mathbf{S}_{ps}$ of singular supermartingale potentials are defined correspondingly. Thus we now have an orthogonal decomposition of each of the lattices $'\mathbf{S}$ and $\mathbf{S}$ into four bands:

$$'\mathbf{S} = {'\mathbf{S}_{mqb}} + {'\mathbf{S}_{ms}} + {'\mathbf{S}_{pqb}} + {'\mathbf{S}_{ps}}, \qquad \mathbf{S} = \mathbf{S}_{mqb} + \mathbf{S}_{ms} + \mathbf{S}_{pqb} + \mathbf{S}_{ps}. \tag{10.1}$$

11. The Orthogonal Decompositions $'\mathbf{S}_m = {'\mathbf{S}_{mqb}} + {'\mathbf{S}_{ms}}$ and $\mathbf{S}_m = \mathbf{S}_{mqb} + \mathbf{S}_{ms}$

In this section we restrict outselves to the continuous parameter context except in the last paragraph where we show how the work can be modified to be applicable to the $('\mathbf{S}, \preceq)$ context. Suppose then (continuous parameter context) that $x(\cdot)$ is an almost surely right continuous positive martingale, that is, $x(\cdot) \in \mathbf{S}_m^+$. Then (Section III.13) $\lim_{t \to \infty} x(t) = x(+\infty)$ exists almost surely.

Case (a): $x(\cdot) \in \mathbf{UI}$. In thise case, equivalently (Theorem 3) if $x(\cdot) \in \mathbf{D}$, equivalently (Theorem 9) if $x(\cdot) \in \mathbf{S}_{qb}$, this martingale is right closable by the random variable $x(+\infty)$ (Section III.14); that is, $x(t) = E\{x(+\infty) | \mathscr{F}(t)\}$ almost surely. Thus $x(+\infty)$ determines $x(\cdot)$ up to an evanescent set.

Case (b): We show that $x(\cdot) \in \mathbf{S}_{ms}$ if and only if $x(+\infty) = 0$ almost surely. If $x(+\infty) = 0$ almost surely, then every bounded specific order minorant of $x(\cdot)$ in $\mathbf{S}^+$ has almost sure limit 0 at $+\infty$; so according to Case (a), the minorant is indistinguishable from the zero process. Hence $x(\cdot) \in \mathbf{S}_{ms}^+$. Conversely, if $x(\cdot) \in \mathbf{S}_{ms}^+$, then for every positive constant c,

$$E\{x(+\infty) \wedge c | \mathscr{F}(\cdot)\} = \lim_{s \to \infty} E\{x(s) \wedge c | \mathscr{F}(\cdot)\} \preceq x(\cdot) \quad \text{a.s.}$$

and the martingale on the left is bounded and so is indistinguishable from the zero process. It follows that $E\{x(+\infty) \wedge c\} = 0$ for every c; so $x(+\infty) = 0$ almost surely.

Case (c): General case. In all cases we can write $x(\cdot)$ in the form

$$x(\cdot) = E\{x(+\infty) | \mathscr{F}(\cdot)\} + [x(\cdot) - E\{x(+\infty) | \mathscr{F}(\cdot)\}] \tag{11.1}$$

and choose the conditional expectations in such a way (Section IV.1) that $E\{x(+\infty)|\mathscr{F}(\cdot)\}$ is an almost surely right continuous martingale. This conditional expectation martingale comes under Case (a). The bracketed difference in (11.1) is a martingale with almost sure limit 0 at $+\infty$ and so comes under Case (b). Thus (11.1) exhibits $x(\cdot)$ as the sum of its quasi-bounded and singular components.

If $x(\cdot)\in{}'\mathbf{S}_m^+$, then $\operatorname{ess\,lim}_{t\uparrow} x(t)$ exists (Section III.13), and Cases (a)–(c) go through as in the continuous parameter context.

12. Local Martingales and Singular Supermartingale Potentials in $(\mathbf{S}, \preceq)$

Local Martingale (Continuous Parameter Context)

A process $x(\cdot)$ in this context is called a *local martingale* if there is an increasing sequence of finite optional times with almost sure limit $+\infty$ such that each process

$$\{x(T_n \wedge t), \mathscr{F}(t), t\in\mathbb{R}^+\} \tag{12.1}$$

is an almost surely right continuous martingale. For example (take $T_n \equiv n$), an almost surely right continuous martingale is a local martingale. Observe that if T_n in (12.1) is replaced by $T_n \wedge n$, that is, if the almost surely right continuous martingale (12.1) is stopped at $t = n$, then the stopped process is a martingale (Section IV.3), uniformly integrable because it is right closed by $x(T_n \wedge n)$. Hence it is no restriction on a local martingale if T_n in (12.1) is supposed bounded and if each martingale (12.1) is supposed uniformly integrable.

It is trivial (calculate expectations or use the fact that $\mathbf{S}_p \perp \mathbf{S}_m$) that an almost surely right continuous supermartingale potential which is a martingale is indistinguishable from the identically zero process. As the following theorem shows the situation is quite different if "martingale" is replaced here by "local martingale."

Theorem. *A process $x(\cdot)$ in $\mathbf{S}_p^+$ is singular if and only if it is a local martingale.*

If $x(\cdot)\in\mathbf{S}_{ps}^+$ and $n\in\mathbb{Z}^+$, define

$$T_n = n \wedge \inf\{t\in\mathbb{R}^+ : x(t)\geq n\}, \qquad y(\cdot) = x(\cdot) - \tau_{T_n}x(\cdot).$$

Then $\lim_{n\to\infty} T_n = +\infty$ because almost every almost surely right continuous supermartingale sample function is bounded on compact intervals. Observe that $\tau_{T_n}x(\cdot)\in\mathbf{S}^+$ with $\tau_{T_n}x(t) < n$ for $t < T_n$ and that $y(\cdot)\in\mathbf{S}^+$ with $y(t) < n$

for $t < T_n$ and $y(t) = 0$ for $t \geq T_n$. Thus $y(\cdot) \preceq x(\cdot)$, and $y(\cdot)$ is bounded; so $y(\cdot) = 0$ quasi everywhere, that is,

$$x(T_n \wedge t) = x(t) = E\{x(T_n)|\mathscr{F}(t)\} \tag{12.2}$$

almost everywhere on the set $\{T_n > t\}$. Since the first and third terms in (12.2) are trivially equal almost everywhere on the set $\{T_n \leq t\}$ [because this set is in $\mathscr{F}(t)$ and the function $x(T_n \wedge t)$ is $\mathscr{F}(t)$ measurable], the process (12.1) is a martingale. Hence $x(\cdot)$ is a local martingale. Conversely, suppose that $x(\cdot) \in \mathbf{S}_p^+$ and that $x(\cdot)$ is a local martingale, so that there is an increasing sequence $T_{\bullet}$ of finite optional times with almost sure limit $+\infty$ such that each process (12.1) is a martingale. If $z(\cdot)$ is in $\mathbf{S}^+$ and is a bounded specific order minorant of $x(\cdot)$, then $z(T_n \wedge \cdot) \preceq x(T_n \wedge \cdot)$; so $z(T_n \wedge \cdot)$ is a martingale with a bound independent of n. The martingale equality $E\{z(0)\} = E\{z(T_n \wedge t)\}$ yields $E\{z(0)\} = E\{z(t)\}$ in the limit, and therefore $z(\cdot) \in \mathbf{S}_m^+$. Since $x(\cdot) \in \mathbf{S}_p^+$, it follows that $z(\cdot) = 0$ quasi everywhere; so $x(\cdot)$ is singular, as was to be proved.

Adaptation to the Parameter Set $\mathbb{Z}^+$

Under the obvious definition of a local martingale with parameter set $\mathbb{Z}^+$ a trivial adaptation of the preceding proof shows that the counterpart of Theorem 12 for the parameter set $\mathbb{Z}^+$ is true.

13. Quasimartingales (Continuous Parameter Context)

In the continuous parameter context (see Section 1) quasimartingales have been defined variously. We shall call an almost surely right continuous adapted process $\{x(\cdot), \mathscr{F}(\cdot)\}$ a quasimartingale if it is L^1 bounded and if there is a constant c for which $0 = t_0 < t_1 < \cdots \to +\infty$ implies that

$$\sum_0^\infty E\{|E\{x(t_k) - x(t_{k+1})|\mathscr{F}(t_k)\}|\} \leq c. \tag{13.1}$$

Observe that, for a given probability space and filtration, the class of quasimartingales is linear and includes L^1 bounded almost surely right continuous martingales (set $c = 0$) and positive almost surely right continuous supermartingales (set $c = E\{x(0)\}$). Hence the members of $\mathbf{S}$ are quasimartingales. According to the following theorem, the converse is also true.

Theorem. *An adapted almost surely right continuous L^1 bounded process $\{x(\cdot), \mathscr{F}(\cdot)\}$ is a quasimartingale if and only if it is the difference between two almost surely right continuous positive supermartingales; that is, $x(\cdot) \in \mathbf{S}$.*

We have already noted that the members of **S** are quasimartingales. Conversely, suppose that $\{x(\cdot), \mathscr{F}(\cdot)\}$ is a quasimartingale satisfying (13.1). A trivial iterated conditional expectation argument shows that if $k \geq j$, the kth term of the sum in (13.1) majorizes the same term with $\mathscr{F}(t_k)$ replaced by $\mathscr{F}(t_j)$. Hence the definition

$$y(t_j) = \sum_{k \geq j} |E\{x(t_k) - x(t_{k+1})|\mathscr{F}(t_j)\}| \tag{13.2}$$

is meaningful, and $E\{y(0)\} \leq c$; the series converges both almost surely and in L^1. Furthermore the L^1 convergence of this series implies that $L^1 \lim_{k\to\infty} E\{x(t_k)|\mathscr{F}(t_j)\}$ exists. Since this is true for all j, so that t_j can be any number in $\mathbb{R}^+$, and since two sequences $t_\bullet$ can be combined into a single one, this L^1 limit property implies that

$$L^1 \lim_{t\to\infty} E\{x(t)|\mathscr{F}(s)\} = x_m(s)$$

exists for all s. Moreover $x_m(s)$ is $\mathscr{F}(s)$ measurable because $\mathscr{F}(0)$ contains the null sets, the process $x_m(\cdot)$ is L^1 bounded because

$$\begin{aligned} E\{|x_m(s)|\} &= E\{|L^1 \lim_{t\to\infty} E\{x(t)|\mathscr{F}(s)\}|\} \leq \lim_{t\to\infty} E\{|E\{x(t)|\mathscr{F}(s)\}|\} \\ &\leq \sup_{t\geq 0} E\{|x(t)|\} < +\infty, \end{aligned} \tag{13.3}$$

and $\{x_m(t_\bullet), \mathscr{F}(t_\bullet)\}$ is a martingale because if $s_1 < s_2$ and if $\Lambda \in \mathscr{F}(s_1)$, then

$$\begin{aligned} \int_\Lambda x_m(s_2)\,dP &= \int_\Lambda L^1 \lim_{t\to\infty} E\{x(t)|\mathscr{F}(s_2)\}\,dP = \lim_{t\to\infty} \int_\Lambda E\{x(t)|\mathscr{F}(s_2)\}\,dP \\ &= \lim_{t\to\infty} \int_\Lambda E\{x(t)|\mathscr{F}(s_1)\}\,dP = \int_\Lambda L^1 \lim_{t\to\infty} E\{x(t)|\mathscr{F}(s_1)\}\,dP \\ &= \int_\Lambda x_m(s_1)\,dP. \end{aligned}$$

Let Q be the set of positive dyadic rational numbers. According to Section IV.1, the right limit process $\{x_{m+}(t), \mathscr{F}(t), t \in \mathbb{R}^+\}$ defined by

$$x_{m+}(t) = \lim_{Q \ni s \downarrow t} x_m(s)$$

[$x_{m+}(t)$ is defined arbitrarily when this right limit does not exist] is an almost surely right continuous martingale and is obviously L^1 bounded. Moreover (Section 4) the process $x_m(\cdot)$ is the difference between two positive almost surely right continuous martingales. At the price of replacing $x(\cdot)$ by $x(\cdot) - x_{m+}(\cdot)$ [which does not change the sum in (13.1) or $y(t_j)$ in (13.2)] we can therefore suppose from now on that $L^1 \lim_{t\to\infty} E\{x(t)|\mathscr{F}(s)\} = 0$. If the absolute value signs in (13.3) are omitted, the sum is

$$x(t_j) - L^1 \lim_{k\to\infty} E\{x(t_k)|\mathscr{F}(t_j)\} = x(t_j) \quad \text{a.s.,}$$

and therefore

$$P\{y(t) \geq x(t)\} = 1, \qquad t \in t_{.}. \tag{13.4}$$

Furthermore, if $j > i$, manipulation of conditional expectations yields

$$\begin{aligned} y(t_i) - E\{y(t_j)|\mathscr{F}(t_i)\} &\geq \sum_{k=i}^{j-1} |E\{x(t_k) - x(t_{k+1})|\mathscr{F}(t_i)\}| \\ &\geq E\{x(t_i) - x(t_j)|\mathscr{F}(t_i)\} \quad \text{a.s.} \end{aligned} \tag{13.5}$$

According to (13.5), the process $\{y(t_.) - x(t_.), \mathscr{F}(t_.)\}$ is a supermartingale. Thus the positive processes

$$\{y(t_.), \mathscr{F}(t_.)\}, \qquad \{y(t_.) - x(t_.), \mathscr{F}(\cdot)\}$$

are supermartingales. We have already noted that $E\{y(0)\} \leq c$. The reader can readily verify that adding points to $t_.$ replaces $y(\cdot)$ by a supermartingale which majorizes $y(\cdot)$ on the original sequence $t_.$. Let $\{y_n(j2^{-n}), \mathscr{F}(j2^{-n}), j \in \mathbb{Z}^+\}$ be, for $n \in \mathbb{Z}^+$, the version of $y(\cdot)$ when $t_. = \{j2^{-n}, j \in \mathbb{Z}^+\}$. For t in Q and n so large that $y_n(t)$ is defined, $y_n(t)$ increases almost surely as n increases. Define $z(\cdot) = \lim_{n\to\infty} y_n(\cdot)$ to obtain positive supermartingales

$$\{z(t), \mathscr{F}(t), t \in Q\}, \qquad \{z(t) - x(t), \mathscr{F}(t), t \in Q\}.$$

Apply Section IV.1 again to find that the positive right limit processes $\underset{+}{z}(\cdot)$ and $\underset{+}{z}(\cdot - x(\cdot)$ are almost surely right continuous supermartingales relative to $\mathscr{F}(\cdot)$ on the parameter set $\mathbb{R}^+$. The desired representation of $x(\cdot)$ is $x(\cdot) = \underset{+}{z}(\cdot) - [\underset{+}{z}(\cdot) - x(\cdot)\}$.

Observation. Theorem 13 asserts that the quasimartingale $x(\cdot)$ has a representation $x(\cdot) = x_1(\cdot) - x_2(\cdot)$, where $x_i(\cdot)$ is a positive almost surely right continuous supermartingale. There is some interest in minimizing $x_1(\cdot)$ and $x_2(\cdot)$. An indirect but interesting way is to observe that the class of pairs $(x_1(\cdot), x_2(\cdot))$ with the stated properties has the property that if $(x_1'(\cdot), x_2'(\cdot))$ and (x_1'', x_2'') are in the class, then $(x_1'(\cdot) \wedge x_1''(\cdot), x_2'(\cdot) \wedge x_2''(\cdot))$ is in the class. Thus the set of first components and the set of second components are both directed downward, and it is not difficult to show that the properly smoothed version of the pair of essential infima of these two sets is the desired minimal pair. A more direct method suggested by the classical discussion of the positive and negative variations of a function of bounded variation is to replace the discussion of $y(\cdot)$ as defined by (13.2) by the processes defined by the two series

$$\sum_{k\geq j}[E\{x(t_k)-x(t_{k+1})|\mathscr{F}(t_j)\}\vee 0], \qquad -\sum_{k\geq j}[E\{x(t_k)-x(t_{k+1})|\mathscr{F}(t_j)\}\wedge 0].$$

The Parameter Set $\mathbb{Z}^+$

A process $\{x(n), \mathscr{F}(n), n\in\mathbb{Z}^+\}$ is called a quasimartingale if it is L^1 bounded and if

$$\sum_0^\infty E\{|E\{x(k)-x(k+1)|\mathscr{F}(k)\}|\} < +\infty.$$

An adapted process is a quasimartingale if and only if it is the difference between two positive supermartingales. The proof follows that in the continuous parameter case, with obvious simplifications.

Chapter VI

Markov Processes

1. The Markov Property

Let $\{x(\cdot), \mathscr{F}(\cdot)\}$ be a stochastic process with state space $(X, \mathscr{X})$ on a filtered probability space $(\Omega, \mathscr{F}, P; \mathscr{F}(t), t \in I)$. The process is called a *Markov process* if when $s < t$ and $A \in \mathscr{X}$, then

$$P\{x(t) \in A \mid \mathscr{F}(s)\} = P\{x(t) \in A \mid x(s)\} \quad \text{a.s.} \tag{1.1}$$

Define $\mathscr{G}(s) = \mathscr{F}\{x(r), r \geq s\}$. It will now be shown that (1.1) implies

$$E\{z \mid \mathscr{F}(s)\} = E\{z \mid x(s)\} \quad \text{a.s.} \tag{1.1'}$$

whenever z is a function from Ω into $\mathbb{R}$ which is $\mathscr{G}(s)$ measurable and is either integrable or positive. If $s < t$ and $z = 1_{\{x(t) \in A\}}$ with A in $\mathscr{X}$, equation (1.1′) reduces to (1.1). The validity of (1.1) or the equivalent (1.1′) will be referred to as the *Markov property*. To prove that (1.1′) is true under (1.1) it is sufficient according to the usual approximation procedure to prove (1.1′) for z the indicator function of a set in $\mathscr{G}(s)$. Since this σ algebra is generated by the algebra of finite disjoint unions of sets of the form $\{x(t_j) \in A_j, j \leq n\}$, with $t_0 = s < \cdots < t_n$ and A_j in $\mathscr{X}$, it is enough to prove (1.1′) for $z = z_0 \cdots z_n$, where $z_j = \phi_j[x(t_j)]$ and ϕ_j is a bounded measurable function from $(X, \mathscr{X})$ into $(\mathbb{R}, \mathscr{B}(\mathbb{R}))$. Equation (1.1′) is trivial when $n = 0$. When $n = 1$,

$$E\{z_0 z_1 \mid \mathscr{F}(s)\} = z_0 E\{z_1 \mid \mathscr{F}(s)\} \quad \text{a.s.,} \tag{1.2}$$

and if $z_1 = 1_A[x(t_1)]$, the right side of (1.2) becomes, using (1.1),

$$z_0 P\{x(t_1) \in A \mid \mathscr{F}(s)\} = z_0 P\{x(t_1) \in A \mid x(s)\} = E\{z_0 z_1 \mid x(s)\} \quad \text{a.s.,} \tag{1.3}$$

so that (1.1′) is true for $z = z_0 z_1$ when z has this special form. Equation (1.1′) for $z_1 = \phi_1[x(t_1)]$ then follows using the usual approximation procedure. We now proceed by induction. If (1.1′) is true for $z = z_0 \cdots z_k$ with an arbitrary choice of $t_0 = s < t_1 < \cdots < t_k$ and functions $\phi_0, \ldots, \phi_k$, for some $k \geq 1$,

$$\begin{aligned}E\{z_0 \cdots z_{k+1}|\mathscr{F}(s)\} &= z_0 E\{E\{z_1 \cdots z_{k+1}|\mathscr{F}(t_1)\}|\mathscr{F}(s)\} \\ &= z_0 E\{E\{z_1 \cdots z_{k+1}|x(t_1)\}|\mathscr{F}(s)\} \\ &= z_0 E\{E\{z_1 \cdots z_{k+1}|x(t_1)\}|x(s)\} \qquad (1.4) \\ &= z_0 E\{E\{z_1 \cdots z_{k+1}|\mathscr{F}(t_1)\}|x(s)\} \\ &= E\{z_0 \cdots z_{k+1}|x(s)\} \quad \text{a.s.,}\end{aligned}$$

as was to be proved.

A manipulation of conditional probabilities which will be left to the reader shows that if the process parameter set is a set of consecutive integers, (1.1) is true in general if true for $t = s + 1$.

The Markov property can be reformulated: $\{x(\cdot), \mathscr{F}(\cdot)\}$ is Markovian if and only if the past and future are independent, given the present, or, in precise form, if $\Lambda \in \mathscr{F}(s)$ and $\mathrm{M} \in \mathscr{G}(s)$, then

$$P\{\Lambda \cap \mathrm{M}|x(s)\} = P\{\Lambda|x(s))P\{\mathrm{M}|x(s)\} \quad \text{a.s.,} \qquad (1.5)$$

or equivalently, if y is $\mathscr{F}(s)$ measurable and z is $\mathscr{G}(s)$ measurable and both are positive or both are bounded,

$$E\{yz|x(s)\} = E\{y|x(s)\}E\{z|x(s)\} \quad \text{a.s.} \qquad (1.5')$$

To derive (1.5′) from the Markov property, suppose that y and z are measurable, as described, and bounded. Using (1.1′),

$$E\{yz|\mathscr{F}(s)\} = yE\{z|\mathscr{F}(s)\} = yE\{z|x(s)\} \quad \text{a.s.;} \qquad (1.6)$$

so performing the operation $E\{-|x(s)\}$ on the first and third terms yields (1.5′). Conversely, under (1.5′)

$$E\{yz|x(s)\} = E\{yE\{z|x(s)\}|x(s)\} \quad \text{a.s.;} \qquad (1.7)$$

so

$$E\{yz\} = E\{yE\{z|x(s)\}\}. \qquad (1.8)$$

If now y is the indicator function of a set Λ in $\mathscr{F}(s)$, equation (1.8) becomes

$$\int_\Lambda z\,dP = \int_\Lambda E\{z|x(s)\}\,dP, \qquad (1.9)$$

which is the integrated version of (1.1′).

The symmetry of (1.5′) implies that if $\{x(\cdot), \mathscr{F}(\cdot)\}$ is Markovian, the process $\{x(\cdot), \mathscr{G}(\cdot)\}$ is Markovian when the parameter order is reversed. Roughly, a Markov process under time reversal is a Markov process.

The Markov Property for Processes with Topological State Spaces

According to the discussion in this section, the Markov property can be stated in the following form: an adapted process $\{x(\cdot), \mathscr{F}(\cdot)\}$ is Markovian if and only if

$$E\{f[x(t)] \mid \mathscr{F}(s)\} = E\{f[x(t)] \mid x(s)\} \quad \text{a.s.} \tag{1.10}$$

when $s < t$ and f is an arbitrary bounded measurable function from the state space into $\mathbb{R}$. We leave it to the reader to verify that if the state space is a Polish space coupled with its Borel sets, the process is Markovian if and only if (1.10) is satisfied whenever f is bounded and continuous, and in particular, if the state space is locally compact and second countable, it is sufficient if (1.10) is satisfied whenever f is continuous with compact support.

Initial Distribution and Transition Function of a Markov Process

Let $\{x(t), \mathscr{F}(t), t \in I\}$ be a Markov process with measurable state space $(X, \mathscr{X})$. If I has a first point t_0, the distribution of $x(t_0)$ is called the *initial distribution of the process*. If there is a stochastic transition function q with parameter set I (Appendix VI.3) such that for s and t in I with $s < t$ and for A in $\mathscr{X}$,

$$P\{x(t) \in A \mid \mathscr{F}(s)\} = q(s, x(s); t, A) \quad \text{a.s.,} \tag{1.11}$$

then $x(\cdot)$ is said to have transition function q. This equation implies, when the operation $E\{- \mid x(s)\}$ is applied to both sides, that the right side is almost surely $P\{x(t) \in A \mid x(s)\}$. Thus (1.11) implies that $\{x(\cdot), \mathscr{F}(\cdot)\}$ is Markovian. Recall that by definition of transition function with parameter set I the Chapman–Kolmogorov equation

$$q(s, \xi; u, A) = \int_X q(t, \eta; u, A) q(s, \xi; t, d\eta) \qquad (s < t < u) \tag{1.12}$$

is satisfied. Observe that in the following equation for iterated conditional expectations (see Section I.4),

$$P\{x(u) \in A \mid \mathscr{F}(s)\} = E\{P\{x(u) \in A \mid \mathscr{F}(t)\} \mid \mathscr{F}(s)\} \quad \text{a.s. } (s < t < u), \tag{1.13}$$

conditioning by $\mathscr{F}(s)$ and $\mathscr{F}(t)$ can be replaced, respectively, by conditioning by $x(s)$ and $x(t)$, in view of (1.1′), and that as so modified (1.13) can be written in the form

$$q(s, x(s); u, A) = \int_X q(t, \eta; u, A) q(s, x(s); t, d\eta) \quad \text{a.s.,} \tag{1.14}$$

which is (1.12) up to the "a.s." in (1.14). Thus the Chapman–Kolmogorov equation amounts to a property of conditional expectations combined with the Markov property.

Absolute Probability Function of a Markov Process

If q is a stochastic transition function with parameter set I and state space $(X, \mathscr{X})$, an *absolute probability function for q* is defined as a function $(t, A) \mapsto \mu(t, A)$ from $I \times \mathscr{X}$ into $[0, 1]$ for which $\mu(t, \cdot)$ is a probability measure and

$$\mu(t, A) = \int_X q(s, \xi; t, A) \mu(s, d\xi) \qquad (s < t). \tag{1.15}$$

If $\{x(\cdot), \mathscr{F}(\cdot)\}$ is a Markov process with transition function q and parameter set I, the function $(t, A) \mapsto \mu(t, A)$ defined by

$$\mu(t, A) = P\{x(t) \in A\} \qquad [(t, A) \in I \times \mathscr{X}] \tag{1.16}$$

is an absolute probability function for q and is called the *absolute probability function of the Markov process*. This absolute probability function and the transition function together determine the finite-dimensional distributions of the process: if $t_1 < \cdots < t_n$ are parameter values,

$$P\{x(t_j) \in A_j, j \le n\} = \int_{A_1} \mu(t_1, d\xi_1) \int_{A_2} q(t_1, \xi_1; t_2, d\xi_2) \cdots \int_{A_n} q(t_{n-1} \xi_{n-1}; t_n, d\xi_n). \tag{1.17}$$

In particular, if I has a first point t_0, and if ν is the distribution of $x(t_0)$,

$$\mu(t, A) = \int_A q(t_0, \xi; t, A) \nu(d\xi) \qquad (t > t_0). \tag{1.18}$$

Recall (Section I.10) that to a prescribed state space and prescribed finite-dimensional distributions correspond a stochastic process with those finite-dimensional distributions if the state space satisfies a certain weak condition (for example, if the state space is a Polish space coupled with its Borel sets) and if the finite-dimensional distributions are consistent. Thus, under the stated restriction on the state space, if I has a first point, to each specified initial distribution and stochastic transition function with parameter set I correspond a Markov process with the specified initial distribution and transition function; if I has no first point, to each stochastic transition function q and absolute probability function for q correspond a Markov process with the specified transition and absolute probability functions. The finite-dimensional distributions of the process are determined by (1.17), and the absolute probability function is determined by the initial distribution if I has a first point.

Notation Involving the Initial Distribution

When I has a first point and the Markov process context requires the identification of the initial distribution, two systems of notation will be used. We shall sometimes identify the initial distribution ν by writing P_ν and E_ν for probabilities and expectations, specializing to P_ξ and E_ξ when ν is supported by $\{\xi\}$; in the latter case the process will be said to have initial point ξ or to be a process from ξ. Alternatively, we may denote probabilities and expectations by P and E but identify the initial distribution by a subscript in the process notation, writing $x_\nu(\cdot)$ for the process and specializing to $x_\xi(\cdot)$ if the process has initial point ξ.

The Stationary Context

If $I = \mathbb{Z}^+$ and if there is a stochastic kernel $(\xi, A) \mapsto p(\xi, A)$ for which

$$q(s, \xi; s+1, A) = p(\xi, A) \qquad (s \in \mathbb{Z}^+),$$

then the transition function value $q(s, \xi; s+t, A)$ does not depend on s for any value of $t = 1, 2, \ldots$, and in fact according to the Chapman–Kolmogorov equation, $(s, A) \mapsto q(s, \xi; s+t, A)$ is the tth kernel iterate of $p(\cdot, \cdot)$. In this case $\{x(\cdot), \mathscr{F}(\cdot)\}$ is said to have stationary probabilities and to have transition function p. If $I = \mathbb{R}^+$ and if q is stationary (Appendix VI.3), that is, if there is a stationary continuous parameter transition function $(t, \xi, A) \mapsto p(t, \xi, A)$ for which

$$q(s, \xi; s+t, A) = p(t, \xi, A) \qquad (s \in \mathbb{R}^+, 0 < t \in \mathbb{R}^+),$$

then $\{x(\cdot), \mathscr{F}(\cdot)\}$ is said to have stationary transition probabilities and to have stationary transition function p. In this context recall that the Chapman–Kolomogorov equation becomes

$$p(s+t, \xi, A) = \int_X p(t, \eta, A) p(s, \xi, d\eta) \qquad (0 < s, t) \tag{1.19}$$

and equation (1.15) linking absolute probability and transition functions becomes

$$\mu(s+t, A) = \int_X p(t, \xi, A)\mu(s, d\xi) \qquad (s \geq 0, t > 0); \tag{1.20}$$

here $s = 0$ yields the version of (1.18) in the present context, since $\mu(0, \cdot)$ is the initial distribution.

2. Choice of Filtration

Let $\{x(\cdot), \mathscr{F}(\cdot)\}$ be a Markov process with an arbitrary linearly ordered parameter set

$\mathscr{F}_0(t) = \mathscr{F}\{x(s), s \leq t\}$,

$\mathscr{F}_0'(t) = \sigma$ algebra generated by $\mathscr{F}_0(t)$ and the null sets,

$\mathscr{F}_0' = \bigvee \mathscr{F}_0'(\cdot)$,

$\mathscr{F}'(t) = \sigma$ algebra generated by $\mathscr{F}(t)$ and the null sets.

The larger the σ algebras of the filtration $\mathscr{F}(\cdot)$ the more significant is the assertion that $\{x(\cdot), \mathscr{F}(\cdot)\}$ has the Markov property. The minimal choice of filtration $\mathscr{F}(\cdot)$ to which $x(\cdot)$ is adapted is $\mathscr{F}_0(\cdot)$; operating on both sides of (1.1) with $E\{-|\mathscr{F}_0(s)\}$ shows that $\{x(\cdot), \mathscr{F}_0(\cdot)\}$ is Markovian. In the other direction it is trivial that $\{x(\cdot), \mathscr{F}'(\cdot)\}$ is Markovian. Thus

$$\mathscr{F}_0(t) \subset \mathscr{F}(t) \subset \mathscr{F}'(t), \mathscr{F}_0'(t) \subset \mathscr{F}'(t), \tag{2.1}$$

and the filtrations $\mathscr{F}_0(\cdot)$, $\mathscr{F}_0'(\cdot)$, $\mathscr{F}(\cdot)$, $\mathscr{F}'(\cdot)$, some of which may be identical, all make $x(\cdot)$ Markovian. The following lemma will be used in Section 7.

Lemma. *If $\mathscr{F}(t) \subset \mathscr{F}_0'$ for some parameter value t, then $\mathscr{F}'(t) \subset \mathscr{F}_0'(t)$.*

To prove the lemma, consider the class Γ of bounded random variables y for which

$$E\{y|\mathscr{F}(t)\} = E\{y|\mathscr{F}_0'\} \quad \text{a.s.} \tag{2.2}$$

If y_1 is bounded and $\mathscr{F}_0'(t)$ measurable, and if y_2 is bounded and is measurable with respect to the σ algebra $\mathscr{F}^t$ generated by the null sets and $\mathscr{F}\{x(s), s \geq t\}$, then $y_1 y_2 \in \Gamma$ because

$$\begin{aligned} E\{y_1 y_2|\mathscr{F}(t)\} &= y_1 E\{y_2|\mathscr{F}(t)\} = y_1 E\{y_2|x(t)\} = y_1 E\{y_2|\mathscr{F}_0'(t)\} \\ &= E\{y_1 y_2|\mathscr{F}_0'(t)\} \quad \text{a.s.} \end{aligned} \tag{2.3}$$

Since Γ is a linear class which is closed under bounded almost everywhere convergence, Γ includes the bounded random variables measurable with respect to $\mathscr{F}_0'(t) \vee \mathscr{F}^t$; that is, Γ includes the bounded $\mathscr{F}_0'$ measurable random variables. In particular, if y is bounded and $\mathscr{F}'(t)$ measurable, the left side of (2.2) is equal almost surely to y, and (2.2) therefore implies that y is $\mathscr{F}_0'(t)$ measurable; that is, $\mathscr{F}'(t) \subset \mathscr{F}_0'(t)$, as was to be proved.

3. Integral Parameter Markov Processes with Stationary Transition Probabilities

Let $(X, \mathscr{X})$ be a measurable space, let μ be a probability measure on $\mathscr{X}$, and let $(\xi, A) \mapsto p(\xi, A)$ be a stochastic kernel with state space $(X, \mathscr{X})$ (Appendix VI.1). Let $\{x(n), \mathscr{F}(n), n \in \mathbb{Z}^+\}$ be a Markov process with initial distribution μ and stationary transition probabilities, with transition function p (Section 1) so that

$$P_\mu\{x(0) \in A\} = \mu(A), \qquad P_\mu\{x(n+1) \in A \mid \mathscr{F}(n)\} = p(x(n), A) \quad \text{a.s.,} \tag{3.1}$$

$$P_\mu\{x(j) \in A_j, j \le n\} = \int_{A_0} \mu(d\xi_0) \int_{A_1} p(\xi_0, d\xi_1) \cdots \int_{A_n} p(\xi_{n-1}, d\xi_n), \tag{3.2}$$

and in particular,

$$P_\xi\{x(j) \in A_j, j \le n\} = 1_{A_0}(\xi) \int_{A_1} p(\xi, d\xi_1) \cdots \int_{A_n} p(\xi_{n-1}, d\xi_n). \tag{3.2'}$$

Abusing language somewhat we shall describe this Markov process from now on as a Markov process with stationary transition function p. The evaluation (3.2) determines the finite-dimensional distributions of the process and thereby determines P_μ on $\mathscr{F}\{x(n), n \in \mathbb{Z}^+\}$, because the latter σ algebra is generated by the algebra $\bigcup_{n=0}^\infty \mathscr{F}\{x(m), m \le n\}$. Conversely, as noted in Section 1, to a specified measurable state space, initial distribution, and transition function corresponds a Markov process $\{\mathring{x}(n), \mathring{\mathscr{F}}(n), n \in \mathbb{Z}^+\}$ with the specified initial distribution and transition function, under a slight restriction on the state space. Here the random variables of the process are defined on the space $\mathring{\Omega}$ of functions from $\mathbb{Z}^+$ into X, $\mathring{x}(n)$ is the function value at n, and (3.2) is now a definition. Since the parameter space is $\mathbb{Z}^+$, the Tulcea generalization of the Kolmogorov extension theorem shows that no restriction need be imposed on the measurable state space. We shall sketch the construction of the process. Each point $\mathring{\omega}$ of $\mathring{\Omega}$ is a sequence $(\xi_0, \xi_1, \ldots)$ of points of X, $\mathring{x}(n, \mathring{\omega}) = \xi_n$, and $\mathring{\mathscr{F}}(n) = \mathscr{F}\{\mathring{x}(0), \ldots, \mathring{x}(n)\}$, $\mathring{\mathscr{F}} = \bigvee_0^\infty \mathring{\mathscr{F}}(n) = \mathscr{F}\{x(n), n \in \mathbb{Z}^+\}$. The measure $\mathring{P}_\mu$ is defined on $\mathring{\mathscr{F}}$ as the extension to $\mathring{\mathscr{F}}$ of the measure on finite-dimensional product sets determined by setting $\mathring{P}\{\mathring{x}(j) \in A_j, j \le n\}$ equal to the right side of (3.2). (The Tulcea proof that the extension exists is omitted.) This construction has the advantage that the space $(\mathring{\Omega}, \mathring{\mathscr{F}})$ is defined without reference to μ or p, an advantage illustrated by the fact that if $\mathscr{X}$ contains the singletons, the function $\xi \mapsto \mathring{P}_\xi\{\mathring{\Lambda}\}$ for $\mathring{\Lambda}$ in $\mathring{\mathscr{F}}$ is $\mathscr{X}$ measurable and for arbitrary μ

$$\mathring{P}_\mu\{\mathring{\Lambda}\} = \int_X \mathring{P}_\xi\{\mathring{\Lambda}\}\mu(d\xi) \tag{3.3}$$

because these assertions are true when $\mathring{\Lambda}$ is a finite-dimensional product set.

If $\{x(\cdot), \mathscr{F}(\cdot)\}$ is an integral parameter Markov process on a probability space $(\Omega, \mathscr{F}, P_\mu)$ with the above state space and transition function, the map $\phi\colon \omega \mapsto [x(0,\omega), x(1,\omega), \ldots]$ is measurable from $(\Omega, \mathscr{F})$ into $(\mathring{\Omega}, \mathring{\mathscr{F}})$, and $P_\mu = \phi^{-1}(\mathring{P}_\mu)$ in the sense that $\phi^{-1}(\mathring{\mathscr{F}}(n)) = \mathscr{F}\{x(j), j \le n\} \subset \mathscr{F}(n)$, $\phi^{-1}(\mathring{\mathscr{F}}) = \mathscr{F}\{x(j), j \in \mathbb{Z}^+\} \subset \mathscr{F}$, and $P_\mu\{\phi^{-1}(\mathring{\Lambda})\} = \mathring{P}_\mu\{\mathring{\Lambda}\}$ for $\mathring{\Lambda} \in \mathring{\mathscr{F}}$. In view of these facts it is usually correct as well as convenient to prove theorems for discrete parameter Markov processes by proving them for the special ones as just defined on $(\mathring{\Omega}, \mathring{\mathscr{F}})$.

Arbitrary Initial Measures

Observe that the preceding discussion remains valid if $\mu(X) \neq 1$, which implies that $P\{\Omega\} \neq 1$. To avoid pathology, we shall always assume that μ is the sum of a sequence of finite measures, but we allow μ to be infinite valued. All that is needed is the acceptance of the idea that in probability theory the measure on the space on which random variables are defined need not be a probability measure; that is, $P\{\Omega\}$ need not be 1! Whenever P may not be a probability measure, this fact will always be mentioned, however.

Substochastic Transition Functions

If p is substochastic, the state space $(X, \mathscr{X})$ can be enlarged (Appendix VI.1) to a state space $(X^\partial, \mathscr{X}^\partial)$ by adjoining an absorbing state ∂, a "trap", to obtain a stochastic transition function extending p to $(X^\partial, \mathscr{X}^\partial)$. The adjoined point ∂ is absorbing in the sense that if a sample path reaches the point, the path stays there from then on. The first time T that ∂ is reached is an optional time, the "lifetime" of the process, equal to $+\infty$ for a path that never reaches ∂. If convenient, the state ∂ can now be dropped, so that to any initial distribution corresponds a process $x(\cdot)$ for which

$$P\{x(n) \text{ is defined}\} = P\{x(n) \in X\} \le 1.$$

In discussing processes with substochastic transition functions it is always to be understood in writing $x(n)$ when the trap point has been dropped that there is an unwritten convention limiting the context to the set on which $x(n)$ is defined, the set on which n is strictly less than the process lifetime.

Theorem (Strong Markov Property in the Discrete Parameter Context). *Let $\{x(n), \mathscr{F}(n), n \in \mathbb{Z}^+\}$ be a Markov process with stationary transition function p, and let S be a finite-valued optional time. Then $\{x(S+n), \mathscr{F}(S+n), n \in \mathbb{Z}^+\}$ is a Markov process with transition function p and initial distribution the distribution of $x(S)$.*

Since $S+n$ is optional as well as S, all that must be proved is that if $A \in \mathscr{X}$, then

$$P\{x(S+1) \in A \mid \mathscr{F}(S)\} = p(x(S), A) \quad \text{a.s.} \tag{3.4}$$

Since $\{S = n\} \in \mathscr{F}(S)$, the left side of (3.4) is equal almost everywhere on $\{S = n\}$ to $P\{x(n+1) \in A \mid x(n)\}$, which itself is almost surely $p(x(n), A)$, as was to be proved.

If S is identically constant in (3.4), this equation is equivalent to the Markov property of $\{x(\cdot), \mathscr{F}(\cdot)\}$ as stated in terms of a transition function. The fact that this relation holds when S is optional is called the *strong Markov property*.

Extension of Theorem 3

A slightly more general version of (3.4) is obtained by observing that the proof of (3.4) also shows more generally that if S is not necessarily finite valued, then (3.4) is true almost everywhere on the set $\{S < +\infty\}$. Still more generally, if p is substochastic, if $x(\cdot)$ has lifetime S', if $A \in \mathscr{X}$, and if $S \le S'$ almost surely, then the proof of (3.4) shows that this relation is true almost everywhere on the set $\{S < S'\}$. A still more general version of Theorem 3 in which $S+1$ is replaced by a random variable measurable relative to $\mathscr{F}(S)$ is useful in the continuous parameter context and in that context will be proved in Section 9. The reader is invited to formulate and prove the counterpart of Theorem 9 in the present context.

4. Application of Martingale Theory to Discrete Parameter Markov Processes

Let $(X, \mathscr{X})$ be a measurable state space, and let $\{x(n), n \in \mathbb{Z}^+\}$ be a Markov process with this state space, from a point of X. It is supposed that the process has stationary transition probabilities, with a transition function p. Call a function u from X into $\overline{\mathbb{R}}$ superharmonic [harmonic] [[subharmonic]] relative to p if u is $\mathscr{X}$ measurable, if $p(\cdot, |u|) < +\infty$, and if $p(\cdot, u) \le u$

$[p(\cdot, u) = u]$ $[[p(\cdot, u) \geq u]]$. A trivial computation shows that the composed process $\{u[x(n)], n \geq k\}$ is a supermartingale [martingale] [[submartingale]] relative to $\mathscr{F}(\cdot)$, with $\mathscr{F}(n) = \mathscr{F}\{x_0, \ldots, x_n\}$ if u is superharmonic [harmonic] [[subharmonic]] and if $E\{|u[x(n)]|\}$ is finite for $n \geq k$. Observe that this expectation is finite when $n = 0$ if u is finite at the initial point of the process and that this expectation is finite by hypothesis when $n = 1$, and therefore if u is superharmonic and positive, this expectation will be finite when $n > 1$. According to the martingale theory convergence Theorem III.13, we conclude that $\lim_{n\to\infty} u[x(n)]$ exists and is finite almost surely if u is positive and superharmonic.

EXAMPLE (Boundary limit theorems in potential theory). Let D be an open Greenian subset of $\mathbb{R}^N$, and for each point ξ of D let $D(\xi)$ be an open relatively compact subset of D containing ξ. Define

$$p(\xi, A) = \mu_{D(\xi)}(\xi, A).$$

Here $\mu_{D(\xi)}$ is harmonic measure; so this measure is supported by $\partial D(\xi)$. It is supposed that $D(\xi)$ depends on ξ in such a way that $p(\cdot, A)$ is Borel measurable when A is a Borel subset of D. The classical superharmonic, harmonic, subharmonic functions are, respectively, superharmonic, harmonic, subharmonic relative to this transition function. The same procedure in the context of parabolic potential theory makes the superparabolic, parabolic, subparabolic functions, respectively, superharmonic, harmonic, subharmonic relative to a transition function defined in terms of parabolic measure. Observe that in this case the sample sequences move in the direction of decreasing ordinate values.

We return to the classical context, and to simplify the situation we make the hypothesis that D is bounded. We shall choose $D(\xi)$ in two ways. First let $D_\bullet$ be an increasing sequence of open relatively compact subsets of D with union D and with $\bar{D}_n \subset D_{n+1}$. Define $D(\xi) = D_k$ for $k = \min\{n : \xi \in D_n\}$. Then p as defined above is a stochastic transition function, and the corresponding Markov process $x(\cdot)$ from a point $x(0)$ of D determines sample sequences which run through successive boundaries of $D_\bullet$ toward ∂D. Furthermore each coordinate function u of $\mathbb{R}^N$ is harmonic and bounded on D; so $u[x(\cdot)]$ is a bounded martingale, and it follows that $\lim_{n\to\infty} u[x(n)]$ exists almost surely. Hence $\lim_{n\to\infty} x(n)$ exists almost surely; that is, almost every $x(\cdot)$ sequence path converges to a boundary point of D. Next we apply this technique to a different choice of $D(\xi)$. This time let ε be a strictly positive number, and choose $D(\xi) = B(\xi, \varepsilon \wedge (|\xi - \partial D|/2))$, so that $p(\xi, \cdot) = \mu_{D(\xi)}(\xi, \cdot)$, that is, $p(\xi, \cdot)$ is the uniform probability distribution on $\partial D(\xi)$. To prove that with this choice of function $p(\cdot, A)$ is Borel measurable when A is a Borel subset of D, it is sufficient to prove that $p(\cdot, f)$ is continuous when f is continuous and bounded on D, and this continuity of $p(\cdot, f)$ follows from the fact that $p(\xi,f)$ is the unweighted average of f on $\partial D(\xi)$ and that the radius of $D(\xi)$ varies continuously with ξ. With this choice of $D(\xi)$ the fact that $\lim_{n\to\infty} = x(\infty)$ exists almost surely is proved as before, and the limit must be on ∂D because

$$|x(n+1) - x(n)| = \varepsilon \wedge \frac{|x(n) - \partial D|}{2} \quad \text{a.s.}$$

The distribution of $x(\infty)$ is a measure $\mu'_D[x(0),\cdot]$ on ∂D. It will be shown later that for both choices of $D(\xi)$ this measure is the harmonic measure $\mu_D[x(0),\cdot]$, and in fact we shall show much more than this, but now we verify this assertion only for ∂D regular. Let ϕ be a continuous function from ∂D into $\mathbb{R}$. Then H_ϕ is a bounded harmonic function on D with boundary limit function ϕ; so $H_\phi[x(\cdot)]$ is a bounded martingale closed (Theorem III.14) by its almost sure limit $H_\phi[x(\infty)]$. Hence

$$\mu'_D[x(0),\phi] = E\{\phi[x(\infty)]\} = H_\phi[x(0)] = \mu_D[x(0),\phi],$$

and since ϕ is arbitrary, $\mu'_D = \mu_D$, as asserted.

For both choices of $D(\xi)$ if u is a positive superharmonic function on D, the composed process $u[x(\cdot)]$ is a positive supermartingale except that the parameter value 0 should be omitted if $u[x(0)] = +\infty$. Thus according to the martingale theory convergence Theorem III.13, a positive superharmonic function has a finite limit on almost every $x(\cdot)$ path to the boundary; this is a Fatou-type boundary limit theorem. Observe that for the second choice of $D(\xi)$ when ε is small, the $x(\cdot)$ paths are sequences whose successive points are close together, and this fact suggests that the above results are valid for suitable continuous paths instead of sequences. This is true: what we have proved is valid when properly interpreted for Brownian motion continuous paths from a point of D to the boundary. In fact for both of the above choices of $D(\xi)$ an $x(\cdot)$ sequence can be identified with a sequence of points tending to a boundary point of D along a Brownian motion path to the point. The probabilistic evaluation of harmonic measure as a Brownian motion hitting distribution, the boundary limit theorem for positive superharmonic functions, and the supermartingale significance of the composition of a positive supermartingale with Brownian motion are parts of Theorem IX.7. The identification of harmonic measure with a Brownian motion hitting distribution is made in Theorem IX.13(a). The latter identification carries with it the fact that the hitting distribution on $\partial D(\xi)$ is $\mu_{D(\xi)}(\xi,\cdot)$ which implies that for each choice of $D(\xi)$ above a sequence $x(\cdot)$ with the assigned distribution can be obtained by choosing $x(1)$ as the first hitting place on $\partial D(x(0))$ by a Brownian motion path from $x(0)$, $x(2)$ as the first hitting place thereafter on $\partial D(x(1))$ and so on.

In the parabolic context the discussion is changed only in that "harmonic" is replaced by "parabolic," and "Brownian motion" by "space-time

Brownian motion." If h [$\dot{h}$] is a strictly positive harmonic [parabolic] function, harmonic [parabolic] measure can be replaced by h-harmonic [$\dot{h}$-parabolic] measure throughout the discussion to derive corresponding results in relative contexts. The continuous path processes for which the results are valid are the conditional [space-time] Brownian motion processes discussed in Chapter X, where the counterparts of the classical context results are proved.

5. Continuous Parameter Markov Processes with Stationary Transition Probabilities

Let $(X, \mathscr{X})$ be a measurable space, let μ be a probability measure on X, and let $(t, \xi, A) \mapsto p(t, \xi, A)$ be a stochastic continuous parameter transition function (Appendix VI.3). Let $\{x(t), \mathscr{F}(t), t \in \mathbb{R}^+\}$ be a Markov process with initial distribution μ and stationary transition probabilities, specifically with transition function p (Section 1), so that p satisfies the Chapman–Kolmogorov equation: for $A \in \mathscr{X}$, $s > 0$, $t > 0$,

$$p(s+t, \xi, A) = \int_X p(t, \eta, A) p(s, \xi, d\eta), \tag{5.1}$$

and for $s \geq 0$, $t > 0$,

$$P_\mu\{x(0) \in A\} = \mu(A), \qquad P_\mu\{x(s+t) \in A \mid \mathscr{F}(s)\} = p(t, x(s), A) \quad \text{a.s.} \tag{5.2}$$

We shall describe this Markov process from now on as a Markov process with stationary transition function p. If $0 = t_0 < \cdots < t_n$ and $A_j \in \mathscr{X}$,

$$\begin{aligned} &P_\mu\{x(t_j) \in A_j, j \leq n\} \\ &\quad = \int_{A_0} \mu(d\xi_0) \int_{A_1} p(t_1, \xi_0, d\xi_1) \cdots \int_{A_n} p(t_n - t_{n-1}, \xi_{n-1}, d\xi_n), \end{aligned} \tag{5.3}$$

and in particular,

$$\begin{aligned} &P_\xi\{x(t_j) \in A_j, j \leq n\} \\ &\quad = 1_{A_0}(\xi) \int_{A_1} p(t_1, \xi, d\xi_1) \cdots \int_{A_n} p(t_n - t_{n-1}, \xi_{n-1}, d\xi_n). \end{aligned} \tag{5.3'}$$

The evaluation (5.3) determines the finite-dimensional distributions of $x(\cdot)$ and thereby determines P_μ on $\mathscr{F}\{x(t), t \in \mathbb{R}^+\}$. Observe that the left side of (5.3′) defines an $\mathscr{X}$ measurable function of ξ whose integral with respect to μ is the left side of (5.3).

Throughout the rest of this section the measurable state space will be

Polish, coupled with its Borel subsets. Given a Polish state space $(X, \mathscr{X})$, a probability measure μ on $\mathscr{X}$, and a stochastic transition function p on $(X, \mathscr{X})$, there is a canonical Markov process with state space $(X, \mathscr{X})$, initial distribution μ, and transition function p. We sketch the definition of this process (cf. Section I.10). The space $\mathring{\Omega}$ is the space of functions $\mathring{\omega}$ from $\mathbb{R}^+$ into X, and $\mathring{x}(t, \mathring{\omega})$ is the value of $\mathring{\omega}$ at t. Define $\mathring{\mathscr{F}}(t) = \mathscr{F}\{\mathring{x}(s), s \leq t\}$, $\mathring{\mathscr{F}} = \bigvee_{t \geq 0} \mathring{\mathscr{F}}(t)$. The measure $\mathring{P}_\mu$ on $\mathring{\mathscr{F}}$ is determined by defining $\mathring{P}_\mu\{\mathring{x}(t_j) \in A_j, j \leq n\}$ by the right side of (5.3), and this definition is extended to $\mathring{\mathscr{F}}$ using Kolmogorov's theorem. The measure $\mathring{P}_\mu$ is then completed, thereby enlarging $\mathring{\mathscr{F}}$ to $\mathring{\mathscr{F}}_\mu$, and finally $\mathring{\mathscr{F}}_\mu(t)$ is defined as the σ algebra generated by $\mathring{\mathscr{F}}(t)$ and the $\mathring{P}_\mu$ null sets. The process $\{\mathring{x}(\cdot), \mathring{\mathscr{F}}_\mu(\cdot)\}$ is the desired canonical process on the probability space $(\mathring{\Omega}, \mathring{\mathscr{F}}_\mu, \mathring{P}_\mu)$. Observe that if $\{x(\cdot), \mathscr{F}(\cdot)\}$ is a Markov process with this state space $(X, \mathscr{X})$ and transition function p, on a probability space $(\Omega_\mu, \mathscr{F}_\mu, P_\mu)$, and if ϕ maps Ω_μ into $\mathring{\Omega}$ by $\phi(\omega) = \mathring{\omega} = x(\cdot, \omega)$, then in view of our convention that process probability measures are complete,

$$\phi^{-1}(\mathring{\mathscr{F}}_\mu) \subset \mathscr{F}, \qquad \phi^{-1}(\mathring{\mathscr{F}}(t)) \subset \mathscr{F}(t),$$

and in addition $\phi^{-1}(\mathring{\mathscr{F}}_\mu(t)) \subset \mathscr{F}(t)$ if $\mathscr{F}(0)$ contains the P_μ null sets. Moreover

$$P_\mu\{\phi^{-1}(\mathring{\Lambda})\} = \mathring{P}_\mu\{\mathring{\Lambda}\} \qquad (\mathring{\Lambda} \in \mathring{\mathscr{F}}_\mu). \tag{5.4}$$

These relations all follow from the fact that for $t_\cdot$ and $A_\cdot$ as above, if $\mathring{\Lambda} = \{\mathring{x}(t_j) \in A_j, j \leq n\}$, then $\phi^{-1}(\mathring{\Lambda}) \in \mathscr{F}(t_n)$, and (5.4) is true for this set $\mathring{\Lambda}$.

If $\mathring{\Lambda} \in \mathring{\mathscr{F}}$, the function $\xi \mapsto \mathring{P}_\xi(\mathring{\Lambda})$ is $\mathscr{X}$ measurable, that is, Borel measurable, and

$$\mathring{P}_\mu\{\mathring{\Lambda}\} = \int_X \mathring{P}_\xi(\mathring{\Lambda})\mu(d\xi) \tag{5.5}$$

because (5.3) is valid for $\mathring{P}_\mu$, and therefore the assertion is true for in $\mathscr{F}\{\mathring{x}(t_j), j \leq n\}$, hence true for $\mathring{\Lambda}$ in the algebra $\bigcup \mathscr{F}\{\mathring{x}(t_j), j \leq n\}$ (where the union is over all finite parameter sets), and finally the assertion is true for $\mathring{\Lambda}$ in $\mathring{\mathscr{F}}$. From the general theory (Appendix VI.2) we know that if for each ξ the measure $\mathring{P}_\xi$ is restricted to the class $\mathscr{U}(\mathring{\mathscr{F}})$ of universally measurable sets over $\mathring{\mathscr{F}}$ or even to the possibly larger class $\mathscr{U}_{\mathring{P}}(\mathring{\mathscr{F}})$ of universally measurable sets over $\mathring{\mathscr{F}}$ relative to the kernel $(\xi, \mathring{\Lambda}) \mapsto \mathring{P}_\xi(\mathring{\Lambda})$ on $\mathscr{X} \times \mathring{\mathscr{F}}$, then the function $\mathring{P}_\cdot\{\mathring{\Lambda}\}$ is universally measurable on X, and (5.5) remains true. In expectation language the function $\mathring{E}_\cdot\{\mathring{x}\}$ is $\mathscr{X}[\mathscr{U}(\mathscr{X})]$ measurable if $\mathring{x}$ is the indicator function of a set in $\mathring{\mathscr{F}}[\mathscr{U}_{\mathring{P}}(\mathring{\mathscr{F}})]$, and therefore by the usual approximation procedure if $\mathring{x}$ is $\mathring{\mathscr{F}}[\mathscr{U}_{\mathring{P}}(\mathring{\mathscr{F}})]$ measurable and is positive or absolutely $\mathring{P}_\mu$ integrable, then $\mathring{E}_\cdot\{\mathring{x}\}$ is $\mathscr{X}[\mathscr{U}(\mathscr{X})]$ measurable and

$$\mathring{E}_\mu\{\mathring{x}\} = \int_X \mathring{E}_\xi\{\mathring{x}\}\mu(d\xi). \tag{5.6}$$

Observe that (5.4) can now be interpreted as an integrated version of

$$P_\mu\{\phi^{-1}(\mathring{\Lambda})|x(0)\} = \mathring{P}_{x(0)}\{\mathring{\Lambda}\} \quad \text{a.s.} \qquad (\mathring{\Lambda}\in\mathscr{U}_{\mathring{P}}(\mathring{\mathscr{F}})), \tag{5.7}$$

and the right side of (1.1′) can be expressed in the form $\mathring{E}_{\mathring{x}(s)}\{z\}$.

In view of (5.4) and the corresponding equation for expectations the relations (5.5) and (5.6) can be applied to noncanonical processes to deduce that

$$P_\mu\{\phi^{-1}(\mathring{\Lambda})\} = \int_X P_\xi\{\phi^{-1}(\mathring{\Lambda})\}\mu(d\xi), \quad E_\mu\{\phi^{-1}(\mathring{x})\} = \int_X E_\xi\{\phi^{-1}(\mathring{x})\}\mu(d\xi), \tag{5.8}$$

where $\mathring{\Lambda}$ and $\mathring{x}$ are as above and $\phi^{-1}(\mathring{x}) = \mathring{x}(\phi)$. Here P_ξ is the probability measure, and E_ξ is the corresponding expectation operator for an arbitrary Markov process with state space $(X, \mathscr{X})$, transition function p, initial point ξ, on a probability space $(\Omega_\xi, \mathscr{F}_\xi, P_\xi)$, and the set Ω_ξ may vary with ξ.

6. Specialization to Right Continuous Processes

Suppose in Section 5 that the transition function p has the property that for each point ξ in the Polish state space X there is a probability space $(\Omega_\xi, \mathscr{F}_\xi, P_\xi)$ on which there is a right continuous process with initial point ξ and transition function p. We can then define the associated right continuous process, in which $\mathring{\Omega}$ is the class of right continuous functions from $\mathbb{R}^+$ into X, $\mathring{x}(t)$, $\mathring{\mathscr{F}}$, $\mathring{\mathscr{F}}(t)$, ϕ are defined as in Section 5 (with this change in the definition of $\mathring{\Omega}$, and $\mathring{P}_\xi\{\mathring{\Lambda}\}$ for $\mathring{\Lambda}$ in $\mathring{\mathscr{F}}$ is defined by

$$\mathring{P}_\xi\{\mathring{\Lambda}\} = P_\xi\{\phi^{-1}(\mathring{\Lambda})\}. \tag{6.1}$$

The measure $\mathring{P}_\xi$ is then completed, thereby enlarging $\mathring{\mathscr{F}}$ to $\mathring{\mathscr{F}}_\xi$. This definition of $\mathring{P}_\xi$ implies that $\mathring{P}_\xi$ satisfies (5.2) (with μ in that equation supported by $\{\xi\}$), and it follows as in Section 5 that the function $\xi \mapsto \mathring{P}_\xi\{\mathring{\Lambda}\}$ is $\mathscr{X}$ measurable when $\mathring{\Lambda}\in\mathring{\mathscr{F}}$. The measure $\mathring{P}_\mu$ defined by (5.4) on $\mathring{\mathscr{F}}$ makes $\{\mathring{x}(\cdot), \mathring{\mathscr{F}}(\cdot)\}$ a right continuous Markov process with initial distribution μ and transition function p. We have thus proved that there is such a process for every μ. It is now clear that everything in Section 5 is true in the present context in which the members of $\mathring{\Omega}$ are right continuous. What is new is that the canonical processes are now right continuous, and this fact induces new process properties. We shall derive properties of the hitting of analytic sets in the present context, keeping the notation of Section 5 but using the fact that the canonical processes are now necessarily progressively measurable.

Suppose that A is an analytic subset of X. We have shown in Section I.9 that

$$\mathring{\Lambda} = \{\mathring{\omega} : \mathring{x}(s, \mathring{\omega}) \in A \text{ for some } s \le t\} \in \mathring{\mathscr{F}}_\mu(t). \tag{6.2}$$

It now follows from (5.4) that for any almost surely right continuous Markov process $x(\cdot)$ on a probability space $(\Omega, \mathscr{F}, P)$ with state space $(X, \mathscr{X})$, initial distribution μ, and transition function p the probability of hitting A at some time $\le t$, that is, the probability that the entry time T of A is at most t, is $P\{T \le t\} = \mathring{P}_\mu\{\phi^{-1}(\mathring{\Lambda})\}$ so that this probability does not depend on the choice of probability space. If μ is supported by $\{\xi\}$, this probability is $\mathring{P}_\xi\{\phi^{-1}(\mathring{\Lambda})\}$ and is universally measurable as a function of ξ. Finally the probability for general μ is the integral $\int_X \mathring{P}_\xi\{\phi^{-1}(\mathring{\Lambda})\}\mu(d\xi)$. The corresponding remarks are valid for hitting and entry time distributions. We conclude that such probabilities and expectations as the following define universally measurable functions of the initial point ξ.

(a) $P_\xi\{\sup_{t \in I} u[x(t)] > c\}$ for I an interval and u a Borel measurable function from X into $\overline{\mathbb{R}}$.
(b) $E_\xi\{\sup_{t \in I} u[x(t)]\}$ if [notation of (a)] $u \ge 0$ or if this expectation with u replaced by $|u|$ is finite for all ξ.
(c) $P_\xi\{\limsup_{t \uparrow T} u[x(t)] > c\}$ for u as in (a) and for T the hitting time of an analytic subset of X by $x(\cdot)$.
(d) $E_\xi\{\limsup_{t \uparrow T} u[x(t)]\}$ if [notation of (c)] $u \ge 0$ or if this expectation with u replaced by $|u|$ is finite for all ξ.

If for every point ξ of the state space there is a continuous process with initial point ξ and the given transition function p, then all the work in this section can also be carried through with $\mathring{\Omega}$ the space of continuous functions from $\mathbb{R}^+$ into X.

The Hitting of an F_σ Set

We now change some of the notation in the right continuous context we are treating. Let A be an analytic subset of X, let I be a compact parameter interval, let $x_\xi(\cdot)$ be a right continuous Markov process from ξ with the given transition function p, and recall that the space on which $x_\xi(\cdot)$ is defined may depend on ξ. Consider the function

$$\xi \mapsto P\{x_\xi(t) \in A \text{ for some } t \text{ in } I\}. \tag{6.3}$$

According to the discussion we have given, the function (6.3) is universally measurable. We shall now show, under further hypotheses imposed on A and the transition function that the function (6.3) is even Borel measurable and that the proof of this Borel measurability does not need the theory of analytic sets. This fact means that some later results in this book can be obtained without the (unpalatable to some readers) invocation of analytic sets. Abstention from analytic set theory makes stochastic process theory much less elegant and sadly incomplete, however. Observe that if $x_\xi(\cdot)$ is as above and if I' is a countable subset of parameter values, the function

$$\xi \mapsto P\{x_\xi(t) \in A \text{ for some } t \text{ in } I'\} \tag{6.4}$$

is Borel measurable. In particular, if A is open and if I' is a countable dense subset of I including both endpoints, the functions in (6.3) and (6.4) are identical. Now suppose that for each ξ the process $x_\xi(\cdot)$ can be supposed continuous. In this context we shall now show that the function (6.3) is Borel measurable when A is an F_σ set. It is sufficient to prove this Borel measurability when A is closed. Let B_n be for $n \geq 1$ the open set of points at distance $< 1/n$ from A. Then A is hit by $x_\xi(\cdot)$ in the parameter interval I if and only if B_n is hit by $x_\xi(\cdot)$ in the parameter interval I for every n. Hence the function (6.3) is the limit as $n \to \infty$ of (6.3) with A replaced by B_n; so the function (6.3) is Borel measurable.

Observe that if $x_\xi(\cdot)$ above is supposed merely almost surely right continuous rather than right continuous, or almost surely continuous rather than continuous, then the conclusions remain valid and the reasoning is unchanged aside from the acknowledgement of exceptional null sets.

7. Continuous Parameter Markov Processes: Lifetimes and Trap Points

Recall that a function defined on a subset A of a measurable space is said to be measurable if A is measurable and if the function is measurable with respect to the class of measurable subsets of A.

Let $(t, \xi, A) \mapsto p(t, \xi, A)$ be a substochastic stationary transition function with state space $(X, \mathscr{X})$. The Chapman–Kolmogorov equation implies that the function $t \mapsto p(t, \xi, X)$ is monotone decreasing on $]0, +\infty[$ for all ξ. We assume throughout this discussion that $\lim_{t \to 0} p(t, \cdot, X) = 1$, which implies that the function $p(\cdot, \xi, X)$ is right continuous on $]0, +\infty[$ for all ξ. Generalizing Section 5 slightly, define a Markov process with state space $(X, \mathscr{X})$, parameter set $\mathbb{R}^+$, initial probability distribution μ, and stationary substochastic transition function p as an adapted process $\{x(t), \mathscr{F}(t), t \in \mathbb{R}^+\}$ satisfying (for $0 \leq s < t$ and $A \in \mathscr{X}$)

$$P\{x(0) \in A\} = \mu(A), \qquad P\{x(t) \in A \mid \mathscr{F}(s)\} = p(t-s, x(s), A) \quad \text{a.s.} \tag{7.1}$$

The process random variables need not be defined on the whole space; so "a.s." means "almost everywhere where $x(s)$ is defined." In any expression involving a random variable of the process the added condition that the random variable is defined is to be understood and will sometimes be written explicitly. Thus the set $\{x(t) \in X\}$ is the domain of definition of $x(t)$. With this convention (5.3) remains valid. According to (7.1) and the Chapman–Kolmogorov equation, the random variable $x(0)$ is defined almost surely, and the function $t \mapsto P\{x(t) \in X\}$, equal to $\int_X p(t, \xi, X)\mu(d\xi)$ when $t > 0$, is monotone decreasing and right continuous on $\mathbb{R}^+$. If $0 < s < t$, the random variable $x(s)$ is defined almost everywhere where $x(t)$ is defined because

$$\begin{aligned} P\{x(s) \in X, x(t) \in X\} &= \int_X \mu(d\xi) \int_X p(t-s, \eta, X) p(s, \xi, d\eta) \\ &= \int_X p(t, \xi, X)\mu(d\xi) = P\{x(t) \in X\}. \end{aligned} \tag{7.2}$$

If there is a positive random variable S ($\leq +\infty$) such that $x(t)$ is defined if and only if $t < S$, the random variable S is called the process *lifetime*. Integral parameter process lifetimes were discussed in Section 3. The existence of a process lifetime is not much of a restriction. In fact without the hypothesis of the existence of a process lifetime define S by

$$S(\omega) = \sup\{r\colon r \text{ is rational}, x(r) \in X\}$$

so that S is almost surely strictly positive and

$$P\{x(t) \in X\} = P\{S \geq t\} = P\{S > t\}.$$

Define $y(t)$ on the set $\{S > t\}$ [which differs from the domain of $x(t)$ by a null set] as $x(t)$ and leave $y(t)$ undefined otherwise. Then the process $\{y(\cdot), \mathscr{F}(\cdot)\}$ is Markovian with transition function p, initial distribution μ, lifetime S and is a standard modification of $\{x(\cdot), \mathscr{F}(\cdot)\}$.

If $\{x(\cdot), \mathscr{F}(\cdot)\}$ is a Markov process with substochastic transition function p and lifetime T, enlarge the state space by adjoining a single point ∂ and extend p to be stochastic on the enlarged space, making ∂ an absorbing point (Appendix VI.3). If now $x(t)$ is defined as ∂ for $t > T$, the extended process is Markovian with transition function the extension of p. Thus results for Markov processes with stochastic transition functions can be applied to those with substochastic transition functions.

Going in the opposite direction, if p is a substochastic transition function with Polish state space X and if μ is a probability distribution on the state space, there is a Markov process with transition function p and initial distribution μ according to the following argument. Adjoin a point ∂ to X as an isolated point of $X \cup \partial$, and extend p by making ∂ an absorbing point. There is then a Markov process $\{\mathring{x}(\cdot), \mathring{\mathscr{F}}(\cdot)\}$ with the extended state space, the extended transition function, and the given initial distribution (supported by X). If $\mathring{x}(t)$ is replaced by its restriction $x(t)$ to the set $\{\mathring{x}(t) \in X\}$, the process $\{x(\cdot), \mathring{\mathscr{F}}(\cdot)\}$ will have the properties described at the beginning of this section and, if desired, can be modified to have a lifetime, as discussed above.

A σ Algebra Equality for Almost Surely Continuous Markov Processes

We shall need the following intuitively obvious fact. Let $x(\cdot)$ be an almost surely continuous Markov process from a point ξ. It is supposed that the state space is a Polish locally compact but not compact space. Let S be the process lifetime. A trap point is adjoined as described above. Let ζ be the adjoined point of the Alexandrov one-point compactification of the state space. It is supposed that $x(S-) = \zeta$ almost surely. For t in $\mathbb{R}^+$ let $\mathscr{F}(t)$ be the σ algebra generated by the null sets of the $x(\cdot)$ probability space and $\mathscr{F}\{x(s), s \leq t\}$, and define $\mathscr{F} = \bigvee_{t \in \mathbb{R}^+} \mathscr{F}(t)$. Let $B_\cdot$ be an increasing sequence of open relatively compact subsets of the state space, containing ξ, with union the state space, and let S_n be the hitting time of ∂B_n by $x(\cdot)$. Then $\lim_{n\to\infty} S_n = S$ almost surely. Define $\mathscr{G} = \bigvee_{n \in \mathbb{Z}^+} \mathscr{F}(S_n)$. Then we show that $\mathscr{F} = \mathscr{G}$. The inclusion $\mathscr{G} \subset \mathscr{F}$ is known from Section I.12. To prove the reverse inclusion, we prove that $x(t)$ is $\mathscr{G}$ measurable for each t. In doing this it is no restriction to identify the process trap point with ζ. Then we need only observe that $x(t \wedge S_n)$ is $\mathscr{F}(t \wedge S_n) \subset \mathscr{F}(S_n) \subset \mathscr{G}$ measurable and that $x(t) = \lim_{n\to\infty} x(t \wedge S_n)$ almost surely.

8. Right Continuity of Markov Process Filtrations; A Zero-One (0-1) Law

In some contexts the desirable right continuity of Markov process filtrations is intrinsic, as illustrated by the following theorem.

Theorem. *Let $\{x(\cdot), \mathscr{F}(\cdot)\}$ be an almost surely right continuous Markov process with Polish state space and with transition function p. Suppose that whenever $t > 0$ and f is a bounded continuous function on the state space, the function $p(t, \cdot, f)$ is continuous, or at least the process $p(t, x(\cdot), f)$ is almost surely right continuous. Then*

(a) *The process $\{x(\cdot), \mathscr{F}^+(\cdot)\}$ is Markovian with transition function p.*
(b) *If for all $t \geq 0$ the σ algebra $\mathscr{F}(t)$ is generated by the null sets and $\mathscr{F}\{x(s), s \leq t\}$, then $\mathscr{F}^+(\cdot) = \mathscr{F}(\cdot)$.*

Observation. If the state space is locally compact and second countable, it will be clear from the proof of the theorem that f in the hypotheses need only be continuous with compact support.

The theorem and the following proof are valid even when p is substochastic. For f as in the theorem the Markov property in the form

$$E\{f[x(s' + t)] | \mathscr{F}(s')\} = p(t, x(s'), f) \quad \text{a.s.} \quad (t > 0) \tag{8.1}$$

implies when $s' \downarrow s > 0$ sequentially that

$$E\{f[x(s+t)]|\mathscr{F}^+(s)\} = p(t, x(s), f) \quad \text{a.s.} \tag{8.2}$$

in view of the hypotheses of the theorem and the dominated convergence theorem for conditional expectations. Thus (a) is true. Under the added hypothesis of (b), Lemma 2 implies that $\mathscr{F}^+(t) \subset \mathscr{F}(t)$; the reverse inclusion is trivial; so (b) is true.

A 0–1 *Law*. Suppose under the hypotheses of (b) in the theorem that the Markov process is one from an initial point. In this case the σ algebra $\mathscr{F}(0) = \mathscr{F}(0+)$ consists of the null sets and their complements. In intuitive phraseology this evaluation of $\mathscr{F}(0+)$ states that every sample function asymptotic property as the parameter value tends to 0 is either almost surely true or almost surely false. For example, if T is the hitting time by $x(\cdot)$ of an analytic subset of the state space, then $P\{T = 0\}$ must be either 0 or 1. See Section VII.6 for a discussion of this 0-1 law and its application in the context of Brownian motion.

9. Strong Markov Property

The following continuous parameter version of the strong Markov property indicates possible strengthening of the discrete parameter version in Section 3, but we shall not need such a strengthening. Recall (from Sections II.1 and II.2) that if T is optional for a filtration $\mathscr{F}(\cdot)$, then $\mathscr{F}^+(T) = \mathscr{F}(T+)$ and that if S is an $\mathscr{F}(T)$ measurable random variable and if $S \geq T$, then S is optional. Recall also that if $x(\cdot)$ is a process with parameter set $\mathbb{R}^+$ and state space the extended reals, and if $0 \leq S \leq +\infty$, then the notation $x(S)$ refers to the function defined on the set $\{S < +\infty\}$ as $x(S)$ and defined as 0 elsewhere. We also adopt the convention that if p is a continuous parameter transition function, then $p(0, \cdot, A) = 1_A$ for A a measurable state space subset.

Theorem. *Let p be a continuous parameter substochastic transition function with a Polish state space, and suppose that p has the following properties*:

(a) *If ξ is a point of the state space, there is a right continuous Markov process $x_\xi(\cdot)$ with transition function p and initial point ξ.*

(b) *If f is a bounded continuous function on the state space (or merely continuous with compact support if the state space is locally compact and second countable) and if $t > 0$, the function $p(t, \cdot, f)$ is continuous, or at least the process $\{p(t, x_\xi(s), f), s \in \mathbb{R}^+\}$ is almost surely right continuous for each $x_\xi(\cdot)$ process in* (a).

Then if $\{x(\cdot), \mathscr{F}(\cdot)\}$ is a right continuous Markov process with transition function p, if S and T are optional times with $S \geq T$, if S is $\mathscr{F}(T)$ measurable, and if A is a Borel subset of the state space,

$$P\{x(S) \in A|\mathscr{F}(T+)\} = p(S-T, x(T), A) \quad \text{a.e. on } \{S < +\infty\}. \tag{9.1}$$

In particular, when $S = T + t$ with t a positive constant,

$$P\{x(T+t)\in A|\mathscr{F}(T+)\} = p(t, x(T), A) \quad \text{a.e. on } \{T < +\infty\}. \tag{9.2}$$

Observation (1). Condition (a) implies that for f as in (b) the function $p(\cdot, \xi, f)$ is right continuous on $\mathbb{R}^+$ because if $x_\xi(\cdot)$ is chosen as in (a), the function $t \mapsto p(t, \xi, f) = E\{f[x_\xi(t)]\}$ is right continuous.

Observation (2). If $\mathscr{F}_T(t) = \mathscr{F}(T+t)$, the theorem implies that the process $\{x(T+\cdot), \mathscr{F}_T^+(\cdot)\}$ is a Markov process on $\{T < +\infty\}$ with transition function p and initial distribution the distribution of $x(T)$. To see this, first note that the process in question is adapted because (Section II.3) the function $x(T+t)$ is $\mathscr{F}(T+t)$ measurable, and second note that if $t_0 \geq 0$ and if the pair $(T, T+t)$ in (9.2) is replaced by $(T+t_0, T+t_0+t)$, then (9.2) becomes the Markov property for the translated process; that is,

$$P\{x(T+t_0+t)\in A|\mathscr{F}_T^+(t_0)\} = p(t, x(T+t_0), A) \quad \text{a.e. on } \{T < +\infty\}. \tag{9.3}$$

Observation (3). If $S = t \vee T$ with t a strictly positive constant, (9.1) yields

$$P\{x(t)\in A|\mathscr{F}(T+)\} = p(t-T, x(T), A) \quad \text{a.e. on } \{T \leq t\}. \tag{9.4}$$

Observation (4). Hypothesis (a) is not weakened if $x_\xi(\cdot)$ is supposed only almost surely right continuous because then there is also a right continuous process with the other stated properties. In the same vein the following is an alternative conclusion to the theorem: if $\{x(\cdot), \mathscr{F}(\cdot)\}$ is an almost surely right continuous Markov process with transition function p, if $\mathscr{F}(0)$ contains the null sets, if S and T are optional times with $S \geq T$, if S is $\mathscr{F}(T)$ measurable, and if A is a Borel subset of the state space, then (9.1) is true. In fact under these hypotheses the process $\{x(\cdot), \mathscr{F}(\cdot)\}$ is indistinguishable from a right continuous process also adapted to $\mathscr{F}(\cdot)$, and the theorem can be applied to the latter process.

In proving the theorem it is no restriction (Section 7) to assume that p is stochastic. We can also assume without loss of generality that $\mathscr{F}(0)$ contains the null sets because (Section I.8) $\mathscr{F}(\cdot)$ can be extended if necessary to make this so. Furthermore it can be assumed that $\mathscr{F}(\cdot)$ is right continuous because (by Theorem 8) the process $\{x(\cdot), \mathscr{F}^+(\cdot)\}$ is Markovian with transition function p.

It will be convenient to prove (9.1) in the equivalent form

$$E\{f[x(S)]|\mathscr{F}(T)\} = p(S-T, x(T), f) \quad \text{a.e. on } \{S < +\infty\} \tag{9.5}$$

when f is as described in (b).

The Special Case (9.2)

If $S = T + t$, equation (9.2) will be proved by proving first that

$$E\{f[x([T]_n + t)] \mid \mathscr{F}([T]_n)\} = p(t, x([T]_n), f) \quad \text{a.e. on } \{[T]_n < +\infty\}. \tag{9.6}$$

That is we prove that for Λ in $\mathscr{F}([T]_n)$

$$\int_{\Lambda \cap \{[T]_n < +\infty\}} f[x([T]_n + t)]\, dP = \int_{\Lambda \cap \{[T]_n < +\infty\}} p(t, x([T]_n), f)\, dP. \tag{9.6'}$$

It is sufficient to prove (9.6′) with the integration set replaced by $\Lambda \cap \{[T]_n = j2^{-n}\}$, a set in $\mathscr{F}(j2^{-n})$, and with this replacement $[T]_n$ can be replaced in the integrands by $j2^{-n}$ so that (9.6′) becomes an immediate consequence of the Markov property of $x(\cdot)$. In particular, (9.6′) is valid when Λ is in $\mathscr{F}(T)$, and with this choice (9.6′) becomes the integrated version of (9.2) when $n \to \infty$.

The Special Case $T = 0$

If z_{jn} is the indicator function of the set $\{[S]_n = j2^{-n}\}$,

$$\begin{aligned} E\{f[x([S]_n)] \mid \mathscr{F}(0)\} &= \sum_{j=0}^{\infty} z_{jn} E\{f[x(j2^{-n})] \mid \mathscr{F}(0)\} \\ &= p([S]_n, x(0), f) \quad \text{a.e. on } \{S < +\infty\}. \end{aligned} \tag{9.7}$$

When $n \to \infty$, the left side of (9.7) becomes $E\{f[x(S)] \mid \mathscr{F}(0)\}$ almost surely because $x(\cdot)$ is right continuous, and the right side has almost sure limit $p(S, x(0), f)$ in view of Observation (1). Thus (9.5) is true when $T = 0$.

General Case

Apply the first special case to find that the process $\{x(T + \cdot), \mathscr{F}_T(\cdot)\}$ is a right continuous Markov process on $\{T < +\infty\}$ with transition function p. Next we apply the second special case. If T is finite valued, the fact that S and T are optional for $\mathscr{F}(\cdot)$ implies (according to Section II.2(h), but the notation there is different) that $S - T$ is optional for $\mathscr{F}_T(\cdot) = \mathscr{F}(T + \cdot)$. Since $S - T$ is $\mathscr{F}_T(0)$ measurable, the second special case can be applied to $\{x(T + \cdot), \mathscr{F}_T(\cdot)\}$ with the pair $(0, S - T)$ of optional times, to show that (9.5) is true. If T is not necessarily finite valued, observe that $S \wedge n$ is $\mathscr{F}(T \wedge n)$ measurable for all n and apply (9.5), but with the pair $(S \wedge n,$

$T \wedge n$) of optional times, to find that for $\Lambda \in \mathscr{F}(T)$, so that

$$\Lambda_n = \Lambda \cap \{S \leq n\} = \Lambda \cap \{T \leq n\} \cap \{S \leq n\} \in \mathscr{F}(n) \cap \mathscr{F}(T) = \mathscr{F}(T \wedge n),$$

$$\int_{\Lambda_n} f[x(S)]\, dP = \int_{\Lambda_n} p(S - T, x(S), f)\, dP.$$

When $n \to \infty$ this equation becomes the integrated version of (9.5).

10. Probabilistic Potential Theory; Excessive Functions

Let p be a continuous parameter substochastic transition function with measurable state space $(X, \mathscr{X})$. G. A. Hunt showed how a far-reaching generalization of classical potential theory can be obtained by defining the analogs of the classical concepts in terms of p. In this theory the role of positive superharmonic functions is taken by the α-excessive functions. Here α is a positive parameter, and an α-excessive function u is by definition a measurable function from X into $\overline{\mathbb{R}}^+$ which satisfies

$$\begin{aligned} &\text{(a)}\quad e^{-\alpha t} p(t, \cdot, u) \leq u, \\ &\text{(b)}\quad \lim_{t \to 0} p(t, \cdot, u) = u. \end{aligned} \tag{10.1}$$

An application of the Chapman–Kolmogorov equation shows that (10.1a) alone implies that the function $t \mapsto e^{-\alpha t} p(t, \cdot, u)$ is monotone decreasing, and it then follows readily that the limit function $\lim_{t \to 0} p(t, \cdot, u)$ is necessarily α-excessive.

Examples. The identically vanishing function is α-excessive for all α. A positive constant function $u = c$ satisfies (10.1a) for all α. Hence, if $\alpha \geq 0$ and if $0 < c < +\infty$, the function u is α-excessive if and only if $\lim_{t \to 0} p(t, \cdot, X) = 1$, and if $c = +\infty$, the function u is α-excessive if and only if $\lim_{t \to 0} p(t, \cdot, X) > 0$. If p is the transition function for Brownian motion in a connected Greenian subset of $\mathbb{R}^N$, it will be shown (Sections IX.6 and IX.8) that a positive function is 0-excessive if and only if it is identically $+\infty$ or is superharmonic. Corresponding results will be proved in Section IX.16 for α-excessive functions ($\alpha > 0$).

The Hunt theory has been presented in varying degrees of generality. For example, the theory includes classical potential theory on Greenian subsets of $\mathbb{R}^N$ with regular boundaries if the following hypotheses are imposed.

H1. The state space is a locally compact Hausdorff space satisfying the second separability condition, coupled with its Borel sets (but sometimes it is necessary for adequate generality to make $\mathscr{X}$ the class of universally measurable sets over the Borel sets).

H2. The transformation $f \mapsto p(t, \cdot, f)$ takes the Banach space of continuous functions f on X with compact support into itself, and this transformation has the identity as strong limit when $t \to 0$.

We shall discuss various aspects of Hunt theory below, as needed. A 0-excessive function is called *excessive*. An α-excessive function is also β-excessive for $\beta > \alpha$. If $p_\alpha(t, \cdot, \cdot) = e^{-\alpha t} p(t, \cdot, \cdot)$, then p_α is a substochastic continuous parameter transition function, and a function is β-excessive relative to p_α if and only if the function is $(\alpha + \beta)$-excessive relative to p. If p is made stochastic by adjunction of a trap point to the state space, as described in Section 7 an α-excessive function before the trap is adjoined remains so if defined as 0 at the trap point. A simple argument shows that the limit of an increasing sequence of α-excessive functions is α-excessive.

If it is supposed that $p(\cdot, \cdot, A)$ is $\mathscr{B}(\mathbb{R}^+) \times \mathscr{X}$ measurable when A is in $\mathscr{X}$, then the role of a Green function is taken by the kernel G_X^α with state space $(X, \mathscr{X})$ and positive parameter α, defined by

$$G_X^\alpha(\xi, A) = \int_0^\infty e^{-\alpha t} p(t, \xi, A) l_1(dt). \tag{10.2}$$

If f is a positive measurable function on X its *α-potential* is the measurable function on X defined by

$$G_X^\alpha(\xi, f) = \int_0^\infty e^{-\alpha t} p(t, \xi, f) l_1(dt). \tag{10.3}$$

This α-potential is α-excessive because

$$p_\alpha(t, \xi, G_X^\alpha(\cdot, f)) = \int_t^\infty e^{-\alpha s} p(s, \xi, f) l_1(ds) \leq G_X^\alpha(\xi, f) \tag{10.4}$$

and there is equality in the limit when $t \to 0$.

It will be shown in Section IX.17 that if the state space is a Greenian subset D of $\mathbb{R}^N$ coupled with its Borel subsets and if p is the transition function of the Brownian motion process in D, then

$$G_D^0(\xi, d\eta) = \text{const } G_D(\xi, \eta) l_N(d\eta).$$

Thus the Hunt potential of a function f becomes the classical potential of the measure $f\, dl_N$ up to a multiplicative constant. A corresponding result will be proved relating parabolic potential theory will space-time Brownian motion.

The analysis of α-excessive functions and α-potentials for the transition function p is the same as the analysis of excessive functions and 0-potentials for the transition function p_α.

Invariant Excessive Functions

If u is not only α-excessive but

$$e^{-\alpha t}p(t,\cdot,u) = u \tag{10.5}$$

for all $t > 0$, then u is called *invariant α-excessive*. For example, if v is α-excessive, the function $\lim_{t\to\infty} e^{-\alpha t}p(t,\cdot,v)$ is invariant α-excessive. If u is α-excessive and if (10.5) is true for some strictly positive t, then u is invariant α-excessive. In fact on the one hand (10.5) is true for $2t$ when true for t because if true for t, then

$$p_\alpha(2t,\cdot,u) = p_\alpha(t,\cdot,p_\alpha(t,\cdot,u)) = p_\alpha(t,\cdot,u) = u,$$

and on the other hand by monotoneity of the function $p_\alpha(\cdot,\xi,u)$ equation (10.5) is true for $t < s$ if true for s.

The concept of an α-excessive function links Hunt potential theory with martingale theory because the condition (10.1a) is precisely the condition that if $\{x(\cdot),\mathscr{F}(\cdot)\}$ is a Markov process with transition function p, then the process $\{e^{-\alpha t}u[x(t)],\mathscr{F}(t),t\in\mathbb{R}^+\}$ is a supermartingale on the parameter set for which the process random variables have finite expectations. The condition (10.5) for an invariant α-excessive function u is the condition that this supermartingale be a martingale on this parameter set.

The following lemma will be useful.

Lemma. *If u is α-excessive and if*

$$u_n = n\int_0^{1/n} e^{-\alpha s}p(s,\cdot,u\wedge n)l_1(ds), \qquad n\in\mathbb{Z}^+,$$

then the sequence $u_\bullet$ is an increasing sequence of bounded α-excessive functions, with limit u.

At the expense of replacing p by p_α we can assume that $\alpha = 0$. Obviously $u_n \le u \wedge n$ and

$$p(t,\cdot,u_n) = n\int_0^{1/n} p(s+t,\cdot,u\wedge n)l_1(ds). \tag{10.6}$$

It is trivial that $u \wedge n$ satisfies (10.1a); so the function $p(\cdot,\xi,u\wedge n)$ is monotone decreasing and furthermore (10.6) implies that this function is continuous. Hence the evaluation (10.6) implies that u_n is excessive. Moreover the monotoneity of $p(\cdot, \xi, u \wedge m)$ implies that the sequence

$$n \mapsto n \int_0^{1/n} p(s, \cdot, u \wedge m) l_1(ds)$$

is monotone increasing for fixed m; so the sequence $u_{\bullet}$ is a monotone increasing sequence. Finally,

$$u \geq \lim_{n\to\infty} u_n \geq \lim_{n\to\infty} n \int_0^{1/n} p(s, \cdot, u \wedge m) l_1(ds) = u \wedge m$$

for all m; so $u_{\bullet}$ has limit u as desired.

Excessive Measures

The dual of an α-excessive function for a transition function p with state space $(X, \mathscr{X})$ is an α-excessive measure, defined as a measure on $\mathscr{X}$ which satisfies, for $A \in \mathscr{X}$,

$$\begin{aligned} &\text{(a)} \quad e^{-\alpha t} \int_X p(t, \xi, A) \mu(d\xi) \leq \mu(A), \\ &\text{(b)} \quad \lim_{t\to 0} \int_X p(t, \xi, A) \mu(d\xi) = \mu(A). \end{aligned} \tag{10.7}$$

Condition (b) is not usually imposed because in most contexts it follows from (a). Further conditions, such as (if X is topological) finiteness of μ on compact sets, are usually imposed. An application of the Chapman–Kolmogorov equation shows that (10.7)(a) alone implies that the function on the left side of (10.7)(a) defines a decreasing function of t. The limit measure when $t \to 0$ necessarily satisfies both conditions in (10.7). A 0-excessive measure is called *excessive*. If μ is not only α-excessive but satisfies, for $A \in \mathscr{X}$,

$$e^{-\alpha t} \int_X p(t, \xi, A) \mu(d\xi) = \mu(A) \tag{10.8}$$

for all $t > 0$, then μ is called *invariant α-excessive*.

The Choice of State Space

In the most common Markov process setting the state space for the transition function p is a topological space coupled with some σ algebra of subsets including the Borel subsets. For example, the state space of Brownian motion in an open subset D of $\mathbb{R}^N$ can be taken as the σ algebra of Borel subsets of D or perhaps more naturally as the σ algebra of l_n measurable subsets of D (the last because the transition distributions are absolutely continuous relative to l_N). At first glance it may seem that whatever the context, it makes no difference which of the possible σ algebras is chosen. But in fact the choice is involved in the definition of excessive functions and measures. In the Brownian motion case it makes no difference in the sense that (Sections IX.6, 8, 18) whichever choice is made of the three possibilities mentioned, a function u is excessive if and only if on each connected component of D this function is either identically $+\infty$ or positive and superharmonic, and a measure is excessive if and only if it is the indefinite integral of an excessive function. For space-time Brownian motion in an open subset $\dot{D}$ of $\dot{\mathbb{R}}^N$, however, the situation is different. It will be seen (Section IX.18) that coupling $\dot{D}$ with its σ algebra of Borel subsets is less suitable than coupling with the much larger σ algebra of those subsets of $\dot{D}$ which meet every hyperplane orthogonal to the ordinate axis in a Borel set. (Replacing "Borel" here by "l_N measurable" would not change the classes of excessive functions and measures.)

When the context is not otherwise unambiguous, we shall use the language "$\mathscr{X}$ measurable excessive function" and "excessive measure on $\mathscr{X}$" to indicate the choice of $\mathscr{X}$.

11. Excessive Functions and Supermartingales

In probabilistic potential theory the state space is usually topological, and enough hypotheses (for example, H1 and H2 in Section 10) are imposed on the given state space and transition function to ensure the following.

(a) For any initial distribution there is a right continuous Markov process $\{x(\cdot), \mathscr{F}(\cdot)\}$ on some probability space, with the specified initial distribution and transition function.

(b) If u is α-excessive, the process $u[x(\cdot)]$ is almost surely right continuous.

If u is a bounded α-excessive function, the process $\{e^{-\alpha t}u[x(t)], \mathscr{F}(t), t \in \mathbb{R}^+\}$ is therefore an almost surely right continuous supermartingale, under the convention that $u[x(t)]$ is defined as 0 at times at least equal to the $x(\cdot)$ process lifetime. Since (Lemma 10) every α-excessive function u is the limit of an increasing sequence of bounded α-excessive functions, Theorem IV.4 implies that for any α-excessive u the process $u[x(\cdot)]$ is almost surely right continuous with left limits. The process $\{e^{-\alpha t}u[x(t)], \mathscr{F}(t), t \in \mathbb{R}^+\}$ is a supermartingale on the parameter set for which the process random variables have finite expectations, and if the expectation is finite at time t_0, the process is a supermartingale on the parameter interval $[t_0, +\infty[$.

12. Excessive Functions and the Hitting Times of Analytic Sets (Notation and Hypotheses of Section 11)

Let p be a substochastic transition function, let A be an analytic subset of the (Polish) state space, and let $\{x_\xi(\cdot), \mathscr{F}_\xi(\cdot)\}$ be an almost surely right continuous Markov process with transition function p and initial point ξ. The probability space on which the process is defined may depend on ξ. Let T_ξ $[T_{t\xi}]$ be the hitting time of A by $x_\xi(\cdot)$, $[x_\xi(t + \cdot)]$. Then $\lim_{t\to 0} T_{t\xi} = T_\xi$ almost surely. Let $\alpha \in \mathbb{R}^+$, let u be an α-excessive function on the state space, and define

$$v(\xi) = E\{e^{-\alpha T_\xi} u[x_\xi(T_\xi)]\}, \tag{12.1}$$

with the convention that the integrand is 0 where $T_\xi = +\infty$. According to Section 6, the function v is universally measurable. If u is bounded, the process $\{e^{-\alpha t}u[x_\xi(t)], \mathscr{F}_\xi(t), t \in \mathbb{R}^+\}$ is a supermartingale, and the supermartingale inequality (Theorem IV.3) at times 0, T_ξ, $t + T_{t\xi}$ yields

$$\begin{aligned} u(\xi) \geq v(\xi) &\geq E\{e^{-\alpha(t+T_{t\xi})} u[x_\xi(t + T_{t\xi})]\} \\ &= e^{-\alpha t} E\{E\{e^{-\alpha T_{t\xi}} u[x_\xi(t + T_{t\xi})] \mid \mathscr{F}(t)\}\} = e^{-\alpha t} E\{v[x_\xi(t)]\} \\ &= e^{-\alpha t} p(t, \xi, v). \end{aligned} \tag{12.2}$$

Moreover, when $t \to 0$ the first expectation tends to $v(\xi)$. Thus v is α-excessive and is majorized by u when u is bounded. If u is not bounded, u is the limit of an increasing sequence of bounded α-excessive functions (Lemma 10); so v is the limit of an increasing sequence of bounded α-excessive functions majorized by u. Hence v is α-excessive and is majorized by u.

If $\alpha = 0$ and $u \equiv 1$, the value $v(\xi)$ is the probability that an $x_\xi(\cdot)$ path hits A at a strictly positive time, and $p(t, \xi, v)$ is the probability that an $x_\xi(\cdot)$ path hits A at some time $> t$. Thus the condition that v be excessive, namely that $p(t, \xi, v)$ increase to $v(\xi)$ as t decreases to 0, has a simple probabilistic interpretation in this special case.

Suppose that $\alpha > 0$, and define the substochastic transition function p_α by $p_\alpha(t, \xi, A) = e^{-\alpha t} p(t, \xi, A)$. Suppose that $u \equiv 1$ so that $v(\xi) = E\{e^{-\alpha T_\xi}\}$. Let S be a positive random variable defined on the same probability space as $x_\xi(\cdot)$, independent of $x_\xi(\cdot)$, with distribution density (relative to l_1) $t \to \alpha e^{-\alpha t}$ on $\mathbb{R}^+$. Then the probability that an $x_\xi(\cdot)$ path hits A at a strictly positive time but before time S is $v(\xi)$. Equivalently, $v(\xi)$ is the probability that a path from ξ of an almost surely right continuous Markov process with transition function p_α hits A at a strictly positive time.

13. Conditioned Markov Processes

Let $(X, \mathscr{X})$ be a measurable space, let p be a stationary substochastic transition function with state space $(X, \mathscr{X})$, and let h be a strictly positive excessive function for p. The hypothesis of strict positivity is unnecessary but avoids some technicalities and is satisfied in the applications to be made. Define $X^h = \{h < +\infty\}$ and

$$p^h(t, \xi, d\eta) = \begin{cases} p(t, \xi, d\eta)\dfrac{h(\eta)}{h(\xi)} & \text{if } \xi \in X^h, \\ 0 & \text{if } \xi \in X - X^h. \end{cases} \tag{13.1}$$

Inequality (10.1a) for excessive functions implies

$$p(t, \xi, X - X^h) = 0 \quad \text{if } \xi \in X^h \tag{13.2}$$

and also implies that p^h is a stationary substochastic transition function with state space $(X, \mathscr{X})$. Moreover, if v is excessive for p, the function v/h redefined as 0 on $X - X^h$ is excessive for p^h. Conversely, if u is excessive for p^h, then $u = 0$ on $X - X^h$, and the function v_0 defined as uh on X^h and $+\infty$ otherwise satisfies the excessive function inequality (10.1a) relative to p. If v is defined as $\lim_{t\to 0} p(t,\cdot, v_0)$, then v is excessive for p, and $u = v/h$ on X^h.

Let $\{x_\xi(\cdot), \mathscr{F}_\xi(\cdot)\}$ be a Markov process from $\xi \in X^h$ with transition function p, lifetime S_ξ, $\mathscr{F}_\xi(s) = \mathscr{F}\{x_\xi(r), r \le s\}$, on a probability space $(\Omega, \mathscr{F}, P)$. Choose $t > 0$, and define a measure P^h on $\mathscr{F}_\xi(t)$ by

$$P^h\{\Lambda\} = \frac{E\{h[x_\xi(t)]; \Lambda\}}{h(\xi)} \quad \text{for } \Lambda \in \mathscr{F}_\xi(t). \tag{13.3}$$

Here as before we adopt the convention that the integral of a function over a set means the integral over the part of the set on which the function is defined. Thus in (13.3) without this convention Λ would be replaced $\Lambda \cap \{S_\xi > t\}$. In particular,

$$P^h\{x_\xi(t) \in X\} = P^h\{x_\xi(t) \in X^h\} = \frac{p(t, \xi, h)}{h(\xi)}. \tag{13.4}$$

A Markov process with transition function p^h is called an *h-path* process. If $x_\xi^h(\cdot)$ is an h-path process from ξ, with lifetime S_ξ^h, the right hand term in (13.4) is the probability that $S_\xi^h \ge t$. More generally, if $0 \le t_1 < \cdots < t_n = t$, the P^h distribution of $x_\xi(t_1), \ldots, x_\xi(t_n)$ is the distribution of $x_\xi^h(t_1), \ldots, x_\xi^h(t_n)$. In particular, if X is a topological space and if $x_\xi(\cdot)$ is almost surely [right] continuous under P, this discussion shows that there is a Markov process on X from ξ, with transition function p^h which is almost surely [right] continuous strictly before time t (on paths for which t is less than the path lifetime), and it follows easily that there is an almost surely [right] continuous process with initial point ξ and transition function p^h.

In probabilistic potential theory hypotheses are imposed on the state space and transition function p ensuring that for any ξ in the state space there is a right continuous Markov process $x_\xi(\cdot)$ from ξ with transition function p and that for such a process the composed process $u[x_\xi(\cdot)]$ is almost surely right continuous whenever u is excessive. In this theory it then follows as sketched above that h-path processes $x_\xi^h(\cdot)$ from points of finiteness of h can be chosen to be right continuous. Moreover, if u is excessive for p, the fact that $u[x_\xi(\cdot)]$ is almost surely right continuous and therefore almost surely right continuous for the measure P^h suggests that $u[x_\xi^h(\cdot)]$ is also almost surely right continuous. Finally the fact that $P^h\{x_\xi(t) \in X - X^h\} = 0$ suggests that almost no h-path from ξ ever hits $X - X^h$. Proofs of these statements and conjectures will be given in more detail in the contexts of Brownian motion and space-time Brownian motion.

14. Tied Down Markov Processes

Let p be a substochastic transition function with state space $(X, \mathfrak{X})$. Suppose that there is a measure l on $\mathfrak{X}$ with respect to which $p(s, \xi, \cdot)$ for $s > 0$ is absolutely continuous with strictly positive density $p'(s, \xi, \cdot)$ and that p' satisfies (identically) the Chapman–Kolmogorov equation for densities:

$$p'(s + t, \xi, \zeta) = \int_X p'(s, \xi, \eta) p'(t, \eta, \zeta) l(d\eta) \tag{14.1}$$

for ξ and η in X. The hypothesis of strict positivity of p' will be satisfied in the applications to be made and avoids technicalities. This hypothesis implies that if h is excessive for p, then $l\{h = +\infty\} = 0$ unless h is identically $+\infty$. Choose $t > 0$, and let ξ, η be any (not necessarily distinct) points of X. If $x_\xi(\cdot)$ is a Markov process with initial point ξ and transition function p, if we consider $x_\xi(\cdot)$ in X, that is, before transition to a trap point, and if $0 < s_1 < \cdots < s_n < t$, the formally calculated joint density of $x_\xi(s_1) = \zeta_1, \ldots, x_\xi(s_n) = \zeta_n$ given that $x_\xi(t) = \eta$ is

$$p'(s_1, \xi, \zeta_1) p'(s_2 - s_1, \zeta_1, \zeta_2) \cdots p'(s_n - s_{n-1}, \zeta_{n-1}, \zeta_n) \frac{p'(t - s_n, \zeta_n, \eta)}{p'(t, \xi, \eta)}. \tag{14.2}$$

This joint density can be interpreted as that of a Markov process $x_{\xi\eta}^t(\cdot)$ on the parameter interval $[0, t[$ with initial value ξ and nonstationary transition density (from ζ_1 at time s_1 to ζ_2 at time s_2 with $0 \le s_1 < s_2 < t$)

$$p'(s_2 - s_1, \zeta_1, \zeta_2) \frac{p'(t - s_2, \zeta_2, \eta)}{p'(t - s_1, \zeta_1, \eta)}. \tag{14.3}$$

The joint density (14.2) is also that of a Markov process reversed in time on the parameter interval $]0, t]$ with initial value η at time t and nonstationary transition density (from ζ_2 at time s_2 to ζ_1 at time s_1 with $0 < s_1 < s_2 \le t$)

$$p'(s_1, \xi, \zeta_1)\frac{p'(s_2 - s_1, \zeta_1, \zeta_2)}{p'(s_2, \xi, \zeta_2)}. \tag{14.4}$$

Probabilities for $x_\xi(\cdot)$ on $[0, t]$ can be obtained from $x^t_{\xi\eta}(\cdot)$ probabilities by integrating out η using as measure the distribution $p(t, \xi, \cdot)$ of $x_\xi(t)$. If h is an excessive function for p and is strictly positive, replacing p by p^h corresponds to replacing $p'(s, \zeta_1, \zeta_2)$ by $p'(s, \zeta_1, \zeta_2)h(\zeta_2)/h(\zeta_1)$, but the densities in (14.3) and (14.4) are unchanged. Thus the probabilities for an h-path process $x^h_\xi(\cdot)$ on $[0, t]$ for ξ in X^h can be obtained from $x^t_{\xi\eta}(\cdot)$ probabilities by integrating out η using as measure the distribution $p^h(t, \xi, \cdot)$ of $x^h_\xi(t)$.

This discussion suggests that many h-path properties, that is, properties of an h-path process $x^h_\xi(\cdot)$, are independent of h. For example, if $x_\xi(\cdot)$ has continuous sample functions, it is plausible that for l almost every η the $x_\xi(\cdot)$ process fixed at η at time t can be chosen to have continuous sample functions, and therefore (as in fact already proved in Section 13) every h-path process can be so chosen.

15. Killed Markov Processes

Let $(t, \xi, A) \mapsto p(t, \xi, A)$ be a continuous parameter stationary transition function with Polish state space X, and suppose that for every point ξ of X there is a right continuous Markov process $x_\xi(\cdot)$ with transition function p and initial point ξ. Let A be an analytic subset of X, and let T_ξ and L_ξ be, respectively, the hitting time and last hitting time of A by $x_\xi(\cdot)$. The function h defined by

$$h(\xi) = P\{T_\xi < +\infty\} = P\{L_\xi > 0\} \tag{15.1}$$

is excessive (Section 12). If A is closed and if $\xi \in X - A$, the process $x_\xi(\cdot)$ killed at time T_ξ, that is, the restriction of $x_\xi(\cdot)$ to $\{T_\xi > t\}$, is Markovian with state space $X - A$ and transition function p_1 given by

$$p_1(t, \eta, B) = P\{x_\eta(t) \in B, T_\eta > t\}. \tag{15.2}$$

Next suppose that A is not necessarily closed but that h is strictly positive, and define $p^h(t, \xi, d\eta) = p(t, \xi, d\eta)h(\eta)/h(\xi)$. Consider $x_\xi(\cdot)$ killed at L_ξ except that the parameter value 0 is dropped; that is, consider $x_{\xi L_\xi}$, defined for each $t > 0$ as the restriction to $\{L_\xi > t\}$ of $x_\xi(t)$. If $0 < t_1 < \cdots < t_n$,

$$P\{x_{\xi L_\xi}(t_j) \in d\eta_j, j \le n\} = P\{x_\xi(t_j) \in d\eta_j, j \le n; L_\xi > t_n\} \\ = p(t_1, \xi, d\eta_1) \cdots p(t_n - t_{n-1}, \eta_{n-1}, d\eta_n)h(\eta_n). \tag{15.3}$$

This evaluation has a simple interpretation: $x_{\xi L_\xi}(\cdot)$ is Markovian with

$$P\{x_{\xi L_\xi}(t) \in d\eta\} = p(t, \xi, d\eta)h(\eta) \tag{15.4}$$

and stationary transition function p^h. That is, $x_{\xi L_\xi}(\cdot)$ probabilities can be computed by computing probabilities for a Markov process from ξ, with transition function p^h, and then multiplying the probabilities obtained by the constant $h(\xi)$. Thus killing a Markov process at the last hitting time of a set leads to a conditioned Markov process in the sense of Section 13. The process killed at L_ξ in the present context is trivially right continuous, verifying a property of conditioned Markov processes derived in Section 13. Observe that p^h may be substochastic even if p is stochastic and that the distribution of $x_{\xi L_\xi}(t)$ may not be a probability measure:

$$P\{x_{\xi L_\xi}(t) \in X\} = p(t, \xi, h); \tag{15.5}$$

when t tends to 0, the right-hand side tends (increasing) to the limit $h(\xi)$.

Chapter VII

Brownian Motion

1. Processes with Independent Increments and State Space $\mathbb{R}^N$

(a) Discrete Parameter Case

Let $y(0), y(1), \ldots$ be mutually independent random variables with distributions $q_0, q_1, \ldots$, respectively, on $\mathbb{R}^N$, and define $x(n) = \sum_0^n y(j)$, $\mathscr{F}(n) = \mathscr{F}\{y(0), \ldots, y(n)\} = \mathscr{F}\{x(0), \ldots, x(n)\}$. If the transition function p_n with state space $\mathbb{R}^N$ is defined by $p_n(\xi, A) = q_{n+1}(A - \xi)$, where A is a Borel subset of $\mathbb{R}^N$ and $A - \xi$ is the translation of A by $-\xi$, the process $\{x(\cdot), \mathscr{F}(\cdot)\}$ is Markovian with initial distribution q_0 and successive transition functions $p_0, p_1, \ldots$. In fact according to the application in Section I.7,

$$\begin{aligned} P\{x(n+1) \in A \,|\, \mathscr{F}(n)\} &= P\{x(n) + y(n+1) \in A \,|\, \mathscr{F}(n)\} \\ &= q_{n+1}(A - x(n)) = p_n(x(n), A) \quad \text{a.s.} \end{aligned} \tag{1.1}$$

If $0 \leq n_1 < \cdots < n_k$, the increments $x(n_2) - x(n_1), \ldots, x(n_k) - x(n_{k-1})$ are independent random variables, and these increments are independent of $x(0)$.

If $y_0, y_1, \ldots$ is an arbitrary sequence of random variables with values in $\mathbb{R}^N$ and if $x_n = \sum_0^n y_j$, the joint distribution of $y_0, y_1, \ldots$ determines that of $x_0, x_1, \ldots$, and conversely. Hence two different distributions of $y_0, y_1, \ldots$ do not induce the same distribution of $x_0, x_1, \ldots$. In the context of the preceding paragraph this one-to-one correspondence between distributions implies that if, for $n \geq 0$, q_n is a probability measure on $\mathbb{R}^N$ and if p_n is defined by $p_n(\xi, A) = q_{n+1}(A - \xi)$, then any Markov process determined by the initial distribution q_0 and these transition functions is a sequence of successive sums of independent random variables, $x_n = \sum_0^n y_j$, for which y_j has distribution q_j.

(b) Continuous Parameter Case

Let $\{x(t), t \in \mathbb{R}^+\}$ be a stochastic process with state space $\mathbb{R}^N$ and with independent increments; that is, $0 \leq t_1 < \cdots t_k$ implies that the increments

$x(t_2) - x(t_1), \ldots, x(t_k) - x(t_{k-1})$ are mutually independent. If $s < t$ and if $q(s, t, \cdot)$ is the distribution of $x(t) - x(s)$, the fact that the distribution of the sum of two independent random variables with state space $\mathbb{R}^N$ is the convolution of their distributions implies that for $0 \le s_1 < s_2 < s_3$,

$$q(s_1, s_3, A) = \int_{\mathbb{R}^N} q(s_2, s_3, A - \eta) q(s_1, s_2, d\eta). \tag{1.2}$$

Suppose, strengthening slightly the condition of independent increments, that $0 < t_1 < \cdots < t_k$ implies that the random variables

$$x(0),\ x(t_1) - x(0),\ \ldots,\ x(t_k) - x(t_{k-1}) \tag{1.3}$$

are mutually independent, so that $x(0)$ is independent of the set of process increments. If now $\mathscr{F}(t) = \mathscr{F}\{x(s), s \le t\}$, the argument used in the discrete parameter case shows that $\{x(\cdot), \mathscr{F}(\cdot)\}$ is a Markov process with initial distribution that of $x(0)$ and transition function given by

$$P\{x(t) \in A \mid \mathscr{F}(s)\} = q(s, t, A - \xi) \quad \text{a.s.} \tag{1.4}$$

Equation (1.2) is the Chapman–Kolmogorov equation for the Markov process transition function. This process is a very special type of Markov process in that as is obvious from the discussion, or as follows readily from (1.4), for $s \in \mathbb{R}^+$ the set of increments $\{x(t) - x(s) : t > s\}$ is independent of the σ algebra $\mathscr{F}(s)$. This independence is equivalent to the mutual independence of the random variables (1.3) for all k and choices of $t_1, \ldots, t_k$. Conversely, if $\{x(\cdot), \mathscr{F}(\cdot)\}$ is a Markov process with state space $\mathbb{R}^N$ and (Section VI.1) stochastic transition function $(s, \xi, t, A) \mapsto q(s, \xi; t, A)$ and if this transition function can be written in the form $(s, \xi, t, A) \mapsto q(s, t, A - \xi)$, where (1.2) is true, then the process $\{x(\cdot), \mathscr{F}(\cdot)\}$ has independent increments in the strengthened sense that for $s \in \mathbb{R}^+$ the set of increments $\{x(t) - x(s): t > s\}$ is independent of $\mathscr{F}(s)$; moreover for $0 \le s < t$ the increment $x(t) - x(s)$ has distribution $q(s, t, \cdot)$. Finally, if v is a probability measure on $\mathscr{B}(\mathbb{R}^N)$, if $q(s, t, \cdot)$ is a probability measure on $\mathscr{B}(\mathbb{R}^N)$, and if (1.2) is satisfied, then there is (Section VI.1) a canonical Markov process with initial distribution v and transition function $(s, \xi, t, A) \mapsto q(s, t, A - \xi)$.

We now summarize the preceding discussion in the special case of continuous parameter processes with stationary independent increments. Let $\{x(t), t \in \mathbb{R}^+\}$ be a stochastic process with state space $\mathbb{R}^N$ for which $0 < t_1 < \cdots t_k$ implies that the random variables (1.3) are mutually independent. Suppose also (increment stationarity) that for $0 \le s_1 < s_2$ the distribution $q(s_1, s_2, \cdot)$ of $x(s_2) - x(s_1)$ depends only on $s_2 - s_1$. If we denote this distribution by $q(s_2 - s_1, \cdot)$, the family $\{q(s, \cdot), s > 0\}$ of probability distributions on $\mathbb{R}^N$ satisfies the convolution equation

$$q(s+t,A)=\int_{\mathbb{R}^N} q(t,A-\xi)q(s,d\xi). \tag{1.5}$$

If we define $\mathscr{F}(t)=\mathscr{F}\{x(s), s\le t\}$, the process $\{x(\cdot),\mathscr{F}(\cdot)\}$ is a Markov process with a stationary transition function given by

$$p(t,\xi,A)=q(t,A-\xi), \tag{1.6}$$

and for $s\in\mathbb{R}^+$ the set of increments $\{x(t)-x(s): t>s\}$ is independent of $\mathscr{F}(s)$. Equation (1.5) is the Chapman–Kolmogorov equation for the transition function p. Conversely, suppose that $\{x(t),\mathscr{F}(t), t\in\mathbb{R}^+\}$ is a Markov process with stationary transition function $(t,\xi,A)\mapsto p(t,\xi,A)$ and suppose that p has the form (1.6). Then the random variables (1.3) are mutually independent, and for $0\le s<t$ the increment $x(t)-x(s)$ has distribution $q(t-s,\cdot)$. Finally to any family $\{q(s,\cdot), s>0\}$ of probability distributions on $\mathbb{R}^N$ satisfying (1.5) along with a further probability distribution ν on $\mathbb{R}^N$ corresponds a Markov process $\{x(\cdot),\mathscr{F}(\cdot)\}$ with initial distribution ν and transition function $(s,\xi,t,A)\mapsto q(t-s,A-\xi)$.

2. Brownian Motion

Define $\mathscr{E}(t,\eta)$ by 1.XV(4.1), but in discussing transition densities it will sometimes be more intuitive to write $\mathscr{E}(t,\xi,\eta)$ for $\mathscr{E}(t,\eta-\xi)$. The variance parameter σ^2 is fixed throughout. A stochastic process $w(\cdot)$ is called an *N-dimensional Brownian motion* or a *Brownian motion in* $\mathbb{R}^N$ if the following four conditions are satisfied:

- BM1. The parameter set is $\mathbb{R}^+$. The state space is $\mathbb{R}^N$.
- BM2. The process has stationary independent increments. If $0\le t_0<t_0+t$, the distribution of the increment $w(t_0+t)-w(t_0)$ has density $\mathscr{E}(t,\cdot)$ relative to l_N.
- BM3. The random variable $w(0)$ is independent of the set of process increments.
- BM4. The process is almost surely continuous.

A Brownian motion in $\mathbb{R}^N$ can also be defined as a Markov process $\{w(t),\mathscr{F}(t), t\in\mathbb{R}^+\}$ satisfying the following three conditions:

- BM1*. BM1.
- BM2*. The (stationary) stochastic transition function has density $\mathscr{E}(t,\xi,\eta)=\mathscr{E}(t,\eta-\xi)$ relative to l_N.
- BM3*. BM4.

Observe that BM1*–BM2* imply that for fixed $s>0$ the set of increments $\{x(t)-x(s): t>s\}$ is independent of $\mathscr{F}(s)$.

The two Brownian motion definitions are equivalent in the following sense. According to Section 1, if $\{w(\cdot), \mathscr{F}(\cdot)\}$ is a Markov process satisfying BM1*–BM3*, then conditions BM1–BM4 are also satisfied, and conversely, if BM1–BM4 are satisfied by a process $w(\cdot)$ and if $\mathscr{F}(t)$ is defined as $\mathscr{F}\{w(s), s \le t\}$, then $\{w(\cdot), \mathscr{F}(\cdot)\}$ satisfies BM1*–BM3*.

Throughout this book a "Brownian motion" specified as $\{w(\cdot), \mathscr{F}(\cdot)\}$ and not otherwise delimited will mean a Markov process satisfying BM1*–BM3* with no further unstated hypotheses imposed on the filtration $\mathscr{F}(\cdot)$.

All sample functions of a Brownian motion can be made continuous without affecting BM1–BM3 or BM1*–BM2* by redefining the discontinuous sample functions to make them continuous. Observe that if $w(\cdot)$ is a Brownian motion from a point, then the N component processes are mutually independent one-dimensional Brownian motions with the same variance parameter as $w(\cdot)$.

If $w(\cdot)$ is a Brownian motion with variance parameter σ^2 and if $t_0 > 0$, the process $\{w(t_0 + t), t \in \mathbb{R}^+\}$ is a Brownian motion with the same variance parameter as $w(\cdot)$ and with initial distribution the distribution of $w(t_0)$. The process $\{w(t/\sigma^2), t \in \mathbb{R}^+\}$ is a Brownian motion with variance parameter 1; the processes $\{[w(t) - w(0)]/\sigma, t \in \mathbb{R}^+\}$ and $\{w(t/\sigma^2) - w(0), t \in \mathbb{R}^+\}$ are both Brownian motions from the origin with variance parameter 1.

Brownian Motion with the Parameter Set $\mathbb{R}$

A Brownian motion with the parameter set $\mathbb{R}$ is defined as a process $\{w(t), \mathscr{F}(t), t \in \mathbb{R}\}$ which satisfies BM1*–BM3* except that the parameter set is $\mathbb{R}$ instead of $\mathbb{R}^+$ and that it is no longer supposed that the filtered measure space on which the process is defined is a probability space or even that the probability value assigned to the whole space is finite. More precisely it is supposed that there is for each t in $\mathbb{R}$ a measure $p(t, \cdot)$ of Borel sets, finite for compact sets, with $0 < p(t, \mathbb{R}^N) \le +\infty$, which is the distribution of $w(t)$, and probabilities are then defined as usual for a Markov process, in terms of p and the transition density assigned in BM2*. Then for $s \in \mathbb{R}$, $t > 0$, $A \in \mathscr{B}(\mathbb{R}^N)$,

$$\int_{\mathbb{R}^N} p(s, d\xi) \int_A \ell(t, \xi, \eta) l_N(d\eta) = p(s + t, A). \tag{2.1}$$

Hence $p(\cdot, \mathbb{R}^N)$ is identically some strictly positive constant β ($\le +\infty$). If s is replaced in (2.1) by $s_0 - t$, we find that $\beta\sigma^{-N}t^{-N/2}l_N(A) \ge p(s_0, A)$, and ($t \to +\infty$) it follows that $\beta = +\infty$. Thus the context imposes the condition that the absolute probability distribution $p(s_0, \cdot)$ be an infinite-valued measure for all s_0. If we define $\dot{u}(\eta, t)$ for $(\eta, t) \in \dot{\mathbb{R}}^N$ by

$$\dot{u}(\eta, t) = \int_{\mathbb{R}^N} \ell(\alpha, \xi, \eta) p(t - \alpha, d\xi) \tag{2.2}$$

with $\alpha > 0$, it follows from (2.1) and the Chapman–Kolmogorov equation that $\dot{u}$ does not depend on the choice of α. The function $\dot{u}(\cdot, t)$ is a version of the density $dp(t, \cdot)/dl_N$ of the $w(t)$ distribution and

$$\int_{\mathbb{R}^N} \dot{u}(\xi, s) l_N(d\xi) = +\infty, \qquad \int_{\mathbb{R}^N} \dot{u}(\xi, s) \ell(t, \xi, \eta) l_N(d\xi) = \dot{u}(\eta, s + t) \tag{2.3}$$

when $t > 0$. Since the function $(\eta, t) \mapsto \ell(t, \xi, \eta) = \ell(t, \xi - \eta)$ is parabolic on $\dot{\mathbb{R}}^N - \{(\xi, 0)\}$, the function $\dot{u}$, which is Borel measurable and is finite valued on a dense subset of $\dot{\mathbb{R}}^N$, has the parabolic average property and so is parabolic on $\dot{\mathbb{R}}^N$. Under our hypotheses when $s \in \mathbb{R}$ the process $\{w(s + t), t \in \mathbb{R}^+\}$ is a Brownian motion with initial distribution of l_N density $\dot{u}(\xi, s)$. The distribution of each increment $x(t_2) - x(t_1)$ with $t_1 \neq t_2$ is an infinite-valued measure. Since the process is Markovian, the reversed time process is also Markovian, and a version of the reverse transition density relative to l_N is easily calculated: the density of a transition from ξ at time s to η at time $t < s$ is

$$\frac{\dot{u}(\eta, t) \ell(s - t, \eta, \xi)}{\dot{u}(\xi, s)}. \tag{2.4}$$

(Recall from Section 1.XVI.8 that a positive parabolic function on $\dot{\mathbb{R}}^N$ is either strictly positive or identically zero.) In terms of ordinary probabilities a version of the distribution of $\{w(s_0 + t), t \in \mathbb{R}^+\}$ conditioned by $w(s_0) = \xi_0$ is that of a Brownian motion from ξ_0, and the distribution of $\{w(s_0 - t), t \in \mathbb{R}^+\}$ under the same condition is that of a Markov process from ξ_0 with transition density determined by (2.4). The special case of stationarity is important: if $p(t, \cdot) = l_N$ for all t, that is, if $\dot{u} \equiv 1$, the process forward and backward transition densities are the same. In this case the processes $\{w(s_0 + t), t \in \mathbb{R}^+\}$ and $\{w(s_0 - t), t \in \mathbb{R}^+\}$ are both Brownian motions with initial distribution l_N.

Existence of Brownian Motions (Parameter Set $\mathbb{R}^+$)

The conditions BM1*–BM2* are conditions for a Markov process with state space $\mathbb{R}^N$ and with a specified transition function. Hence (Section VI.5) such a process exists for an arbitrary choice of initial probability distribution μ, for example, as a canonical process on the space of all functions from $\mathbb{R}^+$ into $\mathbb{R}^N$. A useful additional fact is that since the distribution of the process in question is the integral with respect to $\mu(d\xi)$ of the distribution of the process with initial point ξ, it follows that there is such a process when μ is an arbitrary measure on $\mathbb{R}^N$, finite on compact sets, and not identically 0. In Section 3 we shall show that any process satisfying BM1*–BM2* has a standard modification satisfying BM3*. Thus Brownian motions with parameter set $\mathbb{R}^+$ exist.

Existence of Brownian Motions (Parameter Set $\mathbb{R}$)

We show that if $\dot{u}$ is an arbitrary not identically vanishing positive parabolic function on $\dot{\mathbb{R}}^N$, then there is a Brownian motion with parameter interval $\mathbb{R}$ having $\dot{u}$ as its absolute probability density function. Observe first that $\dot{u}$ satisfies (2.3) by direct calculation, using the representation of $\dot{u}$ in terms of minimal parabolic functions given in Section 1.XVI.8. Observe secondly that if we impose the condition $w(0) = \xi_0$ in addition to the other conditions to be satisfied, the processes $\{w(t), t \in \mathbb{R}^+\}$ and $\{w(t), -t \in \mathbb{R}^+\}$ are to be mutually independent Markov processes whose distributions we have discussed above. Hence (Section VI.5) under the added condition at the parameter value 0 a process $\{w(t), t \in \mathbb{R}\}$ exists as a canonical process on the space of all functions from $\mathbb{R}$ into $\mathbb{R}^N$. The measure so defined on this function space depends on the choice of ξ_0, and we integrate this measure with respect to $\dot{u}(\xi_0, 0) l_N(d\xi_0)$ to get a measure on the function space, and thereby a stochastic process, which has the desired properties aside from sample function continuity. According to Section 3, we can obtain sample function continuity by choosing an appropriate standard modification of the process obtained in this way.

Extension of Brownian Motion for a Finite Interval

Suppose that $b > 0$, that $I = [0, b]$, and that $\{w(t), t \in I\}$ is a process which satisfies the Brownian motion defining conditions BM1–BM4 and BM1*–BM3* except that $\mathbb{R}^+$ is replaced by I as parameter set. Then the process may not be the restriction to I of a Brownian motion process for the parameter set $\mathbb{R}^+$, but if $0 < a < b$, we now exhibit a Brownian motion $\{w(t), t \in \mathbb{R}^+\}$ such that $w'(t) = w(t)$ for $t \leq a$: choose any function f, strictly increasing and continuous, mapping $[a, +\infty[$ onto $[a, b[$, and define

$$w'(t) = \begin{cases} w(t) & \text{if } 0 \leq t \leq a, \\ w(a) + \dfrac{(t-a)^{1/2}[w[f(t)] - w(a)]}{[f(t) - a]^{1/2}} & \text{if } a < t < +\infty. \end{cases}$$

Space-Time Brownian Motion

Let $\dot{\xi}_0 : (\xi_0, s_0)$ be a point of $\dot{\mathbb{R}}^N$, and let $w(\cdot)$ be a Brownian motion in $\mathbb{R}^N$ from ξ_0. The process

$$\{\dot{w}(t), t \in \mathbb{R}^+\} = \{[w(t), s_0 - t], t \in \mathbb{R}^+\}$$

with state space $\dot{\mathbb{R}}^N$ is called a *space-time Brownian motion from* $\dot{\xi}_0$. This process has stationary independent increments and is Markovian relative to the same filtration as $w(\cdot)$. Space-time Brownian motion with an arbitrary initial distribution is defined in the obvious way and has these same properties. In this definition space-time Brownian motion moves downward in $\dot{\mathbb{R}}^N$, that is, in the direction of decreasing ordinate values. The dual motion moving upward will be called *space-cotime Brownian motion.* It will be convenient to introduce the following notation. If $\dot{A}$ is a subset of $\dot{\mathbb{R}}^N$, define $A_t = \{\eta : (\eta, t) \in \dot{A}\}$ and $\dot{A}_t = A_t \times \{t\}$. The transition probability function $(r, (\xi, s), \dot{A}) \mapsto p(r, (\xi, s), \dot{A})$ for space-time Brownian motion has the property that $p(r, (\xi, s), \cdot)$ is a measure supported by $\dot{A}_{s-r}$, and this measure when considered as a measure on $\dot{A}_{s-r}$ is absolutely continuous relative to l_N, with density $\mathscr{G}(r, \xi, \cdot)$. Define

$$\dot{\mathscr{G}}(\dot{\xi}, \dot{\eta}) = \mathscr{G}(s - t, \xi, \eta), \qquad \dot{\xi} = (\xi, s), \qquad \dot{\eta} = (\eta, t)$$

so that $\dot{\mathscr{G}}(\dot{\xi}, \dot{\eta})$ governs the transition density for time $s - t$ when $s > t$,

$$p(r, \dot{\xi}, \dot{A}) = \int_{\dot{A}_{s-r}} \dot{\mathscr{G}}(\dot{\xi}, (\eta, s - r)) l_N(d\eta),$$

and $\dot{\mathscr{G}}(\dot{\xi}, \dot{\eta}) = 0$ when $s \leq t$. This context suggests that an appropriate state space for the transition function p is $\dot{\mathbb{R}}^N$ coupled with the σ algebra of those sets meeting every hyperplane orthogonal to the ordinate axis in a Borel set. (Nothing in the discussion would need any change, however, if "Borel" here is replaced by "l_N measurable.")

3. Continuity of Brownian Paths

In this section it will be shown that a process satisfying conditions BM1–BM3, equivalently, BM1* and BM2*, has a standard modification satisfying BM4. Thus Brownian motions exist. It is sufficient to treat the one-dimensional case.

(a) If x is a Gaussian random variable with mean 0 and variance σ^2 and if $\alpha > 0$, then

$$P\{x \geq \alpha\} < \frac{\sigma}{\alpha} \exp \frac{-\alpha^2}{2\sigma^2}. \tag{3.1}$$

In fact the left side of (3.1) is

$$(2\pi)^{-1/2} \int_{\alpha/\sigma}^{\infty} \exp \frac{-\xi^2}{2} d\xi < \frac{\sigma}{\alpha} \int_{\alpha/\sigma}^{\infty} \xi \exp \frac{-\xi^2}{2} d\xi, \tag{3.2}$$

and the right side is equal to the right side of (3.1).

Recall that a random variable y is said to have a symmetric distribution if $P\{y \geq \alpha\} = P\{y \leq -\alpha\}$ for all α (equivalently, for all positive α). A finite sum of symmetrically distributed mutually independent random variables is itself symmetrically distributed.

(b) If $y(1), \ldots, y(n)$ are mutually independent symmetrically distributed random variables and if $x(k) = \Sigma_1^k y(i)$, then

$$P\{\max_{k\leq n} x(k) \geq \alpha\} \leq 2P\{x(n) \geq \alpha\}. \tag{3.3}$$

Let $T = \min\{j: x(j) \geq \alpha\}$. Then the set $\{T = j\}$ is independent of $x(n) - x(j)$, and since the latter random variable is symmetrically distributed,

$$\begin{aligned} P\{\max_{k\leq n} x(k) \geq \alpha\} &= \sum_1^n P\{T = j\} \leq 2\sum_1^n P\{T = j, x(n) - x(j) \geq 0\} \\ &\leq 2P\{x(n) \geq \alpha\}. \end{aligned} \tag{3.4}$$

(c) If $N = 1$ and if $x(\cdot)$ satisfies the defining conditions BM1–BM3 of a Brownian motion,

$$P\{\sup_{r\leq t} [x(r) - x(0)] \geq \alpha\} \leq 2P\{x(t) - x(0) \geq \alpha\} \qquad (r \text{ rational}). \tag{3.5}$$

According to (b) this inequality is true if the supremum on the left side is replaced by the supremum for only finitely many values of $r \leq t$. The inequality is therefore true as stated except possibly for the countably many values of α which are discontinuities of the distribution function of the supremum on the left. The inequality is true without exception because both sides of (3.5) define left continuous functions of α.

By symmetry, if $x(r) - x(0)$ is replaced in (3.5) by its absolute value, the right side of (3.5) should be doubled.

Theorem. *If $x(\cdot)$ is a process with the finite-dimensional distributions of a Brownian motion, it has a standard modification with continuous sample functions.*

We can assume in the proof that the variance parameter is 1 and that $N = 1$. Applying (c) above, if $m > 0$, $n > 0$ and if r and s are rational,

$$\begin{aligned} P\{\sup_{\substack{0<s-r<1/n \\ s\leq m}} |x(s) - x(r)| \geq 2\alpha\} &\leq \sum_{j=0}^{mn-1} P\{\sup_{0<s-j/n\leq 2/n} |x(s) - x(j/n)| \geq \alpha\} \\ &= mnP\{\sup_{s\leq 2/n} |x(s) - x(0)| \geq \alpha\} \leq 4mnP\{x(2/n) - x(0) \geq \alpha\} \\ &< 8mn^{1/2}\alpha^{-1} \exp\frac{-\alpha^2 n}{4}. \end{aligned} \tag{3.6}$$

Since the last term has limit 0 when $n \to \infty$, the restriction to the rationals of almost every sample function is uniformly continuous on $[0, m[$. In other words the restriction to the rationals of $x(\cdot, \omega)$ for ω not in some null set coincides with a continuous function $x_1(\cdot, \omega)$ on $\mathbb{R}^+$. Since

$$\lim_{t \to s} E\{[x(t) - x(s)]^2\} = 0,$$

$x_1(\cdot)$ is a standard modification of $x(\cdot)$ and has continuous sample functions if $x_1(\cdot)$ is defined, say, to be identically 0 on the exceptional null ω set.

Canonical Brownian Motion in $\mathbb{R}^N$

According to our conventions a Brownian motion in $\mathbb{R}^N$ is any stochastic process satisfying certain distribution and continuity conditions. For some purposes it is convenient to have a canonical Brownian motion, however, for a specified initial distribution and variance parameter. Suppose then that an initial distribution μ and variance parameter σ^2 are specified. Let $\mathring{\Omega}$ be the space of all continuous functions $\mathring{\omega}$ from $\mathbb{R}^+$ into $\mathbb{R}^N$, define $\mathring{w}(t, \mathring{\omega}) = \mathring{\omega}(t)$ and define $\mathring{\mathscr{F}} = \mathscr{F}\{\mathring{w}(t), t \in \mathbb{R}^+\}$. Then (Section I.10) a probability measure $\mathring{p}$ can be defined on $\mathring{\mathscr{F}}$ making $\mathring{w}(\cdot)$ a Brownian motion with initial distribution μ and variance parameter σ^2. The process $\mathring{w}(\cdot)$ will be described as the canonical Brownian motion for the specified μ and σ^2. In the language of Section I.10 $\mathring{w}(\cdot)$ is the canonical process associated with every Brownian motion with the specified initial distribution and variance parameter. In accordance with our conventions, $\mathring{P}$ is to be completed. The measure $\mathring{P}$ is a measure on the space of continuous functions and is commonly described as "Wiener measure" because Wiener was the first to construct this measure rigorously. It is a matter of taste and convenience whether in discussing a canonical Brownian motion from a point ξ of $\mathbb{R}^N$ we further restrict $\mathring{\Omega}$ by the condition $\mathring{\omega}(0) = \xi$; the choice will be specified when it matters. We shall not limit ourselves to canonical Brownian motions because noncanonical ones arise in many contexts. The distributions of the interesting functions defined on a Brownian motion with given initial distribution and variance parameter do not depend on the choice of Brownian motion. For example, the probability that a Brownian path from a point ξ hits a specified analytic set when the parameter varies in a specified interval may depend on the variance parameter but (Section I.12) not on the choice of Brownian motion.

4. Brownian Motion Filtrations

According to Section VI.9, a Brownian motion $\{w(\cdot), \mathscr{F}(\cdot)\}$ has the strong Markov property if $\mathscr{F}(0)$ contains the null sets of the probability space, and this condition is unnecessary if $w(\cdot)$ is continuous instead of merely almost surely continuous. If $\mathscr{F}_1(t)$ is the σ algebra generated by $\mathscr{F}(t)$ and the null sets, the process $\{w(\cdot), \mathscr{F}_1(\cdot)\}$ is still a Brownian motion. Since $s < s' < t$ implies that $w(t) - w(s')$ is independent of $\mathscr{F}_1^+(s)$, the random variable $w(t) - w(s)$ is independent of $\mathscr{F}_1^+(s)$, and this independence implies that the process $\{w(\cdot), \mathscr{F}_1^+(\cdot)\}$ is a Brownian motion. Thus there is no loss of generality in assuming that for a Brownian motion $\{w(\cdot, \mathscr{F}(\cdot)\}$ the σ algebra $\mathscr{F}(0)$ contains the null sets of the probability space and $\mathscr{F}(\cdot) = \mathscr{F}^+(\cdot)$, that is, $\mathscr{F}(\cdot)$ is right continuous. In particular, if $\{w(\cdot), \mathscr{F}(\cdot)\}$ is a Brownian motion with $\mathscr{F}(t) = \mathscr{F}\{w(s), s \le t\}$ and if $\mathscr{F}_1(t)$ is the σ algebra generated by $\mathscr{F}(t)$ and the null sets, then $\{w(\cdot), \mathscr{F}_1(\cdot)\}$ is a Brownian motion, and (by Theorem VI.8) $\mathscr{F}_1(\cdot)$ is right continuous.

In view of the properties of Brownian motions the entry and hitting times of analytic subsets of the state space are optional for a Brownian motion $\{w(\cdot), \mathscr{F}(\cdot)\}$ whenever $\mathscr{F}(0)$ contains the null sets of the probability space and $\mathscr{F}(\cdot)$ is right continuous. The space-time Brownian motions have the strong Markov property and the preceding entry and hitting time property under the stated conditions on the Brownian motions involved.

Theorem. *Let $\{w(\cdot), \mathscr{F}(\cdot)\}$ be a Brownian motion in $\mathbb{R}^N$ with $\mathscr{F}(t)$ generated by $\mathscr{F}\{w(s), s \ge t\}$ and the null sets. Define $\mathscr{F}(+\infty) = \bigvee_{t \in \mathbb{R}^+} \mathscr{F}(t)$.*

(a) *Every $\mathscr{F}(\cdot)$ optional time is predictable.*
(b) *$\mathscr{F}(\cdot)$ is predictable; in fact, if $T_\bullet$ is an increasing sequence of optional times with limit T then $\mathscr{F}(T) = \bigvee_0^\infty \mathscr{F}(T_n)$.*
(c) *Every almost surely right continuous supermartingale $\{x(\cdot), \mathscr{F}(\cdot)\}$ is nearly predictable, almost surely lower semicontinuous, almost surely continuous if the process is a martingale.*

Observation (1). In view of the martingale continuity properties proved in Section IV.1, assertion (c) for martingales is equivalent to the assertion that every martingale $\{x(\cdot), \mathscr{F}(\cdot)\}$ with the filtration $\mathscr{F}(\cdot)$ determined by a Brownian motion as stated in the theorem has an almost surely continuous standard modification.

Observation (2). In (c) for supermartingales if the supermartingale is right closable, Theorem IV.23(c) together with assertions (a) and (b) of the present theorem implies that the $x(\cdot)$ sample function discontinuities can be separated out to make a jump process.

Proof of (c) *in the martingale case.* A trivial argument shows that it is sufficient to prove that ($*$) if x is an integrable $\mathscr{F}(+\infty)$ measurable random variable, then $x(t) = E\{x|\mathscr{F}(t)\}$ can be defined for $t \le +\infty$ to make $x(\cdot)$ a continuous process. In proving ($*$) we shall use repeatedly the fact that if $x_\bullet$ is a sequence of random variables with L^1 limit x and if ($*$) is true for x_n and its martingale $\{x_n(\cdot), \mathscr{F}(\cdot)\}$, then ($*$) is true for x and $x(\cdot)$. In fact, going to a subsequence if necessary, it can be assumed that $E\{|x_{n+1} - x_n|\} < 2^{-n}$ which implies [submartingale maximal inequality applied to $|x_{n+1}(\cdot) - x_n(\cdot)|$]

$$P\{\sup_{t\geq 0}|x_{n+1}(t) - x_n(t)| \geq n^{-2}\} \leq n^2 2^{-n}; \tag{4.1}$$

so (Borel Cantelli theorem) for almost every ω, $\lim_{n\to\infty} x_n(t, \omega)$ exists uniformly in t. The limit process is an almost surely continuous version of $x(\cdot)$ and can then be modified to be continuous. To prove (*), let f_0, f_1 be continuous functions on $\mathbb{R}^N$ with limit 0 at $+\infty$ and define ϕ by $\phi(\alpha, r) = E\{f_1(z + r)\}$, for z an N-dimensional Gaussian random variable whose components are mutually independent with mean 0 and variance α, and for r in $\mathbb{R}$. The function ϕ is continuous. If $0 \leq t_1 < t_2$ and if $x = f_0[w(0)]f_1[w(t_2) - w(t_1)]$, then

$$E\{x|\mathscr{F}(t)\} = \begin{cases} f_0[w(0)]\phi[\sigma^2(t_2 - t_1), 0] & \text{if } t \leq t_1 \\ f_0[w(0)]\phi[\sigma^2(t_2 - t), w(t) - w(t_1)] & \text{if } t_1 < t \leq t_2 \\ f_0[w(0)]\phi[0, w(t_2) - w(t_1)] = x & \text{if } t > t_2, \end{cases} \tag{4.2}$$

neglecting null sets. Thus (*) is true for this choice of x. More generally an only slightly less trivial calculation shows that (*) is true if

$$x = f_0[w(0)]\prod_1^n f_j[w(t_j) - w(t_{j-1})],$$

where f_j is continuous on $\mathbb{R}^N$ with limit 0 at ∞ and $0 \leq t_0 < \cdots < t_n$. It follows (Stone–Weierstrass theorem) that (*) is true for the random variable

$$x = f[w(0), w(t_1) - w(t_0), \ldots, w(t_n) - w(t_{n-1})]$$

for f continuous on $\mathbb{R}^{N(n+1)}$ with limit 0 at ∞ and therefore if x is an integrable random variable which is a Borel measurable function of $w(0)$, $w(t_1) - w(t_0), \ldots, w(t_n) - w(t_{n-1})$, equivalently, if x is an integrable random variable which is a Borel measurable function of $w(0), w(t_0), \ldots, w(t_n)$. To finish the proof, we need only remark that according to Section III.14 every $\mathscr{F}(+\infty)$ measurable and integrable function x can be approximated arbitrarily closely in L^1 by random variables of this special type. □

Proof of (a). To prove that an optional time T is predictable, it suffices to prove that $T \wedge b$ is predictable for every positive constant b, and so it suffices to prove that every bounded optional time T is predictable. Define

$$x(t) = E\{T - T \wedge t|\mathscr{F}(t)\} = E\{T|\mathscr{F}(t)\} - T \wedge t, \qquad 0 \leq t \leq +\infty. \tag{4.3}$$

The process $\{x(\cdot), \mathscr{F}(\cdot)\}$ is a supermartingale, and in view of what we have just proved the conditional expectations in (4.3) can be chosen to make this supermartingale almost surely continuous. Moreover (Theorem IV.2) $x(T) = 0$ almost surely. For $n \geq 1$ define $T_n = \inf\{t\colon x(t) \leq 1/n\}$ to obtain an increasing sequence of optional times for which almost surely $T_n \leq T$, and $T_n < T$ on $\{T > 0\}$ because $x(T_n) = 1/n$ almost everywhere on this set. Furthermore, if $T' = \lim_{n\to\infty} T_n$, then almost everywhere on $\{T > 0\}$

$$0 = x(T') = E\{T - T \wedge T' \mid \mathscr{F}(T')\};$$

so $T = T'$ almost surely. Thus $T_.$ announces T if we ignore exceptional null sets which can be eliminated by trivial modifications of $T_.$. □

Proof of (b). To prove that each set in $\mathscr{F}(T)$ is in $\bigvee_0^\infty \mathscr{F}(T_n)$, let x be the indicator function of a set in $\mathscr{F}(T)$, and recall that we have proved above that we can define $x(t) = E\{x \mid \mathscr{F}(t)\}$ for $t \leq +\infty$ to make $x(\cdot)$ a continuous process. Apply the conditional expectation continuity theorem and the martingale optional sampling Theorem IV.2 to obtain

$$E\{x \mid \bigvee_0^\infty \mathscr{F}(T_n)\} = \lim_{n\to\infty} E\{x \mid \mathscr{F}(T_n)\} = \lim_{n\to\infty} x(T_n) = x \quad \text{a.s.}; \tag{4.4}$$

so x is $\bigvee_0^\infty \mathscr{F}(T_n)$ measurable, as was to be proved. □

Proof of (c) *in the supermartingale case*. According to Theorem IV.23. assertion (c) is implied by (a) and (b). □

5. Elementary Properties of the Brownian Transition Density and Brownian Motion

(a) The transition density $(t, \xi, \eta) \mapsto \ell(t, \xi, \eta) = \ell(t, \eta - \xi)$ satisfies the heat equation

$$\frac{\partial \ell(t, \xi, \eta)}{\partial t} = \frac{\sigma^2}{2} \Delta_\xi \ell(t, \xi, \eta), \tag{5.1}$$

and in fact $\ell(s - t, \xi, \eta) = \dot{G}((\xi, s), (\eta, t))$. This relation between Brownian motion transition density and the parabolic Green function of $\dot{\mathbb{R}}^N$ will be extended to the corresponding relation between the transition density for Brownian motion in an open set D and the parabolic Green function of $D \times \mathbb{R}$ in Section IX.17. As we shall see in cases related to the Dirichlet problem for parabolic functions many Brownian motion probabilities can be evaluated in terms of solutions of the heat equation on the relevant domains with the appropriate boundary conditions or, if there is stationarity in time, can be evaluated in terms of solutions of Laplace's equation with appropriate boundary conditions.

(b) A token of the intimate relation between Brownian motion and classical potential theory is the fact that [see 1.XVII(18.2)] $\int_0^\infty \ell(t, \xi, \eta) l_1(dt)$ is a constant multiple of the Green function of $\mathbb{R}^N$ when $N > 2$. This relation between Brownian transition densities and Green functions will be extended to transition densities for Brownian motions in Green domains in Section IX.17.

(c) A trivial calculation shows that if $w(\cdot)$ is a Brownian motion, then $\lim_{t\to\infty} P\{|w(t)| > c\} = 1$ for every $c > 0$, that is, $\lim_{t\to\infty} |w(t)| = +\infty$ in measure.

(d) The part of almost every Brownian path corresponding to a specified parameter interval has infinite length. To prove this, it is sufficient to show that for one-dimensional Brownian motion $w(\cdot)$ from 0 with variance parameter 1 the part of almost every Brownian sample function corresponding to the parameter interval $[0, \delta]$ is of unbounded variation, for every $\delta > 0$. To see this, define $s_n = \Sigma_1^n |w(j\delta/n) - w((j-1)\delta/n)|$, and note

$$E\{\exp(-s_n)\} = E^n\left\{\exp\left[-\left|w\left(\frac{\delta}{n}\right)\right|\right]\right\} \le E\left\{\exp\left[-n\left|w\left(\frac{\delta}{n}\right)\right|\right]\right\}. \tag{5.2}$$

Since the distribution of $n^{1/2}|w(\delta/n)|$ is independent of n, $\lim_{n\to\infty} n|w(\delta/n)| = +\infty$ in measure, so that $\lim_{n\to\infty} E\{\exp(-s_n)\} = 0$, and since s_n increases with n to the sample function total variation on $[0, \delta]$ when $n \to \infty$ along the integral powers of 2, almost every sample function has infinite total variation on $[0, \delta]$, as asserted.

(e) $(N = 1)$ Let $w(\cdot)$ be a Brownian motion from the origin, with variance parameter σ^2. Then $E\{w(s)w(t)\} = \sigma^2(s \wedge t)$ because if $t > s$,

$$E\{w(s)w(t)\} = E\{[w(t) - w(s)]w(s)\} + E\{w(s)^2\} = \sigma^2 s. \tag{5.3}$$

Conversely, if $w(\cdot)$ is a process with Gaussian finite-dimensional distributions having zero means and if $E\{w(s)w(t)\} = \sigma^2(s \wedge t)$ for some strictly positive constant σ^2, $w(\cdot)$ has the finite-dimensional distributions of a Brownian motion from the origin with variance parameter σ^2.

(f) If $w(\cdot)$ is a Brownian motion from the origin, define

$$y(t) = \begin{cases} tw\left(\dfrac{1}{t}\right) & \text{if } t > 0, \\ 0 & \text{if } t = 0. \end{cases}$$

An application of (e) to the one-dimensional component processes of $y(\cdot)$ shows that $y(\cdot)$ has the same finite-dimensional distributions as $w(\cdot)$. Moreover the $y(\cdot)$ process sample functions are almost all continuous on $]0, +\infty[$, and $\lim_{t\to 0} y(t) = 0$ a.s. if $t \to 0$ along a countable dense set because $w(\cdot)$ has this property. Hence this limit relation is true when $t \to 0$ unrestrictedly. Thus $y(\cdot)$ is Brownian motion, and the relation between $w(\cdot)$ and $y(\cdot)$ shows that properties of Brownian motion for large parameter values can be translated into properties for small parameter values and conversely. For example, continuity of Brownian motion at 0 implies that $\lim_{t\to 0} tw(1/t) = 0$ almost surely, that is,

$$\lim_{t\to\infty} \frac{w(t)}{t} = 0 \quad \text{a.s.} \tag{5.4}$$

This result is a version of the strong law of large numbers in the continuous parameter context, as applied to Brownian motion.

6. The Zero-One Law for Brownian Motion

If the sets of a σ algebra of measurable sets of a measure space are all null sets or the complements of null sets the σ algebra is called *trivial.*

Theorem. *If $w(\cdot)$ is a Brownian motion from a point, the σ algebras $\mathscr{F}_1$ and $\mathscr{F}_2$ generated, respectively, by the null sets and*

$$\bigcap_{t>0} \mathscr{F}\{w(s), s \le t\}, \qquad \bigcap_{t>0} \mathscr{F}\{w(s), s \ge t\} \tag{6.1}$$

are trivial.

It can be assumed that the initial point of the Brownian motion is the origin. If $\Lambda \in \mathscr{F}_1$ and if $t > 0$, the set Λ is independent of the class of differences $\{w(s_2) - w(s_1), t \le s_1 < s_2\}$ and therefore is independent of the σ algebra $\mathscr{F}\{w(s_2) - w(s_1), 0 < s_1 < s_2\}$ which, since $w(0+) = 0$ almost surely, together with the class of null sets generates a σ algebra including $\mathscr{F}_1$. Thus Λ is independent of itself and so has probability either 0 or 1. Apply this result to the Brownian motion $\{tw(1/t), t \in \mathbb{R}^+\}$ [see Section 5(f)] to derive the triviality of $\mathscr{F}_2$.

Observation. Since a Brownian motion process satisfies the conditions of Theorem VI.8(b) after the associated filtration $\mathscr{F}(\cdot)$ is enlarged if necessary to have $\mathscr{F}(0)$ contain the null sets, it follows from Section VI.8 that for a Brownian motion from a point the σ algebra $\mathscr{F}_1$ is trivial. This proof was not used above because a direct proof is more elementary and just as short.

The Zero-One Law for Space-Time Brownian Motion

In view of the definition of space-time Brownian motion in terms of Brownian motion the direct translation of Theorem 6 into the space-time Brownian motion context is true.

Applications. In the following applications $w(\cdot)$ is a Brownian motion in $\mathbb{R}^N$ from the origin, and T^A is the hitting time of a set A by $w(\cdot)$.

(a) $N = 1$. According to Section 5(c), $\lim_{t\to\infty} |w(t)| = +\infty$ in measure; so if $c > 0$, it follows that $P\{w(t) > c\} > 1/4$ for sufficiently large t. Hence

$$P\{w(n) > c \text{ for infinitely many values of } n \text{ in } \mathbb{Z}^+\} \geq \tfrac{1}{4}.$$

Since the condition in the braces defines an $\mathscr{F}_2$ set, the probability in question must be 0 or 1, and so is 1, and therefore $\limsup_{t\to\infty} w(t) = +\infty$ almost surely because c is arbitrary. Similarly or by symmetry the inferior limit is almost surely $-\infty$. Thus almost every $w(\cdot)$ path hits each point of $\mathbb{R}$ at arbitrarily large parameter values. In particular, almost every $w(\cdot)$ path hits 0 at arbitrarily large parameter values and therefore also [Section 5(f)] hits 0 at arbitrarily small strictly positive parameter values. We shall see in Section IX.5 that when $N > 1$, almost no Brownian path in $\mathbb{R}^N$ ever hits a specified point at a strictly positive parameter value.

(b) According to Theorem 6, if A is an analytic subset of $\mathbb{R}^N$, the value of $P\{T^A = 0\}$ must be either 0 or 1. The following examples exploit this fact. The dichotomy suggests that a topology of $\mathbb{R}^N$ [$\dot{\mathbb{R}}^N$] can be defined by making a point a limit point of a set A if the hitting time of every analytic superset A_1 of A by a [space-time] Brownian motion from the point is almost surely 1. We shall see in Section IX.15 that this can be done (even with A_1 open in the Euclidean topology) and that the topology so defined is the [parabolic] fine topology already defined in Section 1.XI.1 [Section 1.XVII.9]. From now on in this section the remarks relating probability results to fine topology results will always be for the classical context, but the corresponding remarks for the parabolic context will also be true.

(c) Let A be an open right circular cone in $\mathbb{R}^N$ with vertex the origin. Then $P\{w(t) \in A\}$ is independent of $t > 0$ and is proportional to the cone central angle. Hence, if $t_\bullet$ is a sequence of strictly positive parameter values with limit 0,

$$P\{w(t_n) \in A \text{ for infinitely many values of } n\} \tag{6.2}$$

is strictly positive, and it follows from (b) that $P\{T^A = 0\} = 1$. Actually Theorem 6 implies the stronger result that the probability in (6.2) is 1. We conclude that if $y(t, \omega)$ is the point of $B(0, 1)$ hit by the ray from the origin through $w(t, \omega)$, then for almost every ω the sequence $y(t_\bullet, \omega)$ is dense in $B(0, 1)$.

Cluster Values along Arcs

Let u be a function from a metric space D into a metric space D' with distance function d'. Recall that if ϕ is a function from an interval $]0, \delta[$ into D, with $C = \phi(]0, \delta[)$, and if the limit $\phi(0+)$ exists, then the closed possibly empty set $\bigcap_{t>0} u(\phi(]0,t[))^-$ is the "cluster set," that is, the set of limiting values, of u at $\phi(0+)$ along C. A point η' is in this cluster set if and only if

$$\liminf_{t\to 0} d'(\eta', u[\phi(t)]) = 0.$$

In (d)–(f) below, u is a Borel measurable function from $\mathbb{R}^N$ into a Polish space D' with metric d'.

(d) If $D' = \bar{\mathbb{R}}$, we weaken the hypotheses on u by demanding only that the set $A_c = \{u > c\}$ be analytic for all c in $\mathbb{R}$. Under this hypothesis

$$\{\limsup_{t\to 0} u[w(t)] > b\} = \bigcup_{r>b} \{T^{A_r} = 0\} \in \mathscr{F}_1 \qquad (r \text{ rational});$$

so $\limsup_{t\to 0} u[w(t)]$ is $\mathscr{F}_1$ measurable and therefore almost surely constant. In particular, if A is an analytic subset of D and if $u = 1_A$, the set A_c is analytic for all c and we find that $P\{T^A = 0\}$ is 0 or 1, as already observed in (b). This special case illustrates the fact that Borel measurability of u is sometimes too strong a hypothesis. If u is supposed Borel measurable, both the above limit superior and the corresponding limit inferior are almost surely constant. We shall see in Section IX.15 that the constant values here are, respectively, the fine topology superior and inferior limits of u at the origin.

(e) A point η' of D' is a cluster value of u either along almost no or along almost every $w(\cdot)$ path back to the origin because according to (d) the probability

$$P\{\liminf_{t\to 0} d'(\eta', u[w(t)]) = 0\}$$

is either 0 or 1. Let A' be the set of points η' for which this probability is 1. We shall see in Section IX.15 that η' is in A' if and only if η' is a fine topology limit value of u at the origin.

(f) The set A' defined in (e) is closed and is the cluster set of u along almost every $w(\cdot)$ path back to the origin and if D' is compact,

$$P\{\lim_{t\to 0} d'(A', w(t)) = 0\} = 1. \tag{6.3}$$

In fact A' is obviously closed so if $\eta' \in D' - A'$, there is a neighborhood B' of η' so small that if $B = u^{-1}(B')$, then $P\{T^{B'} = 0\} = 0$. The set $D' - A'$ is therefore a countable union $\bigcup_j B_j'$ of open sets B_j' with this property. It follows that neglecting a null set of $w(\cdot)$ paths, the cluster set of u along each $w(\cdot)$ path back to the origin is a subset of A'. Furthermore, if A_0' is a countable dense subset of A', the cluster set of u on almost every $w(\cdot)$ path back to the origin includes A_0' and therefore includes A' and so is A'. The set A' may be empty, but if D is compact, A' cannot be empty, and (6.3) is true because if A'' is an open neighborhood of A', a finite subunion of $\bigcup_j B_j'$ covers $D' - A''$.

(g) According to Theorem 6 the value of $P\{\lim_{t\to\infty} u[w(t)]$ exists$\}$ is either 0 or 1. If this probability is 1, then the limit is almost surely constant, that is, a single point of D'. In fact the set A' discussed in (f) is a singleton in this case; so this assertion follows from (f) if D' is compact. If D' is not compact, the assertion follows from the fact that the limit random variable is measurable from a trivial σ algebra into a separable metric space and therefore is almost surely constant. It will be shown in Section IX.15 that $\lim_{t\to 0} u[w(t)] = \eta'$ almost surely if and only if $\text{f}\lim_{\xi\to 0} u(\xi) = \eta'$.

7. Tied Down Brownian Motion

Let $w_\xi(\cdot)$ be a Brownian motion on $\mathbb{R}^N$ from ξ and let η be any point of $\mathbb{R}^N$ (possibly ξ). If $0 < s_1 < \cdots < s_n < t$ the joint density of $w_\xi(s_1), \ldots, w_\xi(s_n)$ given that $w_\xi(t) = \eta$ is

$$\mathscr{E}(s_1, \xi, \zeta_1)\mathscr{E}(s_2 - s_1, \zeta_1, \zeta_2) \cdots \mathscr{E}(s_n - s_{n-1}, \zeta_{n-1}, \zeta_n)\frac{\mathscr{E}(t - s_n, \zeta_n, \eta)}{\mathscr{E}(t, \xi, \eta)}. \tag{7.1}$$

This Gaussian joint density is that of a Markov process $w^t_{\xi\eta}(\cdot)$ on the parameter interval $[0, t]$, with initial value ξ, value η at t, and transition density (from ζ_1 at time s_1 to ζ_2 at time s_2, with $s_1 < s_2 < t$)

$$\mathscr{E}(s_2 - s_1, \zeta_1, \zeta_2)\frac{\mathscr{E}(t - s_2, \zeta_2, \eta)}{\mathscr{E}(t - s_1, \zeta_1, \eta)}. \tag{7.2}$$

The joint density (7.1) is also that of a Markov process reversed in time on $[0, t]$ with initial value η at time t and value ξ at time 0 and transition density (from ζ_2 at time s_2 to ζ_1 at time s_1, with $0 < s_1 < s_2$)

$$\mathscr{E}(s_1, \xi, \zeta_1)\frac{\mathscr{E}(s_2 - s_1, \zeta_1, \zeta_2)}{\mathscr{E}(s_2, \xi, \zeta_2)}. \tag{7.3}$$

Either way the N component processes are mutually independent.

When $N = 1$, the transition density (7.2) is Gaussian with mean and variance, respectively,

$$\frac{\zeta_1(t - s_2) + \eta(s_2 - s_1)}{t - s_1}, \qquad \frac{\sigma^2[(s_2 - s_1)(t - s_2)]}{t - s_1}.$$

The random variable $w^t_{\xi\eta}(s)$ is (for $0 \le s \le t$) Gaussian, with mean and variance, respectively,

$$\frac{\xi(t - s) + \eta s}{t}, \qquad \frac{\sigma^2 s(t - s)}{t}.$$

The joint distribution of $w_{\xi\eta}^t(s_1)$, $w_{\xi\eta}^t(s_2)$ is Gaussian with means and variances as just evaluated and covariance $\sigma^2 s_1(t-s_2)/t$.

Suppose that $N = 1$ and that t is specified, and consider the process defined by $y(s) = w_0(s) - sw_0(t)/t$, $0 \le s \le t$. The $y(\cdot)$ process is almost surely continuous, and its finite-dimensional joint distributions are Gaussian with zero means and

$$E\{y(s_1)y(s_2)\} = \frac{\sigma^2 s_1(t-s_2)}{t}. \tag{7.4}$$

If $y^*(s) = w_{\xi\eta}^t(s) - [\xi(t-s) + \eta s]/t$, $0 \le s \le t$, the $y(\cdot)$ and $y^*(\cdot)$ processes have the same finite-dimensional joint distributions, and it follows that there is a choice of $w_{\xi\eta}^t(\cdot)$, that is, a process with the specified finite-dimensional distributions, with continuous sample functions having values ξ at time 0 and η at time t. In the following the notation $w_{\xi\eta}^t(\cdot)$ will refer to such a process on $\mathbb{R}^N$ except that the continuity condition may be weakened to almost sure continuity. This process is sometimes called a *Brownian bridge*.

Denote expectations and probabilities for $w_{\xi\eta}^t$ by $E_{\xi\eta}^t\{\cdot\}$, $P_{\xi\eta}^t\{\cdot\}$. If ϕ is a bounded Borel measurable function on $(\mathbb{R}^N)^n$ and if $0 \le s_1 < \cdots < s_n < t$, define $x_\xi = \phi[w_\xi(s_1), \ldots, w_\xi(s_n)]$ and $x_{\xi\eta}^t = \phi[w_{\xi\eta}^t(s_1), \ldots, w_{\xi\eta}^t(s_n)]$. These are random variables defined on the probability spaces of $w_\xi(\cdot)$ and $w_{\xi\eta}^t(\cdot)$ processes respectively. Then

$$E_{\xi\eta}^t\{x_{\xi\eta}^t\} = E_\xi\left\{x_\xi \frac{\ell(t-s_n, w_\xi(s_n), \eta)}{\ell(t, \xi, \eta)}\right\}. \tag{7.5}$$

The reader is invited to generalize this equality to a larger class of pairs $(x_\xi, x_{\xi\eta}^t)$.

By definition the $w_{\xi\eta}^t(\cdot)$ finite-dimensional distributions when time is reversed become the $w_{\eta\xi}^t(\cdot)$ distributions. In particular, $w_{\xi\eta}^t(\cdot)$ and $w_{\eta\xi}^t(\cdot)$ probabilities are the same for events that are invariant under time reversal, for example, the event: a sample path enters a specified set at some time in $[0, t]$. That is, $P_{\xi\eta}^t = P_{\eta\xi}^t$ on the class of such symmetric events.

8. André Reflection Principle

Let $\{w(\cdot), \mathscr{F}(\cdot)\}$ be a Brownian motion in $\mathbb{R}$ from the origin. Let α be a strictly positive constant, and let T be the hitting time of α by $w(\cdot)$. Then according to the strong Markov property of Brownian motion [see VI(9.4)] if A is a Borel subset of $\mathbb{R}$ and if A' is the reflection of A in the point α,

$$\begin{aligned} P\{w(t) \in A \mid \mathscr{F}(T)\} &= \int_A \ell(t-T, \alpha, \eta) l_1(d\eta) = \int_{A'} \ell(t-T, \alpha, \eta) l_1(d\eta) \\ &= P\{w(t) \in A' \mid \mathscr{F}(T)\} \quad \text{a.e. on } \{T \le t\}. \end{aligned} \tag{8.1}$$

In particular, if $A = [\alpha, +\infty[$, the value of each integral is $\frac{1}{2}$. Ignore the last two terms in (8.1) in this case and integrate over $\{T \leq t\}$ to obtain

$$P\{w(t) \geq \alpha, T \leq t\} = \frac{P\{T \leq t\}}{2} = \frac{P\{\sup_{s \leq t} w(s) \geq \alpha\}}{2}.$$

Since $P\{w(t) \geq \alpha, T > t\} = 0$, we find that

$$P\{\sup_{s \leq t} w(s) \geq \alpha\} = 2P\{w(t) \geq \alpha\}; \tag{8.2}$$

that is, there is equality in (3.5). Since the left side of (8.2) is $P\{T \leq t\}$, the distribution of T is absolutely continuous relative to l_N, with density given by

$$(\text{Density of the distribution of } T) = (2\pi\sigma^2)^{-1/2}\alpha t^{-3/2} \exp\left(-\frac{\alpha^2}{2\sigma^2 t}\right). \tag{8.3}$$

To derive a refinement of (8.2), suppose that A in (8.1) is a Borel subset of $]-\infty, \alpha[$ and integrate over $\{T \leq t\}$ to obtain

$$P\{w(t) \in A, T \leq t\} = P\{w(t) \in A', T \leq t\}. \tag{8.4}$$

Since the term on the right is trivially equal to $P\{w(t) \in A'\}$, we have proved

$$P\{\sup_{s \leq t} w(s) \geq \alpha, w(t) \in d\eta\} = \ell(t, 2\alpha - \eta) l_1(d\eta) \qquad (\eta < \alpha), \tag{8.5}$$

which implies that

$$P\{\sup_{s \leq t} w(s) < \alpha, w(t) \in d\eta\} = [\ell(t, \eta) - \ell(t, 2\alpha - \eta)] l_1(d\eta) \quad (\eta < \alpha). \tag{8.6}$$

Since the distributions involved here are continuous, "$<\alpha$" and "$\leq\alpha$" are interchangeable in (8.6), and the corresponding remark is valid in the preceding equations.

Intuitive Approach: The André Reflection Principle

The preceding derivations are formally satisfactory but give less insight than the following derivation of corresponding results in $\mathbb{R}^N$, results that could of course also have been derived by the above methods. Let D be a half-space of $\mathbb{R}^N$ whose closure does not contain the origin, and let η' $[A']$ be the reflection of a point η [subset A] of $\mathbb{R}^N$ in the bounding half-plane π of D. Let $\{w(\cdot), \mathscr{F}(\cdot)\}$ be a Brownian motion in $\mathbb{R}^N$ from the origin, and let T be the hitting time of π by $w(\cdot)$. The strong Markov property of Brownian motion implies that the process $\{w(T + \cdot), \mathscr{F}(T + \cdot)\}$ is a Brownian motion independent of $\mathscr{F}(T)$; equivalently, the process $\{w(T+\cdot)-w(T), \mathscr{F}(T+\cdot)\}$ is a Brownian motion from the origin, independent of $\mathscr{F}(T)$. The distribution of the latter Brownian motion is unaffected if the sample paths are reflected in a translation of π containing the origin. (Here we use the fact that the distribution of Brownian motion from the origin is spherically symmetric about the origin and that each coordinate process of this Brownian motion is a Brownian motion in $\mathbb{R}$ from the origin, independent of the other coordinate processes.) Hence (8.4) is true with the present interpretation of $w(\cdot)$ and A', and it follows that

$$P\{T \leq t, w(t) \in d\eta\} = \ell(t, \eta')l_N(d\eta) \qquad (\eta \in D), \tag{8.7}$$

which implies that

$$P\{T > t, w(t) \in d\eta\} = [\ell(t,\eta) - \ell(t,\eta')]l_N(d\eta) \qquad (\eta \in D). \tag{8.8}$$

If $N = 1$ and $\pi = \{\alpha\}$, these results reduce to (8.4)–(8.6). Observe that in the present context if α is the distance from the origin to π, then (8.3) follows because to verify (8.3) we can assume that π is the hyperplane $\{\xi^{(N)} = \alpha\}$, where $\xi^{(N)}$ is the Nth coordinate of ξ, and if $w^{(N)}(t)$ is the Nth coordinate random variable of $w(t)$, then (8.2) follows for $w^{(N)}(\cdot)$ and T is now the hitting time of $\{\alpha\}$ by $w^{(N)}(\cdot)$. We shall interpret (8.8) in Section 9 as the distribution at time t of "Brownian motion in D" from the origin.

9. Brownian Motion in an Open Set ($N \geq 1$)

Let D be a nonempty open subset of $\mathbb{R}^N$. For each point ξ of $\mathbb{R}^N$ denote by $\{w_\xi(\cdot), \mathscr{F}_\xi(\cdot)\}$ a Brownian motion in $\mathbb{R}^N$ from ξ, and let S_ξ be the hitting time of ∂D (Euclidean boundary) by $w_\xi(\cdot)$. Let $\{w(\cdot), \mathscr{F}(\cdot)\}$ be a Brownian motion in $\mathbb{R}^N$ with initial distribution supported by D, and let S be the hitting time of ∂D by $w(\cdot)$. Denote by $w_D(\cdot)$ the process $w(\cdot)$ killed at S, so that $\{w_D(\cdot), \mathscr{F}(\cdot)\}$ is a Markov process with state space D, initial distribution that of $w(\cdot)$, and substochastic transition function p given by

$$p(t, \xi, A) = P\{w_\xi(t) \in A, t < S_\xi\}. \tag{9.1}$$

An almost surely continuous Markov process $\{z(\cdot), \mathscr{G}(\cdot)\}$ with state space D and transition function given by (9.1) will be called a *Brownian motion in* D; the process $\{w_D(\cdot), \mathscr{F}(\cdot)\}$ is a natural example. If $\mathscr{G}(t)$ is the σ algebra generated by the null sets and $\mathscr{F}\{z(s), s \leq t\}$, then $\mathscr{G}(\cdot)$ is right continuous according to Theorem VI.8. Depending on the context, it may or may not be useful to make the transition function stochastic by adjoining a trap point to D.

Since the right side of (9.1) is at most $P\{w_\xi(t) \in A\}$, the measure $p(t, \xi, \cdot)$

is absolutely continuous relative to l_N, determined by a density $\ell_D(t, \xi, \cdot)$. A unique choice of this density will now be made. Let ξ, η be points of D, and let $f(t, \xi, \eta)$ be the probability that a Brownian bridge $w^t_{\xi\eta}(\cdot)$ from ξ to η never hits ∂D. Define ℓ_D as suggested by the definition of the Brownian bridge and the definition of $\ell(t, \xi, \eta)$ for $t \le 0$:

$$\ell_D(t, \xi, \eta) = \begin{cases} \ell(t, \xi, \eta) f(t, \xi, \eta) & \text{if } t > 0, \\ 0 & \text{if } t \le 0. \end{cases} \tag{9.2}$$

In particular, $\ell_D(t, \xi, \eta) = \ell(t, \xi, \eta)$ when $D = \mathbb{R}^N$. By the symmetry remark closing Section 7 the function $\ell_D(t, \cdot, \cdot)$ is symmetric. Equation (7.5) yields

$$\ell_D(t, \xi, \eta) = \ell(t, \xi, \eta) - \lim_{s \uparrow t} E\{1_{\{S_\xi \le s\}} \ell(t - s, w_\xi(s), \eta)\} \tag{9.3}$$

for $t > 0$, and the equation is trivially true for $t \le 0$. The function ℓ_D is Borel measurable on $\mathbb{R} \times D \times D$. The expectation on the right in (9.3) can be evaluated when $0 < s < t$ using the strong Markov property of Brownian motion:

$$\begin{aligned} & E\{E\{1_{\{S_\xi \le s\}} \ell(t - s, w_\xi(s), \eta) | \mathscr{F}_\xi(S_\xi)\}\} \\ &= E\left\{1_{\{S_\xi \le s\}} \int_{\mathbb{R}^N} \ell(t - s, \zeta, \eta) \ell(s - S_\xi, w_\xi(S_\xi), \zeta) l_N(d\zeta)\right\} \\ &= E\{1_{\{S_\xi \le s\}} \ell(t - S_\xi, w_\xi(S_\xi), \eta)\} \end{aligned} \tag{9.4}$$

so that (9.3) reduces to

$$\ell_D(t, \xi, \eta) = \ell(t, \xi, \eta) - E\{\ell(t - S_\xi, w_\xi(S_\xi), \eta)\}. \tag{9.5}$$

This evaluation is valid on $\mathbb{R} \times D \times D$. Since $|w_\xi(S_\xi) - \eta| \ge |\partial D - \eta|$ when $S_\xi < +\infty$, the integrand in (9.5) is bounded uniformly for all t, ξ in D, and η in a compact subset of D. Hence $\ell_D(t, \xi, \cdot)$ is continuous on D and is obviously the only continuous density for the transition probability of Brownian motion in D. The Chapman–Kolmogorov density equation

$$\ell_D(s + t, \xi, \zeta) = \int_D \ell_D(s, \xi, \eta) \ell_D(t, \eta, \zeta) l_N(d\eta) \qquad (s > 0, t > 0) \tag{9.6}$$

is true for all ξ, ζ in D in view of the following facts:

(a) The left side varies continuously with ζ.
(b) The right side varies continuously with ζ, because the integrand does and the integrand is majorized by the integrable function $\ell_D(s, \xi, \cdot)(2\pi\sigma^2 t)^{-N/2}$.

(c) The equation is true for l_N almost every ζ when ξ is fixed because $\not{p}_D$ is a transition density for Brownian motion in D.

In view of the inequality $\not{p}_D \leq \not{p}$ it follows from (9.6) that $\not{p}_D(t, \cdot, \cdot)$ is a continuous bounded function on $D \times D$ for each value of t. Moreover, according to the analysis of the integrand in (9.5), the expectation on the right in (9.5) is for fixed η a bounded function of (ξ, t) on $D \times]0, +\infty[$.

Potential Theory Significance of $\not{p}_D$

Let D be a nonempty open subset of $\mathbb{R}^N$, define $\dot{D} = D \times \mathbb{R}$, and let $\dot{G}_{\dot{D}}$ be the parabolic Green function of $\dot{D}$. The function $((\xi, s), (\eta, t)) \mapsto \dot{G}_{\dot{D}}((\xi, s), (\eta, t))$ is a function of $s - t$, ξ, η, and in Section 1.XVII.18 a function $\not{p}_D$ was defined by

$$\not{p}_D(s - t, \xi, \eta) = \dot{G}_{\dot{D}}((\xi, s), (\eta, t)). \tag{9.7}$$

In Section IX.17 we shall identify $\not{p}_D$ as defined by (9.7) with the Brownian transition density function $\not{p}_D$ defined in this section. The probabilistic evaluation of parabolic measure to be made in Section IX.13 makes the expectation in (9.5) the solution $\dot{H}_f$ of the parabolic Dirichlet problem on $\dot{D}$ with f the restriction to $\partial\dot{D}$ (Euclidean boundary) of $\not{p}(\cdot, \cdot, \eta)$. In this light (9.5) can be interpreted as the Dirichlet solution construction of $\dot{G}_{\dot{D}}$ described in Section 1.XVIII.1.

EXAMPLE. Let D be a half-space of $\mathbb{R}^N$, define $\dot{D} = D \times \mathbb{R}$, and if $\dot{\eta} \in \dot{D}$, denote by $\dot{\eta}'$ the reflection on $\dot{\eta}$ in $\partial\dot{D}$. Then [Section 1.XVII(4.3)] $\dot{G}_{\dot{D}}(\dot{\xi}, \dot{\eta}) = \dot{G}(\dot{\xi}, \dot{\eta}) - \dot{G}(\dot{\xi}, \dot{\eta}')$; that is, if $\not{p}_D$ satisfies (9.7), if $\dot{\xi} = (\xi, s)$ and $\dot{\eta} = (\eta, t)$, and if η' is the reflection in $\mathbb{R}^N$ of η in ∂D,

$$\not{p}_D(t, \xi, \eta) = \not{p}(t, \xi, \eta) - \not{p}(t, \xi, \eta') = \not{p}(t, \eta - \xi) - \not{p}(t, \eta' - \xi). \tag{9.8}$$

Observe that according to (8.8), in which $\xi = 0$, equation (9.8) is true with the probability interpretation of $\not{p}_D$ as a Brownian transition density function.

Canonical Brownian Motion in an Open Set

Let D be a nonempty open subset of $\mathbb{R}^N$. Adjoin a point ∂ to D, defining the topology of $D \cup \partial$ so that $D \cup \partial$ is the one-point compactification of D. Let $\mathring{\Omega}$ be the class of all functions $\mathring{\omega}$ from $\mathbb{R}^+$ into $D \cup \partial$ with the following properties:

$$\mathring{\omega}(s) = \partial \text{ implies that } \mathring{\omega}(t) = \partial \text{ for } t > s.$$

The function $\mathring{\omega}$ is continuous on the interval $[0, \mathring{S}(\mathring{\omega})[$, where $\mathring{S}(\mathring{\omega}) = \inf\{s > 0 : \mathring{\omega}(s) = \partial\}$ and there is a left limit (Euclidean topology of $\mathbb{R}^N$) at $\mathring{S}(\mathring{\omega})$ when $\mathring{S}(\omega) < +\infty$.

Define $\mathring{w}(t, \mathring{\omega}) = \mathring{\omega}(t), \mathring{\mathscr{F}}(t) = \mathscr{F}\{\mathring{w}(s), s \leq t\}, \mathring{\mathscr{F}} = \mathscr{F}\{\mathring{w}(s), s \in \mathbb{R}^+\}$. If $\mathring{P}$ is the completion of a measure on $\mathring{\mathscr{F}}$ under which $\{\mathring{w}(\cdot), \mathring{\mathscr{F}}(\cdot)\}$ is a Brownian motion in D, the process $\{\mathring{w}(\cdot), \mathring{\mathscr{F}}(\cdot)\}$ will be said to be a canonical Brownian motion in D. If $D = \mathbb{R}^N$ this definition yields a canonical Brownian motion in $\mathbb{R}^N$ as already defined in Section 3. One way to obtain a canonical Brownian motion in D is to operate on a canonical Brownian motion in $\mathbb{R}^N$ as follows. If S is the hitting time of the Euclidean boundary of D by the Brownian motion, change every sample function value to ∂ at every parameter value at least equal to S.

10. Space-Time Brownian Motion in an Open Set

If $\dot{D}$ is an open subset of $\dot{\mathbb{R}}^N$, a space-[co]time Brownian motion in $\dot{D}$ is a space-[co]time Brownian motion with initial distribution supported by $\dot{D}$, killed at the hitting time of $\partial\dot{D}$. The killed process is Markovian with substochastic transition function $\dot{p}$ given by

$$\dot{p}(t, \dot{\xi}, \dot{A}) = P\{\dot{w}_{\dot{\xi}}(t) \in \dot{A}, t < \dot{S}_{\dot{\xi}}\}, \qquad \dot{\xi} = (\xi, s), \tag{10.1}$$

where $\dot{w}_{\dot{\xi}}(\cdot)$ is a space-[co]time Brownian motion from $\dot{\xi}$ and $\dot{S}_{\dot{\xi}}$ is the hitting time of $\partial\dot{D}$. From now on only the space-time Brownian motion is considered unless the contrary is stated. We shall use the space-time notation introduced in Section 2. The measure $\dot{p}(t, \dot{\xi}, \cdot)$ is supported by $\dot{D}_{s-t}$, and the set function $A \mapsto \dot{p}(t, \dot{\xi}, A \times \{s - t\})$ for A a Borel subset of D_{s-t} is majorized by

$$A \mapsto \int_A \ell(s - t, \xi, \eta) l_N(d\eta);$$

so the first set function is absolutely continuous relative to l_N and is therefore the integral of a density. A unique choice of this density will now be made. Following the reasoning in the Brownian motion case, let $\dot{\xi} : (\xi, s)$ and $\dot{\eta} : (\eta, t)$ be points of $\dot{D}$, and if $s > t$ let $f(\dot{\xi}, \dot{\eta})$ be the probability that a Brownian bridge $w_{\xi\eta}^{s-t}(\cdot)$ has the property that the space-time bridge

$$\{[w_{\xi\eta}^{s-t}(r), s - r], 0 \leq r \leq t\}$$

from $\dot{\xi}$ to $\dot{\eta}$ never hits $\partial\dot{D}$. Define

$$\dot{\ell}_{\dot{D}}(\dot{\xi}, \dot{\eta}) = \begin{cases} \ell(s - t, \xi, \eta) f(\dot{\xi}, \dot{\eta}) & \text{if } s > t, \\ 0 & \text{if } s \leq t. \end{cases} \tag{10.2}$$

This definition of the transition density (from $\dot{\xi}$ to $\dot{\eta}$ in time $s - t$) leads to the evaluation

$$\dot{\ell}_{\dot{D}}(\dot{\xi}, \dot{\eta}) = \ell(s - t, \xi, \eta) - E\{\ell(s - t - \dot{S}_{\dot{\xi}}, w_{\dot{\xi}}(\dot{S}_{\dot{\xi}}), \eta)\}. \tag{10.3}$$

(See the corresponding transition density derivation for Brownian motion in Section 9). The density $\eta \to \dot{\ell}_{\dot{D}}(\dot{\xi}, \dot{\eta})$ is continuous, and the Chapman–Kolmogorov equation takes the form

$$\dot{\ell}_{\dot{D}}(\dot{\xi}_1, \dot{\xi}_3) = \int_{D_{s_2}} \dot{\ell}_{\dot{D}}(\dot{\xi}_1, \dot{\xi}_2)\dot{\ell}_{\dot{D}}(\dot{\xi}_2, \dot{\xi}_3) l_N(d\xi_2), \qquad \dot{\xi}_i = (\xi_i, s_i), \tag{10.4}$$

satisfied identically for $\dot{\xi}_1$ and $\dot{\xi}_3$ in $\dot{D}$ with $s_1 > s_2 > s_3$.

The condition that an extended real-valued positive universally measurable function $\dot{u}$ on $\dot{D}$ be excessive for space-time Brownian motion in $\dot{D}$ is the validity of the inequality

$$\int_{D_{s_2}} \dot{\ell}_{\dot{D}}(\dot{\xi}_1, \dot{\xi}_2)\dot{u}(\dot{\xi}_2) l_N(d\xi_2) \le \dot{u}(\dot{\xi}_1) \qquad (s_1 > s_2) \tag{10.5}$$

together with equality in the limit when $s_2 \uparrow s_1$. If in addition there is equality in (10.5) whenever $s_1 > s_2$, the function $\dot{u}$ is invariant for space-time Brownian motion.

The transition density $\overset{*}{\ell}_{\dot{D}}$ of space-cotime Brownian motion is defined by dualizing the preceding discussion and satisfies the equation

$$\overset{*}{\ell}_{\dot{D}}(\dot{\eta}, \dot{\xi}) = \dot{\ell}_{\dot{D}}(\dot{\xi}, \dot{\eta}) \tag{10.6}$$

in view of the remarks on time reversal in a Brownian bridge at the end of Section 7.

The Function $\ell_D(\cdot, \cdot, \eta)$ and $\dot{\ell}_{\dot{D}}(\cdot, \dot{\eta})$

Let D be a nonempty open subset of $\mathbb{R}^N$ $(N \ge 1)$, and define $\dot{D} = D \times \mathbb{R}$. Equation (9.6) combined with the inequality $\ell_D \le \ell$ implies that for fixed η in D the function $\ell_D(\cdot, \cdot, \eta)$ is excessive for space-time Brownian motion in $\dot{D}$, invariant excessive for space-time Brownian motion in $\dot{D} - \{(\eta, 0)\}$, continuous and 0 at the points of $\dot{D} - \{(\eta, 0)\}$ with ordinate values ≤ 0. We shall show in Section IX.17 that $\ell_D(s - t, \xi, \eta) = \dot{G}_{\dot{D}}((\xi, s), (\eta, t))$. Even without this identification the fact that $\ell_D(\cdot, \cdot, \eta)$ is superparabolic on $\dot{D}$ and parabolic on $\dot{D} - \{(\eta, 0)\}$ follows from the fact to be proved in Section IX.8 that a space-time Brownian motion excessive function is superparabolic on the set of points strictly below any point of finiteness of the function and that a space-time Brownian motion invariant excessive function is parabolic on the set of points strictly below any point of finiteness of the function. More generally, if $\dot{D}$ is an arbitrary nonempty open subset of $\dot{\mathbb{R}}^N$ and if $\dot{\xi} = (\xi, s)$, $\dot{\eta} = (\eta, t)$ are points of $\dot{D}$, equation (10.4) together with the inequality $\ell_{\dot{D}}(\dot{\xi}, \dot{\eta}) \le \ell(s - t, \xi - \eta)$ implies that the function $\ell_{\dot{D}}(\cdot, \dot{\eta})$ is excessive for space-time Brownian motion in $\dot{D}$, invariant excessive for space-time Brownian motion on $\dot{D} - \{\dot{\eta}\}$, continuous and 0 at the points of $\dot{D} - \{\dot{\eta}\}$ below $\dot{\eta}$. We shall show in Section IX.17 that $\ell_{\dot{D}} = \dot{G}_{\dot{D}}$.

11. Brownian Motion in an Interval

Let I be the one-dimensional interval $]a, b[$, and let ξ be a point of I. Let $w(\cdot)$ be a one-dimensional Brownian motion from the origin, so that $w_\xi(\cdot) = \xi + w(\cdot)$ is a Brownian motion from ξ. Fix $t > 0$, and consider the following events. Each parameter value t_j below depends on the sample function.

- $\langle\!\langle a\rangle\!\rangle'$: $w_\xi(\cdot)$ first meets ∂I at a and does so before time t.
- $\langle\!\langle b\rangle\!\rangle'$: $w_\xi(\cdot)$ first meets ∂I at b and does so before time t.
- $\langle\!\langle b\rangle\!\rangle$: There exists t_1 with $0 < t_1 \le t$ and $w_\xi(t_1) = b$.
- $\langle\!\langle ab\rangle\!\rangle$: There exist t_1, t_2 with $0 < t_1 < t_2 \le t$ and $w_\xi(t_1) = a$, $w_\xi(t_2) = b$.
- $\langle\!\langle bab\rangle\!\rangle$: There exist t_1, t_2, t_3 with $0 < t_1 < t_2 < t_3 \le t$ and $w_\xi(t_1) = b$, $w_\xi(t_2) = a$, $w_\xi(t_3) = b$,

and so on. Then

$$\begin{aligned} \langle\!\langle b\rangle\!\rangle' &= \langle\!\langle b\rangle\!\rangle - \langle\!\langle a\rangle\!\rangle' \cap \langle\!\langle b\rangle\!\rangle, \\ \langle\!\langle a\rangle\!\rangle' \cap \langle\!\langle b\rangle\!\rangle &= \langle\!\langle ab\rangle\!\rangle - \langle\!\langle b\rangle\!\rangle' \cap \langle\!\langle ab\rangle\!\rangle, \\ \langle\!\langle b\rangle\!\rangle' \cap \langle\!\langle ab\rangle\!\rangle &= \langle\!\langle bab\rangle\!\rangle - \langle\!\langle a\rangle\!\rangle' \cap \langle\!\langle bab\rangle\!\rangle, \end{aligned} \tag{11.1}$$

so that

$$\langle\!\langle b\rangle\!\rangle' = \langle\!\langle b\rangle\!\rangle - \langle\!\langle ab\rangle\!\rangle \cup \langle\!\langle bab\rangle\!\rangle - \langle\!\langle abab\rangle\!\rangle \cup \cdots. \tag{11.2}$$

The probabilities of these events are easily calculated using the reflection principle. For example, $P\{\langle\!\langle ab\rangle\!\rangle\}$ is the probability that a $w_\xi(\cdot)$ path reaches b by time t after having hit a, or, reflecting the path in a at the hitting time of a, and setting $c = b - a$

$$\begin{aligned} P\{\langle\!\langle ab\rangle\!\rangle\} &= P\{\min_{s \le t} w_\xi(s) \le a - c\} = P\{\min_{s \le t} w(s) \le a - c - \xi\} \\ &= 2P\{w(t) \ge \xi - b + 2c\}. \end{aligned} \tag{11.3}$$

Computing the probabilities of the sets in (11.2) in this way yields

$$P\{\langle\!\langle b\rangle\!\rangle'\} = 2\sum_0^\infty [P\{w(t) > 2nc + b - \xi\} - P\{w(t) > (2n+2)c - b + \xi\}]. \tag{11.4}$$

Just as (8.2) was strengthened into (8.5), so the condition $w_\xi(t)\in d\eta$ can be imposed on each term of (11.2) to obtain

$$P\{\between b\between', w_\xi(t)\in d\eta\} = \sum_0^\infty [\ell(t, 2nc + b - \xi + |\eta - b|) - \ell(t, (2n+2)c - b + \xi + |\eta - b|)]l_1(d\eta) \qquad (\eta\in\mathbb{R}) \tag{11.5}$$

and similarly

$$P\{\between a\between', w_\xi(t)\in d\eta\} = \sum_0^\infty [\ell(t, 2nc - a + \xi + |a - \eta|) - \ell(t, (2n+2)c + a - \xi + |a - \eta|)]l_1(d\eta) \qquad (\eta\in\mathbb{R}) \tag{11.6}$$

The probability (density) that a Brownian path from ξ will be at η in I in time t without having left I before that time is $\ell(t,\xi,\eta)$ less the sums in (11.5) and (11.6), and trivial manipulation yields

$$\ell_I(t,\xi,\eta) = \sum_{-\infty}^{\infty} [\ell(t, 2nc - \xi + \eta) - \ell(t, 2nc + 2a - \xi - \eta)] \quad (\eta\in I). \tag{11.7}$$

The function ℓ_I is the transition density of Brownian motion in I, defined in Section 9, because $\ell_I(t,\xi,\cdot)$ is continuous on I.

More generally, if D is the interval $I_1\times\cdots\times I_n$ with $I_j =]a_j, b_j[$, it is trivial that the transition density of Brownian motion on D, that is, the probability (density) $\ell_D(t,\xi,\eta)$ that a Brownian path from $\xi: (\xi^{(1)},\ldots,\xi^{(N)})$ will be at $\eta: (\eta^{(1)},\ldots,\eta^{(N)})$ in D at time t, without having left D before that time is

$$\ell_D(t,\xi,\eta) = \prod_1^N \ell_{I_j}(t,\xi^{(j)},\eta^{(j)}). \tag{11.8}$$

If $\dot D = D\times\mathbb{R}$ and if $\dot G_{\dot D}$ is the parabolic Green function of $\dot D$, a comparison of (11.8) with 1.XV(8.5) verifies for D the relation between the transition density ℓ_D and the Green function $\dot G_{\dot D}$ announced in Section 9; that is, the transition density ℓ_D satisfies (9.7).

12. Probabilistic Evaluation of Parabolic Measure for an Interval

Let D be an interval of $\mathbb{R}^N$, and let $]t_1, t_2[$ be an interval of $\mathbb{R}$ so that $\dot D = D\times]t_1,t_2[$ is an interval of $\dot{\mathbb{R}}^N$. It will now be shown that the distribution of the first hitting point of $\partial\dot D$ (Euclidean boundary) by a space-time Brownian motion from $\dot\xi = (\xi, s)$ in $\dot D$ is the parabolic measure $\dot\mu_{\dot D}(\dot\xi,\cdot)$. The discussion will be given for $N = 1$ to simplify the notation. Suppose then that $D =]a, b[$.

Hitting Distribution on the Lower Boundary of $\dot{D}$

According to Section 11, the distribution has density (relative to l_1)

$$\eta \to \ell_D(s - t_1, \xi, \eta) = \dot{G}_{\dot{D}}(\dot{\xi}, (\eta, t_1)),$$

and according to 1.XV(9.1), this density is that of parabolic measure on the lower boundary.

Hitting Distribution on the Lateral Boundary of $\dot{D}$

Only the hitting distribution on the segment of the lateral boundary with abscissa value b will be considered. Let T_b be the first time a Brownian path from ξ reaches b, considering only those paths reaching b before a, so that in the notation of Section 11

$$P\{\emptyset b \emptyset'\} = P\{T_b \leq t\}.$$

Differentiating in (11.4) yields the T_b distribution density:

$$t \mapsto \frac{1}{t} \sum_{-\infty}^{\infty} (2nc + b - \xi)\ell(t, 2nc + b - \xi). \tag{12.1}$$

The first hit of the segment of the lateral boundary with absicca value b is at the point $(b, s - T_b)$; so (12.1) can be interpreted to yield the distribution density of the hitting point, and comparison with equation 1.XV(9.1) shows that this density is the density of parabolic measure, as asserted.

This identification of the parabolic measure $\mu_{\dot{D}}(\dot{\xi}, \cdot)$ with the distribution of the first hitting point of $\partial\dot{D}$ by a space-time Brownian motion from $\dot{\xi}$ will be made for arbitrary open subsets $\dot{D}$ of $\dot{\mathbb{R}}^N$ in Section IX.13. The identification gives an intuitive interpretation of the fact (Section 1.XVIII.1) that the parabolic measure of a boundary subset relative to a reference point vanishes if the set is above the reference point.

13. Probabilistic Significance of the Heat Equation and Its Dual

If $\dot{D}$ is an open subset of $\dot{\mathbb{R}}^N$, we have defined a function $\dot{\ell}_{\dot{D}}$ which determines the space-time Brownian motion transition function, and we have remarked that the identification $\dot{\ell}_{\dot{D}} = \dot{G}_{\dot{D}}$ will be made. In particular, if $\dot{D} = D \times \mathbb{R}$ for D an open subset of $\mathbb{R}^N$, this identification reduces to $\ell_D(s - t, \xi, \eta) = \dot{G}_{\dot{D}}((\xi, s), (\eta, t))$. Since (Section 1.XVIII.1) $\dot{G}_{\dot{D}}(\cdot, (\eta, t))$ is equal to $\dot{G}(\cdot, (\eta, t))$ less the parabolic function Dirichlet solution on $\dot{D}$ for the boundary function

$\dot{G}(\cdot,(\eta,t))|_{\partial\dot{D}}$ (= 0 at ∞), it is sometimes convenient to evaluate $\dot{\ell}_{\dot{D}}$, in particular, to evaluate ℓ_D, by solving the relevant Dirichlet problem. We have already (Section 9, Example) verified the correctness of this evaluation of ℓ_D when D is a half-space and (Section 11) when D is an interval. In these examples the essential point is that the differential equation method of images which leads to the Green function is the counterpart of the probabilistic André reflection principle which leads to the Brownian motion transition function.

In the physical context the function $(\eta,t)\mapsto\dot{\ell}_{\dot{D}}((\xi,s),(\eta,t))$ is the temperature at time t at the point η due to a heat source at ξ activated at time s. At every time t the boundary ∂D_t of the N-dimensional set $\{\eta\colon(\eta,t)\in\dot{D}\}$ is held at temperature 0. Alternatively, the context is that of a substance diffusing from a point source at ξ, starting at time s. At every time t the boundary ∂D_t is an absorbing barrier, and $\dot{\ell}_{\dot{D}}((\xi,s),(\eta,t))$ is the density of the diffusing substance at time t at the point η. In these contexts the derivation of the heat equation is local, and the various physical boundary conditions lead to solutions of the heat equation with corresponding mathematical boundary conditions. For example, suppose that almost every space-time Brownian path from a point of $\dot{D}$ hits $\partial\dot{D}$, as will be true if $\dot{D}$ is bounded. We would expect from the localization just noted that for a sufficiently smooth boundary and for a sufficiently smooth boundary function $\dot{f}$ the expected value of $\dot{f}$ at the first hitting place of a space-time Brownian path from $\dot{\xi}=(\xi,s)$ would define a parabolic function $\dot{u}$ of $\dot{\xi}$. Moreover, since for $\dot{\xi}$ near $\partial\dot{D}$ it is plausible that paths from $\dot{\xi}$ are likely to hit $\partial\dot{D}$ soon, it seems plausible that $\dot{u}$ has boundary limit function $\dot{f}$. Thus it is plausible that $\dot{u}$ as just defined probabilistically is the PWB solution of the Dirichlet problem for parabolic functions whenever $\dot{f}$ is resolutive. In particular, if $\dot{D}=D\times\mathbb{R}^N$ with D a Greenian subset of $\mathbb{R}^N$ and if $(\eta,t)\mapsto\dot{f}(\eta,t)$ is a function of the space variable, say $\dot{f}(\eta,t)=f(\eta)$, the function $\dot{u}$ will be a function of the space variable, say $\dot{u}(\xi,s)=u(\xi)$; so u is harmonic, and it is plausible that u is the PWB solution of the Dirichlet problem for harmonic functions whenever f is resolutive in this context.

The principal tool we shall use in discussing these matters is martingale theory, and the evaluations of Dirichlet solutions will turn out to be applications of the martingale equality. To show how martingale theory arises in a natural way in this context, consider the following problem. Let $w(\cdot)$ be a Brownian motion in $\mathbb{R}^N$ from a point. What functions $\dot{u}$ on $\dot{\mathbb{R}}^N$ have the property that $\{\dot{u}[w(t),t],t\in\mathbb{R}^+\}$ is a martingale, in some local sense at least? If $\delta>0$, a formal application of Taylor's theorem leads to

$$\begin{aligned}0&=E\{\dot{u}[w(t+\delta),t+\delta]\,|\,w(s),s\le t\}-\dot{u}[w(t),t]\\&=\delta\left[\frac{\partial\dot{u}}{\partial t}+\frac{\sigma^2}{2}\Delta\dot{u}\right]+0(\delta^2)\qquad(13.1)\\&=\delta\overset{*}{\Delta}\dot{u}+0(\delta^2),\end{aligned}$$

where the partial derivatives are evaluated at $[w(t), t]$. Thus it is plausible that $\dot{u}$ must be coparabolic and that conversely the process $\dot{u}[w(\cdot), \cdot]$ is a martingale in some local sense if $\dot{u}$ is coparabolic. Equivalently, $\dot{u}$ compounded with space-time [cotime] Brownian motion is a martingale in some local sense if and only if $\dot{u}$ is parabolic [coparabolic]. In particular (if $\dot{u}$ is a function of the space variable), a function u on $\mathbb{R}^N$ composed with Brownian motion is a martingale in some local sense if and only if u is harmonic.

The results suggested in this section will be formulated rigorously and proved in later chapters.

Chapter VIII

The Itô Integral

1. Notation

Let $\{w(\cdot), \mathscr{F}(\cdot)\}$ be a Brownian motion in $\mathbb{R}$, defined on some probability space $(\Omega, \mathscr{F}, P)$. It is supposed that $\mathscr{F}(\cdot)$ is right continuous and that $\mathscr{F}(0)$ contains the null sets. The Itô integral $\int_0^t \phi \, dw$ will be defined for stochastic processes $\phi(\cdot)$ in the space Γ of not necessarily adapted to $\mathscr{F}(\cdot)$ real processes $\{\psi(t), t \in \mathbb{R}^+\}$ with the following property: there is a progressively measurable process $\{\phi(\cdot), \mathscr{F}(\cdot)\}$, depending on $\psi(\cdot)$, for which

$$P\{\omega: \int_0^t |\phi(s, \omega)|^2 \, ds < +\infty\} = 1 \tag{1.1}$$

for all finite t and

$$P\{\phi(t) = \psi(t)\} = 1 \tag{1.2}$$

for l_1 almost every t. In this chapter ds refers to Lebesgue measure l_1. For economy in later references absolute value signs are used in (1.1) and similar contexts because the present discussion will be extended in Section 7 to cover vector processes and processes with complex state spaces. Observe that a process indistinguishable from one in Γ is itself in Γ. Let $l_1 \times P$ be the completed indicated product measure on $\mathbb{R}^+ \times \Omega$, defined on the completion of $\mathscr{B}(\mathbb{R}^+) \times \mathscr{F}$ relative to this product measure. According to Section I.13, if $\{\phi(\cdot), \mathscr{F}(\cdot)\}$ is an extended real-valued adapted process and if the function $\phi(\cdot)$ is $l_1 \times P$ measurable, then this process is in Γ if (1.1) is satisfied.

We metrize Γ by identifying two members $\phi(\cdot)$ and $\psi(\cdot)$ of Γ whenever (1.2) is true for l_1 almost every t, thereby making Γ into a space of equivalence classes, and by defining (abbreviating the notation)

$$\Gamma \operatorname{dist}(\phi, \psi) = \sum_{k=1}^{\infty} 2^{-k} E\left\{1 \wedge \left[\int_0^k |\phi - \psi|^2 \, ds\right]^{1/2}\right\}. \tag{1.3}$$

Here $\phi(\cdot)$ and $\psi(\cdot)$ are progressively measurable, and we are adopting the usual convention: the right-hand side of (1.3) is the distance between the equivalence classes containing $\phi(\cdot)$ and $\psi(\cdot)$. In the following the reader is asked to judge from the context whether notation like $\phi(\cdot)$ refers to an individual process or to an equivalence class. It is also left to the reader to verify the fact, which we shall not use, that Γ is complete in its metric.

The subset of Γ for which (1.1) is true with $t = +\infty$ will be denoted by $\bar{\Gamma}$, and the subset of $\bar{\Gamma}$ for which

$$E\left\{\int_0^\infty |\phi|^2\, ds\right\} < +\infty$$

will be denoted by Γ_2. A metric for $\bar{\Gamma}$ sometimes preferable to that induced by the Γ metric is the $\bar{\Gamma}$ metric in which (1.3) is strengthened to

$$\bar{\Gamma}\operatorname{dist}(\phi, \psi) = E\left\{1 \wedge \left[\int_0^\infty |\phi - \psi|^2\, ds\right]^{1/2}\right\}. \tag{1.4}$$

A Γ_2 metric sometimes preferable to that induced by the Γ or $\bar{\Gamma}$ metric is the L^2 metric in which (1.3) and (1.4) are strengthened to

$$\Gamma_2 \operatorname{dist}(\phi, \psi) = E^{1/2}\left\{\int_0^\infty |\phi - \psi|^2\, ds\right\}. \tag{1.5}$$

The spaces $\bar{\Gamma}$ and Γ_2 are complete in their metrics. There is convergence of a sequence $\{\phi_n(\cdot), n \in \mathbb{Z}^+\}$ to $\phi(\cdot)$ in the Γ metric if and only if

$$\operatorname*{p\,lim}_{n\to\infty} \int_0^t |\phi_n - \phi|^2\, ds = 0 \tag{1.6}$$

for all finite t, and there is convergence in the $\bar{\Gamma}$ metric if and only if (1.6) is true for $t = +\infty$. Here $\phi_n(\cdot)$ and $\phi(\cdot)$ are chosen to be progressively measurable members of their equivalence classes, in Γ or $\bar{\Gamma}$ as required.

There are sometimes formal advantages in having members of Γ defined at the parameter value $+\infty$. This can be effected for our purposes by defining $\phi(+\infty)$ arbitrarily, by defining $\mathscr{F}(+\infty) = \mathscr{F}$, and leaving the Γ, $\bar{\Gamma}$ and Γ_2 metrics unchanged.

Observe that if ψ is a bounded member of Γ, then $\psi(\cdot)\phi(\cdot)$ is in Γ or $\bar{\Gamma}$ or Γ_2 if $\phi(\cdot)$ is. A useful special case is $\psi(\cdot) = 1_{ST}$, defined as the indicator function of the stochastic interval $[\![S, T]\!]$.

As usual the definition of an integral is first given for a simple linear class of integrands. In the present context this class, a subclass of Γ_2, is the class Γ_0 each of whose equivalence classes contains a process $\phi(\cdot)$ defined as follows. There is a finite set $0 < t_1 < \cdots < t_k < +\infty$ and corresponding bounded random variables $f_1, \ldots, f_k$ such that f_j is $\mathscr{F}(t_j)$ measurable and that

$$\phi(t) = \begin{cases} 0 & \text{if } t \le t_1, \\ f_{j-1} & \text{if } t_{j-1} < t \le t_j,\ 2 \le j \le k, \\ 0 & \text{if } t > t_k. \end{cases} \tag{1.7}$$

A function ϕ in Γ_0 has many representations (1.7) because additional partition points can be adjoined without changing $\phi(\cdot)$, adjusting $f_.$ as required. If $\phi(\cdot)$ and $\psi(\cdot)$ in Γ_0 are to be considered simultaneously, partition points can be adjoined if necessary to obtain representations of $\phi(\cdot)$ and $\psi(\cdot)$ with the same partition. Using this fact it is obvious that Γ_0 is an algebra.

Observation on predictability of Itô Integral integrands. The process $\phi(\cdot)$ defined by (1.7) is adapted, left continuous, and therefore predictable. The following lemma states that Γ_0 is dense in the sets Γ, $\bar{\Gamma}$, Γ_2, and a trivial adaptation of the proof shows that each equivalence class of each of these sets contains a predictable member. We shall not use this fact.

2. The Size of Γ_0

Lemma. *The set Γ_0 is dense in Γ, $\bar{\Gamma}$, and Γ_2 in the metrics of these spaces.*

It is sufficient to consider progressively measurable members of these spaces. Suppose first that $\psi(\cdot)$ is a progressively measurable member of Γ, and define

$$\psi_n(t) = \begin{cases} \psi(t) & \text{if } |\psi(t)| \le n \text{ and } t \le n, \\ 0 & \text{otherwise.} \end{cases}$$

Then $\lim_{n\to\infty} \psi_n(\cdot) = \psi(\cdot)$ in the Γ metric. Next define

$$\psi_{nm}(t) = \begin{cases} 0 & \text{if } t \le \dfrac{1}{m}, \\ m \displaystyle\int_{(j-2)/m}^{(j-1)/m} \psi_n\,ds & \text{if } \dfrac{j-1}{m} < t \le \dfrac{j}{m},\ 1 < j \le nm, \\ 0 & \text{if } t > n. \end{cases}$$

The process $\psi_{nm}(\cdot)$ is progressively measurable, $|\psi_{nm}| \le n$, $\psi_{nm}(\cdot) \in \Gamma_0$, and $\lim_{m\to\infty} \psi_{nm}(\cdot) = \psi_n(\cdot)$ in the Γ metric because there is bounded l_1 almost everywhere sample function convergence. This completes the proof of the lemma for the Γ metric and the proof of the lemma is concluded by the observation that for $\psi(\cdot)$ in $\bar{\Gamma}$ or Γ_2 the convergence assertions in the above proof are also valid in the metric of the space $\bar{\Gamma}$ or Γ_2 as the case may be.

Refinement. A trivial computation involving the Schwarz inequality shows that

$$\int_0^t |\psi_{nm}|^2\, ds \le \int_0^t |\psi_n|^2\, ds \le \int_0^t |\psi|^2\, ds \quad \text{a.s.}$$

Thus an arbitrary element $\phi(\cdot)$ of Γ, $\bar{\Gamma}$, or Γ_2 is the limit, in the metric of its space, of a sequence $\{\phi_n(\cdot), n \ge 0\}$ in Γ_0 satisfying almost surely

$$\int_0^t |\phi_n|^2\, ds \le \int_0^t |\phi|^2\, ds \tag{2.1}$$

simultaneously for all t.

3. Properties of the Itô Integral

If $\phi(\cdot) \in \Gamma$ and if $0 \le t < +\infty$ [or $0 \le t \le +\infty$ if $\phi(\cdot) \in \bar{\Gamma}$], the integral $x(t) = \int_0^t \phi\, dw$ will be defined uniquely up to a null set. Moreover it will be shown that a version of $x(t)$ can be chosen for each t in such a way that the map $\phi(\cdot) \mapsto x(\cdot)$ is a linear continuous map from Γ [$\bar{\Gamma}$] [[Γ_2]] into a metric space Γ' [$\bar{\Gamma}'$] [[Γ_2']] of processes defined as follows. The space Γ' is the space of almost surely continuous adapted processes $\{x(\cdot), \mathscr{F}(\cdot)\}$ metrized into a complete metric space (in which indistinguishable processes are identified with each other) by

$$\Gamma' \operatorname{dist}(x(\cdot), y(\cdot)) = \sum_1^\infty 2^{-k} E\{1 \wedge \sup_{t \le k} |x(t) - y(t)|\}. \tag{3.1}$$

The space $\bar{\Gamma}'$ is the subset of Γ' for which almost surely $\lim_{t \to \infty} x(t) = x(+\infty)$ exists and is finite. The $\bar{\Gamma}'$ metric is defined by strengthening (3.1) to

$$\bar{\Gamma}' \operatorname{dist}(x(\cdot), y(\cdot)) = E\{1 \wedge \sup_{t < +\infty} |x(t) - y(t)|\}. \tag{3.2}$$

The space Γ_2' is the subset of $\bar{\Gamma}'$ for which $E\{\sup_{t<+\infty} |x(t)|^2\} < +\infty$. The Γ_2' metric is defined by strengthening (3.2) to

$$\Gamma_2' \operatorname{dist}(x(\cdot), y(\cdot)) = E^{1/2}\{\sup_{t < +\infty} |x(t) - y(t)|^2\}. \tag{3.3}$$

The spaces Γ', $\bar{\Gamma}'$, and Γ_2' are complete in their metrics. When $\phi(\cdot) \in \Gamma$, the integral $\int_0^t \phi\, dw = x(t)$ will be defined for $0 \le t < +\infty$ [or $0 \le t \le +\infty$ if $\phi(\cdot) \in \bar{\Gamma}$] and will have the following properties.

(a) $x(0) = 0$ almost surely, and a version of $x(t)$ can be chosen for each t in such a way that $x(\cdot) \in \Gamma'$. In the following it will be assumed that $x(\cdot)$ is so chosen.

(b) The map $\phi(\cdot) \mapsto x(\cdot)$ is linear and continuous from Γ into Γ', $\bar{\Gamma}$ into $\bar{\Gamma}'$, and Γ_2 into Γ_2' in terms of the metrics of the spaces involved.

(c) If T is a finite optional time, then

$$\int_0^\infty 1_{0T}\phi\, dw = x(T) \quad \text{a.s.} \tag{3.4}$$

The upper limit ∞ is legitimate here because the integrand process is in $\bar{\Gamma}$. Moreover, if $\phi(\cdot) \in \bar{\Gamma}$, then (3.4) is true for an arbitrary optional time T.

Observe that if T is an arbitrary optional time the process $1_{TT}(\cdot)$ is in Γ and is identified with 0 in the Γ metric; therefore if S and T are optional, with $S \le T$, it follows that $1_{0S} + 1_{ST} = 1_{0T}$ under the Γ identifications. Hence if $\phi(\cdot)$ is in Γ and if T is finite valued,

$$\int_0^\infty 1_{ST}\phi\, dw = x(T) - x(S) \quad \text{a.s.} \tag{3.5}$$

If $\phi(\cdot)$ is in $\bar{\Gamma}$, the optional times need not be finite valued. It will sometimes be convenient to use the notation $\int_S^T \phi\, dw$ for the left side of (3.5).

(d) Let S and T be optional times with $S \le T < +\infty$. If f is a bounded $\mathscr{F}(S)$ measurable random variable and if $\phi(\cdot)$ is in Γ, then

$$\int_S^T f\phi\, dw = f\int_S^T \phi\, dw \quad \text{a.s.} \tag{3.6}$$

If $\phi(\cdot)$ is in $\bar{\Gamma}$, the optional times S and T need not be finite valued.

(e) The process $\{x(\cdot), \mathscr{F}(\cdot)\}$ is a local martingale when $\phi(\cdot)$ is in Γ.

(f) Let S and T be optional times with $S \le T \le +\infty$. Suppose that $\phi(\cdot)$ is in Γ and that

$$E\left\{\int_S^T |\phi|^2\, ds\right\} < +\infty. \tag{3.7}$$

Then

$$E\left\{\left|\int_S^T \phi\, dw\right|^2\right\} = \sigma^2 E\left\{\int_S^T |\phi|^2\, ds\right\}, \tag{3.8}$$

and the process $\{x(T \wedge (S + t)), \mathscr{F}(S + t), 0 \le t \le +\infty\}$ is an almost surely continuous martingale. If $\phi(\cdot)$ and $\psi(\cdot)$ in Γ satisfy (3.7), then

$$E\left\{\int_S^T \phi\, dw \int_S^T \psi\, dw\right\} = \sigma^2 E\left\{\int_S^T \phi\psi\, ds\right\}. \tag{3.9}$$

[The integral $\int_S^T \phi\, dw$ is defined in (3.8) and (3.9) even if T is infinite valued because (3.7) implies that the process $1_{ST}(\cdot)\phi(\cdot)$ is in $\bar{\Gamma}$.]

For $\phi(\cdot)$ in Γ define

$$y_\phi(t) = \exp\left\{\int_0^t \phi\, dw - \frac{\sigma^2}{2}\int_0^t |\phi|^2\, ds\right\}, \qquad 0 \le t < +\infty \tag{3.10}$$

and allow $t = +\infty$ when $\phi(\cdot) \in \bar{\Gamma}$. The process $y_\phi(\cdot)$ is almost surely continuous.

(g) For $\phi(\cdot)$ in Γ the process $\{y_\phi(\cdot), \mathscr{F}(\cdot)\}$ is a positive supermartingale on the parameter interval $[0, +\infty]$, with $y_\phi(0) = 1$ almost surely.

(h) If $b > 0$ and $\delta > 0$, then for $\phi(\cdot)$ in Γ,

$$P\left\{\sup_{t \le b}\left|\int_0^t \phi\, dw\right| \ge \delta + \frac{\sigma^2}{2\delta^2}\int_0^b |\phi|^2\, ds\right\} \le 2e^{-1/\delta}. \tag{3.11}$$

(i) If $c \in \mathbb{R}$ and if $p > 1$ then for $\phi(\cdot)$ in Γ

$$E\{|y_\phi(t)|^c\} \le E^{(p-1)/p}\left\{\exp\left[\frac{cp(cp-1)}{p-1}\frac{\sigma^2}{2}\int_0^t |\phi|^2\, ds\right]\right\}; \tag{3.12}$$

if we choose $c > 1$ and if $p = 1 + (c-1)^{1/2}$, then

$$\lim_{c \downarrow 1} \frac{cp(cp-1)}{p-1} = 1. \tag{3.13}$$

(j) For ϕ in Γ the process $\{y_\phi(\cdot), \mathscr{F}(\cdot)\}$ is a local martingale.

(k) If $\phi \in \Gamma$ and if for some $b \le +\infty$,

$$E\left\{\exp\left[\frac{\sigma^2}{2}\int_0^b |\phi|^2\, ds\right]\right\} < +\infty, \tag{3.14}$$

then the process $\{y_\phi(t), \mathscr{F}(t), t \le b\}$ is a martingale.

(l) If $\gamma \in \mathbb{R}$ and if $\phi(\cdot) \in \Gamma_2$, then

$$E\left\{y_\gamma(t)\int_0^t \phi\, dw\right\} = \sigma^2\gamma\int_0^t E\{y_\gamma(s)\phi(s)\}\, ds \tag{3.15}$$

and

$$y_\gamma(t) - 1 = \gamma\int_0^t y_\gamma\, dw \quad \text{a.s.} \tag{3.16}$$

[More generally for $\phi(\cdot)$ in Γ

$$y_\phi(t) - 1 = \int_0^t \phi y_\phi\, dw \quad \text{a.s.},$$

as will be seen in Section 12, Example (b), by means of Itô's formula, but the special case (3.16) will be needed before Itô's formula is derived.]

4. The Stochastic Integral for an Integrand Process in Γ_0

For ϕ in Γ_0, specified by (1.7), the stochastic integral is defined in the obvious way:

$$x(t) = \int_0^t \phi\, dw = \sum_2^k f_{j-1}[w(t_j \wedge t) - w(t_{j-1} \wedge t)], \qquad 0 \le t \le +\infty. \tag{4.1}$$

Under this definition $x(0) = 0$, $x(t)$ is $\mathscr{F}(t)$ measurable, and $x(\cdot)$ is a continuous process, but we allow as other versions of the integral process every process indistinguishable from that in (4.1). Furthermore, if $t_0 = 0$, then almost surely

$$E\{x(t_j)|\mathscr{F}(t_{j-1})\} = \begin{cases} 0 = x(0) & \text{if } j = 1, \\ x(t_{j-1}) + f_{j-1}E\{w(t_j) - w(t_{j-1})|\mathscr{F}(t_{j-1})\} = x(t_{j-1}) & \text{if } j > 1. \end{cases} \tag{4.2}$$

Hence $\{x(t_\bullet), \mathscr{F}(t_\bullet)\}$ is a martingale; so if $0 \le s < t$ and if s, t are in $t_\bullet$, it follows that $E\{x(t)|\mathscr{F}(s)\} = x(s)$ almost surely. Since any pair of parameter values can be adjoined to the set $t_\bullet$, it follows that $\{x(\cdot), \mathscr{F}(\cdot)\}$ is a martingale. This martingale is L^2 bounded, hence has orthogonal increments, and

$$E\{|x(+\infty)|^2\} = \sigma^2 E\left\{\int_0^\infty |\phi(s)|^2\, ds\right\}. \tag{4.3}$$

If $\phi(\cdot)$ is defined by (1.7) and $y_\phi(\cdot)$ by (3.10)

$$E\{y_\phi(t_j)|\mathscr{F}(t_{j-1})\} = \begin{cases} 1 = y_\phi(0) \quad \text{a.s.} & \text{if } j = 1, \\ y_\phi(t_{j-1})E\{\exp[f_{j-1}(w(t_j) & \\ \quad - w(t_{j-1}))]|\mathscr{F}(t_{j-1})\}\exp\left[\dfrac{-\sigma^2}{2}(t_j - t_{j-1})f_{j-1}^2\right] \quad \text{a.s.} & \text{if } j > 1. \end{cases} \tag{4.4}$$

Now f_{j-1} and $w(t_j) - w(t_{j-1})$ are mutually independent, and whenever z is a normally distributed random variable with mean 0 and variance α, $E\{\exp(az)\} = \exp(a^2\alpha/2)$. Hence the right side of (4.4) reduces to $y_\phi(t_{j-1})$. Thus $\{y_\phi(t_\bullet), \mathscr{F}(t_\bullet)\}$ is a martingale, and it follows repeating an argument just used that $\{y_\phi(\cdot), \mathscr{F}(\cdot)\}$ is a martingale.

5. The Stochastic Integral for an Integrand Process in Γ

Suppose that $\phi(\cdot)\in\Gamma_0$. Then (Section 4) the process $\{y_\phi(\cdot),\mathscr{F}(\cdot)\}$ defined by (3.10) is a positive martingale, and if $\delta>0$ and $b>0$, the submartingale maximal inequality III(9.1′) yields

$$P\left\{\sup_{t\le b}\int_0^t \phi\,dw \ge \frac{1}{\delta}+\frac{\sigma^2}{2}\int_0^b \phi^2\,ds\right\} \le P\left\{\sup_{t\le b} y_\phi(t)\ge e^{1/\delta}\right\}\le e^{-1/\delta}. \quad (5.1)$$

If ϕ is replaced by $-\phi$ and the resulting inequality is combined with (5.1), we find

$$P\left\{\sup_{t\le b}\left|\int_0^t \phi\,dw\right| \ge \frac{1}{\delta}+\frac{\sigma^2}{2}\int_0^b \phi^2\,ds\right\}\le 2e^{-1/\delta} \quad (5.2)$$

or, if ϕ is replaced by ϕ/δ^2,

$$P\left\{\sup_{t\le b}\left|\int_0^t \phi\,dw\right| \ge \delta+\frac{\sigma^2}{2\delta^2}\int_0^b \phi^2\,ds\right\}\le 2e^{-1/\delta}. \quad (5.3)$$

We conclude that if $\{\phi_n(\cdot),n\in\mathbb{Z}^+\}$ is a sequence in Γ_0 with limit 0 in the Γ metric, then the sequence $\{\int_0^\bullet \phi_n\,dw, n\in\mathbb{Z}^+\}$ has limit 0 in the Γ' metric. That is, the linear map $\phi\mapsto\int_0^\bullet\phi\,dw$ from Γ_0 in the Γ metric into the metric space Γ' is uniformly continuous. It follows that there is a unique uniformly continuous linear extension of this map, from the Γ metric closure of Γ_0, that is, from Γ, into Γ'. We accept this extension as the definition of the integral $\int_0^\bullet\phi\,dw$; this integral then satisfies Section 3(h).

Furthermore the map $\phi\mapsto\int_0^\bullet\phi\,dw$ takes Γ_0 into $\bar\Gamma'$, and if we apply the reasoning just used but this time with the $\bar\Gamma$ metric on Γ_0 and the $\bar\Gamma'$ metric on the image of Γ_0, we find, setting $b=+\infty$ in (5.2), that the map is uniformly continuous. Hence this map is uniformly continuous from $\bar\Gamma$ into $\bar\Gamma'$.

Finally we consider Γ_0 in the Γ_2 metric. Since $|\int_0^\bullet\phi\,dw|$ is a submartingale for ϕ in Γ_0, the L^2 maximal inequality III(11.1) for the continuous parameter context is applicable and yields

$$E\left\{\sup_{t\ge 0}\left|\int_0^t\phi\,dw\right|^2\right\}\le 4\sigma^2 E\left\{\int_0^\infty |\phi|^2\,ds\right\}, \quad (5.4)$$

and it follows that the map $\phi\mapsto\int_0^\bullet\phi\,dw$ is uniformly continuous from Γ_2 into Γ_2'.

Localization of the Stochastic Integral

We have defined $\int_0^T\phi\,dw$ for T optional as $\int_0^\infty 1_{0T}\phi\,dw$. If ϕ is defined initially only on $]\!]0,T[\![$, define ϕ as 0 on $\mathbb{R}^+\times\Omega-]\!]0,T[\![$, and if the extension is in Γ, define $\int_0^T\phi\,dw$ using the extended integrand.

6. Proofs of the Properties in Section 3

Section 3(a), (b), (h) have already been proved. We prove the remaining properties in the order (cdfegikjl).

Proof of Section 3(c). (c1) In order to prove (3.4) either for $\phi(\cdot)$ in Γ with $T < +\infty$ or for $\phi(\cdot)$ in $\bar{\Gamma}$ with $T \leq +\infty$, it is actually sufficient to prove (3.4) for T bounded. In fact, if (3.4) is true for T bounded, then for general T,

$$\int_0^\infty 1_{0T\wedge n}\phi\,dw = x(T \wedge n) \quad \text{a.s.} \tag{6.1}$$

Now when either $\phi(\cdot)$ is in Γ and $T < +\infty$ or $\phi(\cdot)$ is in $\bar{\Gamma}$ and $T \leq +\infty$, it is true that $\lim_{n\to\infty} x(T \wedge n) = x(T)$ almost surely and the integrand process in (6.1) tends to the process $1_{0T}(\cdot)\phi(\cdot)$ in the $\bar{\Gamma}$ metric so that

$$\begin{aligned} &\operatorname*{p\,lim}_{n\to\infty}\left|\int_0^\infty 1_{0T}\phi\,dw - \int_0^\infty 1_{0T\wedge n}\phi\,dw\right| \\ &\leq \operatorname*{p\,lim}_{n\to\infty}\sup_{t\geq 0}\left|\int_0^t 1_{0T}\phi\,dw - \int_0^t 1_{0T\wedge n}\phi\,dw\right| = 0. \end{aligned} \tag{6.2}$$

Thus (3.4) is true. From now on in the proof of (3.4) we shall suppose that T is bounded and that $\phi(\cdot)$ is in Γ.

(c2) If (3.4) is true for the optional time $[T]_n$, it is true for T because $\lim_{n\to\infty} 1_{0[T]_n}(\cdot)\phi(\cdot) = 1_{0T}(\cdot)\phi(\cdot)$ in the $\bar{\Gamma}$ metric and $\lim_{n\to\infty} x([T]_n) = x(T)$ almost surely.

(c3) Proof of (3.4) for $\phi(\cdot)$ in Γ_0. According to (c2), it is sufficient to prove (3.4) with T replaced by $[T]_n$. At the possible cost of modifying the representation (1.7) of $\phi(\cdot)$ we can suppose that $t_\bullet$ includes the values $\{j2^{-n}, j\in\mathbb{Z}^+\}$ in the interval $[0, t_k]$. Finally by linearity of the map $\phi(\cdot)\mapsto \int_0^\cdot \phi\,dw$ it is then sufficient to prove (3.4) for $\phi(\cdot)$ given by (1.7) with $k = 2$. That is, we now suppose for f the indicator function of the $\mathscr{F}(t_1)$ set $\{[T]_n > t_1\}$ that

$$1_{0[T]_n}(t)\phi(t) = \begin{cases} 0 & \text{if } t \leq t_1, \\ f_1 f & \text{if } t_1 < t \leq t_2, \\ 0 & \text{if } t > t_2. \end{cases} \tag{6.3}$$

This process is in Γ_0, and the definition of the integral for an integrand process in Γ_0 makes (3.4) trivial in this case.

(c4) To prove (3.4) for arbitrary $\phi(\cdot)$ in Γ, let $\{\phi_n(\cdot), n\in\mathbb{Z}^+\}$ be a sequence in Γ_0 with limit $\phi(\cdot)$ in the Γ metric, and define $x_n(t) = \int_0^t \phi_n\,dw$. According to (c3),

$$\int_0^\infty 1_{0T}\phi_n\,dw = x_n(T) \quad \text{a.s.}$$

and (3.4) follows because in view of the boundedness of T,

$$\lim_{n\to\infty} 1_{0T}(\cdot)\phi_n(\cdot) = 1_{0T}(\cdot)\phi(\cdot)$$

in the $\bar{\Gamma}$ metric; so

$$\operatorname*{p\,lim}_{n\to\infty} |x(T) - x_n(T)| \le \operatorname*{p\,lim}_{n\to\infty} \sup_{t\ge 0} \left| \int_0^t 1_{0T}\phi\, dw - \int_0^t 1_{0T}\phi_n\, dw \right| = 0. \quad \square \tag{6.4}$$

Proof of Section 3(d). We can suppose, redefining $\phi(t)$ as 0 for $t \ge T$ if necessary, that $T = +\infty$ and that $\phi(\cdot)$ is in $\bar{\Gamma}$. In view of the metric properties of the map $\phi(\cdot) \mapsto \int_0^\cdot \phi\, dw$ it is sufficient to prove (3.6) with S replaced by $[S]_n$ and with $\phi(\cdot)$ in Γ_0, given by (1.7), and adding points to $t_.$ in (1.7) if necessary, we can suppose that $t_.$ includes the values $\{j2^{-n}, j \in \mathbb{Z}^+\}$ in the interval $[0, t_k]$. Finally in view of the linearity of the above map we can even suppose that $k = 2$. The process $f\phi 1_{[S]_n\infty}(\cdot)$ is now in $\bar{\Gamma}_0$ and has a representation of the form (1.7) with $k = 2$, and the computation verifying (3.6) is trivial. $\square$

Proof of Section 3(f). Suppose first that $S \equiv 0$, $T \equiv +\infty$ and that (3.7) is true. Then $\phi(\cdot)$ is in Γ_2. Let $\{\phi_n(\cdot), n \in \mathbb{Z}^+\}$ be a sequence in Γ_0 with limit $\phi(\cdot)$ in the Γ_2 metric, and define $x_n(t) = \int_0^t \phi_n\, dw$. Then $\lim_{n\to\infty} x_n(\cdot) = x(\cdot)$ in the Γ_2' metric; in particular,

$$\lim_{n\to\infty} E\{|x(t) - x_n(t)|^2\} = 0 \qquad (t \le +\infty).$$

Since the process $\{x_n(t), \mathscr{F}(t), t \le +\infty\}$ is a martingale, it follows that the limit process is a martingale. Moreover

$$E\{|x(+\infty)|^2\} = \sigma^2 E\left\{\int_0^\infty |\phi|^2\, ds\right\}$$

because this equation is valid for $x_n(\cdot)$ and $\phi_n(\cdot)$. In the general case of Section 3(f) if we apply what we have just proved to $1_{ST}(\cdot)\phi(\cdot)$, we find [see (3.5)] that

$$\{x(T \wedge t) - x(S \wedge t), \mathscr{F}(t), t \le +\infty\} = \left\{\int_0^t 1_{ST}\phi\, dw, \mathscr{F}(t), t \le +\infty\right\} \tag{6.5}$$

is a martingale and that (3.8) is satisfied. Equation (3.9) is obtained from (3.8) as usual by polarization. Since the martingale (6.5) is almost surely continuous and right closed, the process remains a martingale when t is replaced by $S + t$; so the process $\{x(T \wedge (S + t)) - x(S), \mathscr{F}(S + t), t \le +\infty\}$ is a martingale; equivalently, $\{x(T \wedge (S + t)), \mathscr{F}(S + t), t \le +\infty\}$ is a martingale. $\square$

Proof of Section 3(e). We prove this property using Section 3(f), just proved. For $\phi(\cdot)$ in Γ and $\alpha > 0$ define

$$T_\alpha = \inf\left\{t: \frac{\sigma^2}{2}\int_0^t |\phi|^2\, ds \geq \alpha\right\}. \tag{6.6}$$

Then T_α is optional, and $(\sigma^2/2)\int_0^{T_\alpha} |\phi|^2\, ds \leq \alpha$. Hence according to Section 3(f), the process $\{x(T_\alpha \wedge t), \mathscr{F}(t), t \leq +\infty\}$ is a right closed and therefore uniformly integrable martingale. Since $\lim_{\alpha\to\infty} T_\alpha = +\infty$ almost surely, it follows that $\{x(\cdot), \mathscr{F}(\cdot)\}$ is a local martingale. □

Proof of Section 3(g). Let $\phi(\cdot)$ be in Γ, and let $\{\phi_n(\cdot), n \in \mathbb{Z}^+\}$ be a sequence in Γ_0 with limit $\phi(\cdot)$ in the Γ metric. Then

$$\lim_{n\to\infty} \int_0^{\bullet} \phi_n\, dw = \int_0^{\bullet} \phi\, dw$$

in the Γ' metric; so $\operatorname{p\,lim}_{n\to\infty} y_{\phi_n}(t) = y(t)$ for each t. The process $\{y_{\phi_n}(\cdot), \mathscr{F}(\cdot)\}$ is a positive martingale, and an application of Fatou's lemma for conditional expectations shows that $\{y_\phi(\cdot), \mathscr{F}(\cdot)\}$ is a supermartingale. □

Proof of Section 3(i). If $\phi(\cdot) \in \Gamma$, $c \in \mathbb{R}$, and if $p > 1$, then

$$y_\phi(t)^c = y_{cp\phi}(t)^{1/p} \exp\left[c(pc-1)\frac{\sigma^2}{2}\int_0^t |\phi|^2\, ds\right], \tag{6.7}$$

from which the inequality (3.12) follows by an application of Hölder's inequality and Section 3(g). The limit relation (3.13) is trivially true for p as stated. □

Proof of Section 3(k). If $b < +\infty$, replace $\phi(\cdot)$ by $1_{0b}(\cdot)\phi(\cdot)$ to obtain an equivalent context in which we can suppose that $b = +\infty$ and $\phi(\cdot) \in \Gamma_2$. We are to prove that $\{y_\phi(t), \mathscr{F}(t), t \leq +\infty\}$ is a martingale; equivalently, since this process is an (almost surely continuous) supermartingale with $E\{y_\phi(0)\} = 1$, we are to prove that $E\{y_\phi(+\infty)\} = 1$. According to Section 2, there is a sequence $\{\phi_n(\cdot), n \in \mathbb{Z}^+\}$ in Γ_0 with limit $\phi(\cdot)$ in the Γ_2 metric and satisfying

$$\int_0^\infty |\phi_n|^2\, ds \leq \int_0^\infty |\phi|^2\, ds, \qquad n \in \mathbb{Z}^+. \tag{6.8}$$

It follows from Section 3(h) that for every real α,

$$\operatorname{p\,lim}_{n\to\infty} \sup_{s\geq 0} |y_{\alpha\phi}(s) - y_{\alpha\phi_n}(s)| = 0.$$

According to Section 4, the process $\{y_{\alpha\phi_n}(t), \mathscr{F}(t), t \le +\infty\}$ is a martingale, with $E\{y_{\alpha\phi_n}(t)\} = 1$, and by Section 3(i) if $0 < \alpha < 1$, the constants c and p can be chosen so that $c > 1, p > 1$, and $cp(cp-1)\alpha^2/(p-1) < 1$, and then in view of (3.12) and (6.8)

$$\sup_{n, t \ge 0} E\{|y_{\alpha\phi_n}(t)|^c\} = \beta_c < +\infty.$$

Hence

$$E\{|y_{\alpha\phi}(t)|^c\} \le \lim_{n\to\infty} E\{|y_{\alpha\phi_n}(t)|^c\} \le \beta_c,$$

$$E\{y_{\alpha\phi}(t)\} = \lim_{n\to\infty} E\{y_{\alpha\phi_n}(t)\} = 1.$$

Thus $\{y_{\alpha\phi}(t), \mathscr{F}(t), t \le +\infty\}$ is an L^c bounded uniformly integrable almost surely continuous martingale. Now define an optional time $S_n(\le +\infty)$ by

$$S_n = \inf\left\{t : \int_0^t \phi\, dw - \sigma^2 \int_0^t |\phi|^2\, ds \le -n\right\}.$$

The martingale sampling theorem implies that $E\{y_{\alpha\phi}(S_n)\} = 1$, that is,

$$\begin{aligned} 1 &= \int_{\{S_n = +\infty\}} y_{\alpha\phi}(+\infty)\, dP + \int_{\{S_n < +\infty\}} y_{\alpha\phi}(S_n)\, dP \\ &= \int_{\{S_n = +\infty\}} y_{\alpha\phi}(+\infty)\, dP + \int_{\{S_n < +\infty\}} \exp\left[\left(\alpha - \frac{\alpha^2}{2}\right)\sigma^2 \int_0^{S_n} |\phi|^2\, ds - \alpha n\right] dP. \end{aligned} \tag{6.9}$$

In view of the definition of S_n the function $\alpha \mapsto e^{\alpha n} y_{\alpha\phi}(\cdot)$ is an increasing function of α when $S_n = +\infty$; furthermore $\alpha \mapsto \alpha - \alpha^2/2$ is an increasing function when $0 < \alpha < 1$. Hence when $\alpha \uparrow 1$ in (6.9), we find

$$1 = \int_{\{S_n = +\infty\}} y_{\phi}(+\infty)\, dP + e^{-n} \int_{\{S_n < +\infty\}} \exp\left[\frac{\sigma^2}{2} \int_0^{S_n} |\phi|^2\, ds\right] dP. \tag{6.10}$$

The sequence $S_.$ is an increasing sequence with $\lim_{n\to\infty} P\{S_n = +\infty\} = 1$ because almost every $y_\phi(\cdot)$ sample function is bounded. Hence when $n \to \infty$, the first integral on the right tends to $E\{y_\phi(+\infty)\}$. The second term on the right is at most

$$e^{-n} E\left\{\exp\left[\frac{\sigma^2}{2} \int_0^\infty |\phi|^2\, ds\right]\right\};$$

so when $n \to \infty$, (6.10) yields $E\{y_\phi(+\infty)\} = 1$, as was to be proved. □

Proof of Section 3(j). If T_α is defined by (6.6) and if $\phi(\cdot)$ is in Γ, then the process $\phi(\cdot)1_{0T_\alpha}(\cdot)$ satisfies the hypotheses of Section 3(k) because

$$\frac{\sigma^2}{2}\int_0^t |\phi 1_{0T_\alpha}|^2\,ds \le \alpha$$

for all t. The process $\{y_\phi(T_\alpha \wedge \cdot), \mathscr{F}(\cdot)\}$ is therefore a martingale, so $\{y_\phi(\cdot), \mathscr{F}(\cdot)\}$ is a local martingale. □

Proof of Section 3(l). It is sufficient to prove (3.15) for $\phi(\cdot)$ in Γ_0 and specified by (1.7) with $k = 2$. In this case when $t_1 < t \le t_2$, the left side of (3.15) can be put in the form

$$E\left\{f_1 \exp\left[\gamma w(t_1) - \frac{\gamma^2\sigma^2 t}{2}\right] E\{[w(t) - w(t_1)]\exp[\gamma(w(t) - w(t_1))] | \mathscr{F}(t_1)\}\right\}. \tag{6.11}$$

In view of the fact that

$$E\{ze^{az}\} = a\alpha e^{a^2\alpha/2}$$

when z is a normally distributed random variable with mean 0 and variance α, the value of the expectation in (6.11) is $\gamma\sigma^2 E\{y_\gamma(t_1)f_1\}(t - t_1)$, as it should be according to (3.15). The verification of (3.15) for other values of t is now trivial. Equation (3.16), an easy consequence of the Itô formula to be proved in Section 12, can be proved at this stage using (3.15) by computing (and finding it to be 0) the expectation of the absolute value of the square of the difference between the two sides of (3.16). □

7. Extension to Vector-Valued and Complex-Valued Integrands

Real Vector-valued Integrands

Suppose that $\{w(\cdot), \mathscr{F}(\cdot)\}$ is a Brownian motion in $\mathbb{R}^N$, with $\mathscr{F}(\cdot)$ right continuous and $\mathscr{F}(0)$ containing the null sets. If β is a vector with N components $\beta^{(1)}, \ldots, \beta^{(N)}$, define $|\beta|^2 = \Sigma_1^N \beta^{(j)^2}$ as usual. Let Γ be the space of processes with state space $\mathbb{R}^N$ whose N component processes are in the space Γ as defined in the one-dimensional case in Section 1, and define the Γ metric by (1.3) with the above interpretation of the notation $|\cdot|$. The corresponding definitions of $\bar{\Gamma}$ and Γ_2 are made. For $\phi(\cdot)$ in Γ define

$$\int_0^t \phi\,dw = \sum_1^N \int_0^t \phi^{(j)}\,dw^{(j)}.$$

The spaces Γ', $\bar{\Gamma}'$, Γ_2' defined in Section 2 are used below, with the same state space $\mathbb{R}$. It is left to the reader to check that at most minor modifications of the proofs in Section 6 show that (a)–(l) in Section 3 are valid for $N \geq 1$ under the following conventions. In (3.9) the integrand $\phi\psi$ is to be interpreted as the inner product $\Sigma_1^N \phi^{(j)}\psi^{(j)}$. In Section 3(l) γ is to be interpreted as a vector; the right side of (3.15) is to be interpreted as the inner product of the vector γ and the vector-valued integral; (3.16) is to be interpreted correspondingly.

Complex Vector-valued Integrands

If the component processes of the integrand processes above have the complex plane as state space, the usual conventions are to be made. That is, $|\beta|^2 = \Sigma_1^N |\beta^{(j)}|^2$; in (3.9) ψ is to be replaced by $\bar{\psi}$, and $\phi\bar{\psi}$ is then the Hermitian symmetric inner product $\Sigma_1^N \phi^{(j)}\bar{\psi}^{(j)}$; ϕ is to be replaced by $\bar{\phi}$ in (3.15) and (3.16). Observe, however, that $y_\phi(t)$ is no longer necessarily a real-valued random variable; so Section 3(g) must be dropped. Furthermore in (3.11) the first term on the right side of the inequality should be changed from δ to 2δ. No other changes are necessary in Section 3(a)–(l). It is easy to deduce from Section 3(g) and the form of $|y_\phi(\cdot)|$ that $\{|y_\phi(\cdot)|, \mathscr{F}(\cdot)\}$ is a supermartingale.

8. Martingales Relative to Brownian Motion Filtrations

If $\{w(\cdot), \mathscr{F}(\cdot)\}$ is a Brownian motion in $\mathbb{R}^N$ and if $\phi(\cdot) \in \Gamma_2$, the process $\{\int_0^\cdot \phi\, dw, \mathscr{F}(\cdot)\}$ is an L^2 bounded almost surely continuous martingale (complex state space) whose random variables have zero expectation. It is a remarkable fact that the converse assertion is true if $\mathscr{F}(\cdot)$ is specified appropriately, as asserted in the following theorem.

Theorem. *Suppose that $\{w(\cdot), \mathscr{F}(\cdot)\}$ is a Brownian motion in $\mathbb{R}^N$, from the origin, and that for all t in $\mathbb{R}^+$ the σ algebra $\mathscr{F}(t)$ is generated by $\mathscr{F}\{w(s), s \leq t\}$ and the null sets.*

(a) *If z is a (complex-valued) $\bigvee_{s \in \mathbb{R}^+} \mathscr{F}(s)$ measurable square integrable random variable, there is a process $\phi(\cdot)$ in Γ_2 for which*

$$z = E\{z\} + \int_0^\infty \phi\, dw \quad \text{a.s.}$$

(b) *If $\{z(\cdot), \mathscr{F}(\cdot)\}$ is an L^2 bounded almost surely continuous martingale, there is a process $\phi(\cdot)$ in Γ_2 for which*

$$z(t) = z(0) + \int_0^t \phi\, dw \quad \text{a.s.} \tag{8.1}$$

In (b) it would be no more general to suppose that $z(\cdot)$ is almost surely right continuous rather than almost surely continuous because Theorem VII.4 implies that an almost surely right continuous martingale relative to a Brownian motion filtration $\mathscr{F}(\cdot)$ as defined in Theorem 8 is almost surely continuous.

Observe that if $0 < \beta < +\infty$ and if z in (a) is $\mathscr{F}(\beta)$ measurable, the upper limit of integration in (a) can be reduced to β. (Apply the operator $E\{-|\mathscr{F}(\beta)\}$ to both sides of the equation for z.) Part (b) is applicable to an L^2 bounded almost surely right continuous martingale $\{z(\cdot), \mathscr{F}(\cdot)\}$ on a finite parameter interval $[0, \beta[$ or $[0, \beta]$ because such a martingale can be extended to one on the parameter interval $\mathbb{R}^+$ to which Theorem 8 is applicable by setting $z(t) = \lim_{s \to \beta} z(s)$ for $t \geq \beta$ in the first case, $z(t) = z(\beta)$ for $t > \beta$ in the second case.

Parts (a) and (b) of the theorem are equivalent. In fact if (a) is true, then under the hypotheses of (b) right close the martingale $z(\cdot)$ by setting $z(+\infty) = \lim_{t \to \infty} z(t)$ and apply (a) to $z(+\infty)$ to obtain, using the fact that $\mathscr{F}(0)$ is trivial,

$$z(t) = E\{z(+\infty)|\mathscr{F}(t)\} = E\{z(0)\} + E\left\{\int_0^\infty \phi\, dw \middle| \mathscr{F}(t)\right\}$$

$$= z(0) + \int_0^t \phi\, dw \quad \text{a.s.}$$

If (b) is true then under the hypotheses of (a) define $z(\cdot)$ as the almost surely continuous martingale $z(\cdot) = E\{z|\mathscr{F}(\cdot)\}$, and then apply (b) and the conditional expectation continuity theorem to find that

$$z = \lim_{t \to \infty} z(t) = E\{z\} + \lim_{t \to \infty} \int_0^t \phi\, dw = E\{z\} + \int_0^\infty \phi\, dw \quad \text{a.s.}$$

To prove (a), suppose to simplify the notation that $N = 1$ and that z is a real random variable. Observe that the class $\{\int_0^\infty \phi\, dw : \phi \in \Gamma_2\}$ ($N = 1$, real context) of random variables is a closed linear subspace of the L^2 space of real random variables; so it is sufficient to prove that if z is $\bigvee_{s \in \mathbb{R}^+} \mathscr{F}(s)$ measurable and is square integrable with $E\{z\} = 0$, and if z is orthogonal to this subspace, then $z = 0$ almost surely.

(a1) We prove first that if $z(t) = E\{z|\mathscr{F}(t)\}$ and if $\phi \in \Gamma_2$, then $\{z(\cdot)\int_0^\cdot \phi\, dw, \mathscr{F}(\cdot)\}$ is a martingale. By the orthogonality hypothesis if $0 \leq s < t$, if f is a bounded $\mathscr{F}(s)$ measurable function, and if ϕ is in Γ_2, then in view of Section 3(d)

$$0 = E\left\{z \int_s^t f\phi\, dw\right\} = E\left\{zf \int_s^t \phi\, dw\right\} = E\left\{z(t) f \int_s^t \phi\, dw\right\};$$

so

$$0 = E\left\{z(t)\int_s^t \phi\, dw \middle| \mathscr{F}(s)\right\} = E\left\{z(t)\int_0^t \phi\, dw \middle| \mathscr{F}(s)\right\} - z(s)\int_0^s \phi\, dw \quad \text{a.s.}$$

This is the desired martingale equality. In particular, it now follows from (3.16) that for every constant γ the process $\{z(\cdot)y_\gamma(\cdot), \mathscr{F}(\cdot)\}$ is a martingale with $E\{z(t)y_\gamma(t)\} = 0$. Observe that ϕ and therefore also γ can be complex here.

(a2) If γ_1 and γ_2 are complex constants and if $0 \le s_1 < s_2$, manipulate conditional expectations to derive the first following equality and apply (a1) to derive the second:

$$\begin{aligned} E\{y_{\gamma_1}(s_1)y_{\gamma_2}(s_2)z\} &= E\{y_{\gamma_1}(s_1)y_{\gamma_2}(s_2)z(s_2)\} = E\{y_{\gamma_1}(s_1)y_{\gamma_2}(s_1)z(s_1)\} \\ &= E\{z(s_1)y_{\gamma_1+\gamma_2}(s_1)\}\exp(\sigma^2\gamma_1\gamma_2 s_1). \end{aligned} \tag{8.2}$$

According to (a1), the last expectation vanishes; so (8.2) yields

$$E\{z\exp[\gamma_1 w(s_1) + \gamma_2 w(s_2)]\} = 0. \tag{8.3}$$

It follows that

$$E\{zf_1[w(s_1)]f_2[w(s_2)]\} = 0 \tag{8.4}$$

whenever f_j is an exponential polynomial of the form $\Sigma_{k=1}^n c_{jk}\exp(i\alpha k\cdot)$ with α real and therefore that (8.4) is true whenever f_j is a continuous periodic function on $\mathbb{R}$. Since any continuous bounded function on $\mathbb{R}$ is the pointwise limit of a bounded sequence of continuous periodic functions (whose periods may become infinite), (8.4) is true whenever f_1 and f_2 are continuous and bounded on $\mathbb{R}$. It then follows that (8.4) is true when f_1 and f_2 are the indicator functions of open subintervals of $\mathbb{R}$, and we conclude that $E\{z|w(s_1), w(s_2)\} = 0$ almost surely. Replace the left side of (8.2) by

$$E\left\{z\prod_1^m y_{\gamma_j}(s_j)\right\}, \; 0 \le s_1 < s_2 < \cdots$$

and proceed as above to find that $E\{z|w(s_.)\} = 0$ almost surely for every finite subset $s_.$ of $\mathbb{R}^+$ and therefore for a countable dense subset of $\mathbb{R}^+$ (conditional expectation continuity theorem). Hence by the almost sure continuity of Brownian motion $E\{z|w(s), s \in \mathbb{R}^+\} = 0$ almost surely, equivalently $E\{z|\bigvee_{s\in\mathbb{R}^+}\mathscr{F}(s)\} = 0$ almost surely, and since z is by hypothesis $\bigvee_{s\in\mathbb{R}^+}\mathscr{F}(s)$ measurable, this vanishing conditional expectation is almost surely z.

9. A Change of Variables

Review of the Inverses of Monotone Functions

Let f be a function from $\mathbb{R}^+$ into $\mathbb{R}^+$ satisfying the following condition.

M. f is monotone increasing, right continuous, and $\lim_{t\to\infty} f(t) = +\infty$.

Define $\hat{f}(t) = \inf\{r : f(r) > t\}$ for $t \in \mathbb{R}^+$. Then $\hat{f}$ satisfies M, and $\hat{f}(t) \le r$ if and only if $f(r+\varepsilon) > t$ whenever $\varepsilon > 0$. Moreover $\hat{\hat{f}} = f$. Finally $\hat{f}[f(r)] = r$ if r is not a point of a constancy interval of f, and $f[\hat{f}(r)] = r$ if r is not a point of a constancy interval of $\hat{f}$, that is, if r is not a discontinuity point of f.

Increasing Families of Optional Times and Their Inverses

Let $\{\mathscr{F}(t), t \in \mathbb{R}^+\}$ be a filtration of a probability space, right continuous and with $\mathscr{F}(0)$ containing the null sets. Let $\{S_t, t \in \mathbb{R}^+\}$ be a right continuous process whose random variables are finite-valued optional times for $\mathscr{F}(\cdot)$ and whose sample functions satisfy condition M. Define an inverse process $\hat{S}_{\bullet}$ by the above procedure, so that

$$\hat{S}_t(\omega) = \inf\{r : S_r(\omega) > t\}, \qquad \{\hat{S}_t \le r\} = \bigcap_{n=1}^{\infty} \{S_{r+1/n} > t\}. \tag{9.1}$$

The process $\{\hat{S}_{\bullet}, \mathscr{F}(\cdot)\}$ is an adapted process whose sample functions satisfy condition M. The filtration $\hat{\mathscr{F}}(\cdot) = \mathscr{F}(S_{\cdot})$ is right continuous, and $\hat{\mathscr{F}}(0)$ contains the null sets. Moreover $\{\hat{S}_t \le r\} \in \hat{\mathscr{F}}(r)$ in view of (9.1) and the right continuity of $S_{\bullet}$. Hence each random variable $\hat{S}_t$ is optional for $\hat{\mathscr{F}}(\cdot)$. Furthermore $\mathscr{F}(t) \subset \hat{\mathscr{F}}(\hat{S}_t)$ for all t; that is, if $\Lambda \in \mathscr{F}(t)$, then $\Lambda \cap \{\hat{S}_t \le \alpha\} \in \hat{\mathscr{F}}(\alpha)$ for all α, equivalently,

$$\Lambda \cap \bigcap_{1}^{\infty} \{S_{\alpha+1/n} > t\} \cap \{S_\alpha \le \beta\} \in \mathscr{F}(\beta) \tag{9.2}$$

for all α and β. In fact (9.2) is trivial if $t > \beta$, whereas if $t \le \beta$, all the sets in (9.2) are in $\mathscr{F}(\beta)$.

A Change of Variables in the Itô Integral

Let $\{w(\cdot), \mathscr{F}(\cdot)\}$ be a one-dimensional Brownian motion with variance parameter σ^2, let $\{\phi(\cdot), \mathscr{F}(\cdot)\}$ be a progressively measurable process in the class Γ (Section 1), and suppose that $\{S_t, t \in \mathbb{R}^+\}$ is a process as above whose sample functions satisfy

$$\int_0^{S_t} |\phi(s)|^2 \, ds = t \tag{9.3}$$

for all t in $\mathbb{R}^+$. Define

$$\hat{w}(t) = \int_0^{S_t} \phi \, dw,$$

and observe that $E\{\hat{w}(t)^2\} = \sigma^2 t$.

Theorem. *The process $\{\hat{w}(\cdot), \hat{\mathscr{F}}(\cdot)\}$ is a one-dimensional Brownian motion with variance parameter σ^2.*

Define $x(\cdot) = \int_0^{\cdot} \phi \, dw$. To prove that $\{\hat{w}(\cdot), \hat{\mathscr{F}}(\cdot)\}$ is a Brownian motion with variance parameter σ^2, it is sufficient to show that for $\alpha < \beta$ the conditional distribution of $x(S_\beta) - x(S_\alpha)$ given $\mathscr{F}(S_\alpha)$ is Gaussian with mean 0 and variance $\sigma^2(\beta - \alpha)$. Thus in view of the characteristic function of the normal distribution it is sufficient to show that for each complex constant γ

$$E\{\exp \gamma[x(S_\beta) - x(S_\alpha)] | \mathscr{F}(S_\alpha)\} = \exp \frac{\gamma^2 \sigma^2 (\beta - \alpha)}{2} \quad \text{a.s.}; \tag{9.4}$$

that is, it is sufficient to show that, in the notation of (3.10),

$$E\{y_{\gamma\phi}(S_\beta) | \mathscr{F}(S_\alpha)\} = y_{\gamma\phi}(S_\alpha) \quad \text{a.s.}$$

In other words the problem is to show that the process $\{y_{\gamma\phi}(S_\cdot), \mathscr{F}(S_\cdot)\}$ is a martingale. Now according to Section 3(k), if $\psi(\cdot) = \gamma\phi 1_{0S_\beta}$, the process

$$\{y_\psi(\cdot), \mathscr{F}(\cdot)\} = \{y_{\gamma\phi}(S_\beta \wedge \cdot), \mathscr{F}(\cdot)\}$$

is a martingale, right closable by $y_{\gamma\phi}(S_\beta)$, and therefore uniformly integrable. Hence (martingale sampling theorem) the martingale equality at times $t = S_\alpha$ and $t = +\infty$ is satisfied; that is, the process $\{y_{\gamma\phi}(S_\cdot), \mathscr{F}(\cdot)\}$ is a martingale, as was to be proved.

Extension to $N > 1$

If $\{w(\cdot), \mathscr{F}(\cdot)\}$ is a Brownian motion in $\mathbb{R}^N$ with variance parameter σ^2 and if ϕ is a progressively measurable process with state space $\mathbb{R}^N$, in the (vector) space Γ, satisfying (9.3), the theorem remains true with no change in proof: the process $\{\hat{w}(\cdot), \hat{\mathscr{F}}(\cdot)\}$ is a one-dimensional Brownian motion with variance parameter σ^2.

Observe that the theorem remains true if null exceptional sets are allowed, that is, if only almost every sample function satisfies condition M and if (9.3) is only true almost everywhere for each t [which implies that (9.3) is true almost surely simultaneously for all t.]

Application (a). If $N \geq 1$ and if $|\phi| = 1$ in Theorem 9, then $S_t = t$, $\hat{\mathscr{F}}(t) = \mathscr{F}(t)$, and $\hat{w}(t) = \int_0^t \phi\, dw$. We leave it to the reader to show that in this case

$$\int_0^t \psi\, d\hat{w} = \int_0^t (\psi\phi)\, dw \quad \text{a.s.} \tag{9.5}$$

whenever $\psi \in \Gamma$ (state space $\mathbb{R}$) and $(\psi\phi)$ is the scalar multiple of the vector ϕ by ψ. More generally, if $\mathbf{M}(t)$ is for each t in $\mathbb{R}^+$ an $N \times N$ orthogonal matrix-valued random variable and if the component processes of $\mathbf{M}(\cdot)$ are in Γ, then the vector process $\{\int_0^\cdot \mathbf{M}\, dw, \mathscr{F}(\cdot)\}$ is an N-dimensional Brownian motion with parameter value σ^2.

Application (b). $(N \geq 1)$ Let $\phi_0(\cdot)$ be a real vector process in Γ, and define

$$\phi(t, \omega) = \begin{cases} \dfrac{\phi_0(t, \omega)}{|\phi_0(t, \omega)|} & \text{if } |\phi_0(t, \omega)| \neq 0, \\ (N^{-1/2}, \ldots, N^{-1/2}) & \text{if } |\phi_0(t, \omega)| = 0, \end{cases}$$

$$\hat{w}(t) = \int_0^t \phi\, dw.$$

Then $|\phi| = 1$; so according to Application (a), the process $\{\hat{w}(\cdot), \mathscr{F}(\cdot)\}$ is a one-dimensional Brownian motion with variance parameter σ^2 and

$$\int_0^t |\phi_0|\, d\hat{w} = \int_0^t \phi_0\, dw \quad \text{a.s.}$$

Thus every vector stochastic integral can be represented as a one-dimensional stochastic integral with a positive integrand.

Application (c). $(N \geq 1)$ Let $\phi(\cdot)$ be a real vector process in Γ, define $x(\cdot) = \int_0^\cdot \phi\, dw$, and define T_α by (6.6). Suppose that T_α is almost surely finite valued for all α. Then according to Theorem 9, the process $\{x(T_\cdot), \mathscr{F}(T_\cdot)\}$ is a Brownian motion. In other words the integral process $x(\cdot)$ is a Brownian motion under a change of time. If T_α is not almost surely finite valued for all α, that is, if $\int_0^\infty |\phi|^2\, ds$ is not almost surely $+\infty$, define

$$\phi_k(t) = \begin{cases} \phi(t) & \text{if } t \leq k, \\ (1, \ldots, 1) & \text{if } t > k, \end{cases}$$

$$T_{k\alpha} = \inf\left\{t : \frac{\sigma^2}{2} \int_0^t |\phi_k|^2\, ds \geq \alpha\right\}, \qquad x_k(t) = \int_0^t \phi_k\, dw.$$

According to these definitions and Theorem 9, the process $\{x_k(T_{k\cdot}), \mathscr{F}(T_{k\cdot})\}$ is a Brownian motion and coincides with $x(T_\cdot)$ on the set $\{T_\alpha < k\}$ for parameter values $< \alpha$.

10. The Role of Brownian Motion Increments

Let $\{w(\cdot), \mathscr{F}(\cdot)\}$ be a Brownian motion in $\mathbb{R}^N$. Fix $t > 0$, define

$$\Delta_{jn}^{(\alpha)} = w^{(\alpha)}(j2^{-n}t) - w^{(\alpha)}((j-1)2^{-n}t),$$

omitting the superscript when $N = 1$, and for f a complex-valued function on $\mathbb{R}^N \times \mathbb{R}^+$ define

$$f_{jn} = f[w(j2^{-n}t), j2^{-n}t],$$

$$f_n(s) = \begin{cases} f_{(j-1)n} & \text{if } (j-1)2^{-n}t < s \leq j2^{-n}t, j \geq 1, \\ f[w(0), 0] & \text{if } s = 0. \end{cases}$$

Lemma. *When f is continuous and finite valued*

$$\operatorname*{p\,lim}_{n\to\infty} \sum_{j=1}^{2^n} f_{(j-1)n} \Delta_{jn}^{(\alpha)} \Delta_{jn}^{(\beta)} = \begin{cases} \sigma^2 \displaystyle\int_0^t f[w(s), s]\, ds & \text{if } \alpha = \beta, \\ 0 & \text{if } \alpha \neq \beta. \end{cases} \tag{10.1}$$

If f is bounded these limits are also L^2 limits.

To simplify the notation, take $\sigma = 1$. Assume first that $|f| \leq c$. Then

$$\begin{aligned} &\lim_{n\to\infty} E\left\{\left|\int_0^t f[w(s), s]\, ds - \sum_{j=1}^{2^n} f_{(j-1)n} 2^{-n} t\right|^2\right\} \\ &= \lim_{n\to\infty} E\left\{\left|\int_0^t [f[w(s), s] - f_n(s)]\, ds\right|^2\right\} \\ &\leq \lim_{n\to\infty} tE\left\{\int_0^t |f[w(s), s] - f_n(s)|^2\, ds\right\} = 0. \end{aligned} \tag{10.2}$$

Thus to prove (10.1) for $\alpha = \beta$ in the L^2 limit sense for f bounded it is sufficient to prove

$$\lim_{n\to\infty} E\left\{\left|\sum_{j=1}^{2^n} f_{(j-1)n}(\Delta_{jn}^{(\alpha)2} - 2^{-n}t)\right|^2\right\} = 0. \tag{10.3}$$

When the indicated square is multiplied out, each term has the form

$$f_{(j-1)n}\bar{f}_{(k-1)n}(\Delta_{jn}^{(\alpha)^2} - 2^{-n}t)(\Delta_{kn}^{(\alpha)^2} - 2^{-n}t), \qquad j \le k.$$

If $j < k$, the last factor is a random variable with expectation 0 and is independent of the other factors. Hence the expectation of the term vanishes. The terms with $j = k$ yield

$$E\left\{\sum_{j=1}^{2^n} |f_{(j-1)n}|^2(\Delta_{jn}^{(\alpha)^2} - 2^{-n}t)^2\right\} \le 2^{-n+1}c^2t^2; \tag{10.4}$$

so (10.3) is true. If f is not bounded, define g_m as any bounded continuous function on $\mathbb{R}^N \times \mathbb{R}^+$, equal to f when $|f| \le m$. Then

$$\lim_{m\to\infty} \int_0^t g_m[w(s), s]\,ds = \int_0^t f[w(s), s]\,ds \quad \text{a.s.}$$

The sum on the left side of (10.1) for $\alpha = \beta$ differs from the sum with f replaced by g_m on a set of small probability when m is large. It follows that (10.1) for $\alpha = \beta$ is true as stated.

To prove (10.1) for $\alpha \ne \beta$ with the limit in the L^2 sense for $|f|$ bounded by c, observe that (for $\sigma = 1$)

$$E\left\{\left|\sum_{j=1}^{2^n} f_{(j-1)n}\Delta_{jn}^{(\alpha)}\Delta_{jn}^{(\beta)}\right|^2\right\} = \sum_{j=1}^{2^n} E\{|f_{(j-1)n}|^2\Delta_{jn}^{(\alpha)^2}\Delta_{jn}^{(\beta)^2}\} \tag{10.5}$$

$$\le c^2 \sum_{j=1}^{2^n} E\{\Delta_{jn}^{(\alpha)^2}\Delta_{jn}^{(\beta)^2}\} = 2^{-n}c^2t^2 \to 0 \qquad (\alpha \ne \beta).$$

The proof for unbounded f follows that when $\alpha = \beta$.

Observation. If $N = 1$ and $f = 1$, the proof of (10.1) for $\alpha = \beta$, dropping some now unnecessary details, yields

$$E\left\{\left|\sum_{j=1}^{2^n} \Delta_{jn}^2 - \sigma^2 t\right|^2\right\} = 2^{-n+1}\sigma^4 t^2. \tag{10.6}$$

An application of the Borel Cantelli theorem yields Lévy's theorem, that is,

$$\lim_{n\to\infty} \sum_{j=1}^{2^n} \Delta_{jn}^2 = \sigma^2 t \quad \text{a.s.} \tag{10.7}$$

11. ($N = 1$) Computation of the Itô Integral by Riemann–Stieltjes Sums

Let $\phi(\cdot)$ be a progressively measurable process in Γ, and choose $b > 0$ and a partition π: $t_0 = 0 < t_1 < \cdots < t_n = b$ of $[0, b]$. Define

$$\delta(\pi) = \max_{j \le n}(t_j - t_{j-1})$$

and

$$\phi_\pi(t) = \begin{cases} \phi(0) & \text{if } t = 0, \\ \phi(t_{j-1}) & \text{if } t_{j-1} < t \le t_j, j \le n, \\ \phi(t) & \text{if } t > b. \end{cases}$$

Then $\phi_\pi(\cdot)$ is in Γ if $\phi(\cdot)$ is almost surely finite valued on π, and

$$\int_0^b \phi_\pi \, dw = \sum_{j=1}^n \phi(t_{j-1})[w(t_j) - w(t_{j-1})]. \tag{11.1}$$

Theorem. (a) *If the restriction to $[0, b]$ of almost every $\phi(\cdot)$ sample function is bounded and l_1 almost everywhere continuous, then*

$$\operatorname*{p\,lim}_{\delta(\pi)\to 0} \int_0^b \phi_\pi \, dw = \int_0^b \phi \, dw. \tag{11.2}$$

(b) *If $E\{|\phi(t)|^2\} < +\infty$ for $t \le b$ and if*

$$\lim_{\substack{t-s\to 0 \\ 0\le s,t\le b}} E\{|\phi(t) - \phi(s)|^2\} = 0, \tag{11.3}$$

then (11.2) *is true as an L^2 limit.*

Proof. (a) Under the hypotheses of (a) the usual reasoning in the discussion of the Riemann integral yields

$$\lim_{\delta(\pi)\to 0} \int_0^b |\phi_\pi - \phi|^2 \, ds = 0 \quad \text{a.s.};$$

so $\lim_{\delta(\pi)\to 0} \Gamma \operatorname{dist}(\phi_\pi, \phi) = 0$, and since the map $\phi(\cdot) \to \int_0^\cdot \phi \, dw$ is continuous in the (Γ, Γ') pair of metrics, it follows that (11.2) is true and even that

$$\operatorname*{p\,lim}_{\delta(\pi)\to 0} \sup_{t \le b} \left| \int_0^t (\phi_\pi - \phi) \, dw \right| = 0. \tag{11.4}$$

(b) Apply Section 3(f) to find

$$E\left\{\left|\int_0^b \phi\,dw - \int_0^b \phi_\pi\,dw\right|^2\right\} = \sigma^2 E\left\{\int_0^b |\phi - \phi_\pi|^2\,ds\right\}$$
$$\leq \sigma^2 b \sup_{\substack{|t-s|\leq\delta(\pi)\\ 0\leq s,t\leq b}} E\{|\phi(t)-\phi(s)|^2\}. \tag{11.5}$$

Therefore (b) is true, and in fact (11.4) is true as an L^2 limit because the map $\phi(\cdot)\mapsto\int_0^{\cdot}\phi\,dw$ is continuous in the (Γ_2,Γ_2') pair of metrics. □

Integration by Parts

If $\phi(\cdot)$ is in Γ and if the restriction to the interval $[0,b]$ of almost every $\phi(\cdot)$ sample function is right continuous and of bounded variation, then

$$\int_0^b \phi\,dw = w(b)\phi(b) - w(a)\phi(a) - \int_0^b w\,d\phi \quad \text{a.s.}, \tag{11.6}$$

where for almost every point ω of the basic measure space the integral on the right is the Riemann–Stieltjes integral $\int_0^b w(s,\omega)\,d_s\phi(s,\omega)$. In fact the hypotheses of Theorem 11 (a) are satisfied, and the sum in (11.1) is equal to

$$w(b)\phi(b) - w(a)\phi(a) - \sum_{j=1}^{n} w(t_j)[\phi(t_j)-\phi(t_{j-1})], \tag{11.7}$$

which almost surely yields the right side of (11.6) when $\delta(\pi)\to 0$.

12. Itô's Lemma

Let $\{w(\cdot),\mathscr{F}(\cdot)\}$ be a Brownian motion in $\mathbb{R}^N$, and let $(\eta,t)\mapsto u(\eta,t)$ be a continuous function on $\mathbb{R}^N\times\mathbb{R}^+$ for which the derivatives

$$u_0 = \frac{\partial u}{\partial t},\qquad u_i = \frac{\partial u}{\partial \eta^{(i)}},\qquad u_{ij} = \frac{\partial^2 u}{\partial\eta^{(i)}\partial\eta^{(j)}}$$

exist and are continuous. Before proving Itô's lemma we prove an important special case: simultaneously for all $t\in\mathbb{R}^+$,

$$u[w(t),t]-u[w(0),0] = \int_0^t u_0\,ds + \sum_{\alpha=1}^{N}\int_0^t u_\alpha\,dw^{(\alpha)}(s) + \frac{\sigma^2}{2}\sum_{\alpha=1}^{N}\int_0^t u_{\alpha\alpha}\,ds$$
$$= \int_0^t \overset{*}{\Delta} u\,ds + \sum_{\alpha=1}^{N}\int_0^t u_\alpha\,dw^{(\alpha)}(s) \quad \text{a.s.} \tag{12.1}$$

The argument in each integrand is $[w(s), s]$. This relation is commonly written in the form

$$du = \mathring{\Delta} u\,dt + \sum_{\alpha=1}^{N} u_\alpha\,dw^{(\alpha)}, \tag{12.1'}$$

and from now on relations like (12.1) will sometimes be written in the corresponding differential form. It is sufficient to prove (12.1) to be valid for fixed t because each side of (12.1) is the tth random variable of an almost surely continuous process. Fix t and denote $j2^{-n}t$ by s_{jn}, so that

$$\begin{aligned} u[w(t), t] - u[w(0), 0] &= \sum_{j=1}^{2^n} \{u[w(s_{jn}), s_{jn}] - u[w(s_{jn}), s_{(j-1)n}]\} \\ &\quad + \sum_{j=1}^{2^n} \{u[w(s_{jn}), s_{(j-1)n}] - u[w(s_{(j-1)n}), s_{(j-1)n}]\}. \end{aligned} \tag{12.2}$$

The jth term in the first sum is $u_0[w(s_{jn}), s_{jn}]2^{-n}t$ up to an error which for each continuous Brownian motion sample function is uniformly $o(2^{-n})$ as j varies and $n \to \infty$. Hence the first sum in (12.2) has the first integral in (12.1) as almost sure limit when $n \to \infty$. The jth term in the second sum in (12.2) is, if $\Delta_{jn}^{(\alpha)} = w^{(\alpha)}(s_{jn}) - w^{(\alpha)}(s_{(j-1)n})$,

$$\sum_{\alpha=1}^{N} u_\alpha[w(s_{(j-1)n}), s_{(j-1)n}]\Delta_{jn}^{(\alpha)} + \frac{1}{2}\sum_{\alpha,\beta=1}^{N} u_{\alpha\beta}[w(s_{(j-1)n}), s_{(j-1)n}]\Delta_{jn}^{(\alpha)}\Delta_{jn}^{(\beta)} \tag{12.3}$$

up to an error which for each continuous Brownian motion sample function is uniformly $o(\Sigma_{\alpha=1}^{N} \Delta_{jn}^{(\alpha)^2})$ as j varies and $n \to \infty$. According to Lemma 10 and Theorem 11, when $n \to \infty$, the sum over $j = 1, \ldots, 2^n$ of the terms in (12.3) has as limit in measure the sum of the second and third sums in (12.1). Finally the summed error $o(\Sigma_{\alpha=1}^{N} \Sigma_{j=1}^{2^n} \Delta_{jn}^{(\alpha)^2})$ is $o(1)$ for almost every Brownian motion sample function because the inside sum has almost sure limit $\sigma^2 t$ when $n \to \infty$ by Lévy's result (10.7).

Localization

If, more specially, u is a function of class $\mathbb{C}^{(2)}$ on $\mathbb{R}^N$ and we consider u composed with $w(\cdot)$, equation (12.1) reduces to

$$u[w(t)] - u[w(0)] = \frac{\sigma^2}{2}\int_0^t \Delta u\,ds + \sum_{\alpha=1}^{N}\int_0^t u_\alpha\,dw^{(\alpha)}(s) \quad \text{a.s.} \tag{12.4}$$

More generally, if u is only defined and of class $\mathbb{C}^{(2)}$ on an open subset D of $\mathbb{R}^N$ and if $P\{w(0) \in D\} = 1$, let S be the hitting time of ∂D by $w(\cdot)$. Then

(12.1) is almost surely valid for $t < S$ in the sense that if D_0 is an arbitrary open relatively compact subset of D and if S_0 is the hitting time of ∂D_0 by $w(\cdot)$, then (12.4) is almost surely true on $\{w(0) \in D_0\}$ if t is replaced by $t \wedge S_0$. To see this, let u_0 be a $\mathbb{C}^{(2)}(\mathbb{R}^N)$ extension of the restriction of u to $\bar{D}_0$. Then an application of (12.4) to u_0 with t replaced by $t \wedge S_0$ gives the desired result.

We now turn to Itô's lemma. Let u and $w(\cdot)$ be defined as at the beginning of this section. Let M be a strictly positive integer, and for $1 \le m \le M$ and $1 \le n \le N$ let ϕ_{mn} and ψ_m be processes in Γ except that the integrability requirement for ψ_m is weakened: it is supposed only that almost every $\psi(\cdot)$ sample function is integrable over finite intervals. Define $x(\cdot) = [x^{(1)}(\cdot), \ldots, x^M(\cdot)]$ by

$$x^{(m)}(t) = x^{(m)}(0) + \sum_{n=1}^{N} \int_0^t \phi_{mn}\, dw^{(n)} + \int_0^t \psi_m\, dt,$$

where $x^{(m)}(0)$ is an arbitrary $\mathscr{F}(0)$ measurable random variable.

Itô's Lemma. *Under the stated conditions, for* $u\colon t \mapsto u[x(t), t]$,

$$du = u_0\, dt + \sum_{m=1}^{M} u_m\, dx^{(m)} + \frac{\sigma^2}{2} \sum_{m,n=1}^{M} u_{mn}\, dx^{(m)}\, dx^{(n)}, \tag{12.5}$$

where $dx^{(m)}\, dx^{(n)}$ *is to be interpreted as the formal product under the convention*

$$(dt)^2 = dt\, dw^{(n)} = 0, \qquad dw^{(m)}\, dw^{(n)} = \sigma^2 \delta_{mn}\, dt$$

so that

$$dx^{(m)}\, dx^{(n)} = \sigma^2 \left(\sum_{\alpha=1}^{N} \phi_{m\alpha} \phi_{n\alpha} \right) dt.$$

To simplify the notation, we give the proof for $M = N = 1$. That is, $dx(t) = \phi\, dw(t) + \psi\, dt$, $u = u[x(t), t]$, and (12.5) reduces to

$$du = \left(u_0 + u_1 \psi + \frac{\sigma^2}{2} u_{11} \phi^2 \right) dt + u_1 \phi\, dw. \tag{12.6}$$

It is sufficient to prove the lemma for ϕ and ψ in Γ_0, given by representations of the form (1.7) with a common partition:

$$\begin{aligned}
&\phi(t) = \psi(t) = 0 && \text{if } t \le t_1, \\
&\phi(t) = f_{j-1}, \quad \psi(t) = g_{j-1} && \text{if } t_{j-1} < t \le t_j,\ 2 \le j \le k, \\
&\phi(t) = \psi(t) = 0 && \text{if } t > t_k.
\end{aligned}$$

In the interval $[0, t_1]$ the evaluation (12.6) is trivial. In the interval $]t_1, t_2]$

$$u[x(t), t] = u[f_1[w(t) - w(t_1)] + g_1(t - t_1), t].$$

A glance at the proof of the special case of Itô's lemma already treated shows that it is applicable with the help of Section 3(d) to yield

$$du = u_0\, dt + u_1 f_1\, dw + u_1 g_1\, dt + \frac{\sigma^2}{2} u_{11} f_1^2\, dt$$

which agrees with (12.6). The verification of (12.6) on the remaining intervals $]t_{j-1}, t_j]$ is similar, and the verification on $]t_k, +\infty[$ is trivial.

The Algebra of Integrals ($N = 1$)

In view of Itô's lemma the integrals of the form

$$\int_0^t \phi\, dw + \int_0^t \psi\, ds$$

with ϕ and ψ as described above form an algebra. In fact, if $dx_i = \phi_i\, dw + \psi_i\, dt$,

$$\begin{aligned} d(x_1 x_2) &= x_1\, dx_2 + x_2\, dx_1 + (\phi_1 \phi_2)\, dt \\ &= (x_1 \phi_2 + x_2 \phi_1)\, dw + (x_1 \psi_2 + x_2 \psi_1 + \phi_1 \phi_2)\, dt. \end{aligned}$$

Example (a) ($N = 1$). If n is a strictly positive integer

$$d(w^n) = n w^{n-1}\, dw + \frac{\sigma^2}{2} n(n-1) w^{n-2}\, dt.$$

Example (b) ($N = 1$). If ϕ is in Γ and if $y_\phi(\cdot)$ is defined by (3.10), then $dy_\phi = \phi y_\phi\, dw$. In particular, as already proved by a different method in Section 6, $dy_\gamma = \gamma y_\gamma\, dw$ for constant γ. Thus the process $y_1(\cdot)$ plays the role of the exponential function in the stochastic calculus based on the Itô integral.

Example (c) ($N \geq 1$). Recall (Section 1.XV.3) that the space-time Hermite polynomials are coparabolic and satisfy 1.XV(3.8). Define $\dot{H}_{m_1 \cdots m_N j}$ as $\dot{H}_{m'_1 \cdots m'_N}$ with

$$m'_i = \begin{cases} m_i - \delta_{ij} & \text{if } m_j > 0, \\ 0 & \text{if } m_j = 0. \end{cases}$$

Then

$$d\dot{H}_{m_1\cdots m_N}[w(t),t] = \sum_{j=1}^{N} m_j \dot{H}_{m_1\cdots m_N j}\, dw_j(t).$$

Thus in the stochastic calculus based on the Itô integral the process $\{\dot{H}_{m_1\cdots m_N}[w(t),t], t\in\mathbb{R}^+\}$ plays the role of the product of powers $m_1,\ldots,m_N$ of the coördinate variables.

13. The Composition of the Basic Functions of Potential Theory with Brownian Motion

Let u be a $\mathbb{C}^{(2)}$ function from $\mathbb{R}^N$ into $\mathbb{R}$. If $\{w(\cdot), \mathscr{F}(\cdot)\}$ is a Brownian motion in $\mathbb{R}^N$ from a point ξ, Itô's lemma yields

$$u[w(t)] - u(\xi) = \int_0^t \langle \operatorname{grad} u, dw\rangle + \frac{\sigma^2}{2}\int_0^t \Delta u\, ds \quad \text{a.s.} \tag{13.1}$$

Hence the process

$$\left\{u[w(t)] - \frac{\sigma^2}{2}\int_0^t \Delta u\, ds; \mathscr{F}(t), t\in\mathbb{R}^+\right\} \tag{13.2}$$

is a local martingale and is a martingale if

$$E\left\{\int_0^t |\operatorname{grad} u|^2\, ds\right\} < +\infty \tag{13.3}$$

for all $t > 0$. In particular, if u is harmonic, the process $\{u[w(\cdot)], \mathscr{F}(\cdot)\}$ is a martingale if (13.3) is satisfied, for example, if u is a harmonic polynomial. More generally, if $\Delta u \le 0$, that is, if u is a $\mathbb{C}^{(2)}$ superharmonic function, and if (13.3) is satisfied, the process $u[w(\cdot)]$ is the sum of a martingale and an adapted process with decreasing sample functions and is therefore a supermartingale if its random variables are integrable. If u is a $\mathbb{C}^{(2)}$ function from an open subset D of $\mathbb{R}^N$ into $\mathbb{R}$, let D_0 be an open relatively compact subset of D, choose ξ in D_0, and let S_0 be the hitting time of ∂D_0 by $w(\cdot)$. There is a $\mathbb{C}^{(2)}$ function u' on $\mathbb{R}^N$, with compact support, extending $u|_{\bar{D}_0}$, and if the preceding argument is applied to u', we find that the process in (13.2) with u replaced by u' is a martingale. This process, stopped at time S_0, is therefore a martingale, and in particular if u is [super]harmonic on D, the process $\{u[w(S_0 \wedge t)]; \mathscr{F}(t), t \le +\infty\}$ is a [super]martingale. In the superharmonic context the supermartingale inequality at times 0, $+\infty$ yields the inequality $u(\xi) \ge E\{u[w(S_0)]\}$, with equality in the harmonic function context. Hence the distribution of $w(S_0)$ is harmonic measure on ∂D_0 relative to ξ. The corresponding discussion in the parabolic context is left to the reader.

The point of the preceding discussion is that the composition of a suitably restricted [super]parabolic function with space-time Brownian motion is a [super]martingale and the composition of a suitably restricted [super]harmonic function with Brownian motion is a [super]martingale. The condition (13.3) and the associated regularity conditions are unnecessarily strong, however. In Chapter IX we shall treat the problems of this section more directly and in more detail, without the use of the Itô integral, and eliminate superfluous hypotheses. This elimination can of course also be effected using the results of this section, but each of the two approaches is too important to omit.

14. The Composition of an Analytic Function with Brownian Motion

Suppose that $N = 2$ and that $\{w(\cdot), \mathscr{F}(\cdot)\}$ is a Brownian motion from a point of $\mathbb{R}^2$, and write $z(t) = w^{(1)}(t) + iw^{(2)}(t)$. Let D be an open subset of the complex plane, and let D_0 be a relatively compact open subset of D. Suppose that $z(0)$ is in D_0 and that f is a complex-valued regular analytic function on D. Then if S_0 is the hitting time of ∂D_0 by $z(\cdot)$, Itô's lemma yields (see the remarks on localization in Section 12)

$$df[z(t)] = f[z(t)]\,dz, \qquad t \le S_0,$$

and $\{f[z(S_0 \wedge t)]; \mathscr{F}(t), t \le +\infty\}$ is a martingale. An adaptation of Application (b) in Section 9 shows that if we define

$$T_\alpha = \inf\left\{t: \int_0^t |f'[z(s)]|^2\,ds = \alpha\right\}, \qquad \beta = \int_0^{S_0} |f'[z(s)]|^2\,ds,$$

then $f[z(T_\cdot \wedge S_0)]$ is a Brownian motion on the complex plane, killed at time β. That is, f maps Brownian paths into Brownian paths with a new time scale.

Chapter IX

Brownian Motion and Martingale Theory

The applications of the Itô integral in Sections VIII.12 to VIII.14 exhibit aspects of the intimate relation between Brownian motion and martingale theory. In the following we shall go from simple examples of this relation to an analysis by means of martingale theory of the composition of the basic functions of the potential theory for Laplace's equation [the heat equation] with Brownian motion [space-time Brownian motion]. This will be effected by a direct method without the use of the Itô integral, but there will be a slight repetition of some of the most elementary topics in Chapter VIII.

1. Elementary Martingale Applications

Let $\{w(\cdot), \mathscr{F}(\cdot)\}$ be a Brownian motion on $\mathbb{R}^N$. The following processes are martingales relative to $\mathscr{F}(\cdot)$ if the initial distribution of the Brownian motion is chosen to make the process random variables have finite expectations, for example, if the initial distribution is supported by a singleton.

M1 $w(\cdot)$

M2 $\{|w(t) - \eta|^2 - N\sigma^2 t, t \in \mathbb{R}^+\}$, η arbitrary

M3 $\{\exp[\langle\gamma, w(t)\rangle - \sigma^2 t \Sigma_1^N \gamma_j^2/2], t \in \mathbb{R}^+\}$, $\gamma = (\gamma_1, \ldots, \gamma_N)$

The first process is a vector martingale (in the obvious definition) because for $s < t$,

$$E\{w(t) - w(s)|\mathscr{F}(s)\} = E\{w(t) - w(s)\} = 0 \quad \text{a.s.} \tag{1.1}$$

since $w(t) - w(s)$ is independent of $\mathscr{F}(s)$ for $s < t$. The second process is a martingale when $N = 1$ and therefore when $N \geq 1$ because $s < t$ implies that

$$\begin{aligned} &E\{[w(t) - \eta]^2|\mathscr{F}(s)\} \\ &= E\{[w(t) - w(s)]^2|\mathscr{F}(s)\} + 2[w(s) - \eta]E\{w(t) - w(s)|\mathscr{F}(s)\} + [w(s) - \eta]^2 \\ &= \sigma^2(t - s) + [w(s) - \eta]^2 \quad \text{a.s.} \end{aligned} \tag{1.2}$$

The third process is a martingale (for any complex vector γ) because if $s < t$

$$E\{\exp\langle\gamma, w(t) - w(s)\rangle \,|\, \mathscr{F}(s)\} = E\{\exp\langle\gamma, w(t) - w(s)\rangle\} \\ = \exp\left[\sigma^2(t-s)\sum_1^N \frac{\gamma_j^2}{2}\right] \quad \text{a.s.} \tag{1.3}$$

Conversely, if $\{w(\cdot), \mathscr{F}(\cdot)\}$ is a stochastic process for which M3 is a martingale whenever the components of γ are purely imaginary, then the process $\{w(\cdot) - w(0), \mathscr{F}(\cdot)\}$ has the Brownian motion finite-dimensional distributions with variance constant σ^2. In fact, if $s < t$, the martingale equality (with $\gamma = i\beta$ and β real)

$$E\left\{\exp\left[i\langle\beta, w(t)\rangle + \sigma^2 t\sum_1^N \frac{\beta_j^2}{2}\right] \Big| \mathscr{F}(s)\right\} = \exp\left[i\langle\beta, w(s)\rangle + \sigma^2 s\sum_1^N \frac{\beta_j^2}{2}\right] \quad \text{a.s.} \tag{1.4}$$

yields

$$E\{\exp i\langle\beta, w(t) - w(s)\rangle \,|\, \mathscr{F}(s)\} = \exp\left[-\sigma^2(t-s)\sum_1^N \frac{\beta_j^2}{2}\right] \quad \text{a.s.} \tag{1.5}$$

The difference $w(t) - w(s)$ is therefore independent of $\mathscr{F}(s)$, and it follows that $w(\cdot)$ has independent increments. Moreover (1.5) implies that $w(t) - w(s)$ is normally distributed with independent components, each of mean 0 and variance $\sigma^2(t-s)$, as was to be proved.

Application of M1. Let $\{w_\xi(\cdot), \mathscr{F}_\xi(\cdot)\}$ be a Brownian motion on $\mathbb{R}^N$ from ξ, let D be a bounded open subset of $\mathbb{R}^N$ containing ξ, and let S_ξ be the hitting time of ∂D, almost surely finite because [Section VII.5(c)] $\lim_{t\to\infty} |w_\xi(t)| = +\infty$ in measure. The function S_ξ is optional for $\mathscr{F}_\xi(\cdot)$ (Section II.4), and therefore (Section IV.3) the stopped process $\{w_\xi(S_\xi \wedge t), \mathscr{F}_\xi(t), t \in \mathbb{R}^+\}$ is a martingale. Equating expectations of the random variables of the stopped process at times 0 and t yields the equation $\xi = E\{w_\xi(S_\xi \wedge t)\}$, which becomes

$$\xi = E\{w_\xi(S_\xi)\} \tag{1.6}$$

when $t \to \infty$. In particular, if $N = 1$ and if D is the interval $]a, b[$, (1.6) becomes

$$\xi = bP\{w_\xi(S_\xi) = b\} + aP\{w_\xi(S_\xi) = a\}, \tag{1.7}$$

which implies

$$P\{w_\xi(S_\xi) = b\} = \frac{\xi - a}{b - a}. \tag{1.8}$$

Thus the probability that a Brownian path from ξ in $]a,b[$ reaches the boundary first at b is the solution of Laplace's equation on $]a,b[$ with boundary limit 1 at b and 0 at a. In other words in this elementary case the distribution of $w_\xi(\cdot)$ at the first hitting place on the boundary is harmonic measure on the boundary relative to the initial point ξ. This evaluation of harmonic measure will be extended to the Borel boundary subsets of an arbitrary Greenian subset D of $\mathbb{R}^N$ in Section 13. That is, if $w_\xi(\cdot)$ is an N-dimensional Brownian motion from ξ in D and if S_ξ is the hitting time of ∂D by $w_\xi(\cdot)$, it will be shown that S_ξ is almost surely finite when $N=2$ and that under the convention $w_\xi(+\infty)=\infty$ when $N>2$ the harmonic measure evaluation (1.8) generalizes to

$$P\{w_\xi(S_\xi)\in A\}=\mu_D(\xi,A). \tag{1.8'}$$

From this it follows that if f is a resolutive boundary function, the PWB solution H_f is given by

$$H_f(\xi)=E\{f[w_\xi(S_\xi)]\}. \tag{1.9}$$

Corresponding results will be proved for PWB^h solutions. The evaluation (1.8) can also be derived by solving the appropriate differential equation problem. The rigorous argument is the following, once it has been shown that the left side of (1.8) defines a harmonic function of ξ, that is, a linear function. Let $w(\cdot)$ be a Brownian motion from 0, so that $\xi+w(\cdot)$ is a Brownian motion from ξ and the measure space does not depend on ξ. A $\xi_0+w(\cdot)$ path (with $a<\xi_0<b$) which hits b before a will do so if the path is translated by $\xi-\xi_0$ with $\xi>\xi_0$; so the function $\xi\mapsto P\{w_\xi(S_\xi)=b\}$ is an increasing function of ξ. Since almost every $w(\cdot)$ path is strictly positive at some parameter values arbitrarily near 0, this increasing function has limit 1 at b, and a similar argument shows that the limit is 0 at a. Hence the function must be given by the right side of (1.8). This trivial example illustrates some of the problems involved in a rigorous justification of the evaluation of Brownian motion probabilities.

Application of M2. Let D be a bounded open subset of $\mathbb{R}^N$, and define $w_\xi(\cdot)$, S_ξ as above. Equating expectations of the random variables of M2 with $\eta=0$ at times 0 and $S_\xi\wedge t$ yields

$$0=E\{|w_\xi(S_\xi\wedge t)|^2-N\sigma^2(S_\xi\wedge t)\}, \tag{1.10}$$

and therefore

$$N\sigma^2E\{S_\xi\}=E\{|w_\xi(S_\xi)|^2\}. \tag{1.11}$$

In the special case $N=1$, $D=]a,b[$ this equation reduces to

$$\sigma^2 E\{S_\xi\} = (\xi - a)(b - \xi), \tag{1.12}$$

in view of the distribution of $w_\xi(S_\xi)$ already evaluated. Thus the function $\xi \mapsto E\{S_\xi\}$ is the solution of the equation $\sigma^2 \Delta u = -2$ with boundary limit function 0. For $N > 1$ and bounded D the martingale equality (1.11) becomes, if we accept (1.8′) (to be proved in Section 13),

$$\sigma^2 N E\{S_\xi\} = \int_{\partial D} |\eta - \xi|^2 \mu_D(\xi, d\eta) = \int_{\partial D} |\eta|^2 \mu_D(\xi, d\eta) - |\xi|^2. \tag{1.12′}$$

As in the case $N = 1$, the function $\xi \mapsto E\{S_\xi\}$ is the solution of the equation $\sigma^2 N \Delta u = -2$ with the boundary limit function 0 in the sense that $\sigma^2 N E\{S_\xi\} + |\xi|^2$ is the harmonic function which is the PWB solution of the harmonic function Dirichlet problem with boundary function $\eta \mapsto |\eta|^2$.

Application of M3. Let $N = 1$, let $w(\cdot)$ be a Brownian motion from the origin, choose $\alpha > 0$, and let T be the hitting time of $\{\alpha\}$ by $w(\cdot)$. The density of the distribution of T is given by VII (8.3) and will now be derived again using the martingale M3. The optional time T is almost surely finite because according to the application of M1 above the probability that a Brownian path from 0 hits α before $-c$ (with $c > 0$) is $c/(c + \alpha)$, which has limit 1 when $c \to +\infty$. Equating expectations of the random variables of M3 at times 0 and $T \wedge t$ yields

$$1 = E\left\{\exp\left[\gamma w(T \wedge t) - \gamma^2 \sigma^2 \frac{T \wedge t}{2}\right]\right\}. \tag{1.13}$$

Take γ real and strictly positive, so that the integrand is at most $\exp(\gamma\alpha)$. When $t \to +\infty$ (1.13) becomes

$$1 = E\left\{\exp\left[\gamma\alpha - \gamma^2 \sigma^2 \frac{T}{2}\right]\right\}, \tag{1.14}$$

that is

$$E\left\{\exp\left[-\gamma^2 \sigma^2 \frac{T}{2}\right]\right\} = \exp(-\gamma\alpha). \tag{1.15}$$

This equation specifies the distribution of T by way of its Laplace transform.

2. Coparabolic Polynomials and Martingale Theory

It was shown in Section VIII.13 that the composition of a [co] parabolic polynomial with space-[co] time Brownian motion from a point is a martingale. We shall establish this result again in this section without using the

Itô integral and incidentally gain insight into the space-time Hermite polynomials. Since every space-time coparabolic polynomial is a linear combination of space-time Hermite polynomials, it is sufficient to consider these as defined in Section 1.XV.3 by

$$e^{\langle \gamma, \eta \rangle - |\gamma|^2 \sigma^2 t/2} = \sum_{n=0}^{\infty} \sum_{m_1 + \cdots + m_N = n} \frac{\gamma^{(1)^{m_1}} \cdots \gamma^{(N)^{m_N}}}{m_1! \cdots m_N!} \dot{H}_{m_1 \cdots m_N}(\dot{\eta}), \qquad \dot{\eta} = (\eta, t) \tag{2.1}$$

and to compose these polynomials with space-cotime Brownian motion from the origin. That is, we consider

$$\{\dot{w}(\cdot), \mathscr{F}(\cdot)\} = \{[w(t), t], \mathscr{F}(t), t \in \mathbb{R}^+\}$$

with $\{w(\cdot), \mathscr{F}(\cdot)\}$ a Brownian motion in $\mathbb{R}^N$ from the origin, and we shall prove that the process $\{\dot{H}_{m_1 \cdots m_N}[\dot{w}(\cdot)], \mathscr{F}(\cdot)\}$ is a martingale.

To prove this fact, define $y_\gamma(t) = \exp[\langle \gamma, w(t) \rangle - |\gamma|^2 \sigma^2 t/2]$ as in VIII (3.10) so that

$$y_\gamma(t) = \sum_{n=0}^{\infty} \sum_{m_1 + \cdots + m_N = n} \frac{\gamma^{(1)^{m_1}} \cdots \gamma^{(N)^{m_N}}}{m_1! \cdots m_N!} \dot{H}_{m_1 \cdots m_N}[\dot{w}(t)] \tag{2.2}$$

and by 1.XV(3.9) (if $n = m_1 + \cdots + m_N$)

$$E\{\dot{H}_{m_1 \cdots m_N}[\dot{w}(t)] \dot{H}_{n_1 \cdots n_N}[\dot{w}(t)]\}$$
$$= \begin{cases} (\sigma^2 t)^{n+N/2} (2\pi)^{N/2} m_1! \cdots m_N! & \text{if } m_. = n_., \\ 0 & \text{otherwise.} \end{cases} \tag{2.3}$$

For fixed γ and t the series in (2.2) is a series of orthogonal random variables with mean square limit sum $y_\gamma(t)$. Thus integration to the limit with respect to the probability measure is legitimate. Now the process $\{\langle \gamma, w(\cdot) \rangle, \mathscr{F}(\cdot)\}$ is a Brownian motion with variance parameter $|\gamma|^2 \sigma^2$; so (Section 1, Example M3) the process $\{y_\gamma(\cdot), \mathscr{F}(\cdot)\}$ is a martingale. The validity of the martingale equality for this process is an identity in γ and implies the validity of the martingale equality for each process $\{\dot{H}_{m_1 \cdots m_N}[\dot{w}(\cdot)], \mathscr{F}(\cdot)\}$, and each of these processes is therefore a martingale. The L^2 martingale maximal inequality (Theorem III.11) together with (2.2) implies that for $b > 0$

$$E\left\{\sup_{t \le b} \left| y_\gamma(t) - \sum_{n=k}^{\infty} \sum_{m_1 + \cdots + m_N = n} \frac{\gamma^{(1)^{m_1}} \cdots \gamma^{(N)^{m_N}}}{m_1! \cdots m_N!} \dot{H}_{m_1 \cdots m_N}[\dot{w}(t)] \right|^2 \right\}$$
$$\le 4(2\pi\sigma^2 t)^{N/2} \sum_{n=k}^{\infty} \frac{(|\gamma|^2 \sigma^2 t)^n}{n!}. \tag{2.4}$$

This inequality implies that almost surely the partial sums in (2.2) converge uniformly for $t \leq b$, $|\gamma| \leq b_1$ for each pair b, b_1.

Harmonic Polynomials

If u is a polynomial on $\mathbb{R}^N$ and is harmonic, u considered as a function on $\dot{\mathbb{R}}^N$ is a coparabolic (and parabolic) polynomial, and therefore the composition of u with a Brownian motion from a point of $\mathbb{R}^N$ is a martingale. This result is generalized in the next section, as is the preceding result for coparabolic polynomials.

3. Superharmonic and Harmonic Functions on $\mathbb{R}^N$ and Supermartingales and Martingales

The following theorem for superharmonic functions on $\mathbb{R}^N$ will be extended in Section 7 to apply to superharmonic functions on Greenian sets but is proved separately because its proof not only is easy but indicates without technical complications why the relation between classical potential theory and martingale theory is so close.

Theorem. *Let $\{w(\cdot), \mathscr{F}(\cdot)\}$ be a Brownian motion in $\mathbb{R}^N$ from ξ. Let u be a superharmonic function on $\mathbb{R}^N$, and suppose that either* (a) *u is bounded below or* (b) *u satisfies for all $t > 0$ the integrability condition*

$$\begin{aligned}\int_{\mathbb{R}^N} &\ell(t, \xi, \eta)|u(\eta)|l_N(d\eta)\\ &= (2\pi\sigma^2 t)^{-N/2}\pi_N \int_0^\infty \exp\frac{-r^2}{2\sigma^2 t} L(|u|, r, \xi) r^{N-1}\,dr < +\infty.\end{aligned} \tag{3.1}$$

Then if $u(\xi) < +\infty$, the process $\{u[w(\cdot)], \mathscr{F}(\cdot)\}$ is a supermartingale, a martingale if u is harmonic. If $u(\xi) = +\infty$, the assertion remains true for the parameter interval $]0, +\infty[$.

If u is a harmonic polynomial, (3.1) is satisfied. Thus this theorem includes the polynomial case noted in Section 2. If u is harmonic, case (a) is trivial because a one-sided bounded harmonic function on $\mathbb{R}^N$ is identically constant according to Section 1.II.2.

To treat case (a), suppose that u is lower bounded and apply the superharmonic function inequality $L(u, r, \cdot) \leq u(\cdot)$ to the equality in (3.1) with $|u|$ replaced by u to obtain

$$\ell(t, \cdot, u) \leq u(\cdot), \tag{3.2}$$

which is precisely the supermartingale inequality, so that $\{u[w(\cdot)], \mathscr{F}(\cdot)\}$ is a supermartingale over the set of parameter values t making $E\{u[w(t)]\}$ finite. If $u(\xi) < +\infty$, the inequality (3.2) at ξ yields $E\{u[w(t)]\} = \ell(t, \xi, u) < +\infty$; so $u[w(\cdot)]$ is a supermartingale on the parameter interval $\mathbb{R}^+$. If $u(\xi) = +\infty$, this argument fails, but the following inequality shows that $E\{u[w(t)]\} < +\infty$ for $t > 0$ so that $\{u[w(t)], \mathscr{F}(t), t > 0\}$ is a supermartingale:

$$\begin{aligned} E\{u[w(t)]\} &\leq (2\pi\sigma^2 t)^{-N/2}\pi_N \int_0^1 L(u, r, \xi) r^{N-1}\, dr + L(u, 1, \xi) \\ &= (2\pi\sigma^2 t)^{-N/2} A(u, 1, \xi) + L(u, 1, \xi) < +\infty. \end{aligned} \tag{3.3}$$

Here we have used the monotoneity of the function $L(u, \cdot, \xi)$.

To treat case (b), suppose that u is superharmonic and satisfies (3.1), which is simply the statement that $E\{|u[w(t)]|\} < +\infty$. Then (3.2) is again valid; so $u[w(\cdot)]$ is a supermartingale as stated, with the parameter value 0 omitted when $u(\xi) = +\infty$. Finally, if u is harmonic and satisfies (3.1), this discussion is applicable to both u and $-u$; so $\{u[w(\cdot)], \mathscr{F}(\cdot)]$ is a martingale.

Application to Potentials

If $N > 2$, a superharmonic potential $G\mu$ of a measure μ satisfies the lower boundedness condition (a). If $N = 2$, a superharmonic potential $G\mu$ satisfies (3.1). In fact $\mu(\mathbb{R}^2) < +\infty$ because $G\mu$ is superharmonic, and the following inequalities (3.4) and (3.5) show that (3.1) is satisfied.

$$\begin{aligned} &\int_{\mathbb{R}^2} \ell(t, \xi, \eta) l_2(d\eta) \int_{\{|\eta-\zeta| \leq 1\}} \log|\eta - \zeta|^{-1} \mu(d\zeta) \\ &\leq (2\pi\sigma^2 t)^{-1} \int_{\mathbb{R}^2} \mu(d\zeta) \int_{\{|\eta-\zeta|\leq 1\}} \log|\eta - \zeta|^{-1} l_2(d\eta) \\ &= (4\sigma^2 t)^{-1} \mu(\mathbb{R}^2) < +\infty. \end{aligned} \tag{3.4}$$

Use this inequality and set $\alpha = (\pi\sigma^2 t/2)^{1/2}$ to derive

$$\begin{aligned} &\int_{\mathbb{R}^2} \ell(t, \xi, \eta) l_2(d\eta) \int_{\{|\eta-\zeta|>1\}} \log|\eta - \zeta| \mu(d\zeta) \\ &\leq \int_{\mathbb{R}^2} \mu(d\zeta) \int_{\mathbb{R}^2} \ell(t, \xi, \eta) \log|\eta - \zeta| l_2(d\eta) + (4\sigma^2 t)^{-1} \mu(\mathbb{R}^2) \\ &\leq \int_{\mathbb{R}^2} \mu(d\zeta) \log\left[\int_{\mathbb{R}^2} \ell(t, \xi, \eta) |\eta - \zeta| l_2(d\eta)\right] + (4\sigma^2 t)^{-1} \mu(\mathbb{R}^2) \end{aligned}$$

$$\le \int_{\mathbb{R}^2} \mu(d\zeta)\log\left[\int_{\mathbb{R}^2} \mathscr{b}(t,\xi,\eta)(|\eta-\xi|+|\xi-\zeta|)l_2(d\eta)\right]+(4\sigma^2 t)^{-1}\mu(\mathbb{R}^2)$$

$$= \int_{\mathbb{R}^2} \log[\alpha+|\xi-\zeta|]\mu(d\zeta)+(4\sigma^2 t)^{-1}\mu(\mathbb{R}^2)$$

$$\le [\log(2\alpha)+(4\sigma^2 t)^{-1}]\mu(\mathbb{R}^2)+\int_{\{|\xi-\zeta|>\alpha\}} \log(2|\xi-\zeta|)\mu(d\zeta); \tag{3.5}$$

the last line is finite because the potential of the projection of μ on the exterior of $B(\xi,\alpha)$ is harmonic on that disk and so is finite at ξ.

Parabolic Context

Theorem 3 is easily translated into the parabolic context. Let $\{\dot{w}(\cdot),\mathscr{F}(\cdot)\}$ be a space-time Brownian motion in $\dot{\mathbb{R}}^N$ from $\dot{\xi}=(\xi,s)$. Let $\dot{u}$ be a superparabolic function on a half-space $\{\operatorname{ord}\dot{\eta}<c\}$ of $\dot{\mathbb{R}}^N$ containing $\dot{\xi}$, and suppose that either (a) $\dot{u}$ is bounded below or (b) $\dot{u}$ satisfies (for all $t>0$)

$$\int_{\mathbb{R}^N} \mathscr{b}(t,\xi,\eta)|\dot{u}(\eta,s-t)|l_N(d\eta)<+\infty. \tag{3.6}$$

Then if $\dot{u}(\dot{\xi})<+\infty$, the process $\{\dot{u}[\dot{w}(\cdot)],\mathscr{F}(\cdot)\}$ is a supermartingale, a martingale if $\dot{u}$ is parabolic. If $\dot{u}(\dot{\xi})=+\infty$, the assertion remains true for the parameter interval $]0,+\infty[$. The proof follows that of Theorem 3.

Application to the Hitting of a Point by a Brownian Path

We have seen in Section VII.6 that almost every path of a Brownian motion in $\mathbb{R}$ hits every point of $\mathbb{R}$ at arbitrarily large parameter values. When $N>1$ we shall now show that almost no path of a Brownian motion in $\mathbb{R}^N$ hits a specified point η of $\mathbb{R}^N$ at a strictly positive parameter value. It is sufficient to consider Brownian motions from a point of $\mathbb{R}^N$. The function $u=G(\eta,\cdot)$ is superharmonic on $\mathbb{R}^N$ with value $+\infty$ at η. This function is positive when $N>2$ and satisfies the integrability condition (3.1) for every point ξ (η not excluded) when $N=2$. If $\{w(\cdot),\mathscr{F}(\cdot)\}$ is a Brownian motion from ξ, the process $\{u[w(\cdot)],\mathscr{F}(\cdot)\}$ is therefore a supermartingale except that the parameter value 0 is to be excluded if $\xi=\eta$. This supermartingale is almost surely continuous, so (Section IV.1) almost every sample function is finite valued. Hence almost no $w(\cdot)$ path ever hits η except when $\xi=\eta$ and the parameter value is 0. It follows that if $N>1$ and if A is a countable subset of $\mathbb{R}^N$, almost no path of a Brownian motion ever hits A at a strictly positive parameter value. This result will be extended to polar sets A in Section 5.

4. Hitting of an F_σ Set

Let A be a subset of $\mathbb{R}^N$, and let $u(\xi, A)$ be the probability (if this probability is defined) that a Brownian motion from ξ hits A at a strictly positive time. The function $u(\cdot, A)$ is defined and Borel measurable if A is an F_σ set (Section VI.6), defined and universally measurable if A is an analytic set according to the same section. We suppose that A is an F_σ set in the present section to show how far one can go using relatively elementary methods. See Section 9 for A analytic; such generality will not be needed in the application to be made in Section 5.

Lemma. *If A is an F_σ subset of $\mathbb{R}^N$ for $N \geq 2$, the function $u(\cdot, A)$ is superharmonic, harmonic on $\mathbb{R}^N - \bar{A}$.*

[If $N = 1$, the function $u(\cdot, A)$ is identically 1 when A is not empty.] If $N > 1$, let B be a ball in $\mathbb{R}^N$ with center ξ, and let T_ξ be the hitting time of ∂B by the Brownian motion $w_\xi(\cdot)$ from ξ. Since Brownian motion is rotationally symmetric about its initial point, the distribution of $w_\xi(T_\xi)$ is the uniform distribution $\mu_B(\xi, \cdot)$ (harmonic measure) on ∂B. To show that $u(\cdot, A)$ is harmonic on $\mathbb{R}^N - \bar{A}$, suppose that $\bar{B} \subset \mathbb{R}^N - \bar{A}$. Apply the strong Markov property of Brownian motion to find that the probability that $w_\xi(\cdot)$ hits A, which is the same as the probability that $w_\xi(\cdot)$ hits A after time T_ξ, is given by

$$u(\xi, A) = E\{u[w_\xi(T_\xi)]\} = \mu_B(\xi, u). \tag{4.1}$$

Thus $u(\cdot, A)$ has the harmonic function average property on $\mathbb{R}^N - \bar{A}$ and hence is harmonic there. To show that u is superharmonic on $\mathbb{R}^N$ observe that if A is an F_σ set, the set $A - B$ is also an F_σ set, and $u(\cdot, A - B)$ is harmonic on B according to what has just been proved. Since $u(\cdot, A - B) \leq u(\cdot, A)$ and there is equality in the limit when B shrinks to its center ξ because (Section 3) $u(\xi, \{\xi\}) = 0$, the function $u(\cdot, A)$ is lower semicontinuous at ξ. Moreover $u(\xi, A)$ is at least equal to the probability that a $w_\xi(\cdot)$ path hits A after time T_ξ; so in the present context (4.1) is modified to the superharmonic function inequality

$$u(\xi, A) \geq E\{u[w_\xi(T_\xi)]\} = \mu_D(\xi, u). \tag{4.2}$$

The function $u(\cdot, A)$ is therefore superharmonic on $\mathbb{R}^N$, as was to be proved.

Parabolic Context

Let $\dot{u}(\xi, \dot{A})$ be the probability, if this probability is defined, that a space-time Brownian motion from ξ hits the set $\dot{A}$ at a strictly positive time. In view of the fact (Section VII.12) that parabolic measure on an interval boundary relative to a point of the interval is the hitting distribution on the boundary of a space-time Brownian motion from the reference point, the method of proof of Lemma 4 with the ball B replaced by an interval shows that when $\dot{A}$ is an F_σ subset of $\dot{\mathbb{R}}^N$ for $N \geq 1$, the function $\dot{u}(\cdot, \dot{A})$ is superparabolic on $\dot{\mathbb{R}}^N$, parabolic on $\dot{\mathbb{R}}^N - \bar{\dot{A}}$. In particular, if $\dot{A} = A \times \mathbb{R}$ with A an F_σ subset of $\mathbb{R}^N$, the set $\dot{A}$ is an F_σ subset of $\dot{\mathbb{R}}^N$, and $\dot{u}((\xi, s), \dot{A}) = u(\xi, A)$ for all s.

5. The Hitting of a Set by Brownian Motion

Theorem. *Let $w(\cdot)$ be a Brownian motion on $\mathbb{R}^N$.*

(a) *If A is a polar subset of $\mathbb{R}^N$, almost no $w(\cdot)$ path meets A at a strictly positive parameter value.*

(b) *If $N > 2$, $\lim_{t\to\infty} w(t) = \infty$ almost surely. Hence almost every Brownian path from a point of a Greenian subset D of $\mathbb{R}^N$ either hits a finite boundary point of D or lies in D and has the boundary point ∞ as a limit.*

(c) *If $N = 2$, almost every $w(\cdot)$ path hits every disk at arbitrarily large parameter values.*

Observation ($N = 2$). According to (a), almost no Brownian path from a point of an open subset D of $\mathbb{R}^2$ hits ∂D if D is not Greenian. Conversely (to be proved in Section 10), almost every Brownian path from a point of D hits ∂D in a finite boundary point if D is Greenian.

Proof of (a). If A is a polar subset of $\mathbb{R}^N$ and $N > 2$, there is a positive superharmonic function u on $\mathbb{R}^N$, identically $+\infty$ on A (Theorem 1.V.2). Choose $\delta > 0$. According to Theorem 3, if $w_\xi(\cdot)$ is a Brownian motion from ξ, the process $\{u[w_\xi(t)], t \geq \delta\}$ is a supermartingale, and according to the submartingale maximal inequality (Section III.9) applied to the negative of this supermartingale, the restriction to the rationals in $[\delta, 1/\delta]$ of almost every $u[w_\xi(\cdot)]$ sample function is bounded. Hence the same assertion is true if $w_\xi(\cdot)$ is replaced by a Brownian motion $w(\cdot)$ with an arbitrary initial distribution. Since u is lower semicontinuous, this function is continuous at each of its infinities, and it follows that almost no $w(\cdot)$ path can hit A at a strictly positive parameter value. If $N = 2$, it will be convenient and sufficient to assume that A is bounded. According to Theorem 1.V.2, there is a measure μ whose potential $u = G\mu$ is identically $+\infty$ on A, and a glance at the proof shows that μ can be chosen with compact support if A is bounded. In view of the application to potentials in Section 3 the function u satisfies (3.1), and the proof of (a) now goes through as in the case $N > 2$. □

Proof of (b). It is sufficient to consider a Brownian motion $w_\xi(\cdot)$ from a point ξ. If $N > 2$ and $u(\eta) = |\eta|^{-N+2}$, the function u satisfies the hypotheses of Theorem 3 so that $u[w_\xi(\cdot)]$ is a positive almost surely continuous supermartingale, omitting the parameter value 0 if $\xi = 0$. Hence (Section III.15) $\lim_{t\to\infty} u[w_\xi(t)]$ exists and is finite almost surely; that is, $\lim_{t\to\infty} |w_\xi(t)|$ exists almost surely, and this limit is $+\infty$ because (Section VII.5(c)) the limit is $+\infty$ in measure. □

Proof of (c). Let B be a disk, and let $u(\xi, B)$ be the probability that a Brownian motion $w_\xi(\cdot)$ from ξ ever hits B. According to Section 4, the positive function $u(\cdot, B)$ is superharmonic on $\mathbb{R}^2$, and it follows (Section 1.II.13) that $u(\cdot, B)$ is identically constant, necessarily identically 1 because this function is 1 on B. Thus almost every Brownian path from ξ hits B at a strictly positive time. If $s > 0$, the process $\{w_\xi(s+t), t \in \mathbb{R}^+\}$ is a Brownian motion with initial distribution the distribution of $w_\xi(s)$, and so almost every Brownian path hits B after time s; that is, almost every Brownian path from a point of $\mathbb{R}^2$ hits B at arbitrarily large parameter values and therefore hits every disk with rational radius and rational center at arbitrarily large parameter values. Part (c) follows. □

Part (c) implies that the cluster set of almost every plane Brownian path as the parameter value becomes infinite is the whole plane. According to Theorem 10 below, (c) is true with "disk" replaced by "analytic nonpolar set."

Theorem 5 exhibits the special nature of dimensionality 2 in classical potential theory. Dimensionality 1 is also special, for example, in that only the empty set of $\mathbb{R}$ is polar; so Theorem 5(a) becomes trivial when $N = 1$. Theorem 5(c) is almost as trivial because almost every Brownian path is unbounded positively and negatively. On the other hand the cases $N = 1, 2$ are not special for the parabolic version of Theorem 5.

Parabolic Context

Theorem 5(a) is true in the parabolic context, that is, almost no space-time Brownian path meets a parabolic-polar set at a strictly positive time. The proof of Theorem 5(a) is applicable, and simpler in the parabolic context in that the proof given above for $N > 2$ is valid in the parabolic context for $N \geq 1$. The parabolic version of Theorem 5(b) is trivial: if $\dot{w}(\cdot)$ is a space-time Brownian motion $\lim_{t\to\infty} \operatorname{ord} \dot{w}(t) = -\infty$. See Section 15 for the hitting of a parabolic-semipolar set by space-time Brownian motion.

6. Superharmonic Functions, Excessive for Brownian Motion

Theorem. *If u is a positive superharmonic function on the open subset D of $\mathbb{R}^N$, then u is excessive for $\mathscr{E}_D$, and $\mathscr{E}_D(t, \cdot, u) < +\infty$ for $t > 0$.*

See Section 8 for the converse theorem. The condition that u be excessive is

$$\begin{aligned} &\text{(a)} \quad \mathscr{E}_D(t, \cdot, u) \le u, \\ &\text{(b)} \quad \lim_{t \to 0} \mathscr{E}_D(t, \cdot, u) = u, \end{aligned} \tag{6.1}$$

and we show first that (6.1b) and the finiteness of $\mathscr{E}_D(t, \cdot, u)$ follow from (6.1a). For each point ξ of D let $\{w_\xi(\cdot), \mathscr{F}_\xi(\cdot)\}$ be a Brownian motion in D from ξ, with lifetime S_ξ (Section VII.9); for example, this process can be a Brownian motion in $\mathbb{R}^N$ killed at the hitting time S_ξ of ∂D. We can assume that $\mathscr{F}_\xi(0)$ contains the null sets and that $\mathscr{F}_\xi(\cdot)$ is right continuous. Since u is lower semicontinuous and positive and $w_\xi(\cdot)$ is almost surely continuous, Fatou's lemma yields

$$\liminf_{t \to 0} \mathscr{E}_D(t, \xi, u) = \liminf_{t \to 0} E\{u[w_\xi(t)]; S_\xi > t\} \ge u(\xi), \tag{6.2}$$

and therefore (6.1a) implies (6.1b). The finiteness of $\mathscr{E}_D(t, \xi, u)$ follows from (6.1a) if $u(\xi) < +\infty$. If $u(\xi) = +\infty$, let B be a ball with center ξ and closure in D. The function $\tau_B u$ is positive and superharmonic on D, finite on B, equal to u on $D - B$. Apply (6.1a) to $\tau_B u$ to find that $\mathscr{E}_D(t, \xi, \tau_B u) < +\infty$ and therefore

$$\mathscr{E}_D(t, \xi, u) \le \mathscr{E}_D(t, \xi, \tau_B u) + \mathscr{E}_D(t, \xi, u 1_B) \le \tau_B u(\xi) + \mathscr{E}(t, \xi, u 1_B) < +\infty \tag{6.3}$$

in view of the fact that u is l_N integrable over B. Thus to prove the theorem, it is sufficient to prove (6.1a).

In the following proof we refine an example given in Section VI.4. Let $B_\varepsilon(\eta)$ be the ball with center η in D and radius $\varepsilon \wedge (|\eta - \partial D|/2)$, in particular, of radius ε if $D = \mathbb{R}^N$. Choose a point η at which u is finite, and define

$$T_0 = 0, \qquad T_{n+1} = \inf\{t > T_n : w_\eta(t) \in B_\varepsilon(w_\eta(T_n))\}, \qquad n > 0.$$

The function T_1 is optional for $\mathscr{F}_\eta(\cdot)$ because T_1 is the hitting time by the almost surely continuous process $\{w_\eta(\cdot), \mathscr{F}_\eta(\cdot)\}$ of a closed set. The function $T_2 - T_1$ is the hitting time of a closed set by the almost surely continuous process

$$\{[w_\eta(T_1 + \cdot), w_\eta(T_1)], \mathscr{F}_\eta(T_1 + \cdot)\}$$

with state space $D \times D$. Hence $T_2 - T_1$ is optional for $\mathscr{F}_\eta(T_1 + \cdot)$, and therefore (Section II.2h) T_2 is optional for $\mathscr{F}_\eta(\cdot)$. Proceeding by induction we find that T_n is optional for $\mathscr{F}_\eta(\cdot)$ for all n. In view of the strong Markov property the discrete parameter process $\{w_\eta(T_n), \mathscr{F}_\eta(T_n), n \in \mathbb{Z}^+\}$ is a Markov process with state space D and stationary transition function q, where

$q(\eta, \cdot)$ is the uniform probability distribution on $\partial B_\varepsilon(\eta)$. Since $q(\eta, u) = L(u, \eta, \varepsilon \wedge (|\eta - \partial D|/2))$, the process $\{u[w_\eta(T_\cdot)], \mathscr{F}_\eta(T_\cdot)\}$ is a supermartingale. Obviously $\lim_{\eta \to \infty} T_n = S_\eta$ almost surely. Choose $t > 0$ and define

$$k = \begin{cases} \min\{n : T_n \geq t\} & \text{if } S_\eta > t \\ +\infty & \text{otherwise.} \end{cases}$$

Then k is optional for $\mathscr{F}_\eta(T_\cdot)$, and $\lim_{\varepsilon \to 0} T_k = t$ almost everywhere where $S_\eta > t$. If $u[w_\eta(T_{+\infty})]$ is defined as 0, the supermartingale inequality at times 0 and k (optional sampling theorem) yields

$$u(\eta) \geq E\{u[w_\eta(T_k)]\}. \tag{6.4}$$

When $\varepsilon \to 0$, this inequality yields, in view of the lower semicontinuity of u and Fatou's lemma,

$$u(\eta) \geq E\{u[w_\eta(t)]; S_\eta > t\} = \mathscr{E}_D(t, \eta, u); \tag{6.5}$$

so (6.1a) is true, as was to be proved.

Probabilistic Interpretation of Theorem 6

In probability language Theorem 6 states the following. Let u be a positive superharmonic function on an open subset D of $\mathbb{R}^N$, let ξ be a point of D, let $\{w_\xi(\cdot), \mathscr{F}_\xi(\cdot)\}$ be a Brownian motion in $\mathbb{R}^N$ from ξ, and let S_ξ be the hitting time of ∂D (Euclidean boundary) by $w_\xi(\cdot)$. Then the process

$$\{u[w_\xi(t)]1_{\{S_\xi > t\}}; \mathscr{F}_\xi(t), t \in \mathbb{R}^+\}$$

is a supermartingale except that the parameter value 0 is to be omitted if $u(\xi) = +\infty$. Equivalently, if $\{w_\xi(\cdot), \mathscr{F}_\xi(\cdot)\}$ is a Brownian motion in D from ξ, with lifetime S_ξ, in which case $w_\xi(t)$ has domain of definition $\{S_\xi > t\}$ when $t > 0$, we can rephrase the result by stating that

$$\{u[w_\xi(t)]; \mathscr{F}_\xi(t), t \in \mathbb{R}^+\}$$

is a supermartingale if defined as 0 for $t \geq S_\xi$, except that the parameter value 0 is to be omitted if $u(\xi) = +\infty$.

Parabolic Context

Theorem 6 and its probabilistic interpretation and proof translate directly into the parabolic context except that balls in the proof of Theorem 6 are to be replaced by intervals.

EXAMPLE. Let $\{w(\cdot), \mathscr{F}(\cdot)\}$ be a Brownian motion in $\mathbb{R}^3$ from the origin, let ξ_0 be a point of $\mathbb{R}^3$ other than the origin, and define $u(\xi) = |\xi - \xi_0|^{-1}$. Then u is a positive superharmonic function on $\mathbb{R}^3$, and (by Theorem 3 or Theorem 6) the process $\{u[w(\cdot)], \mathscr{F}(\cdot)\}$ is an almost surely continuous supermartingale, trivially right closable by 0. This supermartingale has the following properties.

(a) $u[w(\cdot)]$ is L^2 bounded (and therefore uniformly integrable because the function $r \mapsto r^2$ is a uniform integrability test function on $\mathbb{R}^+$). In fact, if $B = B(\xi_0, |\xi_0|/2)$ and if $\ell(t, \xi) = \ell(t, 0, \xi)$, then

$$\int_{\{w(t) \in \mathbb{R}^3 - B\}} u^2[w(t)]\,dP = \int_{\mathbb{R}^3 - B} \ell(t, \xi)|\xi - \xi_0|^{-2} l_3(d\xi) \leq 4|\xi_0|^{-2},$$

$$\int_{\{w(t) \in B\}} u^2[w(t)]\,dP \leq \int_B \ell\left(t, \frac{\xi_0}{2}\right)|\xi - \xi_0|^{-2} l_3(d\xi) = \frac{2\pi}{3}|\xi_0|\ell\left(t, \frac{\xi_0}{2}\right). \tag{6.6}$$

(b) $\lim_{t\to\infty} u[w(t)] = 0$ almost surely because [by Theorem 5(b)] $\lim_{t\to\infty} w(t) = \infty$ almost surely.

(c) Choose $\delta < |\xi_0|$, and let T_δ be the hitting time of $\partial B(\xi_0, \delta)$ by $w(\cdot)$. Then, under the convention that $u[w(+\infty)] = 0$, the process

$$\{u[w(T_\delta \wedge t)]; \mathscr{F}(t), t \in \bar{\mathbb{R}}^+\} \tag{6.7}$$

is a martingale, trivially bounded and therefore uniformly integrable. In fact the process (6.7) is a supermartingale because it can be obtained by stopping at time T_δ the supermartingale $u[w(\cdot)]$ right closed by 0, and the process (6.7) is a submartingale because $2/\delta - u[w(T_\delta \wedge \cdot)]$ is a supermartingale on the parameter set $\bar{\mathbb{R}}^+$. The latter process is a supermartingale because if $w'(\cdot)$ is $w(\cdot)$ killed at $T_{\delta/2}$ to obtain a Brownian motion in $\mathbb{R}^3 - \bar{B}(\xi_0, \delta/2)$ and if u' is the restriction of u to $\mathbb{R}^3 - \bar{B}(\xi_0, \delta/2)$, then the composition $2/\delta - u'[w'(\cdot)]$ of the positive harmonic function $2/\delta - u'$ with $w'(\cdot)$ is a supermartingale which stopped at time T_δ yields $2/\delta - u[w(T_\delta \wedge \cdot)]$. [Assertion (c) is a consequence of Theorem 7(c) below because u' is a bounded harmonic function on $\mathbb{R}^3 - \bar{B}(\xi_0, \delta)$ and is continuous on the closure of its domain, but it seemed more natural to prove (c) at this stage.]

(d) $u(\xi) = E\{u[w(T_\delta)]\}$ because this is the martingale equality for the process (6.7) at times 0, $+\infty$.

(e) $\{u[w(t)]; \mathscr{F}(t), t \in \mathbb{R}^+\}$ is not a martingale because if this uniformly integrable process were a martingale it would be right closable by its limit 0 at $t = +\infty$ (Section III.14) and would therefore almost surely vanish identically.

(f) The process in (e) is a supermartingale potential because it is uniformly integrable; so there is L^1 convergence to 0 in (b).

(g) The process in (e) is a singular potential because $\lim_{\delta\to 0} T_\delta = +\infty$ almost surely since (Section 3 or Theorem 5) almost no $w(\cdot)$ path hits ξ_0, and therefore the process in (e) is a potential which in view of (c) is a local martingale and as such (Theorem V.12) is singular.

(h) The process in (e) is not in the class **D**. In fact, if this process were in **D**, we could go to the limit ($\delta \to 0$) in (d) to find, since

$$\lim_{\delta \to 0} P\{u[w(T_\delta)] = 0\} = 1,$$

the false evaluation $u(\xi_0) = 0$.

7. Preliminary Treatment of the Composition of a Superharmonic Function with Brownian Motion; A Probabilistic Fatou Boundary Limit Theorem

Theorem 6 makes it possible to analyze by martingale theory the composition of a superharmonic function with Brownian motion. "Preliminary" in the section title refers to the fact that although in Theorem 12 below it will be shown that almost every sample function of such a composition is continuous, only right continuity will be proved in this section. Specifically, the process $x_\xi(\cdot)$ in (7.1) below is not only almost surely right continuous, as proved in this section, but is in fact almost surely continuous, and the process $x'_\xi(\cdot)$ in (7.1) is almost surely continuous except for a possible jump discontinuity at the parameter value S_ξ. Theorem 7 is also preliminary in that (Theorem X.8) it is valid in the more general context of relative superharmonic functions and conditional Brownian motion.

In Section 1.XII.19 it was proved that if h is a strictly positive harmonic function on a Greenian subset D of $\mathbb{R}^N$ and if v is a positive superharmonic function on D, then v/h has a minimal-fine limit at M_h almost every point of the Martin boundary of D. We shall see (Theorem 3.III.4) that an equivalent probabilistic formulation of this theorem is Theorem X.8, which does not involve a boundary and which states that v/h has a limit along certain conditional Brownian paths. When $h \equiv 1$, this result is part (a) of the following theorem.

Theorem. *Let D be a Greenian subset of $\mathbb{R}^N$, let ξ be a point of D, and let $\{w_\xi(\cdot), \mathcal{F}_\xi(\cdot)\}$ be a Brownian motion from ξ. It is supposed that $\mathcal{F}_\xi(0)$ contains the null sets. Define*

$$S_\xi = \sup\{t > 0 : w_\xi(s) \in D \text{ for } s < t\},$$

and let u be a positive superharmonic function on D with potential, singular harmonic, and quasi-bounded harmonic components, respectively, u_p, u_{ms}, and u_{mqb}, so that $u = u_p + u_{ms} + u_{mqb}$. Then

(a) $\lim_{t\uparrow S_\xi} u[w_\xi(t)]$ *exists (finite) almost surely.*
Define $\mathscr{F}_\xi(+\infty) = \bigvee_{t\in\mathbb{R}^+} \mathscr{F}_\xi(t)$,

$$\begin{aligned} x'_\xi(t) &= x_\xi(t) = u[w_\xi(t)] && \text{if } t < S_\xi; \\ x'_\xi(t) &= 0, \qquad x_\xi(t) = \lim_{s\uparrow S_\xi} u[w_\xi(s)] && \text{if } S_\xi \le t \le +\infty. \end{aligned} \tag{7.1}$$

(b) *The processes* $\{x'_\xi(\cdot), \mathscr{F}_\xi(\cdot)\}$ *and* $\{x_\xi(\cdot), \mathscr{F}_\xi(\cdot)\}$ *are almost surely right continuous supermartingales except that the parameter value* 0 *is to be omitted if* $u(\xi) = +\infty$. *(The second process will be shown in Section* 11 *to be almost surely continuous.)*

(c) *If* $u = u_p$ *or* $u = u_{ms}$, *the limit in* (a) *vanishes almost surely. If* $u = u_{mqb}$, *the process* $\{x_\xi(\cdot), \mathscr{F}_\xi(\cdot)\}$ *is a uniformly integrable martingale.*

(d) $u(\xi) \ge u_{mqb}(\xi) = E\{\lim_{t\uparrow S_\xi} u[w_\xi(t)]\}$.

Observation. When $S_\xi \le t \le +\infty$, the random variable $x_\xi(t)$ can be defined arbitrarily on the null set on which the limit in (a) does not exist. Note that if $u(\xi) = +\infty$, the function u is necessarily continuous at ξ; so $x'_\xi(\cdot)$ and $x_\xi(\cdot)$ are almost surely continuous at the parameter value 0 in this case, even though this parameter value must be excluded in the supermartingale assertion.

The Optional Time S_ξ

This random variable is the hitting time of the Euclidean boundary by $w_\xi(\cdot)$. The equivalent definition in the theorem was formulated to stress that the theorem does not involve any specific boundary of D. When $N > 2$, the hypothesis that D is Greenian reduces to the hypothesis that D is nonempty and open. In this case S_ξ may be infinite valued with strictly positive probability and, in fact, is almost surely $+\infty$ if $D = \mathbb{R}^N$. When $N = 2$, the hypothesis that D is Greenian, that is, (Theorem 1.V.6) that $\mathbb{R}^2 - D$ is not polar and that D is not empty, is made to avoid trivialities. In fact, if D is not empty and not Greenian, then a positive superharmonic function on D is necessarily identically constant; so the theorem is trivially true. Finally according to Section 10 below, $S_\xi < +\infty$ almost surely when $N = 2$ and D is Greenian.

Proof of (a) *and* (b). The conditional expectation supermartingale inequality for $x'_\xi(\cdot)$ is precisely (6.1a) except when the parameter value $+\infty$ is involved, in which case the inequality becomes trivial because $x'_\xi(+\infty) = 0$ almost surely. Moreover according to Theorem 6,

$$E\{x'_\xi(t)\} = \mathscr{E}_D(t, \xi, u) < +\infty$$

when $t > 0$. It follows that the process $\{x'_\xi(\cdot), \mathscr{F}_\xi(\cdot)\}$ is a supermartingale if the parameter value 0 is omitted when $u(\xi) = +\infty$. Throughout the following proof we assume that $u(\xi) < +\infty$; the modifications to be made in the arguments when $u(\xi) = +\infty$ will be obvious. Let $u_\bullet$ be an increasing sequence of finite-valued positive continuous superharmonic functions on D with limit u (Theorem 1.IV.10). Denote by $x'_{n\xi}(\cdot)$ the primed process (7.1) with u replaced by u_n. According to what we have just proved, the process $\{x'_{n\xi}(\cdot), \mathcal{F}_\xi(\cdot)\}$ is a supermartingale, almost surely right continuous because $w_\xi(\cdot)$ is almost surely continuous. Thus the process $\{x'_\xi(\cdot), \mathcal{F}_\xi(\cdot)\}$ is the limit of an increasing sequence of almost surely right continuous supermartingales, $x'_\xi(t) = \lim_{n\to\infty} x'_{n\xi}(t)$, and is therefore almost surely right continuous (Theorem IV.4). Since an almost surely right continuous supermartingale almost surely has finite left limits (Theorem IV.1), it follows that (a) is true. Thus $x_\xi(\cdot)$ can be defined by (7.1) and is almost surely right continuous. Let $D_\bullet$ be an increasing sequence of open relatively compact subsets of D with union D and with ξ in D_0, and let $S_{n\xi}$ be the hitting time of ∂D_n by $w_\xi(\cdot)$. The $x'_{n\xi}(\cdot)$ process stopped at $S_{n\xi}$, that is, the process

$$\{x'_{n\xi}(t \wedge S_{n\xi}); \mathcal{F}_\xi(t), t \in \mathbb{R}^+\},$$

is an almost surely right continuous supermartingale (Theorem IV.3). Hence

$$u(\xi) \geq E\{x'_{n\xi}(t \wedge S_{n\xi})\}, \quad E\{x'_{n\xi}(t \wedge S_{n\xi}) | \mathcal{F}_\xi(s)\} \leq x'_{n\xi}(s \wedge S_{n\xi}) \quad \text{a.s.} \qquad (7.2)$$

when $0 \leq s < t$. Furthermore, for each parameter value t, $\lim_{n\to\infty} x'_{n\xi}(t \wedge S_{n\xi}) = x_\xi(t)$ almost surely. Apply Fatou's lemma for conditional expectations in (7.2) when $n \to \infty$ to find that the almost surely right continuous process $\{x_\xi(\cdot), \mathcal{F}_\xi(\cdot)\}$ is a supermartingale. (According to Theorem 11 below, this process is actually almost surely continuous.) □

Proof of (c) *and* (d). Denote the expectation in (d) by $u'(\xi)$. The supermartingale inequality for $x_\xi(\cdot)$ at the pair of times 0, $+\infty$ yields $u \geq u'$. The reader is invited to prove, without analytic set theory, that u' is Borel measurable. It follows from the general theory in Section VI.6 that u' is universally measurable, which is all we shall need. If $\bar{B}(\xi, \delta) \subset D$ and if T is the hitting time of $\partial B(\xi, \delta)$ by $w_\xi(\cdot)$, the strong Markov property of Brownian motion implies that $w_\xi(T + \cdot)$ is a Brownian motion with initial distribution the distribution of $w_\xi(T)$, the uniform distribution on $\partial B(\xi, \delta)$ in view of the spherical symmetry of a Brownian motion about its initial point. Hence

$$u'(\xi) = E\left\{E\{\lim_{t\uparrow S_\xi} u[w_\xi(t)] | \mathcal{F}_\xi(T)\}\right\} = L(u', \xi, \delta).$$

Since u' ($\leq u$) is finite l_N almost everywhere on D, this function is harmonic; that is, u' is a harmonic minorant of u. If $u = u_p$ is a potential, then $\mathrm{GM}_D u = 0$ (Section 1.IV.3); so $u = 0$, and the limit in (a) vanishes almost surely. If $u = u_{ms}$ is a singular harmonic function, then for every constant c the function

$u \wedge c$ is a potential (Section 1.IX.10), and therefore again the limit in (a) vanishes almost surely. If u is bounded, say $u \le \alpha$, the inequality $u \ge u'$ can be applied to both u and $\alpha - u$ to find that $u = u'$. Similarly the supermartingale inequality can be applied to both $x_\xi(\cdot)$ and $\alpha - x_\xi(\cdot)$ to find that $\{x_\xi(\cdot), \mathscr{F}_\xi(\cdot)\}$ is a martingale. If $u = u_{mqb}$ is a quasi-bounded harmonic function, the limit of an increasing sequence $u_.$ of positive bounded harmonic functions on D, then the fact that the process $\{x_\xi(\cdot), \mathscr{F}_\xi(\cdot)\}$ in (7.1) with u replaced by u_n is a martingale implies that $(n \to \infty)$ the unprimed process in (7.1) is a martingale, necessarily uniformly integrable because it is right closed [Section III.3(e)]. The equation in (d) is the martingale equality for times $0, +\infty$. □

The Vanishing of a Superharmonic Potential at the Boundary of Its Domain

This vanishing can be phrased in various ways. For example, in view of the Riesz decomposition of a positive superharmonic function u on a Greenian set D the function is a potential if and only if $\mathrm{GM}_D u = 0$; equivalently (Section 1.VIII.11), when $B_.$ is an increasing sequence of open relatively compact subsets of D with union D, the positive superharmonic function u is a potential if and only if

$$\lim_{n\to\infty} \tau_{B_n} u = \lim_{n\to\infty} \mu_{B_n}(\cdot, u) = 0.$$

That is, u is a potential if and only if u has limit 0 at the boundary in the L^1 sense relative to harmonic measure. According to Theorem 13 below, if $T_{\xi n}$ is the hitting time of ∂B_n by a Brownian motion $w_\xi(\cdot)$ from ξ in B_0, then $\mu_{B_n}(\xi, u) = E\{u[w_\xi(T_{\xi n})]\}$; so u is a potential if and only if $\lim_{n\to\infty} E\{u[w_\xi(T_{\xi n})]\} = 0$. This condition is again an L^1 convergence condition. (It is sufficient in the preceding discussion if the L^1 condition is satisfied for a single point ξ in each open connected component of D.) According to Theorem 7(c), a superharmonic potential u has limit 0 along almost every $w_\xi(\cdot)$ path to the boundary. This condition on a positive superharmonic function u, if not reinforced by some kind of L^1 convergence condition, is not sufficient to make u a potential, however. For example, consider for $D = B(0, 1)$ the Poisson kernel function corresponding to the boundary point ζ,

$$u(\xi) = \frac{1 - |\xi|^2}{|\zeta - \xi|^N}.$$

The function u is positive and harmonic on D and has limit 0 at every boundary point except ζ; so although u is not a potential, u has limit 0 along almost every Brownian path from a point of D to the boundary.

Parabolic Context

Theorem 7 and its proof translate directly into the parabolic context. Observe, however, that "right continuity" will not be strengthened below to "continuity" in the parabolic context. That is, if $\dot{u}$ is a positive superparabolic function on an open subset $\dot{D}$ of $\dot{\mathbb{R}}^N$, then the parabolic context analog of the process $x_\xi(\cdot)$ is almost surely right continuous but not necessarily almost surely continuous, as shown by an example in Section 12.

8. Excessive and Invariant Functions for Brownian Motion

Theorem. *If D is an open subset of $\mathbb{R}^N$ and if u is an excessive [invariant excessive] function for ℓ_D, then on each open connected component of D the function u is either superharmonic [harmonic] or identically $+\infty$.*

See Section 6 for the (partial) converse theorem. It can be assumed in the proof that D is connected. Suppose first that u is excessive continuous and bounded on D, and let ξ be a point of D. Then if $\{w_\xi(\cdot), \mathscr{F}_\xi(\cdot)\}$ is a Brownian motion in D from ξ and if $x_\xi'(\cdot)$ is defined by (7.1), the process $\{x_\xi'(\cdot), \mathscr{F}_\xi(\cdot)\}$ is an almost surely right continuous supermartingale. If $\bar{B}(\xi,\delta) \subset D$ and if $T(\delta)$ is the hitting time of $\partial B(\xi,\delta)$ by the Brownian motion, the supermartingale inequality for times 0 and $T(\delta)$ yields the superharmonic function inequality $u(\xi) \geq L(u,\xi,\delta)$ because $w_\xi(T(\delta))$ is uniformly distributed on $\partial B(\xi,\delta)$ (spherical symmetry of Brownian motion about its initial point). Hence u is superharmonic. In the general case define

$$u_n = n\int_0^{1/n} \ell_D(t,\cdot,u \wedge n)l_1(dt). \tag{8.1}$$

According to Section VI.10 the function u_n is excessive for ℓ_D and $u_\cdot$ is an increasing sequence with limit u. Moreover $u_n \leq n$, and in view of the continuity properties of ℓ_D (Section VII.9) the function u_n is continuous. According to the special case of Theorem 8 just treated, the function u_n is superharmonic, and it follows that u, as the limit of an increasing sequence of superharmonic functions, is either superharmonic or identically $+\infty$. If u is invariant excessive and not identically $+\infty$, so that u is superharmonic and positive, we know (by Theorem 6) that $\ell_D(t,\cdot,u) < +\infty$ for $t > 0$. Hence $u = \ell(t,\cdot,u)$ is finite valued, and the process $\{x_\xi'(t), \mathscr{F}_\xi(t), t \in \mathbb{R}^+\}$ is an almost surely right continuous martingale. Note that the parameter set does not include the point $+\infty$. The martingale equality for this process at times 0 and $t \wedge T(\delta)$ is

$$u(\xi) = \int_{\{T(\delta)\leq t\}} u[w_\xi(T(\delta))]\,dP + \int_{\{T(\delta)>t\}} u[w_\xi(t)]\,dP. \tag{8.2}$$

The second integral is at most

$$\int_{\{|\eta-\xi|<\delta\}} u(\eta)\ell_D(t,\xi,\eta)l_N(d\eta) \le (2\pi\sigma^2 t)^{-N/2}\int_{\{|\eta-\xi|<\delta\}} u(\eta)l_N(d\eta)$$

and therefore tends to 0 when $t \to +\infty$. The first integral in (8.2) increases to $L(u,\xi,\delta)$ when $t \to +\infty$, and since u is locally l_N integrable, it follows that u is harmonic, as asserted.

EXAMPLE (a). Let u be a positive superharmonic function on the Greenian set D. Then u is excessive for ℓ_D; so the function $t \mapsto \ell_D(t,\xi,u)$ is monotone decreasing. The function $u_0 = \lim_{t\to\infty} \ell_D(t,\cdot,u)$ is an invariant excessive function (Section VI.10) and is therefore a harmonic minorant of u. A probabilistic interpretation of u_0 will be given in Section X.1.

Parabolic Context

Theorem 8 in the parabolic context becomes: If $\dot{D}$ is an open subset of $\dot{\mathbb{R}}^N$ and if $\dot{u}$ is an excessive [invariant excessive] function for $\dot{\ell}_{\dot{D}}$, then $\dot{u}$ is superparabolic [parabolic] on $\dot{D}$ if each point of $\dot{D}$ is below (relative to $\dot{D}$) some point of finiteness of $\dot{u}$. The proof follows that of Theorem 8 except that balls in $\mathbb{R}^N$ are replaced by intervals in $\dot{\mathbb{R}}^N$. For example, (8.1) becomes in the present context

$$\dot{u}_n(\xi,s) = n\int_{s-1/n}^{s} l_1(dt)\int_{\{\eta:(\eta,t)\in\dot{D}\}} \dot{\ell}_{\dot{D}}((\xi,s),(\eta,t))[\dot{u}(\eta,t)\wedge n]l_N(d\eta), \qquad (8.1')$$

and $\dot{u}_n$ is continuous on $\dot{D}$. Observe that even if $\dot{u}$ is not superparabolic on $\dot{D}$, the function $\dot{u}$ is lower semicontinuous there and is superparabolic on every open set strictly under (relative to $\dot{D}$) a point of finiteness of $\dot{u}$. When $\dot{u}$ is invariant excessive for $\dot{\ell}_{\dot{D}}$, the function $\dot{u}$ is parabolic on every open set strictly under (relative to $\dot{D}$) a point of finiteness of $\dot{u}$.

EXAMPLE (b). If $\dot{D} = \dot{\mathbb{R}}^N$, the function $\dot{u}$ on $\dot{D}$ defined as $+\infty$ or 0 according as the ordinate is strictly positive or not is parabolic excessive but is not superparabolic.

EXAMPLE (c). Let $w(\cdot)$ be a Brownian motion in $\mathbb{R}^N$ with parameter set $\mathbb{R}$, and let $\dot{u}(\cdot,s)$ be the distribution density of $w(s)$ relative to l_N, as defined in Section VII.2. Equation VII(2.3) satisfied by $\dot{u}$ shows that $\dot{u}$ is invariant excessive for space-time Brownian motion; so $\dot{u}$ is parabolic on $\dot{\mathbb{R}}^N$. Conversely, direct calculation shows that a minimal positive parabolic function on $\dot{\mathbb{R}}^N$, given by 1.XVI(8.1), is invariant excessive for space-time Brownian motion. In view of the integral representation 1.XVI(8.2) of an arbitrary positive parabolic function in terms of minimal ones it follows that every positive parabolic function on $\mathbb{R}^N$ is invariant excessive for space-time Brownian motion.

9. Application to Hitting Probabilities and to Parabolicity of Transition Densities

(a) *Hitting Probabilities.* Let D be a Greenian subset of $\mathbb{R}^N$, and let A be an analytic subset of D. Let $w(\cdot)$ be a Brownian motion in D from ξ, and let $u(\xi, A)$ be the probability that a $w(\cdot)$ path hits A at a strictly positive time; that is, $u(\xi, A)$ is the probability that an unkilled Brownian path from ξ hits A before it hits ∂D. Then *the function $u(\cdot, A)$ is superharmonic on D and harmonic on $D - \bar{A}$.* It was shown in Section 4 how this fact can be proved when $D = \mathbb{R}^N$ and A is an F_σ set. The method of proof is applicable in the general case, but we remark that if A is an F_σ set, the theory of analytic sets need not be invoked. At the present stage a second proof that $u(\cdot, A)$ is superharmonic can be given by noting that (Section VI.12) this function is excessive for Brownian motion and so is superharmonic by Theorem 8. It will be shown in Section 14 that $u(\cdot, A) = \underset{+1}{R^A}$. The translation of this discussion to the parabolic context is trivial: "superharmonic" becomes "superparabolic" and so on.

(b) *Parabolicity of Transition Densities.* According to Section VII.10, if D is a nonempty open subset of $\mathbb{R}^N$, if $\dot{D} = D \times \mathbb{R}$, and if η is a point of D, then the function $\ell_D(\cdot, \cdot, \eta)$ is excessive for space-time Brownian motion in $\dot{D}$, invariant excessive for space-time Brownian motion in $\dot{D} - \{(\eta, 0)\}$, and continuous and 0 at the points of $\dot{D} - \{(\eta, 0)\}$ with ordinate values ≤ 0. It follows that $\ell_D(\cdot, \cdot, \eta)$ is superparabolic on $\dot{D}$ and parabolic on $\dot{D} - \{(\eta, 0)\}$. More generally, if $\dot{D}$ is a nonempty open subset of $\dot{\mathbb{R}}^N$ and if $\dot{\eta}$ is a point of $\dot{D}$, the fact (Section VII.10) that $\dot{\ell}_{\dot{D}}(\cdot, \dot{\eta})$ is excessive for space-time Brownian motion in $\dot{D}$, invariant excessive for space-time Brownian motion in $\dot{D} - \{\dot{\eta}\}$, and continuous and 0 at the points of $\dot{D} - \{\dot{\eta}\}$ with ordinate values $\leq \operatorname{ord} \dot{\eta}$, implies that $\dot{\ell}_{\dot{D}}(\cdot, \dot{\eta})$ is superparabolic on $\dot{D}$ and parabolic on $\dot{D} - \{\dot{\eta}\}$. Now apply the Riesz decomposition to find that

$$\dot{\ell}_{\dot{D}}(\cdot, \dot{\eta}) = c\dot{G}_{\dot{D}}(\cdot, \dot{\eta}) + \dot{G}M_{\dot{D}}\dot{\ell}_{\dot{D}}(\cdot, \dot{\eta}). \tag{9.1}$$

Since [notation of VII(10.3)] $|\dot{w}_{\dot{\xi}}(\dot{T}_{\dot{\xi}}) - \dot{\eta}| \geq |\partial\dot{D} - \dot{\eta}|$, the last term in VII(10.3) defines a bounded function of ξ. It follows that $c = 1$ and that the last term in VII(10.3) defines a parabolic function of $\dot{\xi}$ in $\dot{D}$. In Section 17 we shall show that $\dot{\ell}_{\dot{D}} = \dot{G}_{\dot{D}}$, that is, the last term in (9.1) vanishes identically.

10. ($N = 2$). The Hitting of Nonpolar Sets by Brownian Motion

The following theorem strengthens Theorem 5(c).

Theorem. (a) *Let $w(\cdot)$ be a Brownian motion in $\mathbb{R}^2$, and let A be a plane set which is not inner polar (for example, A may be analytic and nonpolar). Then almost every $w(\cdot)$ path hits A at arbitrarily large parameter values.*

(b) *If a plane open set D is Greenian, almost every Brownian path from a point of D hits the boundary. If D is not Greenian, almost no Brownian path from a point of D hits the boundary.*

Proof of (a). It can be assumed that $w(\cdot) = w_\xi(\cdot)$ is a Brownian motion from some point ξ. It can also be assumed that A is compact and nonpolar. Then its complement D is Greenian (Theorem 1.V.6). Choose any point η in the same open connected component of D as ξ. The function $u = G_D(\eta, \cdot) \wedge 1$ is a positive nonconstant continuous superharmonic function on D. According to Section 6, if S_ξ is the hitting time of ∂D, the process $\{u[w_\xi(t)]1_{\{S_\xi > t\}}, t \in \mathbb{R}^+\}$ is a supermartingale. The sample functions of this supermartingale are almost all right continuous, even continuous except for a possible jump at the parameter value S_ξ if $S_\xi < +\infty$. Such a supermartingale has almost surely a limit as the parameter becomes infinite (Section III.13). If $P\{S_\xi = +\infty\} > 0$, there is a $w_\xi(\cdot)$ path with the following properties:

(i) The path never meets A.
(ii) u has a limit c on the path as the parameter becomes infinite.
(iii) The path meets every disk at arbitrarily large parameter values.

But then D is connected, and in view of the continuity of u this function must be identically c, a contradiction. Hence S_ξ is almost surely finite valued; that is, almost every $w_\xi(\cdot)$ path hits A. The proof of (a) is concluded in the same way as the corresponding part of the proof of Theorem 5(c). □

Proof of (b). If D is Greenian, $\mathbb{R}^2 - D$ is not polar (Theorem 1.V.6); so by (a) of the present theorem almost every Brownian path from a point of D hits the boundary. Conversely, if D is not Greenian, $\mathbb{R}^2 - D$ is polar; so by Theorem 5(a) almost no Brownian path from a point of D hits the boundary. □

The Distribution of $w_\xi(S_\xi)$ and Harmonic Measure ($N \geq 2$)

Let D be a Greenian subset of $\mathbb{R}^N$, let ξ be a point of D, let $w_\xi(\cdot)$ be a Brownian motion from ξ, and let S_ξ be the hitting time of ∂D (Euclidean boundary). When $N = 2$, this optional time is almost surely finite according to Theorem 10. When $N > 2$, $\lim_{t\to\infty} w_\xi(t) = \infty$ almost surely, and it is convenient to define $w_\xi(+\infty) = \infty$. It will be shown in Section 13 that for $N \geq 2$ the distribution of $w_\xi(S_\xi)$ is the harmonic measure $\mu_D(\xi, \cdot)$.

11. Continuity of the Composition of a Function with Brownian Motion

In this section all Brownian motions have the same variance parameter.

Theorem. *Let u be a Borel measurable extended real-valued function on an open subset D of* $\mathbb{R}^N$, *and suppose that whenever* $w_\xi(\cdot)$ *is a Brownian motion from* ξ *in D with* S_ξ *the hitting time of* ∂D, *the function* $u[w_\xi(\cdot, \omega)]$ *is for almost every* ω *right continuous on the interval* $[0, S_\xi(\omega)[$. *Then whenever* $w(\cdot)$ *is a Brownian motion in* $\mathbb{R}^N$, *the function* $u[w(\cdot, \omega)]$ *is for almost every* ω *continuous on the intervals on which it is defined.*

According to Section VI.6, if λ is a probability distribution on D and if $w_\lambda(\cdot)$ is a Brownian motion with initial distribution λ, then almost every $u[w_\lambda(\cdot)]$ process sample function is almost surely right continuous up to the hitting time of ∂D because this is true by hypothesis when λ is supported by a singleton. Slightly more generally, if λ is a measure on $\mathbb{R}^N$ and if λ_D is the projection of λ on D, then if $\lambda_D(D) > 0$, we can consider a Brownian motion $w_{\lambda_D}(\cdot)$ with initial distribution λ_D (perhaps not a probability distribution but the generalization to allow this case is trivial) and deduce that almost every sample function of this process is right continuous up to the hitting time of ∂D. Now if $r > 0$, the process $w_\lambda(r + \cdot)$ is a Brownian motion with initial distribution the distribution of $w_\lambda(r)$. It follows from what we have just shown that almost every $u[w_\lambda(r + \cdot)]$ process sample function for which $w_\lambda(r) \in D$, and considered only up to the hitting time by $u[w_\lambda(r + \cdot)]$ of ∂D, is right continuous. The exceptional P_λ null set depends on r, but the union of these null sets as r ranges through the positive rational numbers is a P_λ null sets, and if ω is not in this null set, the sample function $t \mapsto u[w_\lambda(t, \omega)]$, defined on the set of points t with $w_\lambda(t, \omega)$ in D, is right continuous. This argument needs no change if λ is not a probability measure and in particular is valid if $\lambda = l_N$ (see Section VII.2), a choice making the Brownian motion stationary. With this choice suppose that $b > 0$, and consider the process $w'_\lambda(\cdot)$ defined by

$$w'_\lambda(t) = \begin{cases} w_\lambda(b - t) & \text{if } 0 \leq t \leq b \\ w_\lambda(0) + w_\lambda(t) - w_\lambda(b) & \text{if } t > b. \end{cases}$$

This process is a Brownian motion; so almost every $u[w'_\lambda(\cdot)]$ sample function is right continuous where defined. In particular, almost every $u[w_\lambda(b - \cdot)]$ sample function is right continuous, where defined, on the parameter interval

$[0, b[$. Hence (under the present hypothesis that $\lambda = l_N$) for every $b > 0$ almost every $u[w_\lambda(\cdot)]$ process sample function is continuous, where defined, on the parameter interval $[0, b[$ and therefore on $\mathbb{R}^+$. Referring back to Section VI.6 again, it follows that for l_N almost every ξ in $\mathbb{R}^N$ this sample function property holds for $u[w_\lambda(\cdot)]$ when λ is supported by $\{\xi\}$ and therefore holds for $u[w_\lambda(\cdot)]$ when λ is absolutely continuous relative to l_N. If now $w(\cdot)$ is a Brownian motion in $\mathbb{R}^N$ with an arbitrary initial distribution and if $r > 0$, the process $w(r + \cdot)$ is a Brownian motion with an initial distribution absolutely continuous relative to l_N. The theorem follows for parameter values $\geq r$ and therefore as stated.

12. Continuity of Superharmonic Functions on Brownian Motion

Theorem. *If u is a superharmonic function on an open subset D of $\mathbb{R}^N$ and if $w(\cdot)$ is a Brownian motion on $\mathbb{R}^N$, then almost every $u[w(\cdot)]$ sample function (where defined) is finite valued and continuous except for a possible infinity at the parameter value* 0.

It is sufficient to prove the theorem for D a bounded open relatively compact subset of an open set D_1, with u superharmonic on a neighborhood of $\bar{D}$. Adding a constant to u if necessary, it can be assumed that u is positive on D. According to Theorem 7, the hypotheses of Theorem 11 are satisfied by u. Hence the $u[w(\cdot)]$ sample functions are almost surely continuous where defined. These sample functions are almost surely finite valued except possibly at the parameter value 0 because (Theorem 5) almost no Brownian path hits the polar set of infinities of u at a strictly positive time.

Parabolic Context

According to Section 7, if $\dot{x}'_\xi(\cdot)$ is the parabolic context version of the process defined in (7.1), the process $\{\dot{x}'_\xi(\cdot), \dot{\mathscr{F}}_\xi(\cdot)$ is an almost surely right continuous supermartingale except that the parameter value 0 is to be omitted if $\dot{u}(\dot{\xi}) = +\infty$. It follows that except possibly at the parameter value 0, the $\dot{x}'_\xi(\cdot)$ sample functions are almost surely finite valued with finite left limits. It is left to the reader to check that more generally if $\dot{u}$ is any superparabolic function on an open subset of $\dot{\mathbb{R}}^N$ and if $\dot{w}(\cdot)$ is a space-time Brownian motion on $\dot{\mathbb{R}}^N$, then except possibly at the parameter value 0, almost every $\dot{u}[\dot{w}(\cdot)]$ sample function (where defined) is finite valued and right continuous with finite left limits. As the following example shows these sample functions need not be continuous.

Example. Let f be a monotone increasing left continuous function on $\mathbb{R}$, and define $\dot{u}$ on $\dot{\mathbb{R}}^N$ by $\dot{u}(\xi, s) = f(s)$. The function $\dot{u}$ is superparabolic and if $\dot{w}(\cdot)$ is a Brownian motion on $\dot{\mathbb{R}}^N$ the $\dot{u}[\dot{w}(\cdot)]$ sample functions will have jumps if f has discontinuities at values of the argument strictly less than ord $\dot{w}(0)$.

13. Preliminary Probabilistic Solution of the Classical Dirichlet Problem

The following theorem is a special case of the much deeper Theorem 3.II.2 but will be needed before it is possible to prove the latter. Recall that the Euclidean boundary of a Greenian subset of $\mathbb{R}^N$ is PWB resolutive. Although we have not and shall not define a Brownian motion $w(\cdot)$ at the parameter value $+\infty$, we shall occasionally have formulas containing $w(S)$, where S is a possibly infinite valued random variable; if so, $w(+\infty)$ is to be interpreted as ∞.

Theorem. *Let D be a Greenian subset of $\mathbb{R}^N$, let $\{w_\xi(\cdot), \mathscr{F}_\xi(\cdot)\}$ be a Brownian motion from ξ in D, and let S_ξ be the hitting time of the Euclidean boundary ∂D by $w_\xi(\cdot)$.*

(a) *The distribution of $w_\xi(S_\xi)$ on ∂D is the harmonic measure $\mu_D(\xi, \cdot)$.*
(b) *Let f be a* PWB *resolutive boundary function. Then*

$$H_f(\xi) = E\{f[w_\xi(S_\xi)]\}, \tag{13.1}$$

$$\lim_{s \uparrow S_\xi} H_f[w_\xi(s)] = f[w_\xi(S_\xi)] \quad \text{a.s.}, \tag{13.2}$$

and if $x_\xi(t)$ is defined by

$$x_\xi(t) = \begin{cases} H_f[w_\xi(t)] & \text{if } 0 \le t < S_\xi, \\ f[w_\xi(S_\xi)] & \text{if } S_\xi \le t \le +\infty, \end{cases}$$

then the process $\{x_\xi(t), \mathscr{F}_\xi(t), 0 \le t \le +\infty\}$ is an almost surely continuous uniformly integrable martingale.

Observation. The evaluation (13.1) is the martingale equality for the process $x_\xi(\cdot)$ at times $0, +\infty$, equivalently, at times $0, S_\xi$. The representation $H_f = H_{f \vee 0} - H_{(-f) \vee 0}$ of H_f as the difference between two positive harmonic functions makes clear by way of Theorem 7 why the limit in (13.2) exists. We shall not use this argument, however. Equation (13.2) is an elegant justification of the PWB method of solving the Dirichlet problem, in that the solution H_f has the desired boundary limit function on approach along almost every Brownian path. Equation (13.2) will be extended to an arbitrary PWB^h resolutive boundary function, defined on a not necessarily PWB^h resolutive boundary, in Theorem 3.II.2.

Proof of (a). It is sufficient to prove that (13.1) is true when f is a bounded Borel measurable function, say $|f| \leq c$. Let v be a function on D in the upper PWB class for f. Then $v + c$ is a positive superharmonic function on D and has inferior limit at least $f(\eta) + c$ at each boundary point η. Hence Theorem 7(d) with $u = v + c$ yields the inequality $(v + c)(\xi) \geq E\{f[w_\xi(S_\xi)] + c\}$; that is, $v(\xi) \geq E\{f[w_\xi(S_\xi)]\}$. By definition of H_f we conclude that $H_f(\xi) \geq E\{f[w_\xi(S_\xi)]\}$, and (13.1) follows on applying this inequality to both f and $-f$. □

Proof of (b). Equation (13.1) for PWB resolutive f follows from (a) since $H_f = \mu_D(\cdot, f)$. To prove the rest of (b) we first prove that for each $t \geq 0$

$$x_\xi(t) = E\{f[w_\xi(S_\xi)] | \mathscr{F}_\xi(t)\} \quad \text{a.s.} \tag{13.3}$$

This equation is trivial on the set $\{S_\xi \leq t\}$. On the complementary set $\{S_\xi > t\}$ the Markov property of Brownian motion implies that the right-hand side of (13.3) is almost surely $E\{f[w_\xi(S_\xi) | w_\xi(t)\}$. Since the process $w_\xi(t+\cdot)$ is a Brownian motion with initial distribution the distribution of $w_\xi(t)$, this conditional expectation is the expectation of f at the first hit of ∂D by a Brownian motion from $w_\xi(t)$. In view of (a) this expectation is almost surely $\mathrm{H}_f[w_\xi(t)]$. Thus (13.3) is true and shows that $\{x_\xi(\cdot), \mathscr{F}_\xi(\cdot)\}$ is an almost surely right continuous martingale, uniformly integrable because it is right closed, a family of conditional expectations of a given random variable. Since this martingale is almost surely right continuous, it must almost surely have left limits at all points; so the limit in (13.2) exists. (Alternatively apply Theorem 7 as in the above observation to obtain the existence of this limit). To identify the limit as $f[w_\xi(S_\xi)]$, we can apply Theorem VII.4 which states that an almost surely right continuous martingale relative to a suitably defined filtration of a Brownian motion is almost surely continuous, but we can also give the following more elementary direct proof. Observe that in the notation of the proof of Theorem 7 it is clear that $w_\xi(S_{k\xi})$ is $\bigvee_0^\infty \mathscr{F}_\xi(S_{n\xi})$ measurable for all k; so $\lim_{k\to\infty} w_\xi(S_{k\xi}) = w_\xi(S_\xi)$, and therefore also $f[w_\xi(S_\xi)]$ are $\bigvee_0^\infty \mathscr{F}_\xi(S_{n\xi})$ measurable. It now follows from this measurability and the conditional expectation continuity theorem that

$$\lim_{n\to\infty} E\{f[w_\xi(S_\xi)] | \mathscr{F}_\xi(S_{n\xi})\} = E\{f[w_\xi(S_\xi)] | \bigvee_0^\infty \mathscr{F}_\xi(S_{n\xi})\} = f[w_\xi(S_\xi)] \quad \text{a.s.,} \tag{13.4}$$

that is, $\lim_{n\to\infty} H_f[w_\xi(S_{n\xi})] = f[w_\xi(S_\xi)]$ almost surely, and (b) is now completely proved. □

Parabolic Context

Theorem 13 translates directly into the parabolic context with no change in proof.

EXAMPLE. Let A be an analytic subset of $\mathbb{R}^N$, let S_ξ be the hitting time of A by a Brownian motion from ξ, and define $\dot{u}(\xi, s) = P\{S_\xi < s\}$ for $s > 0$. Then $\dot{u}(\xi, s)$ is the probability that a space-time Brownian motion in $\dot{D} = \mathbb{R}^N \times]0, +\infty[$ from (ξ, s) hits $\dot{A} = A \times]0, +\infty[$ during the process lifetime at a strictly positive time. It follows (Section 9) that $\dot{u}$ is superparabolic on $\dot{D}$, parabolic on $\dot{D} - \overline{\dot{A}}$. In particular, if A is closed define $D_0 = \mathbb{R}^N - A$, $\dot{D}_0 = D_0 \times]0, +\infty[$. The parabolic version of Theorem 13 applied to the restriction $\dot{u}_0$ of $\dot{u}$ to $\dot{D}_0$ shows that $\dot{u}_0 = \dot{\mu}_{\dot{D}_0}(\cdot, D_0 \times]0, +\infty[)$.

14. Probabilistic Evaluation of Reductions

Let D be a Greenian subset of $\mathbb{R}^N$, let ξ be a point of D, let $w_\xi(\cdot)$ be a Brownian motion process from ξ, and denote by $T'^A_\xi[T^A_\xi]$ the entry [hitting] time of a set A by $w_\xi(\cdot)$.

Theorem. *If A is an analytic subset of D and if v is a positive superharmonic function on D, then (reductions relative to D)*

$$R^A_v(\xi) = E\{v[w_\xi(T'^A_\xi)];\ T'^A_\xi < T^{\partial D}_\xi\} \tag{14.1}$$

$$\underset{+v}{R^A}(\xi) = E\{v[w_\xi(T^A_\xi)];\ T^A_\xi < T^{\partial D}_\xi\}, \tag{14.1sm}$$

that is,

$$P\{w_\xi(T^A_\xi) \in B, T^A_\xi < T^{\partial D}_\xi\} = \delta^A_D(\xi, B) \tag{14.2}$$

for every Borel set B. In particular,

$$R^A_1(\xi) = P\{T'^A_\xi < T^{\partial D}_\xi\} \tag{14.3}$$

$$\underset{+1}{R^A}(\xi) = P\{T^A_\xi < T^{\partial D}_\xi\}. \tag{14.3sm}$$

Observation (a). Equations (14.1) and (14.1sm) make obvious many of the reduction properties which are not at all obvious from their potential theoretic context, for example, the countable additivity of the function $v \mapsto R^A_v$ [Section 1.VI.3(f)].

Observation (b). The proof of (14.1) and (14.1sm) for compact A given below involves neither analytic set theory nor capacity theory, and since these equations are true for A an F_σ set if true for A compact, it follows that the proof of these equations for A an F_σ set involves neither analytic set theory nor capacity theory.

Observation (c). The right side of (14.3sm) defines a superharmonic function of ξ according to Section 9. More generally the function defined by the right side of (14.1sm) is excessive for Brownian motion (Section VI.12) and is therefore superharmonic or identically $+\infty$ on each open connected component of D. Thus Theorem 14 can be considered as the identification of certain probabilistically defined superharmonic functions as reductions.

It is sufficient to prove (14.1) and (14.1sm) for bounded v (and the boundedness assumption is made tacitly throughout the following proof) because v is the limit of an increasing sequence $n \mapsto v \wedge n$ of bounded positive superharmonic functions.

Proof of (14.1) for A compact. If $\xi \in D - A$, then $R_v^A(\xi) = \underset{+v}{R}^A(\xi)$, $T_\xi'^A = T_\xi^A$, and if A is compact, we have seen (Section 1.VIII.10) that $\underset{+v}{R}^A$ on $D - A$ is the PWB solution on $D - A$ for the boundary function v on $A \cap \partial(D - A)$ and 0 elsewhere. According to Theorem 13 this PWB solution is precisely the right side of (14.1) on $D - A$. On A the two sides of (14.1) are both $v(\xi)$.

Proof of (14.1) and (14.1sm) for analytic A. Fix ξ, v and write $\phi(A)$ for the right side of (14.1) when A is analytic. The function ϕ is an increasing set function because if $x(\cdot)$ is the process $v[w_\xi(\cdot)]$ redefined as 0 for parameter values $\geq T_\xi^{\partial D}$, then $x(\cdot)$ is an almost surely right continuous supermartingale and $\phi(A) = E\{x(T_\xi'^A)\}$; so the fact that $\phi(B) \geq \phi(A)$ when $A \subset B$ is the supermartingale inequality for $x(\cdot)$ at the pair of times $T_\xi'^B$, $T_\xi'^A$. According to Section 1.VI.5, if $\phi_0(A)$ is defined as the left side of (14.1), ϕ_0 is a Choquet capacity on D, that generated by a topological precapacity, the restriction of ϕ_0 to the class Γ of compact subsets of D. Since ϕ is an increasing set function, equal to ϕ_0 on Γ, and since $A_n \uparrow A$ implies that $\phi(A_n) \uparrow \phi(A)$, it follows first that when A is open, $\phi_0(A) = \phi(A)$, second that when A is analytic,

$$\begin{aligned}\phi_0(A) &= \sup\{\phi_0(B): B \subset A, B \text{ compact}\}\\ &= \sup\{\phi(B): B \subset A, B \text{ compact}\} \leq \phi(A),\end{aligned}$$

and finally when A is analytic,

$$\phi_0(A) = \inf\{\phi_0(B): B \supset A, B \text{ open}\} = \inf\{\phi(B): B \supset A, B \text{ open}\} \geq \phi(A).$$

Hence $\phi = \phi_0$ for A analytic, as asserted in (14.1). Equation (14.1sm) is true because (14.1) with A replaced by $A - B(\xi, r)$ becomes (14.1sm) when $r \to 0$.

If (14.1sm) is applied to $v = G_D(\eta, \cdot)$, the right side of (14.1sm) becomes $G_D\lambda(\eta)$, where λ is the distribution of $w_\xi(T_\xi^A)$ for $T_\xi^A < T_\xi^{\partial D}$, so that (14.1sm) becomes the equation defining the measure $\delta_D^A(\eta, \cdot)$. Hence (14.2) is true.

Classical Reductions Composed with Brownian Motion

Let D be a Greenian subset of $\mathbb{R}^N$, and let $\{w_\xi(\cdot), \mathscr{F}_\xi(\cdot)\}$ be a Brownian motion in D from ξ, with lifetime S_ξ. It is supposed that $\mathscr{F}_\xi(0)$ contains the null sets. Let v be a positive superharmonic function on D. Then

$$\lim_{s \uparrow S_\xi} v[w_\xi(s)]$$

exists almost surely, and we define $v[w_\xi(t)]$ for $S_\xi \le t < +\infty$ as this limit, so that (Section 7) $\{v[w_\xi(\cdot)], \mathscr{F}_\xi(\cdot)\}$ is an almost surely continuous supermartingale. Let A be a Borel subset of D, and define $\dot{A} = \{(t, \omega): w_\xi(t, \omega) \in A\}$. The set $\dot{A}$ is nearly predictable because $w_\xi(\cdot)$ is a nearly predictable process. In the classical context $\underset{+v}{R}^A$ is a positive superharmonic function on D, and we define

$$\underset{+v}{R}^A[w_\xi(t)] = \lim_{s \uparrow S_\xi} \underset{+v}{R}^A[w_\xi(s)]$$

for $S_\xi \le t < +\infty$. The supermartingale reduction $\underset{+v[w_\xi(\cdot)]}{R}^{\dot{A}}(\cdot)$ is an almost surely continuous supermartingale, and Theorem 14 combines with Theorem IV.20 to show that this supermartingale is indistinguishable from $\underset{+v}{R}^A[w_\xi(\cdot)]$.

Generalization. Let A and B be analytic subsets of D, and let $T_\xi^{A,B}$ be the hitting time by $w_\xi(\cdot)$ of B after hitting A but before hitting ∂D,

$$T_\xi^{A,B} = \inf\{t: T_\xi^A < t < T_\xi^{\partial D}, w_\xi(t) \in B\}.$$

Then using the notation $\llbracket \;\rrbracket$ for smoothed reductions,

$$\llbracket \llbracket v \rrbracket^B \rrbracket^A(\xi) = E\{v[w_\xi(T_\xi^{A,B})]; T_\xi^{A,B} < T_\xi^{\partial D}\}, \tag{14.4}$$

and in particular,

$$\llbracket \llbracket 1 \rrbracket^B \rrbracket^A(\xi) = P\{T_\xi^{A,B} < T_\xi^{\partial D}\}. \tag{14.5}$$

To prove (14.4), observe that if $z(\cdot)$ is the $w_\xi(\cdot)$ process killed at $T_\xi^{\partial D}$, then (strong Markov property) $z(T_\xi^A + \cdot)$ is a Brownian motion process killed at $T_\xi^{\partial D} - T_\xi^A$ with initial distribution that of $z(T_\xi^A)$, of total value ≤ 1. The hitting time of B by this process is $T_\xi^{A,B} - T_\xi^A$. Thus the right side of (14.4) is

$$E\{E\{v[w_\xi(T_\xi^{A,B})]; T_\xi^{A,B} < T_\xi^{\partial D} | w_\xi(T_\xi^A)\}\} = E\{\llbracket v \rrbracket^B[z(T_\xi^A)]\} = \llbracket \llbracket v \rrbracket^B \rrbracket^A(\xi). \tag{14.6}$$

Parabolic Context. Theorem 14 and the generalization translate directly into the parabolic context, together with their proofs. In particular, in the obvious notation $\dot{\delta}_{\dot{D}}^{\dot{A}}(\dot{\xi}, \dot{B}) = P\{\dot{w}_{\dot{\xi}}(\dot{T}_{\dot{\xi}}^{\dot{A}}) \in \dot{B}; \dot{T}_{\dot{\xi}}^{\dot{A}} < \dot{T}_{\dot{\xi}}^{\partial \dot{D}}\}$.

15. Probabilistic Description of the Fine Topology

Let ξ be a point of $\mathbb{R}^N$, let $w_\xi(\cdot)$ be a Brownian motion from ξ, and for any subset A of $\mathbb{R}^N$ denote by T_ξ^A the hitting time of A by $w_\xi(\cdot)$.

Theorem. *For ξ to be a fine limit point of A it is necessary that*

$$P\{T_\xi^B = 0\} = 1 \tag{15.1}$$

whenever B is an analytic superset of A, and it is sufficient that (15.1) *be true whenever B is an open superset of A.*

According to this theorem, if A is analytic, the point ξ is a fine limit point of A if and only if (15.1) is true when $B = A$; a Borel set A is a fine neighborhood of ξ if and only if almost every $w_\xi(\cdot)$ path lies in A for some parameter interval $[0, t[$ with $t = t(\omega) > 0$. The value t can be taken as the hitting time of $\mathbb{R}^N - A$.

In proving Theorem 15 we can assume that A is a subset of a Greenian set D. Since (Theorem 1.XI.3) ξ is a fine limit point of an analytic set A if and only if $\delta_D^A(\xi, \{\xi\}) = 1$ the theorem for A analytic follows from (14.2) with $B = \{\xi\}$. Since (Section 1.XI.1) ξ is a fine limit point of a set A if and only if ξ is a fine limit point of every open superset of A the theorem follows.

Application to Fine Limits and Cluster Values

Let u be a Borel measurable function from an open subset D of $\mathbb{R}^N$ into a complete separable metric space D', let ξ be a point of D, and let $w_\xi(\cdot)$ be a Brownian motion in $\mathbb{R}^N$ from ξ. An application of Theorem 15 to the analysis of limit values of u along Brownian paths in Section VII.6(d)–(f) yields the following results. If $D' = \overline{\mathbb{R}}$, then

$$\limsup_{t\to 0} u[w_\xi(t)] = \operatorname{f}\limsup_{\eta\to\xi} u(\eta) \quad \text{a.s.}$$

along with the corresponding relation for the inferior limit. For general D' a point η' of D' is a fine cluster value of u at ξ if and only if η' is a cluster value of u along almost every $w_\xi(\cdot)$ path back to ξ, and u has fine limit η' at ξ if and only if $\lim_{t\to 0} u[w_\xi(t)] = \eta'$ almost surely. Recall that a point η' is a cluster value of u along either almost no or almost every $w_\xi(\cdot)$ path back to ξ and that $P\{\lim_{t\to 0} u[w_\xi(t)] \text{ exists}\}$ is either 0 or 1, and in the latter case this limit is almost surely constant, that is, a single point η'.

Parabolic-Fine Topology

In the context of the parabolic-fine topology the counterpart of Theorem 15 with Brownian motion replaced by space-time Brownian motion is true, and the proof is an immediate translation of the proof of Theorem 15 into the parabolic context. The application of Theorem 15 to fine limits and cluster values goes over into the parabolic context with no change.

Application to the Support of a Swept Measure (Notation of Theorem 14)

According to Theorem 14, the distribution of $w_\xi(T_\xi^A)$ for $T_\xi^A < T_\xi^{\partial D}$ is $\delta_D^A(\xi, \cdot)$. Observe that on the one hand for $T_\xi^A < T_\xi^{\partial D}$ the process $w_\xi(T_\xi^A + \cdot)$ is a Brownian motion with initial distribution that of $w_\xi(T_\xi^A)$ and on the other hand for $T_\xi^A < T_\xi^{\partial D}$ and $w_\xi(T_\xi^A)$ not in A the set of strictly positive values of t with $w_\xi(T_\xi^A + t)$ in A must have limit point 0. Hence according to Theorem 15 and the strong Markov property of Brownian motion, the distribution of $w_\xi(T_\xi^A)$ for $T_\xi^A < T_\xi^{\partial D}$, that is, the measure $\delta_D^A(\xi, \cdot)$, is supported by the trace on D of the fine closure $A \cup A^f$ of A. This reasoning in the parabolic context yields the corresponding result for the support of a swept measure in that context. These results have already been derived nonprobabilistically in Sections 1.XI.14 and 1.XVIII.13. In the classical context but not in the parabolic context we can go further. In fact (classical context) almost no $w_\xi(\cdot)$ path hits the polar set $A - A^f$; so $\delta_D^A(\xi, \cdot)$ must be supported by A^f, as proved nonprobabilistically in Section 1.XI.14. A simple time reversal argument shows that $w_\xi(T_\xi^A)$ is almost never a fine interior point of A if $\xi \in D - A^f$; so for such a point ξ the distribution $\delta_D^A(\xi, \cdot)$ is supported by $D \cap \partial^f A^f$, as was proved nonprobabilistically in Section 1.XI.18.

The Hitting of a Parabolic-Semipolar Set by a Space-Time Brownian Motion

Let $\dot{w}(\cdot)$ be a space-time Brownian motion in $\dot{\mathbb{R}}^N$, and let $\dot{A}$ be a parabolic-semipolar subset of $\dot{\mathbb{R}}^N$. We now show that for almost every $\dot{w}(\cdot)$ path the set of parameter values t with $\dot{w}(t) \in \dot{A}$ is at most countable. In view of the structure of a parabolic-semipolar set it is sufficient to prove the assertion for $\dot{A}$ a Borel set parabolic-thin at every point of $\dot{\mathbb{R}}^N$. Moreover it is sufficient to consider only hits of $\dot{A}$ during the parameter interval $]0, c[$ for arbitrary $c > 0$. Let T_1' be the hitting time of $\dot{A}$ by $\dot{w}(\cdot)$. If $\dot{w}(\cdot)$ is a space-time Brownian motion from a point $\dot{\xi}$, then $P\{T_1' = 0\}$ is either 0 or 1 and must be 0 because $\dot{A}$ is parabolic-thin at $\dot{\xi}$. It follows that $P\{T_1' = 0\} = 0$ with no restriction on the distribution of $\dot{w}(0)$. Define $T_1 = T_1' \wedge c$. The process $\dot{w}(T_1 + \cdot)$ is a space-time Brownian motion; so if T_2' is the hitting time of $\dot{A}$ by this process, we conclude that $P\{T_2' = 0\} = 0$. Hence $\dot{w}(T_1 + T_2') \in \dot{A}$ almost everywhere where $T_1 + T_2' < +\infty$. Define $T_2 = (T_1 + T_2') \wedge c$. Continuing in this way we find an increasing sequence $T_\cdot$ of optional times such that almost surely $T_n \leq c$; $T_n < c$ implies that $\dot{w}(T_n) \in \dot{A}$ and $T_n < T_{n+1}$; for $t < T_\beta = \lim_{n\to\infty} T_n$ the inclusion $\dot{w}(t) \in \dot{A}$ is true if and only if $t = T_1$ or $T_2, \ldots$. Here β is the first transfinite ordinal. We can continue by transfinite induction, defining $T_{\beta+1} \le T_{\beta+2} \le \cdots$. Observe that $E\{T_\gamma\} = E\{T_{\gamma+1}\}$ if and only if $T_\gamma = c$ almost surely; so there is a countable ordinal γ such that $T_\gamma = c$ almost surely, and it follows that almost every $\dot{w}(\cdot)$ path meets $\dot{A}$ at most countably often in the parameter interval $]0, c[$, as was to be proved.

Since (Section 1.XVIII.12) a parabolic-semipolar set is also coparabolic semipolar, a parabolic-semipolar set is a countable union of Borel sets which are both parabolic thin and coparabolic thin at every point of $\dot{\mathbb{R}}^N$, so we could have supposed in the preceding proof that $\dot{A}$ has this property. Under this hypothesis on $\dot{A}$ it follows easily that $P\{\bigcup_{n=1}^{\infty} \{T_n = c\}\} = 1$ in the preceding proof; so transfinite induction could have been avoided.

The Iterated Logarithm Law

Let $w(\cdot)$ be a Brownian motion in $\mathbb{R}^N$ from the origin. Then

$$\limsup_{t \to 0} \frac{|w(t)|}{(2\sigma^2 t \log|\log t|)^{1/2}} = 1 \quad \text{a.s.} \tag{15.2}$$

In fact a slightly stronger assertion is true: when $c > 1$,

$$|w(t)|^2 < 2c\sigma^2 t \log|\log t| \tag{15.3}$$

almost surely for all sufficiently small strictly positive values of t, depending on the Brownian path, but when $c = 1$, there are almost surely arbitrarily small strictly positive values of t for which there is equality in (15.3). This result is a restatement in probability language of the significance of Section 1.XVIII.6, Examples (d) and (e). To see this, observe that on the one hand according to Example (d) when $c = 1$, the open set

$$\dot{D} = \{(\xi, s): |\xi|^2 < 2c\sigma^2 |s| \log|\log|s||, -1 < s < 0\}$$

has the origin as a parabolic-regular boundary point, and therefore (Theorem 1.XVIII.7) $\dot{D}$ is not a deleted parabolic-fine neighborhood of the origin. Hence almost every space-time Brownian path from the origin hits $\dot{\mathbb{R}}^N - \dot{D}$ at arbitrarily small strictly positive parameter values; so the limit superior in (15.2) is almost surely at least 1. On the other hand when $c > 1$, according to Example (e), the set $\dot{D}$ has the origin as a parabolic-irregular boundary point, and therefore (Theorem 1.XVIII.7) $\dot{D}$ is a deleted parabolic-fine neighborhood of the origin; so almost every space-time Brownian path from the origin lies in $\dot{D}$ for all sufficiently small strictly positive parameter values, and thus (15.3) is true almost surely for all sufficiently small strictly positive values of t, depending on the path. Hence the limit superior in (15.2) is almost surely at most c and so must be almost surely 1.

16. α-Excessive Functions for Brownian Motion and Their Composition with Brownian Motions

Let D be a connected Greenian subset of $\mathbb{R}^N$. Throughout this section α is a fixed positive number, and if u is a function from D into $\overline{\mathbb{R}}$, the notation $\dot{u}$ refers to the function on $\dot{D} = D \times \mathbb{R}$ defined by $\dot{u}(\eta, t) = e^{\alpha t}u(\eta)$. Define an operator Δ^α with domain the $\mathbb{C}^{(2)}$ class of functions on D by

$$\Delta^\alpha u = \frac{\sigma^2}{2}\Delta u - \alpha u. \tag{16.1}$$

The condition that a universally measurable function u from D into $\overline{\mathbb{R}}^+$ be α-excessive for Brownian motion in D is

$$e^{-\alpha t}b_D(t, \cdot, u) \leq u$$

with equality in the limit when $t \to 0$, and it follows that u is α-excessive if and only if $\dot{u}$ is excessive for space-time Brownian motion in $\dot{D}$. Hence (parabolic context version of Theorem 8) $\dot{u}$ is superparabolic if finite on a dense set, and we can derive the following assertions (α1) and (α2).

(α1) If u is α-excessive, then u is lower semicontinuous and either $u \equiv +\infty$ or $u < +\infty$ quasi everywhere on D.

The finiteness assertion follows from the fact (Section 1.XVIII.11) that a subset A of D is polar if the set $A \times \mathbb{R}$ is parabolic polar.

(α2) If u is a positive function from D into $\mathbb{R}$ of class $\mathbb{C}^{(2)}$, then u is α-excessive if and only if $\Delta^\alpha u \leq 0$, and if u is α-excessive, the Riesz measure associated with $\dot{u}$ (Theorem 1.XVII.6) has l_{N+1} density

$$-\dot{\Delta}\dot{u}(\eta, t) = -e^{\alpha t}\Delta^\alpha u(\eta)$$

at (η, t).

If u is an arbitrary not identically $+\infty$ α-excessive function, the decomposition $\dot{u} = \dot{u}_p + \dot{u}_{mqb} + \dot{u}_{ms}$ of the positive superparabolic function $\dot{u}$ into its potential, quasi-bounded parabolic, and singular parabolic components is invariant under a translation $(\eta, t) \mapsto (\eta, t + c)$. Hence, if $\dot{v}$ is any one of the above three components,

$$\dot{v}(\eta, t) = e^{-\alpha c}\dot{v}(\eta, t + c) = e^{\alpha t}\dot{v}(\eta, 0),$$

and we write $\dot{u}_p(\eta, 0) = u_p(\eta)$, $\dot{u}_{mqb}(\eta, 0) = u_{mqb}(\eta)$, $\dot{u}_{ms}(\eta, 0) = u_{ms}(\eta)$. Then $\Delta^\alpha u_{mqb} = 0$ and $\Delta^\alpha u_{ms} = 0$. Moreover the function $(\eta, t) \mapsto e^{\alpha t}u_p(\eta)$ is a superparabolic potential on $\dot{D}$, and the above translation argument shows that the Riesz measure associated with this potential is a product measure of the form $e^{\alpha t}\mu(d\eta)l_1(dt)$ with μ a measure on D. If u is of class $\mathbb{C}^{(2)}$, we have already evaluated μ: $\mu(d\eta) = -\Delta^\alpha u(\eta)l_N(d\eta)$.

Now let ξ be a point of D, let $\{w_\xi(\cdot), \mathscr{F}_\xi(\cdot)\}$ be a Brownian motion from ξ, and let S_ξ be the hitting time by $w_\xi(\cdot)$ of the Euclidean boundary of D. The definition of S_ξ in Section 7 shows that actually no specific boundary is involved here. In the following u is an arbitrary not identically $+\infty$ α-excessive function on D. According to the parabolic version of Theorem 7 applied to $\dot{u}$, almost every $u[w_\xi(\cdot)]$ sample function, for parameter values $< S_\xi$, is right continuous and is finite except when $u(\xi) = +\infty$ and the parameter value is 0. Apply Theorem 11 to find that "right continuous" here can be replaced by "continuous" and that for an arbitrary Brownian motion $w(\cdot)$ in $\mathbb{R}^N$ almost every $u[w(\cdot)]$ sample function is continuous when the $w(\cdot)$ path is in D and is finite valued except when $w(0) \in D$, $u[w(0)] = +\infty$, and the parameter value is 0. Theorem 7 applied to $\dot{u}$ yields the results (α3)–(α6).

(α3) $\lim_{t \uparrow S_\xi} e^{-\alpha t} u[w_\xi(t)]$ exists ($< +\infty$) almost surely.

Define $\mathscr{F}_\xi(+\infty) = \bigvee_{s \in \mathbb{R}^+} \mathscr{F}_\xi(s)$,

$$x'_\xi(t) = x_\xi(t) = e^{-\alpha t} u[w_\xi(t)] \quad \text{if } t < S_\xi;$$

$$x'_\xi(t) = 0, \qquad x_\xi(t) = \lim_{s \uparrow S_\xi} e^{-\alpha s} u[w_\xi(s)] \quad \text{if } S_\xi \le t \le +\infty.$$

(α4) The processes $\{x'_\xi(\cdot), \mathscr{F}_\xi(\cdot)\}$ and $\{x_\xi(\cdot), \mathscr{F}_\xi(\cdot)\}$ are almost surely continuous supermartingales except that the parameter value 0 is to be omitted if $u(\xi) = +\infty$ and that $x'_\xi(\cdot)$ will have a jump at $t = S_\xi$ when $S_\xi < +\infty$ unless $x'_\xi(S_\xi -) = 0$.

(α5) If $u = u_p$ or $u = u_{ms}$, the limit in (α3) vanishes almost surely. If $u = u_{mqb}$, the process $\{x_\xi(\cdot), \mathscr{F}_\xi(\cdot)\}$ is a uniformly integrable martingale.

(α6) $u(\xi) \ge u_{mqb}(\xi) = E\left\{\lim_{t \uparrow S_\xi} e^{-\alpha t} u[w_\xi(t)]\right\}$.

Example. If u is the sum of a convergent series of positive bounded $\mathbb{C}^{(2)}$ solutions on D of the equation $\Delta^\alpha u = 0$, we now show that $\dot{u} = \dot{u}_{mqb}$; so

$$u(\xi) = E\left\{\lim_{t \uparrow S_\xi} e^{-\alpha t} u[w_\xi(t)]\right\}. \tag{16.2}$$

It is sufficient to prove this assertion when u is bounded. (Observe that boundedness of u does not imply boundedness of $\dot{u}$.) Define

$$\dot{u}_n(\xi, s) = E\left\{n \wedge \lim_{t \uparrow S_\xi} [e^{\alpha(s-t)} u[w_\xi(t)]]\right\}.$$

The parabolic context version of Theorem 7 applied to the positive superparabolic function $\dot{u} \wedge n$ on $\dot{D}$ implies that $\dot{u}_n$ is the quasi-bounded parabolic component of $\dot{u} \wedge n$. Since $\dot{u}_{\cdot}$ is an increasing sequence of bounded parabolic functions with limit $\dot{u}$, the function $\dot{u}$ is quasi bounded.

It is left to the reader to show that if f is a positive Borel measurable function on the Euclidean boundary of D and if we define u on D by

$$u(\xi) = E\{e^{-\alpha S_\xi} f[w_\xi(S_\xi)]\}$$

with the convention that $e^{-\infty} = 0$, then u is either identically $+\infty$ or a $\mathbb{C}^{(2)}$ solution of the equation $\Delta^\alpha u = 0$ and to show that if f is bounded and is continuous at a regular Euclidean boundary point ζ of D and if S_ξ is almost surely finite, then u has limit $f(\zeta)$ at ζ. This result, which can be considerably strengthened, illustrates the possibility of analyzing the Dirichlet problem for solutions of the equation $\Delta^\alpha u = 0$ by means of the corresponding analysis of the Dirichlet problem for parabolic functions. The reader is also invited to generalize the whole discussion of this section by studying the linear space of differences $u_1 - u_2$ of functions on D for which $\dot{u}_1$ and $\dot{u}_2$ are positive and superparabolic on $\dot{D}$. The above restriction to positive functions will thereby be dropped.

α-Excessive Functions for Space-Time Brownian Motion

If $\dot{u}$ is an arbitrary α-excessive function for space-time Brownian motion in an open subset $\dot{D}$ of $\dot{\mathbb{R}}^N$, that is, if the function $\dot{u}_\alpha \colon \dot{\eta} = (\eta, t) \mapsto e^{\alpha t}\dot{u}(\dot{\eta})$ is excessive for space-time Brownian motion on $\dot{D}$, the properties of excessive functions in the parabolic context are immediately applicable to $\dot{u}_\alpha$.

17. Brownian Motion Transition Functions as Green Functions; The Corresponding Backward and Forward Parabolic Equations

Let $\dot{D}$ be an open subset of $\dot{\mathbb{R}}^N$. Throughout this section $\dot{\xi} = (\xi, s)$, $\dot{\eta} = (\eta, t)$, and these points are in $\dot{D}$. Recall that $\ell_{\dot{D}}$ as defined in Section VII.10 is a density determining the transition function of space-time Brownian motion. In particular (Section VII.9), when $\dot{D} = D \times \mathbb{R}$ with D a Greenian subset of $\mathbb{R}^N$, space-time Brownian motion on $\dot{D}$ is governed by a transition density ℓ_D with $\ell_D(t, \xi, \eta) = \ell_{\dot{D}}((\xi, s), (\eta, s - t))$. Furthermore for this choice of $\dot{D}$ the notation ℓ_D was also introduced in Section 1.XVII.18 but was defined there to satisfy (b) of the following theorem. According to (b), these two interpretations of ℓ_D are equal.

Theorem. *Let $\dot{D}$ be a nonempty open subset of $\dot{\mathbb{R}}^N$, and let $\ell_{\dot{D}}$ be the transition density of space-time Brownian motion in $\dot{D}$.*

(a) $\dot{\ell}_{\dot{D}}(\dot{\xi}, \dot{\eta}) = \dot{G}_{\dot{D}}(\dot{\xi}, \dot{\eta})$.

(b) *In particular, if* $\dot{D} = D \times \mathbb{R}$ *with* D *a Greenian subset of* $\mathbb{R}^N$ *and if* ℓ_D *is the transition density of Brownian motion in* D*, then* $\ell_D(t, \xi, \eta) = \dot{G}_{\dot{D}}((\xi, s), (\eta, s - t))$.

Proof. (a) Assertion (a) is true if $\dot{D} = \dot{\mathbb{R}}^N$ because then both sides of the equality in (a) reduce to $\ell(s - t, \xi, \eta)$. In the general case $\dot{\ell}_{\dot{D}}$ is given by VII(10.3), and according to the probabilistic evaluation of parabolic measure in Section 13, equation VII(10.3) means that for fixed $\dot{\eta}$, $\dot{\ell}_{\dot{D}}(\dot{\xi}, \dot{\eta})$ is equal to $\ell(s - t, \xi, \eta)$ less the parabolic Dirichlet solution on $\dot{D}$ at $\dot{\xi}$ for the boundary function $\ell(\cdot - t, \cdot, \eta)$ (defined as 0 at the point ∞ if $\dot{D}$ is unbounded). Since $\dot{G}_{\dot{D}}$ can also be expressed in terms of ℓ in this way (Section 1.XVIII.1), (a) is true.

Assertion (b) is a specialization of (a) requiring no separate proof. □

Observation (a). The transition density $\dot{\ell}_{\dot{D}}$ satisfies two differential equations: $\dot{\ell}_{\dot{D}}(\cdot, \dot{\eta})$ is parabolic on $\dot{D} - \{\dot{\eta}\}$, that is, $\dot{\Delta}_{\dot{\xi}} \dot{\ell}_{\dot{D}}(\dot{\xi}, \dot{\eta}) = 0$ there; $\dot{\ell}_{\dot{D}}(\dot{\xi}, \cdot)$ is coparabolic on $\dot{D} - \dot{\xi}$, that is, $\overset{*}{\Delta}_{\dot{\eta}} \dot{\ell}_{\dot{D}}(\dot{\xi}, \dot{\eta}) = 0$ there. These equations are called, respectively, the backward and forward equations of the space-time Brownian motion because they refer, respectively, to initial and later positions. If $\dot{D} = D \times \mathbb{R}$ with D a Greenian subset of $\mathbb{R}^N$ the transition density ℓ_D also satisfies two differential equations: $\ell_D(t, \xi, \eta)$ defines a parabolic function of (ξ, t) and one of (η, t) for $\xi \neq \eta$. This formulation obscures the difference between the backward and forward equations, however. The point is that $\ell_D(s - t, \xi, \eta)$ is parabolic in (ξ, s) (backward equation) and coparabolic in (η, t) (forward equation) for $(\xi, s) \neq (\eta, t)$.

Observation (b). It was proved in Section 1.XVII.18 that if $\dot{D} = D \times \mathbb{R}$ with D a Greenian subset of $\mathbb{R}^N$, then

$$\int_0^\infty \ell_D(t, \xi, \eta) l_1(dt) = a_N G_D(\xi, \eta) \tag{17.1}$$

for a_N as specified in that section. We shall now sketch how (17.1) can be derived from the probabilistic definition of ℓ_D in Section VII.9 without using the potential theory derivation. Observe first that for $N \geq 3$ equality (17.1) follows from the evaluation of ℓ_D in VII(9.5). In fact (17.1) is true if $D = \mathbb{R}^N$, in which case $\ell_D = \ell$, and on integration VII(9.5) becomes

$$\int_0^\infty \ell_D(t, \xi, \eta) l_1(dt) = a_N[G(\xi, \eta) - \mu_D(\xi, G(\cdot, \eta))] \tag{17.2}$$

in view of the fact that $w_\xi(S_\xi)$ has distribution $\mu_D(\xi, \cdot)$. The bracketed quantity was identified with $G_D(\xi, \eta)$ in Section 1.VIII.3. When $N = 2$, first let D be a half-plane D^+. We have evaluated G_{D+} in 1.XVII(18.3) and in VII(9.8) we have evaluated the Brownian transition density from the origin as initial point for a half-plane not containing the origin. This evaluation is easily adapted to give the expression 1.XVII(18.4) for the transition density of Brownian motion in a half-plane. Equation (17.1) can now be verified for $D = D^+$. When $D \subset D^+$, the evaluation

$$\ell_D(t, \xi, \eta) = \ell_{D^+}(t, \xi, \eta) - E\{\ell_{D^+}(t - S_\xi, w_\xi(S_\xi), \eta)\}, \tag{17.3}$$

derived as VII(9.5) was, is then integrated to prove that (17.1) is true for $D \subset D^+$, and therefore surely if D is an arbitrary nonempty open bounded plane set. Hence, applying this result to each member of an increasing sequence of bounded open subsets of an arbitrary Greenian set D, with union D, yields (17.1) in the general case when $N = 2$. The case $N = 1$ is treated similarly. Observe that the above argument when $N \geq 3$ is simpler than the argument given in Section 1.XVII.18 because parabolic measure is available in the present context.

Application to Energy

In Section 1.XIII.8 it was pointed out that the evaluation of G_D given in Theorem 17 can be used to prove that the energy of a charge is positive.

18. Excessive Measures for Brownian Motion

The symmetry in classical potential theory, expressed, for example, by the facts that the Laplace operator Δ is formally self-adjoint and that $\ell_D(t, \cdot, \cdot)$ and $G_D(\cdot, \cdot)$ are symmetric functions, suggests that the excessive measures for Brownian motion are in some sense the same as their duals, the excessive functions for Brownian motion. The following theorem shows that this identification of ℓ_D-excessive functions with ℓ_D-excessive measures is in fact correct in the most natural sense.

Theorem. *If D is a Greenian subset of $\mathbb{R}^N$, a (totally unrestricted) measure μ of Borel subsets of D is a ℓ_D-excessive measure if and only if there is a ℓ_D-excessive function u such that for every Borel subset A of D*

$$\mu(A) = \int_A u \, dl_N. \tag{18.1}$$

Observe that this theorem means that on a connected open component of D either the measure μ is the indefinite integral of a positive superharmonic function u, or $(u \equiv +\infty)$ $\mu(A)$ is either 0 or $+\infty$ according as A is or is not l_N null.

In the following proof of the theorem we assume to avoid trivialities that D is connected. Recall that by definition a measure μ of Borel subsets of D, not necessarily finite on compact sets, is called excessive if and only if

$$\int_D \mu(d\xi) \int_A \ell_D(t, \xi, \eta) l_N(d\eta) \leq \mu(A) \qquad (t > 0) \tag{18.2}$$

whenever A is a Borel subset of D, with equality in the limit when $t \to 0$. If μ is a ℓ_D-excessive measure, this limiting equality implies that $\mu(A) = 0$ whenever $l_N(A) = 0$. If μ is ℓ_D-excessive and if $\mu(A) = +\infty$ whenever $l_N(A) > 0$, then μ is given by (18.1) with u the identically $+\infty$ excessive function and conversely (18.1) with this excessive function yields an excessive measure. On the other hand, if μ is ℓ_D-excessive and if $\mu(A) < +\infty$ for some not l_N null Borel subset A of D, the finiteness of the right side of (18.2) along with the continuity and strict positivity of ℓ_D for $t > 0$ implies that μ is finite valued on compact sets. Thus in this case μ is absolutely continuous relative to l_N, given by (18.1) with u a Lebesgue measurable function finite l_N almost everywhere on D. Then for fixed $t > 0$,

$$\int_D u(\xi)\ell_D(t, \xi, \eta) l_N(d\xi) \leq u(\eta) \qquad (l_N \text{ a.e. } \eta), \tag{18.3}$$

which in view of the Chapman–Kolmogorov equation and the symmetry of $\ell_D(t, \cdot, \cdot)$ implies that

$$\int_D u(\xi)\ell_D(s + t, \xi, \eta) l_N(d\xi) \leq \int_D u(\xi)\ell_D(s, \xi, \eta) l_N(d\xi) \tag{18.4}$$

for all $s > 0$, $t > 0$, $\eta \in D$. Hence for fixed ξ the left side of (18.3) increases as t decreases. Define $u'(\eta)$ as the limit of this left side when $t \to 0$. Then $u' \leq u$ l_N almost everywhere on D, and there must be equality l_N almost everywhere because the left side of (18.2) has limit $\mu(A)$ when $t \to 0$. The function u is uniquely defined up to an l_N null set, and we now replace u by u'. This change does not affect the left side of (18.3); that is, we now have $u' = u$, and u is obviously excessive. Conversely, if u is a ℓ_D-excessive function and is not identically $+\infty$, that is, if u is a positive superharmonic function on D, then u is locally l_N integrable, and using again the symmetry of $\ell_D(t, \cdot, \cdot)$, we find that

$$\int_D u(\xi)\ell_D(t, \xi, \eta) l_N(d\xi) \leq u(\eta) \tag{18.5}$$

with equality in the (monotone) limit when $t \to 0$. Integrating $\int_A - l_N(d\eta)$ yields the fact that the measure μ given by (18.1) is ℓ_D-excessive.

Space-Time (Parabolic) Context. Let $\dot{D}$ be a nonempty open subset of $\dot{\mathbb{R}}^N$. We use the notation introduced in Section VII.2: if $\dot{A}$ is a subset of $\dot{\mathbb{R}}^N$, define $A_t = \{\eta: (\eta, t) \in \dot{A}\}$ and $\dot{A}_t = A_t \times \{t\}$. Furthermore let $\dot{\mathscr{D}}$ be the σ algebra of subsets $\dot{A}$ of $\dot{D}$ for which $\dot{A} \cap \dot{D}_t$ is a Borel set for all t. A measure $\dot{\mu}$ on the σ algebra $\dot{\mathscr{D}}$ determines a measure $\mu(\cdot, t)$ of Borel subsets of D_t by way of $\mu(A, t) = \dot{\mu}(A \times \{t\})$. Conversely, if for each t there is a measure $\mu(\cdot, t)$ of Borel subsets of D_t, a measure $\dot{\mu}$ on $\dot{\mathscr{D}}$ can be defined by setting

$$\dot{\mu}(\dot{A}) = \sum_{t \in \mathbb{R}} \mu(A_t, t) \tag{18.6}$$

with the obvious convention for uncountable sums. In the following "measure on $\dot{\mathscr{D}}$" refers to a measure defined in this way, but we make no hypotheses on the finiteness of $\dot{\mu}$ or of $\mu(\cdot, t)$ on any class of sets. A measure $\dot{\mu}$ on $\dot{\mathscr{D}}$ is $\ell_{\dot{D}}$-excessive if and only if whenever $\dot{A} \in \dot{D}$,

$$\int_{D_s} \mu(d\xi, s) \int_{A_t} \ell_{\dot{D}}((\xi, s), (\eta, t)) l_N(d\eta) \le \mu(A_t, t) \qquad (s > t), \tag{18.7}$$

and there is equality in the limit when $t \uparrow s$. A $\overset{*}{\ell}_{\dot{D}}$-excessive measure is a measure on $\dot{\mathscr{D}}$ satisfying the obvious dual condition. The counterpart of Theorem 18 in the parabolic context can now be stated: *a measure $\dot{\mu}$ on $\dot{\mathscr{D}}$ is a $\ell_{\dot{D}}$- $[\overset{*}{\ell}_{\dot{D}}$-$]$ excessive measure if and only if there is a $\overset{*}{\ell}_{\dot{D}}$- $[\ell_{\dot{D}}$-$]$ excessive function $\dot{u}$ such that for $\dot{A} \in \dot{\mathscr{D}}$*

$$\mu(A_t, t) = \int_{A_t} \dot{u}(\eta, t) l_N(d\eta). \tag{18.8}$$

In the one direction let $\dot{u}$ be a $\ell_{\dot{D}}$-excessive function, so that $0 \le \dot{u} \le +\infty$, $\dot{u}$ is $\dot{\mathscr{D}}$ measurable, and

$$\int_{D_t} \ell_{\dot{D}}((\xi, s), (\eta, t)) \dot{u}(\eta, t) l_N(d\eta) \le \dot{u}(\xi, s) \qquad (s > t) \tag{18.9}$$

with equality in the limit when $t \uparrow s$. Define a measure $\dot{\mu}$ on $\dot{\mathscr{D}}$ by (18.8) and (18.6). In view of the duality relation VII(10.6) between $\ell_{\dot{D}}$ and $\overset{*}{\ell}_{\dot{D}}$ the inequality (18.9) implies that for $\dot{A} \in \dot{\mathscr{D}}$,

$$\int_{D_t} \mu(d\eta, t) \int_{A_s} \overset{*}{\ell}_{\dot{D}}((\eta, t), (\xi, s)) l_N(d\xi) \le \mu(A_s, s) \tag{18.10}$$

with equality in the (monotone) limit when $t \uparrow s$; so $\dot{\mu}$ is $\overset{*}{\ell}_{\dot{D}}$-excessive. Conversely, if $\dot{\mu}$ is a $\overset{*}{\ell}_{\dot{D}}$-excessive measure, then following the argument for the classical case as given in the first part of this section it is not difficult to see, but more so than in the classical case, and we leave the details to the reader, that there is a $\ell_{\dot{D}}$-excessive function $\dot{u}$ satisfying (18.8).

The dual assertion for $\overset{*}{\ell}_{\dot{D}}$-excessive functions and $\ell_{\dot{D}}$-excessive measures is derived in the same way or reduced to the one just proved by a reflection of $\dot{\mathbb{R}}^N$ in the abscissa hyperplane.

19. Nearly Borel Sets for Brownian Motion

In a general context a set A in the (topological) state space of a stochastic transition function p is called *nearly Borel* if for every initial distribution μ on the state space and an almost surely right continuous Markov process $x_\mu(\cdot)$ with transition function p there are Borel subsets A'_μ and A''_μ of the state space for which $A'_\mu \subset A \subset A''_\mu$ and for which almost no $x_\mu(\cdot)$ path hits $A''_\mu - A'_\mu$ at any strictly positive time. The subtleties of this definition are needed in more general contexts but are quite unnecessary in the Brownian motion context according to the following theorem.

Theorem. *A set A is nearly Borel for Brownian motion in an open subset of $\mathbb{R}^N$ if and only if A differs from a Borel set by a polar set.*

We can suppose that we are dealing with Brownian motions in a connected Greenian set D. If A is nearly Borel and if μ is supported by the singleton $\{\xi\}$, the difference $A''_\mu - A'_\mu$ is a Borel set whose hitting time by a Brownian motion from ξ is almost surely $+\infty$ so that by Theorem 14 the smoothed reduction of the function 1 on the difference set is identically 0. Hence (Theorem 1.V.4) this difference set is polar; so A differs from A'_μ by a polar set. Conversely, if A differs from a Borel set by a polar set, it can be supposed that $A = A_0 \cup A_1$, where A_0 is a Borel set and A_1 is polar, and if so, the sets A'_μ and A''_μ can be chosen, respectively, as A_0 and the union of A_0 with any polar G_δ superset of A_1.

Nearly Borel Sets for Space-Time Brownian Motion. The parabolic context counterpart of Theorem 19 is a direct translation.

20. Brownian Motion into a Set from an Irregular Boundary Point

Let D be a Greenian subset of $\mathbb{R}^N$, and let ∂D be the Euclidean boundary. Let ζ be a finite irregular boundary point of D, let $w(\cdot)$ be a Brownian motion from ζ, and let S be the hitting time of ∂D by $w(\cdot)$. According to the fine topology criterion of irregularity of a boundary point (Theorem 1.XVIII.7) and the corresponding probabilistic criterion by way of Theorem 15, the random variable S is almost surely strictly positive. Let $w'(\cdot)$ be the process $w(\cdot)$ killed at time S. Then $\{w'(t), t > 0\}$ is a Markov process with state space D and transition density ℓ_D. Now let v be a positive superharmonic function on D. A trivial adaptation of Theorem 8 shows that the process $\{v[w'(t)], t > 0\}$, if defined as 0 for $t \geq S$, is almost surely continuous for $t < S$ and is a positive supermartingale if the expectations of the process random variables are finite. If we replace v by $v \wedge c$ if necessary to ensure finiteness of these expectations and then let c tend to $+\infty$, we find by the backward supermartingale convergence theorem that v has a finite or infinite limit along almost every $w(\cdot)$ path back to ζ, that is, v has a fine limit $\leq +\infty$ at ζ. We have thereby obtained a probabilistic proof of the limit result in Section 1.XI.21; a similar argument yields the corresponding parabolic context theorem in Section 1.XVIII.17.

For fixed η in D the function $\dot{\xi} = (\xi, s) \mapsto \ell_D(s, \xi, \eta)$ on $D \times \mathbb{R}$ is superparabolic and positive and is bounded on $\{\dot{\xi}: s > s_0\}$ for each strictly positive s_0; so for $t > 0$ and $\eta \in D$ the limit $\operatorname{pf\,lim}_{\dot{\xi} \to (\zeta, 0)} \ell_D(t + s, \xi, \eta)$ exists and is finite. Denote this limit by $\ell_D(t, \zeta, \eta)$. Equivalently, there is a deleted parabolic-fine neighborhood $\dot{A}$ of $(\zeta, 0)$ such that

$$\lim_{\dot{A} \ni \dot{\xi} \to (\zeta, 0)} \ell_D(t + s, \xi, \eta) = \ell_D(t, \zeta, \eta). \tag{20.1}$$

Here $\dot{A}$ depends on $\dot{\eta} = (\eta, t)$, but in view of the parabolic version of Lemma 1.XI.8 we can choose $\dot{A}$ so that (20.1) is true for $\dot{\eta}$ in a countable dense subset of $D \times]0, +\infty[$. The parabolic context Harnack theorem now implies that (20.1) is true for all $\dot{\eta}$ in this product set and that the convergence is locally uniform so the function $\dot{\eta} \mapsto \ell_D(t, \zeta, \eta)$ is parabolic on this product set. Finally we show that $\ell_D(t, \zeta, \cdot)$ is the distribution density of $w'(t)$. To see this, observe that if f is a continuous function on D with compact support,

$$\lim_{s \to 0} \int_D \ell_D(t - s, w'(s), \eta) f(\eta) l_N(d\eta) = \int_D \ell_D(t, \zeta, \eta) f(\eta) l_N(d\eta) \quad \text{a.s.} \tag{20.2}$$

in view of the fine limit result we have just obtained. The integral on the left is a version of $E\{f[w'(t)] | w'(s)\}$, and in view of the Markov property of Brownian motion this conditional expectation defines a martingale for $0 < s < t$, t fixed, and an application of the conditional expectation continuity theorem and the 0-1 law of Brownian motion shows that when $s \to 0$ sequentially the almost sure limit of this conditional expectation is $E\{f[w'(t)]\}$. This fact combined with (20.2) shows that $\ell_D(t, \zeta, \cdot)$ is the stated density. The corresponding result in the parabolic context is left to the reader.

Chapter X

Conditional Brownian Motion

1. Definition

Let D be an open subset of $\mathbb{R}^N$, and recall that a Brownian motion in D was defined in Section VII.9 as an almost surely continuous Markov process with state space D and transition density ℓ_D. The probability space on which the process is defined may or may not be rich enough to extend the process to be a Brownian motion in $\mathbb{R}^N$.

We now define a Brownian motion in D conditioned by an arbitrary strictly positive superharmonic function h on D. If D is not Greenian, for example, if $N = 2$ and $\mathbb{R}^2 - D$ is polar, the definition yields nothing new because then h is necessarily a constant function and conditional Brownian motion reduces to Brownian motion. We therefore suppose below that D is Greenian. Following Section VI.13, define $D^h = \{h < +\infty\}$, and define

$$\ell_D^h(t, \xi, \eta) = \begin{cases} \ell_D(t, \xi, \eta)\dfrac{h(\eta)}{h(\xi)} & \text{if } (\xi, \eta) \in D^h \times D \\ 0 & \text{if } (\xi, \eta) \in (D - D^h) \times D. \end{cases} \tag{1.1}$$

Observe that the fact that h is ℓ_D-excessive implies that $\ell_D(t, \xi, \cdot) = 0$ l_N almost everywhere on $D - D^h$ when $\xi \in D^h$. An *h-Brownian motion in D* is a Markov process with state space D, transition density ℓ_D^h, and initial distribution supported by D^h. After proving in Section 2 that there is always a continuous process satisfying these conditions, we shall add to the definition the condition that the process be almost surely continuous. It will be shown that then almost no sample path leaves D^h. Thus a 1-Brownian motion in D is what we have defined as a Brownian motion in D.

The Notation P^h and E^h

In previous chapters we have used the notation P and E generically: whenever a probability space was encountered, P and E referred to probabilities and expectations, respectively, on the space, and if several probability spaces were encountered simultaneously, this notation P and E was used for every space. In the first sections of the present chapter this notation could be confusing because we shall deal with h-Brownian motion and Brownian motion simultaneously. To avoid confusion, we shall use P and E for Brownian motion but P^h and E^h for h-Brownian motion. (If h is identically constant, h-Brownian motion reduces to Brownian motion, and so the superscript will be omitted.) After Section 8 of this chapter, however, and throughout later chapters the special notation P^h and E^h will be used only when necessary to avoid ambiguity. Ordinarily the notation for the processes involved, for example, $w^h(\cdot)$ for h-Brownian motion, will warn the reader that P and E on the $w^h(\cdot)$ space refer to h-Brownian motion probabilities and expectations.

Existence and Lifetime of h-Brownian Motion

According to Markov process theory, if $\xi \in D^h$, the transition density $\mathscr{E}_D^h$ determines an h-Brownian motion from ξ. The distribution of the process lifetime S_ξ^h is determined by

$$P^h\{S_\xi^h > t\} = \frac{\mathscr{E}_D(t,\xi,h)}{h(\xi)}, \qquad P^h\{S_\xi^h = +\infty\} = \lim_{t\to\infty} \frac{\mathscr{E}_D(t,\xi,h)}{h(\xi)}. \tag{1.2}$$

The function $h_0 = \lim_{t\to\infty} \mathscr{E}_D(t,\cdot,h)$ is an invariant excessive harmonic minorant of h [Section IX.8, Example (a)]. Thus if h is a potential, the function h_0 vanishes identically; so almost every h-Brownian path has a finite lifetime. If h is a strictly positive superharmonic function on D which is l_N integrable over D, for example, if D is bounded and h is a bounded function, then

$$P^h\{S_\xi^h = +\infty\} \le \frac{(2\pi\sigma^2)^{-N/2}}{h(\xi)} \lim_{t\to\infty} t^{-N/2} \int_D h(\eta) l_N(d\eta) = 0; \tag{1.3}$$

so under these hypotheses almost every h-Brownian path has a finite lifetime. If h is minimal harmonic on D, then $h_0 = ch$ for some constant c, and since h_0 is invariant excessive for $\mathscr{E}_D$,

$$h_0 = \mathscr{E}_D(t,\cdot,h_0) = c\mathscr{E}_D(t,\cdot,h) \to ch_0,$$

so that either $c = 1$, in which case almost every h-Brownian path has an infinite lifetime, or $c = 0$, in which case almost every Brownian path has a finite lifetime.

$\mathscr{E}_D^h$-Excessive Functions

To avoid trivialities, we assume in this paragraph that D is connected. We use (Sections IX.6 and IX.8) the identification of $\mathscr{E}_D$-excessive functions with the positive superharmonic functions together with the identically

$+\infty$ function. According to Section VI.13, if v is a positive superharmonic function on D, the function v/h (defined as 0 on $D - D^h$) is $\mathscr{E}_D^h$-excessive, and conversely, if u is a $\mathscr{E}_D^h$-excessive function, then $u = 0$ on $D - D^h$ and either $u \equiv +\infty$ on D^h or there is a positive superharmonic function v on D such that $u = v/h$ on D^h.

Supermartingales Defined by Composing $\mathscr{E}_D^h$-Excessive Functions with h-Brownian Motion

Let v and h be strictly positive superharmonic functions on D, and let $w_\xi^h(\cdot)$ with lifetime S_ξ^h be an h-Brownian motion from a point ξ of D^h. Define $u = v/h$ on D^h; the values of u on $D - D^h$ will be irrelevant. In view of the preceding paragraph the process $\{u[w_\xi^h(t)], t \in \overline{\mathbb{R}}^+\}$, under the convention that $u[w_\xi^h(t)] = 0$ for $t \geq S_\xi^h$, is a supermartingale if $u(\xi) < +\infty$, that is, if $v(\xi) < +\infty$, just as in the case $h \equiv 1$. Furthermore, as in the case $h \equiv 1$, the fact that (Theorem IX.6) $\mathscr{E}_D(t, \xi, v) < +\infty$ when $t > 0$, even when $u(\xi) = +\infty$, implies that then also $\mathscr{E}_D^h(t, \xi, u) < +\infty$; so if $u(\xi) = +\infty$, the process $\{u[w_\xi^h(t)], 0 < t \leq +\infty\}$ is a supermartingale. We shall see in Section 2 that these supermartingales are almost surely continuous aside from a possible jump discontinuity at the parameter value S_ξ^h, under the condition that $w_\xi^h(\cdot)$ is chosen (as we shall see is possible) to be almost surely continuous.

h-Brownian Motion When h Is Harmonic

In particular, if h is a strictly positive harmonic function on D, the h-Brownian motions play the same role for h-harmonic and h-superharmonic functions that Brownian motions play for harmonic and superharmonic functions. For example, when h is harmonic, a $\mathscr{E}_D^h$-excessive function is, on each connected open component of D, either identically $+\infty$ or h-superharmonic. The fact that almost every path of a Brownian motion $w_\xi(\cdot)$ from a point of D [define $w_\xi(+\infty) = \infty$ if $N > 2$] hits the (Euclidean) boundary implies that almost every path of a Brownian motion in D tends to a boundary point at the path lifetime, but the latter property does not hold for h-Brownian motion in D for general harmonic h. We shall show, however, in Section 4 that when h is harmonic almost every h-Brownian path from a point of D tends to ∂D, although the path cluster set on ∂D may not be a singleton.

Canonical Conditional Brownian Motion

Let D and h be as above. In Section VII.9 we defined a canonical Brownian motion $\mathring{w}(\cdot)$ in D, on a space $\mathring{\Omega}$. A canonical h-Brownian motion is defined similarly, but there is one essential difference: when $h \equiv 1$, almost every path of an h-Brownian motion in D tends to a point of the Euclidean boundary of D at the path lifetime, but this is not true for all choices of h. Hence we must change the definition of $\mathring{\Omega}$ in that we drop the condition in Section VII.9 that the function $\mathring{\omega}$ is to have a left limit in the Euclidean topology of $\mathbb{R}^N$ at the lifetime $\mathring{S}(\mathring{\omega})$. When canonical h-Brownian motion is discussed with no restriction on h, we shall mean the present definition even though the case $h \equiv 1$ is not excluded. In the next section we shall show that h-Brownian motions exist for every choice of h and are almost surely continuous up to their lifetimes. If $w(\cdot)$ is such an h-Brownian motion in D, on a space Ω, the existence of a canonical h-Brownian motion in D with the same initial distribution (supported by D^h) follows easily by mapping. In fact the map $\omega \mapsto w(\cdot, \omega) = \mathring{\omega}$ takes a point ω of Ω into a point $\mathring{\omega}$ of $\mathring{\Omega}$ (we assume that the processes have a common trap point adjoined to D), and the $w(\cdot)$ measure on Ω thereby induces a measure on $\mathring{\Omega}$ under which the coordinate function process $\mathring{w}(\cdot)$ is the desired canonical h-Brownian motion.

Parabolic Context

The translation of the discussion in this section into the parabolic context is left to the reader. Observe, however, that in the parabolic context an example in Section IX.12 shows that the composition of a superparabolic function with space-time Brownian motion, almost surely right continuous with left limits, may have a dense discontinuity set which is the same for every sample function.

2. h-Brownian Motion in Terms of Brownian Motion

In this section we derive relations between h-Brownian motion and Brownian motion in order to reduce h-Brownian motion properties to corresponding Brownian motion properties. Throughout the section D is a Greenian subset of $\mathbb{R}^N$, h is a strictly positive superharmonic function on D, and ξ is a point of D^h. The process $\{w(\cdot), \mathscr{F}(\cdot)\}$ is a Brownian motion in D from ξ with lifetime S defined on a filtered probability space $(\Omega, \mathscr{F}, \mathscr{F}(\cdot), P)$, and the process $\{w^h(\cdot), \mathscr{F}^h(\cdot)\}$ is an h-Brownian motion in D from ξ with lifetime S^h defined on a filtered probability space $(\Omega^h, \mathscr{F}^h, \mathscr{F}^h(\cdot), P^h)$. Observe that

$$\{w(t) \in D\} = \{S > t\}, \qquad \{w^h(t) \in D\} = \{S^h > t\}.$$

A trap state is supposed adjoined to D to make the transition functions stochastic, and h is defined as 0 at the trap state.

We wish to prove the following, in various contexts. Let T be an $\mathscr{F}(\cdot)$ optional time on Ω, and suppose that $\Lambda \in \mathscr{F}(T)$. Let T^h be an $\mathscr{F}^h(\cdot)$ optional time on Ω^h, and suppose that $\Lambda^h \in \mathscr{F}^h(T^h)$. We wish to show that if T and

Λ are defined in the same way in terms of $w(\cdot)$ sample functions as T^h and Λ^h are defined in terms of $w^h(\cdot)$ sample functions, then

$$P^h\{\Lambda^h \cap \{S^h > T^h\}\} = \int_{\Lambda\cap\{S>T\}} h[w(T)]\frac{dP}{h(\xi)}. \tag{2.1}$$

Although we shall only need very special cases of (2.1), its intuitive meaning clarifies the subject so we prove (2.1) under very general hypotheses.

Case (a) of (2.1): $T \equiv t$ for some strictly positive constant t. Define a measure Q^t on the measurable space $(\Omega, \mathscr{F}(t))$ by

$$Q^t(\Lambda) = \int_{\Lambda\cap\{S>t\}} h[w(t)]\frac{dP}{h(\xi)} \tag{2.2}$$

to obtain a measure space on which the process $\{w(s), s \leq t\}_{|\{S>t\}}$ is almost surely continuous. Suppose that $0 < t_1 < \cdots < t_n = t$ and that $A \in \mathscr{B}(D^n)$. Define

$$\mathrm{M}_0 = \{[w(t_1), \ldots, w(t_n)] \in A\}, \qquad \mathrm{M}_0^h = \{[w^h(t_1), \ldots, w^h(t_n)] \in A\}, \tag{2.3}$$

and observe that $\mathrm{M}_0 \subset \{S > t\}$ and $\mathrm{M}_0^h \subset \{S^h > t\}$. In view of the fact that $D - D^h$ is polar and therefore l_N null we can evaluate $Q^t(\mathrm{M}_0)$ in terms of ℓ_D^h:

$$Q^t(\mathrm{M}_0) = \int \cdots_A \int \ell_D^h(t, \xi, \xi_0) \cdots \ell_D^h(t_n - t_{n-1}, \xi_{n-1}, \xi_n) l_N(d\xi_1) \cdots l_N(d\xi_n). \tag{2.4}$$

Thus

$$Q^t(\mathrm{M}) = P^h\{\mathrm{M}^h\} \tag{2.5}$$

for $\mathrm{M} = \mathrm{M}_0$ and $\mathrm{M}^h = \mathrm{M}_0^h$; so (2.1) is true when $T \equiv t$, $T^h \equiv t$, $\Lambda = \mathrm{M}_0$, and $\Lambda^h = \mathrm{M}_0^h$. It follows that (2.1) is true when $T \equiv t$ and $T^h \equiv t$ if Λ $[\Lambda^h]$ is in $\mathscr{F}\{w(s), s \leq t\}_{|\{S>t\}}$ $[\mathscr{F}\{w^h(s), s \leq t\}_{|S^h>t\}}]$ and if Λ and Λ^h are defined in the same way in terms of process sample functions. More precisely, let $\mathring{\Omega}$ be the space of functions $\mathring{\omega}$ from $[0, t]$ into D, and define $\mathring{x}(s, \mathring{\omega}) = \mathring{\omega}(s)$. If $\mathring{\mathrm{M}} \in \mathscr{F}\{\mathring{x}(s), s \leq t\}$, define M as the inverse image of $\mathring{\mathrm{M}}$ under the map $\omega \mapsto w(\cdot, \omega)_{|[0,t]}$ from $\{S > t\}$ into $\mathring{\Omega}$, and define M^h as the inverse image of $\mathring{\mathrm{M}}$ under the map $\omega^h \mapsto w^h(\cdot, \omega^h)_{|[0,t]}$ from $\{S^h > t\}$ into $\mathring{\Omega}$. Then (2.1) is true with $T \equiv T^h \equiv t$, $\Lambda = \mathrm{M}$, and $\Lambda^h = \mathrm{M}^h$. In fact this assertion for the paired sets is true for $\mathring{\mathrm{M}}$ in the algebra of subsets of $\mathring{\Omega}$ of the form $\{[\mathring{w}(t_1), \ldots, \mathring{w}(t_n)] \in A\}$ with $t_{\bullet}$ and A as in (2.3) and therefore for M_0 in the σ algebra $\mathscr{F}\{\mathring{w}(s), s \leq t\}$ generated by this algebra. We leave it to the reader to verify the fact, which we shall not need, that for the class of pairs $(\mathbf{M}, \mathbf{M}^h)$ obtained in this way the class of sets $\mathbf{M}$ is the class of subsets of $\{S > t\}$ in the σ algebra $\mathscr{F}\{w(s), s \le t\}$ and the class of sets $\mathbf{M}^h$ is the class of subsets of $\{S^h > t\}$ in the σ algebra $\mathscr{F}\{w^h(s), s \le t\}$.

Almost Sure Continuity of h-Brownian Motion. Equation (2.5) states that the almost surely continuous process $\{w(s), s \le t\}_{|\{S>t\}}$ under Q^t has the same finite-dimensional distributions as the process $\{w^h(s), s \le t\}_{|\{S^h>t\}}$. It follows that the latter process has an almost surely continuous standard modification and that therefore every h-Brownian motion in D from a point of D^h, and more generally every h-Brownian motion in D with initial distribution supported by D^h, has an almost surely continuous standard modification up to the process lifetime. In more detail, what we have shown implies that when $t > 0$ the restriction to the rationals in $[0, t]$ of almost every $w^h(\cdot)_{|\{S^h>t\}}$ sample function is uniformly continuous and that for fixed s

$$P^h\left\{\lim_{r \to s} w^h(r) \ne w^h(s), s < S^h\right\} = 0 \qquad (r \text{ rational}).$$

Hence, if $w'(s)$ is defined on Ω^h for $s < S^h$ as the limit at s of $w^h(\cdot)$ along the rationals, $w'(\cdot)$ is the required almost surely continuous standard modification of $w^h(\cdot)$ and is itself an h-Brownian motion from ξ. We refine our preliminary definition of h-Brownian motion accordingly: from now on an h-Brownian motion in D is any almost surely continuous up to (but not including) the process lifetime process in D with transition density $\mathscr{p}_D^h$ and with initial distribution supported by D^h. In particular we assume from now on in this section that $w^h(\cdot)$ is almost surely continuous up to but not including the process lifetime.

Killed h-Brownian Motion

Let B be an F_σ subset of D, and let T $[T^h]$ be the hitting time of B by $w(\cdot)$ $[w^h(\cdot)]$. Then if $(\mathbf{M}, \mathbf{M}^h)$ is a pair linked as described above, the pair $(\mathbf{M} \cap \{T > t\}, \mathbf{M}^h \cap \{T^h > t\})$ is a linked pair in the same sense because of the simple expression for the probability of hitting an F_σ set (cf. Section VI.6). Hence Case (a) of (2.1) yields

$$P^h\{\mathbf{M}^h \cap \{S^h \wedge T^h > t\}\} = \int_{\mathbf{M} \cap \{S \wedge T > t\}} h[w(t)] \frac{dP}{h(\xi)}. \tag{2.6}$$

In particular, if D_0 is an open proper subset of D containing ξ, then (2.6) with $B = D - D_0$ shows that $w^h(\cdot)$ killed at T^h has the same finite-dimensional distributions as $h_{|D_0}$-Brownian motion from ξ, up to process lifetimes, that is $w^h(\cdot)$ killed at T^h and then provided with a trap state is an $h_{|D_0}$-Brownian motion in D_0 from ξ.

Observe that since the almost surely continuous process $\{w(s), s \le t\}_{|\{S>t\}}$

under Q^t and the almost surely continuous process $\{w^h(s), s \le t\}_{|\{S^h>t\}}$ have the same finite-dimensional distributions, it follows that these processes have the same probabilities of hitting an analytic subset B of D during a specified parameter interval (Section I.12), and more generally the same argument shows that the same assertion is true if the processes are restricted to sets $\mathbf{M}$ and $\mathbf{M}^h$, respectively, linked as described above. That is, if $T\,[T^h]$ is the hitting time of B by $w(\cdot)\,[w^h(\cdot)]$, then (2.6) remains true.

Representation of an Arbitrary h-Brownian Motion in terms of a Brownian Motion. For $t > 0$ define a measure Q^{th} on $\mathscr{F}^h(t)$ by

$$Q^{th}(\mathbf{M}^h) = \int_{\mathbf{M}^h \cap \{S^h > t\}} \frac{h(\xi)}{h[w^h(t)]} dP^h, \tag{2.7}$$

and note that $Q^{th}(\mathbf{M}^h) = 0$ if and only if $P^h\{\mathbf{M}^h \cap \{S^h > t\}\} = 0$. The process $\{w^h(s), s \le t\}_{|\{S^h>t\}}$ under Q^{th} has the same finite-dimensional distributions as the restricted Brownian motion process $\{w(s), s \le t\}_{|\{S>t\}}$. We can now complete a full circle in the discussion: the integral

$$\int_{\mathbf{M}^h \cap \{S^h > t\}} h[w^h(t)] \frac{dQ^{th}}{h(\xi)},$$

whose value is $P^h\{\mathbf{M}^h \cap \{S^h > t\}\}$ when $\mathbf{M}^h \in \mathscr{F}^h(t)$, expresses $w^h(\cdot)$ process probabilities to time t under the side condition $S^h > t$ in terms of a Brownian motion in D under the corresponding side condition. The important point is that this Brownian motion and the h-Brownian motion involve the same probability space and the same random variables; only the measures are different. In particular, if u is a function on D, then $u[w^h(\cdot)]$ for parameter values $\le t$ is not only a function of $w^h(\cdot)$ under P^h but also under Q^{th}, and we can thereby deduce properties of u on h-Brownian motion from the known properties of u on Brownian motion.

The following assertions are immediate consequences of the representation of h-Brownian motion in terms of Brownian motion that we have obtained.

(a) If A is a polar subset of D, almost no path of an h-Brownian motion in D hits A at a strictly positive parameter value. In particular, almost no such path hits $D - D^h$.

(b) A superharmonic function v on D composed with h-Brownian motion in D yields a process almost surely continuous up to the h-Brownian motion lifetime with finite-valued sample functions except possibly at the parameter value 0.

We have already pointed out in Section 1 that if v is a positive superharmonic function on D and if $u = v/h$ ($= 0$ where $h = +\infty$), then $u[w^h(\cdot)]$ is a supermartingale if defined as 0 for parameter values $\ge S^h$ and with the understanding that the parameter value 0 is to be omitted if $u(\xi) = +\infty$.

We now see from (a) and (b) that this supermartingale is almost surely continuous except for a possible discontinuity at S^h. For almost every sample function the supermartingale right limit exists at S^h trivially and is 0; the left limit at S^h must exist (finite) almost surely because almost surely right continuous supermartingales almost surely have finite left limits at all parameter values, including the parameter value $+\infty$ if the supermartingale is positive.

(c) h-Brownian motion filtrations and the strong Markov property. Theorems VI.8 and VI.9 are not directly applicable to an h-Brownian motion unless h is finite valued, in which case $D^h = D$, but the proofs of these theorems are applicable to the h-Brownian motion context. Hence, if $\mathscr{F}^h(t)$ is generated by the null sets and $\mathscr{F}\{w^h(s), s \leq t\}$, it follows that $\mathscr{F}^h(\cdot)$ is right continuous. Moreover, even without this special choice of filtration, the strong Markov property VI(9.1) holds for $w^h(\cdot)$. These assertions are of course true not only if $w^h(0) = \xi$ but for an arbitrary initial distribution supported by D^h.

(d) The zero-one law for h-Brownian motion. Let $\mathscr{F}_1$ be the σ algebra generated by $\bigcap_{t>0} \mathscr{F}\{w^h(s, s \leq t\}$ and the null sets. Then [assuming as above that $w^h(\cdot)$ is an h-Brownian motion in D from ξ in D^h] $\mathscr{F}_1$ is trivial; that is, it consists of the null sets and their complements. Two proofs of the zero-one law for Brownian motion were given in Section VII.6; the second one is applicable to h-Brownian motion for all h. The zero-one law implies that if B is an arbitrary analytic subset of D and if T is the hitting time of B by $w^h(\cdot)$, then the probability $P\{T=0\}$ must be either 0 or 1. The representation of h-Brownian motion in terms of Brownian motion shows that this probability does not depend on the choice of h, and it follows (Theorem IX.15) that this probability is 1 if and only if ξ is a fine limit point of A. Roughly, the initial character of h-Brownian motion from a point does not depend on the choice of h. Because of this fact, assertions (c)–(f) of Section VII.6 are valid for h-Brownian motion for all h.

h-Brownian Motion at Path Lifetimes [Notation D, ξ, $w(\cdot)$, $w^h(\cdot)$ as Above]

If $w_0(\cdot)$ is a Brownian motion in $\mathbb{R}^N$ from ξ and if S_0 is the hitting time of the Euclidean boundary ∂D by $w_0(\cdot)$, then almost every $w_0(\cdot)$ path killed at S_0 tends to $w_0(S_0)$ at the parameter value S_0. It follows that almost every $w(\cdot)$ path tends to a point of ∂D at the path lifetime, but it does not follow from the relation between Brownian motion and h-Brownian motion that almost every $w^h(\cdot)$ path tends to a point of ∂D at the path lifetime. In fact, when h is a potential, we shall show in Section 4 that almost every h-Brownian path in D tends to a point of D at the path lifetime, and we shall find the distribution of this limit point. When h is harmonic, it will be shown that almost every h-Brownian path in D tends to ∂D but not necessarily to a point of ∂D, at the path lifetime.

3. Contexts for (2.1)

As already explained in Section 2, the first problem in finding a proper context for (2.1) is to find appropriate linked pairs (T, T^h) of optional times and linked pairs (Λ, Λ^h) of sets. The point is that T^h should depend on $w^h(\cdot)$ in the same way T depends on $w(\cdot)$ and Λ^h should be a set depending in the same way on $w^h(\cdot)$ up to time T^h as Λ depends on $w(\cdot)$ up to time T. In Section 2 the situation was simplified by choosing T^h and T to be the same constant function. In this section T^h and T will be nontrivially optional.

Case (b) of (2.1): *$w(\cdot)$ and $w^h(\cdot)$ are canonical, $T = T^h$, $\Lambda = \Lambda^h$.* Let $\{\mathring{w}^h(\cdot), \mathring{\mathscr{F}}^h(\cdot)\}$ be a canonical h-Brownian motion in D from ξ in D^h, on the probability function space $(\mathring{\Omega}^h, \mathring{\mathscr{F}}^h, \mathring{P}^h)$. When $h \equiv 1$, we omit the superscript. Then $\mathring{\Omega} = \mathring{\Omega}^h$, $\mathring{w}(\cdot) = \mathring{w}^h(\cdot)$, and the process lifetimes $\mathring{S}$, $\mathring{S}^h$ are identical. Define

$$\mathring{\mathscr{F}}_0(t) = \mathscr{F}\{\mathring{w}(s), s \le t\} = \mathring{\mathscr{F}}^h(t), \qquad \mathring{\mathscr{F}}_0(+\infty) = \bigvee_{t>0} \mathring{\mathscr{F}}_0(t) = \mathring{\mathscr{F}}_0^h(+\infty).$$

The basic filtration in this case is $\mathring{\mathscr{F}}_0^+(\cdot)$. Let $\mathring{T}$ be an arbitrary $\mathring{\mathscr{F}}_0^+(\cdot)$ optional time and let Λ be an $\mathring{\mathscr{F}}_0^+(\mathring{T})$ set. We make the obvious choices of $\mathring{T}^h$ and $\mathring{\Lambda}^h$, namely $\mathring{T}^h = \mathring{T}$ and $\mathring{\Lambda}^h = \mathring{\Lambda}$, and prove (2.1) in this context. Observe that, with the optional time notation defined in Section II.2, Example (b), and using the fact that

$$\mathring{\Lambda}^h \cap \{\mathring{S}^h > j2^{-n} = [\mathring{T}^h]_n\} \in \mathring{\mathscr{F}}_0(j2^{-n}),$$

(2.6) yields

$$\begin{aligned}
\mathring{P}^h\{\mathring{\Lambda}^h \cap \{\mathring{S}^h > [\mathring{T}^h]_n\}\} &= \sum_{j=0}^{\infty} \mathring{P}^h\{\mathring{\Lambda}^h \cap \{\mathring{S}^h > j2^{-n} = [\mathring{T}^h]_n\}\} \\
&= \sum_{j=0}^{\infty} \int_{\mathring{\Lambda}^h \cap \{\mathring{S} > j2^{-n} = [\mathring{T}]_n\}} h[\mathring{w}(j2^{-n})] \frac{d\mathring{P}}{h(\xi)} \\
&= \int_{\mathring{\Lambda}^h \cap \{\mathring{S} > [\mathring{T}]_n\}} h[\mathring{w}([\mathring{T}]_n)] \frac{d\mathring{P}}{h(\xi)}.
\end{aligned} \tag{3.1}$$

When $n \to \infty$, the left side of (3.1) tends to $\mathring{P}^h\{\mathring{\Lambda}^h \cap \{\mathring{S}^h > \mathring{T}^h\}\}$. The right side is $E\{h[\mathring{w}([\mathring{T}]_n)]; \mathring{\Lambda}^h\}/h(\xi)$. The sequence $\ldots, h[\mathring{w}([\mathring{T}]_1)], h[\mathring{w}([\mathring{T}]_0)]$ ordered as written is a supermartingale, left closable by $h[\mathring{w}(\mathring{T})]$ and (Theorem III.17) uniformly integrable; so there is L^1 as well as almost everywhere convergence to the limit $h[\mathring{w}(\mathring{T})]$. Hence we can integrate to the limit $(n \to \infty)$ in (3.1) to obtain (2.1) in the present context.

Application of Case (b). If $\{w(\cdot), \mathscr{F}(\cdot)\}$ $[\{w^h(\cdot), \mathscr{F}^h(\cdot)\}]$ is a Brownian [h-Brownian] motion in D from ξ on the probability space $(\Omega, \mathscr{F}, P)$

$[(\Omega^h, \mathscr{F}^h, P^h)]$ with lifetime S $[S^h]$, the problem of finding large classes of pairs (T, T^h), (Λ, Λ^h) for which (2.1) is true is partially solved by a reduction to case (b). In fact, if we make the convention that all processes involved have a common trap, the maps $\omega \mapsto w(\cdot, \omega) = \mathring{\omega}$, $\omega^h \mapsto w^h(\cdot, \omega^h) = \mathring{\omega}$ take Ω and Ω^h into $\mathring{\Omega} = \mathring{\Omega}^h$, the space of a canonical Brownian motion in D from ξ. In the notation of case (b), an optional time $\mathring{T}$ on $\mathring{\Omega}$ has as inverse images under these maps an optional time T on Ω and an optional time T^h on Ω^h if we suppose, as we can, that $\mathscr{F}(\cdot)$ and $\mathscr{F}^h(\cdot)$ are right continuous. A set $\mathring{\Lambda}$ in $\mathring{\mathscr{F}}_0^+(\mathring{T})$ has as inverse images a set Λ in $\mathscr{F}(T)$ and a set Λ^h in $\mathscr{F}^h(T^h)$ and (2.1) is true as written if T, Λ, T^h, and Λ^h are obtained in this way because (2.1) is true in the context of case (b). For example, if $\mathring{T}$ is the hitting time by $\mathring{w}(\cdot)$ of a closed subset of D, then T $[T^h]$ is equal almost surely to the hitting time of this set by $w(\cdot)$ $[w^h(\cdot)]$.

Generalization of Case (b) for (2.1)

(Not used below.) It is natural to try to extend Case (b) and thereby the application just noted to allow for the $\mathring{P}$ and $\mathring{P}^h$ null sets. This generalization is needed, for example, if T is to be the hitting time by $\mathring{w}(\cdot)$ of an arbitrary analytic or even merely Borel subset of D. More specifically define $\mathring{\mathscr{F}}(t)$ $[\mathring{\mathscr{F}}^h(t)]$ as the σ algebra generated by $\mathring{\mathscr{F}}_0(t)$ and the $\mathring{P}$ $[\mathring{P}^h]$ null sets. According to Section 2, the filtrations $\mathring{\mathscr{F}}(\cdot)$ and $\mathring{\mathscr{F}}^h(\cdot)$ are right continuous. If $\mathring{T}$ is an $\mathring{\mathscr{F}}(\cdot)$ optional time and if $\mathring{\Lambda} \in \mathring{\mathscr{F}}(\mathring{T})$, there are [from Section II.2(j)] an $\mathring{\mathscr{F}}_0(\cdot)$ optional time $\mathring{T}_0$ and an $\mathring{\mathscr{F}}_0(\mathring{T}_0)$ optional time $\mathring{\Lambda}_0$ such that, up to a $\mathring{P}$ null set, $\mathring{T}_0 = \mathring{T}$ and $\mathring{\Lambda}_0 = \mathring{\Lambda}$. Now according to case (b) of (2.1), an $\mathring{\mathscr{F}}_0^+(\mathring{T}_0)$ subset of $\{\mathring{S} > \mathring{T}_0\}$ is $\mathring{P}$ null if and only if it is $\mathring{P}^h$ null. Hence, up to a $\mathring{P}^h$ null set, $\mathring{T} = \mathring{T}$ and $\mathring{\Lambda}_0 = \mathring{\Lambda}$. Thus if we define $T = \mathring{T}$, $T^h = \mathring{T}_0$, $\Lambda = \mathring{\Lambda}$, $\Lambda^h = \mathring{\Lambda}_0$, equation (2.1) is true in the canonical process context, generalizing case (b). We could equally well have started with an $\mathring{\mathscr{F}}^h(\cdot)$ optional time $\mathring{T}^h$ and an $\mathring{\mathscr{F}}^h(\mathring{T}^h)$ set $\mathring{\Lambda}^h$.

4. Asymptotic Character of h-Brownian Paths at Their Lifetimes

Theorem. *Let D be a Greenian subset of $\mathbb{R}^N$, let h be a strictly positive superharmonic function on D, and let ξ be a point of D^h.*

(a) *If h is harmonic, almost every h-Brownian path from ξ tends to ∂D at the path lifetime.*

(b) *If $h = G_D\mu$ is a superharmonic potential, almost every h-Brownian path from ξ has a finite lifetime and tends to a point ζ of D at its lifetime. The distribution $\nu(\xi, \cdot)$ of ζ is given by $\nu(\xi, d\zeta) = G_D(\xi, \zeta)\mu(d\zeta)/h(\xi)$. If A is polar, $\nu(\xi, A \cap D^h) = 0$.*

Observation (1). Part (a) of the theorem does not state that when h is harmonic, almost every h-Brownian path from ξ tends to a point of ∂D, and in fact it will be proved in Section 3.II.2 that if ∂D is obtained by a metric compactification of D, then almost every h-Brownian path from ξ tends to a point of ∂D if and only if ∂D is h-resolutive, and in that case the distribution of the path limit point on ∂D is $\mu_D^h(\xi, \cdot)$. We can already verify this result in a special case, when $h \equiv 1$ and ∂D is the Euclidean boundary, resolutive according to Theorem 1.VIII.4. In fact, if a Brownian motion in $\mathbb{R}^N$ from ξ in D is killed at the hitting time of the Euclidean boundary ∂D to obtain a Brownian motion in D, almost every path of the Brownian motion in D obtained (without loss of generality for the present purpose) in this way tends to a point of ∂D at the path lifetime, namely to the point at which the original Brownian path first hits ∂D. The distribution of this hitting point on ∂D is $\mu_D(\xi, \cdot)$ according to Theorem IX.13.

Observation (2). We shall prove a stronger result than Theorem 4(b): h-Brownian paths from ξ can be obtained in this case by first choosing the asymptotic endpoint ζ for a path according to the probability distribution $\nu(\xi, \cdot)$ and then choosing a $G_D(\cdot, \zeta)$-Brownian path from ξ (necessarily tending to ζ at the path lifetime). It follows, as stated in (b), that if A is polar, $\nu(\xi, A \cap D^h) = 0$ because (Theorem 1.V.11) $h = G_D\mu = +\infty$ μ almost everywhere on A; so $\mu(A \cap D^h) = 0$.

Proof of (a). Let $w^h(\cdot)$ be an h-Brownian motion in D from ξ, with lifetime S^h, let D_0 be an open relatively compact subset of D containing ξ, and let T^h be the hitting time of ∂D_0 by $w^h(\cdot)$. The superscript h is omitted throughout when $h \equiv 1$. It is trivial that $P\{S > T\} = 1$. According to (2.1) under case (b) as applied in Section 3, if h is harmonic,

$$P^h\{S^h > T^h\} = \int_{\{S>T\}} h[w(T)] \frac{dP}{h(\xi)} = \frac{E\{h[w(T)]\}}{h(\xi)} \\ = \frac{\mu_{D_0}(\xi, h)}{h(\xi)} = 1. \tag{4.1}$$

Thus almost every $w^h(\cdot)$ path hits ∂D_0. To prove (a), we prove that if C is a compact subset of D and if now T^h is the first hitting time by $w^h(\cdot)$ of C after hitting ∂D_0, then $P^h\{S^h > T^h\}$ is arbitrarily small if D_0 is sufficiently large. With this definition of T^h (2.1) yields

$$P^h\{S^h > T^h\} \leq \frac{\left(\sup_C h\right) P\{S > T\}}{h(\xi)}.$$

Now the probability on the right is the probability that a Brownian path in D from ξ hits C after hitting ∂D_0, and this probability is arbitrarily small for sufficiently large D_0 since almost every Brownian path in D from ξ tends to the boundary at the path lifetime. Hence (a) is true. □

Proof of (b). We have already seen in Section 1 that when h is a potential almost every h-Brownian path has a finite lifetime. If D_0 is an open relatively compact subset of D containing ξ and if T^h is the hitting time of ∂D_0 by $w^h(\cdot)$, then (2.1) yields

$$P^h\{S^h > T^h\} = \frac{\mu_{D_0}(\xi, h)}{h(\xi)}$$

as in the proof of (a). Since h is now a potential, the right side of this equality can be made arbitrarily small by choosing D_0 sufficiently large (Section 1.VIII.11); so almost every h-Brownian path from ξ has closure a compact subset of D. Now consider a particular case, $h = G_D(\cdot, \zeta)$ for some point ζ of D other than ξ. We show that in this case almost every $w^h(\cdot)$ path tends to ζ at the path lifetime. To see this, define $D' = D - \{\zeta\}$, and denote by h' the restriction of h to D'. The function h' is harmonic on D', and h'-Brownian motion in D' from ξ can be identified with h-Brownian motion in D from ξ; so it follows from (a) that almost every $w^h(\cdot)$ path tends to $\partial D' = \{\zeta\} \cup \partial D$ at the path lifetime. Since we have just proved that almost every $w^h(\cdot)$ path has closure a compact subset of D, it follows that almost every h-Brownian path from ξ tends to ζ at the path lifetime. More generally, if $h = G_D\mu$,

$$\ell_D^h(t, \xi, \eta) = \int_D \left[\ell_D(t, \xi, \eta)\frac{G_D(\eta, \zeta)}{G_D(\xi, \zeta)}\right] \nu(\xi, d\zeta), \tag{4.2}$$

and if $0 < t_1 < \cdots < t_n$, the joint density of $w^h(t_1), \ldots, w^h(t_n)$ is

$$\int_D \ell_D(t_1, \xi, \xi_1)\ell_D(t_2 - t_1, \xi_1, \xi_2) \cdots \ell_D(t_n - t_{n-1}, \xi_{n-1}, \xi_n)\frac{G_D(\xi_n, \zeta)}{G_D(\xi, \zeta)}\nu(\xi, d\zeta). \tag{4.3}$$

This expression shows that an h-Brownian path from ξ can be obtained by first choosing the asymptotic path endpoint ζ according to the probability distribution $\nu(\xi, d\zeta)$ and then choosing a $G_D(\cdot, \zeta)$-Brownian path from ξ, as stated in Observation (2) above. The formal statement of this construction in the language of conditional probabilities is left to the reader. Part (b) of the theorem follows from this construction. □

h-Brownian Motion for General h

Let h be a strictly positive superharmonic function on D, so that (Riesz decomposition) h is the sum of a positive harmonic function h_1 and a potential

h_2. The evaluation

$$\ell_D^h(t,\xi,\eta) = \frac{h_1(\xi)\ell_D^{h_1}(t,\xi,\eta) + h_2(\xi)\ell_D^{h_2}(t,\xi,\eta)}{h(\xi)} \tag{4.4}$$

of ℓ_D^h in terms of $\ell_D^{h_1}$ and $\ell_D^{h_2}$ can be interpreted as follows: to obtain an h-Brownian path from ξ [when $h(\xi) < +\infty$], choose an h_i-Brownian path from ξ with probability $h_i(\xi)/h(\xi)$. Thus the probability that an h-Brownian path in D from ξ tends to a point of D at the path lifetime is $h_2(\xi)/h(\xi)$, and the distribution of the asymptotic endpoint is determined as described in (b), whereas the probability that an h-Brownian path in D from ξ tends to the boundary of D at the path lifetime is $h_1(\xi)/h(\xi)$. In the latter case the path may or may not almost surely tend to a boundary point, depending on h and the choice of boundary, but the probability that an h-Brownian path in D from ξ has some asymptotic property at the boundary is the product of $h_1(\xi)/h(\xi)$ times the probability that an h_1-Brownian path in D from ξ has this property.

5. h-Brownian Motion from an Infinity of h

If D is a Greenian subset of $\mathbb{R}^N$, if h is a strictly positive superharmonic function on D, and if $h(\xi_0) < +\infty$, an h-Brownian motion in D from ξ_0 is a process $w_{\xi_0}^h(\cdot)$ with the following properties (all densities are relative to l_N).

h-BM(1) $w_{\xi_0}^h(\cdot)$ is an almost surely continuous process from ξ_0 with state space D, except that if the transition density is not stochastic and if a trap ∂ is adjoined to D to obtain a stochastic transition probability, there is a discontinuity at the time of transition to ∂.

h-BM(2) $w_{\xi_0}^h(t)$ has distribution density function $\ell_D^h(t,\xi_0,\cdot)$ when $t > 0$.

h-BM(3) $w_{\xi_0}^h(\cdot)$ is Markovian with transition density function ℓ_D^h on D; equivalently, under h-BM(2), the reverse transition density (transition from η at time t to ξ at time s, with $s < t$) is

$$\frac{\ell_D(s,\xi_0,\xi)\ell_D(t-s,\xi,\eta)}{\ell_D(t,\xi_0,\eta)}. \tag{5.1}$$

Suppose now that all probabilities are multipled by $h(\xi_0)$. This change does not affect the conditional densities in h-BM(3) but replaces h-BM(2) by

h-BM(2′) $w_{\xi_0}^h(t)$ has distribution density function $\ell_D(t,\xi_0,\cdot)h$ when $t > 0$. The fact that the measure space on which the process is defined now has measure $h(\xi_0)$ which may not be 1 causes no difficulty. From now on an *h-Brownian motion from* ξ_0 will mean as before a process satisfying h-BM(1)–(3) when $h(\xi_0) < +\infty$, but satisfying h-BM(1), (2′), (3) when $h(\xi_0) = +\infty$. In the latter case the measure space on which the process is defined necessarily has measure $+\infty$, and in fact

$$P^h\{w^h_{\xi_0}(0) = \xi_0\} = P^h\{w^h_{\xi_0}(t) = \partial\} = +\infty$$

when $t > 0$. We have not yet shown that an h-Brownian motion from an infinity of h exists, and we now proceed to do so. We shall construct the desired process on the space Ω of all functions from $\mathbb{R}^+$ into $D \cup \{\partial\}$ with value ξ_0 at the parameter value 0. Observe first that if a process $w^h_{\xi_0}(\cdot)$ exists, satisfying either h-**BM**(1)–(3) or h-**BM**(1), (2′), (3), then if $\xi' \in D$ and if $s' > 0$, and whether $h(\xi_0)$ is finite or not, the $w^h_{\xi_0}(\cdot)$ process conditioned by $w^h_{\xi_0}(s') = \xi'$ has the following properties: the process on the parameter set $[0, s']$ is independent of the process on the parameter set $[s', +\infty[$; the process on the first parameter set is Markovian, with reverse transition density (5.1); the process on the second parameter set is an h-Brownian motion from ξ'. Now a process $x(\cdot)$ can be constructed on the function space Ω with the finite-dimensional distributions of the so-conditioned process. In fact the standard procedure of Kolmogorov can be used, in which a measure $P^h_{\xi' s'}$ is assigned to Ω making the family of coordinate functions $\{x(t), t \in \mathbb{R}^+\}$ have the specified finite-dimensional distributions for strictly positive parameter values, with $x(0) = \xi_0$. The class of measurable sets is the smallest class making every coordinate function measurable. The probability assigned to Ω itself in this way is 1. If now ξ' is given the distribution on D with density $\ell_D(s', \xi_0, \cdot)h$, the probability measure $P^h_{\xi' s'}$ on Ω becomes a measure

$$P^h_{s'} = \int_D \ell_D(s', \xi_0, \xi')h(\xi')P^h_{\xi' s'} l_N(d\xi')$$

with $P^h_{s'}(\Omega) = \ell_D(s', \xi_0, h)$; the latter value is finite according to Theorem IX.6. The sequence $\{P^h_{1/n}, n \geq 1\}$ of measures on Ω is an increasing sequence with the property that if S is a measurable subset of Ω and if $S' = S \cap \{x(1/n) \in D\}$, then $P^h_{1/n}\{S'\} = P^h_{1/m}\{S'\}$ when $n > m$. The limit of this increasing sequence of measures is a measure P^h on Ω for which $P^h\{\Omega\} = h(\xi_0) \leq +\infty$ and for which the process $x(\cdot)$ is Markovian and satisfies h-**BM**(2′), (3). The measure $P^h_{s'}$ is the restriction of P^h to the class of measurable subsets of $\{x(s') \in D\}$. The measure P^h can be expressed in terms of Brownian motion following Section 2 without the measure normalization used there. That is, if $\{w_{\xi_0}(\cdot), \mathscr{F}(\cdot)\}$ is a Brownian motion in D from ξ_0 with lifetime S, if $s' > 0$, and if $\Lambda \in \mathscr{F}(s')$, then the measure

$$\Lambda \mapsto \int_{\Lambda \cap \{S > s'\}} h[w_{\xi_0}(s')]\, dP$$

assigns a distribution to $\{w_{\xi_0}(s), s \leq s'\}$ which has the same finite-dimensional distributions as $\{x(t), t \leq s'\}$ on the set $\{x(s') \in D\}$ under P^h. It follows, as in the corresponding discussion in Section 2, that $x(\cdot)$ has a stochastic modification $w_{\xi_0}^h(\cdot)$ which is continuous except at the time of transition to ∂; $w_{\xi_0}^h(\cdot)$ satisfies h-BM(1), (2′), (3) as desired.

EXAMPLE. Let $h = G_D(\xi_0, \cdot)$. Then an h-Brownian motion in D from ξ_0 is a process whose paths have finite lifetimes and tend back to ξ_0 at these lifetimes.

6. Brownian Motion under Time Reversal

Let D be a Greenian subset of $\mathbb{R}^N$, and let h be a strictly positive harmonic function on D. Let $w_\xi^h(\cdot)$ be an h-Brownian motion in D from ξ, with lifetime S_ξ^h. It is tempting to conjecture that this process reversed in time, so that the paths tend to ξ at their lifetimes, is a $G_D(\xi, \cdot)$-Brownian motion. The following discussion shows that this conjecture is correct if suitably interpreted.

Denote by $*$ the Alexandrov point in the one-point compactification of D, and define

$$w_\xi^h(t) = \begin{cases} \xi & \text{if } t < 0 \\ * & \text{if } t \geq S_\xi^h. \end{cases}$$

The extended process is a Markov process on the parameter interval $\mathbb{R}$, with nonstationary transition function, and the process reversed in time is therefore also Markovian. This fact is the motivation for the following discussion. It will be useful to reverse the time from a random origin. Let z be a positive random variable, independent of the process $w_\xi^h(\cdot)$, with the distribution l_1. More precisely replace the space Ω on which the given random variables are defined by the product space $\Omega \times \mathbb{R}$, let z be the second coordinate function of this space, and define the measure Q on this product space as the completed product measure of the given Ω measure P^h and l_1. The random variable $w_\xi^h(t)$ becomes a function, also denoted by $w_\xi^h(t)$, on the product space. Define $x(t)$ for $t \in \mathbb{R}$ as $w_\xi^h(z - t)$, so that $x(t)$ is a function defined on a space of infinite measure. An elementary calculation justified by Fubini's theorem shows that the restriction to D of the distribution of $x(s)$ is absolutely continuous relative to l_N with

$$\text{Density at } \zeta \text{ in } D \text{ of the } x(s) \text{ distribution} = \frac{G_D(\xi, \zeta)h(\zeta)}{h(\xi)}. \tag{6.1}$$

Furthermore for $t > 0$ the conditional distribution of $x(s + t)$, given $x(s)$ in $D - \{\xi\}$, can be chosen to be absolutely continuous on D for each value of $x(s)$, with

$$(\text{Density at } \zeta \text{ in } D \text{ of the } x(s+t) \text{ distribution} | x(s))_{|x(s)=\eta} = \frac{\mathscr{p}_D(t, \eta, \zeta) G_D(\xi, \zeta)}{G_D(\xi, \eta)}. \tag{6.2}$$

More generally the conditional density in (6.2) is unaltered if the condition $x(s) = \eta$ is replaced by the more restrictive condition

$$x(s_1) = \eta_1, \qquad \ldots, \qquad x(s_n) = \eta_n, \qquad x(s) = \eta$$

for $s_1 < \cdots < s_n < s$ and $\eta_j \in D$. Thus for any choice of initial parameter value t_0 the restriction of the process $x(t_0 + \cdot)$ to the parameter set $\mathbb{R}^+$ and to the $\Omega \times \mathbb{R}$ set $\{x(t_0) \in D\} = \{0 \le z - t_0 < S_\xi^h\}$ is a Markov process with initial distribution specified by the density (6.1) (which involves h) and transition density (6.2) (which does not involve h) as long as the $x(t_0 + \cdot)$ paths lie in D. More precisely, if the process is killed when the paths reach ξ, so that the process lifetime is $z - t_0$, the so-restricted and killed process $x(t_0 + \cdot)$ is a $G_D(\xi, \cdot)$-Brownian motion, with initial distribution of density (6.1). Furthermore an easy adaptation of Theorem VI.9 (strong Markov property) to the present context shows that if A is an analytic relatively compact subset of D and if T' is the hitting time of A by $x(\cdot)$, then the process $x(T' + \cdot)$, restricted to the set $\{T' < +\infty\}$ and killed when the paths reach ξ, is a $G_D(\xi, \cdot)$-Brownian motion process. More precisely, if we define a filtration $\mathscr{F}'(\cdot)$ of $\Omega \times \mathbb{R}$ by setting $\mathscr{F}'(t)$ for t in $\mathbb{R}$ to be the σ algebra generated by the $P \times l_1$ null sets and $\mathscr{F}\{x(s), -\infty < s \le t\}$, then the process $\{x(T' + t), \mathscr{F}'(T' + t), t \in \mathbb{R}^+\}$, restricted and killed as just described, is a $G_D(\xi, \cdot)$-Brownian motion process. Going back to the original space Ω, define L_ξ^h as the last hitting time of A by $w_\xi^h(\cdot)$, and let $\mathscr{F}(\cdot)$ be the filtration of Ω obtained by defining $\mathscr{F}(t)$ to be the σ algebra of subsets of Ω generated by the P null sets and $\mathscr{F}\{w_\xi^h(L_\xi^h - s), s \le t\}$. Then what we have proved yields in this context that the process $\{w_\xi^h(L_\xi^h - t), \mathscr{F}(t), t \in \mathbb{R}^+\}$ killed at time L_ξ^h is a $G_D(\xi, \cdot)$-Brownian motion, with initial distribution the distribution of $w_\xi^h(L_\xi^h)$. The distribution of $w_\xi^h(L_\xi^h)$ will be evaluated in Section 10.

EXAMPLE (a). Let $D_{\bullet}$ be an increasing sequence of open relatively compact subsets of D with union D, let ξ be a point of D_0, and let $L_{\xi n}^h$ be the last hitting time of D_n by $w_\xi^h(\cdot)$. Then we have proved that the process $w_\xi^h(L_{\xi n}^h - \cdot)$, killed at time $L_{\xi n}^h$, is a $G_D(\xi, \cdot)$-Brownian motion. Observe that the initial distribution of this process is supported by ∂D_n and that $L_{\xi \bullet}^h$ is an increasing sequence, with limit S_ξ^h, the lifetime of the $w_\xi^h(\cdot)$ process. We now show that if S_ξ^h is almost surely finite then the process $w_\xi^h(S_\xi^h - \cdot)$, with lifetime S_ξ^h, is a $G_D(\xi, \cdot)$-Brownian motion on the open parameter interval $]0, +\infty[$. In fact, if f is a positive continuous function on D, with compact support, and if $0 < s < t$, then

$$\begin{aligned} &E^h\{f[w_\xi^h(L_{\xi n}^h - t)] \mid \mathscr{F}(s)\} \\ &= \int_D \ell_D(t - s, w_\xi^h(L_{\xi n}^h - s), \eta) \frac{G_D(\xi, \eta) f(\eta)}{G_D(\xi, w_\xi^h(L_{\xi n}^h - s))} l_N(d\eta) \quad \text{a.s.} \end{aligned} \tag{6.3}$$

(Markov property). When $n \to \infty$, Fatou's lemma yields the inequality

$$E^h\{f[w_\xi^h(S_\xi^h - t)]|\mathscr{F}(s)\} \geq E^h\{f[w_\xi^h(S_\xi^h - t)]|w_\xi^h(S_\xi^h - s)\} \quad \text{a.s.,} \tag{6.4}$$

and there is actually almost sure equality because the expectation of each side is $E^h\{f[w_\xi^h(S_\xi^h - t)]\}$. It follows that the process $w_\xi^h(S_\xi^h - \cdot)$ is a $G_D(\xi, \cdot)$-Brownian motion, as asserted. In particular, suppose that D is provided with a boundary ∂D by a metric compactification with the property that $\lim_{t\uparrow S_\xi^h} w_\xi^h(t)$ exists almost surely. Define $w_\xi^h(S_\xi^h)$ as this limit where the limit exists, and define $w_\xi^h(S_\xi^h)$ arbitrarily elsewhere. This convergence condition is satisfied if the boundary is the Euclidean boundary and $h \equiv 1$ or at least if h has a strictly positive harmonic extension to a neighborhood of $\bar{D}$. When D is connected, we shall show that (Theorem 3.III.2) that this convergence condition is satisfied if and only if the boundary is h-resolutive. If this convergence condition is satisfied, the process $\{w_\xi^h(S_\xi^h - t), t \in \mathbb{R}^+\}$, killed at time S_ξ^h, is an almost surely continuous Markov process whose initial distribution is supported by ∂D. We shall prove in Section 3.III.2 that this initial distribution is the h-harmonic measure $\mu_D^h(\xi, \cdot)$. Observe that, whether or not a boundary is introduced, if $w_\lambda^h(\cdot)$ is an h-Brownian motion in D with initial distribution λ and lifetime S_λ^h, and if S_λ^h is almost surely finite valued, then the process $\{w_\lambda^h(S_\lambda^h - t), t > 0\}$, killed at time S_λ^h, is a $G_D\lambda$-Brownian motion, which can be extended to the parameter set $\mathbb{R}^+$ with initial distribution supported by ∂D if a boundary is introduced with the above stated properties. We shall describe the $w_\lambda^h(S_\lambda^h - \cdot)$ process as a $G_D\lambda$-Brownian motion whether or not the parameter set is extended to $\mathbb{R}^+$.

Example (b). Let ξ_0 and ξ_1 be distinct points of D, and define $B = D - \{\xi_1\}$. If we define h as the restriction to B of $G_D(\xi_1, \cdot)$, the preceding example is applicable to B and h. Observe that (Section 4) almost every h-Brownian motion path in B from ξ_0 has a finite lifetime L and tends to ξ_1 at the path lifetime. According to Example (a), the $G_D(\xi_1, \cdot)$-Brownian motion from ξ_0 reversed at time L is a $G_D(\xi_0, \cdot)$-Brownian motion. Thus, for example, in investigating limits of functions at ξ_1 along Brownian motion paths from ξ_1 we have seen (Section 2) that we can replace the Brownian motion by a $G_D(\xi_0, \cdot)$-Brownian motion from ξ_1 to ξ_0, and we now see that the paths of this conditional Brownian motion can be identified with the paths of a $G_D(\xi_1, \cdot)$-Brownian motion from ξ_0 to ξ_1.

7. Preliminary Probabilistic Solution of the Dirichlet Problem for h-Harmonic Functions; h-Brownian Motion Hitting Probabilities and the Corresponding Generalized Reductions

In this section $w_\xi^h(\cdot)$ is an h-Brownian motion in a Greenian subset D of $\mathbb{R}^N$, from the point ξ of D^h; $w_\xi^h(\cdot)$ has lifetime S_ξ^h, and the entry [hitting] time by $w_\xi^h(\cdot)$ of a subset A of D is $T'^{hA}_\xi[T^{hA}_\xi]$. When $h \equiv 1$, the superscript 1 is dropped from the notation.

Probabilistic Solution of the Dirichlet Problem for h-Harmonic Functions

Suppose here that D is bounded and that h not only is harmonic but has a strictly positive harmonic extension to an open neighborhood D_1 of $\bar{D}$. According to Section 2, an h-Brownian motion in D can be identified with an (extended h)-Brownian motion in D_1 killed at the hitting time of ∂D. Therefore almost every $w_\xi^h(\cdot)$ path tends to a point $w_\xi^h(S_\xi^h-)$ of ∂D at the path lifetime. In particular, if $h \equiv 1$, then according to Theorem IX.13 the harmonic measure $\mu_D(\xi, \cdot)$ is the distribution on ∂D of $w_\xi(S_\xi-)$. We shall now prove the corresponding fact for $\mu_D^h(\xi, \cdot)$. The representations of h-Brownian motion probabilities in terms of Brownian motion probabilities and of h-harmonic measure in terms of harmonic measure (Section 1.VIII.8) yield for B a Borel subset of ∂D,

$$P^h\{w_\xi^h(S_\xi^h-)\in B\} = \int_{\{w_\xi(S_\xi-)\in B\}} h[w_\xi(S_\xi-)]\frac{dP}{h(\xi)} = \frac{\mu_D(\xi, 1_B h)}{h(\xi)} = \mu_D^h(\xi, B). \tag{7.1}$$

That is, extending trivially the meaning of "hitting," the h-harmonic measure $\mu_D^h(\xi, \cdot)$ is the hitting distribution of $w_\xi^h(\cdot)$ on ∂D, as was to be proved. Furthermore Theorem IX.13 can now be translated directly into a theorem on h-Brownian motion, with no change of proof, but recall that D here is bounded and that h here has a harmonic extension to an open neighborhood of $\bar{D}$. The general case, for arbitrary Greenian D and strictly positive harmonic h on D is treated in Section 3.II.2 and is more delicate because even the Euclidean boundary of a Greenian set is not necessarily h-resolutive for every strictly positive harmonic h.

Hitting Probabilities for h-Brownian Motion (h Not Necessarily Harmonic)

Fix A and define

$$u(\xi) = \begin{cases} P^h\{T_\xi^{hA} < S_\xi^h\} & \text{if } \xi \in D^h, \\ 0 & \text{if } \xi \in D - D^h. \end{cases}$$

Since u is $\mathscr{E}_D^h$-excessive (Section VI.12), $u = v/h$ on D^h for some positive superharmonic function v on D. Here $v \le h$ on D^h; so $v \le h$ on D. In particular, if h is harmonic, the function u is h-superharmonic, and the proof given in Section IX.9 for the case $h \equiv 1$ shows that u is h-harmonic on $D - \bar{A}$. The following discussion for general h puts hitting probabilities into the context of reductions.

h-Brownian Motion and Generalized Reductions

Let D be provided with a boundary ∂D by a metric compactification, and let A be a subset of $D \cup \partial D$. Let h be a strictly positive superharmonic function on D, and if u ($\not\equiv +\infty$ on D^h) is a $\mathscr{L}_D^h$-excessive function, define the reduction ${}^hR_u^A$ as the infimum of the class of $\mathscr{L}_D^h$-excessive functions which majorize u both on $A \cap D$ and near $A \cap \partial D$. This reduction has already been defined in Section 1.VIII.1 when h is harmonic, and trivially ${}^1R_u^A = R_u^A$. According to Section 1, every $\mathscr{L}_D^h$-excessive function u vanishes on $D - D^h$, and there is a superharmonic function, which in this section we shall denote by $[uh]$, such that $u = [uh]/h$ on D^h. Obviously ${}^hR_u^A = u = 0$ on $D - D^h$, the reduction ${}^hR_u^A$ is not changed if A is replaced by $A \cap (D^h \cup \partial D)$, and

$$ {}^hR_u^A = \frac{R_{[uh]}^{A\cap(D^h\cup\partial D)}}{h} \quad \text{on } D^h. \tag{7.2}$$

The most natural smoothing of ${}^hR_u^A$ is the $\mathscr{L}_D^h$-excessive function equal to 0 on $D - D^h$ and to

$$\frac{\underset{+[uh]}{R}^{A\cap(D^h\cup\partial D)}}{h}$$

on D^h, and in this section we shall abuse notation by denoting this smoothing by ${}^h\underset{+u}{R}^A$ even though this $\mathscr{L}_D^h$-excessive function is not necessarily the lower semicontinuous smoothing of ${}^hR_u^A$. With this definition, ${}^h\underset{+u}{R}^A \le {}^hR_u^A$ and there is equality quasi everywhere on D; in particular [Section 1.VI.3(b)], there is equality on $D - A$.

Recall from Section IX.14 the reduction evaluations

$$R_v^A(\xi) = E\{v[w_\xi(T_\xi'^A)]\}, \tag{7.3}$$

$$\underset{+v}{R}^A(\xi) = E\{v[w_\xi(T_\xi^A)]\} \tag{7.3sm}$$

for v a positive superharmonic function on D and A an analytic subset of D. The conditions $T_\xi'^A < T_\xi^{\partial D}$ and $T_\xi^A < T_\xi^{\partial D}$ in Section IX.14 are unnecessary here because we have defined $w_\xi(\cdot)$ as a Brownian motion *in D* and v is by convention set equal to 0 at the process trap point. Observe that

$$E\{v[w_\xi(T_\xi^A)]\} = v(\xi)P^v\{T_\xi^A < T_\xi^{\partial D}\}$$

when $v(\xi) < +\infty$. We shall now generalize (7.3) and (7.3sm) to the context of conditional Brownian motion and shall also provide D with a boundary and allow A to contain boundary points. A few preliminary remarks will be needed. In the first place observe that if as above u is a $\mathscr{L}_D^h$-excessive function on D, then $u[w_\xi^h(\cdot)]$ (defined as 0 at parameter values $\ge S_\xi^h$) is an almost surely right continuous supermartingale (Sections 1 and 2) and so has almost sure left limits. In the second place recall that if h_1 is the (Riesz decomposition) harmonic component of h, then $h_1(\xi)/h(\xi)$ is the probability that a

$w_\xi^h(\cdot)$ path tends to ∂D at the path lifetime S_ξ^h. Such a path does not necessarily tend to a single boundary point, however, and for each point ω^h of the probability space on which $w_\xi^h(\cdot)$ is defined we denote by $\Gamma_\xi^h(\omega^h)$ the compact cluster set on ∂D of a path $w_\xi^h(\cdot, \omega^h)$ which tends to ∂D at the path lifetime. To avoid messy typography, we write $I_\xi^h(A)$ for the indicator function of the boundary subset $\Gamma_\xi^h \cap A$ and write 1_h for the indicator function of D^h.

Theorem. *If u is a $\mathscr{E}_D^h$-excessive function ($\not\equiv +\infty$ on D^h) and if A is an analytic subset of $D \cup \partial D$, then for $\xi \in D^h$,*

$$\begin{aligned}{}^hR_u^A(\xi) = E^h\{&u[w_\xi^h(T_\xi'^{hA})]1_{\{T_\xi'^{hA}<S_\xi^h\}}\\ &+ I_\xi^h(A)1_{\{T_\xi'^{hA}\geq S_\xi^h\}} \lim_{t\uparrow S_\xi^h} u[w_\xi^h(t)]\},\end{aligned} \tag{7.4}$$

$$\begin{aligned}{}^h\underset{+u}{R^A}(\xi) = E^h\{&u[w_\xi^h(T_\xi^{hA})]1_{\{T_\xi^{hA}<S_\xi^h\}}\\ &+ I_\xi^h(A)1_{\{T_\xi^{hA}\geq S_\xi^h\}} \lim_{t\uparrow S_\xi^h} u[w_\xi^h(t)]\}.\end{aligned} \tag{7.4sm}$$

Observe that in particular for $\xi \in D^h$,

$${}^hR_{1_h}^A(\xi) = P^h\{T_\xi'^{hA} < S_\xi^h \text{ or } \Gamma_\xi^h \cap A \neq \varnothing\}. \tag{7.5}$$

Replacing $T_\xi'^{hA}$ here by T_ξ^{hA} yields the smoothed reduction.

Since the equalities ${}^hR_u^A(\xi) = {}^h\underset{+u}{R^A}(\xi)$ and $T_\xi'^{hA} = T_\xi^{hA}$ hold when $\xi \in D^h - A$ and since (7.4) is trivial when $\xi \in A$, it follows that (7.4sm) $\Rightarrow$ (7.4). Conversely, (7.4) $\Rightarrow$ (7.4sm) because if $\xi \in D^h \cap A$, an application of (7.4) to $A - \{\xi\}$ yields (7.4sm). From now on we shall assume that u is bounded since the theorem for bounded u can be applied to $u \wedge (n1_h)$ to yield the general result when $n \to +\infty$.

Proof when $A \subset D$. When $A \subset D$ Equations (7.4) and (7.4sm) reduce to

$${}^hR_u^A(\xi) = E^h\{u[w_\xi^h(T_\xi'^{hA})]\}, \tag{7.6}$$

$$\xi \in D^h,$$

$${}^h\underset{+u}{R^A}(\xi) = E^h\{u[w_\xi^h(T_\xi^{hA})]\}, \tag{7.6sm}$$

which are merely reformulations of (7.3) and (7.3sm) by way of (7.2) and the relation between Brownian motion and h-Brownian motion probabilities discussed in Sections 2 and 3.

Proof when $A \cap \partial D$ is a countable union of compact sets. If $A \cap \partial D$ is compact, let $B_\bullet$ be a decreasing sequence of open subsets of $D \cup \partial D$ with intersection $A \cap \partial D$, and define $A_n = (A \cup B_n) \cap D$. By definition of the reduction operation

$$\lim_{n\to\infty} {}^hR_u^{A_n} = {}^hR_u^A.$$

Furthermore (7.4) is true when A is replaced by the subset A_n of D, and

$$\lim_{n\to\infty} E^h\{u[w_\xi^h(T_\xi'^{hA_n})]\}$$

is equal to the right side of (7.4) by the dominated convergence theorem. Since (7.4) is true when $A \cap \partial D$ is compact, this equation is true when $A \cap \partial D$ is a countable union of compact sets. In fact, if $A_\cdot$ is an increasing sequence of analytic subsets of $D \cup \partial D$ with union A, if $A \cap D = A_0 \cap D$, and if each set $A_n \cap \partial D$ is compact, then $\lim_{n\to\infty} {}^hR_u^{A_n} = {}^hR_u^A$ by Section 1.VI.3(e) for $R^{\cdot}_{[uh]}$, and if A_n replaces A on the right side of (7.4), integration to the limit is permissible because only $I_\xi^h(A_n)$ actually changes when n changes, and $I_\xi^h(A_\cdot)$ is a monotone increasing sequence with limit $I_\xi^h(A)$.

Proof in the General Case

Fix $\xi \in D^h$, write $A = B \cup C$, where $B = A \cap D$ and $C = A \cap \partial D$, and consider the function $C \mapsto {}^hR_u^A(\xi)$ for a fixed analytic subset B of D and a varying compact boundary subset C. This set function is strongly subadditive according to Section 1.VI.3(j) for $R^{\cdot}_{[uh]}$. If $C_\cdot$ is a monotone sequence of compact boundary subsets with limit C, then

$$\lim_{n\to\infty} {}^hR_u^{B\cup C_n} = \lim_{n\to\infty} \frac{R_{[uh]}^{B\cup C_n}}{h} = \frac{R_{[uh]}^{B\cup C}}{h} = {}^hR_u^{B\cup C}$$

on D^h. In fact this equation was proved in the preceding paragraph in the increasing case, and this equation is true in the decreasing case by definition of the reduction operation. Thus each side of (7.4) defines a topological precapacity on ∂D for fixed $A \cap D$ and fixed ξ in D^h. Each side is equal to the extension (Appendix II.8) of its precapacity to a Choquet capacity for A analytic because as proved above for each side the value on an F_σ boundary subset is the supremum of the values on compact subsets. Hence (7.4) is true.

Parabolic Context

Theorem 7 and its proof can be translated directly into the parabolic context.

8. Probabilistic Boundary Limit and Internal Limit Theorems for Ratios of Strictly Positive Superharmonic Functions

Let h and v be strictly positive superharmonic functions on a connected Greenian subset D of $\mathbb{R}^N$, and let ν_h and ν_v be the respective associated Riesz measures. According to Theorem 1.XI.4, the function $u = v/h$ has a fine limit at quasi every and $\nu_h + \nu_v$ almost every point of D. According to Theorem 1.XII.19, if h is harmonic, the function u has a minimal-fine limit at M_h almost every Martin boundary point of D. In this section we shall give probabilistic versions of these results. No boundary of D will be involved. The basic theorem is the following, the counterpart for conditional Brownian motion of Theorem IX.7 for Brownian motion. Theorem 8(a) asserts that u has a limit along almost every h-Brownian path at the path lifetime. This result will be applied in the present section, for h a potential in which case the paths tend at their lifetimes to points of D, to derive the internal limit Theorem 1.XII.19. It will be seen in Section 3.III.4 that when h is harmonic, in which case the paths tend at their lifetimes to the one-point Alexandrov boundary, Theorem 8(a) yields the existence of the minimal-fine boundary limit function derived in Theorem 1.XII.19.

Theorem. *Let h and v be strictly positive superharmonic functions on a Greenian subset D of $\mathbb{R}^N$, and let $\{w_\xi^h(\cdot), \mathscr{F}_\xi^h(\cdot)\}$ be an h-Brownian motion in D from a point ξ of $D^h = \{h < +\infty\}$, with lifetime S_ξ^h. It is supposed that $\mathscr{F}_\xi^h(0)$ contains the null sets. Then if $u = v/h$,*

(a) $\lim_{t \uparrow S_\xi^h} u[w_\xi^h(t)]$ *exists (finite) almost surely.*
Define $\mathscr{F}_\xi^h(+\infty) = \bigvee_{t \in \mathbb{R}^+} \mathscr{F}_\xi^h(t)$,

$$\begin{aligned} x_\xi'^h(t) &= x_\xi^h(t) = u[w_\xi^h(t)] \quad \text{if } t < S_\xi^h; \\ x_\xi'^h(t) &= 0, \qquad x_\xi^h(t) = \lim_{s \uparrow S_\xi^h} u[w_\xi^h(s)] \quad \text{if } S_\xi^h \le t \le +\infty. \end{aligned} \tag{8.1}$$

(b) *The processes $\{x_\xi'^h(\cdot), \mathscr{F}_\xi^h(\cdot)\}$ and $\{x_\xi^h(\cdot), \mathscr{F}_\xi^h(\cdot)\}$ are almost surely continuous supermartingales except that the parameter value 0 is to be omitted if $u(\xi) = +\infty$ and that $x_\xi'^h(\cdot)$ may have a jump discontinuity at the parameter value S_ξ^h.*
In (c)–(e) *it is supposed that h is harmonic and that the h-potential, singular h-harmonic, and quasi-bounded h-harmonic components of u (Section 1.IX.11) are denoted by u_p, u_{ms}, and u_{mqb}, respectively.*

(c) *If $u = u_p$ or $u = u_{ms}$, the limit in (a) vanishes almost surely. If $u = u_{mqb}$, the process $\{x_\xi^h(\cdot), \mathscr{F}_\xi^h(\cdot)\}$ is a uniformly integrable martingale.*

(d) $u(\xi) \ge u_{mqb}(\xi) = E^h\{\lim_{s \uparrow S_\xi^h} u[w_\xi^h(s)]\}$.

(e) *If h is a minimal harmonic function the limit in* (a) *is almost surely* $\inf_D u$.

Observation. If h is harmonic, almost every $w_\xi^h(\cdot)$ path tends to the Alexandrov one-point boundary of D at the path lifetime; so (a) can be interpreted as a boundary limit result for every choice of boundary for D. When $S_\xi^h \le t \le +\infty$, the random variable $x_\xi^h(t)$ can be defined arbitrarily on the null set on which the limit in (a) does not exist. Note that if $u(\xi) = +\infty$, the function u is necessarily continuous at ξ; so $x_\xi'^h(\cdot)$ and $x_\xi^h(\cdot)$ are almost surely continuous at the parameter value 0 in this case even though this parameter value must be excluded in the supermartingale assertions.

The proof of (a)–(d) follows that of Theorem IX.7 (the special case $h \equiv 1$) and is therefore omitted. To prove (e), observe that $u = u_p + u_{ms} + u_{qb}$; so in view of (c) it is sufficient to show that u_{mqb} is identically constant, and this constancy follows trivially from the minimality of h.

Extension to Lower-Bounded h-harmonic Functions

If h is harmonic, in Theorem 8 the hypotheses on v can be weakened: instead of positivity it need only be supposed that $v \geq ch$ for some constant c; that is, it need only be supposed that u is lower bounded. This case can be reduced to that of the theorem by replacing v by $v - ch$.
composed process is a martingale.

EXAMPLE. Suppose that D is a Greenian subset of $\mathbb{R}^N$, that v is a positive superharmonic function on D, and that $\zeta \in D$, and define $h = G_D(\zeta, \cdot)$. Since $G_D(\zeta, \cdot)$ is minimal harmonic on $D - \{\zeta\}$ (Section 1.VII.10) and since $G_D(\zeta, \cdot)$-Brownian paths from a point of $D - \{\zeta\}$ almost all tend to ζ at the path lifetimes, the function $u = v/G_D(\zeta, \cdot)$ has $\inf_D u$ as limit along almost every $G_D(\zeta, \cdot)$-Brownian path from any point of $D - \{\zeta\}$ to ζ. According to the symmetry result in Section 6, the function u has this limit at ζ along almost every conditional Brownian path from ζ to a second point of D; equivalently (Section 2), u has this limit at ζ along almost every path of a Brownian motion with initial point ζ. That is (Section IX.15), $\operatorname{f\,lim}_{\eta \to \zeta} u(\eta) = \inf_D u$, as already proved nonprobabilistically [see Theorem 1.XI.4(c)].

Application of Theorem 8 to Derive an Interior Limit Theorem

As noted at the beginning of this section, it was shown in Section 1.XI.4 that v/h has a fine limit at ν_h almost every infinity of h. To derive a probabilistic version of this result, observe first that the harmonic component of h in its Riesz decomposition does not affect the truth of this assertion; so we can assume that h is a potential, $h = G_D \nu_h$. Now according to Section 4, almost every $w_\xi^h(\cdot)$ path tends to a point ζ of D at the path lifetime, the distribution of ζ is $G_D(\xi, \zeta)\nu_h(d\zeta)/h(\xi)$, and the conditional distribution of $w_\xi^h(\cdot)$ given the asymptotic endpoint ζ is that of $G_D(\zeta, \cdot)$-Brownian motion from ξ. Hence according to the preceding example, at $G_D(\xi, \zeta)\nu_h(d\zeta)/h(\xi)$ almost every point ζ of D, equivalently, at ν_h almost every point ζ of D, there is a finite number $c = c(\zeta)$ such that the function v/h has limit ζ along almost every $G_D(\zeta, \cdot)$-Brownian path from an arbitrary point of $D - \{\zeta\}$ to ζ, that is, v/h has fine limit c at ζ. This limit result is trivial unless $v(\zeta) = h(\zeta) = +\infty$ because v and h are continuous in the fine topology. Thus we have proved probabilistically that v/h has a fine limit at ν_h almost every point of D, but further analysis, for example that in Section 1.XI.4, is necessary to identify the limit as $d\nu_v/d\nu_h$ at ν_h almost every infinity of h.

Comparison of the Nonprobabilistic and Probabilistic Fatou Boundary Limit Theorems

If h is harmonic, Theorem 8 states that u has a finite limit along almost every h-Brownian path from a point of D to the one-point boundary of D. On the other hand Theorem 1.XII.19 states that u has a minimal-fine limit at M_h almost every Martin boundary point of D. These two results will be seen to be equivalent in Section 3.III.4 by means of the following reasoning. Let K be a Martin function. It will be shown that almost every $w_\xi^h(\cdot)$ path tends to a minimal Martin boundary point at the path lifetime and that the distribution of $w_\xi^h(S_\xi^h-)$ is the harmonic measure $\mu_D^h(\xi,\cdot)$ on the Martin boundary. In particular, it follows that if ζ is a minimal Martin boundary point, almost every $K(\zeta,\cdot)$-Brownian path from ξ has limit ζ at the path lifetime. Furthermore it will be shown that for arbitrary strictly positive harmonic h the distribution of $w_\xi^h(\cdot)$ can be constructed by choosing a minimal Martin boundary point ζ with the distribution $\mu_D^h(\xi, d\zeta)$ and then choosing $K(\zeta,\cdot)$-Brownian paths from ξ. More precisely the conditional distribution of $w_\xi^h(\cdot)$ given $w_\xi^h(S_\xi^h-)=\zeta$ is the distribution of $K(\zeta,\cdot)$-Brownian motion from ξ. It follows from Theorem 8 that for $\mu_D^h(\xi,\cdot)$ almost every minimal boundary point ζ the positive h-superharmonic function u has a limit $f(\zeta)$ along almost every $K(\zeta,\cdot)$-Brownian path from ξ to ζ. It will be shown that the value $f(\zeta)$ does not depend on the path, does not depend on ξ, and is in fact the minimal-fine limit at ζ whose existence is asserted in Theorem XII.19 of Part 1. The equivalence between the latter theorem and the present theorem in the context of the Martin boundary lies in the facts to be proved in Chapter III of Part 3 that conditional Brownian motion on the Martin space can be generated as described above and that a function has a minimal-fine limit β at a minimal Martin boundary point ζ if and only if the function has limit β along almost every $K(\zeta,\cdot)$-Brownian path from a point of D to ζ. Precise statements will be given in Chapter III of Part 3. In Section 9 it will be seen that the preceding reasoning is easily carried through when D is a ball, in which case the Martin boundary is the Euclidean boundary. It was shown in Sections 1.XII.19 to 1.XII.23 how the fine topology Fatou theorem yields the classical one for a ball or half-space, in which the classical boundary approach is nontangential or normal.

9. Conditional Brownian Motion in a Ball

Let $D = B(0,\delta)$ in $\mathbb{R}^N$, and let K be the ball Poisson kernel,

$$K(\zeta,\eta) = \delta^{N-2}\frac{\delta^2 - |\eta|^2}{|\zeta-\eta|^N}, \qquad |\zeta| = \delta,\ |\eta| < \delta. \tag{9.1}$$

The function $K(\zeta,\cdot)$ is minimal harmonic on D (Section 1.II.16) with value 1 at the origin and limit 0 at every ball boundary point except ζ. Moreover

(Section 1.II.1) $\int_D K(\zeta,\eta) l_N(d\eta) < +\infty$. Fix ζ and consider a $K(\zeta,\cdot)$-Brownian motion from a point ξ of D. According to Section 1, the lifetime of the process is almost surely finite because

$$I(\xi,\zeta,t) = \int_D \ell_D(t,\xi,\eta) K(\zeta,\eta) l_N(d\eta) \le \int_D \ell(t,\xi,\eta) K(\zeta,\eta) l_N(d\eta)$$
$$\le (2\pi\sigma^2 t)^{-N/2} \int_D K(\zeta,\eta) l_N(d\eta) \to 0 \tag{9.2}$$

when $t \to +\infty$. More generally, if h is an arbitrary strictly positive harmonic function with Riesz–Herglotz representation

$$h(\eta) = \int_D K(\zeta,\eta) M_h(d\zeta), \tag{9.3}$$

the lifetime of an h-Brownian motion from ξ is almost surely finite because the probability that the lifetime is at least t is $\int_{\partial D} I(\xi,\zeta,t) M_h(d\zeta)/h(\xi)$, which has limit 0 when $t \to +\infty$. Since the lifetime of an h-Brownian motion when h is a potential is also almost surely finite (Section 1), it follows using the decomposition of conditional Brownian motion in Section 4 that every conditional Brownian motion on a ball has an almost surely finite lifetime.

If ζ is a ball boundary point, almost every $K(\zeta,\cdot)$-Brownian path from ξ tends to the boundary at the path lifetime (Theorem 4), and the function $1/K(\zeta,\cdot)$ has a finite limit along almost every such path (Theorem 8). It follows that almost every $K(\zeta,\cdot)$-Brownian path has limit ζ at the path lifetime. Again let h be an arbitrary strictly positive harmonic function on D with Riesz–Herglotz measure M_h, and recall (from Section 1.VIII.9) that $\mu_D^h(\xi,d\zeta) = K(\zeta,\xi)M_h(d\zeta)/h(\xi)$. If $0 < t_1 < \cdots < t_n$ and if $w_\xi^h(\cdot)$ is an h-Brownian motion in D from ξ with lifetime S_ξ^h, then the joint density of $w_\xi^h(t_1), \ldots, w_\xi^h(t_n)$ in D relative to l_{Nn} is

$$\int_D \ell_D^v(t_1,\xi,\xi_1)\cdots\ell_D^v(t_n - t_{n-1},\xi_{n-1},\xi_n)\mu_D^h(\xi,d\zeta), \qquad v = K(\zeta,\cdot), \tag{9.4}$$

This expression shows that h-Brownian motion paths from ξ can be obtained by first choosing the asymptotic path endpoint ζ according to the distribution $\mu_D^h(\xi,\cdot)$ and then choosing $K(\zeta,\cdot)$-Brownian paths from ξ to ζ. Thus $w_\xi^h(S_\xi^h-)$ exists almost surely and has the distribution $\mu_D^h(\xi,\cdot)$, as already proved (Theorem IX.13) when D is an arbitrary Greenian set, $h \equiv 1$, and ∂D is the Euclidean boundary. In Section 3.II.2 this evaluation of μ_D^h will be extended to every pair $(h, \partial D)$ for which ∂D is h-resolutive, in fact extended in a natural sense to every pair $(h, \partial D)$.

If $h = h_1 + h_2$ is an arbitrary strictly positive superharmonic function on the ball with h_1 positive harmonic and h_2 a potential (Riesz decomposition), the structure of an h-Brownian motion $w_\xi^h(\cdot)$ from ξ with lifetime S_ξ^h is now clear from the decomposition in Section 4. [It is supposed that $h(\xi)$ is finite.] In intuitive language a $w_\xi^h(\cdot)$ path either is an h_1-Brownian path [probability $h_1(\xi)/h(\xi)$] tending to a boundary point $w_\xi^h(S_\xi^h-)$ or is an h_2-Brownian path [probability $h_2(\xi)/h(\xi)$] tending to an interior point $w_\xi^h(S_\xi^h-)$, and the distribution of the asymptotic path endpoint has been found in both cases, respectively, in this section and in Section 4. The extension to the case when $h(\xi) = +\infty$ is left to the reader.

It will be seen in Section 3.III.1 that conditional Brownian motion on an arbitrary Greenian set has a similar structure if the set is provided with the Martin boundary. The process lifetime need not be almost surely finite, however.

The Probabilistic Fatou Theorem for a Ball

Recall that for a ball the Fatou theorems involving radial and nontangential approach to the boundary, and the relations between those theorems and those involving the minimal-fine topology boundary approach, have already been discussed in Sections 1.II.15, 1.XII.19 to 1.XII.23. According to Theorem 8, if v and h are strictly positive superharmonic functions on a ball D and if $w_\xi^h(\cdot)$ is an h-Brownian motion in D from ξ, with lifetime S_ξ^h, then (almost surely) the left limit $v/h[w_\xi^h(S_\xi^h-)]$ exists and is finite. Now suppose that h is harmonic. In view of the structure of h-Brownian motion Theorem 8 in this case is equivalent to the statement that at M_h almost every ball boundary point ζ, that is, at every boundary point up to an h-harmonic measure null set, the function v/h has a finite limit along almost every $K(\zeta,\cdot)$-Brownian path from ξ to ζ. The fact that the existence of a limit in this sense at ζ is equivalent to the existence of a minimal-fine limit at ζ will be proved in Section 3.III.3.

10. Conditional Brownian Motion Last Hitting Distributions; The Capacitary Distribution of a Set in Terms of a Last Hitting Distribution

Let D be a Greenian subset of $\mathbb{R}^N$, let h be a strictly positive superharmonic function on D, let $D^h = \{h < +\infty\}$, and let $w_\xi^h(\cdot)$ be an h-Brownian motion in D, with lifetime S_ξ^h, from ξ in D^h. Let A be an analytic subset of D. Recall (7.5), according to which ${}^hR_{1_h}^A(\xi)$, that is, $\underset{+h}{R}^A(\xi)/h(\xi)$, is the probability that a $w_\xi^h(\cdot)$ path hits A at a strictly positive time; equivalently, if L_ξ^h is the last hitting time of A by $w_\xi^h(\cdot)$,

$$\frac{\underset{+h}{R}^A(\xi)}{h(\xi)} = P\{L_\xi^h > 0\}.$$

Hence almost no $w_\xi^h(\cdot)$ path hits A arbitrarily near ∂D if and only if when B is an open relatively compact subset of D, the value $R_{+h}^{A-B}(\xi)$ can be made arbitrarily small by choosing B sufficiently large. According to Section 1.VIII.11, this condition is satisfied if and only if R_{+h}^A is a potential, as we suppose from now on, $R_{+h}^A = G_D\lambda_A^h$. Under this hypothesis $w_\xi^h(L_\xi^h)$ is well defined aside from the points of the $w_\xi^h(\cdot)$ probability space for which $L_\xi^h = S_\xi^h$, in which case we define $w_\xi^h(L_\xi^h) = w_\xi^h(S_\xi^h-)$, almost surely a point of D. (Recall from Section 1 our convention on the generic use of P and E.)

Theorem. *For $R_{+h}^A = G_D\lambda_A^h$ as just described the distribution of $w_\xi^h(L_\xi^h)$ on $D - \{\xi\}$ is given by*

$$G_D(\xi, \eta)\frac{\lambda_A^h(d\eta)}{h(\xi)}, \tag{10.1}$$

and $P\{w_\xi^h(L_\xi^h) = \xi\} = P\{L_\xi^h = 0\} = 1 - R_{+h}^A(\xi)/h(\xi)$.

According to Section VI.15, on the parameter set $]0, +\infty[$ the $w_\xi^h(\cdot)$ process killed at L_ξ^h becomes an R_{+h}^A-Brownian motion from ξ except that the R_{+h}^A-Brownian motion probabilities are to be multiplied by $R_{+h}^A(\xi)/h(\xi)$. According to Section 4, the distribution of R_{+h}^A-Brownian motion at the path lifetime is $G_D(\xi, \eta)\lambda_A^h(d\eta)/R_{+h}^A(\xi)$, and (10.1) follows. The second assertion is now trivial.

Application to Capacitary Distributions

Since $R_{+1}^A = G_D\lambda_A$ is the capacitary potential of A with respect to D, we find from (10.1) that ($h \equiv 1$) the distribution of $w_\xi(L_\xi)$ on $D - \{\xi\}$ is $G_D(\xi, \eta)\lambda_A(d\eta)$. Thus the last hitting distribution of A by a Brownian motion determines the capacitary distribution λ_A in a very simple way. The first hitting distribution of A by $w_\xi(\cdot)$ leads (Section IX.14) to the sweeping kernel δ_D^A and, in particular, leads to harmonic measure; the last hitting distribution leads to capacitary measure.

11. The Tail σ Algebra of a Conditional Brownian Motion

Let D be a connected Greenian subset of $\mathbb{R}^N$, let h be a strictly positive harmonic function on D, and let $w_\xi^h(\cdot)$ be an h-Brownian motion in D from ξ, with lifetime S_ξ^h. Let $\bar{D}$ be the one-point compactification of D, and make the adjoined point a trap for $w_\xi^h(\cdot)$, so that $w_\xi^h(t)$ is this adjoined point for $S_\xi^h \leq t < +\infty$. Let $\mathscr{F}_\xi^h(t)$ be the σ algebra of subsets of the $w_\xi^h(\cdot)$ probability space generated by the null sets and $\mathscr{F}\{w_\xi^h(s), s \le t\}$. In Section VII.6 it was supposed that $h \equiv 1$ and the asymptotic properties of $w_\xi^h(\cdot)$ paths and of functions on these paths, as the parameter value tends to 0, were investigated. In Section 2 of the present chapter it was shown that the results in Section VII.6 do not depend on the choice of h. In this section we investigate the dual questions, in which the parameter value tends to S_ξ^h instead of 0.

In Sections VII.6 and in Section 2 of the present chapter the key was the initial σ algebra, denoted by $\mathscr{F}_1$ in Section VII.6. In this section the corresponding role is taken by the *tail σ algebra* of sets, determined by $w_\xi^h(\cdot)$ as the parameter tends to S_ξ^h. The natural definition of this σ algebra $\mathscr{G}_\xi^h$ is the following. If $B \subset D$, let S_ξ^{hB} be the hitting time of ∂B by $w_\xi^h(\cdot)$. Define $\mathscr{G}_\xi^h$ as the σ algebra generated by the null sets and the σ algebra

$$\bigcap \{\mathscr{F}\{w_\xi^h(S_\xi^{hB} + t), t \in \mathbb{R}^+\} : B \ni \xi, B \text{ open, relatively compact in } D\}. \quad (11.1)$$

This intersection is unchanged if B is restricted to an increasing sequence of open relatively compact subsets of D with union D. Observe that $\mathscr{G}_\xi^h$ is a σ algebra of subsets of the $w_\xi^h(\cdot)$ probability space, a space which may vary with ξ. Nevertheless the σ algebra (11.1) is defined by conditions on $w_\xi^h(\cdot)$ sample functions which are meaningful for all ξ.

If u is a function from D into $\overline{\mathbb{R}}$, define

$$u_\xi = \limsup_{t \uparrow S_\xi^h} u[w_\xi^h(t)]. \quad (11.2)$$

Let $\mathscr{G}_\xi'^h$ be the smallest σ algebra of subsets of the $w_\xi^h(\cdot)$ probability space containing the null sets and for which u_ξ is measurable whenever u is Borel measurable. The σ algebra $\mathscr{G}_\xi'^h$ is generated by the null sets and the algebra of sets of the form

$$\{[u_{1\xi}, \ldots, u_{n\xi}] \in A'\} \quad (11.3)$$

for $n \ge 1$, A' a Borel subset of $\overline{\mathbb{R}}^n$, and u_i a Borel measurable function from D into $\overline{\mathbb{R}}$. A trivial map shows that $\overline{\mathbb{R}}$ can be replaced here and in the definition of $\mathscr{G}_\xi'^h$ by an arbitrary compact subinterval of $\overline{\mathbb{R}}$.

The following results (a)–(c) on the tail σ algebra and associated concepts will be used in later chapters. A characterization of the tail σ algebra in terms of a suitable boundary of D is given in Section 3.II.4.

(a) Let u be a function from D into $\overline{\mathbb{R}}$ with the property that the set $\{u > c\}$ is analytic for all real c, define u_ξ by (11.2), and define $u'(\xi) = E\{u_\xi\}$ whenever this expectation is meaningful. Then

(a1) u_ξ is $\mathscr{G}_\xi^h$ measurable.
(a2) If u is lower bounded, the function u' is either identically $+\infty$ or h-harmonic and quasi bounded.
(a3) If in (a2) the function u' is h-harmonic, then

$$\lim_{t \uparrow S_\xi^h} u'[w_\xi^h(t)] = u_\xi \quad \text{a.s.} \tag{11.4}$$

for each ξ in D.

(a4) If u is an arbitrary Borel measurable function from D into $\overline{\mathbb{R}}$, then unless $E\{|u_\xi|\} = +\infty$ for all ξ in D, the function u' is h-harmonic and quasi-bounded, and (11.4) is true.

Application. If u is the indicator function of an analytic subset of D and if L_ξ^h is the last hitting time of A by $w_\xi^h(\cdot)$, then (a) implies that the function $\xi \mapsto u'(\xi) = P\{L_\xi^h = S_\xi^h\}$ is h-harmonic and that almost surely $\lim_{t \uparrow S_\xi^h} u'[w_\xi^h(t)]$ exists and is the indicator function of the set $\{L_\xi^h = S_\xi^h\}$. The measurability-type hypothesis imposed on u in (a1)–(a3) was made weaker than Borel measurability to allow for this application.

Proof of (a1). Let $B_\bullet$ be an increasing sequence of open relatively compact subsets of D containing ξ, with union D, and define

$$\Lambda(\xi, m, \alpha) = \{w_\xi^h(\cdot) \text{ hits } (D - \overline{B}_m) \cap \{u > \alpha\}\}.$$

Then up to a $w_\xi^h(\cdot)$ null set $\Lambda(\xi, m, \alpha) \in \mathscr{F}\{w_\xi^h(S_\xi^{hB_m} + t), t \in \mathbb{R}^+\}$ and up to a $w_\xi^h(\cdot)$ null set

$$\{u_\xi > \alpha\} = \bigcup_{k=1}^{\infty} \bigcap_{m=0}^{\infty} \Lambda\left(\xi, m, \alpha + \frac{1}{k}\right);$$

so u_ξ is $\mathscr{G}_\xi^h$ measurable. □

Proof of (a2). Suppose first that u is bounded. Then the strong Markov property of conditional Brownian motion yields

$$E\{u_\xi | \mathscr{F}_\xi^h(S_\xi^{hB_n})\} = E\{u_\xi | w_\xi^h(S_\xi^{hB_n})\} = u'[w_\xi^h(S_\xi^{hB_n})] \quad \text{a.s.} \tag{11.5}$$

Hence (Section 7)

$$u'(\xi) = E\{u_\xi\} = E\{u'[w_\xi^h(S_\xi^{hB_n})]\} = \mu_{B_n}^h(\xi, u'); \tag{11.6}$$

so u' is h-harmonic on B_n for all n and therefore is h-harmonic on D. If u is lower bounded but not necessarily upper bounded, this result applied to $u \wedge n$ implies that the function $\xi \mapsto E\{u_\xi \wedge n\}$ is h-harmonic; so (Section 1.II.3) the function u', the limit of an increasing sequence of bounded h-harmonic functions, is either identically $+\infty$ or h-harmonic and quasi bounded. □

Proof of (a3). We can suppose that u is positive. Then by (a2) either $u' \equiv +\infty$ or u' is h-harmonic and quasi bounded. In the latter case (11.5) is applicable.

Note that u_ξ is $\mathscr{G}_\xi^h$ measurable, that $\mathscr{G}_\xi^h \subset \bigvee_{t\in\mathbb{R}^+}\mathscr{F}(t)$, and that according to Section VI.7 the σ algebra on the right here is $\bigvee_{n\in\mathbb{Z}^+}\mathscr{F}_\xi^h(S_\xi^{hB_n})$; so when $n\to\infty$ in (11.5), the conditional expectation continuity theorem yields

$$u_\xi = E\left\{u_\xi \,\middle|\, \bigvee_{n\in\mathbb{Z}^+}\mathscr{F}_\xi^h(S_\xi^{hB_n})\right\} = \lim_{n\to\infty} u'[w_\xi^h(S_\xi^{hB_n})] \quad \text{a.s.} \tag{11.7}$$

Finally $\lim_{t\uparrow s_\xi^h} u'[w_\xi^h(t)]$ exists almost surely according to Theorem 8, and this limit is almost surely u_ξ according to (11.7). Thus (a3) is true. □

Proof of (a4). If u is Borel measurable, the preceding work is applicable with minor changes when u_ξ is defined by (11.2) with the inferior rather than the superior limit. On applying these results appropriately to $u \vee 0$ and $(-u) \vee 0$ it follows that the function $\xi \mapsto E\{|u_\xi|\}$ on D is either identically $+\infty$ or h-harmonic and quasi bounded, as is then also the function $\xi \mapsto E\{u_\xi\}$, and that in the latter case (11.4) is true. The proof of (a) is now complete. □

(b) $\mathscr{G}_\xi^h = \mathscr{G}_\xi'^h$. Assertion (a1) implies that $\mathscr{G}_\xi'^h \subset \mathscr{G}_\xi^h$. Conversely, fix ξ in D and a set $\Lambda_\xi \in \mathscr{G}_\xi^h$. We shall now show that then $\Lambda_\xi \in \mathscr{G}_\xi'^h$. Define $B_\bullet$ as in the proof of (a1) except that now we also suppose that $\bar{B}_n \subset B_{n+1}$ for all n. According to the Markov property of conditional Brownian motion,

$$P\{\Lambda_\xi|\mathscr{F}(S_\xi^{hB_n})\} = P\{\Lambda_\xi|w_\xi^h(S_\xi^{hB_n})\} \quad \text{a.s.}, \tag{11.8}$$

and (Section I.4) one version of the conditional probability on the right has the form $\phi_n[w_\xi^h(S_\xi^{hB_n})]$, where ϕ_n is a Borel measurable function from ∂B_n into $[0, 1]$. We apply the conditional expectation continuity theorem, again using the fact that

$$\mathscr{G}_\xi^h \subset \bigvee_{t\in\mathbb{R}^+}\mathscr{F}_\xi^h(t) = \bigvee_{n\in\mathbb{R}^+}\mathscr{F}_\xi^h(S_\xi^{hB_n}),$$

and find that

$$1_{\Lambda_\xi} = \lim_{n\to\infty}\phi_n[w_\xi^h(S_\xi^{hB_n})] \quad \text{a.s.} \tag{11.9}$$

If u is the function on D defined for all n as ϕ_n on ∂B_n and defined as 0 elsewhere on D, this function is Borel measurable from D into $[0, 1]$ and

$$1_{\Lambda_\xi} = \limsup_{t\uparrow s_\xi^h}\phi[w_\xi^h(t)] \quad \text{a.s.}$$

Thus $\Lambda_\xi \in \mathscr{G}_\xi'^h$, as was to be proved.

Observation. According to (a) and (b), it is no restriction on $\mathscr{G}_\xi'^h$ if u in (11.2) and the following definition of $\mathscr{G}_\xi'^h$ is bounded and h-harmonic.

(c) If h is minimal harmonic, the following assertions are true.

(c1) The tail σ algebra $\mathscr{G}_\xi^h$ is trivial for every ξ in D.

(c2) If u is an arbitrary Borel measurable function from D into $\overline{\mathbb{R}}$, there is a (not necessarily finite) constant c such that $P\{u_\xi = c\} = 1$ for every ξ in D.

In (c3)–(c5) v is a Borel measurable function from D into a Polish space D', and d' is a metric on D' compatible with the D' topology.

(c3) Let η' be a point of D', and define

$$\mathrm{M}_\xi^h(\eta') = \{\eta' \text{ is a cluster value of } v[w_\xi^h(t)] \text{ when } t \uparrow S_\xi^h\}.$$

Then $\mathrm{M}_\xi^h(\eta') \in \mathscr{G}_\xi^h$, and the function $P\{\mathrm{M}_\cdot^h(\eta')\}$ is either identically 0 or identically 1.

(c4) Let A' be the set of points η' for which the function $P\{\mathrm{M}_\cdot^h(\eta')\}$ is identically 1. The set A' is closed and is the cluster set of almost every $v[w_\xi^h(\cdot)]$ sample function as $t \uparrow S_\xi^h$. If D' is compact,

$$P\{\lim_{t \uparrow S_\xi^h} d'(A', v[w_\xi^h(t)]) = 0\} = 1$$

for every ξ.

(c5) The function $\xi \mapsto P\{\lim_{t \uparrow S_\xi^h} v[w_\xi^h(t)] \text{ exists a.s.}\}$ is either identically 0 or identically 1, and in the latter case there is a point η' of D' such that $\lim_{t \uparrow S_\xi^h} v[w_\xi^h(t)] = \eta'$ almost surely, for every point ξ of D.

Proof of (c1) *and* (c2). We shall prove (c2) which implies (c1). It can be supposed, replacing u by $\arctan u$ if necessary, that u is bounded. The function $\xi \mapsto u'(\xi) = E\{u_\xi\}$ is h-harmonic according to (a2). Since u' is bounded and h is minimal, the function u' is identically constant, $u' \equiv c$, and (11.7) then yields $u_\xi = c$ almost surely, as was to be proved. □

Proof of (c3). A point η' is a cluster value of $v[w_\xi^h(t,\omega)]$ when $t \uparrow S_\xi^h$ if and only if

$$\liminf_{t \uparrow S_\xi^h} d'(\eta', v[w_\xi^h(t,\omega)]) = 0;$$

so if $u(\xi) = -d'(\eta', v(\xi))$, assertions (a) and (c2) combine to imply (c3). □

Proof of (c4). The proof of Section VII.6(e) with obvious changes to the present context yields (c4). □

Proof of (c5). Let Λ_ξ be the $w_\xi^h(\cdot)$ probability space subset for which $v_\xi = \lim_{t \uparrow S_\xi^h} v[w_\xi^h(t)]$ exists. Then $\Lambda_\xi \in \mathscr{G}_\xi^h$ (Section II.5); so $P\{\Lambda_\xi\}$ is either 0 or 1.

The function $\xi \mapsto P\{\Lambda_\xi\}$ is h-harmonic by the same reasoning [see (11.6)] making u' an h-harmonic function. Hence the function $\xi \mapsto P\{\Lambda_\xi\}$ is either identically 0 or identically 1. In the second case fix ξ and observe that v_ξ is $\mathscr{G}_\xi^h$ measurable; so v_ξ is almost surely a point η' of D', that is,

$$\limsup_{t \uparrow s_\xi^h} d'(\eta', v[w_\xi^h(t)]) = 0 \quad \text{a.s.}$$

According to (c2), it follows that this limit relation is true for all ξ; so η' does not depend on ξ. The proof of (c5) is complete. □

The Role of Analytic Set Theory

Although (a)–(c) as treated above involve analytic set theory where, as in (a) and (c2), functions are discussed under very weak hypotheses, the proofs of (b) and (c2) do not involve this theory. Analytic set theory can be avoided in the other parts of (a) and (c) if appropriate restrictions are imposed on the functions under discussion. For example, in (a1) if u is supposed lower semicontinuous, the set $\{u > c\}$ is open, and therefore (Section VI.6) measurability problems for hitting become trivial.

12. Conditional Space-Time Brownian Motion

Let $\dot{D}$ be an open nonempty subset of $\dot{\mathbb{R}}^N$, and let $\dot{h}$ be a positive superparabolic function on $\dot{D}$. The definition and treatment of $\dot{h}$ space-time Brownian motion in $\dot{D}$ follow their counterparts for conditional Brownian motion in a Greenian subset of $\mathbb{R}^N$ except that $\dot{h}$ is allowed to have zeros. We define the transition measure from a zero of $\dot{h}$ to be identically 0. Define $\dot{D}^{\dot{h}} = \{0 < \dot{h} < +\infty\}$. We restrict ourselves to the following remarks. An $\dot{h}$ space-time Brownian motion $\dot{w}_\xi^{\dot{h}}(\cdot)$ from a point ξ of $\dot{D}^{\dot{h}}$ is an almost surely continuous process whose sample functions never leave $\dot{D}^{\dot{h}}$ and proceed downward, that is, in the direction of decreasing ordinate values. If $\dot{v}$ is a positive superparabolic function on $\dot{D}$, the process $(\dot{v}/\dot{h})[\dot{w}_\xi^{\dot{h}}(\cdot)]$ is an almost surely right continuous process with left limits at all parameter values and at the path lifetime and is a supermartingale under the same conventions as in the conditional Brownian motion context. If $\dot{h}$ is parabolic, almost every $\dot{w}_\xi^{\dot{h}}(\cdot)$ path tends to the one-point boundary of $\dot{D}$ at the path lifetime, and $\dot{h}$-parabolic measure $\dot{\mu}_{\dot{B}}^{\dot{h}}(\xi, \cdot)$ on the Euclidean boundary of an open relatively compact subset $\dot{B}$ of $\dot{D}$, containing ξ, is the hitting distribution on this boundary. If $\dot{h} = \dot{G}_{\dot{D}}(\cdot, \dot{\eta})$ for some point $\dot{\eta}$ of $\dot{D}$ and if $\dot{G}_{\dot{D}}(\xi, \dot{\eta}) > 0$, then almost every $\dot{w}_\xi^{\dot{h}}(\cdot)$ path has lifetime $\operatorname{ord}\xi - \operatorname{ord}\dot{\eta}$ and tends to $\dot{\eta}$ at the path lifetime. In general if $\dot{h} = \dot{G}_{\dot{D}}\dot{\mu}$ is a superparabolic potential, almost every $\dot{h}$ space-time Brownian path from a point ξ of $\dot{D}^{\dot{h}}$ has a finite lifetime and tends at the path lifetime to a point $\dot{\eta}$ whose distribution is $\dot{G}_{\dot{D}}(\xi, \dot{\eta})\dot{\mu}(d\dot{\eta})/\dot{h}(\xi)$.

The $\dot{h}$ space-cotime Brownian motion processes are defined in the obvious dual way. When $\dot{G}_{\dot{D}}(\dot{\xi}, \dot{\eta}) > 0$, almost every $\dot{G}_{\dot{D}}(\cdot, \dot{\eta})$ space-time Brownian path from $\dot{\xi}$ tends down to $\dot{\eta}$ at the path lifetime and almost every $\dot{G}_{\dot{D}}(\dot{\xi}, \cdot)$ space-cotime Brownian path from $\dot{\eta}$ tends up to $\dot{\xi}$ at the path lifetime. Moreover a reversal of time in one of these processes yields the other process.

Theorem 8 on the composition with conditional Brownian motion of the ratio of two positive superharmonic functions translates directly into the parabolic context, and its statement is left to the reader. Observe, however, that the sample functions of the composed process, up to the conditional Brownian motion lifetimes, are almost all right continuous but not necessarily continuous, as already noted in Section IX.12 for the special case of the composition with Brownian motion of a superparabolic function. The parabolic counterpart of Theorem 8 can be applied to derive the existence of a coparabolic-fine limit function for the ratio of two positive superparabolic functions (see Theorem 1.XVIII.14 for an exact statement, and see the classical context argument in Section 8).

13. [Space-Time] Brownian Motion in [$\dot{\mathbb{R}}^N$] $\mathbb{R}^N$ with Parameter Set $\mathbb{R}$

All measure densities in this section are relative to l_N.

Brownian motion in $\mathbb{R}^N$ with parameter set $\mathbb{R}$

Let $\{w(t), t \in \mathbb{R}\}$ be a Brownian motion in $\mathbb{R}^N$ with parameter set $\mathbb{R}$. Then (Sections VII.2 and IX.8) there is a positive parabolic function $\dot{u}$ on $\dot{\mathbb{R}}^N$ such that for s in $\mathbb{R}$ the random variable $w(s)$ has distribution density $\dot{u}(\cdot, s)$, with integral $+\infty$ over $\mathbb{R}^N$. The process $w(\cdot)$ has reverse time transition density (transition from ξ at time s to η at time $t < s$)

$$\dot{u}(\eta, t)\ell(s - t, \xi, \eta)/\dot{u}(\xi, s).$$

In particular, suppose that $\dot{u}$ is a minimal parabolic function on $\dot{\mathbb{R}}^N$, so that (Section 1.XVI.8) $\dot{u}$ has the form

$$\dot{u}(\xi, s) = \exp\left[\langle \gamma, \xi \rangle + \sigma^2 |\gamma|^2 \frac{s}{2}\right] \tag{13.1}$$

with γ a point of $\mathbb{R}^N$. Then the above reverse transition density becomes $\ell(s - t, \xi, \eta - \sigma^2(s - t)\gamma)$. This is the transition density of a Brownian motion with drift: conditioned by fixing $w(0) = \xi$, the process

$$\{w(-t) + \sigma^2 t\gamma, t \in \mathbb{R}^+\}$$

is a Brownian motion in $\mathbb{R}^N$ from ξ. According to VII(5.4), it follows that for the conditional and therefore also for the original process,

$$\lim_{t\to-\infty}\frac{w(t)}{t}=\sigma^2\gamma \quad \text{a.s.} \tag{13.2}$$

Observe that if $\gamma\neq 0$, then $\lim_{t\to-\infty} w(t)=\infty$ almost surely. If $\gamma=0$, then (Theorem IX.5) this convergence to ∞ is true if and only if $N>2$. If $\gamma=0$ and $N=2$ $[N=1]$, almost every $w(\cdot)$ sample path hits every disk [point] at arbitrarily large negative parameter values. If $\dot{u}$ is an arbitrary positive parabolic function on $\dot{\mathbb{R}}^N$, then (Section XVI.8) there is a measure $\dot{N}_{\dot{u}}$ on $\mathbb{R}^N$ such that

$$\dot{u}(\xi,s)=\int_{\mathbb{R}^N}\exp\left[\langle\gamma,\xi\rangle+\sigma^2|\gamma|^2\frac{s}{2}\right]\dot{N}_{\dot{u}}(d\gamma), \tag{13.3}$$

and we conclude that if $\dot{u}$ is the absolute probability density function for $\dot{w}(\cdot)$, then $\lim_{t\to-\infty}[w(t)/t]$ exists almost surely and has distribution $\sigma^2\dot{N}_{\dot{u}}(d\gamma)$. Observe that the only condition imposed on $\dot{N}_{\dot{u}}$ is XVI(8.3); so $\dot{N}_{\dot{u}}(\mathbb{R}^N)$ need not be finite. In particular, if $\dot{N}_{\dot{u}}$ is the unit measure supported by the origin, the function $\dot{u}$ is given by (13.1) with $\gamma=0$, and so $\dot{u}\equiv 1$. This is the stationary case. The choice of $\dot{u}$ is the choice of the distribution family $s\mapsto\dot{u}(\xi,s)l_N(d\xi)$; this family is called the *entrance law of the Brownian motion*.

Conditional Brownian Motion in $\mathbb{R}^N$ with Parameter Set $\mathbb{R}$

In the study of h-Brownian motion in $\mathbb{R}^N$ with parameter set $\mathbb{R}$ the function h is a strictly positive superharmonic function on $\mathbb{R}^N$, and the transition density (from ξ at time s to η at time $t>s$) is given by

$$\ell^h(t-s,\xi,\eta)=\ell(t-s,\xi,\eta)\frac{h(\eta)}{h(\xi)}, \tag{13.4}$$

and it is trivial that the absolute probability density function $\dot{u}$ must be replaced by $\dot{u}h$. Observe that if h is harmonic we obtain nothing new because (Section 1.II.2) every positive harmonic function on $\mathbb{R}^N$ is identically constant. Moreover (Section 1.II.13 for $N=2$; the result is trivial when $N=1$) every positive superharmonic function on $\mathbb{R}^N$ is identically constant when $N<3$. Hence in the following we suppose that $N>2$ and that h is a potential. The reverse transition density for (13.4) of an h-Brownian motion $w^h(\cdot)$ with parameter set $\mathbb{R}$ is unchanged by the change from $\dot{u}$ and ℓ to $\dot{u}h$ and ℓ^h aside from the fact that this density is not defined when either ξ or η is in the polar and therefore l_N null set of infinities of h. In particular, suppose that $\dot{u}$ is given by (13.1) and that $h=G(\zeta,\cdot)$ for some point ζ of $\mathbb{R}^N$. Then as t increases from $-\infty$ to the finite process death time, every $w^h(\cdot)$ sample path comes in from the point ∞, along a direction determined by γ if $\gamma \neq 0$, and tends to ζ at the process death time. For general $\dot{u}$ and general h the process is made up of such processes. Thus in general $\lim_{t\to-\infty}[w(t)/t]$ is a random variable with distribution $\sigma^2\dot{N}_{\dot{u}}(d\gamma)$, and (Theorem 4) if $h = G\mu$ and if $h(\xi) < +\infty$, the conditional distribution for $w(s) = \xi$ of the left limit of $w^h(\cdot)$ at the process death time is $G(\xi,\zeta)\mu(d\zeta)/h(\xi)$. Roughly, the choice of $\dot{u}$ determines $w^h(\cdot)$ at the beginning of the process, and the choice of h determines the process at the end.

[Conditional] Space-Time Brownian Motion in $\dot{\mathbb{R}}^N$ with Parameter Set $\mathbb{R}$

A space-time Brownian motion in $\dot{\mathbb{R}}^N$ with parameter set $\mathbb{R}$ is a process of the form $\{(w(t), t_0 - t), t \in \mathbb{R}\}$, where t_0 is an arbitrary constant and $w(\cdot)$ is a Brownian motion in $\mathbb{R}^N$ with parameter set $\mathbb{R}$. A positive parabolic function $\dot{u}$ is then the absolute probability density function for $w(\cdot)$, as explained above. We now study $\dot{h}$ space-time Brownian motion in $\mathbb{R}^N$ with parameter set $\mathbb{R}$, necessarily replacing $\dot{u}$ by $\dot{u}\dot{h}$. In particular, suppose that $\dot{u}$ and $\dot{h}$ are both minimal parabolic functions on $\dot{\mathbb{R}}^N$, given by (3.1) with $\gamma = \gamma_0, \gamma_1$, respectively. Then the process $\{(w(t), \alpha - t)^{\dot{h}}, t \in \mathbb{R}\}$ becomes a space-time process with the original time component. The space component is a process $w'(\cdot)$ with state space $\mathbb{R}^N$ and is Markovian with stationary transition density $\mathscr{E}(t, \xi, \eta - \sigma^2 t\gamma_1)$; so the process $\{w'(t) - \sigma^2 t\gamma_1, t \in \mathbb{R}^+\}$ conditioned by $w'(0) = \xi_0$ is a Brownian motion from ξ_0. The reverse transition density for arbitrary $\dot{h}$ is the same as that for $\dot{h} \equiv 1$. Thus $\lim_{t\to-\infty}[w'(t)/t] = \gamma_0$ almost surely and $\lim_{t\to\infty}[w'(t)/t] = \gamma_1$ almost surely. The character of $w'(\cdot)$ for general $\dot{u}$ and general parabolic $\dot{h}$ can be deduced at once from this special case, and the case when $\dot{h}$ is superparabolic is handled similarly.

GPSR Compliance

The European Union's (EU) General Product Safety Regulation (GPSR) is a set of rules that requires consumer products to be safe and our obligations to ensure this.

If you have any concerns about our products, you can contact us on ProductSafety@springernature.com

In case Publisher is established outside the EU, the EU authorized representative is:

Springer Nature Customer Service Center GmbH
Europaplatz 3
69115 Heidelberg, Germany

Batch number: 09624478

Printed by Printforce, the Netherlands

Classics in Mathematics

Joseph L. Doob

Classical Potential Theory and Its Probabilistic Counterpart

Springer Science+Business Media, LLC

Born in Cincinnati, Ohio on February 27, 1910, Joseph L. Doob studied for both his undergraduate and doctoral degrees at Harvard University. He was appointed to the University of Illinois in 1935 and remained there until his retirement in 1978. Doob worked first in complex variables, then moved to probability under the initial impulse of H. Hotelling, and influenced by A.N Kolmogorov's famous monograph of 1933, as well as by Paul Lévy's work. In his own book *Stochastic Processes* (1953), Doob established martingales as a particularly important type of stochastic process. Kakutani's treatment of the Dirichlet problem in 1944, combining complex variable theory and probability, sparked off Doob's interest in potential theory, which culminated in the present book.

(For more details see:
http://www.dartmouth.edu/~chance/Doob/conversation.html)

Joseph L. Doob

Classical Potential Theory and Its Probabilistic Counterpart

Reprint of the 1984 Edition

Springer

Joseph L. Doob
University of Illinois
Department of Mathematics
101 West Windsor Road #1104
Urbana, IL 61802
USA
e-mail: doob@math.uiuc.edu

Originally published as Vol. 262 of the
Grundlehren der mathematischen Wissenschaften

Cataloging-in-Publication Data applied for

Die Deutsche Bibliothek - CIP-Einheitsaufnahme

Doob, Joseph L.:
Classical potential theory and its probabilistic counterpart / J. L. Doob. - Reprint of the 1984 ed. - Berlin; Heidelberg; New York; Barcelona; Hong Kong; London; Milan; Paris; Singapore; Tokyo: Springer, 2001
(Classics in mathematics)
ISBN 978-3-540-41206-9   ISBN 978-3-642-56573-1 (eBook)
DOI 10.1007/978-3-642-56573-1

Mathematics Subject Classification (2000): 31-XX, 60J45

ISSN 1431-0821
ISBN 978-3-540-41206-9

Originally published by Springer-Verlag Berlin Heidelberg New York in 2001

Printed on acid-free paper   SPIN 10786705   41/3142ck-5 4 3 2 1 0

J. L. Doob

Classical Potential Theory and Its Probabilistic Counterpart

Springer Science+Business Media, LLC

J. L. Doob
Department of Mathematics
University of Illinois
Urbana, IL 61801
U.S.A.

AMS Subject Classificatiens: 31-XX, 60J45

Library of Congress Cataloging in Publication Data
Doob, Joseph L.
Classical potential theory and its probabilistic counterpart.
(Grundlehren der mathematischen Wissenschaften; 262)
Bibliography: p.
1. Potential, Theory of. 2. Harmonic functions. 3. Martingales (Mathematics) I. Title. II. Series.
QA404.7.D66 1983 515.7 83-12446

Originally published by Springer-Verlag New York Inc. in 1984

Typeset by Asco Trade Typesetting Ltd., Hong Kong.

9 8 7 6 5 4 3 2 1

ISBN 978-3-540-41206-9 ISBN 978-3-642-56573-1 (eBook)
DOI 10.1007/978-3-642-56573-1

Contents

Chapter VI

Markov Processes

Chapter VII

Brownian Motion

Chapter X
Conditional Brownian Motion

Part 3

Chapter I
Lattices in Classical Potential Theory and Martingale Theory

Chapter II
Brownian Motion and the PWB Method

Introduction

Potential theory and certain aspects of probability theory are intimately related, perhaps most obviously in that the transition function determining a Markov process can be used to define the Green function of a potential theory. Thus it is possible to define and develop many potential theoretic concepts probabilistically, a procedure potential theorists observe with jaundiced eyes in view of the fact that now as in the past their subject provides the motivation for much of Markov process theory. However that may be it is clear that certain concepts in potential theory correspond closely to concepts in probability theory, specifically to concepts in martingale theory. For example, superharmonic functions correspond to supermartingales. More specifically: the Fatou type boundary limit theorems in potential theory correspond to supermartingale convergence theorems; the limit properties of monotone sequences of superharmonic functions correspond surprisingly closely to limit properties of monotone sequences of supermartingales; certain positive superharmonic functions [supermartingales] are called "potentials," have associated measures in their respective theories and are subject to domination principles (inequalities) involving the supports of those measures; in each theory there is a reduction operation whose properties are the same in the two theories and these reductions induce sweeping (balayage) of the measures associated with potentials, and so on.

The purpose of this book is to develop this correspondence between potential theory and probability theory by examining in detail classical potential theory, that is, the potential theory of Laplace's equation, together with the corresponding probability theory, that is, martingale theory. The joining link which makes this correspondence especially perspicuous is the Brownian motion process, so this process is studied as needed. In order to carry through this program it is necessary to study parabolic potential theory, that is, the potential theory of the heat equation, and the corresponding process of space time Brownian motion. No knowledge of potential theory is presupposed but it is assumed that the reader is familiar with basic probability concepts through conditional expectations. The necessary lattice theory, analytic set theory and capacity theory are covered in the Appendixes.

Thus this book on the one hand contains an introduction to classical and parabolic potential theory and on the other hand contains an introduction to martingale theory, including a smattering of the general theory of stochastic processes and of Markov process theory. There is cross referencing between the nonprobabilistic and probabilistic aspects of the work, and the linking of classical and parabolic potential theory with martingale theory, by Brownian motion and space time Brownian motion, is examined in depth.

One natural criticism of this project is that there is no reason to treat the very special potential theories of the Laplace and heat equations rather than general axiomatic potential theory. Another criticism is that there is no reason to treat potential theory other than as a special subhead of Markov process theory. In the author's opinion, however, classical potential theory is too important to serve merely as a source of illustrations of axiomatic potential theory, which theory in turn is too important in its own right to be left to the probabilists. To learn potential theory from probability is like learning algebraic geometry without the geometry.

It would be quite impossible to cover all those parts of modernized classical potential theory which are relevant to the purpose of this book. Thus there are striking gaps. For example the treatment of energy is skimpy, and Dirichlet spaces and the concept of bounded mean oscillation are not even mentioned in the text. The emphasis is on the Dirichlet problem and related topics; these are treated in considerable depth. The treatments of classical and parabolic potential theories are sometimes separated, sometimes together, but the notation is designed to exhibit the parallelism of the two theories: dots in the notation distinguish parabolic from classical concepts, thereby muddling eyes but saving brains. And the martingale theory notation is designed to point out to readers the corresponding potential theory notation.

Only the part of Markov process theory needed for the relevant discussion of Brownian motion and conditional Brownian motion is covered. In this book a stochastic process is a specified family of random variables, frequently coupled with a filtration to which the family is adapted, but the measure space of the process is left unspecified and there is no translation operator. Thus in a discussion of Brownian motion from a varying initial point the measure space on which the process is defined may vary with the initial point. This definition of a process may not be best for general Markov process theory but is convenient in the special context of this book; it implies for example that no matter how or on what measure space a process is defined, if it has the properties of a Brownian motion (continuous sample functions and the correct distributions of independent increments) then it is a Brownian motion. In a traditional song, a child finds an object which looks smells and tastes like a peanut so the child concludes that the object is a peanut. As stochastic processes are sometimes defined, with special properties demanded of the measure space on which the process random variables are defined, this simple logic is invalid. However the point of view of this book makes it essential in discussing Brownian motion to prove certain invariance properties, for example that two Brownian motion processes in N space, with a common initial point and variance parameter, have the same probability of hitting an analytic set. This fact is not trivial and such questions are treated.

There is nothing very novel in this book. Potential theorists may find the treatment of reductions on boundary sets of interest, as well as the use of iterated reductions to obtain limit theorems. Correspondingly, probabilists may find the new supermartingale crossing inequalities and the technique of iterated reductions of supermartingales of interest. A new domination principle for supermartingales illustrates the fact that classical potential theory still suggests interesting probability results.

The author thanks Bruce Hajek, Naresh Jain and John Taylor for helpful comments on various chapters and, finally, thanks his typist: usually faithful, sometimes accurate.

Notation and Conventions

$\mathbb{R}^N$ is N dimensional Euclidean space, $\mathbb{R} = \mathbb{R}^1$, and $\mathbb{R}^+$ is the set $[0, +\infty[$ of positive reals. $\overline{\mathbb{R}}$ is the set $[-\infty, +\infty]$ of extended reals and $\overline{\mathbb{R}}^+$ is the set $[0, +\infty]$ of positive extended reals.

$\mathbb{Z}$ is the set of integers, $\mathbb{Z}^+$ is the set $0, 1, 2, \ldots$, and $\mathbb{Z}_n^+$ is the set $0, \ldots, n$.

The boundary of an unbounded subset of $\mathbb{R}^N$ contains the adjoined point ∞ of the one point compactification of $\mathbb{R}^N$ unless some other compactification has been specified. This boundary relative to the one point compactification of $\mathbb{R}^N$ will be called the Euclidean boundary of the set.

If ξ is a point of $\mathbb{R}^N$ and A is a subset of $\mathbb{R}^N$ the distance between ξ and A is written $|\xi - A|$.

$B(\xi, \delta)$ is the ball, in whatever metric space is specified, of center ξ and radius δ, specifically in $\mathbb{R}^N$: $B(\xi, \delta) = \{\eta : |\eta - \xi| < \delta\}$.

l_N refers to N dimensional Lebesgue measure.

If A and B are subsets of a space the set of points in A but not in B is denoted by $A - B$.

"Positive" means "≥ 0" and monotone concepts are to be taken in the wide sense, so that for example a constant function from $\mathbb{R}$ into $\mathbb{R}$ is both monotone increasing and monotone decreasing.

If D is an open subset of $\mathbb{R}^N$ the notation $\mathbb{C}^{(k)}(D)$ refers to the class of functions from D into $\mathbb{R}$ which are continuous together with their derivatives of order $\leq k$.

Limit concepts for a function f at a point do not involve the value of f at the point. Thus $\lim_{\eta \to \xi} f(\eta) = \alpha$ means that f is near α in small deleted neighborhoods of ξ.

The notation for a sequence frequently uses a dot for the index set; unless otherwise identified the index set is $\mathbb{Z}^+$, so that $A_{\bullet} = \{A_0, A_1, \ldots\}$.

The set on which a function f satisfies some set S of conditions is frequently denoted by $\{S\}$. Thus if f is a function from $\mathbb{R}$ into $\mathbb{R}$ the positivity set of f is $\{f \geq 0\}$.

The book is divided into three Parts. Section 1.II.3 is Section 3 of Chapter II of Part 1; in any Part, Section II.3 is Section 3 of Chapter II of that part; in any Chapter, Section 3 is Section 3 of that Chapter, and so on.

Part 3

Chapter I

Lattices in Classical Potential Theory and Martingale Theory

1. Correspondence between Classical Potential Theory and Martingale Theory

Submartingales martingales and supermartingales are analogs in the context of martingale theory of subharmonic harmonic and superharmonic functions in the context of classical potential theory. The correspondence between these two contexts has two aspects. In the first place many of the manipulations of supermartingales correspond exactly to manipulations of superharmonic functions. This has been exhibited in previous chapters by the common choice of nomenclature, for example, **D**, **S**, $\mathbf{S}_m$, LM, GM, $\tau_\cdot$, $R^\cdot_\cdot$. In the second place under appropriate hypotheses the composition of a superharmonic function with Brownian motion is a supermartingale; for example, see Section 2.IX.7. In this chapter lattice aspects of classical potential theory and martingale theory will be developed simultaneously.

Throughout this chapter in the *classical potential theory* context D is a fixed connected Greenian subset of $\mathbb{R}^N$, and h is a fixed strictly positive harmonic function on D. The notation $\mathbf{S}^+$, etc., will always refer to classes of positive h-superharmonic functions, etc., on D. The *martingale theory* context is the continuous parameter context as described in Section 2.IV.1; that is, processes with parameter set $\mathbb{R}^+$, on a complete probability measure space $(\Omega, \mathscr{F}, P)$ provided with a right continuous filtration $\mathscr{F}(\cdot)$, are treated. It is supposed that $\mathscr{F}(0)$ contains the null sets. Thus $\mathbf{S}^+, \ldots$ refer to the class of positive almost surely right continuous supermartingales on

$$(\Omega, \mathscr{F}, \mathscr{F}(\cdot), P), \ \ldots .$$

Classes $\mathbf{L}^p$, **D**, **S**, and so on have been defined in both classical potential theory and martingale theory contexts, and it will frequently be possible in the analysis to use the same proof in both contexts, by means of the obvious translation in which "positive h-superharmonic function" becomes "positive almost surely right continuous supermartingale" and operations like $\tau_\cdot$, GM, ... are interpreted according to the context under consideration. It will be convenient in both contexts to abbreviate the notation for the intersection of lattices by writing $\mathbf{S}_{mqb}$ for $\mathbf{S}_m \cap \mathbf{S}_{qb}$, $\mathbf{S}_{ps}$ for $\mathbf{S}_p \cap \mathbf{S}_s$, and so on.

It is trivial that if u is a strictly positive h-superharmonic function, then uh is a strictly positive superharmonic function, and conversely, and that u is in a class $\mathbf{L}^p$, $\mathbf{D}$, $\mathbf{S}$, etc., relative to h if and only if uh is in the corresponding class for $h \equiv 1$. Moreover, when u is a positive h-superharmonic function, $R^A_{uh}/h = {}^hR^A_u$ and $\underset{+uh}{R^A}/h = {}^h\underset{+u}{R^A}$. Thus, although the results in this chapter are stated for general h to facilitate reference, proofs can be given for $h \equiv 1$ without loss of generality.

The Parabolic Context. We have defined the classes mentioned above both in the classical potential theory and the parabolic contexts. In the latter the Greenian subset D of $\mathbb{R}^N$ is replaced by a nonempty open subset $\dot{D}$ of $\dot{\mathbb{R}}^N$, h is replaced by a strictly positive parabolic function $\dot{h}$ on $\dot{D}$, and h-superharmonic functions are replaced by $\dot{h}$-superparabolic functions. In most but not all of the classical potential theory discussion theorems and their proofs will be seen to be directly translatable into the parabolic context, as noted in Section 1.XVIII.19.

The Martingale Theory Lattices $'\mathbf{S}$, *etc.* We have defined and discussed these lattices in Chapter V of Part 2 under very general hypotheses. The most useful case is that with parameter set $\mathbb{Z}^+$, and the reader should have no difficulty in finding the counterparts in this context of the martingale theory continuous context results in the present chapter. Some of the results are not applicable to all linearly ordered parameter sets, however.

2. Relations between Decomposition Components of S in Potential Theory and Martingale Theory

In both the classical potential theory (Chapter IX of Part 1) and martingale theory (Chapter V of Part 2) contexts we have proved that $\mathbf{S}_{mqb}$, $\mathbf{S}_{ms}$, $\mathbf{S}_{pqb}$, $\mathbf{S}_{ps}$ are mutually orthogonal bands in $\mathbf{S}$, with vector sum $\mathbf{S}$. The potential theory proofs are applicable without change to the parabolic context.

In the martingale theory context we have proved (Theorems 2.V.3, 2.V.9) that $\mathbf{S}_m \cap \mathbf{D} = \mathbf{S}_{mqb} = \mathbf{S}_m \cap \mathbf{UI}$. The equality $\mathbf{S}_m \cap \mathbf{D} = \mathbf{S}_{mqb}$ in the potential theory contexts will be proved in Section 5.

In the martingale theory context we have proved (Theorem 2.IV.11) that $\mathbf{S}_p^+ \cap \mathbf{D} = \mathbf{S}_{pqb}^+$. This equality will be proved in the potential theory contexts in Section 9.

3. The Classes $\mathbf{L}^p$ and $\mathbf{D}$

See Sections 1.IX.3, 1.IX.4, 2.II.11, and 2.V.4. In the classical potential theory context we denote these classes more specifically by $\mathbf{L}^p(\mu^h_{D-})$ and $\mathbf{D}(\mu^h_{D-})$ when there is danger of ambiguity. In classical potential theory, in parabolic potential theory, and in martingale theory $\mathbf{S} \subset \mathbf{L}^1$ and for $p > 1$:

The set $\mathbf{S} \cap \mathbf{L}^p$ is a vector sublattice of $(\mathbf{S}, \preceq)$ but not a band; $\mathbf{S} \cap \mathbf{L}^p \subset \mathbf{D}$.

4. PWB-Related Conditions on h-Harmonic Functions and on Martingales

Lemma (Classical Potential Theory Context). *If the h-harmonic function u on D is in* $\mathbf{D}(\mu_{D-}^h)$, *then to each* $\varepsilon > 0$ *and* ξ *in D correspond a lower-bounded h-harmonic pointwise majorant* $u''_{\varepsilon\xi}$ *in* $\mathbf{D}(\mu_{D-}^h)$ *and an upper-bounded h-harmonic pointwise minorant* $u'_{\varepsilon\xi}$ *in* $\mathbf{D}(\mu_{D-}^h)$ *with* $u''_{\varepsilon\xi}(\xi) - u'_{\varepsilon\xi}(\xi) \leq \varepsilon$. *Conversely, if u is an h-harmonic function and if to each* $\varepsilon > 0$ *and* ξ *in D correspond a lower-bounded h-superharmonic pointwise majorant* $u''_{\varepsilon\xi}$ *and an upper-bounded h-subharmonic pointwise minorant* $u'_{\varepsilon\xi}$ *with* $u''_{\varepsilon\xi}(\xi) - u'_{\varepsilon\xi}(\xi) \leq \varepsilon$, *then u is in* $\mathbf{D}(\mu_{D-}^h)$.

(Martingale Theory Context). *If* $x(\cdot) \in \mathbf{D} \cap \mathbf{S}_m$, *that is, if* $\{x(\cdot), \mathscr{F}(\cdot)\}$ *is an almost surely right continuous uniformly integrable martingale, then to each* $\varepsilon > 0$ *correspond a lower-bounded martingale essential majorant* $x''_\varepsilon(\cdot)$ *of* $x(\cdot)$ *in* $\mathbf{D}$ *and an upper-bounded martingale essential minorant* $x'_\varepsilon(\cdot)$ *in* $\mathbf{D}$ *with* $E\{x''_\varepsilon(t) - x'_\varepsilon(t)\} < \varepsilon$. *Conversely, if* $x(\cdot)$ *is an almost surely right continuous martingale and if to each* $\varepsilon > 0$ *correspond a lower-bounded supermartingale essential majorant* $x''_\varepsilon(\cdot)$ *and an upper-bounded submartingale essential minorant* $x'_\varepsilon(\cdot)$ *with*

$$\sup_{t \in \mathbb{R}^+} E\{x''_\varepsilon(t) - x'_\varepsilon(t)\} \leq \varepsilon$$

then $x(\cdot)$ *is in* $\mathbf{D}$.

Proof in the martingale theory context. If $x(\cdot)$ is in $\mathbf{D} \cap \mathbf{S}_m$ apply Section 2.V.2 to find

$$E\{\mathrm{LM}[x(\cdot) \vee n](s)\} - E\{x(s)\} = \sup_{t \in \mathbb{R}^+} E\{x(t) \vee n\} - E\{x(s)\}. \tag{4.1}$$

The difference on the right is constant in s and is at most $\varepsilon/2$ for sufficiently small (negative) n because $x(\cdot)$ is in $\mathbf{D}$. The process $x''_\varepsilon(\cdot)$ is defined as $\mathrm{LM}[x(\cdot) \vee n]$ for such a choice of n, and the process $x'_\varepsilon(\cdot)$ is defined dually. Conversely, if $x''_\varepsilon(\cdot)$ and $x'_\varepsilon(\cdot)$ exist as in the converse assertion, with $x''_\varepsilon(\cdot) \geq -K_\varepsilon$ and $x'_\varepsilon(\cdot) \leq K_\varepsilon$, where $K_\varepsilon > 0$, and if $b > 0$ is chosen so that

$$P\{|x(t)| = b\} = 0,$$

then

$$\begin{aligned} E\{|x(t)|; |x(t)| \geq b\} &\leq E\{x''_\varepsilon(t) - x'_\varepsilon(t)\} - E\{x''_\varepsilon(t); x(t) < -b\} \\ &\quad + E\{x'_\varepsilon(t); x(t) > b\} \\ &\leq \varepsilon + K_\varepsilon P\{|x(t)| > b\}. \end{aligned} \tag{4.2}$$

It follows that $\sup_{t \geq 0} E\{|x(t)|\} < +\infty$ and if K is this supremum, the right side of (4.2) is at most $\varepsilon + K_\varepsilon K/b < 2\varepsilon$ when b is sufficiently large. Hence $x(\cdot) \in \mathbf{D}$. □

The proof in the potential theory context is a translation of that just given.

Observation. The conditions of this lemma have not received much attention in martingale theory, but they are fundamental in potential theory. In fact by their very definition the PWB^h solutions of the Dirichlet problem are in $\mathbf{D}(\mu_{D-}^h)$, according to Lemma 4, for any choice of boundary. Conversely, if the boundary is internally h resolutive (Section 1.VIII.2), there corresponds to every h-harmonic function u in $\mathbf{S}_{mqb}$ [$= \mathbf{S}_m \cap \mathbf{D}(\mu_{D-}^h)$ according to the next section] a PWB^h resolutive boundary function f with $H_f^h = u$.

Parabolic Context

Lemma 4 and its proof need no change in the parabolic context.

5. Class D Property versus Quasi-Boundedness

Theorem. *In the classical and parabolic potential theory contexts and in the martingale theory context*

$$\mathbf{S}_m \cap \mathbf{D} = \mathbf{S}_{mqb}. \tag{5.1}$$

Recall that in the martingale theory context we have proved (Theorems 2.V.3 and 2.V.9) that $\mathbf{S}_m \cap \mathbf{D} = \mathbf{S}_{mqb} = \mathbf{S}_m \cap \mathbf{UI}$; so (5.1) need only be discussed in the potential theory contexts. Actually Lemma 4 yields (5.1) in both contexts. We give the proof of (5.1) in the classical potential theory context in the notation of Lemma 4, with $h \equiv 1$ to simplify the typography. The proof in the martingale theory context, as based on Lemma 4, is essentially the same. If $u \in \mathbf{S}_m^+ \cap \mathbf{D}$, the bounded positive harmonic function $\mathrm{LM}(u'_{\varepsilon\xi} \vee 0)$ is a pointwise majorant of $u'_{\varepsilon\xi}$ and minorant of u; so $u(\xi) - \mathrm{LM}(u'_{\varepsilon\xi} \vee 0) < \varepsilon$. Thus u is the supremum of its positive bounded harmonic minorants and as such is in $\mathbf{S}_{mqb}^+$. Conversely, if $u \in \mathbf{S}_{mqb}^+$, it is trivial from Lemma 4 that $u \in \mathbf{D}$.

Application to $\mathbf{L}^p$ for $p > 1$

In the following discussion we compare certain results for $\mathbf{S}_m \cap \mathbf{L}^p$ in martingale and potential theory contexts. Suppose that $\{x(\cdot), \mathscr{F}(\cdot)\}$ is a uniformly integrable martingale (continuous parameter context), that is, $x(\cdot) \in \mathbf{S}_m \cap \mathbf{D}$. According to the continuous parameter version of Theorem 2.III.14 there is an almost sure limit $x(+\infty) = \lim_{t\to\infty} x(t)$, and $x(t) = E\{x(+\infty)|\mathscr{F}(t)\}$ almost surely, for all t. Moreover according to Theorem

2.III.14, $E\{|x(+\infty)|^p\} < +\infty$ if and only if $x(\cdot)$ is L^p bounded, equivalently (by Section 2.V.4) if and only if $x(\cdot)\in\mathbf{L}^p$. Under the latter condition it follows easily from Section 2.V.2 that

$$E\{|x(+\infty)|^p|\mathscr{F}(\cdot)\} = \mathrm{LM}\,|x(\cdot)|^p \quad \text{a.s.}$$

The following are the classical potential theory counterparts of these martingale theory results. Let D be a connected Greenian subset of $\mathbb{R}^N$, let h be a strictly positive harmonic function on D, and let $u = v/h$ be a positive h-harmonic function on D. According to Theorem 1.XII.10 the Martin boundary is universally resolutive and universally internally resolutive. Hence in the Martin space context the results in the present and preceding sections imply that the class of PWB^h solutions is $\mathbf{S}_m \cap \mathbf{D}(\mu_{D-}^h) = \mathbf{S}_{mqb}$. According to Theorem 1.XII.19, if $u = H_f^h$ then f is the h-harmonic measure almost sure boundary limit function on approach to $\partial^M D$ in the minimal fine topology, and $u = \mu_D^h(\cdot,f)$. Moreover it follows easily from Section 1.IX.2 (the classical potential theory counterpart of Section 2.V.2) that $\mu_D^h(\cdot,|f|^p) < +\infty$ if and only if $u\in\mathbf{L}^p(\mu_{D-}^h)$, and in that event $\mu_D^h(\cdot,|f|^p) = \mathrm{LM}^h\,|u|^p$.

6. A Condition for Quasi-Boundedness

Lemma (Classical Potential Theory Context). *If $u\in\mathbf{S}_{qb}^+$ and $v\in S^+$, then*

$$\lim_{c\to\infty} {}^h\|u\|^{\{v>c\}} = 0 \quad \text{q.e.} \tag{6.1}$$

(Martingale Theory Context). *If $z(\cdot)\in\mathbf{S}_{qb}^+$, $y(\cdot)\in\mathbf{S}^+$, and $\dot{A}_c = \{y(\cdot,\cdot) > c\}$, then*

$$\lim_{n\to\infty} \|z(\cdot)\|^{\dot{A}_n} = 0 \quad \text{q.e.} \quad (n\in\mathbb{Z}^+). \tag{6.2}$$

Observe that (6.1) and (6.2) are identical with the same equations for unsmoothed reductions because $\{v > c\}$ is an open subset of $\mathbb{R}^N$ and $\dot{A}_c$ is a fine-open subset of $\mathbb{R}^+ \times \Omega$. In (6.1) the reduction decreases when c increases. In (6.2) we can only assert that $c < d$ implies that $\|z(\cdot)\|^{\dot{A}_d} \le \|z(\cdot)\|^{\dot{A}_c}$ quasi everywhere, and that is why in (6.2) the limit is sequential. The set $\mathbb{Z}^+$ in (6.2) can of course be replaced by any other unbounded increasing sequence.

Proof in the classical potential theory context. (The martingale theory proof is a direct translation.) If $u = \Sigma_0^\infty u_j$ with u_j in $\mathbf{S}^+$ and bounded, then

$${}^h\|u\|^{\{v>c\}} = \sum_0^\infty {}^h\|u_j\|^{\{v>c\}} \le \sum_0^\infty u_j, \tag{6.3}$$

and (dominated convergence) it is therefore sufficient to show that $\lim_{c\to\infty} {}^h[\![u_j]\!]^{\{v>c\}} = 0$ for all j at the points of finiteness of v. If u_j is bounded by a_j, this desired limit relation follows from

$$ {}^h[\![u_j]\!]^{\{v>c\}} \le \frac{a_j}{c} {}^h[\![c]\!]^{\{v>c\}} \le \frac{a_j v}{c}. \tag{6.4}$$

Observation. The relation (6.1) can also be written in the form

$$\lim_{c\to\infty} [\![uh]\!]^{\{v>c\}} = 0 \quad \text{q.e.} \tag{6.1$'$}$$

Parabolic Context

The lemma and its proof need no change in the parabolic potential theory context.

7. Singularity of an Element of $\mathbf{S}_m^+$

Theorem (Classical Potential Theory Context). *For a function u in $\mathbf{S}_m^+$ to be singular it is sufficient that $u \wedge c \in \mathbf{S}_p^+$ for some strictly positive constant c and necessary that $u \wedge c \in \mathbf{S}_p^+$ for every strictly positive constant c.*

(Martingale Theory Context). (a) *For a process $z(\cdot)$ in $\mathbf{S}_m^+$ to be singular it is sufficient that $z(\cdot) \wedge c \in \mathbf{S}_p^+$ for some strictly positive constant c and necessary that $z(\cdot) \wedge c \in \mathbf{S}_p^+$ for every strictly positive constant c.*

(b) *The process $z(\cdot)$ in $\mathbf{S}_m^+$ is singular if and only if $\lim_{t\to\infty} z(t) = 0$ almost surely.*

The proof of Theorem 7 in the classical potential theory context was given in Section 1.IX.10, and the martingale proof of (a) is a direct translation of that proof. The martingale assertion (b) was proved in Section 2.V.11.

A classical potential theory counterpart of the martingale theory assertion (b) can be formulated in two ways as follows. (Classical potential theory context):

(b1) An h-harmonic function u in $\mathbf{S}_m^+$ is singular if and only if (Theorem 1.XII.19) its minimal-fine boundary limit function on the Martin boundary of D vanishes M_h almost everywhere, that is, up to an h-harmonic measure null set.

(b2) An h-harmonic function u in $\mathbf{S}_m^+$ is singular if and only if (Theorem 2.X.8) when $w_\xi^h(\cdot)$ is an h-Brownian motion in D from ξ, with lifetime S_ξ^h, $\lim_{t\uparrow S_\xi^h} u[w_\xi^h(t)] = 0$ almost surely.

Parabolic Context

Theorem 7 and its proof need no change in the parabolic context. The parabolic criterion corresponding to (b2) does not require a new proof and is of course stated in terms of the limit of a positive h-parabolic function along h space-time Brownian paths.

8. The Singular Component of an Element of $\mathbf{S}^+$

Theorem (Classical Potential Theory Context). *If $u \in \mathbf{S}^+$ and has singular component u_s, then*

$$\lim_{c \to \infty} {}^h\|u\|^{\{u>c\}} = u_s \quad \text{q.e.} \tag{8.1}$$

In particular, u in $\mathbf{S}^+$ is singular if and only if $u = {}^h\|u\|^{\{u>c\}}$ for every (equivalently some) strictly positive constant c.

(Martingale Theory Context). *If $z(\cdot) \in \mathbf{S}^+$ and has singular component $z_s(\cdot)$ and if $A_c = \{z(\cdot, \cdot > c\}$, then*

$$\lim_{n \to \infty} \|z(\cdot)\|^{\dot{A}_n} = z_s(\cdot) \quad \text{q.e. } (n \in \mathbb{Z}^+). \tag{8.2}$$

In particular, $z(\cdot)$ in $\mathbf{S}^+$ is singular if and only if $z(\cdot) = \|z(\cdot)\|^{\dot{A}_c}$ quasi everywhere for every (equivalently some) strictly positive constant c.

Observation. Equations (8.1) and (8.2) are equivalent to the same equations with unsmoothed reductions. See the observation in Section 6 on the corresponding point for Lemma 6, and on the explanation of the sequential convergence ($n \to \infty$) in (8.2) rather than the unrestricted approach ($c \to \infty$) in (8.1).

Proof in the classical potential theory context. (The martingale theory proof is a direct translation.) Suppose first that $u \in \mathbf{S}_s^+$. By Section 1.VI.3(i) as adapted to reductions in the h-superharmonic context,

$$u \preceq {}^h\|u\|^{\{u>c\}} + {}^h\|u\|^{\{u \le c\}}. \tag{8.3}$$

According to the vector lattice decomposition theorem (Theorem 6 of Appendix III), there are members u_1 and u_2 of $\mathbf{S}^+$, specific order minorants, respectively, of the first and second terms on the right, with sum u. Since $\mathbf{S}_s$ is a band the functions u_1 and u_2 must be singular, and since u_2 is bounded by c, $u_2 \in \mathbf{S}_s^+ \cap \mathbf{S}_{qb}$; so $u_2 = 0$. Thus $u = u_1$, and u is the first term on the right in (8.3), as was to be proved. For general u in $\mathbf{S}^+$ write $u = u_{qb} + u_s$, the sum of the quasi-bounded and singular components of u. Then [from Section 1.VI.3(f)]

$$^h\|u\|^{\{u>c\}} = {}^h\|u_{qb}\|^{\{u>c\}} + {}^h\|u_s\|^{\{u>c\}}. \tag{8.4}$$

The first term on the right has limit 0 quasi everywhere when $c \to \infty$ according to Lemma 6, and the second term on the right is u_s for every c because

$$u_s = {}^h\|u_s\|^{\{u_s>c\}} \le {}^h\|u_s\|^{\{u>c\}} \le u_s;$$

so Theorem 8 is true. □

Parabolic Context

Theorem 8 and its proof need no change in the parabolic context.

9. The Class $\mathbf{S}_{pqb}$

Theorem (Classical Potential Theory Context). *The following conditions on a function $u = G_D\mu/h$ in $\mathbf{S}_p^+$ are equivalent:*

PT(a) $u \in \mathbf{S}_{pqb}^+$.
PT(b) $u \in \mathbf{D}(\mu_{D-}^h)$.
PT(c) $\lim_{c\to\infty} {}^h\|u\|^{\{u>c\}} = 0$ q.e.
PT(d) *μ vanishes on polar sets.*

(Martingale Theory Context) *The following conditions on a process $z(\cdot)$ in $\mathbf{S}_p^+$ are equivalent:*

MT(a) $z(\cdot) \in \mathbf{S}_{pqb}^+$
MT(b) $z(\cdot) \in \mathbf{D}$.
MT(c) *If $\dot{A}_c = \{z(\cdot,\cdot) > c\}$, then*

$$\lim_{n\to\infty} \|z(\cdot)\|^{\dot{A}_n} = 0 \quad \text{q.e. } (n \in \mathbb{Z}^+).$$

Observation. PT(c) and MT(c) are equivalent to the same equations with unsmoothed reductions. See the observation in Section 6 on the corresponding point for Theorem 6 and on the explanation of the sequential convergence ($n \to \infty$) in MT(c) rather than the unrestricted approach ($c \to \infty$) in PT(c).

Proof in the classical potential theory context. In clarification of PT(d) recall that (Theorem 1.V.11) a superharmonic potential $G_D\mu$ has the value $+\infty$ at μ almost every point of a polar set. Hence μ vanishes on polar sets if and only if $\mu\{G_D\mu = +\infty\} = 0$. To simplify the notation in the following proof, we take $h \equiv 1$.

PT(a) ⇒ PT(b) If $u = \Sigma_0^\infty u_j$ with each summand in $\mathbf{S}_p^+$ and bounded, let ξ be a point of D, and let B_0 be an open relatively compact subset of D containing ξ. To show that $u \in \mathbf{D}(\mu_{D-})$, we show that the family of pairs

$$\{(u, \mu_B(\xi, \cdot)): B \supset B_0, B \text{ open and relatively compact in } D\} \quad (9.1)$$

is uniformly integrable. Observe that according to the generalized superharmonic function average property in Section 1.VIII.10, the harmonic average $\mu_B(\xi, u)$ increases when B decreases, and therefore attains its maximum value when $B = B_0$. Now define $v_n = \Sigma_{n+1}^\infty u_j$, choose $\varepsilon > 0$, choose n so large that $\mu_{B_0}(\xi, v_n) < \varepsilon/2$, and define $c = \sup_D \Sigma_0^n u_j$. When B is a set in (9.1), the same superharmonic function average inequality yields $\mu_B(\xi, v_n) < \varepsilon/2$, and if A is a Borel subset of ∂B, so small that $\mu_B(\xi, A) < \varepsilon/(2c)$, then

$$\mu_B(\xi, u1_A) \le \mu_B(\xi, v_n) + \mu_B\left(\xi, \sum_0^n u_j 1_A\right) < \varepsilon. \quad (9.2)$$

Hence the family (9.1) is uniformly integrable.

PT(b) ⇒ PT(d) Suppose that for some u in $\mathbf{S}^+$ condition (b) is true but condition (d) is false. Replacing μ by its projection on a suitable compact set if necessary, it can be supposed in deriving a contradiction that μ is supported by a compact polar set A. Let B be an open relatively compact subset of D containing A, let ξ be a point of $B - A$, and let $A_\cdot$ be a decreasing sequence of compact neighborhoods of A, subsets of B, with intersection A. The boundary of $B_n = B - A_n$ consists of ∂B and a subset C_n of ∂A_n. Since u is harmonic on a neighborhood of $\bar{B}_n$,

$$u(\xi) = \mu_{B_n}(\xi, u1_{\partial B}) + \mu_{B_n}(\xi, u1_{C_n}). \quad (9.3)$$

To show that the second term on the right has limit 0 when $n \to \infty$, note first that (Section 1.V.4) there is a function v positive and superharmonic on B, with $v = +\infty$ on A and $v(\xi) < +\infty$. Hence $v \ge 1$ on C_n for large n; so $\limsup_{n\to\infty} \mu_{B_n}(\xi, C_n) \le v(\xi)$, and since for every $\delta > 0$ the function δv satisfies the same conditions as v, it follows that $\lim_{n\to\infty} \mu_{B_n}(\xi, C_n) = 0$. Hence (class $\mathbf{D}$ property) $\lim_{n\to\infty} \mu_{B_n}(\xi, u1_{C_n}) = 0$; so

$$u(\xi) = \lim_{n\to\infty} \mu_{B_n}(\xi, u1_{\partial B}) \le \sup_{\partial B} u. \quad (9.4)$$

Thus u is bounded on $B - A$, and therefore (Section 1.V.5) u has a unique superharmonic extension to B, and this superharmonic extension is harmonic, contrary to fact. It follows that PT(b) ⇒ PT(d).

PT(d) ⇒ PT(a) If $u = G_D\mu$ is a superharmonic potential for which μ vanishes on polar sets, define $A_n = \{n \le u < n+1\}$; and let μ_n be the projection of μ on A_n. Then $\mu = \Sigma_0^\infty \mu_n$ because $\mu\{u = +\infty\} = 0$, and according to the domination principle, $G_D\mu_n \le n+1$. Thus the representation $u = \Sigma_0^\infty G_D\mu_n$ exhibits u as a quasi-bounded potential.

PT(a) $\Leftrightarrow$ PT(c) See Theorem 8.

MT(a) $\Rightarrow$ MT(b) If $z(\cdot) = \Sigma_0^\infty z_j(\cdot)$ with each summand in $\mathbf{S}_p^+$ and bounded, we show that the family $\{z(T): T \text{ optional}\}$ is uniformly integrable. [Define $z(+\infty) = 0$.] Define $y_n(\cdot) = \Sigma_{n+1}^\infty z_j(\cdot)$, choose $\varepsilon > 0$, choose n so large that $E\{y_n(0)\} < \varepsilon/2$, and let c be an upper bound of $\Sigma_0^n z_j(\cdot)$. Then if T is optional, the supermartingale inequality yields

$$E\{z(T)\} \le E\{z(0)\}, \qquad E\{y_n(T)\} \le \frac{\varepsilon}{2};$$

so if $P\{\Lambda\} < \varepsilon/(2c)$,

$$\int_\Lambda z(T)\, dP < E\{y_n(T)\} + \int_\Lambda \sum_0^n z_j(T)\, dP < \varepsilon. \tag{9.5}$$

Hence the family $\{z(T): T \text{ optional}\}$ is uniformly integrable; that is, $z(\cdot) \in \mathbf{D}$.

MT(b) $\Rightarrow$ MT(a) See Theorem 2.IV.11.

MT(a) $\Leftrightarrow$ MT(c) See Theorem 8.

The proof of the theorem is now complete. □

Observation (Discrete Parameter Case). In the martingale theory context $\mathbf{S}_p = \mathbf{S}_{pqb}$ if the parameter set is $\mathbb{Z}^+$ according to Theorem 2.IV.8.

Parabolic Context. The proofs in the classical potential theory context that PT(c) $\Leftrightarrow$ PT(a) $\Rightarrow$ PT(b) $\Rightarrow$ PT(d) need no change in the parabolic context. The above proof that PT(d) $\Rightarrow$ PT(a), however, used the domination principle which is not valid in the parabolic potential theory context according to Section 1.XVII.5. Now define

PT(d′) *μ vanishes on semipolar sets.*

Then PT(d′) $\Leftrightarrow$ PT(d) in the classical potential theory context because semipolar sets are polar in that context and under PT(d′) the domination principle is valid in the parabolic potential theory context (Section 1.XVIII.16); so PT(d′) $\Rightarrow$ PT(a) in the latter context. Thus the parabolic potential theory version of Theorem 9 is slightly weaker than the stated classical potential theory version. It is false that PT(a) $\Rightarrow$ PT(d′) because if $\dot{D} = \dot{\mathbb{R}}^N$ and if $\dot{\mu}$ is supported by the abscissa hyperplane on which $\dot{\mu} = l_N$, then $\dot{\mu}$ is supported by a semipolar set, but $\dot{G}_{\dot{D}}\dot{\mu} \in S_{pqb}^+$.

10. The Class $\mathbf{S}_{ps}$

Theorem (Classical Potential Theory Context). *The following conditions on a function $u = G_D\mu/h$ in $\mathbf{S}_p^+$ are equivalent:*

PT(a) $u \in \mathbf{S}^+_{ps}$.
PT(b) ${}^h\|u\|^{\{u>c\}} = u$ *for every (equivalently some) strictly positive constant* c.
PT(c) μ *is supported by a polar set.*

(Martingale Theory Context). *The following conditions on a supermartingale* $z(\cdot)$ *in* S^+_p *are equivalent*:

MT(a) $z(\cdot) \in \mathbf{S}^+_{ps}$.
MT(b) *If* $A_c = \{z(\cdot,\cdot) > c\}$, *then* $\|z(\cdot)\|^{A_c} = z(\cdot)$ *quasi everywhere for every (equivalently some) strictly positive constant* c.
MT(c) $z(\cdot)$ *is a local martingale.*

PT(a) $\Leftrightarrow$ PT(b) and MT(a) $\Leftrightarrow$ MT(b) according to Theorem 8.

MT(a) $\Leftrightarrow$ MT(c) according to Theorem 2.V.12.

PT(a) $\Leftrightarrow$ PT(c) (Proof for $h \equiv 1$). We use the notation in Section 1.IX.8, in which $\mathbf{M}^+_p$ is the set of measures on D whose potentials are superharmonic. The vector lattice $\mathbf{M}_p = \mathbf{M}^+_p - \mathbf{M}^+_p$ is isomorphic to the vector lattice $(\mathbf{S}_p, \preceq)$ of classical potential theory under the correspondence $\lambda \leftrightarrow G_D\lambda$. The class $\mathbf{M}^+_{pqb}$ of measures in $\mathbf{M}^+_p$ vanishing on polar sets corresponds to the class $\mathbf{S}^+_{pqb}$ according to Theorem 9; so the class $\mathbf{M}^+_{ps}$ of positive components of the class orthogonal to $\mathbf{M}^+_{pqb} - \mathbf{M}^+_{pqb}$, that is, the class of measures in $\mathbf{M}^+_p$ supported by polar sets, corresponds to the class $\mathbf{S}^+_{ps}$, since $\mathbf{S}_s = \mathbf{S}^{\perp}_{qb}$. That is, PT(a) $\Leftrightarrow$ PT(c), and the proof of the theorem is now complete.

Parabolic Context

We shall need a further condition:

PT(c′) μ is supported by a semipolar set.

Then PT(c′) $\Leftrightarrow$ PT(c) in the classical context because in that context semipolar sets are polar. We leave to the reader the easy verification, modifying the discussion in the classical potential theory context, that in the parabolic context PT(a) $\Leftrightarrow$ PT(b) $\Rightarrow$ PT(c′).

11. Lattice Theoretic Analysis of the Composition of an h-Superharmonic Function with an h-Brownian Motion

In this section we use the notation of Section 2.X.8 but restrict the context slightly by assuming that h is harmonic. Thus u is a positive h-superharmonic function on D to which the lattice theoretic discussion of this section is applicable. The restriction to the parameter set $\mathbb{R}^+$ of the process $x^h_\xi(\cdot)$ defined in 2.X(8.1) will be denoted by ${}_0x^h_\xi(\cdot)$. If $u(\xi) < +\infty$, the latter process is an almost surely right continuous supermartingale to which the lattice theoretic discussion is applicable. We use the notation $\mathbf{S}$, $\mathbf{S}_{qb}$, etc., in both potential theory and martingale theory contexts.

(a) *If* $u \in \mathbf{S}^+_{mqb}$, *then* ${}_0x^h_\xi(\cdot) \in \mathbf{S}^+_{mqb}$ *for every point* ξ *of finiteness of* u. This assertion follows trivially from Theorem 2.X.8.

(b) *If* $x^h_\xi(\cdot)$ *is a martingale for some (equivalently every)* ξ, *then* $u \in \mathbf{S}^+_{mqb}$. In fact the martingale equality at times $0, +\infty$ yields equality at ξ in Theorem 2.X.8(d), and therefore there is equality on D. Observe that according to (d) below the present assertion (b) becomes false if $x^h_\xi(\cdot)$ is replaced in the hypothesis by ${}_0x^h_\xi(\cdot)$.

(c) *If* $u \in \mathbf{S}^+_p$, *then* ${}_0x^h_\xi(\cdot) \in \mathbf{S}^+_p$ *for every point* ξ *of finiteness of* u. To see this recall that by 2.X(1.2) if we write $w_\xi(\cdot)$ for a Brownian motion in D from ξ and denote the process lifetime by S_ξ, then

$$E\{{}_0x^h_\xi(t)\} = \int_{\{S_\xi > t\}} \frac{u[w_\xi(t)]h[w_\xi(t)]\,dP}{h(\xi)} = u(\xi)P\{S^{uh}_\xi > t\}. \tag{11.1}$$

Now by hypothesis uh is a superharmonic potential; so (Section 2.X.1) the probability on the right has limit 0 when $t \to \infty$, and (c) follows. Observe that ${}_0x^h_\xi(\cdot) \in \mathbf{S}^+_p$ implies that $x^h_\xi(+\infty) = 0$ almost surely, but that the converse implication is false, although the latter condition is necessary and sufficient that $x^h_\xi(\cdot)$ be in $\mathbf{S}^+_p$. The conclusion of (c) would thus be considerably weakened if ${}_0x^h_\xi(\cdot)$ were replaced by $x^h_\xi(\cdot)$ in the conclusion. This weakening is exhibited explicitly in (d) to be proved next.

(d) *If* $u \in \mathbf{S}^+_{ms}$, *then* ${}_0x^h_\xi(\cdot) \in \mathbf{S}^+_s$, *for every* ξ. [Recall that under the hypothesis on u, $P\{x^h_\xi(+\infty) = 0\} = 1$ for all ξ.] To prove (d), it is sufficient to prove that if $y(\cdot)$ is a bounded supermartingale in S^+ with parameter set $\mathbb{R}^+$ majorized in the specific order by ${}_0x^h_\xi(\cdot)$, then $y(\cdot)$ is a zero process up to a standard modification, equivalently, $E\{y(0)\} = 0$. Now if $D_\cdot$ is an increasing sequence of open relatively compact subsets of D with union D, containing ξ, and if $S^h_{\xi n}$ is the hitting time of ∂D_n by $w^h_\xi(\cdot)$, then the process ${}_0x^h_\xi(\cdot)$ stopped at time $S^h_{\xi n}$ is a martingale, the process $y(\cdot)$ stopped at the same time is a supermartingale, and in view of the order relation $y(\cdot) \preceq {}_0x^h_\xi(\cdot)$ there is another process which added to $y(\cdot)$ yields ${}_0x^h_\xi(\cdot)$ and which when stopped at $S^h_{\xi n}$ is a positive supermartingale. It follows that $y(\cdot)$ stopped at $S^h_{\xi n}$ is a martingale and therefore that

$$E\{y(0)\} = E\{y(S^h_{\xi n})\} \to E\{y\{S^h_\xi -)\} \le E\{{}_0x^h_\xi(S^h_\xi -)\} = E\{x^h_\xi(+\infty)\} = 0, \tag{11.2}$$

as was to be proved.

12. A Decomposition of $\mathbf{S}^+_{ms}$ (Potential Theory Context)

As in the previous sections h is a strictly positive harmonic function on the connected Greenian subset D of $\mathbb{R}^N$. Let $\mathbf{S}^+_{ms\infty}$ be the cone of functions u in $\mathbf{S}^+_{ms}$ for which $\ell^h_D(t, \cdot, u) = u$ for all $t > 0$, that is, the cone of ℓ^h_D-invariant excessive positive singular h-harmonic functions on D, and let $\mathbf{S}_{msf}^+$ be the cone of functions u in $\mathbf{S}_{ms}^+$ for which $\lim_{t\to\infty} \ell_D^h(t,\cdot,u) = 0$. It is easily checked that if u is in one of these classes, its positive specific order minorants are in the same class and that a countable sum of functions in one of these classes is also in that class if the sum is in $\mathbf{S}_m^+$. The classes $\mathbf{S}_{ms\infty} = \mathbf{S}_{ms\infty}^+ - \mathbf{S}_{ms\infty}^+$ and $\mathbf{S}_{msf} = \mathbf{S}_{msf}^+ - \mathbf{S}_{msf}^+$ are therefore orthogonal subbands of $\mathbf{S}_{ms}$. If u is in $\mathbf{S}_{ms}^+$, we have seen in Section 2.IX.8, Example (a) (for $h \equiv 1$ and the general case follows trivially), that $\lim_{t\to\infty} \ell_D^h(t,\cdot,u) = u_0$ is an h-harmonic minorant of u, in $\mathbf{S}_{ms\infty}^+$. Hence $\lim_{t\to\infty} \ell_D^h(t,\cdot,u-u_0) = 0$, that is, $u - u_0 \in \mathbf{S}_{msf}^+$, and therefore $\mathbf{S}_{ms} = \mathbf{S}_{ms\infty} + \mathbf{S}_{msf}$.

The classes $\mathbf{S}_{ms\infty}$ and $\mathbf{S}_{msf}$ have simple probabilistic characterizations. The condition that $u \in \mathbf{S}_{ms\infty}^+$ is the condition that u is in S_{ms}^+ and that the process ${}_0x_\xi^h(\cdot)$ in Section 11 is a martingale for some (equivalently every) ξ, equivalently and more intuitively, that the process $x_\xi'^h(\cdot)$ defined in 2.X(8.1) is a martingale for some (equivalently every) ξ. A second probabilistic interpretation is that (Section 2.X.1) $u \in \mathbf{S}_{ms\infty}^+$ [$u \in \mathbf{S}_{msf}^+$] if and only if the lifetime of uh-Brownian motion in D from some (equivalently every) ξ is almost surely $+\infty$ [almost surely finite]. This characterization inspires the notation.

13. Continuation of Section 11

Section 11(d) is made more precise by (e) and (f) below.

(e) *If* $u \in \mathbf{S}_{ms\infty}^+$, *then* ${}_0x_\xi^h(\cdot) \in \mathbf{S}_{ms}^+$. In fact, as already noted in Section 12, the process ${}_0x_\xi^h(\cdot)$ is a martingale, and this process is singular according to Section 11(d).

(f) *If* $u \in \mathbf{S}_{msf}^+$, *then* ${}_0x_\xi^h(\cdot) \in \mathbf{S}_{ps}^+$ because the equation

$$\lim_{t\to\infty} \ell_D^h(t,\xi,u) = \lim_{t\to\infty} E\{{}_0x_\xi^h(t)\} = 0$$

shows that ${}_0x_\xi^h(\cdot) \in \mathbf{S}_p^+$, and this process is singular according to Section 11(d).

Section 11(c) is made more precise by (g) and (h) below.

(g) *If* $u \in \mathbf{S}_{pqb}^+$, *then* ${}_0x_\xi^h(\cdot) \in \mathbf{S}_{pqb}^+$. In view of Section 11(c) this assertion is trivial.

(h) *If* $u \in \mathbf{S}_{ps}^+$, *then* ${}_0x_\xi^h(\cdot) \in \mathbf{S}_{ps}^+$ *for every* ξ *for which* $u(\xi) < +\infty$. If $u(\xi) < +\infty$, then by Section 11(c) the process ${}_0x_\xi^h(\cdot)$ is in $\mathbf{S}_p^+$. To prove that the process is singular, suppose first that the measure associated with the superharmonic function uh has compact polar support A. (According to Section 10, this measure is supported by a polar set.) Define $D_0 = D - A$, and let u_0 be the restriction of u to D_0. Since almost no $w_\xi^h(\cdot)$ path from the point ξ meets A, the process for u_0 on D_0 analogous to ${}_0x_\xi^h(\cdot)$ for u on D can be taken as this same process ${}_0x_\xi^h(\cdot)$. The fact that ξ may not be in D_0 will not affect the following reasoning. The function u_0 on D_0 is singular as well as h-harmonic because a bounded positive h-harmonic minorant u_1 of u_0 has an h-harmonic extension to D by a trivial generalization of Theorem 1.V.5, and the extension of u_1 is a bounded h-harmonic minorant of u and therefore vanishes identically. Thus ${}_0x_\xi^h(\cdot)\in\mathbf{S}_s^+$ by (e); so ${}_0x_\xi^h(\cdot)\in\mathbf{S}_{ps}^+$ as stated. If, however, μ does not have compact support, the measure is the sum of a sequence of measures with disjoint compact polar supports; so ${}_0x_\xi^h(\cdot)\in\mathbf{S}_{ps}^+$, as was to be proved.

EXAMPLE. Let $N=2$, let D be the upper half-plane, and let $h\equiv 1$. The ordinate function u on D is a singular positive harmonic function, singular because any bounded harmonic minorant of u has limit 0 at every finite boundary point of D and therefore (Theorem 1.V.7) vanishes identically. (It is easy to prove that u is minimal, but this fact will not be needed.) The positive u-harmonic function $1/u$ has a finite limit along almost every u-Brownian path from a point ξ to ∂D so almost every such path tends to ∞. Direct calculation shows that u is subsumed under (e) so that if $w_\xi(\cdot)$ is a Brownian motion in D from ξ with lifetime S_ξ, the process ${}_0x_\xi^h(\cdot)$ with $h\equiv 1$ is a singular martingale. The martingale property amounts to the statement that a one-dimensional Brownian motion from a point $\alpha>0$ killed when it hits the origin is a martingale, a fact easily verified by direct computation.

Chapter II

Brownian Motion and the PWB Method

1. Context of the Problem

Let D be an open nonempty subset of $\mathbb{R}^N$, coupled with a boundary ∂D provided by a metric compactification. To avoid trivial complications, we shall assume that D is connected; if D is disconnected, the results are applicable to each open connected component of D. Let h be a strictly positive harmonic function on D. The PWB method of attacking the first boundary value (Dirichlet) problem for h-harmonic functions on D was detailed in Chapter VIII of Part 1. Recall that the σ algebra of μ_D^h measurable boundary subsets is the σ algebra of boundary subsets A for which the boundary indicator function 1_A is h-resolutive and that $\mu_D^h(\cdot, A) = H_{1_A}^h$. The class of Borel boundary subsets for which 1_A is h-resolutive is a σ algebra, and for each point ξ of D the restriction of $\mu_D^h(\xi, \cdot)$ to this σ algebra, on completion, is the measure $\mu_D^h(\xi, \cdot)$ on the σ algebra of μ_D^h measurable sets. The class of μ_D^h measurable boundary functions f which are $\mu_D^h(\xi, \cdot)$ integrable does not depend on ξ and is the class of h-resolutive boundary functions, and $H_f^h = \mu_D^h(\cdot, f)$. This class of boundary functions will also be described as the class of μ_D^h integrable boundary functions.

The PWB method solves the Dirichlet problem when the boundary is h-regular and the given boundary function is finite valued and continuous, but no relation other than the h-resolutivity itself was given in Chapter VIII of Part 1 to connect an h-resolutive boundary function f on a not necessarily h-resolutive boundary with the solution H_f^h. In Section 2 it will be shown that in the general case h-harmonic measure has a probabilistic evaluation and that H_f^h has f as a boundary limit function along h-Brownian paths. The solution H_f^h, which (by Lemma I.4) is necessarily in the class $\mathbf{D}(\mu_{D-}^h)$, when composed appropriately with h-Brownian motion defines a uniformly integrable martingale, and the evaluation $H_f^h = \mu_D^h(\cdot, f)$ becomes the martingale equality for the parameter values 0, $+\infty$.

Throughout the discussion in Section 2 $\{w_\xi^h(\cdot), \mathscr{F}_\xi^h(\cdot)\}$ is an h-Brownian motion in D from ξ, with lifetime S_ξ^h. With no loss of generality (Section 2.X.2) it can be assumed that $\mathscr{F}_\xi^h(\cdot)$ is right continuous and that $\mathscr{F}_\xi^h(0)$ contains the null sets. Since h is harmonic, almost every $w_\xi^h(\cdot)$ path tends to ∂D at the path lifetime (Section 2.X.4). Let $\Gamma_\xi^h(\omega)$ be the cluster set of $w_\xi^h(\cdot, \omega)$ at the parameter value S_ξ^h. For almost every ω this cluster set is a compact boundary subset. In particular, if $h \equiv 1$ and if ∂D is the Euclidean boundary, the process $w_\xi^h(\cdot)$ can be identified with a Brownian motion in $\mathbb{R}^N$ killed at the hitting time of ∂D, and $\Gamma_\xi^h(\omega)$ is then almost surely a singleton, $\{w_\xi^h[S_\xi^h(\omega)-]\}$. Similarly, if D is bounded, if ∂D is the Euclidean boundary, and if h has a harmonic extension h' to an open neighborhood D' of $\bar{D}$, then $w_\xi^h(\cdot)$ can be identified with an h'-Brownian motion in D' killed at the hitting time of ∂D, and again $\Gamma_\xi^h(\omega)$ is almost surely a singleton, $\{w_\xi^h[S_\xi^h(\omega)-]\}$.

According to the following theorem, whatever the choice of ∂D, the distribution of Γ_ξ^h is the h-harmonic measure $\mu_D^h(\xi,\cdot)$, an h-resolutive function f is for almost every ω a constant function on the cluster set $\Gamma_\xi^h(\omega)$, and H_f^h has the prescribed value of f as limit along almost every $w_\xi^h(\cdot)$ path to the boundary. Furthermore ∂D is h-resolutive if and only if Γ_ξ^h is almost surely a singleton, that is, if and only if the left limit $w_\xi^h(S_\xi^h-)$ exists almost surely. Thus the PWB method generalizes the classical concept of solution of the Dirichlet problem but nevertheless leads to solutions which have the prescribed boundary limit function in a reasonable sense.

The Parabolic Context

The methods used in this chapter are applicable in the parabolic context, and the statements of the results in that context are left to the reader.

2. Probabilistic Analysis of the PWB Method

Theorem. (a) *For every boundary subset A*

$$\bar{H}_{1_A}^h(\xi) = {}^hR_1^A(\xi) = \frac{R_h^A(\xi)}{h(\xi)} \qquad (= P\{\Gamma_\xi^h \cap A \neq \emptyset\} \text{ if } A \text{ is analytic}) \tag{2.1}$$

and

$$\underline{H}_{1_A}^h(\xi) = P\{\Gamma_\xi^h \subset A\} \quad \text{if } \partial D - A \text{ is analytic.} \tag{2.2}$$

(b) *If A is μ_D^h measurable, then*

$$H_{1_A}^h(\xi) = \mu_D^h(\xi, A) = P\{\Gamma_\xi^h \subset A\} = P\{\Gamma_\xi^h \cap A \neq \emptyset\}. \tag{2.3}$$

If f is a μ_D^h measurable boundary function then, for every ξ in D, f is identically constant on almost every cluster set Γ_ξ^h. Conversely if f is a Borel measurable boundary function and if, for some ξ in D, f is identically constant on almost every cluster set Γ_ξ^h, then f is μ_D^h measurable.

(c) *A boundary function f is h resolutive if and only if f is μ_D^h measurable and integrable, and in that case*

$$H_f^h(\xi) = \mu_D^h(\xi, f) = E\{f(\Gamma_\xi^h)\}, \tag{2.4}$$

$$\lim_{t \uparrow S_\xi^h} H_f^h[w_\xi^h(t)] = f(\Gamma_\xi^h) \quad \text{a.s.}, \tag{2.5}$$

and the process $\{x_\xi^h(\cdot), \mathscr{F}_\xi^h(\cdot)\}$ *defined by*

$$x_\xi^h(t) = \begin{cases} H_f^h[w_\xi^h(t)] & \text{if } t < S_\xi^h \\ f(\Gamma_\xi^h) & \text{if } S_\xi^h \le t \le +\infty \end{cases} \tag{2.6}$$

is a uniformly integrable almost surely continuous martingale.

(d) *The boundary is h-resolutive if and only if, for each point* ξ *in* D *(equivalently, for a single point* ξ *in* D*),* Γ_ξ^h *is almost surely a singleton, that is, if and only if the left limit* $w_\xi^h(S_\xi^h-)$ *almost surely exists in the* $D \cup \partial D$ *metric. The distribution of* $w_\xi^h(S_\xi^h-)$ *is then* $\mu_D^h(\xi, \cdot)$.

Observation. The equality $H_f^h(\xi) = E\{f(\Gamma_\xi^h)\}$ is the martingale equality for $x_\xi^h(\cdot)$ at the pair of parameter values 0, $+\infty$.

Proof of (a). The equality $\bar{H}_{1_A}^h = {}^hR_1^A = R_h^A/h$ was pointed out in Section 1.VIII.2. The last equality in (2.1) is a special case of Theorem 2.X.7. An application of (2.1) to $\partial D - A$ yields (2.2). □

Proof of (b). If A is μ_D^h measurable, that is, if 1_A is an h-resolutive boundary function, then if A is a Borel set, the equality of the terms in (2.1) with those in (2.2) yields (2.3). Since every μ_D^h null set is a subset of a Borel μ_D^h null set and since the σ algebra of μ_D^h measurable sets is the completion of the restriction of this σ algebra to the class of its Borel sets, it follows first that (2.3) is trivial when A is a μ_D^h null set and next that (2.3) is true when A is μ_D^h measurable. Conversely, suppose that for some point ξ and Borel boundary subset A the last equation in (2.3) is true. Then the positive harmonic function $\bar{H}_{1_A}^h - \underline{H}_{1_A}^h$ vanishes at ξ and therefore vanishes identically; that is, the set A is μ_D^h measurable. Thus the assertions in (b) for a boundary function f are true when f is the indicator function of a set, and are therefore true without this restriction.

Proof of (c). The h-resolutivity of a boundary function f implies the μ_D^h measurability of f and therefore implies that for each point ξ of D the function f is identically constant on almost every cluster set Γ_ξ^h; that is, $f(\Gamma_\xi^h)$ is almost surely uniquely defined. The first equality in (2.4) was obtained in Section 1.VIII.8, and the last equality follows from the probabilistic evaluation of μ_D^h in (2.3). The following more direct proof of (2.4) also proves (2.5). If u_2 [u_1] is in the upper [lower] PWB^h class for the h-resolutive boundary function f, the h-superharmonic functions $u_2 + c$, $c - u_1$, $u_2 - H_f^h$ are positive if the constant c is sufficiently large. Hence (Theorem 2.10.8) each function u_2, u_1, H_f^h has a finite limit along almost every $w_\xi^h(\cdot)$ path to the boundary, and

$$u_1(\xi) \le E\{\lim_{t\uparrow S_\xi^h} u_1[w_\xi^h(t)]\} \le E\{\lim_{t\uparrow S_\xi^h} H_f^h[w_\xi^h(t)]\} \le E\{\lim_{t\uparrow S_\xi^h} u_2[w_\xi^h(t)]\} \le u_2(\xi). \tag{2.7}$$

Moreover

$$\lim_{t\uparrow S_\xi^h} u_1[w_\xi^h(t)] \le \inf_{\zeta\in\Gamma_\xi^h} f(\zeta) \le \sup_{\zeta\in\Gamma_\xi^h} f(\zeta) \le \lim_{t\uparrow S_\xi^h} u_2[w_\xi^h(t)] \quad \text{a.s.} \tag{2.8}$$

Since u_1 and u_2 can be chosen to make $u_2(\xi) - u_1(\xi)$ arbitrarily small, f must be constant on almost every cluster set Γ_ξ^h, and (2.4) and (2.5) are true.

The martingale assertion in (c) follows from Theorem 2.X.8 and Sections I.4 and I.5 but will be clearer from the following direct proof. Denote the $x_\xi^h(\cdot)$ process in (2.6) by $x_\xi^h(f;\cdot)$. All supermartingales and martingales below are relative to the filtration $\mathscr{F}_\xi^h(\cdot)$. If f is an h-resolutive boundary function, Theorem 2.X.8 implies that the following processes are almost surely continuous positive supermartingales:

$$x_\xi^h(|f|;\cdot), \qquad x_\xi^h(|f|-f;\cdot), \qquad x_\xi^h(|f|\wedge n;\cdot), \qquad x_\xi^h(n-(|f|\wedge n);\cdot). \tag{2.9}$$

The fact that the last two of these processes are supermartingales implies that $x_\xi^h(|f|\wedge n;\cdot)$ is a martingale and therefore $(n\to\infty)$ that $x_\xi^h(|f|,\cdot)$ is a martingale, that is, that (c) is true when f is positive. Since the first two processes in (2.9) are now known to be almost surely continuous martingales, the process $x(f;\cdot)$ is also, as was to be proved. □

Proof of (d). If the boundary is h-resolutive, that is, if the Borel boundary subsets are μ_D^h measurable, choose $\varepsilon > 0$, and let $A_1, \ldots, A_k$ be compact boundary subsets of diameter at most ε and with union ∂D. Since (2.3) implies that Γ_ξ^h is almost surely a subset of any A_j it intersects, the diameter of Γ_ξ^h is almost surely at most ε. Hence Γ_ξ^h is almost surely a singleton; that is, the left limit $w_\xi^h(S_\xi^h-)$ almost surely exists. Conversely, if Γ_ξ^h is almost surely a singleton for some point ξ, the set function $A \mapsto \bar{H}_{1_A}^h(\xi)$ is an additive function of Borel boundary sets by (2.1); so the boundary is h-resolutive (Section 1.VIII.9). □

Special Case: h Is Minimal. If h is minimal, then for every boundary subset A the h-harmonic function $\bar{H}_{1_A}^h = R_h^A/h$ is either identically 1 or identically 0. In fact by minimality $\bar{H}_{1_A}^h$ is a constant function, so that $R_h^A \equiv ch$, and iterating the reduction operation [Section 1.VI.3(h)] yields $ch = cR_A^h = c^2h$; so $c = 0$ or 1. For every boundary point ζ if $R_h^{\{\zeta\}} \equiv 0$, then (since $R_h^{\{\zeta\}}$ is the infimum of the class of reductions on boundary neighborhoods of ζ) $R_h^A \equiv 0$

for A a sufficiently small boundary neighborhood of ζ. The set B of boundary points ζ with $R_h^{\{\zeta\}} \equiv 0$ is therefore open, and $R_h^B \equiv 0$. Thus the set $B' = \partial D - B$ is a compact set with the property that $\Gamma_\xi^h = B'$ almost surely, for each ξ. The class of μ_D^h measurable boundary subsets consists of all boundary subsets which are either supersets of B' or subsets of B. A boundary function f is h-resolutive if and only if it is constant (finite) on B' and the PWB^h solutions are the constant functions. In particular, a boundary is h-resolutive if and only if B' is a singleton $\{\zeta\}$ which supports every h-harmonic measure $\mu_D^h(\xi, \cdot)$, equivalently, if and only if for all ξ in D almost every $w_\xi^h(\cdot)$ path tends to ζ at the path lifetime. See Section 2.X.9 for this situation when D is a ball.

The Role of Capacity Theory

A close examination of the analysis reveals that the theory of analytic sets and Choquet capacities is not needed in the derivation of Theorem 2 if it is supposed that the left limit $w_\xi^h(S_\xi^h-)$ exists almost surely for some (equivalently every) point ξ of D or, going in the other direction, if the boundary is h-resolutive. This hypothesis is satisfied if ∂D is the Euclidean boundary and if either $h \equiv 1$ or D is a relatively compact open subset of an open set on which h has a positive harmonic extension.

3. PWB^h Examples

EXAMPLE (a). The Alexandrov one-point boundary of a Greenian set D is trivially universally resolutive. From the point of view of Theorem 2 universal resolutivity corresponds to the fact that for strictly positive harmonic h on D almost every h-Brownian path from a point of D tends to the one-point boundary at the path lifetime.

EXAMPLE (b). If D is a ball, its Euclidean boundary is universally resolutive according to Section 1.VIII.9; this universal resolutivity corresponds to the fact deduced in Section 2.X.9 that for strictly positive harmonic h on a ball D almost every h-Brownian path from a point of D tends to a point of the Euclidean boundary at the path lifetime.

EXAMPLE (c). Let $N = 2$, let D be a disk, let ζ be a boundary point, and let $h = K(\zeta, \cdot)$ be the minimal harmonic function corresponding to ζ (Section 1.II.1). For ξ in D let $\phi(\xi)$ be the angle between the rays from ζ to ξ and to the disk center, $-\pi/2 < \phi < \pi/2$. Then ϕ is harmonic on D. Define a distance function $d(\cdot, \cdot)$ on $D \times D$ by

$$d(\xi, \eta) = |\xi - \eta| + |\phi(\xi) - \phi(\eta)|. \tag{3.1}$$

If D is provided with a boundary $\partial^r D$ by completion under this distance function, the Euclidean boundary point ζ is ramified into a compact set A of points corresponding to the directions of approach to ζ. In this example h is minimal; so the quasi-bounded h-harmonic functions, in particular the PWB^h solutions, are the constant functions, and the Euclidean boundary singleton $\{\zeta\}$ has h-harmonic measure identically 1. Almost every h-Brownian path from a point ξ of D has limit ζ in the Euclidean metric. We now investigate the cluster sets of these paths on $\partial^r D$, that is, the boundary cluster sets in the d metric defined by (3.1). Recall from Section 1.XII.12 that if ζ is identified with a minimal Martin boundary point of D, no ray from ζ into D is minimal thin at ζ. We shall show in Section III.3 that a Borel subset B of a Greenian subset of $\mathbb{R}^N$ is not minimal thin at a minimal Martin boundary point if and only if almost every conditional Brownian path from a point of the Greenian set to that boundary point hits B arbitrarily near the boundary point. In the present context, using the notation relative to $\partial^r D$, it follows that Γ_ξ^h is almost surely the set A. According to Theorem 2, a function f on $\partial^r D$ is h-resolutive if and only if f is identically constant ($\neq \pm\infty$) on A, and if so, the PWB^h solution for f is identically the constant $f(A)$. The ramified boundary $\partial^r D$ has too many points to be h-resolutive.

Example (d). Let B be an open subset of a Greenian set D, let D be provided with a boundary ∂D by a metric compactification, and let ∂B be the boundary of B relative to this compactification. Then ∂B depends on the choice of ∂D unless B is relatively compact in D. According to Theorem 2, ∂B is h-resolutive if and only if almost every h-Brownian path in D from a point of B either hits $D \cap \partial B$ or tends to a point of $\partial D \cap \partial B$ at the path lifetime. Thus it is sufficient for h-resolutivity of ∂B that ∂D be h-resolutive. This simple resolutivity argument should be compared with the nonprobabilistic argument in Section 1.VIII.8, Example (b). The relation 1.VIII(8.3) connecting μ_D^h and μ_B^h is an immediate consequence of the strong Markov property of conditional Brownian motion.

Example (e). Suppose that the metric compactification $\bar{D} = D \cup \partial D$ of a Greenian set D has the following properties. There is a family $\{\phi_i, i \in I\}$ on D, indexed by some set I, for which

- (e_1) Each function ϕ_i has a continuous extension (also denoted by ϕ_i) to $\bar{D}$.
- (e_2) If $\eta \in \partial D$, a sequence $\eta_.$ in D has limit η if and only if $\lim_{n\to\infty} \phi_i(\eta_n) = \phi_i(\eta)$ for all i; that is, the family $\phi_.$ separates ∂D.
- (e_3) Under these conditions if also for some strictly positive harmonic function h on D $\lim_{t\uparrow S_\xi^h} \phi_i[w_\xi^h(t)]$ exists almost surely simultaneously for all i whenever $w_\xi^h(\cdot)$ is an h-Brownian motion from ξ with lifetime S_ξ^h, it follows that the left limit $w_\xi^h(S_\xi^h -)$ exists almost surely; so ∂D is h-resolutive. Since one standard technique of defining a boundary

for a set D is to choose a compactification of D which makes each function of a specified family of functions on D have a continuous extension to the compactification, this example is frequently encountered. For example, if D is a Greenian subset of $\mathbb{R}^N$ and if ξ_0 is a point of D, the Martin compactification of D (Chapter XII of Part 1) can be defined as the space satisfying (e$_1$) and (e$_2$) for the family of functions $\{\phi_\xi, \xi \in D\}$ defined by $\phi_\xi = G_D(\cdot, \xi)/G_D(\cdot, \xi_0)$. Moreover, if convenient, we can restrict ξ to a countable dense subset of $D - \{\xi_0\}$. A direct argument in Section III.5 will show that (e$_3$) is satisfied for every h and therefore that the Martin boundary is universally resolutive, but in fact this universal resolutivity was derived in Section 1.XII.10 from which it follows (Theorem 2) that almost every h-Brownian path from a point ξ converges to a Martin boundary point and that the distribution of the limit $w_\xi^h(S_\xi^h -)$ is $\mu_D^h(\xi, \cdot)$.

4. Tail σ Algebras in the PWB^h Context

Let D be a connected Greenian subset of $\mathbb{R}^N$, coupled with a boundary ∂D provided by a metric compactification, let h be a strictly positive harmonic function on D, and let $w_\xi^h(\cdot)$ be an h-Brownian motion in D from ξ, with lifetime S_ξ^h. Let $\mathscr{G}_\xi^h$ be the $w_\xi^h(\cdot)$ process tail σ algebra, and let $\Gamma_\xi^h(\omega)$ be the $w_\xi^h(\cdot, \omega)$ cluster set on ∂D. Recall that by definition the σ algebra $\mathscr{G}_\xi^h$ contains the $w_\xi^h(\cdot)$ null sets. Throughout this section ξ and $w_\xi^h(\cdot)$ are fixed.

Theorem. *If ∂D is internally h-resolutive, $\mathscr{G}_\xi^h$ consists [modulo $w_\xi^h(\cdot)$ null sets] of the class of sets of the form*

$$\{\omega\colon \Gamma_\xi^h(\omega) \subset A\} \quad \textit{for Borel } \mu_D^h \textit{ measurable boundary subsets } A. \tag{4.1}$$

In particular, if ∂D is also h-resolutive, $\mathscr{G}_\xi^h$ consists [modulo $w_\xi^h(\cdot)$ null sets] of the sets of the form

$$\{\omega\colon w_\xi^h(S_\xi^h -, \omega) \in A\} \textit{ for Borel boundary subsets } A. \tag{4.2}$$

Observation. Since a μ_D^h measurable boundary subset differs from some Borel and μ_D^h measurable boundary subset by a μ_D^h null set, and since the h-harmonic measure μ_D^h is determined by the distribution of Γ_ξ^h through (2.3), "Borel" can be omitted in (4.1) and replaced by "μ_D^h measurable" in (4.2). Theorem 4 implies that if ∂D is internally h-resolutive, then a $w_\xi^h(\cdot)$ process random variable z is $\mathscr{G}_\xi^h$ measurable if and only if there is a μ_D^h and Borel measurable function g on ∂D such that $z = g(\Gamma_\xi^h)$ almost surely, and if ∂D is also h-resolutive, Γ_ξ^h can be replaced here by $w_\xi^h(S_\xi^h -)$.

Special Case: h Is Minimal. If h is minimal, every bounded h-harmonic function is identically constant; so every choice of boundary for D is internally h-resolutive. The σ algebra $\mathscr{G}_\xi^h$ and the σ algebra of μ_D^h measurable sets are both trivial in this case; that is, their sets are all of measure 0 or 1. If A is a Borel boundary subset of h-harmonic measure 1, the σ algebra of μ_D^h measurable boundary subsets is generated by A and the μ_D^h null sets. If $\Lambda \in \mathscr{G}_\xi^h$ and if Λ is not $w_\xi^h(\cdot)$ null, then $\mathscr{G}_\xi^h$ is the σ algebra generated by Λ and the $w_\xi^h(\cdot)$ null sets. In particular, if ∂D is also h-resolutive, there is a point ζ of ∂D such that $\partial_D^h(\cdot, \{\zeta\}) \equiv 1$. In this case all subsets of ∂D are μ_D^h measurable.

Proof of Theorem 4. If A is μ_D^h measurable, that is, if the boundary function $f = 1_A$ is h-resolutive, then (by Theorem 2)

$$\lim_{t \uparrow s_\xi^h} H_f^h[w_\xi^h(t)] = f(\Gamma_\xi^h) \quad \text{a.s.},$$

and it follows (Section 2.X.11) that $f(\Gamma_\xi^h)$ is $\mathscr{G}_\xi^h$ measurable, that is, $\{\omega\colon \Gamma_\xi^h(\omega) \subset A\} \in \mathscr{G}_\xi^h$. Conversely, if $\Lambda \in \mathscr{G}_\xi^h$, it was shown in Section 2.X.11 that there is a bounded h-harmonic function u such that $\lim_{t \uparrow s_\xi^h} u[w_\xi^h(t)] = 1$ almost surely. Since the boundary is internally h-resolutive, $u = H_f^h$ for some μ_D^h and Borel measurable function f, and u has almost sure boundary limit function f along $w_\xi^h(\cdot)$ paths; that is, $f(\Gamma_\xi^h) = 1_\Lambda$ almost surely. Hence there is a μ_D^h and Borel measurable boundary subset A such that $f = 1_A$ up to a μ_D^h null set, and $\Lambda = \{\omega\colon \Gamma_\xi^h(\omega) \subset A\}$ up to a $w_\xi^h(\cdot)$ null set. If ∂D is also h-resolutive, every Borel measurable boundary subset is μ_D^h measurable, and a μ_D^h measurable boundary subset is a Borel set up to a μ_D^h null set; so the last assertion of the theorem is trivial.

Chapter III

Brownian Motion on the Martin Space

1. The Structure of Brownian Motion on the Martin Space

Let D be a connected Greenian subset of $\mathbb{R}^N$, let K be a Martin function for D, let h be a strictly positive superharmonic function on D, and let $\{w_\xi^h(\cdot), \mathscr{F}_\xi^h(\cdot)\}$ be an h-Brownian motion in D from ξ with lifetime S_ξ^h. For A a subset of D let S_ξ^{hA} and L_ξ^{hA}, respectively, be the hitting and last hitting times of A by $w_\xi^h(\cdot)$. According to Theorem 1.XII.10, if h is harmonic, the Martin boundary is h-resolutive and $\mu_D^h(\xi, d\zeta) = K(\zeta, \xi)M_h(d\zeta)/h(\xi)$, where M_h is the Martin representing measure of h corresponding to K. According to Theorem II.2, the left limit $w_\xi^h(S_\xi^h-)$ exists almost surely and has distribution $\mu_D^h(\xi, \cdot)$ supported (Section 1.XII.7) by the minimal Martin boundary $\partial_1^M D$. In particular, if ζ is a minimal Martin boundary point and if $h = K(\zeta, \cdot)$, then $\mu_D^h(\cdot, \{\zeta\}) = 1$; so $w_\xi^h(S_\xi^h-) = \zeta$ almost surely. With this choice of h we shall sometimes write $w_\xi^\zeta(\cdot)$, $S_\xi^{\zeta A}$, $L_\xi^{\zeta A}$, respectively, for $w_\xi^h(\cdot)$, S_ξ^{hA}, L_ξ^{hA}.

The analysis in Section 2.X.9 of h-Brownian motion in a ball compactified by its Euclidean boundary is applicable to an arbitrary connected Greenian set D compactified by its Martin boundary, with no change except for allowance for the possible presence of nonminimal Martin boundary points. That is, if h is harmonic, an h-Brownian path in D from ξ is obtained by first choosing the minimal asymptotic path endpoint ζ on $\partial_1^M D$ according to the distribution $\mu_D^h(\xi, \cdot)$ and then choosing a $K(\zeta, \cdot)$-Brownian motion path from ξ to ζ. More formally, one version of the conditional distribution of $w_\xi^h(\cdot)$ for $w_\xi^h(S_\xi^h-) = \zeta$ is the distribution of $K(\zeta, \cdot)$-Brownian motion from ξ. If $h = G_D\mu$ is a potential and if $h(\xi) < +\infty$, then (from Section 2.X.4) almost surely S_ξ^h is finite, and the left limit $w_\xi^h(S_\xi^h-)$ exists and is in D; $w_\xi^h(S_\xi^h-)$ has distribution $G_D(\xi, \zeta)\mu(d\zeta)/h(\xi)$. With this choice of h an h-Brownian motion path can be obtained by first choosing the asymptotic path endpoint ζ in D according to the distribution $G_D(\xi, \zeta)\mu(d\zeta)/h(\xi)$ and then choosing a $G_D(\cdot, \zeta)$-Brownian motion path from ξ to ζ; that is, one version of the conditional distribution of $w_\xi^h(\cdot)$ for $w_\xi^h(S_\xi^h-) = \zeta$ is the distribution of $G_D(\cdot, \zeta)$-Brownian motion from ξ. The case $h(\xi) = +\infty$ is treated in Section 2.X.5. If $h = h_1 + h_2$ is an arbitrary strictly positive superharmonic function on D, with h_1 positive harmonic and h_2 a potential (Riesz decomposition), h-Brownian motion from ξ, when $h(\xi) < +\infty$, is h_1-Brownian motion from ξ with probability $h_1(\xi)/h(\xi)$ and is h_2-Brownian motion from ξ with probability $h_2(\xi)/h(\xi)$.

Attainable and Unattainable Martin Boundary Points

According to Section 2.X.1, if ζ is a minimal Martin boundary point, the function $\xi \mapsto P\{S_\xi^\zeta = +\infty\}$ is either identically 0 or identically 1. The point ζ is called *attainable* in the first case and *unattainable* in the second. For example, if D is a ball (Section 2.X.9), its Martin (i.e., Euclidean) boundary points are all minimal and attainable. If D is a half-space its Martin (i.e., Euclidean) boundary points are all minimal, and the finite boundary points are attainable, but the infinite boundary point is not attainable. If $D = \mathbb{R}^N$ with $N > 2$, the minimal harmonic functions are the positive constant functions, and the point ∞ is the only Martin boundary point, is minimal, and is unattainable.

2. Brownian Motions from Martin Boundary Points (Notation of Section 1)

Brownian Motions from Attainable Boundary Sets

Let h be a strictly positive harmonic function on D whose Martin representing measure is supported by the set of attainable minimal Martin boundary points. Then for each point ξ of D the h-harmonic measure $\mu_D^h(\xi, d\zeta) = K(\zeta, \xi)M_h(d\zeta)/h(\xi)$ is also supported by this set. Let $w_\lambda^h(\cdot)$ be an h-Brownian motion in D with initial distribution λ and lifetime S_λ^h; this lifetime is necessarily almost surely finite. We define $w_\lambda^h(S_\lambda^h)$ as $\lim_{t\uparrow S_\lambda^h} w_\lambda^h(t)$, a limit which according to Theorem II.2 exists almost surely and has distribution $\int_D \mu_D^h(\xi, \cdot)\lambda(d\xi)$. According to Section 2.X.6, the process $\{w_\lambda^h(S_\lambda^h - t), t \in \mathbb{R}^+\}$ with lifetime S_λ^h is a $G_D\lambda$-Brownian motion, and obviously λ is the distribution of path endpoints $(t \to S_\lambda^h)$.

Special Case: Brownian Motion from an Attainable Martin Boundary Point to a Point of D

Let ζ be an attainable minimal Martin boundary point, and let ξ be a point of D. Then [particular case of the preceding paragraph with $h = K(\zeta, \cdot)$ and with λ supported by $\{\xi\}$] on the one hand the process $\{w_\xi^\zeta(t), t \in \mathbb{R}^+\}$ is a $K(\zeta, \cdot)$ Brownian motion from ξ, and almost every path tends to ζ at the path lifetime S_ξ^ζ, whereas on the other hand the process $\{w_\xi^\zeta(S_\xi^\zeta - t), t \in \mathbb{R}^+\}$ is a $G_D(\xi, \cdot)$-Brownian motion from ζ, and almost every path tends to ξ at the path lifetime S_ξ^ζ. From now on it will be convenient to write $w_\zeta^\xi(t)$ instead of $w_\xi^\zeta(S_\xi^\zeta - t)$ and denote by ℓ_D^ξ the $G_D(\xi, \cdot)$-Brownian motion transition density.

We now fix ξ and ζ and discuss the absolute probability distributions of the $w_\zeta^\xi(\cdot)$ process: $(t, A) \mapsto P\{w_\zeta^\xi(t) \in A\} = p(t, A)$ for A ranging through the Borel subsets of $D - \{\xi\}$. Since

$$\int_A l_N(d\eta) \int_D \ell_D^\xi(t-s, \eta', \eta) p(s, d\eta') = p(t, A), \qquad 0 < s < t, \tag{2.1}$$

it follows that $p(t, \cdot)$ is absolutely continuous relative to l_N and that one choice of the Radon–Nikodym derivative $dp(t, \cdot)/dl_N$ at η is

$$p_D^\xi(t, \zeta, \eta) = \int_D \ell_D^\xi(t-s, \eta', \eta) p(s, d\eta'). \tag{2.2}$$

Observe that the value of the integral on the right does not depend on the choice of s for $0 < s < t$ because ℓ_D^ξ satisfies the Chapman–Kolmogorov equation; so (2.2) defines $p_D(\cdot, \zeta, \cdot)$ on $]0, +\infty[\times (D - \{\xi\})$. The function $p_D^\xi(t, \zeta, \cdot)$ is the l_N density of the distribution of $w_\zeta^\xi(t)$ in D, and the function $p_D^\xi(\cdot, \zeta, \cdot)$ is $G_D(\xi, \cdot)$-parabolic, that is, $\phi = p_D^\xi(\cdot, \zeta, \cdot) G_D(\xi, \cdot)$ is parabolic, on $]0, +\infty[\times (D - \{\xi\})$ in view of the following properties of ϕ on this product set:

ϕ is Borel measurable.
ϕ is locally l_{N+1} integrable.
ϕ has the parabolic function average property locally, because ℓ_D $(\cdot - s, \eta, \cdot)$ is parabolic on the domain of ϕ.

It is natural to conjecture that when $\eta' \to \zeta$ suitably, the transition density $\ell_D^\xi(t, \eta', \eta)$ tends to $p_D^\xi(t, \zeta, \eta)$, and we now derive one version of such a limit relation: for each strictly positive t,

$$\lim_{s \to 0} \ell_D^\xi(t-s, w_\zeta^\xi(s), \eta) = p_D^\xi(t, \zeta, \eta). \tag{2.3}$$

To see this, observe that the function $(\eta', t') \mapsto \ell_D^\xi(t' \eta', \eta)$ is $G_D(\xi, \cdot)$-parabolic on $(D - \{\xi\}) \times]0, +\infty[$ and, in fact, is invariant excessive for $G_D(\xi, \cdot)$ space-time Brownian motion on this set. Moreover (2.2) implies that

$$E\{\ell_D^\xi(t-s, w_\zeta^\xi(s), \eta)\} = p_D^\xi(t, \zeta, \eta), \qquad 0 < s < t. \tag{2.4}$$

Hence the process $\{\ell_D^\xi(t-s, w_\zeta^\xi(s), \eta), 0 < s < t\}$ is a martingale. This martingale is almost surely continuous and so (from Section 2.III.16) has an almost sure limit as the parameter tends to 0. This limit is a random variable measurable with respect to the $K(\zeta, \cdot)$-Brownian motion tail σ algebra and so (from Section II.4) is almost surely identically constant. Since every martingale is uniformly integrable on a right closed set of parameter values [Section 2.III.3(e)], this constant must be the common expectation in (2.4); that is, (2.3) is true.

Brownian Motion from an Unattainable Minimal Martin Boundary Point to a Point of D

If ζ is an unattainable minimal Martin boundary point of D, that is (Section 2.X.1), if the function $K(\zeta, \cdot)$ is an invariant excessive function for Brownian motion in D, let ξ be a point of D, and let $D_\bullet$ be an increasing sequence of open relatively compact subsets of D, containing ξ, with union D. Let $L^\zeta_{\xi n}$ be the last hitting time of D_n by $w^\xi_\xi(\cdot)$. The function $\|K(\zeta, \cdot)\|^{D_n}$ (reduction relative to D) is the potential of a measure supported by ∂D_n, $\|K(\zeta, \cdot)\|^{D_n} = G_D\mu_n$. According to Section 2.X.10, the distribution of $w^\xi_\xi(L^\zeta_{\xi n})$ is

$$G_D(\xi, \eta)\mu_n(d\eta)/K(\zeta, \xi),$$

and the process $\{w^\xi_\xi(L^\zeta_{\xi n} - t), t \in \mathbb{R}^+\}$ with lifetime $L^\zeta_{\xi n}$ is a $G_D(\xi, \cdot)$-Brownian motion with initial distribution the distribution of $w^\xi_\xi(L^\zeta_{\xi n})$. Observe that $\lim_{n\to\infty} L^\zeta_{\xi n} = +\infty$ almost surely and that $\lim_{n\to\infty} w^\xi_\xi(L^\zeta_{\xi n}) = \zeta$ almost surely.

Example. Denote by d_ξ the Nth coordinate of the point ξ of $\mathbb{R}^N$, and let D be the half-space $\{d_\xi > 0\}$. The Green function G_D was evaluated in Section 1.VIII.9, and in Section 1.XII.3 the Martin boundary was found to be the Euclidean boundary. All boundary points are minimal, and the finite boundary points are attainable. By direct calculation, if ζ is a finite boundary point and if $\sigma^2 = 1$,

$$\lim_{\substack{\eta'\to\zeta \\ s\to 0}} \ell^\xi_D(t - s, \eta', \eta) = \lim_{\substack{\eta'\to\zeta \\ s\to 0}} \frac{\ell_D(t - s, \eta', \eta)G_D(\xi, \eta)}{G_D(\xi, \eta')}$$

$$= \begin{cases} \dfrac{d_\eta|\xi - \zeta|^2}{d_\xi 2\pi t^2} G_D(\xi, \eta) & \text{if } N = 2, \\[2ex] \dfrac{d_\eta|\xi - \zeta|^N}{d_\xi(N - 2)(2\pi)^{N/2} t^{(N+2)/2}} G_D(\xi, \eta) & \text{if } N > 2. \end{cases}$$

According to the results in this section, this limit is $p^\xi_D(t, \zeta, \eta)$. If now ξ tends to a second finite boundary point ζ_1, the transition density b^ξ_D tends to a limit transition density, that of $K(\zeta_1, \cdot)$-Brownian motion, and $p^\xi_D(t, \zeta, \cdot)$ tends to a limit density which can be described as the absolute probability density at time t of a Brownian motion from ζ to ζ_1.

3. The Zero-One Law at a Minimal Martin Boundary Point and the Probabilistic Formulation of the Minimal-Fine Topology (Notation of Section 1)

If ζ is a minimal Martin boundary point, almost every $w^\zeta_\xi(\cdot)$ path tends to ζ at the path lifetime, and (Section 2.X.11) the tail σ algebra of this process is trivial. As detailed in Section 2.X.11 for h-Brownian motion with h a minimal harmonic function, the asymptotic properties 2.VII.6(b)–(g) relating to set and function properties in the neighborhood of the initial point of a Brownian motion correspond to set and function properties near the lifetime of the h-Brownian motion, that is, in the present context, near ζ. For example, Section 2.VII.6(b) corresponds to (b*): if A is an analytic subset of D, the function $\xi \mapsto P\{L_\xi^{\zeta A} = S_\xi^\zeta\}$ is either identically 0 or identically 1.

The probabilistic formulation of the fine topology was given in Section 2.IX.15 [see also Section 2.X.2(d)]. The following theorem describes in probabilistic language the minimal-fine topology at a minimal Martin boundary point, defined nonprobabilistically in Section 1.XII.12.

Theorem. *Let ζ be a minimal Martin boundary point of D. For ζ to be a minimal-fine limit point of the subset A of D it is necessary that for every point ξ of D,*

$$P\{L_\xi^{\zeta B} = S_\xi^\zeta\} = 1 \tag{3.1}$$

whenever B is an analytic superset of A, and it is sufficient that (3.1) *be true for a single point ξ of D whenever B is an open superset of A.*

Thus if A is analytic, the point ζ is a minimal-fine limit point of A if and only if (3.1) is true when $B = A$. A subset A of D for which $D - A$ is analytic, for example, a Borel subset A of D, is a deleted minimal-fine neighborhood of ζ if and only if almost every $w_\xi^\zeta(\cdot)$ path lies in A during some parameter interval $]t, S_\xi^\zeta[$ with $t = t(\omega) < S_\xi^\zeta$; the value of t can be taken as the last hitting time of $D - A$. In both these statements it is necessary that the condition be satisfied for each point ξ of D and sufficient that the condition be satisfied for a single point ξ of D.

Proof of Theorem 3. An equivalent version of (3.1) is

$$P\{\limsup_{t \uparrow S_\xi^\zeta} 1_A[w_\xi^\zeta(t)] = 1\} = 1. \tag{3.1'}$$

According to Section 2.X.11, this condition is satisfied when A is analytic either for every point ξ of D or for no point ξ of D. If B is the trace on D of a neighborhood of ζ, then (Section 2.X.7) when A is analytic,

$$\frac{\|K(\zeta, \cdot)\|^{A \cap B}(\xi)}{K(\zeta, \xi)} = P\{S_\xi^{\zeta A \cap B} < S_\xi^\zeta\}. \tag{3.2}$$

Since the left side is 1 for all B if and only if ζ is a minimal-fine limit point of A and the right side is 1 for all B if and only if (3.1) is satisfied, the theorem is true when A is analytic. Since ζ is a minimal-fine limit point of a subset A of D if and only if ζ is a minimal-fine limit point of every open superset of A (Section 1.XII.12), the theorem is true for every set A.

Application to Function Limits

According to Theorem 3 and the discussion at the beginning of this section, a Borel measurable function from D into a Polish space D' has minimal-fine limit [minimal-fine cluster value] η' at the minimal Martin boundary point ζ if and only if $\lim_{t \uparrow s^\zeta_\xi} u[w^\zeta_\xi(t)] = \eta'$ [$u[w^\zeta_\xi(\cdot)]$ has cluster value η' at ζ along $w^\zeta_\xi(\cdot)$ paths] almost surely, for each point ξ of D, equivalently, for some point ξ of D. In particular, if $D' = \overline{\mathbb{R}}$,

$$\limsup_{t \uparrow s^\zeta_\xi} u[w^\zeta_\xi(t)] = \operatorname{mf}\limsup_{\eta \to \zeta} u(\eta) \quad \text{a.s.}$$

for each point ξ of D, and the corresponding equation for inferior limits is also true.

4. The Probabilistic Fatou Theorem on the Martin Space

Let h be a strictly positive harmonic function on a Greenian subset D of $\mathbb{R}^N$, let v be a positive superharmonic function on D, and define $u = v/h$. According to Theorem 2.X.8,

$$u_\xi = \lim_{t \uparrow s^h_\xi} u[w^h_\xi(t)] \tag{4.1}$$

exists almost surely. In view of the structure of h-Brownian motion discussed in Section 1, the existence of this almost sure limit means that for μ^h_D almost every Martin boundary point ζ (which we can suppose minimal) the function u has a limit along almost every $K(\zeta, \cdot)$-Brownian path to ζ from ξ. According to Section 2.X.11(c), this limit is for each ζ almost surely a number $f(\zeta)$ which does not depend on ξ. In view of Section 3 we have proved that

$$\operatorname{mf}\lim_{\eta \to \zeta} u(\eta) = f(\zeta)$$

exists for μ^h_D almost every ζ, a result we have already proved nonprobabilistically (see Theorem 1.XII.19). Observe that the boundary limit function was identified in Theorem 1.XII.19 as dM_v/dM_h, where M_h [M_v] is the Martin representing measure corresponding to h [to the harmonic component of v in its Riesz decomposition]. The following argument shows how f can be identified in the present context, without invoking Theorem 1.XII.19. It is sufficient to identify f for u an h-potential, u a singular h-harmonic function, and u a quasi-bounded h-harmonic function since (Section 1.IX.11) u is the sum of its components of these types. According to Theorem 2.X.8(c) as now interpreted, the limit function f vanishes μ^h_D almost surely if u is an h-potential or is a singular h-harmonic function. According to Theorem

2.X.8(d), $u(\xi) = E\{u_\xi\} = \mu_D^h(\xi, f)$ when u is quasi bounded and h-harmonic, in view of the expression we have found in Theorem II.2 for h-harmonic measure, and f here is dM_v/dM_h according to Theorem 1.XII.10. Thus we have derived Theorem 1.XII.19 probabilistically.

5. Probabilistic Approach to Theorem 1.XI.4(c) and Its Boundary Counterparts

Throughout this section v is a positive superharmonic function on a connected Greenian subset D of $\mathbb{R}^N$ with $N > 1$. We have already (Sections 1.XII.13 and 1.XII.14) proved the Martin boundary counterparts of Theorem 1.XI.4(c) which states that if ζ is a point of D, the function $v/G_D(\cdot, \zeta)$ has a fine limit at ζ. The first boundary counterpart (Theorem 1.XII.13) of Theorem 1.XI.4(c) asserts that if K is a Martin function for D and if ζ is a minimal Martin boundary point of D, then the function $v/K(\zeta, \cdot)$ has a minimal-fine limit at ζ. In view of the role of a point ζ of D as a minimal Martin boundary point of $D - \{\zeta\}$ [Section 1.XII.12, Example (a)] Theorem 1.XII.13 includes Theorem 1.XI.4(c). As noted in Section 1.XII.19 the boundary limit theorem (Theorem 1.XII.13) is a special case of the Fatou boundary limit theorem for the Martin space of D; so in view of the discussion in Section 4 there is no need to discuss this case probabilistically here except to remark that almost every $K(\zeta, \cdot)$-Brownian path from a point of D to the Martin boundary tends to ζ, and so the probabilistic Fatou boundary limit theorem becomes localized at ζ in this case. The second Martin boundary counterpart (Theorem 1.XII.14) of Theorem 1.XI.4(c) asserts that if $v \not\equiv 0$, the function $v/G_D(\xi, \cdot)$ has a strictly positive perhaps infinite minimal-fine limit at each minimal Martin boundary point ζ, and the probabilistic approach to this result will now be discussed.

It is sufficient to prove the existence and strict positivity of this minimal-fine limit for $v/G_D(\xi, \cdot)$ bounded, at the expense of replacing v if necessary by $v \wedge cG_D(\xi, \cdot)$ for a positive constant c. In the following we therefore suppose that $0 < v/G_D(\xi, \cdot) \leq c$. Let h be a strictly positive harmonic function on D, and let $w_\xi^h(\cdot)$ be an h-Brownian motion in D from ξ with lifetime S_ξ^h. The function h will be chosen to be $K(\zeta, \cdot)$ below, but it is instructive and does not complicate the preliminary work to allow the stated generality of h. If S_ξ^h is almost surely finite, that is, if the Martin representing measure of h is supported by the set of attainable minimal Martin boundary points, the process $\{w_\xi^h(S_\xi^h - t), t \in \mathbb{R}^+\}$ with lifetime S_ξ^h is a $G_D(\xi, \cdot)$-Brownian motion with initial distribution $\mu_D^h(\xi, \cdot)$ (Section 2). Hence the restriction to $D - \{\xi\}$ of the function $v/G_D(\xi, \cdot)$ is excessive for this process; so the composition of $v/G_D(\xi, \cdot)$ with this process, on the parameter set $]0, +\infty[$, defined as 0 for parameter values $\geq S_\xi^h$, is a positive bounded almost surely right continuous supermartingale $x(\cdot)$ and as such has an almost sure limit $x(0+)$, which we denote by $x(0)$. The process $\{x(t), t \in \mathbb{R}^+\}$ is a supermartingale, and $E\{x(0)\}$

> 0 because $v \not\equiv 0$. In particular, if $h = K(\zeta, \cdot)$ for some attainable minimal Martin boundary point ζ, the initial distribution of $w_\zeta^\xi(\cdot) = w_\xi^h(S_\xi^h - \cdot)$ is supported by $\{\zeta\}$; so $v/G_D(\xi, \cdot)$ has a limit at ζ along almost every $G_D(\xi, \cdot)$-Brownian path from ζ to ξ; equivalently (Section 3), $v/G_D(\xi, \cdot)$ has a minimal-fine limit at ζ, equal to $E\{x(0)\}$ and therefore strictly positive, as was to be proved. Observe that without the hypothesis that $h = K(\zeta, \cdot)$ we obtain only the weaker result that for $\mu_D^h(\xi, \cdot)$ almost every minimal Martin boundary point ζ the function $v/G_D(\xi, \cdot)$ has a minimal-fine limit at ζ. In view of this fact we shall assume that $h = K(\zeta, \cdot)$ in discussing limits at unattainable minimal Martin boundary points. Suppose then that ζ is such a point, and let $D_\cdot$ be an increasing sequence of open relatively compact subsets of D with union D, suppose that $\xi \in D_0$, let $L_{\xi n}^\zeta$ be the last hitting time of D_n by $w_\xi^\zeta(\cdot)$, and define the $K(\zeta, \cdot)$-potential $G_D\lambda_n/K(\zeta, \cdot)$ by

$$\frac{G_D\lambda_n}{K(\zeta, \cdot)} = {}^{K(\zeta,\cdot)}R_1^{D_n} = \frac{R_{K(\zeta,\cdot)}^{D_n}}{K(\zeta, \cdot)}.$$

Then (Theorem 2.X.7) $G_D\lambda_n/K(\zeta, \cdot)$ at η is the probability that a $K(\zeta, \cdot)$-Brownian path from η ever hits D_n. According to Section 2.X.10, the distribution of $w_\xi^\zeta(L_{\xi n}^\zeta)$ is $G_D(\xi, \eta)\lambda_n(d\eta)/K(\zeta, \xi)$, and as discussed in Section 2, the process $\{w_\xi^\zeta(L_{\xi n}^\zeta - t), t \in \mathbb{R}^+\}$ is a $G_D(\xi, \cdot)$-Brownian motion with lifetime $L_{\xi n}^\zeta$ and with initial distribution the distribution of $w_\xi^\zeta(L_{\xi n}^\zeta)$. The function $v/G_D(\xi, \cdot)$ is excessive for $G_D(\xi, \cdot)$-Brownian motion on $D - \{\xi\}$; so the process

$$\{x_n(t), t \in \mathbb{R}^+\} = \left\{\frac{v[w_\xi^\zeta(L_{\xi n}^\zeta - t)]}{G_D(\xi, w_\xi^\zeta(L_{\xi n}^\zeta - t)}, t \in \mathbb{R}^+\right\} \tag{5.1}$$

(set equal to 0 at times $\geq L_{\xi n}^\zeta$) is a bounded almost surely right continuous supermartingale. The supermartingale downcrossing inequality is applicable to $x_n(\cdot)$, and we find that the expected number of downcrossings of an interval $[r_1, r_2]$ by $x_n(\cdot)$ is at most $c/(r_2 - r_1)$. That is, the expected number of upcrossings, defined in the obvious way, of $[r_1, r_2]$ by the process

$$\left\{\frac{v[w_\xi^\zeta(t)]}{G_D(\xi, w_\xi^\zeta(t))}, 0 < t \leq L_{\xi n}^\zeta\right\}$$

is at most $c/(r_2 - r_1)$. Since this is true for all n, the same assertion is true for $0 < t < S_\xi^\zeta$, and we deduce just as in the discussion of supermartingale convergence that the function $v/G_D(\xi, \cdot)$ has a limit at ζ along almost every $w_\xi^\zeta(\cdot)$ path; equivalently (Section 3), the function $v/G_D(\xi, \cdot)$ has a minimal-fine limit at ζ. Observe that $E\{x_n(0)\} > 0$ because $v \not\equiv 0$, and observe also that $L_{\xi n+1}^\zeta - L_{\xi n}^\zeta$ is the hitting time of D_n by $x_{n+1}(\cdot)$; so by the supermartingale inequality at optional times $E\{x_{n+1}(0)\} \geq E\{x_n(0)\}$, and therefore $\lim_{n\to\infty} E\{x_n(0)\} > 0$.

Since this limit is the almost sure limit of $v/G_D(\xi,\cdot)$ at ζ along $w_\xi^\zeta(\cdot)$ paths as well as the minimal-fine limit of the function at ζ, this minimal-fine limit must be strictly positive ($\leq +\infty$).

Parabolic Context

The parabolic context version of Theorem 1.XI.4(c) is Theorem 1.XVIII.14(f) together with its dual. Theorem 1.XIX.13 is, together with its dual, the Martin boundary limit counterpart of the latter theorem. The probabilistic approach to these Martin boundary limit theorems is left to the reader.

6. Martin Representation of Harmonic Functions in the Parabolic Context

Let D be a Greenian subset of $\mathbb{R}^N$ ($N \geq 1$), define $\dot{D} = D \times \mathbb{R}$, $\dot{D}^+ = \dot{D} \times]0, +\infty[$, and let h be a positive harmonic function on D. Since h is b_D-excessive, the function $\mathscr{b}_D(\cdot,\xi,h)$ is a finite-valued monotone decreasing function on $]0,+\infty[$. The function $\mathscr{b}_D(\cdot,\cdot,h)$ is parabolic on $\dot{D}^+$ because $\mathscr{b}_D(\cdot,\cdot,\eta)$ is parabolic there for η in D; so $\mathscr{b}_D(\cdot,\cdot,h)$ is obtained by an integral operation on parabolic functions. The fact that h is $\mathscr{b}_D$-excessive implies that $\lim_{t\downarrow 0}\mathscr{b}_D(t,\xi,h) = h(\xi)$ and by Dini's theorem this monotone convergence is locally uniform on D. Thus if $\hat{\mathscr{b}}_D(t,\cdot,h)$ is defined on D as $\mathscr{b}_D(t,\cdot,h)$ for $t > 0$ and as h for $t \leq 0$, the function $\hat{\mathscr{b}}_D(\cdot,\cdot,h)$ is continuous on $\dot{D}$ and in fact is parabolic there because this function is already known to be parabolic on $\dot{D}^+$ and has the parabolic function average property on $\dot{D} - \dot{D}^+$. In the following we write $\mathscr{b}'_D(t,\xi,h)$ for $d\hat{\mathscr{b}}_D(t,\xi,h)/dt$. The function $\mathscr{b}'_D$ is thereby defined, negative, and parabolic on $\dot{D}$, identically 0 on $\dot{D} - \dot{D}^+$. This function vanishes identically if and only if h is $\mathscr{b}_D$ invariant excessive.

(a) The function $-\mathscr{b}'_D(\cdot,\cdot,h)$ is $\mathscr{b}_{\dot{D}}$ invariant excessive on $\dot{D}^+$; that is,

$$-\int_D \mathscr{b}_D(s,\xi,\eta)\mathscr{b}'_D(t,\eta,h)l_N(d\eta) = -\mathscr{b}'_D(s+t,\xi,h) \qquad (s>0, t>0). \tag{6.1}$$

In proving this it will be convenient to denote by $(6.1)^{\leq}$ the relation (6.1) with "$=$" replaced by "$\leq$." Inequality $(6.1)^{\leq}$ is true because $-\mathscr{b}'_D(\cdot,\cdot,h)$ is positive and parabolic and so is $\mathscr{b}_{\dot{D}}$-excessive. If both sides of $(6.1)^{\leq}$ are integrated with respect to $l_1(dt)$ over a compact subinterval of $]0,+\infty[$, we obtain an equality in view of the Chapman–Kolmogorov equation satisfied by $\mathscr{b}_D$. It follows that for fixed (ξ,s) equation (6.1) is true for l_1 almost every t. Furthermore the right side of $(6.1)^{\leq}$ defines a parabolic function of (ξ,s) on $\dot{D}^+$, as does the left side, which is obtained by applying an integral operator to parabolic functions. The difference between the right and left sides thus defines a positive parabolic function on $\dot{D}^+$, necessarily vanishing at all points below a zero. Hence, if $s' > 0$, equation (6.1) is true for $(\xi, s) \in D \times]0, s']$ if t is not in some l_1 null set. When s' runs through the strictly positive integers, we find that there is an l_1 null set A such that (6.1) is true for $(\xi, s) \in \dot{D}^+$ if and only if t is not in A. However, for t not in A and for strictly positive s and δ,

$$\begin{aligned} -\mathscr{E}'_D(\delta + s + t, \zeta, h) &= -\int_D \int_D \mathscr{E}_D(\delta, \zeta, \xi)\mathscr{E}_D(s, \xi, \eta)\mathscr{E}'_D(t, \eta, h) l_N(d\eta) l_N(d\xi) \\ &= -\int_D \mathscr{E}_D(\delta, \zeta, \xi)\mathscr{E}'_D(s + t, \xi, h) l_N(d\xi) \\ &\leq -\mathscr{E}'_D(\delta + s + t, \zeta, h). \end{aligned}$$

Hence there is equality throughout; so $s + t$ is not in A, and it follows that A is empty.

(b) Either $\mathscr{E}'_D(\cdot, \cdot, h) \equiv 0$ (and this is true if and only if h is an invariant excessive harmonic function) or $\mathscr{E}'_D(\cdot, \cdot, h) < 0$ on $\dot{D}^+$ because the positive parabolic function $-\mathscr{E}'_D(\cdot, \cdot, h)$ vanishes identically below each zero and (6.1) implies that this function vanishes identically above each 0 in $\dot{D}^+$.

(c) The function $\mathscr{E}'_D(\cdot, \cdot, h)$ determines h uniquely up to an invariant excessive harmonic function. Since [Section 2.IX.8, Example (a)] the function

$$\lim_{t \to \infty} \mathscr{E}_D(t, \cdot, h) = h_0 \tag{6.2}$$

is an invariant excessive harmonic function majorized by h, we obtain the same class of functions $\mathscr{E}'_D(\cdot, \cdot, h)$ if we restrict h by supposing that $h_0 \equiv 0$.

(d) We have proved that if h is a positive harmonic function on D, the positive parabolic function $\dot{u} = -\mathscr{E}'_D(\cdot, \cdot, h)$ on $\dot{D}$ has the following properties:

(d1) $\dot{u} = 0$ on $\dot{D} - \dot{D}^+$.

(d2) $\displaystyle\int_0^\infty \dot{u}(\xi, s) l_1(ds) < +\infty$.

(d3) $\displaystyle\int_D \mathscr{E}_D(s, \xi, \eta)\dot{u}(\eta, t) l_N(d\eta) = \dot{u}(\xi, s + t) \qquad (s > 0, t > 0)$.

Conversely, we now prove that if $\dot{u}$ is a positive parabolic function on $\dot{D}$ with these three properties, then $\dot{u}$ can be obtained in this way; that is, there is a positive harmonic function h on D such that $\dot{u} = -\mathscr{E}'_D(\cdot, \cdot, h)$. In fact the integral in (d2) defines a positive harmonic function h on D (Section 1.XV.15) and

$$\mathscr{E}_D(t, \xi, h) = \int_t^\infty \dot{u}(\xi, s) l_1(ds);$$

so $-\mathscr{E}'_D(\cdot,\cdot,h) = \dot{u}$. Observe that this procedure yields a function h for which h_0 in (6.2) vanishes identically.

(e) Probabilistic significance of $-\mathscr{E}'_D(\cdot,\cdot,h)$. If h is a strictly positive harmonic function on D and if h_0 in (6.2) vanishes identically, then according to Section 2.X.1 the restriction of $-\mathscr{E}'_D(\cdot,\xi,h)/h(\xi)$ to $\mathbb{R}^+$ is the density relative to l_1 of the distribution of the lifetime of h-Brownian motion in D from ξ.

(f) The positive harmonic function h on D is minimal harmonic if and only if the parabolic function $-\mathscr{E}'_D(\cdot,\cdot,h)$ is minimal parabolic on $\dot{D}$. (It is trivial that a positive parabolic function on $\dot{D}$, vanishing identically off $\dot{D}^+$, is minimal parabolic on $\dot{D}$ if and only if it is minimal parabolic on $\dot{D}^+$.) In fact, if h is minimal harmonic on D and if $\dot{u}_1$ is a positive parabolic minorant of $-\mathscr{E}'_D(\cdot,\cdot,h)$ on $\dot{D}$, then $\dot{u}_1$ satisfies (d1) and (d2). The function $\dot{u}_1$ also satisfies (d3) because both $\dot{u}_1$ and $-\mathscr{E}'_D(\cdot,\cdot,h) - \dot{u}_1$ are positive parabolic functions and so satisfy $(6.1)^{\leq}$ and their sum satisfies (6.1), and so $\dot{u}_1$ must also satisfy (6.1). Thus according to (d) there is a positive harmonic function h_1 on D such that

$$\dot{u}_1 = -\mathscr{E}'_D(\cdot,\cdot,h_1), \quad h_1 = \int_0^\infty \dot{u}_1(\cdot,s)l_1(ds).$$

Hence

$$h_1 \leq \int_0^\infty \dot{u}(\cdot,s)l_1(ds) = h;$$

so $h_1 = ch$, and therefore $\dot{u}_1 = -\mathscr{E}'_D(\cdot,\cdot,h_1) = -c\mathscr{E}'_D(\cdot,\cdot,h)$ on $\dot{D}^+$ and therefore on $\dot{D}$. Thus $-\mathscr{E}'_D(\cdot,\cdot,h)$ is minimal parabolic on $\dot{D}$. Conversely, if $-\mathscr{E}'_D(\cdot,\cdot,h)$ is minimal parabolic on $\dot{D}$ and if h_1 is a positive harmonic minorant of h, then

$$\mathscr{E}_D(s,\xi,h_1) + \mathscr{E}_D(s,\xi,h-h_1) = \mathscr{E}_D(s,\xi,h) = -\int_s^\infty \mathscr{E}'_D(t,\xi,h)l_1(dt).$$

Differentiate to find that $-\mathscr{E}'_D(\cdot,\cdot,h_1) \leq -\mathscr{E}'_D(\cdot,\cdot,h)$; so there is a constant c such that $\mathscr{E}'_D(\cdot,\cdot,h_1) = c\mathscr{E}'_D(\cdot,\cdot,h)$. Integration over $\mathbb{R}^+$ yields $h_1 = ch$, and it follows that h is minimal harmonic on D.

(g) If h is a minimal harmonic function on D but is not $\mathscr{E}_D$ invariant excessive, then h_0 in (6.2) vanishes identically. In this case define the parabolic function $\dot{h}$ on $\dot{D}$ by setting $\dot{h}(\xi,s) = h(\xi)$, and if $\alpha \in \mathbb{R}$, define $\dot{h}_\alpha(\xi,s) = -\mathscr{E}'_D(s-\alpha,\xi,h)$ to obtain a set of minimal parabolic functions on $\dot{D}$. Moreover (from Section 1.XV.17), if $b \in \mathbb{R}$, the restriction of $\dot{h}_\alpha$ to $D \times]-\infty,b[$ is minimal parabolic on this capped cylinder. Aside from normalizations the representation

$$\dot{h} = \int_{-\infty}^\infty \dot{h}_\alpha l_1(d\alpha)$$

is the Martin representation of $\dot{h}$ on $\dot{D}$ in terms of minimal parabolic functions. It is trivial that an $\dot{h}_\alpha$ space-time Brownian motion in $\dot{D}$ from a point (ξ_0, s_0) with $s_0 > \alpha$ has almost sure lifetime $s_0 - \alpha$. As noted in (e) the lifetime of h-Brownian motion in D from ξ_0 has l_1 distribution density $-\ell'_D(\cdot, \xi_0, h)/h(\xi_0)$; equivalently, the lifetime of $\dot{h}$ space-time Brownian motion in $\dot{D}$ from (ξ_0, s_0) has this distribution density. This space-time Brownian motion conditioned by fixing the lifetime to be $s_0 - \alpha$ becomes $\dot{h}_\alpha$ space-time Brownian motion.

Example. If $N = 1$ and if $D =]0, +\infty[$, there are two minimal harmonic functions on D up to multiplicative constants: the function $h \equiv 1$ and the function $\xi \mapsto h(\xi) = \xi$. The first of these leads to the minimal parabolic functions 1.XIX(10.2); the second is ℓ_D invariant excessive.

Appendixes

Appendix I

Analytic Sets

1. Pavings and Algebras of Sets

If X is a space, a *paving* of X is any class $\mathscr{X}$ of subsets of X which includes the empty set. If $\mathscr{X}$ is a paving, $\mathscr{X}_\sigma$ [$\mathscr{X}_\delta$] denotes the class of countable unions [intersections] of $\mathscr{X}$ sets. If a paving includes the complements and the finite [countable] unions of its sets, the paving is called a [σ] *algebra*. The smallest [σ] algebra containing a paving $\mathscr{X}$, that is, the intersection of the collection of all [σ] algebras containing $\mathscr{X}$, is called the [σ] *algebra generated by* $\mathscr{X}$. If $\mathscr{X}$ is an algebra, the σ algebra generated by $\mathscr{X}$ is the smallest collection of sets containing $\mathscr{X}$ and closed with respect to countable monotone unions and intersections.

If $(X, \mathscr{X})$ and $Y, \mathscr{Y})$ are paved spaces, the product paved space is the product space $X \times Y$ together with the *product paving* $\mathscr{X} \times \mathscr{Y}$ consisting of the class of product sets $A \times B$ with A in $\mathscr{X}$ and B in $\mathscr{Y}$. If $\mathscr{X}$ and $\mathscr{Y}$ are σ algebras, $\mathscr{X} \times \mathscr{Y}$ denotes the product σ algebra, that is, the σ algebra generated by the product paving.

2. Suslin Schemes

We use the notation $\mathbb{N} = \mathbb{Z}^+ \times \mathbb{Z}^+ \times \cdots$. Let $(Y, \mathscr{Y})$ be a paved space. A *Suslin scheme* is a map $Y_{\bullet}$ from the set of finite sequences of points of $\mathbb{Z}^+$ into $\mathscr{Y}$, $(n_1, \ldots, n_k) \mapsto Y_{n_1, \ldots, n_k} \in \mathscr{Y}$. To each point $n_{\bullet} = (n_1, n_2, \ldots)$ of $\mathbb{N}$ then corresponds the intersection $Y_{n_1} \cap Y_{n_1 n_2} \cap \cdots$, and the uncountable union $\bigcup_{n \in \mathbb{N}} (Y_{n_1} \cap Y_{n_1 n_2} \cap \cdots)$ is called the *nucleus* of the Suslin scheme and is said to be a set *analytic over* $\mathscr{Y}$. The class of these nuclei will be denoted by $\mathscr{A}(\mathscr{Y})$. If $A \in \mathscr{Y}$ and if $Y_{n_1, \ldots, n_k} = A$ for all k and $n_1, \ldots, n_k$, the nucleus is A; so $\mathscr{Y} \subset \mathscr{A}(\mathscr{Y})$. Furthermore $\mathscr{A}(\mathscr{Y})_\sigma = \mathscr{A}(\mathscr{Y})$ because if $A_j \in \mathscr{A}(\mathscr{Y})$, say

$$A_j = \bigcup_{n \in \mathbb{N}} (Y_{jn_1} \cap Y_{jn_1 n_2} \cap \cdots), \qquad Y_{jn_1, \ldots, n_k} \in \mathscr{Y}, \tag{2.1}$$

let α be a one-to-one map from $\mathbb{Z}^+$ onto $\mathbb{Z}^+ \times \mathbb{Z}^+$ and define

$$Y'_{n_1, \ldots, n_k} = Y_{\alpha(n_1) n_2 \ldots n_k}.$$

Then $\bigcup_0^\infty A_j$ is the nucleus of $Y'_\bullet$. Somewhat less easily $\mathscr{A}(\mathscr{Y})_\delta = \mathscr{A}(\mathscr{Y})$. To prove this, suppose that A_j is given by (2.1). Then $\bigcap_0^\infty A_j$ is an uncountable union whose general summand can be obtained by choosing in succession the first two sets from the union for A_1, then the first from the union for A_2, then the third from the union for A_1, then the second from the union for A_2, then the first from the union for A_3, and so on in the usual diagonal procedure, so that (remembering that the subscripts are dummies)

$$\bigcap_0^\infty A_j = \bigcup_{n \in \mathbb{N}} (Y_{1n_1} \cap Y_{1n_1n_2} \cap Y_{2n_3} \cap Y_{1n_1n_2n_4} \cap Y_{2n_3n_5} \cap \cdots).$$

The left side is the nucleus of $Y''_\bullet$ defined by

$$Y''_{n_1} = Y_{1n_1}, \qquad Y''_{n_1n_2} = Y_{1n_1n_2}, \qquad Y''_{n_1n_2n_3} = Y_{2n_3}, \ldots .$$

The class $\mathscr{A}(\mathscr{Y})$, although closed under countable unions and intersections, is not necessarily a σ algebra because it is not necessarily closed under complementation.

3. Sets Analytic over a Product Paving

Theorem. *Let $(X, \mathscr{X})$ and $(Y, \mathscr{Y})$ be paved spaces. Then $\mathscr{A}(\mathscr{X}) \times \mathscr{A}(\mathscr{Y}) \subset \mathscr{A}(\mathscr{X} \times \mathscr{Y})$. If $\xi \in X$ and $\hat{A} \in \mathscr{A}(\mathscr{X} \times \mathscr{Y})$, then $\{\eta: (\xi, \eta) \in \hat{A}\} \in \mathscr{A}(\mathscr{Y})$.*

To prove the first assertion, suppose that $A \in \mathscr{X}$. Using the obvious notation, it is clear that

$$A \times \mathscr{A}(\mathscr{Y}) \subset \mathscr{A}(\mathscr{X} \times \mathscr{Y}). \tag{3.1}$$

Similarly, if $A \in \mathscr{A}(\mathscr{Y})$,

$$\mathscr{A}(\mathscr{X}) \times A \subset \mathscr{A}(\mathscr{X} \times \mathscr{Y}), \tag{3.2}$$

and this is the desired inclusion relation. To prove the second assertion, suppose that $\hat{A} \in \mathscr{A}(\mathscr{X} \times \mathscr{Y})$, say

$$\hat{A} = \bigcup_{n_\bullet \in \mathbb{N}} (X_{n_1} \times Y_{n_1}) \cap (X_{n_1n_2} \times Y_{n_1n_2}) \cap \cdots .$$

The section of $\hat{A}$ for given ξ is the nucleus of $Z_\bullet(\xi)$ for

$$Z_{n_1 \ldots n_k}(\xi) = \begin{cases} Y_{n_1 \ldots n_k} & \text{if } \xi \in X_{n_1 \ldots n_k}, \\ \varnothing & \text{otherwise.} \end{cases}$$

4. Analytic Extensions versus σ Algebra Extensions of Pavings

Theorem. *If the complement of each set in the paving $\mathscr{Y}$ is in $\mathscr{A}(\mathscr{Y})$, $\mathscr{A}(\mathscr{Y})$ contains the σ algebra generated by $\mathscr{Y}$.*

The algebra $\mathscr{A}_0(\mathscr{Y})$ generated by $\mathscr{Y}$ is the class of finite unions of finite intersections of $\mathscr{Y}$ sets and their complements. Hence $\mathscr{A}_0(\mathscr{Y}) \subset \mathscr{A}(\mathscr{Y})$. Since $\mathscr{A}(\mathscr{Y}) = \mathscr{A}(\mathscr{Y})_\sigma = \mathscr{A}(\mathscr{Y})_\delta$, the class $\mathscr{A}(\mathscr{Y})$ is a monotone class. According to a standard theorem, $\mathscr{A}(\mathscr{Y})$ must therefore contain the σ algebra generated by $\mathscr{A}_0(\mathscr{Y})$, which is the same as that generated by $\mathscr{Y}$.

For example, if $\mathscr{Y}$ is the class of closed [open] subsets of a metric space Y, $\mathscr{A}(\mathscr{Y})$ contains the open [closed] subsets and therefore contains the σ algebra of Borel subsets.

5. Projection Characterization of $\mathscr{A}(\mathscr{Y})$

The following characterization of $\mathscr{A}(\mathscr{Y})$ is important for the application of the Suslin operation to stochastic processes.

Theorem. *Let $(Y, \mathscr{Y})$ be a paved space, and let $(X, \mathscr{X})$ be a topological space paved by its class of compact subsets. Then the projection on Y of a set in $\mathscr{A}(\mathscr{X} \times \mathscr{Y})$ is in $\mathscr{A}(\mathscr{Y})$. For a suitable choice of $(X, \mathscr{X})$, and some suitable choices are compact metric, $\mathscr{A}(\mathscr{Y})$ is the class of projections on Y of the sets in $(\mathscr{X} \times \mathscr{Y})_{\sigma\delta}$.*

Suppose that the paved spaces are as described in the first assertion and that $A \in \mathscr{A}(\mathscr{X} \times \mathscr{Y})$,

$$\begin{aligned} A &= \bigcup_{n_\cdot \in \mathbb{N}} (X_{n_1} \times Y_{n_1}) \cap (X_{n_1 n_2} \times Y_{n_1 n_2}) \cap \cdots \qquad X_{n_1 \ldots n_k} \in \mathscr{X},\ Y_{n_1 \ldots n_k} \in \mathscr{Y} \\ &= \bigcup_{n_\cdot \in \mathbb{N}} (X_{n_1} \cap X_{n_1 n_2} \cap \cdots) \times (Y_{n_1} \cap Y_{n_1 n_2} \cap \cdots). \end{aligned}$$

If $A_{n_1 \ldots n_k}$ is defined by

$$A_{n_1 \ldots n_k} = \begin{cases} Y_{n_1 \ldots n_k} & \text{if } X_{n_1} \cap \cdots \cap X_{n_1 \ldots n_k} \neq \varnothing, \\ \varnothing & \text{otherwise,} \end{cases}$$

the projection of A on Y is the nucleus of $A_\cdot$ and is therefore in $\mathscr{A}(\mathscr{Y})$, as stated in the first assertion of the theorem. To prove the second assertion, define X as the space of all sequences $n_\cdot$ of extended (that is, $\leq +\infty$) positive integers, metrized to be compact by defining the distance between $m_\cdot$ and $n_\cdot$ to be

$$\sum_{j=0}^{\infty} 2^{-j}|\arctan m_j - \arctan n_j|.$$

If $n_1, \ldots, n_k$ are in $\mathbb{Z}^+$, let $X_{n_1 \ldots n_k}$ be the compact set of points of X with first k coordinates $n_1, \ldots, n_k$. The nucleus of the Suslin scheme $Y_.$ on Y is then the projection on Y of

$$\bigcup_{n_. \in \mathbb{N}} (X_{n_1} \cap X_{n_1 n_2} \cap \cdots) \times (Y_{n_1} \cap Y_{n_1 n_2} \cap \cdots)$$
$$= \bigcap_0^{\infty} \bigcup (X_{n_1 \ldots n_k} \times Y_{n_1 \ldots n_k}) \in (\mathscr{X} \times \mathscr{Y})_{\sigma\delta},$$

where for each k the union in the second expression is over all finite-valued k-tuples $n_1 \ldots, n_k$.

6. The Operation $\mathscr{A}(\mathscr{A})$

Theorem. $\mathscr{A}[\mathscr{A}(\mathscr{Y})] = \mathscr{A}(\mathscr{Y})$.

The proof in Section 2 that $\mathscr{A}(\mathscr{Y}) = \mathscr{A}(\mathscr{Y})_\sigma = \mathscr{A}(\mathscr{Y})_\delta$ can be extended to prove this theorem, but the following proof is less tedious. A set in $\mathscr{A}[\mathscr{A}(\mathscr{Y})]$ is the projection on Y of a set in $[\mathscr{X} \times \mathscr{A}(\mathscr{Y})]_{\sigma\delta}$ for a suitable paved space $(X, \mathscr{X})$, as described in Theorem 5. Since (from Sections 2 and 3)

$$[X \times \mathscr{A}(\mathscr{Y})]_{\sigma\delta} \subset [\mathscr{A}(\mathscr{X} \times \mathscr{Y})]_{\sigma\delta} = \mathscr{A}(\mathscr{X} \times \mathscr{Y})$$

and since (by Theorem 5) the projection on Y of a set in the class on the right is in $\mathscr{A}(\mathscr{Y})$, the present theorem is true.

Application. The first assertion in Theorem 3 can now be strengthened. In fact

$$\mathscr{A}[\mathscr{A}(\mathscr{X}) \times \mathscr{A}(\mathscr{Y})] = \mathscr{A}(\mathscr{X} \times \mathscr{Y})$$

because Theorems 3 and 6 combine to yield

$$\mathscr{A}[\mathscr{A}(\mathscr{X}) \times \mathscr{A}(\mathscr{Y})] \subset \mathscr{A}[\mathscr{A}(\mathscr{X} \times \mathscr{Y})] = \mathscr{A}(\mathscr{X} \times \mathscr{Y}).$$

7. Projections of Sets in Product Pavings

Theorem. *Let $(X, \mathscr{X})$ be a locally compact second countable Hausdorff space paved by the class of its compact sets. If $(Y, \mathscr{Y})$ is a measurable space paved by its class of measurable sets, the projection on Y of a set in $\mathscr{A}[\mathscr{A}(\mathscr{X}) \times \mathscr{Y}]$ is in $\mathscr{A}(\mathscr{Y})$.*

In view of Theorem 5 it need only be pointed out that the application in Section 6 implies that $\mathscr{A}(\mathscr{X} \times \mathscr{Y}) = \mathscr{A}[\mathscr{A}(\mathscr{X}) \times \mathscr{Y}]$.

Application. In particular, if $\mathscr{B}(X)$ is the class of Borel subsets of X in Theorem 7, the projection on Y of a set in $\mathscr{A}[\mathscr{B}(X) \times \mathscr{Y}]$ is in $\mathscr{A}(\mathscr{Y})$.

8. Extension of a Measurability Concept to the Analytic Operation Context

Theorem. *If $(X, \mathscr{X})$, $(Y, \mathscr{Y})$ are paved spaces and if f is a function from X into Y with $f^{-1}(\mathscr{Y}) \subset \mathscr{X}$, then $f^{-1}(\mathscr{A}(\mathscr{Y})) \subset \mathscr{A}(\mathscr{X})$.*

This theorem is a trivial consequence of the fact that the inverse image under f of an arbitrary intersection [union] of sets is the intersection [union] of their inverse images.

9. The G_δ Sets of a Complete Metric Space

The following standard topological theorem is proved for orientation.

Theorem. (a) *A G_δ of a complete metric space is homeomorphic with a complete metric space.*

(b) *A set which is a subset of a complete metric space and is homeomorphic with a complete metric space is a G_δ.*

(c) *A complete separable metric space is homeomorphic with a G_δ in a compact metric space.*

Note that since a homeomorph of a separable space is separable, (a) and (b) are trivially specialized to separable spaces. Throughout the following proof, X is a complete metric space with distance function d.

Proof. (a) If $X_1 = \bigcap_0^\infty O_n$ with O_n an open subset of X, define a new distance function d_1 on X_1 by

$$d_1(\xi,\eta) = \sum_0^\infty 2^{-n} \wedge |d^{-1}(\xi, X - O_n) - d^{-1}(\eta, X - O_n)| + d(\xi,\eta)$$

to get a complete metric space (X_1, d_1) homeomorphic with (X_1, d).

(b) Let f map X homeomorphically into a complete metric space X'. Let $\alpha(\xi')$ be the oscillation of the inverse function at ξ' in the closure of $f(X)$, and define α as 0 off the closure of $f(X)$. Then α is upper semicontinuous on X' and vanishes on $f(X)$. Suppose that ξ' is in the closure of $f(X)$ and that $\alpha(\xi') = 0$. Let B_n' be the ball of center ξ' and radius $1/n$, so that $f^{-1}(B_n')$ is open and shrinks ($n \to \infty$) to a point ξ of X. Necessarily $\xi' = f(\xi) \in f(X)$. Let A'_n be the set of points of X' at distance $< 1/n$ from $f(X)$. Then $\bigcap_1^\infty [A'_n \cap \{\alpha < 1/n\}] = f(X)$ is a G_δ, as was to be proved.

(c) Replacing the distance function d on X by $d \wedge 1$ if necessary, it can be supposed that $d \leq 1$. Let $\xi_\bullet$ be a sequence dense in X. Then

$$\xi \mapsto \{d(\xi, \xi_n), n \geq 1\} = f(\xi)$$

is a map of X into the space X' of sequences $b_\bullet$ of positive reals ≤ 1. If the distance between $b_\bullet$ and $b'_\bullet$ in X' is defined as $\Sigma_0^\infty 2^{-n}|b_n - b'_n|$, the space X' is compact, and f is a homeomorphism. According to (b), $f(X)$ is a G_δ. □

10. Polish Spaces

A *Polish space* is defined as a Hausdorff space homeomorphic with a complete separable metric space. Applying Theorem 9 and the fact that a homeomorph of a separable set is separable, the following two assertions are trivial.

(a) A G_δ in a Polish space is a Polish space. In particular, every open subset (or G_δ) of $\mathbb{R}^N$ is a Polish space.

(b) A Polish space can be defined as any homeomorph of a G_δ of a compact metric space or of a G_δ of a complete separable metric space.

If X and Y are complete metric spaces, the product space topology can be metrized to make $X \times Y$ complete metric, and this product space is separable if X and Y are. If X and Y are Polish, the product space with product topology is Polish.

If X is a Hausdorff space which is locally compact and second countable, X is an open subset of its one-point compactification. Since the latter is metrizable, X is Polish.

11. The Baire Null Space

This space, denoted by BN below, is the metric space of sequences $(n_j, j \geq 1\}$ of strictly positive integers with

$$\text{dist}(m_\bullet, n_\bullet) = (\text{first } j \text{ with } m_j \neq n_j)^{-1} \quad \text{if } m_\bullet \neq n_\bullet$$

This space is complete and separable. It is zero dimensional because there is a topological base of clopen sets, the sets obtained by fixing the first k coordinates, $k = 1, 2, \ldots$.

According to Theorem 9, the set of irrationals in $[0, 1]$, a G_δ in the complete metric space $[0, 1]$, is homeomorphic with a complete metric space. In this case the image can be taken as the Baire null space, and the map can be written explicitly by means of the continued fraction representation of an irrational number x in $]0, 1[$,

$$x = \cfrac{1}{n_1 + \cfrac{1}{n_2 + \cdots}}, \qquad n_. \in \mathrm{BN}.$$

It will be convenient to denote by BN^∞ the space of sequences $\{n_j, j \geq 1\}$ for n_j a strictly positive integer or $+\infty$, with metric

$$\operatorname{dist}(m_., n_.) = \sum_1^\infty 2^{-j} |\arctan m_j - \arctan n_j|.$$

The space BN^∞ is compact metric, and the subset of this space with finite coordinates is a G_δ homeomorphic with BN.

Theorem. *Every Polish space is a continuous image of* BN.

Let X be Polish, and assume without loss of generality that X is complete separable metric. Choose closed nonempty sets $X_1, X_2, \ldots$ of diameters ≤ 1 with union X. If $X_{n_1 \ldots n_k}$ has been chosen, choose $X_{n_1 \ldots n_k 1}, X_{n_1 \ldots n_k 2}, \ldots$ closed nonempty sets of diameters $\leq 1/(k+1)$ with union $X_{n_1 \ldots n_k}$. The desired map from BN into X is the one taking $n_.$ into the single point in $X_{n_1} \cap X_{n_1 n_2} \cap \cdots$.

12. Analytic Sets

A set will be called *analytic* if it is a subset of a metrizable space and is analytic over the class of closed subsets of the space. The class of analytic subsets of a metrizable space includes the class of its Borel sets (Section 4). This inclusion is strict in all interesting contexts, for example, if the space is $\mathbb{R}^N$.

Theorem. *The following conditions on a subset A of a Polish space are equivalent.*

(a) *A is analytic.*
(b) *A is the continuous image of a Polish space.*
(c) *A is the continuous image of a G_δ of a compact metric space.*
(d) *A is the continuous image of* BN.
(e) *A is the continuous image of the set of irrationals in* $[0, 1]$.

In view of Theorems 9 and 11 only (a)$\Leftrightarrow$(d) needs proof.

Proof. (a) ⇒ (d) Under (a) there is a complete metric space X such that the set A is the nucleus of a Suslin scheme $X_.: n_1, \ldots, n_k \mapsto X_{n_1 \ldots n_k}$ with each set $X_{n_1 \ldots n_k}$ a closed subset of X. It will be convenient, and irrelevant to the construction of analytic sets, to restrict $n_1, n_2, \ldots$ to be strictly positive. For j and k strictly positive integers choose B_{jk} a closed subset of X, of diameter $< 1/k$, with $\bigcup_{j=1}^{\infty} B_{jk} = X$. Then

$$A = \bigcup (B_{n_1 1} \cap X_{n_2}) \cap (B_{n_3 2} \cap X_{n_2 n_4}) \cap (B_{n_5 3} \cap X_{n_2 n_4 n_6}) \cap \cdots, \quad (12.1)$$

where the union is over all sequences $n_1, n_2, \ldots$ of strictly positive integers. The kth set in parentheses is closed and has diameter $< 1/k$. Let $(a, b) \mapsto \alpha(a, b)$ be a one-to-one map of the set of pairs of strictly positive integers onto the set of strictly positive integers, and define

$$\begin{aligned} X'_{\alpha(n_1, n_2)} &= B_{n_1 1} \cap X_{n_2} \\ X'_{\alpha(n_1, n_2)\alpha(n_3, n_4)} &= (B_{n_1} \cap X_{n_2}) \cap (B_{n_3 2} \cap X_{n_2 n_4}) \\ &\vdots \end{aligned}$$

to get a Suslin scheme $X'_.$ for which, given $n_.$, the set $X'_{n_1 \ldots n_k}$ is closed, has diameter $< 1/k$, decreases as k increases, and either is empty for sufficiently large k or shrinks to a singleton $f(n_.)$. The function f is thereby defined on a subset of BN, is continuous, and maps its domain onto A. To extend f to BN, choose some point ξ of A, and if $X'_{n_1 \ldots n_k}$ is not empty, choose a point $\xi_{n_1 \ldots n_k}$ in this set. If $f(n_.)$ is not defined, either X'_{n_1} is empty, in which case define $f(n_.) = \xi$, or there is a maximal $k \geq 1$ for which $X'_{n_1 \ldots n_k}$ is not empty, in which case define $f(n_.) = \xi_{n_1 \ldots n_k}$. Then f is a continuous map from BN onto A. (Alternatively, it is easy to see that the original domain of f is closed and so is itself a Polish space and as such is the continuous image of BN, which makes A the continuous image of BN.)

(d) ⇒ (a) If (d) is true $A = f(\text{BN})$, where f is continuous from BN into the Polish space containing A. Let $Y_{n_1 \ldots n_k}$ be the set of points of BN with first k coordinates $n_1, \ldots, n_k$. This set is closed and has diameter $1/(k+1)$. Let $X_{n_1 \ldots n_k}$ be the closure of $f(Y_{n_1 \ldots n_k})$. If $n_. \in \text{BN}$,

$$Y_{n_1 \ldots n_k} \downarrow \{n_.\}, \qquad X_{n_1 \ldots n_k} \downarrow f(n_.) \qquad (k \to \infty),$$

and therefore A is the nucleus of $X_.$ and is analytic, as was to be proved. □

13. Analytic Subsets of Polish Spaces

Theorem. *Let X and Y be Polish spaces.*

(a) *If A [B] is an analytic subset of X [Y], $A \times B$ is an analytic subset of $X \times Y$.*

(b) *If $\hat{A}$ is an analytic subset of $X \times Y$ and if $\xi \in X$, the set $\{\eta : (\xi, \eta) \in \hat{A}\}$ is an analytic subset of Y.*
(c) *The projection on Y of an analytic subset of $X \times Y$ is analytic.*
(d) *The image on Y of an analytic subset of X under a Borel measurable map from X into Y is analytic.*

Parts (a) and (b) are special cases of Theorem 3. To prove (c), observe that the projection map is continuous; so in view of Theorem 12(b) the projection on Y of an analytic subset of $X \times Y$ is the continuous image of a continuous image of a Polish space and is therefore analytic. To prove (d), let f be a Borel measurable function from X into Y. The graph of f is a Borel and therefore analytic subset of $X \times Y$. Furthermore, if A is an analytic subset of X, $A \times Y$ is an analytic subset of $X \times Y$; so the intersection $\hat{A}$ of the graph of f with $A \times Y$ is analytic, and its projection on Y, that is to say $f(A)$, is therefore analytic by part (c), as was to be proved.

Appendix II

Capacity Theory

1. Choquet Capacities

If $(X, \mathscr{X})$ is a paved space for which $\mathscr{X}$ is closed under finite unions and intersections, a *Choquet capacity* on $(X, \mathscr{X})$, that is, on X relative to $\mathscr{X}$, is a function I from the class of subsets of X into $\overline{\mathbb{R}}$ with the following properties:

(a) I is increasing; that is, $X_1 \subset X_2$ implies that $I(X_1) \leq I(X_2)$.
(b) $X_n \uparrow X_\infty$ implies that $I(X_n) \to I(X_\infty)$.
(c) $X_n \in \mathscr{X}$ and $X_n \downarrow X_\infty$ implies that $I(X_n) \to I(X_\infty)$.

A subset A of X is called *capacitable* [relative to $(X, \mathscr{X}, I)$] if

$$I(A) = \sup\{I(B)\colon B \subset A, B \in \mathscr{X}_\delta\}. \tag{1.1}$$

It is trivial that the union of a monotone increasing sequence of capacitable sets is capacitable.

2. Sierpinski Lemma

Lemma. *Let $(X, \mathscr{X})$ be a paved space, let A be the nucleus of a Suslin scheme $X_\cdot$ over $\mathscr{X}$, and let $b_1, b_2, \ldots$ be a sequence in $\mathbb{Z}^+$. Define*

$$\begin{aligned} A^k &= \bigcup_{\substack{n_\cdot : n_i \leq b_i \\ i \leq k}} X_{n_1} \cap X_{n_1 n_2} \cap \cdots \\ B^k &= \bigcup_{\substack{n_i \leq b_i \\ i \leq k}} X_{n_1} \cap \cdots \cap X_{n_1 \cdots n_k}, \qquad B = \bigcap_1^\infty B^k. \end{aligned} \tag{2.1}$$

Then $A^k \subset A$, $A^k \subset B^k$, $B^1 \supset B^2 \supset \cdots$, and $B \subset A$.

Only the assertion that $B \subset A$ is not trivial. To prove this relation, suppose that $\xi \in B$, so that for every k there is a k-tuple $(n_1, \ldots, n_k)$ with $n_i \leq b_i$ for which $\xi \in X_{n_1} \cap \cdots \cap X_{n_1 \cdots n_k}$. Call such a k-tuple *admissible*. Then admissibility of $(n_1, \ldots, n_k)$ implies admissibility of $(n_1, \ldots, n_{k-1})$. Since the first integer of an admissible k-tuple is at most b_1, some integer m_1 must be the first integer of an admissible k-tuple for infinitely many values of k. Similarly, for some $m_2 \leq b_2$ the pair (m_1, m_2) must be the first two integers of an admissible k-tuple for infinitely many values of k, and so on, so that there is a sequence $m_{\bullet}$ with $\xi \in X_{m_1} \cap X_{m_1 m_2} \cap \cdots \subset A$, and therefore $B \subset A$, as was to be proved.

Observe that $B \in \mathscr{X}_\delta$ if $\mathscr{X}$ is closed under finite unions and intersections.

The set B increases when members of the sequence $b_{\bullet}$ increase, and it is intuitively reasonable that in some useful sense B can be made close to A by choosing each b_j sufficiently large. The following Choquet Capacity Theorem offers a precise version of this closeness.

3. Choquet Capacity Theorem

Theorem. *Let $(X, \mathscr{X})$ be a paved space for which $\mathscr{X}$ is closed under finite unions and intersections. If I is a Choquet capacity on $(X, \mathscr{X})$, the sets of $\mathscr{A}(\mathscr{X})$ are capacitable.*

If $A \in \mathscr{A}(\mathscr{X})$ and $I(A) = -\infty$, the set A is trivially capacitable because the empty set is in $\mathscr{X}$. If $I(A) > -\infty$, choose $\alpha < I(A)$. It will be proved, in the notation of Sierpinski's lemma, that $I(B) \geq \alpha$ if $b_{\bullet}$ is chosen suitably, and in view of property (c) of a Choquet capacity it is sufficient to choose $b_{\bullet}$ to make $I(B^k) \geq \alpha$ for all k and therefore sufficient to choose $b_{\bullet}$ to make $I(A^k) > \alpha$ for all k. To make such a choice, choose b_1 so large [property (b) of capacity] that $I(A^1) > \alpha$, and if $b_1, \ldots, b_{k-1}$ have been chosen so that $I(A^{k-1}) > \alpha$, choose b_k so large that $I(A^k) > \alpha$, again using property (b).

4. Lusin's Theorem

Theorem. *If $(X, \mathscr{X}, \mu)$ is a complete measure space for which μ is a countable sum of finite measures, for example, if μ is σ finite, then $\mathscr{A}(\mathscr{X}) = \mathscr{X}$.*

It is sufficient to prove the theorem for μ a complete finite measure. Define the outer measure of an arbitrary subset A of X by

$$\mu^*(A) = \inf\{\mu(B) \colon A \subset B \in \mathscr{X}\}.$$

Then μ^* is a Choquet capacity relative to $(X, \mathscr{X})$. In fact the defining property Section 1(a) is trivial, (b) is a standard property of outer measures defined in this way, and (c) is a standard property of finite measures. According to the Choquet capacity theorem, the sets of $\mathscr{A}(\mathscr{X})$ are capacitable, and in the present context capacitability and measurability are equivalent.

5. A Fundamental Example of a Choquet Capacity

Let $(\Omega, \mathscr{F}, P)$ be a complete finite measure space, and define an outer measure P^* on the class of subsets of Ω by

$$P^*(A) = \inf\{P(B): B \supset A, B \in \mathscr{F}\}.$$

Let $\hat{\mathscr{F}}$ be the class of finite unions of subsets of $\mathbb{R}^+ \times \Omega$ of the form $C \times \Lambda$, where C is a compact subset of $\mathbb{R}^+$ and $\Lambda \in \mathscr{F}$. Then $\hat{\mathscr{F}}$ is a paving of $\mathbb{R}^+ \times \Omega$, closed under finite unions and intersections. If $A \subset \mathbb{R}^+ \times \Omega$, let $\pi(A)$ be the projection of A on Ω, and define $I(A) = P^*(\pi(A))$. Observe that

$$\mathscr{A}(\hat{\mathscr{F}}) = \mathscr{A}(\mathscr{B}(\mathbb{R}^+) \times \mathscr{F}) \supset \mathscr{B}(\mathbb{R}^+) \times \mathscr{F},$$

that $\pi(\mathscr{A}(\hat{\mathscr{F}})) = \mathscr{A}(\mathscr{F}) = \mathscr{F}$, and (by Theorem I.5) that $I(A) = P(\pi(A))$ if $A \in \mathscr{A}(\hat{\mathscr{F}})$. An important result in capacity theory is that I so defined is a Choquet capacity on $\mathbb{R}^+ \times \Omega$ relative to $\hat{\mathscr{F}}$, that is, that I satisfies Section 1(a)–(c). Condition (a) that I is monotone increasing is obviously satisfied. The left continuity condition (b) is satisfied because every outer measure defined like P^* in terms of a measure satisfies this condition. To prove that the right continuity condition (c) is satisfied, suppose that $\hat{S}_\bullet$ is a decreasing sequence in $\hat{\mathscr{F}}$ with intersection $\hat{S}$. Then the point ω is in $\pi(\hat{S}_n)$ if and only if the compact set $\{t: (t, \omega) \in \hat{S}_n\}$ is not empty; so $\pi(\hat{S}_\bullet)$ is a decreasing sequence in $\mathscr{F}$ with intersection $\pi(\hat{S})$. Hence

$$\lim_{n \to \infty} I(\hat{S}_n) = \lim_{n \to \infty} P(\pi(\hat{S}_n)) = P(\pi(\hat{S})) = I(\hat{S});$$

that is, (c) is satisfied.

Application. It is trivial that since I is a Choquet capacity relative to $\hat{\mathscr{F}}$, this set function is also a Choquet capacity relative to an arbitrary subpaving of $\hat{\mathscr{F}}$ closed under finite unions and intersections. The Choquet capacity theorem applied to I with various choices of such subpavings of $\hat{\mathscr{F}}$ can be applied (Section 2.II.8) to show that certain subsets of $\mathbb{R}^+ \times \Omega$ contain significant parts of the graphs of suitably defined functions from Ω into $\mathbb{R}^+$.

6. Strongly Subadditive Set Functions

Let $(X, \mathscr{X})$ be a paved space for which $\mathscr{X}$ is closed under finite unions and intersections. A function I from $\mathscr{X}$ into either $]-\infty, +\infty]$ or $[-\infty, +\infty[$ is called *strongly subadditive* on $(X, \mathscr{X})$ if

(a) I is increasing; that is, $A \subset B$ implies that $I(A) \leq I(B)$.
(b) $I(A \cup B) + I(A \cap B) \leq I(A) + I(B)$.

If I is strongly subadditive and α is a constant, $I \wedge \alpha$ and $I + \alpha$ are also strongly subadditive. Under (a), condition (b) is equivalent to

(c) For an arbitrary strictly positive integer n when $A_1, \ldots, A_n$ and $B_1, \ldots, B_n$ are in $\mathscr{X}$ and $A_j \subset B_j$ for $j \leq n$,

$$I\left(\bigcup_1^n B_j\right) + \sum_1^n I(A_j) \leq \sum_1^n I(B_j) + I\left(\bigcup_1^n A_j\right). \tag{6.1}$$

In fact under (a) and under (c) with $n = 2$ inequality (6.1) yields (b) when $A_1 = A \cap B$, $A_2 = A$, $B_1 = B$, $B_2 = A$. Conversely, if (a) and (b) are true, (c) is proved as follows. If $A = B_1$ and $B = A_1 \cup B_2$, conditions (a) and (b) yield

$$I(B_1 \cup B_2) + I(A_1) \leq I(B_1) + I(A_1 \cup B_2). \tag{6.2}$$

If $A = A_1 \cup A_2$ and $B = B_2$, conditions (a) and (b) yield

$$I(A_1 \cup B_2) + I(A_2) \leq I(A_1 \cup A_2) + I(B_2). \tag{6.3}$$

These two inequalities combine to yield (6.1) for $n = 2$, and the general case is proved by induction. If (6.1) is true for a countable sequence (involving countable unions and sums) of pairs (A_j, B_j) in X with $A_j \subset B_j$, the set function I is called *countably strongly subadditive*.

7. Generation of a Choquet Capacity by a Positive Strongly Subadditive Set Function

Let I be a positive strongly subadditive set function on $(X, \mathscr{X})$, and let $\hat{\mathscr{X}}$ be a subclass of $\mathscr{X}_\sigma$, closed under finite intersections and countable unions. Define the function $A \mapsto I^*(A)$ from the class of all subsets of X to $\overline{\mathbb{R}}^+$ by

$$I^*(A) = \sup\{I(B): B \subset A, B \in \mathscr{X}\} \quad \text{if } A \in \hat{\mathscr{X}}, \tag{7.1}$$

and for an arbitrary subset A of X define

$$I^*(A) = \inf\{I^*(B): B \supset A, B \in \hat{\mathscr{X}}\}. \tag{7.2}$$

In (7.2) it is to be understood as usual that the right side is to be $+\infty$ if $\hat{\mathscr{X}}$ contains no superset of A. Equations (7.1) and (7.2) are consistent for A in $\hat{\mathscr{X}}$. Note that I^* need not be equal to I on $\mathscr{X}$. For example, if $\hat{\mathscr{X}}$ is empty, $I^*(A) = +\infty$ for every A.

Theorem. *In the preceding context if*

(a) $I(A_n) \to I(A)$ *when* A_n *and* A *are in* $\mathscr{X}$ *and* $A_n \uparrow A$
(b) $I(A_n) \to I^*(A)$ *when* A_n *is in* $\mathscr{X}$ *and* $A_n \downarrow A$

then I^* is an extension of I, does not depend on the choice of $\hat{\mathscr{X}}$, and is a Choquet capacity on X relative to $\mathscr{X}$, countably strongly subadditive on the class of all subsets of X. If $I^*(A) < +\infty$, the set A is $(X, \mathscr{X})$ capacitable if and only if to each $\varepsilon > 0$ correspond sets A'_ε in $\mathscr{X}_\delta$ and A''_ε in $\hat{\mathscr{X}}$ satisfying

$$A'_\varepsilon \subset A \subset A''_\varepsilon, \qquad I^*(A''_\varepsilon) < I^*(A'_\varepsilon) + \varepsilon. \tag{7.3}$$

It is trivial from (b) with $A_1 = A_2 = \cdots$ that $I^* = I$ on $\mathscr{X}$. The proof that I^* is a countably strongly subadditive Choquet capacity will be carried through in four steps.

(i) *$A_n \uparrow A$ with A_n in $\hat{\mathscr{X}}$ implies that* $I^*(A_n) \to I^*(A)$ because if $A_n = \bigcup_{k=0}^{\infty} A_{nk}$ with A_{nk} in $\mathscr{X}$, if $B_n = \bigcup_{m,k \le n} A_{mk}$, and if B is in $\mathscr{X}$ and is a subset of A, then $B_n \in \mathscr{X}$, $B_n \uparrow A$, $B_n \cap B \uparrow B$, and

$$\lim_{n\to\infty} I^*(A_n) \ge \lim_{n\to\infty} I(B_n \cap B) = I(B). \tag{7.4}$$

The limit on the left is therefore at least (and trivially at most) $I^*(A)$.

(ii) *I^* is strongly subadditive on the class of all subsets of X.* In fact, since the inequality in Section 6(b) is true for A and B in $\mathscr{X}$, increasing sequences of pairs (A, B) yield the same inequality for sets in $\hat{\mathscr{X}}$. If A and B are arbitrary and if either $I^*(A)$ or $I^*(B)$ is infinite, this strong subadditivity inequality is trivially true for I^*. If both values are finite and $A \subset A' \in \hat{\mathscr{X}}$, $B \subset B' \in \hat{\mathscr{X}}$, then

$$I^*(A \cup B) + I^*(A \cap B) \le I^*(A' \cup B') + I^*(A' \cap B') \le I^*(A') + I^*(B'), \tag{7.5}$$

and (ii) is true because the right side can be made arbitrarily close to $I^*(A) + I^*(B)$.

(iii) *$A_n \uparrow A$ implies that $I^*(A_n) \to I^*(A)$.* This conclusion is trivial if $\lim_{n\to\infty} I^*(A_n) = +\infty$. If this limit is finite and if $\varepsilon > 0$, choose A'_n in $\hat{\mathscr{X}}$ in such a way that $A'_n \supset A_n$, $I^*(A'_n) < I^*(A_n) + 2^{-n}\varepsilon$. By strong subadditivity [see (6.1)]

$$I^*\left(\bigcup_1^m A'_n\right) \le I^*\left(\bigcup_1^m A_n\right) + \varepsilon = I^*(A_m) + \varepsilon. \tag{7.6}$$

Hence

$$I^*(A) \le I^*\left(\bigcup_1^\infty A'_n\right) = \lim_{m\to\infty} I^*\left(\bigcup_1^m A'_n\right) \le \lim_{m\to\infty} I^*(A_m) + \varepsilon, \tag{7.7}$$

so that $\lim_{m\to\infty} I^*(A_m)$ is at least (and trivially at most) $I^*(A)$.

(iv) *I^* is countably strongly subadditive* because I^* satisfies (6.1) for arbitrary sets $A_j \subset B_j$, and this inequality yields countable strong subadditivity in view of (iii) when $n \to \infty$.

The conditions (a), (b), and (i)–(iv) imply that I^* is a countably strongly subadditive Choquet capacity relative to $(X, \mathscr{X})$. The last assertion of the theorem is trivial. There remains the proof that I^* does not depend on the choice of $\hat{\mathscr{X}}$. Let I_σ^* be the Choquet capacity obtained with the choice $\hat{\mathscr{X}} = \mathscr{X}_\sigma$, and let I^* be the Choquet capacity obtained with some other choice. It is assumed that (b) in the theorem is true for both choices. It is trivial that $I_\sigma^* \leq I^*$ and that there is equality on $\mathscr{X}$ and therefore also on $\mathscr{X}_\sigma$. Now if A is an arbitrary subset of X,

$$I_\sigma^*(A) = \inf\{I_\sigma^*(B): B \supset A, B \in \mathscr{X}_\sigma\} = \inf\{I^*(B): B \supset A, B \in \mathscr{X}_\sigma\} \geq I^*(A),$$

and therefore $I_\sigma^* = I^*$, as asserted.

8. Topological Precapacities

Let $(X, \mathscr{X})$ be a locally compact second countable space together with its class of compact subsets. Suppose that I is a function from $\mathscr{X}$ into $\bar{\mathbb{R}}^+$ with the following properties:

(a) I is strongly subadditive on $\mathscr{X}$.
(b) If $A_\bullet$ is a monotone sequence of compact sets with compact limit A, then $\lim_{n\to\infty} I(A_n) = I(A)$.

Then I will be called a *topological precapacity* on X. Theorem 7 can be applied with $\hat{\mathscr{X}}$ the class of open subsets of X if $I^*(A) = I(A)$ for A compact, and this is true because if $B_\bullet$ is a decreasing sequence of relatively compact open subsets of X with $\bigcap_0^\infty B_n = \bigcap_0^\infty \bar{B}_n = A$, then in view of (b) and the fact that $I(A) \leq I^*(B_n) \leq I(\bar{B}_n)$,

$$I(A) \leq \lim_{n\to\infty} I^*(B_n) \leq \lim_{n\to\infty} I(\bar{B}_n) = I(A). \tag{8.1}$$

Thus a topological precapacity I has an extension to a countably strongly subadditive Choquet capacity I^* relative to the class of compact sets and

$$I^*(A) = \sup\{I(B): B \text{ compact}, B \subset A\} \qquad (A \text{ open}), \tag{8.2}$$

$$I^*(A) = \inf\{I^*(B): B \text{ open}, A \subset B\} \qquad (A \text{ arbitrary}). \tag{8.3}$$

The class of capacitable sets includes the analytic subsets of X. A set A with $I^*(A)$ finite is capacitable if and only if to each $\varepsilon > 0$ correspond a compact subset A'_ε of A and an open superset A''_ε of A such that $I^*(A''_\varepsilon) < I(A'_\varepsilon) + \varepsilon$. Letting ε run through a sequence with limit 0, it follows that A with $I^*(A)$ finite is capacitable if and only if there is an F_σ subset A' of A and a G_δ superset A'' of A such that $I^*(A') = I^*(A) = I^*(A'')$. This condition is trivially necessary and sufficient for capacitability of A when $I^*(A) = +\infty$.

Note, however, that if A has compact subsets with arbitrarily large values of I, then $I^*(A) = +\infty$, and A is capacitable no matter how complicated its structure.

Example. Let $(X, \mathscr{X})$ be $\mathbb{R}$ together with its compact subsets, and define $I(A)$ for A compact as 0 or 1 according as A is empty or not. Then I is a topological precapacity. Let $\mathscr{X}$ be the class of open subsets of $\mathbb{R}$. The extension I^* is 1 on every nonempty set, and all sets are I^* capacitable. Observe that I^* has the property that there is a decreasing sequence $A_.$ of capacitable sets for which

$$I^*\left(\bigcap_0^\infty A_n\right) = 0 < \lim_{n\to\infty} I^*(A_n) = 1.$$

Application to a Uniqueness Result

Let I be a function from the class of all subsets of a locally compact second countable space into $\bar{\mathbb{R}}^+$. Suppose that I is monotone increasing, that the restriction of I to the class of compact sets is a topological precapacity, and that if B is open

$$I(B) = \sup\{I(A): A \subset B, A \text{ compact}\}.$$

It follows that the extension of this restriction to I^* coincides with I on the class of capacitable sets, in particular, on the analytic sets.

9. Universally Measurable Sets

Let $(X, \mathscr{X})$ be a measurable space, and let μ be a finite measure on $\mathscr{X}$. The class $\mathscr{X}_\mu$ of μ measurable sets obtained by completing μ is the σ algebra of sets generated by $\mathscr{X}$ and the subsets of μ null sets. The intersection $\mathscr{U}(\mathscr{X}) = \bigcap_\mu \mathscr{X}_\mu$ is a σ algebra, the class of *universally measurable* sets. It is easy to verify that $\mathscr{U}(\mathscr{X})$ is unchanged if μ in this definition is allowed to be σ finite or more generally if μ is allowed to be a countable sum of finite measures. Since $\mathscr{U}(\mathscr{X}) \subset \mathscr{X}_\mu$ for every finite measure μ, it follows that $\mathscr{U}(\mathscr{U}(\mathscr{X})) \subset \mathscr{U}(\mathscr{X})$, and the reverse inclusion is trivial.

Apply Lusin's theorem to find that for every finite measure μ on $\mathscr{X}$,

$$\mathscr{U}(\mathscr{X}) \subset \mathscr{A}(\mathscr{U}(\mathscr{X})) \subset \mathscr{A}(\mathscr{X}_\mu) \subset \mathscr{X}_\mu$$

so that

$$\mathscr{U}(\mathscr{X}) \subset A(\mathscr{U}(\mathscr{X})) \subset \mathscr{U}(\mathscr{X}),$$

that is, $\mathscr{A}(\mathscr{U}(\mathscr{X})) = \mathscr{U}(\mathscr{X})$.

If f is a map from the measurable space $(X, \mathscr{X})$ into the measurable space $(Y, \mathscr{Y})$ and if $f^{-1}(\mathscr{Y}) \subset \mathscr{X}$ that is, if f is measurable, a measure μ on X transforms into a measure μ_f on $\mathscr{Y}$ by way of $\mu_f(A) = \mu[f^{-1}(A)]$. Moreover, if μ is complete, this relation remains valid for A in the domain of the completion of μ_f. In particular, this relation is valid for A in $\mathscr{U}(\mathscr{Y})$ and $f^{-1}[\mathscr{U}(\mathscr{Y})] \subset \mathscr{U}(\mathscr{X})$; that is, f is measurable from $(X, \mathscr{U}(\mathscr{X}))$ into $(Y, \mathscr{U}(\mathscr{Y}))$.

Appendix III

Lattice Theory

1. Introduction

The following is an outline of the lattice theory used in this book. Elementary proofs are omitted or merely sketched. Vector spaces are always over the reals. The reader is warned that the nomenclature of this subject has not been standardized and that therefore the definitions given below are not universally accepted.

2. Lattice Definitions

A *lattice* is a class of objects ordered by a transitive reflexive binary relation for which every pair x, y of the objects has a unique order supremum $x \curlyvee y$ and infimum $x \curlywedge y$. The lattice will be called *complete* if every subset has a supremum and infimum. If every upper-bounded subset has a supremum, then every lower-bounded subset has an infimum, the supremum of the lower bounds of the subset, and if every lower-bounded subset has an infimum, then every upper-bounded subset has a supremum, the infimum of the upper bounds of the subset. If these suprema and infima exist, the lattice will be called *conditionally complete*.

A lattice will be said to have the *countability property* if each subset which has a supremum [infimum] contains a countable subset with the same supremum [infimum].

A subset of a lattice is called a *sublattice* if $x \curlyvee y$ and $x \curlywedge y$ are in the subset whenever x and y are.

3. Cones

A cone is a set closed under a commutative addition operation $(x + y)$ and multiplication by positive constants (cx) satisfying the usual associative and distributive laws together with

$$x + y = 0 \Rightarrow x = y = 0,$$

$$x + y = x + y' \Rightarrow y = y' \qquad \text{(unique subtraction)}.$$

When $x + y = z$, the element y will be written $z - x$. According to this definition a cone $\mathcal{N}$ in a vector space $\mathcal{M}$ is a convex set with the property that $x \in \mathcal{N}$ implies that $cx \in \mathcal{N}$ if and only if $c \geq 0$ or $x = 0$. The cone is said to *generate* $\mathcal{M}$ if $\mathcal{M} = \mathcal{N} - \mathcal{N}$, that is, if every element in $\mathcal{M}$ is the difference between two elements of $\mathcal{N}$. An arbitrary cone $\mathcal{N}$ can be immersed in a vector space $\mathcal{M}$ generated by $\mathcal{N}$, as follows. The elements of $\mathcal{M}$ are the pairs x: (x_1, x_2) of elements of $\mathcal{N}$ under the identification $(x_1, x_2) = (y_1, y_2)$ when $x_1 + y_2 = x_2 + y_1$. This identification is a transitive relation because there is unique subtraction in $\mathcal{N}$. Addition of elements of $\mathcal{M}$ and multiplication by real constants are defined by

$$(x_1, x_2) + (y_1, y_2) = (x_1 + y_1, x_2 + y_2)$$

$$c(x_1, x_2) = \begin{cases} (cx_1, cx_2) & \text{if } c \geq 0, \\ ((-c)x_2, (-c)x_1) & \text{if } c < 0. \end{cases}$$

The cone $\mathcal{N}$ is identified with the class of elements of $\mathcal{M}$ having representations of the form $(x, 0)$, and (x_1, x_2) is also written $x_1 - x_2$ when there is no danger of confusion.

4. The Specific Order Generated by a Cone

A cone can be ordered by the convention that $x \precsim y$ if there is a cone element z such that $x + z = y$. This order, called the *specific order*, is transitive and reflexive, and $x \precsim y$ implies that $cx \precsim cy$ when $c \geq 0$ and that $x + z \precsim y + z$ for all cone elements z. Moreover if $x + z \precsim y + z$ for some cone element z, then $x \precsim y$. A supremum or infimum of a set of cone elements is necessarily unique.

If in the specific order every pair of cone elements has a supremum [infimum], then every pair also has an infimum [supremum] and the cone is therefore a lattice. To see this, suppose for example that every pair x, y has a supremum $x \curlyvee y$. Then it will now be shown that $x \curlywedge y$ exists and

$$x \curlyvee y + x \curlywedge y = x + y. \tag{4.1}$$

Since $x + y \succsim x$ and $x + y \succsim y$, it follows that $x + y \succsim x \curlyvee y$. Define $z = x + y - x \curlyvee y$. Then

$$z + x \curlyvee y = x + y \precsim x \curlyvee y + y;$$

so $z \precsim y$ and similarly $z \precsim x$. On the other hand, if $z' \precsim x$ and $z' \precsim y$, then

$$z' + x \curlyvee y = (z' + x) \curlyvee (z' + y) \precsim x + y;$$

so $z' \precsim z$, and it follows that $z = x \curlywedge y$, as asserted.

If $\mathcal{M}$ is a vector space and if $\mathcal{N}$ is a cone in $\mathcal{M}$, the space $\mathcal{M}$ can be given a transitive reflexive order by setting $x \preceq y$ if $y - x \in \mathcal{N}$. The order thereby assigned to $\mathcal{N}$ is the specific order, and $\mathcal{N}$ becomes the set $\mathcal{M}^+$ of positive elements of $\mathcal{M}$, that is, the set of elements $x \succeq 0$. This ordering of $\mathcal{M}$ is called the *specific order relative* to $\mathcal{N}$. It is trivial that in this order the pair of relations $x \preceq y$ and $y \preceq x$ implies that $x = y$. Conversely, let $\mathcal{M}$ be a vector space with a transitive reflexive order in which this implication is valid and in which $x \preceq y$ implies both that $cx \preceq cy$ for $c \in \mathbb{R}^+$ and that $x + z \preceq y + z$ for $z \in \mathcal{M}$. The set $\mathcal{M}^+$ of elements $\succeq 0$ is then a cone, and the $\mathcal{M}$ order is the specific order relative to $\mathcal{M}^+$.

5. Vector Lattices

A "*vector lattice*" $\mathcal{M}$ is defined as a vector space which is a lattice in the specific order determined by some cone $\mathcal{M}^+$ as described in Section 4. The cone $\mathcal{M}^+$ is a sublattice of $\mathcal{M}$ and has the property that to each element x_0 of $\mathcal{M}$ corresponds an element x (for example, $x_0 \curlyvee 0$) of $\mathcal{M}^+$ such that $x \succeq x_0$. Conversely, if $\mathcal{N}$ is a cone in a vector space $\mathcal{M}$, if $\mathcal{N}$ is a lattice in its specific order, and if to each element x_0 of $\mathcal{M}$ corresponds an element x in $\mathcal{N}$ such that $x - x_0$ is in $\mathcal{N}$, then $\mathcal{M}$ is a lattice in the specific order induced by $\mathcal{N}$. To see this, observe that if x_1 and x_2 are in $\mathcal{M}$, there must be an element x in $\mathcal{N}$ such that $x - x_1$ and $x - x_2$ are both in $\mathcal{N}$, and it is easy to verify that the desired supremum $x_1 \curlyvee x_2$ is given by

$$x - [(x - x_1) \curlywedge (x - x_2)]$$

and that then the desired infimum is given by $-[(-x_1) \curlyvee (-x_2)]$. A vector lattice $\mathcal{M}$ is [conditionally] complete if and only if its positive cone is [conditionally] complete in its specific order. For example, if $\mathcal{M}^+$ is conditionally complete, a subset $\{x_\alpha, \alpha \in I\}$ of $\mathcal{M}$ with an upper bound x has a supremum, namely $x - \bigwedge_{\alpha \in I} (x - x_\alpha)$.

If $\mathcal{M}$ is a vector lattice,

$$z + (x \curlyvee y) = (z + x) \curlyvee (z + y), \qquad z + (x \curlywedge y) = (z + x) \curlywedge (z + y) \tag{5.1}$$

and

$$(-x) \curlywedge (-y) = -(x \curlyvee y). \tag{5.2}$$

The property (5.2) can be used to deduce properties involving suprema from dual properties involving infima and conversely. We shall use the notation

$$x^+ = x \curlyvee 0, \qquad x^- = (-x) \curlyvee 0, \qquad |x| = x \curlyvee (-x). \tag{5.3}$$

Applying (5.1) and (5.2),

$$x + y - (x \curlyvee y) = (x + y) + (-x) \curlywedge (-y) = y \curlywedge x;$$

that is, (4.1) is true for vector lattices. We rewrite this result for later reference together with some useful relations easily deducible from it.

$$x \curlyvee y + x \curlywedge y = x + y, \qquad x = x^+ - x^-, \qquad |cx| = |c|\,|x|, \qquad x^+ \curlywedge x^- = 0,$$
$$|x| = x^+ + x^- = x^+ \curlyvee x^-, \qquad |x + y| \preceq |x| + |y|. \tag{5.4}$$

The second relation implies that $\mathscr{M} = \mathscr{M}^+ - \mathscr{M}^+$. If $x = x_1 - x_2$ with $x_i \succeq 0$, then $x_1 \succeq x^+$ and $x_2 \succeq x^-$.

Associative Law. If $\{x_\alpha, \alpha \in I\}$ is a subset of the vector lattice $\mathscr{M}$, if I is partitioned arbitrarily, $I = \bigcup_{\beta \in J} I_\beta$, and if $\bigvee_{\alpha \in I_\beta} x_\alpha = y_\beta$ exists for every β in J, then $\bigvee_{\alpha \in I} x_\alpha$ exists if and only if $\bigvee_{\beta \in J} y_\beta$ exists, in which case the two suprema are the same. This fact is trivial, as is the corresponding dual one for infima.

Distributive Law. Let $\{x_\alpha, \alpha \in I\}$ be a subset of $\mathscr{M}$ for which $\bigvee_{\alpha \in I} x_\alpha$ exists. It follows that $\bigvee_{\alpha \in I} (x_\alpha \curlywedge y)$ exists for y in $\mathscr{M}$ and that

$$\bigvee_{\alpha \in I} (x_\alpha \curlywedge y) = \left(\bigvee_{\alpha \in I} x_\alpha\right) \curlywedge y. \tag{5.5}$$

Dually, the existence of $\bigwedge_{\alpha \in I} x_\alpha$ implies that $\bigwedge_{\alpha \in I} (x_\alpha \curlyvee y)$ exists and that

$$\bigwedge_{\alpha \in I} (x_\alpha \curlyvee y) = \left(\bigwedge_{\alpha \in I} x_\alpha\right) \curlyvee y. \tag{5.6}$$

Only (5.5) will be proved. The right side of (5.5) majorizes $x_\alpha \curlywedge y$ for every α, and conversely, if $x \succeq x_\alpha \curlywedge y$ for every α, then

$$x \succeq x_\alpha \curlywedge y = x_\alpha + y - (x_\alpha \curlyvee y) \succeq x_\alpha + y - \left[\left(\bigvee_{\alpha \in I} x_\alpha\right) \curlyvee y\right];$$

so

$$x \succeq \bigvee_{\alpha \in I} x_\alpha + y - \left[\left(\bigvee_{\alpha \in I} x_\alpha\right) \curlyvee y\right] = \left(\bigvee_{\alpha \in I} x_\alpha\right) \curlywedge y,$$

and therefore (5.5) is true.

6. Decomposition Property of a Vector Lattice

Theorem. *If x, y_1, y_2 are in the positive cone of a vector lattice and if $x \precsim y_1 + y_2$, there are positive elements x_1 and x_2 satisfying*

$$x_1 \precsim y_1, \qquad x_2 \precsim y_2, \qquad x_1 + x_2 = x. \tag{6.1}$$

In fact, if x_1 is defined as $x \curlywedge y_1$ and if $x_2 = x - x_1$, then

$$x_2 = (x - y_1) \curlyvee 0,$$

so that x_2 is positive and is majorized by y_2.

This decomposition theorem implies that if y_1, y_2, y are positive elements of a vector lattice, then

$$(y_1 + y_2) \curlywedge y \precsim y_1 \curlywedge y + y_2 \curlywedge y. \tag{6.2}$$

In fact the left side is majorized by $y_1 + y_2$; so according to the decomposition theorem, there are positive elements x_1, x_2 satisfying

$$x_i \precsim y_i, \qquad x_1 + x_2 = (y_1 + y_2) \curlywedge y.$$

Then $x_i \precsim y$; so $x_i \precsim y_i \curlywedge y$, and it follows that $x_1 + x_2 \precsim y_1 \curlywedge y + y_2 \curlywedge y$, which yields (6.2).

7. Orthogonality in a Vector Lattice

Elements x and y of a vector lattice $\mathscr{M}$ are said to be *orthogonal* if $|x| \curlywedge |y| = 0$, a relation written $x \perp y$. Two subsets of $\mathscr{M}$ are said to be orthogonal if every element of one is orthogonal to every element of the other. The set of elements of $\mathscr{M}$ orthogonal to a set $\mathscr{N}$ is denoted by $\mathscr{N}^\perp$. According to (5.4), $x \perp y$ if and only if $|x| + |y| = |x| \curlyvee |y|$.

8. Bands in a Vector Lattice

A *band* in a vector lattice $\mathscr{M}$ is defined as a vector sublattice $\mathscr{N}$ satisfying the following two conditions:

(a) If a subset of $\mathscr{N}$ has a supremum in $\mathscr{M}$, the supremum is in $\mathscr{N}$.
(b) If $|x| \precsim |y|$ and if $y \in \mathscr{N}$, then $x \in \mathscr{N}$.

Condition (a) implies the validity of the dual condition for infima. If a cone $\mathscr{C}$ of positive elements of $\mathscr{M}$ satisfies conditions (a) and (b), then $\mathscr{C} - \mathscr{C}$ is a band in $\mathscr{M}$. If $\mathscr{N}_0$ is a subset of $\mathscr{M}$, the intersection of all the bands containing $\mathscr{N}_0$ is a band, the band *generated by* $\mathscr{N}_0$.

Theorem. *If $\mathcal{N}$ is a subset of a vector lattice the set $\mathcal{N}^\perp$ is a band.*

To prove linearity of $\mathcal{N}^\perp$, observe that if $x \in \mathcal{N}^\perp$ and $c \in \mathbb{R}$, then $cx \in \mathcal{N}^\perp$ because when y is in $\mathcal{N}$,

$$|cx| \curlywedge |y| \preceq (|c| + 1)(|x| \curlywedge |y|) = 0,$$

and if x_1 and x_2 are in $\mathcal{N}^\perp$, their sum is in $\mathcal{N}^\perp$ because (6.2) implies that when y is in $\mathcal{N}$,

$$|x_1 + x_2| \curlywedge |y| \preceq |x_1| \curlywedge |y| + |x_2| \curlywedge |y| = 0.$$

The band condition (b) is obviously satisfied by $\mathcal{N}^\perp$; so if x and y are in $\mathcal{N}^\perp$, the elements $|x|$, $|y|$, $|x| + |y|$, $x \curlyvee y$, $x \curlywedge y$ are also in $\mathcal{N}^\perp$, which is thereby seen to be a sublattice of the given lattice $\mathcal{M}$. To finish the proof, it will be shown that if $\{x_\alpha, \alpha \in I\}$ is an arbitrary subset of $\mathcal{N}^\perp$ with a supremum $\curlyvee_{\alpha \in I} x_\alpha = x$, then $x \in \mathcal{N}^\perp$. Suppose first that $x_.$ has a smallest element, say $x_\beta = \curlywedge_{\alpha \in I} x_\alpha$, Then $\curlyvee_{\alpha \in I} (x_\alpha - x_\beta) = x - x_\beta$, and if y is in $\mathcal{N}$, an application of the distributive law yields

$$\left[\curlyvee_{\alpha \in I} (x_\alpha - x_\beta)\right] \curlywedge |y| = \curlyvee_{\alpha \in I} [(x_\alpha - x_\beta) \curlywedge |y|] = 0;$$

so $x - x_\beta$, and therefore x are in $\mathcal{N}^\perp$. In the general case choose any index value β, and replace x_α by $x'_\alpha = x_\alpha \curlyvee x_\beta$ for all α. The set $x'_.$ has a smallest element; so its supremum x is in $\mathcal{N}^\perp$, as was to be proved.

9. Projections on Bands

If $\mathcal{N}$ is a band in a conditionally complete vector lattice $\mathcal{M}$, define the map $\pi_\mathcal{N}$ from $\mathcal{M}$ onto $\mathcal{N}$ by

$$\pi_\mathcal{N} x = \curlyvee_{y \in \mathcal{N}} x \curlywedge y \tag{9.1}$$

for $x \succeq 0$, and for arbitrary x define $\pi_\mathcal{N} x = \pi_\mathcal{N} x^+ - \pi_\mathcal{N} x^-$. The map $\pi_\mathcal{N}$ reduces to the identity on $\mathcal{N}$ and $\pi_\mathcal{N} x \preceq x$ for $x \succeq 0$.

Theorem. *If $\mathcal{N}$ is a band in the conditionally complete vector lattice $\mathcal{M}$, then $\mathcal{N}^{\perp\perp} = \mathcal{N}$, and $\mathcal{M}$ is the direct sum of $\mathcal{N}$ and $\mathcal{N}^\perp$: for all $x \in \mathcal{M}$, $x = x_1 + x_2$, where $x_1 = \pi_\mathcal{N} x \in \mathcal{N}$ and $x_2 = \pi_{\mathcal{N}^\perp} \in \mathcal{N}^\perp$. The map $\pi_\mathcal{N}$ is linear, idempotent and order preserving.*

The map $\pi_\mathcal{N}$ is obviously order preserving on $\mathcal{M}^+$. The linearity of $\pi_\mathcal{N}$ to be proved below will therefore imply that $\pi_\mathcal{N}$ is order preserving on $\mathcal{M}$.

If $y \in \mathcal{N}$ and if $x \succeq 0$, then $|y| \wedge (x - \pi_{\mathcal{N}}x) + \pi_{\mathcal{N}}x \preceq x$, and since the left side is a positive element of $\mathcal{N}$, it follows that operating with $\pi_{\mathcal{N}}$ on both sides shows that the left side is majorized by $\pi_{\mathcal{N}}x$. Thus $0 \preceq |y| \wedge (x - \pi_{\mathcal{N}}x) \preceq 0$. The indicated infimum is therefore 0; that is, $x - \pi_{\mathcal{N}}x \in \mathcal{N}^{\perp}$, and x can be written in the form $x = x_1 + x_2$ with $x_1 = \pi_{\mathcal{N}}x \in \mathcal{N}$ and $x_2 = x - \pi_{\mathcal{N}}x \in \mathcal{N}^{\perp}$. This representation of x as the sum of an element of $\mathcal{N}$ and one of $\mathcal{N}^{\perp}$ is trivially extendible to not necessarily positive elements x. Since such a representation must be unique, $\mathcal{M}$ is the direct sum of $\mathcal{N}$ and $\mathcal{N}^{\perp}$, $\mathcal{N}^{\perp\perp} = \mathcal{N}$, and the map $x \mapsto x_1 = \pi_{\mathcal{N}}x$ must be linear. Replace $\mathcal{N}$ by $\mathcal{N}^{\perp}$ above to find that $x_2 = \pi_{\mathcal{N}^{\perp}}x$.

10. The Orthogonal Complement of a Set

Lemma. *Let $\mathcal{N}_0$ be a subset of a conditionally complete vector lattice, and let $\mathcal{N}$ be the band generated by $\mathcal{N}_0$. Then $\mathcal{N}_0^{\perp} = \mathcal{N}^{\perp}$.*

According to Theorem 9, $\mathcal{N}^{\perp\perp} = \mathcal{N}$. Now $\mathcal{N}_0 \subset \mathcal{N}$ implies that $\mathcal{N}^{\perp} \subset \mathcal{N}_0^{\perp}$ and therefore that $\mathcal{N}_0^{\perp\perp} \subset \mathcal{N}^{\perp\perp} = \mathcal{N}$. In the other direction, since $\mathcal{N}_0 \perp \mathcal{N}_0^{\perp}$, it follows that $\mathcal{N}_0 \subset \mathcal{N}_0^{\perp\perp}$; so $\mathcal{N}_0^{\perp\perp}$ is a band containing $\mathcal{N}_0$ and must therefore include $\mathcal{N}$, and so there is equality. Finally $\mathcal{N}_0^{\perp} = \mathcal{N}_0^{\perp\perp\perp} = \mathcal{N}^{\perp}$, as was to be proved.

11. The Band Generated by a Single Element

Let z be a positive element of the conditionally complete vector lattice $\mathcal{M}$, and denote by $\mathcal{N}$ the band generated by z. Call an element y of $\mathcal{M}$ *bounded* (relative to z) if $|y| \preceq cz$ for some constant c. All bounded elements are in $\mathcal{N}$ so any supremum of a set of bounded elements is in $\mathcal{N}$. If B^+ is the class of suprema of sets of positive bounded elements, $B^+ - B^+$ is a band, which must be $\mathcal{N}$; so $B^+ = \mathcal{N}^+$. Actually every element of $\mathcal{N}^+$ is the supremum of an increasing sequence of positive bounded elements of $\mathcal{M}$, and in fact we shall show that

$$\pi_{\mathcal{N}}x = \bigvee_{n \in \mathbb{Z}^+} x \wedge nz \qquad (x \in \mathcal{M}^+). \tag{11.1}$$

To prove (11.1), define πx as the supremum on the right, so that π is a map from $\mathcal{M}^+$ into $\mathcal{N}^+$ and $\pi x \preceq x$. For $k \geq 0$,

$$x \wedge kz = \pi(x \wedge kz) \preceq \pi^2 x \preceq \pi x; \tag{11.2}$$

so $(k \to \infty)$ $\pi x \preceq \pi^2 x \preceq \pi x$, and it follows that $\pi = \pi^2$. The map π is additive because on the one hand using (6.2)

$$\pi(x + y) \preceq \pi x + \pi y \tag{11.3}$$

and on the other hand

$$\pi(x + y) \succeq \pi(x \curlywedge kz + y) \succeq x \curlywedge kz + \bigvee_{n \in \mathbb{Z}^+} [y \curlywedge (n - k)z] = x \curlywedge kz + \pi y; \tag{11.4}$$

so $\pi(x + y) \succeq \pi x + \pi y$, and there must be equality. Moreover $x - \pi x \in \mathcal{N}^\perp$ because $\pi(x - \pi x) = \pi x - \pi^2 x = 0$, and $\pi y = 0$ implies that $y \perp z$, equivalently (by Lemma 10), $y \in \mathcal{N}^\perp$. It follows that $\pi_\mathcal{N}(x - \pi x) = 0$; that is, $\pi_\mathcal{N} x = \pi_\mathcal{N} \pi x = \pi x$, as was to be proved.

12. Order Convergence

Let I be a directed set; that is, I is a set ordered by a transitive reflexive binary relation $\leq$ with the property that each pair of elements has an order upper bound. A subset of I is called *cofinal* if each element of I precedes some element of the subset. The set I has a cofinal countable set if and only if I has a cofinal increasing sequence. Suppose that $x(\cdot)$ is a function from I into a complete lattice L. The inferior and superior limits of $x(\cdot)$ are defined by

$$\liminf_{t\uparrow} x(t) = \bigvee_{t \in I} \bigwedge_{s \geq t} x(s), \qquad \limsup_{t\uparrow} x(t) = \bigwedge_{t \in I} \bigvee_{s \geq t} x(s). \tag{12.1}$$

Then

$$\liminf_{t\uparrow} x(t) \leq \limsup_{t\uparrow} x(t), \tag{12.2}$$

and if these two values are equal, $x(\cdot)$ is said to have this common value as limit. This discussion becomes trivial if I has a last element because then both sides of (12.2) become the value of $x(\cdot)$ at the last element.

The Monotone Case

Suppose that the function $x(\cdot)$ is monotone increasing [decreasing]. Then obviously $x(\cdot)$ has limit $\bigvee_{t \in I} x(t)$ $[\bigwedge_{t \in I} x(t)]$. In particular, suppose that $x(\cdot)$ is monotone increasing and that L has the countability property (Section 2). Define $x = \lim_{t\uparrow} x(t)$. Then we now show that there is an increasing sequence $t_\bullet$ in I such that $\lim_{n\to\infty} x(t_n) = x$ and that $x(t) = x$ whenever t is an upper bound of $t_\bullet$. To see this, let $t'_\bullet$ be a sequence in I for which $\bigvee_{n \geq 0} x(t'_n) = \bigvee_{t \in I} x(t)$. Define $t_0 = t'_0$; if $t_0, \ldots, t_n$ have been defined, define t_{n+1} in I so that $t_{n+1} \geq t_n$ and $t_{n+1} \geq t'_n$. The sequence $t_\bullet$ has the stated properties.

Moreover any increasing sequence $s_.$ in I with $s_n \geq t_n$ for all n has these same properties. Thus if I has a cofinal increasing sequence, we can increase t_n above if necessary to make $t_.$ cofinal in addition to the above stated properties.

EXAMPLE. (Downward- and upward-directed sets). Let L be a partially ordered set, and let L_0 be a subset of L. If L_0 when given the [reverse of the] order inherited from L is a directed set, L is said to be [*downward*] *upward* directed. If L is a lattice and if L_0 has the property that $x \curlyvee y$ [$x \curlywedge y$] is in L_0 when x and y are, then L_0 is a directed set in the order $\preceq$ [$\succeq$] and is directed upward [downward]. In both cases the identity map from L_0 onto itself is a monotone function from L_0 into L; so the order convergence concept is applicable.

13. Order Convergence on a Linearly Ordered Set

Theorem. *Let $x(\cdot)$ be a function from a linearly ordered set I into a complete lattice satisfying the countability condition.*

(a) *If I has no cofinal increasing sequence, there is an element s of I such that $r \geq s$ implies that*

$$\bigcurlywedge_{t \geq r} x(t) = \liminf_{t\uparrow} x(t), \qquad \bigcurlyvee_{t \geq r} x(t) = \limsup_{t\uparrow} x(t). \tag{13.1}$$

(b) *If I has a cofinal increasing sequence, there is a cofinal countable subset J of I such that if $x_J(\cdot)$ is the restriction of $x(\cdot)$ to J, then*

$$\liminf_{t\uparrow} x_J(t) = \liminf_{t\uparrow} x(t), \qquad \limsup_{t\uparrow} x_J(t) = \limsup_{t\uparrow} x(t). \tag{13.2}$$

In view of the discussion of the monotone case in Section 12 there is an increasing sequence $t_.$ in I, cofinal if I has a cofinal increasing sequence, such that

$$\lim_{n\to\infty} \bigcurlywedge_{t \geq t_n} x(t) = \liminf_{t\uparrow} x(t), \qquad \lim_{n\to\infty} \bigcurlyvee_{t \geq t_n} x(t) = \limsup_{t\uparrow} x(t)$$

and that if s is an upper bound of $t_.$, then (13.1) is true. If I has no cofinal increasing sequence, there are (infinitely many) such points s, and we are done. If I has a cofinal increasing sequence, the sequence $t_.$ is cofinal, and we define $I_n = \{t\colon t_n \leq t < t_{n+1}\}$. Let I_n' be a countable subset of I_n, chosen so that $x(\cdot)$ has the same infimum and supremum on I_n' as on I_n. The set $J = \bigcup_0^\infty I_n'$ satisfies (13.2).

Appendix IV

Lattice Theoretic Concepts in Measure Theory

1. Lattices of Set Algebras

Recall that a [σ] algebra of subsets of a set X is a nonempty class of subsets containing complements, finite [countable] unions, and finite [countable] intersections of its members. The intersection of an arbitrary collection of [σ] algebras is a [σ] algebra. The smallest [σ] algebra containing a class Γ of subsets of X, that is, the intersection of the collection of all [σ] algebras containing Γ, is the [σ] algebra generated by Γ. In particular, the generated [σ] algebra is called *countably generated* if Γ is countable. If to each point α of some set I corresponds a collection $\mathscr{F}_\alpha$ of subsets of X, we denote by $\mathscr{F}\{\mathscr{F}_\alpha, \alpha \in I\}$ or $\mathscr{F}\{\bigcup_{\alpha\in I}\mathscr{F}_\alpha\}$ the σ algebra generated by $\bigcup_{\alpha\in I}\mathscr{F}_\alpha$. It will frequently be true in this book that each collection $\mathscr{F}_\alpha$ is a σ algebra and that $\mathscr{F}_.$ is directed upwards in the sense that if α and β are in I, there is a point γ in I such that $\mathscr{F}_\alpha \cup \mathscr{F}_\beta \subset \mathscr{F}_\gamma$. In this case $\bigcup_{\alpha\in I}\mathscr{F}_\alpha$ is an algebra but need not be a σ algebra.

The class $\mathscr{L}$ of σ algebras of subsets of X ordered by inclusion ($\subset$) is a complete lattice in which if $\{\mathscr{F}_\alpha, \alpha\in I\}$ is a collection of these σ algebras,

$$\bigvee_{\alpha\in I}\mathscr{F}_\alpha = \mathscr{F}\{\mathscr{F}_\alpha, \alpha\in I\}, \qquad \bigwedge_{\alpha\in I}\mathscr{F}_\alpha = \bigcap_{\alpha\in I}\mathscr{F}_\alpha.$$

Observe that to each set A in $\bigvee_{\alpha\in I}\mathscr{F}_\alpha$ corresponds a countable subset J of I, depending on A, such that $A \in \bigvee_{\alpha\in J}\mathscr{F}_\alpha$, because the class of sets A with this property is a σ algebra containing $\bigcup_{\alpha\in I}\mathscr{F}_\alpha$ but contained in and therefore equal to $\bigvee_{\alpha\in I}\mathscr{F}_\alpha$. Observe also that the class of sub σ algebras of a given σ algebra is a complete sublattice of $\mathscr{L}$. We omit the corresponding remarks on the lattice of algebras of subsets of X.

2. Measurable Spaces and Measurable Functions

Recall that a measurable space is a set X coupled with a distinguished σ algebra $\mathscr{X}$ of its subsets. In this book if X is endowed with a topology, then unless some other choice is specified the class $\mathscr{X}$ will be the class $\mathscr{B}(\mathscr{X})$ of

Borel subsets of X, that is, the σ algebra generated by the class of open subsets of X. If $(X, \mathscr{X})$ and $(Y, \mathscr{Y})$ are measurable spaces, a function ϕ from X into Y is defined as measurable if $\phi^{-1}(\mathscr{Y}) \subset \mathscr{X}$, and for this it is sufficient that $\phi^{-1}(A) \in \mathscr{X}$ for each set A in a subclass of $\mathscr{Y}$ which generates the σ algebra $\mathscr{Y}$. If ϕ_1 is a measurable function from $(X, \mathscr{X})$ into $(Y, \mathscr{Y})$ and if ϕ_2 is a measurable function from $(Y, \mathscr{Y})$ into a third measurable space $(Z, \mathscr{Z})$, then the composed function $\phi_2(\phi_1)$ is measurable from $(X, \mathscr{X})$ into $(Z, \mathscr{Z})$.

If X is a set, if $\mathscr{F}_\alpha$ is a σ algebra of subsets of X for each point α of some index set I, and if ϕ is a measurable function from $(X, \bigvee_{\alpha \in I} \mathscr{F}_\alpha)$ into a measurable space $(Y, \mathscr{X})$, then if $\mathscr{Y}$ is countably generated, there is a countable subset J of I such that ϕ is measurable from $(X, \bigvee_{\alpha \in J} \mathscr{F}_\alpha)$ into $(Y, \mathscr{Y})$. In fact it is sufficient to show that for each set A in a countable generating class for Y the set $\phi^{-1}(A)$ is in the σ algebra generated by a countable subclass of $\mathscr{F}_\bullet$, and we have already remarked in Section 1 that this is true.

Suppose that to each point t of a set I corresponds a measurable space $(X_t, \mathscr{X}_t)$. The product σ algebra $\times_{t \in I} \mathscr{X}_t$ is defined as the σ algebra of subsets of the product space $\times_{t \in I} X_t$ generated by the class of product sets $\times_{t \in I} A_t$ with $A_t = X_t$ for every value of t with one possible exception and for that value $A_t \in \mathscr{X}_t$. The product measurable space is defined as $(\times_{t \in I} X_t, \times_{t \in I} \mathscr{X}_t)$. If $(X, \mathscr{X})$ is a measurable space and if $(X_t, \mathscr{X}_t)$ is a replica of $(X, \mathscr{X})$ for every t in a set I, we denote the product measurable space by $(X^I, \mathscr{X}^I)$, or by $(X^k, \mathscr{X}^k)$ if I has cardinality k.

3. Composition of Functions

If y is a function from a space X into a measurable space $(Y, \mathscr{Y})$, the smallest σ algebra $\mathscr{F}$ of subsets of X making y measurable from $(X, \mathscr{F})$ into $(Y, \mathscr{Y})$ will be denoted by $\mathscr{F}\{y\}$. That is, $\mathscr{F}\{y\} = y^{-1}(\mathscr{Y})$ is the class of subsets of X of the form $\{y \in A\}$ with A in $\mathscr{Y}$. If ϕ is a measurable function from $(Y, \mathscr{Y})$ into a measurable space $(Z, \mathscr{Z})$, the composed function $z = \phi(y)$ is measurable from $(X, \mathscr{F}\{y\})$ into $(Z, \mathscr{Z})$. Conversely, if z is a measurable function from $(X, \mathscr{F}\{y\})$ into $(Z, \mathscr{Z})$, it is a useful fact that z must have this form $\phi(y)$ under suitable restrictions on the measurable space $(Z, \mathscr{Z})$. The following lemma, stated in the preceding context, gives one such restriction.

Lemma. *Let $(Z, \mathscr{Z})$ be a Polish space coupled with its class of Borel sets. Then if z is a measurable function from $(X, \mathscr{F}\{y\})$ into $(Z, \mathscr{Z})$, there is a measurable function ϕ from $(Y, \mathscr{Y})$ into $(Z, \mathscr{Z})$ such that $z = \phi(y)$.*

We can suppose that Z has been assigned a metric making the space complete and separable. To prove the lemma, suppose first that z has as range a sequence $\zeta_\bullet$ of points of Z. Then $z^{-1}(\zeta_i) \in \mathscr{F}\{y\}$; so there is a (perhaps not uniquely determined) set B_i in $\mathscr{Y}$ such that $z^{-1}(\zeta_i) = y^{-1}(B_i)$. If we replace $B_0, B_1, \ldots$ by $B_0, B_1 - B_0, B_1 - (B_0 \cup B_1), \ldots$ if necessary, we can suppose that the sequence $B_{\cdot}$ is disjoint, and we define ϕ on Y as ζ_i on B_i and $\phi \equiv$ const (an arbitrary point of Z) on the remaining set $Y - \bigcup_{i=0}^{\infty} B_i$. This function ϕ has the desired properties. For arbitrary z there is a convergent sequence $z_{\cdot}$ of countably valued measurable functions from $(X, \mathscr{F}\{y\})$ into $(Z, \mathscr{Z})$ with limit z. For example, for each positive integer n choose a partition $B_{n0}, B_{n1}, \ldots$ of Z into disjoint nonempty Borel sets of diameter $\leq 2^{-n}$, choose a point ζ_{nj} in B_{nj}, and define z_n on X as ζ_{nj} on $z^{-1}(B_{nj})$. Then by the lemma in the special case just treated there is a measurable function ϕ_n from $(Y, \mathscr{Y})$ into $(Z, \mathscr{Z})$ such that $z_n = \phi_n(y)$, and it follows that the sequence $\phi_{\cdot}$ is convergent on the range of y. Since the convergence set of $\phi_{\cdot}$ is in $\mathscr{Y}$, if we redefine $\phi_n \equiv$ const on the divergence set (the same point of Z for all n), the sequence $\phi_{\cdot}$ is convergent on Y to a function ϕ with the required properties.

Application to Product Spaces. (Notation of the Beginning of This Section.) If k is a strictly positive integer and if $y_k = [y(1), \ldots, y(k)]$ is a function from X into Y^k, then $\mathscr{F}\{y_k\}$, usually written $\mathscr{F}\{y(1), \ldots, y(k)\}$, becomes the class of subsets of X of the form $\{[y(1), \ldots, y(k)] \in A_k\}$ with $A_k \in \mathscr{Y}^k$. Observe that $\mathscr{F}\{y_k\}$ is the smallest σ algebra $\mathscr{F}$ of subsets of X making every function $y(i)$ measurable from $(X, \mathscr{X})$ into $(Y, \mathscr{Y})$. More generally, if I is an arbitrary nonempty set and if $\{y(t), t \in I\}$ is a family of functions from X into Y, indexed by I, let $y(t, \xi)$ be the value at ξ in X of the function $y(t)$. The function $y_I : \xi \mapsto y(\cdot, \xi)$ is a measurable function from $(X, \mathscr{F}\{y_I\})$ into $(Y^I, \mathscr{Y}^I)$, and $\mathscr{F}\{y_I\}$, usually written $\mathscr{F}\{y(t), t \in I\}$, can also be described as the smallest σ algebra $\mathscr{F}$ of subsets of X making every function $y(t)$ measurable from $(X, \mathscr{F})$ into $(Y, \mathscr{Y})$; that is,

$$\mathscr{F}\{y(t), t \in I\} = \bigvee_{t \in I} \mathscr{F}\{y(t)\}.$$

According to Section 1, a set A in $\mathscr{F}\{y(t), t \in I\}$ is in $\mathscr{F}\{y(t), t \in J\}$ for some countable subset J of I, depending on A. It follows that if Y is countably generated, a measurable function y from $(X, \mathscr{F}\{y(t), t \in I\})$ into $(Y, \mathscr{Y})$ is measurable from $(X, \mathscr{F}\{y(t), t \in J\})$ into $(Y, \mathscr{Y})$ for some countable subset J of I depending on y.

4. The Measure Lattice of a Measurable Space

Let $(X, \mathscr{X})$ be a measurable space, and let $\mathscr{M}^+$ be the class of measures on $\mathscr{X}$, ordered by setting $\mu \precsim \nu$ if $\mu \leq \nu$ on $\mathscr{X}$, that is, if $\mu(A) \leq \nu(A)$ for every A in $\mathscr{X}$. From now on all subsets of X mentioned are in $\mathscr{X}$, and this qualification will no longer be stated. The space $\mathscr{M}^+$ in this order is a lattice with

$$\begin{aligned}(\mu \curlyvee \nu)(A) &= \sup\{\mu(B) + \nu(A - B) : B \subset A\} \\ (\mu \curlywedge \nu)(A) &= \inf\{\mu(B) + \nu(A - B) : B \subset A\}.\end{aligned} \tag{4.1}$$

The only nontrivial part of this assertion is the countable additivity of the set functions defined in (4.1). If $\lambda(A)$ is the right-hand side of the first line and if $A = \bigcup_{j=0}^{\infty} A_j$ (disjoint union),

$$\begin{aligned}\lambda(A) &= \sup\left\{\sum_{j=1}^{\infty} [\mu(B_j) + \nu(A_j - B_j): B \subset A, B_j = B \cap A_j\right\} \\ &= \sum_{j=0}^{\infty} \sup\{\mu(B_j) + \nu(A_j - B_j): B_j \subset A_j\} = \sum_{j=0}^{\infty} \lambda(A_j).\end{aligned} \tag{4.2}$$

Thus λ is a measure, and an analogous argument shows that the second line of (4.1) defines a measure.

The class $\mathcal{M}^+$ has the identically vanishing measure as order infimum and the measure assigning $+\infty$ to each nonempty set as order supremum. Moreover the lattice $\mathcal{M}^+$ is a complete lattice. In fact we now prove that an arbitrary subset $\{\mu_\alpha, \alpha \in I\}$ of $\mathcal{M}^+$ has an order supremum (and therefore also an order infimum). In proving that the supremum exists it can be supposed that $\mu_.$ contains every supremum of finitely many of its elements because adjoining these suprema does not change the conditions for an overall supremum. Define μ on $\mathfrak{X}$ by $\mu(A) = \sup_{\alpha \in I} \mu_\alpha(A)$. It is trivial that $\mu = \bigvee_{\alpha \in I} \mu_\alpha$ once it is shown that μ is a measure. If $A = \bigcup_0^\infty A_j$ (disjoint union),

$$\mu_\alpha(A) = \sum_{j=0}^{\infty} \mu_\alpha(A_j) \leq \sum_{j=0}^{\infty} \mu(A_j) \tag{4.3}$$

for all α; so μ is countably subadditive. Finite additivity is trivial and implies

$$\mu(A) = \sum_{j=0}^{n} \mu(A_j) + \mu\left(\bigcup_{j=n+1}^{\infty} A_j\right) \geq \sum_{j=0}^{n} \mu(A_j); \tag{4.4}$$

so $(n \to \infty)$ μ is countably superadditive as well as subadditive and therefore is a measure.

Support of a Measure

A support of a measure μ on $(X, \mathfrak{X})$ is a set with the property that its complement is μ null. In some contexts uniqueness can be attained by a further condition on a support. For example, if μ is a measure of Borel subsets of an open subset X of $\mathbb{R}^N$, finite on compact subsets, then μ has a smallest support closed relative to X.

Orthogonal Measures

If μ and ν are measures on $(X, \mathfrak{X})$, they are *orthogonal* if $\mu \curlywedge \nu = 0$, that is, if $(\mu \curlywedge \nu)(X) = 0$. This condition is satisfied if and only if the measures have disjoint supports. In fact the latter condition is obviously sufficient for orthogonality. Conversely, if the measures are orthogonal, it follows from (4.1) with $A = X$ that for each positive integer n there is a set B_n such that $\mu(B_n) + \nu(X - B_n) < 2^{-n}$. If $A = \limsup_{n\to\infty} B_n$ $(= \bigcap_{n=1}^{\infty} \bigcup_{m=n}^{\infty} B_m)$, then $\mu(A) = \nu(X - A) = 0$; so μ and ν have the respective supports $X - A$ and A.

Almost Everywhere–Almost Surely

If $\lambda \in \mathcal{M}^+$, it will sometimes be convenient to follow probability usage and write "λ almost surely" instead of "λ almost everywhere."

5. The σ Finite Measure Lattice of a Measurable Space (Notation of Section 4)

Let $\mathcal{M}_\sigma^+$ be the class of σ finite measures on $(X, \mathfrak{X})$, in the order inherited from $\mathcal{M}^+$. Then $\mathcal{M}_\sigma^+$ is a cone, ordered in its specific order, and is a conditionally complete lattice, a sublattice of $\mathcal{M}^+$.

Theorem. *The lattice $\mathcal{M}_\sigma^+$ has the countability property; that is if $\{\mu_\alpha, \alpha \in I\}$ is an upper-bounded subset of $\mathcal{M}_\sigma^+$, there is a countable subset J of I such that $\bigvee_{\alpha\in J} \mu_\alpha = \bigvee_{\alpha\in I} \mu_\alpha$.*

It can be assumed in the proof that $\mu_.$ contains every supremum of finitely many of its elements since these suprema can be adjoined to $\mu_.$ if necessary to achieve this, without altering the generality of the context. It can also be assumed that $\mu_.(X)$ is bounded because if ν is an upper bound of $\mu_.$ in $\mathcal{M}_\sigma^+$, there is a disjoint sequence $A_.$ of sets of finite ν measure, with union X, so that $\mu_.(A_j) \le \nu(A_j) < +\infty$, and if the theorem is known to be true for the projections on A_j of the measures involved, the theorem follows as stated. Under these assumptions choose index values $\alpha_1, \alpha_2, \ldots$ in such a way that $\mu_{\alpha_1} \le \mu_{\alpha_2} \le \cdots$ and that $\lim_{n\to\infty} \mu_{\alpha_n}(X) = \sup_{\alpha\in I} \mu_\alpha(X)$. Define $\mu = \bigvee_{n\ge1} \mu_{\alpha_n}$; that is, $\mu(A) = \lim_{n\to\infty} \mu_{\alpha_n}(A)$ for every set A in $\mathfrak{X}$. Then $\mu \le \bigvee_{\alpha\in I} \mu_\alpha$. In the other direction observe first that for all α,

$$\mu \vee \mu_\alpha = \left(\bigvee_{n\ge1} \mu_{\alpha_n}\right) \vee \mu_\alpha = \bigvee_{n\ge1} (\mu_{\alpha_n} \vee \mu_\alpha).$$

Hence by definition of μ,

$$(\mu \vee \mu_\alpha)(X) = \lim_{n\to\infty} (\mu_{\alpha_n} \vee \mu_\alpha)(X) = \mu(X).$$

The measure $\mu \vee \mu_\alpha - \mu$ has value 0 on X and is therefore the zero measure; that is, $\mu \ge \mu_\alpha$, and so $\mu \ge \bigvee_{\alpha\in I} \mu_\alpha$. Hence there is equality, and we have proved that the supremum in question is the supremum of the sequence $\mu_{\alpha_.}$.

6. The Hahn and Jordan Decompositions

Let λ be a signed measure on a measurable space $(X, \mathscr{X})$. In the following all sets mentioned are in $\mathscr{X}$. A Hahn decomposition of X for λ is a partition of X into a *positivity* set A^+ on whose subsets $\lambda \geq 0$ and a *negativity* set $A^- = X - A^+$ on whose subsets $\lambda \leq 0$. The signed measure λ can be expressed as the difference between two measures (Jordan decomposition). For example,

$$\lambda = \mu - v, \qquad \mu(A) = \lambda(A \cap A^+), \qquad v(A) = -\lambda(A \cap A^-).$$

This particular Jordan decomposition is minimal in the sense that $\mu \leq \mu'$ and $v \leq v'$ whenever $\mu' - v'$ is a Jordan decomposition of λ. Conversely, if μ and v are measures on X, their difference $\mu - v$ need not be a signed measure, but if both μ and v are σ finite, there is an increasing sequence $B_{\bullet}$ of sets with union X such that both μ and v are finite on every set B_j; so the restrictions of μ and v to the class of subsets of B_j are finite, and we can therefore build a set A^+ from the positivity sets of these restrictions such that $\mu \geq v$ on the subsets of A^+ and $\mu \leq v$ on the subsets of $A^- = X - A^+$. In this way we have found what amounts to a Hahn decomposition of X for $\mu - v$ and thereby found a Jordan decomposition of this difference, although the difference may not be defined on all $\mathscr{X}$ sets. In Section 7 such differences will be rigorously defined and ordered to obtain a vector lattice.

7. The Vector Lattice $\mathscr{M}_\sigma$

According to Sections III.3 to III.5, the cone $\mathscr{M}_\sigma^+$ can be identified with the positive cone of a conditionally complete vector lattice $\mathscr{M}_\sigma$ consisting of pairs (μ_1, μ_2) of members of $\mathscr{M}_\sigma^+$, with the convention that $(\mu_1, \mu_2) = (v_1, v_2)$ when $\mu_1 + v_2 = \mu_2 + v_1$. and with the ordering $(\mu_1, \mu_2) \succeq (v_1, v_2)$ when $\mu_1 + v_2 \geq \mu_2 + v_1$. Each member μ of $\mathscr{M}_\sigma$ has a minimal representation (μ^+, μ^-) in which the component measures are orthogonal, that is, have disjoint supports, and are minimal in the sense that if (μ_1, μ_2) is a representation of μ, then $\mu_1 \geq \mu^+$ and $\mu_2 \geq \mu^-$. If μ^+ or μ^- is a finite measure, then μ can be identified with the signed measure $\mu^+ - \mu^-$. A set A is a *support* for the element $\mu = (\mu_1, \mu_2)$ of $\mathscr{M}_\sigma$ if there is a representation of μ with both components supported by A, equivalently, if A is a support for the measure $\mu_1 + \mu_2$. It follows that two elements of $\mathscr{M}_\sigma$ are orthogonal if and only if they have disjoint supports.

Charges

Whenever we deal with an extended real-valued countably additive set function on a measurable space $(X, \mathscr{X})$, it is to be understood when X has a topology that $\mathscr{X}$ is either the class $\mathscr{B}(\mathscr{X})$ of Borel subsets of X or the perhaps partial completion of $\mathscr{B}(\mathscr{X})$ relative to some class of measures. Suppose now that X is a locally compact second countable Hausdorff space. (In this book X will in fact usually be an open subset of $\mathbb{R}^N$.) In this context it is to be understood in the absence of a contrary statement that the set functions are finite valued on compact sets. A member μ of $\mathscr{M}_\sigma$ for which $|\mu| = \mu^+ + \mu^-$ is finite valued on compact sets will be called a *charge*. If either μ^+ or μ^- is finite valued, the charge can be identified with the signed measure $\mu^+ - \mu^-$, in particular with the measure μ^+ if $\mu^- \equiv 0$. We shall not use the term "Radon measure" in this book, in which "measure" always implies positivity. The class of charges is a sublattice of $\mathscr{M}_\sigma$ and is itself a conditionally complete vector lattice with the countability property.

In the following sections it is to be understood that the measures and charges discussed are defined on some measurable space, specified further as needed by the context.

8. Absolute Continuity and Singularity

If $\lambda \in \mathscr{M}_\sigma^+$, an element μ of $\mathscr{M}_\sigma$ will be called *absolutely continuous relative to* λ if μ has a representation (μ_1, μ_2) for which λ null sets are also $\mu_1 + \mu_2$ null sets; that is, in the usual terminology the measures μ_1 and μ_2 are absolutely continuous relative to λ. If so, μ_1 and μ_2 can be taken as the minimal choices $\mu_1 = \mu^+$, $\mu_2 = \mu^-$. For any acceptable choice (μ_1, μ_2) there are unique up to λ null sets positive finite-valued measurable functions, which we shall denote by $d\mu_1/d\lambda$, $d\mu_2/d\lambda$, for which

$$\mu_i(A) = \int_A \frac{d\mu_i}{d\lambda}\, d\lambda.$$

Conversely, if f_1 and f_2 are positive λ almost everywhere finite-valued measurable functions and if μ_i is defined by

$$\mu_i(A) = \int_A f_i\, d\lambda, \tag{8.1}$$

then (μ_1, μ_2) is a λ absolutely continuous member of $\mathscr{M}_\sigma$. We shall write $d\mu_i = f_i\, d\lambda$ for (8.1). If $\nu = (\nu_1, \nu_2)$ with $d\nu_i = g_i\, d\lambda$ is a second λ absolutely continuous member of $\mathscr{M}_\sigma$, then

$$\begin{aligned} \mu \curlyvee \nu &= (\mu_1', \mu') \quad \text{with } d\mu_1' = (f_1 + g_2) \vee (f_2 + g_1)\, d\lambda, \quad d\mu' = (f_2 + g_2)\, d\lambda, \\ \mu \curlywedge \nu &= (\mu_2', \mu') \quad \text{with } d\mu_2' = (f_1 + g_2) \wedge (f_2 + g_1)\, d\lambda. \end{aligned} \tag{8.2}$$

In particular, if μ and ν are positive, we can choose $f_2 \equiv g_2 \equiv 0$.

An element μ of $\mathcal{M}_\sigma$ will be called *singular relative to* λ if $\mu \perp \lambda$, that is, if μ and λ have a pair of disjoint supports. The Lebesgue decomposition theorem implies that each element of $\mathcal{M}_\sigma$ has a unique representation as the sum of a λ absolutely continuous element and a λ-singular element of $\mathcal{M}_\sigma$. According to Section III.11 the positive elements in the band generated by λ are the elements of $\mathcal{M}_\sigma^+$ of the form $\bigvee_0^\infty (v \curlywedge n\lambda)$, with v in $\mathcal{M}_\sigma^+$. The measure $v \curlywedge n\lambda$ is majorized by $n\lambda$ and is therefore λ absolutely continuous, as is the limit measure μ obtained when $n \to \infty$. Conversely, if μ is a λ absolutely continuous measure, the measure $\mu \curlywedge n\lambda$ has the Radon–Nikodym density $(d\mu/d\lambda) \wedge n$; so $\mu = \bigvee_0^\infty (\mu \curlywedge n\lambda)$, and μ is in the band generated by λ. Hence the band generated by λ is the set of λ absolutely continuous members of $\mathcal{M}_\sigma$ and the Lebesgue decomposition is precisely the decomposition of $\mathcal{M}_\sigma$ into the orthogonal bands of λ absolutely continuous and λ-singular elements. If $(X, \mathscr{X})$ is a locally compact second countable Hausdorff space coupled with its class of Borel sets, the lattice $\mathcal{M}_\sigma$ can be replaced in this discussion by the lattice of charges.

9. Lattices of Measurable Functions on a Measure Space

Every concept in this section is relative to a specified complete measure λ in $\mathcal{M}_\sigma^+$. That is, we are discussing functions on a complete σ finite measure space $(X, \mathscr{X}, \lambda)$. We conform to the usual custom that in discussing a function from a measurable space into $\overline{\mathbb{R}}$ "measurable" means that the range measurable space is $(\overline{\mathbb{R}}, \mathscr{B}(\overline{\mathbb{R}}))$. Let $\mathcal{M}^*$ be the class of measurable functions from $(X, \mathscr{X})$ into $\overline{\mathbb{R}}$ with the convention that a function need be defined only λ almost everywhere and that two functions are identified if they are equal λ almost everywhere. Strictly speaking $\mathcal{M}^*$ is a space of equivalence classes of functions, but we shall adhere to the usual convenient carelessness of analysts in which "function" and "member of $\mathcal{M}^*$" may be either a function or an equivalence class, as indicated by the context. The space $\mathcal{M}^*$ is a lattice under the order $\leq$ to be interpreted as follows: if f and g are measurable functions and if $\mathbf{f}$ and $\mathbf{g}$ are the respective equivalence classes containing these functions, the inequality $\mathbf{f} \leq \mathbf{g}$ means that $f \leq g$ up to a λ null set. Similarly $\mathbf{f} \vee \mathbf{g}$ [$\mathbf{f} \wedge \mathbf{g}$] is the equivalence class containing $f \vee g$ [$f \wedge g$]. As already noted, we shall blur the distinction between function and equivalence class.

The subclass $\mathcal{M}_\sigma^*$ of λ almost everywhere finite members of $\mathcal{M}^*$ is a vector sublattice of $\mathcal{M}^*$. The set of positive members of $\mathcal{M}^*$ is not a cone, but the set $\mathcal{M}_\sigma^{*+}$ of positive members of $\mathcal{M}_\sigma^*$ is a cone and $\mathcal{M}_\sigma^* = \mathcal{M}_\sigma^{*+} - \mathcal{M}_\sigma^{*+}$.

Theorem. *The lattice $\mathcal{M}^*$ is complete, and $\mathcal{M}_\sigma^*$ is conditionally complete. Both lattices have the countability property that the supremum [infimum] of a set, if it exists, is the supremum [infimum] of a countable subset.*

Since the transformation $f \mapsto \arctan f$ is order preserving from $\mathcal{M}^*$ into $\mathcal{M}_\sigma^*$, it is sufficient to prove the assertion of the theorem for $\mathcal{M}_\sigma^*$ and therefore for $\mathcal{M}_\sigma^{*+}$. Each element f of $\mathcal{M}_\sigma^{*+}$ determines an element μ_f of $\mathcal{M}_\sigma^+$ by way of $d\mu_f = f\,d\lambda$, and the map $f \mapsto \mu_f$ is an order isomorphism between $\mathcal{M}_\sigma^*$ and the lattice of λ absolutely continuous elements of $\mathcal{M}_\sigma^+$. The theorem now follows from the conditional completeness of $\mathcal{M}_\sigma^+$ and Theorem 5.

Essential Order Terminology

The order of $\mathcal{M}^*$ is called the *essential order*. If $\{f_\alpha, \alpha \in I\}$ is a subset of $\mathcal{M}^*$, the elements $\bigvee_{\alpha \in I} f_\alpha$ and $\bigwedge_{\alpha \in I} f_\alpha$ are denoted, respectively, by $\operatorname{ess\,sup}_{\alpha \in I} f_\alpha$ and $\operatorname{ess\,inf}_{\alpha \in I} f_\alpha$. Thus according to Theorem 9, there is a countable subset J of I such that $\sup_{\beta \in J} f_\beta \geq f_\alpha$ for every α neglecting a λ null set depending on α. If I is directed, $\operatorname{ess\,lim\,sup}_{\alpha\uparrow} f_\alpha$ is defined (Section III.12) as

$$\operatorname*{ess\,inf}_{\beta \in I} \big[\operatorname*{ess\,sup}_{\alpha \geq \beta} f_\alpha\big],$$

and the essential limit infimum is defined dually. A function f is called the *essential limit* of $f_.$ if both the essential limit superior and essential limit inferior are equal λ almost everywhere to f.

10. Order Convergence of Families of Measurable Functions

Theorem. *Let I be a linearly ordered set with no last element and with a cofinal increasing sequence. Let $x(\cdot)$ be a function from I into the space $\mathcal{M}_\sigma^*$ based on a complete σ finite measure space $(X, \mathfrak{X}, \lambda)$. Then there is a cofinal increasing sequence $t_.$ for which*

$$\liminf_{n\to\infty} x(t_n) = \operatorname*{ess\,lim\,inf}_{t\uparrow} x(t), \qquad \limsup_{n\to\infty} x(t_n) = \operatorname*{ess\,lim\,sup}_{t\uparrow} x(t) \quad \text{a.s.} \tag{10.1}$$

This theorem with I an open interval of $\mathbb{R}$ is useful in stochastic process theory. To prove the theorem, assume [if necessary, replacing the given measure λ by a finite measure with the same null sets and replacing $x(t)$ by $\arctan x(t)$] that λ is finite and that $|x(\cdot)| \leq \text{const}\,(< +\infty)$. Let $t'_.$ be a cofinal increasing sequence in I, define $I'_n = \{t : t'_n \leq t \leq t'_{n+1}\}$, and choose $t'_{n0}, \ldots, t'_{na_n}$ in I'_n in such a way that the functions $\min_{j \leq a_n} x(t'_{nj})$ and $\max_{j \leq a_n} x(t'_{nj})$ are, respectively, at L^1 distance at most 2^{-n} from $\operatorname{ess\,inf}_{t \in I'_n} x(t)$ and $\operatorname{ess\,sup}_{t \in I'_n} x(t)$. Then

$$\lim_{n\to\infty} \big[\min_{j \leq a_n} x(t'_{nj}) - \operatorname*{ess\,inf}_{t \in I'_n} x(t)\big] = \lim_{n\to\infty} \big[\operatorname*{ess\,sup}_{t \in I'_n} x(t) - \max_{j \leq a_n} x(t'_{nj})\big] = 0 \quad \text{a.s.}$$

because the sums of the integrals of the bracketed λ almost everywhere positive functions converge. Hence (10.1) is true with $t_.$ the set $\{t_{nj}: n \geq 0, j \leq a_n\}$ arranged in increasing order.

Generalization

Let S be a compact metric space, let $\hat{S}$ be the space of measurable functions from $(X, \mathscr{X}, \lambda)$ into $(S, \mathscr{B}(S))$, and let $y(\cdot)$ be a function from I into $\hat{S}$. Theorem 10 can be applied to $x(t) = f[y(t)]$ for f a real finite-valued continuous function on S. In fact a slight extension of the proof of the theorem yields the fact that there is a cofinal increasing sequence $t_.$ in I for which

$$\limsup_{n\to\infty} f[y(t_n)] = \operatorname{ess}\limsup_{t\uparrow} f[y(t)] \quad \text{a.s.}$$

simultaneously for every function f in a countable dense subset of $\mathbb{C}(S)$, and therefore simultaneously for every function in $\mathbb{C}(S)$. In particular, if I is itself countable, "ess" can be omitted in (10.2), and it is then easy to conclude that, neglecting a λ null set, $y(t)$ has the same cluster values when t increases through $t_.$ as through I. This result is important in discussing stochastic process separability. If it is known that $\lim_{n\to\infty} x(t_n)$ exists λ almost surely whenever $t_.$ is a cofinal increasing sequence in I, it follows that $\operatorname{ess}\lim_{t\uparrow} x(t)$ exists and is λ almost surely this sequential limit, which obviously does not depend, neglecting null sets, on the choice of $t_.$. If I is countable, it follows that $\lim_{t\uparrow} x(t)$ exists almost surely.

Fatou's Lemma and the Lebesgue Dominated Convergence Theorem

Let $\{x(t), t \in I\}$ be an upper-directed (in the essential order) family of measurable functions from the σ finite measure space $(X, \mathscr{X}, \lambda)$ into $\overline{\mathbb{R}}$. The ordering of the functions defines an ordering of I, and

$$\operatorname{ess}\lim_{t\uparrow} x(t) = \operatorname{ess}\sup_{t\in I} x(t) = \sup_{t\in J} x(t) \quad \text{a.s.,}$$

where (by Theorem 9) J is some countable subset of I and (from Section III.12) we can choose J to be an increasing sequence $t_.$, cofinal if I has a countable cofinal set, so that

$$\operatorname{ess}\sup_{t\in I} x(t) = \lim_{n\to\infty} x(t_n) \quad \text{a.s.}$$

[If I does not have a countable cofinal set, this result is trivial because according to Theorem III.12 there is a point t' of I such that $x(t) = x(t')$ almost surely whenever $t \geq t'$.] If $\operatorname{ess}\inf_{t\in I} x(t) \geq 0$, then

$$\int_X \operatorname*{ess\,sup}_{t\in I} x(t)\,d\lambda = \lim_{t\uparrow} \int_X x(t)\,d\lambda.$$

Now suppose that $\beta \mapsto f_\beta$ is a function from a directed set I into the space of measurable functions from $(X, \mathscr{X}, \lambda)$ into $\bar{\mathbb{R}}$. Then if $\operatorname{ess\,inf}_{\beta\in I} f_\beta \geq 0$, the preceding monotone limit result yields Fatou's lemma in the present context:

$$\int_X \operatorname*{ess\,lim\,inf}_{\beta\uparrow} f_\beta\,d\lambda = \lim_{\alpha\uparrow} \int_X \operatorname*{ess\,inf}_{\beta\geq\alpha} f_\beta\,d\lambda \leq \liminf_{\alpha\uparrow} \int_X f_\alpha\,d\lambda.$$

Drop the positivity hypothesis and apply the usual argument using Fatou's lemma to prove the Lebesgue dominated convergence theorem in the present context: if $\operatorname{ess\,lim}_{\beta\uparrow} f_\beta = f$ exists and if $\operatorname{ess\,sup}_{\beta\in I} |f_\beta|$ is λ integrable, then

$$\int_X f\,d\lambda = \lim_{\beta\uparrow} \int_X f_\beta\,d\lambda.$$

11. Measures on Polish Spaces

Whenever a measure on a Polish space X is discussed in this book, it is to be understood that the domain of the measure is $\mathscr{X} = \mathscr{B}(X)$ or, if so stated, the completion of $\mathscr{B}(X)$ with respect to a specified measure or family of measures. It is an important property of Polish spaces that a finite measure λ on $\mathscr{X}$ (and therefore also a σ finite measure on $\mathscr{X}$) is inner regular in the sense that for A in $\mathscr{X}$

$$\lambda(A) = \sup\{\lambda(B): B \subset A, B \text{ compact}\}. \tag{11.1}$$

We sketch the proof for λ a finite measure. In the first place (11.1) is true if "compact" is replaced by "closed" because the class of Borel sets A for which (11.1) as so modified is true includes the closed sets and is closed under countable unions and intersections, and is therefore $\mathscr{B}(X)$. Thus it is sufficient to verify (11.1) as stated for closed sets A. Let A be a closed set, and let ε be a strictly positive number. In some metric for X let $A_.$ be a sequence of nonempty closed subsets of A, of diameter < 1, with union A. Define $B_0 = A$ and $B_1 = \bigcup_{j=0}^n A_j$, where n is so large that $\lambda(B_1) > \lambda(B_0) - \varepsilon$. Continue by induction: if B_k has been defined as a closed subset of B_{k-1} with $\lambda(B_k) > \lambda(A) - \varepsilon$, we can use the preceding argument to find B_{k+1}, a closed subset of B_k, the union of finitely many closed sets of diameter $< 1/(k+1)$, with $\lambda(B_{k+1}) > \lambda(A) - \varepsilon$. The set $B = \bigcap_{k=0}^\infty B_k$ is a closed set which can be covered by finitely many sets of diameter $< 1/k$ for every $k > 0$; so B is compact. Moreover $\lambda(B) \geq \lambda(A) - \varepsilon$. Hence (11.1) is true.

This theorem implies that for any σ finite Borel measure μ on a Polish space the class of continuous functions with compact support on the space is a dense subset of $L^1(\mu)$.

12. Derivates of Measures

Let μ and ν be charges on $\mathbb{R}^N$ with $\nu \geq 0$, and let ζ be a point of $\mathbb{R}^N$. If S is a closed convex subset of $\mathbb{R}^N$ containing ζ, let $d'(S)$ be the distance from ζ to ∂S, and let $d''(S)$ be the diameter of S. A number q will be called the *convex derivate* at ζ of μ with respect to ν if whenever $S_\bullet$ is a sequence of closed convex sets containing ζ for which

$$d''(S_n) > 0, \qquad \lim_{n\to\infty} d''(S_n) = 0, \qquad \liminf_{n\to\infty} \frac{d'(S_n)}{d''(S_n)} > 0, \tag{12.1}$$

it follows that $\nu(S_n) > 0$ for large n and that

$$\lim_{n\to\infty} \frac{\mu(S_n)}{\nu(S_n)} = q. \tag{12.2}$$

In particular, if (12.2) is required only when each set S_n is a closed ball with center ζ, the derivate is called the *symmetric derivate*. The convex derivate exists at ν almost every point ζ, and the derivate function is ν almost surely the Radon Nikodym derivative $d\mu/d\nu$. The latter notation will also be used for the convex and symmetric derivates, with the intended interpretation always specified.

If there is a finite number q such that

$$\frac{d|\mu - q\nu|}{d\nu}(\zeta) = 0,$$

where the derivate is in the convex [symmetric] sense, then $(d\mu/d\nu)(\zeta) = q$ in the same sense, and we shall then say that the derivate is q in the convex [symmetric] variational sense. We outline the proof, standard when $N = 1$ and $\nu = l_1$, that $d\mu/d\nu$ exists ν almost everywhere on $\mathbb{R}^N$ in the convex variational sense. If the function $d\mu/d\nu$ (convex derivate) is denoted by f and if q is any real number,

$$\frac{d|\mu - q\nu|}{d\nu} = |f - q| \tag{12.3}$$

ν almost surely; so (12.3) is true ν almost surely simultaneously for all rational q, and it follows from an easy continuity argument that (12.3) is true almost surely simultaneously for all q. If ζ is not in the exceptional set for the last statement, set $q = f(\zeta)$ to complete the proof.

Appendix V

Uniform Integrability

This Appendix, consisting of only one section, lists for easy reference what is needed in this book of the uniform integrability concept.

It will be convenient to call a function Φ from $\mathbb{R}^+$ to $\mathbb{R}^+$ a *uniform integrability test function* if Φ is monotone increasing and convex and if $\lim_{s\to\infty} [\Phi(s)/s] = +\infty$.

Let $(X, \mathscr{X}, \lambda)$ be a measure space with $\lambda(X) < +\infty$. All functions on X in the discussion below are measurable from $(X, \mathscr{X})$ into $(\overline{\mathbb{R}}, \mathscr{B}(\overline{\mathbb{R}}))$ unless specified otherwise. Recall that a family $\{f_t, t \in I\}$ of such functions is said to be uniformly integrable (relative to λ) if

$$\lim_{n\to\infty} \sup_{t\in I} \int_{\{|f_t|>n\}} |f_t| \, d\lambda = 0.$$

The following uniform integrability properties are used in this book. Except in (d), all functions of the family are in $L^1(\lambda)$.

(a) The family $f_.$ is uniformly integrable if and only if it is L^1 bounded and if the set function family $\{A \mapsto \int_A |f_t| \, d\lambda, t \in I\}$ is uniformly absolutely continuous relative to λ; more formally $f_.$ is uniformly integrable if and only if

$$\sup_{t\in I} \int_X |f_t| \, d\lambda < +\infty, \quad \text{and} \quad \lim_{\lambda(A)\to 0} \sup_{t\in I} \int_A |f_t| \, d\lambda = 0.$$

(b) The family $f_.$ is uniformly integrable if and only if there is a uniform integrability test function Φ such that $\sup_{t\in I} \int_X \Phi(|f_t|) \, d\lambda < +\infty$.

(c) A λ almost everywhere convergent sequence of functions on X is L^1 convergent if and only if the sequence is uniformly integrable.

(d) If $f_.$ is a sequence of positive functions on X, then (by Fatou's lemma)

$$\liminf_{n\to\infty} \int_X f_n \, d\lambda \geq \int_X \liminf_{n\to\infty} f_n \, d\lambda.$$

In particular, if $\lim_{n\to\infty} f_n = f$ λ almost surely, if each f_n is integrable, and if

$$\lim_{n\to\infty} \int_X f_n \, d\lambda = \int_X f \, d\lambda < +\infty,$$

then the sequence $f_\bullet$ is uniformly integrable.

(e) A uniformly integrable sequence $f_\bullet$ on X is weakly sequentially compact; that is, there is a subsequence $f_{a_\bullet}$ and a λ integrable function f on X such that

$$\lim_{n\to\infty} \int_X f_{a_n} g \, d\lambda = \int_X fg \, d\lambda$$

for every bounded function g on X.

Generalization

Let $(X, \mathscr{X})$ be a measurable space, and let I be an arbitrary nonempty set. Suppose that for each t in I there is a function f_t on X and a finite measure λ_t on $\mathscr{X}$. The family is said to be uniformly integrable relative to $\lambda_\bullet$ if

$$\sup_{t\in I} \mu_t(X) < +\infty, \quad \text{and} \quad \lim_{n\to\infty} \sup_{t\in I} \int_{\{|f_t|>n\}} |f_t| \, d\lambda_t = 0.$$

The obvious translations of (a) and (b) into the present context are valid.

Appendix VI

Kernels and Transition Functions

1. Kernels

Let $(X, \mathscr{X})$ and $(Y, \mathscr{Y})$ be measurable spaces. A *kernel* from the first space into the second is defined as a function from $X \times \mathscr{Y}$ into $\overline{\mathbb{R}}^+$ with the following properties:

(K1) $p(\cdot, A)$ is $\mathscr{X}$ measurable when $A \in \mathscr{Y}$.
(K2) $p(\xi, \cdot)$ is a measure on $\mathscr{Y}$ when $\xi \in X$.

The kernel is called *substochastic* if $p(\cdot, \mathscr{Y}) \leq 1$ and *stochastic* if $p(\cdot, \mathscr{Y}) \equiv 1$. If $(X, \mathscr{X}) = (Y, \mathscr{Y})$, the kernel is said to have state space $(X, \mathscr{X})$.

Kernels Given by Densities

If q is a measurable function from $(X \times Y, \mathscr{X} \times \mathscr{Y})$ into $(\overline{\mathbb{R}}^+, \mathscr{B}(\overline{\mathbb{R}}^+))$ and if λ is a measure on Y, define

$$p(\xi, A) = \int_A q(\xi, \eta)\lambda(d\eta) \qquad (\xi \in X, A \in \mathscr{Y})$$

to obtain a kernel p with density q relative to λ.

Notation

If p is a kernel from $(X, \mathscr{X})$ into $(Y, \mathscr{Y})$ and if f is a measurable function from $(Y, \mathscr{Y})$ into $(\overline{\mathbb{R}}, \mathscr{B}(\overline{\mathbb{R}}))$, we write $p(\xi, f)$ for $\int_Y f(\eta) p(\xi, d\eta)$ when this integral is well defined, for example, when $f \geq 0$ and f is $\mathscr{Y}$ measurable. If p has density q relative to a specified measure, we also write $q(\cdot, f)$ for $p(\cdot, f)$.

Extension of a Substochastic Kernel to a Stochastic Kernel

Let p be a substochastic kernel with state space $(X, \mathscr{X})$. This kernel can be extended to be stochastic as follows. Adjoin a *trap point* (also called a *cemetery*) ∂ to obtain an enlarged space $X' = X \cup \{\partial\}$, and define $\mathscr{X}'$ as the

σ algebra of subsets of X' generated by $\mathscr{X}$ and the singleton $\{\partial\}$. A stochastic kernel p' with state space $(X', \mathscr{X}')$ is then determined by setting $p' = p$ on $X \times \mathscr{X}$ and

$$p'(\xi, \{\partial\}) = \begin{cases} 1 - p(\xi, X) & \text{if } \xi \in X, \\ 1 & \text{if } \xi = \partial. \end{cases}$$

This kernel is the desired extension of p.

2. Universally Measurable Extension of a Kernel

If p is a kernel from $(X, \mathscr{X})$ into $(Y, \mathscr{Y})$ and if the function $p(\cdot, X)$ is bounded, we can extend each measure $p(\xi, \cdot)$ by partial completion to the σ algebra $\mathscr{U}(\mathscr{Y})$ of universally measurable sets over $\mathscr{Y}$. Under this extension p becomes a kernel from $(X, \mathscr{U}(\mathscr{X}))$ into $(Y, \mathscr{U}(\mathscr{Y}))$. To see this, let μ be the completion of a finite measure on $\mathscr{X}$, suppose that $A \in \mathscr{U}(\mathscr{Y})$, define a measure ν on $\mathscr{X}$ by $\nu(\cdot) = \int_X p(\xi, \cdot)\mu(d\xi)$, and extend ν by partial completion to the σ algebra $\mathscr{U}(\mathscr{Y})$. Since A is ν measurable, there are sets A_1, A_2 in $\mathscr{Y}$ with $A_1 \subset A \subset A_2$ and $\nu(A_1) = \nu(A_2)$. This equality implies that $p(\cdot, A_1) = p(\cdot, A) = p(\cdot, A_2)$ μ almost everywhere on X; that is, $p(\cdot, A)$ is μ measurable for every μ, as was to be proved. This argument shows that the definition of $\nu(A)$ for A in $\mathscr{Y}$ remains valid for A in $\mathscr{U}(\mathscr{Y})$.

It is important for the application to probability that we have actually proved more than was asserted. In fact let $\mathscr{U}_p(\mathscr{Y})$ be the σ algebra of those subsets of Y which, for every choice of μ, are in the completion of the measure ν defined above. This σ algebra includes $\mathscr{U}(\mathscr{Y})$, and its sets will be called the *universally measurable sets over* $\mathscr{Y}$ *relative to* p. What we have proved is that if $p(\xi, \cdot)$ is extended by partial completion to this σ algebra, then $p(\cdot, A)$ is universally measurable, and the definition of ν remains valid.

3. Transition Functions

Let I be a linearly ordered set, let $(X, \mathscr{X})$ be a measurable space, and suppose that to each pair (r, s) of points of I with $r < s$ corresponds a stochastic kernel $p(r, \cdot, s, \cdot)$ with state space $(X, \mathscr{X})$ such that for A in $\mathscr{X}$ and $r_1 < r_2 < r_3$ in I the Chapman–Kolmogorov equation

$$p(r_1, \xi, r_2, A) = \int_X p(r_2, \eta, r_3, A)p(r_1, \xi, r_2, d\eta) \tag{3.1}$$

is satisfied. Such a function p is called a *Markov transition function*, or simply a *transition function*. In some contexts it is necessary to allow the kernels to be substochastic or even infinite valued, but any divergence from the above definition will be specified explicitly. The parameter set is usually either the set of strictly positive integers *discrete parameter case* or the interval $]0, +\infty[$ (*continuous parameter case*).

Discrete Parameter Stationary Case

Let $(X, \mathscr{X})$ be a measurable space, let $p(1, \cdot, \cdot)$ be a kernel with this state space, and define inductively a sequence of kernels with this state space by

$$p(n+1, \xi, A) = \int_X p(1, \eta, A)p(n, \xi, d\eta), \qquad n \geq 1. \tag{3.2}$$

Then

$$p(r+s, \xi, A) = \int_X p(r, \eta, A)p(s, \xi, d\eta). \tag{3.3}$$

If $p(1, \cdot, \cdot)$ is a stochastic kernel, the function $(m, \xi, n, A) \mapsto p(n-m, \xi, A)$ for $0 < m < n$ is a transition function with parameter set the set of strictly positive integers, but in this context it should cause no confusion if the function $(n, \xi, A) \mapsto p(n, \xi, A)$ for $n > 0$ is called a *stationary transition function* or *temporally homogeneous transition function* and if (3.3) is identified as the relevant Chapman–Kolmogorov equation.

If the given kernel $p(1, \cdot, \cdot)$ is substochastic, (3.3) yields a substochastic transition function p. Let $p'(1, \cdot, \cdot)$ be the extension of the substochastic kernel $p(1, \cdot, \cdot)$ to be stochastic by way of a trap point. Then the inductive definition (3.2) applied to $p'(1, \cdot, \cdot)$ leads to a (stochastic) transition function p', and $p'(n, \cdot, \cdot)$ is for each strictly positive integer n the extension of $p(n, \cdot, \cdot)$ to be stochastic by way of the same trap point.

Continuous Parameter Stationary Case

Let $(X, \mathscr{X})$ be a measurable space, and let p be a function from $]0, +\infty[\times X \times \mathscr{X}$ into $\mathbb{R}^+$ with the following properties:

(a) For each t in $]0, +\infty[$ the function $p(t, \cdot, \cdot)$ is a kernel with state space $(X, \mathscr{X})$.

(b) The Chapman–Kolmogorov equation (3.3) (with r and s now in $]0, +\infty[$) is satisfied.

If each kernel $p(t, \cdot, \cdot)$ is stochastic, the function $(r, \xi, s, A) \mapsto p(s-r, \xi, A)$ for $0 < r < s$ is a transition function with parameter set $]0, +\infty[$, but as in the discrete parameter case the function p itself is called a stationary or temporally homogeneous transition function. If the kernels are allowed to be substochastic, in which case p is a substochastic transition function, a single trap point can be adjoined to X to make each kernel $p(t, \cdot, \cdot)$ stochastic and thereby extend p to be a (stochastic) transition function.

Universally Measurable Extensions

In the discrete parameter stationary case it is trivial that if $p(1, \cdot, X)$ in (3.2) is bounded and if $p_u(1, \cdot, \cdot)$ is the universally measurable extension of $p(1, \cdot, \cdot)$, that is (Section 2), the extension of this kernel to the state space $(X, \mathscr{U}(\mathscr{X}))$, then the nth iterate $p_u(n, \cdot, \cdot)$ of $p_u(1, \cdot, \cdot)$ obtained in (3.2) when $p(1, \cdot, \cdot)$ is replaced by $p_u(1, \cdot, \cdot)$ is the extension of $p(n, \cdot, \cdot)$ to the extended state space. The corresponding result in the continuous parameter stationary case will now be proved. That is, in this case we prove that if $p(t, \cdot, X)$ is bounded for each $t > 0$ and if $p_u(t, \cdot, \cdot)$ is the extension of the kernel $p(t, \cdot, \cdot)$ to the state space $(X, \mathscr{U}(\mathscr{X}))$, then the Chapman–Kolmogorov equation (3.3) is satisfied by p_u. To verify (3.3) for p_u with A in $\mathscr{U}(\mathscr{X})$, fix r, s, and ξ, and choose A_1 and A_2 in $\mathscr{X}$ to satisfy

$$A_1 \subset A \subset A_2, \qquad p(r + s, \xi, A_1) = p(r + s, \xi, A_2).$$

Then

$$p_u(r + s, \xi, A) = p_u(r + s, \xi, A_i) = \int_X p_u(s, \eta, A_i) p_u(r, \xi, d\eta), \qquad i = 1, 2; \tag{3.4}$$

so (3.4) is valid when A_i is replaced in the integrand by A, as was to be proved. In this context if we strengthen the hypotheses on p by supposing that for A in $\mathscr{X}$ the function $p(t, \cdot, A)$ is $\mathscr{B}(]0, +\infty[) \times \mathscr{X}$ measurable, then if f is a Lebesgue measurable function from $\mathbb{R}^+$ into $\overline{\mathbb{R}}^+$, the function

$$(\xi, A) \mapsto \int_0^\infty f(t) p_u(t, \xi, A) l_1(dt)$$

is a kernel with state space $(X, \mathscr{X})$ and so (Section 2) when considered on $X \times \mathscr{U}(\mathscr{X})$ is a kernel with that state space, and therefore the above integral defines a $\mathscr{U}(\mathscr{X})$ measurable function for fixed A in $\mathscr{U}(\mathscr{X})$. Functions defined by integrals of this form are basic in probabilistic potential theory.

Appendix VII

Integral Limit Theorems

1. An Elementary Limit Theorem

Let X be a metric space, let λ be a completed measure of Borel subsets of X, and let $k_.$ be a sequence of λ measurable functions from X into $\overline{\mathbb{R}}$. In many contexts there is a point ζ of X with the property that $\lambda(\{\zeta\}) = 0$ and that for f a suitably restricted function from X into $\overline{\mathbb{R}}$, continuous at ζ,

$$\lim_{n\to\infty} \int_X k_n f\, d\lambda = f(\zeta). \tag{1.1}$$

The following theorem gives conditions that such a limit relation be true. The conditions are phrased for ease in application rather than for maximal generality.

Theorem. *Suppose that $k_.$ and f satisfy the following conditions:*

(a) *Each function k_n is positive, with $\int_X k_n\, d\lambda = 1$.*
(b) *The function f is λ measurable, and if A is an open neighborhood of ζ, then $\lim_{n\to\infty} \int_{X-A} k_n |f|\, d\lambda = \lim_{n\to\infty} \int_{X-A} k_n\, d\lambda = 0$.*

Under these conditions

$$\limsup_{n\to\infty} \int_X k_n f\, d\lambda \le \limsup_{\xi\to\zeta} f(\xi). \tag{1.2}$$

If in addition to (b) *the function f is continuous at ζ, then* (1.1) *is true.*

If the right side of (1.2) is $+\infty$, inequality (1.2) becomes trivial. Thus to prove (1.2), which implies the last assertion of the theorem, it is sufficient to show that if $f \le \alpha < +\infty$ on a deleted open neighborhood A of ζ, then the left side of (1.2) is at most α. We can suppose, adding a constant to f if necessary, that $\alpha \ge 0$. Then

$$\limsup_{n\to\infty} \int_X k_n f\, d\lambda \le \alpha + \limsup_{n\to\infty} \int_{X-A} k_n |f|\, d\lambda = \alpha,$$

as was to be proved.

2. Ratio Integral Limit Theorems

For each positive integer n let $k(n, \cdot)$ be a positive Borel measurable function on $\mathbb{R}^N$. Let H and U be, respectively, a measure and a charge on $\mathbb{R}^N$. Define

$$a_n = \int_{\mathbb{R}^N} k(n, \eta) U(d\eta), \qquad b_n = \int_{\mathbb{R}^N} k(n, \eta) H(d\eta). \tag{2.1}$$

In many contexts $k(n, \eta)$ is defined in such a way that

$$\lim_{n \to \infty} \frac{a_n}{b_n} = \frac{dU}{dH}(0). \tag{2.2}$$

We shall derive conditions on $k(n, \cdot)$ for (2.2) to be true. The following remark will be useful. Suppose that H is fixed and that $0 < b_n < +\infty$ for all n. The charge U will be allowed to vary in some specified linear class Γ of charges, including H and including $|U|$ with U, for which

$$\int_{\mathbb{R}^N} k(n, \eta) |U|(d\eta) < +\infty.$$

Suppose it has been proved that (2.2) is true under the following special conditions: $U \in \Gamma$, $U \geq 0$, and the derivate in (2.2) is the symmetric derivate and vanishes. It then follows that (2.2) is true for U in Γ whenever the derivate on the right exists as a symmetric variational derivate (see Section IV.12 for the derivate definitions). In fact if q is the value of this derivate,

$$|a_n - qb_n| \leq \int_{\mathbb{R}^N} k(n, \eta) |U - qH|(d\eta),$$

and by hypothesis $|U - qH| \in \Gamma$, and by definition of the symmetric variational derivate the symmetric derivate $d|U - qH|/dH$ exists and vanishes at the origin.

The preceding remark is of course also valid if the derivate in question is the convex instead of the symmetric derivate.

In some contexts ratio integral limit theorems of the type described are easily reduced to rather simple ratio integral limit theorems in one dimension. Such a one-dimensional ratio integral limit theorem is proved in the next section.

3. A One-Dimensional Ratio Integral Limit Theorem

The following ratio integral limit theorem will be used to derive the desired ratio limit theorem for the Poisson integral for harmonic functions on a ball. Let $k_{\bullet}$ be a sequence of positive Borel measurable functions on $\mathbb{R}^+$, and let

α, β be monotone increasing functions on $\mathbb{R}^+$, vanishing at 0, strictly positive on $]0, +\infty[$, and right continuous except possibly at 0. Define

$$a_n = \int_0^\infty k_n \, d\alpha, \qquad b_n = \int_0^\infty k_n \, d\beta. \tag{3.1}$$

The following theorem stating conditions which imply that

$$\lim_{n\to\infty} \frac{a_n}{b_n} = 0 \tag{3.2}$$

is phrased for ease in application rather than generality.

Theorem. *Suppose that a_n, b_n as defined by* (3.1) *are finite for all n and that the following conditions, in which p is some strictly positive integer, are satisfied.*

(a) *k_n is positive and monotone decreasing and has a continuous derivative k_n'.*

(b) *If $a > 0$,*

$$\int_0^a \liminf_{n\to\infty} \frac{k_n(r)r^{p-1}}{k_n(a)} l_1(dr) = +\infty.$$

(c) *There is a strictly positive function f on $\mathbb{R}^+$ such that the function k_n/f is monotone decreasing for large n and that $\int_0^\infty f \, d\alpha < +\infty$.*

(d) $$\lim_{r\to 0} \frac{\alpha(r)}{\beta(r)} = 0, \qquad \limsup_{r\to 0} \frac{r^p}{\beta(r)} < +\infty.$$

Then (3.2) *is true.*

To prove the theorem, choose ε strictly positive and so small that the limit superior in (d) is strictly less than $1/\varepsilon$. If a is so small that $\alpha \le \varepsilon\beta$ on $[0, a]$, then

$$\begin{aligned}\int_0^a k_n \, d\alpha &= \alpha(a)k_n(a) - \int_0^a \alpha k_n' \, dl_1 \le \alpha(a)k_n(a) - \varepsilon \int_0^a \beta k_n' \, dl_1 \\ &= k_n(a)[\alpha(a) - \varepsilon\beta(a)] + \varepsilon \int_0^a k_n \, d\beta \le \varepsilon b_n.\end{aligned} \tag{3.3}$$

If a is so small that $\varepsilon r^p \le \beta(r)$ for r in $[0, a]$, then

$$\begin{aligned}b_n \ge \int_0^a k_n \, d\beta &= \beta(a)k_n(a) - \int_0^a \beta k_n' \, dl_1 \ge \beta(a)k_n(a) - \varepsilon \int_0^a r^p k_n'(r) l_1(dr) \\ &= k_n(a)[\beta(a) - \varepsilon a^p] + \varepsilon p \int_0^a k_n(r) r^{p-1} l_1(dr) \ge \varepsilon \int_0^a k_n(r) r^{p-1} l_1(dr).\end{aligned} \tag{3.4}$$

Combining these two inequalities, we find that for sufficiently small a,

$$\limsup_{n\to\infty}\frac{a_n}{b_n}\le\varepsilon+\limsup_{n\to\infty}\frac{\displaystyle\int_a^\infty k_n\,d\alpha}{\varepsilon\displaystyle\int_0^a k_n(r)r^{p-1}l_1(dr)}. \tag{3.5}$$

The numerator in the fraction on the right is at most

$$\int_a^\infty\frac{k_n}{f}f\,d\alpha\le\frac{k_n(a)}{f(a)}\int_0^\infty f\,d\alpha.$$

[Incidentally this inequality shows that condition (c) implies the finiteness of a_n.] In view of condition (b) the limit superior in (3.5) is 0, and therefore (3.2) is true.

4. A Ratio Integral Limit Theorem Involving Convex Variational Derivates

In the following theorem, phrased to be applicable to the desired ratio boundary limit theorem for parabolic functions on a half-space, $K(s,\cdot)$ is for each strictly positive s a positive Borel measurable function on $\mathbb{R}^+$, U is a charge on $\mathbb{R}^N$, H is a measure on $\mathbb{R}^N$, and u and h are defined on $\mathbb{R}^N\times\,]0,+\infty[$ by

$$u(\xi,s)=\int_{\mathbb{R}^N}K(s,|\xi-\eta|)U(d\eta),\qquad h(\xi,s)=\int_{\mathbb{R}^N}K(s,|\xi-\eta|)H(d\eta). \tag{4.1}$$

Theorem. *Suppose that the following conditions are satisfied.*

(a) *For each $s>0$ the function u, with U replaced by $|U|$, and the function h are locally bounded on $\mathbb{R}^N$.*

(b) *For each $s>0$ the function $K(s,\cdot)$ is a positive monotone decreasing function on $\mathbb{R}^+$, with a continuous derivative, and $K(s,0)=1$.*

(c) *There is a strictly positive function $\delta(\cdot)$ on $]0,+\infty[$ satisfying*

$$\lim_{s\to0}\delta(s)=0,\qquad\liminf_{s\to0}K(s,\delta(s))>0.$$

(d) *If $a>0$, there is a strictly positive number $s_0=s_0(a)$ for which, uniformly for $r\ge a$,*

$$\lim_{s\to0}\frac{K(s,r)}{K(s_0,r)\delta(s)^N}=0.$$

Then if at a point ζ of $\mathbb{R}^N$ the derivate dl_N/dH exists as a finite convex derivate and dU/dH exists as a finite convex variational derivate, and if $0 < c < 1$, it follows that

$$\lim_{s\to 0} \sup_{|\xi-\zeta|\le c\delta(s)} \left|\frac{u(\xi,s)}{h(\xi,s)} - \frac{dU}{dH}(\zeta)\right| = 0. \tag{4.2}$$

Observe that (Section IV.12) dU/dH and dl_N/dH exist and are finite as convex variational derivates at H almost every point ζ of $\mathbb{R}^N$.

For fixed H the class of charges U satisfying (a) is linear, contains $|U|$ with U, and contains H. Hence according to the argument in Section 2, it is sufficient to prove that (4.2) is true for ζ when U is positive and both dl_N/dH and dU/dH exist as finite convex derivates with $(dU/dH)(\zeta) = 0$. To simplify the notation, we assume that $\zeta = 0$. For $r > 0$ define $\alpha(\xi, r)$ and $\beta(\xi, r)$, respectively, as the U and H measures of $\bar{B}(\xi, r)$, and define $\alpha(\xi, 0) = \beta(\xi, 0) = 0$. By definition of the convex derivate

$$\begin{aligned} &\lim_{r\to 0}\sup\left\{\frac{\alpha(\xi,r)}{\beta(\xi,r)} : |\xi| \le cr\right\} = 0,\\ &\lim_{r\to 0}\sup\left\{\frac{r^N}{\beta(\xi,r)} : |\xi| \le cr\right\} < +\infty. \end{aligned} \tag{4.3}$$

To obtain a minorant of h, observe that

$$h(\xi,s) \ge \int_{[0,\delta(s)]} K(s,r)\,d\beta(\xi,r) \ge K(s,\delta(s))\beta(\xi,\delta(s)) \tag{4.4}$$

so that

$$\int_{[0,\delta(s)]} K(s,r)\frac{d\alpha(\xi,r)}{h(\xi,s)} \le \frac{\alpha(\xi,\delta(s))}{K(s,\delta(s))\beta(\xi,\delta(s))}. \tag{4.5}$$

If $a > 0$ and if s is so small that $\delta(s) < a$,

$$\begin{aligned} \int_{]\delta(s),a]} K(s,r)\,d\alpha(\xi,r) &= K(s,a)\alpha(\xi,a) - K(s,\delta(s))\alpha(\xi,\delta(s))\\ &\quad - \int_{]\delta(s),a]} K'(s,r)\alpha(\xi,r)l_1(dr). \end{aligned} \tag{4.6}$$

If now $\varepsilon > 0$ and if a is so small that $\alpha(\xi,r) \le \varepsilon\beta(\xi,r)$ when $|\xi| \le cr$ and $r \le a$, then the right side of (4.6) is majorized when $K'\alpha$ is replaced by $\varepsilon K'\beta$, and integration by parts yields

$$
\begin{aligned}
\int_{]\delta(s),a]} K(s,r)\,d\alpha(\xi,r) &\le K(s,a)[\alpha(\xi,a)-\varepsilon\beta(\xi,a)] \\
&\quad - K(s,\delta(s))[\alpha(\xi,\delta(s))-\varepsilon\beta(\xi,\delta(s))] \\
&\quad + \varepsilon\int_{]\delta(s),a]} K(s,r)\,d\beta(\xi,r) \\
&\le \varepsilon\beta(\xi,\delta(s)) + \varepsilon h(\xi,s).
\end{aligned}
\tag{4.7}
$$

Thus for sufficiently small a and $|\xi| \le c\delta(s)$,

$$
\int_{]\delta(s),a]} K(s,r)\frac{d\alpha(\xi,r)}{h(\xi,s)} \le \frac{\varepsilon}{K(s,\delta(s))} + \varepsilon. \tag{4.8}
$$

Finally, if s_0 is chosen as in condition (d) of the theorem, if $\varepsilon' > 0$, and then if s is sufficiently small,

$$
\begin{aligned}
\int_{]a,+\infty[} K(s,r)\frac{d\alpha(\xi,r)}{h(\xi,s)} &\le \varepsilon'\int_{]a,+\infty[} K(s_0,r)\,d\alpha(\xi,r)\frac{\delta(s)^N}{h(\xi,s)} \\
&\le \frac{\varepsilon' u(\xi,s_0)\delta(s)^N}{K(s,\delta(s))\beta(\xi,\delta(s))}.
\end{aligned}
\tag{4.9}
$$

Now u/h is the sum of the integrals on the left in (4.5), (4.8), and (4.9). The first of these integrals tends to 0 with s when $|\xi| \le c\delta(s)$ by (4.3). The second can be made arbitrarily small in the limit ($s \to 0$) by choosing ε small and then choosing a small. For such a choice of a the right side of (4.9) is arbitrarily small with ε' when $|\xi| \le c\delta(s)$. Hence Theorem 4 is true.

Appendix VIII

Lower Semicontinuous Functions

1. The Lower Semicontinuous Smoothing of a Function

Let H be a Hausdorff space, and if $\xi \in H$, let $N(\xi)$ be the set of neighborhoods of ξ. Recall that a function u from H into $\overline{\mathbb{R}}$ is lower semicontinuous if and only if the set $\{u > c\}$ is open for c in $\mathbb{R}$. If u is not necessarily lower semicontinuous, the lower function $\underset{+}{u}$ of u, defined by

$$\underset{+}{u}(\xi) = \sup_{A \in N(\xi)} \inf_{\eta \in A} u(\eta) = u(\xi) \wedge \liminf_{\eta \to \xi} u(\eta), \tag{1.1}$$

is lower semicontinuous, is a minorant of u, and majorizes every lower semicontinuous minorant of u. The function $\underset{+}{u}$ will be called the *lower semicontinuous smoothing of* u in this book. Finally recall that the supremum u of a family $\{u_\beta, \beta \in I\}$ of lower semicontinuous functions is lower semicontinuous because

$$\{u > c\} = \bigcup_{\beta \in I} \{u_\beta > c\} \tag{1.2}$$

and the sets on the right are open.

2. Suprema of Families of Lower Semicontinuous Functions

Theorem. *Let $\{u_\beta, \beta \in I\}$ be a family of lower semicontinuous functions from a second countable Hausdorff space into $\overline{\mathbb{R}}$, and if $J \subset I$, define $u^J = \sup_{\beta \in J} u_\beta$. Then u^I is lower semicontinuous, and there is a countable subset J of I such that $u^J \equiv u^I$. In particular, if $u_\bullet$ is directed upward, there is an increasing sequence u_β with limit u^I.*

We have already noted in Section 1 that u^I is lower semicontinuous. Since the domain of definition of the functions is second countable, (1.2) implies that there is a countable subset J of I such that

$$\{u^I > c\} = \bigcup_{\beta \in I} \{u_\beta > c\}$$

simultaneously for every rational c and therefore for all c. Then u^J is lower semicontinuous, $u^J \leq u^I$, and $\{u^J > c\} = \{u^I > c\}$ for all c. Hence $\{u^J \geq c\} = \{u^I \geq c\}$ also, and subtraction yields $\{u^J = c\} = \{u^I = c\}$; so $u^J \equiv u^I$. The second assertion of the theorem is left to the reader.

3. Choquet Topological Lemma

Lemma. *Let $\{u_\beta, \beta \in I\}$ be a family of functions from a second countable Hausdorff space into $\overline{\mathbb{R}}$, and if $J \subset I$, define $u^J = \inf_{\beta \in J} u_\beta$. Then there is a countable subset J of I such that $\underset{+}{u}^J = \underset{+}{u}^I$. In particular, if I is directed downward, there is a decreasing sequence $u_{\beta_\cdot}$ with limit v such that $\underset{+}{v} = \underset{+}{u}^I$.*

Let H be the given second countable Hausdorff space, and let $H_\cdot$ be a sequence of open subsets of H containing each set of a countable base for the topology of H infinitely often. After replacing u_β by $\arctan u_\beta$ if necessary, we can assume that the family $u_\cdot$ is uniformly bounded. For $n \geq 1$ choose ξ_n in H_n to satisfy $u^I(\xi_n) - \inf_{\eta \in H_n} u^I(\eta) \leq 1/n$. There is a point β_n in I for which $u_{\beta_n}(\xi_n) \leq u^I(\xi_n) + 1/n$. Set $J = \beta_\cdot$. Then

$$\inf_{\eta \in H_n} u^J(\eta) \leq \inf_{\eta \in H_n} u^I(\eta) + \frac{2}{n},$$

and it follows that $\underset{+}{u}^J \leq \underset{+}{u}^I$. The reverse inequality is trivial; so there is equality. The second assertion of the lemma is left to the reader.

Historical Notes

The modernization of classical potential theory during the last fifty years has proceeded for the most part in small steps, many of which were more and more rigorous derivations in more and more general contexts of facts stated years earlier. Thus it is certain that the following notes, which give reference to the "originators" of the more important advances, will appear to knowledgeable readers to be inaccurate and unfair. The historical notes to the probability discussion are only slightly more reliable. It is trivial for research mathematicians to discover mistakes in historical accounts of their own subjects, especially when the accounts rely on publication dates and on the references in research papers. Thus no one should have much confidence in a detailed history of the development of a mathematical theory. The difficulty is compounded by the fact that many theorems are proved by several authors in more or less overlapping formulations and that (see the many papers on the Martin boundary) authors are frequently unwilling to read their colleagues' papers carefully enough to state for the record the exact differences between the scopes of their papers. Thus the reader is requested to read the following notes with a grain of salt and a lump of tolerance.

Part 1

See Burkhardt and Meyer [1, 1900] and Lichtenstein [1, 1919] for reviews of nineteenth and early twentieth century potential theory, and see Kellogg [2, 1929] for potential theory just before its renaissance. See Brelot [11, 1952] and [17, 1972] for more modern reviews. The books of Tsuji [1, 1959], Helms [1, 1969], Brelot [15, 1969], Constantinescu and Cornea [1, 1972], Landkof [1, 1972], and Wermer [1, 1974] on potential theory contain much of the material in Part 1 but are organized differently with emphasis on different aspects of the subject. The books of Murali Rao [4, 1977], Port and Stone [1, 1978], and Chung [2, 1981] derive their potential theory as a by-product of probability.

Chapter 1.I

It would be both futile and difficult to give a detailed historical background to the work in Chapter I of Part 1, but note that the development in Section 8 culminating in (8.3) follows [Green 1, 1828].

Chapter 1.II

Section 1. What was essentially the Poisson integral for $N = 2$ was displayed by Poisson in 1820. In [Poisson 1, 1823], he displayed his integral for $N = 3$ in its present form and tried to prove the boundary limit part of Theorem 1 for continuous boundary functions. Schwarz [1, 1872] gave the first rigorous proof of Theorem 1 (for continuous boundary functions).

Sections 2 and 3. See Harnack [1, 1887] for his inequality and convergence theorem, the latter of course only for monotone sequences. The fact that a one-sided bounded harmonic function on $\mathbb{R}^N$ is identically constant is due to Bôcher [1, 1903].

Section 4. F. Riesz [2, 1926; 3, 1930] inaugurated the systematic study of superharmonic and subharmonic functions; the properties and applications in this chapter are almost all due to him. See [Radó 1, 1937] for early references and further details on these functions.

Section 10. Theorem 10 on the concavity of spherical averages of superharmonic functions was proved in [Riesz 2, 1926] in the dual form for convexity of spherical averages of subharmonic functions.

Section 12. Lord Kelvin (= W. Thomson) introduced his transformation in [Thomson 1, 1847].

Section 14. The theorem that a positive harmonic function on a ball is the Poisson integral of a measure has been called the "Herglotz theorem" with a reference to [Herglotz 1, 1911], but this theorem was proved first by F. Riesz [1, 1911], and in his own paper Herglotz refers to the Riesz paper for this result.

Section 15. The essential content of Theorem 15 is that u/h has nontangential limit function dM_u/dM_h almost everywhere (M_h measure) on the ball boundary. Here the derivate dM_u/dM_h has different possible definitions, and the exceptional M_h null boundary subset depends on the definition chosen. No distinction is made between these definitions in the following remarks. Fatou [1, 1906] proved Theorem 15 for the case $N = 2$, $h \equiv 1$, and proofs were given later for $N > 2$ with $h \equiv 1$ by other mathematicians. The theorem was proved by Doob [12, 1959] in the general case.

Section 16. Minimal harmonic functions were introduced into potential theory by Martin [1, 1941].

Chapter 1.III

Section 1. F. Riesz [2, 1926] introduced least harmonic majorants of subharmonic functions.

Section 3. See these notes to Chapter VI for the history of the Fundamental Convergence Theorem.

Section 4. The reduction concept developed from the dual concept of "extremization." If D is a Greenian subset of $\mathbb{R}^N$, if u is a negative subharmonic function on D, and if A is a subset of D, the extremization of u on A was defined as the supremum of the class of negative subharmonic functions majorized quasi everywhere on $D - A$ by u. See Brelot [8, 1945] for further details and references to earlier related work by him and A. F. Monna. Brelot [13, 1956] extended work of Martin [1, 1941] by treating reductions of positive superharmonic functions on certain subsets of ∂D. The treatment of reductions in this book extends the standard treatment by allowing reductions on an arbitrary subset of the closure of D in an arbitrary compactification.

Section 7. The natural order decomposition is apparently due to Mokobodzki. See Mokobodzki and Sibony [1, 1968] for the relation between the natural order decomposition theorem and additivity of reductions in a very general context.

Chapter 1.IV

Cartan's paper [2, 1945] on potentials on special sets had an important influence in the modernization of potential theory. See references below to key results in this paper.

Sections 7 and 8. F. Riesz [3, 1930] derived the existence of the measure associated with a superharmonic function and proved what is now known as the Riesz Decomposition Theorem.

Chapter 1.V

Section 1. Brelot [5, 1941] introduced polar sets to take the place of sets of inner capacity 0 (called inner polar sets in this book) as the negligible sets of classical potential theory. Cartan proved in [2, 1945], although he announced the result in [1, 1942], that the polar sets are the sets of capacity 0 in the capacity definition of Chapter 1.XIII.

Section 2. Choquet [2, 1957] showed, in a trivially different context, that if A is a polar G_δ subset of a Greenian subset D of $\mathbb{R}^N$, then there is a measure μ supported by A such that A is the set of infinities of $G_D\mu$. G. C. Evans had proved the existence of such a potential for A polar and compact.

Section 5. Theorem 5 is usually stated for A a set closed in D, but the formulation in Theorem 5 is sometimes more convenient. The fact that a bounded harmonic function on an open set D less a compact polar subset has a harmonic extension to D was proved by Bouligand [1, 1926]. Vasilesco [2, 1937] proved Theorem 5 for A compact; Brelot [5, 1941], for A closed in D. The fact that a one sided-bounded harmonic function defined on an open deleted neighborhood of a point ξ must have the form $cG(\xi, \cdot) + h$ with h harmonic on the full neighborhood goes back at least as far as [Bôcher 1, 1903]. The less general fact that a bounded harmonic function defined on an open deleted neighborhood of a point can be extended to be harmonic on the full neighborhood was proved by Schwarz [1, 1872].

Section 6. Szegö [1, 1924] proved that in $\mathbb{R}^2$ the complement of a compact set A has a Green function if and only if the Robin constant of A (see Section XIII.18 for the Robin constant) is strictly positive. This result is equivalent to Theorem 6.

Section 8. The Evans–Vasilesco theorem references are Evans [2, 1935] and Vasilesco [1, 1935].

Section 10. The principle of the maximum is sometimes called the *Maria–Frostman* principle of the maximum because Maria [1, 1934] and Frostman [1, 1935] proved this special case of the domination principle (with $D = \mathbb{R}^N$ and $N > 2$).

Chapter 1.VI

Section 1. Szpilrajn [1, 1933] proved the Fundamental Convergence Theorem in the weak version of Theorem III.3 in the context of a locally L^1 convergent sequence of superharmonic functions. Radó [1, 1937] remarked that Szpilrajn's result implied Theorem III.3 in the context of a monotone decreasing locally lower bounded sequence of superharmonic functions. (Both mathematicians actually phrased their results in the dual context, for sequences of subharmonic functions.) Brelot [2, 1938] strengthened the Szpilrajn–Radó result by proving that the exceptional set $\{\underset{+}{u} < u\}$ is inner polar. Cartan [2, 1945], but he announced the result in [1, 1942], proved that this exceptional set is polar.

Section 2. In a series of notes in the *Paris Comptes Rendus* leading up to his definitive "Theory of capacities" [1, 1955] Choquet analyzed set functions satisfying strong subadditivity and related conditions and obtained his capacitability theorem (Theorem 3 of Appendix II), which makes easy the proof that an analytic inner polar set is polar.

Section 3. The reduction properties listed go back to Brelot's early work except for the properties involving sets meeting ∂D and for property (o) which corresponds to certain crossing inequalities in martingale theory (see Section 2.III.22).

Chapter 1.VII

Writers who describe G_D as "the Green's function" should be condemned to differentiate the Lebesgue's measure using the Radon–Nikodym's theorem.

Section 1. The older mathematical literature contains many attempts to show that open sets of various types have Green functions. The modern approach removes the difficulty by generalizing the definition of a Green function. This trick shifts the problem to that of showing that the Green function has certain desired properties, but some of the old smoothness demands have disappeared. For example, it is no longer necessary to define harmonic measure by means of a Green function, as was done in Section I.8.

Chapter 1.VIII

See Vasilesco [3, 1938] for a survey of progress in the Dirichlet problem just before Brelot's [3, 1939] definitive study of what is now called the PWB method.

Section 1. The fact that relativization of a family of functions having an average property (1.1) yields a family with a corresponding average property (1.3) was pointed out in [Doob 9, 1958]. In a Markov process probabilistic context this relativization corresponds to a modification of transition probabilities (Section 2.VI.13) conditioning paths to go where the conditioning function is large. The analysis of the Dirichlet problem on a Greenian subset of $\mathbb{R}^N$ provided with its Martin boundary in [Brelot 13, 1956] was essentially, although not explicitly, for relative harmonic functions.

The Dirichlet problem for harmonic functions has been attacked in several ways, but it was finally seen that each way involves two distinct problems. (i) The first problem is to find, for a specified suitably restricted boundary function f, certainly for an arbitrary finite-valued continuous boundary function f, a perhaps generalized "solution" u_f. The function u_f is to be harmonic, and if the given domain and the given boundary function are sufficiently smooth, the function u_f is to have boundary limit function f; that is, u_f is the solution of the Dirichlet problem in the classical sense. (ii) If u_f exists and if a domain boundary point ζ is specified, the second problem is to find conditions on the domain and boundary function ensuring that u_f has limit $f(\zeta)$ at ζ. The simple Zaremba [1, 1911] example [Section 2, Example (a)] and the more discouraging Lebesgue [1, 1912] spine example (Section 15) showed that the Dirichlet problem may not be solvable in the classical sense even when the boundary is the Euclidean boundary and the boundary function is finite valued and continuous. Poincaré [1, 1890; 2, 1899], however, had shown that for a sufficiently smooth Euclidean boundary the Dirichlet problem has a solution in the classical sense for every finite-valued continuous boundary function. The problem of finding Dirichlet solutions in more general contexts was solved by generalizing the definition of "Dirichlet solution." The PWB method of finding (generalized) Dirichlet solutions is a development of a method devised by Perron [1, 1923] (independently and almost simultaneously by Remak [1, 1924]) and perfected by Brelot [3, 1939]. Wiener [2, 1924; 3, 1924; 4, 1925] was the first to assign a generalized Dirichlet solution to an arbitrary finite-valued continuous Euclidean boundary function. He used two methods of which one was an elaboration of the Perron method, but he did not obtain Brelot's later definitive results because a miscalculation in [4, 1925] gave him a supposed counterexample which prevented him from proceeding to the class of bounded Borel measurable boundary functions. He showed, however, that for the Euclidean boundary the finite-valued continuous boundary functions are resolutive.

Lebesgue was perhaps the first to see clearly the distinction between problems (i) and (ii) above; so it is not surprising that he coined the term "regular boundary point." Kellogg [1, 1928] for $N = 2$ and Evans [1, 1933] for $N \geq 2$ showed that every subset of strictly positive capacity of the Euclidean boundary contains a regular boundary point; that is, the set of irregular boundary points is inner polar. This fact implies that the F_σ set of irregular Euclidean boundary points is polar.

Sections 12 to 14. Poincaré [1, 1890] used the idea of what is now called a barrier. Lebesgue [2, 1912] coined the term and used the existence of barriers for a regular region to prove the convergence to the Dirichlet solution (for a finite-valued continuous boundary function) of a certain sequence of approximants. Lebesgue [3, 1924] found that for the existence of a barrier at a Euclidean boundary point it is necessary and sufficient that the point be regular. Bouligand [1, 1926] showed that it is sufficient for regularity that the barrier be a weak one and proved the related result that a Euclidean boundary point is regular if and only if the Green function $G_D(\xi, \cdot)$ of the given domain D has limit 0 at the boundary point for every pole ξ (equivalently, a single pole if D is connected).

Section 15. In the years when potential theory was finally conquering the Dirichlet problem the cone regularity condition was ascribed to both Poincaré and Zaremba. Only Clio now remembers why Zaremba was later given the sole credit; the condition will be given both names in this book.

Section 18. The extension of G_D in Section 18 was analyzed by Brelot [6, 1944]. See [Brelot 1, 1938], [M. Riesz 1, 1938], [Frostman 2, 1938], and [Vallée Poussin 4, 1938] for earlier discussions of the relation between $\mathrm{GM}_D v$ (for v superharmonic on a neighborhood of $\bar{D}$) and $\mu_D(\cdot, v)$, of the kernel δ_D^A, and of related questions.

Section 19. The proof of Theorem 19(b4) follows that in [Murali Rao 3, 1974]. The result was proved by Brelot [6, 1944] in the course of his treatment of potential theory on $\mathbb{R}^N$ extended by a point at infinity.

Chapter 1.IX

Everything in this chapter is in the folk lore of potential theory, and much of it has appeared explicitly.

Sections 9 and 10. The classes of quasi-bounded and singular harmonic functions were first discussed by Parreau [1, 1951].

Chapter 1.X

The concept of sweeping (balayage) goes back at least as far as Gauss [1, 1840]. There are two approaches to the subject, one by operations on functions, now treated by Brelot's reduction technique, the other by orthogonality-minimization techniques originated by Gauss but first made rigorous by Frostman [1, 1935; 2, 1938] and Cartan [2, 1945; 3, 1946]. The first approach is stressed in this book. The following outline of Poincaré's method of solving the Dirichlet problem [1, 1890; 2, 1899] shows why his method is identified with sweeping, but see ⟨below⟩ Vallée Poussin's remark about Poincaré's work. Let D be a bounded open subset of $\mathbb{R}^N$, with a smooth boundary. For u a $\mathbb{C}^{(2)}$ function on a neighborhood of $\bar{D}$, with $\Delta u < 0$ (so that u is superharmonic on a neighborhood of $\bar{D}$), Poincaré evaluated what we now identify as $GM_D u$ by the method of proof of Theorem III.1 and proved that under his smoothness hypotheses on D the function $GM_D u$ was the desired Dirichlet solution for the boundary function $u_{|\partial D}$. If Δu is not supposed strictly negative, Poincaré reduced the problem to the case already treated by writing u as the difference between two functions with the above properties of u. Using this solution and an approximation procedure Poincaré solved the Dirichlet problem for an arbitrary continuous boundary function. Thus the basis of Poincaré's method was the construction of $GM_D u$ for u superharmonic on D. His construction, in the language of the proof of Theorem III.1, was to apply operators of the form τ_B with B a ball relatively compact in D. For u superharmonic on D the operation $u \mapsto \tau_B u$ sweeps the Riesz measure for u out of B (observe that $\tau_B u = R_u^{D-B}$ when u is positive on D, so that reduction notation is applicable), but Poincaré did not go on to analyze this aspect of his method. Vallée Poussin in a series of papers including [1, 1932; 2, 1937; 3, 1938; 4, 1938] developed sweeping further and in fact in [3, 1938] he asserted that he had coined the word "balayage" and that (in spite of his own title for [1, 1932]) Poincaré's work had nothing to do with sweeping. Vallée Poussin concentrated on the swept measures; so it was he who developed and applied the interpretation of the sweeping kernel as harmonic measure. He applied sweeping to analyze capacity, the role of sets of inner capacity 0, and many other topics, not always including full details, obtaining among other results one equivalent to Theorem XI.13. The full use of the reduction technique on functions had to wait for the final form of the Fundamental Convergence Theorem, which was quickly applied to reductions by Brelot [8, 1945]. Most of Chapter X first appeared in the latter paper.

Chapter 1.XI

See [Brelot 16, 1971] for further elaboration of the fine topology and its applications.

Section 1. Brelot [4, 1940] defined thinness of a set at a point by XI(2.1). Cartan pointed out in a letter to him that thinness could be interpreted by means of what is now called the fine topology. For some years thereafter, however, the fine topology was merely a tool for phrasing results elegantly; in fact topological language was not commonly used in discussions involving thinness. Cartan [3, 1946] noted that the fine topology is the restriction of a certain topology of measures to the class of unit measures supported by singletons. According to Constantinescu and Cornea [1, 1972], the Baire property of the fine topology was noted by Cornea in 1966.

Section 4. The existence of the fine limits in Theorem 4(a) was proved by Doob [8, 1957] by probabilistic and [11, 1959] by nonprobabilistic methods. The present proof is new. The identification of these limits could easily (see the application in Section XII.19) have been made in these earlier papers but was not.

Section 5. See [Brelot 6, 1944] for classical potential theory extended to $\mathbb{R}^N \cup \{\infty\}$.

Section 6. Theorem 6 is due to Brelot [7, 1944].

Section 9. Theorem 9 is due to Cartan [see Deny 1, 1950, p. 171] for fine limits and to Doob [11, 1959] for fine cluster values (in the Martin boundary context).

Section 11. Theorem 11 is due to Doob [17, 1966].

Section 12. Theorem 12 is due to Brelot [4, 1940].

Sections 14 and 15. Theorems 14 and 15 are due to Brelot [8, 1945].

Section 18. Theorem 18 is due to Brelot [10, 1948], but see also Vallée Poussin 4, 1938].

Section 19. See [Fuglede 1, 1972] for a discussion of classical potential theory based on fine superharmonic and fine harmonic functions.

Sections 21 and 22. The material in these sections is taken from [Brelot 9, 1946].

Section 23. See the note to Section XVIII.16 on the parabolic context domination principle.

Chapter 1.XII

Sections 1 to 9. Martin [1, 1941] defined what is now called the Martin boundary and proved what is now called the Martin representation theorem (Theorem 9). By now the Martin boundary of a set D on which harmonic functions in some generalized sense are defined is a rather vague concept. Roughly any set of points ζ, to each of which corresponds a minimal harmonic function $K(\zeta, \cdot)$, which is adjoined to D along with possible other points to form a topological space and which then gives as a representation of an arbitrary positive harmonic function u an integral $\int K(\zeta, \cdot)M_u(d\zeta)$ is called a Martin boundary for D. The treatment in Sections 1 to 9 follows Martin as modernized by Brelot [13, 1956] and put into the context of h-harmonic functions. The significance for its Martin representation of quasi boundedness of a harmonic function was pointed out by Parreau [1, 1951].

Section 10. Theorem 10 is due to Brelot [13, 1956]. Heins [1, 1959] defined a "generalized harmonic measure" on a Riemann surface D as a positive harmonic function u for which $0 \le u \le 1$ and $\mathrm{GM}_D[u \wedge (1-u)] = 0$. According to the remark "Intrinsic definition of h-harmonic measure," such a function u is actually the harmonic measure of a Martin boundary subset. (The Martin boundary discussion is applicable to Riemann surfaces.)

Sections 11 to 14. Naïm [1, 1957] defined the minimal-fine topology on a Martin space (in a more satisfying way than that in Section 12, where her elegant approach is considerably abbreviated) and proved Theorems 11, 13, and 14. Theorems 13(a) and 14 had already been proved in different terminology for D a half-space by Lelong [1, 1949].

Section 16. Naïm [1, 1957] proved the part of Theorem 16 involving minimal-fine limits. The cluster set part of Theorem 16 is taken from [Doob 11, 1959].

Section 18. Theorem 18 is due to Naïm [1, 1957].

Section 19. Theorem 19 was proved by Doob [8, 1957] using probabilistic and [11, 1959] nonprobabilistic methods. This theorem has been generalized to axiomatic potential theory contexts by Gowrisankaran (to Brelot harmonic spaces) and by Sibony (to a context including Bauer harmonic spaces). See [Taylor 1, 1980] for a simple version of Sibony's result with references to previous work.

Section 21. Theorem 21 was proved by Brelot and Doob [1, 1963] along with further theorems on the relations between nontangential and minimal-fine boundary limits of functions defined on a half-space. See the references there to work of Constantinescu and Cornea and others on the relations between nontangential and minimal-fine cluster values of functions defined on a ball and see [Brossard 1, 1978] for such relations derived in probabilistic language.

Sections 22 and 23. Theorem 23(a) for $n = 2$, in the disk rather than the half-plane context, was proved by Littlewood [1, 1928] long before the fine topology was invented. His result was generalized to the N-dimensional context by Privalov [1, 1938]. The proofs of Lemma 22 and Theorem 23 are taken (corrected) from [Doob 16, 1965], an unreadably garbled paper.

Fine topology confusion. Let D be a half-space, let ζ be a finite boundary point of D, and let A be a subset of D. The relations between thinness of A at ζ, minimal thinness of A at ζ, and the relations between these two concepts when A is in a cone with vertex ζ are not obvious and have sometimes been misstated in the literature. The situation has been complicated by the habit of Doob and some other authors of omitting the qualifier "minimal" in discussing minimal-fine limits and cluster values at Martin boundary points. Jackson [1, 1970] surveyed this situation and made the necessary corrections.

Chapter 1.XIII

Section 1. In a pioneering paper Gauss [1, 1840] considered measures ν supported by a smooth surface A in $\mathbb{R}^3$. For f a given function on A he studied the properties of ν chosen to minimize $\int_A (G\nu - 2f)\,d\nu$ for specified $\nu(A)$; the measure ν was by hypothesis given by a smooth density relative to surface area. He took it for granted that a minimizing measure μ existed and showed that then $G\mu - f$ is identically constant on the support of μ. In particular ($f =$ const), Gauss thereby obtained an equilibrium distribution for A and, if f is the restriction to A of the potential of a measure λ, Gauss obtained the sweeping of λ onto A. Finally Gauss showed how to manipulate his result to solve the Dirichlet problem for the domain bounded by A. His work was of course not rigorous but offered ideas not fully digested for a century. Careful formulations suggested by some of his ideas are presented in Section 14. The first rigorous derivation of an equilibrium distribution for an arbitrary compact subset of $\mathbb{R}^3$ was carried through by Frostman [1, 1935], whose basic tool was a modernization of Gauss's technique of minimization of the energy integral. The capacity of an arbitrary compact subset A of $\mathbb{R}^3$ was defined by Wiener [2, 1924] by solving the (generalized) Dirichlet problem for the unbounded open connected component of $\mathbb{R}^3 - A$ with boundary function 1 on ∂A and 0 at the point ∞. This solution can be identified with the restriction to D of the potential $G\mu$ of a measure on A, and Wiener defined the capacity of A as $\mu(A)$. [Since $G\mu$ is in fact the smoothed reduction $\hat{R}^A_{+1}$ (relative to $\mathbb{R}^3$), the potential $G\mu$ is the capacitary potential of A.] This capacity definition was extended by Vallée Poussin [1, 1932] who defined the capacity of an arbitrary subset A of $\mathbb{R}^3$ as the supremum of $\nu(\mathbb{R}^3)$ for measures ν supported by compact subsets of A with potentials $G\nu \leq 1$, and he showed that this definition coincided with Wiener's for compact sets. The Vallée Poussin capacity is the inner capacity as defined in Section 11.

The problem of finding an equilibrium distribution for a set is sometimes called the *Robin problem* in view of [Robin 1, 1886] in which Robin showed that the problem for a smooth surface in $\mathbb{R}^3$ is equivalent to the problem of solving a certain integral equation.

Section 6. Cartan [2, 1945] showed that the space $\mathscr{E}^+$ is complete. (He worked with what we have called "special" domains.) See [Brelot 12, 1953–1954] for a discussion of BLD functions.

Sections 7 and 8. Frostman [1, 1935; 2, 1938] and M. Riesz [1, 1938] gave the first rigorous proofs of Theorem 7.

Section 10. Choquet was led to his theory of capacities by potential theory problems. He proved Theorem 10 in [1, 1955].

Section 17. The Wiener thinness criterion was derived in [Wiener 3, 1924] in the context of boundary point regularity for the Dirichlet problem. Brelot [4, 1940] put it into the equivalent thinness context.

Section 18. See [Choquet 3, 1958] for an approach to logarithmic capacity somewhat different from that in Section 18, but observe that he does not use the standard terminology. H. Jackson kindly showed the author an example demonstrating that the logarithmic capacity set function is not subadditive.

Chapter 1.XIV

The material in this chapter is well known and mostly elementary. It is a fine example of material that is easy to develop, useful to have available for reference, and for which it is as difficult as it is pointless to find the original source.

Chapter 1.XV

Section 1. Parabolic potential theory has not been treated systematically in the literature. Petrowsky [2, 1934–1935] is a valuable source for the Dirichlet problem. Tychonoff [1, 1935] has many useful results. Widder's book [2, 1975] is useful for the solutions of the heat equation on a slab. Some work in axiomatic potential theory has included applications to parabolic potential theory. See, for example, [Constantinescu and Cornea 1, 1972], a book which covers contexts far more general than parabolic potential theory with explicit applications to parabolic potential theory. Probabilists have invoked parabolic potential theory as a convenient tool in studying Brownian motion in a potential theoretic context. See, for example, [Doob 6, 1955], but the reader should be warned that the statement there that for an open set bounded by hyperplanes the Euclidean boundary is regular for the heat equation is false.

Section 5. The proof of Theorem 5 is taken from [Tychonoff 1, 1935].

Section 11. See [Moser 1, 1964] for a more precise Harnack-type inequality than that given in Section 11.

Section 16. The Appell transformation dates back to [Appell 1, 1892].

Chapter 1.XVI

Sections 1 to 6. The essential result in these sections is the equivalence of Theorem 6 parts (a1) and (a2), due to Widder [1, 1944]. The preliminary results leading up to Theorem 6 are mostly taken from Widder's paper and [Tychonoff 1, 1935].

Section 7. Theorem 7 with $\dot{h} \equiv 1$ is old as a special case of a general integral limit theorem; see, for example, [Titchmarsh 1, 1937]. Theorem 7 as stated was first proved in [Doob 13, 1960] along with a general ratio integral limit theorem. See these notes to Appendix VII for references to such ratio integral limit theorems. See [Hartman and Wintner 1, 1950] for a Fatou boundary limit theorem for parabolic functions on an interval.

Chapter 1. XVII

Many results in Chapters XVII and XVIII of Part 1 can be obtained by translating classical results and their proofs into the parabolic context. Such results will not be referenced in these notes. Some of the classical reduction property proofs are no longer valid in the parabolic context, however. Where essentially different proofs are given, they are due to Boboc, Constantinescu or Cornea, separately or together, and have been taken from [Constantinescu and Cornea 1, 1972], which contains proofs valid in a very general context along with detailed references.

Section 13. The direct half of Theorem 13 (in a general axiomatic context) is due to Brelot [14, 1962].

Section 18. Theorem 18 is due to Hunt [1, 1956], who proved it in a probabilistic context.

Chapter 1.XVIII

Section 1. Sternberg [1, 1929] was apparently the first to observe that the Perron method of attacking the Dirichlet problem could be applied in the parabolic context, although he restricted himself to functions on very special domains.

Section 4. Babuška and Vyborný [1, 1962] showed that a finite Euclidean boundary point ξ of a subset D of $\mathbb{R}^N$ is regular for Laplace's equation if and only if every Euclidean boundary point (ξ, s) of $D \times \mathbb{R}$ is regular for the heat equation.

Section 6. The iterated logarithm criterion of Example (d) is due to Petrowsky [2, 1934–1935] who knew that this criterion corresponded to the Khintchine iterated logarithm law for Brownian motion (Section 2.IX.15).

Sections 11 and 12. Meyer [1, 1962] showed that in a very general potential theoretic context, defined probabilistically, a set is polar [semipolar] if and only if it is copolar [cosemipolar].

Section 14. Theorem 14 and Applications (a) and (b) appear to be new. Applications (c) and (d) in a more general (probabilistic) context are exercises in Blumenthal and Getoor's [1, 1968] and are proved in their [2, 1970].

Section 15. See [Doob 13, 1960] for a probabilistic version of Theorem 15, and see Section XIX.15 for a proof of Theorem 15 by Martin boundary technique. It is natural to conjecture that when $\dot{h} \equiv 1$, then (15.1) is true as a normal limit (that is, as a limit along the normal line to the boundary point) as well as a coparabolic-fine limit, but in fact Kaufman and Wu [1, 1982] have exhibited a measure ν on the upper half-space $\dot{D}$ of $\dot{\mathbb{R}}$ such that $\limsup_{s \to 0} \dot{G}_{\dot{D}} \nu(\xi, s) = +\infty$ for $0 < \xi < 1$.

According to Theorem 15, a positive parabolic function $\dot{v}$ on a slab $\mathbb{R}^N \times]0, \delta[$ has coparabolic-fine limit function $d\dot{N}_{\dot{v}}/dl_N$ at l_N almost every point of the abscissa hyperplane. Koranyi and Taylor [1, 1983] have shown that this result can be used to prove Theorem XVI.7 in the special case $\dot{h} \equiv 1$, but their method fails for general $\dot{h}$. That is, the reasoning used in proving Theorem XII.21 to go from minimal-fine boundary limits to non-tangential boundary limits does not seem to have a counterpart in the parabolic context.

Section 16. In a general axiomatic setting including parabolic potential theory Janssen [2, 1975] proved that (notation of Theorem 16) $\dot{h} \leq \dot{v}$ if $\text{p*f}\limsup_{\dot{\eta} \to \zeta} \dot{v}(\dot{\eta})/\dot{h}(\dot{\eta}) \geq 1$ for $\dot{\nu}_{\dot{h}}$ almost every ζ. This result becomes Theorem 16 with the help of Theorem 14. Blumenthal and Getoor state that $\dot{h} \leq \dot{v}$ under condition (c′) in a more general context but with $\dot{v} \equiv$ const as an exercise in their [1, 1968] and give the proof in their [2, 1970].

Chapter 1.XIX

Various versions of a parabolic context Martin boundary have appeared in the literature but that given here, especially that relative to a Martin point set pair, seems closest to Martin's ideas and the easiest for calculational purposes. See Janssen [1, 1971], for example, for the Martin boundary obtained by convex set methods in a context including parabolic potential theory.

Section 15. In the following discussion of functions on the right half-plane $]0, +\infty[\times \mathbb{R}$ we shall write that a function on this set has a [one-sided] parabolic limit α at a boundary point $(0, \tau)$ if the function has the limit α as $(\xi, s) \to (0, \tau)$ with $[\beta_1 \xi^2 < s - \tau < \beta_2 \xi^2]$ $|\tau - s| < \beta\xi^2$ for each strictly positive choice of $[\beta_1, \beta_2$ with $\beta_1 < \beta_2]$ β. According to Section 10, if h is a strictly positive parabolic function on the right half-plane, it can be written in the form

$$\dot{h}(\dot{\xi}) = \int_{-\infty}^{+\infty} \hat{K}(\tau', \dot{\xi}) \dot{N}_h'(d\tau') + \int_{-\infty}^{0} \hat{K}(\gamma'', \dot{\xi}) \dot{N}_h''(d\gamma''),$$

where $\dot{N}_h'$ is a uniquely determined measure on $\mathbb{R}$ and $\dot{N}_h''$ is a uniquely determined measure on $-\mathbb{R}^+$. Let $\dot{v}$ be a positive superparabolic function on the right half-plane with parabolic component (Riesz decomposition) determined by measures $\dot{N}_{\dot{v}}'$ and $\dot{N}_{\dot{v}}''$. According to Section 15, the function $\dot{v}/\dot{h}$ has coparabolic-fine limit $d\dot{N}_{\dot{v}}'/d\dot{N}_h'$ at $\dot{N}_h'$ almost every point of the ordinate axis. This result should be compared with the following results. Kemper [1, 1972] proved that if $\dot{h} \equiv 1$, that is, if $\dot{N}_h' = l_1$ and $\dot{N}_h'' \equiv 0$, and if $\dot{v}$ is parabolic, then $\dot{v}/\dot{h}$ has the stated limit at l_1 almost every point of the ordinate axis in the sense of parabolic approach to the point. On the other hand, Kaufman and Wu [1, 1982] have exhibited an example of a positive parabolic function $\dot{h}$ on the right half-plane specified by a continuous singular relative to l_1 measure $\dot{N}_h'$ and by $\dot{N}_h'' \equiv 0$ such that $\liminf_{\xi \to 0} \dot{h}(\xi, s) = 0$ for all ξ. Thus $1/\dot{h}$ does not give the limit $dl_1/d\dot{N}_h'$ along normal approach to the ordinate boundary, but Kaufman and Wu showed that if $\dot{v}$ and $\dot{h}$ are positive and parabolic on the right half-plane, then $\dot{v}/\dot{h}$ has the limit $d\dot{N}_{\dot{v}}'/d\dot{N}_h'$ at $\dot{N}_h'$ almost every point of the ordinate axis in the sense of one-sided parabolic approach to the point. On the other hand, if $\dot{v}$ is a superparabolic potential on the right half-plane, Wu [1, 1979] has shown that $\lim_{\xi \to 0} \dot{v}(\xi, s) = 0$ for l_1 almost every s. One-sided parabolic approach does not yield this zero limit, for example, if $\dot{v}$ is the superparabolic potential of a measure assigning a strictly positive value to each point of a countable dense subset of the right half-plane.

Part 2

The most useful reference books for Part 2 are [Meyer 6, 1966], [Dellacherie and Meyer 1, 1975; 2, 1980], and [Dellacherie 2, 1972]. For Markov processes see [Dynkin 1, 1960; 2, 1965] and [Blumenthal and Getoor 1, 1968]. [Meyer 8, 1968] contains a useful summary of key ideas in the general theory of stochastic processes, including classification of optional times, classification of types of stochastic process measurability, and a discussion of section theorems. The books of K. M. Rao [4, 1977], Port and Stone [1, 1978], and Chung [2, 1981] are useful references for the relations between probability and classical potential theory.

Chapter 2.I

Section 1. The concept of a filtered measure space always underlay probabilistic analysis involving the Markov property, and the concept was finally formalized in [Doob 4, 1953] to deal with martingale theory. In the schools of Dynkin and Meyer the concept has become the base of all analysis involving Markov process theory, martingale theory, and stochastic integration. The Dynkin approach to Markov processes in his [2, 1965] leads to a further refinement of a filtered space which is not needed in this book.

Section 2. Dynkin defined process measurability, in the context of Markov processes, in his [1, 1960] essentially as what is now called progressive measurability. Meyer defined and applied progressive measurability in [2, 1962–1963]. This measurability was rediscovered and given its present name in [Chung and Doob 1, 1965], written in ignorance of the earlier work.

Section 5. Let $\ldots, x_0, x_1, \ldots$ be mutually independent random variables, each uniformly distributed on the interval $[0, 1]$, and define $\mathscr{F}(n) = \mathscr{F}\{\ldots, x_n\}$. In what amounts to this context Theorem 5 was proved by Jessen [1, 1934] and for $n \to +\infty$ by Lévy [1, 1935], who supposed x in Theorem 5 to be the indicator function of a set. It is clear in Lévy's discussion, however, that he does not use the special independence context, and in fact this special context is not presupposed in his treatment of the same convergence result in his [2, 1937]. Theorem 5 as stated is in [Doob 1, 1940].

Section 9. The first rigorous treatment in a general continuous parameter context of the hitting of a set by a process was given by Hunt [2, 1957]. Meyer [2, 1962–1963] was the first to discuss the hitting of progressively measurable sets.

Section 10. Kolmogorov [1, 1933] gave the measure theoretic basis for probability, including the definitions of expectation and conditional expectation and a construction of measures on infinite-dimensional product spaces. The latter construction is what is needed for canonical processes.

Section 13. For variations of Theorem 13 see [Chung and Doob 1, 1965], [Meyer 6, 1966], and [Dellacherie and Meyer 1, 1975].

Section 14. See [Dellacherie 2, 1972] for a full discussion of the important σ algebras of subsets of $\mathbb{R}^+ \times \Omega$, first defined by Meyer. See [Meyer 6, 1966; 8, 1968] for earlier discussions.

Chapter 2.II

Section 1. Like the Molière character who spoke in prose without realizing it, mathematicians had dealt with optional times for centuries without finding it necessary to define them precisely. Precise definitions were finally introduced, along with σ algebra filtrations, to answer the needs of Markov process and martingale theories. Meyer [6, 1966; 8, 1968] first classified optional times.

Section 4. Hunt [2, 1957] was the first to show that in an appropriate context the hitting time of an analytic set is optional. Doob [5, 1954] had shown that the hitting time by Brownian motion of a capacitable set is optional, but in 1954 it was not known that all Borel sets are capacitable in his context. Meyer [2, 1962–1963] proved that the hitting time of a progressively measurable set is optional.

Section 7. Observe that predictable optional times are defined slightly differently in [Dellacherie and Meyer 1, 1975].

Section 8. Section theorems were introduced into probability theory by Meyer [2, 1962–1963], more successfully in [6, 1966]. See Dellacherie [1, 1972], from which the proof of Theorem 8 is taken, for an elegant unified approach to them.

Section 9. The proof of Theorem 9 is taken from the correction sheet in [Dellacherie and Meyer 1, 1975].

Section 10. Dellacherie [1, 1969] proved that, in the notation of this section, a predictable subset $\dot{A}$ of $\mathbb{R}^+ \times \Omega$ is semipolar if for almost every ω the set $\{t\colon (t,\omega)\in \dot{A}\}$ is countable.

Chapter 2.III

Before martingales had been formally christened, Lévy [1, 1935; 2, 1937], Bernstein [1, 1937], and other mathematicians had analyzed some of their properties in special contexts; usually the martingales in question arose as partial sums $n \mapsto \Sigma_0^n y_j$ of a sequence $y_.$ of random variables under the condition $E\{y_j|y_0, \ldots, y_{j-1}\} = 0$ so that the sums arose as generalizations of sums of independent random variables with zero means. Ville [1, 1939] defined a martingale very nearly as a positive martingale is now defined but tied it to a sequence of independent random variables under analysis. His fundamental tool, a fact he proved, was that almost every sample sequence of a positive martingale sequence is bounded (see Theorem 9). Doob [1, 1940] discussed martingales and proved the basic martingale convergence theorems under the name "family of random variables with the property *E*." ("*E*" was chosen not as the initial letter of "expectation" but as the first letter in the alphabet following D.) Under the respective names "semimartingale" and "lower semimartingale," submartingales and supermartingales were introduced in [Snell 1, 1952] and [Doob 4, 1953]. This obviously inappropriate nomenclature was chosen under the malign influence of the noise level of radio's SUPERman program, a favorite supper-time program of Doob's son during the writing of [Doob 4, 1953]. For further work in martingale theory see [Neveu 1, 1972] (discrete parameter context) and [Dellacherie and Meyer 2, 1980] (continuous parameter context).

Section 4. The trivial but useful map in this section appears to be new.

Section 6. Theorem 6 and its generalizations to various forms of optional sampling in Chapters III and IV are adapted from [Doob 1, 1940; 4, 1953] with improvements from [Meyer 6, 1966].

Section 9. Theorem 9 is taken from [Doob 4, 1953]. In the martingale case or the |martingale| case Theorem 9(a) was proved by Lévy [2, 1937], Bernstein [1, 1937], and Ville [1, 1939].

Section 11. Theorem 11 is taken from [Doob 4, 1953].

Section 12. Inequality (12.2) is new, as are (12.3) and (12.4) except when $y(\cdot) \equiv 1$. Crossings (implicit in [Doob 1, 1940]) were formally introduced into martingale theory in Doob [3, 1951], where a variant of (12.3) was obtained for $x(\cdot)$ a martingale and $y(\cdot) \equiv 1$. Snell in [1, 1952] extended crossing inequalities to submartingales and supermartingales. Dubins [1, 1966] obtained a variant of (12.4) with $y(\cdot) \equiv 1$.

Sections 13 to 17. These results are taken from [Doob 4, 1953]. In the martingale case Theorems 13 and 17 were in [Doob 1, 1940].

Section 19. This may be the first time that the natural decomposition theorem appears in print in the martingale theory context, but the result is merely a special application of the general theory.

Section 22. The first application of reduction theory to martingale theory was made by Snell [1, 1952]. See Neveu [1, 1972] for further discussion of related problems. The infimum of the class of supermartingales majorizing a given process under appropriate side conditions is called the *Snell envelope* of the given process.

Chapter 2.IV

Section 1. The study of the continuity properties of continuous parameter martingale sample functions was initiated in [Doob 1, 1940], completed in [Doob 3, 1951], and extended to supermartingales in [Doob 4, 1953]. Perhaps the first sophisticated use of optional times appeared in the first of these papers to obtain martingale right continuity properties (without crossing inequalities, unknown at that time).

Sections 2 and 3. Theorems 2 and 3 are adapted from [Doob 4, 1953] with improvements in formulation from [Meyer 6, 1966].

Section 4. Theorem 4, the counterpart of, but much deeper than, the classical potential theory rather elementary fact (Section 1.II.4) that the limit of an increasing sequence of superharmonic functions on a connected open subset of $\mathbb{R}^N$ is either superharmonic or identically $+\infty$, is due to Meyer [6, 1966]. The elementary right continuity proof is new.

Section 5. This theorem appears to be new, but see related work of Mertens [1, 1972]. As Mertens' work indicates, the choice of standard continuous parameter supermartingales need not be the almost surely right continuous ones. For some purposes the right continuity hypothesis can be advantageously replaced by a combination of an appropriate type of measurability together with invariance of the supermartingale inequality under optional sampling.

Sections 6 to 12. Doob [4, 1953] proved the rather trivial discrete parameter decomposition Theorem 8 and after searching for several years found a mathematician who could prove the continuous parameter version Theorem 11 [Meyer 4, 1962; 5, 1963]. See the appendix to Meyer [6, 1966] for earlier decomposition results due to Volkonski, Šur, and Meyer which are for supermartingales associated with Markov processes. Perhaps as tribute to Doob's persistent search for this holy grail, the Meyer decomposition is sometimes inappropriately called the Doob–Meyer or even the Doob decomposition. Meyer proved Theorem 11 with the monotone increasing process $A(\cdot)$ natural in the sense of Section 7. The equivalence between natural and nearly predictable [Theorem 7(a3)] was proved by Doléans [1, 1967]. Rao's [1, 1969] proof of Theorem 11 is followed in the text. See [Dellacherie and Meyer 2, 1980] for extensive applications of the Meyer decomposition.

Section 13. The supermartingale domination principle appears to be new. This example of the potential theory of supermartingales indicates how closely such a theory parallels classical potential theory. Unfortunately, it was impractical to include the elegant Airault–Föllmer [1, 1974] results or the Azéma [1, 1972] and Azéma–Jeulin [1, 1976] results in this direction. Among many other things these papers contain a counterpart for supermartingales of conditional Markov processes. The most glaring probabilistic potential theory omission from this book, however, is that of additive functionals of Markov processes (see, for example, [Dynkin 2, 1963] and [Blumenthal and Getoor 1, 1968]). The author has three convincing excuses for this omission, of increasing cogency: (1) the existence of these books, (2) lack of space, (3) ignorance of the subject.

Sections 17 to 20. Reduction theory in this form seems to be new, but see also related work involving Snell envelopes by Mertens [1, 1972], Azéma [1, 1972], and Azéma and Jeulin [1, 1976]. Results in the latter paper together with Theorem 13 suggest that in some contexts $R^{\dot{\Lambda}}_{z(\cdot)}$ in Section 17 should be defined as the infimum of the class of almost surely right continuous positive supermartingales whose left limit processes majorize that of $z(\cdot)$ on $\dot{\Lambda}$. This definition is also suggested by the awkward extra hypotheses in Section 18(m).

Section 21. This energy treatment is due to Meyer [3, 1962–1963].

Chapter 2.V

Sections 2 and 4. Krickeberg [1, 1956] introduced lattice ideas into martingale theory. In particular, he discussed the LM operator and proved Theorem 4(c) for $p = 1$, the "Krickeberg decomposition" of a martingale. See [Dellacherie and Meyer 2, 1980] for further elaboration of this decomposition.

Section 12. Local martingales were introduced into martingale theory by Itô and S. Watanabe [1, 1965].

Section 13. Quasimartingales were introduced by Fisk [1, 1965]. Theorem 13 is due to K. M. Rao [2, 1969]. See his paper and that of Dellacherie and Meyer [2, 1980] for further discussion of the decomposition of a quasimartingale. See also [Föllmer 1, 1973] for a correspondence, unfortunately omitted from this book, between quasimartingales relative to a filtered probability space and measures on the σ algebra of predictable sets determined by the filtration.

Chapter 2.VI

This chapter treats only those aspects of Markov process theory needed in this book. Readers interested in going further into probabilistic potential theory, which is based on Markov processes, may consult [Dynkin 2, 1965], [Blumenthal and Getoor 1, 1968], and [Chung 2, 1981]. Knowledgeable readers will observe that translation operators are not mentioned in this chapter. They are not needed in this book.

Section 1. A. A. Markov introduced what are now called Markov processes into probability theory in [1, 1906].

Section 3. The strong Markov property in the discrete parameter case is trivial to prove, but the realization that such a property is relevant and requires proof had to await the sophistication not present before the property was needed in the technically more difficult continuous parameter case. Even now rather than explicitly recalling and applying this property, it is frequently easier, when the discrete parameter strong Markov property is relevant in a computation, to make the computation detailed enough to avoid explicit statement and use of the property.

Section 4. Kakutani [3, 1945] defined random walks on domains of $\mathbb{R}^2$ as in the example in this section. He proved, omitting some details, what amounts to the fact that the distribution of $x(n)$ in the example tends to harmonic measure relative to the initial point.

Section 6. Hunt [2, 1957] was the first to consider in depth the hitting probabilities of sets by Markov process trajectories.

Section 8. See Meyer [7, 1967] for a more complete analysis of the filtrations generated by Markov processes. The 0-1 law in this section is known as the Blumenthal 0-1 law in view of [Blumenthal 1, 1957]. Blumenthal's paper is probably the first paper to treat a Markov process explicitly as a process adapted to a general filtration and having the Markov property relative to the filtration.

Section 9. The first version of the strong Markov property in a continuous parameter context was proved by Doob [2, 1945] in the context of a countable state space. Hunt [1, 1956] proved a version for processes with independent increments. The first generally applicable versions were proved independently by Dynkin and Yushkevich [1, 1956] and Blumenthal [1, 1957].

Sections 10 to 12. Hunt [2, 1957] introduced excessive functions and measures into Markov process theory as part of his fundamental development [1, 1957; 2, 1957; 3, 1958] of probabilistic potential theory based on Markov processes.

Section 13. Conditioned Markov processes were introduced in [Doob 8, 1957] in the context of Brownian motion. See [Dynkin 2, 1963] for conditioned Markov processes in the context of multiplicative linear functionals.

Section 15. See [Meyer, Smythe, and Walsh 1, 1972] for a deep discussion of Markov processes killed at various kinds of random times.

Chapter 2.VII

The Brownian motion process is the link between classical potential theory and martingale theory, and only those Brownian motion properties relevant to this linkage are discussed in this book. For more information on Brownian motion see, for example, [Lévy 4, 1948], [Ciesielski 1, 1966], [Freedman 1, 1971], [Itô and McKean 1, 1974], [Knight 1, 1981], and, especially for the physical significance of Brownian motion, [Nelson 1, 1967]. The process takes its name from the English botanist Brown who in 1827 observed irregular motion of pollen particles in a liquid. Einstein in 1905 obtained from physical considerations the fact that the mean square displacement of a Brownian particle is a multiple of the displacement time and evaluated this multiple. Bachelier [1, 1900; 2, 1901] and later papers derived many distributions involving Brownian motion considered as a limit of random walks and saw the connection with the heat equation. The first rigorous construction of a Brownian motion process was given by Wiener [1, 1923], but his work remained unknown to or at least ignored by other probabilists for about 15 years. Thus probabilists were at a disadvantage in treating sample functions of Brownian motion; although they realized that in some sense these sample functions were continuous, it was not clear how to formulate this property without Wiener's work. It was intuitively clear, however, that many distributions involving Brownian motion from a point of a domain involved a process whose paths were ordinary Brownian motion paths until they hit the boundary of the domain, where they were absorbed or reflected in some way so that the transition probabilities involved, like the transition densities of Brownian motion in $\mathbb{R}^N$, were governed by the heat equation in the domain. The conduct of the paths at the boundary then determined corresponding heat equation boundary conditions.

Sections 2 and 3. The definition of a Brownian motion process used here implies that one example is the canonical example: a measure is assigned to the space of continuous functions from $\mathbb{R}^+$ into $\mathbb{R}^N$ making the coordinate functions random variables with the properties BM1–BM4. This measure is known as *Wiener measure*, and Brownian motion is sometimes defined as this canonical example. Under this definition a process satisfying BM1–BM4 is not necessarily a Brownian motion.

Section 4. Theorem 4 is a special case of a theorem of Meyer [7, 1967]. The proof in the text follows that of Chung and Walsh [2, 1974].

Section 6. The zero-one law for Brownian motion can be considered as a special case of a general zero-one law of Kolmogorov [1, 1933].

Section 8. The André reflection principle goes back to [André 1, 1887], where a reflection idea is used in a ballot counting problem. The distributions in Section 8 were derived by Bachelier (see the above notes to this chapter), but his work was ignored and rediscovered by many others.

Section 9. Hunt [1, 1956] derived the basic properties of the transition density for Brownian motion in an open set.

Section 11. The transition density for Brownian motion in an interval was found by Bachelier (see the above notes to this chapter).

Chapter 2.VIII

The integral $\int_I f\,dw$ with f a function from I into $\mathbb{R}$ and $w(\cdot)$ a Brownian motion was first discussed by Wiener [1, 1923]. Itô [1, 1944] allowed the integrand to depend on the Brownian paths, thereby inaugurating a far-reaching further development. See McKean [1, 1969] for many applications of Itô's integral, and see [Dellacherie and Meyer 2, 1980] for stochastic integrals with differentials more general than Brownian motion differentials.

Section 3. The problem of finding conditions that y_ϕ as defined by (3.10) be a martingale is a special case of a problem proposed by Girsanov in 1960 in which $w(\cdot)$ is a local martingale. Various sufficient conditions have been found; for example, Novikov [1, 1972] found a condition reducing to (k) in our context.

Section 8. Theorem 8 is a consequence of Itô's [3, 1951] construction of a complete orthonormal system on a Brownian motion measure space. See also [Kunita and S. Watanabe 1, 1967] for a systematic study of general stochastic integrals in L^2 contexts. The proof in Section 8 is taken from [Parthasarathy 1, 1978]. Dudley [1, 1977] proved that every finite-valued $\mathscr{F}\{w(t), s \in \mathbb{R}^+\}$ measurable function can be expressed as an Itô integral with integrand in Γ.

Section 10. Lévy [3, 1940] proved that if $t_{\bullet}$ is a dense sequence in $[0, b]$ and if $t_0^{(n)}, \ldots, t_n^{(n)}$ are $t_0, \ldots, t_n$ in increasing order, then

$$\lim_{n\to\infty} \sum_1^n [w(t_j^{(n)}) - w(t_{j-1}^{(n)})]^2 = \sigma^2 b \quad \text{a.s.}$$

See [Doob 10, 1953] for a martingale proof of this limit theorem.

Section 12. Itô's lemma, also called *Itô's formula*, was proved in [2, 1951].

Section 14. The fact that the composition of an analytic function with plane Brownian motion is Brownian motion with a new time scale is due to Lévy [4, 1948].

Chapter 2.IX

The martingale theory properties of the composition of superharmonic and harmonic functions with Brownian motion were treated in [Doob 5, 1954], and the corresponding discussion in the parabolic context was given in [Doob 6, 1955]. Nonroutine results in Chapter IX, aside from results specifically attributed otherwise, are taken from these papers, although most of the proofs are different and some of the results are slightly refined.

Section 5. Lévy [3, 1940] proved Theorem 5(c) and stated Theorem 5(a) for A a singleton. Kakutani [1, 1944] proved Theorem 5(b) and [2, 1944] sketched a proof of Theorem 5(a) for $N = 2$.

Section 10. Theorem 10(a) was indicated without proof details in [Kakutani 2, 1944].

Section 11. Theorem 11 in its classical and parabolic versions [Doob 5, 1954; 6, 1955] was generalized by Hunt [2, 1957], Dynkin [2, 1963], and Meyer [7, 1967] to yield various continuity properties of the composition of excessive functions with their corresponding Markov processes.

Section 13. The idea that Theorem 13(a) and its parabolic counterpart must be true is old, but a rigorous proof had to await the rigorous development of harmonic measure and of Brownian motion. Courant, Friedrichs, and Lewy [1, 1928] obtained harmonic function Dirichlet solutions for smooth domains by solving a corresponding problem for difference equations and going to the obvious limit. They refer to the probabilistic interpretation of their difference equation. Petrowsky [1, 1933–1934] gave a similar discussion, and Khintchine [1, 1933] in a corresponding discussion for both harmonic and parabolic functions mentions the Brownian motion interpretation of the limiting case. Such ideas were common at the time, but Wiener's [1, 1923] treatment of Brownian motion was unknown or unappreciated, and, for example, Khintchine's Brownian motion interpretation was therefore somewhat artificial. Kakutani [2, 1944; 3, 1945] stated Theorem 3(a) with indications of a proof.

Section 14. Theorem 14 (in a much more general context) is due to Hunt [2, 1957].

Section 15. The dichotomy in Theorem 5 in the classical and parabolic contexts [Doob 5, 1954; 6, 1955] has been generalized to define fine and cofine topologies in the probabilistic potential theory of Markov processes. Khintchine [1, 1933] proved the iterated logarithm law for Brownian motion, aside from the fact that he could not quite define the probabilities involved without a rigorous definition of Brownian motion.

Section 17. Theorem 17 is part of the folk lore of the subject but is new as stated.

Chapter 2.X

Section 6. See [Chung and Walsh 1, 1969] for a general discussion of reversing a Markov process.

Section 8. The application of martingale theory to derive a probability version of Theorem 1.XI.4 is taken from [Doob 8, 1957]. See [Weil 1, 1969] and [Airault 1, 1973] for such work in a more general context. Airault shows that the specification of the limit in Theorem 1.XI.4(b) is possible because if the Riesz measure associated with a positive superharmonic function h is supported by a polar set, the lifetime of an h Brownian motion process is predictable. Unfortunately the stated hypotheses underlying most probabilistic analysis of this sort, including the above references, are too strong to cover the parabolic context.

Section 10. The evaluation of the capacitary distribution in terms of last hitting distributions (in a more general context) is due to Chung [1, 1973].

Part 3

Chapter 3.I

The lattice theoretic results in this chapter are mostly routine or in the folklore; so few references to their origin will be given.

Section 4. Lemma 4, the key to the connection between uniform integrability and the PWB method, is taken from [Doob 7, 1956], where the potential theory version is proved in an axiomatic potential theory context.

Section 8. See [Arsove and Leutwiler 1, 1974] for a general approach to Theorem 8.

Section 12. Lamb [1, 1971] pointed out the decomposition $\mathbf{S}_{ms} = \mathbf{S}_{ms\infty} + \mathbf{S}_{msf}$.

Chapter 3.II

Section 2. The delicate parts of Theorem 2 not treated in previous chapters are taken from [Doob 7, 1956; 9, 1958].

Chapter 3.III

Section 1. See [Kunita and T. Watanabe 1, 1965; 2, 1966], [Meyer 9, 1968], and [Dynkin 3, 1969; 4, 1971] for probability-potential theory approaches to the Martin space and process paths on it, carried out in Markov process contexts at various levels of generality. The reader will observe that this chapter is rather skimpy and that a parabolic context counterpart is missing. The point is that the author is a firm believer in finite-valued stopping times.

Section 4. Observe that, as pointed out in the notes to Section XI.4, the probabilistic proof of the Fatou boundary limit theorem for the Martin space preceded the nonprobabilistic proof.

Section 6. The example is taken from [Doob 10, 1958], which contains several examples of conditional Brownian motions in a half-space, involving paths to and from boundary points.

Appendixes

Appendixes I and II

The operation constructing a nucleus from a Suslin scheme was devised by Suslin in [1, 1917]. Apparently Lusin coined the name "analytic sets." Choquet created his capacity theory in [1, 1955; 4, 1959]. Theorem II.5, which makes analytic set theory conveniently applicable in many probabilistic contexts, is taken from [Meyer 6, 1966]. See [Dellacherie and Meyer 1, 1975; 2, 1980] for more on analytic sets and capacities and their application to probabilistic potential theory.

Appendixes III to VI

The material in these appendixes, except possibly for Section 12 of Appendix IV for which see [Besicovitch 1, 1946], is traditional and routine and is assembled for the reader's convenience. [Peressini 1, 1967] is a useful source for a reader who wants enough but not too much vector lattice theory.

Appendix VII

See [Doob 13, 1960; 14, 1960–1961] for ratio integral limit theorems; those in Appendix VII are adaptations to fit the needs of this book.

Bibliography

Hélène Airault: [1] "Théorème de Fatou et frontière de Martin." *J. Funct. Anal.* **12** (1973), 418–455.

Hélène Airault and Hans Föllmer: [1] "Relative densities of semimartingales." *Inventiones Math.* **27** (1974), 299–327.

Desiré André: [1] Solution directe d'un problème résolu par M. Bertrand. *C.R. Acad. Sci. Paris* **105** (1887), 436–437.

P. Appell: [1] Sur l'équation $\partial^2 z/\partial x^2 - \partial z/\partial y = 0$ et la théorie du chaleur. *J. Math. Pures. Appl.* (4) **8** (1892), 187–216.

Maynard Arsove and Heinz Leutwiler: [1] Quasi bounded and singular functions. *Trans. Amer. Math. Soc.* **189** (1974), 275–302.

Jacques Azéma: [1] Quelques applications de la théorie générale des processus. I. *Inventiones Math.* **18** (1972), 293–336.

J. Azéma and T. Jeulin: [1] Précisions sur la measure de Föllmer. *Ann. Inst. Henri Poincaré* **12** (1976), 257–283.

I. Babuška and R. Výborný: [1] Reguläre und stabile Randpunkte für das Problem der Wärmeleitungsgleichung. *Ann. Polon. Math.* **12** (1962), 91–104.

Louis Bachelier: [1] Théorie de la spéculation. *Ann. Sci. Ecole Norm. Sup.* (3) **17** (1900), 21–86.
[2] Théorie mathématique du jeu. *Ann. Sci. Ecole Norm. Sup.* (3) **18** (1901), 143–210.

Serge Bernstein: [1] On some transformations of the Chebychev inequality. (Russian) *Dokl. Akad. Nauk SSSR* **17** (1937), 275–277.

A. S. Besicovitch: [1] "A general form of the covering principle and relative differentiation of additive functions II." *Proc. Cambridge Phil. Soc.* **42** (1946), 1–10.

R. M. Blumenthal: [1] An extended Markov property. *Trans. Amer. Math. Soc.* **85** (1957), 52–72.

R. M. Blumenthal and R. K. Getoor: [1] *Markov Processes and Potential Theory*. New York, Academic, 1968.
[2] "Dual processes and potential theory." *Proc. Twelfth Biennial Sem. Can. Math. Congr.* 1970, 137–156.

Maxime Bôcher: [1] "Singular points of functions which satisfy partial differential equations of the elliptic type." *Bull. Amer. Math. Soc.* **9** (1903), 455–465.

Georges Bouligand: [1] "Sur le problème de Dirichlet." *Ann. Soc. Polon. Math.* **4** (1926), 59–112.

M. Brelot: [1] "Fonctions sous-harmoniques et balayage I, II." *Acad. Roy. Belgique, Bull. Cl. Sci.* (5) **24** (1938), 301–312, 421–436.
[2] "Sur le potentiel et les suites de fonctions surharmoniques." *C. R. Acad. Sci. Paris* **207** (1938), 836–838.
[3] "Familles de Perron et problème de Dirichlet." *Acta Litt. Sci. Szeged* **9** (1939), 133–153.
[4] "Points irréguliers et transformations continues en théorie du potentiel." *J. Math. Pures. Appl.* **19** (1940), 319–337.
[5] "Sur la théorie autonome des fonctions sousharmoniques." *Bull. Sci. Math.* **65** (1941), 72–98.
[6] "Sur le rôle du point à l'infini dans la théorie des fonctions harmoniques." *Ann. Sci. Ecole Norm. Sup.* **61** (1944), 301–332.
[7] "Sur les ensembles effilés." *Bull. Sci. Math.* **68** (1944), 12–36.
[8] "Minorantes sous-harmoniques, extrémales et capacités." *J. Math. Pures. Appl.* **24** (1945), 1–32.

[9] "Etude générale des fonctions harmoniques ou surharmoniques positive au voisinage d'un point-frontière irregulier." *Ann. Univ. Grenoble* **22** (1946), 201–219.
[10] "Quelques propriétés et applications du balayage." *C. R. Acad. Sci. Paris* **227** (1948), 19–21.
[11] "La théorie moderne du potentiel." *Ann. Inst. Fourier Grenoble* **4** (1952), 113–140 (1954).
[12] "Etude et extensions du principe de Dirichlet." *Ann. Inst. Fourier Grenoble* **5** (1953–1954), 371–419.
[13] "Le problème de Dirichlet. Axiomatique et frontière de Martin." *J. Math. Pures. Appl.* **35** (1956), 297–335.
[14] "Quelques propriétés et applications nouvelles de l'effilement." *Sem. (Brelot-Choquet-Deny) Théorie du Potentiel* **6** (1961–1962), 1-27–1-40, 1962.
[15] *Eléments de la Théorie Classique du Potentiel*, 4th ed. Centre du Documentation Universitaire Paris, 1969.
[16] *On Topologies and Boundaries in Potential Theory*. Lecture Notes in Mathematics 175, Berlin, Springer-Verlag, 1971.
[17] "Les étapes et les aspects multiples de la théorie du potentiel," *L'Enseignment Math. Ser.* II **18** (1972), 1–36.

M. Brelot and J. L. Doob: [1] "Limites angulaires et limites fines." *Ann. Inst. Fourier Grenoble* **13** (1963), 395–415.

Jean Brossard: [1] Comportement "non-tangentiel" et comportement "Brownien" des fonctions harmoniques dans un demi-espace. Démonstration probabiliste d'un théoreme de Calderon et Stein. *Sém. Prob. XII 1976/77*, Lecture Notes in Mathematics 649, Berlin, Springer-Verlag, 1978, pp. 378–397.

Heinrich Burkhardt and W. Franz Meyer: [1] "Potentialtheorie (Theorie der Laplace-Poissonschen Differentialgleichung)." *Enzyc. Math. Wiss.* IIA7b (1900), 464–503.

Henri Cartan: [1] "Capacité extérieure et suites convergentes." *C. R. Acad. Sci. Paris* **214** (1942), 944–946.
[2] "Théorie du potentiel newtonien: énergie, capacité, suites de potentiels." *Bull. Soc. Math. Fr.* **73** (1945), 74–106.
[3] "Théorie générale du balayage en potentiel newtonien." *Ann. Inst. Fourier Grenoble* **22** (1946), 221–280.

Gustave Choquet: [1] "Theory of capacities." *Ann. Inst. Fourier Grenoble* **5** (1953–1954), 131–295 (1955).
[2] "Potentiels sur un ensemble de capacité nulle. Suites de potentiels." *C. R. Acad. Sci. Paris* **244** (1957), 1707–1710.
[3] "Capacitabilité en potentiel logarithmique." *Acad. Roy. Belg. Bull. Cl. Sci.* (5) **44** (1958), 321–326.
[4] "Forme abstraite du théoreme de capacitabilité." *Ann. Inst. Fourier Grenoble* **9** (1959), 83–89.

Kai Lai Chung: [1] "Probabilistic approach in potential theory to the equilibrium problem." *Ann. Inst. Fourier Grenoble* **23/3** (1973), 313–322.
[2] *Lectures from Markov Processes to Brownian Motion*. Berlin, Springer-Verlag, 1981.

Kai Lai Chung and J. L. Doob: [1] "Fields, optionality and measurability." *Amer. J. Math.* **87** (1965), 397–424.

Kai Lai Chung and John B. Walsh: [1] "To reverse a Markov process." *Acta Math.* **123** (1969), 225–251.
[2] "Meyer's theorem on predictability." *Ztschr. Wahrscheinlichkeitstheorie verw. Geb.* **29** (1974), 253–256.

Z. Ciesielski: [1] *Lectures on Brownian motion, heat conduction and potential theory*. Aarhus, Denmark: Aarhus Univ., 1966.

Corneliu Constantinescu and Aurel Cornea: [1] *Potential theory of harmonic spaces*. Berlin, Springer-Verlag, 1972.

R. Courant, K. Friedrichs, and H. Lewy: [1] "Über die partiellen Differenzengleichungen der mathematischen Physik." *Math. Ann.* **100** (1928), 32–74.

Claude Dellacherie: [1] Ensembles aléatoires I, II. *Sem. Prob. III 1967–8*, Lecture Notes in Mathematics 88. Berlin, Springer-Verlag, 1969.
[2] *Capacités et Processus Stochastiques*. Erg. Math. u. ihrer Grenzgebiete 67. Berlin, Springer-Verlag, 1972.

Claude Dellacherie and Paul-André Meyer: [1] Probabilités et potentiel. Chapters I–IV. *Act. Sci. Ind.* **1372** (1975), Paris, Hermann.
[2] Chapters V–VIII *Act. Sci. Ind.* **1385** (1980), Paris, Hermann.

Jacques Deny: [1] "Les potentiels d'énergie finie." *Acta Math.* **82** (1950), 107–183.
C. Doléans (= C. Doléans-Dade): [1] "Processus croissant naturels et processus croissant tres-bien-measurables." *C. R. Acad. Sci. Paris* **264** (1967), 874–876.
J. L. Doob: [1] "Regularity properties of certain families of chance variables." *Trans. Amer. Math. Soc.* **47** (1940), 455–486.
[2] "Markoff chains—denumerable case." *Trans. Amer. Math. Soc.* **58** (1945), 455–473.
[3] "Continuous parameter martingales." *Proc. Sec. Berkeley Symp. Math. Statistics Prob.* 1950, Berkeley, 1951, pp. 269–277.
[4] *Stochastic Processes.* New York, Wiley, 1953.
[5] "Semimartingales and subharmonic functions." *Trans. Amer. Math. Soc.* **77** (1954), 86–121.
[6] "A probability approach to the heat equation." *Trans. Amer. Math. Soc.* **80** (1955), 216–280.
[7] "Probability methods applied to the first boundary value problem." *Proc. Third Berkeley Symp. Math. Statistics and Prob. 1954/5* Vol. 2, Berkeley, 1956, pp. 49–80.
[8] "Conditional Brownian motion and the boundary limits of harmonic functions." *Bull. Soc. Math. Fr.* **85** (1957), 431–458.
[9] "Probability theory and the first boundary value problem." *Ill. J. Math.* **2** (1958), 19–36.
[10] "Boundary limit theorems for a halfspace." *J. Math. Pures. Appl.* (9) **37** (1958), 385–392.
[11] "A non-probabilistic proof of the relative Fatou theorem." *Ann. Inst. Fourier Grenoble* **9** (1959), 293–300.
[12] "A relativized Fatou theorem." *Proc. Nat. Acad. Sci. USA* **45** (1959), 215–222.
[13] "A relative limit theorem for parabolic functions." *Trans. Second Prague Conference on Information Theory, Statistical Decision Functions, Random Processes.* Prague, Czechoslovak Acad. Sci., 1960, pp. 61–70.
[14] "Relative theorems in analysis." *J. Anal. Math.* **8** (1960–1961), 289–306.
[15] "Conformally invariant cluster value theory." *Ill. J. Math.* **5** (1961), 521–549.
[16] "Some classical function theory theorems and their modern versions." *Ann. Inst. Fourier Grenoble* **15** (1965), 113–136.
[17] "Applications to analysis of a topological definition of smallness of a set." *Bull. Amer. Math. Soc.* **72** (1966), 579–600.
Lester E. Dubins: [1] "A note on upcrossings of semimartingales." *Ann. Math. Stat.* **37** (1966), 728.
R. M. Dudley: [1] "Wiener functions as Itô integrals." *Ann. Prob.* **5** (1977), 140–141.
E. B. Dynkin: [1] *Foundations of the Theory of Markov Processes* (translation of his 1959 Russian book: *Основания Теории Марковских Проиессов*). Oxford, Pergamon, 1960.
[2] *Markov Processes* (translation of his 1963 Russian book: *Марковские Процессы*). Berlin, Springer-Verlag, 1965.
[3] "The space of exits of a Markov process." *Russ. Math. Surv.* **24/4** (1969), 89–157.
[4] "Initial and final behavior of Markov process trajectories." *Russ. Math. Surv.* **26** (1971), 165–185.
E. B. Dynkin and A. A. Yushkevich: [1] Strong Markov processes. *Theory Prob. Appl.* **1** (1956), 134–139.
G. C. Evans: [1] "Application of Poincaré's sweeping-out process." *Proc. Nat. Acad. Sci. USA* **19** (1933), 457–461.
[2] "On potentials of positive mass." *Trans. Amer. Math. Soc.* **37** (1935), 226–253.
P. Fatou: [1] "Séries trigonométriques et séries de Taylor." *Acta Math.* **30** (1906), 335–400.
D. L. Fisk: [1] "Quasi-martingales". *Trans. Amer. Math. Soc.* **120** (1965), 369–389.
Hans Föllmer: [1] "On the representation of semimartingales." *Ann. Prob.* **1** (1973), 580–589.
David Freedman: [1] *Brownian motion and diffusion.* San Francisco, Holden-Day, 1971.
Otto Frostman: [1] "Potentiel d'équilibre et capacité des ensembles avec quelques applications à la théorie des fonctions." *Meddel. Lunds Univ. Mat. Sem.* **3** (1935), 1–118.
[2] "Sur le balayage des masses." *Acta Litt. Sci. Univ. Szeged, Sec. Sci. Math.* **9** (1938), 43–51.
Bent Fuglede: [1] *Finely Harmonic Functions.* Lecture Notes in Mathematics 289. Berlin, Springer-Verlag, 1972.
C. F. Gauss: [1] Allgemeine Lehrsätze in Beziehung auf die im vehrkehrten Verhältnisse des Quadrats der Entfernung wirkenden Anziehungs- und Abstossungs-Kräfte. Gauss Werke 5, pp. 197–242, 1840, Göttingen, 1867.
George Green: [1] An essay on the application of mathematical analysis to the theories of

electricity and magnetism. Nottingham 1828. *Math. Papers*. London, Macmillan, 1871, pp. 9–41.
A. Harnack: [1] *Die Grundlagen der Theorie des logarithmischen Potentials und der eindeutigen Potentialfunktion*. Leipzig, Teubner, 1887.
Philip Hartman, Aurel Wintner: [1] "On the solutions of the equation of heat conduction." *Amer J. Math.* **72** (1950), 367–395.
Maurice Heins: [1] On the principle of harmonic measure. *Comment. Math. Helv.* **33** (1959), 47–58.
L. L. Helms: [1] *Introduction to Potential Theory*. New York, Wiley, 1969.
G. Herglotz: [1] "Über Potenzreihen mit positivem reellen Teil im Einheitskreis." *Ber. Verhandl. Sächs. Akad. Wiss. Leipzig Math.-Phys. Kl.* **63** (1911), 501–511.
G. A. Hunt: [1] "Some theorems concerning Brownian motion." *Trans. Amer. Math. Soc.* **81** (1956), 294–319.
[2] "Markoff processes and potentials I." *Ill. J. Math.* **1** (1957), 44–93.
[3] "Markoff processes and potentials II." *Ill. J. Math.* **1** (1957), 316–369.
[4] "Markoff processes and potentials III." *Ill. J. Math.* **2** (1958), 151–213.
Kiyosi Itô: [1] "Stochastic integral." *Proc. Imp. Acad. Tokyo* **20** (1944), 519–524.
[2] "On a formula concerning stochastic differentials." *Nagoya Math. J.* **3** (1951), 55–65.
[3] "Multiple Wiener integral." *J. Math. Soc. Japan* **3** (1951), 157–169.
K. Itô and H. P. McKean, Jr.: [1] *Diffusion processes and their sample paths*. Berlin, Springer-Verlag, 1974.
K. Itô and S. Watanabe: [1] "Transformation of Markov processes by multiplicative functionals." *Ann. Inst. Fourier Grenoble* **15**, 1, (1965), 15–30.
H. L. Jackson: [1] "Some results on thin sets in a halfplane." *Ann Inst. Fourier Grenoble* **20** (1970), 201–218.
Klaus Janssen: [1] *Martin Boundary and H^p Theory of Harmonic Spaces*. Lecture Notes in Mathematics 226. Berlin, Springer-Verlag, 1971, pp. 102–151.
[2] "A co-fine domination principle for harmonic spaces." *Math. Ztschr.* **141** (1975), 185–191.
Børge Jessen: [1] "The theory of integration in a space of an infinite number of dimensions." *Acta Math.* **63** (1934), 249–323.
Shizuo Kakutani: [1] "On Brownian motions in *n*-space." *Proc. Imp. Acad. Tokyo* **20** (1944), 648–652.
[2] "Two dimensional Brownian motion and harmonic functions." *Proc. Imp. Acad. Tokyo* **20** (1944), 706–714.
[3] "Markoff process and the Dirichlet problem." *Proc. Japan Acad.* **21** (1945), 227–233 (1949).
Robert Kaufman and Jang-Mei Wu: [1] "Parabolic potential theory." *J. Differential Equations* **43** (1982), 204–234.
O. D. Kellogg: [1] "Unicité des fonctions harmoniques." *C. R. Acad. Sci. Paris* **187** (1928), 526–527.
[2] *Foundations of Potential Theory*. Berlin, Springer-Verlag, 1929.
John T. Kemper: [1] "Temperatures in several variables. Kernel functions, representations, and parabolic boundary values." *Trans. Amer. Math. Soc.* **167** (1972), 243–262.
A. Khintchine: [1] "Asymptotische Gesetze der Wahrscheinlichkeitsrechnung." *Erg. Math. Grenzgebiete* **2/4** (1933), Berlin, Springer-Verlag.
Frank B. Knight: [1] Essentials of Brownian motion and diffusion. *Math. Surveys 18*. Providence, Amer. Math. Soc., 1981.
A. Kolmogoroff: [1] "Grundbegriffe der Wahrschleinlichkeitsrechnung." *Erg. Math. Grenzgebiete* **2/3** (1933), Berlin, Springer-Verlag.
Adam Koranyi and J. C. Taylor: [1] "Fine convergence and parabolic convergence for the Helmholtz equation and the heat equation." *Ill. J. Math.*
K. Krickeberg: [1] "Convergence of martingales with a directed index set." *Trans. Amer. Math. Soc.* **83** (1956), 313–357.
Hiroshi Kunita and Shinzo Watanabe: [1] "On square integrable martingales." *Nagoya Math. J.* **30** (1967), 209–245.
Hiroshi Kunita and Takesi Watanabe: [1] "Markov processes and Martin boundaries Part 1." *Ill. J. Math.* **9** (1965), 485–526.
[2] "On certain reversed processes and their applications to potential theory and boundary theory." *J. Math. Mech.* **15** (1966). 393–434.
Charles W. Lamb: [1] "A note on harmonic functions and martingales." *Ann. Math. Stat.*

42 (1971), 2044–2049.
N. S. Landkof: [1] *Foundations of modern potential theory* (translation from his Russian book of 1966: Основы современной теории потенциала) Berlin, Springer-Verlag, 1972.
H. Lebesgue: [1] "Sur les cas d'impossibilité du problème de Dirichlet." *C. R. Séances Soc. Math. Fr.* (1912).
[2] "Sur le problème de Dirichlet." *C. R. Acad. Sci. Paris* **154** (1912), 335–337.
[3] "Conditions de régularité, conditions d'irrégularité, conditions d'impossibilité dans le problème de Dirichlet." *C. R. Acad. Sci. Paris* **178** (1924), 349–354.
J. Lelong: [1] "Etude au voisinage de la frontière des fonctions surharmoniques positive dans un demi-espace." *Ann. Sci. Ecole Norm. Sup.* (3) **66** (1949), 125–159.
Paul Lévy: [1] "Propriétés asymptotiques des sommes de variables aléatoires enchainées." *Bull. Soc. Math. Fr.* **59** (1935), 1–32.
[2] *Théorie de l'Addition des Variables Aléatoires.* Paris, Gauthier-Villars, 1937.
[3] "Le mouvement brownien plan." *Amer. J. Math.* **62** (1940), 487–550.
[4] *Processes Stochastiques et Mouvement Brownien.* Paris, Gauthier-Villars, 1948 (2d. ed., 1965).
L. Lichtenstein: [1] "Neuere Entwicklung des Potentialtheorie. Konforme Abbildung." *Encykl. Math. Wiss.* IIC3 (1919), 177–377.
J. E. Littlewood: [1] "Mathematical Notes (8): On functions subharmonic in a circle (II)." *Proc. London Math. Soc.* (2) **28** (1928), 383–394.
Alfred J. Maria: [1] "The potential of a positive mass and the weight function of Wiener." *Proc. Nat. Acad. Sci. USA* **20** (1934), 485–489.
A. A. Markov: [1] "Extension of the law of large numbers to dependent events" (Russian). *Bull. Soc. Phys. Math. Kazan* (2) **15** (1906), 135–156.
R. S. Martin: [1] "Minimal positive harmonic functions." *Trans. Amer. Math. Soc.* **49** (1941), 137–172.
H. P. McKean, Jr.: [1] *Stochastic Integrals.* New York, Academic, 1969.
Jean-Francois Mertens: [1] "Théorie des processus stochastiques généraux. Applications aux surmartingales." *Ztschr. Wahrscheinlichkeitstheorie* **22** (1972), 54–68.
Paul-André Meyer: [1] "Fonctions multiplicatives et additives de Markov." *Ann. Inst. Fourier Grenoble* **12** (1962), 125–230.
[2] "Une présentation de la théorie des ensembles sousliniens. Applications aux processus stochastiques." *Sém. Brelot-Choquet-Deny (Théorie du potentiel)* **7** (1962–1963).
[3] "Interprétation probabiliste de la notion d'energie." *Sém. Brelot-Choquet-Deny (Théorie du Potentiel)* **7** (1962–1963).
[4] "A decomposition theorem for supermartingales." *Ill. J. Math.* **6** (1962), 193–205.
[5] "Decomposition of supermartingales: the uniqueness theorem." *Ill. J. Math.* **7** (1963), 1–17.
[6] *Probability and Potentials.* Waltham, Blaisdell, 1966.
[7] *Processus de Markov.* Lecture Notes in Mathematics 26. Berlin, Springer-Verlag, 1967.
[8] Guide détaillé de la théorie générale des processus. *Sém. Prob. II.* Lecture Notes in Mathematics 51. Berlin, Springer-Verlag, 1968.
[9] *Processus de Markov: la frontière de Martin.* Lecture Notes in Mathematics 77. Berlin, Springer-Verlag, 1968.
P. A. Meyer, R. T. Smythe, and J. B. Walsh: Birth and death of Markov processes. *Proc. Sixth Berkeley Symp. Math. Stat. Prob. 1970/71* Vol. III. Berkeley, Univ. of California, 1972, pp. 295–305.
Gabriel Mokobodzki and Daniel Sibony: "Sur une propriété characteristique des cones de potentiels." *C. R. Acad. Sci. Paris* **266** (1968), 215–218.
Jürgen Moser: [1] "A Harnack inequality for parabolic differential equations." *Comm. Pure Appl. Math.* **17** (1964), 101–134.
Linda Naïm (= Linda Lumer-Naïm): [1] "Sur le rôle de la frontière de R. S. Martin dans la théorie du potential." *Ann. Inst. Fourier Grenoble* **7** (1957), 183–281.
Edward Nelson: [1] *Dynamical Theories of Brownian Motion.* Math. Notes Princeton U. Press, Princeton, 1967.
J. Neveu: [1] *Martingales à Temps Discret.* Paris, Masson, 1972.
A. A. Novikov: [1] "On an identity for stochastic integrals." *Theory Prob. Appl.* **17** (1972), 717–720.
M. Parreau: [1] "Sur les moyennes des fonctions harmoniques et analytiques et la classification des surfaces de Riemann." *Ann. Inst. Fourier Grenoble* **3** (1951), 103–197.

K. R. Parthasarathy: [1] "Square integrable martingales orthogonal to every stochastic integral." *Stochastic Proc. Appl.* **7** (1978), 1–7.

Anthony L. Peressini: [1] *Ordered Topological Vector Spaces*. New York, Harper, 1967.

O. Perron: [1] "Eine neue Behandlung der ersten Randwertaufgabe fur $\Delta u = 0$." *Math. Ztschr.* **18** (1923), 42–54.

I. Petrowsky: [1] "Über das Irrfahrtproblem." *Math. Ann.* **109** (1933–1934), 425–444.

[2] "Zur ersten Randwertaufgabe der Wärmeleitungsgleichung." *Comp. Math.* **1** (1934–1935), 383–419.

H. Poincaré: [1] "Sur les équations aux dérivées partielles de la physique mathématique." *Amer. J. Math.* **12** (1890), 211–294.

[2] *Théorie du Potentiel Newtonien*. Paris, Gauthier-Villars, 1899.

S. D. Poisson: [1] "Addition au mémoire précédent et au mémoire sur la manière d'exprimer les fonctions par les séries de quantités périodiques." *J. Ecole Roy. Polytechnique 19 cahier* **12** (1823), 145–162.

Sidney C. Port, Charles J. Stone: [1] *Brownian Motion and Classical Potential Theory*. New York, Academic, 1978.

N. Privalov: [1] "Boundary problems of the theory of harmonic and subharmonic functions in space" (Russian). *Mat. Sbornik* **3** (1938), 3–25.

Tibor Radó: [1] "Subharmonic functions." *Erg. Math. Grenzgebiete* **5** (1), Berlin, Springer-Verlag, 1937. (Reprinted New York, Chelsea, 1949.)

K. Murali Rao: [1] "On decomposition theorems of Meyer." *Math. Scand.* **24** (1969), 66–78.

[2] "Quasi martingales." *Math. Scand.* **24** (1969), 79–92.

[3] "On Green functions in $\mathbb{R}^2$." *Israel J. Math.* **19** (1974), 313–328.

[4] Brownian motion and classical potential theory. *Math Inst.*, Aarhus Univ., 1977.

Robert Remak: [1] "Über potentialkonvexe Funktionen." *Math. Ztschr.* **20** (1924), 126–130.

F. Riesz: [1] "Sur certains systèmes singuliers d'équations intégrales." *Ann. Sci. Ecole Norm. Sup.* (3) **28** (1911), 33–62.

[2] "Sur les fonctions subharmoniques et leur rapport a la théorie du potentiel I." *Acta Math.* **48** (1926), 329–343.

[3]II. *Acta Math.* **54** (1930), 321–360.

M. Riesz: [1] "Intégrales de Riemann-Liouville et potentiels." *Acta Litt. Sci. Univ. Szeged* **9** (1) (1938), 1–42.

G. Robin: [1] "Sur la distribution de l'électricité à la surface des conducteurs fermés et des conducteurs ouverts." *Ann. Sci. Ecole Norm. Sup.* (3) **3** (1886), Supp. 1–58.

H. A. Schwarz: [1] "Zur Integration der partiellen Differentialgleichung $\partial^2 u/\partial x^2 + \partial^2 u/\partial y^2 = 0$." *J. Reine Angew. Math.* **74** (1872), 218–253.

J. L. Snell: [1] "Application of martingale system theorems." *Trans. Amer. Math. Soc.* **73** (1952), 293–312.

M. Souslin: [1] "Sur une définition des ensembles mesurables B sans nombres transfinis." *C. R. Acad. Sci. Paris* **164** (1917), 88–91.

W. Sternberg: [1] "Über die Gleichung der Wärmeleitung." *Math. Ann.* **101** (1929), 394–398.

G. Szegö: [1] "Bemerkung zu einer Arbeit von Herrn M. Fekete: Über die Verteilung der Wurzeln bei gewissen algebraischen Gleichungen mit ganzzahligen Koeffizienten." *Math. Ztschr.* **21** (1924), 203–208.

E. Szpilrajn (= E. Marczewski): [1] "Remarques sur les fonctions sousharmoniques." *Ann. Math.* (2) **34** (1933), 588–594.

J. C. Taylor: [1] An elementary proof of the theorem of Fatou-Naïm-Doob. *1980 Sem. on Harmonic Analysis*, Montreal, 1980, pp. 153–163.

William Thomson (= Lord Kelvin): "Extraits de deux lettres addressées à M. Liouville." *J. Math. Pures. Appl.* (1) **12** (1847), 256–264.

E. C. Titchmarsh: [1] *Introduction to the Theory of Fourier Integrals*. Oxford, Clarendon Press, 1937.

M. Tsuji: [1] *Potential Theory in Modern Function Theory*. Tokyo, Maruzen, 1959.

A. Tychonoff: [1] "Théorème d'unicité pour l'équation de la chaleur." *Mat. Sbornik* **42** (1935), 199–216.

C. de la Vallée Poussin: [1] "Extension de la méthode du balayage de Poincaré et problème de Dirichlet." *Ann. Inst H. Poincaré* **2** (1932), 169–232.

[2] "Les nouvelles méthodes de la théorie du potentiel et le problème généralisé de Dirichlet." *Act. Sci. Ind.* **578** (1937), Paris, Hermann.

[3] "Potentiel et problème généralisé de Dirichlet." *Math. Gazette* **22** (1938), 17–36.

[4] "Points irrégulier. Détermination des masses par les potentiels." *Acad. Belg. Bull. Cl. Sci.* (5) **24** (1938), 368–384.

Florin Vasilesco: [1] "Sur la continuité du potentiel à travers les masses, et la démonstration d'un lemme de Kellogg." *C. R. Acad. Sci. Paris* **200** (1935), 1173–1174.

[2] "Sur une application des familles normales de distributions de masse." *C. R. Acad. Sci. Paris* **205** (1937), 212–215.

[3] La notion de point irrégulier dans le problème de Dirichlet. *Act. Sci. Ind.* **660** (1938), Paris, Masson.

Jean Ville: [1] *Etude Critique de la Notion de Collectif*. Paris, Gauthier-Villars, 1939.

Michel Weil: [1] "Propriétés de continuité fine des fonctions coexcessives." *Ztschr. Wahrscheinlichkeitstheorie verw. Geb.* **12** (1969), 75–86.

John Wermer: [1] *Potential Theory*. Lecture Notes in Mathematics 408. Berlin, Springer-Verlag, 1974.

D. V. Widder: [1] "Positive temperatures on an infinite rod." *Trans. Amer. Math. Soc.* **55** (1944), 85–95.

[2] *The Heat Equation*. New York, Academic, 1975.

Norbert Wiener: [1] "Differential space." *J. Math. Phys. Mass. Inst. Tech.* **2** (1923), 131–174.

[2] "Certain notions in potential theory." *J. Math. Phys. Mass. Inst. Tech.* **3** (1924), 24–51.

[3] "The Dirichlet problem." *J. Math. Phys. Mass. Inst. Tech.* **3** (1924), 127–146.

[4] "Note on a paper by O. Perron." *J. Math. Phys. Mass. Inst. Tech.* **4** (1925), 21–32.

Jang-Mei G. Wu: [1] "On parabolic measures and subparabolic functions." *Trans. Amer. Math. Soc.* **25** (1979), 171–185.

S. Zaremba: [1] "Sur le principe de Dirichlet." *Acta Math.* **34** (1911), 293–316.

Notation Index

Index

References are to sections, say 1.II.3 or, for the Appendixes, A.I.2. The index covers neither the Historical Notes nor the Bibliography.

CIM

M. Aigner Combinatorial Theory ISBN 978-3-540-61787-7

A. L. Besse Einstein Manifolds ISBN 978-3-540-74120-6

N. P. Bhatia, G. P. Szegő Stability Theory of Dynamical Systems ISBN 978-3-540-42748-3

J. W. S. Cassels An Introduction to the Geometry of Numbers ISBN 978-3-540-61788-4

R. Courant, F. John Introduction to Calculus and Analysis I ISBN 978-3-540-65058-4

R. Courant, F. John Introduction to Calculus and Analysis II/1 ISBN 978-3-540-66569-4

R. Courant, F. John Introduction to Calculus and Analysis II/2 ISBN 978-3-540-66570-0

P. Dembowski Finite Geometries ISBN 978-3-540-61786-0

A. Dold Lectures on Algebraic Topology ISBN 978-3-540-58660-9

J. L. Doob Classical Potential Theory and Its Probabilistic Counterpart ISBN 978-3-540-41206-9

R. S. Ellis Entropy, Large Deviations, and Statistical Mechanics ISBN 978-3-540-29059-9

H. Federer Geometric Measure Theory ISBN 978-3-540-60656-7

S. Flügge Practical Quantum Mechanics ISBN 978-3-540-65035-5

L. D. Faddeev, L. A. Takhtajan Hamiltonian Methods in the Theory of Solitons ISBN 978-3-540-69843-2

I. I. Gikhman, A. V. Skorokhod The Theory of Stochastic Processes I ISBN 978-3-540-20284-4

I. I. Gikhman, A. V. Skorokhod The Theory of Stochastic Processes II ISBN 978-3-540-20285-1

I. I. Gikhman, A. V. Skorokhod The Theory of Stochastic Processes III ISBN 978-3-540-49940-4

D. Gilbarg, N. S. Trudinger Elliptic Partial Differential Equations of Second Order ISBN 978-3-540-41160-4

H. Grauert, R. Remmert Theory of Stein Spaces ISBN 978-3-540-00373-1

H. Hasse Number Theory ISBN 978-3-540-42749-0

F. Hirzebruch Topological Methods in Algebraic Geometry ISBN 978-3-540-58663-0

L. Hörmander The Analysis of Linear Partial Differential Operators I – Distribution Theory and Fourier Analysis ISBN 978-3-540-00662-6

L. Hörmander The Analysis of Linear Partial Differential Operators II – Differential Operators with Constant Coefficients ISBN 978-3-540-22516-4

L. Hörmander The Analysis of Linear Partial Differential Operators III – Pseudo-Differential Operators ISBN 978-3-540-49937-4

L. Hörmander The Analysis of Linear Partial Differential Operators IV – Fourier Integral Operators ISBN 978-3-642-00117-8

K. Itô, H. P. McKean, Jr. Diffusion Processes and Their Sample Paths ISBN 978-3-540-60629-1

T. Kato Perturbation Theory for Linear Operators ISBN 978-3-540-58661-6

S. Kobayashi Transformation Groups in Differential Geometry ISBN 978-3-540-58659-3

K. Kodaira Complex Manifolds and Deformation of Complex Structures ISBN 978-3-540-22614-7

Th. M. Liggett Interacting Particle Systems ISBN 978-3-540-22617-8

J. Lindenstrauss, L. Tzafriri Classical Banach Spaces I and II ISBN 978-3-540-60628-4

R. C. Lyndon, P. E Schupp Combinatorial Group Theory ISBN 978-3-540-41158-1

S. Mac Lane Homology ISBN 978-3-540-58662-3

C. B. Morrey Jr. Multiple Integrals in the Calculus of Variations ISBN 978-3-540-69915-6

D. Mumford Algebraic Geometry I – Complex Projective Varieties ISBN 978-3-540-58657-9

O. T. O'Meara Introduction to Quadratic Forms ISBN 978-3-540-66564-9

G. Pólya, G. Szegő Problems and Theorems in Analysis I – Series. Integral Calculus. Theory of Functions ISBN 978-3-540-63640-3

G. Pólya, G. Szegő Problems and Theorems in Analysis II – Theory of Functions. Zeros. Polynomials. Determinants. Number Theory. Geometry ISBN 978-3-540-63686-1

W. Rudin Function Theory in the Unit Ball of $\mathbb{C}^n$ ISBN 978-3-540-68272-1

S. Sakai C*-Algebras and W*-Algebras ISBN 978-3-540-63633-5

C. L. Siegel, J. K. Moser Lectures on Celestial Mechanics ISBN 978-3-540-58656-2

T. A. Springer Jordan Algebras and Algebraic Groups ISBN 978-3-540-63632-8

D. W. Stroock, S. R. S. Varadhan Multidimensional Diffusion Processes ISBN 978-3-540-28998-2

R. R. Switzer Algebraic Topology: Homology and Homotopy ISBN 978-3-540-42750-6

A. Weil Basic Number Theory ISBN 978-3-540-58655-5

A. Weil Elliptic Functions According to Eisenstein and Kronecker ISBN 978-3-540-65036-2

K. Yosida Functional Analysis ISBN 978-3-540-58654-8

O. Zariski Algebraic Surfaces ISBN 978-3-540-58658-6

GPSR Compliance

The European Union's (EU) General Product Safety Regulation (GPSR) is a set of rules that requires consumer products to be safe and our obligations to ensure this.

If you have any concerns about our products, you can contact us on ProductSafety@springernature.com

In case Publisher is established outside the EU, the EU authorized representative is:

Springer Nature Customer Service Center GmbH
Europaplatz 3
69115 Heidelberg, Germany

Batch number: 09624486

Printed by Printforce, the Netherlands